WATER TREATMENT PLANT OPERATION

Volume II

Fifth Edition

A Field Study Training Program

prepared by

Office of Water Programs
College of Engineering and Computer Science
California State University, Sacramento

≈ ≈ ≈

Kenneth D. Kerri, Project Director

≈ ≈ ≈

2006

Funding for the production of this operator training manual was provided by University Enterprises, Inc., California State University, Sacramento. Mention of trade names or commercial products does not constitute endorsement or recommendation for use by the project director; the Office of Water Programs; or California State University, Sacramento.

Copyright © 2006 by
University Enterprises, Inc.

First edition published 1983. Fifth edition 2006.

All Rights Reserved.
Printed in the United States of America

12 11 10 09 08 07 06   1 2 3 4 5 6

ISBN
978-1-59371-036-1
1-59371-036-4

http://www.owp.csus.edu

OFFICE OF WATER PROGRAMS

The Office of Water Programs is a nonprofit organization operating under University Enterprises, Inc., California State University, Sacramento, to provide distance learning courses for persons interested in the operation and maintenance of drinking water and wastewater facilities. These training programs were developed by people who explain, through the use of our manuals, how they operate and maintain their facilities. The university, fully accredited by the Western Association of Schools and Colleges, administers and monitors these training programs, under the direction of Dr. Ramzi J. Mahmood.

Our training group develops and implements programs and publishes manuals for operators of water treatment plants, water distribution systems, wastewater collection systems, and municipal and industrial wastewater treatment and reclamation facilities. We also offer programs and materials for pretreatment facility inspectors, environmental compliance inspectors, and utility managers. All training is offered as distance learning, using correspondence, video, or computer-based formats with opportunities for continuing education and contact hours for operators, supervisors, managers, and administrators.

Materials and opportunities available from our office include manuals in print, CD, DVD, or video formats, and enrollments for courses providing CEU (Continuing Education Unit) contact hours. Here is a sample:

- Water Treatment Plant Operation, 2 volumes (print, course enrollment)
- Water Distribution System Operation and Maintenance (print, CD, course enrollment, five on-line courses)
- Small Water System Operation and Maintenance (print, video, CD, course enrollment, six on-line courses)
- Water Systems Operation and Maintenance Video Training Series (print, video, DVD, course enrollment)
- Utility Management (print, course enrollment)
- Manage for Success (print, course enrollment)
- and more

These and other materials may be ordered from:

Office of Water Programs
California State University, Sacramento
6000 J Street
Sacramento, CA 95819-6025
(916) 278-6142 – phone
(916) 278-5959 – FAX

or

visit us on the web at http://www.owp.csus.edu

ADDITIONAL VOLUMES OF INTEREST

Water Treatment Plant Operation, Volume I

The Water Treatment Plant Operator
Water Sources and Treatment
Reservoir Management and Intake Structures
Coagulation and Flocculation
Sedimentation
Filtration
Disinfection
Corrosion Control
Taste and Odor Control
Plant Operation
Laboratory Procedures

Water Distribution System Operation and Maintenance

The Water Distribution System Operator
Storage Facilities
Distribution System Facilities
Water Quality Considerations in Distribution Systems
Distribution System Operation and Maintenance
Disinfection
Safety
Distribution System Administration

Small Water System Operation and Maintenance

The Small Water System Operator
Water Sources and Treatment
Wells
Small Water Treatment Plants
Disinfection
Safety
Laboratory Procedures
Setting Water Rates for Small Water Utilities

Manage for Success

Supervising
Communicating
Human Relations
Planning and Organizing
Training and Teaching Skills
Problem-Solving Skills (Looking for Opportunities)
Decision Making
Technical Issues and Regulatory Compliance
Financial Management
Computers in Managing a Utility
Emergency Planning
Health and Safety Programs
Community Relations
Personal and Professional Skills

PREFACE
VOLUME II

Volume II is a continuation of Volume I. In Volume I, the emphasis was on the knowledge and skills needed by operators of conventional surface water treatment plants. Volume II stresses information needed by those operators, but also includes information on specialized water treatment processes for iron and manganese control, fluoridation, softening, arsenic removal, trihalomethanes, membrane filtration, demineralization, and the handling and disposal of process wastes. Topics of importance to the operators of all water treatment plants include maintenance, instrumentation, safety, advanced laboratory procedures, water quality regulations, administration, and how to solve water treatment plant arithmetic problems. The material in this manual will be easier to understand if you know and understand the meanings of words used by water treatment plant operators. Words specific to each chapter will be defined at the beginning of the chapter as well as in footnotes on the pages where the words are used in text. All of the words you will need to know are defined in the Appendix at the end of this manual.

You may wish to concentrate your studies on those chapters that apply to your water treatment plant. Upon successful completion of both volumes, you will have gained a broad and comprehensive knowledge of the water treatment field.

The Project Director is indebted to the many operators and other persons who contributed to this manual. Every effort was made to acknowledge material from the many excellent references in the water treatment field. Reviewers Leonard Ainsworth, Jack Rossum, and Joe Monscvitz deserve special recognition for their extremely thorough review and helpful suggestions. John Trax, Chet Pauls, and Ken Hay, Office of Drinking Water, US Environmental Protection Agency, and John Gaston, Bill MacPherson, Bert Ellsworth, Clarence Young, Ted Bakker, and Beverlie Vandre, Sanitary Engineering Branch, California Department of Health Services, all performed outstanding jobs as resource persons, consultants, and advisors. Larry Hannah served as Education Consultant. Illustrations were drawn by Martin Garrity. Charlene Arora helped type the field test and final manuscript for printing. Special thanks are well deserved by the Program Administrator, Gay Kornweibel, who typed, administered the field test, managed the office, administered the budget, and did everything else that had to be done to complete this project successfully.

                Kenneth D. Kerri
                Office of Water Programs
                California State University, Sacramento
                6000 J Street
                Sacramento, CA 95819-6025
                (916) 278-6142 – phone
                wateroffice@csus.edu – e-mail

2006

OBJECTIVES OF THIS MANUAL

Proper installation, inspection, operation, maintenance, repair, and management of water treatment plants have a significant impact on the operation and maintenance costs and effectiveness of the plants. The objective of this manual is to provide water treatment plant operators with the knowledge and skills required to operate and maintain water treatment plants effectively, thus eliminating or reducing the following problems.

1. Health hazards created by the production or output of unsafe water from the plant;

2. System failures that result from the lack of proper installation, inspection, preventive maintenance, surveillance, and repair programs designed to protect the public's investment in the plant;

3. Taste and odor complaints from consumers;

4. Turbid or colored waters that are unacceptable to consumers;

5. Corrosion damages to pipes, equipment, tanks, and structures at the water treatment plant and in the distribution system;

6. Complaints from the public or local officials due to the unreliability or failure of the water treatment plant to perform as designed; and

7. Fire damage caused by insufficient water at a time of need.

SCOPE OF THIS MANUAL

This manual on water treatment plant operation is divided into two volumes. Volume I stresses the knowledge and skills needed by an operator working in a conventional water treatment plant used for treating surface waters. Volume II emphasizes material needed by operators to control iron and manganese, soften hard waters, remove arsenic, and control trihalomethanes. Also contained in Volume II is information needed by all operators responsible for the administration and management of a water treatment plant, such as maintenance, instrumentation, safety, and laboratory procedures.

Volume I contains information on:

1. What water treatment plant operators do;
2. How to manage reservoirs and intake structures;
3. How to operate and maintain coagulation, flocculation, sedimentation, and filtration water treatment processes;
4. Disinfection of water;
5. Procedures for controlling corrosion;
6. Techniques for identifying the causes of taste and odor problems and suggestions for correcting such problems;
7. Procedures for operating, maintaining, and administering a water treatment plant; and
8. Basic laboratory procedures.

Volume II contains information on:

1. How to control iron and manganese;
2. Procedures for fluoridating water;
3. Techniques for softening water;
4. How to control trihalomethanes;
5. How to remove arsenic;
6. Techniques for treating suspended and dissolved solids in water;
7. Handling and disposal of process wastes;
8. Procedures for maintaining processes, equipment, and facilities;
9. How to maintain and troubleshoot instrumentation;
10. Techniques for recognizing hazards and developing safe procedures and safety programs;
11. Advanced laboratory procedures for analyzing samples of water;
12. Water quality regulations; and
13. Administrative considerations for supervisors and managers.

Material in this manual furnishes you with information concerning situations encountered by most water treatment plant operators in most areas. These materials provide you with an understanding of the basic operational and maintenance concepts for water treatment plants and with an ability to analyze and solve problems when they occur. Operation and maintenance programs for water treatment plants will vary with the age of the plant, the extent and effectiveness of previous programs, and local conditions. You will have to adapt the information and procedures in this manual to your particular situation.

Technology is advancing very rapidly in the field of operation and maintenance of water treatment plants. To keep pace with scientific advances, the material in this program must be periodically revised and updated. This means that you, the water treatment plant operator, must be aware of new advances and recognize the need for continuous personal training reaching beyond this program. Training opportunities exist in your daily work experience, from your associates, and from attending meetings, workshops, conferences, and classes.

USES OF THIS MANUAL

Originally, this manual was developed to serve as a home-study course for operators in remote areas or persons unable to attend formal classes either due to shift work, personal reasons, or the unavailability of suitable classes. This home-study training program uses the concepts of self-paced instruction where you are your own instructor and work at your own speed. In order to certify that a person has successfully completed this program, objective tests and special answer sheets for each chapter are provided when a person enrolls in this course.

Once operators started using this manual for home study, they realized that it could serve effectively as a textbook in the classroom. Many colleges and universities have used the manual as a text in formal classes often taught by operators. In areas where colleges are not available or are unable to offer classes in the operation of water treatment plants, operators and utility agencies can join together to offer their own courses using the manual.

Occasionally, a utility agency has enrolled from three to over 300 of its operators in this training program. A manual is purchased for each operator. A senior operator or a group of operators are designated as instructors. These operators help answer questions when the persons in the training program have questions or need assistance. The instructors grade the objective tests, record scores, and notify California State University, Sacramento, of the scores when a person successfully completes this program. This approach eliminates any waiting while papers are being graded and returned by the university.

This manual was prepared to help operators run their water treatment plants. Please feel free to use it in the manner that best fits your training needs and the needs of other operators. We will be happy to work with you to assist you in developing your training program. Please feel free to contact:

Project Director
Office of Water Programs
California State University, Sacramento
6000 J Street
Sacramento, CA 95819-6025
(916) 278-6142 – phone
(916) 278-5959 – FAX
wateroffice@csus.edu – e-mail

INSTRUCTIONS TO PARTICIPANTS IN HOME-STUDY COURSE

Procedures for reading the lessons and answering the questions are contained in this section.

To progress steadily through this program, you should establish a regular study schedule. For example, many operators set aside two hours during two evenings a week for study.

The study material is contained in 12 chapters. Some chapters are longer and more difficult than others. For this reason, many of the chapters are divided into two or more lessons. The time required to complete a lesson will depend on your background and experience. Some people might require an hour to complete a lesson and some might require three hours; but that is perfectly all right. The important thing is that you understand the material in the lesson.

Each lesson is arranged for you to read a short section, write the answers to the questions at the end of the section, check your answers against suggested answers; and then *YOU* decide if you understand the material sufficiently to continue or whether you should read the section again. You will find that this procedure is slower than reading a typical textbook, but you will remember much more when you have finished the lesson.

Some discussion and review questions are provided following each lesson in the chapters. These questions review the important points you have covered in the lesson. Write the answers to the discussion and review questions in your notebook.

In the appendix at the end of this manual, you will find some comprehensive review questions and suggested answers. These questions and answers are provided as a way for you to review how well you remember the material. You may wish to review the entire manual before you attempt to answer the questions. Some of the questions are essay-type questions, which are used by some states for higher-level certification examinations. After you have answered all the questions, check your answers with those provided and determine the areas in which you might need additional review before your next certification or civil service examination. Please do not send your answers to California State University, Sacramento.

You are your own teacher in this program. You could merely look up the suggested answers at the end of the chapters or comprehensive review questions or copy them from someone else, but you would not understand the material. Consequently, you would not be able to apply the material to the operation of your plant or recall it during an examination for certification or a civil service position.

You will get out of this program what you put into it.

SUMMARY OF PROCEDURE

OPERATOR (YOU)

1. Read what you are expected to learn in each chapter; the major topics are listed at the beginning of the chapter.
2. Read sections in the lesson.
3. Write your answers to questions at the end of each section in your notebook. You should write the answers to the questions just as you would if these were questions on a test.
4. Check your answers with the suggested answers.
5. Decide whether to reread the section or to continue with the next section.
6. Write your answers to the discussion and review questions at the end of each lesson in your notebook.

ORDER OF WORKING LESSONS

To complete this program, you will have to work all of the lessons. You may proceed in numerical order, or you may wish to work some lessons sooner.

Safety is a very important topic. Everyone working in a water treatment plant must always be safety conscious. Operators daily encounter situations and equipment that can cause a serious disabling injury or illness if the operator is not aware of the potential danger and does not exercise adequate precautions. For these reasons you may decide to work on the chapter on Safety early in your studies. In each chapter, safe procedures are always stressed.

TECHNICAL CONSULTANTS

John Brady
Gerald Davidson
Larry Hannah

Jim Sequeira
Susuma Kawamura
Mike Young

NATIONAL ENVIRONMENTAL TRAINING ASSOCIATION REVIEWERS

George Kinias, Project Coordinator

E. E. "Skeet" Arasmith
Terry Engelhardt
Dempsey Hall
Jerry Higgins

Andrew Holtan
Deborah Horton
Kirk Laflin
Rich Metcalf

William Redman
Kenneth Walimaa
Anthony Zigment

PROJECT REVIEWERS

Leonard Ainsworth
Ted Bakker
Jo Boyd
Dean Chausee
Walter Cockrell
Fred Fahlen
David Fitch
Richard Haberman
Lee Harry
Jerry Hayes
Ed Henley
Charles Jeffs

Chet Latif
Frank Lewis
Perry Libby
D. Mackay
William Maguire
Nancy McTigue
Joe Monscvitz
Angela Moore
Harold Mowry
Theron Palmer
Eugene Parham
Catherine Perman

David Rexing
Jack Rossum
William Ruff
Gerald Samuel
Carl Schwing
David Sorenson
Russell Sutphen
Robert Wentzel
James Wright
Mike Yee
Clarence Young

WATER TREATMENT PLANT OPERATION
COURSE OUTLINE

VOLUME II, FIFTH EDITION

Chapter	Topic	Page
12	IRON AND MANGANESE CONTROL Jack Rossum Special Section by Gerald Davidson	1
13	FLUORIDATION Harry Tracy	27
14	SOFTENING Don Gibson Marty Reynolds	69
15	SPECIALIZED TREATMENT PROCESSES Mike McGuire Chet Anderson	121
16	MEMBRANE TREATMENT PROCESSES (MEMBRANE FILTRATION AND DEMINERALIZATION) Dave Argo Ken Kerri Revised by Warren Casey	155
17	HANDLING AND DISPOSAL OF PROCESS WASTES George Uyeno	227
18	MAINTENANCE Parker Robinson	255
19	INSTRUMENTATION AND CONTROL SYSTEMS Leonard Ainsworth Revised by William H. Hendrix	385
20	SAFETY Joe Monscvitz	451
21	ADVANCED LABORATORY PROCEDURES Jim Sequeira	505
22	DRINKING WATER REGULATIONS Tim Gannon Revised by Jim Sequeira and Ken Kerri	549
23	ADMINISTRATION Lorene Lindsay Portions by Tim Gannon and Jim Sequeira	609

Appendix COMPREHENSIVE REVIEW QUESTIONS AND SUGGESTED ANSWERS 703
HOW TO SOLVE WATER TREATMENT PLANT ARITHMETIC PROBLEMS 729
ABBREVIATIONS 770
WATER WORDS 771
SUBJECT INDEX 847

CHAPTER 12

IRON AND MANGANESE CONTROL

by

Jack Rossum

With a Special Section by

Gerald Davidson

TABLE OF CONTENTS
Chapter 12. IRON AND MANGANESE CONTROL

			Page
OBJECTIVES			4
WORDS			5
12.0	NEED TO CONTROL IRON AND MANGANESE		7
12.1	MEASUREMENT OF IRON AND MANGANESE		7
	12.10	Occurrence of Iron and Manganese	7
	12.11	Collection of Iron and Manganese Samples	8
	12.12	Analysis for Iron and Manganese	8
12.2	REMEDIAL ACTION		8
	12.20	Alternate Source	8
	12.21	Phosphate Treatment	10
	12.22	Removal by Ion Exchange	12
	12.23	Oxidation by Aeration	12
	12.24	Oxidation with Chlorine	14
	12.25	Oxidation with Permanganate	14
	12.26	Operation of Filters	15
	12.27	Proprietary Processes by Bill Hoyer	15
	12.28	Monitoring of Treated Water	16
	12.29	Summary	16
12.3	OPERATION OF AN IRON AND MANGANESE REMOVAL PLANT by Gerald Davidson		18
	12.30	Description of Equipment and Process	18
	12.31	Regeneration of Manganese Greensand	20
	12.32	Troubleshooting	20
12.4	MAINTENANCE OF A CHEMICAL FEEDER		20
12.5	TROUBLESHOOTING RED WATER PROBLEMS		22
12.6	ARITHMETIC ASSIGNMENT		23
12.7	ADDITIONAL READING		23
12.8	ACKNOWLEDGMENT		23
DISCUSSION AND REVIEW QUESTIONS			23
SUGGESTED ANSWERS			24

OBJECTIVES
Chapter 12. IRON AND MANGANESE CONTROL

Following completion of Chapter 12, you should be able to:

1. Identify and describe the various processes used to control iron and manganese.

2. Collect samples for analysis of iron and manganese.

3. Safely operate and maintain the following iron and manganese control processes:

    a. Phosphate treatment
    b. Ion exchange
    c. Oxidation by aeration
    d. Oxidation with chlorine
    e. Oxidation with permanganate
    f. Greensand
    g. Proprietary processes

4. Troubleshoot red water problems.

WORDS
Chapter 12. IRON AND MANGANESE CONTROL

ACIDIFICATION (uh-SID-uh-fuh-KAY-shun)   ACIDIFICATION

The addition of an acid (usually nitric or sulfuric) to a sample to lower the pH below 2.0. The purpose of acidification is to fix a sample so it will not change until it is analyzed.

AQUIFER (ACK-wi-fer)   AQUIFER

A natural, underground layer of porous, water-bearing materials (sand, gravel) usually capable of yielding a large amount or supply of water.

BACKFLOW   BACKFLOW

A reverse flow condition, created by a difference in water pressures, that causes water to flow back into the distribution pipes of a potable water supply from any source or sources other than an intended source. Also see BACKSIPHONAGE.

BACKSIPHONAGE   BACKSIPHONAGE

A form of backflow caused by a negative or below atmospheric pressure within a water system. Also see BACKFLOW.

BENCH-SCALE ANALYSIS (TEST)   BENCH-SCALE ANALYSIS (TEST)

A method of studying different ways or chemical doses for treating water or wastewater and solids on a small scale in a laboratory. Also see JAR TEST.

BREAKPOINT CHLORINATION   BREAKPOINT CHLORINATION

Addition of chlorine to water or wastewater until the chlorine demand has been satisfied. At this point, further additions of chlorine will result in a free chlorine residual that is directly proportional to the amount of chlorine added beyond the breakpoint.

CHELATION (key-LAY-shun)   CHELATION

A chemical complexing (forming or joining together) of metallic cations (such as copper) with certain organic compounds, such as EDTA (ethylene diamine tetracetic acid). Chelation is used to prevent the precipitation of metals (copper). Also see SEQUESTRATION.

COLLOIDS (KALL-loids)   COLLOIDS

Very small, finely divided solids (particles that do not dissolve) that remain dispersed in a liquid for a long time due to their small size and electrical charge. When most of the particles in water have a negative electrical charge, they tend to repel each other. This repulsion prevents the particles from clumping together, becoming heavier, and settling out.

DIVALENT (dye-VAY-lent)   DIVALENT

Having a valence of two, such as the ferrous ion, Fe^{2+}. Also called bivalent.

GREENSAND   GREENSAND

A mineral (glauconite) material that looks like ordinary filter sand except that it is green in color. Greensand is a natural ion exchange material that is capable of softening water. Greensand that has been treated with potassium permanganate ($KMnO_4$) is called manganese greensand; this product is used to remove iron, manganese, and hydrogen sulfide from groundwaters.

INSOLUBLE (in-SAWL-yoo-bull)   INSOLUBLE

Something that cannot be dissolved.

ION EXCHANGE

A water or wastewater treatment process involving the reversible interchange (switching) of ions between the water being treated and the solid resin contained within an ion exchange unit. Undesirable ions are exchanged with acceptable ions on the resin or recoverable ions in the water being treated are exchanged with other acceptable ions on the resin.

ION EXCHANGE RESINS

Insoluble polymers, used in water or wastewater treatment, that are capable of exchanging (switching or giving) acceptable cations or anions to the water being treated for less desirable ions or for ions to be recovered.

RESINS

See ION EXCHANGE RESINS.

SEQUESTRATION (SEE-kwes-TRAY-shun)

A chemical complexing (forming or joining together) of metallic cations (such as iron) with certain inorganic compounds, such as phosphate. Sequestration prevents the precipitation of the metals (iron). Also see CHELATION.

ZEOLITE

A type of ion exchange material used to soften water. Natural zeolites are siliceous compounds (made of silica) that remove calcium and magnesium from hard water and replace them with sodium. Synthetic or organic zeolites are ion exchange materials that remove calcium or magnesium and replace them with either sodium or hydrogen. Manganese zeolites are used to remove iron and manganese from water.

CHAPTER 12. IRON AND MANGANESE CONTROL

12.0 NEED TO CONTROL IRON AND MANGANESE

Like the cities of Minneapolis and St. Paul, iron and manganese are referred to as a pair. They are, in fact, two distinct elements and are often found in water separately. Neither of them has any direct adverse health effects. Indeed, both are essential to the growth of many plants and animals, including humans; however, the iron and manganese found in drinking water have no nutrient value for humans. Even if they were available in beneficial amounts, the presence of iron and manganese in drinking water would still be objectionable.

Clothes laundered in water containing iron and manganese above certain levels come out stained. When bleach is added to remove the stains, they are only intensified and become fixed so that no amount of further washing with iron-free water will remove the stains. They can be removed by treatment with oxalic acid, but this is rather hard on fabrics, or by the use of commercial rust removers. Excessive amounts of iron and manganese are also objectionable because they stain plumbing fixtures, bathtubs, and sinks.

Perhaps the most troublesome consequence of iron and manganese in the water is that they promote the growth of a group of microorganisms known as iron bacteria. These organisms obtain energy for their growth from the chemical reaction that spontaneously occurs between iron and manganese and dissolved oxygen. These bacteria form thick slimes on the walls of the distribution system mains. Such slimes are rust colored from iron and black from manganese. Variations in flow cause these slimes to come loose and create dirty water (a big source of consumer complaints). Furthermore, these slimes will cause foul tastes and odors in the water.

The growth of iron bacteria is controlled by chlorination. However, when water containing iron is chlorinated, the iron is converted into rust particles, and manganese is converted into a jet-black compound called manganese dioxide. These materials form a loosely adherent coating on the pipe walls. Pieces of this coating will break loose from the pipe walls when there are changes or reversals of flow in the distribution system.

Iron and manganese in water can be easily detected by observing the color of the inside walls of filters and the filter media. If the raw water is prechlorinated, there will be black stains on the walls below the water level and a black coating over the top portion of the sand filter bed. This black color will usually indicate a high level of manganese in the raw water while a brownish-black stain indicates the presence of both iron and manganese.

The generally acceptable limit for iron in drinking water is 0.3 mg/L and that for manganese is 0.05 mg/L. However, if the water contains more than 0.02 mg/L of manganese, the operator should initiate an effective flushing program to avoid complaints. By regularly flushing the water mains, the buildup of black manganese dioxide can be prevented.

QUESTIONS

Write your answers in a notebook and then compare your answers with those on page 24.

12.0A   What problems are caused by iron and manganese in drinking water?

12.0B   How can the growth of iron bacteria be controlled?

12.0C   What are the generally acceptable limits for iron and manganese in drinking water?

12.1 MEASUREMENT OF IRON AND MANGANESE

12.10 Occurrence of Iron and Manganese

Because both iron and manganese react with dissolved oxygen to form INSOLUBLE COMPOUNDS,[1] they are not found in high concentrations in waters containing dissolved oxygen except as COLLOIDAL SUSPENSIONS[2] of the oxides. Accordingly, surface waters are generally free from both iron and manganese. One exception to this rule is that manganese up to one mg/L or higher may be found in shallow reservoirs and may come and go several times a year.

Iron or manganese is most frequently found in water systems supplied by wells and springs. Horizontal wells under rivers are notoriously prone to produce water containing iron. Bacteria will reduce iron oxides in soil to the soluble, DIVALENT[3] form of iron (Fe^{2+}), which will produce groundwater with a high iron content.

Iron bacteria can make use of the ferrous ion (Fe^{2+}). These bacteria will oxidize the iron and use the energy for reducing

---

[1] *Insoluble* (in-SAWL-yoo-bull).   Something that cannot be dissolved.

[2] *Colloids* (KALL-loids).   Very small, finely divided solids (particles that do not dissolve) that remain dispersed in a liquid for a long time due to their small size and electrical charge. When most of the particles in water have a negative electrical charge, they tend to repel each other. This repulsion prevents the particles from clumping together, becoming heavier, and settling out.

[3] *Divalent* (dye-VAY-lent).   Having a valence of two, such as the ferrous ion, Fe^{2+}. Also called bivalent.

carbon dioxide to organic forms (slimes). The manganous ion (Mn^{2+}) is used in a similar fashion by certain bacteria. Very small concentrations of iron and manganese in water can cause problems because bacteria obtain the nutrients (iron and manganese) from water in order to grow, even when the concentrations are very low.

Iron bacteria are found nearly everywhere. They are frequently found in iron water pipes and everywhere else that a combination of dissolved oxygen and dissolved iron is usually or frequently present. Only one cell of iron bacteria is needed to start an infestation of iron bacteria in a well or a distribution system. Unfortunately, it is almost impossible to drill a well and maintain sterile conditions to prevent the introduction of iron bacteria.

12.11 Collection of Iron and Manganese Samples

The best way to determine if there is an iron and manganese problem in a water supply is to look at the plumbing fixtures in a couple of houses. If the fixtures are stained, then there is a problem. Determination of the concentrations of iron and manganese in water is useful when evaluating well waters for use and treated waters for effectiveness of treatment processes.

The results of tests for iron and manganese are wrong more often than they are right. This is because samples for these substances are difficult to collect. Both iron and manganese form loosely adherent (not firmly attached) scales on pipe walls, including the sample lines. When the sample tap is opened, particles of scale may be dislodged and enter the sample bottle. If many of these particles enter the sample bottle, the error can become very large. Furthermore, unless the sample is acidified (enough nitric acid added to drop the pH to less than 2), both iron and manganese tend to form an adherent scale on the walls of the sample bottle in the few days that sometimes elapse before the analysis is started. When the sample is poured from the bottle for testing, most of the iron and manganese will then remain inside the sample bottle.

To avoid this situation, samples should be taken from a plastic sample line located as close to the well or other source as possible. Open the sampling tap slowly so that the flow rate is suitable for filling the sample bottle. Allow the sample water to flow for at least one minute for each 10 feet (3 m) of sample line before the sample is collected.

Samples for iron and manganese should be tested within 48 hours unless they have been acidified. If the sample contains any clay or if any particles of rust are picked up from a steel pipe or fitting, an acidified sample will dissolve the iron in these substances and the results will be too high. If clay or rust particles are observed in a sample, do not acidify and ask the laboratory to analyze the sample immediately. Furthermore, many laboratories fail to be sure that iron and manganese are in the divalent form (Fe^{2+} or Mn^{2+}) by adding enough nitric acid prior to the tests to lower the pH to less than 2, so laboratory errors may be even greater than sampling errors.

12.12 Analysis for Iron and Manganese

The preferred method of testing for iron and manganese is atomic absorption, but for the small plant the equipment is too expensive. With careful attention to laboratory procedures, colorimetric methods (comparing colors of unknowns with known standards) can provide sufficient accuracy in most instances. Colorimetric methods use either a spectrophotometer, a filter photometer, or the less satisfactory set of matched Nessler tubes with standards. Good results have been obtained by the use of properly calibrated colorimeters (Figure 12.1). For detailed procedures on how to use a spectrophotometer to measure iron and manganese, see Chapter 21, "Advanced Laboratory Procedures."

QUESTIONS

Write your answers in a notebook and then compare your answers with those on page 24.

12.1A How do iron and manganese form insoluble compounds?

12.1B Why must iron and manganese samples be acidified when they are collected?

12.1C Where should a sample for iron and manganese testing be collected?

12.2 REMEDIAL ACTION

Several methods are available to control iron and manganese in water. This section discusses how to operate the most common treatment processes.

12.20 Alternate Source

The construction of a plant to remove iron and manganese will cost as much as or more than a new well so it pays to investigate the possibility of obtaining an alternate supply of water that is free from iron and manganese. This investigation should include samples from nearby private wells, discussions with well drillers who have been active in the locality, and discussions with engineers in the state agency responsible for the regulation of well drilling.

If the water produced by the well contains dissolved oxygen along with iron and manganese, this is an indication that water is being drawn from more than one *AQUIFER*.[4] One or more of the aquifers must be producing water containing dissolved oxygen but is free of iron and manganese since oxygen reacts with both elements to form insoluble compounds. Furthermore, it is highly probable that the iron- or manganese-bearing water is from deeper aquifers. It may be possible to cure the problem simply by sealing off these deeper aquifers.

---

[4] *Aquifer* (ACK-wi-fer). A natural, underground layer of porous, water-bearing materials (sand, gravel) usually capable of yielding a large amount or supply of water.

Iron and Manganese 9

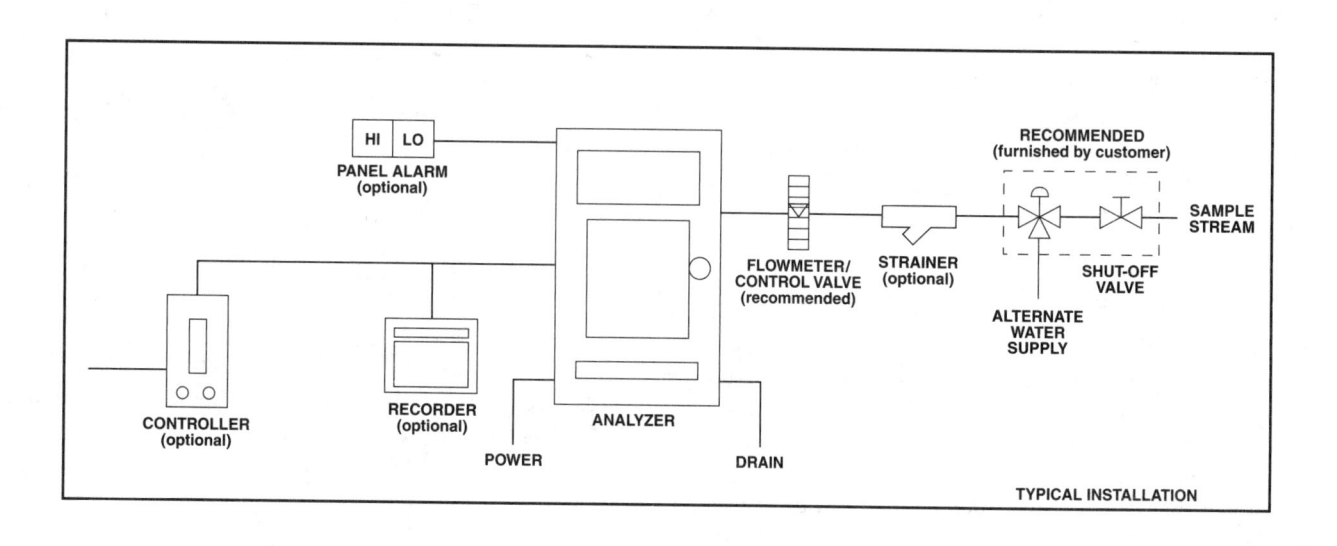

*Fig 12.1 Typical continuous on-line pump colorimeter analyzer
for iron, manganese, or permanganate*
(Permission of HACH)

12.21 Phosphate Treatment

If the water contains manganese up to 0.3 mg/L and less than 0.1 mg/L of iron, an inexpensive and reasonably effective control can be achieved by feeding the water with one of the three polyphosphates mentioned below. Chlorine usually must be fed along with the polyphosphate to prevent the growth of iron bacteria. The effect of the polyphosphate is to delay the precipitation of oxidized manganese for a few days so that the scale that builds up on the pipe walls is greatly reduced.

The chlorine dose for phosphate treatment should be sufficient to produce a free chlorine residual of approximately 0.25 mg/L after a five-minute contact time (a higher chlorine dose may be required with some water to maintain a free chlorine residual of at least 0.2 mg/L throughout the distribution system).

Any of the three polyphosphates (pyrophosphate, tripolyphosphate, and metaphosphate) can be used, but sodium metaphosphate is effective in lower concentrations than the others. The proper phosphate dose is determined by laboratory BENCH-SCALE TESTS[5] in the following manner.

1. Treat a series of samples with a standard chlorine solution to determine the chlorine dose required to produce the desired chlorine residual.

2. Prepare a standard polyphosphate solution by dissolving 1.0 gram of polyphosphate in a liter of distilled water.

3. Treat another series of samples with varying amounts of polyphosphate solution. One mL of the standard polyphosphate solution (0.1% solution) in a liter sample is equivalent to 8.34 pounds of polyphosphate per million gallons (see Examples 2 and 3 on page 17). Stir to ensure that the polyphosphate has been well mixed, and continue stirring while adding the previously determined chlorine dose so as to minimize the creation of high concentrations of chlorine.

4. Observe the samples daily against a white background, noting the amount of discoloration. The proper polyphosphate dose is the lowest dose that delays noticeable discoloration for a period of four days.

Samples for the above bench test should be as fresh as possible and should be kept away from direct sunlight to avoid heating.

Polyphosphate treatment to control iron and manganese is usually most effective when the polyphosphate is added upstream from the chlorine, but satisfactory results may be obtained by feeding them together. The chlorine should never be fed ahead of the polyphosphate because the chlorine will oxidize the iron and manganese (cause insoluble precipitates to form too soon).

If you are able to install a one-half inch (12 mm) polyethylene hose in the well so that it discharges a few inches below the suction screen, you can construct a very satisfactory semiautomatic feed system. Use a gased chlorinator whose water supply is obtained from the well discharge downstream from the check valve (Figure 12.2). In this way, the chlorinator operates only when the pump is running. The chlorine solution is fed down the polyethylene tube. Polyphosphate is fed down the same tube by means of a plastic tee. The phosphate is fed by means of an electrically operated solution feeder connected so as to run when the well pump runs.

The chlorine solution flowing through the polyethylene tube is extremely corrosive. If the tube does not discharge into flowing water, the corrosive effect of the solution on a metal surface can be disastrous. Wells have been destroyed by corrosion from this cause. The following simple test should be made at least once every three months.

EXAMPLE 1

1. Calculate the time required for water to flow from the pump suction to the pump discharge.

   a. Record the following information:

      (1) Length of pump column from suction to discharge in feet, 324 ft.

      (2) Diameter of pump column in inches, 8 in.

      (3) Discharge rate from pump in gallons per minute, 423 GPM.

   b. Calculate the volume of the pump column in cubic inches.

   $$\text{Volume, cu in} = (0.785)(\text{Diameter, in})^2(\text{Length, ft})(12 \text{ in/ft})$$

   $$= (0.785)(8 \text{ in})^2(324 \text{ ft})(12 \text{ in/ft})$$

   $$= 195{,}333 \text{ cubic inches}$$

   c. Convert the pump column volume from cubic inches to gallons.

   $$\text{Volume, gal} = \frac{\text{Volume, cu in}}{231 \text{ cu in/gal}}$$

   $$= \frac{195{,}333 \text{ cu in}}{231 \text{ cu in/gal}}$$

   $$= 846 \text{ gallons}$$

   d. Determine the time required for the water to flow from the pump suction to the pump discharge.

   $$\text{Time, min} = \frac{\text{Volume, gallons}}{\text{Pump Discharge, GPM}}$$

   $$= \frac{846 \text{ gallons}}{423 \text{ gal/min}}$$

   $$= 2.0 \text{ minutes}$$

---

[5] *Bench-Scale Analysis (Test).* A method of studying different ways or chemical doses for treating water or wastewater and solids on a small scale in a laboratory. Also see JAR TEST.

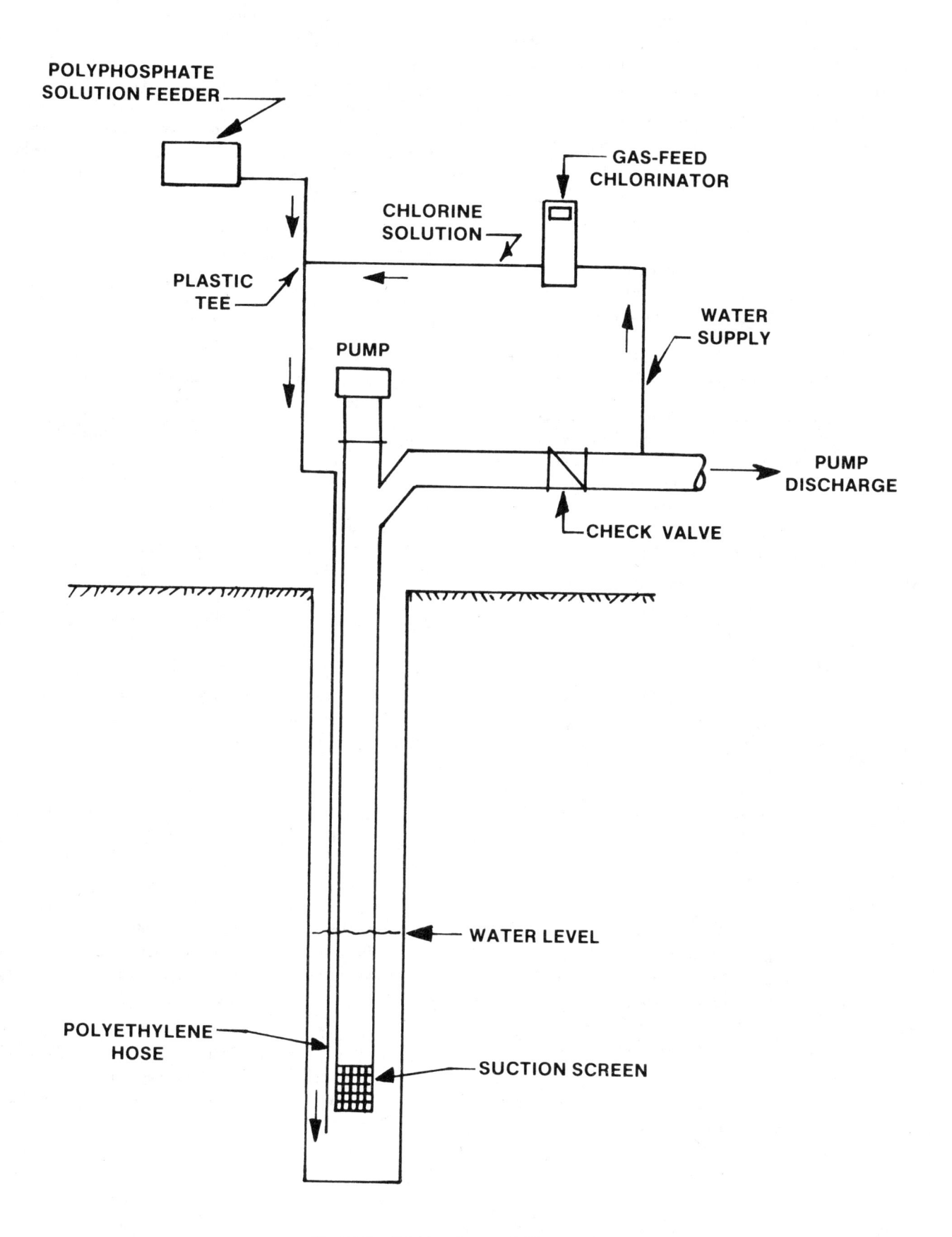

Fig. 12.2 Polyphosphate and chlorine system

2. Turn off the water supply to the chlorinator. Since some of the chlorine in the feed line will drain into the well, it may take several minutes for the chlorine residual to disappear. Check to be sure there is no chlorine residual in the water.

3. Turn the chlorinator back on, noting the exact time. Take samples for chlorine residual every 15 seconds. If the chlorine residual has reached its proper value within 30 seconds of the calculated time (two minutes in this example), you can be sure that the polyethylene tube is properly positioned below the suction screen.

Solutions of polyphosphate containing more than one-half pound per gallon (60 gm/L) may be very viscous (thick like molasses), depending upon which of the polyphosphates is used. Do not use a solution much over 48 hours old because the polyphosphates react slowly with water to form orthophosphates, which are much less effective in preventing manganese deposits.

There are some reports in the literature of the successful use of tetrasodium pyrophosphate on iron-bearing waters, but many attempts to control iron have failed. The available information is not clear as to whether the process works only under special conditions or whether the reports of success are in error.

QUESTIONS

Write your answers in a notebook and then compare your answers with those on page 24.

12.2A How could you find out if nearby wells produce water containing iron and manganese?

12.2B If a well produces water containing dissolved oxygen as well as iron and manganese, how would you attempt to solve the iron and manganese problem?

12.2C What are bench-scale tests?

12.2D Why should polyphosphate solutions over 48 hours old not be used?

12.22 Removal by Ion Exchange[6]

Iron and manganese ion exchange units are similar to a downflow pressure filter. The water to be treated enters the unit through an inlet distributor located in the top. The water is forced (usually by a pump) down through the ion exchange resin into an underdrain structure. From the underdrain structure, the treated water flows out of the unit to the next treatment process, to storage, or into the distribution system.

The actual location of the *ION EXCHANGE RESINS*[7] with respect to other water treatment processes will depend on the raw water quality and the design engineer. If the water to be treated contains no oxygen, both iron and manganese may be removed by ion exchange using the same resins that are used for water softening. If the water being treated contains any dissolved oxygen, the resin becomes fouled with iron rust or insoluble manganese dioxide. The resin can be cleaned but this is expensive and the capacity is reduced. Well water may contain no oxygen in normal operating conditions except immediately after the well is first turned on. If this is the case, provisions should be made to run the well to waste until the oxygen is no longer present.

In one eastern city, an ion exchange plant had operated for seven years, reducing iron from 52 mg/L to 0.1 mg/L and manganese from 1.3 mg/L to zero. When a pump was repaired, a gasket on the suction side of the pump was improperly installed, allowing air to enter the raw water. Within three months the resin was fouled by iron oxide.

The main advantage of ion exchange for iron and manganese removal is that the plant requires little attention. The disadvantages are the danger of fouling the ion exchange resin with oxides and high initial cost.

To operate an ion exchange unit, try to operate as close as possible to design flows. Monitor the treated water for iron and manganese on a daily basis. When iron and manganese start to appear in the treated water, the unit must be regenerated. Regenerate with a brine solution that is treated with 0.01 pound of sodium bisulfite per gallon (1.2 gm/L) of brine to remove oxygen present. After regeneration is complete, dispose of the brine in an approved manner.

See Chapter 14, "Softening," for procedures on how to calculate the frequency for regenerating the unit. Details are given in Chapter 17, "Handling and Disposal of Process Wastes," on how to properly dispose of brine solutions.

12.23 Oxidation by Aeration

Iron can be oxidized by aerating the water to form insoluble ferric hydroxide. As shown in Figure 12.3, this reaction is accelerated by an increase in pH. The rates indicated in Figure 12.3 were determined at 25°C (77°F) under laboratory conditions. If the water contains any organic substances, the rates will be significantly lower; reduced temperatures will also lower the rates. The oxidation of manganese by aeration is so slow that this process is not used on waters with high manganese concentrations.

Since pH is increased by the removal of carbon dioxide, it is important that the aeration (which removes carbon dioxide) be as efficient as possible. Lime is sometimes added to the water to increase the pH as well as remove carbon dioxide. The higher the pH, the shorter the time required for aeration, as shown in Figure 12.3.

---

[6] *Ion Exchange.* A water or wastewater treatment process involving the reversible interchange (switching) of ions between the water being treated and the solid resin contained within an ion exchange unit. Undesirable ions are exchanged with acceptable ions on the resin or recoverable ions in the water being treated are exchanged with other acceptable ions on the resin.

[7] *Ion Exchange Resins.* Insoluble polymers, used in water or wastewater treatment, that are capable of exchanging (switching or giving) acceptable cations or anions to the water being treated for less desirable ions or for ions to be recovered.

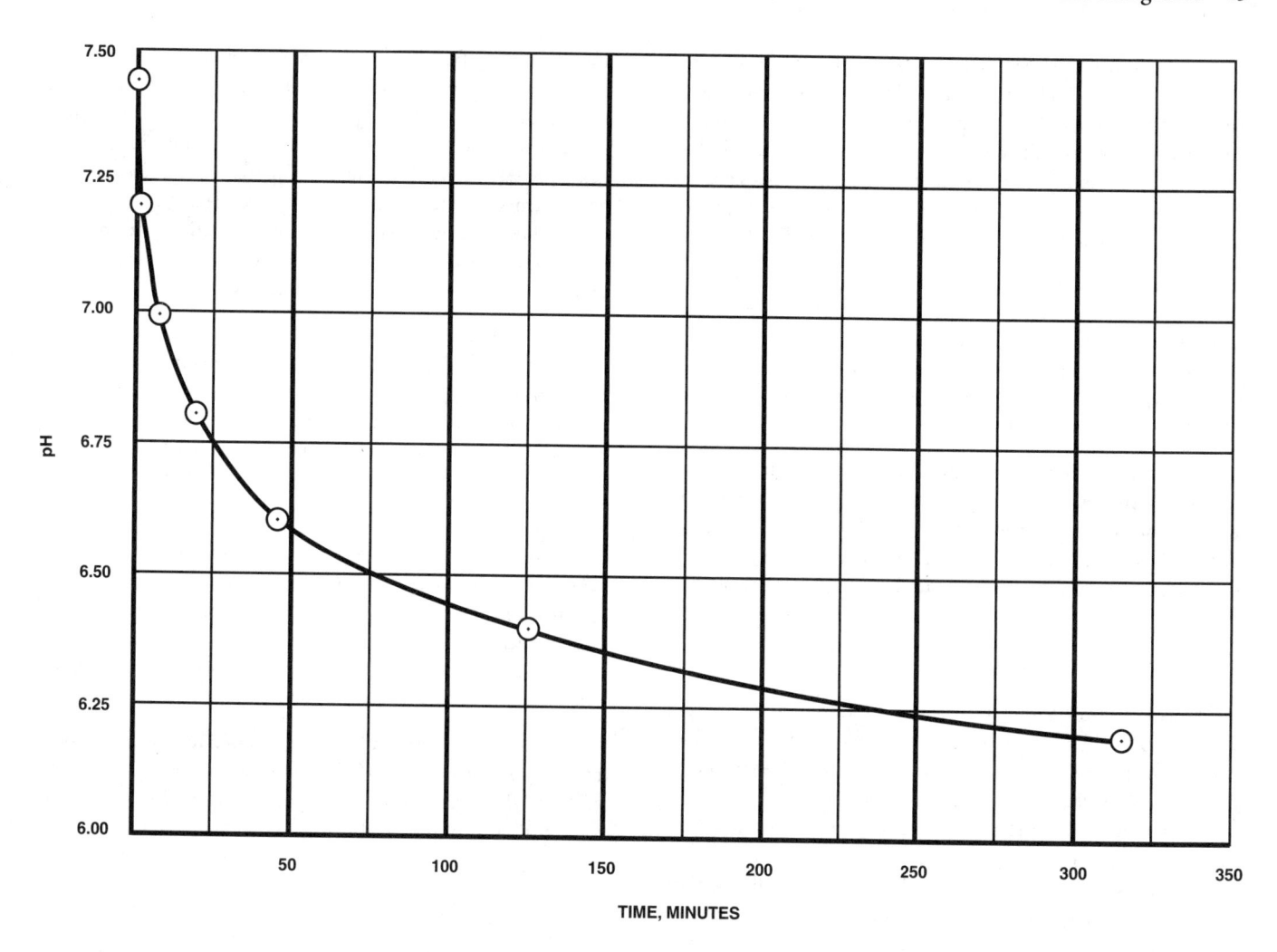

Fig. 12.3 pH versus time to oxidize 99 percent iron

Operation of the aeration process to remove iron and manganese requires careful control of the flow through the process. If the flow becomes too great, not enough time will be available for the reactions to occur. Flows are controlled by the use of variable speed pumps or the selection of the proper number or combinations of pumps. Carefully monitor the iron and manganese content of the treated water. If iron is detected, the flows may have to be reduced.

There are several methods of providing aeration. Either the water being treated is dispersed (scattered) into the air or air is bubbled through the water. Aeration may be achieved by the use of compressed air that passes through diffusers in the water. These diffusers produce many small bubbles that allow the transfer of oxygen in the air to dissolved oxygen in the water.

Other aeration techniques include forced draft, multiple trays, cascades, and sprays. These methods may cause slime growths to develop on surfaces or coatings on media. Slime growths and coatings on media should be controlled to prevent the development of tastes and odors in the product water and the sloughing off of the slimes. Chlorination may be used to control slime growths and coatings. Regularly inspect aeration equipment for the development of anything unusual.

A reaction basin (sometimes called a collection or detention basin) follows the aeration process. The purpose of the reaction basin is to allow time for the oxidation reactions to take place. The aeration process should produce sufficient dissolved oxygen for the iron to be oxidized to insoluble ferric hydroxide. A minimum recommended detention time is 20 minutes with desirable detention times ranging from 30 to 60 minutes (see Example 4, page 17). As shown in Figure 12.3, the pH of the water strongly influences the time for the reaction to take place. Sometimes chlorine is added before the reaction basin.

The reaction basin may be a cylindrically (circular) shaped basin similar to a clarifier. Often the basin is baffled to prevent short-circuiting and the deposition of solids. Since there are no provisions for sludge removal, the basin must be drained and cleaned regularly. If the basins are not cleaned, slugs of deposits or sludge or also mosquito or fly larvae (young of any insect) could reach the filters in the next process and cause them to plug.

Operators of reaction basins must always be on the alert for potential sources of contamination. These basins should have covers and protective lids to keep out rain, stormwater runoff, rodents, and insects. All vents must be properly screened. The outlet to the drain must not be connected directly to a sewer or stormwater drain. There must be an air gap or some other protective device to prevent contamination from *BACKFLOW*.[8]

After the ferric hydroxide is formed, it is removed by sedimentation or by filtration alone. If only filtration is used, water from the reaction basin is usually pumped to pressure filters for filtration. The water may also be pumped or flow by gravity to rapid sand filters. For details on how to operate and maintain filters, see Chapter 6, "Filtration."

The main advantage of this method is that no chemicals are required; however, lime may be added to increase the pH. The major disadvantage is that small changes in raw surface water quality may affect the pH and soluble organics level and slow the oxidation rates to a point where the capacity of the plant is reduced.

12.24 Oxidation with Chlorine

Chlorine will oxidize manganese to the insoluble manganese dioxide and will oxidize iron to insoluble ferric hydroxide, which can then be removed by filtration. The higher the chlorine residual, the faster this reaction occurs. Some very compact plants have been constructed by treating the water to a free chlorine residual of 5 to 10 mg/L, filtering, and dechlorinating to a residual suitable for domestic use. Do not use high doses of chlorine if the water contains a high level of organic color because excessive concentrations of total trihalomethanes (TTHMs) could develop. The water is dechlorinated by the use of reducing agents such as sulfur dioxide (SO_2), sodium bisulfite ($NaHSO_3$), and sodium sulfite (Na_2SO_3). Bisulfite is commonly used because it is cheaper and more stable than sulfite. When dechlorinating with reducing agents, be very careful not to overdose because inadequate disinfection could result (no chlorine residual remains) and if the dissolved oxygen level in the water is depleted, fish kills could occur in home aquariums. Frequently, a reaction basin (as described in Section 12.23, "Oxidation by Aeration") is installed between the chlorination processes to allow time for the reactions to occur.

12.25 Oxidation with Permanganate

Potassium permanganate oxidizes iron and manganese to insoluble oxides, and can be used to remove these elements in the same way chlorine is used. The dose of potassium permanganate must be exact. Bench-scale tests are required to determine the proper dosage. Too small a dose will not oxidize all the manganese in the water and too large a dose will allow permanganate to enter the system and may produce a pink color in the water. Actual observations of the water being treated will tell you if any adjustments of the chemical feeder are necessary. Most well waters have relatively consistent concentrations of iron and manganese. Therefore, once you set the chemical feeder, adjustments of the dosage usually are not necessary.

Experience from many water treatment plants has shown that a regular filter bed (a rapid sand filter or a dual media filter bed) can remove manganese as long as iron and manganese concentrations are both less than one milligram per liter. These plants use either chlorine or permanganate to oxidize the iron and manganese before the water being treated flows through the filter bed.

Potassium permanganate is often used with manganese *ZEOLITE*[9] or manganese *GREENSAND*.[10] Greensand is a granular material. After the greensand has been treated with potassium permanganate, it can oxidize both iron and manganese to their insoluble oxides. The greensand also acts as a filter and must be backwashed to remove the insoluble oxides.

Manganese greensand filters can be operated in three modes: continuous regeneration (CR), intermittent regeneration (IR), and catalytic regeneration. The method used will depend on the concentrations of iron and manganese in the water and the pH of the water.

The manganese greensand continuous regeneration process can be used for waters containing iron concentrations as high as 15 mg/L or more, but with such high concentrations, frequent backwashing will be necessary. Generally, waters having iron concentrations in the range of 0.5 to 3.0 mg/L can be treated with more acceptable run lengths of 18 to 36 hours before backwashing.

In the CR process, chlorine and potassium permanganate are typically added to the raw water ahead of the manganese greensand bed. Chlorine is added first to oxidize most of the iron and any sulfide. A slight excess of potassium permanganate is then added to oxidize the remaining iron and soluble manganese. This reaction produces insoluble oxides. When the raw water passes through the manganese greensand bed, two things occur: (1) the insoluble iron oxide particles are filtered out, and (2) any remaining permanganate is reduced to manganese oxides by the greensand. These manganese oxides attach to the grains of greensand, thereby continuously regenerating the manganese

---

[8] *Backflow.* A reverse flow condition, created by a difference in water pressures, that causes water to flow back into the distribution pipes of a potable water supply from any source or sources other than an intended source. Also see BACKSIPHONAGE.

[9] *Zeolite.* A type of ion exchange material used to soften water. Natural zeolites are siliceous compounds (made of silica) that remove calcium and magnesium from hard water and replace them with sodium. Synthetic or organic zeolites are ion exchange materials that remove calcium or magnesium and replace them with either sodium or hydrogen. Manganese zeolites are used to remove iron and manganese from water.

[10] *Greensand.* A mineral (glauconite) material that looks like ordinary filter sand except that it is green in color. Greensand is a natural ion exchange material that is capable of softening water. Greensand that has been treated with potassium permanganate ($KMnO_4$) is called manganese greensand; this product is used to remove iron, manganese, and hydrogen sulfide from groundwaters.

greensand. As the run progresses, the filter bed becomes clogged with insoluble oxides and the differential pressure increases. When head loss reaches a predetermined point or a certain number of gallons of water have been treated, the filter must be backwashed to remove the filtered particulates.

The intermittent regeneration (IR) process is suitable for raw water containing only manganese or mostly manganese with small amounts of iron. The raw water flows through a manganese greensand bed where oxidation of manganese occurs directly on the grains of greensand. Some iron will also be oxidized directly on the manganese greensand but if the iron concentrations are high, iron oxides will quickly coat or foul the media. To prevent fouling of the media, the iron is sometimes converted to its insoluble form before the water enters the greensand bed by adding chlorine ahead of the filter or aerating the water before it enters the greensand bed. After treating a certain number of gallons of water or when head loss reaches a predetermined point, the capacity of the greensand to oxidize manganese and iron is used up and the media must be backwashed and regenerated. Regeneration consists of the downflow passage of a dilute potassium permanganate solution through the bed using 1.5 ounces (43 milligrams) of $KMnO_4$ per cubic foot of media, followed by thorough rinsing of the media.

When the well water contains low concentrations of iron and manganese (less than 1.0 mg/L) and the pH is greater than 7.0, the catalytic regeneration mode of operation may be a suitable method for removing iron and manganese. In this process, sufficient chlorine is applied ahead of the filter to maintain a chlorine residual of 0.5 to 1.0 mg/L through a bed containing a specially refined grade of greensand. The chlorine keeps the greensand continuously regenerated. Manganese in the raw water can usually be oxidized directly by the manganese oxide coating on the greensand, provided that the pH is neutral and adequate detention time is provided. The advantages of catalytic oxidation are that the filter run lengths are longer than for other process modes, higher filtration rates can be used, the backwash waste suspended solids are low because the manganese is oxidized on the manganese greensand grains, and the chemical operating costs are minimal because only chlorine is required. For more information on the manganese greensand process, see Section 12.3, "Operation of an Iron and Manganese Removal Plant."

12.26 Operation of Filters

When iron and manganese are oxidized to insoluble forms by aeration, chlorination, or permanganate, the oxidation processes are usually followed by filters to remove the insoluble material. In addition to the procedures for operating and maintaining filters that were outlined in Chapter 6, "Filtration," the procedures discussed in this section apply to filters used to remove iron and manganese.

Iron tests should be made monthly on the water entering a filter to be sure the iron is in the ferric (Fe^{3+}) state. Collect a sample of the water and pass the water through a filter paper. Run an iron test on the water that has passed through the filter. If the iron is still in the soluble ferrous (Fe^{2+}) state, there will be iron in the water. If aeration is being used to oxidize the iron from the soluble ferrous to the insoluble ferric state and iron is still present in the soluble state in the water entering the filter, try adding chlorine or potassium permanganate. If chlorine or potassium permanganate is being used and soluble iron is in the water entering the filter, try increasing the chemical dose. If potassium permanganate is being used, the sand may be replaced by greensand to improve the efficiency of the process.

If oxidation is being accomplished by either aeration or chlorination, a free chlorine residual must be maintained in the effluent of the filter to prevent the insoluble ferric iron from returning to the soluble ferrous form and passing through the filter.

Most iron removal treatment plants are designed so that the filters are backwashed according to head loss. If iron breakthrough is a problem, filters should be backwashed when breakthrough occurs or just before breakthrough is expected. Accurate records can reveal when breakthrough occurs and also when breakthrough can be expected.

12.27 Proprietary Processes
by Bill Hoyer

Several patented processes are available for iron and manganese control. The best way to learn about the effectiveness and maintenance requirements of these processes is to contact someone who has one. Once you are operating one of these processes, the manufacturer is a good source of help when troubleshooting. Remember that various sources of raw water are different and what works for one operator may not work at your water treatment plant.

Electromedia iron and manganese removal systems are generally used on groundwater supplies at individual well sites because of their compactness and simplicity of treatment. The system uses reaction vessels where chemical reactions take place and an adsorptive media that requires no regeneration by special chemicals. Chlorine is used as the oxidizing chemical because of its cost and efficiency. (Any suitable oxidizing chemical can be used.) Almost 30 percent less chlorine, pound for pound, is required to perform the same amount of oxidation as potassium permanganate.

After oxidation with chlorine, a small dose of sulfur dioxide (0.25 to 0.50 mg/L) is introduced prior to the second reaction vessel. This dosage is factory set according to the general mineral analysis of the raw water. Dosage should not be altered. The sulfur dioxide is used to accelerate the oxidation of any sulfur compounds in the water to form compounds having no objectionable taste or odor.

The water is then sent to a filter operating at a preset rate of up to 15 gallons per minute per square foot (10 liters per second per square meter or 10 millimeters per second). In the filter vessel, the iron and manganese are adsorbed on the surface of the media until backwashing. The media can withstand a very high backwash rate (20 gallons per minute per square foot, 13.6 L/sec/sq m or 13.6 mm/sec) and requires only a four-minute backwash to obtain thorough cleaning.

The filter effluent can be sampled by a continuously monitoring analyzer that drives a 30-day strip-chart recorder. The recorder may have a color-coded indicating strip to direct the operator in the proper chemical dosage. If the recorder trace falls out of the green, the operator increases the chlorine dosage. The dosage is adjusted by turning one knob and can be read immediately. The effect of the change can be seen on the chart trace in five minutes and will reach a steady state within 10 minutes. Thus the operator can quickly determine the proper dosage. With the chemical dosage set properly, a free chlorine residual exists in the filter effluent providing the required disinfection in the distribution system. Variations in water quality are quickly reflected in the chart tracing. Since no permanganate is used, there are no "black water" or "pink water" complaints from accidental underdosage or overdosage of chemicals.

The process uses *BREAKPOINT CHLORINATION*[11] and the very effective adsorptive qualities of the media. Each system is provided with an automatic control panel that permits adjustment of any of the filter cycles simply by rotating a timer knob. Status of the system is displayed on the front panel with pilot lights for easy viewing. The automatic control panel operates the manually set chemical feed system using gaseous chlorine and gaseous sulfur dioxide. Backwash is accomplished automatically by using a process signal and filtration timer with a differential pressure override.

Maintenance on the system is quite limited. Most systems are built with an automatic standby for the chemicals that will switch from an empty container to a full container. Table 12.1 lists the recommended maintenance.

TABLE 12.1 RECOMMENDED MAINTENANCE FOR THE ELECTROMEDIA PROCESS

	Daily	Weekly	Monthly
1. Inspection of chart paper for proper chemical dosage	X		
2. Free chlorine residual test	X		
3. Total chlorine residual test	X		
4. Addition of buffer solution in the analyzer		X	
5. Colorimetric analysis of the influent and effluent for iron concentration		X	
6. Laboratory tests for analysis of influent and effluent for iron and manganese concentration			X
7. Changing of chart paper			X
8. Routine maintenance checks associated with valves, pipes, and pumps			X

Each application for iron and manganese removal is based on the general mineral analysis of the raw water. The required chemical treatment is provided for iron, manganese, and sulfide treatment. Additional equipment may be provided where corrosivity or chelating compounds are found to be present. Aeration may precede the process where methane extraction or carbon dioxide removal is required. Plant operators are directed to the operation and maintenance instructions provided with the equipment for additional details.

12.28 Monitoring of Treated Water

When controlling iron and manganese by aeration or with chemicals, the product water must be monitored closely. If laboratory facilities are available, the treated water can be analyzed for iron and manganese to be sure treatment is adequate. A quick way to monitor treated water is to collect a sample and add a dose of chlorine. If a brown- or rust-colored floc develops, then the treatment is inadequate. You will either have to increase the chemical doses or reduce the flows. If a pink color appears in the product water when using permanganate, then the dose is too high and must be reduced until the pink color is no longer visible.

12.29 Summary

Small iron and manganese water treatment plants can be very difficult to operate. If your plant is not operating as desired, talk to other operators in your area and see if they have any suggestions. If you have problems, you will have to try different chemical doses and procedures. Keep accurate records so you can evaluate the effectiveness of your efforts.

A lot of "iron complaints" in drinking water are caused by old steel or cast-iron water mains. A possible solution to this problem is to inject polyphosphates directly into the distribution mains. See Section 12.5, "Troubleshooting Red Water Problems," for additional ideas on how to solve problems.

---

[11] *Breakpoint Chlorination.* Addition of chlorine to water or wastewater until the chlorine demand has been satisfied. At this point, further additions of chlorine will result in a free chlorine residual that is directly proportional to the amount of chlorine added beyond the breakpoint.

FORMULAS

A standard polyphosphate solution is prepared by mixing and dissolving a known amount of polyphosphate in a container and adding distilled water to the one-liter mark. To determine the settings on polyphosphate chemical feeders:

1. Prepare a series of samples and test with polyphosphate.

2. Select the optimum dosage in mg/L.

3. Calculate the chemical feeder setting in pounds of polyphosphate per day.

$$\text{Stock Solution, mg/mL} = \frac{(\text{Polyphosphate, grams})(1{,}000 \text{ mg/gram})}{(\text{Solution, liter})(1{,}000 \text{ mL/L})}$$

$$\text{Dose, mg/L} = \frac{(\text{Stock Solution, mg/mL})(\text{Volume Added, mL})}{\text{Sample Volume, L}}$$

$$\text{Dose, lbs/MG} = \frac{(\text{Dose, mg/L})(3.785 \text{ L/gal})(1{,}000{,}000/\text{Million})}{(1{,}000 \text{ mg/gm})(454 \text{ gm/lb})}$$

$$\text{Chemical Feeder, lbs/day} = (\text{Flow, MGD})(\text{Dose, mg/L})(8.34 \text{ lbs/gal})$$

EXAMPLE 2

A standard polyphosphate solution is prepared by mixing and dissolving 1.0 gram of polyphosphate in a container and adding distilled water to the one-liter mark. Determine the concentration of the stock solution in milligrams per milliliter. If 5 milliliters of the stock solution are added to a one-liter sample, what is the polyphosphate dose in milligrams per liter and pounds per million gallons?

Known	Unknown
Polyphosphate, gm = 1.0 gm	1. Stock Solution, mg/L
Solution, L = 1.0 L	2. Dose, mg/L
Stock Solution, mL = 5 mL	3. Dose, lbs/MG
Sample, L = 1 L	

1. Calculate the concentration of the stock solution in milligrams per milliliter.

$$\text{Stock Solution, mg/mL} = \frac{(\text{Polyphosphate, gm})(1{,}000 \text{ mg/gm})}{(\text{Solution, L})(1{,}000 \text{ mL/L})}$$

$$= \frac{(1.0 \text{ gm})(1{,}000 \text{ mg/gm})}{(1.0 \text{ L})(1{,}000 \text{ mL/L})}$$

$$= 1.0 \text{ mg/mL}$$

2. Determine the polyphosphate dose in the sample in milligrams per liter.

$$\text{Dose, mg/L} = \frac{(\text{Stock Solution, mg/mL})(\text{Vol Added, mL})}{\text{Sample Volume, L}}$$

$$= \frac{(1.0 \text{ mg/mL})(5 \text{ mL})}{1 \text{ L}}$$

$$= 5.0 \text{ mg/L}$$

3. Determine the polyphosphate dose in the sample in pounds of polyphosphate per million gallons.

$$\text{Dose, lbs/MG} = \frac{(\text{Dose, mg/L})(3.785 \text{ L/gal})(1{,}000{,}000/\text{Million})}{(1{,}000 \text{ mg/gm})(454 \text{ gm/lb})}$$

$$= \frac{(5.0 \text{ mg/L})(3.785 \text{ L/gal})(1{,}000{,}000/\text{Million})}{(1{,}000 \text{ mg/gm})(454 \text{ gm/lb})}$$

$$= 42 \text{ lbs/MG}$$

EXAMPLE 3

Determine the chemical feeder setting in pounds of polyphosphate per day if 0.4 MGD is treated with a dose of 5 mg/L.

Known	Unknown
Flow, MGD = 0.4 MGD	Chemical Feeder, lbs/day
Dose, mg/L = 5 mg/L	

Determine the chemical feeder setting in pounds per day.

$$\text{Chemical Feeder, lbs/day} = (\text{Flow, MGD})(\text{Dose, mg/L})(8.34 \text{ lbs/gal})$$

$$= (0.4 \text{ MGD})(5 \text{ mg/L})(8.34 \text{ lbs/gal})$$

$$= 17 \text{ lbs/day}$$

FORMULAS

To calculate the average detention time for a reaction basin:

1. Determine the dimensions of the basin.

2. Measure and record the flow of water being treated.

$$\text{Basin Vol, cu ft} = (0.785)(\text{Diameter, ft})^2(\text{Depth, ft})$$

$$\text{Basin Vol, gal} = (\text{Basin Vol, cu ft})(7.48 \text{ gal/cu ft})$$

$$\text{Detention Time, min} = \frac{(\text{Basin Vol, gal})(24 \text{ hr/day})(60 \text{ min/hr})}{\text{Flow, gal/day}}$$

EXAMPLE 4

A reaction basin 12 feet in diameter and 5 feet deep treats a flow of 200,000 gallons per day. What is the average detention time in minutes?

Known	Unknown
Diameter, ft = 12 ft	Detention Time, min
Depth, ft = 5 ft	
Flow, GPD = 200,000 GPD	

1. Calculate the basin volume in cubic feet.

$$\text{Basin Vol, cu ft} = (0.785)(\text{Diameter, ft})^2(\text{Depth, ft})$$

$$= (0.785)(12 \text{ ft})^2(5 \text{ ft})$$

$$= 565 \text{ cu ft}$$

2. Convert the basin volume from cubic feet to gallons.

$$\text{Basin Vol, gal} = (\text{Basin Vol, cu ft})(7.48 \text{ gal/cu ft})$$

$$= (565 \text{ cu ft})(7.48 \text{ gal/cu ft})$$

$$= 4{,}226 \text{ gal}$$

3. Determine the average detention time in minutes for the reaction basin.

$$\text{Detention Time, min} = \frac{(\text{Basin Vol, gal})(24 \text{ hr/day})(60 \text{ min/hr})}{\text{Flow, gal/day}}$$

$$= \frac{(4,226 \text{ gal})(24 \text{ hr/day})(60 \text{ min/hr})}{200,000 \text{ gal/day}}$$

$$= 30 \text{ minutes}$$

QUESTIONS

Write your answers in a notebook and then compare your answers with those on page 24.

12.2E What happens if water being treated for iron and manganese by ion exchange contains any dissolved oxygen?

12.2F How does the pH of the water influence the rate of oxidation of iron to insoluble ferric hydroxide?

12.2G What is the purpose of a reaction basin following the aeration process?

12.2H After chlorine has been added to oxidize iron and manganese, how is the water dechlorinated?

12.2I What are the three modes of operation for a manganese greensand process called?

12.3 OPERATION OF AN IRON AND MANGANESE REMOVAL PLANT
by Gerald Davidson

The very low recommended limits for iron (0.3 mg/L) and manganese (0.05 mg/L) in water make these contaminants difficult to treat and sometimes the processes are expensive. Because of this, operators should know how the processes work and what to check for when something goes wrong and the limits are exceeded. This section describes the operation of an iron and manganese removal plant using the continuously regenerated (CR) manganese greensand process.

12.30 Description of Equipment and Process

The filter is the most important piece of treatment equipment in an iron and manganese removal plant. Figure 12.4 illustrates a typical filter consisting of layers of gravel, manganese greensand, and anthracite coal. Some of the differences between greensand filters and conventional filters are: (1) the greensand is finer than filter sand; (2) the filtration rate is slower (2 to 5 GPM/sq ft (1.3 to 3.4 liters per sec/sq m or 1.3 to 3.4 mm/sec)); and (3) the backwash rate is lower (12 GPM/sq ft at 55°F versus 15 GPM/sq ft for sand/anthracite filters.

Figure 12.5 shows a typical flow diagram for a continuous regeneration (CR) manganese greensand plant. Treatment with manganese greensand can remove 95 percent of both iron and manganese. However, if the iron concentration is above 10 mg/L or manganese is above 5 mg/L, the efficiency of the greensand (or any other type of filter) drops very quickly as insoluble oxide particles accumulate on the grains of greensand and within the media bed. Pretreatment may be required to oxidize and remove a major portion of the iron and any sulfide present.

At the plant shown in Figure 12.5, raw water being pumped from a 50-foot (15-m) deep well contains 3.0 mg/L iron and 0.75 mg/L manganese (see Chapter 21 for test procedures) and it has a pH of 6.2. The pH of the water has a significant effect on iron removal. The oxidation potential of chlorine and potassium permanganate decreases as the pH increases, although the rate of reaction increases significantly with the increase in pH. Because the pH of the water in this example is 6.2, which is the minimum recommended pH for use of the CR process, the pH is adjusted to within the range of 6.5 to 6.8 by adding sodium carbonate or sodium hydroxide or by passing the water through a calcite neutralizing filter.

Next, the water is injected with chlorine, flash mixed, and flocculated for a period of ten minutes in order to oxidize most of the iron. The chlorine dosage can be calculated using the following formula:

Cl_2 Required, mg/L = 1 × Fe Conc, mg/L

Therefore,

Cl_2 Required, mg/L = 1 × 3.0 mg/L

= 3 mg/L

After chlorination, the water is injected with potassium permanganate ($KMnO_4$) to complete the oxidation of any

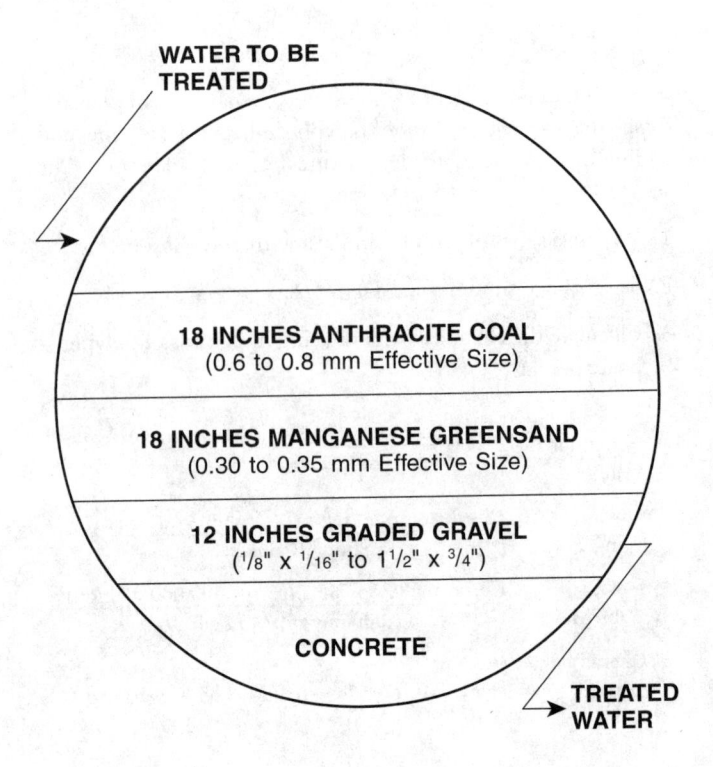

Fig. 12.4 Multi-media manganese greensand filter (horizontal)

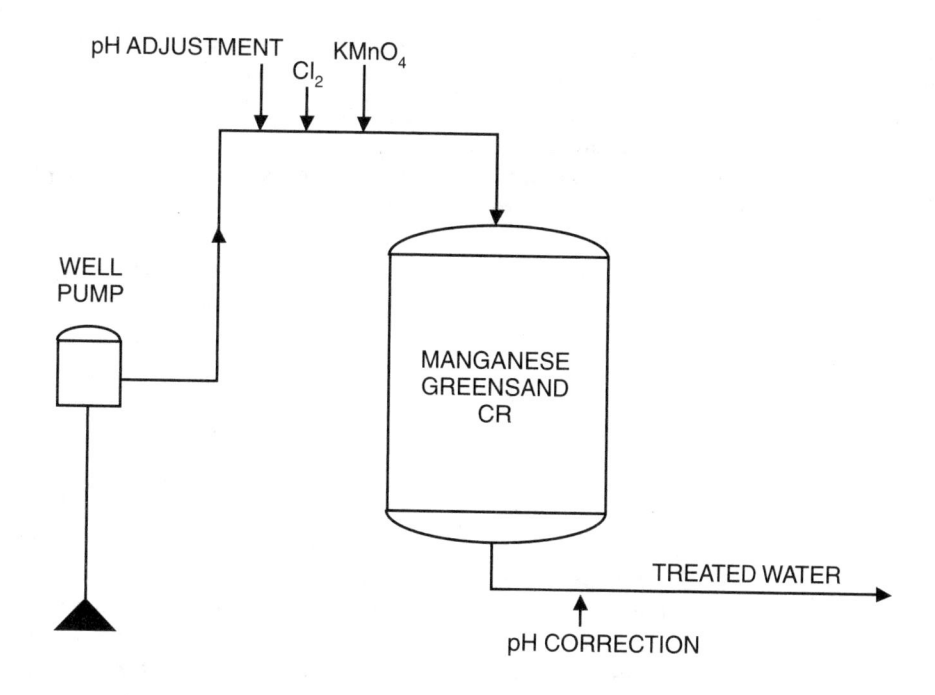

Fig. 12.5 Typical flow diagram of a continuous regeneration (CR) greensand process
(Permission of Hungerford & Terry, Inc.)

remaining iron and soluble manganese. Sodium hydroxide (30 mg/L)[12] may also be added if needed for corrosion control (pH correction, Figure 12.5). The KMnO$_4$ dosage can be calculated using the following formula:

KMnO$_4$ Required, mg/L = (0.2 × Fe Conc, mg/L) + (2 × Mn Conc, mg/L)

Therefore,

KMnO$_4$ Required, mg/L = (0.2 × 3.0 mg/L) + (2 × 0.75 mg/L)
= 2.1 mg/L

NOTE: The calculated 2.1 mg/L potassium permanganate dose is the minimum dose. This dose assumes there are no other oxidizable compounds in the raw water. However, typical oxidizable compounds usually found include organic color, bacteria, and even hydrogen sulfide (H$_2$S). Therefore, the actual dose may be higher. A bench-scale test should be performed to determine the required dose.

To prepare a stock solution of potassium permanganate, mix 2 to 4 ounces (56 to 113 mg) of potassium permanganate chemical per gallon of water (15 to 30 mg/L) in a solution tank. Mix thoroughly to make sure the chemical dissolves completely. KMnO$_4$ is soluble up to 8.3 ounces (235 mg) per gallon (62 mg/L) of water at 68°F so you should have little trouble dissolving the chemical.

Problems will develop in the iron and manganese removal process using manganese greensand if you have too short a detention time for the chemicals to react. That is, it takes a little time for the chemicals to start working. If the plant you are operating does not allow sufficient detention time for the chemical reactions to take place, you should perform jar tests to see if a flash mix will improve performance. In some plants, the injection of the potassium permanganate solution in the volute of the pump will produce complete mixing of potassium permanganate.

The potassium permanganate feed system in our example plant consists of a 50-gallon (190-L) polyethylene solution vat, two ¼-HP mixers, liquid level switches, and a metering pump. The mixers do not have to run continuously because of the solubility of potassium permanganate in water.

---

[12] *NOTE:* Addition of 30 mg/L of sodium hydroxide (NaOH) will increase the sodium content of the water by 17 mg/L. If you are trying to keep the sodium level below 20 mg/L, then the sodium in the raw water must be below 3 mg/L.

The influent flow to the manganese greensand/anthracite filter is approximately 1,500 to 2,000 GPM. If the chemical dosage rate for $KMnO_4$ has been correctly calculated, the filter influent will have a slight pink/orange color due to the presence of a slight excess of permanganate. As the water is treated in the filter, this color will disappear because the permanganate will be reduced to manganese oxide by the manganese greensand. The manganese oxides will then precipitate on the grains, thus continuously regenerating the manganese greensand. It is important not to underfeed permanganate for an extended period of time. Without adequate permanganate, you will eventually exhaust the oxidative capacity of the media. If this happens, iron and manganese may pass through the filter into the distribution system and you will have to recharge the greensand (see Section 12.31).

As the filter run progresses and the amount of insoluble precipitates builds up in the media, the pressure drop across the filter will gradually increase. When the head loss reaches 10 psi (0.07 pascals), or after treating a predetermined number of gallons, the bed must be backwashed to remove the filtered particulates. Backwashing usually takes 10 to 15 minutes. Since the media is continuously regenerated during operation, there is no need to regenerate the greensand after backwashing. The used backwash water can be collected and reused after allowing the precipitates to settle out. The settled precipitates may be disposed of in a sanitary sewer or sent to a drying bed.

In addition to backwashing, some plants use an air washing system to more thoroughly clean the manganese greensand grains on a weekly basis. Air wash is particularly useful for plants treating water that has a relatively high concentration of manganese. Manganese hydroxide oxidation products tend to be sticky or gummy and may cause the grains of greensand to stick together in clumps, especially in long filter runs. Backwashing alone may not always separate these grains and, over a period of time, "mudballs" may form. Agitating the greensand with an air wash (or better still an air-water wash) helps to keep the filter media in a free, loose, fluid condition. This will result in cleaner media and longer runs by minimizing cementation of the media bed.

Daily tests the operator should perform include iron, manganese, pH, and chlorine residual. The iron and manganese test tells the operator if the treatment plant is working and meeting state and federal water quality requirements. The pH test is also very important because of the relationship between pH and the corrosivity of water and because effective iron and manganese removal by the CR manganese greensand process requires a pH of 6.2 to 6.8.

12.31 Regeneration of Manganese Greensand

If the potassium permanganate charge is lost in the filter bed due to underdosing of potassium permanganate, the operator must regenerate the greensand. There are two ways to regenerate the bed:

1. Shut down and pour a saturated solution of potassium permanganate (about 5 percent) into the filters; let the saturated solution sit for approximately 24 hours. After 24 hours, backwash the filters at a normal rate to flush out the excess potassium permanganate.

2. Recharge the greensand by increasing the potassium permanganate dosage until pink water flows out of the greensand media. Then, decrease the potassium permanganate until you have a slight pink color before filtration. There should be no pink water after filtration when the water is being pumped into the distribution system. If there is still pink water after filtration, keep decreasing the potassium permanganate dose until no pink water is present in the water after filtration. The pink color is the best indication that the greensand is regenerated or recharged.

One problem with this method is that you might reduce the permanganate level too far and pass water with iron and manganese. This could cause red-colored water in the distribution system and stain clothes or bathroom fixtures. Because of the importance of the potassium permanganate in the greensand process, it is highly recommended that some type of fail-safe system be installed to prevent filtering water in the event the potassium permanganate solution vat reaches a low level. Without permanganate, the greensand could lose its charge and iron and manganese will enter the distribution system. When a low level is reached, the plant should be automatically shut down. Typical fail-safe systems include low-level alarms in the vat or an automatic switchover system to another vat when the level drops too low in the vat in use.

Some operators find that method No. 1 is more effective. Since all treatment plants are unique in some respect, one method or the other or some modification may work best for your plant. Therefore, procedures and methods should be developed through actual experience. These methods should be adopted only if they provide the desired results without eliminating any concepts of design or of good operating practice.

12.32 Troubleshooting

See Table 12.2, "Troubleshooting Manganese Greensand Systems," for a listing of troubles, causes, and remedies.

> ## QUESTIONS
>
> Write your answers in a notebook and then compare your answers with those on page 25.
>
> 12.3A How does a water's pH affect the manganese greensand process?
>
> 12.3B How can an operator quickly tell whether enough potassium permanganate is being fed in a continuous regeneration manganese greensand process?
>
> 12.3C Why should a greensand plant be shut down when the permanganate solution level in the vat gets low?

12.4 MAINTENANCE OF A CHEMICAL FEEDER

In small water treatment plants that remove iron and manganese, a hypochlorite solution may be used to provide chlorine instead of using chlorine gas. Commercial sodium hypochlorite solutions (such as Clorox) contain an excess of caustic (sodium hydroxide, $NaOH$). When the solution is diluted with water

Iron and Manganese

TABLE 12.2 TROUBLESHOOTING MANGANESE GREENSAND SYSTEMS

Trouble	Cause	Remedy
SYSTEM TYPE: CR/IR[a]		
1. Filter effluent clear, iron low, manganese higher than raw water.	Manganese being leached from greensand grains; bed is insufficiently regenerated.	Increase frequency of regeneration. Regenerate bed with sufficient $KMnO_4$ (1½ oz/cu ft) so heavy purple color comes through bed. Make sure proper influent chemicals (Cl_2 or $KMnO_4$) are being continuously fed.
2. Filter effluent turbid with yellow to brownish color. Iron and manganese high.	Too much alkali being fed ahead of filter.	Reduce alkali feed. Maintain correct pH prior to filter at 6.2–6.5. Post pH correct if higher pH required in system.
	Polyphosphate being fed ahead of filter.	Discontinue polyphosphate feed.
	Channeling through filters.	Check bed surface for mounds, pockets, and channeling. Backwash and air-scrub if possible.
	Iron organically bound: reactions with oxidizing agent produce a nonfilterable colloid.	Feed alum or other coagulant prior to filter. Amount determined in field.
3. Excessive pressure drop across bed immediately after backwashing.	Accumulation of greensand fines at surface of bed.	Remove fines by scraping after backwashing. In severe cases, bed replacement may be required.
		Check maximum pressure differential recommended and do not exceed 10 psi ΔP as additional fines may be formed.
	Backwash rate too low.	Increase backwash rate to 10–12 GPM/sq ft.
	Filter bed cemented, as evidenced by mounding around periphery of vessel.	Break up cemented areas with air-water wash combination. Bed replacement may be necessary.
	Well throwing fine sand, silt, and colloidal clay.	Check well supply, especially immediately after pump start-up. Allow well to pump overboard at start of pumping cycle.
4. On multiple unit installations, water quality good on some units, bad on others.	Unequal distribution of prefeed chemicals.	Inject chemical at a point where thorough mixing of chemicals with raw water occurs before diversion to the various filters.
SYSTEM TYPE: CR[a]		
5. Iron breakthrough before recommended maximum ΔP of 10 psi is reached.	Some iron waters filter in depth and do not build up head loss.	Backwash should be initiated by total number of gallons treated rather than by head loss. Use ΔP as a backup to prevent exceeding ΔP of 10 psi.
6. Faint pink color in filter effluent.	Permanganate feed rate is too high.	Operate filter for 1–2 hours with $KMnO_4$ feed off. Then reset feeder at slightly lower setting.
SYSTEM TYPE: IR[a]		
7. Low capacity.	Manganese oxide coating stripped from greensand grains due to insufficient regeneration. May be especially troublesome with high sulfide water.	Increase frequency of regeneration. Prefeed Cl_2 with sulfide waters. Replace bed if required.
	Manganese greensand heavily iron-fouled.	Use CR method with dual-media anthracite/manganese greensand bed to prevent iron fouling.
	Excessive grain growth due to high manganese oxide buildup.	Increase frequency of regeneration. Bed replacement may eventually be required.

[a] CR—Continuous Regeneration
  IR—Intermittent Regeneration

(Reprinted with permission of Hungerford & Terry, Inc., Clayton, NJ)

containing carbonate alkalinity,[13] the resulting solution becomes supersaturated with calcium carbonate. This calcium carbonate tends to form a coating on the poppet valves in the solution feeder. The coated valves do not seal properly and the feeder fails to pump the hypochlorite solution properly.

Calcium carbonate scale can be removed from poppet valves by using the following procedure:

1. Fill a one-quart (946-mL) Mason jar half full of tap water.
2. Add one fluid ounce (44 mL) of 30- to 37-percent hydrochloric acid (HCl) to the Mason jar.
3. Finish filling the jar to the top with tap water.
4. Place the suction hose of the hypochlorinator in the jar and pump the entire contents of the jar through the system.
5. Return the suction hose to the solution tank and resume normal operation.

The hydrochloric acid (HCl), also called muriatic acid, can be obtained from stores selling swimming pool supplies.

One way to avoid the formation of calcium carbonate coatings is to obtain the dilution water for the hypochlorite from an ordinary home water softener.

For additional information on the operation and maintenance of various types of chemical feeders, see Chapter 13, "Fluoridation."

QUESTIONS

Write your answers in a notebook and then compare your answers with those on page 25.

12.4A What problems may develop in a chemical feeder pumping sodium hypochlorite?

12.4B How can the problems caused by calcium carbonate scale on a hypochlorinator's poppet valves be solved?

12.5 TROUBLESHOOTING RED WATER PROBLEMS

The first step when troubleshooting red or dirty water complaints is to be sure the iron and manganese treatment processes are working properly. If the iron and manganese are being removed by the treatment processes, investigate the distribution system for sources of the problem.

Red water or dirty water problems may be caused by iron or manganese in the water, corrosive water, or iron bacteria in the distribution system. When an unstable or corrosive water (see Chapter 8, "Corrosion Control") is pumped into the distribution system, the water attacks cast-iron pipes and metal service lines, picks up iron, and causes red water complaints. All water treatment plants should run a Marble Test (see page 528 in Chapter 21). If the test indicates that the water is corrosive, the addition of caustic (sodium hydroxide, NaOH) to the water to increase the pH will help. When the water becomes stable (according to the Marble Test), some of the red water complaints could be eliminated.

The growth of iron bacteria inside water mains causes one of the most troublesome and most difficult to eliminate red water problems. These bacteria are not harmful. They live and accumulate in the iron in the water flowing through the distribution system. As the bacterial growths increase, slimes will build up in the mains and eventually slough off into the water. When these slimes come out a consumer's water tap, you can expect complaints of red water and slimes.

Slime growths can be controlled by maintaining a free chlorine residual throughout the distribution system. Sometimes the residual is very difficult to maintain. If bacterial growths have been in the distribution system for a long time and are flourishing, it is very difficult to maintain a free chlorine residual at the extremes or in dead ends of the system. Also, if the water has a natural high chlorine demand, your chlorination equipment may not be capable of feeding enough chlorine to maintain a free chlorine residual. Remember that frequently when consumers complain about a strong chlorine taste or swimming pool-tasting water, the solution is to add more chlorine in order to get past the breakpoint.

One way to rid a distribution system of iron bacteria is to develop a flushing program.[14] Flushing should start at the location where the water enters the distribution system, such as an elevated tank. Flush the water mains by working toward the extremes or most distant points of the distribution system. Usually only one portion of the distribution system is flushed, followed by another portion until the entire system has been flushed.

A common practice is to open a hydrant at the extreme end of the system at the start of the flushing job to be sure the water being flushed will carry the sediment and insoluble precipitates in the desired direction and out of the system. Flushing is often done late at night when water demands are low so facilities will not be overworked and consumers will not be inconvenienced.

Valves will have to be opened and closed in the proper sequences to be sure the desired mains are being flushed and that no one will be without water. Hydrants that are opened to allow flushing must be of sufficient size to produce flushing velocities (2.5 up to 5.0 ft/sec preferred or 0.75 to 1.5 m/sec) in the mains. Also, the mains providing the flushing flows must have sufficient capacity to deliver the desired flows.

When flushing a system, be sure the pressure in the distribution system does not drop below 20 psi (1.4 kg/sq cm or 138 kPa). If a four-inch (100-mm) water main is flushed using a six-inch (150-mm) hydrant, the water pressure in the main downstream from the hydrant could become dangerously low. When this happens, the distribution system could be subject to contamination by *BACKSIPHONAGE*.[15] Never allow a backsiphon condition to develop in a distribution system.

---

[13] See Chapter 14, "Softening," for a discussion of carbonate alkalinity.
[14] See *WATER DISTRIBUTION SYSTEM OPERATION AND MAINTENANCE* in this series of manuals, Chapter 5, Section 5.706, "Pipe Flushing."
[15] *Backsiphonage.* A form of backflow caused by a negative or below atmospheric pressure within a water system. Also see BACKFLOW.

In summary, to minimize red water or dirty water problems and complaints, you must provide adequate treatment to control iron and manganese. This is necessary to ensure that the water pumped into the distribution system contains little or no iron and manganese. The water must be stable (noncorrosive) so that iron will not be picked up in the distribution system. Corrosion control treatment processes can produce a stable water. If bacterial growths are a problem, a free chlorine residual must be maintained in all water throughout the distribution system. If red or dirty water problems exist in a distribution system, a thorough flushing program can be very helpful.

12.6 ARITHMETIC ASSIGNMENT

Turn to the Arithmetic Appendix at the back of this manual. Read and work the problems in Section A.30, "Iron and Manganese Control." Check the arithmetic in this section using an electronic calculator. You should be able to get the same answers. Section A.50 contains similar problems using metric units.

12.7 ADDITIONAL READING

1. *NEW YORK MANUAL,* Chapter 13,* "Iron and Manganese."

2. *TEXAS MANUAL,* Chapter 11,* "Special Water Treatment (Iron and Manganese Removal)."

* Depends on edition.

NOTE: *A MANUAL OF INSTRUCTION FOR WATER TREATMENT PLANT OPERATORS,* also called *NEW YORK MANUAL,* is available from Health Education Services, PO Box 7126, Albany, NY 12224. Price, $36.00, includes cost of shipping and handling.

MANUAL OF WATER UTILITIES OPERATIONS, also called *TEXAS MANUAL,* is available from Texas Water Utilities Association, 1106 Clayton Lane, Suite 101 East, Austin, TX 78723-1093. Price to members, $25.00; nonmembers, $35.00; plus $5.00 shipping and handling.

12.8 ACKNOWLEDGMENT

Portions of this chapter were revised by Raymond Ellis, Hungerford & Terry, Inc.

QUESTIONS

Write your answers in a notebook and then compare your answers with those on page 25.

12.5A List the possible causes of red or dirty water complaints.

12.5B How can slime growths be controlled in water distribution systems?

Please answer the discussion and review questions next.

DISCUSSION AND REVIEW QUESTIONS
Chapter 12. IRON AND MANGANESE CONTROL

The purpose of these questions is to indicate to you how well you understand the material in the chapter. Write the answers to these questions in your notebook.

1. Why should iron and manganese be controlled in drinking water?
2. Why are accurate results of tests for iron and manganese difficult to obtain?
3. Why is chlorine usually fed with polyphosphates when controlling iron and manganese?
4. How do polyphosphates control manganese?
5. How is the proper polyphosphate dose determined?
6. What happens when an ion exchange resin becomes fouled with iron rust or insoluble manganese dioxide?
7. When should iron and manganese ion exchange units be regenerated?
8. How would you determine whether or not to adjust the flows to an oxidation by aeration process to remove iron?
9. Why should reaction basins be drained and cleaned?
10. What are the advantages and disadvantages of the oxidation by aeration process to remove iron?
11. What happens if the dose of potassium permanganate to remove iron and manganese is not exact?
12. How does an operator decide when to backwash a manganese greensand filter?
13. What must you do if the potassium permanganate charge is lost in the filter bed?
14. How would you attempt to prevent red or dirty water complaints?

SUGGESTED ANSWERS
Chapter 12. IRON AND MANGANESE CONTROL

Answers to questions on page 7.

12.0A When clothes are washed in water containing iron and manganese, they will come out stained. Excessive iron and manganese in water can also stain plumbing fixtures. Iron bacteria will cause thick slimes to form on the walls of water mains. These slimes are rust colored from iron and black from manganese. Variations in flow cause these slimes to slough, which results in dirty water. Furthermore, these slimes will cause foul tastes and odors in the water.

12.0B The growth of iron bacteria is easily controlled by chlorination. However, when water containing iron is chlorinated, the iron is converted into rust particles, and manganese is converted into a jet-black compound called manganese dioxide.

12.0C The generally acceptable limit for iron in drinking water is 0.3 mg/L and that for manganese is 0.05 mg/L.

Answers to questions on page 8.

12.1A Iron and manganese react with dissolved oxygen to form insoluble compounds.

12.1B Iron and manganese samples are acidified when they are collected to prevent the formation of iron and manganese scales on the walls of the sample bottles.

12.1C Samples for iron and manganese testing should be collected as close to the well or source of water as possible.

Answers to questions on page 12.

12.2A To determine if nearby wells contain iron and manganese, samples could be collected and analyzed from nearby private wells. Also, discussions with well drillers who have been active in the locality and with engineers in the state agency responsible for the regulation of well drilling will be helpful.

12.2B If a well produces water containing dissolved oxygen as well as iron and manganese, the iron and manganese are probably coming from the deeper aquifers. Try to seal off the deeper aquifers.

12.2C Bench-scale tests are a method of studying different ways or chemical doses for treating water or wastewater and solids on a small scale in a laboratory.

12.2D If polyphosphate solutions are much over 48 hours old, they will react slowly with water to form orthophosphates, which are much less effective in preventing manganese deposits.

Answers to questions on page 18.

12.2E If water being treated for iron and manganese by ion exchange contains any dissolved oxygen, the resin becomes fouled with iron rust or insoluble manganese dioxide.

12.2F The higher the pH, the faster the rate of oxidation of iron to insoluble ferric hydroxide.

12.2G The purpose of the reaction basin is to allow time for the oxidation reactions to take place. The aeration process should produce sufficient dissolved oxygen for the iron to be oxidized to insoluble ferric hydroxide.

12.2H Water can be dechlorinated by the use of reducing agents such as sulfur dioxide (SO_2), sodium bisulfite ($NaHSO_3$), and sodium sulfite (Na_2SO_3). Bisulfite is commonly used because it is cheaper and more stable than sulfite.

12.2I The three process modes for manganese greensand are continuous regeneration, intermittent regeneration, and catalytic regeneration.

Answers to questions on page 20.

12.3A   The oxidation potential of chlorine and potassium permanganate used in the manganese greensand process decreases as the pH increases. To operate efficiently, the CR manganese greensand process requires a pH range of 6.2 to 6.8.

12.3B   When adequate potassium permanganate is being fed in a manganese greensand process, the filter influent will have a slight pink/orange color.

12.3C   A greensand plant should be shut down when the permanganate solution level in the vat gets low because of the importance of the permanganate in the process. Without permanganate, the greensand could lose its charge and iron and manganese will enter the distribution system.

Answers to questions on page 22.

12.4A   When a chemical feeder pumps sodium hypochlorite, calcium carbonate coatings may develop on the poppet valves if the dilution water contains carbonate alkalinity. Coated valves do not seal properly and the feeder fails to pump the hypochlorite solution properly.

12.4B   The problems caused by calcium carbonate scale on a hypochlorinator's poppet valves can be solved in two ways:

1. A hydrochloric acid solution can be pumped through the system.
2. The dilution water for the hypochlorite can be obtained from an ordinary home water softener.

Answers to questions on page 23.

12.5A   Red or dirty water complaints may be caused by:

1. Iron or manganese in the water
2. Corrosive waters
3. Iron bacteria in the distribution system

12.5B   Slime growths in distribution systems can be controlled by maintaining a free chlorine residual throughout the distribution system and by a distribution system flushing program.

CHAPTER 13

FLUORIDATION

by

Harry Tracy

TABLE OF CONTENTS
Chapter 13. FLUORIDATION

		Page
OBJECTIVES		31
WORDS		32
13.0	IMPORTANCE OF FLUORIDATION	33
13.1	FLUORIDATION PROGRAMS	33
13.2	COMPOUNDS USED TO FURNISH FLUORIDE ION	34
13.3	FLUORIDATION SYSTEMS	35
	13.30 Chemical Feeders	35
	13.31 Saturators	35
	13.32 Downflow Saturators	45
	13.33 Upflow Saturators	45
	13.34 Large Hydrofluosilicic Acid Systems	47
13.4	FINAL CHECKUP OF EQUIPMENT	47
	13.40 Avoid Overfeeding	47
	13.41 Review of Designs and Specifications	47
13.5	CHEMICAL FEEDER START-UP	49
13.6	CHEMICAL FEEDER OPERATION	49
	13.60 Fine-Tuning	49
	13.61 Preparation of Fluoride Solution	49
	13.62 Fluoridation Log Sheets	52
	13.620 Hydrofluosilicic Acid	52
	13.621 Sodium Silicofluoride	54
	13.63 Equipment Check Procedures	54
13.7	PREVENTION OF OVERFEEDING	54
13.8	UNDERFEEDING	58
13.9	SHUTTING DOWN CHEMICAL SYSTEMS	58
13.10	MAINTENANCE	58
13.11	SAFETY IN HANDLING FLUORIDE COMPOUNDS	59
	13.110 Avoid Overexposure	59
	13.111 Symptoms of Fluoride Poisoning	59

13.112	Basic First Aid	60
13.113	Protecting Yourself and Your Family	60
13.114	Training	60
13.12	**CALCULATING FLUORIDE DOSAGES**	60
13.13	**ARITHMETIC ASSIGNMENT**	65
13.14	**ADDITIONAL READING**	65
13.15	**ACKNOWLEDGMENTS**	65
DISCUSSION AND REVIEW QUESTIONS		66
SUGGESTED ANSWERS		66

OBJECTIVES
Chapter 13. FLUORIDATION

Following completion of Chapter 13, you should be able to:

1. Explain the reason for fluoridating drinking water.
2. Describe how fluoridation programs are implemented.
3. List the compounds used to furnish fluoride ion.
4. Review designs and specifications of fluoridation equipment.
5. Inspect fluoridation equipment.
6. Start up a chemical feeder.
7. Operate and maintain a chemical feeder.
8. Calculate fluoride dosages.
9. Prepare fluoride solutions.
10. Develop and keep accurate fluoride log sheets.
11. Prevent overfeeding of fluoride.
12. Shut down chemical feed systems.
13. Safely handle fluoride compounds.

WORDS
Chapter 13. FLUORIDATION

BATCH PROCESS

A treatment process in which a tank or reactor is filled, the water (or wastewater or other solution) is treated or a chemical solution is prepared, and the tank is emptied. The tank may then be filled and the process repeated. Batch processes are also used to cleanse, stabilize, or condition chemical solutions for use in industrial manufacturing and treatment processes.

DAY TANK

A tank used to store a chemical solution of known concentration for feed to a chemical feeder. A day tank usually stores sufficient chemical solution to properly treat the water being treated for at least one day. Also called an AGE TANK.

ENDEMIC (en-DEM-ick)

Something peculiar to a particular people or locality, such as a disease that is always present in the population.

FLUORIDATION (floor-uh-DAY-shun)

The addition of a chemical to increase the concentration of fluoride ions in drinking water to a predetermined optimum limit to reduce the incidence (number) of dental caries (tooth decay) in children. Defluoridation is the removal of excess fluoride in drinking water to prevent the mottling (brown stains) of teeth.

GRAVIMETRIC FEEDER

A dry chemical feeder that delivers a measured weight of chemical during a specific time period.

POSITIVE DISPLACEMENT PUMP

A type of piston, diaphragm, gear, or screw pump that delivers a constant volume with each stroke. Positive displacement pumps are used as chemical solution feeders.

SATURATOR (SAT-yoo-ray-tor)

A device that produces a fluoride solution for the fluoridation process. The device is usually a cylindrical container with granular sodium fluoride on the bottom. Water flows either upward or downward through the sodium fluoride to produce the fluoride solution.

VOLUMETRIC FEEDER

A dry chemical feeder that delivers a measured volume of chemical during a specific time period.

CHAPTER 13. FLUORIDATION

13.0 IMPORTANCE OF FLUORIDATION

During the period 1902 to 1931, Dr. Frederick S. McKay, a dentist practicing in Colorado Springs, noted what seemed an *ENDEMIC*[1] brown stain on the teeth of his patients. McKay devoted much of his time researching the case of mottled (brown, chalky deposits) tooth enamel but it was not until 1931 that the cause was found to be excessive fluoride in the water supplies (2 to 13 mg/L). During this period, McKay had also noted that the mottled teeth seemed to have fewer dental caries (decay or cavities).

The next logical step was to add fluoride to waters that were deficient in fluoride and to discover if children drinking water treated with fluoride had fewer cavities. In 1945, controlled fluoridation was started in the cities of Grand Rapids, Michigan, and Newburgh, New York, with control cities of Muskegon and Kingston.

Finally, in 1955, the results were in and they showed a 60 percent reduction of caries in the children who drank fluoridated water compared to the children in the control cities.

The progress of fluoridation did not go smoothly. Antifluoridationists became increasingly vocal and were able to stop fluoridation in many cities through action in the political arena. Although dentists practicing in areas with naturally high fluoride waters noted that their patients had remarkably few cavities, there were still those disfiguring brown stains. Mottling of the teeth occurs when the fluoride level exceeds about 1.5 mg/L. Fluoride concentrations in excess of 1,000 mg/L have been found in waters from volcanic regions. Waters with fluoride concentrations more than 1.4 to 2.4 mg/L should be treated to reduce the level to approximately one milligram per liter. Recommended control limits for fluoride depend on the annual average of maximum daily air temperatures (Table 13.1). The maximum contaminant level (MCL) is 4.0 mg/L.

QUESTIONS

Write your answers in a notebook and then compare your answers with those on page 66.

13.0A What happens if a person drinks water with an excessive concentration of fluoride?

13.0B What happens if children drink a recommended dose of fluoride?

13.1 FLUORIDATION PROGRAMS

Generally speaking, fluoridation programs start with citizens' inquiries about fluoridation of their water supplies and encouragement of the local dental society. These requests for the addition of fluoride to prevent dental caries are passed along to appropriate governing agencies. The governing body will usually rely upon a vote of the people, or it may be forced into a vote by threat of a referendum. If the decision is made to fluoridate, the water department or water company will almost always make the final decisions as to types of fluoride chemicals and feeding equipment to be used.

TABLE 13.1 PRIMARY DRINKING WATER REGULATIONS FOR FLUORIDE AND RECOMMENDED LEVELS

Annual Average Maximum Daily Air Temperatures[a]		Recommended Control Limits of Fluoride Levels, mg/L			Maximum Contaminant Level, mg/L
°F	°C	Lower	Optimum	Upper	
53.7 & Below	12.0 & Below	0.9	1.2	1.7	4.0
53.8 to 58.3	12.1 to 14.6	0.8	1.1	1.5	4.0
58.4 to 63.8	14.7 to 17.6	0.8	1.0	1.3	4.0
63.9 to 70.6	17.7 to 21.4	0.7	0.9	1.2	4.0
70.7 to 79.2	21.5 to 26.2	0.7	0.8	1.0	4.0
79.3 to 90.5	26.3 to 32.5	0.6	0.7	0.8	4.0

[a] Contact your local Weather Service Office to determine the "Annual Average Maximum Daily Air Temperature" for your service area.

---

[1] *Endemic* (en-DEM-ick). Something peculiar to a particular people or locality, such as a disease that is always present in the population.

13.2 COMPOUNDS USED TO FURNISH FLUORIDE ION

The most commonly used compounds to fluoridate water systems are sodium fluoride, sodium fluorosilicate, and hydrofluosilicic (HI-dro-FLOO-o-suh-LIS-ick) acid. There are also a few systems using such compounds as hydrofluoric acid and ammonium silicofluoride. All of these chemicals are refined from minerals found in nature and they yield fluoride ions identical to those found in natural waters. Hydrofluosilicic acid (also called fluorosilicic acid) is the compound most commonly used in several states (California, Florida, and Illinois).

The American Water Works Association has developed standards for the purity and composition of the fluoride compounds most commonly used. In order for you to be confident of the fluoride compound you are using, insist that your supplier furnish only compounds meeting the appropriate AWWA specifications.

The plant should have as part of its records several copies of the appropriate AWWA standard for reference. Standards can be purchased from American Water Works Association, Bookstore, 6666 West Quincy Avenue, Denver, CO 80235.

Standard	Chemical	Price[a] Members	Nonmembers	Order No.
B701-06	Sodium Fluoride	$37.50	$51.50	42701
B702-06	Sodium Fluorosilicate	$37.50	$51.50	42702
B703-06	Fluorosilicic Acid	$37.50	$51.50	42703

[a] Price includes cost of shipping and handling.

Also, NSF International's[2] "Drinking Water Additives Standards Set," NSF Standard 60, "Drinking Water Treatment Chemicals - Health Effects," most current edition, lists companies by chemical name that supply chemicals approved for use in drinking waters. Fluoridation drinking water treatment chemicals approved by NSF include:

1. Fluosilicic acid
2. Hydrofluosilicic acid
3. Sodium fluoride

Only the fluoride ion in these compounds is of any importance in the fluoridation of water; therefore, pound for pound, each compound will provide a different final fluoride level. If you switch from one type of fluoride compound to another, you will have to make separate calculations for each type. Table 13.2 summarizes the important properties of fluoride compounds.

When selecting a fluoridation chemical, several important factors must be considered. If powdered or crystal forms are used, solubility of the chemical in water will determine how readily it dissolves in water and how well it remains in solution. Operator safety and ease of handling must be given serious consideration. Storage and feeding requirements as well as costs must also be studied when selecting any chemical.

TABLE 13.2 FLUORIDE COMPOUNDS

	Sodium Silicofluoride Na_2SiF_6	Sodium Fluoride NaF	Hydrofluosilicic Acid H_2SiF_6
1. Form	Powder	Powder or Crystal	Liquid
2. Molecular Weight	188.1	42.0	144.1
3. Commercial Purity, %	98–99	95–98	22–30 by Weight
4. Fluoride Ion, % (Purity, %)	60.7 (100%) 59.8 (98.5%)	45.3 (100%) 43.4 (96%)	79.2 (100%) 23.8 (30%)
5. Density	55–72 lbs/cu ft	65–90 lbs/cu ft	10.5 (30%) lbs/gal
6. Solubility in Water, % (gram/100 mL water at 77°F or 25°C)	0.76	4.05	100[a]
7. pH of Saturated Solution	3.5	7.6	1.2 (1% Solution)

[a] Infinite because we are dealing with a liquid.

REMEMBER, All fluoridation chemicals are POISONOUS at HIGH LEVELS.

Hydrofluosilicic acid is usually the easiest fluoridation chemical to feed. However, hydrofluosilicic acid produces poisonous fumes that must be vented and is very irritating to your skin. Sodium fluoride is easier to feed than the other powdered fluoridation chemicals because it is more soluble in water.

Operators can receive instructions from the manufacturer on how to make the chemical solutions and how much solution to meter per million gallons. See Section 13.61, "Preparation of Fluoride Solution," for calculations and procedure details.

Prior to fluoridation, the water should be checked for its natural fluoride level. If there is natural fluoride in the water, it is only necessary to add enough to bring the total to the desired level recommended by the local health authorities.

QUESTIONS

Write your answers in a notebook and then compare your answers with those on page 66.

13.1A Who makes the final decisions as to types of fluoride chemicals and feeding equipment to be used?

13.2A List the three compounds most commonly used to fluoridate water.

---

[2] NSF International, PO Box 130140, 789 North Dixboro Road, Ann Arbor MI, 48113-1040. Phone 1-800-NSF-MARK (800-673-6275), website: http://www.nsf.org.

13.3 FLUORIDATION SYSTEMS

Drinking waters may come to contain fluoride ions by three different types of situations. First, the raw water source may have adequate or excessive fluoride ions naturally present. Second, sometimes two water sources are blended together to produce an acceptable level of fluoride ions. This can occur when one source has a higher than acceptable level of fluoride ions and the other is below the desired level. This chapter is mainly concerned with the third situation in which fluoride ions must be added to the water to achieve an acceptable level of fluoride ions.

13.30 Chemical Feeders

Fluoride ions are added to water by either chemical solution feeders or dry feeders. Solution feeders are *POSITIVE DISPLACEMENT*[3] diaphragm pumps (Figure 13.1), peristaltic pumps (Figure 13.2), or electronic pumps (Figure 13.3), that deliver a fixed amount of liquid fluoridation chemical with each stroke or pulse. The dry feeders are either *VOLUMETRIC*[4] or *GRAVIMETRIC*[5] types of chemical feeders. Volumetric feeders (Figures 13.4 and 13.5) are usually simpler, less expensive, less accurate, and feed smaller amounts of chemicals than gravimetric feeders. Gravimetric feeders (Figure 13.6) are usually more accurate than volumetric feeders; however, they are more expensive and require more space for installation. The amount fed is measured on the basis of the weight of chemical to be fed to the system. Fluoride chemical feeders must be very accurate.

Whatever the type of feeding system or chemical used, the design should be planned by an engineer experienced in developing feeding systems (Figure 13.7). The design must incorporate means to prevent both overfeeding and backsiphonage along with a means to monitor the amount of chemical used. It is also desirable to incorporate some means of feeding fluoride that is adjusted (paced) according to the plant flow rate. Also, a means to continuously measure the finished water's fluoride ion content with an adjustable high-fluoride alarm is desirable. Fluoride doses must never be metered against a negative or suction head.

13.31 Saturators[6]

The saturator (Figure 13.8) is a special application of a solution feeder. A small pump delivers a saturated solution of sodium fluoride into the water system. The principle of a saturator is that a saturated solution will result if water is allowed to trickle through a bed containing sodium fluoride.

Although saturated solutions of sodium fluoride can be manually prepared, generally the easiest and best way is an automatic feed device. Only crystal-grade (20 to 60 mesh) sodium fluoride should be fed with a saturator. Sodium fluoride has a nearly constant solubility at normal temperature ranges and thus produces a fluoride solution of uniform strength. Sodium silicofluoride is not recommended to feed through the saturator because of its very low solubility in water (see Table 13.2). Maintain a depth of six to ten inches (150 to 250 mm) in the sodium fluoride bed of the saturator. Saturators should be stirred every day to prevent fluoride solids from building up on the bottom. The use of powdered sodium fluoride is not recommended because this chemical will clog the saturator.

Many dry chemical feed systems include a mixer, dissolving tank, and *DAY TANK*[7] (Figure 13.9). The mixer mixes a known amount of chemicals with a measured amount of water in a dissolving tank or solution chamber. This is a "batch mixed" process because the chemicals are mixed with a specified amount of water, rather than being mixed in flowing water. The dissolving tank allows the chemicals to become dissolved in water. The solution is then either continuously applied to the water being treated or is stored in a day tank or storage tank. The day tank usually stores enough chemical solution to properly treat the plant flow for at least one day. The chemical solution is fed from the day tank to the water being treated by a chemical feeder (feed pump) whose feed rate continuously adjusts to the flow being treated (flow-paced).

When working with fluoridation systems using a sodium fluoride solution, the hardness of the water is very important. Hard water can produce problems in systems using saturators and dissolving tanks through the formation of low-solubility deposits of calcium and magnesium fluoride compounds. If the dilution water has a hardness of less than 10 mg/L hardness as $CaCO_3$, there will be no problem. If the hardness range is above 10 mg/L and below 75 mg/L, scaling will continue to occur and the cost of ongoing cleaning and maintenance must be considered. Above a hardness of 75 mg/L, a softened water must be used for the water to prepare a fluoride solution in order to prevent severe scaling in the equipment.

Water can be softened prior to use with fluoridation equipment by the use of zeolite ion exchangers. As an alternative to the zeolite water softener, polyphosphates (at 7 to 15 mg/L) may be used to prevent plugging by scale in the feed system. The polyphosphates are added to the day tank to prevent scale deposits. If neither a zeolite softener nor polyphosphates are used, plugging may occur at any point in the feed system, including valves,

---

[3] *Positive Displacement Pump.* A type of piston, diaphragm, gear, or screw pump that delivers a constant volume with each stroke. Positive displacement pumps are used as chemical solution feeders.
[4] *Volumetric Feeder.* A dry chemical feeder that delivers a measured volume of chemical during a specific time period.
[5] *Gravimetric Feeder.* A dry chemical feeder that delivers a measured weight of chemical during a specific time period.
[6] *Saturator* (SAT-yoo-ray-tor). A device that produces a fluoride solution for the fluoridation process. The device is usually a cylindrical container with granular sodium fluoride on the bottom. Water flows either upward or downward through the sodium fluoride to produce the fluoride solution.
[7] *Day Tank.* A tank used to store a chemical solution of known concentration for feed to a chemical feeder. A day tank usually stores sufficient chemical solution to properly treat the water being treated for at least one day. Also called an AGE TANK.

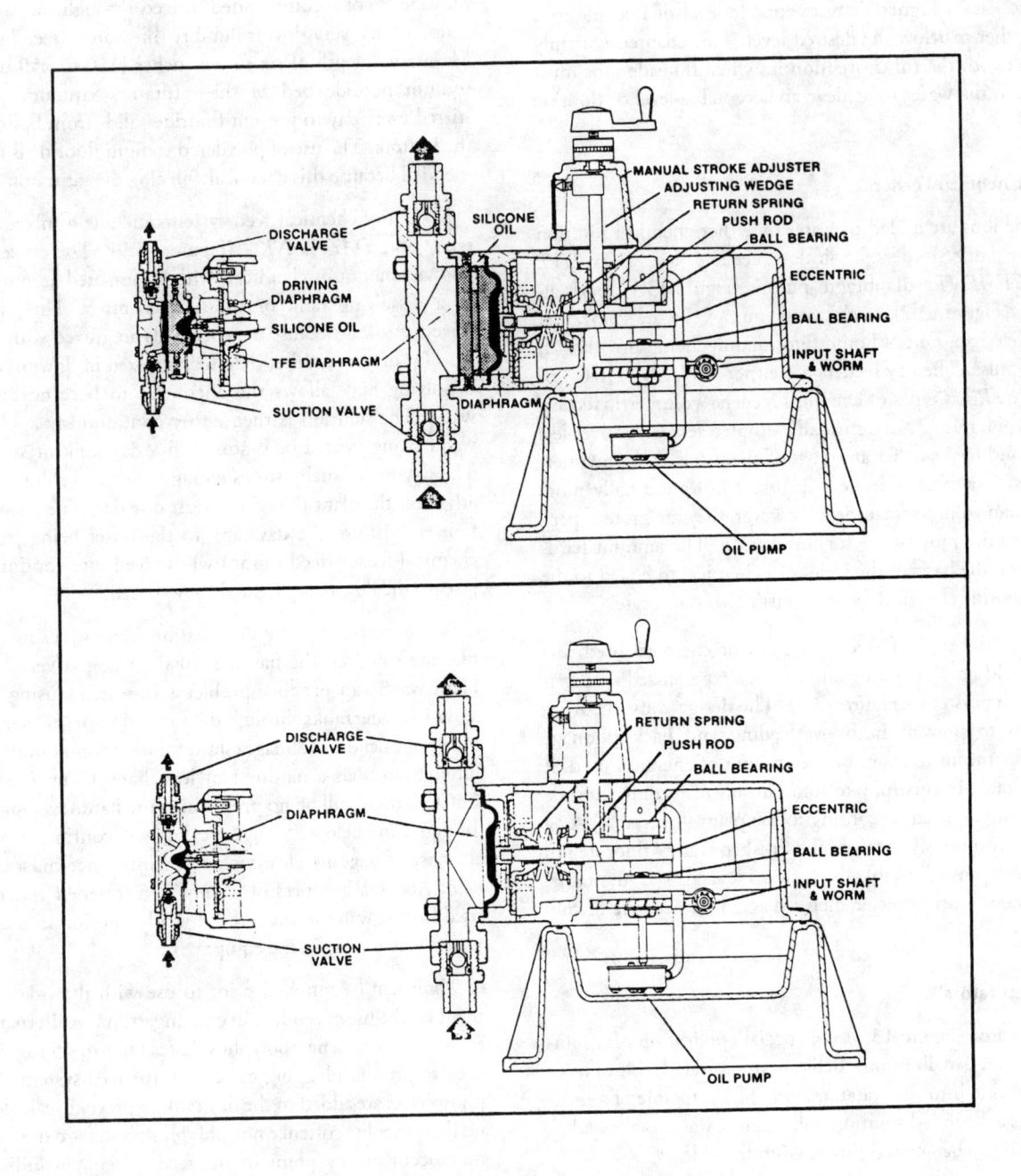

Fig. 13.1 Positive displacement diaphragm pumps
(Permission of Wallace & Tiernan, Inc.)

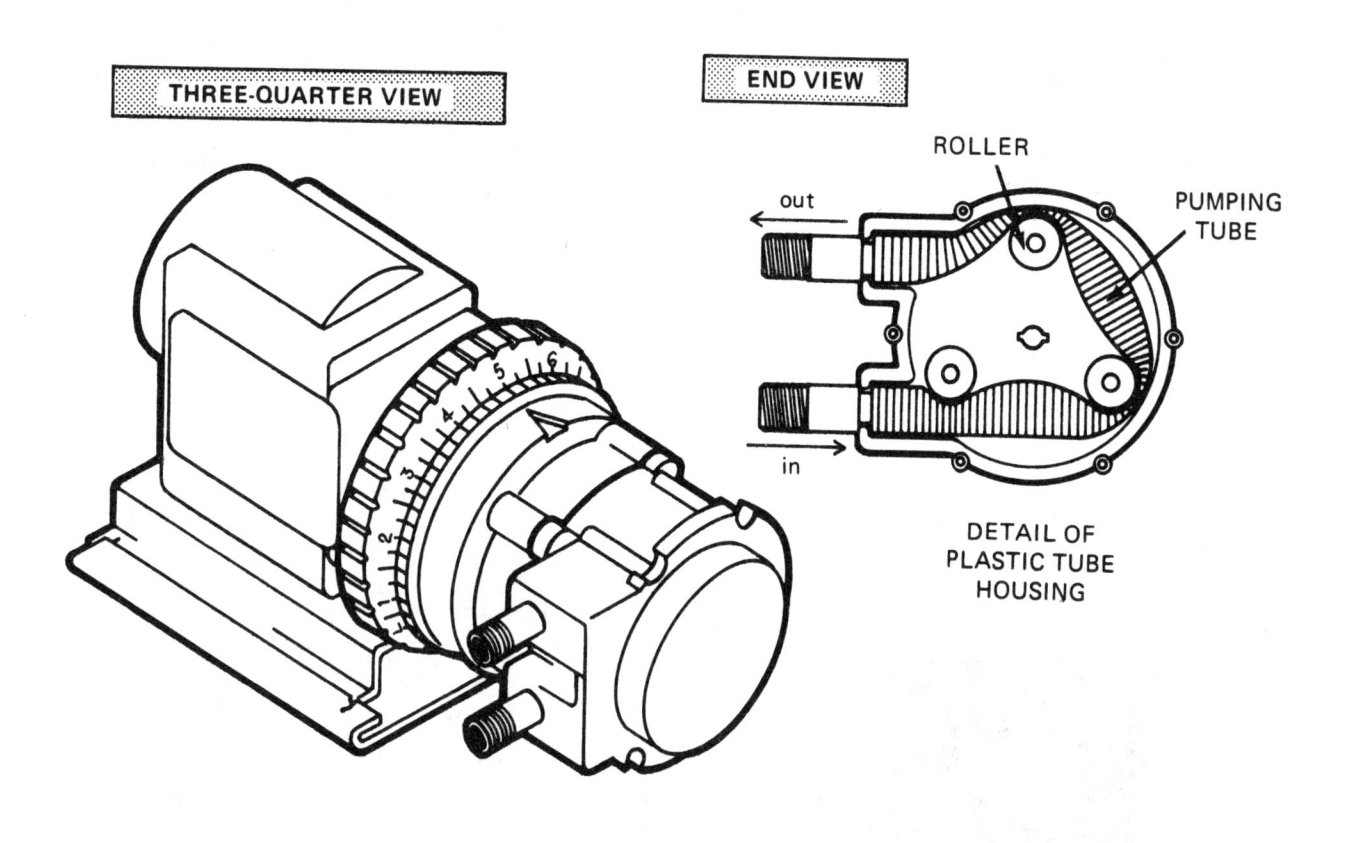

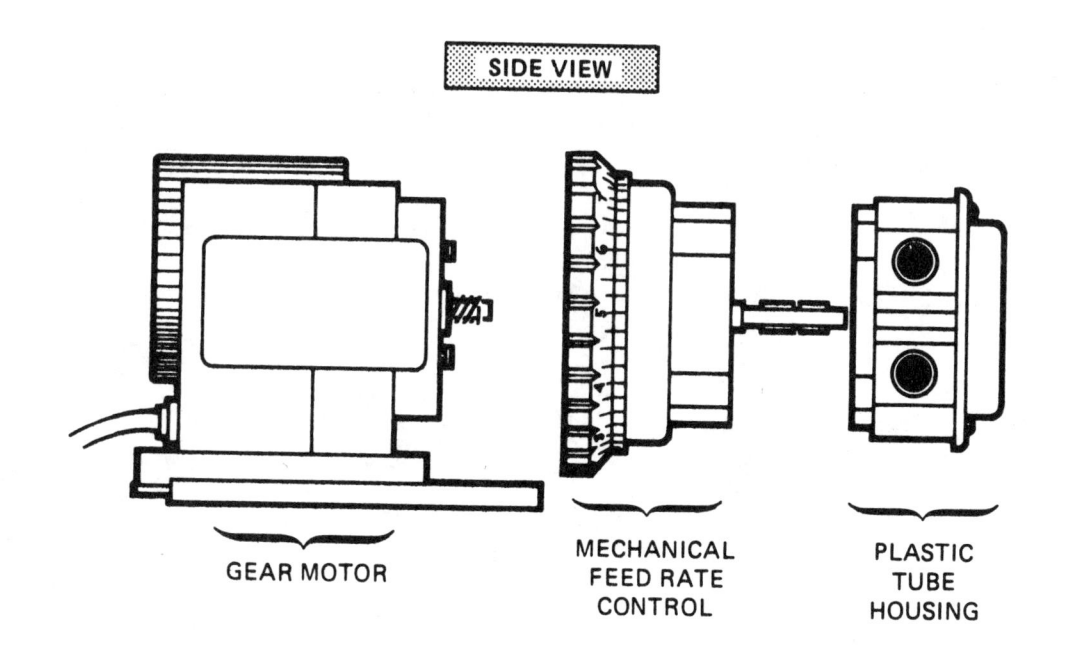

Fig. 13.2 Peristaltic feeder
(Reproduced from *WATER FLUORIDATION*, A Training Course Manual for Engineers
and Technicians, by permission of the Dental Disease Prevention Activity,
US Public Health Service)

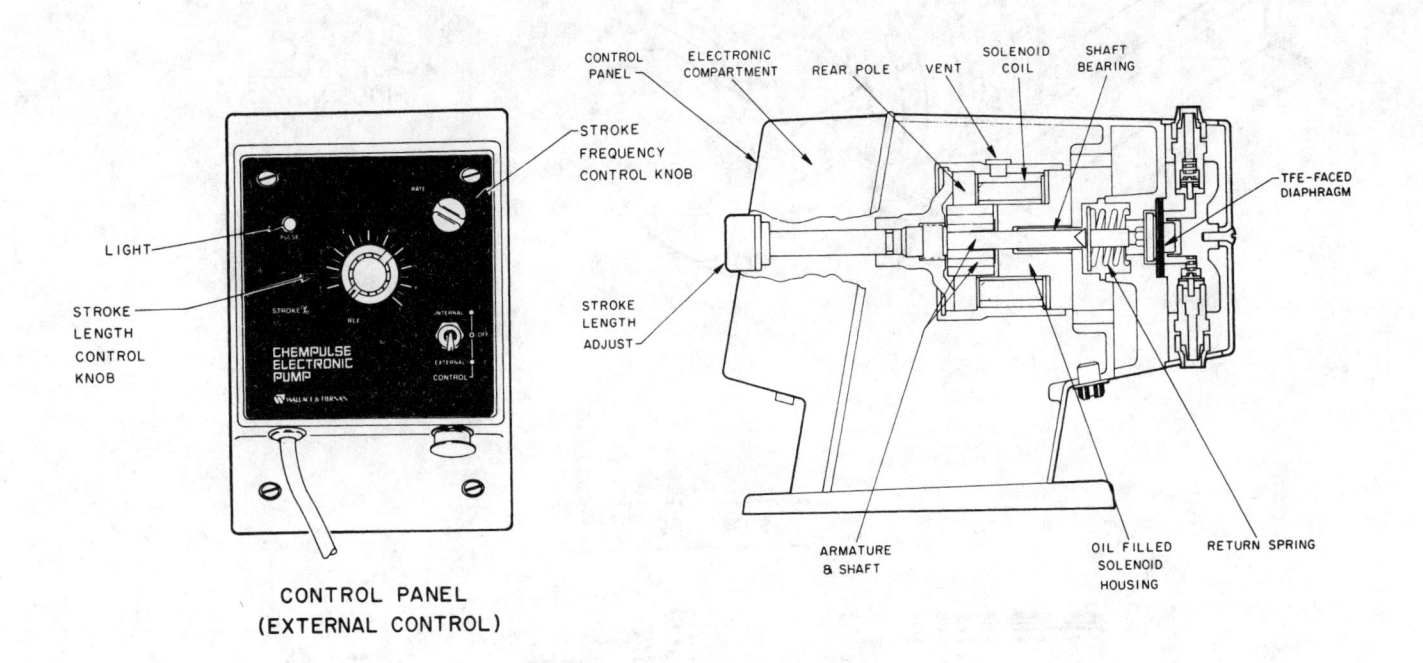

Fig. 13.3 Electronic feeder
(Permission of Wallace & Tiernan, Inc.)

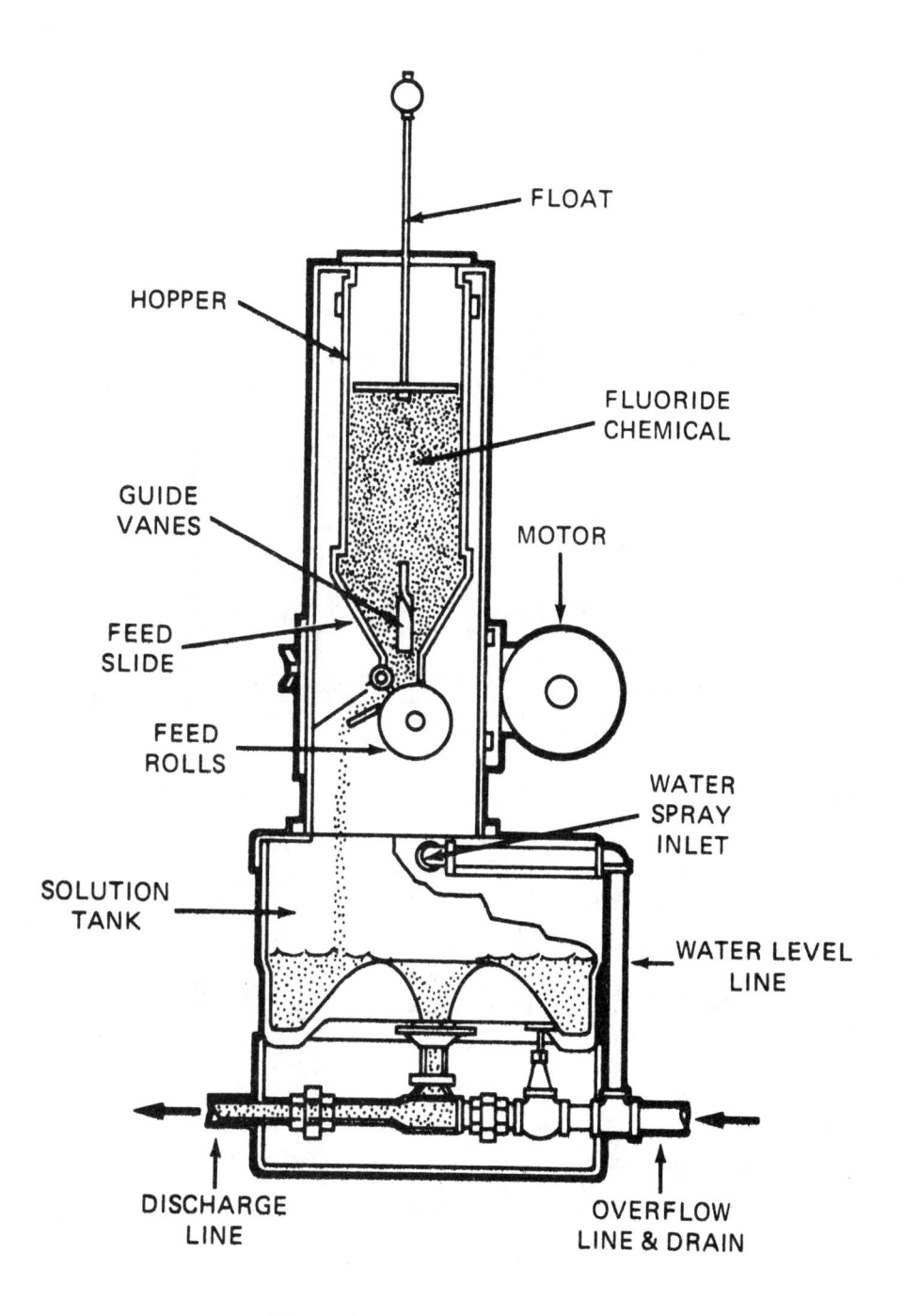

Fig. 13.4 Volumetric feeder, roll-type
(Reproduced from *WATER FLUORIDATION,* A Training Course Manual for Engineers and Technicians, by permission of the Dental Disease Prevention Activity, US Public Health Service)

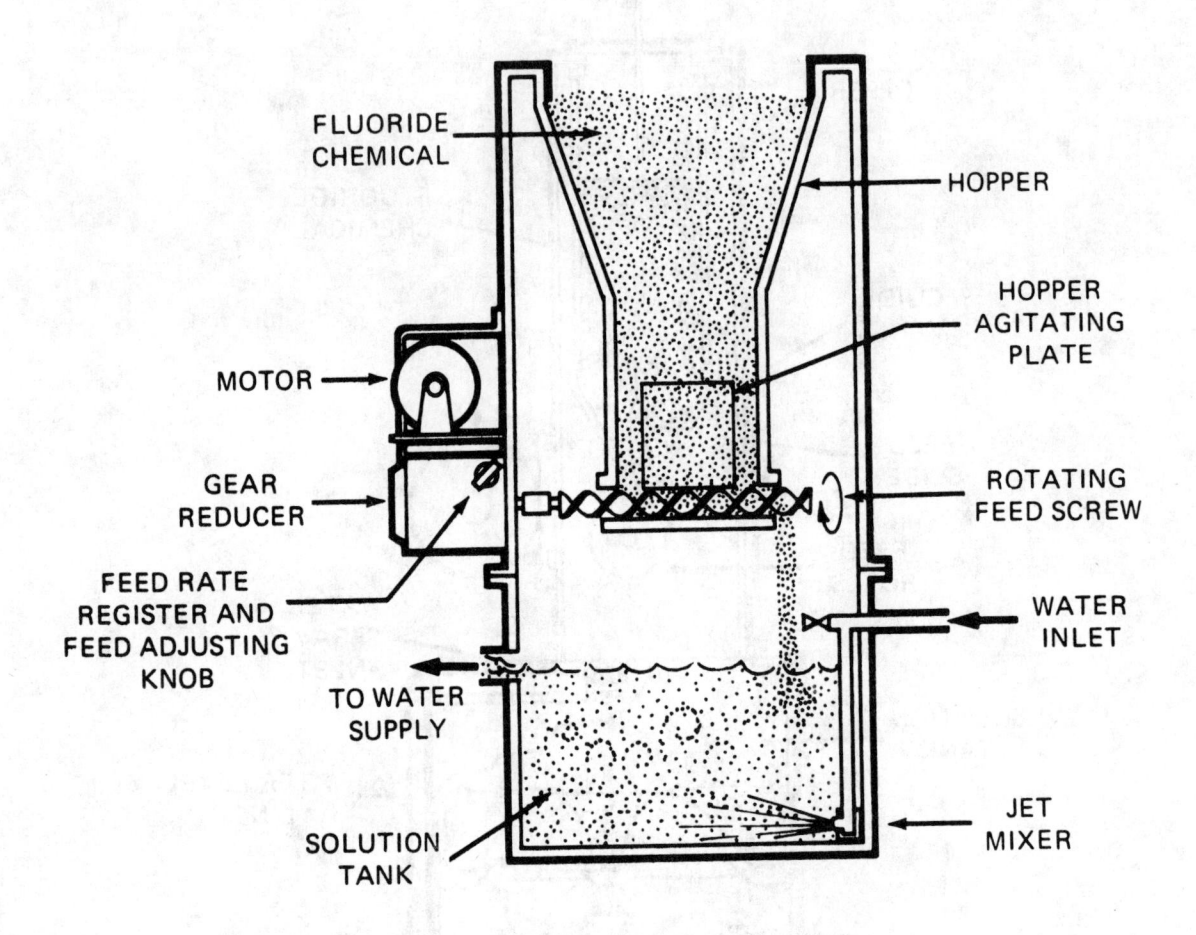

Fig. 13.5 Volumetric feeder, screw-type
(Reproduced from WATER FLUORIDATION, A Training Course Manual for Engineers
and Technicians, by permission of the Dental Disease Prevention Activity,
US Public Health Service)

Fluoridation 41

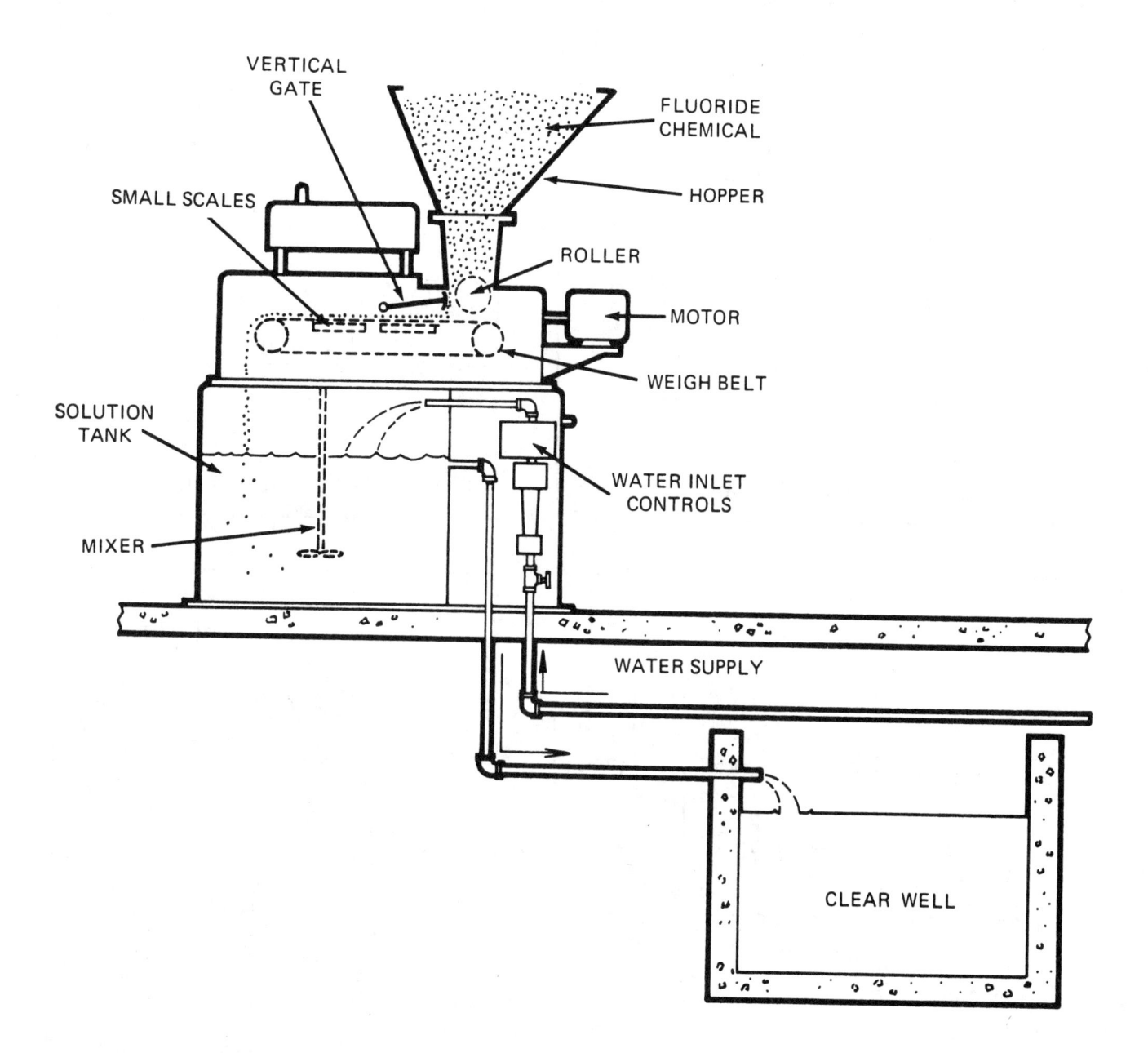

Fig. 13.6 Gravimetric feeder, belt-type
(Reproduced from *WATER FLUORIDATION*, A Training Course Manual for Engineers and Technicians, by permission of the Dental Disease Prevention Activity, US Public Health Service)

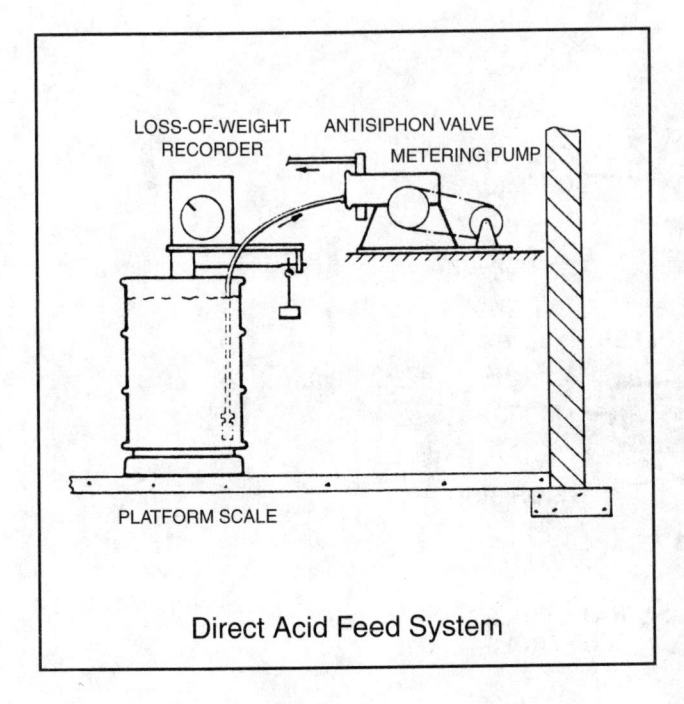

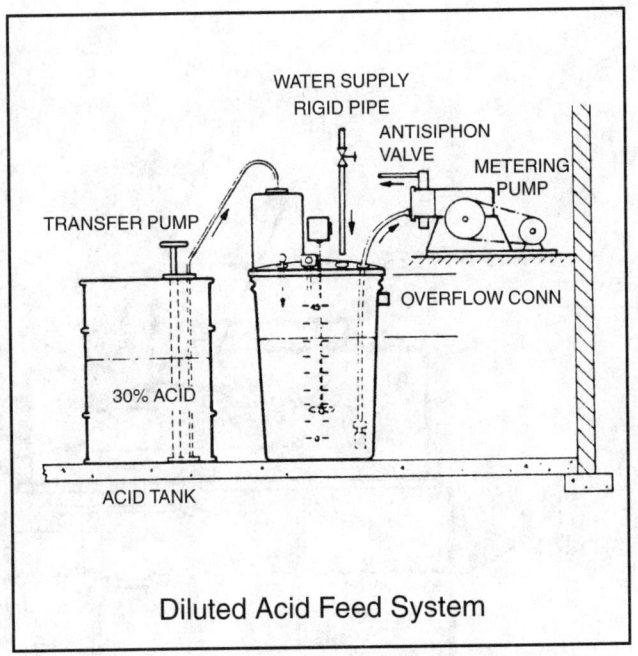

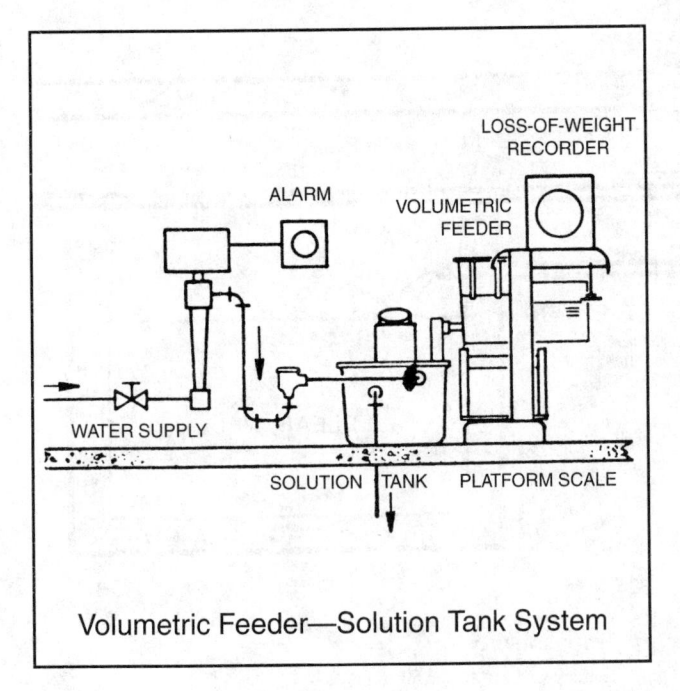

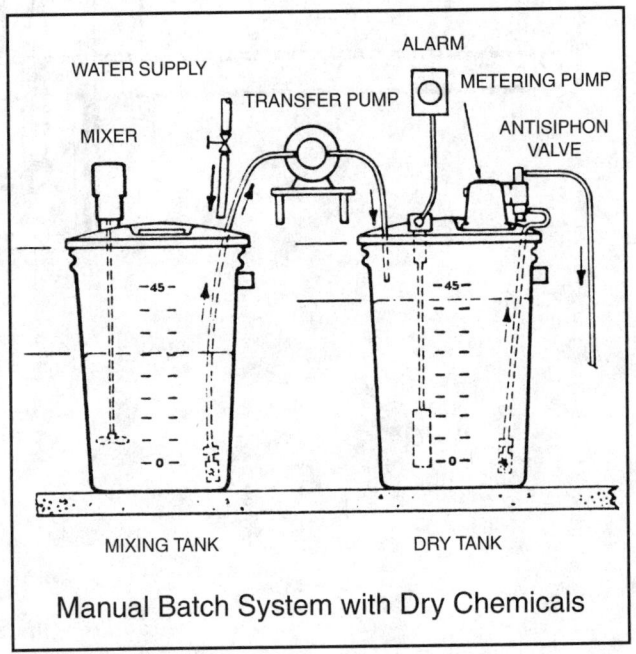

Fig. 13.7 Typical fluoridation systems
(Permission of Wallace & Tiernan, Inc.)

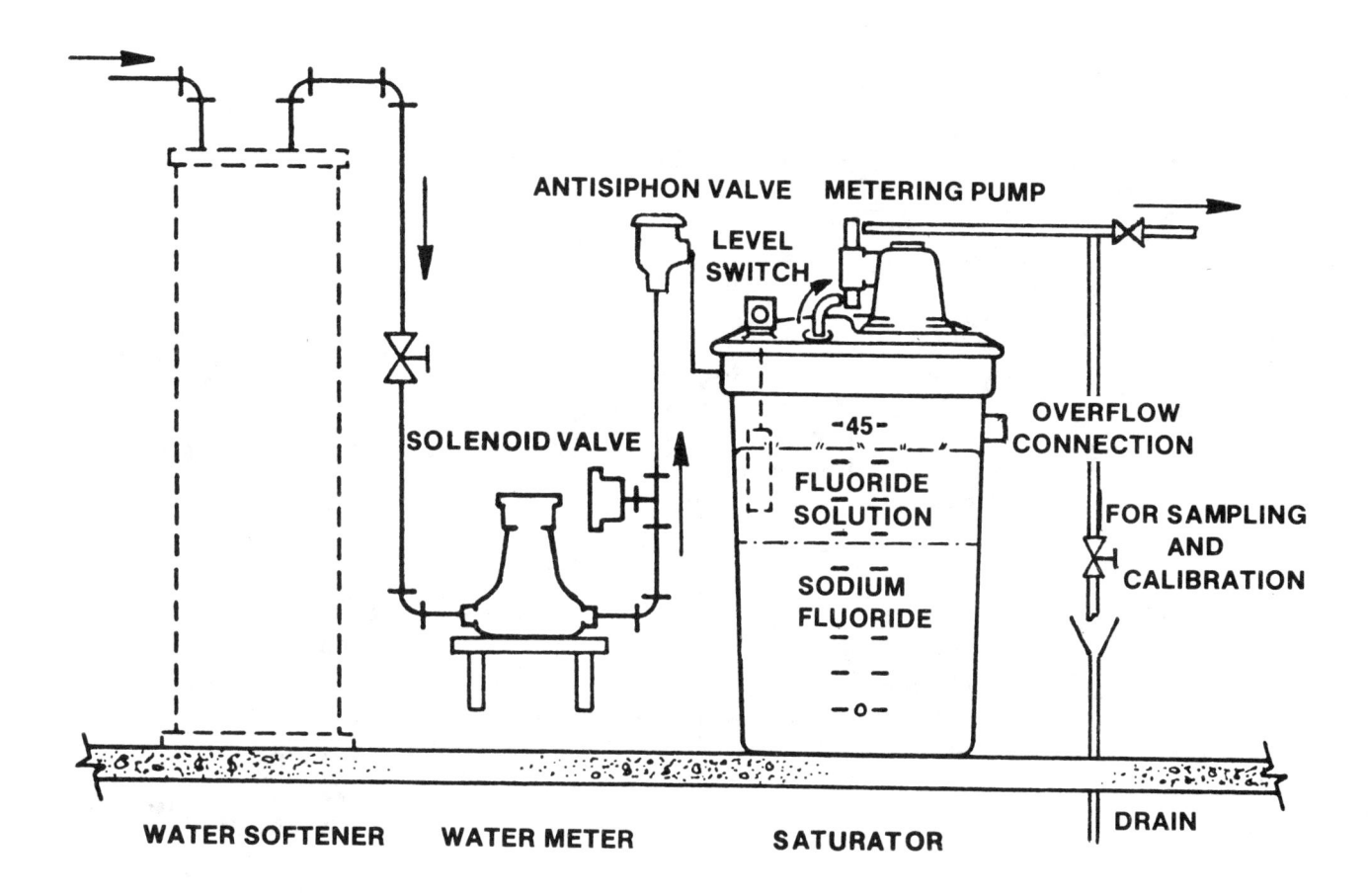

Fig. 13.8 Fluoride saturator
(Permission of Wallace & Tiernan, Inc.)

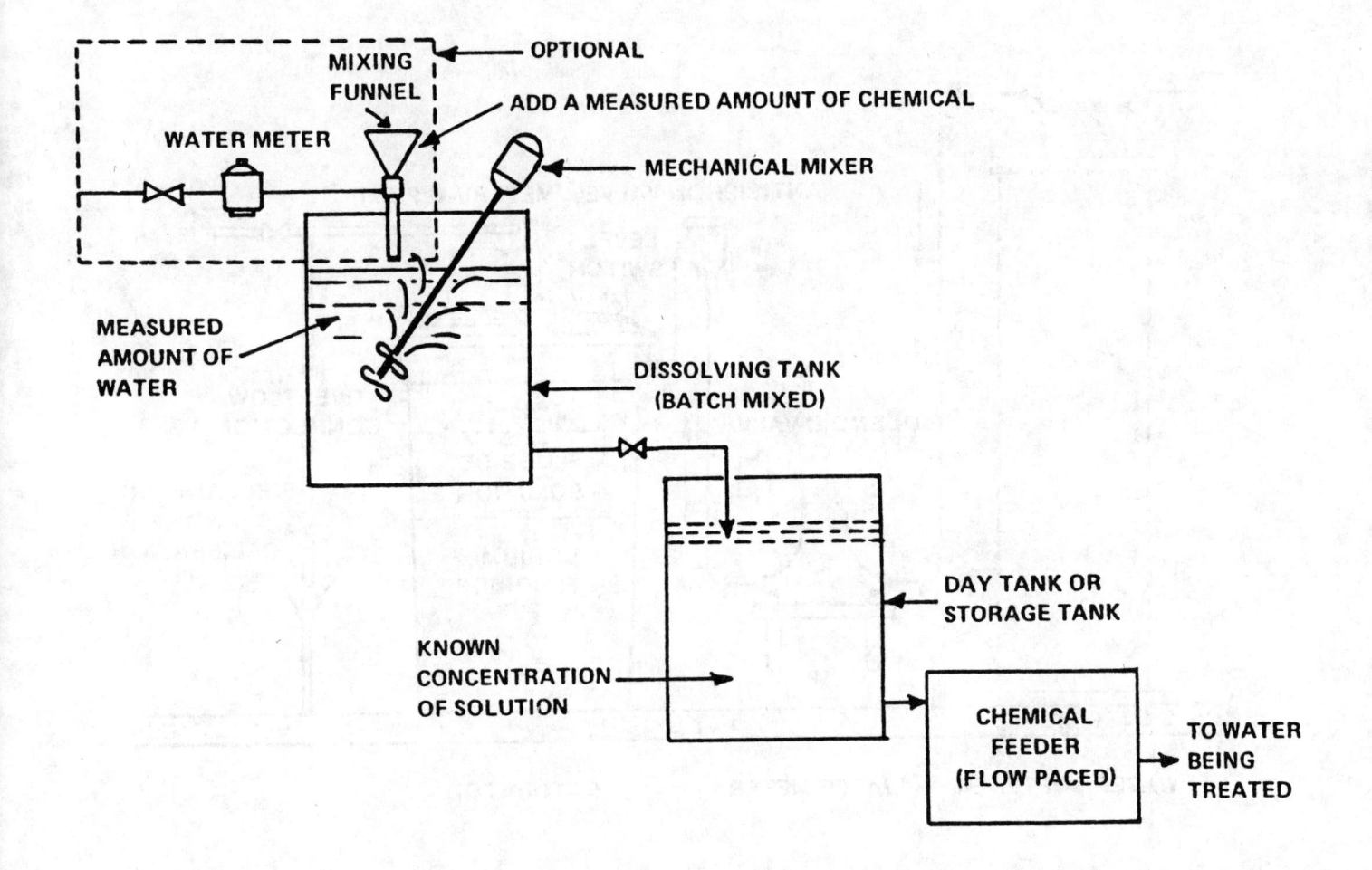

Fig. 13.9 Dry chemical dissolver, day tank, and feeder

saturator bed, and injection point. Remove these hardness deposits by flushing the system with vinegar or a five-percent solution of hydrochloric acid (muriatic acid). The saturator beds also may require the removal of water hardness deposits.

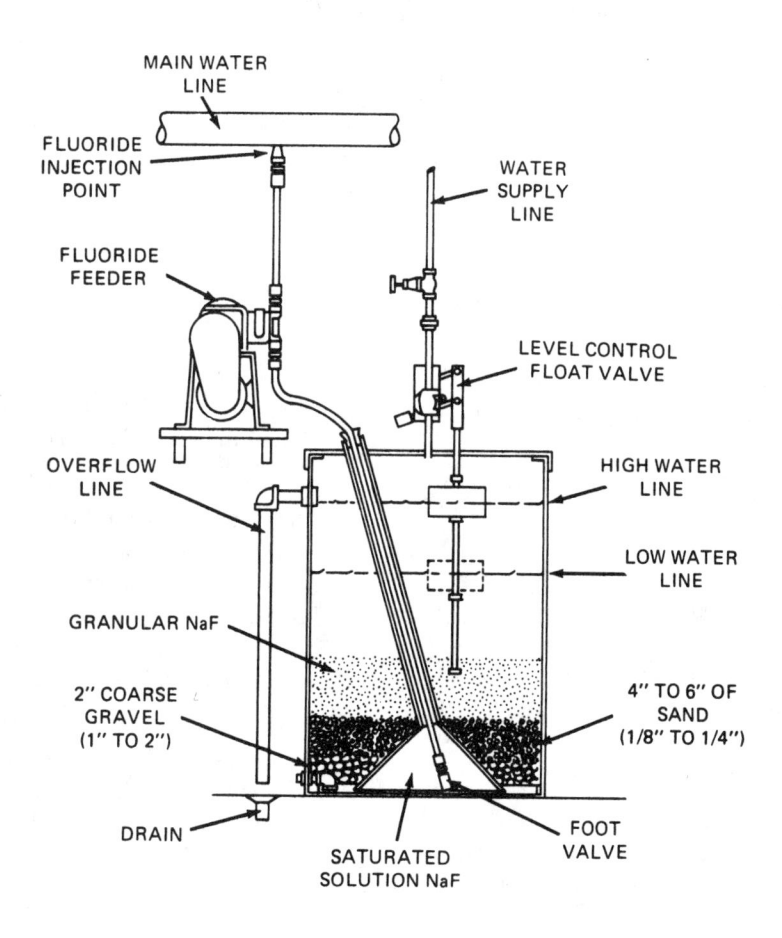

13.32 Downflow Saturators (Figure 13.10)

There are two kinds of saturators, the upflow saturator and the downflow saturator. In the downflow saturator, the solid sodium fluoride is held in a plastic drum or barrel and is isolated from the prepared solution by a plastic cone or a pipe manifold. A filtration barrier is provided by layers of sand and gravel to prevent particles of undissolved sodium fluoride from infiltrating the solution area under the cone or within the pipe manifold. The feeder pump draws the solution from within the cone or manifold at the bottom of the plastic drum. Downflow saturators require clean gravel and sand. In some systems, the gravel and sand must be cleaned every day or two.

13.33 Upflow Saturators (Figure 13.11)

In an upflow saturator, undissolved sodium fluoride forms its own bed below which water is forced upward under pressure. No barrier is used since the water comes up through the bed of sodium fluoride and the specific gravity of the solid material keeps it from rising into the area of the clear solution above. A spider-type water distributor located at the bottom of the tank contains hundreds of very small slits. Water, forced under pressure through these slits, flows upward through the sodium fluoride bed at a controlled rate to ensure the desired four-percent solution. The feeder pump intake line floats on top of the solution in order to avoid withdrawal of undissolved sodium fluoride. The water pressure requirements are 20 psi (138 kPa) minimum to 125 psi (862 kPa) maximum and the flow is regulated at 4 GPM (0.25 L/sec). Since introduction of water to the bottom of the saturator constitutes a definite cross connection, a mechanical siphon-breaker must be incorporated into the water line; or better still, the saturator can be factory modified to include an air gap and a water feed pump.

Figures 13.8 and 13.11 show two configurations of upflow-type saturators that feed and prepare constant-strength fluoride solution from granular sodium fluoride. The upflow type is the preferred type over the downflow saturator, as it is much easier to clean and maintain. Under normal conditions, it

Fig. 13.10 Downflow saturator
(Reproduced from *WATER FLUORIDATION*, A Training Course Manual for Engineers and Technicians, by permission of the Dental Disease Prevention Activity, US Public Health Service)

46 Water Treatment

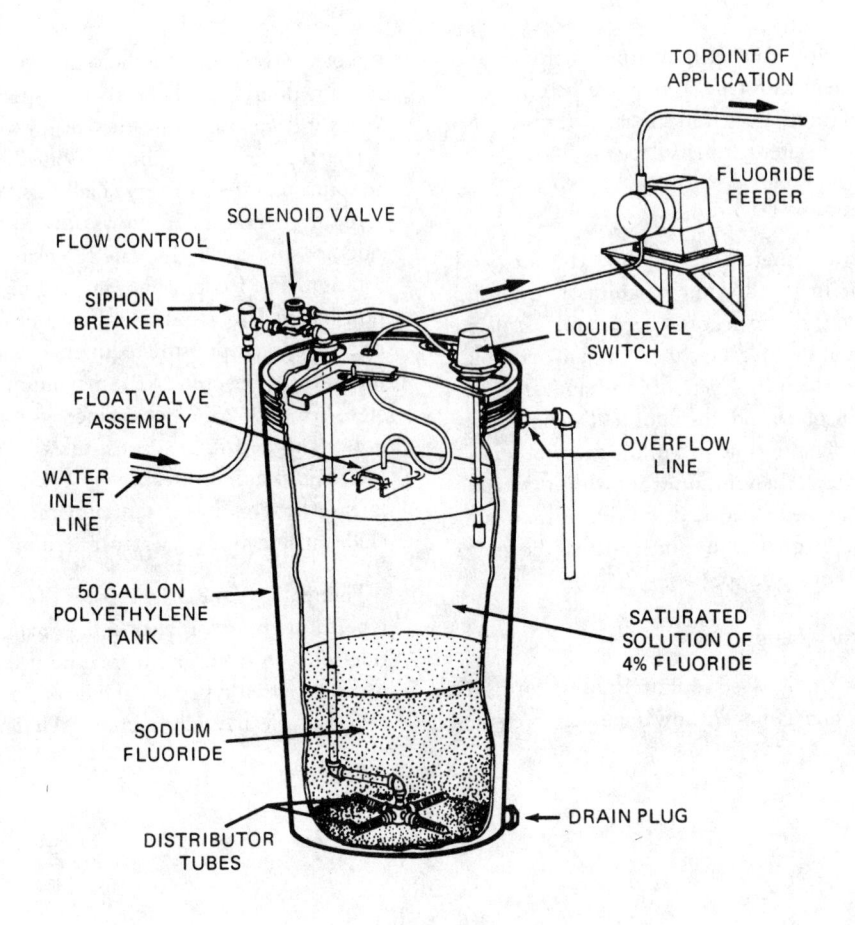

Fig. 13.11 Upflow saturator
(Reproduced from *WATER FLUORIDATION,* A Training Course Manual for Engineers
and Technicians, by permission of the Dental Disease Prevention Activity,
US Public Health Service)

should need cleaning only once a year. For these reasons, the construction and use of an upflow saturator is discussed in some detail.

To prepare an upflow saturator for use, the following steps should be taken.

1. With the distributor tubes in place, and the floating suction device removed, add 200 to 300 pounds (91 to 136 kg) of sodium fluoride directly to the tank. Any type of sodium fluoride can be used, from coarse crystal to fine crystal, but fine crystal will dissolve better than coarse material. If crystal is not available, powder can be used, but it is not as desirable as a crystal form of sodium fluoride.

2. Connect the solenoid water valve to an electric outlet and turn on the water supply. The water level should be slightly below the overflow; if it is not, the liquid level switch should be adjusted.

3. Replace the intake float and connect it to the feeder intake line. The saturator is now ready to use.

4. By looking through the translucent wall of the saturator tank, you should be able to see the level of undissolved sodium fluoride. Whenever the level is low enough, add another 100 pounds (45 kg) of fluoride.

5. The water distributor slits are supposed to be essentially self-cleaning. The accumulation of insolubles and precipitates is not as serious a problem as it is in a downflow saturator; however, a periodic cleaning is still required. Frequency of cleaning is dictated by the severity of use and the rate of accumulation of debris.

6. Because of the thicker bed of sodium fluoride attainable in an upflow saturator, higher withdrawal rates are possible. With 300 pounds (136 kg) of sodium fluoride in the saturator tank, more than 15 gallons per hour (58 L/hr) of saturated solution can be fed. This rate is sufficient to treat about 5,000 gallons per minute (135 L/sec) of water to a fluoride level of 1.0 mg/L.

7. The fixed water inlet rate of 4 GPM (0.25 L/sec) should register satisfactorily on a ⅝-inch (16-mm) meter.

From a financial point of view, many water agencies will want to design their fluoridation plants for unattended operation. In such cases, the system design will include provisions for automatic shutdown and an alarm system. For the sake of the operator who has to respond at all hours, alarm lights should be wired to indicate reasons for plant shutdown. For example, alarms could be set to indicate low water flow in the main pipeline, high fluoride flow, high or low fluoride levels, low injection water pressure, power outage, and a running time meter to indicate downtime. These warning systems are particularly helpful in large systems.

13.34 Large Hydrofluosilicic Acid Systems
(Figure 13.12)

A more complicated system for fluoridation is a closed-loop control feeding system using hydrofluosilicic acid. This system finds use in large installations where the hydrofluosilicic acid can be delivered by tanker truck of around 4,000 gallons (15,140 L), although, of course, smaller amounts can be purchased. The installation depicted in Figure 13.12 can treat up to 285 million gallons per day (1,079 ML/day). The advantages in this system are the elimination of dust hazards and minimal labor requirements.

The storage tanks are made of filament-wound fiberglass. The interior lining consists of a surfacing fabric with a veil-type covering serving as a final barrier. Steel tanks lined with at least 3/32 inch (2.34 mm) polyvinyl chloride sheet or neoprene sheet secured to the metal surface with adhesive can also be used. Hydrofluosilicic acid storage tanks constructed of plastic should be housed in enclosures to protect the tanks from vandalism. Leaks could be dangerous to passersby and will kill surrounding vegetation. The tanks must be vented to the outside as the fumes from the acid are highly corrosive.

Inspection of internal conditions of the hydrofluosilicic tanks should be made on two-year intervals as some lining deterioration can be expected over a period of time. Should small leaks occur in the PVC piping, repairs should be made at once; they will only become worse and any acid dripping on concrete surfaces will dissolve the surface fairly quickly.

The use of a closed-loop control system in an unattended plant utilizing a fluoride analyzer as one of the controls is not recommended. The problem is that if the analyzer goes haywire and incorrectly indicates low fluoride levels, the system will try to correct itself and increase the addition of fluoride chemical. The net result will be to actually overfluoridate the water supply.

QUESTIONS

Write your answers in a notebook and then compare your answers with those on page 66.

13.3A List the three different types of situations whereby drinking waters may contain fluoride ions.

13.3B List three important design features of fluoridation systems.

13.3C What is a saturator?

13.3D What problems can be created by hard water in systems using saturators and dissolving tanks?

13.4 FINAL CHECKUP OF EQUIPMENT

13.40 Avoid Overfeeding

The operator must be certain that there will be no overfeed of fluoride ions. A gross overfeed can cause illness and bad public relations. Of all the chemicals used in the water treatment plant, fluoride concentrations probably require the closest attention to maximum dosages.

13.41 Review of Designs and Specifications

When reviewing fluoride feeding system designs and specifications, the operator should check the items listed below.

1. Review the results of predesign tests to determine the fluoride rate for both the present and future. The fluoridator should be sized to handle the full range of chemical doses or provisions should be made for future expansion.

2. Determine if sampling points are provided to measure chemical feeder output.

3. Examine plans for valving to allow flushing the system with water before removing from service.

4. Be sure corrosion-resistant drip pans and drains are provided to prevent chemical leaks from reaching the floor, for example, drips from pump packing.

5. Check that all piping, valves, and fittings are made of corrosion-resistant materials such as PVC or Stainless Steel Type 316.

6. Determine the amount of maintenance required. The system should require a minimum of maintenance. Equipment should be standard, with replacement parts readily available.

7. Consider the effect of changing head conditions (both feeder suction and discharge head conditions) on the chemical feeder output. Changing head conditions will not affect the output if the proper chemical feeder has been specified and installed.

8. Determine whether locations for monitoring readouts and dosage controls are convenient to the operation center and easy to read and record.

9. Any switches that throw the equipment from automatic into a hand or manual mode should be equipped with a color-coded red warning light to indicate that the equipment is on "hand" or "manual." This can easily be accomplished by a double-throw, double-pole toggle switch. Lights with different colors can be used to indicate normal or automatic operation as well as *ON* or *OFF* in order to avoid confusion.

48 Water Treatment

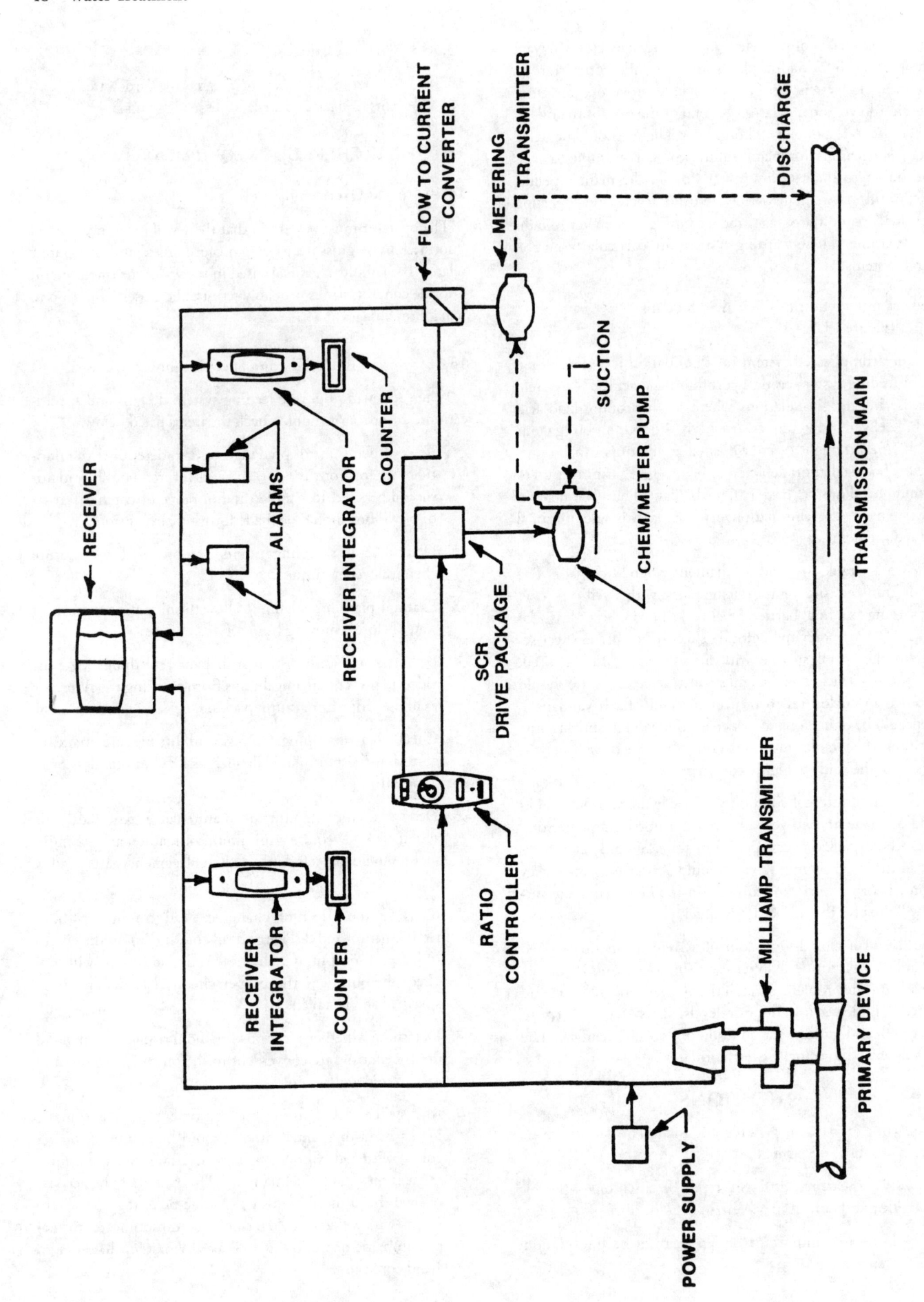

Fig. 13.12 *Large automatic hydrofluosilicic acid system*

10. The location where fluoride is added to the water should be where there will be the least possible removal of fluoride by other chemicals added to the water (after filtration and before the clear well).

11. Be sure the chemical hoppers are located where there is plenty of room so they can be conveniently and safely filled with the fluoride chemical.

12. Dust exhaust systems should be installed where substantial amounts of dry chemicals are handled.

13. In any fluoridation system, except the sodium fluoride saturator, scales are necessary for weighing the quantity of chemical (including solution) fed per day.

14. Alarms are important to signal and prevent both the loss of feed and overfeeding.

15. Any potable water line connected to a chemical feeder unit must be provided with a vacuum breaker or an air gap to prevent a cross connection.

13.5 CHEMICAL FEEDER START-UP

After the chemical feed system has been purchased and installed, carefully check the system out before start-up. Even if the contractor who installed the system is responsible for ensuring that the equipment operates as designed, the operation by plant personnel, the functioning of the equipment, and the results from the process are the responsibility of the chief operator. Therefore, before start-up, check the items listed below.

1. Inspect the electrical system for proper voltage, for properly sized overload protection, for proper operation of control lights on the control panel, for proper safety lock-out switches and operation, and for proper equipment rotation.

2. Confirm that the manufacturer's lubrication and start-up procedures are being followed. Equipment may be damaged in minutes if it is run without lubrication.

3. Examine all fittings, inspection plates, and drains to ensure that they will not leak when placed into service.

4. Determine the proper positions for all valves. A positive displacement pump will damage itself or rupture lines in seconds if allowed to run against a closed valve or system.

5. Be sure that the type of fluoride to be fed is available and in the hopper or feeder. A progressive cavity pump will be damaged in minutes if it is allowed to run dry.

6. Inspect all equipment for binding or rubbing.

7. Confirm that safety guards are in place.

8. Examine the operation of all auxiliary equipment including the dust collectors, fans, cooling water, mixing water, and safety equipment.

9. Check the operation of alarms and safety shutoffs. If it is possible, operate these devices by manually tripping each one. Examples of these devices are alarms and shutoffs for high water, low water, high temperature, high pressure, and high chemical levels.

10. Be sure that safety equipment such as eye wash stations, drench showers, dust masks, face shields, gloves, and vent fans are in place and functional.

11. Record all important nameplate data and place it in the plant files for future reference.

QUESTIONS

Write your answers in a notebook and then compare your answers with those on pages 66.

13.4A Why must overfeeding of fluoridation chemicals be prevented?

13.4B What should be the capacity or size of the fluoridator?

13.5A What items should be considered when inspecting the fluoridation electrical system?

13.5B List the safety equipment that should be available near a fluoridation system.

13.6 CHEMICAL FEEDER OPERATION

13.60 Fine-Tuning

Once the chemical feed equipment is in operation and the major bugs are worked out, the feeder will need to be fine-tuned. To aid in fine-tuning and build confidence in the entire chemical feed system, the operator must maintain accurate records (see Section 13.62, "Fluoridation Log Sheets").

A comment or remarks section should be used to note abnormal conditions, such as a feeder plugged for a short time, related equipment that malfunctions, and other problems. Daily logs should be summarized into a form that operators can use as a future reference.

13.61 Preparation of Fluoride Solution

To learn how to make up a fluoride solution, let us assume a hypothetical case using the following data:

1. Flow to be treated is 10 million gallons per day.

2. Hydrofluosilicic acid 20 percent is the chemical to be used.

3. The unfluoridated water contains 0.05 mg/L (ppm) fluoride ion (F^-).

4. The desired fluoride concentration in the treated water is 1.0 mg/L.

What should the feed rate be?

A treatment chart, such as those on pages 50 and 51, can be used to determine the feed rate. For example, using "Treatment Chart I, Hydrofluosilicic Acid," on page 50, locate the 10 MGD flow on the left-hand scale of the graph and follow that line to the right until it intersects the 20 percent diagonal line. Project this point down vertically to the intersection of bottom line indicating gallons per day (or gallons per hour) required to

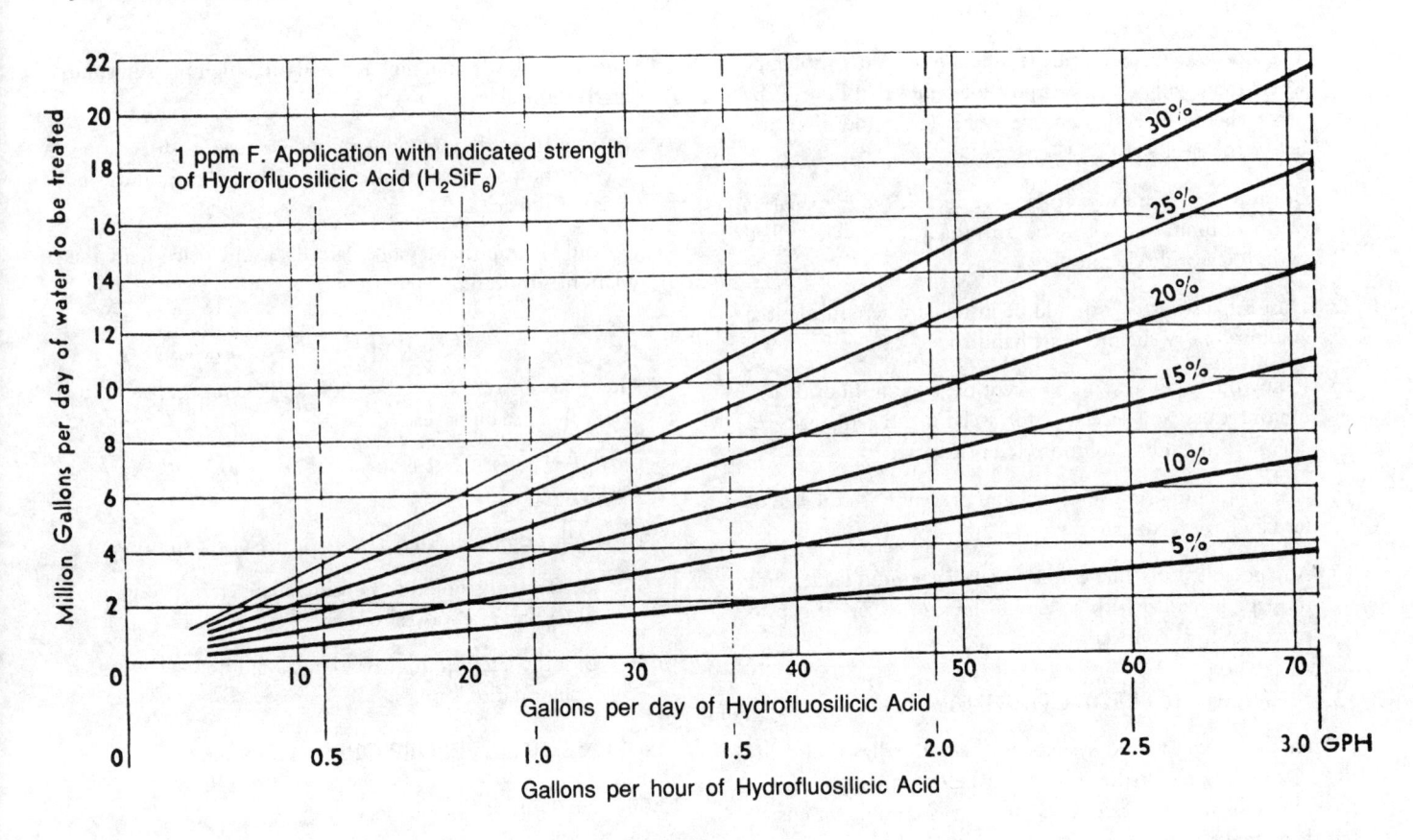

TREATMENT CHART I
Hydrofluosilicic Acid

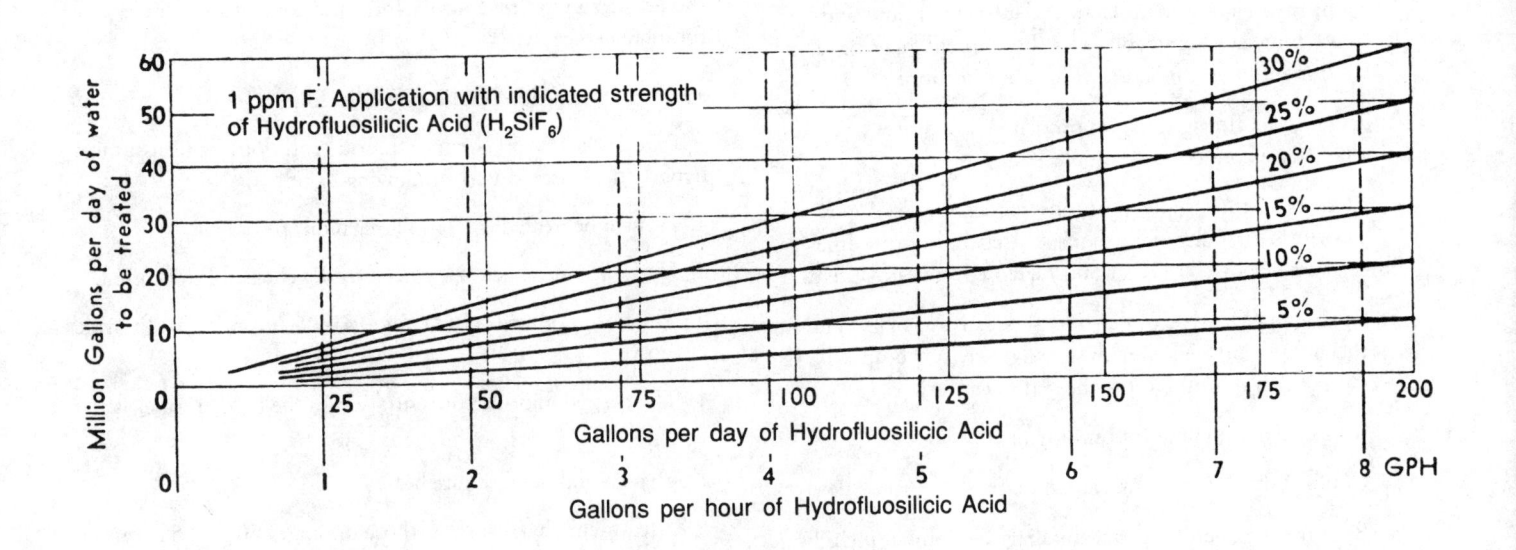

TREATMENT CHART II
Hydrofluosilicic Acid

(Permission of Wallace & Tiernan, Inc.)

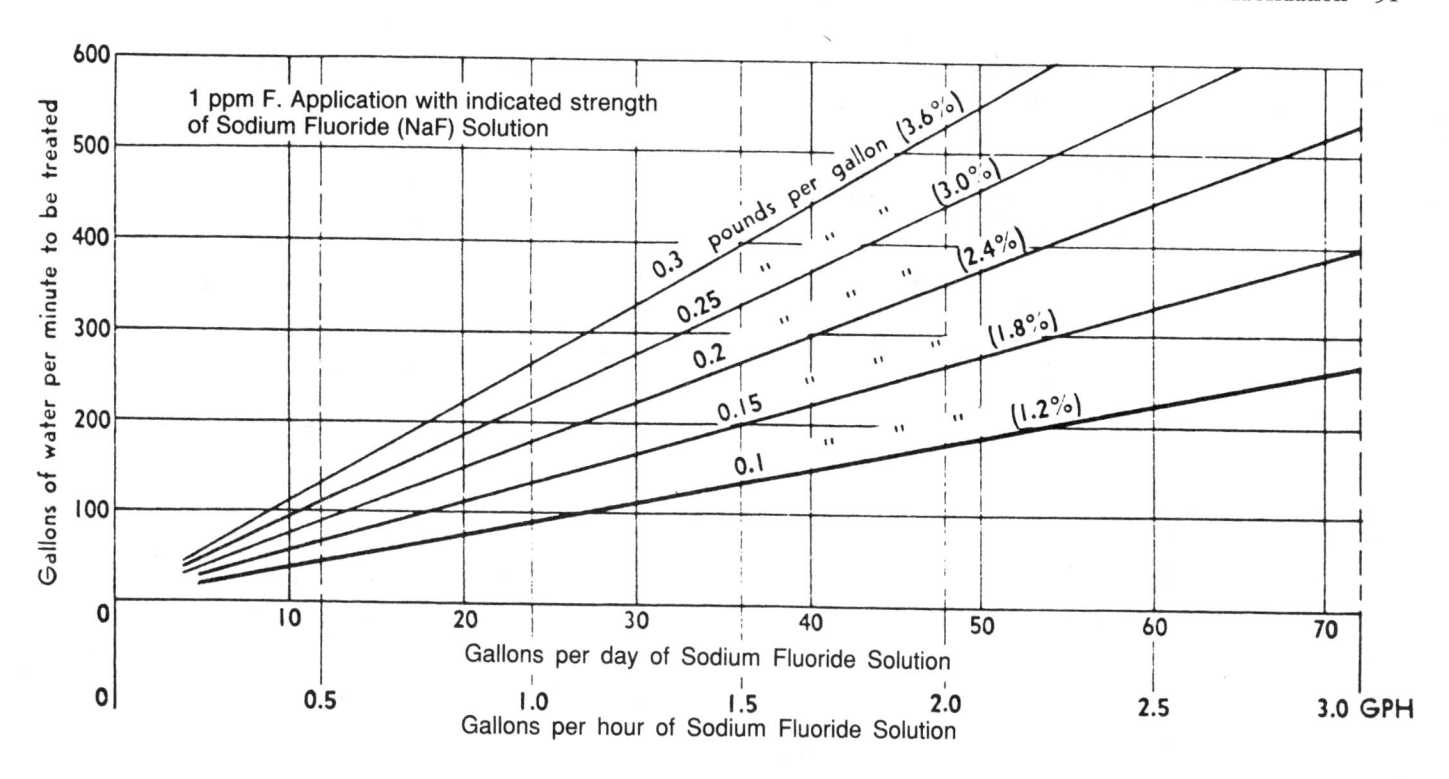

TREATMENT CHART III
Sodium Fluoride

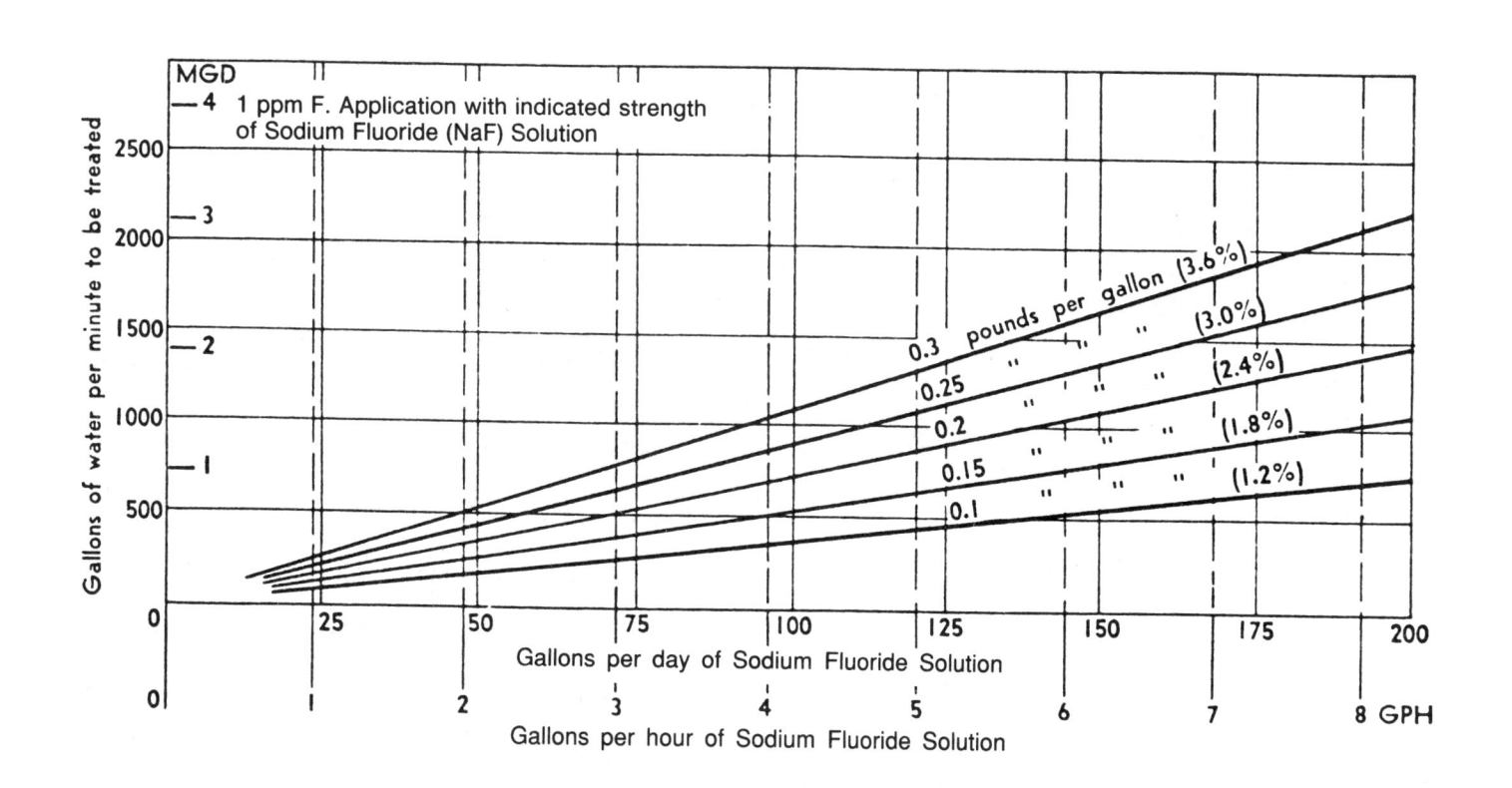

TREATMENT CHART IV
Sodium Fluoride

(Permission of Wallace & Tiernan, Inc.)

produce a one-mg/L (ppm) dose of fluoride (F). The answer is 50 gallons per day or a little less than 2.1 GPH (gallons per hour). Multiply the 50 by (1.00 − 0.05) to give the needed treatment of 47.5 gallons per day or 2 GPH. The 1.00 is the desired dose of 1.00 mg/L and the 0.05 is the actual fluoride concentration of 0.05 mg/L in the untreated water.

In some cases, it might be desirable to use a weaker acid solution to avoid feed rates below the minimum capacity of the metering pump. Dilution then is in order. The concentration may be reduced by volumetric proportions, for example one gallon of 20-percent acid plus one gallon of water results in two gallons of 10-percent acid. If possible, try to avoid having to dilute acid because of potential errors and problems, especially with hard water. Peristaltic and electronic feeder pumps (Figures 13.12 and 13.3) may be used when the feed rates are low.

For more practice using the treatment charts, see Section 13.12, "Calculating Fluoride Dosages," for eleven example problems.

13.62 Fluoridation Log Sheets

You will probably want to design your own log sheets so they will be consistent with the installation features at your plant. Sample log sheets are shown on Figures 13.13, 13.14, and 13.15 (see pages 53, 55, and 56).

13.620 Hydrofluosilicic Acid

Figure 13.13 shows a typical log sheet from a hydrofluosilicic acid station. An explanation of the various columns is given below.

1. "Date" refers to calendar date when readings were logged or the date a shipment of fluoride was received.
2. "Time" refers to time event happened.
3. "Tank" that is supplying the feeder is circled. "Gals." refers to the gauge reading of the amount of acid in the tank.
4. "% & Sp. Gr." Each delivery is accompanied by a vendor's laboratory analysis. The specific gravity is not measured until the tank is ready to be placed in service. When mixing acids of varying strengths, the end percentage must be calculated and entered in the proper column. See Example 10 in Section 13.12, "Calculating Fluoride Dosages."
5. For each tank follow the directions given in Step 4.
6. "Tank Loss Gals." refers to the amount of feed during the reporting period. In the sample 2,930 − 2,600 = 330 gallons. The feeding equipment should be equipped with an acid totalizer readout.
7. The "ratio" column indicates the feed setting computed using the acid strength, specific gravity, and required dosage. The following steps illustrate how to calculate the feed setting for a specific piece of equipment.
    (a) (Sp. Gr.)(lbs/gal water)(% H_2SiF_6)(% F^- (in H_2SiF_6)) = lbs F^-/gal.
    (b) Substituting figures in the above formula.
    (c) (1.226)(8.34)(0.229)(0.791) = 1.85 lbs F^-/gal.
    (d) Dosage: 8.34 ÷ 1.85 = 4.51 gallons acid/MG water.

Explanation:

1 liter weighs 1 million milligrams.

$$\text{Dosage, } \frac{mg}{L} = \frac{mg}{1\text{ M mg}} = \frac{lbs}{1\text{ M lbs}} = \frac{gal}{1\text{ M Gal}}$$

$$\text{or } 1\text{ mg}/L = \frac{8.34\text{ lbs}}{1\text{ M Gal}}$$

$$= \frac{8.34\text{ lbs/M Gal Water}}{1.85\text{ lbs/gal Acid}}$$

$$= 4.51\text{ gal Acid/M Gal Water}$$

(e) In order to compensate for the .05 mg/L F^- in the raw water supply, the above figure of 4.51 should be reduced by 5%, which is the relationship of the desired level of say 1 mg/L F^- to the raw water level of .05 mg/L F^-.

(f) 4.51 − (.05 × 4.51) = 4.51 − .23 = 4.28 gal/MG.

(g) Ratio setting therefore is 4.28 ÷ 4.80 or 0.89.

(h) The flow capacity of the pipeline water meter at 100% is 300 MGD.

(i) The flow capacity of the acid feed pump is 1,440 gallons of H_2SiF_6/day.

The ratio of the above two 100% capacities is 1,440 ÷ 300 or 4.8 gal/MG.

Note the difference of the setting of 0.88 and the calculated figure of 0.89. This adjustment is made in order that the fluoride dosage will agree with the laboratory results. In all instances, the laboratory results should determine the feed settings.

The small difference in calculated setting and actual setting can also result from accumulated errors in the control equipment, that is, flow transmitter, square root ($\sqrt{}$) extractor, and ratio controller.

8. The H_2SiF_6 column records the totalizer values.
9. "H_2SiF_6 Gals." is the actual amount of acid fed into the system and is derived as follows:

885,005.50 − 884,676.08 = 329.42 gallons

This figure should be fairly close to the reading obtained at Step 6. If it is not, look for errors in readings, leaks, or equipment malfunctioning.

10. "Water Meter Totalizer" is the cumulative total of the amount of water being treated measured by a Venturi or some other type of primary water meter.
11. "Water M/Gals." is the actual amount of water passing through the water meter for the time period involved and again is derived by simple subtraction:

268.00 − 191.01 = 76.99 Million Gallons.

BYPASS TUNNEL FLUORIDE STATION

Week Ending _December 11, 2005_

DATE	TIME	ACID STORAGE									TANK LOSS GALS.	RATIO	H2SIF6 TOTALIZER	H2SIF6 GALS.	WATER METER TOTALIZER	WATER M/GALS.	ACID GAL/MG	F RESID. PPM	DOWN TIME	OBS. BY
		TANK 1		TANK 2		TANK 3		TANK 4											
		GALS.	% & SP. GR.	GALS.	% & SP. GR.	GALS.	% & SP. GR.	GALS.	% & SP. GR.										
END OF PREVIOUS WEEK		2930	1.226 22.9	1370	23.0	120	23.9	2315	23.1		0.88	884676.08		191.01			0.98		
12/5	11:00A	2600								330	0.88	885005.50	329.42	268.00	76.99	4.28	0.97		TM

BIF METER FLOW F & P METER

REMARKS:

Fig. 13.13 Log sheet for bypass tunnel fluoride station

12. "Acid Gal/MG" is the rate of treatment for hydrofluosilicic acid and is obtained by dividing the figures from Step 9 by the figure from Step 11.

    329.42 ÷ 76.99 = 4.28

13. "F Resid. PPM" is actual fluoride content in the treated water as read by continuous flow fluoride ion analyzer.

14. "Down Time." The equipment should be equipped with a running time meter reading in seconds that begins to operate any time the plant shuts down. This will give reasons for low feed as indicated by the readings in Step 12. The operator should know why this deviation occurred.

15. "Obs. By." This will be the operator's initials.

13.621 Sodium Silicofluoride

1. Figure 13.14 is a typical log sheet for a gravimetric feeder feeding powdered sodium silicofluoride.

2. "Date" and "Time" are entered in the first two columns.

3. "Totalizer Reading (lbs.)." This reading indicates a cumulative reading of the amount of silicofluoride that has been fed by the machine.

4. "Weight Loss per 24 hrs. (lbs.)" is the amount of silicofluoride actually fed during the time frame and is determined from the readings observed in Step 3 as follows:

    44,165.5 − 43,276.9 = 888.6.

5. "Mach. Feed Setting" is the feed rate being used. This rate may vary from machine to machine depending upon gear ratios and other devices used to control the rate of chemical feed. If the laboratory tests indicate low or high fluoride ion level, the adjustment is made with this setting on a percentage basis. If the laboratory fluoride ion level is 10 percent low, then this setting should be raised 10 percent.

6. "Chem. Added to Bin (lbs.)" is the amount of chemical taken from storage and dumped into the feeder hopper.

7. "Chemical Left in Storage (lbs.)" is the amount of fluoride in the bulk storage. This is useful in programming supply orders and in checking the accuracy of the feeder over a period of time. To check accuracy, compare this amount with the amount indicated in Step 4 over a six-month or one-year period.

8. "Pump Operating" is useful if several pumps are available to inject the dissolved sodium silicofluoride into the water main.

9. "Water Meter Reading (m.g.)" is the reading from the main line water meter.

10. "Water Treated (m.g.)" is the amount of water actually treated and is derived from Step 9 data as follows:

    39,829.73 − 39,762.41 = 67.32.

11. "Dosage (lbs. per m.g.)" represents the actual dosage of fluoride in the water. This figure should be constant, barring downtime, changes in machine feed setting, Step 5, or equipment malfunctioning. The value is derived from the weight loss (Step 4) divided by the water treated (Step 10) as follows:

    888.6 ÷ 67.32 = 13.2.

12. "Plant Down Time (Hrs.)" is the period of time fluoride was not being fed and the plant was shut down because of power failure or automatic shutdown. The installation should be equipped with a resettable running time meter reading hours and tenths.

13. Figure 13.15 is a typical small plant log sheet used in plants utilizing sodium silicofluoride. This log sheet is provided through the courtesy of the City of Palo Alto, California.

13.63 Equipment Check Procedures

Figure 13.16 is a flowchart indicating the procedures an operator could follow when starting water fluoridation equipment. The flowchart procedures offer potential ways of correcting problems that may develop and help operators achieve a system with a good operating efficiency.

QUESTIONS

Write your answers in a notebook and then compare your answers with those on page 67.

13.6A What should be the feed rate in gallons per day for treating 6 MGD with hydrofluosilicic acid 20 percent if the desired fluoride ion concentration is 1.2 mg/L? Assume the raw water does not contain any fluoride ion.

13.6B What could be the causes of differences between the recorded volume of acid used from a storage tank and the volume of acid actually fed into the system as measured by a flowmeter?

13.7 PREVENTION OF OVERFEEDING

1. Operators must ensure that no overfeeding occurs because no additional benefits result from overfeeding and there is a waste of chemicals and money. Excessive overfeeding could be harmful to consumers. Overfeeding can be prevented by proper operation and continuous monitoring of the product water.

2. If the size of the installation warrants, a continuous fluoride ion analyzer should be installed in the treated water line located downstream a sufficient distance so that adequate mixing is ensured.

3. In a large plant involving shift operation, grab samples can be analyzed for the fluoride level during each shift; otherwise, once-a-day checks will suffice.

Fluoridation 55

FLUORIDE STATION REPORT

Station _Sunset Supply Line_   Week Ending _December 11, 2005_

Date	Time	Totalizer Reading (lbs.)	Weight Loss per 24 hrs. (lbs.)	Mach. Feed Setting	Chem. Added to Bin (lbs.)	Chemical Left in Storage (lbs.)	Pump Operating	Water Meter Reading (m.g.)	Water Treated (m.g.)	Dosage (lbs. per m.g.)	Plant Down Time (Hrs.)	Feeder Time Lapse (sec.)	Observ.
END OF PREVIOUS WEEK		43276.9											
12/5	10:20A	44165.5	888.6	52.8		42,000	1A	39762.41		13.2			
				52.8	400	41,600	1A	39829.73	67.32	13.2		12735	✓

Fig. 13.14 *Log sheet for fluoride station reports*

City of Palo Alto **Water Division**

WEEKLY WATER PRODUCTION AND TREATMENT LOG

Week Ending Wed.________________, 20___. Station:________________________

	Sacks on hand	FLUORIDE Feeder Lbs. Start	Add	Final	Used	Feed Setting	Calc mg/l	Timer setting
Total								
Wed.								
Tues.								
Mon.								
Fri.								
Thurs.								
Previous								

WATER (CCF) Meter Record		Consumption	SAND (ml)

Pump A	No. motor starts	___ + ___ + ___ = ___
	No. running hours	___ – ___ = ___
Lead Pump ↕		
Pump B	No. motor starts	___ + ___ + ___ = ___
	No. running hours	___ – ___ = ___

OTHER METER READINGS			
Meter Function			
Present Reading			
Previous Reading			
Difference			

	CHLORINE Cylinders on hand	Cylinder, Lbs. Gross	Tare	Net #1	Net #2	Used	Feed Rate	OTO resid. mg/l
Previous								
Thurs.								
Fri.								
Mon.								
Tues.								
Wed.								
Total								

MISCELLANEOUS MEASUREMENTS			
Well static level, ft.			
Well pumping level, ft.			
Head against well, pump, psi			
Amp draw reading	L	C	R

REMARKS

Fig. 13.15 Weekly water production and treatment log

Fluoridation 57

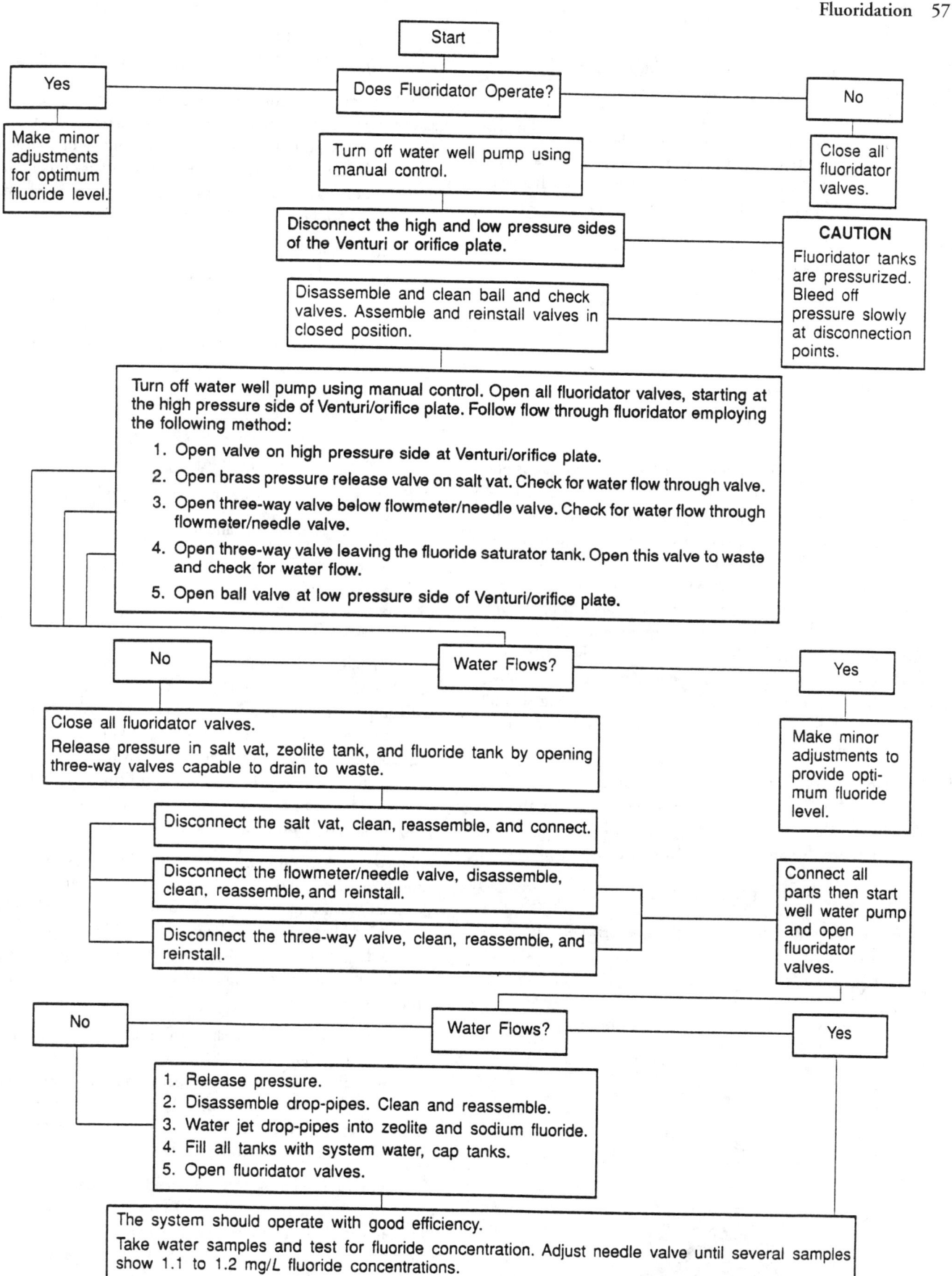

```
                              Start
                                │
                                ▼
        ┌─── Yes ───   Does Fluoridator Operate?   ─── No ───┐
        │                                                     │
        ▼                       ▼                             ▼
   Make minor          Turn off water well pump          Close all
   adjustments         using manual control.             fluoridator
   for optimum                  │                        valves.
   fluoride level.              ▼
                       Disconnect the high and low             
                       pressure sides of the              CAUTION
                       Venturi or orifice plate.          Fluoridator tanks
                                │                         are pressurized.
                                ▼                         Bleed off
                       Disassemble and clean ball         pressure slowly
                       and check valves. Assemble         at disconnection
                       and reinstall valves in            points.
                       closed position.
                                │
                                ▼
```

Turn off water well pump using manual control. Open all fluoridator valves, starting at the high pressure side of Venturi/orifice plate. Follow flow through fluoridator employing the following method:

1. Open valve on high pressure side at Venturi/orifice plate.
2. Open brass pressure release valve on salt vat. Check for water flow through valve.
3. Open three-way valve below flowmeter/needle valve. Check for water flow through flowmeter/needle valve.
4. Open three-way valve leaving the fluoride saturator tank. Open this valve to waste and check for water flow.
5. Open ball valve at low pressure side of Venturi/orifice plate.

```
        ┌─── No ───      Water Flows?      ─── Yes ───┐
        │                                              │
        ▼                                              ▼
```

Close all fluoridator valves.
Release pressure in salt vat, zeolite tank, and fluoride tank by opening three-way valves capable to drain to waste.

Make minor adjustments to provide optimum fluoride level.

Disconnect the salt vat, clean, reassemble, and connect.

Disconnect the flowmeter/needle valve, disassemble, clean, reassemble, and reinstall.

Disconnect the three-way valve, clean, reassemble, and reinstall.

Connect all parts then start well water pump and open fluoridator valves.

```
        ┌─── No ───      Water Flows?      ─── Yes ───┐
        │                                              │
        ▼                                              │
```

1. Release pressure.
2. Disassemble drop-pipes. Clean and reassemble.
3. Water jet drop-pipes into zeolite and sodium fluoride.
4. Fill all tanks with system water, cap tanks.
5. Open fluoridator valves.

The system should operate with good efficiency.
Take water samples and test for fluoride concentration. Adjust needle valve until several samples show 1.1 to 1.2 mg/L fluoride concentrations.

Fig. 13.16 Water fluoridation equipment check procedures
(Reproduced with permission from *OPERATOR'S DRUMBEAT*, Volume 1, Number 6, March–April 1989. Albuquerque Area Indian Health Service, J. R. Olguin, Publisher)

4. If the plant uses one of the solid fluoride compounds and the operator questions whether there is total solubility, the fluoride feeder can be shut down and the lack of fluoride traced out in the distribution system. There should be a sudden drop to zero fluoride or to the background level if total solubility is not being achieved (the undissolved solid fluoride compound will settle out).

5. All liquid systems should be checked for positive protection against backsiphonage from fluoride storage tanks.

6. Shut down the plant if there is any significant overfeeding. Start flushing the affected mains and notify the local and state health departments. The water department and the health departments will then decide if public notification should be undertaken.

13.8 UNDERFEEDING

In contrast to the chlorination process where continuous operation must be ensured, fluoridation does not have to be continuous. Shutdowns for cleaning, adjustments, or due to safety controls can be tolerated for short time periods. This does not mean that sloppy operation and maintenance is desirable. Every attempt should be made to maintain constant feeding. For example, the installation of standby electrical generating equipment just to maintain fluoridation equipment in operation would not be warranted. If the standby generator had to be purchased for other reasons, then the emergency circuit may also include the fluoride feeding equipment. Underfeeding should not be allowed because this results in a very significant reduction of the benefits of fluoridation.

Daily inspection of the fluoridation equipment, fluoride tests on the treated water, and calculation of the dosage from water treatment and chemical use data can greatly minimize the possibility of both overfeeding and underfeeding.

QUESTIONS

Write your answers in a notebook and then compare your answers with those on page 67.

13.7A   Why should overfeeding be prevented?

13.7B   What should be done if significant overfeeding occurs?

13.8A   Why might a fluoridation operation be shut down?

13.9 SHUTTING DOWN CHEMICAL SYSTEMS

If the fluoridation equipment is going to be shut down for an extended length of time, it should be cleaned out to prevent corrosion or the solidifying of the chemical. Lines and equipment could be damaged when restarted if chemicals left in them solidify. Operators could be seriously injured if they open a chemical line that has not been properly flushed out.

The following items should be included in your checklist for shutting down the chemical system:

1. Flush out the chemical supply with water.

2. Run dry chemicals completely out of the equipment and clean equipment by using a vacuum cleaner.

3. Flush out all the solution lines with water until the lines are clean.

4. Shut off the electrical power.

5. Shut off the water supply and protect from freezing.

6. Drain and clean the mix and feed tanks.

7. Padlock (lock out) the main electric switch box to the fluoride equipment.

13.10 MAINTENANCE

Maintenance should follow the same routine as with any similar chemical feeder, including regular cleanup and painting of the equipment and appurtenant metal piping and conduits. In order to give the plant a fresher look and hold down on painting, consider using all plastic piping even though it is needed only for the water supply. Conduit and fittings should also be plastic for the same reason. Vacuum any gears and other similar parts to remove fluoride dust.

Since fluoride solutions are extremely corrosive, be constantly on the lookout for drips or leaks and any other evidence of corrosion. Repair these conditions as quickly as possible. Also look for the buildup of insoluble deposits in feed lines and equipment. Schedule the removal of insoluble deposits on a regular basis to prevent buildups from creating any problems.

All containers of fluoride chemicals must be disposed of in an acceptable manner. Thoroughly rinse all containers with water to remove all traces of chemicals before allowing containers to leave your plant. You may burn the containers if a nuisance will not be created. Remember that fluoride fumes can kill vegetation and are harmful to people.

You do not need to be too concerned about checking the feed rate by catching a given amount of fluoride over a time period. The log will show long-period discrepancies and the daily laboratory tests will indicate any drifting from the desired fluoride concentration in the treated water.

Either you or the laboratory personnel must analyze the fluoridated water daily. Check the results for any deviations from the norm and take corrective action. Hand-held colorimeters are available for measuring fluoride in water. See Chapter 21, "Advanced Laboratory Procedures," for details on how to analyze samples for the fluoride ion.

An important part of your maintenance program is the prevention of any sanitary defects that could adversely affect the safety or quality of your treated drinking water. Sanitary defects that could develop in fluoridation systems include:

1. Lack of or inadequate start-stop controls
2. Inadequate feed rate control equipment
3. No analyzer to measure fluoride ion levels in treated water
4. Lack of or inadequate backflow safeguards
5. Fluoridation chemical not meeting AWWA specifications
6. Inadequate free chlorine residual in treated water

13.11 SAFETY IN HANDLING FLUORIDE COMPOUNDS[8]

From the operator's viewpoint, fluoride chemicals have one thing in common with all other chemicals found in treatment plants: Fluoride chemicals can seriously injure or kill the careless or untrained operator. Safety should be of special concern to you because it is your own health that is at stake.

13.110 Avoid Overexposure

One of the major causes of overexposure is the inhalation of fluoride dust. This usually occurs while a dry feeder or saturator is being loaded. To protect yourself from the dust of dry fluoride compounds, be sure the dust collector system works properly. Even with the use of dust collector systems, dust will circulate in the air. Always use approved respirators equipped with cartridges for organic dusts and vapors, protective coveralls, and gloves when emptying sacks or cleaning up equipment and plant surfaces.

When loading a saturator, dust will be minimized if crystalline sodium fluoride is fed instead of powdered sodium fluoride. When loading a dry feeder, you should wear a mask, apron, and rubber gloves to minimize exposure.

When the protection gear is removed, the remaining small traces of chemical should also be removed from your body. Some large water plants have dust collection systems that use a partial vacuum to draw dust from your body and vent it to the outside air after filtering.

Care should be taken when emptying bags of chemicals into a feeder hopper. The bags should be opened carefully at the top and the contents poured gently to minimize dust. Care should also be taken during storage of the bags. Bags should be stored in a dry place, preferably off the floor. If bags are stacked too high there is the possibility of them falling and breaking open.

If a saturator is used, you should be cautious about allowing the solution to come in contact with skin and clothing. If this does happen, the affected area should be washed immediately with water. This also applies to other fluoride solutions (such as the dissolving water used in a dry feeder).

If a fluoride acid is being fed, extra precaution must be taken. Fluoride acid is probably the most corrosive chemical found in a water plant. The pH of fluoride acid is approximately 1.2 and it will eat through glass faster than chlorine. Special care should be taken to keep fumes to a minimum. If the acid does come in contact with your skin, you may not be able to wash it off fast enough to prevent a burn. If this happens, standard first aid should be administered as soon as possible.

A good pair of safety goggles should be worn at all times when working around fluoridation equipment where there is any possibility of splashing fluoride solutions. Be especially cautious around the fluoride acids as the concentrated acid can dissolve the whites of one's eyes in addition to the usual burns associated with acids. Another "must" is a safety shower. This must be located within easy access to both the unloading operation and points of liquid usage.

Another safety precaution that should be followed is the labeling of all feeders and solution tanks. Proper labeling will help prevent placement of chemicals in the wrong feeder. If possible, fluoride chemical should be tinted blue to differentiate it from other water treatment chemicals.

13.111 Symptoms of Fluoride Poisoning

In the event that someone is poisoned, it is vitally important to recognize the early symptoms.

Some of the obvious signs of poisoning are vomiting, stomach cramps, and diarrhea. Usually, the person will become very weak, have trouble speaking, be very thirsty, and have poor color vision. In cases of extreme poisoning, there are strong, jerky muscle contractions in the arms and legs leading to convulsions. If poisoning is not treated immediately, the person may die. Fatal doses range from 4 to 5 gm, or about a tablespoon. This equals about 2,000 times the amount of fluoride swallowed by a person from a water supply.

If a person is poisoned by inhaling fluoride, the first symptoms will be a sharp, biting pain in the nose followed by a runny nose or nosebleed. It is doubtful that a person could inhale enough fluoride to produce the same effects as encountered from drinking a large amount of fluoride. However, the sudden presence of bad stomach cramps and pains in the nose and eyes should not be ignored.

---

[8] Portions of the material in this section were adapted from "Safety Procedures Necessary During Fluoridation Process," by Ed Hansen. Reproduced from *OPFLOW*, Volume 9, No. 7, (July 1983) by permission. Copyright 1983, The American Water Works Association.

The victim should see a doctor immediately, and the water treatment practices should be checked to determine the source of the fluoride poisoning. It is probably a good idea to check out treatment practices occasionally.

The importance of quick treatment for fluoride poisoning cannot be overemphasized. In such cases, a doctor should be called immediately, and if the poisoning is severe, an ambulance should be called.

13.112 Basic First Aid

Once it is established that fluoride is the cause of the poisoning, first aid should be started while waiting for medical help. The following are recommended first-aid procedures:

1. Move the person away from any contact with fluoride and keep warm.

2. Inhalation: Remove exposed person to an uncontaminated area immediately. If breathing has stopped, start artificial respiration at once. Oxygen should be provided for an exposed person having difficulty breathing (but only by an authorized person) until exposed person is able to breathe easily unassisted. Exposed person should be examined by a physician.

3. Eye Contact: Flush eyes for at least 15 minutes with large amounts of water. Eyelids should be held apart during the flushing to ensure contact of water with all accessible tissue of the eyes and lids. Medical attention should be given as soon as possible.

4. Skin Contact: Exposed person should be removed to an uncontaminated area and subjected immediately to a drenching shower of water for a minimum of 15 to 20 minutes. Remove all contaminated clothing while under shower. Medical attention should be given as soon as possible for all burns, regardless of how minor they seem.

5. Ingestion: If conscious, give the exposed person large quantities of water immediately to dilute the acid. *DO NOT* induce vomiting. If hydrofluosilicic acid is swallowed, vomiting may cause the acid to go the wrong way into the lungs and cause more serious problems. Milk may be given for its soothing effect. If sodium fluoride is swallowed, vomiting may be induced. A physician should be contacted immediately.

6. Take the person to the hospital as soon as possible.

If common sense and good safety practice are used, the hazard to the water plant operator should be as small as the hazard to the water consumer.

Fluoridation chemicals are poisonous. Protect yourself from these toxic chemicals.

13.113 Protecting Yourself and Your Family

Avoid swallowing fluoridation chemicals. Do not eat, drink, or smoke in or around chemical storage or feed areas. Do not inhale chemical dusts or vapors. Wear a respirator. Be sure exhaust fans and dust collectors are operating properly. Prevent hydrofluosilicic acid from coming in contact with your skin or eyes because hydrofluosilicic acid is very corrosive. If any hydrofluosilicic acid touches you, flood the contact area with plenty of water. If you are acutely poisoned by a fluoride chemical, you may be thirsty, vomit, and have stomach cramps, diarrhea, difficulty in speaking, and disturbed color vision. If any of these symptoms occur, consult a physician immediately.

When leaving the fluoride plant, wash your hands and change coveralls so that fluoride dust is not carried home.

13.114 Training

Special safety training must be given to all operators who will handle fluoride compounds. Training must include how to safely receive compounds from supplier, store until needed, prepare solutions, load feeders, and dose water being treated.

QUESTIONS

Write your answers in a notebook and then compare your answers with those on page 67.

13.9A Why should fluoridation equipment be cleaned out if the equipment is going to be shut down for an extended length of time?

13.10A How can fluoride dust be removed from gears?

13.10B During maintenance of fluoridation equipment, how concerned should you be about checking the feed rate by catching a given amount of fluoride over a time period?

13.11A What are the symptoms of acute fluoride poisoning?

13.12 CALCULATING FLUORIDE DOSAGES

FORMULAS

1. Treatment charts, such as the ones located on pages 50 and 51, can be used to determine feed rates. The feed rate is usually based on a dose of one mg/L; therefore actual feed rates must be adjusted.

$$\text{Actual Feed Rate, GPD} = \frac{(\text{Chart Feed Rate, GPD})(\text{Actual Dose, mg}/L)}{1 \text{ mg}/L}$$

2. Feed rates may be calculated on the basis of pounds per day or gallons per day. Consideration must be given to the pounds of fluoride ion per pound of commercial chemical.

$$\text{Feed Rate, lbs/day} = \frac{(\text{Flow, MGD})(\text{Dose, mg}/L)(8.34 \text{ lbs/gal})(100\%)}{\text{Solution, \% F}}$$

or

$$\text{Feed Rate, lbs/day} = \frac{\text{Feed Rate, lbs F/day}}{\text{lbs F/lb Commercial Chemical}}$$

or

$$\text{Feed Rate, gal/day} = \frac{\text{Feed Rate, lbs/day}}{\text{Chemical Solution, lbs/gal}}$$

3. If the water being treated contains some fluoride ion, but not sufficient, then a feed dose must be calculated.

   Feed Dose, mg/L = Desired Dose, mg/L – Actual Concentration, mg/L

4. Commercial chemicals usually are not 100 percent pure. Also, the chemical only contains a portion of the ion of concern (fluoride ion in this chapter).

   $$\text{Portion F} = \frac{(\text{Commercial Purity, \%})(\text{Fluoride Ion, \%})}{(100\%)(100\%)}$$

   The portion F is the pounds of F per pound of commercial chemical. For example, 0.6 pound F per one pound of commercial sodium silicofluoride.

5. To calculate the fluoride dosage, or any chemical dosage, you need to know the pounds of chemical and volume of water in million gallons.

   $$\text{Dosage, mg/}L = \frac{\text{Chemical, lbs}}{(\text{Water, M Gal})(8.34 \text{ lbs/gal})}$$

   $$= \frac{\text{lbs Chemical}}{\text{Million lbs Water}}$$

   If we substitute milligrams for pounds, we get

   $$= \frac{\text{mg Chemical}}{\text{Million mg Water}}$$

   One million milligrams of water occupy a volume of one liter.

   $$= \frac{\text{mg Chemical}}{\text{Liter of Water}}$$

   $$= \text{mg/}L$$

6. To determine the amount of feed solution in either gallons or gallons per day to treat a water, you need to know the amount of water to be treated in gallons or gallons per day, the feed dose in milligrams per liter, and the feed solution in milligrams per liter.

   $$\text{Feed Solution, gal} = \frac{(\text{Flow, gal})(\text{Feed Dose, mg/}L)}{\text{Feed Solution, mg/}L}$$

   NOTE: If the "Feed Solution" is in gallons per day instead of gallons, then the "Flow" must be in gallons per day also instead of gallons.

7. When mixing the same two acids or chemicals, but of different strengths, the volumes or flows of the chemicals and their strengths must be known.

   $$\text{Mixture Strength, \%} = \frac{(\text{Volume 1, gal})(\text{Strength 1, \%}) + (\text{Volume 2, gal})(\text{Strength 2, \%})}{\text{Volume 1, gal} + \text{Volume 2, gal}}$$

   NOTE: The "Volumes" may be in gallons or treated as flows in GPD or MGD. The "Strengths" may be in percentages or concentrations such as mg/L.

8. When using chemicals for fluoridation, we need to know the percentage fluoride ion purity. This information will allow us to convert the pounds of chemical dosage to pounds of fluoride ion available.

   $$\text{Fluoride Ion Purity, \%} = \frac{(\text{Molecular Weight of Fluoride})(100\%)}{\text{Molecular Weight of Chemical}}$$

EXAMPLE 1

A flow of 4 MGD is to be treated with a 20-percent solution of hydrofluosilicic acid (H_2SiF_6). The water to be treated contains no fluoride and the desired fluoride concentration is 1.8 mg/L. What should be the feed rate of hydrofluosilicic acid? Use the treatment charts.

Known	Unknown
Flow, MGD = 4 MGD	1. Feed Rate, gal/day
Acid Solution, % = 20%	2. Feed Rate, gal/hr
Desired F, mg/L = 1.8 mg/L	

1. Use "Treatment Chart I" on page 50 because we are treating a relatively small flow (4 MGD).

2. Start on the left side at the 4 MGD value and move horizontally to the right to the 20 percent diagonal line.

3. At this point, drop vertically downward to the bottom lines and read the feed rates for one mg/L (ppm).

   a. Feed Rate, gallons per day = 19 gal/day
   b. Feed Rate, gallons per hour = 0.8 gal/hr

4. Calculate the feed rate to produce the desired fluoride concentration of 1.8 mg/L.

   a. $$\text{Feed Rate, GPD} = \frac{(\text{Feed Rate, GPD})(\text{Desired F, mg/}L)}{1 \text{ mg/}L}$$

   $$= \frac{(19 \text{ GPD})(1.8 \text{ mg/}L)}{1 \text{ mg/}L}$$

   $$= 34.2 \text{ gal/day}$$

   b. $$\text{Feed Rate, gal/hr} = \frac{(\text{Feed Rate, gal/hr})(\text{Desired F, mg/}L)}{1 \text{ mg/}L}$$

   $$= \frac{(0.8 \text{ gal/hr})(1.8 \text{ mg/}L)}{1 \text{ mg/}L}$$

   $$= 1.44 \text{ gal/hr}$$

62 Water Treatment

EXAMPLE 2

A flow of 4 MGD is to be treated with a 20-percent solution of hydrofluosilicic acid (H_2SiF_6) that contains a fluoride purity of 79.2%. The water to be treated contains no fluoride and the desired fluoride concentration is 1.8 mg/L. Assume the hydrofluosilicic acid weighs 9.8 pounds per gallon. What should be the feed rate of hydrofluosilicic acid? Calculate the feed rate.

Known	Unknown
Flow, MGD = 4 MGD	1. Feed Rate, gal/day
Acid Solution, % = 20%	2. Feed Rate, gal/hr
Acid lbs/gal = 9.8 lbs/gal	
Purity, % = 79.2%	
Desired F, mg/L = 1.8 mg/L	

1. Calculate the hydrofluosilicic acid feed rate in pounds per day.

$$\text{Feed Rate, lbs/day} = \frac{(\text{Flow, MGD})(\text{Desired F, mg/}L)(8.34 \text{ lbs/gal})(100\%)(100\%)}{(\text{Acid Solution, \%})(\text{Purity, \%})}$$

$$= \frac{(4 \text{ MGD})(1.8 \text{ mg/}L)(8.34 \text{ lbs/gal})(100\%)(100\%)}{(20\%)(79.2\%)}$$

$$= 379 \text{ lbs Acid/day}$$

2. Determine the feed rate of the acid in gallons per day.

$$\text{Feed Rate, gal/day} = \frac{\text{Feed Rate, lbs/day}}{\text{Acid, lbs/gal}}$$

$$= \frac{379 \text{ lbs Acid/day}}{9.8 \text{ lbs Acid/gal}}$$

$$= 39 \text{ gal Acid/day}$$

NOTE: We obtained a feed rate of 34 gallons of acid per day from "Treatment Chart I." The differences result from the problems of drawing and reading the chart accurately.

3. Calculate the feed rate in gallons of acid per hour.

$$\text{Feed Rate, gal/hr} = \frac{\text{Feed Rate, gal/day}}{24 \text{ hr/day}}$$

$$= \frac{39 \text{ gal Acid/day}}{24 \text{ hr/day}}$$

$$= 1.6 \text{ gal Acid/hr}$$

EXAMPLE 3

A flow of 200 GPM is to be treated with a 2.4-percent (0.2 lbs/gallon) solution of sodium fluoride (NaF). The water to be treated contains 0.7 mg/L of fluoride ion and the desired fluoride ion concentration is 1.6 mg/L. What should be the feed rate of sodium fluoride? Use the treatment charts.

Known	Unknown
Flow, GPM = 200 GPM	1. Feed Rate, gal/day
NaF Solution, % = 2.4%	2. Feed Rate, gal/hr
Desired F, mg/L = 1.6 mg/L	
Actual F, mg/L = 0.7 mg/L	

1. Use "Treatment Chart III" on page 51 because we are treating 200 GPM.
2. Start on the left side at the 200 GPM value and move horizontally to the right to the 2.4 percent diagonal line.
3. At this point, drop vertically downward to the bottom lines and read the feed rates for one mg/L (ppm).

   a. Feed Rate, gallons per day = 26.5 gal/day
   b. Feed Rate, gallons per hour = 1.1 gal/hr

4. Calculate the feed rate to produce the desired fluoride concentration of 1.6 mg/L.

$$\text{Feed Dose, mg/}L = \text{Desired F, mg/}L - \text{Actual F, mg/}L$$

$$= 1.6 \text{ mg/}L - 0.7 \text{ mg/}L$$

$$= 0.9 \text{ mg/}L$$

$$\text{Feed Rate, GPD} = \frac{(\text{Feed Rate, GPD})(\text{Feed Dose, mg/}L)}{1 \text{ mg/}L}$$

$$= \frac{(26.5 \text{ gal/day})(0.9 \text{ mg/}L)}{1 \text{ mg/}L}$$

$$= 23.8 \text{ GPD}$$

$$\text{Feed Rate, gal/hr} = \frac{(\text{Feed Rate, gal/hr})(\text{Feed Dose, mg/}L)}{1 \text{ mg/}L}$$

$$= \frac{(1.1 \text{ gal/hr})(0.9 \text{ mg/}L)}{1 \text{ mg/}L}$$

$$= 0.99 \text{ gal/hr or 1 gal/hr}$$

EXAMPLE 4

A flow of 200 GPM is to be treated with a 2.4-percent (0.2 pound per gallon) solution of sodium fluoride (NaF). The water to be treated contains 0.7 mg/L of fluoride ion and the desired fluoride ion concentration is 1.6 mg/L. What should be the feed rate of sodium fluoride? Calculate the feed rate. Assume the sodium fluoride has a fluoride ion purity of 45.3 percent.

Known	Unknown
Flow, GPM = 200 GPM	1. Feed Rate, gal/day
NaF Solution, % = 2.4%	2. Feed Rate, gal/hr
NaF Solution, lbs/gal = 0.2 lb/gal	
Desired F, mg/L = 1.6 mg/L	
Actual F, mg/L = 0.7 mg/L	
Purity, % = 45.3%	

1. Convert flow from gallons per minute to million gallons per day.

$$\text{Flow, MGD} = \frac{(\text{Flow, gal/min})(60 \text{ min/hr})(24 \text{ hr/day})}{1,000,000/\text{Million}}$$

$$= \frac{(200 \text{ gal/min})(60 \text{ min/hr})(24 \text{ hr/day})}{1,000,000/\text{Million}}$$

$$= 0.288 \text{ MGD}$$

2. Determine the fluoride feed dose in milligrams per liter.

   Feed Dose, mg/L = Desired F, mg/L − Actual F, mg/L

   = 1.6 mg/L − 0.7 mg/L

   = 0.9 mg/L

3. Calculate the feed rate in pounds of fluoride ion per day.

   Feed Rate, lbs F/day = (Flow, MGD)(Feed Dose, mg/L)(8.34 lbs/gal)

   = (0.288 MGD)(0.9 mg/L)(8.34 lbs/gal)

   = 2.16 lbs F/day

4. Convert the feed rate from pounds of fluoride per day to gallons of sodium fluoride solution per day.

   Feed Rate, gal/day = $\dfrac{\text{(Feed Rate, lbs F/day)(100\%)}}{\text{(NaF Solution, lbs F/gallon)(Purity, \%)}}$

   = $\dfrac{(2.16 \text{ lbs F/day})(100\%)}{(0.2 \text{ lb F/gal})(45.3\%)}$

   = 23.8 gal/day

   NOTE: We obtained a feed rate of 23.8 gal/day in both Examples 3 and 4. Sometimes you may get slightly different results using these two methods of calculating the feed rate. The differences could result from inaccurately preparing and reading the treatment chart as well as the assumed purity of fluoride ion in the sodium fluoride.

5. Convert the feed rate from gallons per day to gallons per hour.

   Feed Rate, gal/hr = $\dfrac{\text{Feed Rate, gal/day}}{24 \text{ hr/day}}$

   = $\dfrac{23.8 \text{ gal/day}}{24 \text{ hr/day}}$

   = 1.0 gal/hr

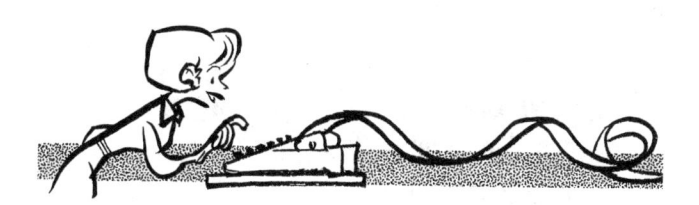

EXAMPLE 5

A flow of 1 MGD is treated with sodium silicofluoride (Na$_2$SiF$_6$) to provide a fluoride ion dose of 1.4 mg/L. What is the feed rate in pounds per day? Commercial sodium silicofluoride has a purity of 98.5 percent and the fluoride ion purity of sodium silicofluoride is 60.7 percent.

Known		Unknown
Flow, MGD	= 1 MGD	Feed Rate, lbs/day
Dose, mg/L	= 1.4 mg/L	
Na$_2$SiF$_6$ Purity, %	= 98.5%	
Fluoride Ion Purity, %	= 60.7%	

1. Calculate the portion of fluoride ion in the commercial sodium silicofluoride.

   Portion F = $\dfrac{(\text{Na}_2\text{SiF}_6 \text{ Purity, \%})(\text{Fluoride Ion Purity, \%})}{(100\%)(100\%)}$

   = $\dfrac{(98.5\%)(60.7\%)}{(100\%)(100\%)}$

   = 0.598 lb F/lb Commercial Na$_2$SiF$_6$

   This says that there is 0.598 pound of fluoride ion in a pound of commercial sodium silicofluoride.

2. Calculate the pounds of fluoride required per day.

   Fluoride, lbs/day = (Flow, MGD)(Dose, mg/L)(8.34 lbs/gal)

   = (1 MGD)(1.4 mg/L)(8.34 lbs/gal)

   = 11.7 lbs F/day

3. Determine the chemical feed rate for the commercial sodium silicofluoride in pounds per day.

   Feed Rate, lbs/day = $\dfrac{\text{Fluoride, lbs/day}}{\text{Fluoride, lbs/lb Commercial Na}_2\text{SiF}_6}$

   = $\dfrac{11.7 \text{ lbs F/day}}{0.598 \text{ lb F/lb Commercial Na}_2\text{SiF}_6}$

   = 19.6 lbs/day Commercial Na$_2$SiF$_6$

EXAMPLE 6

A flow of 1.4 MGD is treated with sodium silicofluoride. The raw water contains 0.4 mg/L of fluoride ion and the desired fluoride ion concentration is 1.6 mg/L. What should be the chemical feed rate in pounds per day? Assume each pound of commercial sodium silicofluoride (Na$_2$SiF$_6$) contains 0.6 pound of fluoride ion.

Known		Unknown
Flow, MGD	= 1.4 MGD	Feed Rate, lbs/day
Raw Water F, mg/L	= 0.4 mg/L	
Desired F, mg/L	= 1.6 mg/L	
Chemical, lbs F/lb	= 0.6 lb F/lb	

1. Determine the fluoride feed dose in milligrams per liter.

   Feed Dose, mg/L = Desired F, mg/L − Raw Water F, mg/L

   = 1.6 mg/L − 0.4 mg/L

   = 1.2 mg/L

2. Calculate the fluoride feed rate in pounds per day.

   Feed Rate, lbs F/day = (Flow, MGD)(Feed Dose, mg/L)(8.34 lbs/gal)

   = (1.4 MGD)(1.2 mg/L)(8.34 lbs/gal)

   = 14.0 lbs F/day

3. Determine the chemical feed rate in pounds of commercial sodium silicofluoride per day.

$$\text{Feed Rate, lbs/day} = \frac{\text{Feed Rate, lbs F/day}}{\text{lbs F/lb Commercial Na}_2\text{SiF}_6}$$

$$= \frac{14.0 \text{ lbs F/day}}{0.6 \text{ lb F/lb Commercial Na}_2\text{SiF}_6}$$

$$= 23.3 \text{ lbs/day Commercial Na}_2\text{SiF}_6$$

EXAMPLE 7

The totalizer for a water treatment plant indicated that a total of 100,000 gallons of water had been treated with three pounds of 98 percent pure sodium fluoride (NaF). The fluoride ion purity for sodium fluoride is 45.3 percent. What was the added fluoride ion dosage in milligrams per liter?

Known	Unknown
Water Treated, MG = 0.1 M Gal	Fluoride Dosage, mg/L
NaF, lbs = 3 lbs	
NaF Purity, % = 98%	
F Ion Purity, % = 45.3%	

1. Calculate the portion of fluoride ion in the commercial sodium fluoride.

$$\text{Portion F} = \frac{(\text{NaF Purity, \%})(\text{Fluoride Ion Purity, \%})}{(100\%)(100\%)}$$

$$= \frac{(98\%)(45.3\%)}{(100\%)(100\%)}$$

$$= 0.444$$

or = 0.444 lb F/lb commercial NaF

2. Calculate the pounds of fluoride used.

Fluoride, lbs = (Commercial NaF, lbs)(0.444 lb F/lb Comm NaF)

= (3 lbs Comm NaF)(0.444 lb F/lb Comm NaF)

= 1.33 lbs F

3. Calculate the fluoride dosage in milligrams per liter.

$$\text{Fluoride Dosage, mg/}L = \frac{\text{Fluoride, lbs F}}{(\text{Water Treated, M Gal})(8.34 \text{ lbs/gal})}$$

$$= \frac{1.33 \text{ lbs F}}{(0.1 \text{ M Gal})(8.34 \text{ lbs/gal})}$$

$$= \frac{1.33 \text{ lbs F}}{0.834 \text{ Million lbs Water}}$$

$$= \frac{1.6 \text{ lbs F}}{1 \text{ M lbs Water}}$$

$$= 1.6 \text{ mg/}L$$

EXAMPLE 8

Determine the percentage of fluoride ion in the feed solution from a saturator. The saturator contains 95 percent pure sodium fluoride, the maximum water solubility for sodium fluoride is four percent, and sodium fluoride is 45.3 percent fluoride ion.

Known	Unknown
Commercial NaF Purity, % = 95%	Solution, % F
NaF Solubility, % = 4%	
F Ion Purity, % = 45.3%	

Calculate the percentage of fluoride ion in the feed solution.

$$\text{Solution, \% F} = \frac{(\text{NaF Solubility, \%})(\text{F Ion Purity, \%})}{100\%}$$

$$= \frac{(4\%)(45.3\%)}{100\%}$$

$$= 1.8\%$$

NOTE: In a saturator, the commercial NaF purity of 95 percent does not enter into the calculations because the four percent solubility is all NaF.

EXAMPLE 9

The feed solution from a saturator containing 1.8 percent fluoride ion is used to treat a total flow of 400,000 gallons of water. The raw water has a fluoride ion content of 0.5 mg/L and the desired fluoride in the finished water is 1.8 mg/L. How many gallons of feed solution are needed?

Known	Unknown
Flow Vol, gal = 400,000 gal	Feed Solution, gallons
Raw Water F, mg/L = 0.5 mg/L	
Desired F, mg/L = 1.8 mg/L	
Feed Solution, % F = 1.8% F	

1. Convert the feed solution from a percentage fluoride ion to milligrams fluoride ion per liter of water.

1.0% F = 10,000 mg F/L

$$\text{Feed Solution, mg/}L = \frac{(\text{Feed Solution, \%})(10,000 \text{ mg/}L)}{1\%}$$

$$= \frac{(1.8\%)(10,000 \text{ mg/}L)}{1\%}$$

$$= 18,000 \text{ mg/}L$$

2. Determine the fluoride feed dose in milligrams per liter.

Feed Dose, mg/L = Desired F, mg/L − Raw Water F, mg/L

= 1.8 mg/L − 0.5 mg/L

= 1.3 mg/L

3. Calculate the gallons of feed solution needed.

$$\text{Feed Solution, gal} = \frac{(\text{Flow Vol, gal})(\text{Feed Dose, mg}/L)}{\text{Feed Solution, mg}/L}$$

$$= \frac{(400{,}000 \text{ gal})(1.3 \text{ mg}/L)}{18{,}000 \text{ mg}/L}$$

$$= 28.9 \text{ gallons}$$

EXAMPLE 10

A hydrofluosilicic acid (H_2SiF_6) tank contains 300 gallons of acid with a strength of 18 percent. A commercial vendor delivers 2,000 gallons of acid with a strength of 20 percent to the tank. What is the resulting strength of the mixture as a percentage?

Known	Unknown
Tank Contents, gal = 300 gal	Mixture Strength, %
Tank Strength, % = 18%	
Vendor, gal = 2,000 gal	
Vendor Strength, % = 20%	

Calculate the strength of the mixture as a percentage.

$$\text{Mixture Strength, \%} = \frac{(\text{Tank, gal})(\text{Tank, \%}) + (\text{Vendor, gal})(\text{Vendor, \%})}{(\text{Tank, gal} + \text{Vendor, gal})}$$

$$= \frac{(300 \text{ gal})(18\%) + (2{,}000 \text{ gal})(20\%)}{(300 \text{ gal} + 2{,}000 \text{ gal})}$$

$$= \frac{5{,}400 + 40{,}000}{2{,}300}$$

$$= \frac{45{,}400}{2{,}300}$$

$$= 19.7\%$$

EXAMPLE 11

Sodium silicofluoride (Na_2SiF_6) is used as the chemical to fluoridate a water supply. What is the fluoride ion purity as a percentage?

Known	Unknown
Atomic Weights	Fluoride Ion Purity, %
Na = 22.99	
Si = 28.09	
F = 19.00	

1. Determine the molecular weight of the fluoridation chemical, sodium silicofluoride, Na_2SiF_6.

Symbol	No. Atoms	×	Atomic Wt.[a]	=	Molecular Wt.
Na_2	2	×	22.99	=	45.98
Si	1	×	28.09	=	28.09
F_6	6	×	19.00	=	114.00
	Molecular Wt of Chemical			=	188.07

[a] Atomic weight values can be obtained from a chemistry book.

2. Calculate the fluoride ion purity as a percentage.

$$\text{Fluoride Ion Purity, \%} = \frac{(\text{Molecular Weight of Fluoride})(100\%)}{\text{Molecular Weight of Chemical}}$$

$$= \frac{(114.00)(100\%)}{188.07}$$

$$= 60.62\%$$

This means that there is 0.6062 pound of fluoride ion in every pound of sodium silicofluoride.

13.13 ARITHMETIC ASSIGNMENT

Turn to the Arithmetic Appendix at the back of this manual. Read and work the problems in Section A.31, "Fluoridation." Check the arithmetic in this section using an electronic calculator. You should be able to get the same answers. Section A.51 contains similar problems using metric units.

13.14 ADDITIONAL READING

1. *NEW YORK MANUAL*, Chapter 16,* "Fluoridation."

2. *TEXAS MANUAL*, Chapter 11,* "Special Water Treatment (Fluoridation)."

3. *WATER FLUORIDATION PRINCIPLES AND PRACTICES* (M4). Obtain from American Water Works Association (AWWA), Bookstore, 6666 West Quincy Avenue, Denver, CO 80235. Order No. 30004. ISBN 1-58321-311-2. Price to members, $67.50; nonmembers, $97.50; price includes cost of shipping and handling.

4. *WATER FLUORIDATION*, A Manual for Engineers and Technicians. Available from Centers for Disease Control and Prevention, Division of Oral Health, 4770 Buford Highway, Atlanta, GA 30341. No longer available.

* Depends on edition.

13.15 ACKNOWLEDGMENTS

The author wishes to acknowledge the assistance graciously given by Robert A. Hewitt, Assistant Water Quality Engineer, San Francisco Water Department, San Francisco, California, and Tom Reeves, Centers for Disease Control and Prevention, Center for Prevention Services, Dental Disease Prevention Activity, Atlanta, Georgia.

Please answer the discussion and review questions next.

DISCUSSION AND REVIEW QUESTIONS
Chapter 13. FLUORIDATION

The purpose of these questions is to indicate to you how well you understand the material in the chapter. Write the answers to these questions in your notebook.

1. Why are drinking waters fluoridated?
2. What factors would you consider when selecting a fluoridation chemical?
3. How can water be softened prior to use with fluoridation equipment?
4. What items should be considered when reviewing plans and specifications for the location of fluoride chemical hoppers?
5. Why is it important for operators to ensure that no overfeeding of fluoridation chemicals occurs? How can overfeeding be prevented?
6. What should be done if significant overfeeding occurs?
7. What should be done when fluoridation equipment is going to be shut down for an extended length of time?
8. How would you dispose of fluoride chemical containers?
9. How would you protect yourself from the dust of dry fluoride compounds?

SUGGESTED ANSWERS
Chapter 13. FLUORIDATION

Answers to questions on page 33.

13.0A If a person drinks water with an excessive amount of fluoride, the teeth become mottled (brown, chalky deposits).

13.0B Children who drink a recommended dose of fluoride have fewer dental caries (decay or cavities).

Answers to questions on page 34.

13.1A The water department or water company makes the final decisions as to types of fluoride chemicals and feeding equipment to be used.

13.2A The three compounds most commonly used to fluoridate water are sodium fluoride, sodium fluorosilicate, and hydrofluosilicic acid.

Answers to questions on page 47.

13.3A Drinking waters may come to contain fluoride ions by three different types of situations:
  1. Raw water source may have adequate or excessive fluoride ions naturally present.
  2. Two water sources may be blended together (one higher and one lower than acceptable level) to produce an acceptable level.
  3. Fluoride ions must be added to the water to achieve an acceptable level.

13.3B Fluoridation systems must incorporate means to prevent both overfeeding and backsiphonage along with means to monitor the amount of chemical used.

13.3C A saturator is a device that produces a fluoride solution for the fluoridation process. The device is usually a cylindrical container with granular sodium fluoride on the bottom. Water flows either upward or downward through the sodium fluoride to produce the fluoride solution.

13.3D Hard water can produce problems in systems using saturators and dissolving tanks through the formation of low-solubility deposits of calcium and magnesium fluoride compounds.

Answers to questions on page 49.

13.4A Overfeeding of fluoridation chemicals must be prevented to avoid illness and bad public relations.

13.4B The fluoridator should be sized to handle the full range of both present and future doses or provisions should be made for future expansion.

13.5A When inspecting the fluoridation electrical system, inspect the system for: (1) proper voltage, (2) properly sized overload protection, (3) proper operation of control lights on the control panel, (4) proper safety lockout switches and operation, and (5) proper equipment rotation.

13.5B Safety equipment that should be available near a fluoridation system include an eye wash station, drench showers, dust masks, face shields, gloves, and vent fans.

Answers to questions on page 54.

13.6A What should be the feed rate in gallons per day for treating 6 MGD with hydrofluosilicic acid 20 percent if the desired fluoride ion concentration is 1.2 mg/L? Assume the raw water does not contain any fluoride ion.

Known	Unknown
Flow, MGD = 6 MGD	Feed Rate, gal/day
Conc Fluoride, mg/L = 1.2 mg/L	
Hydrofluosilicic Acid, % = 20%	

1. Use "Treatment Chart I, Hydrofluosilicic Acid." Start at the left side with 6 MGD and move horizontally to the right to the intersection of the 20% diagonal line.
2. Drop down vertically to the chemical feed rate of 30 gallons per day, for 1 mg/L fluoride application.
3. Adjust the flow rate for a dose of 1.2 mg/L.

$$\text{Flow Rate, gal/day} = \frac{(\text{Flow Rate from Chart, gal/day})(\text{Conc Fluoride, mg/}L)}{1 \text{ mg/}L}$$

$$= \frac{(30 \text{ gal/day})(1.2 \text{ mg/}L)}{1 \text{ mg/}L}$$

$$= 36 \text{ gal/day}$$

13.6B Differences between the volume of acid used from a storage tank and the volume actually fed into the system could be caused by errors in readings, leaks, or equipment malfunctioning.

Answers to questions on page 58.

13.7A Overfeeding should be prevented because no additional benefits result from overfeeding and there is a waste of chemicals and money. Excessive overfeeding could be harmful to consumers.

13.7B If significant overfeeding occurs, the plant should be shut down. The affected mains should be flushed and the local and state health departments notified.

13.8A A fluoridation operation could be shut down for cleaning, adjustments, or due to safety controls.

Answers to questions on page 60.

13.9A If fluoridation equipment is going to be shut down for an extended length of time, it should be cleaned out to prevent corrosion or the solidifying of the chemical. Lines and equipment could be damaged when re-started if chemicals left in them solidify.

13.10A Fluoride dust can be removed from gears by the use of a vacuum cleaner.

13.10B During maintenance of fluoridation equipment, you do not need to be too concerned about checking the feed rate by catching a given amount of fluoride over a time period. The log will show long-period discrepancies and the daily laboratory tests will indicate any drifting from the desired fluoride concentration in the treated water. Analyze the fluoridated water daily. Check the results for any deviations from the norm and take corrective action. Hand-held colorimeters are available for measuring fluoride in water.

13.11A If you are acutely poisoned by a fluoride chemical, you may be thirsty, vomit, and have stomach cramps, diarrhea, difficulty in speaking, and disturbed color vision.

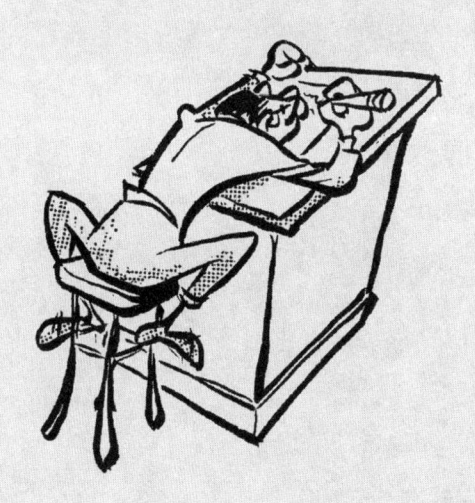

CHAPTER 14

SOFTENING

by

Don Gibson

and

Marty Reynolds

TABLE OF CONTENTS
Chapter 14. SOFTENING

	Page
OBJECTIVES	73
WORDS	74

LIME–SODA ASH SOFTENING by Don Gibson

LESSON 1

14.0	WHAT MAKES WATER HARD?		77
14.1	WHY SOFTEN WATER?		78
14.2	CHEMISTRY OF SOFTENING		79
	14.20	Hardness	79
	14.21	pH	80
	14.22	Alkalinity	80
14.3	HOW WATER IS SOFTENED		82
	14.30	Basic Methods of Softening	82
	14.31	Chemical Reactions	83
		14.310 Lime	83
		14.311 Removal of Carbon Dioxide	83
		14.312 Removal of Carbonate Hardness	83
		14.313 Removal of Noncarbonate Hardness	84
		14.314 Stability	84
		14.315 Caustic Soda Softening	84
		14.316 Calculation of Chemical Dosages	85
	14.32	Lime Softening	86
	14.33	Split Lime Treatment	86
	14.34	Lime–Soda Ash Softening	87
	14.35	Caustic Soda Softening	87
	14.36	Handling, Application, and Storage of Lime	89
14.4	INTERACTIONS WITH COAGULANTS		90
14.5	STABILITY		90
14.6	SAFETY		91
14.7	SLUDGE RECIRCULATION AND DISPOSAL		92

14.8	RECORDS		93
14.9	JAR TESTS		93
	14.90	Typical Procedures	93
	14.91	Examples	93
	14.92	Calculation of Chemical Feeder Settings	97
DISCUSSION AND REVIEW QUESTIONS			98

ION EXCHANGE SOFTENING by Marty Reynolds

LESSON 2

14.10	DESCRIPTION OF ION EXCHANGE SOFTENING PROCESS		99
14.11	OPERATIONS		102
	14.110	Service	102
	14.111	Backwash	104
	14.112	Brine	104
	14.113	Rinse	105
14.12	CONTROL TESTING OF ION EXCHANGE SOFTENERS		105
14.13	LIMITATIONS CAUSED BY IRON AND MANGANESE		105
14.14	DISPOSAL OF SPENT BRINE		106
14.15	MAINTENANCE		106
14.16	TROUBLESHOOTING		107
	14.160	Test Units	107
	14.161	Service Stage	107
	14.162	Backwash Stage	107
	14.163	Brine Injection Stage	108
	14.164	Rinse Stage	108
14.17	START-UP AND SHUTDOWN OF UNIT		108
14.18	ION EXCHANGE ARITHMETIC		110
14.19	BLENDING		112
14.20	RECORDKEEPING		114
14.21	ARITHMETIC ASSIGNMENT		114
14.22	ADDITIONAL READING		114
14.23	ACKNOWLEDGMENTS		114
DISCUSSION AND REVIEW QUESTIONS			115
SUGGESTED ANSWERS			115

OBJECTIVES
Chapter 14. SOFTENING

Following completion of Chapter 14, you should be able to:

1. Explain what makes water hard and the advantages of softening.
2. Describe the processes used to soften water.
3. Prepare chemical doses to soften water with considerations given to coagulants and stability.
4. Safely handle softening chemicals.
5. Dispose of process sludges and brines.
6. Keep neat and accurate softening records.
7. Perform jar tests and apply results.
8. Operate and maintain chemical precipitation and ion exchange softening processes.
9. Start up and shut down water softening units.
10. Blend softened waters with unsoftened waters (split treatment) for delivery to consumers.

WORDS
Chapter 14. SOFTENING

ALKALINITY (AL-kuh-LIN-it-tee)
The capacity of water or wastewater to neutralize acids. This capacity is caused by the water's content of carbonate, bicarbonate, hydroxide, and occasionally borate, silicate, and phosphate. Alkalinity is expressed in milligrams per liter of equivalent calcium carbonate. Alkalinity is not the same as pH because water does not have to be strongly basic (high pH) to have a high alkalinity. Alkalinity is a measure of how much acid must be added to a liquid to lower the pH to 4.5.

ANION (AN-EYE-en)
A negatively charged ion in an electrolyte solution, attracted to the anode under the influence of a difference in electrical potential. Chloride ion (Cl^-) is an anion.

CALCIUM CARBONATE EQUILIBRIUM
A water is considered stable when it is just saturated with calcium carbonate. In this condition, the water will neither dissolve nor deposit calcium carbonate. Thus, in this water the calcium carbonate is in equilibrium with the hydrogen ion concentration.

CALCIUM CARBONATE ($CaCO_3$) EQUIVALENT
An expression of the concentration of specified constituents in water in terms of their equivalent value to calcium carbonate. For example, the hardness in water that is caused by calcium, magnesium, and other ions is usually described as calcium carbonate equivalent. Alkalinity test results are usually reported as mg/L $CaCO_3$ equivalents. To convert chloride to $CaCO_3$ equivalents, multiply the concentration of chloride ions in mg/L by 1.41, and for sulfate, multiply by 1.04.

CATION (KAT-EYE-en)
A positively charged ion in an electrolyte solution, attracted to the cathode under the influence of a difference in electrical potential. Sodium ion (Na^+) is a cation.

DIVALENT (dye-VAY-lent)
Having a valence of two, such as the ferrous ion, Fe^{2+}. Also called bivalent.

EQUIVALENT WEIGHT
That weight that will react with, displace, or is equivalent to one gram atom of hydrogen.

GREENSAND
A mineral (glauconite) material that looks like ordinary filter sand except that it is green in color. Greensand is a natural ion exchange material that is capable of softening water. Greensand that has been treated with potassium permanganate ($KMnO_4$) is called manganese greensand; this product is used to remove iron, manganese, and hydrogen sulfide from groundwaters.

HARD WATER
Water having a high concentration of calcium and magnesium ions. A water may be considered hard if it has a hardness greater than the typical hardness of water from the region. Some textbooks define hard water as a water with a hardness of more than 100 mg/L as calcium carbonate.

HARDNESS, WATER

A characteristic of water caused mainly by the salts of calcium and magnesium, such as bicarbonate, carbonate, sulfate, chloride, and nitrate. Excessive hardness in water is undesirable because it causes the formation of soap curds, increased use of soap, deposition of scale in boilers, damage in some industrial processes, and sometimes causes objectionable tastes in drinking water.

HYDRATED LIME

Limestone that has been burned and treated with water under controlled conditions until the calcium oxide portion has been converted to calcium hydroxide ($Ca(OH)_2$). Hydrated lime is quicklime combined with water. $CaO + H_2O \rightarrow Ca(OH)_2$. Also called slaked lime. Also see QUICKLIME.

INSOLUBLE (in-SAWL-yoo-bull)

Something that cannot be dissolved.

ION

An electrically charged atom, radical (such as SO_4^{2-}), or molecule formed by the loss or gain of one or more electrons.

ION EXCHANGE

A water or wastewater treatment process involving the reversible interchange (switching) of ions between the water being treated and the solid resin contained within an ion exchange unit. Undesirable ions are exchanged with acceptable ions on the resin or recoverable ions in the water being treated are exchanged with other acceptable ions on the resin.

ION EXCHANGE RESINS

Insoluble polymers, used in water or wastewater treatment, that are capable of exchanging (switching or giving) acceptable cations or anions to the water being treated for less desirable ions or for ions to be recovered.

METHYL ORANGE ALKALINITY

A measure of the total alkalinity in a water sample. The alkalinity is measured by the amount of standard sulfuric acid required to lower the pH of the water to a pH level of 4.5, as indicated by the change in color of methyl orange from orange to pink. Methyl orange alkalinity is expressed as milligrams per liter equivalent calcium carbonate.

NPDES PERMIT

National Pollutant Discharge Elimination System permit is the regulatory agency document issued by either a federal or state agency that is designed to control all discharges of potential pollutants from point sources and stormwater runoff into US waterways. NPDES permits regulate discharges into US waterways from all point sources of pollution, including industries, municipal wastewater treatment plants, sanitary landfills, large animal feedlots, and return irrigation flows.

PHENOLPHTHALEIN (FEE-nol-THAY-leen) ALKALINITY

The alkalinity in a water sample measured by the amount of standard acid required to lower the pH to a level of 8.3, as indicated by the change in color of phenolphthalein from pink to clear. Phenolphthalein alkalinity is expressed as milligrams per liter of equivalent calcium carbonate.

PRECIPITATE (pre-SIP-uh-TATE)

(1) An insoluble, finely divided substance that is a product of a chemical reaction within a liquid.

(2) The separation from solution of an insoluble substance.

QUICKLIME

A material that is mostly calcium oxide (CaO) or calcium oxide in natural association with a lesser amount of magnesium oxide. Quicklime is capable of combining with water, that is, becoming slaked. Also see HYDRATED LIME.

RECARBONATION (re-kar-bun-NAY-shun)

A process in which carbon dioxide is bubbled into the water being treated to lower the pH.

RESINS

See ION EXCHANGE RESINS.

SATURATION

The condition of a liquid (water) when it has taken into solution the maximum possible quantity of a given substance at a given temperature and pressure.

SLAKE

To mix with water so that a true chemical combination (hydration) takes place, such as in the slaking of lime.

SLAKED LIME

See HYDRATED LIME.

SUPERSATURATED

An unstable condition of a solution (water) in which the solution contains a substance at a concentration greater than the saturation concentration for the substance.

TITRATE (TIE-trate)

To titrate a sample, a chemical solution of known strength is added drop by drop until a certain color change, precipitate, or pH change in the sample is observed (end point). Titration is the process of adding the chemical reagent in small increments (0.1–1.0 milliliter) until completion of the reaction, as signaled by the end point.

CHAPTER 14. SOFTENING

Lime–Soda Ash Softening by Don Gibson

(Lesson 1 of 2 Lessons)

14.0 WHAT MAKES WATER HARD?[1]

Water hardness is a measure of the soap- or detergent-consuming power of water. Technically, hardness is caused by *DIVALENT*[2] metallic cations that are capable of reacting: (1) with soap (detergent) to form precipitates, and (2) with certain anions present in water to form scale.

Cations Causing Hardness	Most Common Anions
Calcium, Ca^{2+}	Bicarbonate, HCO_3^-
Magnesium, Mg^{2+}	Sulfate, SO_4^{2-}
Strontium, Sr^{2+}	Chloride, Cl^-
Iron, Fe^{2+}	Nitrate, NO_3^-
Manganese, Mn^{2+}	Silicate, SiO_3^{2-}

Calcium and magnesium are usually the only cations that are present in significant concentrations. Therefore, hardness is generally considered to be an expression of the total concentration of the calcium and magnesium ions that are present in the water. However, if any of the other cations listed are present in significant amounts, they should be included in the hardness determination.

Table 14.1 describes various levels of hardness. Different textbooks will use similar classifications. Hardness levels in source waters, local conditions, and local usage will influence consumers' attitudes toward the hardness of their water.

To help you understand this chapter on water softening, some of the terms used are defined below.

HARD WATER is a water having a high concentration of calcium and magnesium ions. A water may be considered hard if it has a hardness greater than the typical hardness of water from the region. Some textbooks define hard water as a water with a hardness of more than 100 mg/L as calcium carbonate.

HARDNESS is a characteristic of water caused mainly by the salts of calcium and magnesium, such as bicarbonate, carbonate, sulfate, chloride, and nitrate. Excessive hardness in water is undesirable because it causes the formation of soap curds, increased use of soap, deposition of scale in boilers, damage in some industrial processes, and it sometimes causes objectionable tastes in drinking water.

TABLE 14.1 DESCRIPTION OF VARIOUS LEVELS OF HARDNESS[a]

Description	Hardness in Terms of mg/L as Calcium Carbonate
1. Extremely soft to soft	0–45
2. Soft to moderately hard	46–90
3. Moderately hard to hard	91–130
4. Hard to very hard	131–170
5. Very hard to excessively hard	171–250
6. Too hard for ordinary domestic use	OVER 250

[a] From L. A. Lipe and M. D. Curry, "Ion Exchange Water Softening," a discussion for water treatment plant operators, 1974–75 seminar series sponsored by Illinois Environmental Protection Agency.

CALCIUM HARDNESS is caused by calcium ions (Ca^{2+}).

MAGNESIUM HARDNESS is caused by magnesium ions (Mg^{2+}).

TOTAL HARDNESS is the sum of the hardness caused by both calcium and magnesium ions.

CARBONATE HARDNESS is caused by the alkalinity present in water up to the total hardness. This value is usually less than the total hardness.

NONCARBONATE HARDNESS is that portion of the total hardness in excess of the alkalinity.

ALKALINITY (AL-kuh-LIN-it-tee) is the capacity of water or wastewater to neutralize acids. This capacity is caused by the water's content of carbonate, bicarbonate, hydroxide, and occasionally borate, silicate, and phosphate. Alkalinity is expressed in milligrams per liter of equivalent calcium carbonate. Alkalinity is not the same as pH because water does not have to be strongly basic (high pH) to have a high alkalinity. Alkalinity is a measure of how much acid must be added to a liquid to lower the pH to 4.5.

---

[1] Portions of the material covered in the first three sections of this chapter were provided by Don Gibson, Marty Reynolds, Susumu Kawamura, Terry Engelhardt, Jack Rossum, and Mike Curry.
[2] *Divalent* (dye-VAY-lent). Having a valence of two, such as the ferrous ion, Fe^{2+}. Also called bivalent.

CALCIUM CARBONATE (CaCO₃) EQUIVALENT is an expression of the concentration of specified constituents in water in terms of their equivalent value to calcium carbonate. For example, the hardness in water that is caused by calcium, magnesium, and other ions is usually described as calcium carbonate equivalent. Alkalinity test results are usually reported as mg/L CaCO₃ equivalents. To convert chloride to CaCO₃ equivalents, multiply the concentration of chloride ions in mg/L by 1.41, and for sulfate, multiply by 1.04.

14.1 WHY SOFTEN WATER?

The dissolved minerals (calcium and magnesium ions) in water cause difficulties in doing the laundry and in dishwashing in the household. These ions also cause a coating to form inside the hot water heater similar to that in a tea kettle after repeated use.

Hardness, in addition to inhibiting the cleaning action of soaps, will tend to shorten the life of fabrics that are washed in hard water. The scum or curds may become lodged in the fibers of the fabric and cause them to lose their softness and elasticity.

In industry, hardness can cause even greater problems. Many processes are affected by the hardness content of the water used. Industrial plants using boilers for processing steam or heat must remove the hardness from their makeup water, even beyond what a water treatment plant would do. The reason for this is that the minerals will plate out on the boiler tubes and form a scale. This scale forms an insulation barrier that prevents proper heat transfer, thus causing excessive energy requirements to fire the boilers. The problems associated with process water softening are too numerous to go into; however, everything from food processing to intricate manufacturing processes is affected by the hardness of water.

In addition to the removal of hardness from water, some other benefits of softening include:

1. Removal of iron and manganese
2. Control of corrosion when proper stabilization of water is achieved
3. Disinfection due to high pH values when using lime (especially the excess lime softening process)
4. Sometimes a reduction in tastes and odors
5. Reduction of some total solids content by the lime treatment process
6. Removal of radioactivity

Possible limitations of softening might include:

1. Free chlorine residual is predominantly hypochlorite at pH levels above 7.5 and is a less powerful disinfectant.
2. Costs and benefits must be carefully weighed to justify softening.
3. Ultimate disposal of process wastes.
4. At the pH levels associated with softening chemical precipitation, the trihalomethane fraction in the treated water may increase (depends on several other factors).
5. Production of an "aggressive" water that would tend to corrode metal ions from the distribution system piping. Hard waters usually do not corrode pipe. However, excessively hard water can cause scaling on the inside of pipes and thereby restrict flow.

In many cases, the decision to soften the water is left up to each community as softening is done mostly as a customer service. Hard water does not have an adverse effect on health, but can create several unwanted side effects, some of which are:

1. Over a period of time, the detergent-consuming power of hard water can be very costly.
2. Scale problems on fixtures will be more noticeable.
3. The lifetime of several types of fabrics will be reduced with repeated washing in hard water. Also, a residue can be left in clothing, creating a dirty appearance.

Once the decision is made to soften, a method must be selected. The two most common methods used to soften water are chemical precipitation (lime–soda ash) and *ION EXCHANGE*.[3] Ion exchange softening can best be applied to waters high in noncarbonate hardness and where the total hardness does not exceed 350 mg/L. This method of softening can produce a water of zero hardness, as opposed to lime softening where zero hardness cannot be reached.

Ion exchange softening will also remove noncarbonate hardness without the addition of soda ash, which is required with lime softening. Ion exchange is a nonselective method of softening. This means it will remove total hardness (the sum of carbonate and noncarbonate hardness) making it a very desirable means of water softening.

Limitations of the ion exchange softening process include an increase in the sodium content of the softened water if the ion exchanger is regenerated with sodium chloride. The sodium level should not exceed 20 mg/L in treated water because of the potentially harmful effect on persons susceptible to hypertension. Also, the ultimate disposal of spent brine and rinse waters from softeners can be a major problem for many installations.

QUESTIONS

Write your answers in a notebook and then compare your answers with those on page 115

14.0A What causes hardness in water?

14.0B Why is excessive hardness undesirable in a domestic water supply?

14.1A What are some of the limitations of the ion exchange softening process?

---

[3] *Ion Exchange.* A water or wastewater treatment process involving the reversible interchange (switching) of ions between the water being treated and the solid resin contained within an ion exchange unit. Undesirable ions are exchanged with acceptable ions on the resin or recoverable ions in the water being treated are exchanged with other acceptable ions on the resin.

14.2 CHEMISTRY OF SOFTENING

To understand how water hardness is described and also how hardness is removed from water by softening processes, operators need to have an idea of the chemical reactions that take place in water. In this section, hardness, pH, and alkalinity reactions in water will be discussed.

14.20 Hardness

Hardness is due to the presence of divalent metallic cations in water, but the Twenty-First Edition of *STANDARD METHODS*[4] identifies only calcium and magnesium as hardness constituents. Hardness is a factor commonly measured by *TITRATION*[5] as described in Volume I (see Chapter 11, Section 11.3, 5. "Hardness"). Individual divalent cations may be measured in the laboratory using an atomic absorption (AA) spectrophotometer for very accurate work.

Hardness is usually reported as *CALCIUM CARBONATE (CaCO₃) EQUIVALENT*.[6] This procedure allows us to combine or add up the hardness caused by both calcium and magnesium and report the results as total hardness.

$$\text{Calcium Hardness, mg}/L \text{ as CaCO}_3 = (\text{Calcium, mg}/L)\left(\frac{\text{Equivalent Weight of CaCO}_3}{\text{Equivalent Weight of Calcium}}\right)$$

$$= (\text{Ca, mg}/L)\left(\frac{50}{20}\right)$$

$$= 2.50(\text{Ca, mg}/L)$$

This equation indicates that if the calcium concentration in milligrams per liter is multiplied by 2.50, the result is the calcium hardness in milligrams per liter as calcium carbonate. The *EQUIVALENT WEIGHT*[7] of most elements or chemical radicals (SO_4^{2-} is a radical) can be obtained by dividing the molecular weight by the valence.

$$\text{Equivalent Weight of Calcium} = \frac{\text{Atomic Weight}}{\text{Valence}}$$

$$= \frac{40}{2}$$

$$= 20$$

To express the magnesium hardness of water as calcium carbonate equivalent, use the following formula.

$$\text{Magnesium Hardness, mg}/L \text{ as CaCO}_3 = (\text{Magnesium, mg}/L)\left(\frac{\text{Equivalent Weight of CaCO}_3}{\text{Equivalent Weight of Magnesium}}\right)$$

$$= (\text{Mg, mg}/L)\left(\frac{50}{12.15}\right)$$

$$= 4.12(\text{Mg, mg}/L)$$

The total hardness of water is the sum of the calcium and magnesium hardness as $CaCO_3$.

$$\text{Total Hardness, mg}/L \text{ as CaCO}_3 = \text{Calcium Hardness, mg}/L \text{ as CaCO}_3 + \text{Magnesium Hardness, mg}/L \text{ as CaCO}_3$$

EXAMPLE 1

Determine the total hardness as $CaCO_3$ for a sample of water with a calcium content of 30 mg/L and a magnesium content of 20 mg/L.

Known	Unknown
Calcium, mg/L = 30 mg/L	Total Hardness,
Magnesium, mg/L = 20 mg/L	mg/L as CaCO₃

Calculate the total hardness as milligrams per liter of calcium carbonate equivalent.

$$\text{Total Hardness, mg}/L \text{ as CaCO}_3 = \text{Calcium Hardness, mg}/L \text{ as CaCO}_3 + \text{Magnesium Hardness, mg}/L \text{ as CaCO}_3$$

$$= 2.50(\text{Ca, mg}/L) + 4.12(\text{Mg, mg}/L)$$

$$= 2.50(30 \text{ mg}/L) + 4.12(20 \text{ mg}/L)$$

$$= 75 \text{ mg}/L + 82.4 \text{ mg}/L$$

$$= 157.4 \text{ mg}/L \text{ as CaCO}_3$$

Total hardness is also described as the sum of the carbonate hardness (temporary hardness) and noncarbonate hardness (permanent hardness).

$$\text{Total Hardness, mg}/L \text{ as CaCO}_3 = \text{Carbonate Hardness, mg}/L \text{ as CaCO}_3 + \text{Noncarbonate Hardness, mg}/L \text{ as CaCO}_3$$

---

[4] *STANDARD METHODS FOR THE EXAMINATION OF WATER AND WASTEWATER*, 21st Edition. Obtain from American Water Works Association (AWWA), Bookstore, 6666 West Quincy Avenue, Denver, CO 80235. Order No. 10084. ISBN 0-87553-047-8. Price to members, $198.50; nonmembers, $266.00; price includes cost of shipping and handling.

[5] *Titrate* (TIE-trate). To titrate a sample, a chemical solution of known strength is added drop by drop until a certain color change, precipitate, or pH change in the sample is observed (end point). Titration is the process of adding the chemical reagent in small increments (0.1–1.0 milliliter) until completion of the reaction, as signaled by the end point.

[6] *Calcium Carbonate Equilibrium.* A water is considered stable when it is just saturated with calcium carbonate. In this condition, the water will neither dissolve nor deposit calcium carbonate. Thus, in this water the calcium carbonate is in equilibrium with the hydrogen ion concentration.

[7] *Equivalent Weight.* That weight that will react with, displace, or is equivalent to one gram atom of hydrogen. The equivalent weight of an element (such as Ca^{2+}) is equal to the atomic weight divided by the valence.

$$\text{Equivalent Weight of CaCO}_3 = \frac{\text{Molecular Weight}}{\text{Valence}}$$

$$= \frac{100}{2}$$

$$= 50$$

The amount of carbonate and noncarbonate hardness depends on the alkalinity of the water. This relationship can be described as follows.

1. When the alkalinity (expressed as calcium carbonate equivalent) is greater than the total hardness, all the hardness is in the carbonate form.

$$\text{Carbonate Hardness, mg}/L \text{ as } CaCO_3 = \text{Total Hardness, mg}/L \text{ as } CaCO_3$$

2. When the total hardness is greater than the alkalinity, the alkalinity is carbonate hardness, and noncarbonate hardness is the difference between total hardness and alkalinity.

$$\text{Carbonate Hardness, mg}/L \text{ as } CaCO_3 = \text{Alkalinity, mg}/L \text{ as } CaCO_3$$

$$\text{Noncarbonate Hardness, mg}/L \text{ as } CaCO_3 = \text{Total Hardness, mg}/L \text{ as } CaCO_3 - \text{Alkalinity, mg}/L \text{ as } CaCO_3$$

14.21 pH

pH is an expression of the intensity of the basic or acidic condition of a liquid. Mathematically, pH is the logarithm (base 10) of the reciprocal of the hydrogen ion activity.

$$pH = \text{Log} \frac{1}{\{H^+\}}$$

If $\{H^+\} = 10^{-6.5}$, then pH = 6.5. The pH may range from 0 to 14, where 0 is most acidic, 14 most basic, and 7 neutral. Natural waters usually have a pH between 6.5 and 8.5.

Table 14.2 shows the relationship between pH and hydrogen and hydroxide ions.

TABLE 14.2 RELATIONSHIP BETWEEN pH AND HYDROGEN AND HYDROXIDE IONS

pH	Hydrogen Ion (H^+), Moles/Liter	Hydroxide Ion (OH^-), Moles/Liter
0	1.0	0.000 000 000 000 01
1	0.1	0.000 000 000 000 1
2	0.01	0.000 000 000 001
3	0.001	0.000 000 000 01
4	0.000 1	0.000 000 000 1
5	0.000 01	0.000 000 001
6	0.000 001	0.000 000 01
7	0.000 000 1	0.000 000 1
8	0.000 000 01	0.000 001
9	0.000 000 001	0.000 01
10	0.000 000 000 1	0.000 1
11	0.000 000 000 01	0.001
12	0.000 000 000 001	0.01
13	0.000 000 000 000 1	0.1
14	0.000 000 000 000 01	1.0

When treating waters, the pH is very important. The pH of water may be increased or decreased by the addition of certain chemicals used to treat water (Table 14.3). In many instances, the effect on pH of adding one chemical is neutralized by the addition of another chemical. When softening water by chemical precipitation processes (lime–soda softening for example), the pH must be raised to 11 for the desired chemical reactions to occur. The levels of carbon dioxide, bicarbonate ion, and carbonate ion in waters are very sensitive to pH.

TABLE 14.3 INFLUENCE OF WATER TREATMENT CHEMICALS ON pH

Lowers pH	Increases pH
Aluminum Sulfate (Alum), $Al_2(SO_4)_3 \cdot 18H_2O$	Calcium Hypochlorite, $Ca(OCl)_2$
Carbon Dioxide, CO_2	Caustic Soda, NaOH
Chlorine, Cl_2	Hydrated Lime, $Ca(OH)_2$
Ferric Chloride, $FeCl_3$	Soda Ash, Na_2CO_3
Hydrofluosilicic Acid, H_2SiF_6	Sodium Aluminate, $NaAlO_2$
Sulfuric Acid, H_2SO_4	Sodium Hydrochlorite, NaOCl

The stability of treated water is determined by measuring the pH and calculating the Langelier Index (see Volume I, Chapter 8, Section 8.322, "Calcium Carbonate Saturation"). This index reflects the equilibrium pH of a water with respect to calcium and alkalinity.

Langelier Index (L.I.) = $pH - pH_s$

where pH = actual pH of the water

pH_s = pH at which water having the same alkalinity and calcium content is just saturated with calcium carbonate

A negative Langelier Index indicates that the water is corrosive and a positive index indicates that the water is scale forming. After water has been softened, the treated water distributed to consumers must be stable (neither corrosive nor scale forming).

14.22 Alkalinity

Alkalinity is the capacity of water or wastewater to neutralize acids. This capacity is caused by the water's content of carbonate, bicarbonate, hydroxide, and occasionally borate, silicate, and phosphate. Alkalinity is expressed in milligrams per liter of equivalent calcium carbonate. Alkalinity is not the same as pH because water does not have to be strongly basic (high pH) to have a high alkalinity. Alkalinity is a measure of how much acid must be added to a liquid to lower the pH to 4.5.

Alkalinity is measured in the laboratory by the addition of color indicator solutions and the alkalinity is then determined by the amount of acid required to reach a titration end point (specific color change) (see Volume I, Chapter 11, "Laboratory

Procedures," Section 11.3, 1. "Alkalinity"). The P (phenolphthalein) end point is at pH 8.3. When the pH is below 8.3, there is no P alkalinity present. When the pH is above 8.3, P alkalinity is present. No carbon dioxide is present when the pH is above 8.3, so there is no carbon dioxide in the water when P alkalinity is present. Also, hydroxide and carbonate alkalinity are not present when pH is below 8.3.

The relationship between the various alkalinity constituents (bicarbonate (HCO_3^-), carbonate (CO_3^{2-}), and hydroxide (OH^-)) can be based on the P (phenolphthalein) and T (total or methyl orange) alkalinity as shown in Table 14.4 and Figure 14.1.

When the pH is less than 8.3, all alkalinity is in the bicarbonate form and is commonly referred to as natural alkalinity. When the pH is above 8.3, the alkalinity may consist of bicarbonate, carbonate, and hydroxide. As the pH increases, the alkalinity progressively shifts to carbonate and hydroxide forms.

Total alkalinity is the sum of the bicarbonate, carbonate, and hydroxide. Each of these values can be determined by measuring the P and T alkalinity in the laboratory and referring to Table 14.4. Alkalinity is expressed in milligrams per liter as calcium carbonate equivalents. Alkalinity is influenced by chemicals used to treat water as shown in Table 14.5.

TABLE 14.4 ALKALINITY CONSTITUENTS

Alkalinity, mg/L as $CaCO_3$

Titration Result	Bicarbonate	Carbonate	Hydroxide
P = 0	T	0	0
P is less than ½T	T−2P	2P	0
P = ½T	0	2P	0
P is greater than ½T	0	2T−2P	2P−T
P = T	0	0	T

where P = phenolphthalein alkalinity
T = total alkalinity

TABLE 14.5 INFLUENCE OF WATER TREATMENT CHEMICALS ON ALKALINITY

Lowers Alkalinity	Increases Alkalinity
Aluminum Sulfate (Alum), $Al_2(SO_4)_3 \cdot 18H_2O$	Calcium Hypochlorite, $Ca(OCl)_2$
Carbon Dioxide, CO_2	Caustic Soda, NaOH
Chlorine Gas, Cl_2	Hydrated Lime, $Ca(OH)_2$
Ferric Chloride, $FeCl_3$	Soda Ash, Na_2CO_3
Ferric Sulfate, $Fe_2(SO_4)_3$	Sodium Aluminate, $NaAlO_2$
Sulfuric Acid, H_2SO_4	

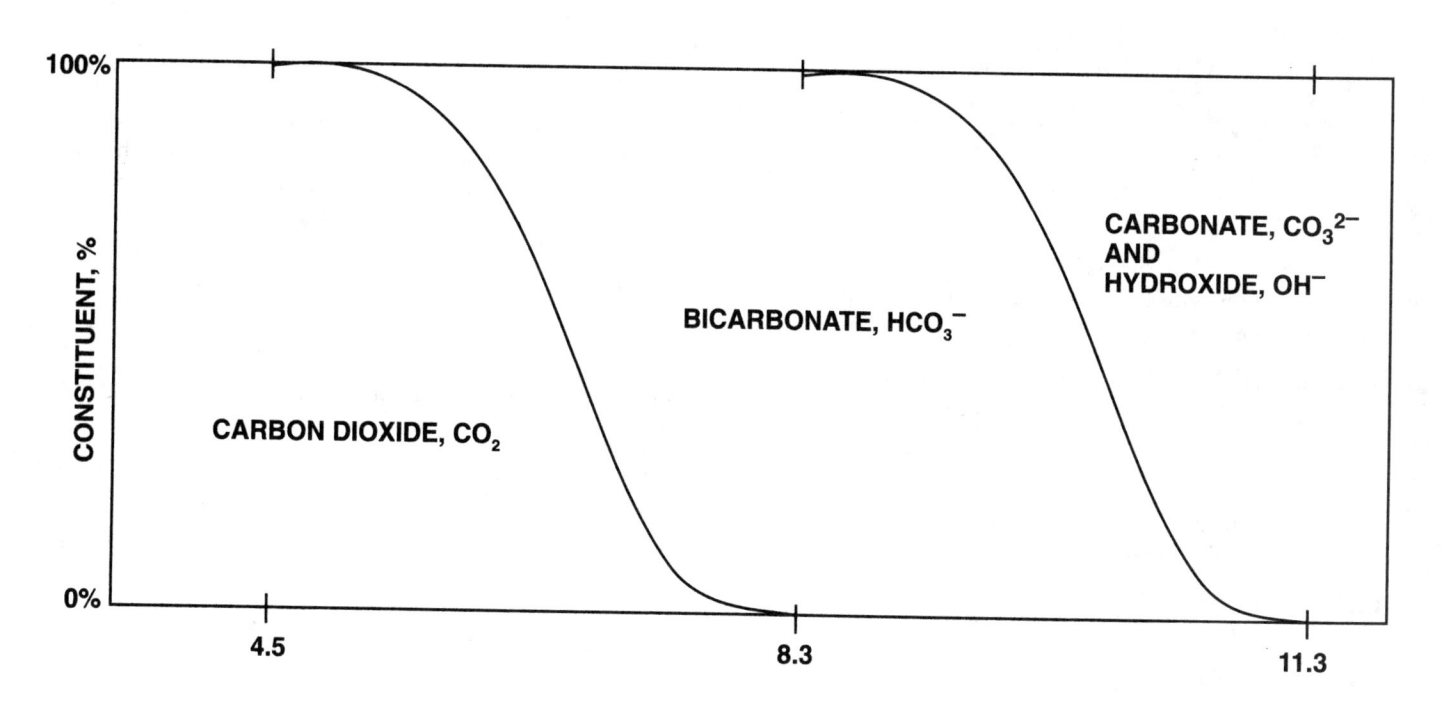

Fig. 14.1 Relationship between pH and alkalinity constituents (CO_2, HCO_3^-, CO_3^{2-}, and OH^-)

EXAMPLE 2

Results from alkalinity titrations on a raw water sample were as follows:

Known

Sample Size, mL	= 100 mL
mL Titrant Used to pH 8.3, A	= 0 mL
Total mL of Titrant Used, B	= 8.2 mL
Acid Normality, N	= 0.02 N H_2SO_4

Unknown

1. Total Alkalinity, mg/L as $CaCO_3$
2. Bicarbonate Alkalinity, mg/L as $CaCO_3$
3. Carbonate Alkalinity, mg/L as $CaCO_3$
4. Hydroxide Alkalinity, mg/L as $CaCO_3$

See Volume I, Chapter 11, "Laboratory Procedures," Section 11.3, 1. "Alkalinity" for details and formulas.

1. Calculate the phenolphthalein alkalinity in mg/L as $CaCO_3$.

$$\text{Phenolphthalein Alkalinity, mg/}L \text{ as } CaCO_3 = \frac{A \times N \times 50{,}000}{mL \text{ of Sample}}$$

$$= \frac{(0 \text{ }mL) \times (0.02 \text{ }N) \times (50{,}000)}{100 \text{ }mL}$$

$$= 0 \text{ mg/}L \text{ as } CaCO_3$$

2. Calculate the total alkalinity in mg/L as $CaCO_3$.

$$\text{Total Alkalinity, mg/}L \text{ as } CaCO_3 = \frac{B \times N \times 50{,}000}{mL \text{ of Sample}}$$

$$= \frac{(8.2 \text{ }mL) \times (0.02 \text{ }N) \times (50{,}000)}{100 \text{ }mL}$$

$$= 82 \text{ mg/}L \text{ as } CaCO_3$$

3. Refer to Table 14.4 for alkalinity constituents. The first row indicates that since P = 0, the total alkalinity is equal to the bicarbonate alkalinity.

$$\text{Bicarbonate Alkalinity, mg/}L \text{ as } CaCO_3 = \text{Total Alkalinity, mg/}L \text{ as } CaCO_3$$

$$= 82 \text{ mg/}L \text{ as } CaCO_3$$

The first row also indicates that since P = 0, the carbonate and hydroxide alkalinities are also zero.

$$\text{Carbonate Alkalinity, mg/}L \text{ as } CaCO_3 = 0 \text{ mg/}L \text{ as } CaCO_3$$

$$\text{Hydroxide Alkalinity, mg/}L \text{ as } CaCO_3 = 0 \text{ mg/}L \text{ as } CaCO_3$$

QUESTIONS

Write your answers in a notebook and then compare your answers with those on pages 115 and 116.

14.2A What laboratory procedures are used to measure hardness?

14.2B Determine the total hardness as $CaCO_3$ for a sample of water with a calcium content of 25 mg/L and a magnesium content of 14 mg/L.

14.2C Which water treatment chemicals lower the pH when added to water?

14.2D Results from alkalinity titrations on a sample of water were as follows: sample size, 100 mL; mL titrant used to pH 8.3, 1.2 mL; total mL of titrant used, 5.6 mL; and the acid normality was 0.02 N H_2SO_4. Calculate the total, bicarbonate, carbonate, and hydroxide alkalinity as $CaCO_3$.

14.3 HOW WATER IS SOFTENED

14.30 Basic Methods of Softening

The two basic methods of softening a municipal water supply are chemical precipitation and ion exchange. Ion exchange will be discussed in the second portion of this chapter in Sections 14.10 through 14.21, pages 99 through 114. We will begin here with the chemical precipitation methods, mainly lime–soda ash softening and variations of this process.

Hardness is not completely removed by the chemical precipitation methods used in water treatment plants. That is, hardness is not reduced to zero. Water having a hardness of 150 mg/L as $CaCO_3$ or more is usually treated to reduce the hardness to 80 to 90 mg/L when softening is chosen as a water treatment option.

The minimum hardness that can be achieved by the lime–soda ash process is around 30 to 40 mg/L as $CaCO_3$. The effluent from an ion exchange softener could contain almost zero hardness. Regardless of the method used to soften water, consumers usually receive a blended water with a hardness of around 80 to 90 mg/L as $CaCO_3$ when softening is used in water treatment plants.

Lime–soda softening may produce benefits in addition to the softening of water. These advantages include:

1. Removal of iron and manganese
2. Reduction of solids
3. Removal and inactivation of bacteria and viruses due to high pH
4. Control of corrosion and scale formation with proper stabilization of treated water
5. Removal of excess fluoride

Limitations of the lime–soda softening process include:

1. Inability to remove all hardness
2. A high degree of operator control must be exercised for maximum efficiency in cost, hardness removal, and water stability
3. Color removal may be complicated by the softening process due to high pH levels
4. Large quantities of sludge are created that must be handled and disposed of in an acceptable manner

QUESTIONS

Write your answers in a notebook and then compare your answers with those on page 116.

14.3A   What is the minimum hardness that can be achieved by the lime–soda ash process?

14.3B   List some of the benefits that could result from the lime–soda softening process in addition to softening the water.

14.31 Chemical Reactions

In the chemical precipitation process, the hardness-causing ions are converted from soluble to insoluble forms. Calcium and magnesium become less soluble as the pH increases. Therefore, calcium and magnesium can be removed from water as insoluble precipitates at high pH levels.

Addition of lime to water increases the hydroxide concentrations, thus increasing the pH. Addition of lime to water also converts alkalinity from the bicarbonate form to the carbonate form, which causes the calcium to be precipitated as calcium carbonate ($CaCO_3$). As additional lime is added to the water, the phenolphthalein (P) alkalinity increases to a level where hydroxide becomes present (excess causticity) allowing magnesium to precipitate as magnesium hydroxide.

Following the chemical softening process, the pH is high and the water is *SUPERSATURATED*[8] with excess caustic alkalinity in either the hydroxide or carbonate form. Carbon dioxide can be used to decrease the causticity and scale-forming tendencies of the water prior to filtration.

The chemical reactions that take place in water during the chemical precipitation process are described in the remainder of this section. The procedures for softening water depend on whether the hardness to be removed is carbonate or noncarbonate hardness. Carbonate hardness (also called "temporary hardness") can be removed by the use of lime only. Removal of noncarbonate hardness (also called "permanent hardness") requires both lime and soda.

14.310   Lime

The lime used in the chemical precipitation softening process may be from either *HYDRATED LIME*[9] ($Ca(OH)_2$, calcium hydroxide, or "slaked" lime) or calcium oxide (CaO, *QUICKLIME*[10] or "unslaked" lime). The hydrated lime may be used directly. The calcium oxide or quicklime must first be *SLAKED*.[11] This involves adding the calcium oxide (CaO) pellets to water and heating to cause "slaking" (the formation of calcium hydroxide ($Ca(OH)_2$)) before use. Small facilities commonly use hydrated lime ($Ca(OH)_2$). Large facilities may find it more economical to use quicklime (CaO) and slake it on site.

14.311   *Removal of Carbon Dioxide*

The application of lime for the removal of carbonate hardness also removes carbon dioxide. Carbon dioxide does not contribute to hardness and therefore does not need to be removed. However, carbon dioxide will consume a portion of the lime to be used and therefore must be considered. Equation (1) describes the reaction of carbon dioxide with lime.

(1) Carbon Dioxide + Lime → Calcium Carbonate ↓ + Water

$$CO_2 + Ca(OH)_2 \rightarrow CaCO_3 \downarrow + H_2O$$

14.312   *Removal of Carbonate Hardness*

The equations below describe the removal of carbonate hardness.

(2) Calcium Bicarbonate + Lime → Calcium Carbonate ↓ + Water

$$Ca(HCO_3)_2 + Ca(OH)_2 \rightarrow 2CaCO_3 \downarrow + 2H_2O$$

(3) Magnesium Bicarbonate + Lime → Calcium Carbonate ↓ + Magnesium Carbonate + Water

$$Mg(HCO_3)_2 + Ca(OH)_2 \rightarrow CaCO_3 \downarrow + MgCO_3 + H_2O$$

(4) Magnesium Carbonate + Lime → Calcium Carbonate ↓ + Magnesium Hydroxide ↓

$$MgCO_3 + Ca(OH)_2 \rightarrow CaCO_3 \downarrow + Mg(OH)_2 \downarrow$$

When lime is added to water, any carbon dioxide present is converted to calcium carbonate if enough lime is added (Equation (1)). With the addition of more lime, the calcium bicarbonate will be precipitated as calcium carbonate. To remove both the calcium and magnesium bicarbonate, an excess of lime must be used.

---

[8] *Supersaturated.* An unstable condition of a solution (water) in which the solution contains a substance at a concentration greater than the saturation concentration for the substance.

[9] *Hydrated Lime.* Limestone that has been burned and treated with water under controlled conditions until the calcium oxide portion has been converted to calcium hydroxide ($Ca(OH)_2$). Hydrated lime is quicklime combined with water. $CaO + H_2O \rightarrow Ca(OH)_2$. Also called slaked lime. Also see QUICKLIME.

[10] *Quicklime.* A material that is mostly calcium oxide (CaO) or calcium oxide in natural association with a lesser amount of magnesium oxide. Quicklime is capable of combining with water, that is, becoming slaked. Also see HYDRATED LIME.

[11] *Slake.* To mix with water so that a true chemical combination (hydration) takes place, such as in the slaking of lime.

14.313 Removal of Noncarbonate Hardness

Magnesium noncarbonate hardness requires the addition of both lime and soda ash (sodium carbonate, Na_2CO_3).

(5) Magnesium Sulfate + Lime → Magnesium Hydroxide↓ + Calcium Sulfate

$$MgSO_4 + Ca(OH)_2 \rightarrow Mg(OH)_2\downarrow + CaSO_4$$

(6) Calcium Sulfate + Soda Ash → Calcium Carbonate↓ + Sodium Sulfate

$$CaSO_4 + Na_2CO_3 \rightarrow CaCO_3\downarrow + Na_2SO_4$$

Equation (6) is also one of the equations for the removal of calcium noncarbonate hardness. Similar equations can be written for the removal of noncarbonate hardness caused by calcium and magnesium chloride.

14.314 Stability

The main chemical reaction products from the lime–soda softening process are $CaCO_3\downarrow$ and $Mg(OH)_2\downarrow$. The water thus treated has been chemically changed and is no longer stable because of pH and alkalinity changes. Lime–soda softened water is usually supersaturated with calcium carbonate ($CaCO_3$). The degree of instability and excess calcium carbonate depends on the degree to which the water is softened. Calcium carbonate hardness is removed at a lower pH than magnesium carbonate hardness. If maximum carbonate hardness removal is practiced (thus requiring a high pH to remove the magnesium carbonate hardness), the water will be supersaturated with calcium carbonate and magnesium hydroxide. Under these conditions, deposition of precipitates will occur in filters and pipelines.

Excess lime addition to remove magnesium carbonate hardness results in supersaturated conditions and a residual of lime, which will produce a pH of about 10.9. The excess lime is called caustic alkalinity since it raises the pH. If the pH is then lowered, better precipitation of calcium carbonate and magnesium hydroxide will occur. Alkalinity will be lowered also. This is usually accomplished by pumping carbon dioxide (CO_2) gas into the water. This addition of carbon dioxide to the treated water is called RECARBONATION.[12]

Recarbonation may be carried out in two steps. The first addition of carbon dioxide would follow excess lime addition to lower the pH to about 10.4 and encourage the precipitation of calcium carbonate and magnesium hydroxide. A second addition of carbon dioxide after treatment removes noncarbonate hardness. This would again lower the pH to about 9.8 and would encourage precipitation. By carrying out recarbonation prior to filtration, the buildup of excess lime and also calcium carbonate and magnesium hydroxide precipitates in the filters will be prevented or minimized. The recarbonation reaction for excess lime removal is shown below.

(7) Calcium Hydroxide + Carbon Dioxide → Calcium Carbonate↓ + Water

$$Ca(OH)_2 + CO_2 \rightarrow CaCO_3\downarrow + H_2O$$

Care must be exercised when using recarbonation. Feeding excess carbon dioxide may result in no lowering of the hardness by causing calcium carbonate precipitates to go back into solution and cause carbonate hardness.

(8) Calcium Carbonate + Carbon Dioxide + Water → Calcium Bicarbonate

$$CaCO_3 + CO_2 + H_2O \rightarrow Ca(HCO_3)_2$$

14.315 Caustic Soda Softening

An alternative method in the lime–soda softening process is the use of sodium hydroxide (NaOH, often called caustic soda) in place of soda ash. The chemical reactions of sodium hydroxide with carbonate and noncarbonate hardness are listed below.

(9) Carbon Dioxide + Sodium Hydroxide → Sodium Carbonate + Water

$$CO_2 + 2NaOH \rightarrow Na_2CO_3 + H_2O$$

(10) Calcium Bicarbonate + Sodium Hydroxide → Calcium Carbonate↓ + Sodium Carbonate + Water

$$Ca(HCO_3)_2 + 2NaOH \rightarrow CaCO_3\downarrow + Na_2CO_3 + 2H_2O$$

(11) Magnesium Bicarbonate + Sodium Hydroxide → Magnesium Hydroxide↓ + Sodium Carbonate + Water

$$Mg(HCO_3)_2 + 4NaOH \rightarrow Mg(OH)_2\downarrow + 2Na_2CO_3 + 2H_2O$$

(12) Magnesium Sulfate + Sodium Hydroxide → Magnesium Hydroxide↓ + Sodium Sulfate

$$MgSO_4 + 2NaOH \rightarrow Mg(OH)_2\downarrow + Na_2SO_4$$

These chemical reactions show that in removing carbon dioxide and carbonate hardness, sodium carbonate (Na_2CO_3, soda ash) is formed, which will react to remove the noncarbonate hardness. Not only will sodium hydroxide substitute for soda ash, but it may replace all or part of the lime ($Ca(OH)_2$) requirement for removal of the carbonate hardness. The use of caustic soda (usually as a 50-percent solution) may have several advantages:

1. Stability in storage
2. Less sludge is formed
3. Ease of handling and storage

Safe handling procedures for caustic soda must be used at all times. A 50-percent caustic solution is very dangerous. Caustic soda is a strong base and will attack fabrics and leather and cause severe burns to the skin. Rubber gloves, respirator, safety goggles, and a rubber apron must be worn when handling caustic soda. A safety shower and an emergency eye wash station must be readily available at all times.

The decision to use caustic soda rather than soda ash depends on the quality of the source water and the delivered costs of the various chemicals.

QUESTIONS

Write your answers in a notebook and then compare your answers with those on pages 116 and 117.

14.3C What causes the pH to increase during the lime–soda softening process?

14.3D How can the scale-forming tendencies be decreased in water after the chemical softening process?

---

[12] *Recarbonation* (re-kar-bun-NAY-shun). A process in which carbon dioxide is bubbled into the water being treated to lower the pH. The pH may also be lowered by the addition of acid. Recarbonation is the final stage in the lime–soda ash softening process. This process converts carbonate ions to bicarbonate ions and stabilizes the solution against the precipitation of carbonate compounds.

14.3E What is recarbonation?

14.3F Under what conditions might caustic soda softening be used?

14.316 Calculation of Chemical Dosages

There are several different approaches to calculating chemical doses for the lime–soda softening process. This section illustrates one step-by-step procedure. To use this procedure you need to obtain a chemical analysis of the water you are softening. From this analysis obtain the known values for your water similar to the "Knowns" listed in Example 3. Then calculate the dosages of chemicals for your water by following the steps in the example.

To help you understand where some of the numbers come from in the formulas, we have listed the molecular weights of the major chemical components involved in the chemical precipitation softening process.

Quicklime, CaO = 56
Hydrated Lime, $Ca(OH)_2$ = 74
Magnesium, Mg^{2+} = 24.3
Carbon Dioxide, CO_2 = 44
Magnesium Hydroxide, $Mg(OH)_2$ = 58.3
Soda Ash, Na_2CO_3 = 106
Alkalinity, as $CaCO_3$ = 100
Hardness, as $CaCO_3$ = 100

FORMULAS

1. The lime dosage for softening can be estimated by using the following formula:

$$\frac{\text{Quicklime (CaO)}}{\text{Feed, mg/}L} = \frac{(A + B + C + D)1.15}{\text{Purity of Lime, as a decimal}}$$

Where A = CO_2 in source water
(mg/L as CO_2)(56/44)

B = Bicarbonate alkalinity removed in softening
(mg/L as $CaCO_3$)(56/100)

C = Hydroxide alkalinity in softener effluent
(mg/L as $CaCO_3$)(56/100)

D = Magnesium removed in softening
(mg/L as Mg^{2+})(56/24.3)

1.15 = Excess lime dosage
(using a 15 percent excess)

NOTE: If hydrated lime $(Ca(OH)_2)$ is used instead of quicklime, substitute 74 for 56 in A, B, C, and D.

2. The soda ash dosage to remove noncarbonate hardness can be estimated by using the formula below.

$$\frac{\text{Soda Ash (Na}_2\text{CO}_3\text{)}}{\text{Feed, mg/}L} = (\text{Noncarbonate Hardness, mg/}L \text{ as } CaCO_3)(106/100)$$

3. The dosage of carbon dioxide required for recarbonation can be estimated using the formula below.

$$\frac{\text{Total CO}_2}{\text{Feed, mg/}L} = (Ca(OH)_2 \text{ excess, mg/}L)(44/74)$$
$$+ (Mg^{2+} \text{ residual, mg/}L)(44/24.3)$$

EXAMPLE 3

Calculate the hydrated lime $(Ca(OH)_2)$ with 90 percent purity, soda ash, and carbon dioxide dose requirements in milligrams per liter for the water shown below.

Known

Constituents	Source Water	Softened Water After Recarbonation and Filtration
CO_2, mg/L	= 6 mg/L	= 0 mg/L
Total Alkalinity, mg/L	= 170 mg/L as $CaCO_3$	= 30 mg/L as $CaCO_3$
Total Hardness, mg/L	= 280 mg/L as $CaCO_3$	= 70 mg/L as $CaCO_3$
Mg^{2+}, mg/L	= 21 mg/L	= 3 mg/L
pH	= 7.5	= 8.8
Lime Purity, %	= 90%	

Unknown

1. Hydrated Lime, mg/L
2. Soda Ash, mg/L
3. Carbon Dioxide, mg/L

1. Calculate the hydrated lime $(Ca(OH)_2)$ required in milligrams per liter.

A = $(CO_2, \text{mg/}L)(74/44)$
 = (6 mg/L)(74/44)
 = 10 mg/L

B = (Alkalinity, mg/L)(74/100)
 = (170 mg/L − 30 mg/L)(74/100)
 = 104 mg/L

C = 0   Hydroxide Alkalinity = 0

D = $(Mg^{2+}, \text{mg/}L)(74/24.3)$
 = (21 mg/L − 3 mg/L)(74/24.3)
 = 55 mg/L

$$\frac{\text{Hydroxide Lime}}{(Ca(OH)_2) \text{ Feed, mg/}L} = \frac{(A + B + C + D)1.15}{\text{Purity of Lime, as a decimal}}$$

$$= \frac{(10 \text{ mg/}L + 104 \text{ mg/}L + 0 + 55 \text{ mg/}L)1.15}{0.90}$$

$$= \frac{(169 \text{ mg/}L)(1.15)}{0.90}$$

= 216 mg/L

2. Calculate the soda ash required in milligrams per liter.

$$\begin{aligned}\text{Total Hardness Removed,} \\ \text{mg}/L \text{ as CaCO}_3\end{aligned} = \begin{aligned}\text{Total Hardness,} \\ \text{mg}/L \text{ as CaCO}_3\end{aligned} - \begin{aligned}\text{Total Hardness Remaining,} \\ \text{mg}/L \text{ as CaCO}_3\end{aligned}$$

$$= 280 \text{ mg}/L - 70 \text{ mg}/L$$

$$= 210 \text{ mg}/L \text{ as CaCO}_3$$

$$\begin{aligned}\text{Noncarbonate} \\ \text{Hardness,} \\ \text{mg}/L \text{ as} \\ \text{CaCO}_3\end{aligned} = \begin{aligned}\text{Total Hardness} \\ \text{Removed,} \\ \text{mg}/L \text{ as CaCO}_3\end{aligned} - \left(\begin{aligned}\text{Carbonate} \\ \text{Hardness,} \\ \text{mg}/L \text{ as CaCO}_3\end{aligned} - \begin{aligned}\text{Carbonate} \\ \text{Hardness} \\ \text{Remaining,} \\ \text{mg}/L \text{ as CaCO}_3\end{aligned}\right)$$

$$= 210 \text{ mg}/L - (170 \text{ mg}/L - 30 \text{ mg}/L)$$

$$= 70 \text{ mg}/L \text{ as CaCO}_3$$

$$\begin{aligned}\text{Soda Ash (Na}_2\text{CO}_3) \\ \text{Feed, mg}/L\end{aligned} = \left(\begin{aligned}\text{Noncarbonate Hardness,} \\ \text{mg}/L \text{ as CaCO}_3\end{aligned}\right)(106/100)$$

$$= (70 \text{ mg}/L)(106/100)$$

$$= 74 \text{ mg}/L$$

3. Calculate the dosage of carbon dioxide required for recarbonation.

$$\text{Excess Lime, mg}/L = (A + B + C + D)(0.15)$$

$$= (10 \text{ mg}/L + 104 \text{ mg}/L + 0 + 55 \text{ mg}/L)(0.15)$$

$$= (169 \text{ mg}/L)(0.15)$$

$$= 25 \text{ mg}/L$$

$$\begin{aligned}\text{Total CO}_2 \\ \text{Feed, mg}/L\end{aligned} = (\text{Ca(OH)}_2 \text{ excess, mg}/L)(44/74) \\ + (\text{Mg}^{2+} \text{ residual, mg}/L)(44/24.3)$$

$$= (25 \text{ mg}/L)(44/74) + (3 \text{ mg}/L)(44/24.3)$$

$$= 15 \text{ mg}/L + 6 \text{ mg}/L$$

$$= 21 \text{ mg}/L$$

QUESTION

Write your answer in a notebook and then compare your answer with the one on page 117.

14.3G Calculate the hydrated lime (Ca(OH)$_2$) with 90 percent purity, soda ash, and carbon dioxide dose requirements in milligrams per liter for the water shown below.

Constituents	Source Water	Softened Water After Recarbonation and Filtration
CO$_2$, mg/L	= 5 mg/L	= 0 mg/L
Total Alkalinity, mg/L	= 150 mg/L as CaCO$_3$	= 20 mg/L as CaCO$_3$
Total Hardness, mg/L	= 240 mg/L as CaCO$_3$	= 50 mg/L as CaCO$_3$
Mg^{2+}, mg/L	= 16 mg/L	= 2 mg/L
pH	= 7.4	= 8.8
Lime Purity, %	= 90%	

14.32 Lime Softening (Figure 14.2)

Water having hardness caused by calcium and magnesium bicarbonate (carbonate hardness) can usually be softened to an acceptable level using only lime. The lime reacts with the bicarbonate to form calcium carbonate, which will precipitate and settle out (convert from soluble to insoluble form) at a pH above 10, and magnesium carbonate, which will remain in solution. The magnesium carbonate reacts with additional lime at a pH above 11 to form magnesium hydroxide, which will precipitate.

In practice, if enough hardness can be removed by reacting lime with the calcium bicarbonate, softening can be accomplished at less expense. This procedure is sometimes called partial lime softening (no magnesium removal), which removes hardness caused by calcium ions. This may be referred to as calcium hardness. On the other hand, if some of the magnesium is to be removed, additional lime will be required.

Figure 14.2 is a flow diagram of a typical straight lime softening treatment plant. Settling should be provided after the addition of carbon dioxide (recarbonation) to ease the load on the filters. Recarbonation is used to lower the pH of the water. When properly recarbonated, the water is still supersaturated with calcium carbonate (CaCO$_3$). If the pH is much above 9, the water will usually cause scale to form. By recarbonation, the pH can be lowered to a range between 8.8 and 8.4 and the Langelier Index will still be positive; therefore there will be little or no corrosion. A polyphosphate is sometimes added to the water to prevent excessively heavy calcium carbonate scale deposits from forming. A polyphosphate may not be necessary if recarbonation is properly controlled. Addition of acid will not accomplish the same things as recarbonation and the addition of a polyphosphate; however, adding acid will lower the pH.

14.33 Split Lime Treatment

The amount of calcium and magnesium in source waters may vary. When the water contains a high level of magnesium, a method known as split lime treatment may be used (Figure 14.3). Split lime treatment can be used in lime treatment only or lime–soda ash treatment. In split lime treatment, a portion of the water (say 80 percent) is treated with an excess amount of lime to remove the magnesium at a pH of over 11. Then, source water (the other 20 percent) is added in the next basin to neutralize (lower the pH) the excess-lime-treated portion. The percentages will vary depending upon the water hardness, treatment layout, and desired results.

Split lime treatment softening can eliminate the need for recarbonation as well as offer a significant savings in lime feed. Since the fraction of the water that is treated has a high lime dose, magnesium is almost completely removed from this portion. When this water is mixed with the unsoftened water, the carbon dioxide and bicarbonate in the unsoftened fraction of the water tend to recarbonate in the final blend or mix of the treated water (effluent).

If the water shown in Figure 14.4 was treated by conventional treatment (not split treatment), it would require a lime dose of 400 mg/L as CaCO$_3$ (which is 25 percent higher) and a carbon

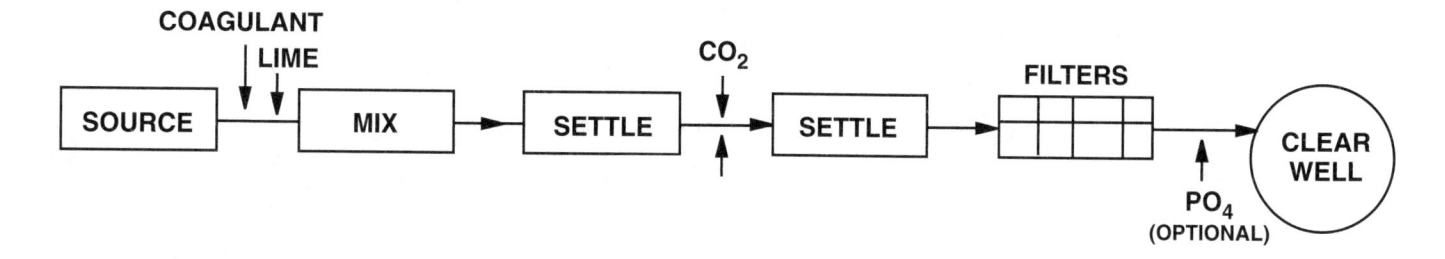

Fig. 14.2 *Straight lime treatment*

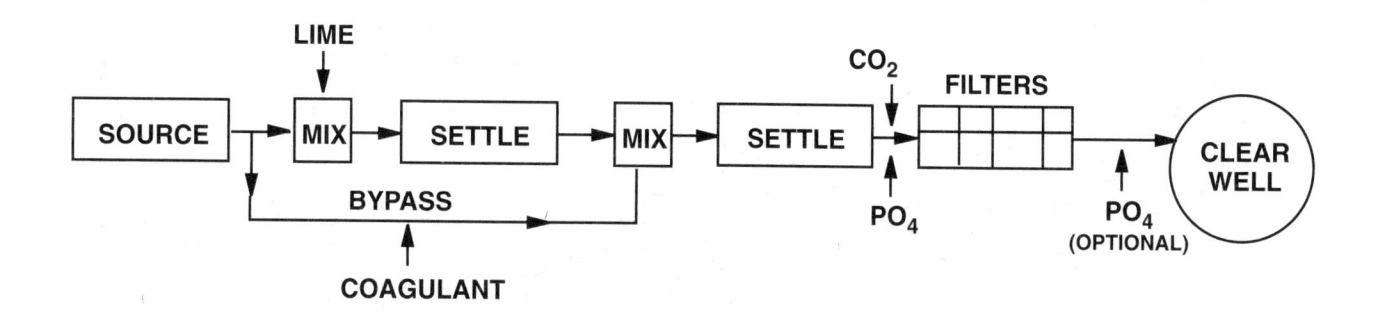

Fig. 14.3 *Split lime treatment*

dioxide dose of 145 mg/L as $CaCO_3$ to produce a water having a hardness of 61 mg/L as $CaCO_3$ and a pH of 8.63.

While split treatment may be used in the lime–soda process, it is often advantageous to use a lime–ion exchange process (see Section 14.10). The salt used to remove noncarbonate hardness in the ion exchange process is much less expensive than the soda ash required in the lime–soda ash process.

The curves shown in Figure 14.4 assume that carbonate equilibrium has been achieved. In practice, it is not possible to attain equilibrium, but if the reactions take place in solids contact units, the results are very close to carbonate equilibrium.

The proper fraction of water to bypass is rather critically dependent on the lime dose and chemical composition of the unsoftened water. The proper fraction may be calculated, but the calculations are very complex. An experienced water chemist can perform the calculations.

QUESTIONS

Write your answers in a notebook and then compare your answers with those on page 117.

14.3H What compounds are formed when calcium and magnesium are precipitated out of water in the lime softening process?

14.3I What hardness is removed by partial lime softening (no magnesium removal)?

14.3J What is split lime treatment?

14.34 Lime–Soda Ash Softening

Let us look now at hardness requiring lime–soda ash treatment for removal (Figure 14.5).

When water cannot be softened to the desired level with lime only, it no doubt contains noncarbonate hardness. Noncarbonate hardness requires the addition of a compound that increases carbonate concentration, usually soda ash (sodium carbonate).

A water could contain only calcium hardness, yet require both lime and soda ash treatment. This would occur if the hardness were only calcium bicarbonate, sulfate, or chloride. In other words, all of the hardness is calcium carbonate and calcium noncarbonate hardness. This would not require split treatment (Figure 14.6).

14.35 Caustic Soda Softening

An alternative to the lime–soda ash process is the use of caustic soda (sodium hydroxide, NaOH) instead of soda ash.

The reactions of caustic soda with the carbonate and noncarbonate hardness are given in Section 14.315. Recall that caustic soda reacts with the carbonate hardness to form soda ash (sodium carbonate), which will react with calcium sulfate to form calcium carbonate ($CaCO_3$), as shown previously.

The advantages of using liquid caustic soda include ease of handling and feeding, lack of deterioration in storage, and less calcium carbonate sludge to handle and dispose of. Caustic soda is capable of removing both carbonate and noncarbonate hardness. Therefore, caustic soda may be used instead of soda ash,

88 Water Treatment

UNSOFTENED WATER

TEMPERATURE	= 25°C	Ca	= 150 mg/L as CaCO$_3$
TDS	= 400 mg/L	Mg	= 100 mg/L as CaCO$_3$
pH	= 7.2	Alk	= 250 mg/L as CaCO$_3$

DOSAGE

LIME = 320 mg/L as CaCO$_3$

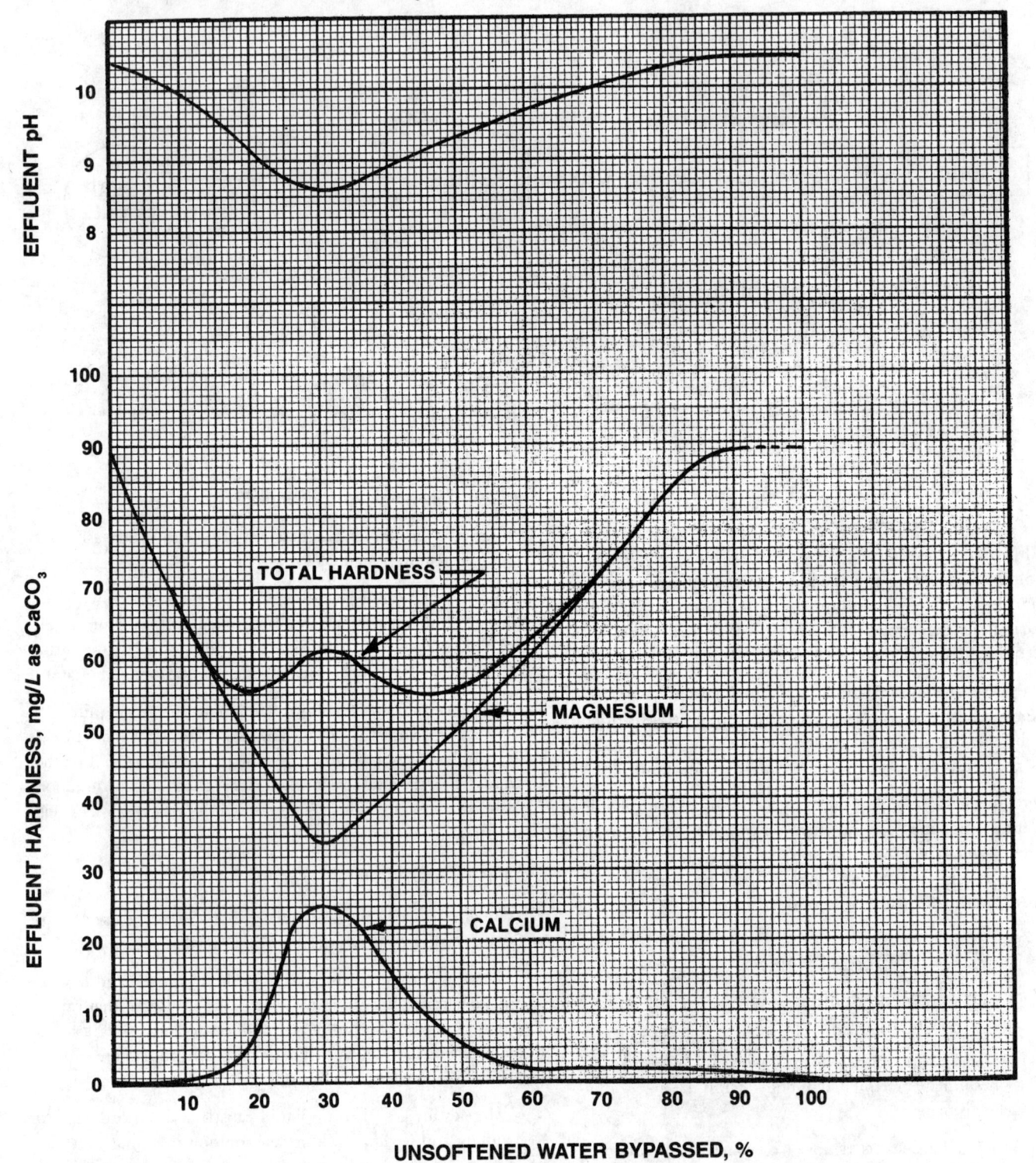

Fig. 14.4 Split treatment softening

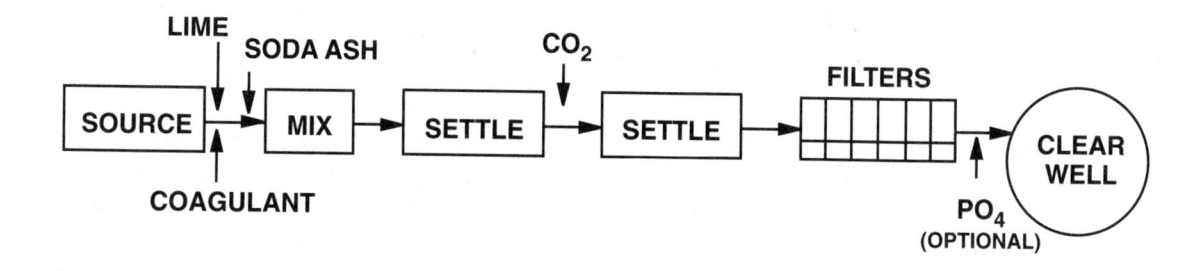

Fig. 14.5 Lime–soda ash treatment

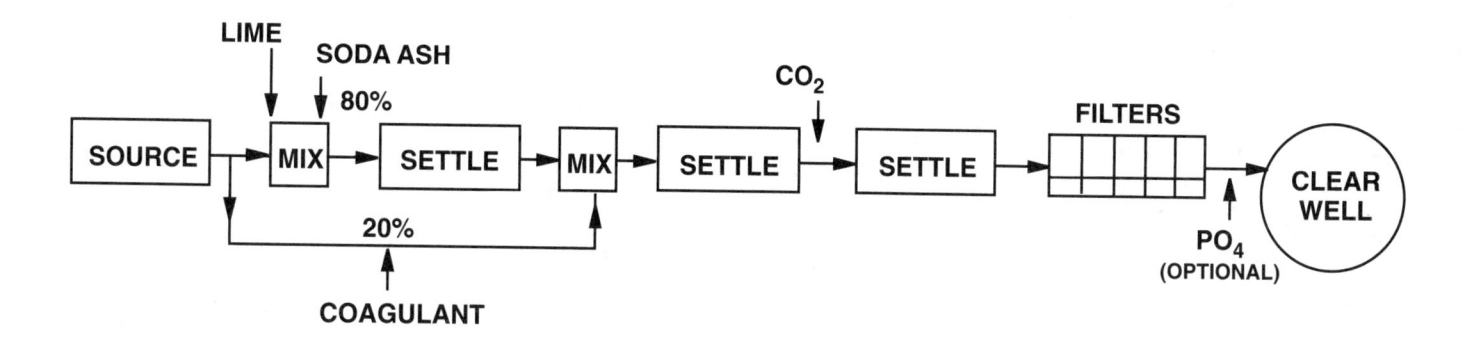

Fig. 14.6 Lime–soda ash split treatment

but also in place of part or all of the lime requirement. The use of caustic soda depends on a comparison of the costs of caustic soda, lime, and soda ash, and the characteristics of the source water.

Two points to observe are that sodium does not contribute to hardness, thus all the reactions having sodium compounds as an end product are nonhardness-producing compounds. However, sodium levels in drinking water should be less than 20 mg/L. The second point is that the precipitated compounds ($CaCO_3\downarrow$ and $Mg(OH)_2\downarrow$) are the desired end products whether lime or lime–soda ash or caustic soda treatment is used.

QUESTIONS

Write your answers in a notebook and then compare your answers with those on page 118.

14.3K Under what conditions would lime–soda ash softening be used?

14.3L What chemical is used to remove noncarbonate hardness in the chemical precipitation softening process?

14.36 Handling, Application, and Storage of Lime

Where the daily requirements for lime are small, lime is usually delivered to the water treatment plant in bags. At larger treatment plants either quicklime (CaO) or hydrated lime ($Ca(OH)_2$) is delivered in bulk quantities. Truckloads of lime are commonly transferred to weathertight bins or silos by mechanical or pneumatic conveying systems.

Storage areas for bagged lime must be covered to prevent rain from wetting the bags. Bagged quicklime (CaO or calcium oxide) should never be stored close to combustible materials because considerable heat will be generated if the lime accidentally gets wet. Quicklime may be stored as long as six months, but in general should not be stored over three months. Hydrated lime should not be stored for more than three months before use.

Lime may be applied by dry feeding techniques using volumetric or gravimetric feeders. Lime is too insoluble to make "solution feeding" by pump feeders practical because of the accumulation of carbonate precipitation. See Chapter 13, "Fluoridation," for additional details and pictures of the various types of chemical feeders.

Operator safety must be considered before attempting to work with lime. A properly designed lime feeding system can minimize or eliminate lime dust problems. If lime dust is a problem, operators must wear protective clothing to avoid burns from contact with lime. Protective clothing includes long-sleeved shirt with sleeves and collar buttoned, trousers with legs down over tops of shoes or boots, head protection, and gloves. Clothing should not fit too tightly around your neck, wrists, or ankles. A protective cream should be applied to exposed parts of the body, especially your neck, face, and wrists. You should wear a lightweight filter mask and tight-fitting safety glasses with side shields to protect yourself from the lime dust.

If lime comes in contact with your skin or eyes, immediately flush the affected areas with water and consult a physician if necessary. Do not rub your eyes if they are irritated with lime dust because rubbing will make the irritation worse. Keep any lime burns covered with a bandage during healing to prevent infection.

After handling lime, you should take a shower. If your clothes are covered with dust, or splattered with a lime slurry, take them off and have them washed. If possible, wear clean clothes on every shift.

For additional information regarding lime, contact the National Lime Association, 200 N. Glebe Road, Suite 800, Arlington, VA 22203-3728, and request a copy of their publication, *LIME HANDLING, APPLICATION, AND STORAGE IN TREATMENT PROCESSES,* 7th Edition, 1995. Order No. 213. Price, $19.95, plus shipping and handling. Lime, as well as other water treatment chemicals, should comply with the Standards of the American Water Works Association.

QUESTIONS

Write your answers in a notebook and then compare your answers with those on page 118.

14.3M   How is lime delivered to plants where the daily requirements are small?

14.3N   Why should quicklime be kept dry?

14.3O   What types of chemical feeders are used to apply dry lime?

14.4 INTERACTIONS WITH COAGULANTS

Coagulation is discussed in detail in Volume I, Chapter 4, "Coagulation and Flocculation." However, the interactions of lime and soda ash with metallic coagulants such as alum, iron salts (ferric chloride, ferric sulfate, and ferrous sulfate), sodium aluminate, and many polymers are important.

Alum and iron salts are acidic and react with the alkalinity in water to cause a demand the same way that free carbon dioxide will. Therefore, this acidic condition must be met before softening can occur. In other words, extra lime will be required as the alum or iron feed rate goes up and, therefore, less lime will be required as the alum or iron feed rate is reduced.

Cationic polymers are not very pH sensitive and are often used as coagulant aids in softening plants rather than alum or iron salts. On the other hand, when sodium aluminate (a basic rather than an acidic compound) is the coagulant, the lime required to achieve a specific hardness reduction will be less, and will vary the opposite of alum or iron salts. The proportion of lime required in either instance is directly related to the coagulant dosage as well as the hardness removal desired. Approximate relationships can be calculated; however experimentation

is in order since plant equipment and source water variations are primary factors in the efficiencies of each waterworks. Jar tests are discussed later in Section 14.9.

If you are treating highly colored waters, these waters must be coagulated for color removal at low pH values. Alum is a good coagulant under these conditions. Ozone, permanganate, and chlorine may be tried along with alum to oxidize color. The high pH values required during softening tend to set the color, which then becomes very difficult to remove.

QUESTIONS

Write your answers in a notebook and then compare your answers with those on page 118.

14.4A   What happens to the lime dose when the alum dose is increased for coagulation?

14.4B   How can color be removed from water?

14.5 STABILITY

In nature, most waters are more or less stable. That is, they are in chemical balance. If a water is not stable, either: (1) scale will form on the filter sand, distribution mains, and household plumbing, or (2) the water will be corrosive. When lime is added, the chemical balance is changed. The calcium carbonate ($CaCO_3$) formed in lime treatment is scale forming unless the exact chemical balance is achieved, which is seldom the case.

Under most conditions, a slight excess of lime is fed to cause a caustic condition to ensure complete reactions and achieve the desired results. In order to prevent scale formation on the filter sand, distribution mains, and household plumbing, the excess caustic and unprecipitated carbonate ions (pin floc) must be converted to soluble forms. Recarbonation is the most common way to do this. Again, as with all chemical treatment, recarbonation must be controlled to achieve the desired results.

Recarbonation lowers the pH to about 8.8 and thus converts some of the carbonate (CO_3^{2-}) back to the original bicarbonate (HCO_3^-) that existed in the source water. Recarbonation can be accomplished, to a degree, by using the source water in the split treatment mode discussed earlier. Usually, this is not adequate so further recarbonation is required. One reason for using source water as a neutralizing agent is that the recarbonation process is much less costly than if a high caustic water (high pH) is neutralized by chemical addition.

Use of carbon dioxide gas is the most common method of recarbonation. The reactions are:

1. $Ca(OH)_2 + CO_2 + H_2O \rightarrow Ca(HCO_3)_2$

2. $CaCO_3 + CO_2 + H_2O \rightarrow Ca(HCO_3)_2$

These reactions may be looked at as lime softening in reverse and will increase the hardness slightly. In these reactions, you are producing bicarbonate ions, which were removed in softening as carbonate hardness. This process tends to move the water back toward its original state, thus rendering it more stable.

The use of acids such as sulfuric or hydrochloric instead of recarbonation with carbon dioxide (CO_2) does not produce the same results. When carbon dioxide is added to a water containing calcium ions (Ca^{2+}) and hydroxide ions (OH^-), a calcium carbonate ($CaCO_3$) precipitate will form and the water will be saturated (or supersaturated) with calcium carbonate. If a strong acid is added to neutralize the softened water, which is highly basic, these reactions will not take place.

The marble test is the simplest method of measuring stability in the laboratory. Run the marble test[13] as outlined below:

1. Collect a sample of tap water that has been softened and stopper the sample bottle (avoid splashing into the flask).

2. To an identical sample, add one gram of powdered calcium carbonate. Mix and let stand for an hour or so.

3. Filter both samples (so they are both exposed to the same conditions).

4. Run pH and alkalinity tests on both samples.

5. The goal is to have the sample of softened tap water as nearly matched to the softened sample treated with calcium carbonate as possible. The plant treatment must be controlled to permit this condition to exist. If the pH and alkalinity in the softened sample are higher than in the softened sample treated with calcium carbonate, you are probably overtreating your supply and have scale-forming water. But, if the pH and alkalinity in your untreated softened sample are lower than in the treated one (calcium carbonate added), you are undertreating your supply. If they are similar, then stability is near.

Another way to check your water is to suspend a couple of nails on strings in your filter. Observe the nails occasionally to see if they are rusting or scaling up. To further protect the distribution system as well as prevent scale formation in the filter bed, 0.7 to 1.0 mg/L polyphosphate could be fed far enough ahead of the filters to allow mixing before the water reaches the filters. Addition of polyphosphate can prevent the formation of scale on filter media and in distribution system mains, but polyphosphate does not prevent corrosion. The Langelier Index (see Volume I, Chapter 8, Section 8.322, "Calcium Carbonate Saturation") is another approach to determining the corrosivity of water.

Caution should be exercised when using polyphosphate compounds. If they are converted to the orthophosphate form, they will lose their effectiveness. With the addition of phosphorus to water, there could be an increase in bacterial growths in the distribution system. Also, some wastewater treatment plants have phosphorus discharge limitations and polyphosphates added to drinking water can cause wastewater treatment plants to violate their discharge requirements.

QUESTIONS

Write your answers in a notebook and then compare your answers with those on page 118.

14.5A What problems are sometimes created when a slight excess of lime is fed during softening to cause a caustic condition to ensure complete reactions?

14.5B How can excess caustic and unprecipitated carbonate ions (pin floc) be removed from softened water?

14.5C What test is used to determine if a water is stable?

14.5D How can nails be used to determine if a water is stable?

14.6 SAFETY

When quicklime reacts with water in the slaking process (Figure 14.7), it gets hot enough to cause serious burns. Also, being caustic in nature, it can harm your eyes and skin. *ALWAYS* wear goggles or a face shield when working with lime that has been or is in the process of slaking. Flush with water if exposed to lime. Seek medical attention if it gets in your eyes. As for hands or face burns, immediately wash the affected areas and consult a physician if the burns appear serious.

Feeding equipment has moving parts. All moving machinery is a potential safety hazard. A paste-type slaker is particularly dangerous. This type of slaker will "eat you alive." Never put your hand in or near the slaker paddles while the slaker is running. Use wooden paddles as cleaning tools on any slaker in operation. A metal tool will damage the slaker and could even injure the operator if dropped by accident. However, a wooden paddle will likely be broken up with no damage to the equipment or the operator.

Types of equipment vary greatly. Usually, the operator has little or no input in this area. Engineers usually design a plant and specify the type of chemical feed equipment.

Equipment suppliers are usually quite cooperative in advising any operator in the use and care of their equipment in your treatment plant. Detailed start-up, shutdown, and maintenance procedures are available in the equipment manuals supplied by the manufacturer of the equipment.

Another important safety precaution is to avoid using the same conveyor or bin for alternately handling both quicklime and one of the coagulants containing water, such as alum, ferric sulfate, or copperas. This water may be withdrawn by the quicklime and could generate enough heat to cause a fire. Explosions have been reported to have been caused by lime–alum mixtures in enclosed bins. Therefore, always clean facilities before switching from one chemical to another.

---

[13] For additional information on the marble test, see Chapter 21, "Advanced Laboratory Procedures," Test Procedure 9, "Marble Test (Calcium Carbonate Saturation Test)."

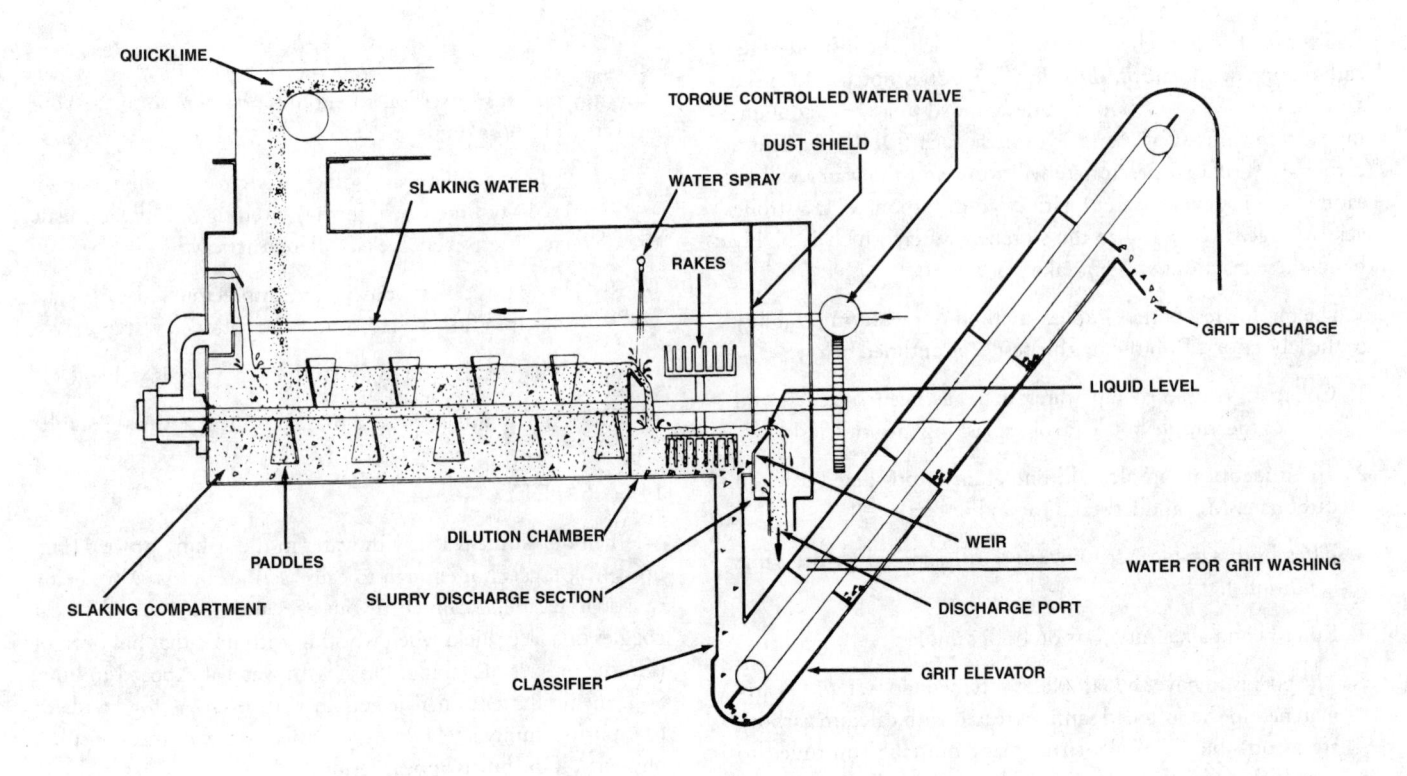

Fig. 14.7 Lime-slaking system
(Permission of Wallace & Tiernan, Inc.)

QUESTIONS

Write your answers in a notebook and then compare your answers with those on page 118.

14.6A Why should wooden paddles be used as cleaning tools on any slaker in operation?

14.6B Where would you look for information on how to safely maintain equipment?

14.7 SLUDGE RECIRCULATION AND DISPOSAL

Considerable sludge may be produced by the lime and lime–soda softening processes. When calcium and magnesium hardness are converted from soluble forms to insoluble precipitates (calcium carbonate and magnesium hydroxide), these precipitates form sludge. This sludge is removed from the bottom of settling basins and may be recirculated or must be disposed of by an acceptable procedure.

In some instances, sludge is recirculated back into the primary mix area of conventional plants to help seed the process. The advantages are: (1) recirculation speeds up the precipitation process, and (2) some reduction of chemical requirements may result. One disadvantage is that an increase in magnesium could result. Only trial and error will really determine if sludge recirculation will serve a useful purpose in your plant.

Sludge disposal is a problem everywhere. Perhaps the most common method is landfill disposal. This is accomplished by dewatering the sludge (drying beds or mechanical means) and then hauling the sludge to landfill sites developed solely for sludge disposal or sanitary landfills. Generally, a sludge with a Ca:Mg ratio of less than 2:1 will be difficult to dewater, whereas a sludge with a Ca:Mg ratio of greater than 5:1 will dewater relatively easily. The less water in the sludge, the less volume to transport to the disposal site and the less space required in the landfill.

To a lesser degree, sanitary sewer disposal is sometimes used. This only moves the sludge to another location for someone else to deal with. Some work has also been done with land application as a substitute for agricultural lime to increase the pH of highly acidic soils. The lime sludge is applied at a rate that will produce the optimum soil pH for the crops to be planted. For additional information on sludge disposal, see Chapter 17, "Handling and Disposal of Process Wastes."

14.8 RECORDS

Records should be kept on the amounts of chemicals ordered and the amounts fed. Laboratory results should be recorded in a permanent laboratory book. See Chapter 18, "Maintenance," for details on how to keep equipment maintenance schedules and records. Again, every plant is different, but accurate records will help a good operator be a better operator.

QUESTIONS

Write your answers in a notebook and then compare your answers with those on page 118.

14.7A   What is a disadvantage of recirculating sludge back to the primary mix area?

14.7B   How could you determine if sludge recirculation will serve a useful purpose in your plant?

14.8A   What types of records should be kept regarding treatment plant chemicals?

14.9 JAR TESTS

14.90 Typical Procedures

Approximate amounts of chemicals required can be calculated (see Section 14.316, "Calculation of Chemical Dosages"); however, the best method of determining the proper dosages is by the use of the jar test. See Volume I, Chapter 11, "Laboratory Procedures," for details on the equipment and procedures required to run jar tests.

Chemical reagents may be made up by adding one gram of reagent to a liter of water.[14] This will produce a 0.1-percent chemical solution. When 1.0 gram/liter of lime, soda ash, or coagulant is made up as the stock solution, one mL of the stock solution in a one-liter sample of water equals one-mg/L dosage. If large doses of lime are required, add 10 grams of lime per liter. With this stock solution, one mL of stock solution in a one-liter sample of water equals ten-mg/L dosage.

Set up 6 samples and estimate the dosage required by adding varying amounts to each sample. Trial and error will put you in the ballpark. To refine the dosage, pick the best looking sample from the settling properties of the floc to establish the optimum lime dose. Then, do the same with varying amounts of soda ash, leaving the lime dosage constant. By running pH, alkalinity, and hardness tests, you can find the optimum dosages that give the desired softened water results.

The exact procedures used to soften water by chemical precipitation using the lime–soda ash process depend on the hardness and other chemical characteristics of the water being treated. A series of jar tests are commonly used to determine optimum dosages. In many cases, the feed rates determined by jar tests do not produce the exact same results in an actual plant. This is because of differences in water temperature, size and shape of jars as compared with plant basins, mixing equipment, and influence of coagulant (a heavy alum feed will neutralize more of the lime). You must remember that jar test results are a starting point. You may have to make additional adjustments to the chemical feeders in your plant based on actual analyses of the treated water.

Convert the jar test results to plant feed rates. Always feed enough chemical to achieve the desired results, but do not overfeed. Overtreating is a waste of money and quality control will suffer.

14.91 Examples

Let us set up some jar tests to determine the optimum dosages for lime or lime–soda treatment to remove hardness from a municipal water supply. To get started, add 10.0 grams of hydrated lime to a one-liter graduated cylinder or flask and fill to the one-liter mark with tap water. Thoroughly mix this stock solution in order to thoroughly suspend all of the lime. One mL of this solution (which has been thoroughly mixed) in a liter of water is the same as a lime dose of ten mg/L, or 0.5 mL in 500 mL is still the same as a ten-mg/L lime dose.

Set up a series of hardness tests by adding 5.0 mL, 10.0 mL, 15.0 mL, 20.0 mL, 25.0 mL, 30.0 mL, 35.0 mL, and 40.0 mL to one-liter (1,000-mL) beakers or jars. Fill the beakers to the 1,000-mL mark with the water being tested. Mix thoroughly for as long as normal mixing will occur in your plant. Allow the precipitate to settle (20 minutes if this is the settling time in your plant) and measure the hardness remaining in the water above the precipitate. A plot of the hardness remaining against the lime dosage will reveal the optimum dosage. Examination of Figures 14.8, 14.9, and 14.10[15] reveals that the water of all three cities responded differently to the increasing lime dosage. City 1 (Figure 14.8) should be providing a lime dose of 100 mg/L. The cost of increasing the dosage to 150 mg/L is not worth the slight reduction in hardness from 110 to 100 mg/L as $CaCO_3$. Note that an overfeed of lime will actually increase the hardness.

City 2 (Figure 14.9) should be providing a lime dose of 200 mg/L. A dose of 300 mg/L will reduce hardness, but the increase in lime costs is too great. City 3 (Figure 14.10) should be dosing lime between 200 and 250 mg/L. Note that the greater the lime dose, the less the hardness, but the greater the quantities of sludge that must be handled and disposed of.

---

[14] Some operators add 10 grams of reagent to a liter of water. This will produce a one-percent solution. One mL of the stock solution in one liter will produce a ten-mg/L dosage.

[15] These figures were adapted from an article titled, "Use of Softening Curve for Lime Dosage Control," by Michael D. Curry, P.E., which appeared in THE DIGESTER/OVER THE SPILLWAY, published by the Illinois Environmental Protection Agency.

94 Water Treatment

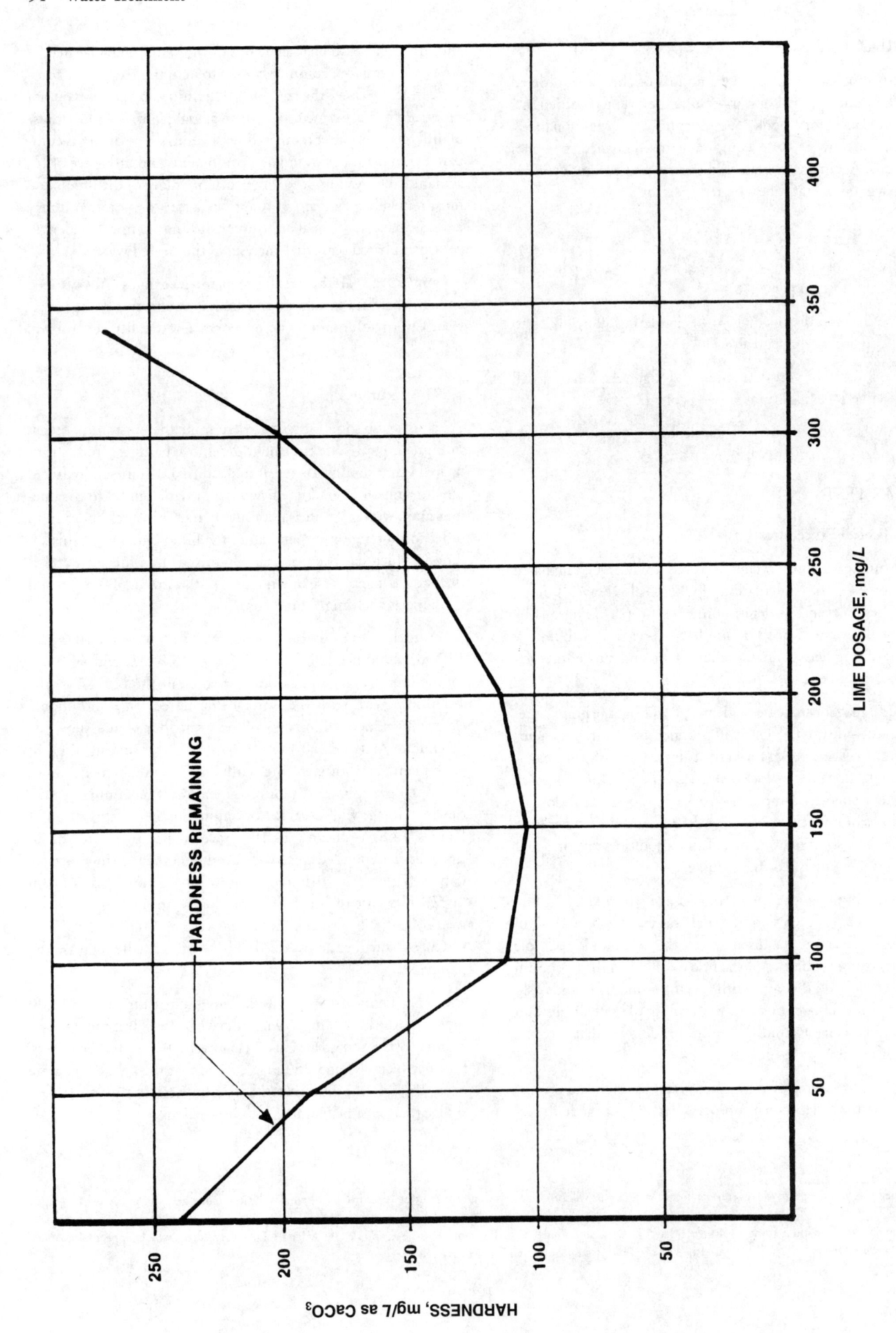

Fig. 14.8 Softening curve for City 1

Softening 95

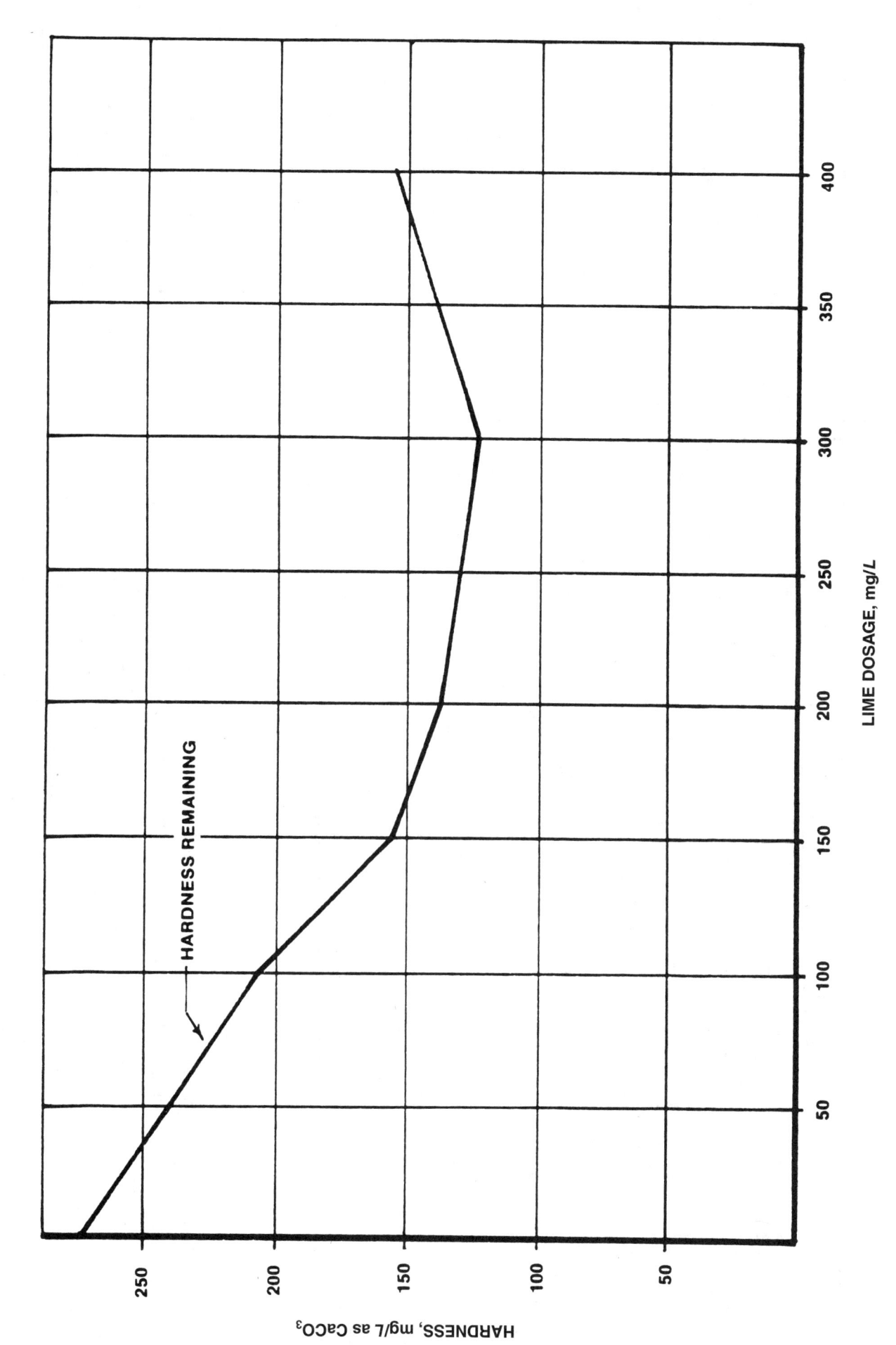

Fig. 14.9 Softening curve for City 2

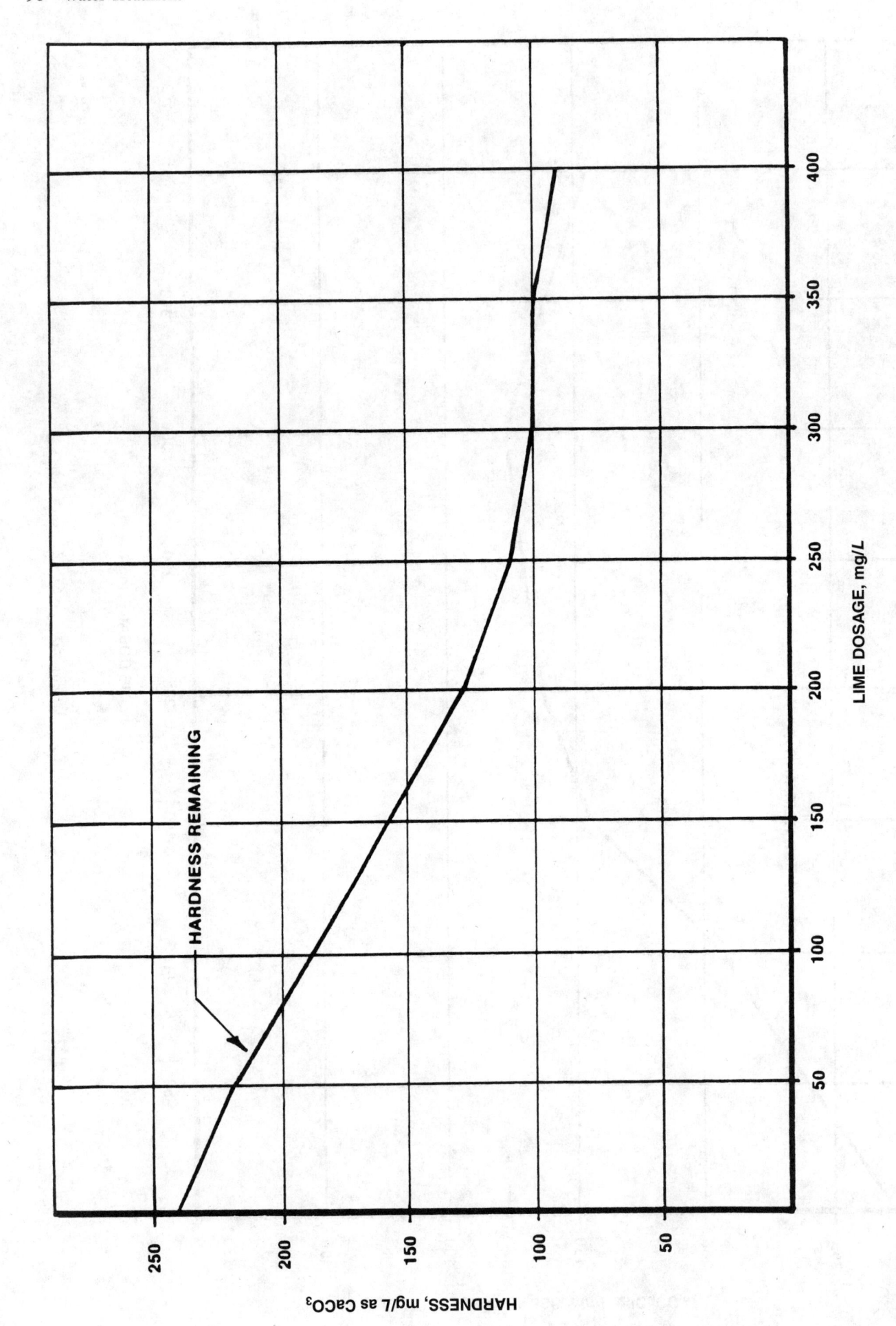

Fig. 14.10 Softening curve for City 3

If lime added to the water does not remove sufficient hardness, select the optimum lime dose and then add varying amounts of soda ash. From Figure 14.9, we found that the optimum lime dose was 200 mg/L (300 mg/L would have reduced the hardness slightly). Let us take six one-liter containers and add 20 mL of our lime stock solution (a dosage of 200 mg/L). Prepare a stock solution of soda ash similar to our lime solution by adding 10 grams of soda ash to a one-liter container, fill with distilled water and mix thoroughly. Add zero, 2.5 mL (25-mg/L dose), 5 mL, 7.5 mL, 10 mL, and 12.5 mL to the one-liter containers. Mix thoroughly, allow the precipitate to settle, and measure the hardness remaining in the water above the precipitate. A plot of hardness remaining against the soda ash dosage will reveal the desired dosage. We would like the final hardness to be in the 80- to 90-mg/L as $CaCO_3$ range in this example.

To select the optimum doses of lime and soda ash, consider the items discussed below.

1. Optimum dosage of lime was based on increments of 50 mg/L. You should refine this test by trying at least two 10-mg/L increments above and below the optimum dose. From Figure 14.8 we found that 100 mg/L was the optimum dose. Try lime doses of 80, 90, 100, 110, and 120 mg/L.

2. Optimum dosage of soda ash can be refined by trying smaller increments also.

3. Try slightly increasing the actual lime dose in your plant to see if there is any decrease in the remaining hardness. Is the decrease in hardness worth the increase in lime costs?

4. Try slightly increasing and decreasing both lime and soda ash dosages at your plant one at a time, and evaluate the results.

5. If you are treating well water or a water of constant quality, all you have to do to maintain proper treatment is to make minor adjustments to keep the system fine-tuned.

6. If you are treating water from a lake or a river and the water quality (including temperature) changes, you will have to repeat these procedures whenever the raw water quality changes. Water quality changes of concern include raw water hardness, alkalinity, pH, turbidity, and temperature.

7. Remember, you do not want to produce water of zero hardness. If you can get the hardness down to around 80 to 90 mg/L, that usually will be low enough for most domestic consumers. When selecting a target hardness level for a water softening plant, consider the uses of the softened water and the cost of softening.

QUESTIONS

Write your answers in a notebook and then compare your answers with those on page 118.

14.9A If lime added to water does not reduce the hardness of a water sufficiently, what would you do?

14.9B What items should be considered when selecting a target hardness level for a water softening plant?

14.92 Calculation of Chemical Feeder Settings

After chemical doses have been calculated or determined from jar tests, convert the results to plant chemical feed rates. Depending on the type of chemical feeder, you may have to calculate the feed rates in pounds per day, pounds per hour, or pounds per minute. Always feed enough chemical to achieve the desired results, but do not overfeed. Overtreating is a waste of money and quality control will suffer.

EXAMPLE 4

The optimum lime dosage from the jar tests is 230 mg/L. If the flow to be treated is 6 MGD, what is the feeder setting in pounds per day and the feed rate in pounds per minute?

Known	Unknown
Lime Dose, mg/L = 230 mg/L	1. Feeder Setting, lbs/day
Flow, MGD = 6 MGD	2. Feed Rate, lbs/min

1. Calculate the feeder setting in pounds per day.

$$\text{Feeder Setting, lbs/day} = (\text{Flow, MGD})(\text{Lime, mg}/L)(8.34 \text{ lbs/gal})$$

$$= (6 \text{ MGD})(230 \text{ mg}/L)(8.34 \text{ lbs/gal})$$

$$= 11{,}509 \text{ lbs/day}$$

2. Calculate the feed rate in pounds per minute.

$$\text{Feed Rate, lbs/min} = \frac{\text{Feeder Setting, lbs/day}}{(60 \text{ min/hr})(24 \text{ hr/day})}$$

$$= \frac{11{,}509 \text{ lbs/day}}{(60 \text{ min/hr})(24 \text{ hr/day})}$$

$$= 8.0 \text{ lbs/min}$$

When the calculated feed rate of eight pounds of lime per minute is put into the plant process, observations and tests will determine if optimum levels are met. In many instances, jar tests and actual plant feed rates do not agree exactly. This is because of temperature, size and shape of jars versus size and shape of plant facilities, mixing time, and influence of the coagulant (a heavy alum feed would neutralize more of the lime). Jar tests are merely indicators or a point at which to begin.

If underfeeding results, reactions will not be complete and the results will be undertreated water having a hardness higher than desired.

If overfeeding results, chemicals are being wasted. Also, it is quite possible to have excessive calcium in the water. This results in unstable conditions, which cause buildup on filter media sand grains and the interior of the water mains. This is where the stability test enters the picture (refer to Section 14.5, "Stability").

The above discussion has dealt with establishing the proper lime feed. The same process would be used to determine the soda ash requirements if you are removing noncarbonate hardness. Set up the lime feed jar tests as discussed. Pick the optimum dosage. Then set up another series of jars using the same lime feed rate in all jars. Now, vary the soda ash feed rate.

EXAMPLE 5

How much soda ash is required (pounds per day and pounds per minute) to remove 50 mg/L noncarbonate hardness as $CaCO_3$ from a flow of 6 MGD?

Known		Unknown
Noncarbonate Hardness Removed, mg/L as $CaCO_3$ = 50 mg/L		1. Feeder Setting, lbs/day
Flow, MGD = 6 MGD		2. Feed Rate, lbs/min

1. Calculate the soda ash dose in milligrams per liter. (See Section 14.316, "Calculation of Chemical Dosages" for an explanation of the following formula.)

$$\text{Soda Ash, mg}/L = \left(\begin{array}{c}\text{Noncarbonate Hardness,}\\ \text{mg}/L \text{ as } CaCO_3\end{array}\right)(106/100)$$

$$= (50 \text{ mg}/L)(106/100)$$

$$= 53 \text{ mg}/L$$

2. Determine the feeder setting in pounds per day.

$$\text{Feeder Setting, lbs/day} = (\text{Flow, MGD})(\text{Soda Ash, mg}/L)(8.34 \text{ lbs/gal})$$

$$= (6 \text{ MGD})(53 \text{ mg}/L)(8.34 \text{ lbs/gal})$$

$$= 2{,}652 \text{ lbs/day}$$

3. Calculate the soda ash feed rate in pounds per minute.

$$\text{Feed Rate, lbs/min} = \frac{\text{Feeder Setting, lbs/day}}{(60 \text{ min/hr})(24 \text{ hr/day})}$$

$$= \frac{2{,}652 \text{ lbs/day}}{(60 \text{ min/hr})(24 \text{ hr/day})}$$

$$= 1.8 \text{ lbs/min}$$

After you determine the proper feed rates and implement them, if you are treating well water or other water of constant quality, all you have to do to maintain proper treatment is make minor adjustments to keep the system fine-tuned.

On the other hand, if you treat a river or lake supply subject to constant or frequent changes in water quality, it is an entirely different set of circumstances. Until you learn from experience to judge the chemical changes necessary by the fluctuations in raw water hardness, alkalinity, and turbidity, you almost have to check yourself daily by the jar test method.

When your treatment process does not work properly, the first thing to check is whether or not the feeder is feeding properly. If it is, the next step is to check your source water quality. Generally, one of these two will be the cause of your problem.

QUESTIONS

Write your answers in a notebook and then compare your answers with those on pages 118 and 119.

14.9C Why should overfeeding of chemicals be avoided?

14.9D What should be the lime feeder setting in pounds per day to treat a flow of 2 MGD when the optimum lime dose is 160 mg/L?

14.9E How much soda ash is required in pounds per day to remove 40 mg/L of noncarbonate hardness from a flow of 2 MGD?

END OF LESSON 1 OF 2 LESSONS
on
SOFTENING

Please answer the discussion and review questions next.

DISCUSSION AND REVIEW QUESTIONS

Chapter 14. SOFTENING

(Lesson 1 of 2 Lessons)

At the end of each lesson in this chapter you will find some discussion and review questions. The purpose of these questions is to indicate to you how well you understand the material in the lesson. Write the answers to these questions in your notebook.

1. Why should water be softened?
2. What are the benefits of softening water in addition to hardness removal?
3. In the two-step recarbonation process, what is accomplished by a second addition of carbon dioxide after treatment?
4. What are the advantages of using liquid caustic soda to soften water?
5. Why should quicklime never be stored close to combustible materials?
6. How can operators protect themselves from lime?
7. Why is the stability of a water important?
8. Why is a paste-type slaker dangerous?
9. Why should you avoid using the same conveyor or bin for alternately handling both quicklime and one of the coagulants containing water, such as alum?
10. Why is sludge sometimes recirculated back into the primary mix area of conventional plants?
11. When running jar tests, how would you determine the optimum chemical dosages for the lime–soda softening process?
12. When your lime–soda ash softening treatment process does not work properly, what is the first thing you should check?

CHAPTER 14. SOFTENING

Ion Exchange Softening by Marty Reynolds

(Lesson 2 of 2 Lessons)

14.10 DESCRIPTION OF ION EXCHANGE SOFTENING PROCESS

The term "Zeolite" is most often associated with sodium ion exchangers and should be considered to mean the same as the term ion exchange. Most ion exchange units in use today use sulfonated polystyrene resins as the exchange media. Ion exchange softening can be defined as exchanging hardness-causing ions (calcium and magnesium) for the sodium ions that are attached to the ion exchange resins to create a soft water.

The treatment plant operator should be aware of the three basic types of softeners on the market.

1. An upflow unit in which the water enters from the bottom and flows up through the ion exchange bed and out the top.
2. A unit that is constructed and operated like a gravity rapid sand filter. The water enters the top, flows down through the ion exchange bed, and out the bottom.
3. The pressure downflow ion exchange softener, which is the most common, will be discussed in this lesson. See Figures 14.11 and 14.12. Pressure filters may be either horizontal or vertical units. Vertical units are preferred because there is less chance of short-circuiting.

To help explain the construction and activity that occurs in an ion exchanger, let us compare it to a pressure filter. The water enters the unit through an inlet distributor located in the top; it is forced (usually pumped) down through a bed of some type of media into an underdrain structure. From the underdrain structure, the treated water flows out of the unit and into storage or into the distribution system.

The flow pattern through a filter and softener are similar, the key difference being the action that takes place in the media or bed of each unit. The filter bed may be considered an adsorption and mechanical straining device used to remove suspended solids from the water. The bed usually consists of sand, anthracite (crushed coal), or a combination of both. Once the bed becomes saturated with the insoluble material (usually clay, suspended solids and iron, manganese hydroxide), the filter is taken out of service, backwashed, and returned to service. This pressure filter will continue to operate until the condition reoccurs and the procedure is repeated.

The bed, media, or resin in an ion exchange softener, however, is much more complex. This resin serves as a medium in which an ion exchange takes place. As hard water is passed through the resin, the sodium ions on the resin are exchanged for the calcium and magnesium ions (in the case of sodium exchange resins). The sodium ions are released from the exchange resin and remain in the water, which flows out of the softener. The calcium and magnesium ions, however, are retained by the resin. The softener effluent is free from calcium and magnesium ions and, therefore, is softened (Figure 14.13).

Once a softener has exchanged all of its sodium ions and the resin is saturated with calcium and magnesium, it will no longer produce soft water. At this time, the unit must be taken out of service and the calcium and magnesium must be removed from the resin by exchange with sodium ions. This process is referred to as a regeneration cycle.

In a regeneration cycle, the calcium and magnesium ions that have been retained by the resin must be removed and the sodium ions restored. In order for the exchange to take place, the resin must hold all ions loosely. If the calcium and magnesium ions cannot be removed, the resin will not accept the addition of new sodium ions that are necessary for additional softening.

Salt, in the form of a concentrated brine solution, is used to regenerate (recharge) the ion exchange resin. When salt is added to water it changes into or ionizes to form sodium cation (Na^+) and chloride anions (Cl^-). When the brine solution is fed into the resin, the sodium cations are exchanged for calcium and magnesium cations. As the brine solution travels down through the resin, the sodium cations are attached to the resin while the calcium, magnesium, and chloride (from the salt) ions flow to waste. After the regeneration has taken place, the bed is ready to be placed in service again to remove calcium and magnesium by ion exchange.

100 Water Treatment

VERTICAL SOFTENER TANK

____" DIAMETER
____" SIDE SHELL HEIGHT
____ P.S.I. W.P.

— AIR RELEASE LINE
BAFFLE
FREEBOARD
____" ZEOLITE = ____ CU. FT.

____" FILTER SAND, SIZE ____ TO ____ MM.
____" GRADED GRAVEL, SIZE 1/4" x #10
____" GRADED GRAVEL, SIZE 1/2" x 1/4"
____" GRADED GRAVEL, SIZE 3/4" x 1/2"
____" GRADED GRAVEL, SIZE 1 1/2" x 3/4"

GRAVEL BED
UNDERDRAIN

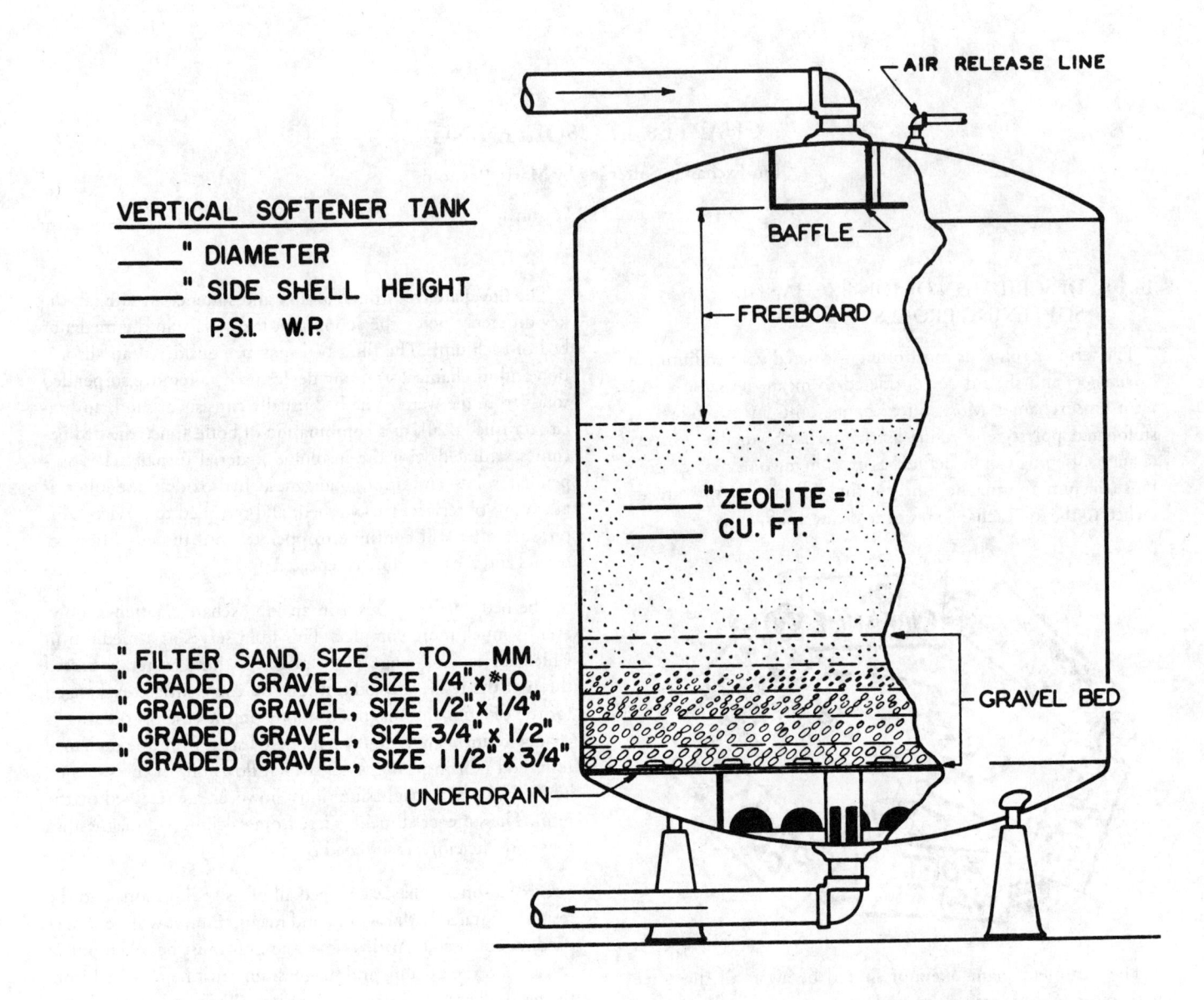

STANDARD SOFTENER

UNDERDRAIN:-
RIGIDLY SUPPORTED PLATE OVER 100% FILTER AREA WITH STAINLESS STEEL BAFFLE ASSEMBLIES.

ZEOLITE:- ____ CUBIC FEET.
SUPPORTING GRAVEL:- ____" GRADED.

STAINLESS STEEL BAFFLE ASSEMBLY

Fig. 14.11 Pressure downflow ion exchange softener
(Permission of General Filter Company)

Softening 101

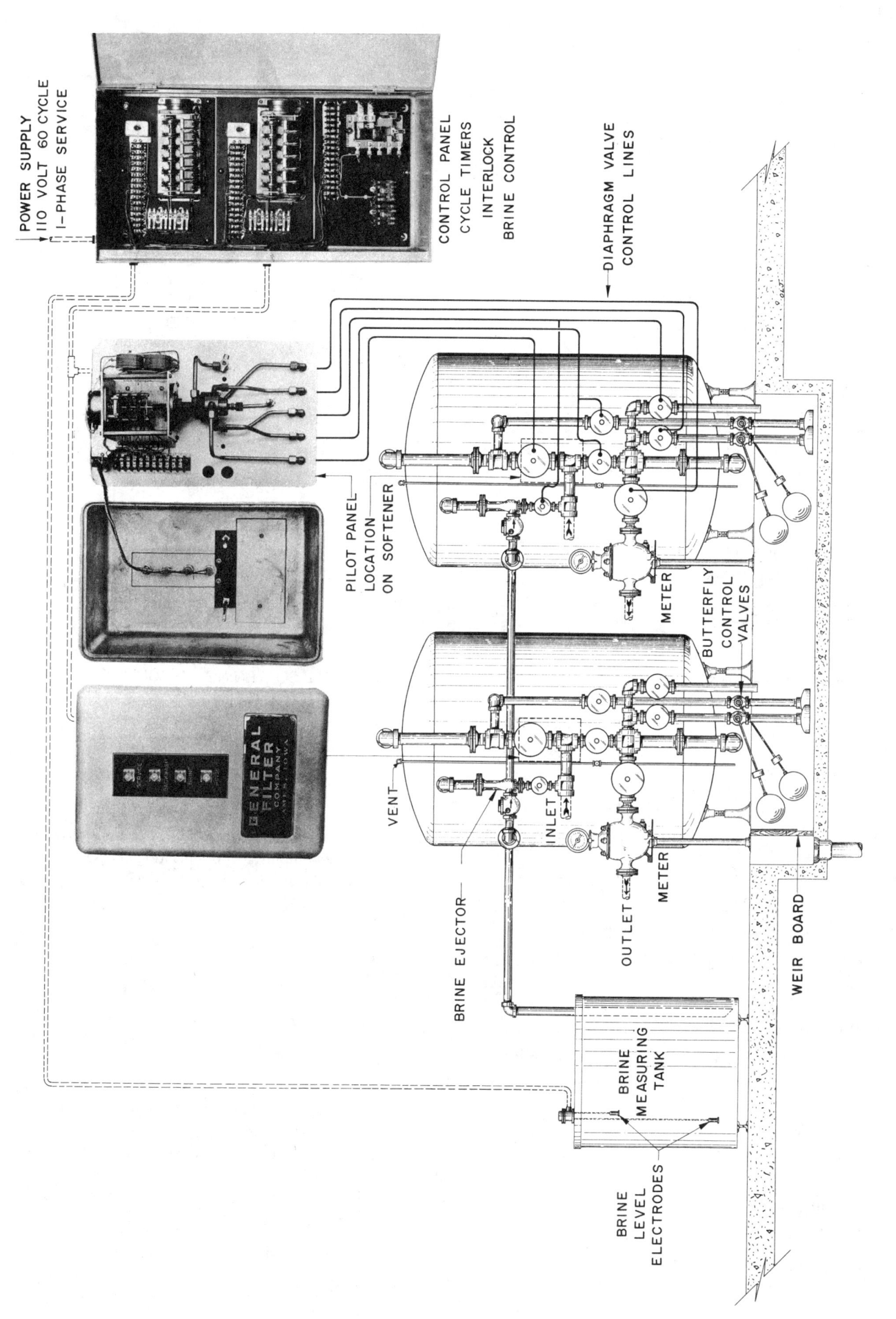

Fig. 14.12 *Semiautomatic controls for ion exchanger*
(Permission of General Filter Company)

102 Water Treatment

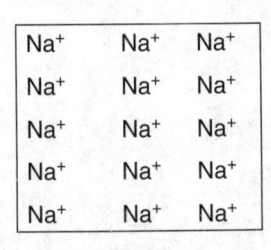

Na⁺	Na⁺	Na⁺
Na⁺	Na⁺	Na⁺
Na⁺	Na⁺	Na⁺
Na⁺	Na⁺	Na⁺
Na⁺	Na⁺	Na⁺

[1]
EXCHANGE RESIN
PRIOR TO SOFTENING
(AFTER REGENERATION)

PLUG FLOW-PISTON EFFECT
HARDNESS APPLIED

Ca⁺⁺ Mg⁺⁺
↓ ↓ ↓ ↓

Ca⁺⁺	Mg⁺⁺	Ca⁺⁺
Ca⁺⁺	Ca⁺⁺	Mg⁺⁺
Mg⁺⁺	Ca⁺⁺	Ca⁺⁺
Ca⁺⁺	Na⁺	Ca⁺⁺
Na⁺	Na⁺	Na⁺

↓ ↓ ↓ ↓

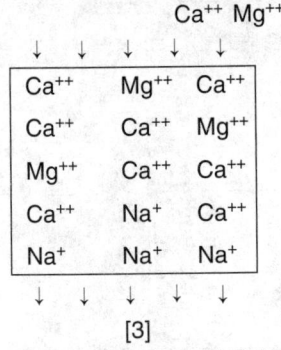

[3]
Na⁺ AND SOFT WATER EXCHANGE
RESIN DURING SOFTENING CYCLE

PLUG FLOW-PISTON EFFECT
HARDNESS APPLIED

Ca⁺⁺ Mg⁺⁺
↓ ↓ ↓ ↓

Ca⁺⁺	Mg⁺⁺	Ca⁺⁺
Na⁺	Na⁺	Na⁺
Na⁺	Na⁺	Na⁺
Na⁺	Na⁺	Na⁺
Na⁺	Na⁺	Na⁺

↓ ↓ ↓ ↓

[2]
Na⁺ AND SOFT WATER EXCHANGE
RESIN AFTER START OF SOFTENING CYCLE

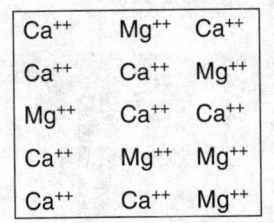

Ca⁺⁺	Mg⁺⁺	Ca⁺⁺
Ca⁺⁺	Ca⁺⁺	Mg⁺⁺
Mg⁺⁺	Ca⁺⁺	Ca⁺⁺
Ca⁺⁺	Mg⁺⁺	Mg⁺⁺
Ca⁺⁺	Ca⁺⁺	Mg⁺⁺

[4]
EXCHANGE RESIN EXHAUSTED
(ALL SOFTENING CAPACITY LOST—
READY TO REGENERATE, SEE [1])

Fig. 14.13 Ion exchange resin condition during softening cycle

QUESTIONS

Write your answers in a notebook and then compare your answers with those on page 119.

14.10A List the three basic types of softeners on the market.

14.10B What happens during the regeneration cycle of an ion exchange softener?

14.11 OPERATIONS

Many factors influence the procedures used to operate an ion exchange unit and the efficiency of the softening process. These factors include:

1. Characteristics of the ion exchange resin
2. Quality of the source water
3. Rate of flow applied to the softener
4. Salt dosage during regeneration
5. Brine concentration
6. Brine contact time

Each ion exchange softener, regardless of manufacturer, will have at least four common stages of operation. These stages are listed below and will be explained as each occurs in the softener operation (see Figure 14.14).

1. Service
2. Backwash
3. Brine
4. Rinse

14.110 Service

The service stage of each unit is where the actual softening of the water occurs. Hard water is forced into the top of the unit and allowed to flow down through the exchange resin. As this takes place, the calcium and magnesium ions exchange with sodium on the resin. The sodium ions are released into the water and the exchange capacity of the unit is slowly exhausted.

The length of each service stage is dependent on several factors; source water hardness is a main consideration. The harder the water, the more calcium and magnesium must be removed to

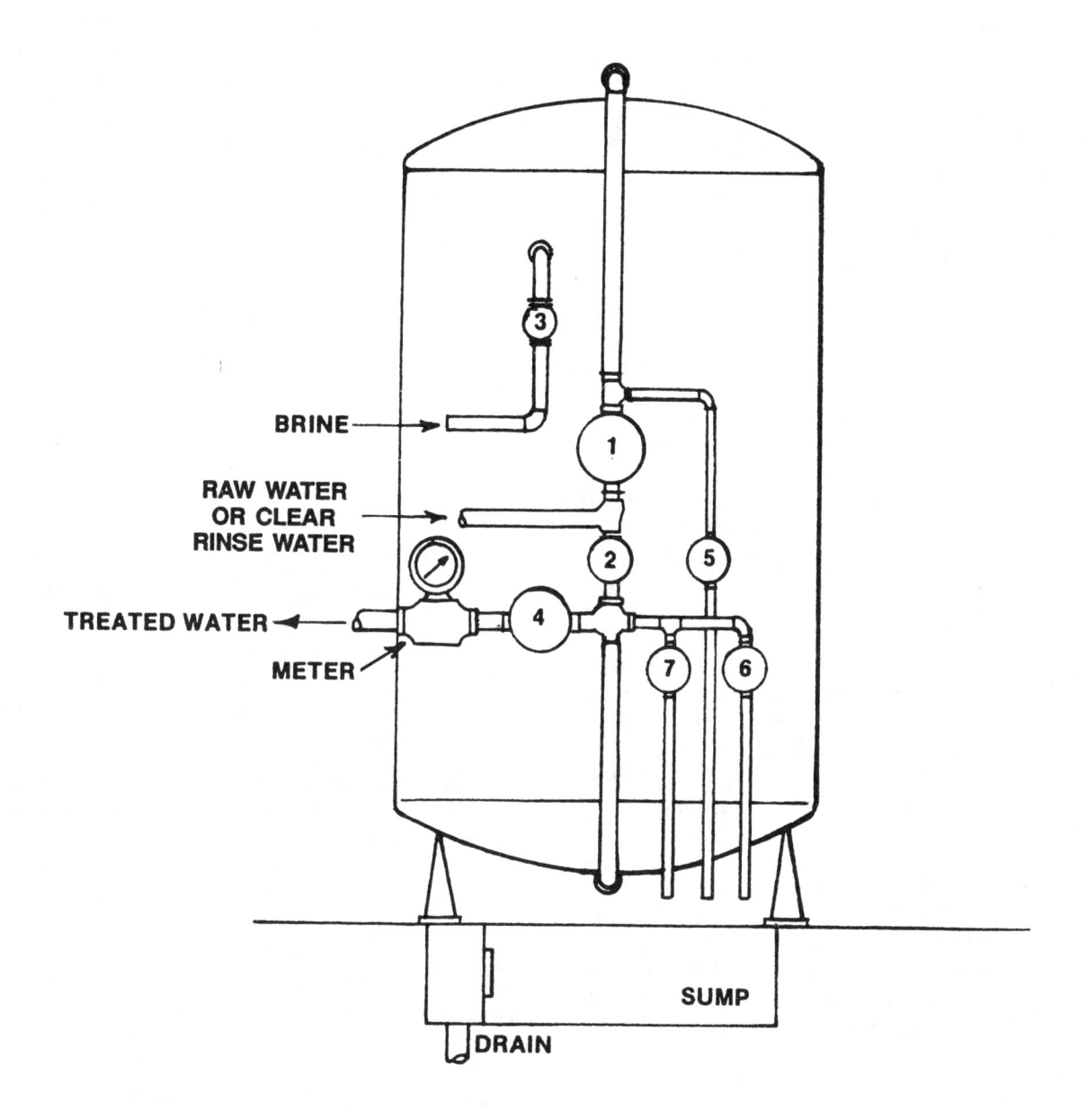

OPERATION	VALVE NUMBER						
	1	2	3	4	5	6	7
SERVICE	OPEN	CLOSE	CLOSE	OPEN	CLOSE	CLOSE	CLOSE
BACKWASH	CLOSE	OPEN	CLOSE	CLOSE	OPEN	CLOSE	CLOSE
BRINE	CLOSE	CLOSE	OPEN	CLOSE	CLOSE	OPEN	CLOSE
RINSE	OPEN	CLOSE	CLOSE	CLOSE	CLOSE	CLOSE	OPEN

Fig. 14.14 Valve positions for each stage of ion exchange softener operation
(Permission of General Filter Company)

reach a level of zero hardness. Simply stated, the harder the water, the less water you can treat before the resin becomes exhausted. As long as the design flow for the ion exchange unit is not exceeded, changes in the hardness of the source water may be automatically adjusted for in the ion exchange unit. The effluent from the unit usually will have zero hardness until the unit needs regeneration. If the total dissolved solids (TDS) in the water supply is fairly high (above 500 mg/L), there may be some leakage. If a high TDS water has a high sodium content, the sodium may hinder the process by causing a local exchange on the media of calcium and magnesium (hardness leakage) for some of the sodium. The amount of hardness leakage depends on the TDS and the salt dosage (percent salt) used for regeneration.

Other factors involved are the size of the softener and the exchange capacity of the resin. Softeners can vary in size from a few cubic feet to several hundred cubic feet. The size of the unit will generally be consistent with regard to the overall treatment plant design. In other words, the softener should be capable of producing enough softened water so that the mix or blend of softened and unsoftened water will produce a treated water with the desired level of hardness.

The exchange resin will also vary in its removal capacity. There are many types of strong acid cation exchange resins on the market today. Most will range in capacity from 20,000 to 30,000 grains of hardness removal per cubic foot (0.011 to 0.016 kg/cu m) of resin. The removal ability of the resin is usually expressed in grains of hardness removal per cubic foot of material or resin.

The source water hardness, the size of the unit, and the removal capacity of the resin will determine the amount of water that can be treated before the softener must be regenerated. With a few simple calculations, an operator can determine the softening capacity of the units. Calculations and examples will be given at the end of the lesson. See Example 8 in Section 14.18, "Ion Exchange Arithmetic."

14.111 Backwash

The second stage of the ion exchange softener process is the backwash. In this stage, the unit is taken out of service and the flow pattern through the unit is reversed. The purpose of this is to expand and clean the resin particles and also to free any material such as iron, manganese, and particulates that might have been removed during the softening stage. The backwash water entering the softener at the beginning of this stage should be applied at a slow, steady rate. If the water enters the unit too quickly, it could create a surge in the resin and wash it out of the unit with the water going to waste.

Ideal bed expansion during the softener backwash should be 75 to 100 percent. In other words, when the unit is backwashed, the resin should expand to occupy a volume from 75 to 100 percent greater than when in normal service. An example of this would be an ion exchange softener with 24 inches (60 cm) of resin while in service. When the unit is backwashed, the resin should expand to 48 inches (120 cm) for a 100-percent expansion of the bed. As the bed expands, a shearing action due to the backwash water and some scrubbing action will free any material that might have formed on the resin particles during the softening stage.

During the backwash, a small amount of resin could be lost. This amount, however, should be minimal and you should check the backwash effluent at different intervals to ensure that the resin is not being lost. A glass beaker can be used to catch a sample of the effluent while the unit is backwashing. A trace amount of resin should cause no alarm, but a steady loss of resin could indicate a problem in the unit and the cause should be located and corrected as soon as possible. Too much loss of resin may be caused by an improper freeboard on the tank or wash troughs. The backwash duration and flow rate will vary depending on the manufacturer, the type and size of resin used, and the water temperature.

14.112 Brine

The third stage is most often termed the regeneration or brine stage. At this point, the sodium ion concentration of the resin is recharged by pumping a concentrated brine solution onto the resin. The solution is allowed to circulate through the unit and displace all water from the resin in order to provide full contact between the brine solution and the resin.

Most treatment plants use a brine solution to regenerate their softening units. The optimum brine concentration coming in contact with the ion exchange resin is around 10- to 14-percent sodium chloride solution. Concentrated brine is only used when the water within the softener tank serves as the dilution water. A 26-percent brine solution (fully concentrated or saturated) causes too great of an osmotic shock on the ion exchange resin and can cause it to break up. The salt dosage used to prepare the brine solution is one of the most important factors affecting the ion exchange capacity and ranges from 5 to 15 pounds of salt per cubic foot (80 to 240 kg/cu m) of resin. See Example 10 on page 112 for procedures on how to calculate the salt dosage and gallons of brine solution required. Brine concentrations less than saturated require longer contact time and more solution must be applied to the unit to achieve a successful regeneration.

The regeneration stage of the softener is very important and the operator should be certain it is properly carried out. In the regeneration stage, the sodium ions present in the brine solution are exchanged with the calcium and magnesium ions on the resin. The ions on the resin were exchanged during the service or softening stage. The regeneration rate is usually one to two GPM per cubic foot (2.2 to 4.4 liters per second per cubic meter) of resin for the first 55 minutes and then three to five GPM per cubic foot (6.6 to 11 liters per second per cubic meter) for the last five minutes of fast drain. If the regeneration process is performed correctly, the result is a bed that is completely recharged with sodium ions and will again soften water when the unit is returned to service.

14.113 Rinse

The fourth and final stage of softener operations is the rinse stage. After adequate contact time has been allowed between the brine solution and resin, a clear rinse is applied from the top of the unit to remove the waste products and excess brine solution from the softener. The flow pattern is very similar to the service stage except that the softener effluent goes to waste instead of storage. The waste discharge contains high concentrations of calcium and magnesium chloride. Most rinse stages will last between 20 and 40 minutes, depending on the size of the unit and the manufacturer. See Section 14.14, "Disposal of Spent Brine," for procedures on how to dispose of the waste discharge.

Again, the operator should pay close attention to the softener while it rinses. The rinse must be long enough to remove the heavy concentration of waste from the unit. If the rinse is not of the correct length and the unit returns to service, a salty taste will be very noticeable in the softener effluent. Taste the waste effluent near the end of the rinse stage to determine if the majority of chloride ions have been removed. The chloride ion concentration may also be measured by titration as outlined in Chapter 21, "Advanced Laboratory Procedures," or by measuring the conductivity of the water. If the water still has a strong salty taste or excessive chloride ions are present, check the rinse rate and timer settings. The unit may need adjustment to increase the duration of the rinse stage.

QUESTIONS

Write your answers in a notebook and then compare your answers with those on page 119.

14.11A What is the main consideration in determining the length of the service stage of an ion exchange softener?

14.11B What is the purpose of the backwash stage of an ion exchange softener?

14.11C How are ion exchange softeners regenerated?

14.11D Where does the softener effluent go during the rinse stage?

14.12 CONTROL TESTING OF ION EXCHANGE SOFTENERS

In most small treatment plants, the operator has to perform many jobs and may not always have time to monitor the softening units as they should be monitored. If a few simple test procedures are learned and carried out on a regular basis, the operator can feel confident the ion exchange units are operating properly. Control tests the operator should perform are listed in this section.

1. Softener Influent

   Be aware of the iron and manganese levels entering the softener. These levels should be kept to a minimum to prevent fouling of the media bed as the unit will remove a certain amount of iron and manganese before becoming plugged. Insoluble particles of iron and manganese will plug the filter media. Soluble ionic iron (Fe^{2+}) and manganese (Mn^{2+}) will exchange onto the media and will not be fully removed by regeneration. If the source water entering the plant is high in iron and manganese, proper oxidation and filtration of the water before the softener should reduce the levels and prevent problems from developing in the softeners.

   Monitor source water hardness on a routine basis. Generally, hardness will not vary, but if it changes, you will need to adjust the amount of water treated by each softener before the media becomes exhausted and the unit must be regenerated.

2. Softener Effluent

   At the end of a regeneration stage, as the unit goes back into service, check the effluent for hardness. This one test will tell you if the regeneration of the softener has been properly conducted. Allow a few minutes to ensure that all of the rinse water in the unit has been purged (removed). Run a hardness test on the effluent side of the unit. The results should indicate a water of zero hardness. Several test kits are available on the market today that are fairly quick and simple to use to measure water hardness.

14.13 LIMITATIONS CAUSED BY IRON AND MANGANESE

Ion exchange units are very versatile. The primary purpose of the unit is to remove calcium and magnesium from the water thus making the water soft. Ion exchange softeners, however, will also remove iron and manganese in either the soluble or precipitated form. If this occurs, the iron and manganese will seriously affect the life of the exchange capacity of the resin. If water high in iron and manganese is applied to the ion exchange resin for very long, iron fouling or the loss of exchange capacity will result.

When the softeners remove iron in the ferrous (soluble) or ferric (solid) form, the two problems discussed below could result.

1. If water with iron in the ferrous form is applied to the softener, the resin will remove the iron from the water. The iron can be retained on the surface of the resin or is sometimes captured deep inside the resin itself. As this happens, the resin or bed will develop an orange or rusty appearance. If the resin becomes iron coated, the efficiency of the softener will be reduced greatly.

2. The second problem associated with high iron levels is a plugging or clogging of the resin bed. When water containing iron in the ferric form (solid) is applied to the unit, it will act like a filter and strain the iron from the water, leaving the iron trapped in the bed. If high iron loadings continue, the upper layer of the bed could become plugged, forcing the water to channel or short-circuit through the bed. The result is incomplete contact between the water and media, thus creating hardness leakage and loss of softening efficiency.

Iron and manganese must be reduced to their lowest possible limits before applying water to the softener. Oxidation of iron and manganese (see Chapter 12, "Iron and Manganese Control") before applying water to ion exchange units is very helpful. You should also be aware of the chlorine levels applied to the softening units. Normal chlorine dosages will not present a problem, but high residuals could damage the resin and reduce its life span.

14.14 DISPOSAL OF SPENT BRINE

One of the largest problems associated with the design and operation of ion exchange softening plants is the disposal of the softener waste.

The waste discharge from softeners consists mostly of calcium, magnesium, and sodium chlorides. These by-products are corrosive to material they contact and possess varying toxic levels in relationship to the environment.

Many water treatment plants discharge spent brine into nearby sewers. This procedure may be approved if the downstream wastewater treatment plant and receiving waters can handle the brine. Usually, the water treatment plant must have some type of holding tank to store the spent brine. The brine is slowly discharged into the sewer at a rate that will not upset (or be toxic to) the biological treatment processes at the wastewater treatment plant. Also, the salt level in the effluent from the wastewater treatment plant must not adversely impact the aquatic life in the receiving waters or cause a violation of the wastewater treatment plant's *NPDES PERMIT*.[16]

Some water treatment plants may be issued an NPDES Permit to discharge spent brine into receiving waters. This could happen only if the flow in the receiving waters was very high (plenty of dilution) and the flow of spent brine was very low. A holding tank would be needed for the spent brine and the brine could be discharged very slowly. The receiving waters would have to be monitored to be sure that the discharge of brine will not cause a significant increase in the level of brine.

Sanitary landfills also may be an acceptable means of disposing of spent brine. See Chapter 17, "Handling and Disposal of Process Wastes," for additional information.

Each ion exchange treatment plant probably has only one approved method of waste disposal. Very few options are available to plants discharging this type of waste. Alternate waste disposal methods available for spent brine are discussed in Chapter 17, "Handling and Disposal of Process Wastes."

The operator needs to be aware of the seriousness involved with softener waste. If a problem develops at the treatment plant, the operator should be working with the agency in the area that governs waste disposal as several considerations must be studied when changing a disposal method.

QUESTIONS

Write your answers in a notebook and then compare your answers with those on page 119.

14.12A Which water quality indicators should be monitored in the effluent of an ion exchange softener?

14.13A What happens when high chlorine residual levels are applied to the softening units?

14.14A Why is the disposal of spent brine a problem?

14.15 MAINTENANCE

Most of the ion exchange water softening equipment on the market today is fully automated (Figure 14.12). The reason for most of this automation is to reduce the time an operator must spend with each unit. Automation is fine for operational control, but it does not mean a unit is maintenance free. Systems like this have a tendency to lead operators astray. A small routine maintenance item can go unnoticed until it becomes a full-scale problem if the operator does not run a regular maintenance schedule on the equipment. For example, most valves on ion exchange units are pneumatic or are equipped with some type of self-operating device (valve actuators). This does not mean, however, the valve will operate each time it is required to do so without a regular examination and overhaul. The operator must check the equipment to ensure it is always in proper working order. One valve that fails to open or to close during a regeneration stage could mean trouble (a storage tank full of salty water or no brine at all).

The components of an ion exchange softening system that should receive constant attention are the brine pumps and piping. A saturated brine solution is very corrosive and will attack any unprotected metallic surface it comes in contact with. Try to keep the system as tight as possible. An uncontained brine leak will only get worse.

If you must change the pipe work in the brine system, give serious consideration to installing PVC pipe. The material is much cheaper than bronze and will outlast steel or galvanized pipe when properly installed and supported. Future repairs are also much easier to make if PVC pipe is used.

The pump on the brine system is most often made of brass, which offers some additional protection from the brine solution. The impeller should be bronze and the shaft stainless steel. A strainer or screen device should be installed ahead of the pump on the suction side.

---

[16] *NPDES Permit.* National Pollutant Discharge Elimination System permit is the regulatory agency document issued by either a federal or state agency that is designed to control all discharges of potential pollutants from point sources and stormwater runoff into US waterways. NPDES permits regulate discharges into US waterways from all point sources of pollution, including industries, municipal wastewater treatment plants, sanitary landfills, large animal feedlots, and return irrigation flows.

Most treatment plants buy salt in bulk to make their brine solution. Regardless of the salt supplier, a certain amount of insoluble material will accompany each bulk delivery. This insoluble material will consist of rocks, coal, sand, and other particles that can clog or destroy an impeller if they reach the pump. Check the strainer assembly on the brine pumps quite often and keep spare parts on hand in case replacement is required.

The use of packing on the pumps is recommended over mechanical seals. Regardless of how well the strainers perform, a small amount of sand will usually end up in the pump. The combination of sand and mechanical seals will most often result in high repair and maintenance costs. Packing is cheaper and easier to install and maintain than mechanical seals and packing will usually outlast mechanical seals in this type of installation.

The brine pump motor should have a heavy-duty rating and a body made of cast iron is preferable. Aluminum or mild steel motor housings do not hold up as well as cast iron when subjected to the corrosive environment around the brine pumping station.

An area most often neglected until problems arise is the bulk brine storage area of the treatment plant. Most storage areas are underground pits equipped with rock or gravel strainers above some type of underdrain collection system. Over a period of time, the strainers will silt in with sand and impurities received with salt deliveries. The best way to prevent this from occurring is to regularly shut down, drain, and replace the strainer systems in the pit. This is a great deal of work, but it is a necessity if the brine system is to stay in operation.

Some brine storage areas have become so clogged that the brine solution could not penetrate the strainer media and reach the underdrain system. Like the head loss on a filter, the strainers can become so clogged that the solution cannot seep through to the underdrain system. If this happens and the system cannot be shut down for cleaning, a pipe can be driven down through the sand and impurities into the gravel layers. If enough holes are driven through the zones of impurities, the solution will eventually seep into the underdrains and can be pumped into the softeners. This is a temporary repair measure only and the storage area should be cleaned as soon as possible.

Inspect the brine solution makeup water line while the storage area is shut down. This line must be kept in good working order because it provides potable water to the salt supply. This water makes the saturated brine solution that is used to regenerate the softener. PVC pipe would provide excellent service in a corrosive environment such as a storage area.

Wet salt brine storage tanks are another location at a water treatment plant where sanitary defects may develop. The makeup water line must have a free fall or air gap someplace in the system to prevent the backflow of a brine solution into the potable water supply. The brine tanks must be protected from contamination just like any other water storage facility. The cover and access hatches must be of the raised-lip, overlapping-cover type. All vents must be properly screened to keep out insects, birds, and rodents.

Not every area of maintenance in an ion exchange softening plant can be covered here because each plant will differ with the type of equipment used and its method of operation. Set up a maintenance routine that is appropriate for your treatment plant. The objective of the maintenance routine must be to keep the plant operating and hold repair costs to a minimum.

QUESTIONS

Write your answers in a notebook and then compare your answers with those on page 119.

14.15A   What could happen if a valve fails to open or close during a regeneration stage?

14.15B   Why should the brine pumps and piping receive constant attention?

14.15C   Why is packing on brine pumps recommended over mechanical seals?

14.16 TROUBLESHOOTING

14.160 Test Units

Ion exchange softeners, if properly operated and maintained, will usually provide years of trouble-free service. If problems arise, however, the operator should be able to identify and correct the situation without a great deal of difficulty.

The best way to ensure that a softener will continue to operate properly is to occasionally test the unit during various stages of operation. Learn to recognize minor problems as they develop and make the necessary repairs before a full-scale problem exists.

Items an operator should check on in each stage are discussed in this section.

14.161 Service Stage

In the service stage, while the unit is producing soft water, hardness tests should be run on the softener effluent to ensure the water has a hardness of zero. One grain hardness per gallon of water (17.1 mg/L) showing up in the effluent should not cause alarm, but concentrations of hardness higher than one grain hardness per gallon signal the need to investigate the softener's operation more closely.

14.162 Backwash Stage

Check the backwash stage for adequate flow rates and for sufficient time to complete the stage. Unless the flow rate is high enough to remove trapped turbidity particles and other insoluble material that is trapped in the resin, a loss of softener efficiency could result. Check the timer on the backwash stage to make sure the unit is washed for the required length of time.

If high iron concentrations have been applied to the softener, check the condition of the resin by visually inspecting the top layer of the bed. Color is a key factor to watch for in units beginning to show signs of an iron-fouled resin. The resin will be an orange, rusty color, while the backwash effluent will appear a light orange at the end of the backwash stage. Also, the head loss on the unit will run higher than normal as the bed becomes plugged with iron.

If iron fouling appears to be a problem, the length of the backwash stage should be increased to wash as much of the particulate matter from the resin as possible. A means of surface washing the resin must be provided for this procedure to be effective. Avoid exceeding recommended manufacturer's flow rates to prevent washing resin from the unit.

A chemical cleaner can be used to remove heavy iron coatings from the resin itself. These cleaners are mostly sodium hydrosulfite and can be mixed in solution form and poured into the softener. The hydrosulfite could also be added to the resin during the regeneration stage by dumping a concentrated powder form in with the brine solution. Consult the resin supplier or manufacturer before using any cleaner on the resin.

14.163 Brine Injection Stage

The brine injection stage of the softener sequence must be correctly applied or the unit will not perform satisfactorily when it returns to service. If the resin is not regenerated during this stage, there will be no sodium ions to exchange with the hardness ions in the water when the water is applied to the unit.

The brine storage area should always contain enough salt to provide a brine solution when makeup water is added.

Also, check the amount of salt solution that is pumped into the softener. This is usually done with a meter that is preset to deliver the exact amount of brine solution required to regenerate the softener. If a brine solution is less than saturated, longer contact time is required between the media and the solution.

The required amount of solution must be delivered consistently to achieve a successful regeneration of the unit. If hardness leakage appears early in the service or softening stage, check the amount and saturation of brine solution in the brine system since these are the main reasons for hardness leakage.

If hardness leakage is excessive immediately following a regeneration stage, shut the unit down and check the media level. The bed could be disrupted from excessive backwash or rinse rates. Iron fouling could also cause a channeling condition to occur and cause the water to short-circuit through the media without contacting the complete bed volume.

14.164 Rinse Stage

The rinse stage of the softener should be checked when tests indicate problems are developing. The rinse rate is a key factor in keeping the softener functioning properly. If the rinse starts too soon, the brine solution could be forced out of the unit before adequate contact time has elapsed. If the rinse rate is too low, all the waste material might not be removed from the unit before it goes into the service stage.

The rate settings on the unit should be compared, on all stages, against the actual manufacturer's recommended settings. As equipment ages, it wears. Over a period of time, valves might need adjustment to keep the unit operating within the manufacturer's guidelines.

QUESTIONS

Write your answers in a notebook and then compare your answers with those on page 119.

14.16A What is the maximum hardness level expected in the effluent of a properly operating ion exchange softener before the operator should investigate the operation more closely?

14.16B What is the purpose of the backwash stage of an ion exchange softener operation?

14.16C What would you do if hardness leakage is excessive immediately following a regeneration stage?

14.16D What problems may occur if the rinse stage starts too soon or the rinse rate is too low?

14.17 START-UP AND SHUTDOWN OF UNIT

At times, it becomes necessary to shut a unit down and take it out of service for repairs or inspection. If the operator will follow common-sense guidelines, no problems should arise from unit shutdowns.

When dealing with these units, drain and fill them slowly. This will prevent surging of the media, which will either wash it out of the unit or disrupt it, thus making the media uneven and creating channeling problems. If you suspect problems with a unit, the last thing you want to do is make the situation worse by rapidly backfilling or backwashing the softener.

Most units are equipped with automatic air release valves (Figure 14.15). Be certain these valves are operating properly, because the venting of trapped air is an important step in filling a unit after it has been shut down.

During unit shutdown, make a complete visual inspection. Now is the time to detect and correct minor problems that might otherwise develop into bigger ones. Check the brine inlet distributors while the unit is down. They should be visible from the top hatch on most softeners.

Make sure the pipe work is level and the nozzles or openings are not clogged. The distributors play an important role in the regeneration stage by applying the brine solution evenly to the top of the resin bed. If the pipe work is deteriorating from the brine solution, PVC pipe should be used as a replacement.

If the resin and gravel support material are removed from the softener, check the underdrain structures in the unit and repair any problems you discover. In filling the unit with gravel and resin, each zone of the bed should be leveled and sized according to manufacturer's specifications.

The procedure for filling a unit with water after a total shutdown is very important. The flow into the unit should be from the bottom at a slow, controlled rate. This is done by putting the unit in the backwash position, running water into the unit from the bottom and out the backwash effluent valve. The purpose of this procedure is to fill the unit with water and purge the air that was trapped in the resin and softener during the replacement process.

Softening 109

AIR RELEASE ASSEMBLY

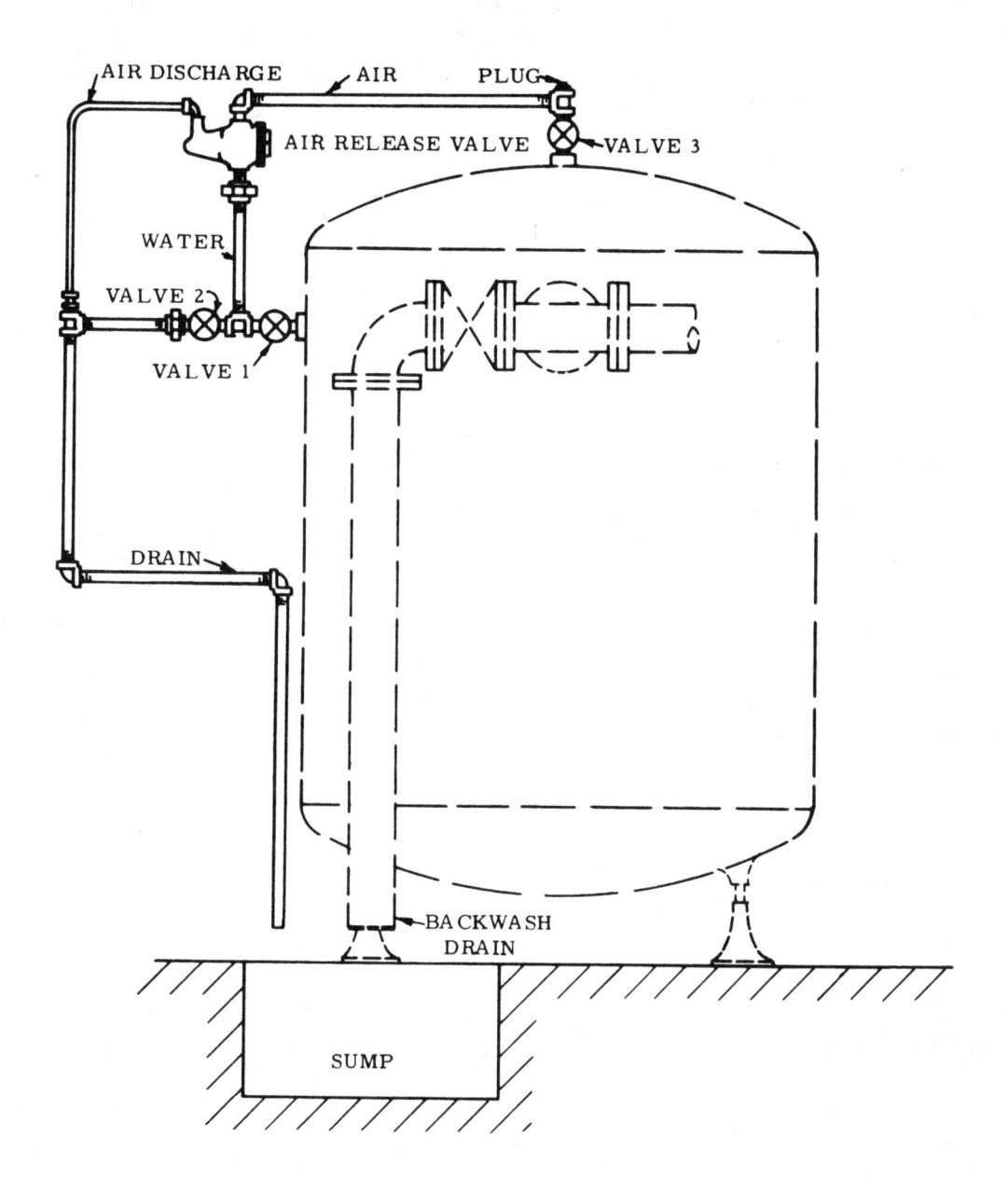

Pressure aeration and pressure filtration type filter plants require an automatic air release assembly to prevent accumulation of an excessive volume of air in the pressure filter tanks. This air release assembly consists of an automatic air release valve, and necessary pipe, valves and fittings to install on filter tank. The air release valve is a float operated type and must be installed with center line of valve level with or above top of filter tank. Air from top of filter enters air release valve at top connection and water from filter inlet pipe enters valve at bottom connection. Excessive air from filter fills valve body with air forcing water level down and thereby allowing float to drop. Downward movement of float allows excessive air to escape through the needle valve until an air-water pressure balance is restored.

In normal operation valves 1 and 3 are open and valve 2 is closed. To flush air release valve close valve 1 and open valve 2 which allows water from top of filter to flush down through the valve to the drain. Valve 3 is left open at all times unless it is necessary to remove air release valve.

Fig. 14.15 Air release assembly
(Permission of General Filter Company)

After the unit is filled, the backwash rate should be increased to normal and continued until the effluent is clear. Again, care should be taken when bringing the rates up to the manufacturer's recommendations to prevent disrupting or displacing resin from the bed.

Once the unit has been satisfactorily backwashed, the bed should be regenerated. This can be accomplished by running the softener through a normal brine and rinse procedure before it is returned to service. Run a hardness test on the effluent to ensure all stages have performed correctly and the unit is softening water.

QUESTIONS

Write your answers in a notebook and then compare your answers with those on page 119.

14.17A Why must ion exchange softeners be drained and filled slowly during start-up and shutdown?

14.17B What should be done if the pipe work in an ion exchange softener is deteriorating from the brine solution?

14.18 ION EXCHANGE ARITHMETIC

Hardness is usually expressed as mg/L of $CaCO_3$. In ion exchange softening, however, hardness is most often expressed in terms of grains per gallon or grains of hardness removed from the water being treated.

The exchange capacity of most softeners is expressed as kilograins (1,000 grains) of hardness removed per cubic foot of resin.

Salt, in solution form, is used to regenerate ion exchange softeners. The theoretical salt requirement is 0.17 pound (0.08 kg) of salt for 1,000 grains of hardness removed. Most regenerations, however, require 0.3 to 0.5 pound (0.14 to 0.23 kg) of salt per 1,000 grains of hardness removal.

In this section, you will learn how to calculate the volume of the brine solution required to regenerate the softening unit as well as the pounds of salt required for regeneration. The concentration of brine solution used at each treatment plant may vary. Table 14.6 lists the pounds of salt present in the percentage of brine solution being used.

FORMULAS AND CONVERSION FACTORS

Hardness is usually expressed as milligrams of hardness per liter of water as $CaCO_3$.

Treatment for hardness is often discussed as grains of hardness per gallon of water.

1 grain per gallon = 17.1 milligrams per liter

or   1 gpg = 17.1 mg/L

7,000 grains = 1 pound

To convert grains per gallon to milligrams per liter.

$$\text{Hardness, mg}/L = \frac{(\text{Hardness, grains/gallon})(17.1 \text{ mg}/L)}{1 \text{ gpg}}$$

TABLE 14.6 SALT SOLUTION CHARACTERISTICS

Percent NaCl or grams per 100 grams of solution	Specific Gravity at 15°C or 59°F	Salameter Degree	Lbs NaCl per US Gal	Lbs NaCl per Cu Ft
1.0	1.0073	4	0.084	0.63
2.0	1.0145	8	0.169	1.27
3.0	1.0217	11	0.255	1.91
4.0	1.0290	15	0.343	2.57
5.0	1.0362	19	0.432	3.23
6.0	1.0437	23	0.522	3.90
7.0	1.0511	27	0.612	4.59
8.0	1.0585	30	0.705	5.28
9.0	1.0659	34	0.799	5.98
10.0	1.0734	38	0.874	6.69
11.0	1.0810	42	0.990	7.41
12.0	1.0885	45	1.09	8.14
13.0	1.0962	49	1.19	8.83
14.0	1.1038	53	1.29	9.63
15.0	1.1115	67	1.39	10.4
16.0	1.1194	60	1.49	11.2
17.0	1.1273	65	1.60	12.0
18.0	1.1352	68	1.70	12.7
19.0	1.1432	72	1.81	13.5
20.0	1.1511	76	1.92	14.4
21.0	1.1593	80	2.03	15.2
22.0	1.1676	84	2.14	16.0
23.0	1.1758	87	2.25	16.9
24.0	1.1840	91	2.37	17.7
25.0	1.1923	95	2.48	18.6
26.0	1.2010	99	2.60	19.5
26.4	1.2040	100	2.65	19.8

To convert milligrams per liter to grains per gallon.

$$\text{Hardness, grains/gallon} = \frac{(\text{Hardness, mg}/L)(1 \text{ gpg})}{17.1 \text{ mg}/L}$$

To find the exchange capacity (removal capacity) of a softener, you need to know the removal capacity of the softener in grains per cubic foot of resin or in kilograins per cubic foot of resin and the volume of the resin in cubic feet.

$$\text{Exchange Capacity, kilograins} = (\text{Removal Capacity, kilograins/cu ft})(\text{Media Vol, cu ft})$$

$$\text{Water Treated, gal} = \frac{\text{Exchange Capacity, kilograins}}{\text{Hardness Removed, kilograins/gal}}$$

$$\text{Operating Time, hr (At a Given Flow Rate Before Regeneration)} = \frac{(\text{Water Treated, gal})(24 \text{ hr/day})}{\text{Avg Daily Flow, gal/day}}$$

To determine the amount of salt required for regeneration, you need to know the pounds of salt per 1,000 grains required for regeneration. To calculate the gallons of brine required for regeneration, you need to know the percent brine solution or the pounds of salt per gallon of brine.

$$\text{Salt Needed, lbs} = (\text{Salt Required, lbs}/1{,}000 \text{ gr})(\text{Hardness Removed, gr})$$

$$\text{Brine, gallons} = \frac{\text{Salt Needed, lbs}}{\text{Salt Solution, lbs Salt/gallon of Brine}}$$

Softening 111

EXAMPLE 6

How many milligrams of hardness per liter are there in a water with 16 grains of hardness per gallon of water?

Known	Unknown
Hardness, gpg = 16 gpg	Hardness, mg/L

Calculate the hardness of the water in milligrams per liter.

$$\text{Hardness, mg}/L = \frac{(\text{Hardness, grains/gallon})(17.1 \text{ mg}/L)}{1 \text{ grain/gallon}}$$

$$= \frac{(16 \text{ grains/gallon})(17.1 \text{ mg}/L)}{1 \text{ grain/gallon}}$$

$$= 274 \text{ mg}/L$$

EXAMPLE 7

Convert the hardness of a water at 290 mg/L to grains per gallon.

Known	Unknown
Hardness, mg/L = 290 mg/L	Hardness, grains/gallon

Convert the hardness from milligrams per liter to grains per gallon.

$$\text{Hardness, grains/gallon} = \frac{(\text{Hardness, mg}/L)(1 \text{ grain/gallon})}{17.1 \text{ mg}/L}$$

$$= \frac{(290 \text{ mg}/L)(1 \text{ grain/gallon})}{17.1 \text{ mg}/L}$$

$$= 17 \text{ grains/gallon}$$

EXAMPLE 8

An ion exchange softener contains 50 cubic feet of resin with a hardness removal capacity of 20 kilograins per cubic foot of resin. The water being treated has a hardness of 300 mg/L as $CaCO_3$. How many gallons of water can be softened before the softener will require regeneration?

Known	Unknown
Resin Volume, cu ft = 50 cu ft	Water Treated, gal
Removal Capacity, gr/cu ft = 20,000 grains/cu ft	
Hardness, mg/L = 300 mg/L	

1. Convert the hardness from mg/L to grains per gallon.

$$\text{Hardness, grains/gallon} = \frac{(\text{Hardness, mg}/L)(1 \text{ grain/gallon})}{17.1 \text{ mg}/L}$$

$$= \frac{(300 \text{ mg}/L)(1 \text{ grain/gallon})}{17.1 \text{ mg}/L}$$

$$= 17.5 \text{ grains/gallon}$$

2. Calculate the exchange capacity of the softener in grains.

$$\text{Exchange Capacity, grains} = (\text{Resin Vol, cu ft})(\text{Removal Capacity, grains/cu ft})$$

$$= (50 \text{ cu ft})(20,000 \text{ grains/cu ft})$$

$$= 1,000,000 \text{ grains of Hardness Removal Capacity}$$

3. Calculate the volume of water in gallons that may be treated before regeneration.

$$\text{Water Treated, gal} = \frac{\text{Exchange Capacity, grains}}{\text{Hardness, grains/gallon}}$$

$$= \frac{1,000,000 \text{ grains}}{17.5 \text{ grains/gallon}}$$

$$= 57,143 \text{ gallons}$$

Therefore, 57,000 gallons of water with 17.5 grains of hardness per gallon of water can be treated before the resin becomes exhausted.

EXAMPLE 9

An ion exchange softening plant has two softeners that are eight feet in diameter and the units have a resin depth of six feet. The resin has a 20 kilograin removal ability. How many gallons of water can be treated if the hardness is 14 grains per gallon? If the flow rate to the softeners is 500 gallons per minute, how long will they operate before regeneration is required?

Known	Unknown
Number of Softeners = 2 Softeners	1. Water Treated, gal
Diameter, ft = 8 ft	2. Operating Time, hr
Resin Depth, ft = 6 ft	
Removal Capacity, gr/cu ft = 20,000 grains/cu ft	
Hardness, grains/gallon = 14 grains/gallon	
Flow, gallons/min = 500 gallons/min	

1. Calculate the total volume of softener media.

$$\text{Resin Vol, cu ft} = (0.785)(\text{Diameter, ft})^2(\text{Depth, ft})(\text{No. Softeners})$$

$$= (0.785)(8 \text{ ft})^2(6 \text{ ft})(2 \text{ Softeners})$$

$$= 603 \text{ cu ft}$$

2. Calculate the total exchange capacity of the two softeners in grains.

$$\text{Exchange Capacity, grains} = (\text{Resin Vol, cu ft})(\text{Removal Capacity, grains/cu ft})$$

$$= (603 \text{ cu ft})(20,000 \text{ grains/cu ft})$$

$$= 12,060,000 \text{ grains of Removal Capacity of the Beds}$$

3. Calculate the volume of water in gallons that may be treated before the resin is exhausted.

$$\text{Water Treated, gal} = \frac{\text{Exchange Capacity, grains}}{\text{Hardness, grains/gallon}}$$

$$= \frac{12{,}060{,}000 \text{ grains}}{14 \text{ grains/gallon}}$$

$$= 861{,}429 \text{ gallons Can Be Treated Before Resin is Exhausted}$$

4. Find the length of time the softeners can run before requiring regeneration.

$$\text{Operating Time, hr} = \frac{\text{Water Treated, gal}}{(\text{Avg Daily Flow, gal/min})(60 \text{ min/hr})}$$

$$= \frac{861{,}429 \text{ gal}}{(500 \text{ gal/min})(60 \text{ min/hr})}$$

$$= 28.7 \text{ hours of Operation Before Regeneration is Required}$$

EXAMPLE 10

An ion exchange softener will remove 1,000,000 grains of hardness before the resin becomes exhausted. If 0.3 pound of salt is required per 1,000 grains of hardness, how many pounds of salt are needed? If a 15-percent salt solution is used to regenerate the unit, how many gallons of brine are required? Table 14.6 indicates that 1.39 pounds of salt is present in each gallon of 15-percent brine solution.

Known		Unknown
Hardness Removal, grains	= 1,000,000 grains	1. Salt Needed, lbs
Salt Required, lbs/1,000 gr	= 0.3 lb/1,000 gr	2. Brine, gal
Salt Solution, lbs/gal	= 1.39 lbs/gal	

1. Determine the pounds of salt needed for regeneration.

$$\text{Salt Needed, lbs} = (\text{Salt Required, lbs/1,000 gr})(\text{Hardness Removed, gr})$$

$$= \frac{(0.3 \text{ lb Salt})(1{,}000{,}000 \text{ grains})}{(1{,}000 \text{ grains})}$$

$$= 300 \text{ lbs of Salt}$$

2. Find the gallons of brine solution required.

$$\text{Brine, gal} = \frac{\text{Salt Needed, lbs}}{\text{Salt Solution, lbs/gallon of Brine}}$$

$$= \frac{300 \text{ lbs of Salt}}{1.39 \text{ lbs of Salt/gallon of Brine}}$$

$$= 216 \text{ gallons of Brine (15-percent Salt Solution)}$$

EXAMPLE 11

Use the same information as in Example 10, except use a 12-percent brine solution. Table 14.6 indicates that 1.09 pounds of salt are present in each gallon of 12-percent brine solution. Three hundred pounds of salt are needed for regeneration. How many gallons of 12-percent brine solution are required?

Known		Unknown
Salt Needed, lbs	= 300 lbs	Brine, gal
Salt Solution, lbs/gal	= 1.09 lbs/gal	

Find the gallons of brine solution required.

$$\text{Brine, gal} = \frac{\text{Salt Needed, lbs}}{\text{Salt Solution, lbs/gal of Brine}}$$

$$= \frac{300 \text{ lbs}}{1.09 \text{ lbs/gal}}$$

$$= 275 \text{ gallons of 12-percent Brine Solution}$$

NOTE: More gallons of brine solution are required when using a 12-percent brine solution than when using a 15-percent solution. The weaker concentration requires more gallons to achieve the same results.

QUESTIONS

Write your answers in a notebook and then compare your answers with those on page 120.

14.18A A source water has a hardness of 150 mg/L as $CaCO_3$. What is the hardness in grains per gallon?

14.18B An ion exchange softener contains 60 cubic feet of resin with a hardness removal capacity of 25 kilograins per cubic foot of resin. The water being treated has a hardness of 250 mg/L as $CaCO_3$. How many gallons of water can be softened before the softener will require regeneration?

14.19 BLENDING

Ion exchange softeners will produce a water with zero hardness. Water with zero hardness must not be sent into a distribution system. Water with zero hardness is very corrosive and, over a period of time, will attack steel pipes in the system and create red water problems. Also, providing a water supply with zero hardness water would be very expensive.

At most softening plants, the zero hardness effluent from the softeners is mixed with filtered water having a known hardness concentration. In other words, a certain amount of water the treatment plant produces will bypass the softening units (split treatment). This water has a known hardness concentration and is mixed in various proportions with the softener effluent to arrive at a desired level of hardness in the finished water (Figure 14.16).

An example would be a treatment plant that has a filtered water hardness of 16 grains/gallon. If the desired plant effluent hardness is 8 grains/gallon, fifty percent of the plant influent must be softened and the other fifty percent would be filtered water mixed together with the softener effluent. The result would be water that has a hardness of 8 grains per gallon.

Softening

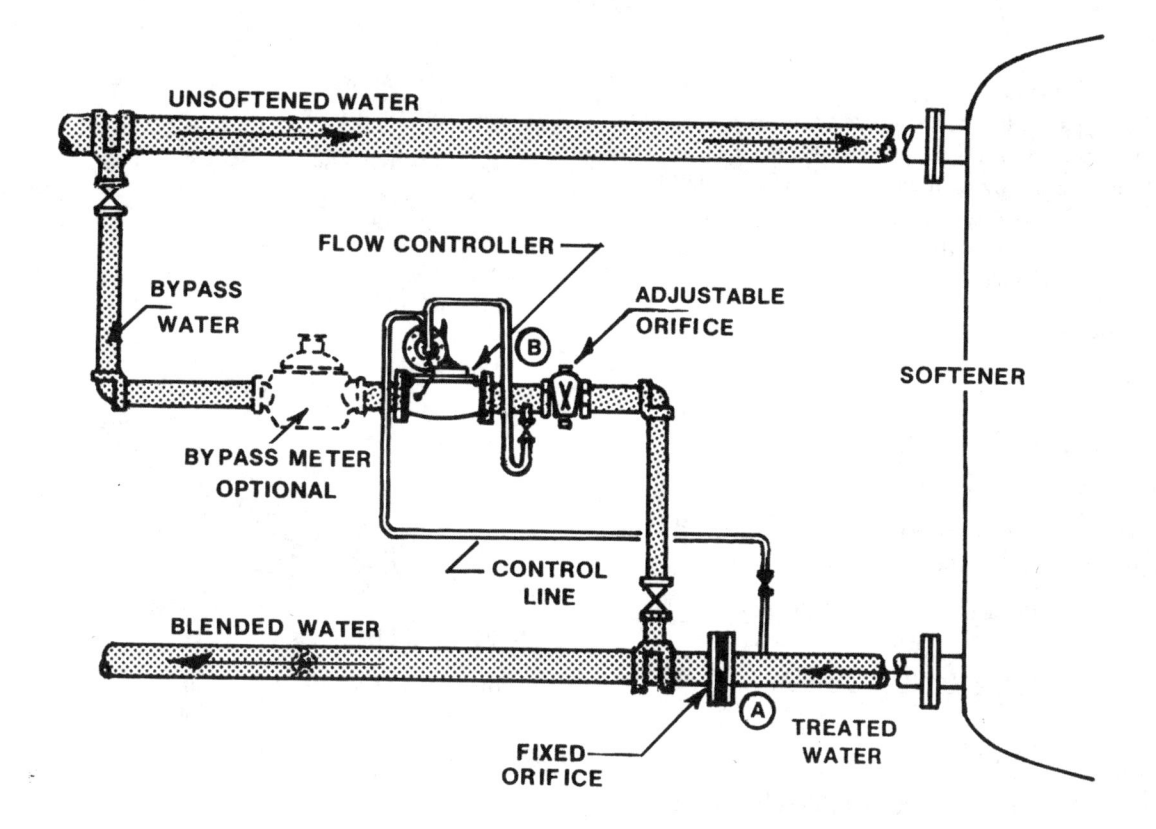

Fig. 14.16 Automatic softener bypass
(Permission of General Filter Company)

The blending of water is very simple and is usually controlled by a valve and meter. The operator adjusts the exact gallons per minute bypassing the softener to produce the desired hardness.

FORMULAS

To calculate the bypass flow in gallons per day to blend water, determine the total flow, the source water hardness, and the desired effluent hardness.

The softener capacity in gallons and both the softener and bypass flows in gallons per day are needed to determine the volume of bypass water.

The total flow produced by the plant before regeneration is the sum of the flows through the softener and the bypass flow.

$$\text{Bypass Flow, GPD} = \frac{(\text{Total Flow, GPD})(\text{Plant Effl Hardness, gpg})}{\text{Source Water Hardness, gpg}}$$

$$\text{Bypass Water, gal} = \frac{(\text{Softener Capacity, gal})(\text{Bypass Flow, GPD})}{\text{Softener Flow, GPD}}$$

Total Flow, gal = Softener Capacity, gal + Bypass Water, gal

EXAMPLE 12

A softener plant treats 120,000 gallons per day. The source water has a hardness of 15 grains per gallon (256 mg/L) and the desired hardness in the plant effluent is 5 grains per gallon (86 mg/L). How much water in gallons per day must bypass the softener to produce the desired level of hardness?

Known	Unknown
Total Flow, GPD = 120,000 GPD	Bypass Flow, GPD
Source Hardness, gpg = 15 gpg	
Effl Hardness, gpg = 5 gpg	

Calculate the bypass flow in gallons per day.

$$\text{Bypass Flow, GPD} = \frac{(\text{Total Flow, GPD})(\text{Plant Effl Hardness, gpg})}{\text{Source Water Hardness, gpg}}$$

$$= \frac{(120,000 \text{ GPD})(5 \text{ gpg})}{15 \text{ gpg}}$$

$$= 40,000 \text{ GPD}$$

EXAMPLE 13

Using the information in Example 12, how many gallons of water will be bypassed before the softener requires regeneration? The softener has the capacity to treat 105,000 gallons. From Example 12, the bypass flow is 40,000 GPD and the total flow is 120,000 GPD. Therefore, the softener flow is 80,000 GPD (120,000 GPD − 40,000 GPD). What is the total flow produced by the plant per regeneration?

Known	Unknown
Softener Capacity, gal = 105,000 gal	1. Bypass Water, gal
Softener Flow, GPD = 80,000 GPD	2. Total Flow, gal
Bypass Flow, GPD = 40,000 GPD	

1. Calculate the gallons of water that will be bypassed before the softener requires regeneration.

$$\text{Bypass Water, gal} = \frac{(\text{Softener Capacity, gal})(\text{Bypass Flow, GPD})}{\text{Softener Flow, GPD}}$$

$$= \frac{(105{,}000 \text{ gal})(40{,}000 \text{ GPD})}{80{,}000 \text{ GPD}}$$

$$= 52{,}500 \text{ gallons}$$

2. Determine the total flow produced by the plant per regeneration.

$$\text{Total Flow, gal} = \text{Softener Capacity, gal} + \text{Bypass Water, gal}$$

$$= 105{,}000 \text{ gal} + 52{,}500 \text{ gal}$$

$$= 157{,}500 \text{ gallons}$$

14.20 RECORDKEEPING

Keeping correct and up-to-date records is as important as performing scheduled maintenance on a regular routine. The recordkeeping system should be set up to record data on a daily basis. Record the total flow through the softener each day, along with the blend rates and gallons that have by-passed the unit. The total gallons of brine used each day, along with the pounds of salt used to keep the ion exchange softener in good working order should be recorded. Records of the results of tests performed on the softeners (source water, softener effluent, and blended water) should be kept up to date in order to warn the operator of any problems that might be developing with the softening unit. Good records are an important part of a successful treatment plant operation. Many problems can be avoided or solved with an adequate recordkeeping system if you review your daily records and compare them with the normal records to determine operating problems.

QUESTIONS

Write your answers in a notebook and then compare your answers with those on page 120.

14.19A Why is the zero hardness effluent from the ion exchange softener mixed (blended) with filtered water having a known hardness concentration?

14.20A What records should be kept by the operator of an ion exchange softening plant?

14.21 ARITHMETIC ASSIGNMENT

Turn to the Arithmetic Appendix at the back of this manual. Read and work the problems in Section A.32, "Softening." Check the arithmetic in this section using an electronic calculator. You should be able to get the same answers. Section A.52 contains similar problems using metric units.

14.22 ADDITIONAL READING

1. *NEW YORK MANUAL,* Chapter 11,* "Softening."
2. *TEXAS MANUAL,* Chapter 11,* "Special Water Treatment (Softening and Ion Exchange)."

* Depends on edition.

14.23 ACKNOWLEDGMENTS

Portions of the material discussed on ion exchange softening came from the following sources:

1. Bowers, Eugene, "Ion Exchange Softening" in *WATER QUALITY AND TREATMENT,* 4th Ed., published by American Water Works Association (AWWA), 6666 West Quincy Avenue, Denver, CO 80235.

2. Lipe, L. A. and M. D. Curry, "Ion Exchange Water Softening," a discussion for water treatment plant operators, 1974–75 seminar series sponsored by Illinois Environmental Protection Agency.

3. Hoover, Charles P., *HOOVER'S WATER SUPPLY AND TREATMENT,* revised by Merrill L. Riehl, 1994. Available from National Lime Association, 200 N. Glebe Road, Suite 800, Arlington, VA 22203-3728. Order No. 211. Price, $39.95, plus shipping and handling.

END OF LESSON 2 OF 2 LESSONS

on

SOFTENING

Please answer the discussion and review questions next.

DISCUSSION AND REVIEW QUESTIONS
Chapter 14. SOFTENING
(Lesson 2 of 2 Lessons)

Write the answers to these questions in your notebook. The question numbering continues from Lesson 1.

13. What happens in the resin or media in an ion exchange softener during the softening stage?

14. How would you ensure that large amounts of resin are not being lost during the backwash stage?

15. How would you determine if an ion exchange softener rinse stage has been successful?

16. What happens if an ion exchange softener removes iron in the ferrous (soluble) or ferric (solid) form?

17. What types of insoluble material may be found in salt? What problems can be caused by this material and how can these problems be prevented?

18. How would you prevent the strainers under the bulk brine storage area from silting in with sand and impurities?

19. How would you determine if iron has fouled the resin of an ion exchange softener?

20. How are ion exchange units filled with water after total shutdown?

21. Why is water with zero hardness not delivered to consumers?

SUGGESTED ANSWERS
Chapter 14. SOFTENING

ANSWERS TO QUESTIONS IN LESSON 1

Answers to questions on page 78.

14.0A Hardness is caused mainly by the calcium and magnesium ions in water.

14.0B Excessive hardness is undesirable because it causes the formation of soap curds, increased use of soap, deposition of scale in boilers, damage in some industrial processes, and it sometimes causes objectionable tastes in drinking water.

14.1A Limitations of the ion exchange softening process include: (1) an increase in the sodium content of the softened water, and (2) ultimate disposal of spent brine and rinse waters from softeners can be a major problem for many installations.

Answers to questions on page 82.

14.2A Hardness is commonly measured by titration. Individual divalent cations may be measured in the laboratory by using an atomic absorption (AA) spectrophotometer.

14.2B Determine the total hardness as $CaCO_3$ for a sample of water with a calcium content of 25 mg/L and a magnesium content of 14 mg/L.

Known	Unknown
Calcium, mg/L = 25 mg/L	Total Hardness,
Magnesium, mg/L = 14 mg/L	mg/L as $CaCO_3$

Calculate the total hardness as milligrams per liter of calcium carbonate equivalent.

$$\text{Total Hardness, mg/}L \text{ as } CaCO_3 = \text{Calcium Hardness, mg/}L \text{ as } CaCO_3 + \text{Magnesium Hardness, mg/}L \text{ as } CaCO_3$$

$$= 2.50(Ca, \text{mg/}L) + 4.12(Mg, \text{mg/}L)$$

$$= 2.50(25 \text{ mg/}L) + 4.12(14 \text{ mg/}L)$$

$$= 62.5 \text{ mg/}L + 57.7 \text{ mg/}L$$

$$= 120.2 \text{ mg/}L \text{ as } CaCO_3$$

116 Water Treatment

14.2C Water treatment chemicals that lower the pH when added to water include alum, carbon dioxide, chlorine (Cl_2), ferric chloride, hydrofluosilicic acid, and sulfuric acid.

14.2D Results from alkalinity titrations on a sample of water were as follows: sample size, 100 mL; mL titrant used to pH 8.3, 1.2 mL; total mL of titrant used, 5.6 mL; and the acid normality was 0.02 N H_2SO_4. Calculate the total, bicarbonate, carbonate, and hydroxide alkalinity as $CaCO_3$.

Known

Sample Size, mL	= 100 mL
mL Titrant Used to pH 8.3, A	= 1.2 mL
Total mL of Titrant Used, B	= 5.6 mL
Acid Normality, N	= 0.02 N H_2SO_4

Unknown

1. Total Alkalinity, mg/L as $CaCO_3$
2. Bicarbonate Alkalinity, mg/L as $CaCO_3$
3. Carbonate Alkalinity, mg/L as $CaCO_3$
4. Hydroxide Alkalinity, mg/L as $CaCO_3$

1. Calculate the phenolphthalein alkalinity in mg/L as $CaCO_3$.

$$\text{Phenolphthalein Alkalinity, mg}/L \text{ as } CaCO_3 = \frac{A \times N \times 50{,}000}{\text{m}L \text{ of Sample}}$$

$$= \frac{(1.2 \text{ m}L) \times (0.02 \text{ }N) \times 50{,}000}{100 \text{ m}L}$$

$$= 12 \text{ mg}/L \text{ as } CaCO_3$$

2. Calculate the total alkalinity in mg/L as $CaCO_3$.

$$\text{Total Alkalinity, mg}/L \text{ as } CaCO_3 = \frac{B \times N \times 50{,}000}{\text{m}L \text{ of Sample}}$$

$$= \frac{(5.6 \text{ m}L) \times (0.02 \text{ }N) \times 50{,}000}{100 \text{ m}L}$$

$$= 56 \text{ mg}/L \text{ as } CaCO_3$$

3. Refer to Table 14.4 for alkalinity constituents.

From Table 14.4 we want the second row because P = 12 mg/L, which is less than ½T [(½)(56 mg/L) = 28 mg/L]. Therefore,

a. Bicarbonate Alkalinity, mg/L as $CaCO_3$
$= T - 2P$
$= 56 \text{ mg}/L - 2(12 \text{ mg}/L)$
$= 32 \text{ mg}/L$ as $CaCO_3$

b. Carbonate Alkalinity, mg/L as $CaCO_3$
$= 2P$
$= 2(12 \text{ mg}/L)$
$= 24 \text{ mg}/L$ as $CaCO_3$

c. Hydroxide Alkalinity, mg/L as $CaCO_3$
$= 0 \text{ mg}/L$ as $CaCO_3$

Answers to questions on page 83.

14.3A The minimum hardness that can be achieved by the lime–soda ash process is around 30 to 40 mg/L as $CaCO_3$.

14.3B Benefits that could result from the lime–soda softening process in addition to softening include:

1. Removal of iron and manganese
2. Reduction of solids
3. Removal and inactivation of bacteria and viruses due to high pH
4. Control of corrosion and scale formation with proper stabilization of treated water
5. Removal of excess fluoride

Answers to questions on pages 84 and 85.

14.3C The pH increases during the lime–soda softening process when lime is added to water, which increases the hydroxide concentrations.

14.3D After the chemical softening process, the scale-forming tendencies of water can be decreased by the use of carbon dioxide.

14.3E Recarbonation is a process in which carbon dioxide is bubbled into the water being treated to lower the pH. Recarbonation may be carried out in two steps. The first addition of carbon dioxide would follow excess lime addition to lower the pH to about 10.4 and encourage the precipitation of calcium carbonate and magnesium hydroxide. A second addition of carbon dioxide after treatment removes noncarbonate hardness. By carrying out recarbonation prior to filtration, the buildup of excess lime and also calcium carbonate and magnesium hydroxide precipitates in the filters will be prevented or minimized.

14.3F Caustic soda softening might be used in place of soda ash. The decision to use caustic soda rather than soda ash depends on the quality of the source water and the delivered costs of various chemicals.

Answer to question on page 86.

14.3G Calculate the hydrated lime (Ca(OH)$_2$) with 90 percent purity, soda ash, and carbon dioxide dose requirements in milligrams per liter for the water shown below.

Known

Constituents	Source Water	Softened Water After Recarbonation and Filtration
CO$_2$, mg/L	= 5 mg/L	= 0 mg/L
Total Alkalinity, mg/L	= 150 mg/L as CaCO$_3$	= 20 mg/L as CaCO$_3$
Total Hardness, mg/L	= 240 mg/L as CaCO$_3$	= 50 mg/L as CaCO$_3$
Mg^{2+}, mg/L	= 16 mg/L	= 2 mg/L
pH	= 7.4	= 8.8
Lime Purity, %	= 90%	

Unknown

1. Hydrated Lime, mg/L
2. Soda Ash, mg/L
3. Carbon Dioxide, mg/L

1. Calculate the hydrated lime (Ca(OH)$_2$) required in milligrams per liter.

$$A = (CO_2, mg/L)(74/44)$$
$$= (5 \text{ mg}/L)(74/44)$$
$$= 8 \text{ mg}/L$$

$$B = (\text{Alkalinity, mg}/L)(74/100)$$
$$= (150 \text{ mg}/L - 20 \text{ mg}/L)(74/100)$$
$$= 96 \text{ mg}/L$$

$$C = 0 \quad \text{Hydroxide Alkalinity} = 0$$

$$D = (Mg^{2+}, mg/L)(74/24.3)$$
$$= (16 \text{ mg}/L - 2 \text{ mg}/L)(74/24.3)$$
$$= 43 \text{ mg}/L$$

$$\text{Hydrated Lime (Ca(OH)}_2\text{) Feed, mg}/L = \frac{(A + B + C + D)1.15}{\text{Purity of Lime, as a decimal}}$$

$$= \frac{(8 \text{ mg}/L + 96 \text{ mg}/L + 0 + 43 \text{ mg}/L)1.15}{0.90}$$

$$= \frac{(147 \text{ mg}/L)(1.15)}{0.90}$$

$$= 188 \text{ mg}/L$$

2. Calculate the soda ash required in milligrams per liter.

$$\text{Total Hardness Removed, mg}/L \text{ as CaCO}_3 = \text{Total Hardness, mg}/L \text{ as CaCO}_3 - \text{Total Hardness Remaining, mg}/L \text{ as CaCO}_3$$

$$= 240 \text{ mg}/L - 50 \text{ mg}/L$$

$$= 190 \text{ mg}/L \text{ as CaCO}_3$$

$$\text{Noncarbonate Hardness, mg}/L \text{ as CaCO}_3 = \text{Total Hardness Removed, mg}/L \text{ as CaCO}_3 - (\text{Carbonate Hardness, mg}/L \text{ as CaCO}_3 - \text{Carbonate Hardness Remaining, mg}/L \text{ as CaCO}_3)$$

$$= 190 \text{ mg}/L - (150 \text{ mg}/L - 20 \text{ mg}/L)$$

$$= 60 \text{ mg}/L \text{ as CaCO}_3$$

$$\text{Soda Ash (Na}_2\text{CO}_3\text{) Feed, mg}/L = \left(\text{Noncarbonate Hardness, mg}/L \text{ as CaCO}_3\right)(106/100)$$

$$= (60 \text{ mg}/L)(106/100)$$

$$= 63.6 \text{ mg}/L$$

3. Calculate the dosage of carbon dioxide required for recarbonation.

$$\text{Excess Lime, mg}/L = (A + B + C + D)(0.15)$$
$$= (8 \text{ mg}/L + 96 \text{ mg}/L + 0 + 43 \text{ mg}/L)(0.15)$$
$$= (147 \text{ mg}/L)(0.15)$$
$$= 22 \text{ mg}/L$$

$$\text{Total CO}_2 \text{ Feed, mg}/L = (\text{Ca(OH)}_2 \text{ excess, mg}/L)(44/74) + (Mg^{2+} \text{ residual, mg}/L)(44/24.3)$$

$$= (22 \text{ mg}/L)(44/74) + (2 \text{ mg}/L)(44/24.3)$$

$$= 13 \text{ mg}/L + 4 \text{ mg}/L$$

$$= 15 \text{ mg}/L$$

Answers to questions on page 87.

14.3H In the lime softening process, calcium is precipitated out as calcium carbonate and magnesium as magnesium hydroxide.

14.3I Partial lime softening (no magnesium removal) removes hardness caused by calcium ions. This may be referred to as calcium hardness.

14.3J In split lime treatment, a portion of the water is treated with excess lime to remove the magnesium at a high pH. Then, source water (the remaining portion) is added in the next basin to neutralize (lower the pH) the excess-lime-treated portion.

Answers to questions on page 89.

14.3K   Lime–soda ash softening is used when lime alone will not remove enough hardness.

14.3L   Noncarbonate hardness is removed by the addition of soda ash (sodium carbonate) in the chemical precipitation softening process.

Answers to questions on page 90.

14.3M   Where the daily requirements for lime are small, lime is usually delivered to the water treatment plant in bags.

14.3N   Considerable heat is generated if quicklime accidentally gets wet.

14.3O   Lime may be applied by dry feeding techniques using volumetric or gravimetric feeders.

Answers to questions on page 90.

14.4A   When the alum dose increases for coagulation, the lime dose must be increased also.

14.4B   Color can be removed from water by coagulation with alum at low pH values. The high pH values required during softening tend to "set" the color, which then becomes very difficult to remove.

Answers to questions on page 91.

14.5A   A slight excess of lime can cause a scale to form on the filter sand, distribution mains, and household plumbing.

14.5B   Excess caustic and unprecipitated carbonate ions (pin floc) can be removed from softened water by recarbonation. Recarbonation is the bubbling of carbon dioxide through the water being treated to lower the pH. Recarbonation can be accomplished, to a degree, by using source water in the split treatment mode.

14.5C   The marble test is used to determine if a water is stable. The Langelier Index is also used to determine the corrosivity of water.

14.5D   Suspending a couple of nails on strings in a filter can indicate if the water is stable. If the nails are rusting, the water is corrosive. If a scale forms on the nails, then scale is forming on your filter media and in your distribution system.

Answers to questions on page 92.

14.6A   Wooden paddles should be used as cleaning tools on any slaker in operation. A metal tool will damage the slaker and could even injure the operator if dropped by accident. However, a wooden paddle will likely be broken up with no damage to the equipment or the operator.

14.6B   Information on how to safely maintain equipment may be found in equipment manuals provided by equipment suppliers and manufacturers.

Answers to questions on page 93.

14.7A   A disadvantage of recirculating sludge back to the primary mix area is that an increase in magnesium could result.

14.7B   Only trial and error will really determine if sludge recirculation will serve a useful purpose in your plant.

14.8A   Records should be kept on the amounts of treatment plant chemicals ordered and the amounts fed.

Answers to questions on page 97.

14.9A   If lime added to water does not reduce the hardness of a water sufficiently, use the optimum lime dose and run jar tests with varying soda ash doses. Select the soda ash dose that will produce a water with a hardness of around 80 to 90 mg/L.

14.9B   When selecting a target hardness level for a water softening plant, consider the uses of the softened water and the cost of softening.

Answers to questions on page 98.

14.9C   The overfeeding of chemicals is a waste of money and quality control will suffer.

14.9D   What should be the lime feeder setting in pounds per day to treat a flow of 2 MGD when the optimum lime dose is 160 mg/L?

Known	Unknown
Flow, MGD = 2 MGD	Feeder Setting, lbs/day
Lime Dose, mg/L = 160 mg/L	

Calculate the lime feeder setting in pounds per day.

$$\text{Feeder Setting, lbs/day} = (\text{Flow, MGD})(\text{Lime, mg}/L)(8.34 \text{ lbs/gal})$$

$$= (2 \text{ MGD})(160 \text{ mg}/L)(8.34 \text{ lbs/gal})$$

$$= 2{,}669 \text{ lbs/day}$$

14.9E   How much soda ash is required in pounds per day to remove 40 mg/L noncarbonate hardness from a flow of 2 MGD?

Known	Unknown
Flow, MGD = 2 MGD	Feeder Setting, lbs/day
Hardness Removed, mg/L = 40 mg/L	

1. Calculate the soda ash dose in milligrams per liter.

$$\text{Soda Ash, mg}/L = (1.06)(\text{Hardness Removed, mg}/L)$$

$$= (1.06)(40 \text{ mg}/L)$$

$$= 42.4 \text{ mg}/L$$

2. Determine the soda ash feeder setting in pounds per day.

$$\text{Feeder Setting, lbs/day} = (\text{Flow, MGD})(\text{Soda Ash, mg/}L)(8.34 \text{ lbs/gal})$$

$$= (2 \text{ MGD})(42.4 \text{ mg/}L)(8.34 \text{ lbs/gal})$$

$$= 707 \text{ lbs/day}$$

ANSWERS TO QUESTIONS IN LESSON 2

Answers to questions on page 102.

14.10A The three basic types of softeners on the market are:
1. An upflow unit
2. A gravity flow unit
3. A pressure downflow unit (the most common)

14.10B During the regeneration cycle, the softener is taken out of service. Salt, in the form of a concentrated brine solution, is used to regenerate (recharge) the ion exchange resin. When the brine solution is fed into the resin, the sodium cations are exchanged for calcium and magnesium cations. As the brine solution travels down through the resin, the sodium cations are attached to the resin while the calcium, magnesium, and chloride (from the salt) ions flow to waste. After the regeneration has taken place, the bed is ready to be placed in service again to remove calcium and magnesium by ion exchange.

Answers to questions on page 105.

14.11A The source water hardness is the main consideration in determining the length of the service stage of an ion exchange softener.

14.11B The purpose of the backwash stage is to expand and clean the media or resin particles and also to free any material such as iron, manganese, and particulates that might have been removed during the softening stage.

14.11C Ion exchange softeners are regenerated by the use of a concentrated brine solution.

14.11D During the rinse stage, the softener effluent goes to waste.

Answers to questions on page 106.

14.12A Hardness should be monitored in the effluent of an ion exchange softener.

14.13A High chlorine residual levels applied to softening units could damage the resin and reduce its life span.

14.14A The disposal of spent brine is a problem because the brine is very corrosive and toxic to many living things in the environment.

Answers to questions on page 107.

14.15A One valve that fails to open or close during a regeneration stage could mean a storage tank full of salty water or no brine.

14.15B Brine pumps and piping must receive constant attention because a saturated brine solution is very corrosive and will attack any unprotected metallic surface it comes in contact with. Try to keep the system as tight as possible. An uncontained brine leak will only get worse.

14.15C Packing is recommended over mechanical seals on brine pumps because, regardless of how well the strainers perform, a small amount of sand will usually end up in the pump. The combination of sand and mechanical seals will most often result in high repair and maintenance costs. Packing is cheaper and easier to install and maintain than mechanical seals and packing will usually outlast mechanical seals in this type of installation.

Answers to questions on page 108.

14.16A The maximum expected hardness level in the effluent from an ion exchange softener should not exceed one grain hardness per gallon (17.1 mg/L). Concentrations of hardness higher than one grain hardness per gallon signal the need to investigate the softener's operation more closely.

14.16B The purpose of the backwash stage is to remove trapped turbidity particles and other insoluble material that is trapped in the resin.

14.16C If hardness leakage is excessive immediately following a regeneration stage, shut the unit down and check the media level. The bed could be disrupted from excessive backwash or rinse rates. Iron fouling could also cause a channeling condition to occur and cause the water to short-circuit through the media without contacting the complete bed volume.

14.16D If the rinse stage starts too soon, the brine solution could be forced out of the unit before adequate contact time has elapsed. If the rinse rate is too low, all the waste material might not be removed from the unit before it goes into the service stage.

Answers to questions on page 110.

14.17A Ion exchange softeners must be drained and filled slowly during start-up and shutdown to prevent surging of the media, which will either wash it out of the unit or disrupt it, thus making the media uneven and creating channeling problems.

14.17B If the pipe work in an ion exchange softener is deteriorating from the brine solution, PVC pipe should be used as a replacement.

Answers to questions on page 112.

14.18A A source water has a hardness of 150 mg/L as $CaCO_3$. What is the hardness in grains per gallon?

Known	Unknown
Hardness, mg/L = 150 mg/L	Hardness, grains/gallon

Calculate the source water hardness in grains per gallon.

$$\text{Hardness, grains/gallon} = \frac{(\text{Hardness, mg}/L)(1 \text{ gpg})}{17.1 \text{ mg}/L}$$

$$= \frac{(150 \text{ mg}/L)(1 \text{ gpg})}{17.1 \text{ mg}/L}$$

$$= 8.8 \text{ grains/gal}$$

14.18B An ion exchange softener contains 60 cubic feet of resin with a hardness removal capacity of 25 kilograins per cubic foot of resin. The water being treated has a hardness of 250 mg/L as $CaCO_3$. How many gallons of water can be softened before the softener will require regeneration?

Known	Unknown
Resin Vol, cu ft = 60 cu ft	Water Treated, gal
Removal Capacity, gr/cu ft = 25,000 gr/cu ft
Hardness, mg/L = 250 mg/L

1. Convert the hardness from mg/L to grains per gallon.

$$\text{Hardness, grains/gallon} = \frac{(\text{Hardness, mg}/L)(1 \text{ grain/gal})}{17.1 \text{ mg}/L}$$

$$= \frac{(250 \text{ mg}/L)(1 \text{ grain/gal})}{17.1 \text{ mg}/L}$$

$$= 14.6 \text{ grains/gal}$$

2. Calculate the exchange capacity of the softener in grains.

$$\text{Exchange Capacity, grains} = (\text{Resin Vol, cu ft})(\text{Removal Capacity, gr/cu ft})$$

$$= (60 \text{ cu ft})(25,000 \text{ grains/cu ft})$$

$$= 1,500,000 \text{ grains of Hardness Removal Capacity}$$

3. Calculate the volume of water in gallons that may be treated before regeneration.

$$\text{Water Treated, gal} = \frac{\text{Exchange Capacity, grains}}{\text{Hardness, grains/gallon}}$$

$$= \frac{1,500,000 \text{ grains}}{14.6 \text{ grains/gal}}$$

$$= 102,700 \text{ gal}$$

Therefore, 102,700 gallons of water with 14.6 grains of hardness per gallon of water can be treated before the resin becomes exhausted.

Answers to questions on page 114.

14.19A Zero hardness effluent from the ion exchange softener is mixed (blended) with filtered water having a known hardness concentration to achieve a desired level of hardness in the finished water. Water with zero hardness must not be sent into a distribution system. Water with zero hardness is very corrosive and, over a period of time, will attack steel pipes in the system and create red water problems. Also, providing a water supply with zero hardness water would be very expensive.

14.20A Records that should be kept by the operator of an ion exchange softening plant include:

1. Total daily flow through unit
2. Blend rates
3. Total daily gallons that have bypassed unit
4. Gallons of brine used each day
5. Pounds of salt used each day
6. Results of tests performed on source water, softener effluent, and blended water

CHAPTER 15

SPECIALIZED TREATMENT PROCESSES

by

Mike McGuire

and

Chet Anderson

NOTICE

Keep in contact with your state drinking water agency to obtain the rules and regulations that currently apply to your water utility. The EPA's toll-free Safe Drinking Water Hotline at (800) 426-4791 is also an excellent source of information about drinking water regulations.

TABLE OF CONTENTS

Chapter 15. SPECIALIZED TREATMENT PROCESSES

		Page
OBJECTIVES		125
WORDS		126

TRIHALOMETHANES by Mike McGuire

LESSON 1

15.0	THE TRIHALOMETHANE (THM) PROBLEM		129
15.1	FEASIBILITY ANALYSIS PROCESS		131
15.2	PROBLEM DEFINITION		131
	15.20	Sampling	131
	15.21	THM Calculations	132
	15.22	Chemistry of THM Formation	133
15.3	CONTROL STRATEGIES		134
15.4	EXISTING TREATMENT PROCESSES		135
15.5	TREATMENT PROCESS RESEARCH STUDY RESULTS		135
	15.50	Consider Options	135
	15.51	Remove THMs After They Are Formed	135
	15.52	Remove THM Precursors	136
	15.53	Alternative Disinfectants	138
15.6	SELECTION AND IMPLEMENTATION OF A COST-EFFECTIVE ALTERNATIVE		139
15.7	REGULATORY UPDATE		139
15.8	SUMMARY AND CONCLUSIONS		140
15.9	ADDITIONAL READING		140
DISCUSSION AND REVIEW QUESTIONS			141

ARSENIC by Chet Anderson

LESSON 2

15.10	THE ARSENIC PROBLEM		142
	15.100	Why Are We Concerned About Arsenic?	142
	15.101	What Are the Sources of Arsenic?	142
	15.102	Chemistry of Arsenic	142
15.11	TREATMENT FOR REDUCTION OR REMOVAL OF ARSENIC		142
	15.110	New Source Alternative to Treatment	142
	15.111	Summary of Arsenic Treatment Options	143

15.112	Engineered Blending	143
15.113	Ion Exchange (IX)	144
15.114	Activated Alumina (AA)	146
15.115	Oxidation-Filtration and Iron-Based Adsorption	146
15.116	Point-Of-Use (POU) and Point-Of-Entry (POE) Devices	146
15.117	Proprietary Media	146

15.12 TYPICAL ARSENIC TREATMENT PLANT 148

15.120	Plant Start-Up and Shutdown	148
15.121	Operation	148
15.122	Maintenance	149
15.123	Troubleshooting	149
15.124	Safety and Security Issues	149
15.125	Review of Plans and Specifications	149

15.13 WASTEWATER AND RESIDUALS 149

15.14 MONITORING 150

15.140	Analysis of Arsenic	150
15.141	Types of Arsenic Sampling/Monitoring	150
15.142	Monitoring for Compliance	150
15.143	Monitoring for Process Control	151

15.15 RECORDKEEPING AND REPORTING 151

15.150	Records	151
15.151	Reporting	151

15.16 ARITHMETIC ASSIGNMENT 152

15.17 ADDITIONAL READING 152

DISCUSSION AND REVIEW QUESTIONS 152

SUGGESTED ANSWERS 153

OBJECTIVES

Chapter 15. SPECIALIZED TREATMENT PROCESSES

Following completion of Chapter 15, you should be able to:

1. Describe how trihalomethanes are formed.
2. Explain why trihalomethanes are a problem in drinking water.
3. Collect samples for trihalomethane analysis.
4. Identify control strategies for trihalomethanes.
5. Describe treatment processes capable of controlling trihalomethanes.
6. Select and implement a cost-effective means of controlling trihalomethanes.
7. Describe why we are concerned about arsenic.
8. Review the various water treatment processes used for reduction or removal of arsenic.
9. Safely operate, maintain, and troubleshoot a typical arsenic treatment plant.
10. Review the plans and specifications for an arsenic treatment plant.
11. Properly dispose of wastewater and residuals from an arsenic treatment plant.
12. Monitor an arsenic treatment plant for process control and compliance.
13. Properly perform recordkeeping and reporting tasks for an arsenic treatment plant.

WORDS
Chapter 15. SPECIALIZED TREATMENT PROCESSES

ABSORPTION (ab-SORP-shun)
The taking in or soaking up of one substance into the body of another by molecular or chemical action (as tree roots absorb dissolved nutrients in the soil).

ACTIVATED ALUMINA
A charged form of aluminum, used with a synthetic, porous media in an ion exchange adsorption process to remove charged contaminants.

ADSORPTION (add-SORP-shun)
The gathering of a gas, liquid, or dissolved substance on the surface or interface zone of another material.

CARCINOGEN (kar-SIN-o-jen)
Any substance that tends to produce cancer in an organism.

COPRECIPITATION
A treatment process that occurs when ferrous iron is added to water or metallic wastestreams and subsequently oxidized in an aerator. The oxidized iron, which is insoluble, precipitates along with other metallic contaminants present in the water or wastestream, thereby enhancing metals removal.

EMPTY BED CONTACT TIME (EBCT)
A measure of the time during which a water to be treated is in contact with the treatment medium in a contact vessel, assuming that all liquid passes through the vessel at the same velocity. EBCT is equal to the volume of the empty bed divided by the flow rate.

ENDOCRINE (EN-doe-krin) EFFECTS
An altering of the organs in the human body responsible for secreting hormones into the bloodstream. Endocrine glands include the thyroid gland, the pancreas, and the adrenal glands.

HALOACETIC (HAL-o-uh-SEE-tick) ACID (HAA)
A class of disinfection by-products, formed mainly during the chlorination of water, containing natural organic matter. HAA5 is the sum of the concentrations, in milligrams per liter, of five haloacetic acid compounds.

MCL
Maximum Contaminant Level. The largest allowable amount. MCLs for various water quality indicators are specified in the National Primary Drinking Water Regulations (NPDWR).

NSF
NSF International is a noncommercial, not-for-profit organization concerned with public health safety and environmental protection. NSF Standard 60 lists certified drinking water chemicals and NSF Standard 61 lists certified drinking water system components.

NONVOLATILE MATTER
Material such as sand, salt, iron, calcium, and other mineral materials that are only slightly affected by the actions of organisms and are not lost on ignition of the dry solids at 550°C (1,022°F). Volatile materials are chemical substances usually of animal or plant origin. Also see INORGANIC WASTE and VOLATILE SOLIDS.

PRECURSOR, THM (PRE-curse-or)

Natural, organic compounds found in all surface and groundwaters, which may react with halogens (such as chlorine) to form trihalomethanes (THMs); they must be present in order for THMs to form.

REPRESENTATIVE SAMPLE

A sample portion of material, water, or wastestream that is as nearly identical in content and consistency as possible to that in the larger body being sampled.

SPLIT SAMPLE

A single grab sample that is separated into at least two parts such that each part is representative of the original sample. Often used to compare test results between field kits and laboratories or between two laboratories.

TRIHALOMETHANES (THMs) (tri-HAL-o-METH-hanes)

Derivatives of methane, CH_4, in which three halogen atoms (chlorine or bromine) are substituted for three of the hydrogen atoms. Often formed during chlorination by reactions with natural organic materials in the water. The resulting compounds (THMs) are suspected of causing cancer.

VISCOSITY (vis-KOSS-uh-tee)

A property of water, or any other fluid, that resists efforts to change its shape or flow. Syrup is more viscous (has a higher viscosity) than water. The viscosity of water increases significantly as temperatures decrease. Motor oil is rated by how thick (viscous) it is; 20 weight oil is considered relatively thin while 50 weight oil is relatively thick or viscous.

VOLATILE (VOL-uh-tull)

(1) A volatile substance is one that is capable of being evaporated or changed to a vapor at relatively low temperatures. Volatile substances can be partially removed from water or wastewater by the air stripping process.

(2) In terms of solids analysis, volatile refers to materials lost (including most organic matter) upon ignition in a muffle furnace for 60 minutes at 550°C (1,022°F). Natural volatile materials are chemical substances usually of animal or plant origin. Manufactured or synthetic volatile materials, such as plastics, ether, acetone, and carbon tetrachloride, are highly volatile and not of plant or animal origin. Also see NONVOLATILE MATTER.

CHAPTER 15. SPECIALIZED TREATMENT PROCESSES
Trihalomethanes by Mike McGuire

(Lesson 1 of 2 Lessons)

15.0 THE TRIHALOMETHANE (THM) PROBLEM

For the past few decades, water utilities have been concerned about the presence of organic compounds in drinking water. The analytical methods for detecting inorganic compounds such as calcium, magnesium, and iron have been known for many decades. However, the ability to analyze for organic compounds in water has been developed only recently. What are organic compounds? Organic compounds are defined as those compounds that contain a carbon atom. Carbon is one of the basic chemical elements. Examples of organic compounds include: proteins, carbohydrates, fats, vitamins, and a wide variety of compounds that modern technology has created.

In 1974, researchers with the US Environmental Protection Agency (EPA) and in the Netherlands published their findings that trihalomethanes are formed in drinking water when free chlorine comes in contact with naturally occurring organic compounds (*THM PRECURSORS* [1]). Trihalomethanes are a class of organic compounds in which there has been a replacement of the three hydrogen atoms in the methane molecule with three halogen atoms (chlorine or bromine). The four most commonly found THMs are chloroform, bromodichloromethane, dibromochloromethane, and bromoform (Figure 15.1). While it is theoretically possible to form iodine-substituted THMs, they are rarely found in treated water and they are not regulated at this time. In general, methane is not involved in the THM reaction. The production of THMs can generally be shown as:

EQUATION 1

Free Chlorine + Natural Organics (precursor) + Bromide → THMs + Other Products

Free chlorine is added to drinking water as a disinfectant. The naturally occurring organics get into water when the water partially dissolves organic materials from algae, leaves, bark, wood, soil, and other similar materials. This dissolving action is similar to what happens when a teabag is placed in hot water; the water dissolves those parts of the tea leaves that are soluble organic and inorganic compounds. While it is possible to form THMs by reactions between chlorine and industrial organic chemicals, the overwhelming bulk of THM precursors in water are from natural organic compounds.

One source of bromide is seawater. Water agencies whose supplies are subject to seawater intrusion can expect THMs in their treated water to have high levels of bromide. Bromide reaction products can be found in most surface waters, even where bromide concentrations are low. The "other products" formed in this reaction are very poorly understood and are not regulated at this time.

After THMs were discovered in drinking waters around the country, several studies were made of the possible health effects of THMs in general and chloroform (a THM) in particular. These tests indicated that chloroform caused cancer in laboratory animals (rats and mice) and was suspected of causing cancer in humans. Further studies comparing people who used different sources of drinking water suggested that there may be a link between the presence of manmade organic compounds like THMs and increased levels of cancer. Animal feeding experiments and population studies are not definite proof that THMs in drinking water cause cancer. Under the Safe Drinking Water Act, EPA may pass a regulation for any contaminant that MAY HAVE any adverse health effect.

On November 29, 1979, the THM regulations were published in the *FEDERAL REGISTER*; they were amended on February 28, 1983 (see Section 15.7, "Regulatory Update"). The general aspects of the regulation are outlined below:

MAXIMUM CONTAMINANT LEVEL: 0.080 mg/L total trihalomethanes (TTHMs)—sum of the concentrations of chloroform, bromodichloromethane, dibromochloromethane, and bromoform. (See *NOTICE* on page 131.)

APPLIES TO: All community water systems that add a disinfectant to their water supply and serve a population greater than 10,000 persons.

MONITORING REQUIREMENTS: Monitoring compliance is based on an annual running TTHM (total trihalomethane) average of four quarters of data. Schedule, locations, and numbers of samples depend on system size and should be worked out with the state or EPA.

ENSURING MICROBIOLOGICAL QUALITY: State or EPA must be notified of significant modifications to treatment processes to remove TTHMs in order to ensure microbiological quality of the treated water.

---

[1] *Precursor, THM* (PRE-curse-or). Natural, organic compounds found in all surface and groundwaters, which may react with halogens (such as chlorine) to form trihalomethanes (THMs); they must be present in order for THMs to form.

Methane, CH₄

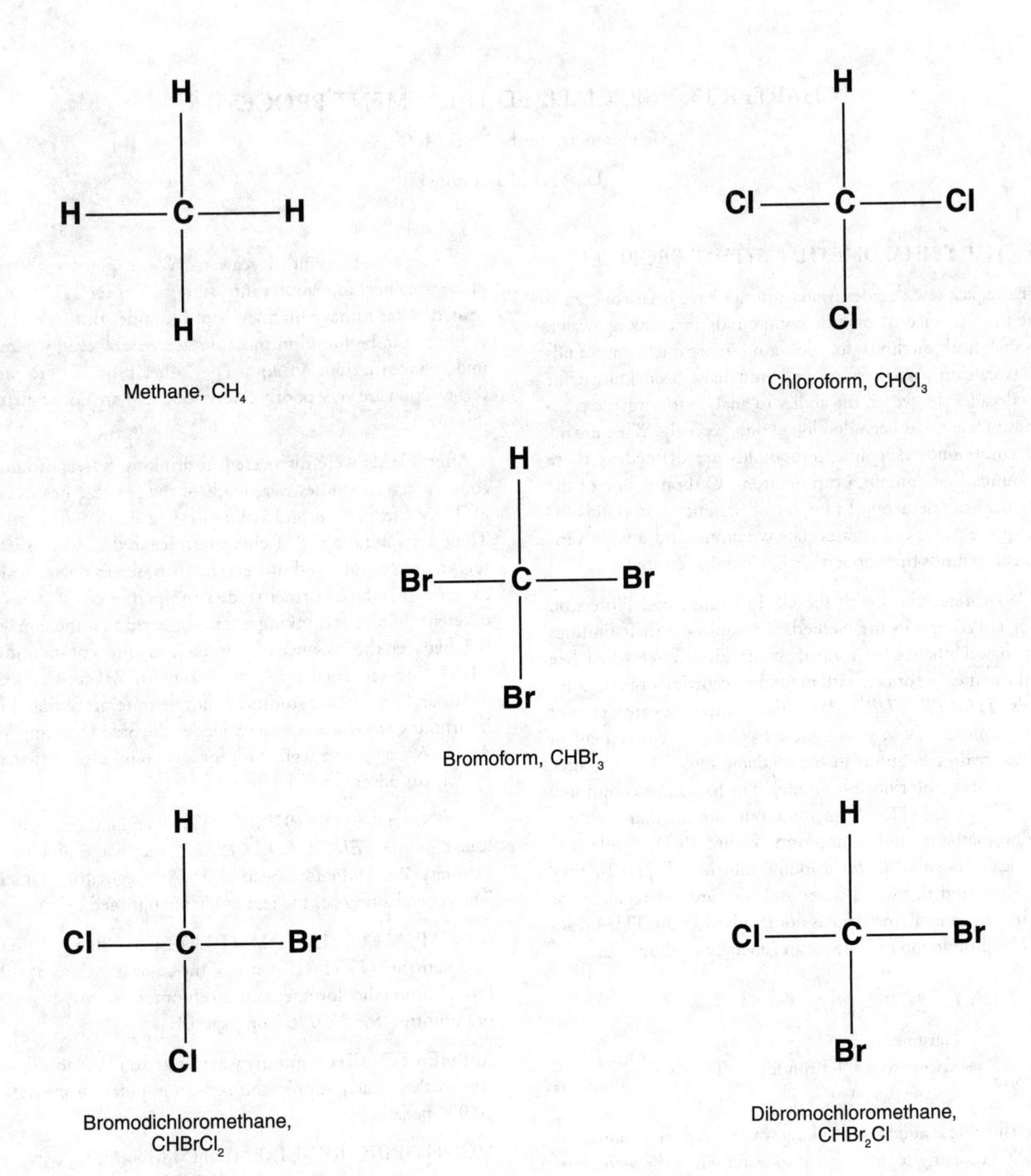

Chloroform, CHCl₃

Bromoform, CHBr₃

Bromodichloromethane, CHBrCl₂

Dibromochloromethane, CHBr₂Cl

Fig. 15.1 Methane and THMs

The MCL for TTHMs was not established on the basis of the health-effects data, but was set as a feasible level for compliance.

NOTICE

Operators should be aware that the Disinfectants/Disinfection By-Products Rule (D/DBP), which was promulgated on December 16, 1998, will significantly affect both large and small drinking water utilities. The D/DBP Rule lowered the MCL for total trihalomethanes (TTHMs) from 0.10 mg/L to 0.080 mg/L and set new limits for various disinfectants and disinfection by-products. See Chapter 22, "Drinking Water Regulations," Section 22.23, "Disinfectants and Disinfection By-Products (D/DBPs)," for additional details.

The rest of this lesson is devoted to a discussion of how to collect samples for THM analysis and how a utility can evaluate the many alternatives available to control THMs in its system. This discussion is presented in outline form. A much more detailed treatment of control techniques for THMs is presented in an EPA publication entitled "Treatment Techniques for Controlling Trihalomethanes in Drinking Water," by J. M. Symons, et al., September 1981.[2] A large part of this lesson is a summary of material from that source.

QUESTIONS

Write your answers in a notebook and then compare your answers with those on page 153.

15.0A How are trihalomethanes formed in drinking water?

15.0B What are THM precursors?

15.0C Why is free chlorine added to drinking water?

15.0D What is one source of bromide in drinking water?

15.1 FEASIBILITY ANALYSIS PROCESS

In any problem-solving process, it is useful to follow a series of prescribed steps that will lead you to the most cost-effective solution. Table 15.1 lists the stages of a feasibility analysis process that has been used to solve a water utility's THM problem. However, the process outlined in Table 15.1 is very general and can also be applied to solving other treatment or operational problems.

15.2 PROBLEM DEFINITION

In order to determine the extent of a THM problem in a system, a reliable analytical technique must be used. THM analytical services may be developed by the utility or may be purchased

TABLE 15.1 FEASIBILITY ANALYSIS PROCESS

1. Determine the extent of the THM problem
   a. Monitor THM levels
   b. THM chemistry (time of formation)
2. Evaluate control strategies
   a. Change sources of supply
   b. Treatment options
      (1) Remove THMs
      (2) Remove precursors
      (3) Adjust or modify the chlorine application points
      (4) Use alternative disinfectants
3. Evaluate existing treatment processes
4. Examine studies of proposed treatment processes (Bench-, pilot-, and full-scale)
5. Select a cost-effective option
6. Implement the chosen option

from a contract laboratory. A discussion of the THM analytical methods will not be made here; rather, the reader is referred to the *FEDERAL REGISTER* publication of the regulation or to the EPA document prepared by J. M. Symons, which was previously discussed.

15.20 Sampling

To determine the extent of the THM problem, collect *REPRESENTATIVE SAMPLES*[3] from the distribution system of the water utility and analyze them according to an approved method to determine if the utility is in compliance with the THM regulation. Of course, four quarters of data are needed to make a definite judgment on the MCL. However, even one quarter of data can show how close the system will be to complying with the MCL. See Section 15.21, "THM Calculations," Examples 1 and 2, for procedures on how to calculate quarterly average TTHM levels and annual TTHM running averages.

The following THM sampling protocol was developed from 141.132 CFR and applies to drinking water distribution systems serving at least 10,000 people. Refer to 141.132 CFR for THM sampling protocols for smaller distribution systems.

---

[2] Available from National Technical Information Service (NTIS), 5285 Port Royal Road, Springfield, VA 22161. Order No. PB82-163197. Price, $81.00, plus $5.00 shipping and handling per order.

[3] *Representative Sample.* A sample portion of material, water, or wastestream that is as nearly identical in content and consistency as possible to that in the larger body being sampled.

To collect samples to determine THM levels, use the following procedures:

1. A minimum of four samples per quarter (every three months) must be taken on the same day for each treatment plant in the distribution system.
2. Twenty-five percent of these samples must be collected from the extremities of the distribution system (the points farthest from each treatment plant).
3. Seventy-five percent of all the four samples must be representative of the population served by the distribution system.

Do not collect samples from swivel faucets, faucets with aerators, or faucets with hoses because of the possibility of contaminating the sample or loss of THMs.

To collect samples for THM analysis, use a narrow-mouth, screw-cap, glass sample bottle that can hold at least 25 mL of water. Use a polytetrafluorethylene (PTFE)-faced silicon-septia bottle-cap liner to provide an airtight seal over the sample bottle. The bottle caps must screw tightly on the sample bottles.

Some sample bottles will contain a small amount of a chemical reducing agent (usually sodium thiosulfate or sodium sulfite). The reducing agent will stop the chemical reaction that occurs between chlorine and the THM precursors (humic and fluvic acids). By stopping this chemical reaction, THMs will not continue to form in the sample after it has been collected from the distribution system. When using sample bottles that contain a reducing agent, do not rinse out the reducing agent before collecting the sample.

Some sample bottles will not contain a reducing agent. Water samples from these bottles will be tested for the maximum concentration of TTHMs that can form over an extended period of time. These tests cannot be performed if a reducing agent has been added to the sample. When using sample bottles that do not contain a reducing agent, do not add any chemicals to the bottles.

When collecting water samples for THM analysis, use the following procedures:

1. Turn on the sampling tap.
2. Allow sufficient time (about five minutes) for the water temperature to become constant.
3. Fill the sample bottle until it begins to overflow.
4. Set the bottle on a level surface and place the bottle-cap liner on top of the bottle.
5. Screw the bottle cap tightly on the bottle and turn the bottle upside down.
6. The sample is properly sealed if no air bubbles are present.
7. If air bubbles are present, remove the bottle cap and bottle-cap liner, turn on the sampling tap and add a small amount of water to the sample in the bottle. Repeat steps 4 through 6.

A good practice is to collect two samples at each location. This procedure allows the laboratory to double-check test results and, if a sample bottle is broken, there will be another sample available for testing.

Each sample bottle must include a label on which important information is recorded. Be sure to write on the label the sample location, date, and name of person collecting the sample.

Samples should be sent to the laboratory immediately after they are collected and should be analyzed within 14 days. When sending samples to the laboratory, be sure to include the complete name and address of the person to whom the test results are to be returned. Samples must be chilled to 4°C immediately after collection and kept at 4°C during storage. Do not use dry ice when shipping or storing samples because the water in the bottles may freeze and break the sample bottles.

15.21 THM Calculations

FORMULAS

In order to calculate the average of a group of measurements, sum up the measurements and divide the total by the number of measurements.

$$\text{Average} = \frac{\text{Sum of Measurements}}{\text{Number of Measurements}}$$

To calculate the running annual average, sum up the average measurements for each quarter and divide the total by the number of quarters.

$$\text{Running Average} = \frac{\text{Sum of Averages for Each Quarter}}{\text{Number of Quarters}}$$

Whenever data for a new quarter become available, the newest quarterly average replaces the oldest quarterly average and the running annual average is recalculated.

EXAMPLE 1

A water utility collected and analyzed eight samples from a water distribution system on the same day for TTHMs. The results are shown below.

Sample No.	1	2	3	4	5	6	7	8
TTHM, µg/L	80	50	70	110	90	120	80	90

What was the average TTHM for the day?

Known	Unknown
Results from analyses of 8 TTHM samples	Average TTHM level for the day

Calculate the average TTHM level in micrograms per liter.

$$\text{Avg TTHM, µg}/L = \frac{\text{Sum of Measurements, µg}/L}{\text{Number of Measurements}}$$

$$= \frac{80\ \text{µg}/L + 50\ \text{µg}/L + 70\ \text{µg}/L + 110\ \text{µg}/L + 90\ \text{µg}/L + 120\ \text{µg}/L + 80\ \text{µg}/L + 90\ \text{µg}/L}{8\ \text{Measurements}}$$

$$= \frac{690\ \text{µg}/L}{8}$$

$$= 86\ \text{µg}/L$$

EXAMPLE 2

The results of the quarterly average TTHM measurements for two years are given below. Calculate the running annual average of the four quarterly measurements in micrograms per liter.

Quarter	1	2	3	4	1	2	3	4
Avg Quarterly TTHM, µg/L	87	72	99	82	62	111	138	89

Known	Unknown
Results from analyses of 2 years of TTHM samples	Running annual average of quarterly TTHM measurements

Calculate the running annual average of the quarterly TTHM measurements.

$$\text{Annual Running TTHM Average, µg}/L = \frac{\text{Sum of Average TTHM for Four Quarters}}{\text{Number of Quarters}}$$

Quarters 1, 2, 3, and 4

$$\text{Annual Running TTHM Average, µg}/L = \frac{87\ \text{µg}/L + 72\ \text{µg}/L + 99\ \text{µg}/L + 82\ \text{µg}/L}{4}$$

$$= \frac{340\ \text{µg}/L}{4}$$

$$= 85\ \text{µg}/L$$

Quarters 2, 3, 4, and 1

$$\text{Annual Running TTHM Average, µg}/L = \frac{72\ \text{µg}/L + 99\ \text{µg}/L + 82\ \text{µg}/L + 62\ \text{µg}/L}{4}$$

$$= \frac{315\ \text{µg}/L}{4}$$

$$= 79\ \text{µg}/L$$

Quarters 3, 4, 1, and 2

$$\text{Annual Running TTHM Average, µg}/L = \frac{99\ \text{µg}/L + 82\ \text{µg}/L + 62\ \text{µg}/L + 111\ \text{µg}/L}{4}$$

$$= \frac{354\ \text{µg}/L}{4}$$

$$= 89\ \text{µg}/L$$

Quarters 4, 1, 2, and 3

$$\text{Annual Running TTHM Average, µg}/L = \frac{82\ \text{µg}/L + 62\ \text{µg}/L + 111\ \text{µg}/L + 138\ \text{µg}/L}{4}$$

$$= \frac{393\ \text{µg}/L}{4}$$

$$= 98\ \text{µg}/L$$

Quarters 1, 2, 3, and 4

$$\text{Annual Running TTHM Average, µg}/L = \frac{62\ \text{µg}/L + 111\ \text{µg}/L + 138\ \text{µg}/L + 89\ \text{µg}/L}{4}$$

$$= \frac{400\ \text{µg}/L}{4}$$

$$= 100\ \text{µg}/L$$

SUMMARY OF RESULTS

Quarter	1	2	3	4	1	2	3	4
Avg Quarterly TTHM, µg/L	87	72	99	82	62	111	138	89
Annual Running TTHM Avg, µg/L				85	79	89	98	100

15.22 Chemistry of THM Formation

An understanding of the chemistry of THM formation is crucial if a water utility is to solve a THM problem. Equation 1 shown in Section 15.0, "The Trihalomethane (THM) Problem," describes the overall mechanism. Very little is known about the specific reactions that free chlorine and natural organics (precursors) undergo. In general, the effects of time, temperature, pH, and concentrations of the chemicals on the production of THMs have been studied by various investigators and are fairly well understood.

Depending on the type of natural organics present in the water, the time it takes for 0.080 mg/L (80 µg/L) of THMs to form may range from minutes to days. Set up a THM monitoring program on the source water(s) of the utility to measure the production of THMs over an appropriate time period (time from when chlorine is first added to water until water is consumed). A plot of the THMs produced against time will give you an idea of the TTHM (Total Trihalomethanes) formation potential (TTHMFP) of each source water. For many systems, a large part of the production of THMs will take place after the water leaves the treatment plant.

The higher the temperature, the faster the THMs will be produced. As might be expected, a dependence on temperature will probably show up as a seasonal effect—higher THM levels in the summer than in the winter. Temperature may not be the only controlling factor, however; higher THM levels may show up in the winter, as they have in California.

The higher the pH of the water, the faster the production of THMs. For most water utilities this will not be a concern; however, utilities raising the pH of treated water by caustic soda or by lime for corrosion control (Langelier Index) or using lime softening should be aware that free chlorine in contact with natural organics at a pH of 10.5 or higher will produce THMs much faster than if the pH was near 7.0.

The higher the concentrations of free chlorine and natural organics in the water, the more THMs will be produced. In the past, the amount of free chlorine that utilities used was only limited by economics and possible taste and odor complaints from consumers. Careful use of chlorine may help a utility to lower the THMs in its system. However, because of the danger of using too little chlorine (inadequate disinfection) in a system, the THM regulation specifically requires state or EPA approval of major treatment changes to meet the regulation.

The concentration of precursors in water is as important as the types of precursors that are found in water. Some naturally occurring organic compounds can produce 10 or 100 times more THMs on an equivalent basis than organics from another source. Also, some types of precursors will produce THMs faster than others. For these reasons it is important to evaluate the TTHMFP of each source of supply as a possible THM control measure.

The effect of higher bromide concentrations on THM production is not as clear as the effects of temperature and pH. The more bromide present, the more bromide-containing THMs will be formed. Free chlorine selectively attacks the bromide ion and changes it to bromine, which reacts quickly with precursors to form bromoform, dibromochloromethane, and bromodichloromethane. The usual result of high bromide levels is higher THM levels because more molecules are available (chlorine + bromine) to participate in THM-forming chemical reactions.

Now that some of the basics of THM chemistry are understood and a THM problem can be properly defined, it is time to look at some of the possible control strategies.

QUESTIONS

Write your answers in a notebook and then compare your answers with those on page 153.

15.1A   List the major steps that a water utility could take to solve a THM problem.

15.1B   List the possible control strategies that could be evaluated to control a THM problem.

15.2A   What important factors influence the production of trihalomethanes?

15.2B   How does lime used for softening influence the production of THMs?

15.3 CONTROL STRATEGIES

Assuming that a utility discovers a THM problem in its system, there are two ways to control it: change the source of supply or provide some type of treatment. Changing the source of supply can consist of an entire range of alternatives such as shifting between wells of different water quality, drawing water from different levels in a reservoir, or abandoning a surface water supply altogether during part of the year. Since most utilities do not have the flexibility to abandon a source of supply, this alternative will have limited application.

The three treatment options available to control THMs are as follows:

1. Remove THMs after they are formed.
2. Remove THM precursors before chlorine is added.
3. Use a disinfectant other than free chlorine.

A later section will examine each of these options and the processes associated with them. At this point, it is useful to discuss overall treatment strategies. The general equation for forming THMs illustrates how each of the three options can work.

EQUATION 1

$$\text{Free Chlorine} + \text{Natural Organics (precursor)} + \text{Bromide} \rightarrow \text{THMs} + \text{Other Products}$$

Removing THMs after they are formed is generally not the strategy of choice unless there is a particular circumstance at the utility that warrants its evaluation. Since precursors are not necessarily removed when THMs are removed, there is the problem of continued THM formation, especially in the distribution system.

Removing precursors before free chlorine is added has some major advantages, particularly if the precursors can be removed by a fairly inexpensive process. Also, removing precursors allows the continued use of free chlorine as a disinfectant, which has been proven to be an effective barrier against disease for many decades. As Equation 1 shows, fewer precursors also means the formation of fewer "other products." These other products consist of high-molecular-weight organic compounds that contain chlorine and bromine. The health significance of these other products is not always known, but concern has been raised by regulatory agencies.

Using a disinfectant other than free chlorine has a number of advantages and disadvantages that must be evaluated case by case. Abandoning free chlorine is a serious move in view of its superior performance as a disinfectant. However, if any of the alternative disinfectants are lower cost alternatives, they must be given careful consideration.

15.4 EXISTING TREATMENT PROCESSES

Before beginning a complex, expensive research program, it is valuable to examine how well existing treatment processes can control the formation of THMs. The following sections cover the potential of individual processes for THM control; however, some generalizations can be made with regard to existing unit processes. Aeration unit processes are sometimes available in water treatment plants to control tastes and odors. The same process may show measurable removals of THMs after they are formed. Oxidation of tastes and odors with chlorine dioxide (ClO$_2$) and potassium permanganate are common unit processes available in water treatment plants. Chlorine dioxide does not form THMs. Permanganate sometimes can be used to oxidize THM precursors if they are affected by this kind of treatment.

Coagulation/sedimentation/filtration and softening processes can remove THM precursors depending on the types that are present in the water supply. Powdered activated carbon and granular activated carbon used for taste and odor control can have a limited impact on the removal of both THMs and THM precursors.

QUESTIONS

Write your answers in a notebook and then compare your answers with those on page 153.

15.3A If a utility discovers a THM problem, what are two ways to control the problem?

15.3B Why is abandoning the use of free chlorine considered a serious move?

15.4A List the water treatment processes that can be used to control THMs.

15.5 TREATMENT PROCESS RESEARCH STUDY RESULTS

15.50 Consider Options

There is a long list of treatment options that can be investigated for the control of THMs. Since a large number of them have already been studied and reported on, it is not necessary for every utility to repeat this work. A careful evaluation of the results published by the US EPA will help a utility focus on the treatment processes that should be looked at on a bench-, pilot-, or full-scale basis.

Most feasible treatment options include the removal of precursor materials prior to the formation of a THM, the avoidance of generation of a THM by use of an alternate disinfectant, or the actual removal of a THM by means of aeration or carbon adsorption. Also, the geographic and climatological conditions can have a very important influence on the choice of a process. For example, aeration is not a desirable method of treatment where severe cold weather is common.

15.51 Remove THMs After They Are Formed

There are three treatment processes available to remove THMs after they have been formed:

1. Oxidation

   a. Ozone

   b. Chlorine dioxide

   c. Ozone/ultraviolet light

2. Aeration

   a. Open storage

   b. Diffused air

   c. Towers

3. Adsorption

   a. Powdered activated carbon

   b. Synthetic resins

OXIDATION. Oxidation of THMs using any one of the three oxidants listed above has not been very successful. The combination of ozone/ultraviolet light showed some promise; however, the cost-effectiveness of the process has yet to be demonstrated.

AERATION. In contrast, aeration is an effective process for removing THMs from water, although the individual THMs are removed at different efficiencies. THM removal efficiencies by aeration, ranging from the easiest to most difficult, are from chloroform to bromodichloromethane to dibromochloromethane and to bromoform. Allowing water containing THMs to stand uncovered will ultimately result in the THMs leaving the water, since they are *VOLATILE*[4] compounds that are poorly soluble in water. In other words, THMs have a natural tendency to migrate from water into the atmosphere if given the chance. Because of this tendency, THM reductions may be noticeable in effluents from uncovered finished water reservoirs after a significant detention time (days). Note: The Interim Enhanced Surface Water Treatment Rule (IESWTR) prohibits construction of new uncovered finished water facilities after February 1999.)

More efficient removal of THMs can be accomplished if energy is put into the aeration process. A convenient way to put energy into aeration is by bubbling air into water. Many water

---

[4] *Volatile* (VOL-uh-tull). (1) A volatile substance is one that is capable of being evaporated or changed to a vapor at relatively low temperatures. Volatile substances can be partially removed from water or wastewater by the air stripping process. (2) In terms of solids analysis, volatile refers to materials lost (including most organic matter) upon ignition in a muffle furnace for 60 minutes at 550°C (1,022°F). Natural volatile materials are chemical substances usually of animal or plant origin. Manufactured or synthetic volatile materials, such as plastics, ether, acetone, and carbon tetrachloride, are highly volatile and not of plant or animal origin. Also see NONVOLATILE MATTER.

treatment plants currently have an aeration process of some kind to help control tastes and odors in the source water. The efficiencies of these existing processes would not be expected to be very great for THM removal.

Operators should realize that aeration of treated water can cause a significant amount of contamination. Air in many areas may contain large amounts of dust, dirt, bacteria, and other pollutants that can contaminate treated water and also lead to operation and maintenance problems.

In the research results that are currently available, countercurrent tower aeration (Figure 15.2) has produced the highest removals of THMs with air-to-water ratios (the ratio of the volume of air added to the volume of water treated) in the 20 to 1 to 50 to 1 range. Treatment efficiencies greater than 90 percent removal have been demonstrated with some aeration towers on some types of water.

Countercurrent aeration towers are designed so that the water and air pass over a packing material countercurrent to each other (in opposite directions). A significant amount of theoretical work has been done on the possible tower designs for any given set of treatment conditions. Pilot-scale testing is usually recommended before a full-scale plant is constructed.

Aeration is most effective on the more volatile chemicals. Chloroform is the most volatile of the THMs and is generally the most easily removed by aeration. Bromoform, on the other hand, is the least volatile THM and, consequently, is the hardest to remove by aeration. If the TTHM content of the water contains a significant amount of bromoform, aeration may not be the most desirable technique to investigate.

ADSORPTION. THMs can be removed by a wide variety of activated carbons and synthetic resins. The adsorption process involves the individual THM compounds leaving the water and becoming attached to the surface of the carbon or resin. THMs are generally considered difficult to adsorb on any surface. The efficiency by adsorption from easiest to most difficult is bromoform, dibromochloromethane, bromodichlromethane, and chloroform.

Powdered activated carbon (PAC) is usually added as a treatment chemical in the rapid-mix process or in the sedimentation basin effluent. PAC is normally used in water treatment for taste and odor control at dosages of less than 20 mg/L. Studies have shown that PAC dosages of 100 mg/L or more are necessary to get significant removals of THMs. Chloroform is particularly difficult to remove with PAC.

Synthetic resins such as XE-340 have been demonstrated to be effective in removing THMs from water; however, economics must be taken into consideration, since the cost of the resins is high in comparison with other alternatives. Regeneration of the resins has not been worked out satisfactorily. Pilot-scale studies show some promise. Resin manufacturers are continually developing new processes to improve the performance of their products.

QUESTIONS

Write your answers in a notebook and then compare your answers with those on page 153.

15.5A Which is the better process for removing THMs after they are formed, oxidation or aeration?

15.5B How does the storage of water in uncovered reservoirs affect THM levels?

15.5C How does the adsorption process work?

15.52 Remove THM Precursors

A variety of treatment processes have been investigated to remove THM precursors before they come in contact with chlorine. These include:

1. Aeration
2. Oxidation
    a. Ozone (prior to coagulation and clarification)
    b. Chlorine dioxide
    c. Permanganate
    d. Ozone/ultraviolet light
    e. Hydrogen peroxide
3. Clarification
    a. Coagulation/sedimentation/filtration
    b. Softening
4. Adsorption
    a. Powdered activated carbon
    b. Granular activated carbon
    c. Synthetic resins
5. Ion Exchange

AERATION. Since THM precursors are not volatile compounds, it is not surprising that aeration is ineffective in removing them from water.

OXIDATION. All of the oxidants listed above have some effect on removing or modifying THM precursors. Since THM precursors vary so much between locations, it is difficult to generalize on the effectiveness of any of the oxidants. In fact, some studies have demonstrated that the formation potential for THMs can *increase* with the application of certain dosages of ozone and potassium permanganate. In general, it is necessary to perform bench- or pilot-scale studies on the water in question before the usefulness of any of these oxidants can be considered. The US EPA is also concerned with the production of potentially harmful by-products that could result from the use of any of these oxidants. Once again, studies on the water to be treated are necessary to determine whether or not this is a problem.

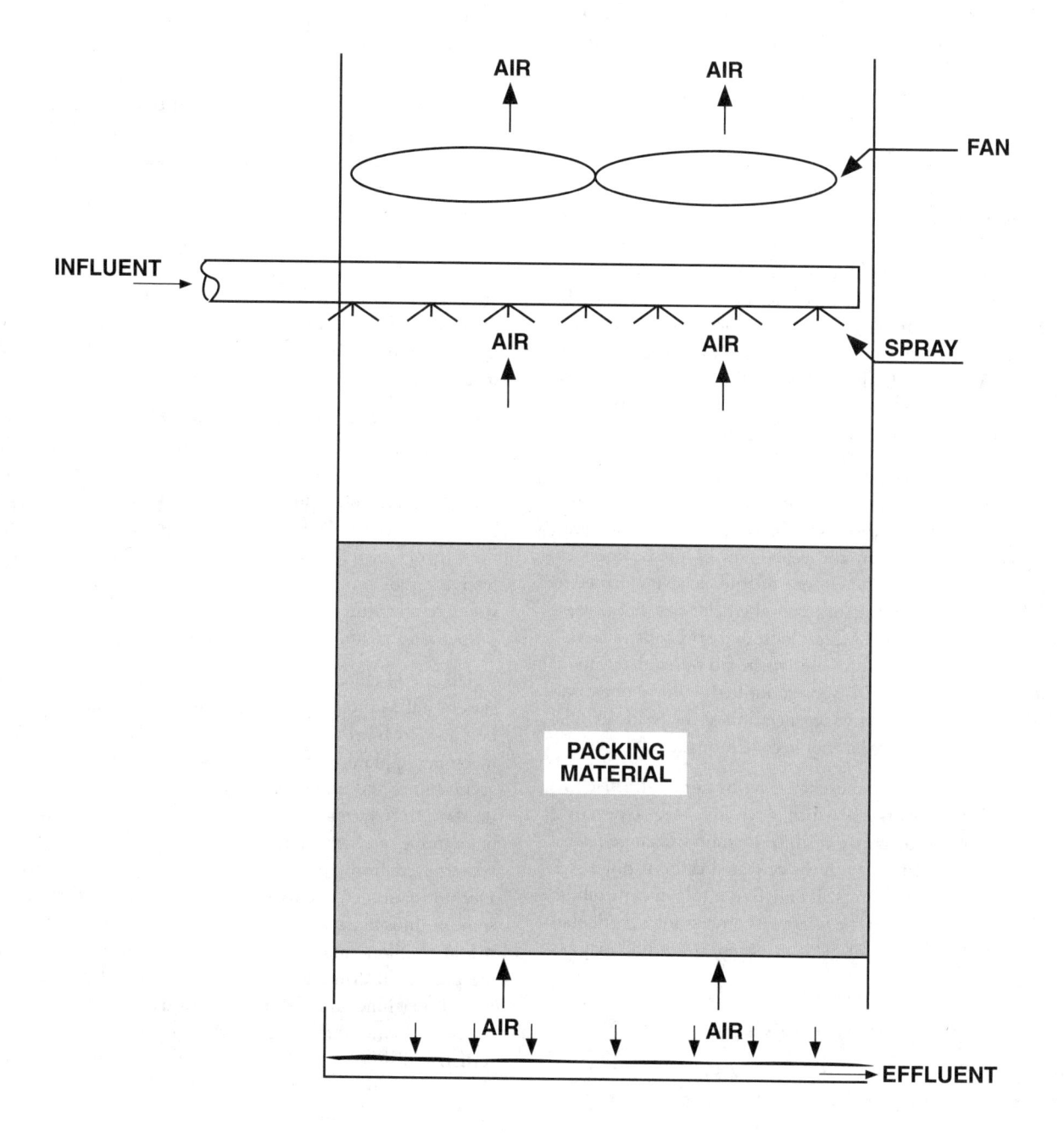

Fig. 15.2 Countercurrent aeration tower

In the southeastern United States and in other locations, water supplies are relatively warm and contain high levels of organic matter. Under these conditions, controlled oxidation with small doses of ozone can actually coagulate organic material and make conventional sedimentation more effective. Too much ozone can break the organics down and make them more reactive with chlorine. However, small, controlled doses of ozone may be an effective microflocculant and may be added to conventional water treatment plants to improve the physical removal of THM precursors to the point that prechlorination disinfection is possible.

CLARIFICATION. The clarification process used in water treatment plants has the potential for removing significant amounts of THM precursors. Dozens of studies by the US EPA have demonstrated widely varying removal efficiencies (0 to 100 percent) because of the highly variable nature of THM precursors from place to place. The use of this process, which is available in most water treatment plants, to remove THM precursors holds great promise for an economical solution to any THM problem. Moving the addition of free chlorine to a point following the clarification process is the key to success for this approach. Many water utilities have adopted this approach to solve their problem.

ADSORPTION. The use of powdered activated carbon (PAC) and granular activated carbon (GAC) are effective in removing THM precursors; however, the economics of these processes must be carefully evaluated. Dozens of studies have reported a wide variety of THM precursor removal efficiencies. Because of the high cost of PAC and GAC, their use as THM control methods will be restricted to those cases where no other alternatives are available. Synthetic resins showed limited removal potential for THM precursors. Effective regeneration of the resins for additional precursor removal has not been demonstrated.

ION EXCHANGE. Anion exchange resins can be effective for removing THM precursors, which generally have a negative charge. Both strong-base and weak-base anion exchange resins have been investigated. As with the activated carbons discussed above, anion exchange resins will only find a role in controlling THMs if the economics of the treatment process for a particular site are favorable. Disposal of the spent regenerant liquid may be a problem.

QUESTIONS

Write your answers in a notebook and then compare your answers with those on pages 153 and 154.

15.5D List the major treatment processes that have been investigated to remove THM precursors before they come in contact with chlorine.

15.5E What is the key to success in using clarification to remove THM precursors from the water being treated?

15.53 Alternative Disinfectants

Removing free chlorine from the chlorine/bromide/precursor reaction will stop the formation of any significant amount of THMs. However, free chlorine has been an effective barrier between people and disease-causing bacteria since the beginning of this century, and abandoning its use is a very serious step. There are other disinfectants that can be used instead of free chlorine, but the advantages and disadvantages of each alternative must be carefully evaluated.

The most commonly considered alternative disinfectants are ozone, chlorine dioxide, and chloramines. Ozone is a gas that is produced by passing oxygen through an electrical discharge. While ozone is a highly effective disinfectant, it is very expensive, it must be generated on site, and it does not leave a residual in the treated water. Chlorine dioxide is a gas produced by the reaction of free chlorine and sodium chlorite. Chlorine dioxide is a very effective disinfectant that does leave a residual in the treated water; however, there are some concerns regarding the health implications of the inorganic breakdown products, chlorite and chlorate. The THM regulation recommended a 0.5 mg/L limit for the total concentration of chlorine dioxide, chlorite, and chlorate in water after chlorine-dioxide treatment.

Chloramines are produced in water by the reaction between free chlorine and ammonia. Chloramines are weaker disinfectants than free chlorine, ozone, or chlorine dioxide, but the residuals remain much longer than free chlorine and they have been used successfully by dozens of water utilities. For example, the City of Denver has used chloramines for many years. The effectiveness of monochloramines as a disinfectant depends on water temperature, pH, and biological quality, as well as the proper ratio of ammonia to chlorine.

The use of chloramines can also cause problems in a utility's system unless proper precautions are taken. Chloramines must be removed from the water before it is used in kidney dialysis machines. Chloramines in water can pass through kidney dialysis machines and into a patient's blood where the ammonia will decrease the oxygen-carrying capacity of the blood. In addition, chloramines are toxic to fish in home aquariums, and they must be removed from water before it comes in contact with fish. Dechlorination of water with activated carbon, ascorbic acid, or sodium thiosulfate will prevent any of these problems if the removal of chloramines is properly controlled. Any oxidants that are present in drinking water can cause problems with kidney dialysis machines and fish in home aquariums. However, chloramines are somewhat more difficult to remove than the other alternative disinfectants. For additional information, see Section 7.26, "Chloramination," in *WATER TREATMENT PLANT OPERATION*, Volume I.

Before an alternative disinfectant is applied to any system, the source of the water supply, water quality, and treatment effectiveness for bacteriological control must be evaluated. For example, the use of a weaker disinfectant such as chloramines may not be appropriate for a surface water supply that is highly contaminated with discharges from municipal and industrial wastewater treatment plants unless an extra high dosage and a long contact time are provided. Also, many of the conventional water treatment processes are capable of removing bacteria, viruses, and protozoa from the water (for example, softening and coagulation/sedimentation/filtration). These conventional processes may help to provide the required disinfection barrier between a contaminated supply and the population served, which could allow the use of a less potent disinfectant in the distribution system.

Upgraded monitoring (more samples and tests) of the distribution system before and after a disinfectant change must be provided by the utility. The THM regulation specifies guidelines that the states must use in establishing such a monitoring program. Guidelines describing coliforms, standard plate count, turbidity, and nutrients are included in the suggested monitoring list. With "before-and-after" monitoring by the water utility, it will be possible to determine if there is any significant degradation of the bacteriological quality in the distribution system. Control of THMs must not be accomplished at the expense of a higher risk of bacterial and viral diseases among the population that is being served. Therefore, a decision to use a disinfectant other than free chlorine must be based on a carefully considered plan. A utility that rushes into the use of an alternative disinfectant without the required studies is likely to experience many problems that are easily avoided with proper planning.

QUESTIONS

Write your answers in a notebook and then compare your answers with those on page 154.

15.5F What items must be considered before an alternative disinfectant is applied to any system?

15.5G What type of distribution system monitoring must be provided by a utility before and after a disinfectant change?

15.5H What water quality indicators should be monitored before and after a disinfectant change?

15.6 SELECTION AND IMPLEMENTATION OF A COST-EFFECTIVE ALTERNATIVE

A detailed evaluation of the comparative economics of the many treatment processes described above is outside the scope of this lesson. In many cases, a utility will commission a special cost-effectiveness study that will be accomplished in-house or by an outside consultant. The US EPA THM treatment manual presents a detailed look at cost estimates for various alternatives with equivalent THM control levels. However, these data are not current and should be updated to reflect current economic conditions whenever cost studies are conducted.

If existing processes are not capable of solving a utility's THM problem, the least-cost solution will probably be an alternative disinfectant. While no statistics are currently available, evidence from discussions with consultants and utility managers suggests that alternative disinfectants, especially chloramines, are the overwhelming least-cost solution for water utilities with a THM problem. However, the use of chloramines may cause problems for persons using kidney dialysis machines.

Implementation of a THM control strategy requires a number of well-defined steps:

1. Full-scale design
2. Construction
3. Start-up
4. Operation

The length of time required to complete these steps will depend on the complexity of the control strategy chosen and the availability of engineering services to complete the assigned tasks. Throughout the implementation phase, it is important that the bench-, pilot-, and full-scale tests initiated in the feasibility analysis phase be continued so that the chosen strategy can be refined and optimized. For example, a pilot plant can be used to train treatment plant operators to use the new technology that will soon be on line.

QUESTIONS

Write your answers in a notebook and then compare your answers with those on page 154.

15.6A If existing water treatment processes are not capable of solving a THM problem, what is the most likely least-cost solution?

15.6B What appears to be the most popular alternative disinfectant?

15.7 REGULATORY UPDATE

On February 28, 1983, the US EPA published in the *FEDERAL REGISTER* (page 8406) an amendment to the THM regulation originally published on November 29, 1979. The amendment specifies the treatment alternatives that a utility must consider or investigate in detail before it can apply for and receive a variance from meeting the MCL as defined under the regulation. The US Environmental Protection Agency or the state may require a community water system to use a "generally available" technology before granting a variance. "Generally available" or Group 1 treatment techniques are:

1. Use of chloramines or chlorine dioxide as an alternative or supplement to chlorine for oxidation and disinfection.

2. Use of chloramines, chlorine dioxide, or potassium permanganate as an alternative to chlorine for preoxidation.

3. Moving the point of chlorination in order to reduce THM formation.

4. Improvement of existing clarification.

5. Use of powdered activated carbon (PAC), intermittently, as necessary, to reduce TTHM or THM precursors. The dosage of PAC is not to exceed an annual average of 10 mg/L.

Any of these technologies may be required in the variance unless the regulatory agency, US EPA, or the state determines that "such treatment method ... is not available and effective for TTHM control for the system." The rule allows exemption from the use of a technique if the method would not be technically appropriate and technically feasible for the system or if the method would result in only a marginal reduction in TTHM.

The rule also allows the regulator to require the study of Group 2 technologies by water systems where Group 1 technologies are not appropriate or sufficient in meeting the MCL. If a Group 2 technology would be technically feasible and economically reasonable and result in significant TTHM reductions in line with the cost of treatment, then the regulator can require the use of a Group 2 technology.

The listed Group 2 technologies are introduction of off-line water storage, aeration, introduction of clarification, alternative sources of raw water, and the use of ozone as an alternative or supplement to chlorine for disinfection or oxidation.

The February 28, 1983, amendment to the THM regulations did not mention granular activated carbon (GAC) or biological activated carbon (BAC) as treatment alternatives that must be considered. These two treatment methods were judged to be too expensive and are not widely enough used in the US to warrant their evaluation for THM control. In general, the amendment was designed to reduce the economic impact of the THM regulation on those utilities that have THM problems and limited resources to drastically modify their treatment procedures.

The Disinfectants/Disinfection By-Products Rule (D/DBP), promulgated by the EPA on December 16, 1998, further regulates not only THMs but also other disinfection by-products and residual concentrations of some disinfectants. The goal of the rule is a balance between projected microbiological and chemical risks. See Chapter 22, "Drinking Water Regulations," Section 22.23, "Disinfectants and Disinfection By-Products (D/DBPs)," for additional information about specific limits, effective dates of the regulations, and types of systems required to comply with the D/DBP regulations. Also, see the poster provided with this manual, Stages 1 and 2 Disinfectants and Disinfection By-Products Rule (D/DBPR), Total Trihalomethanes (TTHMs), and Haloacetic Acids (HAA5).

Utilities that may be affected by THM regulations are advised to follow future developments in the *FEDERAL REGISTER*.

15.8 SUMMARY AND CONCLUSIONS

1. Trihalomethanes are produced when free chlorine, which is added as a disinfectant, reacts with naturally occurring bromide and organic compounds.

2. A trihalomethane regulation is now in effect that has established a 0.080 mg/L maximum contaminant level and monitoring requirements.

3. A feasibility analysis process is a series of logical steps to arrive at a cost-effective solution to a THM problem:

    a. Determine the extent of THM problem

       (1) Monitor

       (2) Examine THM chemistry

    b. Evaluate control strategies

       (1) Change sources of supply

       (2) Treatment options

          (A) Remove THMs

          (B) Remove precursors

          (C) Use alternative disinfectants

    c. Evaluate existing treatment processes

    d. Research studies of treatment processes

       (1) Bench-, pilot-, and full-scale

    e. Select a cost-effective option

    f. Implement chosen option

4. There are three treatment options available to control THMs:

    a. Remove THMs after they are formed

    b. Remove THM precursors before chlorine is added

    c. Use a disinfectant other than free chlorine

5. A water utility must not create a possible health problem by ignoring bacteriological safeguards in an attempt to solve a THM problem.

6. An amendment to the THM regulation specifies treatment techniques that must be evaluated before a utility may receive a variance. Since this amendment affects a utility's feasibility analysis procedure, the steps outlined in the amendment should be followed.

QUESTIONS

Write your answers in a notebook and then compare your answers with those on page 154.

15.7A What treatment processes must utilities evaluate before applying for and receiving a variance?

15.7B If treatment processes are not technically feasible and economically reasonable, then what should utilities consider?

15.9 ADDITIONAL READING

1. "Treatment Techniques for Controlling Trihalomethanes in Drinking Water," US Environmental Protection Agency. EPA No. 600-2-81-156. Obtain from National Technical Information Service (NTIS), 5285 Port Royal Road, Springfield, VA 22161. Order No. PB82-163197. Price, $81.00, plus $5.00 shipping and handling per order.

END OF LESSON 1 OF 2 LESSONS

on

SPECIALIZED TREATMENT PROCESSES

Please answer the discussion and review questions next.

DISCUSSION AND REVIEW QUESTIONS

Chapter 15. SPECIALIZED TREATMENT PROCESSES

(Lesson 1 of 2 Lessons)

At the end of each lesson in this chapter you will find some discussion and review questions. The purpose of these questions is to indicate to you how well you understand the material in the lesson. Write the answers to these questions in your notebook.

1. How are trihalomethanes formed in drinking water?

2. On what basis was the Maximum Contaminant Level (MCL) for total trihalomethanes (TTHMs) established?

3. What is the influence of higher temperatures and pH on the production of trihalomethanes (THMs)?

4. What are some of the options if a water utility decides to investigate changing the source of the water supply?

5. What are the advantages of removing precursors before free chlorine is added?

6. Under what conditions might ozone be used prior to clarification and filtration?

7. List the three alternative disinfectants to free chlorine and the advantages and limitations of each one.

8. What items must be considered before an alternative disinfectant is applied to any system?

9. What are the advantages of using a pilot plant in the implementation of a THM control strategy?

CHAPTER 15. SPECIALIZED TREATMENT PROCESSES

Arsenic by Chet Anderson

(Lesson 2 of 2 Lessons)

15.10 THE ARSENIC PROBLEM

15.100 Why Are We Concerned About Arsenic?

Arsenic is a natural element that is found in water at low concentrations. Ingestion of very high levels of arsenic can cause acute (short-term) adverse health effects. High levels of arsenic are typically not found in drinking water sources. Low levels of arsenic consumed over a long period of time can cause cancer and noncancerous toxic effects on humans. The types of cancer related to chronic (long-term) exposure to arsenic in drinking water include skin, bladder, lung, kidney, nasal passages, liver, and prostate cancer. Noncarcinogenic (not cancer-causing) effects include cardiovascular, pulmonary, immunological, neurological, and *ENDOCRINE EFFECTS*.[5]

The US EPA MCL for arsenic is 0.010 mg/L (or 10 ppb). States must adopt an MCL at least as stringent as 10 ppb.

15.101 What Are the Sources of Arsenic?

Arsenic in drinking water sources can come from either natural occurrences (such as volcanic action, erosion of rock, or a result of forest fires) or human activities. Arsenic is most commonly found in water sources as a result of natural leaching from soils. Arsenic in water sources also can result from mining, agricultural, or manufacturing operations. Approximately 90 percent of the arsenic used by industry in the US is currently used for wood-preservative purposes. Arsenic is also found in paints, dyes, metals, and other manufactured products.

The occurrence of high levels of arsenic in drinking water is mainly in groundwater, although there are a few exceptions in surface waters. Western states in the US have a disproportionate number of drinking water systems with arsenic problems. The source of this problem is primarily related to naturally occurring arsenic in soil and is normally not related to mining wastes.

15.102 Chemistry of Arsenic

Arsenic is an element with the chemical symbol As. Arsenic is normally a cation with a valence (charge) of either +3 or +5. However, when arsenic is dissolved in water it will be present as part of a complex that may have either a negative charge or a neutral charge. Arsenic can be either in the trivalent (+3 or III) arsenite form or the pentavalent (+5 or V) arsenate form, depending on the level of oxidation. The oxidation state of arsenic is very important when treating arsenic because the arsenic (III) form is not removed by common treatment processes. In water, the arsenic (III) is part of a neutrally charged complex while arsenic (V) is part of a negatively charged ion. The arsenic (V) can be removed by either ion exchange, activated alumina, or by the coagulation-filtration systems using ferric salt coagulants. The pH of the water is important in determining the arsenic speciation (arsenic (III) or arsenic (V) form).

Both forms of arsenic commonly occur in groundwater: As(III) and As(V). NP4, an arsenic-reducing microbe from the genus *Sulfurospirillum*, can transform As(V) to As(III), the more toxic form of arsenic. Also, naturally occurring bacteria can increase arsenic concentrations in groundwater. Members of the genus *Geobacter* and other iron-reducing bacteria grow in the absence of oxygen. They can transform solid-phase iron [Fe(III)] into water-soluble Fe(II). Solid-phase Fe(III) can trap arsenic on its surface, but if transformed into Fe(II), it would release the arsenic into groundwater.

> ### QUESTIONS
>
> Write your answers in a notebook and then compare your answers with those on page 154.
>
> 15.10A What are the potential sources of arsenic in drinking water?
>
> 15.10B Why is the oxidation state of arsenic very important when treating arsenic?

15.11 TREATMENT FOR REDUCTION OR REMOVAL OF ARSENIC

15.110 New Source Alternative to Treatment

Before considering treatment of a water source to reduce or remove arsenic, water utilities should investigate a new source as an alternative water supply. Construction of new, low arsenic sources may be the most economical solution for some water systems. However, this alternative will likely be restricted to very

---

[5] *Endocrine* (EN-doe-krin) *Effects.* An altering of the organs in the human body responsible for secreting hormones into the bloodstream. Endocrine glands include the thyroid gland, the pancreas, and the adrenal glands.

small systems where drilling of a single well may create the desired solution. This solution assumes that low arsenic groundwater is available and can be recovered for drinking purposes. There are costs and risks associated with verifying that an alternative groundwater of acceptable water quality is available. Changing to a surface water supply means a tradeoff between building an arsenic removal plant or a surface water treatment plant.

15.111 Summary of Arsenic Treatment Options

There are a number of treatment methods that can be used to remove arsenic from drinking water sources. These methods can be classified into three general treatment categories: *ADSORPTION*[6] (A), *COPRECIPITATION*[7] (C), and physical separation (P). Those treatment technologies that are considered feasible from a technical and economic viewpoint are described as Best Available Technology (BAT).

BAT processes available for treating arsenic include the following technologies:

1. Engineered blending
2. *ACTIVATED ALUMINA*[8] (C)
3. Coagulation/filtration (C)
4. Ion exchange (A)
5. Lime softening (C)
6. Reverse osmosis (RO) (P)
7. Electrodialysis removal (EDR) (P)
8. Oxidation/filtration (P)
9. Point-of-use (POU) and point-of-entry (POE) devices (A), (C), or (P)

Almost all arsenic treatment processes (except reverse osmosis and blending) depend on the electrostatic interactions of a negatively charged As(V) ion with the treatment media (or chemical). For this reason, it is necessary to add an oxidation agent to ensure that all of the neutral (uncharged) As(III) is changed (oxidized) to the negatively charged anionic As(V) state. This change can be achieved by using chlorine or potassium permanganate ($KMnO_4$). Chlorine is the most common and readily available oxidant. Ozone and hydrogen peroxide could also be effective.

15.112 Engineered Blending

Blending of higher arsenic level sources with low arsenic level sources is an acceptable arsenic treatment option as long as the blending is properly engineered, operated, and controlled. Because the arsenic levels in both surface and groundwaters tend to be quite stable, blending can be effectively controlled. Variations in source water arsenic levels must be thoroughly evaluated before considering blending as an option.

The key elements that must be provided for in order for blending to be successful include:

1. Low arsenic source(s) with sufficient quantity and reliability.
2. The high and low arsenic sources must be located in such proximity that they can be mixed before entering the distribution system.
3. Mixing must be thorough and proven (for example, the use of a storage or detention tank, if available, provides for adequate mixing and a buffer in the event of unexpected changes in raw water arsenic levels).
4. The system must be able to function with both sources out of operation since failure of the low arsenic source will render both sources unusable.
5. There must be adequate monitoring, operation, and record-keeping.

Before a blending operation can be implemented, theoretical calculations must be made to determine if blending is feasible and will be an effective solution. Factors that must be known in order to make the required blending calculations include the following:

Q_1 = Pumping capacity of Source #1 (usually in gallons per minute (GPM))

Q_2 = Pumping capacity of Source #2 (GPM)

Q_b = Combined flow of the two sources (GPM)

C_1 = Arsenic level of Source #1 (in mg/L or ppm)

C_2 = Arsenic level of Source #2 (mg/L)

C_b = Arsenic level in blended water (mg/L)

The theoretical resulting arsenic level in the blended water (C_b) is calculated using the following equation:

$$Q_1 \times C_1 + Q_2 \times C_2 = C_b (Q_1 + Q_2)$$

or

$$C_b = \frac{Q_1 C_1 + Q_2 C_2}{Q_b}$$

Where

$$Q_b = Q_1 + Q_2$$

---

[6] *Adsorption* (add-SORP-shun). The gathering of a gas, liquid, or dissolved substance on the surface or interface zone of another material.
[7] *Coprecipitation.* A treatment process that occurs when ferrous iron is added to water or metallic wastestreams and subsequently oxidized in an aerator. The oxidized iron, which is insoluble, precipitates along with other metallic contaminants present in the water or wastestream, thereby enhancing metals removal.
[8] *Activated Alumina.* A charged form of aluminum, used with a synthetic, porous media in an ion exchange adsorption process to remove charged contaminants.

EXAMPLE 1

$Q_1 = 200$ GPM

$Q_2 = 300$ GPM

$C_1 = 0.020$ mg/L

$C_2 = 0.000$ mg/L

$Q_b = Q_1 + Q_2 = 200$ GPM $+ 300$ GPM $= 500$ GPM

$$C_b = \frac{Q_1 C_1 + Q_2 C_2}{Q_b}$$

$$= \frac{200 \text{ GPM} \times 0.020 \text{ mg}/L + 300 \text{ GPM} \times 0.000 \text{ mg}/L}{500 \text{ GPM}}$$

$$= \frac{4 \text{ mg}/L}{500 \text{ GPM}}$$

$$= 0.008 \text{ mg}/L$$

As part of the operational requirements for approval of blending operations, the regulatory agency will probably require frequent, perhaps daily, calculation of the theoretical blended water arsenic levels to provide verification of laboratory analyses and operational procedures.

Some issues that are related to blending that require consideration include:

1. Blending processes may be complex in some situations.
2. Well failure/maintenance failure will affect the availability of low arsenic sources.
3. Water quality could potentially change over time.
4. System may need water distribution system modeling to ensure reliable blended arsenic levels in the drinking water.

15.113 Ion Exchange (IX) (Figure 15.3)

The basics for using ion exchange for removing arsenic from drinking water sources are similar to any ion exchange treatment process using an anion-based resin.

Some specific concerns related to the use of ion exchange to remove arsenic include:

1. **Effect of pH.** The ion exchange of the arsenate ion, with chloride on the chloride-form, strong-base resins that are most likely to be used, occurs effectively in the pH range of about 6.5 to 9.0. Typically, most groundwaters are in this pH range. Therefore, adjustment of pH prior to ion exchange is rarely needed.
2. **Effect of competing ions.** Because of the relative affinity that the strong-base resins have for different anions, the source raw water quality is very important. Sulfate and nitrate are both preferentially adsorbed instead of arsenic on the ion exchange resin, so the levels of these water quality constituents are critical. If nitrate removal is also needed and is being done, sulfate is especially critical. An effect known as "chromatographic" peaking can occur when sulfate can displace previously adsorbed arsenate ions on the resin and cause higher levels of arsenic in the effluent than in the raw water. The resin bed must be monitored and regenerated well before this "peaking" situation occurs. Also, if the sulfate levels are above 120 to 150 mg/L, ion exchange costs are prohibitive.
3. **Effect of arsenic concentration.** Because arsenic is present in groundwater in very low concentrations relative to other competing ions, such as sulfate, and is not the preferred ion by the exchange resin, the concentration of arsenic (whether at 10 ppb or 100 ppb) will not dramatically affect the length of ion exchange run before leakage occurs. For example, the time for the length of ion exchange run could be half the run length time if the arsenic concentration is increased ten times.
4. **Resin type.** Strong-base resins are normally used for arsenic removal. However, resin selection should be a site-specific decision based on the results of pilot tests.
5. **Configuration of ion exchange columns or vessels.** Use of multiple ion exchange beds can help address issues of arsenic breakthrough and the chromatographic effect.
6. **Direction of flow.** The usual procedure is to have the raw water and the regenerant brine both flow through the resin bed in the downward direction (also called cocurrent flow). If the regenerant flow is in the opposite direction (upward flow or countercurrent flow), potential advantages include minimizing the leakage of arsenate ions from the vessel into the treated water and, possibly, the reduction of volume and concentration of brine required.
7. **Number of ion exchange units.** The use of multiple ion exchange units may have very significant benefits in terms of the removal of arsenic because of the breakthrough potential and the competing water quality indicator issues.

TYPICAL ION EXCHANGE ELEMENTS FOR ARSENIC REMOVAL

- Strong-base anion exchange resin
- Empty bed contact time (EBCT)[9] = 1.5 minutes
- Typical bed volumes treated = 400
- Regenerant, salt
- Salt use, 1,850 lbs salt per million gallons treated (222 kilograms salt per million liters treated)
- Recycle brine, up to 36 times
- Precipitate arsenic in brine with ferric chloride before reuse

NOTE: Each treatment situation will be different because of unique water quality factors and local conditions.

---

[9] *Empty Bed Contact Time (EBCT).* A measure of the time during which a water to be treated is in contact with the treatment medium in a contact vessel, assuming that all liquid passes through the vessel at the same velocity. EBCT is equal to the volume of the empty bed divided by the flow rate.

Specialized Treatment 145

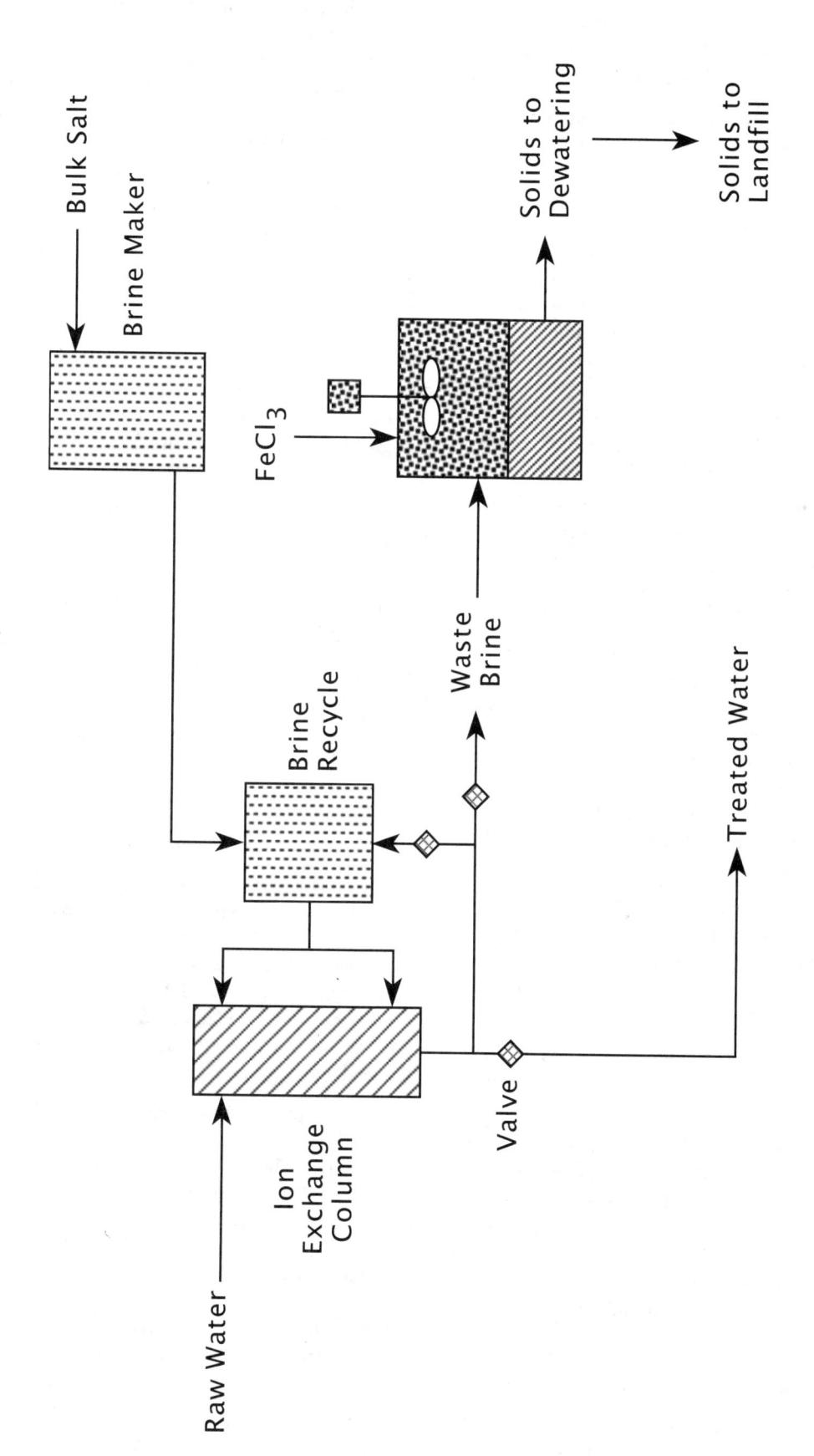

Fig. 15.3 Ion exchange arsenic removal treatment process

15.114 Activated Alumina (AA) (Figure 15.4)

Activated alumina is an adsorption process used to remove arsenic from drinking water. Critical elements of an activated alumina treatment process are described in this section.

Importance of pH. The activated alumina adsorption process for removal of arsenic from drinking water is highly dependent on pH. This adsorption process may not perform efficiently unless the pH is lowered to the range of about 5.5 to 6.0. pH reduction requires the addition of acid (usually sulfuric acid) to lower the pH of most groundwaters. After arsenic has been adsorbed on the activated alumina, the treated water will require the pH to be raised to 7.5 or higher by adding caustic (sodium hydroxide) to control corrosion. Operation of this process will require operators to have advanced skills. The use and storage requirements for pH adjustment chemicals, the costs of the chemicals, and the safe handling and storage of chemicals may make activated alumina an unacceptable process for many small- to medium-sized water systems.

Water quality concerns. Activated alumina can be effective when the groundwater contains high levels of total dissolved solids (TDS). Water quality indicators that compete with arsenic ions (arsenate) for adsorptions sites on the media include selenium, fluoride, chloride, and sulfate (at high levels) ions.

Spent media. Activated alumina media may be regenerated or discarded. Regeneration, however, requires the addition of strong-base and acid chemicals and also is not complete. For these reasons, on-site regeneration of activated alumina media is not likely to be feasible. Shipping the media to an off-site location for regeneration is typically not feasible either. The spent media is likely to be nonhazardous and can be disposed of in a landfill. If the media cannot be discarded, activated alumina will probably not be feasible for smaller systems.

Wastewater disposal. Regeneration of the activated alumina media produces a highly concentrated brine wastestream, which may create disposal problems. Metering the waste brine stream as a low flow into a sanitary sewer system may be possible in some locations.

Media availability. Activated alumina production is limited at this time and media may not be readily available in some locations and at some points in time in the future.

Configuration. Typically, two activated alumina media columns are operated in sequence to allow efficient operation, process control, and safety benefits. The media in the first column is run to exhaustion (unable to remove any more arsenic) and the second column provides a polishing step. Then, the media in the first column is replaced and the sequence is reversed (the second column removes arsenic and the first column provides the polishing step).

15.115 Oxidation-Filtration and Iron-Based Adsorption

There are several treatment systems that have been used successfully for removing iron and manganese (such as greensand filters) that may provide for co-removal of arsenic. The effectiveness of these processes in removing arsenic depends on the presence of sufficient iron in the raw water. To be effective, the iron/arsenic ratio must be at least 20 parts iron to 1 part arsenic in order to obtain 80 percent arsenic removal efficiency.

15.116 Point-Of-Use (POU) and Point-Of-Entry (POE) Devices

Point-of-use (POU) and point-of-entry (POE) devices are technologies that could be effective in removing arsenic from drinking water, especially for small systems. The devices generally use one or more BAT treatment processes.

Critical elements of any point-of-use or point-of-entry system should include the following items:

1. All POU and POE units must be owned, controlled, and maintained by the public water system (PWS) or a contractor hired by the PWS to ensure proper operation and maintenance of the units.

2. POU and POE units must have mechanical warnings to automatically notify consumers of operational problems.

3. The public water system (PWS) must be able to obtain regular access to the units to perform necessary maintenance and monitoring.

4. A rigorous preventive maintenance program and frequent sampling is required.

Technologies used by POU and POE devices to remove arsenic from drinking water include activated alumina, iron-based sorbents, and reverse osmosis.

15.117 Proprietary Media

There are several other iron-based technologies that use adsorption as the mechanism for arsenic removal. These newer technologies include granular ferric hydroxide (GFH) and several new proprietary media. Operators and water utility agencies need to visit existing sites and conduct pilot tests at their site to evaluate the feasibility and performance of the newer, proprietary media, which are typically continually being improved by the vendor.

The GFH media systems remove arsenic, chromium, lead, selenium, antimony, uranium, and other heavy metals from groundwater. The system is operated as a fixed bed adsorber. Typical installation is in pressure vessels to allow a single pumping stage for the treatment system. Water continuously passes through the media bed where arsenic is adsorbed from the water onto the media. The vessels are arranged in parallel or in series arrangement with an empty bed contact time of five minutes. The pressure vessels are backwashed once every two to six weeks to prevent bed compaction and remove trapped particulates.

Once the GFH media is exhausted, it is removed from the vessels and new media is installed. Exhausted media in most instances can be disposed of in a nonhazardous landfill.

Specialized Treatment 147

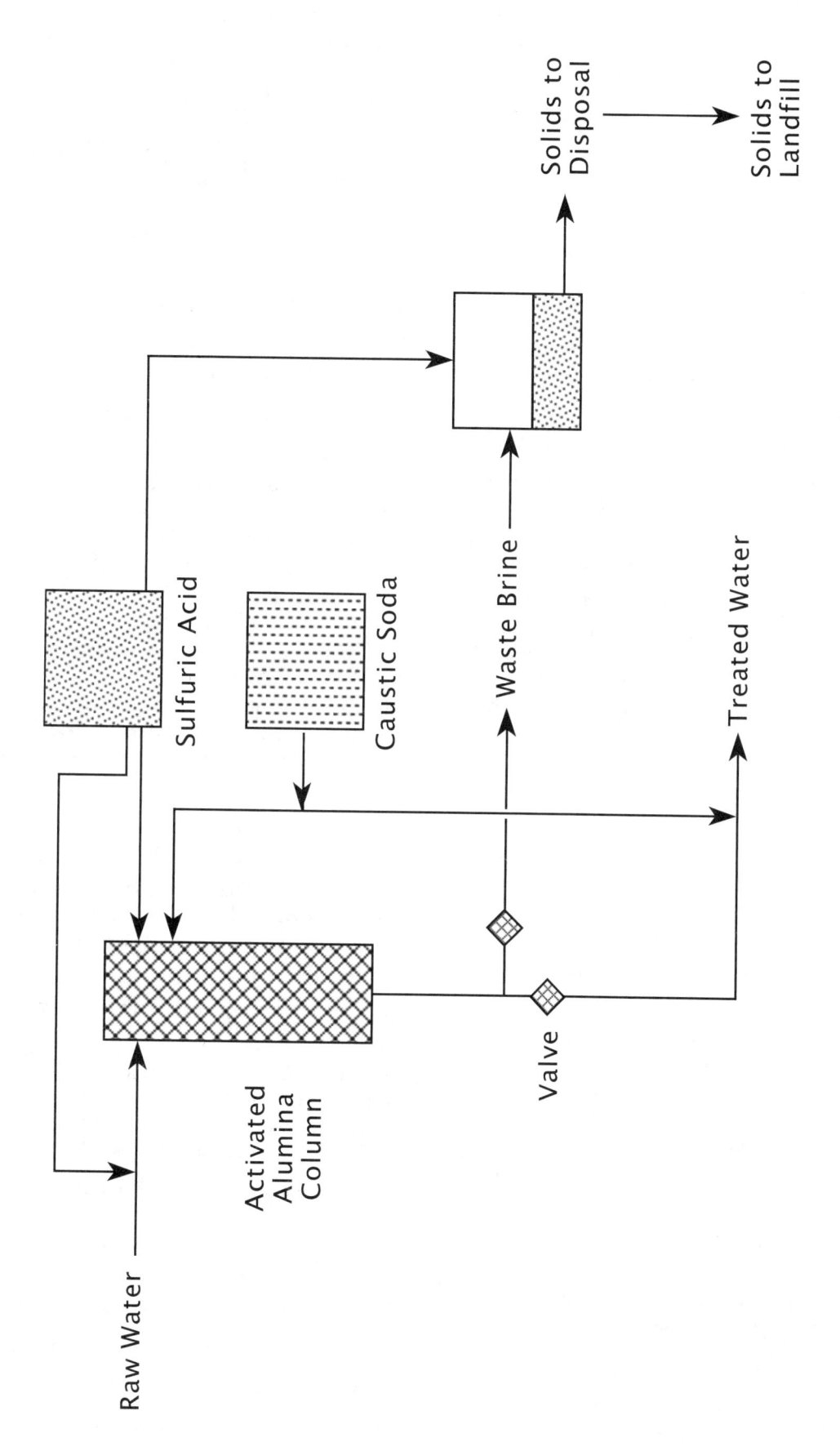

Fig. 15.4 Activated alumina arsenic removal treatment process

Parallel or series operation depends on the required removal concentrations. If a consistent 90-percent reduction of arsenic is needed across the system, the series design is used. However, if the required percentage removal is less than 90 percent, then the parallel design is typically used.

QUESTIONS

Write your answers in a notebook and then compare your answers with those on page 154.

15.11A   What are the three general treatment categories used to remove arsenic from drinking sources?

15.11B   Which tasks must be performed for blending to be an acceptable arsenic treatment option?

15.11C   Which ions in groundwater are preferentially adsorbed instead of arsenic on an ion exchange resin?

15.11D   Which technologies are used by POU and POE devices to remove arsenic from drinking water?

15.12 TYPICAL ARSENIC TREATMENT PLANT

Most arsenic removal treatment plants use either ion exchange or some type of adsorption media. The operation and maintenance (O & M) of these processes are similar. This section will emphasize the O & M of an ion exchange unit. If your plant uses another adsorption media, the O & M procedures will be similar.

15.120 Plant Start-Up and Shutdown

See Section 14.17, page 108, "Start-Up and Shutdown of Unit," for an explanation of these procedures for an ion exchange unit.

15.121 Operation

Many factors influence the procedures used to operate an ion exchange unit and the efficiency of the arsenic removal process. These factors include:

1. Characteristics of the ion exchange resin
2. Quality of the source water
3. Rate of flow applied to the ion exchange unit
4. Salt dosage during regeneration
5. Brine concentration
6. Brine contact time

Every water treatment plant should have an approved operating plan. The plant should always be operated in accordance with the approved plan. The detailed steps for normal operation, abnormal operation, troubleshooting, maintenance, monitoring, records, and reporting events are spelled out in the operating plan. The operating plan is the operator's bible for the plant.

Initial operating plans are frequently prepared by the design engineer who cannot anticipate all of the actual operating conditions. Therefore, it is essential that the operator modify the plan routinely, as needed, to reflect real-world conditions. The operator is responsible for working with others in the agency to make sure the operating plan is effective and up to date.

Each ion exchange unit, regardless of manufacturer, will have at least four common stages of operation. These stages are listed below and explained in previous sections (see Figure 14.14, page 103).

1. "Service," see Section 14.110, page 102
2. "Backwash," see Section 14.111, page 104
3. "Brine," see Section 14.112, page 104
4. "Rinse," see Section 14.113, page 105

Most ion exchange arsenic removal plants are fully automated, but must be designed to permit manual override upon demand. Automated valves should be hydropneumatically controlled valves actuated by signals through a computer system such as a Programmable Logic Controller (PLC). Each vessel containing the ion exchange resin may have six or more valves to control raw water, brine, rinse water, and backwash water flows in and out of the vessel. In addition, there will be main line raw water, backwash water, wastewater, and other valves servicing all the vessels.

A detailed valve sequencing chart or list is needed showing the exact sequence and timing of operation of each valve or valves on an ion exchange vessel. This chart or list will be specific for a given water treatment plant.

Daily operation of an arsenic removal water treatment plant typically includes:

1. Reading and recording all flowmeter values and recording instantaneous rate-of-flow and cumulative volumes. Unusual readings need to be investigated and corrective action taken, if necessary.

2. Checking salt system and brine levels and verifying flow rates and salt usage.

3. Reading all on-line instruments for water quality indicators; noting set points for alarms or automatic shutdown; and recording and taking any corrective action needed. Calibrating the instruments, as needed.

4. Collecting grab samples of raw and treated water, as required, and performing analyses or arranging shipment to an approved laboratory. Conducting any calibrations needed based on these results.

5. Checking air compressor and automatic valves.

6. Checking all valve control positions by computer or visually.

7. Logging all data in records.

8. Reporting to superintendent, manager, or regulatory agency, as needed.

Operators must be sure to use only *NSF*[10] certified media and chemicals.

15.122 Maintenance

See Section 14.15, "Maintenance," page 106, which contains information on how to maintain ion exchange units.

15.123 Troubleshooting

See Section 14.16, "Troubleshooting," page 107, which contains information on how to troubleshoot the ion exchange units, the service stage, the backwash stage, the brine injection stage, and the rinse stage.

15.124 Safety and Security Issues

Operation of ion exchange and other arsenic removal treatment systems does not present any unique or unusual safety or security issues that are not present in any typical water treatment plant. Safety issues of concern for operators include working with electrical systems and mechanical equipment, and the storage and handling of chemicals and water. Operators are expected to be properly trained to be familiar with and adhere to good safety and security procedures.

15.125 Review of Plans and Specifications

For most treatment plants, it is very beneficial to the design engineer, to the operator, and to the facility for the operator to review the plans and specifications of an expanded or new arsenic removal plant prior to the completion of the plans and specifications. The design review helps the design engineer know what details to look for to make operation and maintenance easier and to anticipate problems that might otherwise require design modifications after construction is completed. Without the operator's assistance, modification might later be necessary because someone forgot or did not have the knowledge to recommend specific details for better operational control of the arsenic removal process. Operators should very carefully review the location of valves and the ease of adjusting and maintaining valves on the ion exchange unit.

The chemical storage and supply building should be equipped with emergency eye wash stations and deluge showers and floor washdown equipment. Electrical control panels and chemical feed pumps should be protected from accidental spills or splashing during washdown. If polymers are used, all possible locations of polymer spillage should have a false grated floor, with washdown drain, to allow the spilled material to pass to a point where its *VISCOUS*,[11] slippery nature is not a threat to operator safety. The ultimate destination of floor drains should always comply with sewer-use ordinances or discharge permits.

15.13 WASTEWATER AND RESIDUALS

Operators of arsenic removal treatment plants must have facilities capable of handling various types of wastestream liquids, sludges, and solids. The first step is to identify the sources of the liquids, sludges, and solids and then ensure that the treatment plant properly collects, treats, and disposes of these wastestreams. The sources of liquids, sludges, and solids include:

1. Liquids: Backwash water, regenerant solutions (brines, caustic), and reject water

2. Sludges (semiliquid residuals): Sludge from coagulation/precipitation, and iron/manganese sludge (sludge may be subjected to added treatment to concentrate and dewater)

3. Solids: Spent filter/adsorption media, resins, and concentrated salts from evaporation ponds

Alternative methods of ultimately disposing of liquids, sludges, and solids from arsenic removal treatment processes include the following:

1. Discharge to receiving stream (if allowed, pretreatment will be required)

2. Discharge to sanitary sewer that flows to wastewater treatment plant (POTW) (may require pretreatment)

3. Land disposal

4. Evaporation pond, with solids to landfill

5. Sanitary landfill (sludge or solids to landfill and brine to a POTW)

6. Hazardous waste facility

Contaminated brine streams may need pretreatment prior to discharge either to a receiving stream, sanitary sewer, or land disposal. In addition to the presence of arsenic in the brine stream, total dissolved solids (TDS) may cause a disposal problem by increasing the salinity in the receiving water (surface water or groundwater).

Another concern is the possibility that the contaminated brine stream may be classified as a hazardous waste. The determination of whether the wastestream is hazardous or not may depend on whether federal procedures or local procedures are used to make the hazardous waste determination. This determination may be a critical factor in both the selection of a treatment process and the operation of the facility. If a treatment residual is deemed hazardous, the cost of treating the hazardous waste may be prohibitive.

---

[10] *NSF.* NSF International is a noncommercial, not-for-profit organization concerned with public health safety and environmental protection. NSF Standard 60 lists certified drinking water chemicals and NSF Standard 61 lists certified drinking water system components.

[11] *Viscosity* (vis-KOSS-uh-tee). A property of water, or any other fluid, that resists efforts to change its shape or flow. Syrup is more viscous (has a higher viscosity) than water. The viscosity of water increases significantly as temperatures decrease. Motor oil is rated by how thick (viscous) it is; 20 weight oil is considered relatively thin while 50 weight oil is relatively thick or viscous.

> **QUESTIONS**
>
> Write your answers in a notebook and then compare your answers with those on page 154.
>
> 15.12A   What are the common stages of operation of an ion exchange unit?
>
> 15.12B   What are the safety issues of concern for operators of an ion exchange unit?
>
> 15.13A   What are the sources of waste liquids, sludges, and solids from an arsenic removal treatment plant?

15.14 MONITORING

15.140 Analysis of Arsenic

Currently, there are four US EPA-approved laboratory methods of analyzing for arsenic in drinking water at levels below the maximum contaminant level (MCL). These methods use mass spectrometry (MS) or atomic adsorption (AA). Most commercial water laboratories can do these analyses.

For arsenic removal process control, there are a number of commercial test kits available that can measure arsenic levels below the MCL. The accuracy and precision of these test kits vary and their reliability in producing data for controlling treatment operations must be verified regularly by comparing results of *SPLIT SAMPLES*[12] obtained by a certified/approved laboratory. The frequency of analyzing verification test samples should depend on experience with previous results.

Currently, there are no readily available continuous, on-line, arsenic measuring systems to monitor arsenic removal treatment processes. Potential indicator (surrogate) measurements such as pH or conductivity may provide an on-line indication of the process efficiency in removing arsenic.

Field test kits for arsenic determination are available from the following companies:

1. Hach Company, PO Box 389, Loveland, CO 80539-0389; phone, (800) 227-4224

2. Industrial Test Systems, Inc., 1875 Langston Street, Rock Hill, SC 29730; phone, (800) 861-9712

15.141 Types of Arsenic Sampling/Monitoring

There are several types of arsenic sampling/monitoring that are required.

1. Compliance: Compliance monitoring samples are the mandated samples that must be analyzed by an approved laboratory and reported to the regulatory agency. The results include all samples required and used to calculate conformance with the MCL.

2. Process Control: These samples are the essential samples needed on an ongoing basis to verify proper operation of the arsenic removal treatment plant. Many of theses samples are "mandated" and "reportable" to the regulatory agency as part of an approved operations plan.

3. Calibration/Verification: These samples are the essential samples taken to calibrate instruments, compare results with laboratory data, and verify settings on process controls.

4. Special: Additional samples may be needed to troubleshoot or investigate problems.

Arsenic at the relatively low levels likely to be encountered in drinking water sources does not pose an acute health effect, if the human exposure is short term (hours or days rather than years or a lifetime). The lack of acute health effects minimizes the need for continuous, on-line monitoring to prevent any instantaneous exceeding of the MCL or of a treatment target. This fact is recognized in the regulatory scheme for arsenic that allows the averaging of monitoring data over quarters for a year.

15.142 Monitoring for Compliance

Monitoring for determining compliance with the arsenic MCL is mandated by the US EPA regulation and is usually also spelled out by the state/province regulatory agency. The compliance monitoring is based on sampling each "point of entry" (POE) of water into the distribution system. This means that each source that pumps directly to the system will need to be sampled separately. Similarly, if multiple sources are mixed before their mutual POE, the sampling is performed on the mixed waters. Also, if one or more wells are treated before delivery to the system, the combined treated water will need to be monitored for compliance.

Each POE must be sampled initially once after the effective date (January 23, 2006) and then every three years for wells and every year for surface waters. However, if the initial sample exceeds the MCL, then quarterly monitoring must be conducted thereafter until the system is reliably and consistently below the MCL. The quarterly results are averaged (on a running annual average basis) to determine compliance with the MCL. If the MCL were exceeded, treatment requirements would be imposed on the water system.

Water utilities must do intensive monitoring to determine which sources exceed the arsenic MCL. If sources exceed the MCL, the water utilities will need to plan, fund, design, and install treatment facilities to remove arsenic or to blend waters to meet the MCL. With the approval and permitting of each water treatment plant, the monitoring for determining compliance with the arsenic MCL will be specified. Water utilities operating arsenic removal and blending treatment facilities have stringent requirements for meeting the arsenic MCL and some will have even lower operational targets on an ongoing basis.

---

[12] *Split Sample.* A single grab sample that is separated into at least two parts such that each part is representative of the original sample. Often used to compare test results between field kits and laboratories or between two laboratories.

15.143 Monitoring for Process Control

Water utilities should have a detailed operations plan and a monitoring program that will greatly minimize the likelihood of MCL exceedances occurring. Continuous monitoring of some indicators of arsenic treatment efficiency such as pH or conductivity may be desirable to provide better operational oversight and control.

Regular and frequent samples must be taken of raw water, plant effluent, and, if applicable, blended water. A minimum of daily grab samples from each source location may be required, at least initially. Test kits are useful for analyzing samples for arsenic, provided regular split-sample analyses are performed by a certified laboratory. A frequency of performing at least one split sample per week is typical. If the raw water is demonstrated to be very consistent, less frequent split samples may be acceptable. In most groundwater sources, arsenic levels do not vary dramatically either seasonally or over time, but regular monitoring is needed to document the lack of variation for each source.

Measuring electrical conductivity (EC), a surrogate for total dissolved solids (TDS), and pH can be very helpful process-control guidelines in processes such as ion exchange. However, operators must realize that EC cannot usually be correlated directly with arsenic reduction since other commonly present ions such as sulfate are also reflected by EC tests and are removed at different rates and preferences than arsenic.

Other water quality sampling and analyses may be needed depending on the specific site situation. These other analyses may include pH, sulfate, and bicarbonate. Control testing of the chlorination or other disinfection processes used is essential.

Special sampling of residuals and wastestreams is required for arsenic removal facilities. This special sampling may include determination of whether the waste is hazardous under federal and state/province regulatory definitions. Monitoring of waste discharges is required by water pollution control agencies or local public sewering agencies.

15.15 RECORDKEEPING AND REPORTING

Each water treatment plant should have an operations plan that will list the required records and reporting.

15.150 Records

Detailed plant operating records are essential to enable operators to effectively troubleshoot problems, evaluate long-term trends and subtle changes in treatment capacity or efficiency, assess operating costs, plan for major maintenance, and demonstrate compliance and treatment effectiveness. The records may be very important in the unlikely case of litigation against the water utility.

There are no standard recordkeeping forms that work for all water treatment plants; however, the following records need to be kept for all plants on some type of form.

1. Daily operations log: This information could be kept in a book in which the operator notes the times and types of inspections and also records any unusual events and information.

2. Daily inspection form: This form will provide a consistent, organized form for recording all meter readings, laboratory results, notations of valve operation, tank levels, and other important information. The form is essential to record critical information.

3. Monthly summary form: This form will provide an overview for management and may be a required regulatory report summarizing the overall operation and the actual performance of the plant.

4. Blending calculations record: This form simplifies the regular calculation of theoretical blends and enables the operator to compare and verify test results. Detailed records are required if engineered blending is used for arsenic compliance.

15.151 Reporting

There are at least four types of reports that are required in the operation of ion exchange arsenic removal water treatment plants. These types of reports are similar to the types of reports required for any of the arsenic removal treatment plants.

1. Routine: Monthly summary reports are generally required to be submitted to the regulatory agency. These reports are due by the tenth of the following month and must be signed by the responsible utility official. The signature carries with it the responsibility and liability for ensuring that the report is accurate. Annual summary reports also may be required by either management or the regulatory agency.

2. Special: These special reports include appropriate notification of other officials in the organization (other operators, superintendent, manager) of special or unusual conditions that may affect the overall system operation or to alert them of the need for materials, supplies, maintenance, or equipment replacement.

3. Violations: Regulatory law typically has provisions for special notification of the regulatory agency if a violation occurs. The nature and extent of the violation dictates the immediacy and method of the notification required. Often the operator will have taken the necessary corrective action immediately and no further action is required by the utility agency beyond making the required report.

4. Emergency: *NOTE:* Any failure of an arsenic removal treatment plant or exceeding the arsenic MCL probably will *NOT* necessitate an emergency warning. This is because the arsenic levels in the water supplied to the consumers probably will not reach the arsenic level that a short-term exposure would cause adverse health effects. However, if the water quality test results of the situation indicate an emergency, an emergency notice may be necessary. This could involve an emergency notice to all consumers using the most expeditious means available (radio, TV, or door-to-door notification). Usually, the decision for such a notice is made by management in cooperation with the regulatory agency. The initial report of an emergency by the operator must be done

without delay. In a small water system, the operator may have the responsibility, in the absence of others, to make the decision that there is an emergency and to initiate an immediate notice.

QUESTIONS

Write your answers in a notebook and then compare your answers with those on page 154.

15.14A How can an operator verify the precision and accuracy of commercial field arsenic test kits?

15.15A What types of records should be kept by the operator of an ion exchange arsenic removal water treatment plant?

15.15B What types of reports are required in the operation of an ion exchange arsenic removal water treatment plant?

15.16 ARITHMETIC ASSIGNMENT

Turn to the Arithmetic Appendix at the back of this manual. Read and work the problems in Section A.33, "Specialized Treatment Processes." Check the arithmetic in this section using an electronic calculator. You should be able to get the same answers.

15.17 ADDITIONAL READING

1. "Arsenic Treatment Technology Evaluation Handbook for Small Systems," US Environmental Protection Agency, Technical Report, July 2003. EPA No. 816-R-03-014. Obtain from National Technical Information Service (NTIS), 5285 Port Royal Road, Springfield, VA 22161. Order No. PB-2003-106891. Price, $47.50, plus $5.00 shipping and handling per order. Also available at www.epa.gov/safewater.

2. "Arsenic Removal from Drinking Water by Ion Exchange and Activated Alumina Plants," US Environmental Protection Agency, Technical Report, October 2000. EPA No. 600-R-00-088. Obtain from National Technical Information Service (NTIS), 5285 Port Royal Road, Springfield, VA 22161. Order No. PB-2001-104343. Price, $59.50, plus $5.00 shipping and handling per order.

END OF LESSON 2 OF 2 LESSONS

on

SPECIALIZED TREATMENT PROCESSES

Please answer the discussion and review questions next.

DISCUSSION AND REVIEW QUESTIONS

Chapter 15. SPECIALIZED TREATMENT PROCESSES

(Lesson 2 of 2 Lessons)

Write your answers to these questions in your notebook. The question numbering continues from Lesson 1.

10. What are the potential health effects of the consumption of low levels of arsenic over a long period of time?

11. What should a water utility do before considering treatment of a water source to reduce or remove arsenic?

12. Which issues related to blending must be considered before a blending program is implemented?

13. Which factors influence the procedures used to operate an ion exchange unit and the efficiency of the arsenic removal process?

14. Which items should an operator consider when reviewing the plans and specifications for an ion exchange unit to remove arsenic from drinking water?

15. What are the required types of arsenic sampling/monitoring?

16. Why are detailed arsenic removal treatment plant operating records essential?

SUGGESTED ANSWERS
Chapter 15. SPECIALIZED TREATMENT PROCESSES

ANSWERS TO QUESTIONS IN LESSON 1

Answers to questions on page 131.

15.0A   Trihalomethanes are formed in drinking water when free chlorine comes in contact with naturally occurring organic compounds (THM precursors). Trihalomethanes are a class of organic compounds in which there has been a replacement of the three hydrogen atoms in the methane molecule with three halogen atoms (chlorine or bromine).

15.0B   THM precursors are defined as natural, organic compounds found in all surface and groundwaters, which may react with halogens (such as chlorine) to form trihalomethanes (THMs); they must be present in order for THMs to form.

15.0C   Free chlorine is added to drinking water as a disinfectant.

15.0D   One source of bromide in drinking water is seawater.

Answers to questions on page 134.

15.1A   The major steps that a water utility could take to solve a THM problem include:

1. Determine the extent of the THM problem
2. Evaluate control strategies
3. Evaluate existing treatment processes
4. Examine studies of proposed treatment processes
5. Select a cost-effective option
6. Implement the chosen option

15.1B   Control strategies that could be evaluated to control a THM problem include:

1. Change sources of supply
2. Treatment options
   (a) Remove THMs
   (b) Remove precursors
   (c) Adjust or modify the chlorine application points
   (d) Use alternative disinfectants

15.2A   Important factors that influence the production of trihalomethanes include time, temperature, pH, and concentrations of chemicals.

15.2B   Those utilities that use lime softening should be aware that free chlorine in contact with natural organics at a pH of 10.5 or higher will produce THMs faster than if the pH was near 7.0.

Answers to questions on page 135.

15.3A   The two methods of controlling a THM problem are: (1) change the source of supply, or (2) provide some type of treatment.

15.3B   Abandoning the use of free chlorine is a serious move in view of its superior performance as a disinfectant.

15.4A   Water treatment processes that can be used to control THMs include:

1. Aeration
2. Oxidation with chlorine dioxide and potassium permanganate
3. Coagulation, sedimentation, and filtration
4. Softening processes
5. Powdered activated carbon and granular activated carbon applications

Answers to questions on page 136.

15.5A   Aeration is a much more effective process than oxidation for removing THMs after they have been formed.

15.5B   THM concentrations should be lower in waters that have been stored in uncovered reservoirs because of loss to the atmosphere.

15.5C   The adsorption process involves the individual THM compounds leaving the water and becoming attached to the surface of the carbon or resin.

Answers to questions on page 138.

15.5D   The major treatment processes that have been investigated to remove THM precursors before they come in contact with chlorine include:

1. Aeration
2. Oxidation (including ozone oxidation prior to coagulation and clarification)
3. Clarification
4. Adsorption
5. Ion exchange

15.5E Moving the addition of free chlorine to a point following the clarification process is the key to success when using clarification to remove THM precursors.

Answers to questions on page 139.

15.5F Before an alternative disinfectant is applied to any system, the source of the water supply, water quality, and treatment effectiveness for bacteriological control must be evaluated.

15.5G Upgraded monitoring (more samples and tests) of the distribution system before and after a disinfectant change must be provided by a utility.

15.5H Before and after a disinfectant change, the distribution system monitoring program should include coliforms, standard plate count, turbidity, and nutrients.

Answers to questions on page 139.

15.6A If existing water treatment processes are not capable of solving a THM problem, the least-cost solution will probably be an alternative disinfectant.

15.6B Apparently, the most popular alternative disinfectants are chloramines.

Answers to questions on page 140.

15.7A Before applying for and receiving a variance, utilities must evaluate the following treatment processes:

1. Use of chloramines or chlorine dioxide as an alternative or supplement to chlorine for oxidation and disinfection.
2. Use of chloramines, chlorine dioxide, or potassium permanganate as an alternative to chlorine for preoxidation.
3. Moving the point of chlorination in order to reduce THM formation.
4. Improvement of existing clarification.
5. Use of powdered activated carbon (PAC), intermittently, as necessary, to reduce TTHM or THM precursors. The dosage of PAC is not to exceed an annual average of 10 mg/L.

15.7B If treatment processes are not technically feasible and economically reasonable, then utilities must consider:

1. Off-line water storage
2. Aeration
3. Introduction of clarification
4. Alternative sources of raw water
5. Use of ozone as an alternative or supplement to chlorine for disinfection or oxidation

ANSWERS TO QUESTIONS IN LESSON 2

Answers to questions on page 142.

15.10A Potential sources of arsenic in drinking water include either natural occurrences (such as volcanic action, erosion of rock, or a result of forest fires) or from human activities.

15.10B The oxidation state of arsenic is very important when treating arsenic because the arsenic (III) form is not removed by common treatment processes.

Answers to questions on page 148.

15.11A The three general treatment categories used to remove arsenic from drinking water sources are adsorption, coprecipitation, and physical separation.

15.11B Blending of higher arsenic level sources with low arsenic level sources is an acceptable arsenic treatment option as long as the blending is properly engineered, operated, and controlled.

15.11C Sulfate and nitrate are both preferentially adsorbed instead of arsenic on an ion exchange resin.

15.11D Technologies used by POU and POE devices to remove arsenic from drinking water include activated alumina, iron-based sorbents, and reverse osmosis.

Answers to questions on page 150.

15.12A The common stages of operation of an ion exchange unit include: (1) service, (2) backwash, (3) brine, and (4) rinse.

15.12B Safety issues of concern for operators of ion exchange units include working with electrical systems and mechanical equipment, and the storage and handling of chemicals and water.

15.13A Sources of waste liquids, sludges, and solids from an arsenic removal treatment plant include:

1. Liquids: Backwash water, regenerant solutions (brines, caustic), and reject water
2. Sludges (semiliquid residuals): Sludge from coagulation/precipitation; and iron/manganese sludge (sludge may be subjected to added treatment to concentrate and dewater)
3. Solids: Spent filter/adsorption media, resins, and concentrated salts from evaporation ponds

Answers to questions on page 152.

15.14A Operators can verify the precision and accuracy of commercial field arsenic test kits by comparing results of split samples obtained by a certified/approved laboratory.

15.15A Types of records that should be kept by the operator of an ion exchange arsenic removal water treatment plant include a daily operations log, daily inspection form, monthly summary form, and a blending calculations record, if appropriate.

15.15B Types of reports that are required in the operation of an ion exchange arsenic removal water treatment plant include routine, special, violations, and emergency.

CHAPTER 16

MEMBRANE TREATMENT PROCESSES
(MEMBRANE FILTRATION AND DEMINERALIZATION)

by

Dave Argo

and

Ken Kerri

Revised by

Warren Casey

TABLE OF CONTENTS
Chapter 16. MEMBRANE TREATMENT PROCESSES
(MEMBRANE FILTRATION AND DEMINERALIZATION)

		Page
OBJECTIVES		160
WORDS		161

LESSON 1

			Page
16.0	MEMBRANE TREATMENT TECHNOLOGIES		165
16.1	DESCRIPTION OF MEMBRANE FILTRATION UNITS		166
	16.10	Pressure Vessels or Submerged Flow	166
	16.11	Membrane Flow Types	166
	16.12	Membrane Fouling	167
	16.13	Pretreatment	167
		16.130 Operation Experience	167
		16.131 Summary of Experiences	174
		16.132 Need for Bench- or Pilot-Scale Studies	174
16.2	OPERATION AND MAINTENANCE		174
	16.20	SCADA System	174
	16.21	Components of a Typical Plant	175
	16.22	Operational Procedures	177
	16.23	Membrane Performance Monitoring	177
	16.24	Troubleshooting Equipment and Process Failures	177
16.3	RECORDKEEPING		180
	16.30	Records	180
	16.31	Summary of Records	180
		16.310 Water Quality Monitoring Records	180
		16.311 Plant Operation and Water Production	181
		16.312 Troubleshooting Records	181
		16.313 Membrane Filter Maintenance and Inspection	181
16.4	MAINTENANCE PROGRAM		181
	16.40	Types of Maintenance	181
	16.41	Routine Maintenance	181

	16.42	Preventive Maintenance	181
	16.43	Breakdown or Corrective Maintenance	182
	16.44	Acknowledgment	182

DISCUSSION AND REVIEW QUESTIONS ... 182

LESSON 2

16.5	SOURCES OF MINERALIZED WATER	183
16.6	DEMINERALIZING PROCESSES	184
16.7	REVERSE OSMOSIS	186
	16.70 What Is Reverse Osmosis?	186
	16.71 Reverse Osmosis Membrane Structure and Composition	186
	16.72 Membrane Performance and Properties	188
	16.73 Definition of Flux	189
	16.74 Mineral Rejection	189
	16.75 Effects of Feedwater Temperature and pH on Membrane Performance	194
	16.76 Recovery	194

DISCUSSION AND REVIEW QUESTIONS ... 198

LESSON 3

16.8	COMPONENTS OF A REVERSE OSMOSIS UNIT	199
	16.80 Pressurization Pump	199
	16.81 Piping	199
	16.82 Pressure Vessel Housings	199
	16.83 Concentrate Control Valve	199
	16.84 Sample Valves	202
	16.85 Flush Connections	202
	16.86 Cleaning Connections	202
	16.87 Permeate Rinse	202
	16.88 Permeate Drawback Tank	202
	16.89 Energy Recovery Devices	202
	16.810 Membranes	202
16.9	OPERATION	205
	16.90 Pretreatment	205
	16.91 Removal of Turbidity and Suspended Solids	205
	16.92 pH and Temperature Control	206
	16.93 Other Potential Scalants	206
	16.94 Microbiological Organisms	206
	16.95 RO Plant Operation	207
	16.96 Typical RO Plant Operations Checklist	207

		16.97	Membrane Cleaning	211
		16.98	Safety	211
			16.980 Use of Proper Procedures	211
			16.981 Chemical Safety	211
			16.982 Hydraulic Safety	212
			16.983 Electrical Safety	212

DISCUSSION AND REVIEW QUESTIONS 212

LESSON 4

16.10 ELECTRODIALYSIS 213

16.11 PRINCIPLES OF ELECTRODIALYSIS 214

    16.110 Anions and Cations in Water 214

    16.111 Effect of Direct Current (DC) Potential on Ions 215

    16.112 Anion- and Cation-Permeable Membranes and Three-Cell Unit 215

    16.113 Multicompartment Unit 215

16.12 PARTS OF AN ELECTRODIALYSIS UNIT 215

    16.120 Flow Diagram 215

    16.121 Pretreatment 215

    16.122 Pumping Equipment and Piping 220

    16.123 DC Power Supply 220

    16.124 Membrane Stack 220

    16.125 Chemical Flush System 220

16.13 ROUTINE OPERATING PROCEDURES 220

    16.130 Design Specifications for Feedwater 220

    16.131 Detailed Operating Procedures 221

16.14 SAFETY PRECAUTIONS 221

16.15 ARITHMETIC ASSIGNMENT 223

16.16 ADDITIONAL READING 223

DISCUSSION AND REVIEW QUESTIONS 223

SUGGESTED ANSWERS 224

OBJECTIVES

Chapter 16. MEMBRANE TREATMENT PROCESSES (MEMBRANE FILTRATION AND DEMINERALIZATION)

Following completion of Chapter 16, you should be able to:

1. Describe various membrane technologies used to treat water.
2. Explain various types of membrane filtration processes.
3. Evaluate potential membrane pretreatment alternatives.
4. Operate and maintain a membrane filtration treatment plant.
5. Keep accurate and appropriate records for a membrane filtration plant.
6. Develop, implement, and maintain a membrane filtration plant maintenance program.
7. Describe the various demineralizing processes.
8. Explain how the reverse osmosis process works.
9. Operate and maintain a reverse osmosis demineralization plant.
10. Explain the principles of electrodialysis.
11. Identify and describe the parts of an electrodialysis plant.
12. Operate and maintain an electrodialysis plant.
13. Safely perform your duties around reverse osmosis and electrodialysis plants.

WORDS

Chapter 16. MEMBRANE TREATMENT PROCESSES
(MEMBRANE FILTRATION AND DEMINERALIZATION)

atm
The abbreviation for atmosphere. One atmosphere is equal to 14.7 psi or 100 kPa.

ANGSTROM (ANG-strem)
A unit of length equal to one-tenth of a nanometer or one-tenbillionth of a meter (1 Angstrom = 0.000 000 000 1 meter). One Angstrom is the approximate diameter of an atom.

BENCH-SCALE ANALYSIS (TEST)
A method of studying different ways or chemical doses for treating water or wastewater and solids on a small scale in a laboratory. Also see JAR TEST.

CHELATION (key-LAY-shun)
A chemical complexing (forming or joining together) of metallic cations (such as copper) with certain organic compounds, such as EDTA (ethylene diamine tetracetic acid). Chelation is used to prevent the precipitation of metals (copper). Also see SEQUESTRATION.

COLLOIDS (KALL-loids)
Very small, finely divided solids (particles that do not dissolve) that remain dispersed in a liquid for a long time due to their small size and electrical charge. When most of the particles in water have a negative electrical charge, they tend to repel each other. This repulsion prevents the particles from clumping together, becoming heavier, and settling out.

CONCENTRATION POLARIZATION
(1) A buildup of retained particles on the membrane surface due to dewatering of the feed closest to the membrane. The thickness of the concentration polarization layer is controlled by the flow velocity across the membrane.
(2) Used in corrosion studies to indicate a depletion of ions near an electrode.
(3) The basis for chemical analysis by a polarograph.

CONDUCTANCE
A rapid method of estimating the dissolved solids content of a water supply. The measurement indicates the capacity of a sample of water to carry an electric current, which is related to the concentration of ionized substances in the water. Also called SPECIFIC CONDUCTANCE.

CROSS-FLOW FILTRATION
A type of membrane filtration where the water being filtered flows across the surface of the membrane to keep the particle buildup and fouling to a minimum. The flow that is not filtered becomes concentrated and flows out the end of the membrane fiber as a wastestream.

DEAD-END FILTRATION
A type of membrane filtration where the water being filtered flows through the membrane, but there is no wastestream from the system. All solids accumulate on the membrane during filtration and are removed during backwash.

DEMINERALIZATION (DEE-min-er-al-uh-ZAY-shun)
A treatment process that removes dissolved minerals (salts) from water.

DISINFECTION BY-PRODUCT (DBP)

A contaminant formed by the reaction of disinfection chemicals (such as chlorine) with other substances in the water being disinfected.

DISSOLVED ORGANIC CARBON (DOC)

That portion of the organic carbon in water that passes through a 0.45 μm pore-diameter filter.

DISSOLVED ORGANIC MATTER (DOM)

That portion of the organic matter in water that passes through a 0.45 μm pore-diameter filter.

ENZYMES (EN-zimes)

Organic or biochemical substances that cause or speed up chemical reactions.

FEEDWATER

The water that is fed to a treatment process; the water that is going to be treated.

FLUX

A flowing or flow.

FULVIC ACID

A complex organic compound that can be derived either from soil or water. Aquatic fulvic acids are major precursors of disinfection by-products.

HUMIC ACID

A polymeric constituent of soils, lignite, and peat. Aquatic humic acids are major precursors of disinfection by-products.

HUMIC SUBSTANCES

Natural organic matter resulting from partial decomposition of plant or animal matter and forming the organic portion of soil.

HYDROLYSIS (hi-DROLL-uh-sis)

(1) A chemical reaction in which a compound is converted into another compound by taking up water.
(2) Usually a chemical degradation of organic matter.

MATERIAL SAFETY DATA SHEET (MSDS)

A document that provides pertinent information and a profile of a particular hazardous substance or mixture. An MSDS is normally developed by the manufacturer or formulator of the hazardous substance or mixture. The MSDS is required to be made available to employees and operators or inspectors whenever there is the likelihood of the hazardous substance or mixture being introduced into the workplace. Some manufacturers are preparing MSDSs for products that are not considered to be hazardous to show that the product or substance is not hazardous.

MEMBRANE FOULING

The cause of a loss of flow through the membrane as a result of material being retained on the surface of the membrane or within the membrane pores. Membrane fouling may be reversible or irreversible. Loss of flow through the membrane by reversible fouling can be recovered by regular backwashing of the membrane surface. Flow loss by irreversible fouling cannot be recovered.

MICROFILTRATION (MF)

A pressure-driven membrane filtration process that separates particles down to approximately 0.1 μm diameter from influent water using a sieving process.

NANOFILTRATION (NF)

A pressure-driven membrane filtration process that separates particles down to approximately 0.002 to 0.005 μm diameter from influent water using a sieving process.

NATURAL ORGANIC MATTER (NOM)

Humic substances composed of humic and fulvic acids that come from decayed vegetation.

OSMOSIS (oz-MOE-sis)

The passage of a liquid from a weak solution to a more concentrated solution across a semipermeable membrane. The membrane allows the passage of the water (solvent) but not the dissolved solids (solutes). This process tends to equalize the conditions on either side of the membrane.

PERMEATE (PURR-me-ate)

(1) To penetrate and pass through, as water penetrates and passes through soil and other porous materials.

(2) The liquid (demineralized water) produced from the reverse osmosis process that contains a low concentration of dissolved solids.

PILOT-SCALE STUDY

A method of studying different ways of treating water or wastewater and solids or to obtain design criteria on a small scale in the field.

POINT-OF-USE/POINT-OF-ENTRY (POU/POE)

Point-of-use applications refer to a water treatment device that treats water at the location of the end user. Point-of-entry application refers to a water treatment device that is located at the inlet to an entire building or facility.

REVERSE OSMOSIS (oz-MOE-sis) (RO)

The application of pressure to a concentrated solution, which causes the passage of a liquid from the concentrated solution to a weaker solution across a semipermeable membrane. The membrane allows the passage of the water (solvent) but not the dissolved solids (solutes). In the reverse osmosis process, two liquids are produced: (1) the reject (containing high concentrations of dissolved solids), and (2) the permeate (containing low concentrations). The clean water (permeate) is not always considered to be demineralized. Also see OSMOSIS.

SCADA (SKAY-dah) SYSTEM

Supervisory Control And Data Acquisition system. A computer-monitored alarm, response, control, and data acquisition system used to monitor and adjust treatment processes and facilities.

SALINITY

(1) The relative concentration of dissolved salts, usually sodium chloride, in a given water.

(2) A measure of the concentration of dissolved mineral substances in water.

SEQUESTRATION (SEE-kwes-TRAY-shun)

A chemical complexing (forming or joining together) of metallic cations (such as iron) with certain inorganic compounds, such as phosphate. Sequestration prevents the precipitation of the metals (iron). Also see CHELATION.

SPECIFIC CONDUCTANCE

A rapid method of estimating the total dissolved solids content of a water supply. The measurement indicates the capacity of a sample of water to carry an electric current, which is related to the concentration of ionized substances in the water. Also called CONDUCTANCE.

TOTAL DISSOLVED SOLIDS (TDS)

All of the dissolved solids in a water. TDS is measured on a sample of water that has passed through a very fine mesh filter to remove suspended solids. The water passing through the filter is evaporated and the residue represents the total dissolved solids. Also see SPECIFIC CONDUCTANCE.

ULTRAFILTRATION (UF)

A pressure-driven membrane filtration process that separates particles down to approximately 0.01 μm diameter from influent water using a sieving process.

CHAPTER 16. MEMBRANE TREATMENT PROCESSES
(MEMBRANE FILTRATION AND DEMINERALIZATION)

(Lesson 1 of 4 Lessons)

16.0 MEMBRANE TREATMENT TECHNOLOGIES

Membrane treatment technologies are advancing rapidly in the water treatment field. Membranes are being used in municipal water treatment plants, in home *POINT-OF-USE OR POINT-OF-ENTRY (POU/POE)*[1] applications, in reclamation facilities, and wastewater treatment plants to remove both suspended and dissolved minerals from water.

Membranes contain very fine pore openings that allow water to pass through and block the passage of any contaminant larger than the pore diameter. Membranes used in water treatment are classified by their pore diameter. Classifications, from largest pore diameters to smallest, are *MICROFILTRATION (MF)*,[2] *ULTRAFILTRATION (UF)*,[3] *NANOFILTRATION (NF)*,[4] and *REVERSE OSMOSIS (RO)*.[5] The size ranges for each classification are presented in Figure 16.1, along with the relative sizes of materials commonly present in water.

Microfiltration (MF) and ultrafiltration (UF) are used in water treatment for particle, sediment, algae, bacteria, and virus removal. MF and UF are effective in the removal of *Giardia* and *Cryptosporidium*. Reverse osmosis (RO) membranes are used for desalination/demineralization and in home drinking water units. Reverse osmosis (RO) and nanofiltration (NF) membranes are used to remove *DISSOLVED ORGANIC MATTER (DOM)*[6] and dissolved contaminants such as arsenic, nitrate, pesticides, and radionuclides. Also, these membranes can remove ions such as calcium and magnesium (for softening) and also sodium and chloride (for deionization, desalination, or demineralization of brackish waters). NF can be used to reduce the concentration of *NATURAL ORGANIC MATTER (NOM)*[7] to control the formation of *DISINFECTION BY-PRODUCTS (DBPs)*.[8]

The type of membrane used depends on the constituents to be removed from the water being treated. During water treatment, water is typically pumped against the surface of the membrane, but may be pulled through the membrane by a vacuum (submerged membrane). The water pressure forces water through the membrane and the constituents that do not pass through form a wastestream that may require treatment and proper disposal.

> ### QUESTIONS
> Write your answers in a notebook and then compare your answers with those on page 224.
>
> 16.0A Microfiltration (MF) and ultrafiltration (UF) are used in water treatment for the removal of which constituents?
>
> 16.0B Reverse osmosis (RO) and nanofiltration (NF) membranes are used to remove which constituents from water?

---

[1] *Point-Of-Use/Point-Of-Entry (POU/POE).* Point-of-use applications refer to a water treatment device that treats water at the location of the end user. Point-of-entry application refers to a water treatment device that is located at the inlet to an entire building or facility.

[2] *Microfiltration (MF).* A pressure-driven membrane filtration process that separates particles down to approximately 0.1 μm diameter from influent water using a sieving process.

[3] *Ultrafiltration (UF).* A pressure-driven membrane filtration process that separates particles down to approximately 0.01 μm diameter from influent water using a sieving process.

[4] *Nanofiltration (NF).* A pressure-driven membrane filtration process that separates particles down to approximately 0.002 to 0.005 μm diameter from influent water using a sieving process.

[5] *Reverse Osmosis* (oz-MOE-sis) *(RO).* The application of pressure to a concentrated solution, which causes the passage of a liquid from the concentrated solution to a weaker solution across a semipermeable membrane. The membrane allows the passage of the water (solvent) but not the dissolved solids (solutes). In the reverse osmosis process, two liquids are produced: (1) the reject (containing high concentrations of dissolved solids), and (2) the permeate (containing low concentrations). The clean water (permeate) is not always considered to be demineralized. Also see OSMOSIS.

[6] *Dissolved Organic Carbon (DOC).* That portion of the organic carbon in water that passes through a 0.45 μm pore-diameter filter.

[7] *Natural Organic Matter (NOM).* Humic substances composed of humic and fulvic acids that come from decayed vegetation.

[8] *Disinfection By-Product (DBP).* A contaminant formed by the reaction of disinfection chemicals (such as chlorine) with other substances in the water being disinfected.

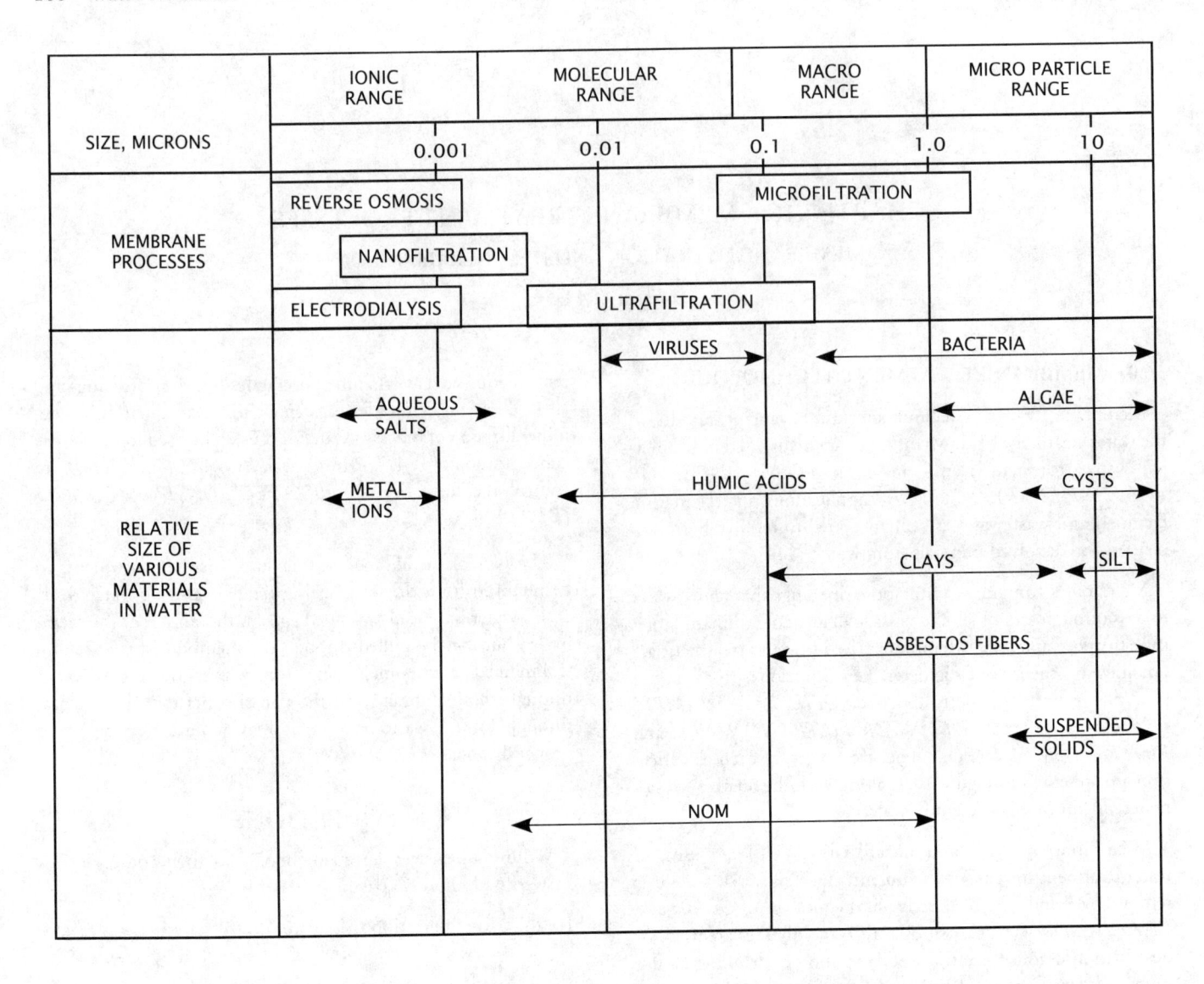

Fig. 16.1 Size ranges of membrane processes and contaminant
(Adapted from AWWA, *WATER QUALITY AND TREATMENT: A HANDBOOK OF COMMUNITY WATER SUPPLIES*, 5th edition, McGraw-Hill, 1999)

16.1 DESCRIPTION OF MEMBRANE FILTRATION UNITS

16.10 Pressure Vessels or Submerged Flow

Typical water treatment membrane filtration units are installed either in pressure vessels or submerged in tanks. The membranes are hollow fibers or threads (Figures 16.2 and 16.3) with an outside diameter ranging from 0.5 to 2 mm and a wall thickness (membrane thickness) of 0.07 to 0.6 mm. In the pressure vessel installation, thousands of membrane fibers or threads (Figure 16.3) are arranged in racks or skids, with each pressure vessel from 4 to 12 inches (100 to 300 mm) in diameter and 3 to 18 feet (0.9 to 5.5 m) long (Figure 16.4). Filtration flow may be from outside to inside or from inside to outside of the hollow fiber membrane (Figures 16.5, 16.6, and 16.7).

Submerged or immersed membrane filtration systems (Figures 16.8 and 16.9) are membrane modules suspended in basins containing the water to be treated. The treated water may be pulled through the membrane by vacuum. Some treatment plants have removed sand from the sand filters and installed submerged membrane modules in the old filter basin.

16.11 Membrane Flow Types

There are two different types of membrane filtration feedwater flows: the *CROSS-FLOW FILTRATION*[9] system (Figure 16.7)

---

[9] *Cross-Flow Filtration.* A type of membrane filtration where the water being filtered flows across the surface of the membrane to keep the particle buildup and fouling to a minimum. The flow that is not filtered becomes concentrated and flows out the end of the membrane fiber as a wastestream.

Fig. 16.2   One hollow fiber membrane

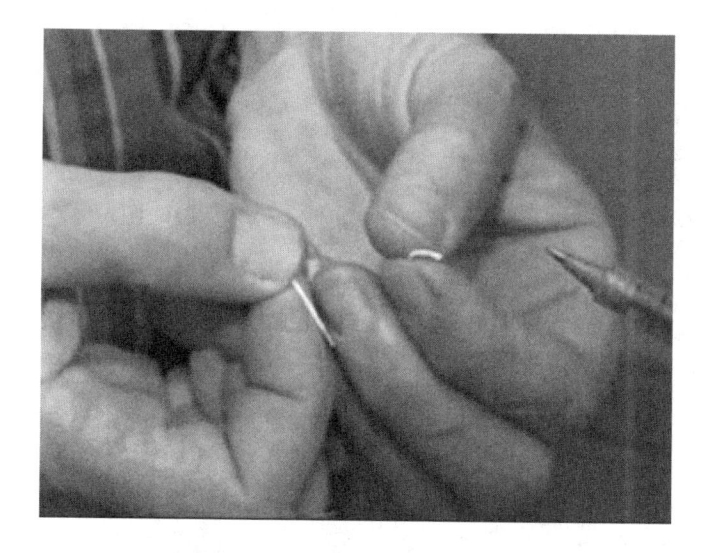

Fig. 16.3   Broken membrane tube with fibers showing

and the DEAD-END FILTRATION[10] system (Figure 16.6). In the cross-flow filtration system, the flow is from the inside of the membrane, through the membrane, and the filtered water flows out of the system. The flow inside the membrane flows along the inside surface of the membrane, becomes concentrated, and flows out of the end of the membrane fiber as a wastestream. In the dead-end filtration system, the water being filtered may flow either from the outside into the hollow fiber or from the inside to the outside, but there is no wastestream from the system. All solids accumulate on the membrane during filtration and are removed during backwash.

16.12   Membrane Fouling[11]

Membrane fouling can be a serious problem when operating membrane filtration processes. Membrane fouling can be described by whether the cause of the fouling can be removed (reversible or irreversible), by the material causing the fouling (biological, organic, particulate, or dissolved), and by the means of fouling (cake formation or membrane pore blockage).

Whether the cause of the fouling can be removed (reversible or irreversible) depends on the type of membrane used and the constituents in the source water. During continued operation, the flow through the membrane may decrease, but the flow may be recovered by backwashing and cleaning.

The constituents in the water being filtered may cause fouling. During membrane filtration, microorganisms are transported to the membrane surface where biofouling may occur.

These microorganisms may not be removed by backwashing, but may be controlled by chlorine. However, some membrane materials (cellulose acetate (CA) or polypropylene (PP)) may be damaged by chlorine. Membrane manufacturers are now tending to use materials that are not damaged by chlorine.

Dissolved organic matter (DOM) may cause membrane fouling. The extent of the fouling problem depends on the characteristics of the dissolved organic matter, the membrane material, and the characteristics of the water being filtered.

During membrane filtration, particulate matter from the water being filtered collects on the membrane surface in a porous mat called a filter cake. Particulate fouling is usually reversible during periodic backwashing.

Natural organic matter (NOM) can be the most common form of membrane fouling. Dissolved organic matter (DOM) includes wastes and portions of aquatic plants and animals as well as organic matter washed into surface water from land. Sources of DOM include organic chemicals found in biological systems and also dissolved organic chemicals from industrial and commercial wastes. Fouling depends on the characteristics of the DOM, the membrane material, and the source water.

16.13   Pretreatment

16.130   Operation Experience

When operators first started using membrane filtration (microfiltration (MF) and ultrafiltration (UF)), most water

---

[10] *Dead-End Filtration.*   A type of membrane filtration where the water being filtered flows through the membrane, but there is no wastestream from the system. All solids accumulate on the membrane during filtration and are removed during backwash.

[11] *Membrane Fouling.*   The cause of a loss of flow through the membrane as a result of material being retained on the surface of the membrane or within the membrane pores. Membrane fouling may be reversible or irreversible. Loss of flow through the membrane by reversible fouling can be recovered by regular backwashing of the membrane surface. Flow loss by irreversible fouling cannot be recovered.

168 Water Treatment

Fig. 16.4   Racks of membrane pressure vessels

Membrane Treatment Processes 169

Fig. 16.5 Membrane filtration—outside into membrane flow

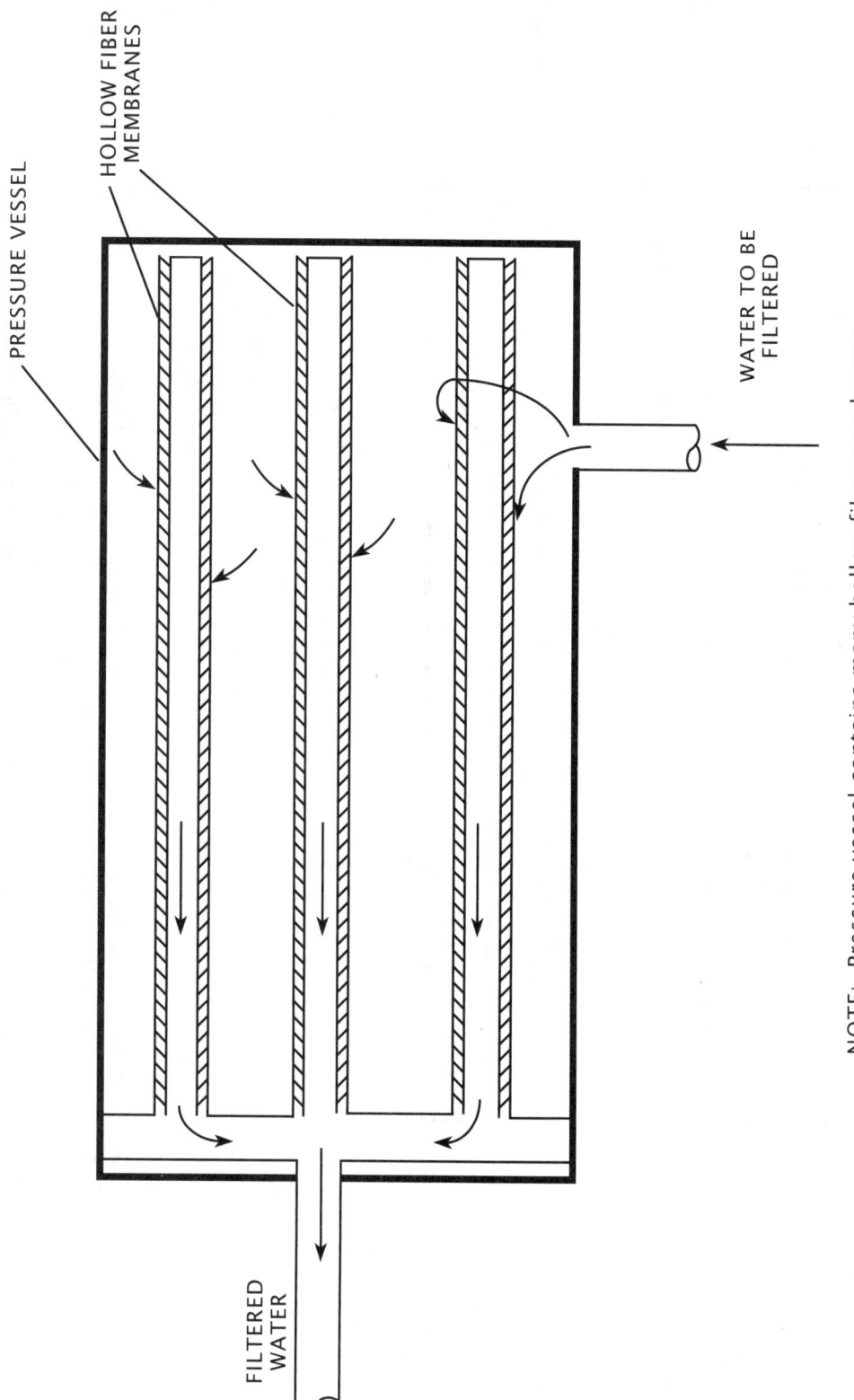

NOTE: Pressure vessel contains many hollow fiber membranes packed closely together.

170 Water Treatment

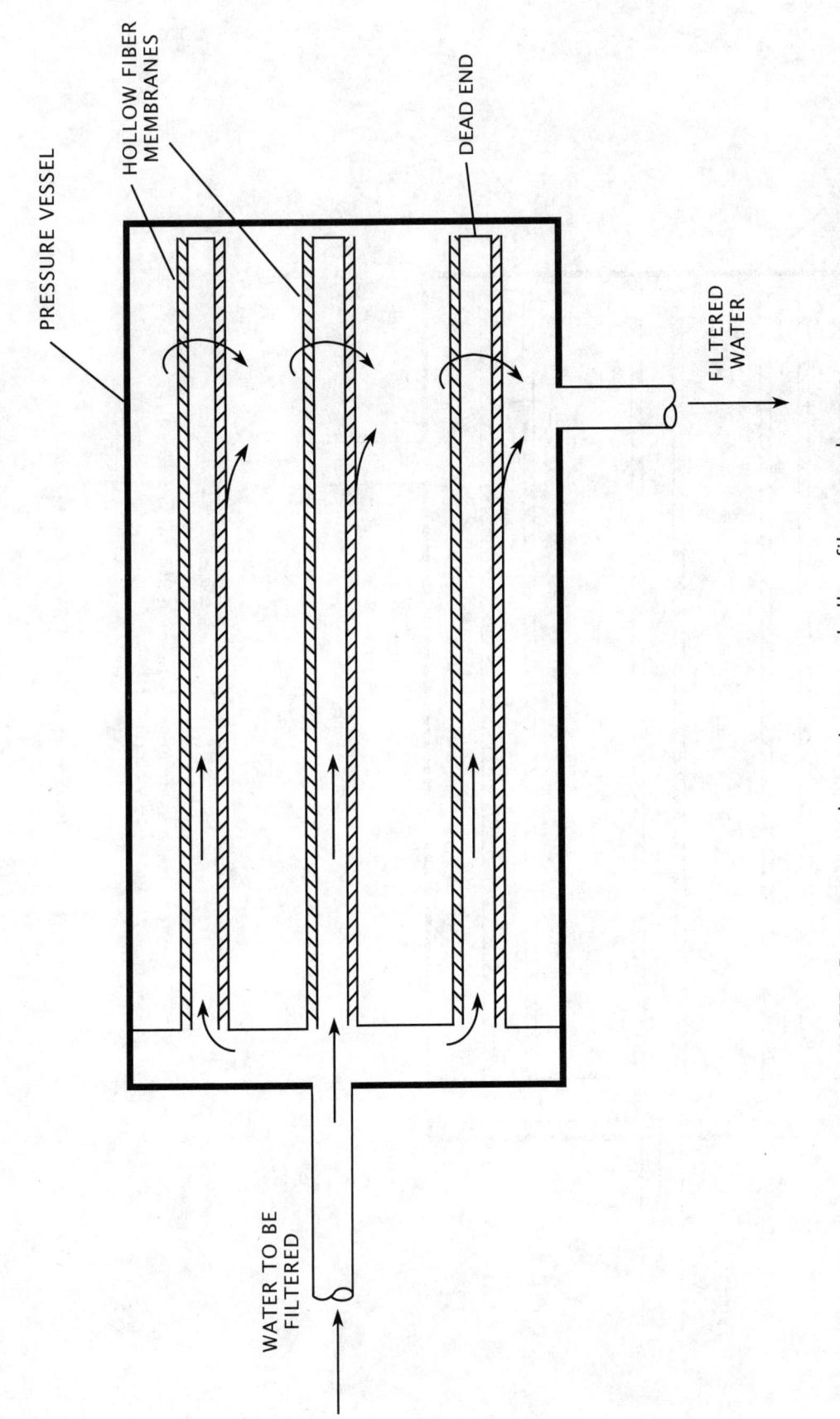

NOTE: Pressure vessel contains many hollow fiber membranes packed closely together.

Fig. 16.6 Dead-end membrane filtration—inside membrane flow to outside of membrane

Membrane Treatment Processes 171

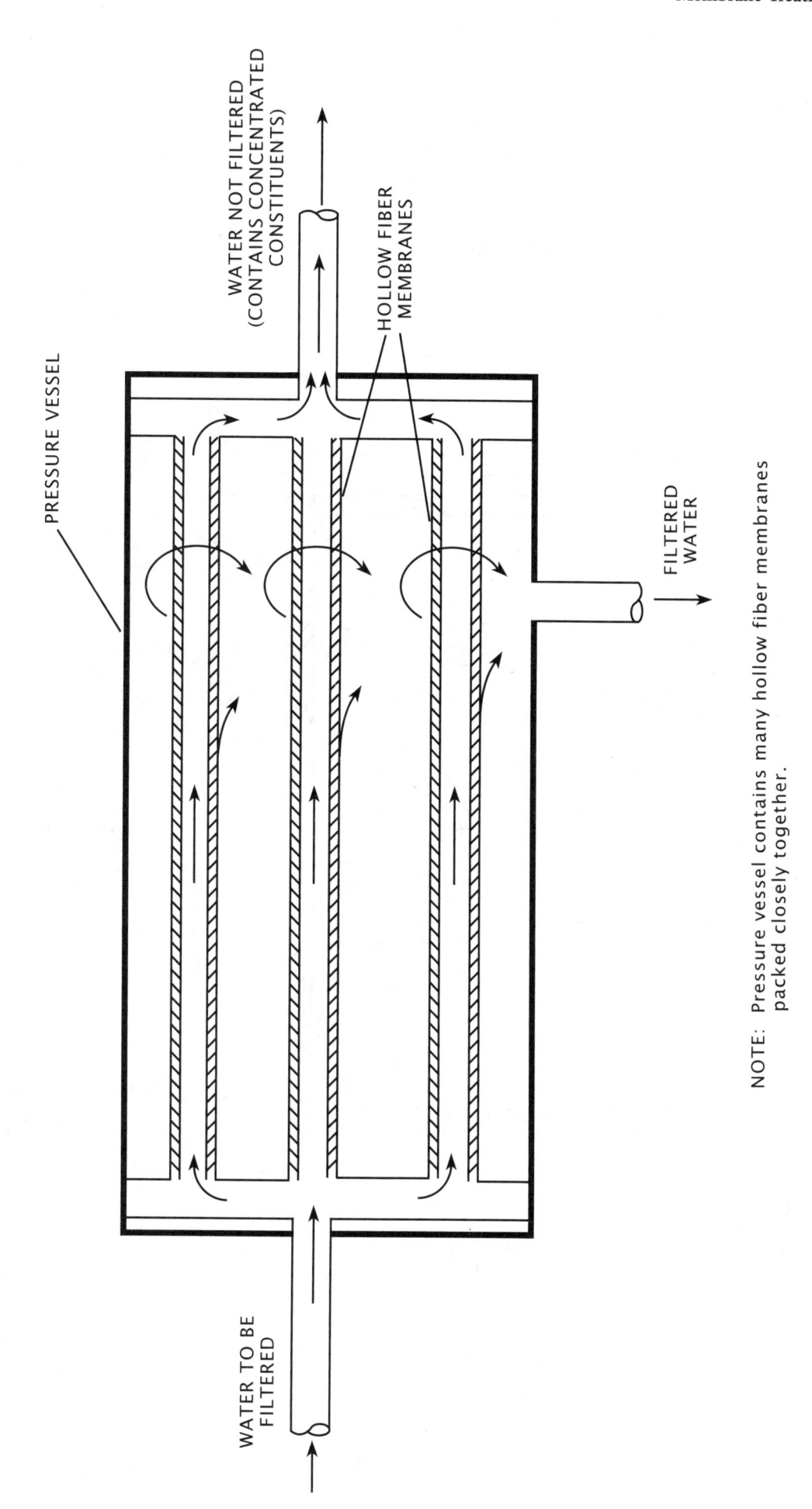

Fig. 16.7 *Cross-flow membrane filtration—inside membrane flow to outside of membrane, across membrane surface*

NOTE: Pressure vessel contains many hollow fiber membranes packed closely together.

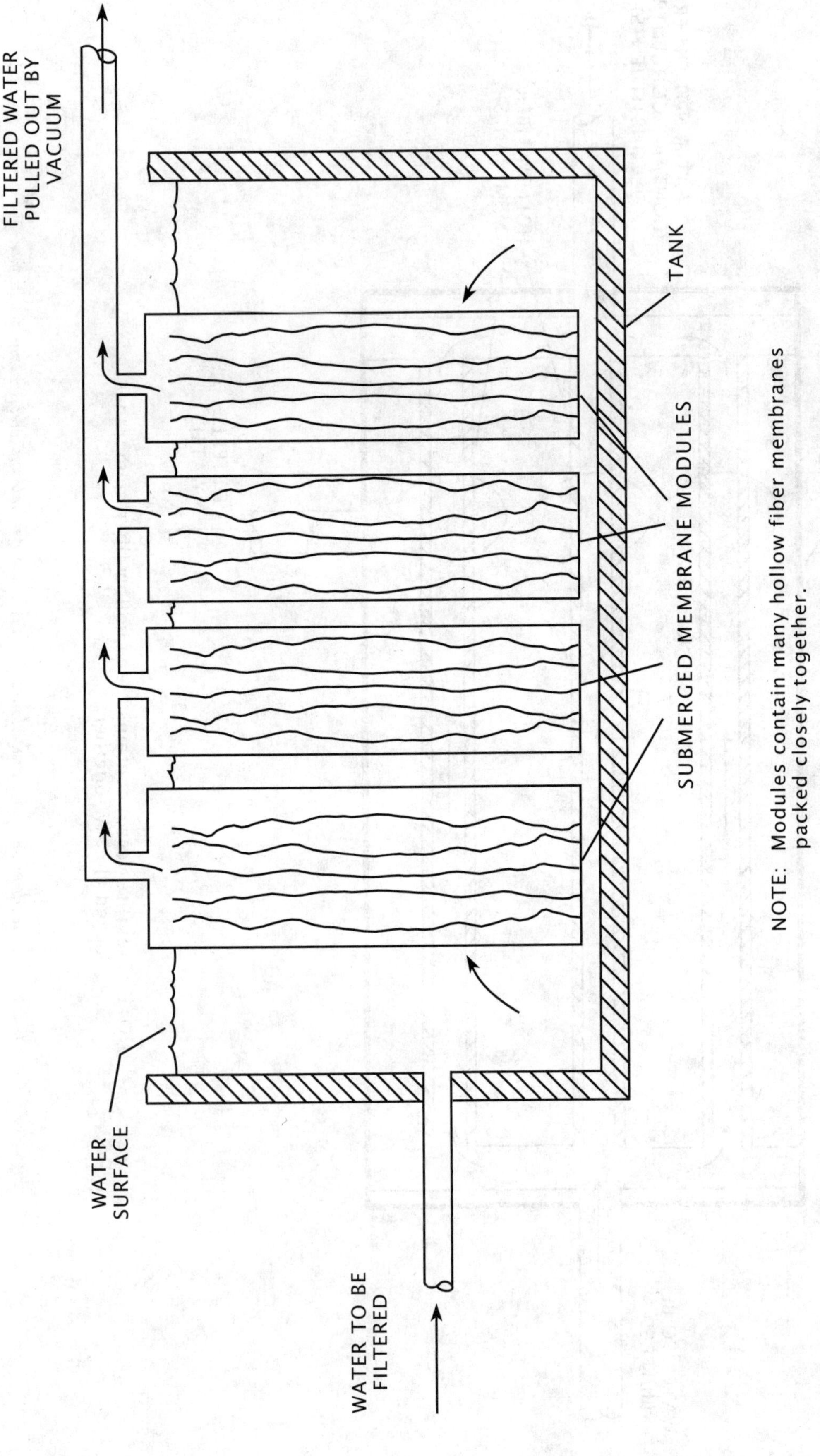
FILTERED WATER
PULLED OUT BY
VACUUM
TANK
SUBMERGED MEMBRANE MODULES
WATER
SURFACE
WATER TO BE
FILTERED
NOTE: Modules contain many hollow fiber membranes
packed closely together.

Fig. 16.8 Schematic of submerged membrane filtration system

Fig. 16.9 Submerged membrane filtration system

sources used were of high-quality, low-turbidity surface waters that required minimum treatment. Membrane filtration effectively removed any turbidity present in the water. The effectiveness of membrane filtration in treating low-turbidity water encouraged operators to consider using coagulation pretreatment before membrane filtration.

Operators have learned that the interactions among coagulants, the various constituents in natural waters, and the membrane materials are very complex, making the effect of coagulation on membrane performance difficult to predict. In some treatment plants, the flow through the membrane increases with coagulation pretreatment and in other plants there is a reduction in flow.

When using coagulation pretreatment before membrane filtration, operators have experienced all of the following, sometimes contradictory, situations:

- When using river water, cellulose acetate (CA) ultrafiltration membranes fouled in two days, but pretreatment with 21 mg/L of ferric chloride essentially eliminated fouling.

- After several months of operation, a plant using lake water and cellulose acetate (CA) ultrafiltration membranes experienced rapid fouling during a period of lake turnover. Pretreatment with 14 mg/L ferric chloride restored the system to normal operation.

- Some water treatment plants have experienced improved membrane performance (flows through the membranes) after pretreatment with alum or ferric salts.

- Coagulation with either ferric chloride or a commercial polyaluminum coagulant slowed short-term reversible fouling, but did not reduce the rate or extent of irreversible fouling.

- Coagulation can be effective for removing particulate matter, but may have no effect on fouling by *DISSOLVED ORGANIC CARBON (DOC)*.[12]

- Some plants have found that coagulation makes fouling worse.

- Coagulation with ferric chloride significantly increased fouling of microfiltration and ultrafiltration membranes during

---

[12] *Dissolved Organic Carbon (DOC).* That portion of the organic carbon in water that passes through a 0.45 μm pore-diameter filter.

filtration of river water containing humic and fulvic acid solutions.

- Low alum doses increased the rate of fouling, but high doses decreased the fouling of 10 mg/L solutions of fulvic acids from swamp waters filtered through polysulfone membranes. Also, a similar reduction in membrane performance was observed at low ferric chloride doses, but overall membrane performance improved at higher doses.

Membrane configuration (layout), membrane materials, and coagulation conditions may also affect whether coagulation improves or decreases membrane performance:

- Coagulation pretreatment caused a decline in performance of a pressurized membrane system, but an improvement in performance of a submerged membrane system.
- Differences in membrane performance can be caused by changes in cake resistance. A reduction in fouling can be obtained by the removal of dissolved organic carbon (DOC).

16.131 Summary of Experiences

A summary of these operator experiences indicates that pretreatment coagulation before membrane filtration has produced inconsistent results regarding membrane fouling and a decline of flows through the membrane. These inconsistencies apparently result from the fact that both coagulation and membrane filtration may be performed under a wide variety of operating conditions. For example, coagulation may be performed with or without flocculation and with or without sedimentation before membrane filtration. Membrane filtration may be performed with submerged or pressurized membranes and with constant pressure or constant flow conditions. Studies conducted on a specific source water with an individual coagulant and a set coagulant dose cannot be easily compared with other source waters, coagulants, and doses.

Another important consideration in the relationship between coagulation and membrane performance is that coagulation affects both particulate matter and dissolved organic carbon (DOC), both of which can affect membrane performance. Coagulation collects particles into layer masses and, if settling is practiced, removes particles from solution, which may alter membrane fouling because of cake resistance. Coagulation also removes DOC from the water, which may alter membrane fouling caused by adsorption.

16.132 Need for Bench- or Pilot-Scale Studies[13, 14]

In view of these conflicting experiences with pretreatment and membrane filtration, what can an operator do to optimize membrane performance? A suggested approach is to work with the membrane manufacturers. Invite the manufacturers to your facility and ask them to perform bench- or pilot-scale studies. Indicate that the results of the studies should provide you with the following information:

1. The selection of the most appropriate coagulant for your specific application. This selection is dependent on interactions between the coagulant, source water, and membrane material.

2. The selection of the most appropriate coagulant dose. Coagulant dose is an important factor in determining whether membrane performance improves or decreases. For example, at low alum doses, membrane fouling is frequently worse than with no coagulant, but at the dose for enhanced coagulation, membrane performance is usually better.

3. The coagulant selection should include consideration of removing DOC. The coagulant type (for example, alum, ferric sulfate, or PACL) that tends to be most effective at removing DOC in a particular source water is typically the most effective at improving membrane performance.

4. A cost comparison for each alternative investigated.

> ## QUESTIONS
>
> Write your answers in a notebook and then compare your answers with those on page 224.
>
> 16.1A Where are typical water treatment membrane filtration units installed?
>
> 16.1B How can a decrease in flow through a membrane caused by fouling be recovered?
>
> 16.1C Why do operators consider using coagulation pretreatment before membrane filtration?
>
> 16.1D Why should bench- or pilot-scale studies be conducted to optimize membrane performance?

16.2 OPERATION AND MAINTENANCE

16.20 SCADA System[15] (Figures 16.10 and 16.11)

The operation of membrane filtration water treatment processes is typically almost fully automated. These systems usually have a Supervisory Control And Data Acquisition (SCADA) system and are designed to run unattended. The SCADA system is used to monitor operational conditions and to make process adjustments within selected set points. SCADA monitors alarm conditions and follows protocols for alarm notification and plant shutdown.

Typically, operators maintain a laptop computer equipped with full plant operation compatibility using a dial-in connection. When the plant is unattended, the on-call operator can use

---

[13] *Bench-Scale Analysis (Test).* A method of studying different ways or chemical doses for treating water or wastewater and solids on a small scale in a laboratory. Also see JAR TEST.

[14] *Pilot-Scale Study.* A method of studying different ways of treating water or wastewater and solids or to obtain design criteria on a small scale in the field.

[15] *SCADA* (SKAY-dah) *System.* Supervisory Control And Data Acquisition system. A computer-monitored alarm, response, control, and data acquisition system used to monitor and adjust treatment processes and facilities.

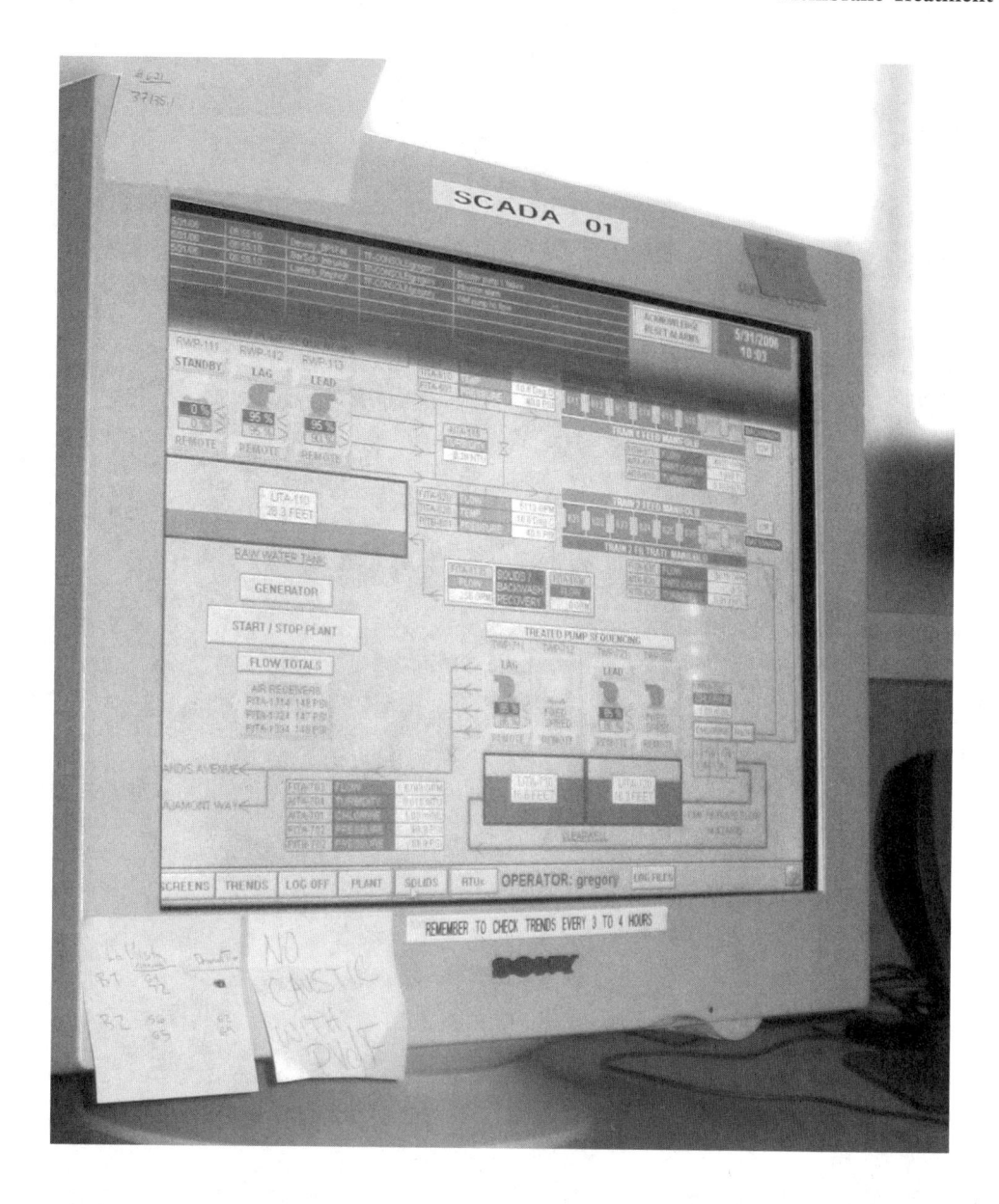

Fig. 16.10 SCADA system (plant control room)

a laptop computer to confirm plant process performance and make process and equipment adjustments prior to returning to the plant.

At the beginning and end of a shift, the operator should review the SCADA information and walk through the water treatment plant. The plant walk-through includes looking and listening for anything unusual.

16.21 Components of a Typical Plant

The operation and maintenance of a typical membrane filtration water treatment plant requires the operator to be familiar with the purposes of the following units.

1. Raw water source. The raw water sources for most membrane plants are surface waters, including high-quality streams and lakes.

2. Raw water pump station. The raw water pump station diverts water from the source into the treatment units.

3. Pretreatment facilities, if applicable. Pretreatment facilities consist of chemical treatment processes to remove turbidity prior to filtration.

4. Membrane filters. The membrane filters are used to remove particles, sediment, algae, bacteria, and viruses from the water being treated by water filtration through membranes.

5. Backwashing. Backwashing may occur on the basis of three possible conditions:

   a. Elapsed time since the last backwash

   b. Total water produced since the last backwash

   c. Pressure increase since the last backwash

Fig. 16.11   SCADA system (controls at membrane rack)

Backwashing may be by air or by water, depending on the plant. Over time, the membranes require a more aggressive cleaning to remove accumulations and this is accomplished with the use of a proprietary chemical and caustic solution through a clean-in-place (CIP) process. This process results in the discharge of a wastestream that requires proper disposal.

6. Backwash water treatment (Figure 16.12). Water from backwash cleaning requires treatment and, depending on treatment, may be disinfected and discharged to the distribution or storage system or returned to the raw water intake, blended with raw water, and re-treated by pretreatment (if necessary) and membrane filtration.

7. Clean-in-place membrane regeneration system (Figure 16.13). The clean-in-place membrane regeneration system uses aggressive chemicals based on the material of the membrane.

8. Chemical feed systems. Chemical feed systems feed chlorine for the disinfection of the treated water. Caustic is fed after chlorination to increase the pH for corrosion control.

9. CT chamber. The chlorine contact chamber (CT chamber) provides the required contact time to achieve pathogen inactivation.

10. Clear well. The clear well provides storage for treated water before the water is pumped into the distribution system.

11. Treated water pump station. Treated water is pumped from the clear well into the distribution system. The pump station discharge is monitored for flow, pressure, turbidity, chlorine residual, and particle count.

12. Standby engine generator. The plant should be equipped with a standby emergency engine generator to provide power in the event of a power failure.

16.22 Operational Procedures

Daily Operations

1. Inspect intake structure and security
2. Inspect building security
3. Inspect and maintain:
   - Pumps, raw water, and treated water
   - Membrane filters, flows, turbidity, and particle count monitoring
   - Chemical feed system, disinfection, and pH (corrosion control)
   - CT chamber or tank
   - Clear well
   - Treated water pump station
   - Standby engine generator

4. Pretreatment facilities (if applicable), inspect and maintain:
   - Chemicals, storage, leaks, quantities (need to order more?)
   - Chemical feeders
   - Chemical doses

Weekly Operations

1. Particle count measurement on each membrane skid effluent

Monthly Operations

1. Raw water and treated water pump stations
   - Check mechanical seals
   - Turbidimeter calibration
2. Membrane filters
   - Turbidimeter calibration
3. Standby engine generator
   - Run/Test

Yearly Operations

1. Raw water and treated water pump stations
   - Check motor efficiency, electrical phase balance
   - Calibrate flowmeters
   - Calibrate pressure transducers
   - Change packing and oil on top bearing and grease bearings of pumps
2. Membrane filters
   - Calibrate flowmeters
   - Calibrate pressure transducers
3. Chemical feed system
   - Change hydraulic fluid in feed pumps
4. CT chamber and clear well
   - Dive inspection

16.23 Membrane Performance Monitoring

The operation of the membranes includes monitoring and testing for membrane filtration rate and membrane integrity. This is accomplished by pressure decay testing and sonic testing. When broken or damaged membrane fibers are discovered, they are repaired or replaced.

16.24 Troubleshooting Equipment and Process Failures

When the membrane filtration water treatment plant experiences an equipment or process failure, the operators troubleshoot to determine the cause of the problem. When the cause is determined, they will complete any repairs they are capable of performing. Any parts or outside services that are required will

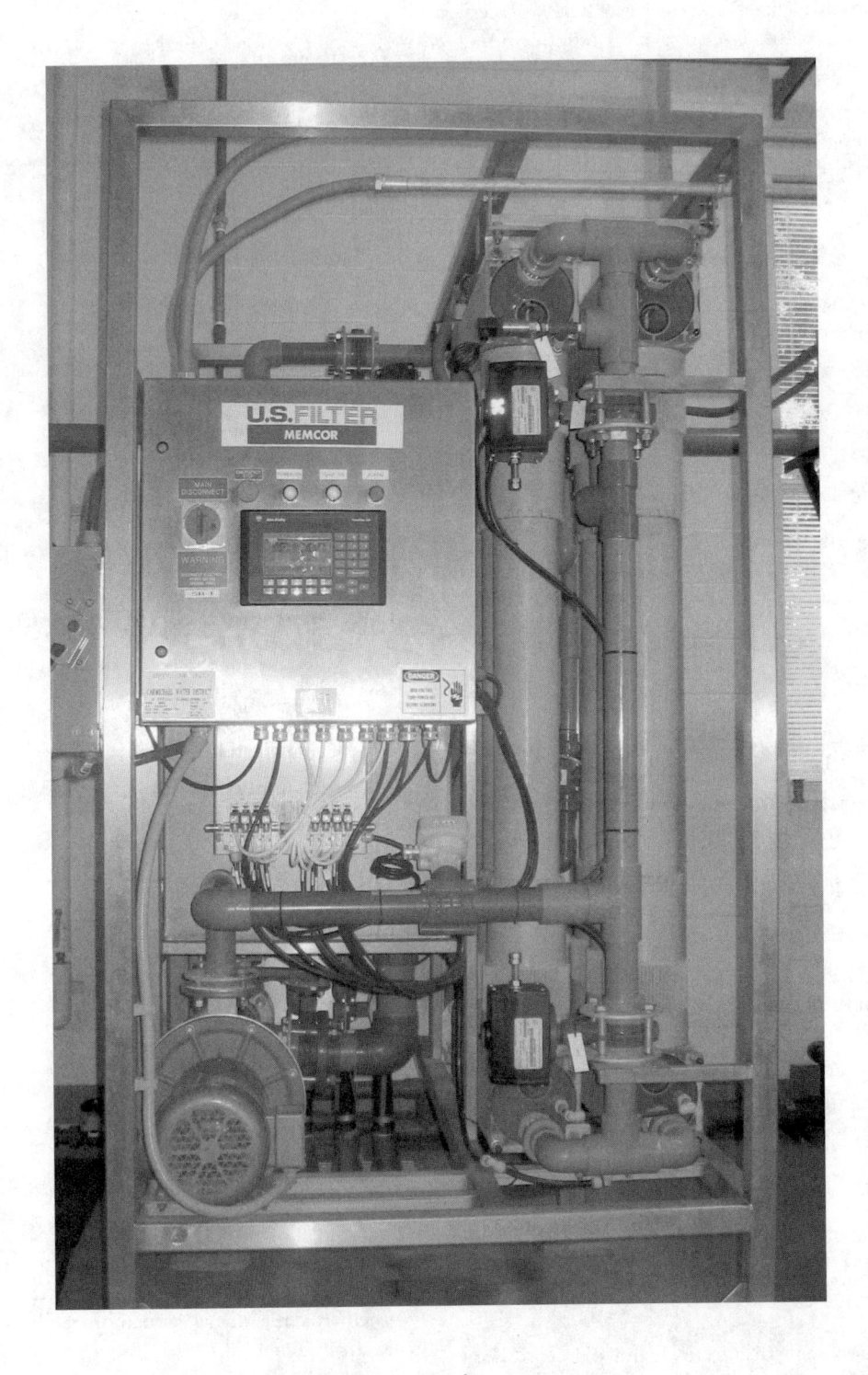

Fig. 16.12   Backwash water treatment

Fig. 16.13   Clean-in-place chemical preparation

be ordered or scheduled. The incident is recorded in the operator logbook. This logbook is maintained daily and includes daily and critical events.

Typical problems that may be encountered when operating a membrane filtration water treatment plant are described in Table 16.1 along with the cause and the corrective action.

> ## QUESTIONS
>
> Write your answers in a notebook and then compare your answers with those on page 224.
>
> 16.2A How can an on-call operator use a laptop computer?
>
> 16.2B Backwashing membrane filters may occur under what kinds of possible conditions?
>
> 16.2C How is membrane performance monitored?
>
> 16.2D What kinds of problems may an operator encounter when operating a membrane filtration water treatment plant?

16.3 RECORDKEEPING

16.30 Records

Membrane filtration water treatment plants typically use a SCADA system for water quality and water production monitoring and recordkeeping. These records include influent and effluent water quality, the hours of plant operation, plant flow rates, water production and distribution, process monitoring records, chemical dose rates and residual trends, alarms, and many other important conditions. In addition, the operations staff maintains handwritten records of maintenance, meter readings, chemical storage levels, and other process data that is read during the operator's walk-throughs at the plant.

16.31 Summary of Records

Membrane filtration water treatment plant records are usually maintained both electronically by the SCADA system and manually on hard copies. The records include water quality monitoring data, water production data, maintenance information and tasks performed, calibration dates and procedures, cleaning dates and procedures, testing methods, and *MATERIAL SAFETY DATA SHEETS (MSDSs)*.[16]

16.310 Water Quality Monitoring Records

The membrane filtration water treatment plant uses the SCADA system to continuously monitor turbidity and chlorine residual. The turbidity monitoring is of the raw, filtered, treated, and backwash returned water. Required disinfection levels are achieved by providing enough contact time and residual chlorine concentration.

TABLE 16.1 TROUBLESHOOTING MEMBRANE EQUIPMENT AND PROCESS PROBLEMS

Problem	Cause	Corrective Action
Air compressor failure	Mechanical failure inside unit.	Rebuild failed compressor.
	Belt failure.	Replace belt.
	Oil pump leak.	Repair leak.
		Use a portable air compressor.
Raw water pump station failure	Electronic failure of the uninterrupted power supply (UPS).	Bypass the UPS.
		Conduct routine monthly checks of all plant UPS units.
Computer lockup	Central SCADA computer failure.	Operate plant with only local control over plant operation.
		Hire an outside contractor to bring computer back to normal operating conditions.

---

[16] *Material Safety Data Sheet (MSDS).* A document that provides pertinent information and a profile of a particular hazardous substance or mixture. An MSDS is normally developed by the manufacturer or formulator of the hazardous substance or mixture. The MSDS is required to be made available to employees and operators or inspectors whenever there is the likelihood of the hazardous substance or mixture being introduced into the workplace. Some manufacturers are preparing MSDSs for products that are not considered to be hazardous to show that the product or substance is not hazardous.

16.311 Plant Operation and Water Production

The membrane filtration water treatment plant operation and water production records are maintained by the SCADA system. This includes water levels, flow rates, hours of operations, and quality of water produced.

16.312 Troubleshooting Records

Always record the conditions and causes of problems and failures. Also, record the corrective action or solution taken and the effectiveness of the corrective action. This information is very helpful in preventing future problems and failures and in evaluating potential solutions to failures.

16.313 Membrane Filter Maintenance and Inspection

Filter maintenance and inspection records for the membranes consist of the following:

- Transmembrane pressure (as indication of need for CIP (clean-in-place))
- Filtration production by skid
- Backwash occurrence including water production, transmembrane pressure, and run time for each period prior to backwash
- Membrane integrity testing using pressure decay testing
- Membrane sonic testing
- Membrane repair testing and membrane fiber repairs

QUESTIONS

Write your answers in a notebook and then compare your answers with those on page 224.

16.3A Membrane filtration water treatment plants continuously monitor which water quality indicators?

16.3B Which membrane filtration water treatment plant operation and water production records are maintained in a SCADA system?

16.4 MAINTENANCE PROGRAM

16.40 Types of Maintenance

The maintenance program at a membrane filtration water treatment plant includes different levels of activity performed throughout the year. There are three main types of equipment maintenance activities:

1. Routine maintenance
2. Preventive maintenance
3. Breakdown or corrective maintenance

16.41 Routine Maintenance

Routine maintenance is the first level of maintenance performed at a water treatment plant. The operator conducts routine maintenance on a daily basis. The operator observes each piece of equipment and checks for noise, heat, odors, vibration, and leaks. Depending on the skill level required, the operator may investigate the problems encountered and conduct on-site repair or schedule a contract service to maintain the equipment. A daily checklist is filled out to confirm that all equipment is running properly or to report problems encountered.

16.42 Preventive Maintenance

The preventive maintenance program (PMP) (also known as preventive maintenance (PM) program) includes scheduled maintenance based on operating time. To prevent catastrophic equipment failure events and to eliminate or minimize conditions that would result in plant downtime requires an effective PMP. The frequency of PMP activity is set to comply with the equipment manufacturer's recommendations and historical operations and maintenance experience.

Typical PMP activities at a water treatment plant include replacing lubricants, inspecting impellers, replacing bearings, and tensioning drive belts. Equipment at the plant should be monitored continuously by the SCADA system to ensure an effective PMP. The information maintained in the SCADA system associated with the PMP log includes the following items:

- Description of the equipment, equipment number, and location within the plant
- Manufacturer's address, phone number, and person to contact
- Supplier or local representative with address, phone number, and date of purchase
- Size, model type, and serial numbers
- Electrical, mechanical, and other pertinent performance data
- Type and frequency of preventive maintenance tasks to be accomplished
- Space to note when preventive maintenance was performed and by whom
- Lubrication schedule, denoting frequency of oil changes, greasing and general lubrication, types of lubricants to be used, and guidance on disposal of old oil
- Location of drawings and operation manuals
- Spare parts available from stock
- Safety considerations
- Work hours required, costs, and material or supplies consumed

The preventive maintenance activity is planned such that replacement parts are available in order to reduce equipment downtime. The skills required to perform preventive maintenance are different than those required for routine maintenance. Preventive maintenance is often carried out by outside contractors.

16.43 Breakdown or Corrective Maintenance

Breakdown or corrective maintenance relates to repairs to equipment that has failed in service. Due to normal redundancy built into the plant design, breakdown maintenance generally does not interrupt treatment and allows time for the plant repairs or replacement to be undertaken and completed. On completion of the repair task, the action is recorded in the breakdown maintenance log.

16.44 Acknowledgment

Material in this lesson is from the Carmichael Water District's "Bajamont Way Membrane Filtration Plant Operations Plan" prepared by Alex Peterson, P.E., Kennedy/Jenks Consultants. Steven Nugent is the General Manager of the Carmichael Water District. The cooperation of Alex Peterson and Steven Nugent is greatly appreciated.

QUESTIONS

Write your answers in a notebook and then compare your answers with those on page 224.

16.4A How frequently does an operator conduct routine maintenance?

16.4B What is the preventive maintenance program (PMP) scheduled maintenance based on?

16.4C What does breakdown maintenance relate to?

END OF LESSON 1 OF 4 LESSONS
on
MEMBRANE TREATMENT PROCESSES

Please answer the discussion and review questions next.

DISCUSSION AND REVIEW QUESTIONS
Chapter 16. MEMBRANE TREATMENT PROCESSES
(MEMBRANE FILTRATION AND DEMINERALIZATION)
(Lesson 1 of 4 Lessons)

At the end of each lesson in this chapter you will find some discussion and review questions. The purpose of these questions is to indicate to you how well you understand the material in the lesson. Write the answers to these questions in your notebook.

1. How can membrane fouling be described?
2. What should be the frequency of performing operational procedures at a membrane filtration plant?
3. How are membrane filtration water treatment plant records usually maintained?
4. What are the main types of equipment maintenance activities?

CHAPTER 16. MEMBRANE TREATMENT PROCESSES
(MEMBRANE FILTRATION AND DEMINERALIZATION)
(Lesson 2 of 4 Lessons)

16.5 SOURCES OF MINERALIZED WATER

As our country's population continues to grow, so does our demand for more water resources. Traditionally, water supplies have been obtained from "fresh water" sources. This constantly increasing need for water has started to deplete the available fresh water supplies in some areas of the country.

Faced with potential shortages, water planners must now consider new treatment technologies that, until recently, were not considered to be economically feasible. Since most of the earth's water supplies are saline (the ocean is high in dissolved minerals) rather than fresh, these impurities must be removed. One process receiving considerable attention is demineralization. Demineralization is the process that removes dissolved minerals (salts) from water.

All available water supplies can be classified according to their mineral quality. All waters contain various amounts of *TOTAL DISSOLVED SOLIDS (TDS)*,[17] including fresh water. A majority of the dissolved materials are inorganic minerals (salts). Minerals are compounds commonly found in nature that consist of positive metallic ions (such as calcium, sodium) bonded to negative ions (such as chloride, sulfate, carbonate). Many of these compounds are soluble in water and come from the weathering and erosion of the earth's surface.

Fresh water supplies, which have been the major sources of water developed in the past, usually contain less than 1,000 mg/L of total dissolved solids. Secondary drinking water standards recommend 500 mg/L TDS as the limit. Waters containing slightly higher concentrations can be used without adverse effects.

Brackish water contains from 1,000 to 10,000 mg/L TDS (seawater has 35,000 mg/L TDS). Most brackish water is found in groundwater. Figure 16.14 shows that over one half of the United States overlays groundwater containing TDS levels ranging from 1,000 to 3,000 mg/L. To date, brackish water has not been widely used for municipal drinking water supplies because of its highly mineralized taste and associated problems such as scaling in pipes. With the advent of new treatment technologies, however, demineralization of brackish waters (including reuse of wastewater) has great potential for further development.

The largest available source of water in terms of quantity is classified as seawater, which usually contains more than 35,000 mg/L TDS. While seawater is becoming an important future water resource because of its seemingly unlimited availability in coastal areas, it is more expensive to treat than brackish water because of its greater TDS concentration.

The purpose of this chapter is to introduce and familiarize the water treatment plant operator with the newer treatment processes that have been developed to remove the dissolved minerals (TDS) from water. The development of the membrane demineralization processes has significantly reduced the cost of demineralization. This savings, combined with diminished fresh water supplies, will increase the use of demineralization treatment processes. The large quantities of mineralized groundwater and the increased *SALINITY*[18] of many rivers and lakes due to waste discharges, agricultural runoff, and other uses will increase the need for demineralization.

Some areas of the United States, such as the Florida Gulf Coast, are already turning to demineralization. The worldwide capacity of brackish water demineralization plants is increasing every year. Thus, it is important that the water treatment plant operator become more knowledgeable concerning the methods used to demineralize water.

QUESTIONS

Write your answers in a notebook and then compare your answers with those on page 224.

16.5A What is demineralization?

16.5B Why is seawater more expensive to treat than brackish water?

---

[17] *Total Dissolved Solids (TDS).* All of the dissolved solids in a water. TDS is measured on a sample of water that has passed through a very fine mesh filter to remove suspended solids. The water passing through the filter is evaporated and the residue represents the total dissolved solids. Also see SPECIFIC CONDUCTANCE.

[18] *Salinity.* (1) The relative concentration of dissolved salts, usually sodium chloride, in a given water. (2) A measure of the concentration of dissolved mineral substances in water.

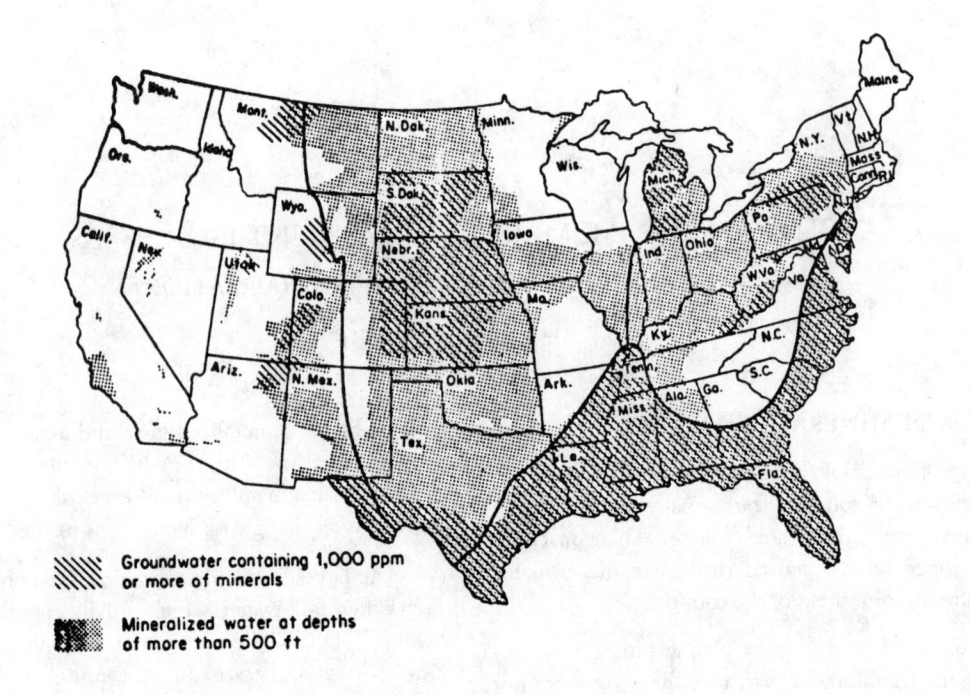

Fig. 16.14 Map of the conterminous United States showing depth to and quality of shallowest groundwater containing more than 1,000 mg/L dissolved solids
(From a paper by Bill Katz, "Treating Brackish Water for Community Supplies," published in Proceedings in "Role of Desalting Technology," a series of Technology Transfer Workshops presented by the Office of Water Research and Technology)

16.6 DEMINERALIZING PROCESSES

Methods of removing minerals from water can be divided into two classes: (1) those that use a phase change such as freezing or distillation, and (2) nonphase change methods such as reverse osmosis, electrodialysis, and ion exchange.

Demineralizing processes have primarily been used to remove dissolved inorganic material (TDS) from industrial water and wastewater, municipal water and wastewater, and seawater. However, some processes will also remove suspended material, organic material, bacteria, and viruses. Application of the various demineralizing processes is partially dependent upon the total dissolved solids (TDS) concentration of the water to be treated. Figure 16.15 illustrates the approximate TDS range for use of two phase change processes (distillation and freezing) and three nonphase change processes (reverse osmosis, electrodialysis, and ion exchange).

The selection of a demineralizing process for a particular application depends on several factors including:

1. Mineral concentration in *FEEDWATER*[19] (brackish water supply)

2. Product water quality required

3. Brine disposal facilities

4. Pretreatment required

5. Need to remove other material such as bacteria and viruses

6. Availability of energy and chemicals required for the process

7. Cost

The basic system is similar for all demineralizing processes and includes the processes shown in Figure 16.16.

---

[19] *Feedwater.* The water that is fed to a treatment process; the water that is going to be treated.

Membrane Treatment Processes   185

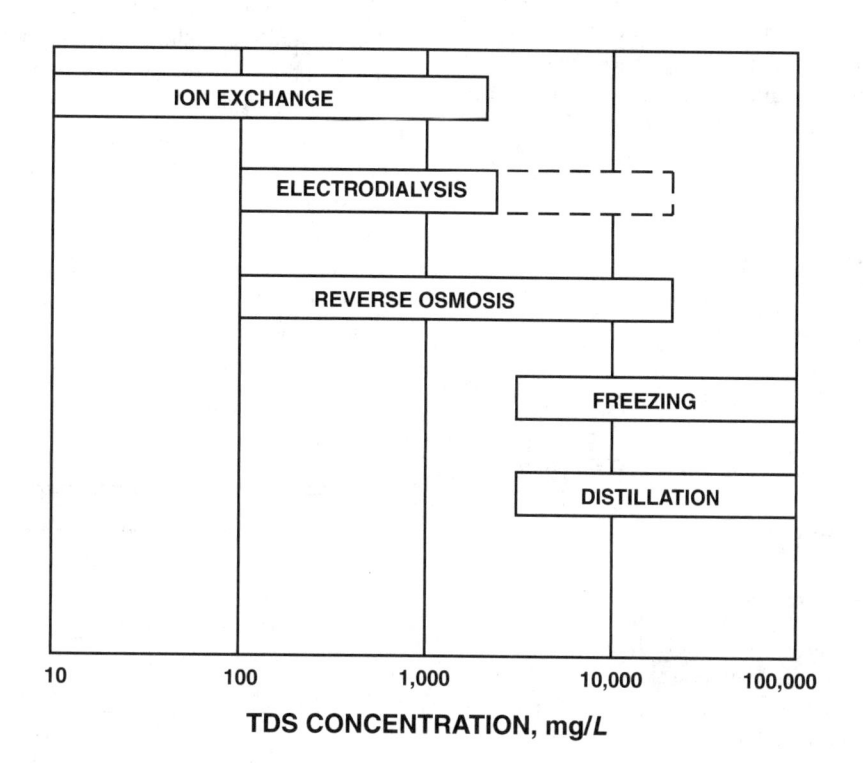

NOTE: The dashed lines indicate a feasible range of operation, but not typical range.

Fig. 16.15   Demineralization processes versus feedwater TDS concentrations

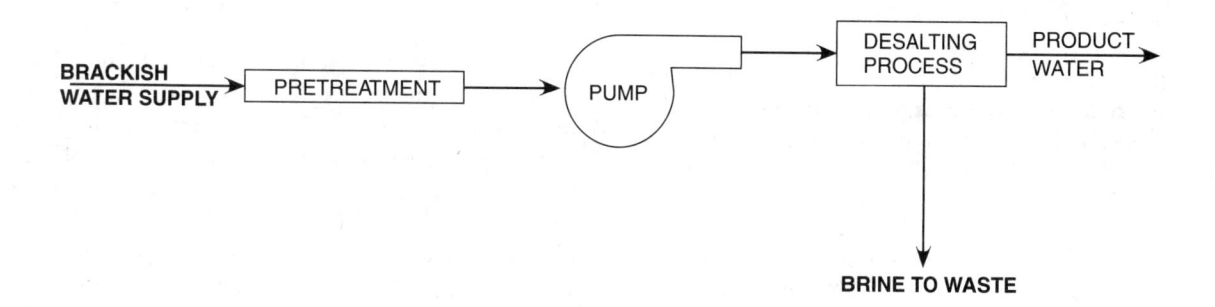

Fig. 16.16   Basic system for demineralizing processes

Since freezing and distillation apply primarily to seawater demineralization and their widespread use seems unlikely, these processes will not be discussed. Neither will ion exchange because of its limited use for brackish water (1,000 to 3,000 mg/L TDS). Currently, activity within the water industry is focused primarily on the membrane demineralizing processes known as reverse osmosis, electrodialysis, and nanofiltration. Also see *ADVANCED WASTE TREATMENT*, Chapter 4, "Solids Removal From Secondary Effluents," Section 4.5, "Cross Flow Membrane Filtration."

QUESTIONS

Write your answers in a notebook and then compare your answers with those on page 224.

16.6A List the two classes of methods of removing minerals from water.

16.6B List the common membrane demineralizing processes.

16.7 REVERSE OSMOSIS

16.70 What Is Reverse Osmosis?

Osmosis can be defined as the passage of a liquid from a weak solution to a more concentrated solution across a semipermeable membrane. The membrane allows the passage of the water (solvent) but not the dissolved solids (solutes).

Osmosis plays a vital role in many biological processes. Nutrient and waste minerals are transported by osmosis through the cells of animal tissues, which show varying degrees of permeability to different dissolved solids. A striking example of a natural osmotic process is the behavior of blood cells placed in pure water. Water passes through the cell walls to dilute the solution inside the cell. The cell swells and eventually bursts, releasing its red pigment. If the blood cells are placed in a concentrated sugar solution, the reverse process occurs; the cells shrink and shrivel up as water moves out into the sugar solution.

The top half of Figure 16.17 illustrates osmosis. The transfer of the water (solvent) from the freshwater side of the membrane continues until the level (shown in shaded area) rises and the head or pressure is large enough to prevent any net transfer of the solvent (water) to the more concentrated solution. At equilibrium, the quantity of water passing in either direction is equal; the difference in water level between the two sides of the membrane is defined as the osmotic pressure of the solution.

If a piston is placed on the more-concentrated solution side of the semipermeable membrane (Figure 16.17) and a pressure, P, is applied that is greater than the osmotic pressure, water flows from the more-concentrated solution to the freshwater side of the membrane. This condition illustrates the process of reverse osmosis.

16.71 Reverse Osmosis Membrane Structure and Composition

The two types of semipermeable membranes that are used most often for demineralization are cellulose acetate and thin film composites. Cellulose acetate, the first commercially available membrane (since the mid-1960s) is made by casting a cellulose acetate/solvent solution onto a porous support material. After quenching and annealing operations, the cellulose acetate membrane, often referred to as CA, exhibits the structure shown in Figure 16.18. The cellulose acetate membrane is asymmetric, meaning that one side is different from the other. The total cellulose acetate layer is 50 to 100 microns thick; however, a thin, dense layer approximately 0.2 micron thick exists at the surface. This thin, dense layer serves as the rejecting barrier of the membrane. The characteristics of a CA membrane are given in Table 16.2.

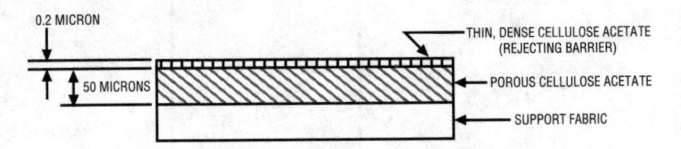

Fig. 16.18 Cellulose acetate membrane cross section

TABLE 16.2 CELLULOSE ACETATE MEMBRANE TYPICAL CHARACTERISTICS

Net Driving Pressure:	400 psi
NaCl Rejection:	92–97%
Flux Rate at 400 psi, 77°F:	25 GFD[a]
Operating pH range:	4.0–6.0
Cleaning pH range:	3.0–6.0
Cost relative to thin film composite membrane:	lower
Allowable feedwater chlorine concentration:	1.0 mg/L
Maximum operating temperature:	104°F (40°C)

Subject to biological attack
Subject to hydrolysis
Salt passage rate typically doubles after three years
Most suitable for treatment of municipal wastes and some heavily pretreated surface supplies (due to lower fouling rate vs. thin film)

[a] Gallons of flux per square foot per day.

During the 1970s, researchers realized the need for a membrane with better *FLUX*[20] and rejection characteristics than those of cellulose acetate. The basic approach to developing a better membrane was to improve the efficiencies of the thin rejecting layer and the porous substrate by casting each layer separately during the manufacturing process, hence the term "thin film composite." A typical thin film composite membrane is shown in Figure 16.19. The first step in preparing a thin film composite membrane is the casting of the porous support, usually a polysulfone solution, onto the support fabric. The next step is to contact the composite support with the polymers that will actually form the semipermeable thin film rejecting barrier.

---

[20] *Flux.* A flowing or flow.

Membrane Treatment Processes   187

OSMOSIS—NORMAL FLOW FROM LOW TO HIGH CONCENTRATION

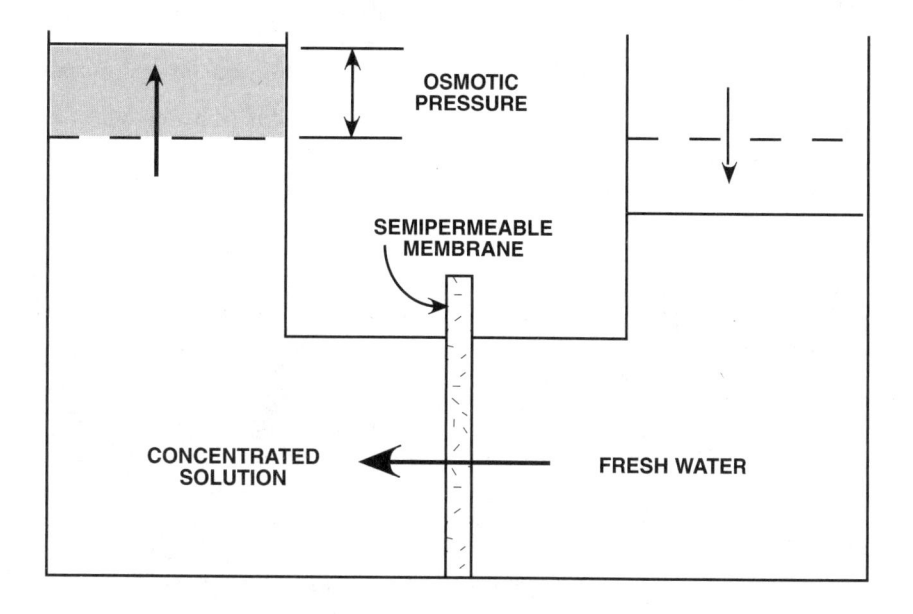

REVERSE OSMOSIS—FLOW REVERSED BY APPLICATION OF PRESSURE TO HIGH CONCENTRATION SOLUTION

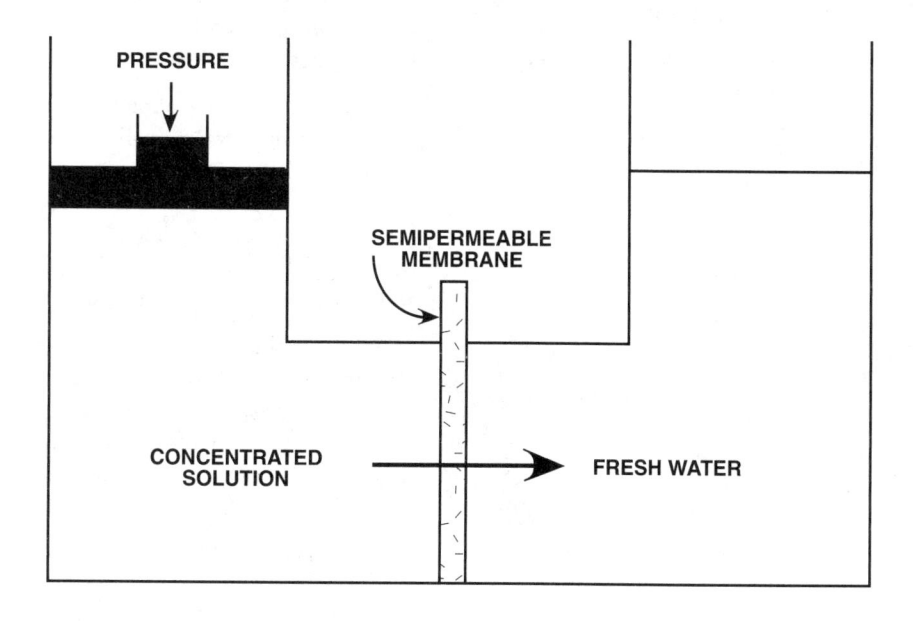

Fig. 16.17   Flows through a semipermeable membrane

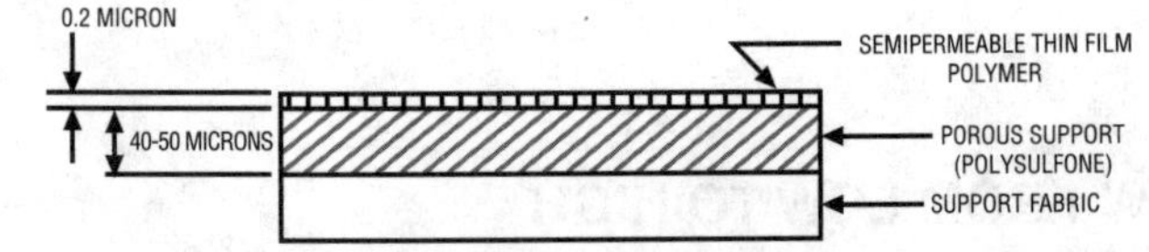

Fig. 16.19 Thin film composite membrane cross section

The thin film rejecting barrier is formed *in situ*, or in position, by interfacial polymerization. It is the ability to form the semipermeable membrane separate from the support layers that has enabled membrane manufacturers to select polymers that will produce membranes with optimum dissolved solids rejection and water flux rates. The characteristics of thin film composite membranes are given in Table 16.3.

TABLE 16.3 THIN FILM COMPOSITE MEMBRANE TYPICAL CHARACTERISTICS

Net Driving Pressure:	200 psi
NaCl Rejection:	98–99%
Flux Rate at 200 psi, 77°F:	25–30 GFD[a]
Operating pH range:	3.0–10.0
Cleaning pH range:	2.0–12.0
Cost relative to cellulose acetate membrane:	higher
Allowable feedwater chlorine concentration:	none
Maximum operating temperature:	113°F (45°C)
Salt passage increase after three years:	≤30%

Not subject to biological attack, hydrolysis, or compaction
Higher rejection and flux rates than cellulose acetate membrane
Higher fouling rates than cellulose acetate on certain waste and surface water supplies
Sensitive to oxidants in feedwater

---
[a] Gallons of flux per square foot per day.

16.72 Membrane Performance and Properties

The basic behavior of semipermeable cellulose acetate reverse osmosis membranes can be described by two equations. The product water flow through a semipermeable membrane can be expressed as shown in Equation 1.

EQUATION 1

$$F_w = A(\Delta P - \Delta \pi)$$

Where

F_w = Water flux (gm/sq cm-sec)

A = Water permeability constant (gm/sq cm-sec atm[21])

ΔP = Pressure differential applied across the membrane (atm)

$\Delta \pi$ = Osmotic pressure differential across the membrane (atm)

Note that the water flux is the flow of water in grams per second through a membrane area of one square centimeter. Think of this as similar to the flow through a rapid sand filter in gallons per minute through a filter area of one square foot (GPM/sq ft).

The mineral (salt) flux (mineral passage) through the membrane can be expressed as shown in Equation 2.

EQUATION 2

$$F_s = B(C_1 - C_2)$$

Where

F_s = Mineral flux (gm/sq cm-sec)

B = Mineral permeability constant (cm/sec)

$C_1 - C_2$ = Concentration gradient across the membrane (gm/cu cm)

The water permeability (A) and mineral permeability (B) constants are characteristics of the particular membrane that is used and the processing it has received.

An examination of Equations 1 and 2 shows that the water flux (the rate of water flow through the membrane) is dependent on the applied pressure, while the mineral flux is *not* dependent on pressure. As the pressure of the feedwater is increased, the flow of water through the membrane increases while the flow of minerals remains essentially constant. Therefore, both the quantity and the quality of the purified product (the *PERMEATE*[22]) should increase with increased pressure. This occurs because there is more water to dilute the same amount of mineral.

The water flux (F_w) *decreases* as the mineral content of the feed increases because the osmotic pressure contribution ($\Delta \pi$) increases with increasing mineral content. In other words, since $\Delta \pi$ increases, the term ($\Delta P - \Delta \pi$) decreases, which results in a decrease in F_w, the water flux. Further, as more and more feedwater passes through the membrane, the mineral content of the feedwater becomes higher and higher (more concentrated). The osmotic pressure contribution ($\Delta \pi$) of the concentrate increases, resulting in a lower water flux.

Finally, since the membrane rejects a constant percentage of mineral, product water quality decreases with increased feedwater concentration. Also, note that Equation 2 reveals that the greater the concentration gradient ($C_1 - C_2$) across the membrane, the greater the mineral flux (mineral flow). Therefore, the greater the feed concentration, the greater the mineral flux and also mineral concentration in the product water.

Water treatment plant operators must have a basic understanding of these mathematical relationships that describe RO (reverse osmosis) membrane performance. To help develop a better understanding of the interrelationships of flux, rejection, time, temperature, pH, and recovery, further explanation of these variables continues in the next section.

EXAMPLE 1

Convert a water flux of 5×10^{-4} gm/sq cm-sec[23] to gallons per day per square foot.

---
[21] *atm.* The abbreviation for atmosphere. One atmosphere is equal to 14.7 psi or 100 kPa.
[22] *Permeate* (PURR-me-ate). (1) To penetrate and pass through, as water penetrates and passes through soil and other porous materials. (2) The liquid (demineralized water) produced from the reverse osmosis process that contains a low concentration of dissolved solids.
[23] 5×10^{-4} is the same as 0.0005.

Known	Unknown
Water Flux, gm/sq cm-sec = 5×10^{-4} gm/sq cm-sec	Flow, GPD/sq ft

Convert the water flux from gm/sq cm-sec to flow in GPD/sq ft.

$$\text{Flow, GPD/sq ft} = \frac{(\text{Water Flux, gm/sq cm-sec})(2.54 \text{ cm/in})^2(12 \text{ in/ft})^2(60 \text{ sec/min})(60 \text{ min/hr})(24 \text{ hr/day})}{(1{,}000 \text{ gm}/L)(3.785 \, L/\text{gal})}$$

$$= \frac{(0.0005 \text{ gm/sq cm-sec})(2.54 \text{ cm/in})^2(12 \text{ in/ft})^2(60 \text{ sec/min})(60 \text{ min/hr})(24 \text{ hr/day})}{(1{,}000 \text{ gm}/L)(3.785 \, L/\text{gal})}$$

$$= 10.6 \text{ GPD/sq ft}$$

QUESTIONS

Write your answers in a notebook and then compare your answers with those on page 225.

16.7A What is the osmotic pressure of a solution?

16.7B What are the two types of semipermeable membranes that are used most often for demineralization?

16.7C What is the meaning of water flux and of mineral flux? What units are used to express measurement of these quantities?

16.7D When additional pressure is applied to the side of a membrane with a concentrated solution, what happens?

16.7E When higher mineral concentrations occur in the feedwater, what happens to the permeate (or product water)?

16.73 Definition of Flux

The term flux is the expression used to describe the rate of water flow through the semipermeable membrane. Flux is usually expressed in gallons per day per square foot of membrane surface or in grams per second per square centimeter.

The average membrane flux rate of a reverse osmosis system is an important operating guideline. In practice, most reverse osmosis systems will require periodic cleaning. It has been demonstrated that the cleaning frequency can be dependent on the average membrane flux rate of the system. Too high a flux rate may result in excessive fouling rates requiring frequent cleaning. Some general industry guidelines for acceptable flux rates are:

Feedwater Source	Flux Rate, GFD
Industrial/Municipal Waste	8–12
Surface (river, lake, ocean)	8–14
Well	14–20

EXAMPLE 2

The permeate flow through an arrangement (or array) of RO membrane pressure vessels is 1,330,000 gallons per day (GPD). Feedwater first flows to 33 vessels operating in parallel. The concentrate from the 33 first-pass vessels is combined and sent to a set of 11 second-pass vessels. Each vessel (or tube) contains six membrane elements. Each element contains many fiber membranes, which provide a total of 325 square feet of membrane surface area per element. Calculate the average membrane flux rate for the system in gallons per day per square foot (GFD).

Known	Unknown
Permeate Flow, GPD = 1,330,000 GPD	Average Flux Rate, GFD
No. of Vessels = 44 (33 + 11)	
No. of Elements per Vessel = 6	
Membrane Area per Element, sq ft = 325 sq ft	

1. Determine the total membrane area in the system.

$$\text{Membrane Area, sq ft} = (\text{No. Vessels})(\text{No. Elements/vessel})(\text{Surface Area/element})$$

$$= (44 \text{ vessels})(6 \text{ elements/vessel})(325 \text{ sq ft/element})$$

$$= 85{,}800 \text{ sq ft}$$

2. Calculate the average membrane flux rate for the system.

$$\text{Average Flux Rate, GFD} = \frac{\text{Permeate Flow, GFD}}{\text{Membrane Area, sq ft}}$$

$$= \frac{1{,}330{,}000 \text{ GPD}}{85{,}800 \text{ sq ft}}$$

$$= 15.5 \text{ GFD}$$

Even under ideal conditions (pure feedwater and no fouling of the membrane surface), there is a decline in water flux with time. This decrease in flux is due to membrane compaction. This phenomenon is considered comparable to "creep" observed in other plastics or even metals when subjected to compressing stresses (pressure).

The term "flux decline" is used to describe the loss of water flow through the membrane due to compaction plus fouling. In the real world, feedwaters are never pure and contain suspended solids, dissolved organics and inorganics, bacteria, algae, and other potential foulants. These impurities can be deposited or grow on the membrane surface, thus hindering the flow of water through the membrane.

16.74 Mineral Rejection

The purpose of demineralization is to separate minerals from water; the ability of the membrane to reject minerals is called the mineral rejection. Mineral rejection is defined as:

EQUATION 3

$$\text{Rejection, \%} = \left(1 - \frac{\text{Product Concentration}}{\text{Feedwater Concentration}}\right) \times 100\%$$

Mineral rejections can be determined by measuring the TDS and using Equation 3. Rejections also may be calculated for individual constituents in the solution by using their concentrations.

The basic equations that describe the performance of a reverse osmosis membrane indicate that rejection decreases as feedwater mineral concentration increases. Remember, this is because the higher mineral concentration increases the osmotic pressure. Figure 16.20 illustrates the rejection performance for a typical RO (reverse osmosis) membrane operating on three different feedwater solutions. This figure shows that as feed mineral concentration increases (TDS in mg/L), rejection decreases at a given feed pressure. Notice also that rejection improves as feed pressure increases.

Typical rejection for most commonly encountered dissolved inorganics is usually between 92 and 99 percent. Divalent ions such as calcium and sulfate are better rejected than monovalent ions such as sodium or chloride. Table 16.4 lists the typical rejection of an RO membrane operating on a brackish feedwater.

TABLE 16.4 TYPICAL REVERSE OSMOSIS REJECTIONS OF COMMON CONSTITUENTS FOUND IN BRACKISH WATER

Contaminant	Units	Feedwater Concentration	Percent Removal
EC[a]	µmhos	1,400	92
TDS[a]	mg/L	900	92
Calcium	mg/L	100	99
Chloride	mg/L	120	92
Sulfate	mg/L	338	99
Sodium	mg/L	158	92
Ammonia	mg/L	22.5	94
Nitrate	mg/L	2.9	55
COD[a]	mg/L	12.5	95
TOC[a]	mg/L	6.0	88
Silver	µg/L	1.2	88
Arsenic	µg/L	<5.0	—
Aluminum	µg/L	71.0	93
Barium	µg/L	24.0	96
Beryllium	µg/L	<1.0	—
Cadmium	µg/L	3.4	98
Cobalt	µg/L	4.6	>90
Chromium	µg/L	3.6	80
Copper	µg/L	12.7	63
Iron	µg/L	24.0	91
Mercury	µg/L	0.8	41
Manganese	µg/L	1.0	85
Nickel	µg/L	2.5	88
Lead	µg/L	<1.0	—
Selenium	µg/L	<5.0	—
Zinc	µg/L	<100.0	—

[a] EC, Electrical Conductivity; TDS, Total Dissolved Solids; COD, Chemical Oxygen Demand; and TOC, Total Organic Carbon.

EXAMPLE 3

Estimate the ability of a reverse osmosis plant to reject minerals by calculating the mineral rejection as a percent. The feedwater contains 1,500 mg/L TDS and the product water TDS is 150 mg/L.

Known	Unknown
Feedwater TDS, mg/L = 1,500 mg/L	Mineral Rejection, %
Product Water TDS, mg/L = 150 mg/L	

Calculate the mineral rejection as a percent.

$$\text{Mineral Rejection, \%} = \left(1 - \frac{\text{Product TDS, mg}/L}{\text{Feed TDS, mg}/L}\right)(100\%)$$

$$= \left(1 - \frac{150 \text{ mg}/L}{1,500 \text{ mg}/L}\right)(100\%)$$

$$= (1 - 0.1)(100\%)$$

$$= 90\%$$

While most demineralization applications require the use of a membrane with high rejection rates (greater than 95 percent), some applications can use a membrane with lower rejection rates (80 percent) and lower operating pressures (less than 150 psi). The membranes that fit this classification are commonly referred to as softening or nanofiltration membranes. These membranes produce the same quantity of water as standard RO membranes but at lower operating pressures. A comparison of standard and softening RO membranes is given below:

	Standard	Softening (Nanofiltration)
Flux	25–30 GFD	25–30 GFD
Applied Pressure	225 psi	150 psi
Minimum Salt Rejection	97–98%	75–80%
Hardness Rejection	>99%	>95%

Softening or nanofiltration membranes are seeing widespread use for demineralization of municipal water supplies that require high rejection rates for hardness and THM formation potential, and moderate TDS rejection.

Reverse osmosis membrane manufacturers provide computer software to project the permeate water quality of a reverse osmosis system. Figures 16.21 and 16.22 are examples of computer printouts of permeate projections showing the expected initial performance of a thin film composite membrane (Figure 16.21) and a cellulose acetate membrane (Figure 16.22). The same feedwater was used for both projections.

QUESTIONS

Write your answers in a notebook and then compare your answers with those on page 225.

16.7F Water flux is usually expressed in what units?

16.7G What is "flux decline"?

16.7H How is mineral rejection measured?

16.7I What are the two major differences between standard RO membranes and softening membranes?

Membrane Treatment Processes 191

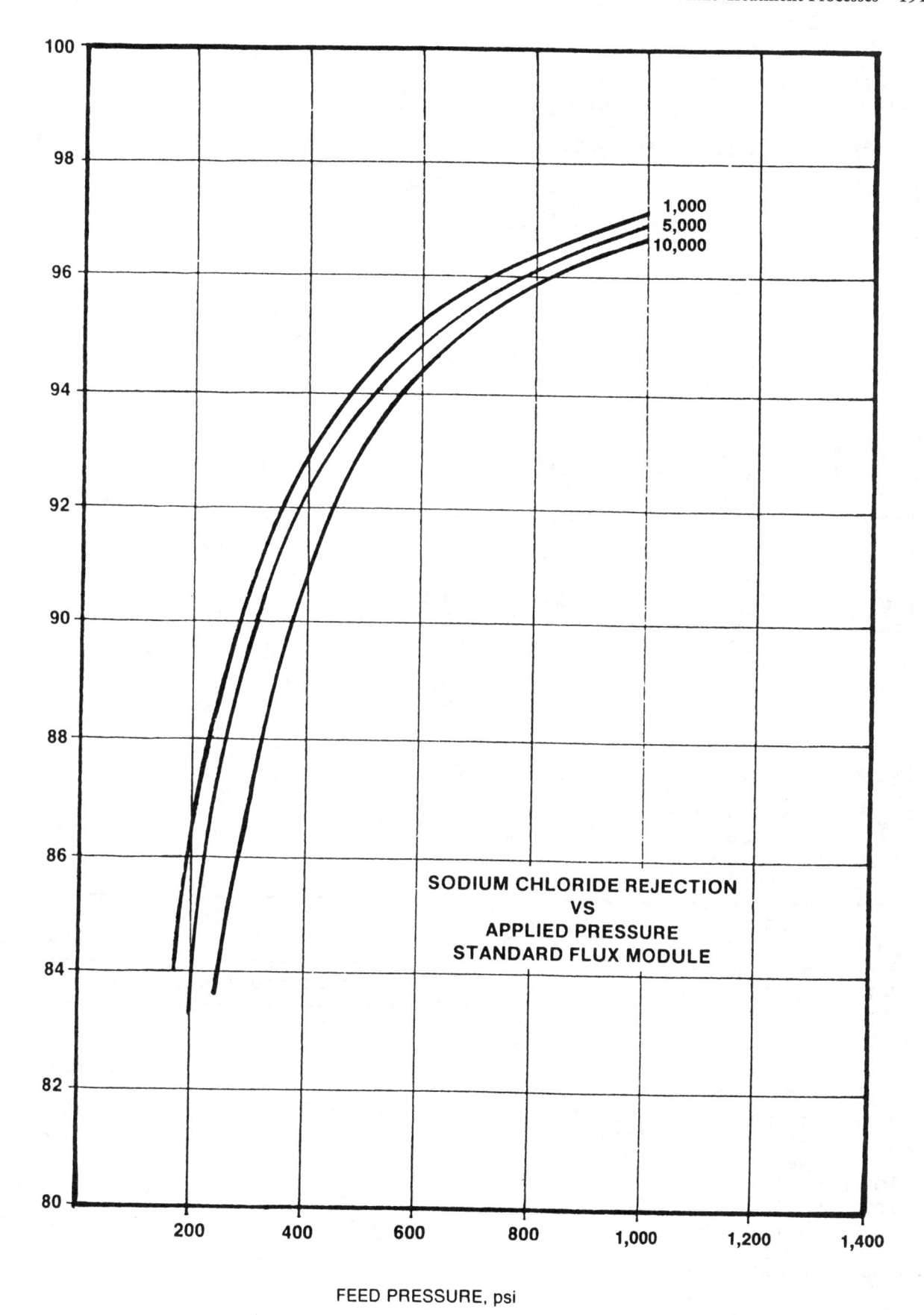

Fig. 16.20 *Typical RO rejection for three different feedwater concentrations of TDS in mg/L*
(Source: *REVERSE OSMOSIS PRINCIPLES AND APPLICATIONS*,
by Fluids Systems, Division of UOP, October 1970)

192 Water Treatment

```
HYDRANAUTICS DESIGN PROGRAM -  VERSION 4.01 (1991)           05-26-06
Calculation was made by: CWC

Project name : SANTA ANA WATERSHED           Permeate flow :  1330000 GPD

Feedwater temperature :      21.0 C     Recovery :       77.0%
Raw water pH :               7.20       Element age :    0.0 years
Acid dosage, ppm(100%):      32.7 H2SO4  Flux decline coefficient :      -0.025
Acidified feed CO2,ppm :     66.7       3-yr salt passage increase factor :1.2


Feed Pressure : 157.4 psi                Concentrate Pressure : 118.1 psi

Pass   Feed Flow     Conc.  Flow    Beta   Conc.    Element      Element  Array
       Total Vessel  Total  Vessel         Press.   Type         No.
       gpm    gpm    gpm    gpm            psi

 1    1199.5  36.3   454.7  13.8    1.22   142.7   8040-LSY-CPA2   198    33x6
 2     454.7  41.3   275.9  25.1    1.08   118.1   8040-LSY-CPA2    66    11x6
```

	Raw water		Feed water		Permeate		Concentrate	
Ion	mg/l	ppm*	mg/l	ppm*	mg/l	ppm*	mg/l	ppm*
Ca	140.0	349.1	140.0	349.1	1.8	4.6	602.5	1502.5
Mg	42.0	172.8	42.0	172.8	0.6	2.3	180.8	743.8
Na	168.0	365.2	168.0	365.2	10.5	22.8	695.4	1511.7
K	3.8	4.9	3.8	4.9	0.3	0.4	15.5	19.9
NH4	0.0	0.0	0.0	0.0	0.0	0.0	0.0	0.0
Ba	0.0	0.0	0.0	0.0	0.0	0.0	0.0	0.0
Sr	0.0	0.0	0.0	0.0	0.0	0.0	0.0	0.0
CO3	0.3	0.5	0.1	0.2	0.0	0.0	2.1	3.5
HCO3	367.0	300.8	326.6	267.7	18.8	15.4	1357.2	1112.5
SO4	243.0	253.1	275.1	286.5	2.3	2.3	1188.4	1237.9
Cl	162.0	228.5	162.0	228.5	5.2	7.4	686.8	968.7
F	0.4	1.1	0.4	1.1	0.0	0.1	1.7	4.4
NO3	93.0	75.0	93.0	75.0	5.9	4.8	384.5	310.1
SiO2	40.0		40.0		0.7		171.7	
TDS	1259.5		1251.0		46.1		5286.5	
pH	7.2		6.9		5.7		7.5	

Notes: *ppm as CaCO3.   Calculated concentrations are accurate to +/- 10%

	Raw water	Feed water	Concentrate
CaSO4/Ksp*100,%	7.9	8.8	55.2
SrSO4/Ksp*100,%	0.0	0.0	0.0
BaSO4/Ksp*100,%	0.0	0.0	0.0
SiO2 sat.,%	34.2	34.2	146.8
Langelier ind.	0.21	-0.14	1.69
Stiff & Davis ind.	0.20	-0.15	1.44
Ionic strength	0.03	0.03	0.11
Osmotic press.,psi	9.2	9.1	38.5

Fig. 16.21 Computer printout of permeate projections for a thin film composite membrane
(Source: Hydranautics)

```
HYDRANAUTICS DESIGN PROGRAM -   VERSION 4.01 (1991)           05-26-06
Calculation was made by: CWC

Project name : SANTA ANA WATERSHED            Permeate flow :    1330000 GPD

Feedwater temperature :     21.0 C      Recovery :      77.0%
Raw water pH :               7.20       Element age :    0.0 years
Acid dosage, ppm(100%):      0.0 H2SO4  Flux decline coefficient :    -0.035
Acidified feed CO2,ppm :    37.9        3-yr salt passage increase factor :2.0


Feed Pressure : 297.9 psi                    Concentrate Pressure : 257.5 psi

Pass    Feed Flow      Conc. Flow     Beta    Conc.    Element      Element  Array
     Total  Vessel   Total  Vessel            Press.   Type         No.
      gpm    gpm      gpm    gpm               psi

 1   1199.5  36.3    480.9   14.6     1.21    282.8   8040-MSY-CAB1   198    33x6
 2    480.9  43.7    275.9   25.1     1.11    257.5   8040-MSY-CAB1    66    11x6
```

Ion	Raw water mg/l	Raw water ppm*	Feed water mg/l	Feed water ppm*	Permeate mg/l	Permeate ppm*	Concentrate mg/l	Concentrate ppm*
Ca	140.0	349.1	140.0	349.1	2.5	6.2	600.4	1497.3
Mg	42.0	172.8	42.0	172.8	0.7	3.1	180.1	741.2
Na	168.0	365.2	168.0	365.2	27.9	60.6	637.2	1385.1
K	3.8	4.9	3.8	4.9	0.8	1.0	13.8	17.7
NH4	0.0	0.0	0.0	0.0	0.0	0.0	0.0	0.0
Ba	0.0	0.0	0.0	0.0	0.0	0.0	0.0	0.0
Sr	0.0	0.0	0.0	0.0	0.0	0.0	0.0	0.0
CO3	0.3	0.5	0.3	0.5	0.0	0.0	4.6	7.7
HCO3	367.0	300.8	367.0	300.8	21.3	17.5	1524.3	1249.4
SO4	243.0	253.1	243.0	253.1	1.8	1.9	1050.5	1094.2
Cl	162.0	228.5	162.0	228.5	22.6	31.8	628.8	886.9
F	0.4	1.1	0.4	1.1	0.0	0.1	1.6	4.3
NO3	93.0	75.0	93.0	75.0	24.3	19.6	323.1	260.6
SiO2	40.0		40.0		10.4		139.0	
TDS	1259.5		1259.5		112.3		5103.5	
pH	7.2		7.2		6.0		7.8	

Notes: *ppm as CaCO3.   Calculated concentrations are accurate to +/- 10%

	Raw water	Feed water	Concentrate
CaSO4/Ksp*100,%	7.9	7.9	50.1
SrSO4/Ksp*100,%	0.0	0.0	0.0
BaSO4/Ksp*100,%	0.0	0.0	0.0
SiO2 sat.,%	34.2	34.2	118.8
Langelier ind.	0.21	0.21	2.04
Stiff & Davis ind.	0.20	0.20	1.79
Ionic strength	0.03	0.03	0.11
Osmotic press.,psi	9.2	9.2	37.0

Fig. 16.22 Computer printout of permeate projections for a cellulose acetate membrane
(Source: Hydranautics)

16.75 Effects of Feedwater Temperature and pH on Membrane Performance

In reverse osmosis operation, feedwater temperature has a significant effect on membrane performance and must therefore be taken into account in system design and operation. Essentially, the value of the water permeation constant is only constant for a given temperature. As the temperature of the feedwater increases, flux increases. Usually, flux is reported at some standard temperature reference condition, such as 25°C. Figure 16.23 illustrates the increase in flux for a standard RO module over a range of operating temperatures when 400 psi (2,758 kPa or 28 kg/sq cm) net operating pressure is applied.

Cellulose acetate membrane is subject to long-term *HYDROLYSIS*.[24] Hydrolysis results in a lessening of mineral rejection capability. The rate of hydrolysis is accelerated by increased temperature, and is also a function of feed pH (Figure 16.24). Slightly acidic pH values (5 to 6) ensure a lower hydrolysis rate, as do cooler temperatures. Therefore, to ensure the longest possible lifetime of the cellulose acetate membrane and to slow hydrolysis, acid is added as a pretreatment step before demineralization. Table 16.5 indicates the relative time for mineral passage to increase 200 percent at different feedwater pH levels. Thin film composite membranes are not subject to hydrolysis but pH adjustment of the feedwater may be required for scale control.

TABLE 16.5 TIME REQUIRED TO ACHIEVE A 200 PERCENT INCREASE IN MINERAL PASSAGE AT 23°C AT VARIOUS pH LEVELS

pH	Time
5.0	6 years
6.0	3.8 years
7.0	1 year
8.0	0.14 year = 51 days
9.0	0.01 year = 3.6 days

16.76 Recovery

Recovery is defined as the percentage of feed flow that is recovered as product water. Expressed mathematically, recovery can be determined by Equation 4.

EQUATION 4

$$\text{Recovery, \%} = \left(\frac{\text{Product Flow}}{\text{Feed Flow}}\right)(100\%)$$

The recovery rate is usually determined or limited by two considerations. The first is the desired product water quality. Since the amount of mineral passing through the membrane is influenced by the concentration differential between the brine and product, there is a possibility of exceeding product quality criteria with excessive recovery. The second consideration concerns the solubility limits of minerals in the brine. One should not concentrate the brine to a degree that would precipitate minerals on the membrane. This effect is commonly referred to as *CONCENTRATION POLARIZATION*.[25]

The most common and serious problem resulting from concentration polarization is the increasing tendency for precipitation of sparingly soluble salts and the deposition of particulate matter on the membrane surface.

In any flowing hydraulic system, the fluid near a solid surface travels more slowly than the main stream of the fluid. In other words, there is a liquid boundary layer at the solid surface. This is also true at the surface of the membrane in a spiral-wound element or in any other membrane packaging configuration. Since water is transmitted through the membrane at a much more rapid rate than minerals, the concentration of the minerals builds up in the boundary layer (concentration polarization) and it is necessary for the minerals to diffuse back into the flowing stream. Polarization will reduce both the flux and rejection of a reverse osmosis system. Since it is impractical to totally eliminate the polarization effect, it is necessary to minimize it by good design and operation.

The boundary layer effect can be minimized by increased water flow velocity and by promoting turbulence within the RO elements. Brine flow rates can be kept high as product water is removed by staging (reducing) the module pressure vessels. This is popularly referred to as a "Christmas Tree" arrangement. Typical flow arrangements such as 4 units - 2 units - 1 unit (85 percent recovery) or 2 units - 1 unit (75 percent recovery) are used most often (Figure 16.25).

These configurations consist of feeding water to a series of pressure vessels in parallel where about 50 percent of the water is separated by the membrane as product water and 50 percent of the water is rejected. The reject is then fed to half as many vessels in parallel where again about 50 percent is product water and 50 percent rejected. This reject becomes the feed for the next set of vessels. By arranging the pressure vessels in the 4-2-1 arrangement, it is possible to recover over 85 percent of the feedwater as product water and to maintain adequate flow rates across the membrane surface to minimize polarization. For example, a

---

[24] *Hydrolysis* (hi-DROLL-uh-sis). (1) A chemical reaction in which a compound is converted into another compound by taking up water. (2) Usually a chemical degradation of organic matter.

[25] *Concentration Polarization.* (1) A buildup of retained particles on the membrane surface due to dewatering of the feed closest to the membrane. The thickness of the concentration polarization layer is controlled by the flow velocity across the membrane. (2) Used in corrosion studies to indicate a depletion of ions near an electrode. (3) The basis for chemical analysis by a polarograph.

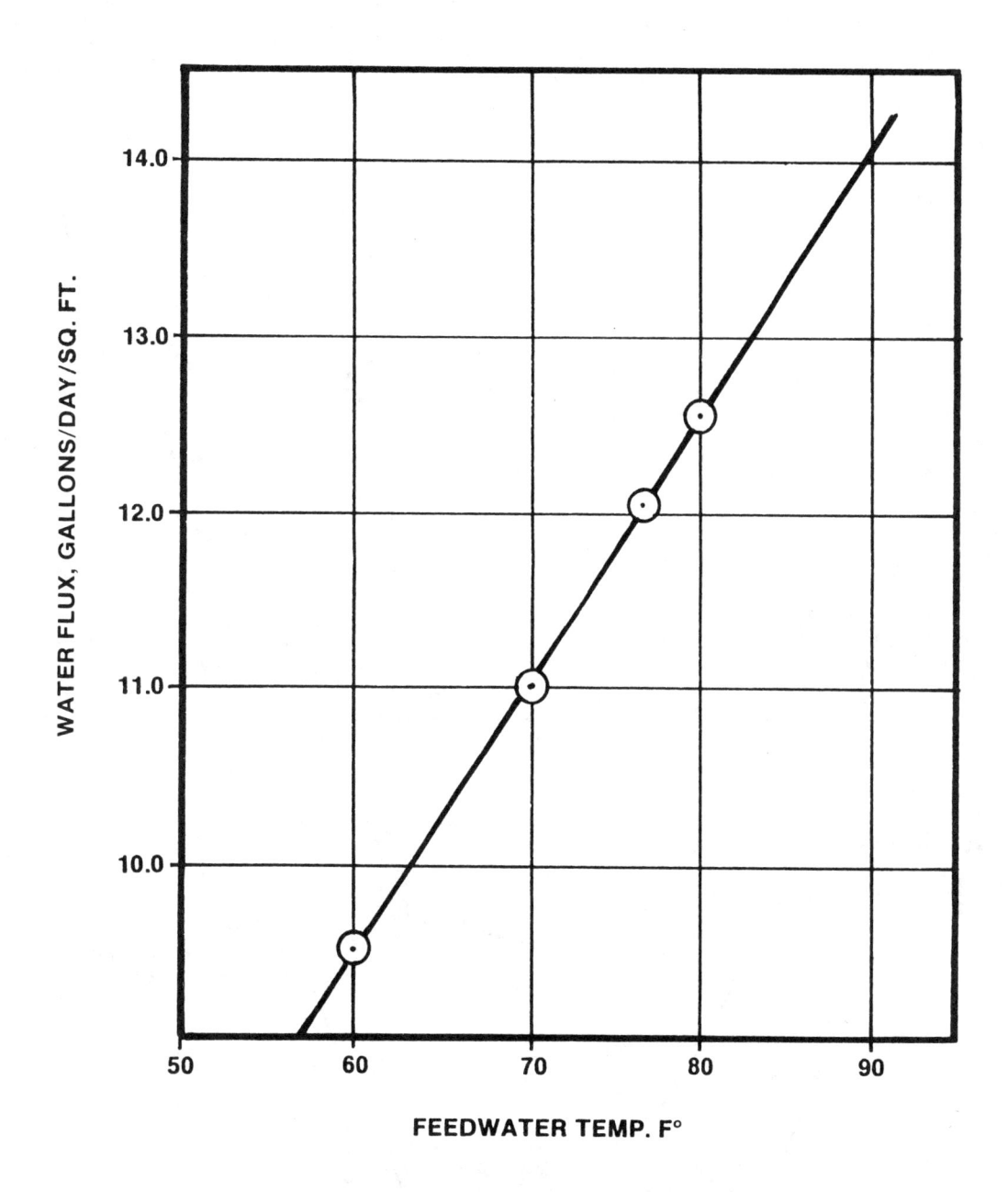

Fig. 16.23 *Effect of temperature on water flux rate, cellulose acetate membrane operating pressure at 400 psi (2,758 kPa or 28 kg/sq cm) net*
(Source: REVERSE OSMOSIS PRINCIPLES AND APPLICATIONS, by Fluids Systems, Division of UOP, October 1970)

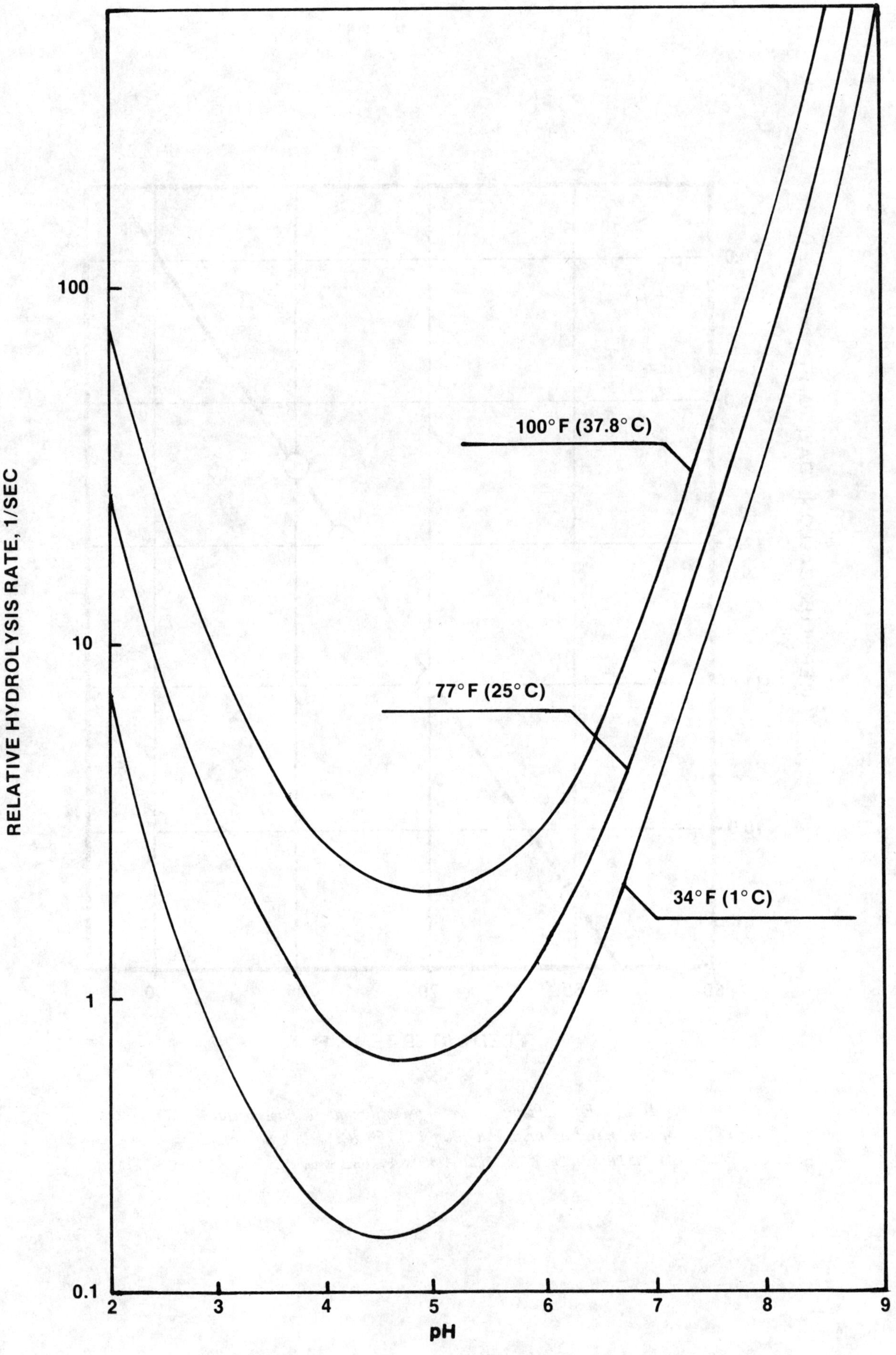

Fig. 16.24 Effect of temperature and pH on hydrolysis rate for cellulose acetate membrane

FOUR PRESSURE VESSELS **TWO PRESSURE VESSELS** **ONE PRESSURE VESSEL**

FEEDWATER → → BRINE TO WASTE

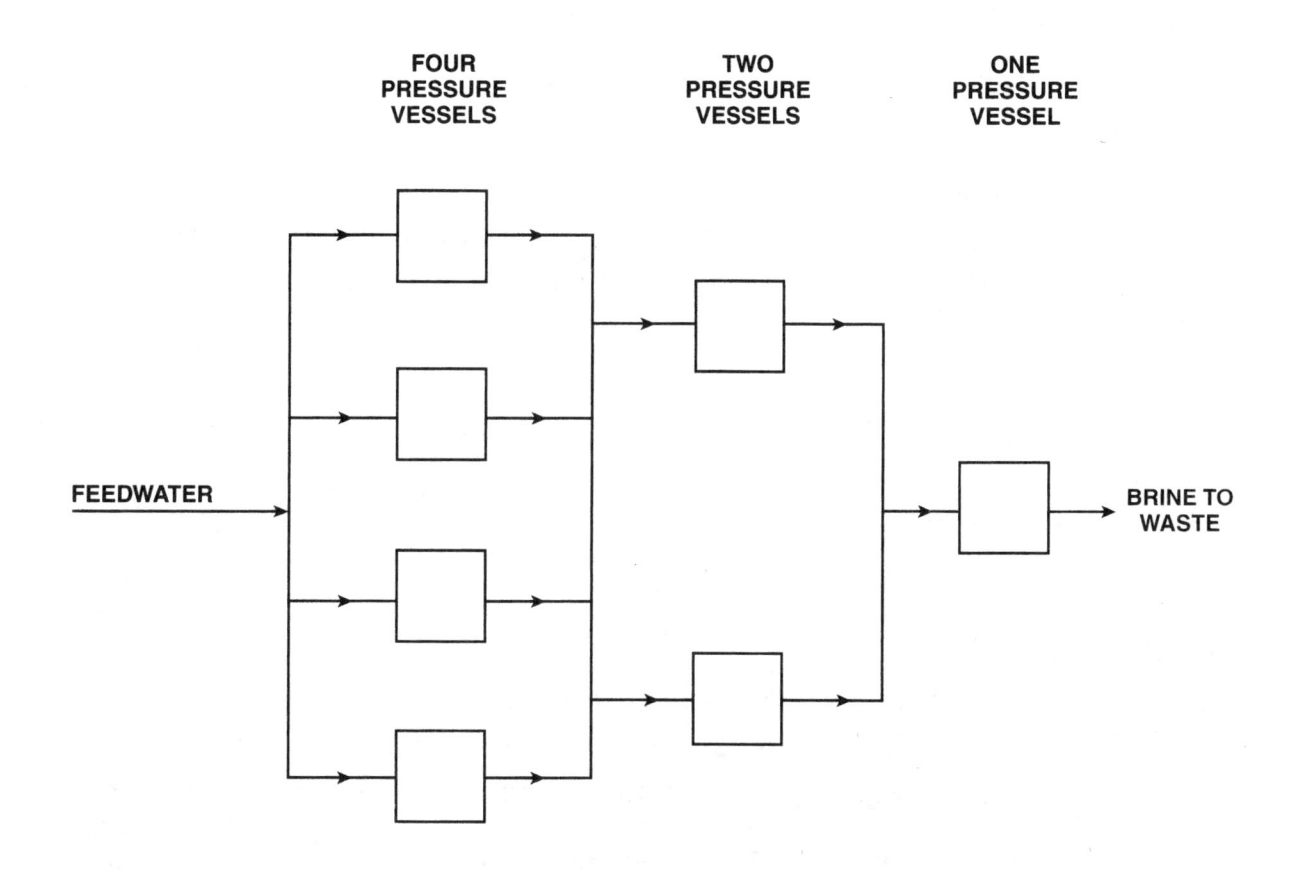

NOTES:
1. BRINE FLOWS OUT OF PRESSURE VESSELS TO NEXT VESSEL.
2. PRODUCT WATER IS NOT SHOWN. PRODUCT WATER FLOWS OUT OF EACH VESSEL INTO A COMMON HEADER.

Fig. 16.25 Typical 4-2-1 "Christmas Tree" arrangement

system consisting of a total of 35 vessels would have a 20-10-5 pressure vessel arrangement for an 85 percent recovery.

EXAMPLE 4

Estimate the percent recovery of a reverse osmosis unit with a 4-2-1 arrangement if the feed flow is 5.88 MGD and the product flow is 5.0 MGD.

Known	Unknown
Product Flow, MGD = 5.0 MGD	Recovery, %
Feed Flow, MGD = 5.88 MGD	

Calculate the recovery as a percent.

$$\text{Recovery, \%} = \frac{(\text{Product Flow, MGD})(100\%)}{\text{Feed Flow, MGD}}$$

$$= \frac{(5.0 \text{ MGD})(100\%)}{5.88 \text{ MGD}}$$

$$= 85\%$$

QUESTIONS

Write your answers in a notebook and then compare your answers with those on page 225.

16.7J   How will an increase in feedwater temperature influence the water flux?

16.7K   How does hydrolysis influence the mineral rejection capability of a membrane?

16.7L   How is recovery defined?

16.7M   Recovery rate is usually limited by what two considerations?

16.7N   Define concentration polarization.

END OF LESSON 2 OF 4 LESSONS

on

MEMBRANE TREATMENT PROCESSES

Please answer the discussion and review questions next.

DISCUSSION AND REVIEW QUESTIONS
CHAPTER 16.   MEMBRANE TREATMENT PROCESSES
(MEMBRANE FILTRATION AND DEMINERALIZATION)
(Lesson 2 of 4 Lessons)

Write the answers to these questions in your notebook. The question numbering continues from Lesson 1.

5. Why has brackish water not been widely used for municipal drinking water supplies?

6. What is reverse osmosis?

7. Indicate what will happen to both the water flux and mineral flux when:

   a. Pressure differential applied across the membrane (ΔP) increases

   b. Osmotic pressure differential across the membrane ($\Delta \pi$) increases

   c. Concentration gradient across the membrane ($C_1 - C_2$) increases

8. What usually happens to water flux with time? Explain.

9. How does fouling develop on membranes?

10. What factors influence the rate of hydrolysis of a membrane? Explain.

11. What is the most common and serious problem resulting from concentration polarization?

12. Why do demineralization plants use a pressure vessel Christmas Tree configuration?

CHAPTER 16. MEMBRANE TREATMENT PROCESSES
(MEMBRANE FILTRATION AND DEMINERALIZATION)
(Lesson 3 of 4 Lessons)

16.8 COMPONENTS OF A REVERSE OSMOSIS UNIT

A reverse osmosis unit is shown in Figure 16.26. Several RO units, along with the necessary auxiliary equipment for a large reverse osmosis demineralization plant, are shown in Figure 16.27. The major components are shown in the schematic, Figure 16.28.

16.80 Pressurization Pump

The pressures required for reverse osmosis can range from 100 to 1,200 psi. Typically, the pressure ranges can be broken down as follows:

Application	Pressure Range (psi)
Softening	100–200
Brackish	200–500
Brackish/Seawater Mixture, Industrial Concentrating	500–800
Seawater	800–1,200

Two basic types of pumps are used for pressurizing the feedwater: centrifugal and positive displacement. Important characteristics of each type of pump relating to its use in RO applications are listed below.

Centrifugal Pumps

1. Typically used for applications less than 500 psi.
2. Most cost effective for applications below 500 psi.
3. Single impellers geared to operate higher than motor speed create excessive noise.
4. Multistage centrifugal pumps are more costly than single-stage but more efficient.

Positive Displacement Pumps

1. Typically used for applications greater than 500 psi.
2. Very efficient for seawater (800 to 1,200 psi).
3. Flow pulsations require use of pulsation dampener for velocities greater than 2 FPS (feet per second).

The output of a centrifugal pump may be throttled by use of a multiturn throttling valve. This is often done for new systems, or after a successful membrane cleaning.

The output of a positive displacement pump may not be throttled. The pump discharge line should contain a pressure relief mechanism. Optional items would be a bypass valve to control flow to the membrane section and a pulsation dampener.

16.81 Piping

The selection of piping material depends on the water salinity and pressure. Seawater reverse osmosis requires the use of high-grade stainless steels for high-pressure lines. The most common types of materials used to date are 316L and 317L. These have been chosen for the high-molybdenum content. Brackish water plants typically use 304 and 316 stainless steel.

Low-pressure piping is typically polyvinyl chloride (PVC) or fiber-reinforced plastic (FRP). Some exotic materials such as 316SS and polyvinylidene fluoride (PVDF) are used in high-purity applications such as for semiconductor rinse water.

16.82 Pressure Vessel Housings

Several spiral-wound membrane elements (described later in Section 16.810) are connected in series and contained in pressure vessels. For most applications, a maximum of six 40-inch long spiral-wound elements are contained in a single vessel. Due to improvements in the hydraulics of the spiral-wound design, seven 40-inch long elements have been placed in one vessel. The standard material of construction is fiber-reinforced polyester (FRP). The pressure vessels are available in 200, 400, 600, 1,000, and 1,200 psi ratings. Some manufacturers can provide vessels constructed and stamped according to ASME Code—Section X.

Hollow fiber bundles (described later in Section 16.810) are packaged in individual fiberglass housings. For seawater desalination, these housings can be rated up to 1,200 psi.

16.83 Concentrate Control Valve

A regulating valve located in the concentrate line provides a means of applying a back pressure to the membrane. Positioning this valve in conjunction with the pump discharge valve (bypass valve for positive displacement pumps) will set the concentrate and permeate flow rates.

Fig. 16.26   Reverse osmosis unit

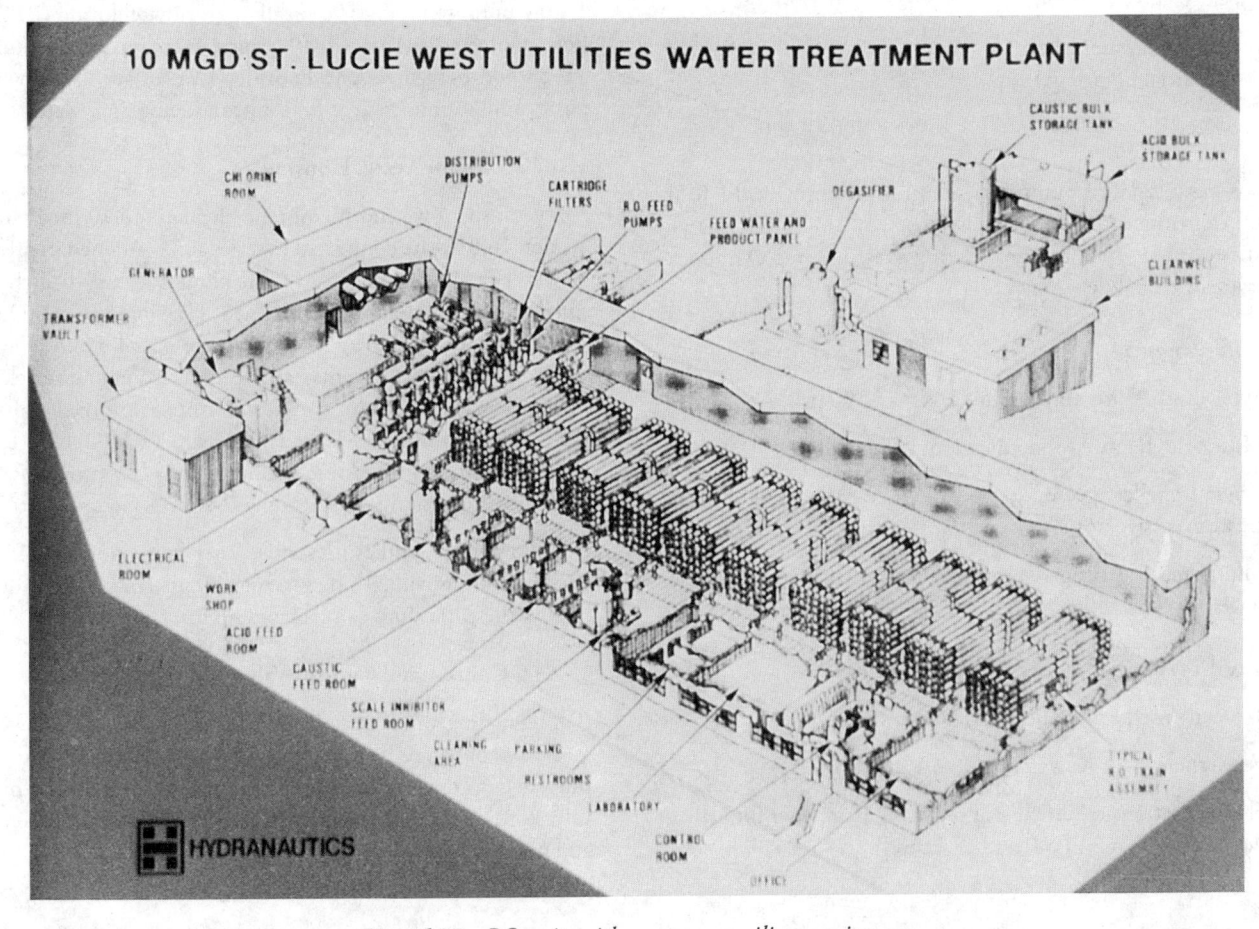

Fig. 16.27   RO units with necessary auxiliary equipment

Membrane Treatment Processes 201

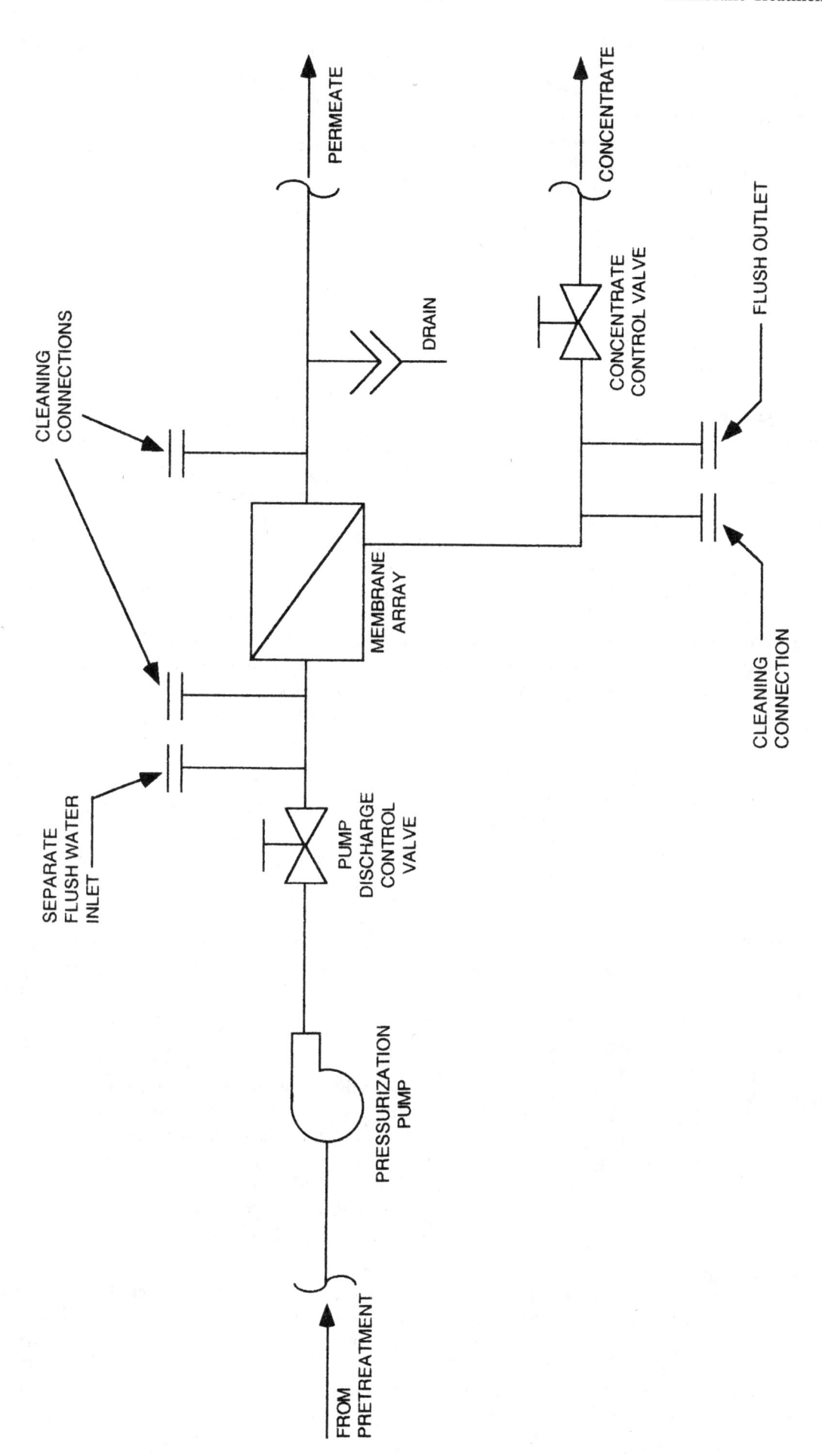

Fig. 16.28 Schematic of reverse osmosis unit

16.84 Sample Valves

Sample valves should be located on the feed, permeate, and concentrate lines. Locations should be such that samples can be taken during all modes of operation such as servicing, flushing, cleaning, and rinsing. Sample valves should also be located in the permeate line of each permeator or pressure vessel.

16.85 Flush Connections

Provisions should be made for flushing the unit for certain applications. Examples would be seawater or brackish waters with high organic content. The flush water could be acidified feed or permeate. If permeate is used, a separate inlet would be required. For all units that require flushing, a separate outlet in the concentrate line upstream of the concentrate control valve should be provided. The concentrate control valve would restrict flush flow, which is usually greater than the design concentrate flow.

16.86 Cleaning Connections

All units should have cleaning connections for each bank of permeators or pressure vessels connected in parallel. Isolation valves for each bank would allow for one bank to soak while the upstream or downstream bank is being cleaned. On large systems with many vessels or permeators connected in parallel, the cleaning system is sized for economic reasons to clean only a portion of the bank. In this case, valves are required to isolate the specific amount of vessels that can be cleaned at one time.

16.87 Permeate Rinse

It is useful to have provisions for sending the permeate from one bank or unit to drain. Some processes require that the permeate achieve quality by rinsing to drain after a shutdown period. Also, for certain troubleshooting procedures, poor quality permeate can be directed to drain while individual vessels are checked for poor quality.

16.88 Permeate Drawback Tank

For seawater applications, a permeate drawback tank may be provided. The purpose of the drawback tank is to provide a supply of water for an off-line unit that is subject to osmosis. Upon shutdown with removal of applied pressure, reverse osmosis ceases and osmosis begins. During osmosis, flow will occur from permeate to the feed-concentrate side of the membrane. Flushing the feed-concentrate channels with permeate after shutdown should prevent natural osmosis but, as a precaution, drawback banks may be provided to prevent dehydration of the membrane.

16.89 Energy Recovery Devices

Energy recovery devices installed in the high-pressure concentrate line are used in some seawater reverse osmosis plants. The basic principle of operation is the conversion of the potential energy of the high-pressure concentrate into kinetic energy. A nozzle directs the concentrate flow toward a rotor with dished vanes. The flow strikes the vanes and turns the rotor. The shaft of the rotor is connected to a pump that prepressurizes the seawater feed. Due to lack of cost effectiveness, corrosion problems, and size restrictions, energy recovery turbines of this type have seen limited use.

16.810 Membranes

Operating plants use the RO principle in several different membrane configurations. Three types of commercially available membrane configurations that have been used in operating plants include:

1. Spiral wound
2. Hollow fine fiber
3. Tubular

The spiral-wound RO module was developed by Gulf Environmental Systems Company (now Fluid Systems Division, UOP) under contract to the US Office of Saline Water. This RO unit was conceived as a method of obtaining a relatively high ratio of membrane area to pressure vessel volume. The membrane is supported on both sides of a backing material and sealed with glue on three of the four edges of the laminate. The laminate is also sealed to a central tube that has been drilled to allow the demineralized water to enter. The membrane surfaces are separated by a screen material that acts as a brine spacer. The entire package is then rolled into a spiral configuration and wrapped in a cylindrical form. The membrane modules are loaded, end to end, into a pressure vessel as shown in Figure 16.29. Feed flow is parallel to the central tube while permeate flows through the membrane toward the central tube. Plants using this type of system include the brackish water demineralizing plants at Key Largo, Florida, and Kashima, Japan; the wastewater demineralizing plants in California; and the seawater demineralizing plant at Jeddah in Saudi Arabia. Spiral-wound membrane elements are pictured in Figure 16.29.

The hollow fiber type of membrane was developed by DuPont and Dow Chemical. The membranes manufactured by DuPont are made of aromatic polyamide fibers about the size of a human hair with an inside diameter of about 0.0016 inch (0.04 mm). In these very small diameters, fibers can withstand high pressures. In an operating process, the fibers are placed in a pressure vessel; one end of each fiber is sealed and the other end protrudes outside the vessel. The brackish water is under pressure on the outside of the fibers and product water flows inside of the fiber to the open end. A DuPont module is illustrated in Figure 16.30. For operating plants, the membrane modules are assembled in a configuration similar to the spiral-wound unit. Municipal demineralizing plants (manufactured by DuPont) in Greenfield, Iowa, and in Florida and seawater demineralizing plants in the Middle East use this type of membrane.

Tubular membrane processes operate on much the same principle as the hollow fine fiber except that the tubes are much larger in diameter, on the order of 0.5 inch (12 mm). Use of this type of membrane system is usually limited to special situations such as for wastewater with a high suspended solids concentration. The tubular membrane process is not economically competitive with other available systems for treatment of most waters.

Membrane Treatment Processes 203

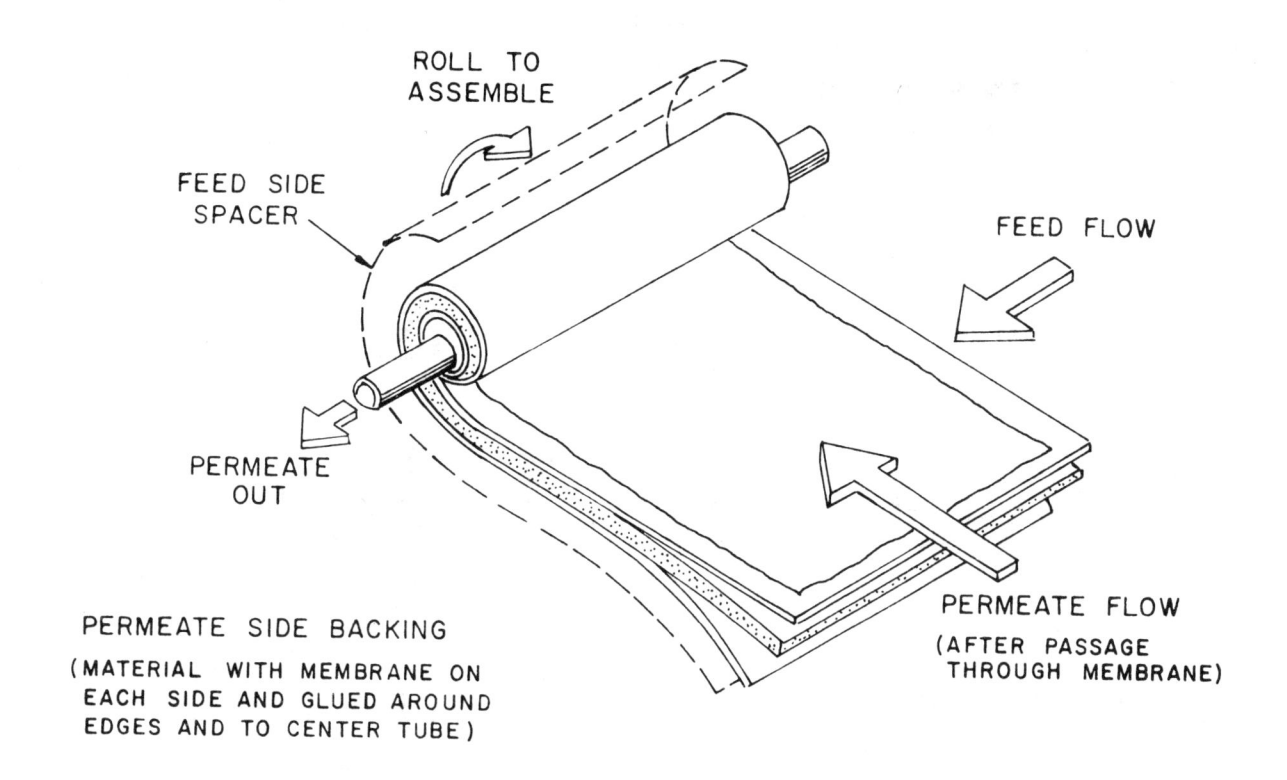

SPIRAL-WOUND REVERSE OSMOSIS MODULE

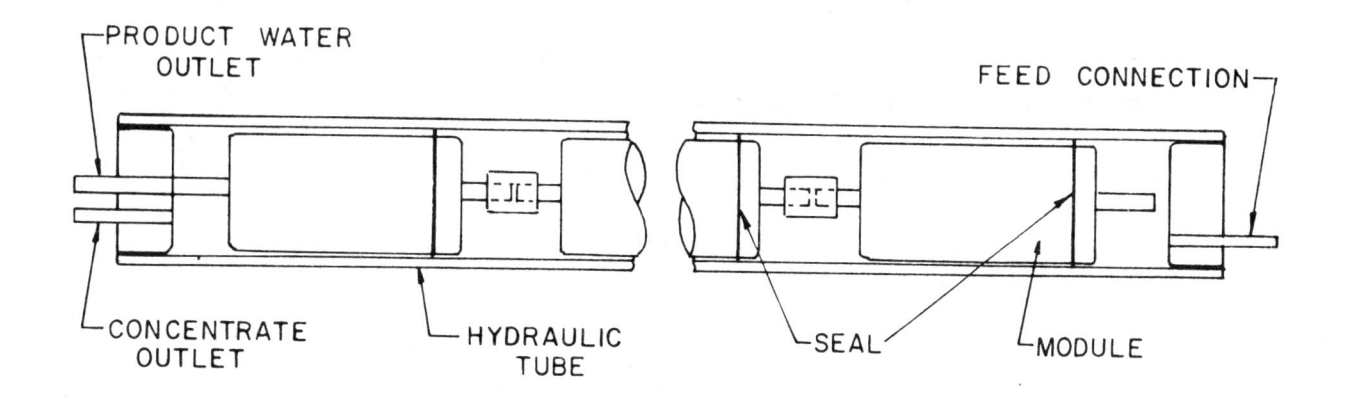

PRESSURE VESSEL ASSEMBLY

Fig. 16.29 Spiral-wound reverse osmosis module (as manufactured by UOP)
(From a paper by Mack Wesner, "Desalting Process and Pretreatment," published in Proceedings on "Role of Desalting Technology,"
a series of Technology Transfer Workshops presented by the Office of Water Research and Technology)

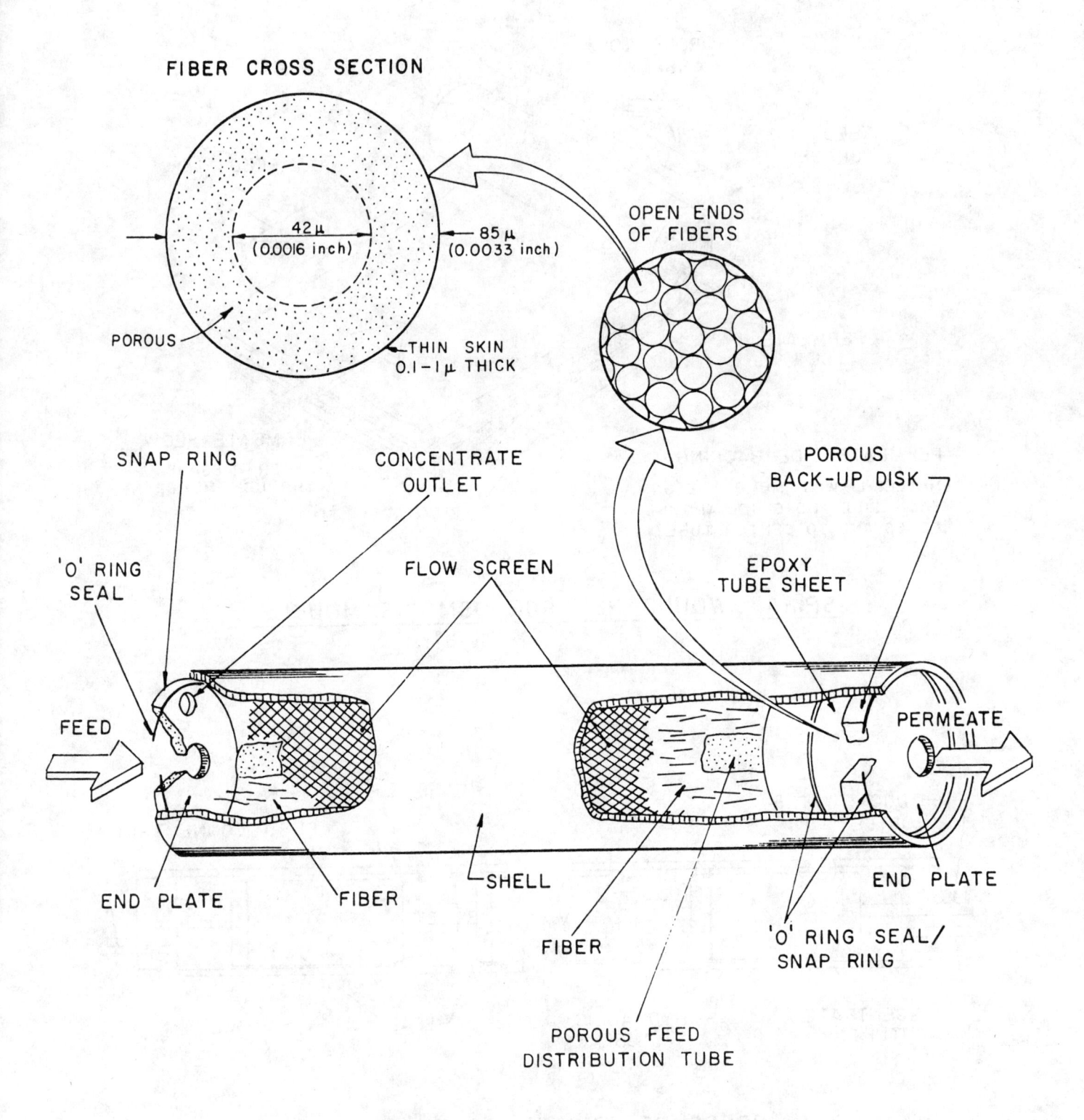

Fig. 16.30 Hollow fiber reverse osmosis module (as manufactured by DuPont)
(From a paper by Mack Wesner, "Desalting Process and Pretreatment," published in Proceedings on "Role of Desalting Technology,"
a series of Technology Transfer Workshops presented by the Office of Water Research and Technology)

> **QUESTIONS**
>
> Write your answers in a notebook and then compare your answers with those on page 225.
>
> 16.8A List the three types of commercially available membrane configurations that have been used in operating plants.
>
> 16.8B What type of membrane process is used to treat wastewater with a high suspended solids concentration?

16.9 OPERATION

16.90 Pretreatment

Water to be demineralized always contains impurities that should be removed by pretreatment to protect the membrane and to ensure maximum efficiency of the reverse osmosis process. Depending on the water to be demineralized, it is usually necessary to treat the feedwater to remove materials and conditions potentially harmful to the RO process such as:

1. Remove turbidity/suspended solids
2. Adjust pH and temperature
3. Remove materials to prevent scaling or fouling
4. Disinfect to prevent biological growth

Table 16.6 provides a summary of pretreatment requirements for various feedwater constituents.

16.91 Removal of Turbidity and Suspended Solids

In general, the feedwater should be filtered to protect the reverse osmosis system and its accessory equipment. When the water source is a groundwater or previously treated municipal or industrial supply, this may be accomplished by a simple screening procedure. However, such a procedure may not be adequate

TABLE 16.6 RO PRETREATMENT REQUIREMENTS

Constituent	Problem	Treatment
Fine suspended solids (that is, causing turbidity)	Fouling. Collects on face of element and on the feed concentrate mesh spacer.	Clarification and/or coagulant addition followed by media filtration.
Gross suspended solids and particulates (that is, sand, debris)	Blockage of lead element face and entrapment within feed/concentrate channels.	Media filtration or cartridge filtration. Cartridge filters usually required to protect high-pressure pumps as well.
Oil and Grease	Fouls membrane surface. Difficult to remove by cleaning.	Ultrafiltration. Batch separators, skimmers.
Microbiological matter	Fouling of the feed/concentrate mesh spacer. Adheres to some membrane surfaces.	Addition of oxidizing agent. In limited cases, ultraviolet sterilization.
Oxidizing agents (that is, chlorine)	Causes membrane degradation for some types of membranes.	Activated carbon or addition of reducing agents (sodium bisulfite).
Carbonates (CO_3, HCO_3)	Scaling. Combines with excess calcium to form calcium carbonate scale on membranes' surfaces.	Acid addition to convert alkalinity to carbon dioxide. Softening to remove calcium. Scale inhibitor in some cases.
Sulfate (SO_4)	Scaling. Combines with calcium, barium, strontium, or aluminum to form scale on membranes' surfaces.	Scale inhibitor addition. Softening to remove cations that could combine with sulfate. Reduce recovery of RO unit.
Silica	Scaling. Forms insoluble gel on membrane surface when supersaturated.	Lime softening. Reduce recovery of RO unit.
Iron (Ferric, +3)	Scaling and fouling. Forms iron oxides. Iron in the ferrous (+2) form is acceptable.	For groundwaters, no aeration or addition of oxidizing agents prior to RO unit. Greensand filters for removal of all forms of iron prior to RO unit.
Hydrogen Sulfide (H_2S)	Scaling and fouling (due to formation of elemental sulfur) when H_2S is aerated or oxidized prior to RO unit.	Best treatment technique is to avoid aeration or oxidation prior to RO unit and to remove H_2S from permeate by degasification.

when the source is an untreated surface water. The amount of suspended matter in surface waters may vary by several orders of magnitude and may change radically in character and composition in a very short time. In such cases, in addition to the mechanical action of the filter, the operator may have to introduce chemicals for coagulation and flocculation and use filtration equipment in which the media can be washed or renewed at low cost. Pressure and gravity sand filters and diatomaceous earth filters may be required, particularly for large installations. When the particulates approach or are COLLOIDAL,[26] chemical treatment and filtration are almost essential.

Cartridge filters function as a particle safeguard and not as a primary particle removal device. In general, the influent turbidity to a cartridge filter should be less than one TU. Typical cartridge filter sizes range from 5 to 20 microns and loading rates vary from 2 to 4 GPM/sq ft (1.4 to 2.8 mm/sec).

16.92 pH and Temperature Control

As previously discussed, an important limiting factor in the life of cellulose acetate membranes in reverse osmosis is the rate of membrane hydrolysis. Cellulose acetate will break down (hydrolyze) to cellulose and acetic acid. The rate at which this hydrolysis occurs is a function of feedwater or source water pH and temperature. As the membrane hydrolyzes, both the amount of water and the amount of solute that permeate the membrane increase and the quality of the product water deteriorates. The rate of hydrolysis is at a minimum at a pH of about 4.7, and it increases with both increasing and decreasing pH. Thus, it is standard practice to inject acid, usually sulfuric acid, to adjust feedwater pH to 5.5. Not only does pH adjustment minimize the effect of hydrolysis, but it is also essential in controlling precipitation of scale-forming or membrane-fouling minerals.

Calcium carbonate and calcium sulfate are probably the most common scaling salts encountered in natural waters and are certainly the most common cause of scale in reverse osmosis systems. The addition of a small amount of acid can reduce the pH to a point where the alkalinity is reduced; this shifts the equilibrium to the point where calcium bicarbonate, which is much more soluble, is present at all points within the reverse osmosis loop. Neutralization of 75 percent of the total alkalinity usually provides sufficient pH adjustment to achieve calcium carbonate scale control and bring the membrane into a reasonable part of the hydrolysis curve. The pH reached by 75-percent neutralization is about 5.7. Calcium carbonate precipitation is also inhibited by the control procedure used for calcium sulfate.

Calcium sulfate is relatively soluble in water in comparison to calcium carbonate. Again, however, as "pure" or product water is removed from a feed solution containing calcium and sulfate, these chemicals become further concentrated in the feedwater. When the limits of saturation are eventually exceeded, precipitation of calcium sulfate will occur. Since calcium sulfate solubility occurs over a wide pH range, the scale control method used to inhibit calcium sulfate precipitation is a threshold treatment[27] with sodium hexametaphosphate (SHMP). This precipitation inhibitor represses both calcium carbonate and calcium sulfate by interfering with the crystal formation process. Other polyphosphates may also be used but are not as effective as the hexametaphosphate. Generally 2 to 5 mg/L of SHMP are added to the feedwater to decrease precipitation of calcium sulfate.

16.93 Other Potential Scalants

The oxides or hydroxides most commonly found in water are those of iron, manganese, and silica. The oxidized and precipitated forms of iron, manganese, and silica can be a serious problem to any demineralization scheme because they can coat the reverse osmosis membrane with a tenacious (difficult to remove) film, which will affect performance. The scale inhibitor most frequently used is sodium hexametaphosphate.

16.94 Microbiological Organisms

Reverse osmosis modules provide a large surface area for the attachment and growth of bacterial slimes and molds. These organisms may cause membrane fouling or even module plugging. There is also some evidence that occasionally the enzyme systems of some organisms will attack the cellulose acetate membrane. The continuous application of chlorine to produce a 1 to 2 mg/L chlorine residual will help inhibit or retard the growth of most of the organisms encountered. However, caution must be exercised since continuous exposure of the membrane to higher chlorine residuals will impair membrane efficiency. Shock concentrations of up to 10 mg/L of chlorine are applied from time to time. When an oxidant-intolerant polyamide-type membrane is used, chlorination must be followed with dechlorination. One of the dechlorination agents, sodium bisulfite, is also known to be a disinfectant. Another disinfection option is the use of ultraviolet disinfection, which leaves no oxidant residual in the water.

QUESTIONS

Write your answers in a notebook and then compare your answers with those on page 225.

16.9A How are turbidity and suspended solids removed from feedwater to the reverse osmosis system?

16.9B How are colloidal particulates removed from feedwater to the reverse osmosis system?

---

[26] *Colloids* (KALL-loids). Very small, finely divided solids (particles that do not dissolve) that remain dispersed in a liquid for a long time due to their small size and electrical charge. When most of the particles in water have a negative electrical charge, they tend to repel each other. This repulsion prevents the particles from clumping together, becoming heavier, and settling out.

[27] Threshold treatment refers to the practice of using the least amount of chemical to produce the desired effect.

16.9C What happens to the product water as an acetate membrane hydrolyzes?

16.9D How is the precipitation of calcium sulfate prevented?

16.9E How is biological fouling on membranes controlled?

16.95 RO Plant Operation

Following proper pretreatment, the water to be demineralized is pressurized by high-pressure feed pumps and delivered to the RO pressure vessel membrane assemblies. An example of a typical RO plant layout is given in Figure 16.31. The membrane assemblies consist of a series of pressure vessels (usually fiber-reinforced plastic) arranged in the "Christmas Tree" layout depending on the desired recovery. Typical operating pressure for brackish water demineralizing varies from 150 to 400 psi (1,050 to 2,760 kPa or 10.5 to 28 kg/sq cm). A control valve on the influent manifold regulates the operating pressure. The volumes of feed flow and of product water are also monitored. The demineralized water is usually called permeate, and the reject, concentrate (brine). The recovery rate is controlled by increasing feed flow (increase operating pressure) and controlling the concentrate (brine) or reject with a preset brine control valve.

The operator must properly maintain and control all flows and recovery rates to avoid possible damage to the membranes from scaling.

You must remember that the

BRINE FLOW VALVES ARE NEVER TO BE FULLY CLOSED.

Should they be accidentally closed during operation, 100 percent recovery will result in almost certain damage to the membranes due to the precipitation of inorganic salts ($CaSO_4$). Product or permeate flow is not regulated and varies as feedwater pressure and temperature change as previously discussed.

Most RO systems are designed to operate automatically and require a minimum of operator attention. However, the continuous monitoring of system performance is an important aspect of operation. An example of a typical operation log for monitoring the Orange County Water District's 5 MGD (19 MLD) RO plant is given in Figure 16.32.

QUESTIONS

Write your answers in a notebook and then compare your answers with those on page 225.

16.9F How is the operating pressure on a reverse osmosis unit regulated?

16.9G The demineralized water is usually called _____, the reject, _____.

16.9H How does the product or permeate flow vary or change?

16.96 Typical RO Plant Operations Checklist

1. Check cartridge filters. Properly installed filters ensure additional removal of suspended solids that could damage either the high-pressure feed pumps or foul the membrane elements. Cartridge filters should be replaced whenever the head loss exceeds the manufacturer's recommendation or effluent turbidity exceeds one TU.

2. Start up and check scale inhibitor feeding equipment and adjust feed rate to desired dose (2 to 5 mg/L). Most RO systems should not be operated without the addition of a scale inhibitor to protect membranes from precipitation of calcium sulfate or other inorganics. The scale inhibitor most frequently used is sodium hexametaphosphate.

3. If chlorine is used to prevent biological fouling, start chlorine feed and adjust dose to produce a chlorine residual of between 1 and 2 mg/L.

4. Start up and adjust acid feed system to correct feedwater pH to a level between 5.0 and 6.0 to protect membranes from possible damage due to hydrolysis. Note, feedwater should always be bypassed until the pH is properly adjusted.

5. Most RO systems are designed with automatic controls and various shutdown alarms. These alarms prevent start-up or running of the unit until proper operating conditions are reached. After satisfying these conditions, high-pressure feed pumps can be started and water delivered to the RO units. A control valve is used to regulate feedwater pressure. Typical operating pressures vary from 150 to 400 psi (1,050 to 2,760 kPa or 10.5 to 28 kg/sq cm).

6. Adjust permeate and concentrate flow to establish the desired recovery rate.

7. Once flow has been established, check the differential pressure (ΔP) across the RO unit (ΔP = feed pressure − concentrate pressure), which is usually indicated by a meter and recorded. The importance of ΔP relates to cleaning. When the elements become fouled, ΔP usually increases, thus indicating the need for cleaning. The ΔP should not exceed 60 psi per 6-element pressure vessel (414 kPa or 4.1 kg/sq cm) because of possible damage to the RO modules.

8. With the system on line, monitor the performance. Rely on flow measurements, product water quality, and various pressure indications. A sample of a typical log sheet is shown in Figure 16.32.

208 Water Treatment

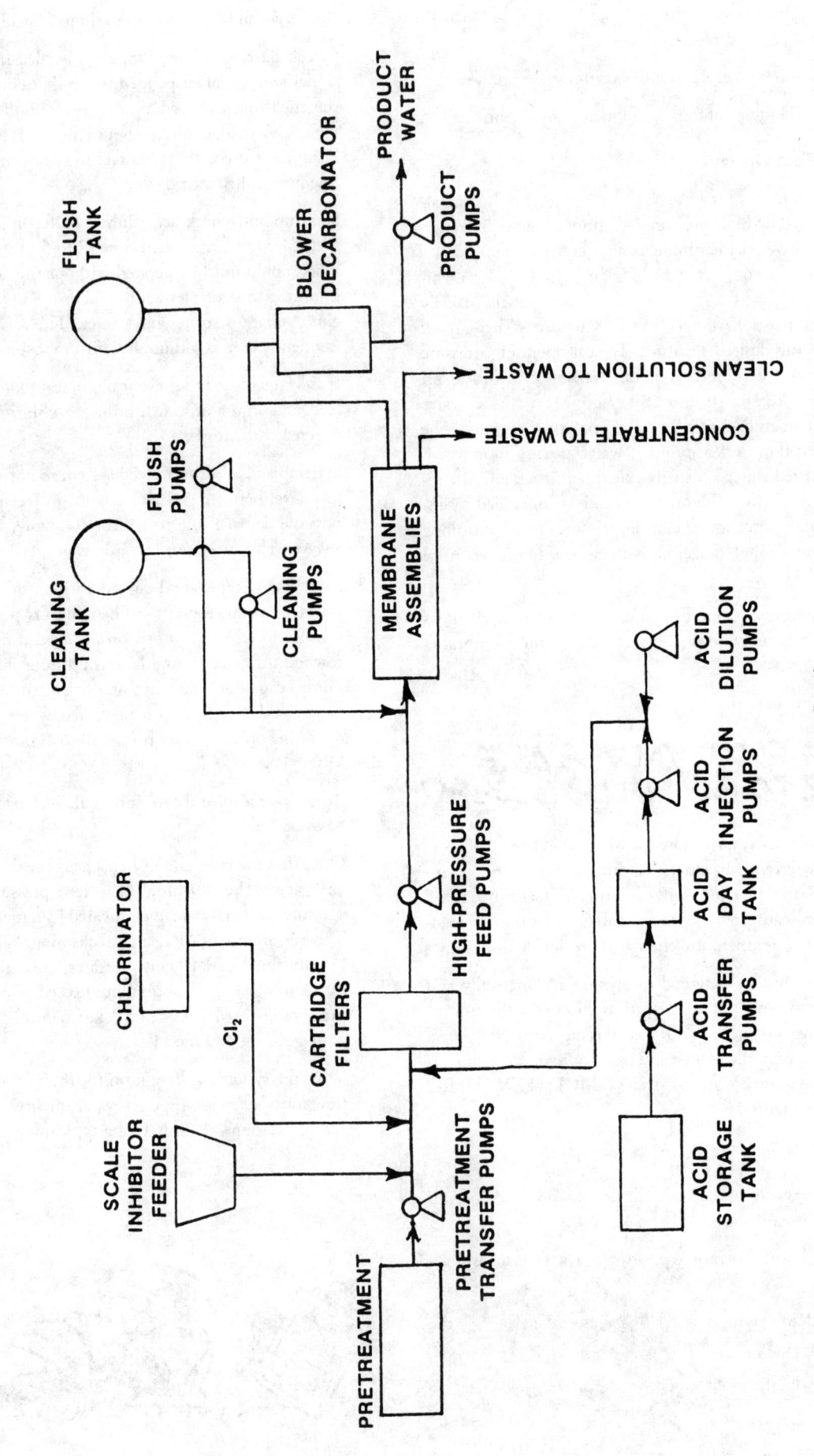

Fig. 16.31 *RO flow diagram*

TIME	Pretreatment											
	Cl mg/ℓ	Cartridge Filters				Turb NTU	Temp °F	Feed Cond μmhos/cm	pH	Pump-2 Disch.		
		#1 psig	#2 psig	#3 psig	#4 psig					A psig	B psig	C psig
0400												
1200												
2000												

TIME	RO Unit 1			RO Unit 2			Total Brine				Total Product		
	Feed psig	Feed Flow MGD	Product Cond. (μ)	Feed psig	Feed Flow MGD	Product Cond. (μ)	Flow MGD	pH	Conductivity μmhos/cm	KWH Ton	pH	Conductivity μmhos/cm	Flow MGD
0400													
1200													
2000													

TIME	RO Unit 1																	
	Section 1A-Product EC						Section 1B-Product EC						Section 1C-Product EC					
	Feed psig	ΔP			Product gpm	Brine MGD	Feed psig	ΔP			Product gpm	Brine MGD	Feed psig	ΔP			Product gpm	Brine MGD
		1st	2nd	3rd				1st	2nd	3rd				1st	2nd	3rd		
0400																		
1200																		
2000																		

TIME	RO Unit 2																	
	Section 2A-Product EC						Section 2B-Product EC						Section 2C-Product EC					
	Feed psig	ΔP			Product gpm	Brine MGD	Feed psig	ΔP			Product gpm	Brine MGD	Feed psig	ΔP			Product gpm	Brine MGD
		1st	2nd	3rd				1st	2nd	3rd				1st	2nd	3rd		
0400																		
1200																		
2000																		

Fig. 16.32 Data sheet, Orange County Water District 5.0 MGD (19 MLD) reverse osmosis system

210 Water Treatment

Shift	Operator

24 Hour Totalizer

TIME	Feed Flow MGD	Bypass Flow MGD	Total Product MGD	Total Brine MGD
(A) 2400 (II)				
(B) 2400 (I)				

Cl_2	SHMP	ACID	ELAPSED TIME		POWER
2400 (II)			RO #1	RO #2	2400 (II)
2400 (I)	lbs	gal			2400 (I)

REMARKS: _____

Date _____

Fig. 16.32 Data sheet, Orange County Water District 5.0 MGD (19 MLD) reverse osmosis system (continued)

16.97 Membrane Cleaning

Periodically, the performance of the RO system will decline. This is usually observed when either the product water flow rate (flux) decreases or salt removal (rejection) decreases. Table 16.7 summarizes common causes of membrane damage or loss of performance. Note that in Cases III and IV the corrective action requires cleaning of the element. Provisions for the periodic cleaning of the reverse osmosis elements are usually included in the system design. This makes it possible to clean impurities off the membrane surfaces and restore normal flow rates without removing the elements from the pressure vessels. Element cleaning should be performed at regular intervals to ensure as low an operating pressure as practical. The elements should be cleaned when the pressure required to maintain the rated capacity has either been increased by 15 percent (or a 15-percent decrease in product water flow has occurred at constant pressure), or a rise of 15 percent in the system differential pressure has been observed.

Most RO systems are provided with in-place cleaning systems. This includes tanks, pumps, valves, and piping for mixing and pumping cleaning solutions through the membrane elements. For cleaning, the unit is shut down and cleaning solutions are pumped through the vessels in a manner similar to feedwater. Typically, cleaning solutions are passed through the pressure vessels at low pressure and at flow rates where the ΔP does not exceed 60 psi (414 kPa or 4.1 kg/sq cm) to avoid damaging the elements. The cleaning solutions are returned to clean tanks at the end of a cleaning cycle, which usually lasts about one hour. Different cleaning solutions are available for use depending upon the type of fouling. Membranes are typically cleaned for approximately 45 minutes after which the cleaning solution is spent.

To remove inorganic precipitates, use an acid flush of citric acid. For biological or organic fouling, various solutions of detergents, sequestrants, chelating agents, bactericides, and enzymes are available. Examples include sodium tripolyphosphate, B13, Triton X-100, and EDTA.

To improve the long-term performance of an RO system, the membranes should be flushed with flush water during periods of shutdown to remove raw feedwater and concentrate. If raw water is allowed to remain in the unit, precipitation may occur. Flushing is also done after cleaning to remove the cleaning solution prior to system start-up. In some cases, where the system is shut down for long periods of time, formaldehyde may be added to the flush water to inhibit biological growth.

> **QUESTIONS**
>
> Write your answers in a notebook and then compare your answers with those on pages 225 and 226.
>
> 16.9I Why is chlorine added to the feedwater to a reverse osmosis unit?
>
> 16.9J Why must the operator check the differential pressure (ΔP) across the RO unit?
>
> 16.9K When should the reverse osmosis elements be cleaned?

16.98 Safety

16.980 Use of Proper Procedures

As in any water treatment plant, there are forces and chemicals used in a reverse osmosis plant that must be handled properly to ensure the safety and protection of personnel. Safety needs for demineralization plants can be divided into three general groups consisting of chemicals, electrical, and hydraulics.

16.981 Chemical Safety

Operation of an RO plant requires the use of a wide variety of chemicals. Whenever you must handle chemicals, follow the

---

TABLE 16.7 SUMMARY OF COMMON CAUSES OF MEMBRANE DAMAGE

	Symptom	Cause	Restoration Procedure
Case I[a]	1. Lower product water flow rate 2. Higher salt rejection	Membrane compaction[b] accelerated by operating pressure greater than 500 psi (3,450 kPa or 35 kg/sq cm).	None. Required element replacement when product water flow rate reaches an unacceptable level.
Case II[a]	1. Higher product water flow rate 2. Lower salt rejection	Membrane hydrolysis 1. pH outside operating limits. 2. Bacterial degradation. 3. Temperature outside operating limits.	Injection of chemical or element replacement.
Case III[c]	1. Lower product water flow rate 2. Lower salt rejection	Membrane fouling.	Element cleaning.
Case IV[c]	1. Lower product water flow rate 2. High ΔP	Membrane fouling.	Element cleaning.

[a] CA–cellulose acetate membranes.
[b] Membrane Compaction. Product water flow rate declines with operational time in addition to fouling of the membrane surface due to other factors. Water flow rate plotted versus time on log-log paper will yield a straight line (flow rate decline).
[c] All types of membranes.

proper procedures for each chemical. The manufacturer's recommendations for use of each chemical must be observed. Chemicals that are commonly used in an RO plant operation and that require special handling include:

1. Acid
2. Chlorine
3. Sodium hexametaphosphate
4. Formaldehyde
5. Citric acid
6. Numerous cleaning agents

See Chapter 20, "Safety," for more detailed procedures on the safe use of hazardous chemicals.

16.982 Hydraulic Safety

For the reverse osmosis process to function properly, hydraulic pressure in excess of the solution's average osmotic pressure (π) is required. Within the plant, therefore, most of the pipes, tubing, vessels, and their associated equipment, along with the substances inside these items, operate under varying levels of hydraulic pressure (150 to 400 psi (1,050 to 2,760 kPa or 10.5 to 28 kg/sq cm)). Therefore, prior to any repairs, modifications, or work of any kind, no matter how minor, know the substances contained, isolate the piece of equipment, and equalize pressure levels to atmospheric pressure.

After being repaired, any piece of equipment should be purged of all foreign substances before being restarted. When bringing a piece or pieces of equipment on line, increase the hydraulic pressures slowly. Keep all personnel in a safe area to maximize their personal safety.

16.983 Electrical Safety

An RO plant consists of a series of electrically powered pumps and mechanical equipment. Electric shocks due to the use of electrical equipment occur without warning and are usually serious. The average individual thinks of the hazards of electric shock in terms of high voltage and does not always realize that it is primarily the current that kills, not the voltage. Consequently, persons who work around low-voltage equipment do not always have the same healthy respect for current as they do for high voltage. Whenever working around electrically operated equipment, strictly observe all applicable rules of the 2007 National Electrical Safety Code.[28]

QUESTIONS

Write your answers in a notebook and then compare your answers with those on page 226.

16.9L List the three general groups of safety needs for a demineralization plant.

16.9M What type of electrical equipment is used around reverse osmosis plants?

END OF LESSON 3 OF 4 LESSONS

on

MEMBRANE TREATMENT PROCESSES

Please answer the discussion and review questions next.

DISCUSSION AND REVIEW QUESTIONS

CHAPTER 16. MEMBRANE TREATMENT PROCESSES

(MEMBRANE FILTRATION AND DEMINERALIZATION)

(Lesson 3 of 4 Lessons)

Write the answers to these questions in your notebook. The question numbering continues from Lesson 2.

13. Why does water to be demineralized require pretreatment?
14. What problems are created for demineralization processes by the oxidized and precipitated forms of iron, manganese, and silica?
15. How does the operator of a reverse osmosis plant avoid possible damage to the membrane from scaling?
16. What will happen in a reverse osmosis plant if the brine flow valves are accidentally closed during operation?
17. What is the purpose of cartridge filters and when should they be replaced?
18. How can the operator determine if the performance of the RO system is declining?
19. What does hydraulic safety consist of around a reverse osmosis process?

---

[28] *2007 NATIONAL ELECTRICAL SAFETY CODE.* Obtain from Institute of Electrical and Electronic Engineers, Inc., IEEE Customer Service Center, 445 Hoes Lane, PO Box 1331, Piscataway, NJ 08855-1331. Order No. SH95514. ISBN 0-7381-4893-8. Price to members, $105.00; nonmembers, $130.00; plus $7.95 shipping and handling.

CHAPTER 16. MEMBRANE TREATMENT PROCESSES
(MEMBRANE FILTRATION AND DEMINERALIZATION)
(Lesson 4 of 4 Lessons)

16.10 ELECTRODIALYSIS

Electrodialysis (ED) is a well-developed process with a history of many years of operation on brackish well water supplies. A 650,000 GPD (2.5 MLD) ED plant, manufactured by Ionics, Inc., Watertown, Massachusetts, began operation on well water at Buckeye, Arizona, in September 1972 and has been in continuous operation to date. ED plants are also in operation demineralizing municipal water supplies in Siesta Key, Florida; Sanibel Island, Florida; Sorrento Shores, Florida; and at the Foss Reservoir in Oklahoma. The process is also used for industrial water demineralizing.

Typical removals of inorganic salts from brackish water by ED range from 25 to 40 percent of dissolved solids per stage of treatment. Higher removals require treatment by multiple stages in series. Less than 20 percent of the organics remaining in activated carbon-treated secondary effluent are removed by electrodialysis. Energy required for ED is about 0.2 to 0.4 kilowatt-hour per 1,000 gallons (kWh/1,000 gal) for each 100 mg/L dissolved solids removed, plus 2 to 3 kWh/1,000 gal for pumping feedwater and brine. Advantages of the ED process include: (1) well-developed technology, including equipment and membranes; (2) efficient removal of most inorganic constituents; and (3) waste brine contains only salts removed plus a small amount of acid used for pH control in some ED applications.

In the ED process, brackish water flows between alternating cation-permeable and anion-permeable membranes as illustrated in Figure 16.33. A direct electric current provides the motive force to cause ions to migrate through the membranes. Many alternating cation- and anion-permeable membranes, each separated by a plastic spacer, are assembled into membrane stacks. The spacers (about 0.04-inch (1-mm) thick) contain the water streams within the stack and direct the flow of water through a tortuous path across the exposed faces of the membranes. Membrane thicknesses generally range between 0.005 and 0.025 inch (0.125 and 0.625 mm).

Physically, the equipment takes the form of a plate-and-frame assembly similar to that of a filter press. The spacers determine the thickness of the solution compartments and also define the flow paths of the water over the membrane surface. Several hundred membranes and their separating spacers are usually assembled between a single set of electrodes to form a membrane stack. End plates and tie rods complete the assembly. When a membrane is placed between two salt solutions and subjected to the passage of a direct electric current, most of the current will be carried through the membrane by ions, hence the membrane is said to be ion selective. Typical selectivities are greater than 90 percent. When the passage of current is continued for a sufficient length of time, the solution on the side of the membrane that is furnishing the ions becomes partially desalted, and the solution adjacent to the other side of the membrane becomes more concentrated. These desalting and concentrating phenomena occur in thin layers of solution immediately adjacent to the membrane, resulting in the desalting of the bulk of the solution.

Passage of water between the membranes of a single stack, or stage, usually requires 10 to 20 seconds, during which time the entering minerals in the feedwater are removed. The actual percentage removal that is achieved varies with water temperature, type and amounts of ions present, flow rate of the water, and stack design. Typical removals per stage range from 25 to 40 percent and systems use one to six stages. An ED system will operate at temperatures up to 110 degrees Fahrenheit (110°F (43°C)) and the removal efficiency increases with increasing temperature. Ion-selective membranes in commercial electrodialysis equipment are commonly guaranteed for as long as 5 years and experience has demonstrated an effective life of over 10 years.

The most commonly encountered problem in ED operation is scaling (or fouling) of the membranes by both organic and inorganic materials. Alkaline scales are troublesome in the concentrating compartments when the diffusion of ions to the surface of the anion membrane in the diluting cell is insufficient to carry the current. Water is then electrolyzed and hydroxide ions pass through the membrane and raise the pH in the cell. This increase is often sufficient to cause precipitation of materials such as magnesium hydroxide or calcium carbonate. The accumulation of particulate matter increases the electrical resistance of the membrane; this may damage or destroy the membranes. This condition can be offset by feeding acid to the concentrate water stream to maintain a negative Langelier Index to ensure scale-free operation.

Ionics, Inc., has developed a type of ED unit that does not require the addition of acid or other chemicals for scale control. This system reverses the DC current direction and the flow path of the dilution and concentrating streams every 15 minutes. The electrodes reverse by switching the polarity of the cathodes and anodes. The stream flow paths also exchange their source every 15 minutes. Motor-operated valves controlled by timers

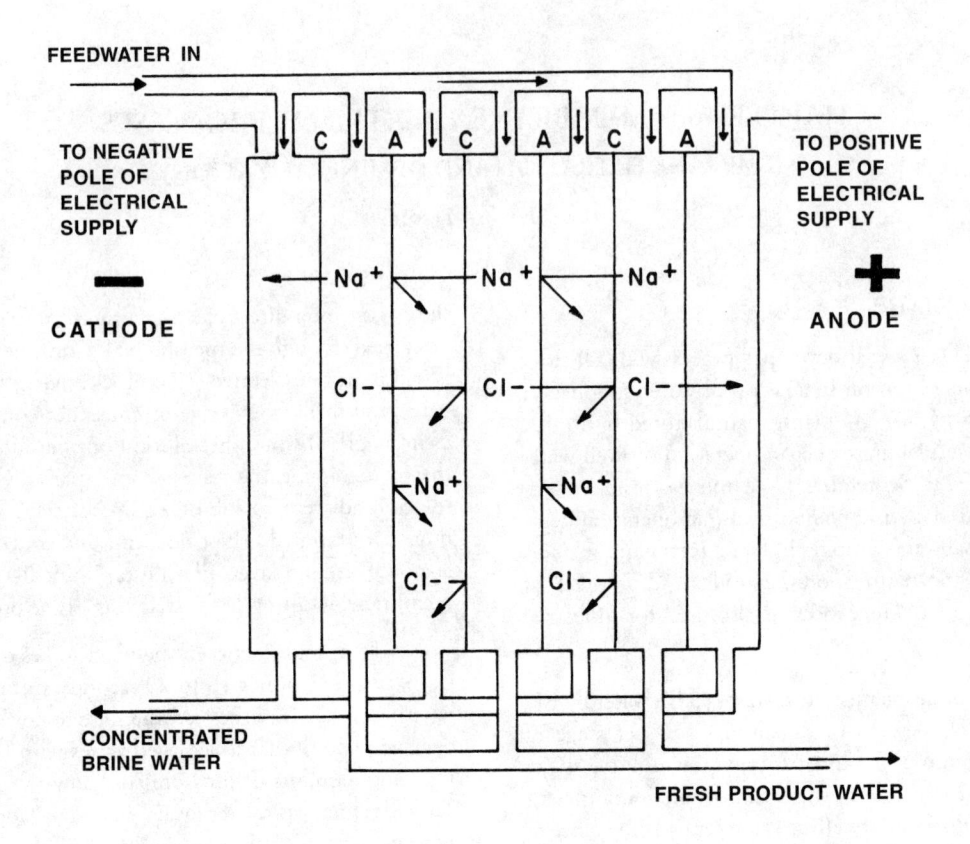

Fig. 16.33 Electrodialysis demineralization process
(From *STANDARD OPERATION INSTRUCTION PLAN FOR ELECTRODIALYSIS*, prepared by Ionics, Inc.)

switch the streams so that the flow path that was previously the diluting stream becomes the concentrating stream and the flow path that was previously the concentrating stream becomes the diluting stream. This reversing polarity system is commonly referred to as electrodialysis polarity reversal (EDR).

QUESTIONS

Write your answers in a notebook and then compare your answers with those on page 226.

16.10A What are the typical removals of inorganic salts from brackish water by electrodialysis (ED) per stage of treatment?

16.10B What is a membrane stack in an electrodialysis unit?

16.10C What is the most commonly encountered problem in ED operation?

16.11 PRINCIPLES OF ELECTRODIALYSIS

16.110 Anions and Cations in Water

When most common salts, minerals, acids, and alkalis are dissolved in water, each molecule splits into two oppositely charged particles called "ions." All positively charged ions are known as "cations" and all negatively charged ions, as "anions." For instance, when common table salt (sodium chloride or NaCl) is dissolved in water, it separates into positive sodium ions (Na$^+$) and negative chloride ions (Cl$^-$). The following ions are found in seawater or brackish water in appreciable quantities.

Cations		Anions	
Sodium	Na$^+$	Chloride	Cl$^-$
Calcium	Ca^{2+}	Bicarbonate	HCO$_3^-$
Magnesium	Mg^{2+}	Sulfate	SO$_4^{2-}$

16.111 Effect of Direct Current (DC) Potential on Ions

If a DC potential is applied across a solution of salt in water by means of insertion of two electrodes in the solution, the cations will move toward a negative electrode, which is known as the "cathode," and the anions will move toward the positive electrode, which is known as the "anode." In Figure 16.34 (A) we have a solution of sodium chloride in water. The cations (Na^+) and anions (Cl^-) are moving about at random. In Figure 16.34 (B) a DC potential has been introduced in the solution and the anions move toward the positive electrode and the cations move toward the negative electrode.

16.112 Anion- and Cation-Permeable Membranes and Three-Cell Unit

Advantage could be taken of this movement of ions if proper barriers were available to isolate the purified zone in Figure 16.34 (B) so as to prevent remixing. There are two types of membranes that can be used as such barriers:

1. Cation-Permeable Membranes—Permit only the passage of cations (positively charged ions).

2. Anion-Permeable Membranes—Permit only the passage of anions (negatively charged ions).

Introduction of a cation-permeable membrane and anion-permeable membrane into a salt solution to form three watertight compartments (Figure 16.34 (C)) followed by a direct electric current into the water (Figure 16.34 (D)) will result in the demineralization of the central compartment.

In the three-cell unit shown in Figure 16.34 (C) and (D), "1" is the anode (positive electrode), "2" is the anion-permeable membrane, "3" is the cation-permeable membrane, and "4" is the cathode (negative electrode). In Figure 16.34 (C) there is no electric flow so the ions move at random in their respective compartments. In Figure 16.34 (D) the introduction of a DC potential gives these ions direction: the cations (Na^+) move toward the cathode and the anions (Cl^-) toward the anode. The following occurs:

1. Na^+ from compartment A cannot pass through anion-permeable membrane (2) into compartment B

2. Cl^- from compartment A reacts at the anode (1) to give off chlorine gas

3. Na^+ from compartment B passes through cation-permeable membrane (3) into compartment C

4. Cl^- from compartment B passes through anion-permeable membrane (2) into compartment A

5. Na^+ from compartment C reacts at the cathode to give off hydrogen gas and hydroxyl ions (OH^-)

6. Cl^- from compartment C cannot pass through cation-permeable membrane (3) into compartment B

This description indicates how the overall effect has produced a demineralization of the central compartment.

16.113 Multicompartment Unit

Figure 16.35 presents a multicompartment unit similar in principle to a stack. Letter "A" designates the anion-permeable membranes; letter "C," the cation-permeable membranes; the "+" sign, the anode; the "−" sign, the cathode. A salt solution of Na^+ and Cl^- ions flows between the membranes. On application of the DC potential, the overall effect will be as shown, a movement of ions from the compartments bounded by an anion-permeable membrane on the left and a cation-permeable membrane on the right into the adjacent compartments. The compartments losing salt are labeled "dilute" and those receiving the transferred salt are labeled "brine." Two electrode compartments are also found in the drawing. Each is bordered by a cation-permeable membrane and the electrode. At the anode, a reaction takes place evolving chlorine and oxygen gases; at the cathode, hydrogen gas is produced and hydroxyl ions (OH^-) are left in the solution. Hydroxyl ions are alkaline.

> **QUESTIONS**
>
> Write your answers in a notebook and then compare your answers with those on page 226.
>
> 16.11A What happens if a DC potential is applied across a solution of salt in water by means of insertion of two electrodes in the solution?
>
> 16.11B What type of ions can pass through cation-permeable membranes?
>
> 16.11C In a multicompartment ED unit, the compartments losing salt are labeled _____ and those receiving the transferred salt, _____.

16.12 PARTS OF AN ELECTRODIALYSIS UNIT

16.120 Flow Diagram

The basic electrodialysis unit consists of:

1. Pretreatment equipment

2. Pumping equipment (feed, brine, and recirculation)

3. DC power supply

4. Membrane stack and electrodes

5. In-place cleaning system

Figure 16.36 shows a typical flow diagram and Figure 16.37 a photo of an electrodialysis unit.

16.121 Pretreatment

A certain degree of pretreatment of the feedwater supply is necessary in order to prepare it for demineralization in the stacks. Pretreatment depends on the specific water being treated, but it usually includes the removal of suspended or dissolved solids that could adversely affect the surface of the membranes or mechanically block the narrow passageways in the individual cells. Cartridge filters are used as a particle safeguard before the

216 Water Treatment

A

B

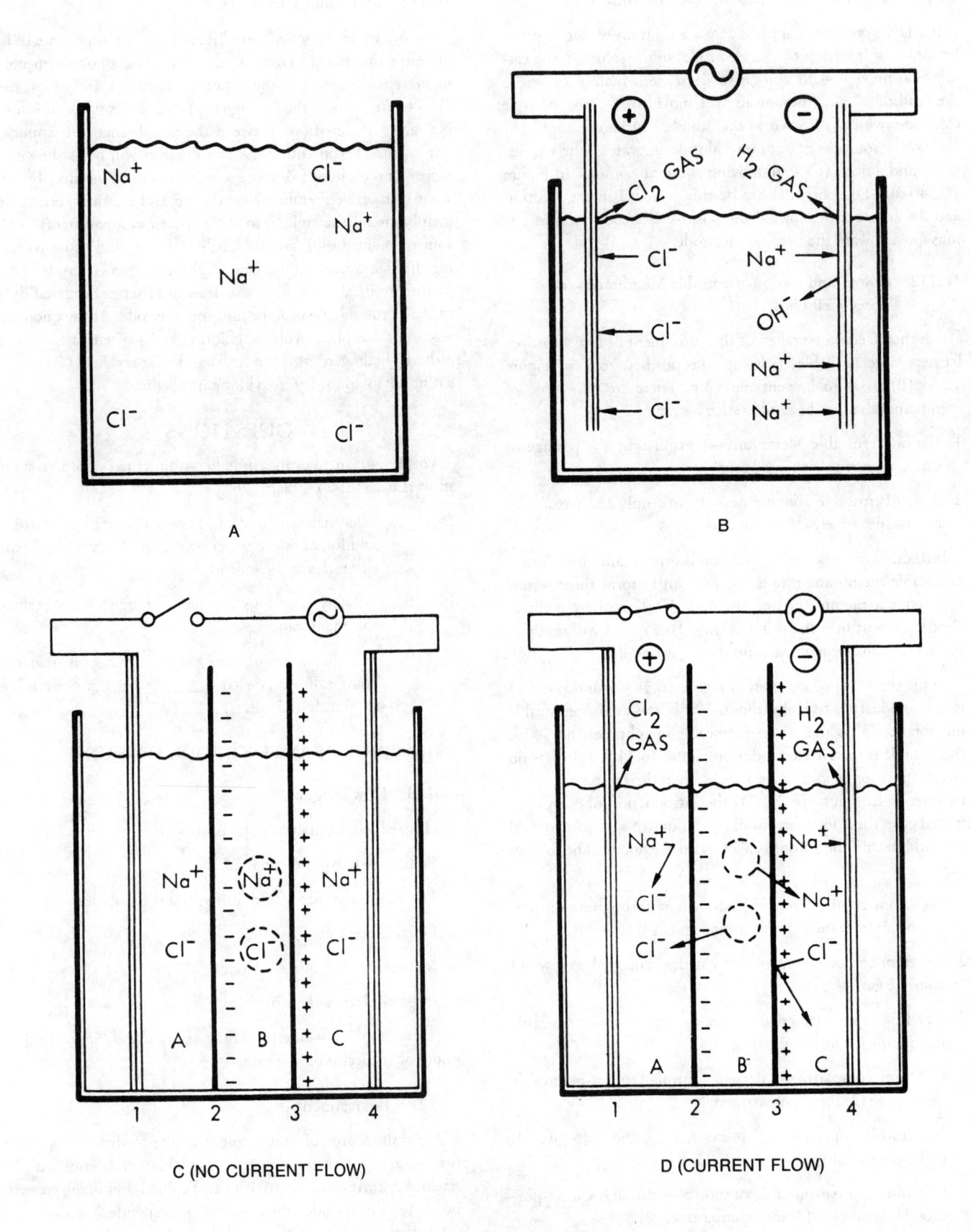

C (NO CURRENT FLOW)

D (CURRENT FLOW)

Fig. 16.34 Influence of current flow
(From *STANDARD OPERATION INSTRUCTION PLAN FOR ELECTRODIALYSIS*, prepared by Ionics, Inc.)

Fig. 16.35 Multicompartment ED stack
(From *STANDARD INSTRUCTION PLAN FOR ELECTRODIALYSIS*, prepared by Ionics, Inc.)

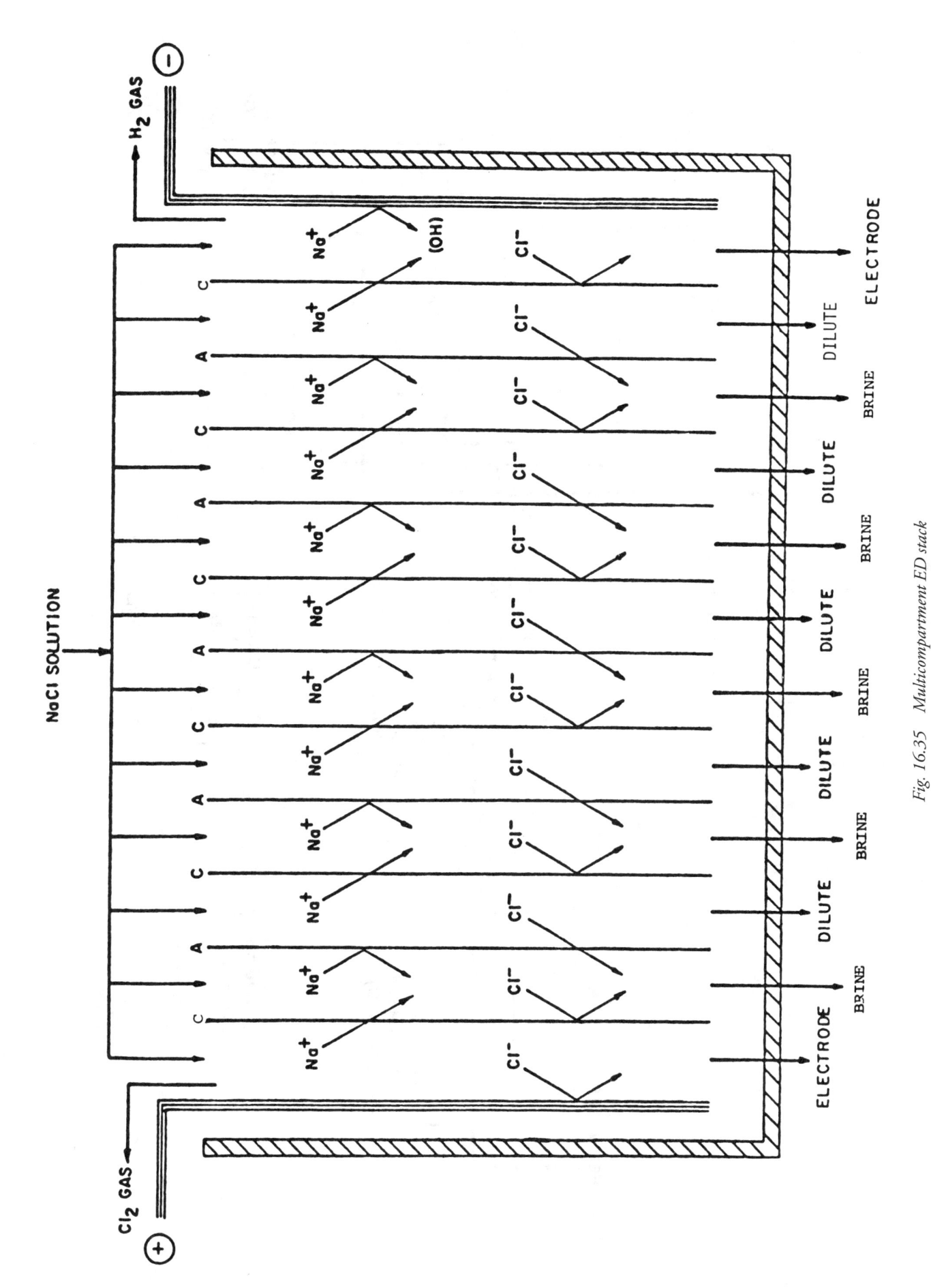

218 Water Treatment

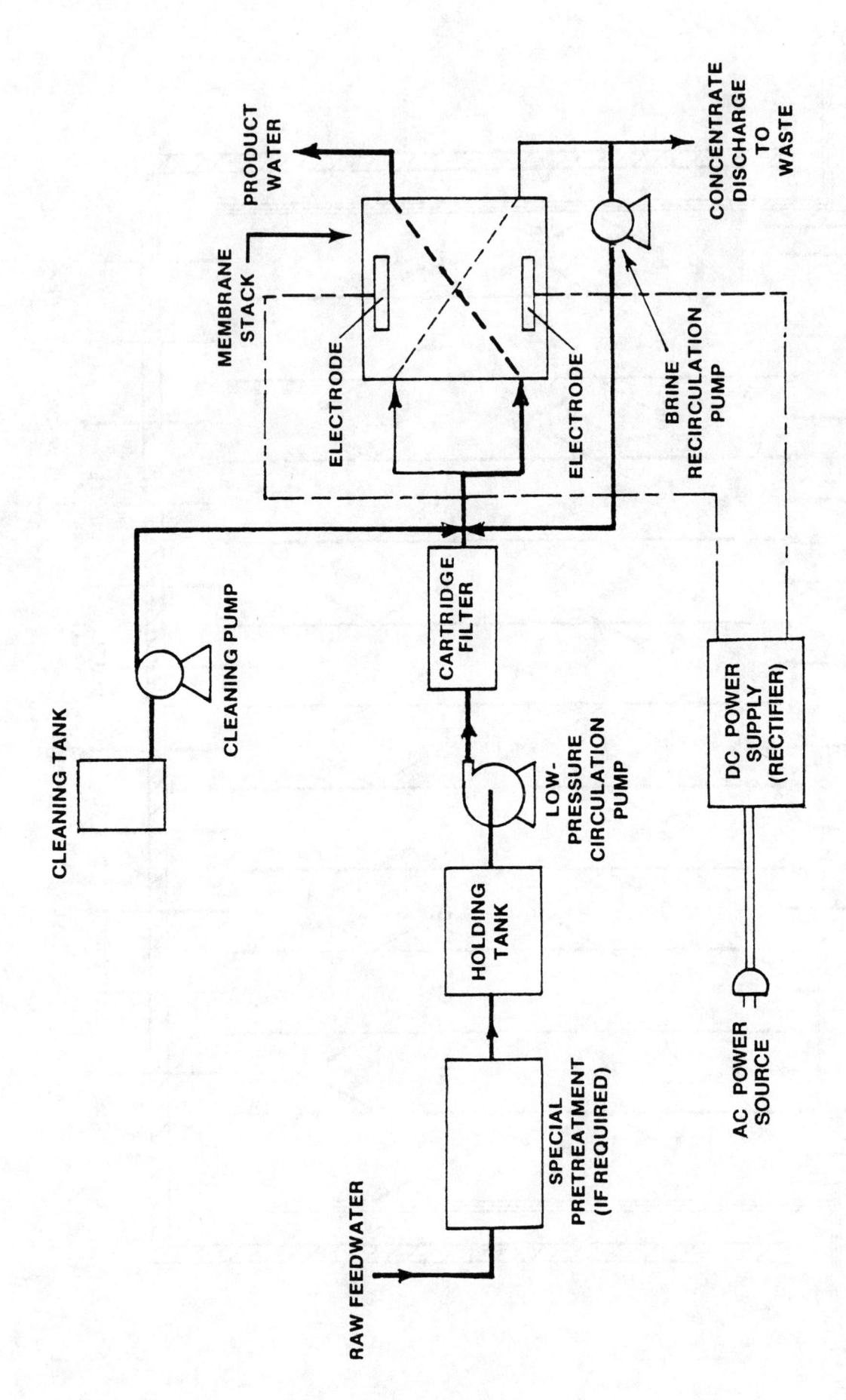

Fig. 16.36 Typical electrodialysis flow diagram

Membrane Treatment Processes 219

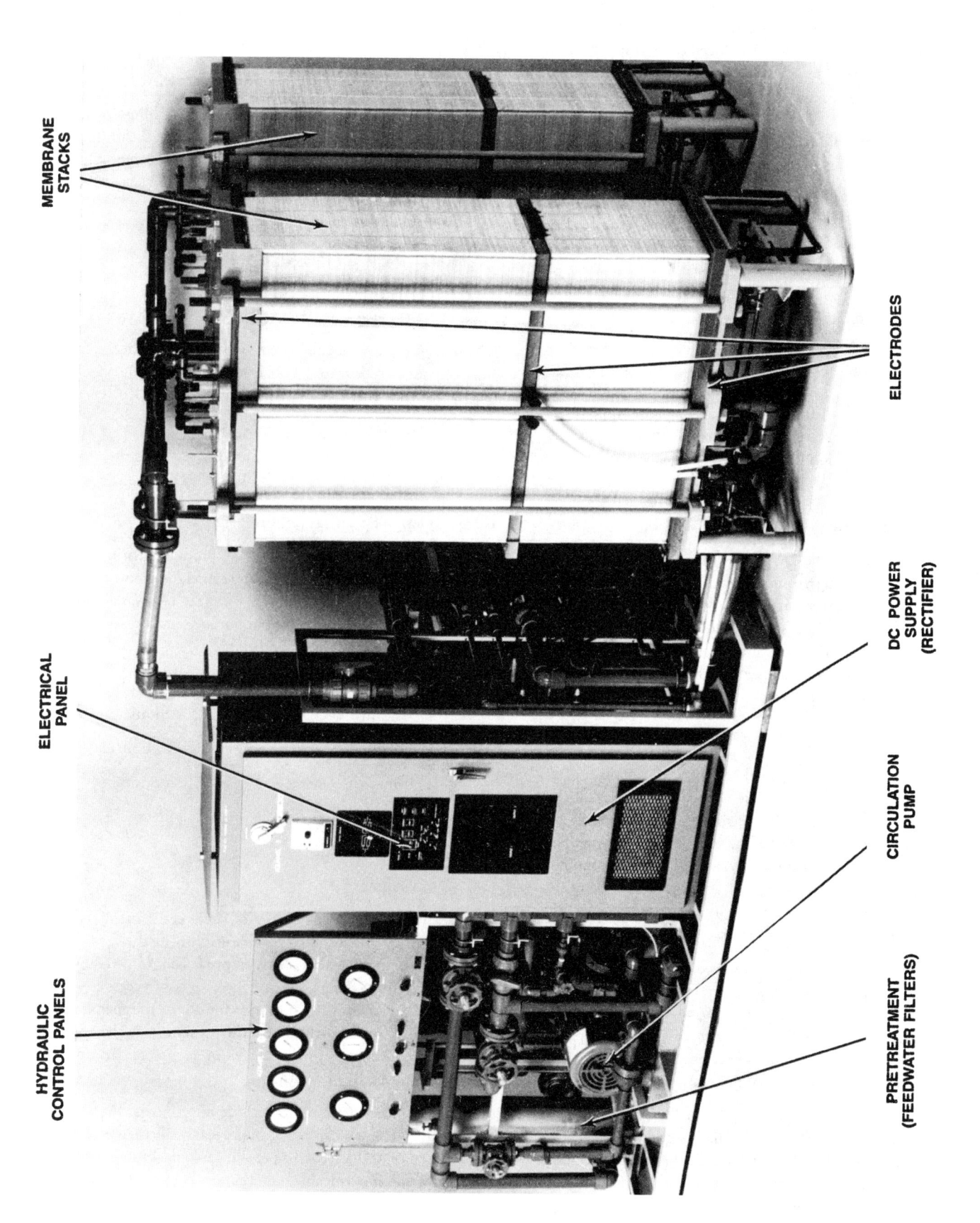

Fig. 16.37 Basic parts of an electrodialysis unit
(Courtesy of Ionics, Inc.)

ED unit. Before development of the electrodialysis reversal (EDR) unit, acid addition to prevent carbonate scaling was always practiced. With the electrodialysis reversal process, the requirement for acid addition is reduced or eliminated. Removal of specific materials such as iron, manganese, or chlorine residual, if required, is included in pretreatment.

16.122 Pumping Equipment and Piping

In the electrodialysis process, the water pump(s) is used only for circulation of the water through the stack. The head loss for this circulation varies with the construction of the stacks, number of stages, stacks, and piping, but generally a pumping pressure of only about 50 to 75 psi (345 to 517 kPa or 3.5 to 5.2 kg/sq cm) is needed.

Electrodialysis (ED) systems can be constructed with common materials found in most water treatment applications because only low operating pressures are required, as compared to reverse osmosis (RO) systems. This has allowed the use of a great deal of standard plastic pipe and fittings. The use of plastic pipe produces benefits including lower cost (compared to stainless steel), high resistance to corrosion in a saline environment, and ease of construction.

16.123 DC Power Supply

The rectifier provides the DC power to the membrane stack assembly. The input (alternating current (AC)) is converted by the rectifier to direct current, which is applied to the electrodes on each side of the membrane stack to remove the ions from the feed stream. This equipment also includes a control module for periodic reversal of the current every 15 to 30 minutes on all new electrodialysis polarity reversal (EDR) models.

16.124 Membrane Stack

The membrane "stack" is so called because it is composed of a large number of stacked pieces, like a deck of cards. Half of these pieces are spacers and half are membranes, which alternate from the bottom to the top of the stack. In other words, in examining any portion of the stack, you will find a membrane above and below every spacer (except at the electrodes) and a spacer above and below every membrane. Two membranes or two spacers should never occur together.

Each membrane stack constitutes one stage of demineralization and is a separate hydraulic and electrical stage. The total number of stacks in the unit will be arranged in either one line or two lines running in parallel (each with an equal number of stacks). Since all the stacks in a line are connected in a series, the number of stacks per line will equal the number of stages of demineralization.

The membranes and spacers in the main section of the stack make up the number of cell pairs noted in the stack specifications. A cell pair consists of one anion-permeable membrane, one cation-permeable membrane, and two intermembrane spacers and is the basic demineralizing element. The metal electrodes located at the ends of the stack apply the DC electric power required for demineralization.

16.125 Chemical Flush System

ED units are equipped with a clean-in-place (CIP) flush system to allow periodic flushing of the membrane stacks and associated piping with acid solutions down to pH 1 or with brine solutions up to 10 percent sodium chloride.

The two chemical solutions that are used most often for stack cleaning are a five-percent solution of hydrochloric acid (for removal of scale and normal cleaning), and a five-percent salt solution that has caustic soda added to adjust the pH to between 12 and 13 (for removal of organic fouling or slime).

QUESTIONS

Write your answers in a notebook and then compare your answers with those on page 226.

16.12A   What must be removed by pretreatment of the feedwater supply to the electrodialysis unit?

16.12B   What is the purpose of the rectifier in an electrodialysis unit?

16.13 ROUTINE OPERATING PROCEDURES

16.130 Design Specifications for Feedwater

The electrodialysis desalting unit will produce demineralized water at a rate dependent on water temperature and mineral composition of feedwater. The quality of the feedwater and its ionic composition are extremely important and the design of the ED unit is based on these conditions.

The ions most often encountered in feedwater are:

Cations	Anions
1. Calcium	1. Bicarbonate
2. Iron	2. Chloride
3. Magnesium	3. Sulfate
4. Silica	
5. Sodium	

An excessive concentration of any of these constituents could lead to chemical fouling due to scaling. Iron in the feedwater will cause certain process problems; above 0.1 mg/L certain precautions have to be taken. One of the effects of excess iron in feedwater is the deposit of an orange film onto the membrane surface, which increases the electrical resistance of the membrane stack. Concentrations of iron in excess of 0.3 mg/L should be removed by pretreatment.

There are other important considerations regarding feedwater quality. These include pH, biological quality, and bacteriological quality of the feed. To prevent biological fouling of the cation- and anion-permeable membranes, the feedwater should be free of bacteria. Proper control of feedwater pH is also important, particularly in terms of corrosion control in piping and plumbing equipment. Because chlorine attacks the ED membrane, the feedwater cannot contain any chlorine residual. If prechlorination is practiced, the feedwater must be dechlorinated before entering the ED unit. Generally, the unit should not be operated when the feedwater contains any of the following:

1. Chlorine residual of any concentration
2. Hydrogen sulfide of any concentration
3. Calgon or other hexametaphosphates in excess of 10 mg/L
4. Manganese in excess of 0.1 mg/L
5. Iron in excess of 0.3 mg/L

16.131 Detailed Operating Procedures

Detailed operating procedures vary from one system to the next. Most ED or EDR units come designed with fully automatic control systems. A typical operating log used to monitor an ED system is given in Figure 16.38.

The detailed specifications for any plant will give the proper settings for the various controls on the unit. These control settings should be checked and recorded at least once every 24 hours using the sample log sheet given in Figure 16.38. Any action needed to keep the plant running according to the specifications should be taken immediately.

In addition to checking the specifications, the routine maintenance schedule outlined below should be followed closely in order to reduce the risk of lengthy and expensive downtimes. Any process problems discovered must be acted upon immediately.

Daily

1. Fill out log sheet.
2. Verify that electrodes are bumping and flowing properly.
3. Inspect stacks for excess external leakage (greater than 10 gallons per hour or 38 liters per hour per stack).
4. Check the pressure drop across the cartridge filter and change the cartridges whenever the pressure drop reaches 10 psi (69 kPa or 0.7 kg/sq cm).

Weekly

1. Voltage probe the membrane stacks.
2. Check the oil level on pumps fitted with automatic oilers.
3. Inspect all piping and skid components for leaks.
4. Twice per week, measure all electrode waste flows.

QUESTIONS

Write your answers in a notebook and then compare your answers with those on page 226.

16.13A List the ions most often encountered in the feedwater to an electrodialysis unit.

16.13B What items must be considered to prevent biological fouling of the cation- and anion-permeable membranes?

16.13C Generally, the electrodialysis unit should not be operated when the feedwater contains_____. (List the appropriate water quality constituents.)

16.13D List the recommended daily activities for the operator of an electrodialysis unit.

16.14 SAFETY PRECAUTIONS

1. Grounding. The entire unit, including the stacks, must be connected to an electrical ground of each potential. At the time of installation, it is necessary to ground the skid or the control panel cabinet, either by a metal conduit or a separate grounding wire.

   Each time the unit is moved or dismantled, check the ground connections before turning on the power. The skid, power supply cabinet, and stack(s) must always be firmly connected to the building ground or other suitable ground.

2. Check the electrode tab connecting bolts and be sure these are tight and there is no corrosion. Loose connections at these points will cause overheating, which could result in serious damage to the membrane stack.

3. Do not touch wet stack sides or electrode tabs when the DC power is on.

4. Always wear rubber gloves when voltage probing the membrane stack.

5. When washing down the area, never direct a hose on the membrane stack when the DC power is on.

6. Never operate a dry centrifugal pump, even when checking rotation.

7. Never apply DC voltage to the membrane stack without water flowing through the stack.

8. Expect the DC amperage to drop when the feedwater temperature drops. Never increase the DC stack voltage as the water temperature drops in an attempt to raise currents to those recorded at the higher temperatures unless you have received specific instructions to do so from the manufacturer.

9. Expect the DC amperage to rise when the feedwater temperature rises. As this happens, the DC stack voltages must be lowered until the DC amperage returns to the normal setting. This conserves power and prevents damage to the stack.

10. Never allow oil, organic solutions, solvents, detergents, wastewater, chlorine, nitric acid, strong bleach, or other oxidizing agents to come in contact with the membranes and spacers unless directed to do so by the manufacturer. Membranes can be damaged by a feedwater containing even 0.1 mg/L free chlorine.

11. Always keep the membranes wet. Store in the membrane tube supplied or in the original plastic bags provided the seals are not broken.

Date										
Polarity										
Feed Temp (°F)										
Feed TDS (mg/L)										
Product TDS (mg/L)										
Product Conductivity										
Dilute Flowrate (GPM)										
Brine Make-up (GPM)										
PRESSURES	Stack Inlet									
	Stack Outlet									
	Differential In									
	Differential Out									
	Before Filter									
	After Filter									
	Electrode Inlet									
Stage 1 Volts	Line 1									
	Line 2									
Stage 1 Amps	Line 1									
	Line 2									
Stage 2 Volts	Line 1									
	Line 2									
Stage 2 Amps	Line 1									
	Line 2									
Stage 3 Volts	Line 1									
	Line 2									
Stage 3 Amps	Line 1									
	Line 2									
Stage 4 Volts	Line 1									
	Line 2									
Stage 4 Amps	Line 1									
	Line 2									
Stage 5 Volts	Line 1									
	Line 2									
Stage 5 Amps	Line 1									
	Line 2									
Stage 6 Volts	Line 1									
	Line 2									
Stage 6 Amps	Line 1									
	Line 2									

Fig. 16.38   Typical operating log sheet for ED unit

12. Do not smoke or use exposed flames or sparks in the gas separator tank area due to the presence of potentially explosive gases.

13. Do not service the gas separator tank when the unit is in operation. Especially avoid the vent lines where toxic and explosive gases can be present. If it is necessary to service the tank, operate the unit for 30 minutes without DC power, then wait an additional hour before beginning work or ventilate with fans to ensure complete dispersion of dangerous gases.

14. If it is necessary to troubleshoot any of the electrical panels, be extremely careful of the live panel voltages. This maintenance should be done only by someone familiar with the circuits and wiring. The unit should never be operated with the panel doors open, except for maintenance purposes, and only by experienced personnel.

15. Should shorting occur from a metal end plate across the plastic end block to the electrode, immediately turn off the rectifier. Try to eliminate the cause of the shorting by wiping excess moisture off the block. Also, be sure to completely remove the black carbon that has formed at the point of shorting. If this is not effectively done, the shorting will recur when the rectifier is turned back on.

16. Feedwater containing Calgon or other hexametaphosphates will cause high membrane stack resistance. Avoid operation when these are present.

17. Red warning lamps are mounted on the wire way for the stack power connections. The lamps are lit when the DC power is applied to the stacks.

18. When the plant is on automatic, the plant is controlled by the product water tank's level switch. Therefore, when working on the equipment, the plant should be switched to manual operation and locked out, thus avoiding the possibility of an unexpected start-up.

19. Use of the "STOP" switch or "STOP/START" switch activates an automatic flushing cycle and therefore does not immediately stop operation of all components of the unit. If the operation of the entire unit must be stopped immediately, the main breaker should be switched off.

QUESTIONS

Write your answers in a notebook and then compare your answers with those on page 226.

16.14A What problems can be created by loose connections at the electrode tab connecting bolts?

16.14B What happens to the DC amperage when the feedwater temperature drops?

16.14C How can shorting be prevented from the metal end plate across the plastic end block to the electrode?

16.14D How can the operation of the entire electrodialysis unit be stopped immediately?

16.15 ARITHMETIC ASSIGNMENT

Turn to the Arithmetic Appendix at the back of this manual. Read and work the problems in Section A.34, "Membrane Treatment Processes." Check the arithmetic in this section using an electronic calculator. You should be able to get the same answers. Section A.54 contains similar problems using metric units.

16.16 ADDITIONAL READING

1. *TEXAS MANUAL,* Chapter 11,* "Special Water Treatment (Desalting)."

* Depends on edition.

END OF LESSON 4 OF 4 LESSONS

on

MEMBRANE TREATMENT PROCESSES

Please answer the discussion and review questions next.

DISCUSSION AND REVIEW QUESTIONS
CHAPTER 16. MEMBRANE TREATMENT PROCESSES
(MEMBRANE FILTRATION AND DEMINERALIZATION)
(Lesson 4 of 4 Lessons)

Write the answers to these questions in your notebook. The question numbering continues from Lesson 3.

20. How does an electrodialysis unit demineralize (desalt) brackish water?

21. What are the parts of a basic electrodialysis unit?

22. What are the benefits of using plastic pipe in an electrodialysis plant?

23. What is the purpose of the chemical flush system in an electrodialysis unit?

24. An excessive concentration of any specific ion in the feedwater to an electrodialysis unit can cause what problem?

25. When should you check to be sure that an electrodialysis unit is properly grounded?

SUGGESTED ANSWERS
Chapter 16. MEMBRANE TREATMENT PROCESSES
(MEMBRANE FILTRATION AND DEMINERALIZATION)

ANSWERS TO QUESTIONS IN LESSON 1

Answers to questions on page 165.

16.0A Microfiltration (MF) and ultrafiltration (UF) are used in water treatment for particle, sediment, algae, bacteria, and virus removal as well as in the removal of *Giardia* and *Cryptosporidium*.

16.0B Reverse osmosis (RO) and nanofiltration (NF) membranes are used to remove dissolved organic matter and dissolved contaminants such as arsenic, nitrate, pesticides, and radionuclides. Also, these membranes can remove ions such as calcium and magnesium (for softening) and also sodium and chloride (for deionization, desalination, or demineralization of brackish waters).

Answers to questions on page 174.

16.1A Typical water treatment membrane filtration units are installed either in pressure vessels or submerged in tanks.

16.1B A decrease in flow through a membrane can be recovered by backwashing and cleaning.

16.1C Operators consider using coagulation pretreatment before membrane filtration because of the effectiveness of membrane filtration in treating low-turbidity water.

16.1D Bench- or pilot-scale studies should be conducted to:

1. Select the most appropriate coagulant
2. Select the most appropriate coagulant dose
3. Consider the effectiveness of removing DOC
4. Compare costs of alternatives investigated

Answers to questions on page 180.

16.2A When the plant is unattended, the on-call operator can use a laptop computer to confirm plant process performance and make process and equipment adjustments prior to returning to the plant.

16.2B Backwashing membrane filters may occur on the basis of three possible conditions:

1. Elapsed time since the last backwash
2. Total water produced since the last backwash
3. Pressure increase since the last backwash

16.2C The operation of membranes includes monitoring and testing for membrane filtration rate and membrane integrity.

16.2D Problems that an operator may encounter when operating a membrane filtration water treatment plant include: (1) air compressor failure, (2) raw water pump station failure (electronic failure of the uninterrupted power supply (UPS)), and (3) computer lockup.

Answers to questions on page 181.

16.3A Membrane filtration water treatment plants continuously monitor turbidity and chlorine residual.

16.3B Membrane filtration water treatment plant operation and water production records maintained in a SCADA system include water levels, flow rates, hours of operations, and quality of water produced.

Answers to questions on page 182.

16.4A The operator conducts routine maintenance on a daily basis.

16.4B The preventive maintenance program (PMP) includes scheduled maintenance based on operating time.

16.4C Breakdown maintenance relates to repairs to equipment that has failed in service.

ANSWERS TO QUESTIONS IN LESSON 2

Answers to questions on page 183.

16.5A Demineralization is the process that removes dissolved minerals (salts) from water.

16.5B Seawater is more expensive to treat than brackish water because of its greater TDS concentration.

Answers to questions on page 186.

16.6A Methods of removing minerals from water can be divided into two classes:

1. Those that use a phase change such as freezing or distillation
2. Nonphase change methods such as reverse osmosis, electrodialysis, and ion exchange

16.6B The common membrane demineralizing processes are reverse osmosis, electrodialysis, and nanofiltration.

Answers to questions on page 189.

16.7A The osmotic pressure of a solution is the difference in water level between the two sides of a membrane.

16.7B The two types of semipermeable membranes that are used most often for demineralization are cellulose acetate and thin film composites.

16.7C The water flux is the flow of water in grams per second through a membrane area of one square centimeter (or gallons per day per square foot) while the mineral flux is the flow of minerals in grams per second through a membrane area of one square centimeter.

16.7D When additional pressure is applied to the side of a membrane with a concentrated solution, the water flux (rate of water flow through the membrane) will increase, but the mineral flux (rate of flow of minerals) will remain constant.

16.7E When higher mineral concentrations occur in the feedwater, the mineral concentrations will increase in the permeate (or product water).

Answers to questions on page 190.

16.7F Water flux is usually expressed in gallons per day per square foot (or grams per second per square centimeter) of membrane surface.

16.7G The term "flux decline" is used to describe the loss of water flow through the membrane due to compaction plus fouling.

16.7H Mineral rejection is defined as:

$$\text{Rejection, \%} = \left(1 - \frac{\text{Product Concentration}}{\text{Feedwater Concentration}}\right) \times 100\%$$

Mineral rejection can be determined by measuring the TDS and using the above equation. Rejections also may be calculated for individual constituents in the solution by using their concentrations.

16.7I Softening membranes operate at lower pressures and lower rejection rates than standard reverse osmosis membranes.

Answers to questions on page 198.

16.7J An increase in feedwater temperature will increase the water flux.

16.7K Hydrolysis of a membrane results in a lessening of mineral rejection capability.

16.7L Recovery is defined as the percentage of feed flow that is recovered as product water.

$$\text{Recovery, \%} = \left(\frac{\text{Product Flow}}{\text{Feed Flow}}\right)(100\%)$$

16.7M Recovery rate is usually limited by: (1) desired product water quality, and (2) the solubility limits of minerals in the brine.

16.7N Concentration polarization is a buildup of retained particles on the membrane surface due to dewatering of the feed closest to the membrane. The thickness of the concentration polarization layer is controlled by the flow velocity across the membrane.

ANSWERS TO QUESTIONS IN LESSON 3

Answers to questions on page 205.

16.8A The three types of commercially available membrane configurations that have been used in operating plants are: (1) spiral wound, (2) hollow fine fiber, and (3) tubular.

16.8B The tubular membrane process is used to treat wastewater with a high suspended solids concentration.

Answers to questions on pages 206 and 207.

16.9A To protect the reverse osmosis system and its accessory equipment, the feedwater should be filtered. When the water source is a groundwater or a previously treated municipal or industrial supply, filtration may be accomplished by a simple screening procedure. An untreated surface water will probably require coagulation, flocculation, sedimentation, and filtration.

16.9B Colloidal particulates are removed from feedwater by chemical treatment and filtration.

16.9C As an acetate membrane hydrolyzes, both the amount of water and the amount of solute that permeate the membrane increase and the quality of the product water deteriorates.

16.9D The scale control method that is used to inhibit calcium sulfate precipitation is a threshold treatment with 2 to 5 mg/L of sodium hexametaphosphate (SHMP).

16.9E A 1 to 2 mg/L chlorine residual is maintained to control biological fouling.

Answers to questions on page 207.

16.9F Operating pressure on a reverse osmosis unit is regulated by a control valve on the influent manifold.

16.9G The demineralized water is usually called permeate, the reject, concentrate.

16.9H Product or permeate flow is not regulated and varies as feedwater pressure and temperature change.

Answers to questions on page 211.

16.9I Chlorine is added to the feedwater to prevent biological fouling.

16.9J The operator must check the differential pressure across the RO unit to know when to clean the elements. When the elements become fouled, ΔP usually increases, thus indicating the need for cleaning.

16.9K   The reverse osmosis elements should be cleaned when the operator observes: (1) lower product water flow rate, (2) lower salt rejection, (3) higher differential pressure (ΔP), and (4) higher operating pressure.

Answers to questions on page 212.

16.9L   Safety needs for demineralization plants can be divided into three general groups consisting of chemicals, electrical, and hydraulics.

16.9M   Electrical equipment used around reverse osmosis plants consists of a series of electrically powered pumps.

ANSWERS TO QUESTIONS IN LESSON 4

Answers to questions on page 214.

16.10A   Typical removals of inorganic salts from brackish water by ED range from 25 to 40 percent of dissolved solids per stage of treatment.

16.10B   A membrane stack in an electrodialysis unit consists of several hundred membranes and their separating spacers assembled between a single set of electrodes. End plates and tie rods complete the assembly.

16.10C   The most commonly encountered problem in ED operation is scaling (or fouling) of the membranes by both organic and inorganic materials. Alkaline scales are troublesome in the concentrating compartments when the diffusion of ions to the surface of the anion membrane in the diluting cell is insufficient to carry the current.

Answers to questions on page 215.

16.11A   If a DC potential is applied across a solution of salt in water by means of insertion of two electrodes in the solution, the cations will move toward a negative electrode, which is known as the "cathode," and the anions will move toward the positive electrode, which is known as the "anode."

16.11B   Only cations (positively charged ions) can pass through cation-permeable membranes.

16.11C   In a multicompartment ED unit, the compartments losing salt are labeled "dilute" and those receiving the transferred salt, "brine."

Answers to questions on page 220.

16.12A   Iron, manganese, and chlorine residual must be removed from the feedwater supply to the electrodialysis unit.

16.12B   The rectifier provides the DC power to the membrane stack assembly. The input (alternating current (AC)) is converted by the rectifier to DC, which is applied to the electrodes on each side of the membrane stack to remove the ions from the feed stream.

Answers to questions on page 221.

16.13A   The ions most often encountered in the feedwater to an electrodialysis unit include:

Cations	Anions
1. Calcium	1. Bicarbonate
2. Iron	2. Chloride
3. Magnesium	3. Sulfate
4. Silica	
5. Sodium	

16.13B   To prevent biological fouling of the cation- and anion-permeable membranes, the operator must control feedwater pH, biological quality, and bacteriological quality.

16.13C   Generally, the electrodialysis unit should not be operated when the feedwater contains any of the following:

1. Chlorine residual of any concentration
2. Hydrogen sulfide of any concentration
3. Calgon or other hexametaphosphates in excess of 10 mg/L
4. Manganese in excess of 0.1 mg/L
5. Iron in excess of 0.3 mg/L

16.13D   The recommended daily activities for the operator of an electrodialysis unit include:

1. Fill out log sheet.
2. Verify that electrodes are bumping and flowing properly.
3. Inspect stacks for excess external leakage.
4. Check the pressure drop across the cartridge filter and change the cartridges whenever the pressure drop reaches 10 psi.

Answers to questions on page 223.

16.14A   Loose connections at the electrode tab connecting bolts will cause overheating, which could result in serious damage to the membrane stack.

16.14B   Expect the DC amperage to drop when the feedwater temperature drops.

16.14C   Should shorting occur from the metal end plate across the plastic end block to the electrode, immediately turn off the rectifier. Try to eliminate the cause of the shorting by wiping excess moisture off the block. Also, be sure to completely remove the black carbon that has formed at the point of shorting.

16.14D   If the operation of the entire unit must be stopped immediately, the main breaker should be switched off.

CHAPTER 17

HANDLING AND DISPOSAL OF PROCESS WASTES

by

George Uyeno

TABLE OF CONTENTS

Chapter 17. HANDLING AND DISPOSAL OF PROCESS WASTES

		Page
OBJECTIVES		230
WORDS		231
17.0	NEED FOR HANDLING AND DISPOSAL OF PROCESS WASTES	233
17.1	SOURCES OF TREATMENT PROCESS WASTES	233
17.2	PROCESS SLUDGE VOLUMES	234
17.3	METHODS OF HANDLING AND DISPOSING OF PROCESS WASTES	235
17.4	DRAINING AND CLEANING OF TANKS	235
17.5	BACKWASH RECOVERY PONDS (SOLAR LAGOONS)	237
17.6	SLUDGE DEWATERING PROCESSES	240
	17.60 Solar Drying Lagoons	240
	17.61 Sand Drying Beds	240
	17.62 Belt Filter Presses	241
	17.63 Centrifuges	244
	17.64 Filter Presses	245
	17.65 Vacuum Filters	245
17.7	DISCHARGE INTO COLLECTION SYSTEMS (SEWERS)	248
17.8	DISPOSAL OF SLUDGE	249
17.9	EQUIPMENT	251
	17.90 Vacuum Trucks	251
	17.91 Sludge Pumps	251
17.10	PLANT DRAINAGE WATERS	251
17.11	MONITORING AND REPORTING	252
DISCUSSION AND REVIEW QUESTIONS		252
SUGGESTED ANSWERS		253

OBJECTIVES

Chapter 17. HANDLING AND DISPOSAL OF PROCESS WASTES

Following completion of Chapter 17, you should be able to:

1. Explain why the operators of water treatment plants should be concerned about the handling and disposal of process wastes.

2. Identify the sources of water treatment plant wastes.

3. Drain and clean sedimentation tanks.

4. Discharge process wastes to collection systems (sewers).

5. Operate and maintain backwash recovery ponds (lagoons) and sludge drying beds.

6. Dispose of process wastes.

7. Safely operate and maintain sludge handling and disposal equipment.

8. Monitor and report on the disposal of process wastes.

WORDS

Chapter 17. HANDLING AND DISPOSAL OF PROCESS WASTES

CENTRIFUGE
A mechanical device that uses centrifugal or rotational forces to separate solids from liquids.

CONDITIONING
Pretreatment of sludge to facilitate removal of water in subsequent treatment processes.

DECANT (de-KANT)
To draw off the upper layer of liquid (water) after the heavier material (a solid or another liquid) has settled.

DEWATER
(1) To remove or separate a portion of the water present in a sludge or slurry. To dry sludge so it can be handled and disposed of.

(2) To remove or drain the water from a tank or a trench. A structure may be dewatered so that it can be inspected or repaired.

SEPTIC (SEP-tick) or SEPTICITY
A condition produced by bacteria when all oxygen supplies are depleted. If severe, the bottom deposits produce hydrogen sulfide, the deposits and water turn black, give off foul odors, and the water has a greatly increased oxygen and chlorine demand.

SLUDGE (SLUJ)
(1) The settleable solids separated from liquids during processing.

(2) The deposits of foreign materials on the bottoms of streams or other bodies of water or on the bottoms and edges of wastewater collection lines and appurtenances.

SUPERNATANT (soo-per-NAY-tent)
Liquid removed from settled sludge. Supernatant commonly refers to the liquid between the sludge on the bottom and the scum on the surface.

THICKENING
Treatment to remove water from the sludge mass to reduce the volume that must be handled.

CHAPTER 17. HANDLING AND DISPOSAL OF PROCESS WASTES

17.0 NEED FOR HANDLING AND DISPOSAL OF PROCESS WASTES

The need for handling and disposal of potable water treatment plant wastes is a problem that must be faced by all plant operators. Many articles and books have been published on potable water treatment processes. Their emphasis is usually on producing wholesome and pure water for human consumption in compliance with EPA and state and local health department regulations, but very few mention sludge handling and disposal in any great detail. In response to a growing population and increasing concern about pollution of natural water sources, pollution control agencies, health departments, and fish and game departments established programs to enforce rules to prevent any waste discharge that would tend to discolor, pollute, or generally be harmful to aquatic or plant life or the environment.

The law that restricts or prohibits the discharge of process wastes from water treatment plants is Public Law 92-500, the Water Pollution Control Act Amendments of 1972. This Act clearly includes treatment plant wastes such as sludge from a water treatment plant. These wastes are considered an industrial waste that requires compliance with the provisions of the Act. Under the National Pollutant Discharge Elimination System (NPDES) provisions, a permit must be obtained in order to discharge wastes from a water treatment plant. Water treatment plants are classified into three categories.

Category 1   Plants that use one of the following three processes: (1) coagulation, (2) oxidative iron and manganese removal, or (3) direct filtration.

Category 2   Plants that use only chemical softening processes.

Category 3   Plants that use combinations of coagulation and chemical softening, or oxidative iron and manganese removal and chemical softening.

Enforcement of PL 92-500 is the responsibility of each state. Many NPDES permits have been issued by the states to water treatment plants using state standards applicable to the local conditions at the time the permits were issued. Water quality indicators for which waste discharge limitations have been issued include pH, total suspended solids, settleable solids, total iron and manganese, flow rate, total dissolved solids (TDS), BOD, turbidity, total residual chlorine, temperature, floating solids, and visible forms of waste.

Water treatment plants can no longer simply discharge dirty backwash water or settled sludge into lakes, rivers, streams, or tributaries as was done in the past. Current regulations require daily monitoring of any discharge and analysis of such water quality indicators as pH, turbidity, TDS, settleable solids, or other harmful materials. The results of the analyses must be logged and reported frequently to the proper authorities and must conform to their rigid standards. For these reasons, it is absolutely necessary to make provisions for facilities to handle these wastes on a routine basis. The most important part of an operator's job is still the end product, good potable water, but an operator's duties are not complete until all by-products and wastes are disposed of in an acceptable and documented manner.

QUESTIONS

Write your answers in a notebook and then compare your answers with those on page 253.

17.0A   Why are strict laws needed regarding the disposal of process wastes?

17.0B   If a discharge results from the disposal of process wastes, what water quality indicators require daily monitoring?

17.1 SOURCES OF TREATMENT PROCESS WASTES

Although there are many types of water treatment plants and methods for treating water, most of them probably operate in the following general manner. Alum or polymers are applied to the water in a rapid-mix chamber. The mixture is agitated by mechanical means or through a cylinder designed for hydraulic flash mixing for coagulation.

Following coagulation, the water passes through mechanical flocculators or a series of baffles for flocculation. The water then moves into the sedimentation tank where the floc is allowed to settle out before the water moves to the filters. The sedimentation tanks may be of various shapes and depths; however, they are most commonly rectangular or circular. Many large plants are equipped with either mechanical rakes or scrapers that periodically remove sludge from a hopper, or with a vacuum-type sludge removal device. The sludge is continuously scraped into the hopper. The hopper is emptied from one to three times per day for 20 to 30 minutes each time depending on the size of the hopper and the density of the sludge. Sludge is then usually moved to drying beds. The smaller and older plants may not have these sludge handling facilities available. Many new water treatment plants are equipped with sludge collection headers with squeegees. This system does not need any sludge hoppers. The collection headers are supported by a travelling bridge or floats. The sludge is pumped out of the bottom of the basin and into a sludge channel on the walkway level. This system is described in Chapter 5, "Sedimentation."

Another type of plant similar to the one above contains an upflow solids-contact unit with clarifiers. The clarifiers are usually circular and have sludge draw-off lines that must be monitored; the solids are then drawn off periodically as sludge.

For small plants and in areas where water must be pumped, pressure filters may be used and the coagulant is applied directly to the filter. Sedimentation tanks or clarifiers may occasionally accompany the use of pressure filters but this is not usually the case if the quality of the source water is good.

In another type of plant layout (not too commonly used), the sedimentation tank also functions as the backwash recovery area. In this case, the backwash wastewater is pumped back to the head of the plant. Most of the solids will settle out when the water flows through the sedimentation basin. This method does eliminate the need for backwash recovery ponds or lagoons.

Diatomaceous earth filtration is different from all other types of filtration in its method of operation. There is usually no pretreatment of the water. Disposal of backwash wastes is still a problem. Table 17.1 summarizes the various sources of treatment process wastes and the methods of collecting, handling, and disposing of these wastes.

TABLE 17.1 COLLECTION, HANDLING, AND DISPOSAL OF PROCESS WASTES

SOURCES OF WASTES

1. Trash racks
2. Grit basins
3. Alum, ferric hydroxide, or polymer sludges from sedimentation basins
4. Filter backwash
5. Lime–soda softening
6. Ion exchange brine

COLLECTION OF SLUDGES

1. Mechanical scrapers or vacuum devices
2. Manual (hoses and squeegees)
3. Pumps (into tank trucks or dewatering facilities)
4. Floating headers

DEWATERING OF SLUDGES

1. Solar drying lagoons
2. Sand drying beds
3. Centrifuges[a]
4. Belt presses
5. Filter presses
6. Vacuum filters

DISPOSAL OF SLUDGES AND BRINES

1. Wastewater collection systems (sewers)
2. Landfills (usually dewatered sludges)
3. Spread on land

---

[a] Mechanical devices that use centrifugal or rotational forces to separate solids from liquids (sludge from water).

QUESTIONS

Write your answers in a notebook and then compare your answers with those on page 253.

17.1A How is sludge removed from sedimentation tanks?

17.1B How is sludge removed from an upflow solids-contact type of unit?

17.2 PROCESS SLUDGE VOLUMES

All of the different methods of sludge collection require some type of facilities for handling processed waste. The possibilities here include sedimentation tanks, backwash recovery ponds, drying beds, lagoons, ponds, holding tanks, adequate land, access to sewer systems, or vacuum trucks or similar equipment for removal and disposal of the waste material.

The amount of sludge accumulation at a typical water treatment plant depends on the type and amount of suspended matter in the source water being treated as well as on the level of dosage and the type of coagulant used. As an example, let us examine sludge production at two 5-MGD (19-MLD) plants. The annual water production at each was 800 to 900 million gallons (3,000 to 3,400 megaliters). Plant One had no source water stabilizing reservoir and the raw water turbidity ranged from 3 units during the summer months to over 100 units during the winter, with an annual average alum dosage of 11 mg/L. Yearly sludge accumulation was approximately 500,000 gallons (1.9 megaliters). Plant Two, with a 15-million gallon (56.8-megaliters) source water stabilizing reservoir, treated raw water that never exceeded 20 turbidity units at the intake with an average alum dosage of 8 mg/L. The annual sludge accumulation was approximately 300,000 gallons (1.14 megaliters). In both cases, nonionic polymer was used for filter aid at approximately 15 ppb. The source water stabilizing reservoir reduced the turbidity level of the water to be treated and provided a more constant quality of water that required less alum and produced less sludge.

Organic polymers may be used instead of alum to reduce the quantity of sludge produced. Polymer sludges are relatively denser and easier to dewater for subsequent handling and disposal. Not all waters can be treated by using polymers instead of alum.

For plants without sludge collection devices, the volume of sludge produced and the frequency of cleaning the sedimentation tank are affected by several factors. Items to consider include:

1. Water demand
2. Suspended solids loads and when peak demands occur
3. Water temperature (as the temperature of the water increases, the settling rate of the solids will increase)
4. Detention time (as the detention time increases, the amount of solids that settle out will increase)
5. Volume of sludge deposited in the basin (as the volume of sludge increases, the detention time decreases as well as the efficiency of the basin)

6. Volume of treated water storage for the system (the greater the volume of treated water storage, the more time is available for sludge removal)

7. Time required to clean and make any necessary repairs during the shutdown

8. Availability of adequate drying beds, lagoons, landfill, a vacuum tank truck, pumps or equipment, and adequate help with all necessary safety equipment and procedures

Sedimentation tanks should be drained and cleaned at least twice a year and more often if the sludge buildup interferes with the treatment processes (filtration and disinfection). Alum or polymer sludge solids content is only 0.5 to 1 percent for continuous sludge removal and 2 to 4 percent when the sludge is allowed to accumulate and compact. Therefore, the sludge can flow readily in pipes or be pumped, especially with wastewater-type pumps.

QUESTIONS

Write your answers in a notebook and then compare your answers with those on page 253.

17.2A How can a source water stabilizing reservoir reduce the volume of sludge handled?

17.2B How often should sedimentation tanks be drained and cleaned?

17.3 METHODS OF HANDLING AND DISPOSING OF PROCESS WASTES (Figure 17.1)

Various methods are used to handle and dispose of process wastes. The disposal facilities at your plant will depend on when the plant was built, the region where the plant is located (topography and climate), the sources of sludge, and the methods of ultimate disposal.

An effective method of handling sludge is to regularly during the day remove sludge from sedimentation tanks to a drying bed. When one drying bed is full of sludge, the sludge is allowed to dry while the other drying beds are being filled. A key to speedy drying is the regular removal of the water on top of the sludge.

Some plants require that portions of the facilities be shut down twice a year, the tanks drained, and the sludge removed. This is an excellent time to inspect the tanks and equipment and perform any necessary maintenance and repairs.

Backwash recovery ponds or lagoons are used to separate the water from the solids after the filters have been backwashed. The water is usually returned or recycled to the plant headworks for treatment with the source water. These ponds also may be used to concentrate or thicken sludges from sedimentation tanks. Sludges from the lime–soda softening process are usually stored in lagoons. The drainage water is removed and the sludge may be covered or hauled off to a disposal site. Lime softening sludges may be applied to agricultural lands to achieve the best soil pH for optimum crop yields.

Larger plants or plants that produce large volumes of sludge may use THICKENING,[1] CONDITIONING,[2] and DEWATERING[3] processes to reduce the volume of sludge that must be handled and ultimately disposed of. Sometimes, polymers are added to sludges for conditioning prior to dewatering. Belt filter presses, centrifuges, filter presses, vacuum filters, solar lagoons, and sand drying beds are some of the processes used to dewater sludges.

Ultimately, process wastes such as trash, grit, sludge, and brine must be disposed of in a manner that will not harm the environment. Trash and grit may be disposed of in landfills. Sometimes, sludge and brine are discharged into wastewater collection systems (sewers); however, this procedure may cause operational problems for the wastewater treatment plant operator. To avoid upsetting wastewater treatment plants, discharges to sewers must be made very slowly to take advantage of the dilution provided by the wastewater. Sludges are commonly disposed of by spreading on land or dumping in landfills. The method used will depend on the volume of the sludge, sludge moisture content, land available, and distance from the plant to the ultimate disposal site.

The remainder of this chapter will discuss the detailed operational procedures that an operator must consider when handling and disposing of process wastes. Sections are also provided on equipment operation and maintenance as well as on monitoring and reporting.

QUESTIONS

Write your answers in a notebook and then compare your answers with those on page 253.

17.3A When should sedimentation tanks be inspected and repaired?

17.3B List the methods that may be used to dewater sludge.

17.4 DRAINING AND CLEANING OF TANKS

Plants without mechanical or vacuum-type sludge collectors will require the use of manual labor to remove the sludge once or twice a year. When two or more sedimentation tanks are designed into a plant, the job of cleaning is made easier. While one sedimentation tank is down, the other(s) can remain in operation. The cleaning of sedimentation tanks should be done prior to or after peak demand months. Generally, early spring and the fall of the year are the better times to take some facilities out of service for cleaning.

---

[1] *Thickening.* Treatment to remove water from the sludge mass to reduce the volume that must be handled.
[2] *Conditioning.* Pretreatment of sludge to facilitate removal of water in subsequent treatment processes.
[3] *Dewater.* (1) To remove or separate a portion of the water present in a sludge or slurry. To dry sludge so it can be handled and disposed of. (2) To remove or drain the water from a tank or a trench. A structure may be dewatered so that it can be inspected or repaired.

236 Water Treatment

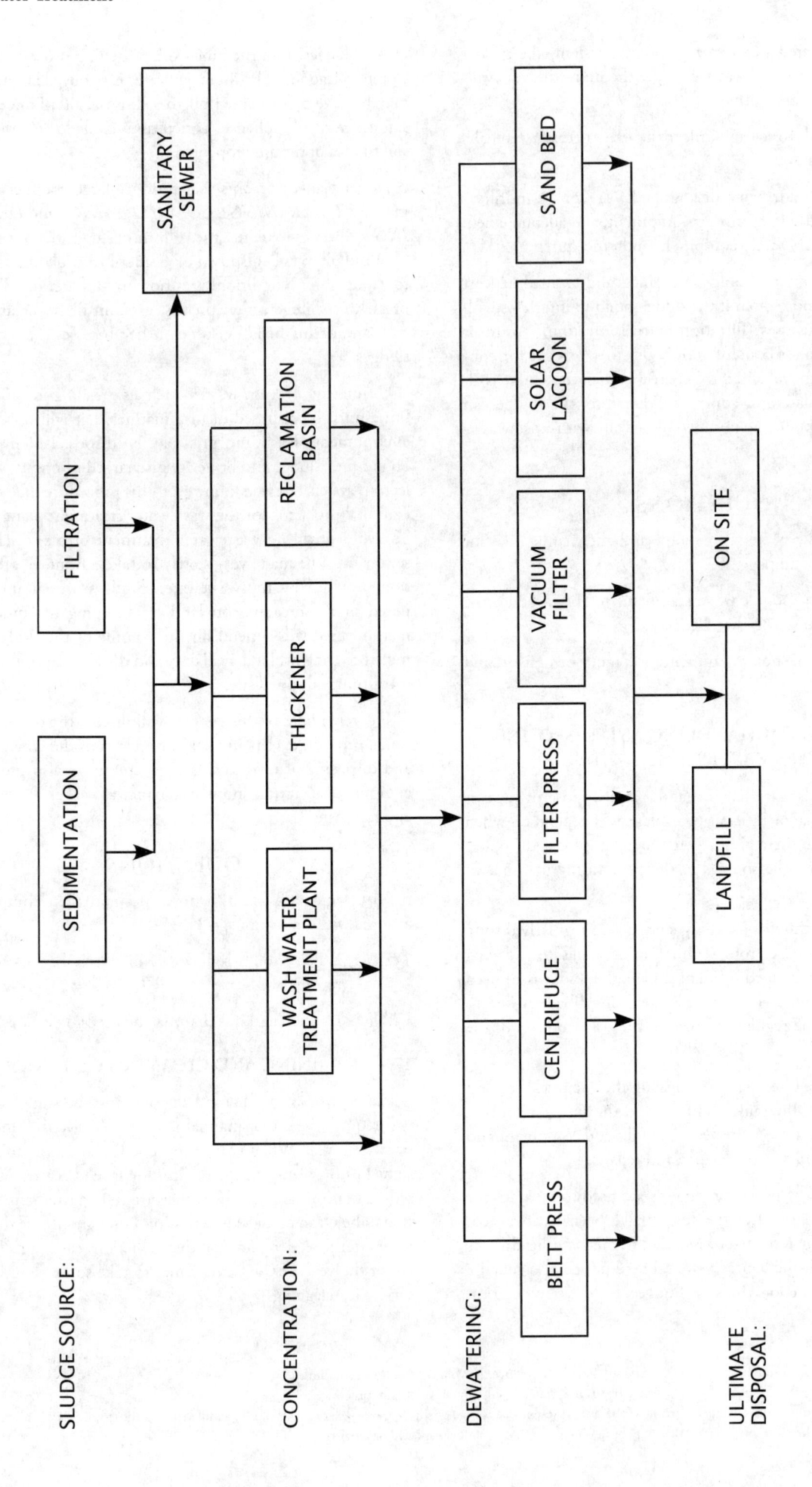

Fig. 17.1 Sludge processing alternatives

Before draining any in-ground tank, always determine the level of the water table. If the water table is high, an empty tank could float like a cork on the water surface and cause considerable damage to the tank and piping. A properly designed tank will have provisions to drain high water tables or will contain other protective features (bottom pressure-relief disks to let groundwater into the tank to prevent damage).

After any necessary intake valve(s) or gate(s) changes are made, drain the water down to the sludge blanket by partially opening the drain valve from the sedimentation tank into the lagoon or drying bed. After the first few minutes (if the valve is not wide open), the water will become clear. This portion of the water can be diverted from the lagoon or drying bed, if proper plumbing is available, and returned to the source to be reprocessed. Most drying beds will not handle this great a volume of water unless this draining process is extended over a period of a few days. Pump(s) can be used to transfer the settled water to another sedimentation tank that is still in operation.

As the water gets down to the sludge, fully open the drain valve into the drying bed(s). A large quantity of the sludge will drain by itself. As shown in Figure 17.2, the tank wall is 10 feet (3 m) high with the sludge level showing about five feet (1.5 m) from the top. When the level drops down to about two feet (0.6 m) of depth by the drain opening, the sludge will have to be assisted by an operator with a squeegee (Figure 17.3).

During the draining stages, the walls and all the equipment should be completely hosed down and inspected for damage. All necessary repairs should be made at this time. Once sludge dries on any coated surface, it is difficult to remove so it is important to hose everything down during the process of draining and while the sludge is still wet. All gears, sprockets, and moving parts should be lubricated immediately after hosing down to prevent "freeze up" resulting from exposure to the air during inspection and repair. By using drying beds and drying bed *DECANT*[4] pumps, ample amounts of water may be used for cleaning and assisting draining of the sludge. Under these conditions, two to three operators can clean out one sedimentation tank for a plant of 5 to 10 MGD (19 to 38 MLD) in one day. Additional time is required for initial drawdown, gathering up of tools and equipment, final cleanup, and any repairs that may be needed.

Sludge that settles out near the entrance to the sedimentation tank is more dense, especially when a polymer is used for flocculation aid. Therefore, the drain should be located in the headworks area. Once the sludge ceases to flow freely, even with the dilution water, then operators will have to push it toward the drain with squeegees (Figure 17.4).

The volume of sludge can vary with the size of the basin or clarifier, the quality of the source water being treated, the use of alum, polymer or combinations of both, and the frequency of cleaning. This volume may range from 100,000 to 200,000 gallons (0.38 to 0.76 ML), depending on the size of the basin and how long the sludge has accumulated in the basin.

The following precautions must be exercised whenever operators are in any closed tank (confined space):

1. Do not operate gasoline engines in the tank.

2. Test the atmosphere in the tank for oxygen deficiency/enrichment, flammable or explosive gases, and toxic gases, and provide adequate ventilation of clean air at all times.

3. Provide a source of water to clean off boots and tools where the operators come out of the tank.

4. Use the buddy system. At least one person must be outside the tank and watching anyone inside the tank, and another person must be nearby and available to help in an emergency.

Before filling the tank, thoroughly inspect and repair all equipment and valves. Wash everything down with clean water or a solution of 200 mg/L chlorine to disinfect the basin. If a chlorine wash solution is not used, fill the tank 10 percent full with a 50-mg/L chlorine solution and then finish filling it with clean water from the plant. The final free chlorine residual should not be so high that water with a free chlorine residual greater than 0.5 mg/L reaches the consumers.

Although manually draining and cleaning tanks requires more operator hours and plant downtime than mechanical sludge removal, it does have its advantages. A more sanitary condition in the tank is obtained by cleaning up algae buildups or other deposits that are not picked up by mechanical collectors and regular inspection of equipment can eliminate many potential breakdown conditions. Even basins with continuous sludge collection systems should be drained once a year for inspection and maintenance.

QUESTIONS

Write your answers in a notebook and then compare your answers with those on page 253.

17.4A   How can sludge be removed from tanks without mechanical or vacuum-type sludge collectors?

17.4B   When draining a sedimentation tank, what should be done with the settled water above the sludge?

17.4C   What precautions must be exercised whenever an operator enters a closed tank (confined space)?

17.5 BACKWASH RECOVERY PONDS (SOLAR LAGOONS)

Because of water pollution control legislation enacted since the 1960s, many water treatment plants have backwash recovery ponds (Figure 17.5). In many instances, these ponds can serve a dual purpose. In addition to their primary function as backwash recovery ponds, they can also be used to collect the sludge from

---

[4] *Decant* (de-KANT).   To draw off the upper layer of liquid (water) after the heavier material (a solid or another liquid) has settled.

Fig. 17.2   Sludge being drained from a clarifier

Fig. 17.3   Operator with a squeegee assisting sludge out of clarifier

Fig. 17.4 Operators pushing sludge toward drain with squeegees and vacuum truck removing sludge

Fig. 17.5 Backwash recovery pond

sedimentation tanks and clarifiers with a few modifications. While these modified ponds are capable of performing both functions at the same time, it is critical to time these operations so that they do not overlap. You want to avoid having to backwash the filters while you are draining a sedimentation tank. Water for hosing down the sedimentation basin and assisting the flow of sludge should be used sparingly. Also, the backwash recovery pump suction pipe intake should be floated near the surface, by use of a flexible hose and tire tube or any similar float, so that any excess water can be recycled without also drawing out sludge. This will be very important if the filters must be backwashed at the same time sludge is being cleaned out of the backwash recovery ponds.

A vacuum tank truck will be needed to move the wet sludge (a vacuum tank truck is shown in Figure 17.4). The capacity of the vacuum truck's tank in the picture is 5,000 gallons (19 cu m) and it has a 6-inch (150-mm) suction hose. About 15 minutes are required to fill the tank if sludge is fed to the suction end constantly without breaking the vacuum. A lift of 12 feet (3.6 m) can be obtained without too much difficulty.

Sludge is sometimes applied to land as a soil conditioner. Polymer sludges are suitable as a soil conditioner. Sludges produced by direct filtration, without coagulants, usually make excellent soil conditioners both with and without polymers. The sludge may be applied either wet or dry. Because commercial soil conditioner is becoming more expensive, sale of sludge as a soil conditioner can help to offset sludge handling and disposal costs.

Most plants that use the lime–soda ash softening process collect the sludge and dewater it in a lagoon. A variable length riser or discharge pipe is used to draw off the water that is separated from the sludge. When the lagoon is full and the sludge is dried, the surface may be covered with soil as in a landfill operation. In some plants where space is scarce, the dried sludge is hauled off to a landfill and the lagoon refilled. Lime softening sludges also can be disposed of by application to agricultural soils to adjust the pH for optimum crop yields.

Discharge of lime–soda sludges to the wastewater collection system (sewers) is a poor practice because: (1) the sewers could become plugged regularly, and (2) the operator at the wastewater treatment plant will have to handle and dispose of the sludge. However, the lime–soda sludge may help the wastewater treatment plant operator by: (1) adjusting the pH, or (2) serving as a coagulant aid in treating the incoming wastewater.

QUESTIONS

Write your answers in a notebook and then compare your answers with those on page 253.

17.5A Why is timing critical if backwash recovery ponds are used to handle sludge from sedimentation basins?

17.5B Why should the suction pipe for the backwash recovery pump be floated near the surface of the pond?

17.5C How can lime–soda softening sludge be disposed of ultimately?

17.6 SLUDGE DEWATERING PROCESSES[5]

17.60 Solar Drying Lagoons

Solar drying lagoons are shallow, small-volume storage ponds in which treatment process sludge (sometimes concentrated) is stored for extended time periods. Sludge solids settle to the bottom of the lagoon by plain sedimentation (gravity settling) and the clear *SUPERNATANT*[6] water is skimmed off the top with the aid of an outlet structure that drains the clear surface waters. Evaporation removes additional water and the solar drying process proceeds until the sludge reaches a concentration of from 30 to 50 percent solids. At this point, the sludge can be disposed of on site or at a sanitary landfill. Obviously, the solar drying process is dependent on environmental conditions (weather) and may take many months to complete. For this reason, several lagoons should be provided (a minimum of three) so that sludge loading and drying can be rotated from one lagoon to another.

17.61 Sand Drying Beds

Sand drying beds have been used extensively in municipal wastewater treatment where high solids volumes are handled. Sand drying beds are similar in construction to a sand filter, and consist of a layer of sand, a support gravel layer, an underdrain system, and some means for manual or mechanical removal of the sludge (see Figures 17.6, 17.7, and 17.8). They are built with underdrains covered with gradations of aggregate and sand. The drains discharge into a sump where recovery pumps can return the water drained from the sludge back to the plant to be reprocessed. Frequently, three beds are used so one can be dried out while one is being filled from the drawdown of wet sludge from the sedimentation tanks. The third bed contains dried sludge that is being hauled out.

The efficiency of the sand drying bed dewatering process can be greatly improved by preconditioning the process sludge with chemical coagulants. The drying time can vary from days to

---

[5] Portions of this section were prepared by Jim Beard.
[6] *Supernatant* (soo-per-NAY-tent). Liquid removed from settled sludge. Supernatant commonly refers to the liquid between the sludge on the bottom and the scum on the surface.

Fig. 17.6 Sludge drying beds

weeks, depending on weather conditions and the degree of preconditioning of the sludge. The frequency of removal of dried sludge will vary with different plants depending on the volume of sludge produced, size of drying beds, and drying conditions (weather).

Sludge has a unique characteristic about it that once it has even partially dried, it will not expand; therefore, layer after layer of wet sludge can be added over a period of time. This procedure will work as long as the solids content of the applied sludge is at least 2 to 3 percent. When drying this type of sludge, large cracks and checks will develop on the surface and extend down through the sludge to the sand. The proper time to remove this dried sludge is when no more than one foot (0.3 m) has accumulated and dried into a checkered pattern. A piece of dry sludge can then be picked up off the sand. The dried sludge can easily be removed with a front-end loader onto a dump truck and be hauled off to a landfill. However, the operator must exercise extreme caution so that only the dried sludge is picked up with the minimum possible disturbance to the sand and aggregate. The loader bucket should be operated carefully since there may be only about ten inches (36 cm) of sand cover over the underdrains. The loader bucket capacity should be limited to one or two cubic yards of sludge. Concrete tracks should be provided for larger equipment to collect the dried sludge.

QUESTIONS

Write your answers in a notebook and then compare your answers with those on page 254.

17.6A What is the minimum recommended number of solar drying lagoons?

17.6B Describe a typical sludge drying bed.

17.6C When is the proper time to remove dried sludge from the drying bed?

17.6D What precautions must be exercised when operating a front-end loader to remove sludge from a drying bed?

17.62 Belt Filter Presses

Continuous-belt filter presses are popular because of their relative ease of operation, low energy consumption, small land requirements, and their ability to produce a relatively dry filter cake (material removed from the filter press has about 35 to 40 percent solids). There are two primary mechanisms by which free water is separated from the sludge solids in a belt press:

1. Gravity drainage

2. Pressure dewatering

242 Water Treatment

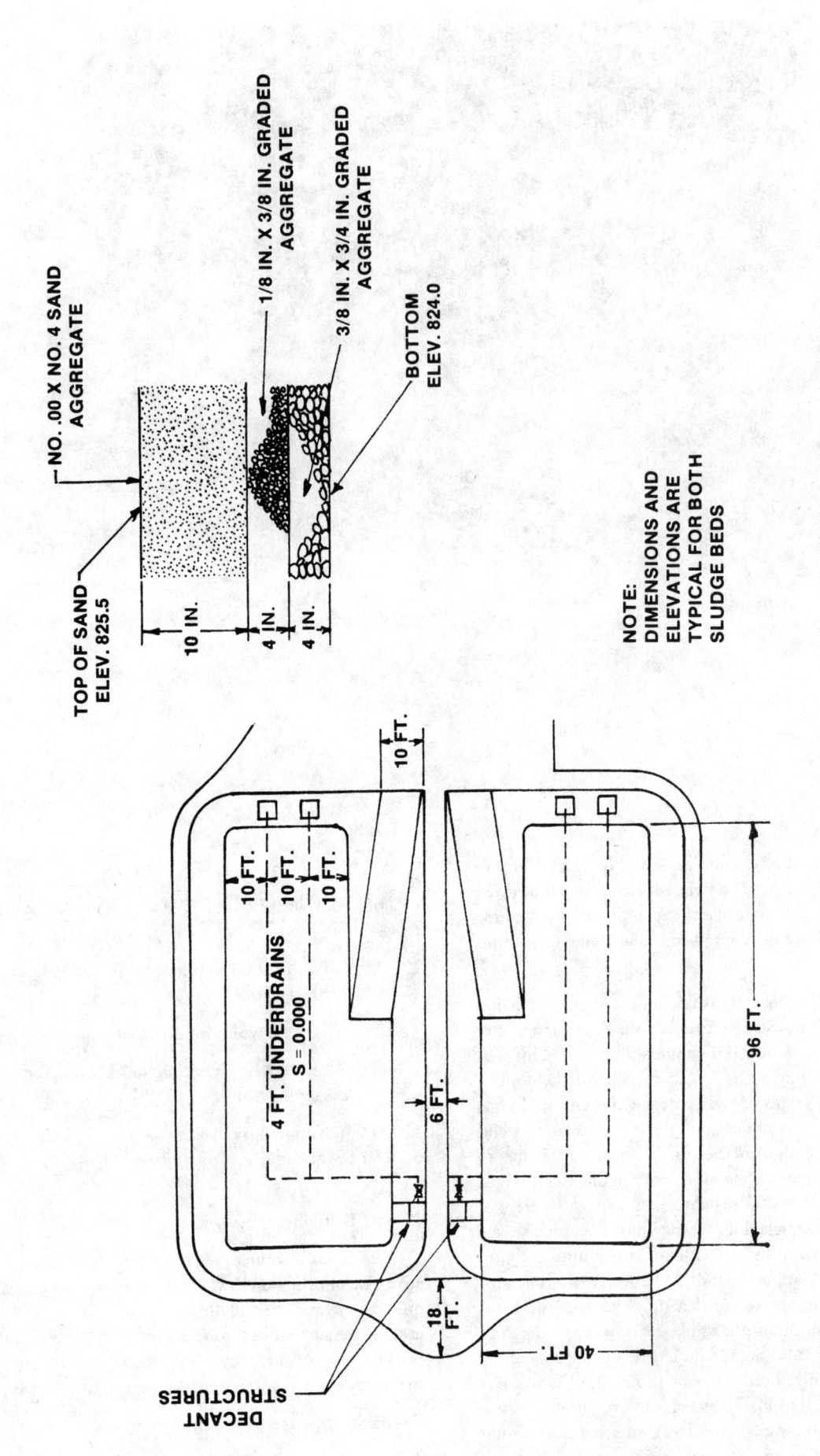

Fig. 17.7 Sludge drying beds

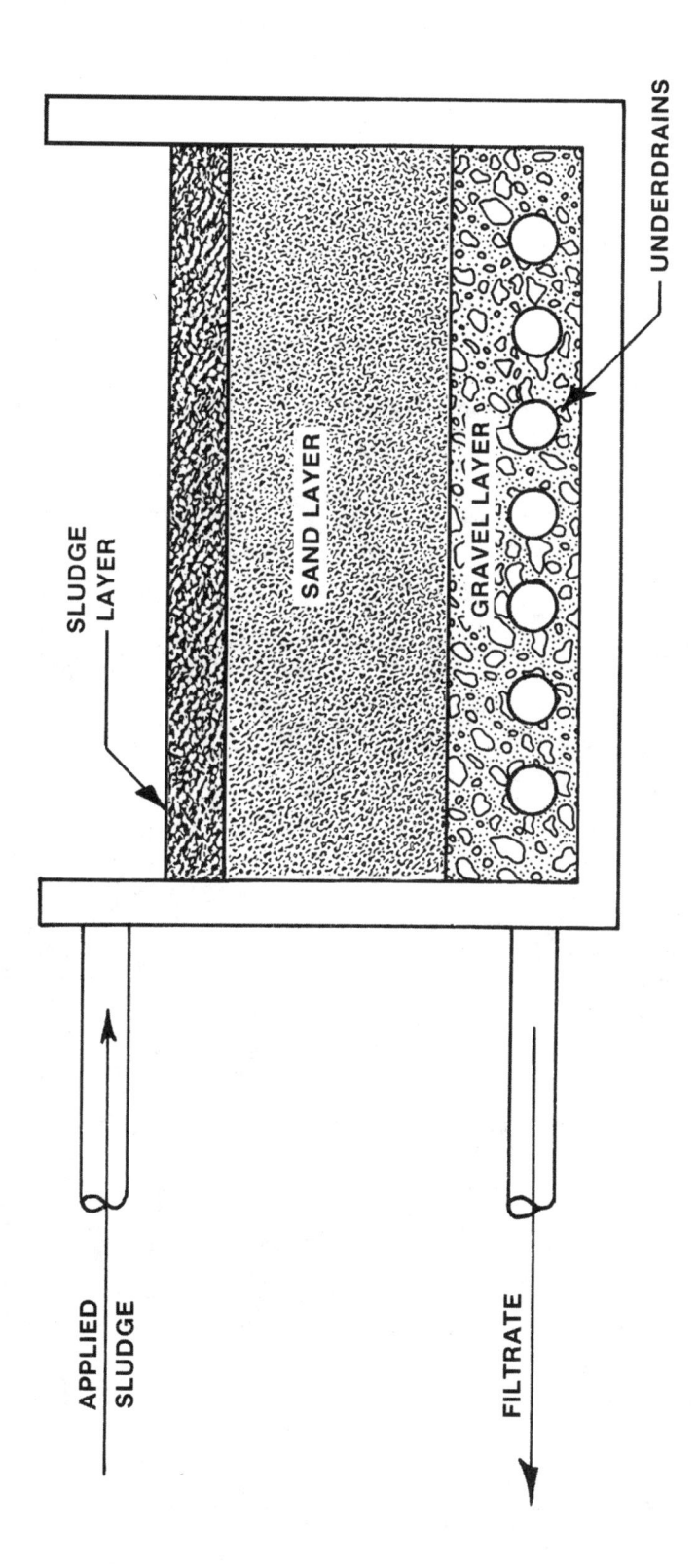

Fig. 17.8 Sectional view of sludge drying bed

Sludge is conveyed and dewatered between two endless belts (Figure 17.9). After the sludge is initially mixed with a polymer in a rotary drum conditioner, it is dewatered in three distinct zones:

1. A horizontal zone for gravity drainage
2. A vertical sandwich draining zone
3. A final dewatering zone containing an arrangement of staggered rollers that produce a multiple-shear force action that squeezes out the remaining free water

Each belt is washed with a high-pressure/low-volume water spray.

17.63 Centrifuges

Centrifuges have been used to dewater municipal sludges for some time. Problems with the earlier units included erosion of surfaces hit by high-speed particles, and poor performance capacity. Design improvements and the use of polymers have generally eliminated these problems. The principal advantage of this dewatering technique is that the density of the sludge cake can be varied from a thickened liquid slurry to a dry cake. The major limitation of using centrifuges is high energy consumption.

There are two basic types of centrifuges, the scroll type and the basket type. The scroll centrifuge operates continuously, while the basket centrifuge is a batch-type unit. Solids capture is generally greater with the basket centrifuge.

In the scroll centrifuge (Figure 17.10), solids are introduced horizontally into the center of the unit. The spinning action forces the solids against the outer wall of the bowl, where they are transported to the discharge end by a rotating screw conveyor. Clear supernatant liquid is discharged over an adjustable weir on the opposite end of the unit.

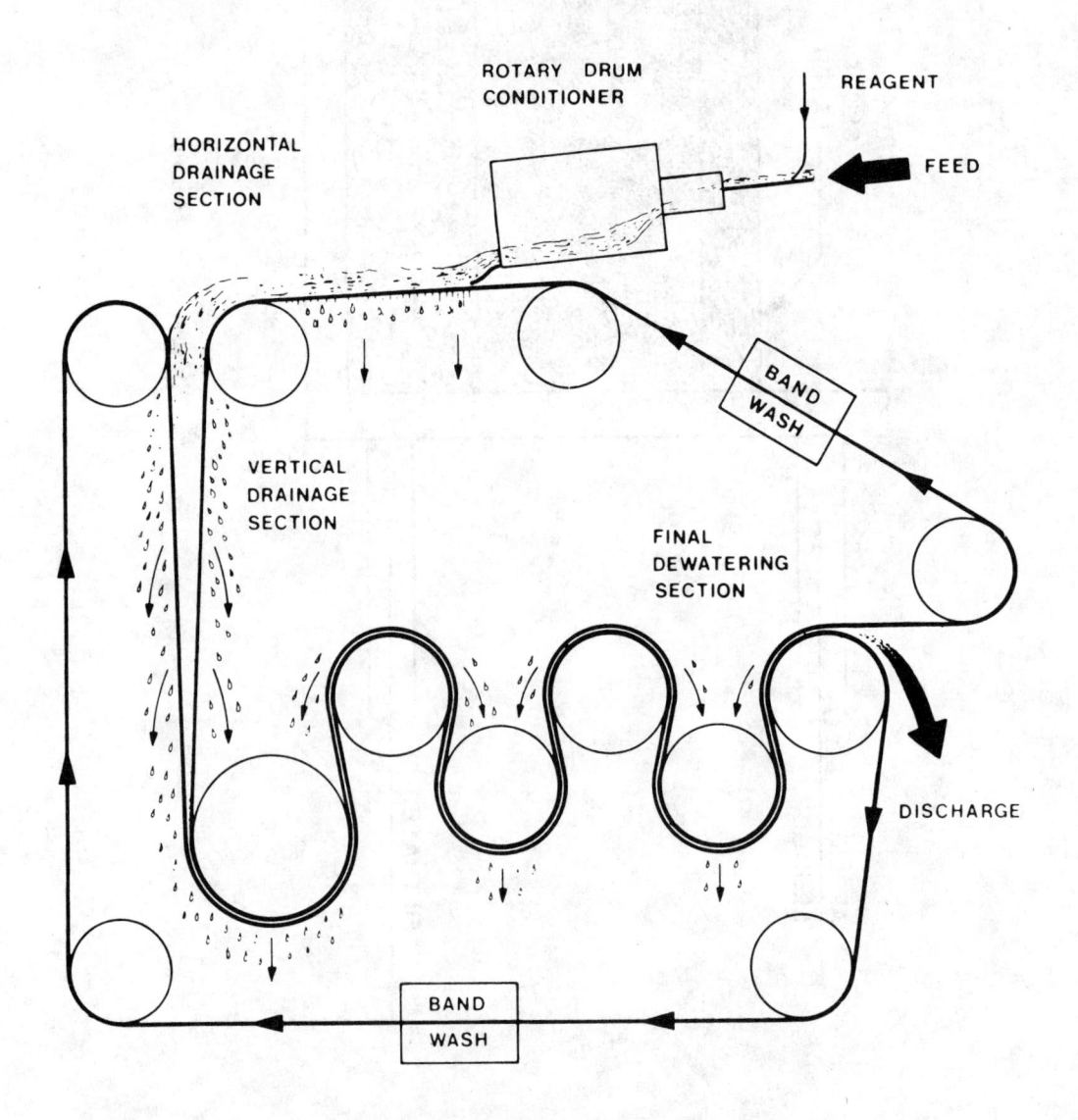

Fig. 17.9 Flow diagram of Winklepress (a continuous-belt filter press)
(Permission of Ashbrook-Simon-Hartley)

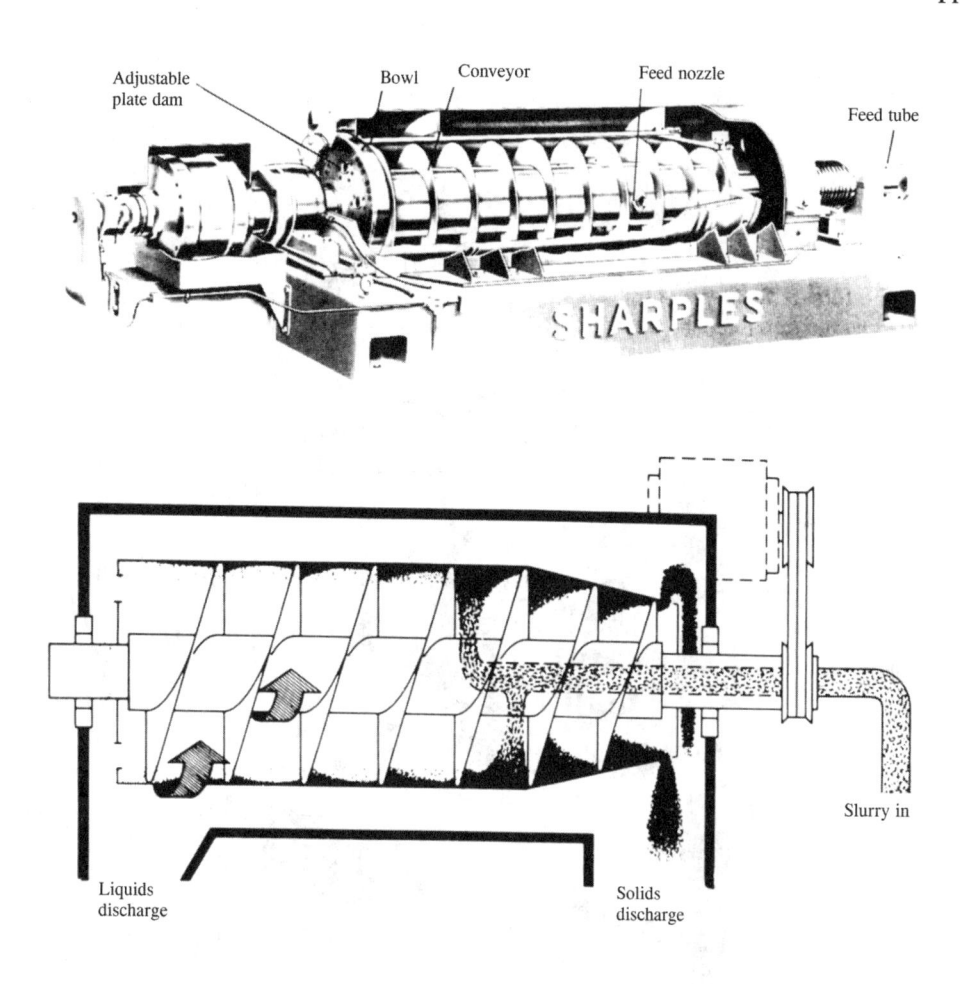

Fig. 17.10 Solid-bowl scroll centrifuge
(Permission of Sharples-Stokes Division, Penwalt Corporation)

In the basket centrifuge (Figure 17.11), sludge is introduced vertically into the bottom of the bowl and the supernatant is discharged over a weir at the top of the bowl. When the solids concentration in the supernatant becomes too high, the operation is stopped and the dense solids cake is removed by a knife unloader.

17.64 Filter Presses

Filter presses have been successfully used to process difficult-to-dewater sludges (alum sludges). These machines are best suited for sludges with a high specific resistance (the internal resistance of a sludge cake to the passage of water). Filter presses produce very dry cakes, a clear filtrate, and have a very high solids capture.

A filter press consists of a series of vertical plates covered with cloth that supports and retains the filter cake (Figure 17.12). These plates are rigidly held in a frame. Sludge is fed into the press at increasing pressures for about half an hour. The plates are then pressed together for one to four hours at pressures as high as 225 psi (15.8 kg/sq cm or 1,551 kiloPascals). Water passes through the cloth while the solids are retained, forming a cake that is removed when the press is depressurized.

17.65 Vacuum Filters

Vacuum filtration was once the main chemical sludge dewatering process. However, its use has declined due to development of devices such as the belt press, which consumes less energy, is less sensitive to polymer dosage, and does not require use of a precoat (a substance applied to the filter before applying sludge for dewatering).

A vacuum filter consists of a cylindrical drum that rotates, partially submerged, in a tank of chemically conditioned sludge (Figure 17.13). As the drum slowly rotates, a vacuum is applied under the filter medium (belt) to form a cake on the surface. As the belt rotates, suction is maintained to promote additional dewatering. As the belt passes the top of the drum, it separates from the drum and passes over a small-diameter roller for discharge of the cake. The belt is then washed before it re-enters the vat. A precoat of diatomaceous earth is required to dewater gelatinous (jelly-like) alum sludge.

246   Water Treatment

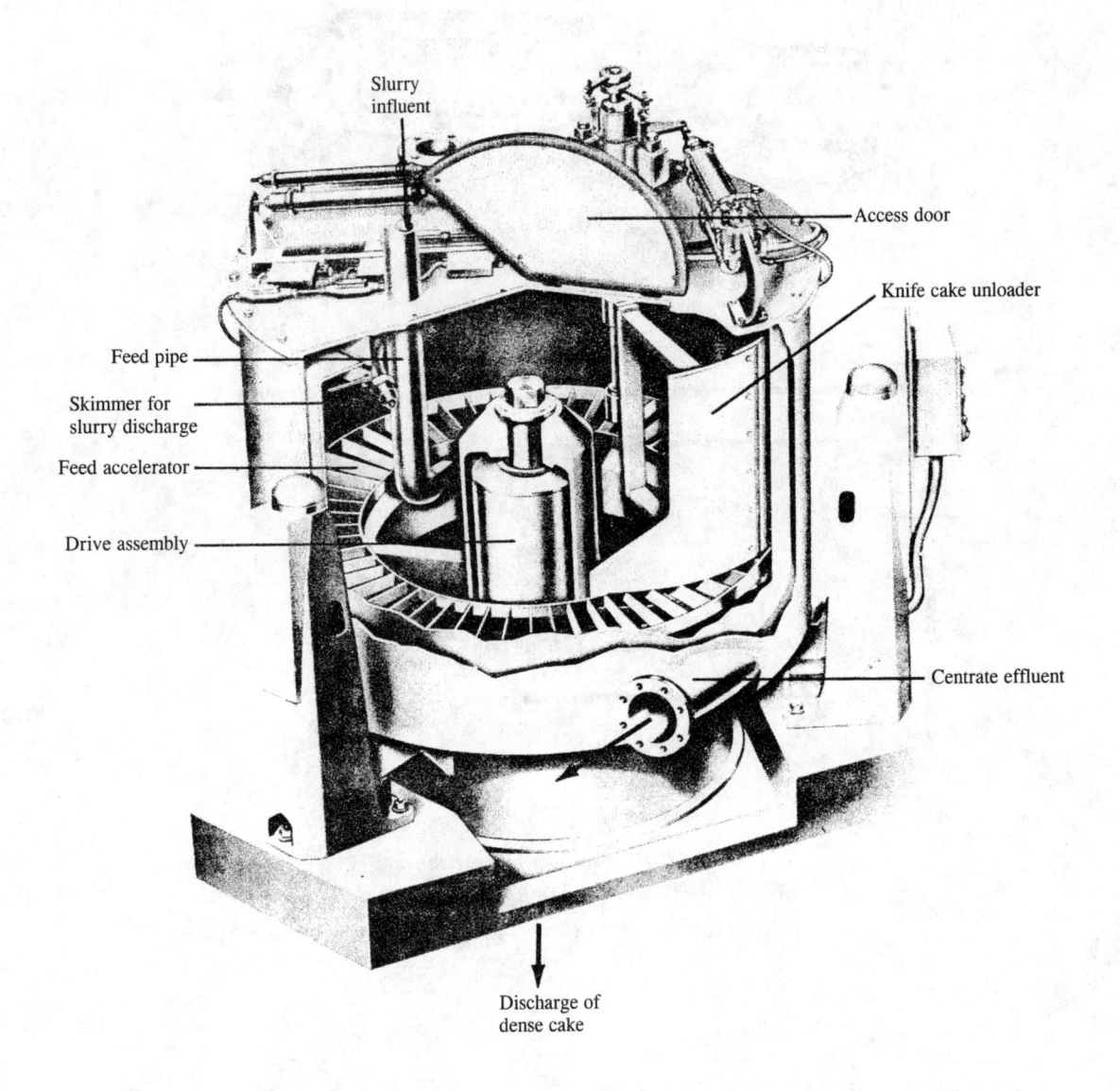

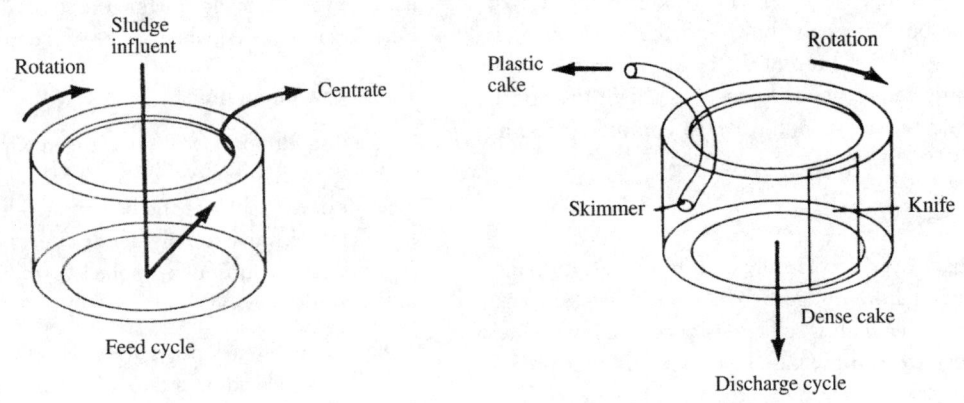

Fig. 17.11   Imperforate basket centrifuge
(Permission of Sharples-Stokes-Division, Penwalt Corporation)

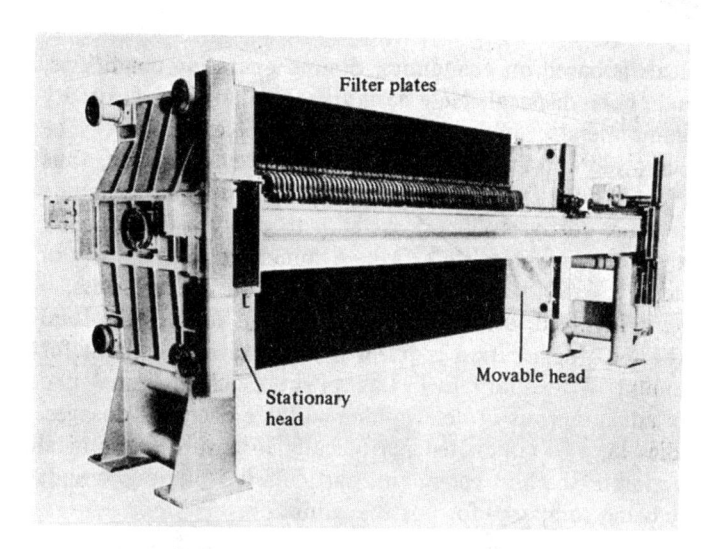

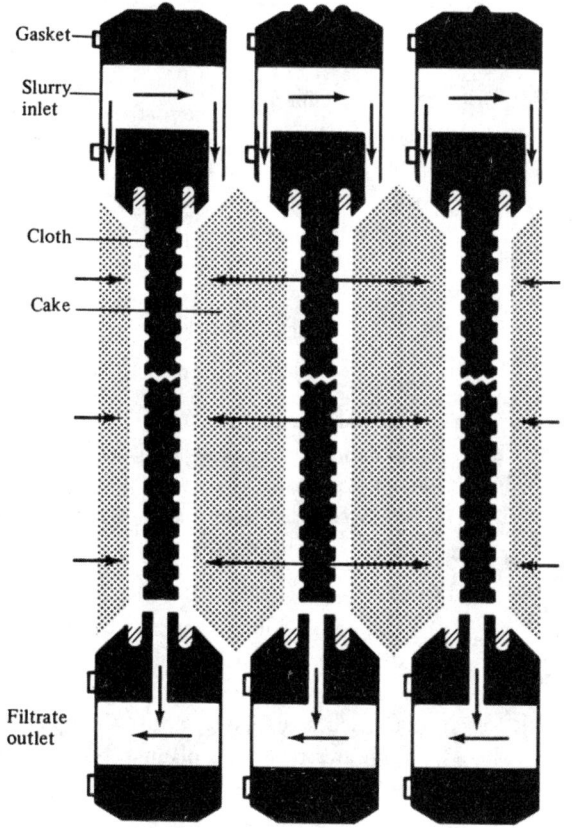

Fig. 17.12 Filter press
(Permission of Metropolitan Water District of Southern California)

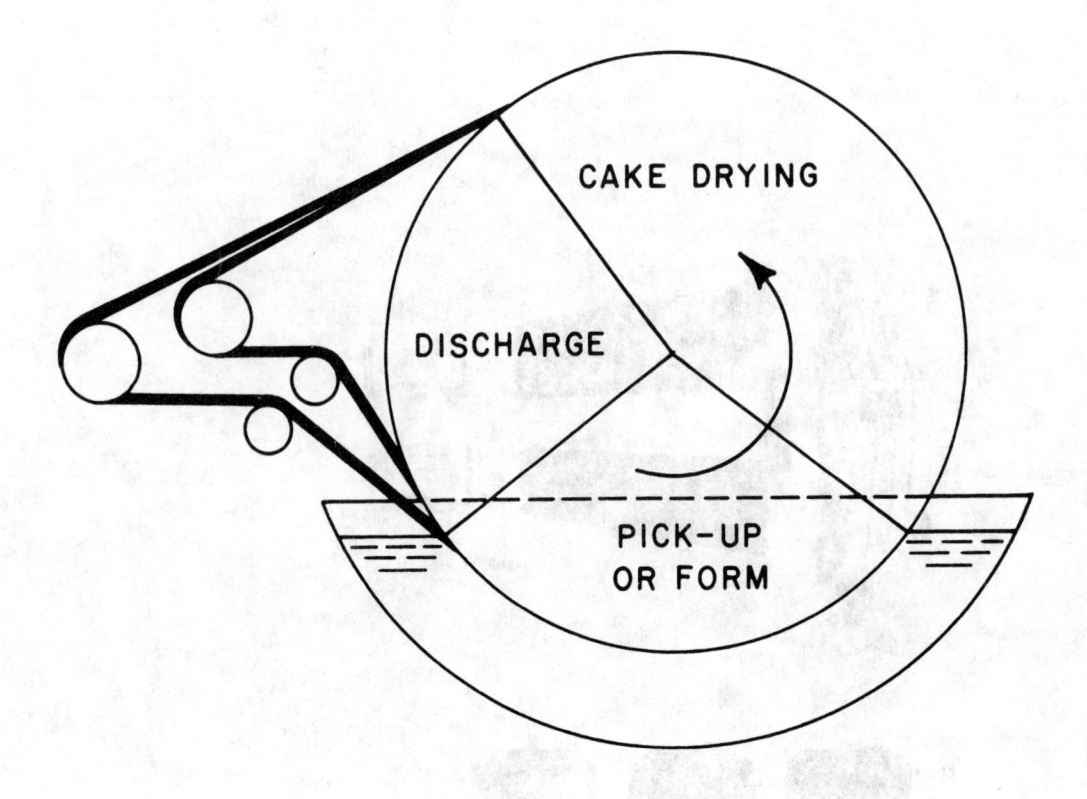

Fig. 17.13 Operating zones of a vacuum filter
(Permission of Metropolitan Water District of Southern California)

QUESTIONS

Write your answers in a notebook and then compare your answers with those on page 254.

17.6E List the methods available for dewatering sludges.

17.6F List the major advantages and limitations of using centrifuges to dewater sludges.

17.6G When is a precoat of diatomaceous earth required on vacuum filters?

17.7 DISCHARGE INTO COLLECTION SYSTEMS (SEWERS)

The easiest method of sludge disposal would be to send the sludge down the wastewater collection (sewer) system. This does create some complications even if the wastewater treatment plant has the capacity to handle the load. Sludge discharged to sewers could cause sewer blockages. Also, the fees charged by the wastewater treatment plant to process and dispose of the sludge could be prohibitive. The charges are usually based upon annual flow, chemical or biochemical oxygen demand, suspended solids, and peak and average discharge. There are also increased monitoring requirements and costs associated with a sewer discharge. The water treatment plant must have a holding tank so that the sludge can be released at a uniform rate throughout the day or released only during the wastewater treatment plant's low-flow period.

Brine from ion exchange units may be discharged into wastewater collection systems. Usually, the brine is discharged during the day to take advantage of high flows for dilution. When you plan such a discharge, notify the operator of the downstream wastewater treatment plant to be sure you will not create any unnecessary problems.

QUESTIONS

Write your answers in a notebook and then compare your answers with those on page 254.

17.7A What are the complications of discharging sludge to sewers?

17.7B When is brine from ion exchange units usually discharged into wastewater collection systems?

17.8 DISPOSAL OF SLUDGE[7]

Sludge is commonly disposed of in sanitary landfills. Other methods of ultimate disposal include lagoons, land application, and sanitary sewers. The method of disposal depends on the source and type of sludge, as well as economic, environmental, and community awareness/sensitivity considerations.

Water treatment plant lime sludge is an excellent liming agent for agricultural purposes. Lime sludge must be applied at a rate to achieve the best soil pH for optimum crop yield. Optimum levels of nitrogen and phosphorus are also important to achieve high crop yields.

The application of nitrogen fertilizers causes a reduction in soil pH. If optimum soil pH conditions do not exist, crop yields will be reduced. Therefore, sufficient quantities of lime must be applied as a means of counteracting the fertilizer applications.

Lime softening sludges can also aid in the reclamation of spoiled lands by neutralizing acid soils. Disposal of lime softening sludge on strip mine land will help minimize the discharge of acidic compounds and low pH drainage waters.

Although most lime softening sludges are an excellent liming agent for agricultural and land reclamation purposes, some lime softening sludges must be disposed of in a sanitary landfill due to the lack of availability of agricultural land or excessive costs. Landfilling of lime softening sludge is a practical alternative where this method is cost effective (minimum cost of disposal).

Alum sludge has a tendency to cause soils to harden and does not provide any beneficial value. For this reason, water treatment plant alum sludge must not be applied to agricultural land. The sludge may be applied to a dedicated land disposal site. The sludge is applied to the land and plowed into the soil. Landfilling is another method of ultimate disposal of alum sludge.

Some water treatment plants use a slow sand filter or settling pond for the treatment of iron filter backwash wastewater. The slow sand filter must be cleaned occasionally by removing the top 2 to 3 inches (50 to 75 mm) of sand and iron sludge. The material removed must be disposed of in a sanitary landfill.

Water treatment plants that soften water by the ion exchange (zeolite) softening method produce a wastewater that has high concentrations of total dissolved solids and chloride compounds. Ion exchange softening wastes should not be discharged untreated into low-flow streams. The wastewater treatment processes capable of reducing the total dissolved solids and chloride concentrations to acceptable levels are very energy consumptive and expensive. Ion exchange softener regeneration wastewater may be very carefully and slowly discharged into a sanitary sewer system. The operator at the wastewater treatment plant must be notified in advance.

If a sanitary sewer system is not available for the disposal of ion exchange softening wastes, holding tanks should be installed to store the liquid from the regeneration and rinse cycles. This liquid should be ultimately disposed of in a sanitary landfill.

Filter backwash wastewater must be recycled through the water treatment plant or placed in wastewater storage ponds for additional treatment and disposal. The remainder of this section discusses some of the procedures used by operators to dispose of sludges.

Wet sludge can be disposed of in an open field by use of spray bars or dumped out for landfill (Figure 17.14). When releasing the wet sludge at one spot, the back of the truck should face downhill so it will drain faster and be emptied out completely (Figure 17.15). This practice will also reduce the chances of the truck getting stuck in the sludge. Usually, it takes about 10 minutes to empty a truck.

Sometimes, individuals will request the sludge for fill and some contractors have used the sludge to mix with decomposed granite (DG) (a type of rock found in some regions) for fill purposes.

Sludge drying beds for sedimentation tank wastes also can be used to dry sludge from nearby backwash recovery ponds, but this requires the sludge to be handled for a second time. An open field spray bar application is one method for disposing of backwash recovery sludges because after a few weeks, the residual is hardly noticeable. PVC pipe may be used to dispose of backwash recovery sludges instead of using a spray bar.

Where sludge is repeatedly spread in a single landfill site, be prepared to mix the sludge in with the native soil because it is unsightly. Sludge should be disposed of as close as possible to the water treatment plant to reduce hauling costs. Process wastes also can be dried in lagoons and covered over permanently.

In plants with a size range from 5 to 10 MGD (19 to 38 MLD), it will take four operators approximately two days to complete the job of draining and cleaning a sedimentation tank or a backwash recovery pond and disposing of the sludge.

In plants where a backwash pond is not available, the sludge can be moved to a sump. A smaller suction hose must be used to empty small sumps; instead of the 6-inch (150-mm) hose, use either a 3- or 4-inch (75- or 100-mm) hose. With a smaller suction hose, it will take 25 minutes or more to fill the vacuum tank. Five or six operators will be needed to keep the sump filled. Of course, during the time that the truck is on the road to the dump site and back again, the operators are standing by and the sludge cannot be moved. If two trucks were used, this disadvantage could be overcome, but the cost may also increase. Therefore, the larger the sump and the truck's tank, the cheaper the operation from the standpoint of labor costs.

---

[7] Portions of this section were obtained from *ILLINOIS EPA SLUDGE REGULATION GUIDANCE DOCUMENT,* Illinois Environmental Protection Agency, 1021 North Grand Avenue East, PO Box 19276, Springfield, IL 62794-9276.

Fig. 17.14   Use of sludge for a landfill

Fig. 17.15   Emptying sludge from a truck

Plants of one MGD (3.8 M*L*D) capacity may be able to use some of the local septic tank pumpers or vacuum trucks to an advantage. Such companies usually have made arrangements for the use of dump sites that may be used to dispose of the sludge they pump.

> ## QUESTIONS
>
> Write your answers in a notebook and then compare your answers with those on page 254.
>
> 17.8A List the methods of ultimate sludge disposal.
>
> 17.8B Why should sludge be disposed of as close as possible to the water treatment plant?

17.9 EQUIPMENT

17.90 Vacuum Trucks

The use of a vacuum truck is highly recommended because these trucks develop the fewest problems with clogging. The larger the suction pipe the better. Any object smaller than the hose size can be readily sucked through unless there are too many objects in the sludge. Wedging of several rocks or sticks can occur, slowing up the process of sludge removal. Any object that can be seen should be removed by hand and discarded. Any well-operated plant should be void of these solids, but they are sometimes accidentally dropped in or thrown in by vandals. Leaves and other small objects that may get by the plant inlet screens are not too much of a problem unless an excessive amount exists. Under these conditions, the heavily accumulated portions should be scooped out to prevent any clogging of drain pipes or suction hoses.

17.91 Sludge Pumps

Many small treatment plants exist, especially in rural areas, where disposal of sludge would appear less troublesome. A couple of thousand gallons of sludge can be moved by gravity or pumped out to an open field and mixed (disked) into the soil. Even a quarter of an acre can handle many years of dried sludge from a small plant. Unfortunately, most of these plants do not have the land available, so the sludge in its wet form must be hauled out to a disposal site or a small lagoon or pond must be excavated for sludge collection. If sludge must be moved wet, use a sludge pump[8] to pump it into a tank or hire a septic tank pumper. Septic tank pumps, especially the suction hose, must be thoroughly rinsed and disinfected before use in any water treatment plant facilities. This applies also to all tools and equipment used.

A sludge pump can be an advantage over a self-priming centrifugal pump because of its large suction and discharge hose (usually three inches (75 mm) in diameter). These pumps are rated at about 60 GPM (3.8 *L*/sec) and can handle more solids. However, there are also self-priming types of wastewater pumps available that are designed so that solids do not actually pass through the impellers. These pumps may be used to pump sludges from water treatment plants. In either case, an excessive amount of foreign solids other than sludge itself can cause some pumping problems.

Wet sludge from a sedimentation tank will flow through pipes by gravity even if there are some ups and downs in the line provided the sludge is under some head. If difficulty arises, add some water for dilution or raise the end of the suction hose closer to the surface of the sump where the sludge is more diluted.

Because of the differences in volume between wet and dried sludge, it is always preferable to contain the wet sludge at the plant for drying. If the plant water source is a canal or reservoir close to the site, construct a small pond parallel to it and return the supernatant to the source for recycling by removing baffle boards. Only fresh backwash wastewater should be recycled. Water separated or drained from old sludge can cause serious problems if recycled. This water is likely to be *SEPTIC*[9] and could cause taste and odor problems. Also, this water will contain millions of bacteria and microorganisms that should not be recycled through a water treatment plant.

If necessary, semidry or dried sludge can be hauled out of the pond.

17.10 PLANT DRAINAGE WATERS

There are several sources of drainage waters in a water treatment plant that must be properly handled and disposed of. These sources include the laboratory, shops, and plant drainage water from leaks and spills. If continuous sampling pumps provide the lab with continuously flowing water from various plant processes, this water could be discharged to a sewer. Any reagents, toxics, or potentially pathogenic wastes from the lab must be properly treated and packaged before ultimate disposal in landfills. Drainage waters from leaks and other sources in the plant may be discharged to sewers. These drainage waters may

---

[8] Types of sludge pumps include diaphragm, nonclog, and progressive cavity pumps.

[9] *Septic* (SEP-tick) or *Septicity*. A condition produced by bacteria when all oxygen supplies are depleted. If severe, the bottom deposits produce hydrogen sulfide, the deposits and water turn black, give off foul odors, and the water has a greatly increased oxygen and chlorine demand.

be recycled through the plant; however, extreme caution must be exercised at all times to avoid contributing to taste and odor problems, health hazards, or operational problems.

17.11 MONITORING AND REPORTING

The location of the water treatment plant and the methods used to ultimately dispose of the process wastes will dictate the monitoring and reporting requirements. These reporting requirements may be established by local or state health or pollution control agencies. Monitoring and reporting will usually involve measuring and recording volumes of sludges or brines, percent solids, and other measurements that will prove that these processes are not creating any adverse environmental impacts.

QUESTIONS

Write your answers in a notebook and then compare your answers with those on page 254.

17.9A How do objects that plug sludge suction hoses get into water treatment plants?

17.9B What types of pumps can be used to pump wet sludge into a tank on a truck?

17.10A List the sources of plant drainage waters.

17.11A What factors will dictate the monitoring and reporting requirements for your sludge disposal program?

Please answer the discussion and review questions next.

DISCUSSION AND REVIEW QUESTIONS
Chapter 17. HANDLING AND DISPOSAL OF PROCESS WASTES

The purpose of these questions is to indicate to you how well you understand the material in the chapter. Write the answers to these questions in your notebook.

1. The amount of sludge accumulation at a typical water treatment plant depends on what factors?

2. What items should be considered when determining the frequency of cleaning a sedimentation basin?

3. What happens to the water separated from sludges in backwash recovery ponds or lagoons?

4. What precaution must be taken before draining any in-ground tank?

5. What duties should be performed by operators as the sludge is being drained from a sedimentation tank?

6. How would you fill a sedimentation tank after the tank has been emptied, inspected, and the necessary repairs completed?

7. How is sludge from the lime–soda ash softening process dewatered?

8. Why are three sand drying beds frequently installed at water treatment plants?

9. Why should the back of the sludge truck face downhill when releasing wet sludge at one point?

10. What is the purpose of monitoring and reporting for a sludge disposal program?

SUGGESTED ANSWERS
Chapter 17. HANDLING AND DISPOSAL OF PROCESS WASTES

Answers to questions on page 233.

17.0A Strict laws are needed regarding the disposal of process wastes to prevent rivers and streams from becoming more polluted. These laws are designed to prevent any waste discharge that could discolor, pollute, or generally be harmful to aquatic or plant life or the environment.

17.0B Current regulations require daily monitoring of any discharge and analysis of such water quality indicators as pH, turbidity, TDS, settleable solids, or other harmful materials.

Answers to questions on page 234.

17.1A Sludge is removed from sedimentation tanks by mechanical rakes or scrapers that periodically draw out sludge from a hopper, or a vacuum-type sludge removal device may be used.

17.1B Sludge is removed from upflow solids-contact units through sludge draw-off lines that must be monitored.

Answers to questions on page 235.

17.2A A source water stabilizing reservoir can reduce the volume of sludge handled by reducing the turbidity in the water being treated. Lower turbidities reduce the amount of alum required and thus the volume of sludge that settles out.

17.2B Sedimentation tanks should be drained and cleaned at least twice a year and more often if the sludge buildup interferes with the treatment processes.

Answers to questions on page 235.

17.3A Sedimentation tanks should be inspected and repaired (if necessary) when the tanks are drained and cleaned.

17.3B Sludge may be dewatered by the use of belt filter presses, centrifuges, filter presses, vacuum filters, solar lagoons, and sand drying beds.

Answers to questions on page 237.

17.4A Sludge must be removed manually from tanks without mechanical or vacuum-type sludge collectors. The tanks must be drained and then operators must push the sludge to the drain lines with squeegees or the sludge must be pumped out into tank trucks.

17.4B When draining a sedimentation tank, after any necessary intake valve(s) or gate(s) changes are made, drain the water down to the sludge blanket by partially opening the drain valve from the sedimentation tank into the lagoon or drying bed. After the first few minutes (if the valve is not wide open), the water will become clear. This portion of the water can be diverted from the lagoon or drying bed, if proper plumbing is available, and returned to the source to be reprocessed. Most drying beds will not handle this great a volume of water unless this draining process is extended over a period of a few days. Pump(s) can be used to transfer the settled water to another sedimentation tank that is still in operation.

17.4C The following precautions must be exercised whenever operators are in any closed tank (confined space):

1. Do not operate gasoline engines in the tank.
2. Test the atmosphere in the tank for oxygen deficiency/enrichment, flammable or explosive gases, and toxic gases, and provide adequate ventilation of clean air at all times.
3. Provide a source of water to clean off boots and tools where the operators come out of the tank.
4. Use the buddy system. At least one person must be outside the tank and watching anyone inside the tank, and another person must be nearby and available to help in an emergency.

Answers to questions on page 240.

17.5A Timing is critical if backwash recovery ponds are used to handle sludge from sedimentation basins because you want to avoid having to backwash the filters while you are draining a sedimentation tank.

17.5B The suction pipe for the backwash recovery pump should be floated near the surface of the pond so that any excess water can be recycled without also drawing out sludge. This will be very important if the filters must be backwashed at the same time sludge is being cleaned out of the backwash recovery ponds.

17.5C Lime–soda softening sludge can be disposed of ultimately by:

1. Covering the lagoon with soil
2. Hauling the dried sludge to a landfill
3. Spreading on agricultural soils to adjust the pH for optimum crop yields

254 Water Treatment

Answers to questions on page 241.

17.6A   The minimum recommended number of solar drying lagoons is three.

17.6B   Sludge drying beds are made with underdrains covered with gradations of aggregate and sand. The drains discharge into a sump where recovery pumps can return the water drained from the sludge back to the plant to be reprocessed.

17.6C   The proper time to remove sludge from the drying bed is when no more than one foot of sludge has accumulated and dried into a checkered pattern. A piece of dry sludge can then be picked up off the sand.

17.6D   When operating a front-end loader to remove sludge from a drying bed, be careful so only the dried sludge is picked up with the minimum possible disturbance to the sand and aggregate. The loader bucket capacity should be limited to one or two cubic yards of sludge.

Answers to questions on page 248.

17.6E   Sludges may be dewatered using: (1) solar drying lagoons, (2) sand drying beds, (3) belt filter presses, (4) centrifuges, (5) filter presses, and (6) vacuum filters.

17.6F   The principal advantage of using centrifuges to dewater sludges is that the density of the sludge cake can be varied from a thickened liquid slurry to a dry cake. The major limitation of using centrifuges is high energy consumption.

17.6G   A precoat of diatomaceous earth is required to dewater gelatinous alum sludge.

Answers to questions on page 248.

17.7A   The complications of discharging sludge to sewers include:
1. Possibility of causing sewer blockages
2. Wastewater treatment plant would have to process and dispose of the sludge
3. Fees charged by the wastewater treatment plant could be prohibitive
4. Increased monitoring requirements and costs associated with a sewer discharge
5. Water treatment plant must have a holding tank so that the sludge can be released at a uniform rate throughout the day

17.7B   Brine from ion exchange units is usually discharged into wastewater collection systems during the day to take advantage of high flows for dilution.

Answers to questions on page 251.

17.8A   Methods of ultimate sludge disposal include disposal in sanitary landfills, lagoons, land application, and discharge to sanitary sewers.

17.8B   Sludge should be disposed of as close as possible to the water treatment plant to reduce hauling costs.

Answers to questions on page 252.

17.9A   Objects that plug sludge suction hoses get into water treatment plants by being accidentally dropped in or thrown in by vandals.

17.9B   If sludge must be moved wet, use a sludge pump to pump it into a tank. Types of sludge pumps include diaphragm, nonclog, and progressive cavity pumps.

17.10A Sources of plant drainage waters include the laboratory, shops, and plant drainage water from leaks and spills.

17.11A Monitoring and reporting requirements for a sludge disposal program are dictated by the location of the water treatment plant and the methods used to ultimately dispose of the process wastes.

CHAPTER 18

MAINTENANCE

by

Parker Robinson

TABLE OF CONTENTS
Chapter 18. MAINTENANCE

	Page
OBJECTIVES	262
WORDS	263

LESSON 1

- **18.0 TREATMENT PLANT MAINTENANCE—GENERAL PROGRAM** ... 267
  - 18.00 Preventive Maintenance Records ... 267
  - 18.01 Library of Manufacturers' Operation and Parts Manuals ... 269
  - 18.02 Emergencies ... 269
  - 18.03 Lockout/Tagout Procedure ... 269
- **18.1 ELECTRICAL EQUIPMENT** ... 271
  - 18.10 Beware of Electricity ... 271
    - 18.100 Attention ... 271
    - 18.101 Recognize Your Limitations ... 271
  - 18.11 Electrical Fundamentals ... 272
    - 18.110 Introduction ... 272
    - 18.111 Volts ... 272
    - 18.112 Direct Current (DC) ... 272
    - 18.113 Alternating Current (AC) ... 272
    - 18.114 Amps ... 273
    - 18.115 Watts ... 274
    - 18.116 Power Requirements ... 274
    - 18.117 Conductors and Insulators ... 274
  - 18.12 Tools, Meters, and Testers ... 274
    - 18.120 Voltage Testing ... 274
    - 18.121 Ammeter ... 277
    - 18.122 Megger ... 280
    - 18.123 Ohmmeters ... 281
  - 18.13 Switch Gear ... 281
    - 18.130 Equipment Protective Devices ... 281
    - 18.131 Fuses ... 281
    - 18.132 Circuit Breakers ... 281

	18.133	Overload Relays	282
	18.134	Motor Starters	282
18.14	Electric Motors		284
	18.140	Classifications	284
	18.141	Troubleshooting	287
		18.1410 Step-By-Step Procedures	287
		18.1411 Troubleshooting Guide for Electric Motors	288
		18.1412 Troubleshooting Guide for Magnetic Starters	290
		18.1413 Trouble/Remedy Procedures for Induction Motors	292
	18.142	Recordkeeping	293
18.15	Auxiliary Electrical Power		293
	18.150	Safety First	293
	18.151	Standby Power Generation	293
	18.152	Emergency Lighting	296
	18.153	Batteries	297
18.16	High Voltage		298
	18.160	Transmission	298
	18.161	Switch Gear	298
	18.162	Power Distribution Transformers	298
18.17	Electrical Safety Checklist		299
18.18	Additional Reading		299
DISCUSSION AND REVIEW QUESTIONS			300

LESSON 2

18.2	**MECHANICAL EQUIPMENT**		301
	18.20	Repair Shop	301
	18.21	Pumps	301
		18.210 Centrifugal Pumps	301
		18.211 Let's Build a Pump!	301
		18.212 Vertical Centrifugal Pumps	310
		18.213 Horizontal Centrifugal Pumps	310
		18.214 Reciprocating or Piston Pumps	310
		18.215 Progressive Cavity (Screw-Flow) Pumps	312
		18.216 Chemical Metering Pumps	312
	18.22	Lubrication	312
		18.220 Purpose of Lubrication	312
		18.221 Properties of Lubricants	312
		18.222 Lubrication Schedule	315
		18.223 Precautions	315

		18.224	Pump Lubrication	315
		18.225	Equipment Lubrication	316

DISCUSSION AND REVIEW QUESTIONS ... 317

LESSON 3

	18.23	Pump Maintenance	318
		18.230 Section Format	318
		18.231 Preventive Maintenance	318
		1. Pumps, General	318
		2. Reciprocating Pumps, General	325
		3. Propeller Pumps, General	326
		4. Progressive Cavity Pumps, General	326
		5. Pump Controls	327
		6. Electric Motors	327
		7. Belt Drives	330
		8. Chain Drives	331
		9. Variable Speed Belt Drives	332
		10. Couplings	332
		11. Shear Pins	334
	18.24	Pump Operation	334
		18.240 Starting a New Pump	334
		18.241 Pump Shutdown	336
		18.242 Pump-Driving Equipment	336
		18.243 Electrical Controls	336
		18.244 Operating Troubles	336
		18.245 Starting and Stopping Pumps	338
		18.2450 Centrifugal Pumps	338
		18.2451 Positive Displacement Pumps	340

DISCUSSION AND REVIEW QUESTIONS ... 341

LESSON 4

	18.25	Compressors	342
	18.26	Valves	344
		18.260 Uses of Valves	344
		18.261 Gate Valves	344
		18.262 Maintenance of Gate Valves	344
		12. Gate Valves	344
		18.263 Globe Valves	346
		18.264 Eccentric Valves	347
		18.265 Butterfly Valves	347

		18.266	Check Valves	347
		18.267	Maintenance of Check Valves	351
			13. Check Valves	351
		18.268	Automatic Valves	360

DISCUSSION AND REVIEW QUESTIONS ... 360

LESSON 5

18.3 INTERNAL COMBUSTION ENGINES ... 362

    18.30 Gasoline Engines ... 362
        18.300 Need to Maintain Gasoline Engines ... 362
        18.301 Maintenance ... 362
        18.302 Starting Problems ... 362
        18.303 Running Problems ... 362
        18.304 How to Start a Gasoline Engine ... 363
            18.3040 Small Engines ... 363
            18.3041 Large Engines ... 363
    18.31 Diesel Engines ... 364
        18.310 How Diesel Engines Work ... 364
        18.311 Operation ... 364
        18.312 Fuel System ... 364
        18.313 Water-Cooled Diesel Engines ... 367
        18.314 Air-Cooled Diesel Engines ... 368
        18.315 How to Start Diesel Engines ... 368
        18.316 Maintenance and Troubleshooting ... 368
    18.32 Cooling Systems ... 368
    18.33 Fuel Storage ... 370
        18.330 Code Requirements ... 370
        18.331 Diesel Fuel ... 370
        18.332 Gasoline ... 370
        18.333 Liquified Petroleum Gas (LPG) ... 370
        18.334 Natural Gas ... 371
    18.34 Standby Engines ... 371

18.4 CHEMICAL STORAGE AND FEEDERS ... 371

    18.40 Chemical Storage ... 371
    18.41 Drainage from Chemical Storage and Feeders ... 372
    18.42 Use of Feeder Manufacturer's Manual ... 372
    18.43 Solid Feeders ... 372
    18.44 Liquid Feeders ... 372
    18.45 Gas Feeders ... 372

	18.46	Calibration of Chemical Feeders	372
		18.460 Large-Volume Metering Pumps	372
		18.461 Small-Volume Metering Pumps	372
		18.462 Dry-Chemical Systems	372
	18.47	Chlorinators	375

18.5 TANKS AND RESERVOIRS 376
    18.50 Scheduling Inspections 376
    18.51 Steel Tanks 376
    18.52 Cathodic Protection 377
    18.53 Concrete Tanks 377

18.6 BUILDING MAINTENANCE 377

18.7 ARITHMETIC ASSIGNMENT 378

18.8 ADDITIONAL READING 378

18.9 ACKNOWLEDGMENTS 378

DISCUSSION AND REVIEW QUESTIONS 378

SUGGESTED ANSWERS 379

OBJECTIVES
Chapter 18. MAINTENANCE

Following completion of Chapter 18, you should be able to:

1. Develop a maintenance program for your plant, including equipment, buildings, grounds, channels, and tanks.

2. Start a maintenance recordkeeping system that will provide you with information to protect equipment warranties, to prepare budgets, and to satisfy regulatory agencies.

3. Schedule maintenance of equipment at proper time intervals.

4. Perform maintenance as directed by manufacturers.

5. Recognize symptoms that indicate equipment is not performing properly, identify the source of the problem, and take corrective action.

6. Recognize the serious consequences that could occur when inexperienced, unqualified, or unauthorized persons attempt to troubleshoot or repair electrical panels, controls, circuits, wiring, or equipment.

7. Communicate with electricians by indicating possible causes of problems in electrical panels, controls, circuits, wiring, and motors.

8. Properly select and use the following pieces of equipment (if qualified and authorized):

    a. Multimeter

    b. Ammeter

    c. Megger

    d. Ohmmeter

9. Safely operate and maintain auxiliary electrical equipment, including during standby and emergency situations.

10. Describe how a pump is put together.

11. Discuss the application or use of different types of pumps.

12. Start and stop pumps.

13. Maintain the various types of pumps.

14. Operate and maintain a compressor.

15. Develop and conduct an equipment lubrication program.

16. Start up, operate, maintain, and shut down gasoline engines, diesel engines, heating, ventilating, and air conditioning systems.

NOTE: Special maintenance information is given in the previous chapters on treatment processes, where appropriate.

WORDS
Chapter 18. Maintenance

AIR GAP

An open, vertical drop, or vertical empty space, between a drinking (potable) water supply and potentially contaminated water. This gap prevents the contamination of drinking water by backsiphonage because there is no way potentially contaminated water can reach the drinking water supply.

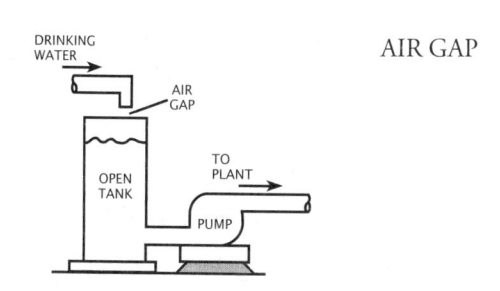

ALTERNATING CURRENT (AC)

An electric current that reverses its direction (positive/negative values) at regular intervals.

AMPERAGE (AM-purr-age)

The strength of an electric current measured in amperes. The amount of electric current flow, similar to the flow of water in gallons per minute.

AMPERE (AM-peer)

The unit used to measure current strength. The current produced by an electromotive force of one volt acting through a resistance of one ohm.

AMPLITUDE

The maximum strength of an alternating current during its cycle, as distinguished from the mean or effective strength.

AXIAL TO IMPELLER

The direction in which material being pumped flows around the impeller or flows parallel to the impeller shaft.

AXIS OF IMPELLER

An imaginary line running along the center of a shaft (such as an impeller shaft).

BRINELLING (bruh-NEL-ing)

Tiny indentations (dents) high on the shoulder of the bearing race or bearing. A type of bearing failure.

CATHODIC (kath-ODD-ick) PROTECTION

An electrical system for prevention of rust, corrosion, and pitting of metal surfaces that are in contact with water, wastewater, or soil. A low-voltage current is made to flow through a liquid (water) or a soil in contact with the metal in such a manner that the external electromotive force renders the metal structure cathodic. This concentrates corrosion on auxiliary anodic parts, which are deliberately allowed to corrode instead of letting the structure corrode.

CAVITATION (kav-uh-TAY-shun)

The formation and collapse of a gas pocket or bubble on the blade of an impeller or the gate of a valve. The collapse of this gas pocket or bubble drives water into the impeller or gate with a terrific force that can knock metal particles off and cause pitting on the impeller or gate surface. Cavitation is accompanied by loud noises that sound like someone is pounding on the impeller or gate with a hammer.

CIRCUIT

The complete path of an electric current, including the generating apparatus or other source; or, a specific segment or section of the complete path.

CIRCUIT BREAKER

A safety device in an electric circuit that automatically shuts off the circuit when it becomes overloaded. The device can be manually reset.

CONDUCTOR

(1) A pipe that carries a liquid load from one point to another point. In a wastewater collection system, a conductor is often a large pipe with no service connections. Also called a CONDUIT. Also see INTERCEPTOR (INTERCEPTING) SEWER or INTERCONNECTOR.

(2) In plumbing, a pipe installed to drain water from the roof gutters or roof catchment to the storm drain or other means of disposal. Also called a DOWNSPOUT.

(3) In electricity, a substance, body, device, or wire that readily conducts or carries electric current. Also called a CONDUIT.

COULOMB (KOO-lahm)

A measurement of the amount of electrical charge carried by an electric current of one ampere in one second. One coulomb equals about 6.25×10^{18} electrons (6,250,000,000,000,000,000 electrons).

CROSS CONNECTION

(1) A connection between drinking (potable) water and an unapproved water supply.

(2) A connection between a storm drain system and a sanitary collection system.

(3) Less frequently used to mean a connection between two sections of a collection system to handle anticipated overloads of one system.

CURRENT

A movement or flow of electricity. Electric current is measured by the number of coulombs per second flowing past a certain point in a conductor. A coulomb is equal to about 6.25×10^{18} electrons (6,250,000,000,000,000,000 electrons). A flow of one coulomb per second is called one ampere, the unit of the rate of flow of current.

CYCLE

A complete alternation of voltage or current in an alternating current (AC) circuit.

DATEOMETER (day-TOM-uh-ter)

A small calendar disk attached to motors and equipment to indicate the year in which the last maintenance service was performed.

DIRECT CURRENT (DC)

Electric current flowing in one direction only and essentially free from pulsation. Also see ALTERNATING CURRENT.

ELECTROLYTE (ee-LECK-tro-lite)

A substance that dissociates (separates) into two or more ions when it is dissolved in water.

ELECTROMOTIVE FORCE (EMF)

The electrical pressure available to cause a flow of current (amperage) when an electric circuit is closed. Also called VOLTAGE.

ELECTRON

(1) A very small, negatively charged particle that is practically weightless. According to the electron theory, all electrical and electronic effects are caused either by the movement of electrons from place to place or because there is an excess or lack of electrons at a particular place.

(2) The part of an atom that determines its chemical properties.

END BELLS
Devices used to hold the rotor and stator of a motor in position.

FUSE
A protective device having a strip or wire of fusible metal that, when placed in a circuit, will melt and break the electric circuit if heated too much. High temperatures will develop in the fuse when a current flows through the fuse in excess of that which the circuit will carry safely.

GROUND
An expression representing an electrical connection to earth or a large conductor that is at the earth's potential or neutral voltage.

HERTZ (Hz)
The number of complete electromagnetic cycles or waves in one second of an electric or electronic circuit. Also called the frequency of the current.

HYGROSCOPIC (hi-grow-SKAWP-ick)
Absorbing or attracting moisture from the air.

JOGGING
The frequent starting and stopping of an electric motor.

LEAD (LEED)
A wire or conductor that can carry electric current.

MANDREL (MAN-drill)
(1) A special tool used to push bearings in or to pull sleeves out.
(2) A testing device used to measure for excessive deflection in a flexible conduit.

MEG
(1) Abbreviation of MEGOHM.
(2) A procedure used for checking the insulation resistance on motors, feeders, bus bar systems, grounds, and branch circuit wiring. Also see MEGGER.

MEGGER (from megohm)
An instrument used for checking the insulation resistance on motors, feeders, bus bar systems, grounds, and branch circuit wiring. A megger reads in millions of ohms. Also see MEG.

MEGOHM (MEG-ome)
Millions of ohms. Mega- is a prefix meaning one million, so 5 megohms means 5 million ohms.

MULTISTAGE PUMP
A pump that has more than one impeller. A single-stage pump has one impeller.

NAMEPLATE
A durable, metal plate found on equipment that lists critical operating conditions for the equipment.

OSHA (O-shuh)
The Williams-Steiger Occupational Safety and Health Act of 1970 (OSHA) is a federal law designed to protect the health and safety of workers, including the operators of water supply and treatment systems and wastewater collection and treatment systems. The Act regulates the design, construction, operation, and maintenance of water and wastewater systems. OSHA regulations require employers to obtain and make available to workers the Material Safety Data Sheets (MSDSs) for chemicals used at industrial facilities and treatment plants. OSHA also refers to the federal and state agencies that administer the OSHA regulations.

OHM
The unit of electrical resistance. The resistance of a conductor in which one volt produces a current of one ampere.

POLE SHADER
A copper bar circling the laminated iron core inside the coil of a magnetic starter.

POWER FACTOR
The ratio of the true power passing through an electric circuit to the product of the voltage and amperage in the circuit. This is a measure of the lag or lead of the current with respect to the voltage. In alternating current, the voltage and amperes are not always in phase; therefore, the true power may be slightly less than that determined by the direct product.

PRUSSIAN BLUE
A blue paste or liquid (often on a paper like carbon paper) used to show a contact area. Used to determine if gate valve seats fit properly.

RADIAL TO IMPELLER
Perpendicular to the impeller shaft. Material being pumped flows at a right angle to the impeller.

RESISTANCE
That property of a conductor or wire that opposes the passage of a current, thus causing electric energy to be transformed into heat.

ROTOR
The rotating part of a machine. The rotor is surrounded by the stationary (nonmoving) parts (stator) of the machine.

SEIZING or SEIZE UP
Seizing occurs when an engine overheats and a part expands to the point where the engine will not run. Also called freezing.

SHEAVE
V-belt drive pulley, which is commonly made of cast iron or steel.

SHIM
Thin metal sheets that are inserted between two surfaces to align or space the surfaces correctly. Shims can be used anywhere a spacer is needed. Usually shims are 0.001 to 0.020 inch (0.025 to 0.50 mm) thick.

SINGLE-STAGE PUMP
A pump that has only one impeller. A multistage pump has more than one impeller.

STATOR
That portion of a machine that contains the stationary (nonmoving) parts that surround the moving parts (rotor).

STETHOSCOPE
An instrument used to magnify sounds and carry them to the ear.

VOLTAGE
The electrical pressure available to cause a flow of current (amperage) when an electric circuit is closed. Also called ELECTROMOTIVE FORCE.

WATER HAMMER
The sound like someone hammering on a pipe that occurs when a valve is opened or closed very rapidly. When a valve position is changed quickly, the water pressure in a pipe will increase and decrease back and forth very quickly. This rise and fall in pressures can cause serious damage to the system.

CHAPTER 18. MAINTENANCE

(Lesson 1 of 5 Lessons)

18.0 TREATMENT PLANT MAINTENANCE—GENERAL PROGRAM

A water treatment plant operator has many duties. Most of them have to do with the efficient operation of the plant. An operator has the responsibility to produce a water that will meet all the requirements established for the plant. By doing this, the operator develops a good working relationship with the regulatory agencies, water users, and plant neighbors.

Another duty an operator has is that of plant maintenance. A good maintenance program is a must in order to maintain successful operation of the plant. A successful maintenance program will include everything from mechanical equipment to the care of the plant grounds, buildings, and structures.

Mechanical maintenance is of prime importance as the equipment must be kept in good operating condition in order for the plant to maintain peak performance. Manufacturers provide information on the mechanical maintenance of their equipment. You should thoroughly read their literature on your plant equipment and understand the procedures. Contact the manufacturer or the local representative if you have any questions. Follow the instructions very carefully when performing maintenance on equipment. You also must recognize tasks that may be beyond your capabilities or repair facilities, and you should request assistance when needed.

For a successful maintenance program, your supervisors must understand the need for and benefits from equipment that operates continuously as intended. Disabled or improperly working equipment is a threat to the quality of the plant output, and repair costs for poorly maintained equipment usually exceed the cost of maintenance.

18.00 Preventive Maintenance Records

Preventive maintenance programs help operating personnel keep equipment in satisfactory operating condition and aid in detecting and correcting malfunctions before they develop into major problems.

A frequent occurrence in a preventive maintenance program is the failure of the operator to record the work after it is completed. When this happens, the operator must rely on memory to know when to perform each preventive maintenance function. As days pass into weeks and months, the preventive maintenance program is lost in the turmoil of everyday operation.

The only way an operator can keep track of a preventive maintenance program is by good recordkeeping. A good recordkeeping system tells when maintenance is due and also provides a record of equipment performance. Poor performance is a good justification for replacement or new equipment. Good records also help keep your warranty in force. Whatever recordkeeping system is used, it should be kept up to date on a daily basis and not left to memory for some other time. Equipment service cards and service record cards (Figure 18.1) are easy to set up and require little time to keep up to date.

An *EQUIPMENT SERVICE CARD* (master card) should be filled out for each piece of equipment in the plant. Each card should have the equipment name on it, such as Raw Water Intake Pump No. 1.

1. List each required maintenance service with an item number.

2. List maintenance services in order of frequency of performance. For instance, show daily service as items 1, 2, and 3 on the card; weekly items as 4 and 5; monthly items as 6, 7, 8, and 9; and so on.

3. Describe each type of service under work to be done.

Make sure all necessary inspections and services are shown. For reference data, list paragraph or section numbers as shown in the pump maintenance section of this lesson (Section 18.23, page 318). Also, list frequency of service as shown in the time schedule columns of the same section. Under time, enter day or month service is due. Service card information may be changed to fit the needs of your plant or particular equipment as recommended by the equipment manufacturer. Be sure the information on the cards is complete and correct.

The *SERVICE RECORD CARD* should have the date and work done, listed by item number and signed by the operator who performed the service. Some operators prefer to keep both cards clipped together, while others place the service record card near the equipment.

When the service record is filled, it should be filed for future reference and a new card attached to the master card. The equipment service card tells what should be done and when, while the service record card is a record of what you did and when you did it. Many operators keep this information on a computer.

268 Water Treatment

EQUIPMENT SERVICE CARD					
EQUIPMENT: #1 Raw Water Intake Pump					
Item No.	Work To Be Done	Reference[a]	Frequency	Time	
1	Check water seal and packing gland	Par. 1	Daily	
2	Listen for unusual noises	Par. 6	Daily	
3	Operate pump alternately	Par. 1	Weekly	Monday
4	Inspect pump assembly	Par. 1	Weekly	Wednesday
5	Inspect and lube bearings	Par. 1	Quarterly	1-4-7-10[b]
6	Check operating temperature of bearings	Par. 1	Quarterly	1-4-7-10[b]
7	Check alignment of pump and motor	Par. 1	Semiann.	4 & 10
8	Inspect and service pumps	Par. 1	Semiann.	4 & 10
9	Drain pump before shutdown	Par. 1		

SERVICE RECORD CARD					
EQUIPMENT: #1 Raw Water Intake Pump					
Date	Work Done (Item No.)	Signed	Date	Work Done (Item No.)	Signed
1-5-06	1-2-3	J.B.			
1-6-06	1-2	J.B.			
1-7-06	1-2-4-5-6	R.W.			

[a] Par. 1 refers to Paragraph 1 in Section 18.23 of this manual. Par. 6 is also in Section 18.23.
[b] 1-4-7-10 represent the months of the year when the equipment should be serviced—1-January, 4-April, 7-July, and 10-October.

Fig. 18.1   Equipment service card and service record card

18.01 Library of Manufacturers' Operation and Parts Manuals

A plant library can contain helpful information to assist in plant operation. Material in the library should be cataloged and filed for easy use. Items in the library should include:

1. Plant operation and maintenance instruction manuals
2. Plant plans and specifications
3. Manufacturers' instructions
4. Reference books on water treatment
5. Operator training manual on water treatment
6. Professional journals and publications
7. First-aid book
8. Reports from other plants
9. A dictionary

18.02 Emergencies

If your plant has not developed procedures for handling potential emergencies, do it now. Emergency procedures must be established for operators to follow when emergencies are caused by the release of chlorine or hazardous or toxic chemicals into the raw water supply, power outages, broken transmission lines or distribution mains, and attacks by terrorists. These procedures should include a list of emergency phone numbers located near all telephones, especially one that is unlikely to be affected by the emergency. Update the phone numbers annually.

1. Police
2. Fire
3. Hospital or Physician
4. Responsible Plant Officials
5. Local Emergency Disaster Office
6. *CHEMTREC*, (800) 424-9300
7. Emergency Team (if your plant has one)

The *CHEMTREC* toll-free number may be called at any time. Personnel at this number will give information on how to handle emergencies created by hazardous materials and will notify appropriate emergency personnel.

An emergency team for your plant may be trained and assigned the task of responding to specific emergencies such as chlorine leaks. This emergency team must meet the following strict specifications at all times.

1. Team personnel must be physically and mentally qualified.
2. Proper equipment must be available at all times, including:
   a. Protective equipment, including self-contained breathing apparatus
   b. Repair kits
   c. Repair tools
   d. First-aid supplies
3. Proper training must take place on a regular basis and include instruction about:
   a. Properties and detection of hazardous chemicals
   b. Safe procedures for handling and storage of chemicals
   c. Types of containers, safe procedures for shipping containers, and container safety devices
   d. Installation of repair devices
   e. First-aid procedures
4. Team members must be exposed regularly to simulated field emergencies or practice drills. Team response must be carefully evaluated and any errors or weaknesses corrected.
5. Emergency team performance must be reviewed annually on a specified date. Review must include:
   a. Training program
   b. Response to actual emergencies
   c. Team physical and mental examinations

WARNING. One person should never be permitted to attempt an emergency repair alone. Always wait for trained assistance. Valuable time, needed to correct a serious emergency, could be lost rescuing a foolish individual.

For additional information on emergencies, see Chapter 7, "Disinfection," Section 7.52, "Chlorine Leaks," Chapter 10, "Plant Operation," Section 10.9, "Emergency Conditions and Procedures," and Chapter 23, "Administration," Section 23.10, "Emergency Response." Chapter 23 contains information on what to do if a toxic substance gets into your water supply.

18.03 Lockout/Tagout Procedure

OSHA standards require that all equipment that could unexpectedly start up or release stored energy must be locked out and tagged whenever it is being worked on. Common forms of stored energy are electrical energy, spring-loaded equipment, hydraulic pressure, and compressed gases that are stored under pressure.

The operator who will be performing the work on the equipment is the person who installs a lock (either key or combination) on the energy isolating device (switch, valve) to positively lock the switch or valve in a safe position, preventing the equipment from starting or moving. At the same time, a tag such as the one shown in Figure 18.2 is installed at the lock indicating why the equipment is locked out and who is working on it. No other person is authorized to remove the tag or lockout device unless the employer has provided specific procedures and training for removal by others.

Even after equipment is locked out and tagged, it still may not be safe to work on. Stored energy in gas, air, water, steam,

270 Water Treatment

DANGER

OPERATOR WORKING ON LINE

DO NOT CLOSE THIS SWITCH WHILE THIS TAG IS DISPLAYED

TIME OFF: _____

DATE: _____

SIGNATURE: _____

This is the ONLY person authorized to remove this tag.

INDUSTRIAL INDEMNITY/INDUSTRIAL UNDERWRITERS/
INSURANCE COMPANIES

4E210—R66

Fig. 18.2 Typical warning tag
(Source: Industrial Indemnity/Industrial Underwriters/Insurance Companies)

and hydraulic systems (such as water pumping systems) must be drained or bled down to prevent its sudden release, which could injure an operator working on the system or equipment. Elevated machine members, flywheels, and springs should be physically blocked or secured in place to prevent movement.

Before bleeding down pressurized systems, think about the pressures involved and what will be discharged to the atmosphere and work area (toxic or explosive gases, corrosive chemicals). Will other safety precautions have to be taken? What volume of bleed down material will there be and where will it go? If dealing with chemicals, greases, or lubricants, how can they be cleaned up or contained? Many times an operator has removed a pump volute cleanout only to find that the discharge check valve was not seated or one of the isolation valves was not fully seated. Once the pump was open, however, the pump intake structure drained into the pump room through the opened pump.

All rotating mechanical equipment must have guards installed to protect operators from becoming entangled or caught up in belts, pulleys, drive lines, flywheels, and couplings. Keep the guards installed even when no one is working on the equipment.

Various pieces of machinery are equipped with travel limit switches, pressure sensors, pressure reliefs, shear pins, and stall or torque switches to ensure proper and safe operation of the equipment. Never disconnect a device, or install larger shear pins than those specified on the original design, or modify pressure or temperature settings.

The basic elements of a proper lockout/tagout procedure are as follows:

1. Notify all affected employees that a lockout or tagged system is going to be used and the reason why. The authorized employee shall know the type and level of energy that the equipment uses and shall understand the hazard presented by the equipment.

2. If the equipment is operating, shut it down using the procedures established for the equipment.

3. Operate the switch, valve, or other energy isolating device(s) so that the equipment is isolated from its energy source(s). Stored energy such as that in springs, elevated machine parts, rotating flywheels, hydraulic systems, and air, gas, steam, or water pressure must be released or restrained by methods such as repositioning, blocking, or bleeding down.

4. Lock out or tag out the energy isolating device with your assigned individual lock or tag.

5. After ensuring that no personnel are exposed, and as a check that the energy source is disconnected, operate the pushbutton or other normal operating controls to make certain the equipment will not operate. A common problem when

motor control centers (MCCs) contain many breakers is to lock out the wrong equipment (locking pump #3 and thinking it is #2). Always confirm the dead circuit. *CAUTION!* Return operating controls to the neutral or OFF position after the test.

6. The equipment is now locked out or tagged out and work on the equipment may begin.

7. After the work on the equipment is complete, all tools have been removed, guards have been reinstalled, and employees are in the clear, remove all lockout or tagout devices. Operate the energy isolating devices to restore energy to the equipment.

8. Notify affected employees that the lockout or tagout device(s) has been removed before starting the equipment.

QUESTIONS

Write your answers in a notebook and then compare your answers with those on page 379.

18.0A Why should you plan a good maintenance program for your treatment plant?

18.0B What general items would you include in your maintenance program?

18.0C Why should you have a good recordkeeping system for your maintenance program?

18.0D What is the difference between an equipment service card and a service record card?

18.0E Prepare a list of emergency phone numbers for your treatment plant.

18.0F Why is it necessary to bleed down pressurized systems before making repairs?

18.1 ELECTRICAL EQUIPMENT

18.10 Beware of Electricity

18.100 Attention

A. *DO NOT ATTEMPT TO INSTALL, TROUBLESHOOT, MAINTAIN, REPAIR, OR REPLACE ELECTRICAL EQUIPMENT, PANELS, CONTROLS, WIRING, OR CIRCUITS UNLESS YOU:*

1. *KNOW WHAT YOU ARE DOING*

2. *ARE QUALIFIED*

3. *ARE AUTHORIZED*

Section 18.11, "Electrical Fundamentals," is presented to provide you with an understanding and awareness of electricity. The purpose of the section is to help you provide electricians with the information they will need when you contact them and request their assistance. You must be extremely familiar with electricity before attempting any major repairs.

B. Due to the wide variety of equipment and manufacturers in the water treatment field, detailed procedures for the maintenance of some types of equipment were very difficult to include in this chapter. Also, manufacturers are continually improving their products and some details would soon be out of date. For details concerning the operation, maintenance, and repair of a particular piece of equipment, refer to the O & M instruction manual or contact the manufacturer.

C. Effective equipment maintenance is the key to successful system performance. The better your maintenance, the better your facilities will perform. Abuse your equipment and facilities and they will abuse you. Everyone must realize that if the equipment cannot work, no one can work.

18.101 Recognize Your Limitations

In the water departments of all cities, there is a need for maintenance operators to know something about electricity. Duties could range from repairing a taillight on a trailer or vehicle to repairing complex pump controls and motors. Very few maintenance operators do the actual electrical repairs or troubleshooting because this is a highly specialized field and unqualified people can seriously injure themselves and damage costly equipment. For these reasons, you must be familiar with electricity, know the hazards, and recognize your own limitations when you must work with electrical equipment.

Most municipalities employ electricians or contract with a commercial electrical company that they call when major problems occur. However, the maintenance operator should be able to explain how the equipment is supposed to work and what it is doing or is not doing when it fails. After studying this section, you should be able to tell an electrician what appears to be the problem with electrical panels, controls, circuits, and equipment.

The need for safety should be apparent. If proper safe procedures are not followed in operating and maintaining the various electrical equipment used in water treatment facilities, accidents can happen that cause injuries, permanent disability, or loss of life. Some of the serious accidents that have happened and could have been avoided occurred when machinery was not shut off, locked out, and tagged properly (Figure 18.2). Possible accidents include:

1. Maintenance operator could be cleaning pump and have it start, thus losing an arm, hand, or finger.

2. Electric motors or controls not properly grounded could lead to possible severe shock, paralysis, or death.

3. Improper circuits such as a wrong connection, safety devices jumped, wrong fuses, or improper wiring can cause fires or injuries due to incorrect operation of machinery.

Another consideration for having a basic working knowledge of electricity is to prevent financial losses resulting from motors burning out and from damage to equipment, machinery, and control circuits. Additional costs result when damages have to be repaired, including payments for outside labor.

> **WARNING**
>
> Never work in electrical panels or on electrical controls, circuits, wiring, or equipment unless you are qualified and authorized. By the time you find out what you do not know about electricity, you could find yourself too dead to use the knowledge.

QUESTIONS

Write your answers in a notebook and then compare your answers with those on page 379.

18.10A Why must unqualified or inexperienced people be extremely careful when attempting to troubleshoot or repair electrical equipment?

18.10B What could happen when machinery is not shut off, locked out, and tagged properly?

18.11 Electrical Fundamentals

18.110 Introduction

This section contains a basic introduction to electrical terms and information plus directions on how to troubleshoot problems with electrical equipment.

Most electrical equipment used in water treatment plants comes with an instruction manual and is labeled with nameplate information indicating the proper voltage and allowable current in amps.

18.111 Volts

Voltage (E) is also known as Electromotive Force (EMF), and is the electrical pressure available to cause a flow of current (amperage) when an electric circuit is closed.[1] This pressure can be compared with the pressure or force that causes water to flow in a pipe. Some pressure in a water pipe is required to make the water move. The same is true of electricity. A force is necessary to push electricity or electric current through a wire. This force is called voltage. There are two types of current: Direct Current (DC) and Alternating Current (AC).

18.112 Direct Current (DC)

Direct current (DC) flows in one direction only and is essentially free from pulsation. Direct current is seldom used in water treatment plants except in electronic equipment, some control components of pump drives, and standby lighting. Direct current is used exclusively in automotive equipment, certain types of welding equipment, and a variety of portable equipment. Direct current is found in various voltages such as 6 volts, 12 volts, 24 volts, 48 volts, and 110 volts. All batteries are direct current. DC voltage can be measured by holding the positive and negative leads of a DC multimeter on the corresponding terminals of the DC device such as a battery. Direct current usually is not found in higher voltages (over 24 volts) around plants except in motor generator sets. Care must be taken when installing battery cables and wiring that Positive (+) and Negative (−) poles are connected properly to wires marked (+) and (−). If not properly connected, you could get an arc of electricity across the unit that could cause an explosion.

18.113 Alternating Current (AC)

An alternating current circuit is one in which the voltage and current periodically change direction and *AMPLITUDE*.[2] In other words, the current goes from zero to maximum strength, back to zero, and to the same strength in the opposite direction. Most AC circuits have a frequency of 60 *CYCLES*[3] per second. "Hertz" is the term we use to describe the frequency of cycles completed per second so our AC voltage would be 60 Hertz (Hz).

Alternating current is classified as:

1. Single phase
2. Two phase
3. Three phase, or polyphase

The most common of these are single phase and three phase. The various voltages you probably will find on your job are 110 volts, 120 volts, 208 volts, 220 volts, 240 volts, 277 volts, 440 volts, 480 volts, 2,400 volts, and 4,160 volts.

Single-phase power is found in lighting systems, small pump motors, various portable tools, and throughout our homes. This power is usually 120 volts and sometimes 240 volts. Single phase means that only one phase of power is supplied to the main electrical panel at 240 volts and has three wires or leads. Two of these leads have 120 volts each, the other lead is neutral and usually is coded white. The neutral lead is grounded. Many appliances and power tools have an extra ground (commonly a green wire) on the case for additional protection.

Three-phase power is generally used with motors and transformers found in water treatment plants, and usually is 208, 220, 240 volts, or 440, 460, 480, and 550 volts. Higher voltages are used in some pump stations. Three-phase power is used when higher power requirements or larger motors are used because efficiency is usually higher and motors require less maintenance. Generally speaking, all motors above two horsepower are

---

[1] Electricians often talk about closing an electric circuit. This means they are closing a switch that actually connects circuits together so electricity can flow through the circuit. Closing an electric circuit is like opening a valve on a water pipe.

[2] *Amplitude.* The maximum strength of an alternating current during its cycle, as distinguished from the mean or effective strength.

[3] *Cycle.* A complete alternation of voltage or current in an alternating current (AC) circuit.

three phase unless there is a problem with the power company getting three-phase power to the installations. Three-phase power usually is brought in to the point of use with three leads. There is power on all three leads and the fuse switches will generally appear as shown in Figure 18.3.

When making voltage measurements on three-phase power circuits, take three readings: (1) between lead 1 and lead 2, (2) between 1 and 3, and (3) between 2 and 3. The imbalance between readings should not exceed five percent of the average of the three readings and the average should not be below the nominal voltage (208, 220, 240, 460) nor should it exceed the nominal voltage by more than five percent. Voltages that do not meet these limits will place undue stress on electrical equipment, especially motors.

When there is power in three leads and a fourth lead is brought in, it is a neutral lead. Incoming power goes through a meter and then some type of disconnecting switch. This switch could be a fuse switch or a circuit breaker. The purpose of the disconnect switch is to open whenever a short or fault occurs and thus protect both the electric circuits and electrical equipment.

Circuit breakers (Figure 18.4) are used to protect electric circuits from overloads. Most circuit breakers are metal conductors that de-energize the main circuit when excess current passes through a metal strip causing it to overheat and open the main circuit.

Two-phase systems will not be discussed because they are seldom found in water treatment facilities.

18.114 Amps

An ampere (I) is the practical unit of electric current. This is the current produced by a pressure of one volt in a circuit having a resistance of one ohm. Amperage is the measurement of current or electron flow and is an indication of work being done or "how hard the electricity is working."

In order to understand amperage, one more term must be explained. The ohm is the practical unit of electrical resistance (R). "Ohm's Law" states that in a given electrical circuit the amount of current (I) in amperes is equal to the pressure in volts (E) divided by the resistance (R) in ohms. The following three formulas are given to provide you with an indication of the relationships among current, resistance, and EMF (electromotive force).

$$\text{Current, amps} = \frac{\text{EMF, volts}}{\text{Resistance, ohms}} \qquad I = \frac{E}{R}$$

$$\text{EMF, volts} = (\text{Current, amps})(\text{Resistance, ohms}) \qquad E = IR$$

$$\text{Resistance, ohms} = \frac{\text{EMF, volts}}{\text{Current, amps}} \qquad R = \frac{E}{I}$$

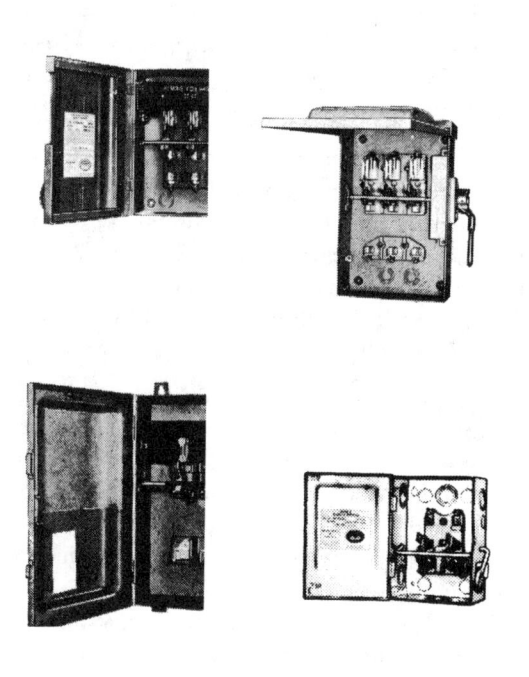

Fig. 18.3 Fuse switches
(Courtesy of Consolidated Electrical Distributors, Inc.)

Fig. 18.4 Circuit breakers
(Courtesy of Consolidated Electrical Distributors, Inc.)

These equations are used by electrical engineers for calculating circuit characteristics. If you memorize the following relationship, you can always figure out the correct formula.

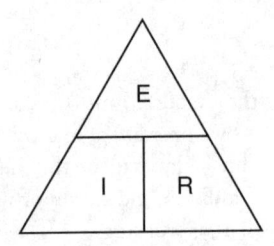

To use the above triangle you cover up with your finger the term you do not know or are trying to find out. The relationship between the other two known terms will indicate how to calculate the unknown. For example, if you are trying to calculate the current, cover up I. The two knowns (E and R) are shown in the triangle as E/R. Therefore, I = E/R. The same procedure can be used to find E when I and R are known or to find R when E and I are known.

18.115 Watts

Watts (W) and kilowatts (kW) are the units of measurement of the rate at which power is being used or generated. In DC circuits, watts (W) equal the voltage (E) multiplied by the current (I).

Power, watts = (Electromotive Force, volts)(Current, amps)

or P, watts = (E, volts)(I, amps)

In AC polyphase circuits the formula becomes more complicated because of the inclusion of two additional factors. First, there is the square root of 3, for three-phase circuits, which is equal to 1.73. Secondly, there is the power factor, which is the ratio of the true or actual power passing through an electric circuit to the product of the voltage times the amperage in the circuit. For standard three-phase induction motors, the power factor will be somewhere near 0.9. The formula for power input to a three-phase motor is:

$$\text{Power, kilowatts} = \frac{(E, \text{volts})(I, \text{amps})(\text{Power Factor})(1.73)}{1,000 \text{ watts/kilowatt}}$$

Since 0.746 kilowatt equals 1.0 horsepower, then the power output of a motor is:

$$\frac{\text{Power Output,}}{\text{horsepower}} = \frac{(\text{Power Input, kilowatts})(\text{Efficiency, \%})}{(0.746 \text{ kilowatt/horsepower})(100\%)}$$

18.116 Power Requirements

Power requirements (PR) are expressed in kilowatt hours. 500 watts for two hours or one watt for 1,000 hours equals one kilowatt hour. The power company charges so many cents per kilowatt hour.

Power Req, kW-hr = (Power, kilowatts)(Time, hours)

PR, kW-hr = (P, kW)(T, hr)

18.117 Conductors and Insulators

A material, like copper, that permits the flow of electric current is called a conductor. Material that will not permit the flow of electricity, like rubber, is called an insulator. Such material when wrapped or cast around a wire is called insulation. Insulation is commonly used to prevent the loss of electrical flow by two conductors coming into contact with each other.

> **QUESTIONS**
>
> Write your answers in a notebook and then compare your answers with those on page 379.
>
> 18.11A How can you determine the proper voltage and allowable current in amps for a piece of equipment?
>
> 18.11B What are two types of current?
>
> 18.11C Amperage is a measurement of what?

18.12 Tools, Meters, and Testers

> **WARNING**
>
> Never enter any electrical panel or attempt to troubleshoot or repair any piece of electrical equipment or any electric circuit unless you are qualified and authorized.

18.120 Voltage Testing

In order to maintain, repair, and troubleshoot electrical equipment and circuits, the proper tools are required. You will need a *MULTIMETER* to check for voltage. There are several types on the market and all of them work. They are designed to be used on energized circuits and care must be exercised when testing. By holding one lead on ground and the other on a power lead, you can determine if the circuit is energized.

Be sure the multimeter that you are using has sufficient range to measure the voltage you would expect to find. In other words, do not use a multimeter with a limit of 600 volts on a circuit that normally is energized at 2,400 volts. With the multimeter, you can tell if the current is AC or DC and the intensity or voltage, which will probably be one of the following: 120, 208, 240, 480, 2,400, or 4,160.

Do not work on any electric circuits unless you are qualified and authorized. Use a multimeter and other circuit testers to determine if a circuit is energized, or if all voltage is off. This should be done after the main switch is turned off to make sure it is safe to work inside the electrical panel. Always be aware of the possibility that even if the disconnect to the unit you are working on is off, the control circuit may still be energized if the circuit originates at a different distribution panel. Also, a capacitor in the unit may have sufficient energy stored to cause considerable harm to an operator, such as a power factor correction capacitor on a motor. Check with a multimeter before and during the time the switch is pulled off to have a double-check. This procedure ensures that the multimeter is working and that you have good continuity to your tester. Use circuit testers to measure voltage or current characteristics to a given piece of equipment and to make sure that you have or do not have a "live" circuit.

In addition to checking for power, a multimeter can be used to test for open circuits, blown fuses, single phasing of motors, grounds, and many other uses. Some examples are illustrated in the following paragraphs.

In the circuit shown below (Figure 18.5), test for power by holding one lead of the multimeter on point "A" and the other at point "B." If no power is indicated, the switch is open or faulty. The sketch shows the switch in the "open" position.

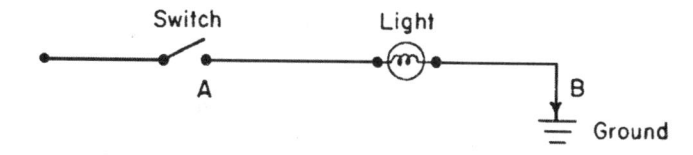

Fig. 18.5 Single-phase circuit (switch in open position)

To test for power at Point "A" and Point "B" in Figure 18.6, open the switches as shown. Using a multimeter with clamp-on leads, clamp a lead on Line 1 and a lead on Line 2, at points A and B, between the fuses and the load. Bring the multimeter and leads out of the panel and close the panel door as far as possible without cutting or damaging the meter leads. Some switches cannot be closed if the panel door is open. The panel door is closed when testing because hot copper sparks could seriously injure you when the circuit is energized and the voltage is high. Close the switches.

1. Multimeter should register at 220 volts. If there is no reading at points "A" and "B," the fuse or fuses could be "blown."

2. Move multimeter to L1 and L2. If there is still no reading on the multimeter, check for an open switch in another location, or call the power company to find out if power is out.

3. If a 220-volt reading is registered at L1 and L2, move the test leads to point "A" and the neutral lead. If a reading of 110 volts is observed, the fuse on line "A" is OK. If there is no voltage reading, the fuse on line "A" could be blown. Move the lead from line "A" to line "B." Observe the reading. If 110-volt power is not observed, the fuse on line "B" could be blown. Another possibility to consider is that the neutral line could be broken.

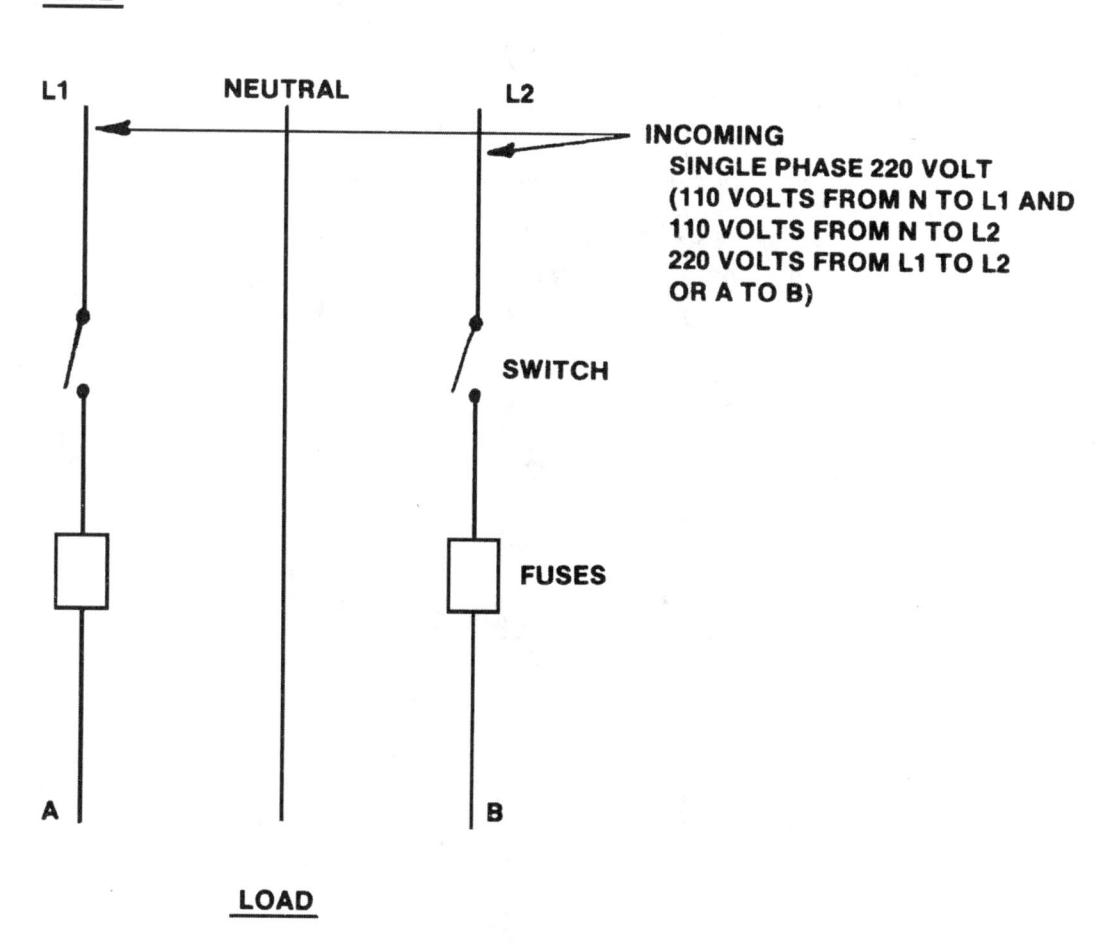

Fig. 18.6 Single-phase, three-lead circuit

276 Water Treatment

> **WARNING**
>
> Turn off power and be sure that there is no voltage in either power line before changing fuses. Use a fuse puller. Test circuit again in the same manner to make sure fuses or circuit breakers are OK. 220 volts power or voltage should be present between points "A" and "B." If fuse or circuit breaker trips again, shut off and determine the source of the problem.

Referring to Figure 18.7, test for voltage in three-phase circuits as follows—with the switches closed and the load disconnected, check for voltage (probably either 220 or 440) between points A3–B3, A3–C3, and between B3–C3. A zero voltage reading on any or all of the three tests indicates a problem ahead of this location that could be at another switch or a power company problem. Assuming normal readings were found at A3, B3, and C3, repeat the three readings at points A2, B2, and C2 with the switches closed. Any zero voltage readings are an indication of a defective switch. Assuming normal readings were found at A2, B2, and C2, repeat the three readings at points A1, B1, and C1 with the switches closed. If any two voltage readings are zero, one fuse is blown and it will be the one in the line that was common to the two zero readings. If all three voltage readings are zero, either two or three fuses are blown. To determine which fuses are blown, refer to Table 18.1. Note that a zero voltage reading indicates a blown fuse.

TABLE 18.1 LOCATING A BLOWN FUSE

Blown Fuse in Line	Use Either Test
L1	A1–B2 or A1–C2
L2	B1–A2 or B1–C2
L3	C1–A2 or C1–B2

Another way of checking the multimeter with the load connected on this three-phase circuit would be to take your multimeter and place one lead on the bottom and one lead on the top of each fuse. You should not get a voltage reading on the multimeter. This is because electricity takes the path of least resistance. If you get a reading across any of the fuses (top to bottom), that fuse is bad.

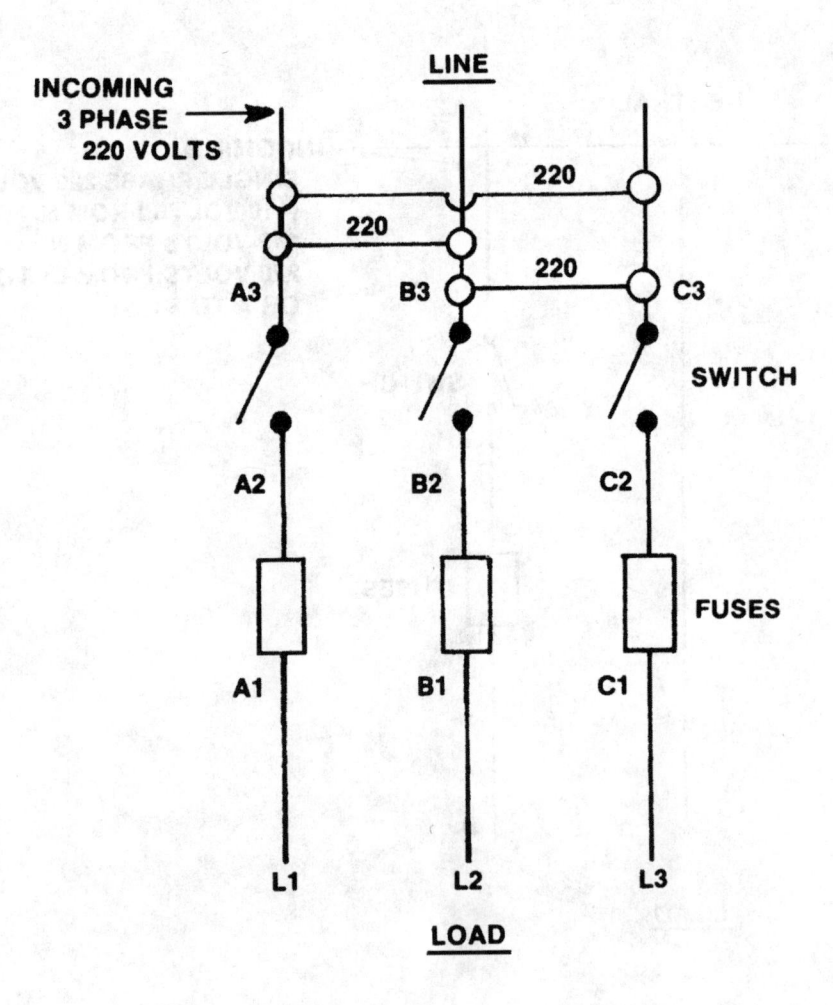

Fig. 18.7   Three-phase circuit, 220 volts

Always make sure that when you use a multimeter it is set for the proper voltage. If voltage is unknown and the meter has different scales that are manually set, always start with the highest voltage range and work down. Otherwise, the multimeter could be damaged. Look at the equipment instruction manual or nameplate for the expected voltage. Actual voltage should not be much higher than given unless someone goofed when the equipment was wired and inspected.

18.121 *Ammeter*

Another meter used in electrical maintenance and testing is the *AMMETER*. The ammeter records the current or "amps" flowing in the circuit. There are several types of ammeters, but only two will be discussed in this section. The ammeter generally used for testing is called a clamp-on type. The term "clamp-on" means that it can be clamped around a wire supplying a motor, and no direct electrical connection needs to be made. Each "leg" or lead on a three-phase motor must be individually checked.

The first step should be to read the motor nameplate data and find what the amperage reading should be for the particular motor or device you are testing. After you have this information, set the ammeter to the proper scale. Set it on a higher scale than necessary if the expected reading is close to the top of the meter scale. Place the clamp around one lead at a time. Record each reading and compare with the nameplate rating. If the readings are not similar to the nameplate rating, find the cause, such as low voltage, bad bearings, poor connections, or excessive load. If the ammeter readings are higher than expected, the high current could produce overheating and damage the equipment. Try to find the problem and correct it.

Current imbalance is undesirable because it causes uneven heating in a motor that can shorten the life expectancy of the insulation. However, a small amount of current imbalance is to be expected in the leads to a three-phase motor. This imbalance can be caused by either peculiarities in the motor or by a power company imbalance. To isolate the cause, make the following test. Note that this test should be done by a qualified electrician. Refer to Figure 18.8.

If the current readings on Lines L1, L2, and L3 are about the same both before and after the wiring change, this is an

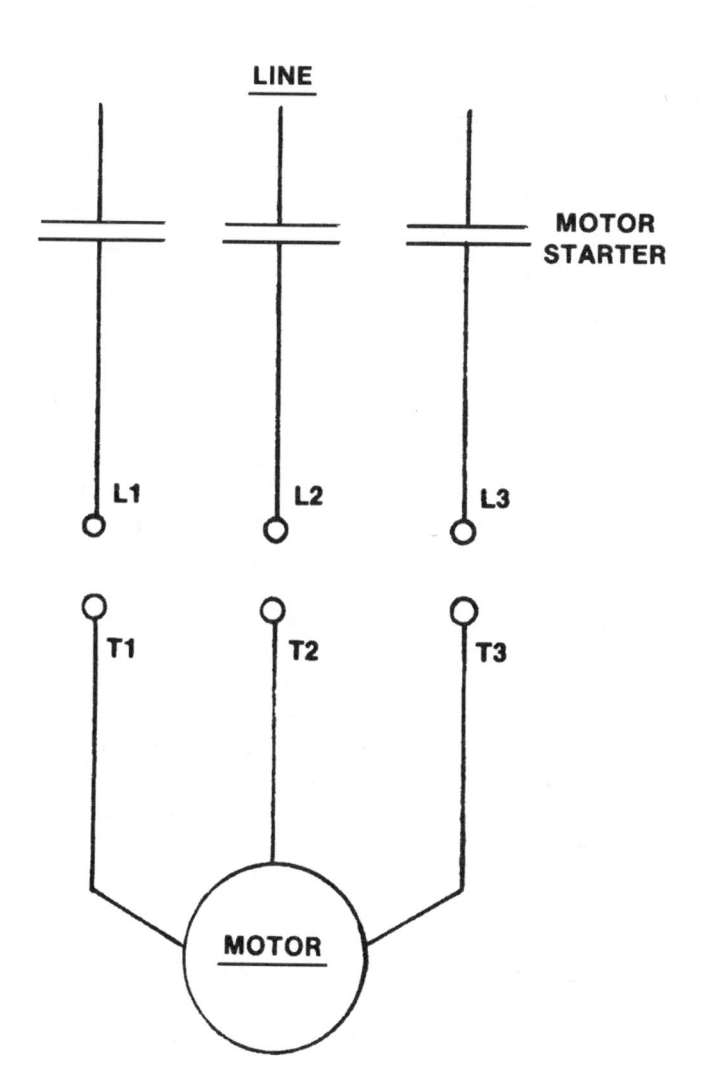

1. With the motor wired to its starter, L1 to T1, L2 to T2, and L3 to T3, measure and record the amperage on L1, L2, and L3.

2. De-energize the circuit and reconnect the motor as follows: L1–T3, L2–T1, L3–T2. This wiring change will not change the direction of the rotation of the motor.

3. Start up the motor and again measure and record the amperage on L1, L2, and L3.

Fig. 18.8 Determination of current imbalance

indication that the imbalance is being caused by the power company and they should be asked to make adjustments to correct the condition. However, if the current readings followed the motor terminal (T) numbers rather than the power line (L) numbers, the problem is within the motor and there is not much that can be done except contact the motor manufacturer for a possible exchange.

When using a clamp-on ammeter, be sure to set the meter on a high enough range or scale for the starting current if you are testing during start-up. Starting currents range from 500 to 700 percent higher than running currents and using too low a range can ruin an expensive and delicate instrument. Newer clamp-on ammeters automatically adjust to the proper range and can measure both starting or peak current and normal running current.

Another type of ammeter is one that is connected in line with the power lead or leads. Generally, they are not portable and are usually installed in a panel or piece of equipment. They require physical connections to put them in series with the motor or apparatus being tested. Current transformers (CT) are commonly used with this type of ammeter so that the meter does not have to conduct the full motor current. These ammeters are usually more accurate than the clamp-on type and are used in motor control centers and pump panels.

PROBLEM: Voltage imbalance. A common problem found in pump stations with a high rate of motor failure is voltage imbalance or unbalance. Unlike a single-phase condition, all three phases are present but the phase-to-phase voltage is not equal in each phase.

Voltage imbalance can occur in either the utility side or the pump station electrical system. For example, the utility company may have large, single-phase loads (such as residential services) that reduce the voltage on a single phase. This same condition can occur in the pump station if a large number of 120/220 volt loads are present. Slight differences in voltage can cause disproportional current imbalance; this may be six to ten times as large as the voltage imbalance. For example, a two-percent voltage imbalance can result in a 20-percent current imbalance. A 4.5-percent voltage imbalance will reduce the insulation life to 50 percent of the normal life. This is the reason a dependable voltage supply at the motor terminals is critical. Even relatively slight variations can greatly increase the motor operating temperatures and burn out the insulation.

It is common practice for electrical utility companies to furnish power to three-phase customers in open delta or wye configurations. An open delta or wye system is a two-transformer bank that is a suitable configuration where lighting loads are *large* and three-phase loads are *light*. This is the exact opposite of the configuration needed by most pumping facilities where three-phase loads are *large*. (Examples of three-transformer banks include Y-delta, delta-Y, and Y-Y.) In most cases, three-phase motors should be fed from three-transformer banks for proper balance. The capacity of a two-transformer bank is only 57 percent of the capacity of a three-transformer bank. The two-transformer configuration can cause one leg of the three-phase current to furnish higher amperage to one leg of the motor, which will greatly shorten its life.

Operators should acquaint themselves with the configuration of their electric power supply. When an open delta or wye configuration is used, operators should calculate the degree of current imbalance existing between legs of their polyphase motors. If you are unsure about how to determine the configuration of your system or how to calculate the percentage of current imbalance, always consult a qualified electrician. *Current imbalance between legs should never exceed 5 percent under normal operating conditions (NEMA Standards MGI-14.35).*

Loose connections will also cause voltage imbalance as will high-resistance contacts, circuit breakers, or motor starters.

Another serious consideration for operators is voltage fluctuation caused by neighborhood demands. A pump motor in near perfect balance (for example, 3 percent unbalance) at 9:00 am could be as much as 17 percent unbalanced by 4:00 pm on a hot day due to the use of air conditioners by customers on the same grid. Also, the hookup of a small market or a new home to the power grid can cause a significant change in the degree of current unbalance in other parts of the power grid. Because energy demands are constantly changing, water system operators should have a qualified electrician check the current balances between legs of their three-phase motors at least once a year.

SOLUTION: Motor connections at the circuit box should be checked frequently (semiannually or annually) to ensure that the connections are tight and that vibration has not caused the insulation on the conductors to wear away. Measure the voltage at the motor terminals and calculate the percentage imbalance (if any) using the procedures below.

Do not rely entirely on the power company to detect unbalanced current. Complaints of suspected power problems are frequently met with the explanation that all voltages are within the percentages allowed by law and no mention is made of the percentage of current unbalance, which can be a major source of problems with three-phase motors. A little research of your own can pay large benefits. For example, a small water company in Central California configured with an open delta system (and running three-phase unbalances as high as 17 percent as a result) was routinely spending $14,000 a year for energy and burning out a 10-HP motor on the average of every 1.5 years (six 10-HP motors in 9 years). After consultation, the local power utility agreed to add a third transformer to each power board to bring the system into better balance. Pump drop leads were then rotated, bringing overall current unbalances down to an average of 3 percent; heavy-duty three-phase capacitors were added to absorb the prevalent voltage surges in the area; and computerized controls were added to the pumps to shut them off when pumping volumes got too low. These modifications resulted in a saving in energy costs the first year alone of $5,500.00.

FORMULAS

Percentage of current unbalance can be calculated by using the following formulas and procedures:

$$\text{Average Current} = \frac{\text{Total of Current Value Measured on Each Leg}}{3}$$

$$\%\text{ Current Unbalance} = \frac{\text{Greatest Amp Difference from the Average}}{\text{Average Current}} \times 100\%$$

PROCEDURES

1. Measure and record current readings in amps for each leg. (Hookup 1.) Disconnect power.

2. Shift or roll the motor leads from left to right so the drop cable lead that was on terminal 1 is now on 2, lead on 2 is now on 3, and lead on 3 is now on 1. (Hookup 2.) Rolling the motor leads in this manner will not reverse the motor rotation. Start the motor, measure and record current reading on each leg. Disconnect power.

3. Again shift drop cable leads from left to right so the lead on terminal 1 goes to 2, 2 goes to 3, and 3 to 1. (Hookup 3.) Start pump, measure and record current reading on each leg. Disconnect power.

4. Add the values for each hookup.

5. Divide the total by 3 to obtain the average.

6. Compare each single leg reading to the average current amount to obtain the greatest amp difference from the average.

7. Divide this difference by the average to obtain the percentage of unbalance.

8. Use the wiring hookup that provides the lowest percentage of unbalance.

CORRECTING THE THREE-PHASE POWER UNBALANCE

Example: Check for current unbalance for a 230-volt, 3-phase, 60-Hz submersible pump motor, 18.6 full load amps.

Solution: Steps 1 to 3 measure and record amps on each motor drop lead for Hookups 1, 2, and 3 (Figure 18.9).

	Step 1 (Hookup 1)	Step 2 (Hookup 2)	Step 3 (Hookup 3)
(T_1)	DL_1 = 25.5 amps	DL_3 = 25 amps	DL_2 = 25.0 amps
(T_2)	DL_2 = 23.0 amps	DL_1 = 24 amps	DL_3 = 24.5 amps
(T_3)	DL_3 = 26.5 amps	DL_2 = 26 amps	DL_1 = 25.5 amps
Step 4	Total = 75 amps	Total = 75 amps	Total = 75 amps
Step 5	Average Current = $\frac{\text{Total Current}}{3 \text{ readings}}$ = $\frac{75}{3}$ = 25 amps		
Step 6	Greatest amp difference from the average:	(Hookup 1) = 25 − 23 = 2 (Hookup 2) = 26 − 25 = 1 (Hookup 3) = 25.5 − 25 = .5	
Step 7	% Unbalance	(Hookup 1) = 2/25 × 100 = 8 (Hookup 2) = 1/25 × 100 = 4 (Hookup 3) = 0.5/25 × 100 = 2	

As can be seen, Hookup 3 should be used since it shows the least amount of current unbalance. Therefore, the motor will operate at maximum efficiency and reliability on Hookup 3.

By comparing the current values recorded on each leg, you will note the highest value was always on the same leg, L_3. This indicates the unbalance is in the power source. If the high current values were on a different leg each time the leads were changed, the unbalance would be caused by the motor or a poor connection.

If the current unbalance is greater than 5 percent, contact your power company for help.

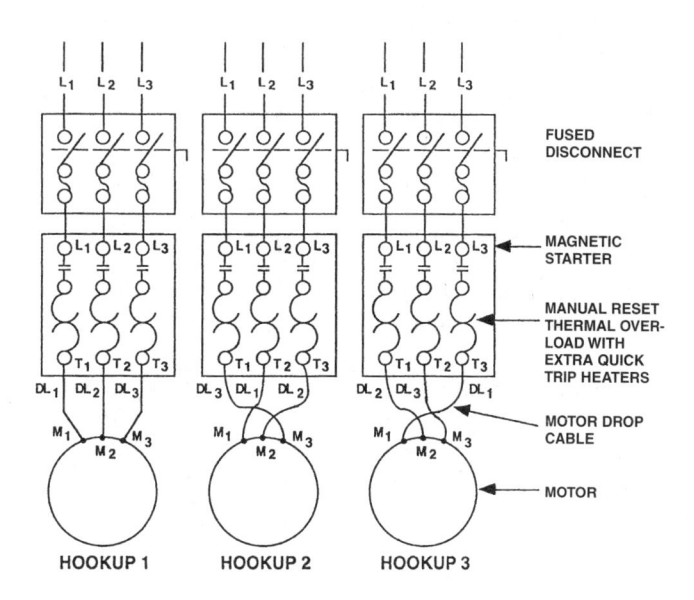

Fig. 18.9 Three hookups used to check for current unbalance

Acknowledgment

Material on unbalanced current was provided by James W. Cannell, President, Canyon Meadows Mutual Water Company, Inc., Bodfish, California. His contribution is greatly appreciated.

18.122 Megger

A *MEGGER* is a device used for checking the insulation resistance on motors, feeders, bus bar systems, grounds, and branch circuit wiring.

> **WARNING**
>
> Use a megger only on de-energized circuits or motors.

There are three general types of meggers: crank operated, battery operated, and instrument. There are two leads to connect. One lead is clamped to a ground lead and the other to the lead you are testing. The readings on the megger will range from "0" (ground) to infinity (perfect), depending on the condition of your circuit.

The megger is usually connected to a motor terminal at the starter, and the other lead to the ground lead. Results of this test indicate if the insulation is deteriorating or cut.

Insulation resistance of electrical equipment is affected by many variables such as the equipment design, the type of insulating material used, including binders and impregnating compounds, the thickness of the insulation and its area, cleanliness (or uncleanliness), moisture, and temperature. For insulation resistance measurements to be conclusive in analyzing the condition of equipment being tested, these variables must be taken into consideration.

Such factors as the design of the equipment, the kind of insulating material used, and its thickness and area cease to be variables after the equipment has been put into service, and minimum insulation resistance values can be established within reasonable tolerances. The variables that must be considered after the equipment has been put into service, and at the time that the insulation resistance measurements are being made, are uncleanliness, moisture, temperature, and damage such as fractures.

The most important requirements in the reliable operation of electrical equipment are cleanliness and the elimination of moisture penetration into the insulation. This is merely good housekeeping but it is essential in the maintenance of all types of electrical equipment. The very fact that insulation resistance is affected by moisture and dirt, with due allowances for temperature, makes the megger insulation test the valuable tool that it is in electrical maintenance. The test is an indication of cleanliness and good housekeeping as well as a detector of deterioration and impending trouble.

Suggested safe levels or minimum values of insulation resistance have been developed. These values should be provided by the equipment manufacturer and should serve as a guide for equipment in service. However, periodic tests on equipment in service will usually reveal readings considerably higher than the suggested minimum safe values. Records of periodic tests must be kept because persistent downward trends in insulation resistance usually give fair warning of impending trouble, even though the actual values may be *higher* than the suggested minimum safe values.

Also, allowances must be made for equipment in service showing periodic test values *lower* than the suggested minimum safe values, so long as the values remain stable or consistent. In such cases, after due consideration has been given to temperature and humidity conditions at the time of the test, there may be no need for concern. This condition may be caused by uniformly distributed leakages of a harmless nature, and may not be the result of a dangerous localized weakness. Here again, records of insulation resistance tests over a period of time reveal changes that may justify investigation. The trend of the curve may be more significant than the numerical values themselves.

For many years, one *MEGOHM*[4] has been widely used as a fair allowable lower limit for insulation resistance of ordinary industrial electrical equipment rated up to 1,000 volts. This value is still recommended for those who may not be too familiar with insulation resistance testing practices, or who may not wish to approach the problem from a more technical point of view.

For equipment rated above 1,000 volts, the "one megohm" rule is usually stated: "A minimum of one megohm per thousand volts." Although this rule is somewhat arbitrary, and may be criticized as lacking an engineering foundation, it has stood the test of a good many years of practical experience. This rule gives some assurance that equipment is not too wet or not too dry and has saved many an unnecessary breakdown.

More recent studies of the problem, however, have resulted in formulas for minimum values of insulation resistance that are based on the kind of insulating material used and the electrical and physical dimensions of the types of equipment under consideration.[5]

Motors and wirings should be megged at least once a year, and twice a year, if possible. The readings taken should be recorded and plotted in some manner so that you can determine when insulation is breaking down. Meg motors and wirings after a pump station has been flooded. If insulation is wet, excessive current could be drawn and cause pump motors to "kick out."

---

[4] *Megohm* (MEG-ome). Millions of ohms. Mega- is a prefix meaning one million, so 5 megohms means 5 million ohms.

[5] Portions of the preceding paragraphs were taken from INSTRUCTION MANUAL FOR MEGGER INSULATION TESTERS, No. 21-J, pages 42 and 43, published by Biddle Instruments, no longer in print. For additional information, see *A STITCH IN TIME: THE COMPLETE GUIDE TO ELECTRICAL INSULATION TESTING*. Available from AVO Training Institute, Technical Resource Center, Electrical Training Services, 4271 Bronze Way, Dallas, TX 75237-1019. Order No. AVOB001. Price, $7.00.

18.123 Ohmmeters

OHMMETERS, sometimes called circuit testers, are valuable tools used for checking electrical circuits. An ohmmeter is used only when the electrical circuit is OFF, or de-energized. The ohmmeter supplies its own power by using batteries. An ohmmeter is used to measure the resistance (ohms) in a circuit. These are most often used in testing the control circuit components such as coils, fuses, relays, resistors, and switches. They are used also to check for continuity.

An ohmmeter has several scales that can be used. Typical scales are: R × 1, R × 10, R × 1,000, and R × 10,000. Each scale has a level of sensitivity for measuring different resistances. To use an ohmmeter, set the scale, start at the low point (R × 1), and put the two leads across the part of the circuit to be tested such as a coil or resistor and read the resistance in ohms. A reading of infinity would indicate an open circuit, and a zero would read no resistance. These usually would be used only by skilled technicians because they are very delicate instruments.

All meters should be kept in good working order and calibrated periodically. They are very delicate, susceptible to damage, and should be well protected during transportation. When readings are taken, they should always be recorded on a machinery history card for future reference. Meters are a good way to determine pump and equipment performance. *CAUTION: Never use a meter unless qualified and authorized.*

QUESTIONS

Write your answers in a notebook and then compare your answers with those on page 379.

18.12A How can you determine if there is voltage in a circuit?

18.12B What are some of the uses of a multimeter?

18.12C What precautions should be taken before attempting to change fuses?

18.12D How do you test for voltage with a multimeter when the voltage is unknown?

18.12E What could be the cause of amp readings different from the nameplate rating?

18.12F How often should motors and wirings be megged?

18.12G An ohmmeter is used to check the ohms of resistance in what control circuit components?

18.13 Switch Gear

18.130 Equipment Protective Devices

Operators need safety devices to protect themselves and plant equipment from the destructive effects of electricity. Water systems have pressure valves, pop-offs, and different safety equipment to protect the pipes and equipment. So must electricity have safety devices to contain the voltage and amperage that come in contact with the wiring and equipment. The first piece of equipment that must be protected is the main electrical panel or control unit where the power enters. This protection is provided by either fuses or a circuit breaker.

18.131 Fuses

The power company has installed fuses on their power poles to protect their equipment from damage. We also must install something to protect the main control panel and wiring from damage due to excessive voltage or amperage.

A fuse is a protective device having a strip or wire of fusible metal that, when placed in a circuit, will melt and break the electrical circuit when subjected to an excessive temperature. This temperature will develop in the fuse when a current flows through the fuse in excess of what the circuit will carry safely. This means that the fuse must be capable of de-energizing the circuit before any damage is done to the wiring it is safely protecting. Fuses are used to protect operators, main circuits, branch circuits, heaters, motors, and various other electrical equipment.

There are several types of fuses, each being used for a certain type of protection. Some of these are:

1. Current-Limiting Fuses: These fuses open so quickly while clearing a short-circuit current that the potential fault current is not allowed to reach its peak. They are used to protect power distribution circuits.

2. Dual-Element Fuses: These fuses provide a time delay in the low overload range and a fast-acting element for short-circuit protection. These fuses are used for motor protection circuits.

There are many other types of fuses used for special applications, but the above are the most common.

A fuse must never be bypassed or jumped. This is the only protection the circuit has; without it, serious damage to equipment and possible injury to operators can occur. Make sure that all fuses are replaced with the proper size and type indicated for that circuit. If you have any doubt, check the electrical blueprints or contact your electrical engineer.

18.132 Circuit Breakers

The circuit breaker (Figure 18.4) is another safety device and is used in the same place as a fuse. Most circuit breakers consist of a switch that opens automatically when the current or the voltage exceeds or falls below a certain limit. Unlike a fuse that has to be replaced each time it blows, a circuit breaker can be reset after a short delay to allow time for cooling. This is done by moving the handle to the OFF position or slightly past, and then moving it back to the ON position. Also, unlike a fuse, a circuit breaker can be visually inspected to find out if it has been tripped. The handle will be at the mid position between ON and OFF. Several different types of circuit breakers are being used today and each one is selected for a special protective purpose.

18.133 Overload Relays

Three-phase motors are usually protected by overload relays. This is accomplished by having heater strips, bimetal, or solder pots that open on current rise (overheating), and open the control circuit. This, in turn, opens the power control circuit, which de-energizes the starter and stops power to the motor. Such relays are also known as heaters or "thermal overloads." Sizing of these overloads is very critical and should coincide with the nameplate rating on the motor. Sizing depends on the service factor of the electric motor. Usually, they range from 100 to 110 percent of the motor nameplate ratings and should never exceed 125 percent (usually 115 percent) of the motor rating. For example, if the motor is rated for 10 amps, the overloads should be sized from 10 to 11 amps.

Again, *never increase the rating of the overload heaters because of tripping. You should find the problem and repair it.* There are many other protective devices for electricity such as motor winding thermostats, phase protectors, low voltage protectors, and ground fault protectors. Each has its own special applications and should never be tampered with or jammed.

"Ground" is an expression representing an electrical connection to earth or a large conductor that is at the earth's potential or neutral voltage. Motor frames and all electrical tools and equipment enclosures should be connected to ground. This is generally referred to simply as grounding, or equipment ground.

The third prong on cords from electric hand tools is the equipment ground and must never be removed. When an adapter is used with a two-prong receptacle, the green wire on the adapter should be connected under the center screw on the receptacle cover plate. Many times, equipment grounding, especially at home, is achieved by connecting onto a water pipe or drain rather than a rod driven into the ground. This practice generally is not recommended when plastic pipes and other nonconducting pipe materials are used unless it is known that the piping is all metal and not interrupted. Also, corrosion can be accelerated if pipes of different metals are used. A rod driven into dry ground is not very effective as a ground.

18.134 Motor Starters

A motor starter is a device or group of devices that are used to connect the electrical power to a motor. These starters can be either manually or automatically controlled.

Manual and magnetic starters range in complexity from a single ON/OFF switch to a sophisticated automatic device using timers and coils. The simplest motor starter is used on single-phase motors where a circuit breaker is manually turned on and the motor starts. This type of starter also is used on three-phase motors of smaller horsepower, and on fan motors, machinery motors, and several other applications where it is not necessary to have automatic control.

Magnetic starters (Figures 18.10 and 18.11) are commonly used to start pumps, compressors, blowers, and anything where automatic or remote control is desired. They permit low power circuits to energize the starter of equipment at a remote location or to start larger starters (Figure 18.12). A magnetic starter is operated by electromagnetic action. This starter has contactors and they operate by energizing a coil that closes the contact, thus starting the motor. The circuit that energizes the starter is called the control circuit and it may operate on a lower voltage (115 volts) than the motor. Whenever a starter is used as a part of an integrated circuit (such as for flow, pressure, or temperature control), a magnetic starter or controller is necessary.

Magnetic starters are sized for their voltage and horsepower ratings. These are divided into classes. The most common starter is Class A. A Class A starter is an alternating current, air-break and oil-immersed, manual or magnetic controller for service on 600 volts or less. It is capable of interrupting operating overloads up to and including 10 times their normal motor rating, but not short circuits or faults beyond operating overloads. Additional class information can be found in electrical catalogs, manuals, and manufacturers' brochures.

There are a number of different types of three-phase magnetic motor starters available. The simplest and most common is the across-the-line full-voltage starter. This starter consists of three contacts, a magnetic actuating device, and overload detection. This starter subjects the power system to the full surge current on start-up and may cause the lights in the treatment plant to dim momentarily.

To reduce the inrush current when starting polyphase motors, a number of other types of starters are available.

1. Auto-Transformer Type Reduced-Voltage Starters. These begin the motor start sequence by applying a reduced voltage to the motor for a few seconds. The voltage is controlled by a time-delay relay within the starter. The reduced voltage is obtained from transformers that are a part of the starter. These transformers are designed to operate for only a few seconds at a time and can easily be burned out if the motor is started too frequently.

2. Solid-State Reduced-Voltage Starters. These starters do the same job as the auto-transformer type reduced-voltage starters but they do not need transformers because the voltage and current are electrically controlled.

3. Part Winding Starters. These starters are used with special motors that have two separate sets of windings on the same motor frame. By energizing the windings about one second apart, the inrush current is limited to about half that of a normal motor with a full-voltage starter.

4. Wye-Delta Starters. These starters are used with motors that have all leads brought out to the terminal box. The motor is first started with wye-connected coils and switched over to a delta connection for running. The result is the same as if you used a reduced-voltage starter.

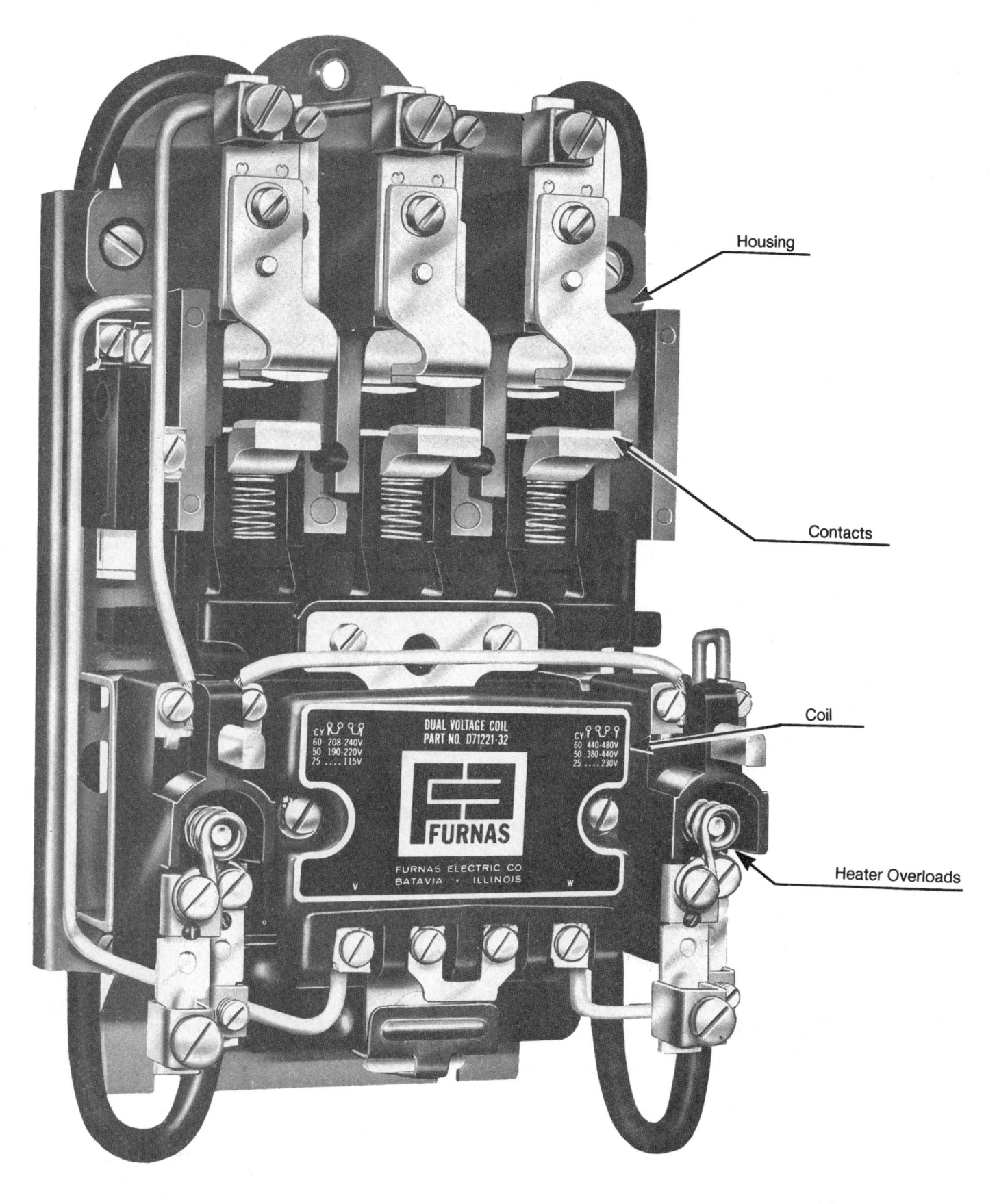

Fig. 18.10 Three-phase magnetic starter
(Courtesy of Furnas Electric Company)

284 Water Treatment

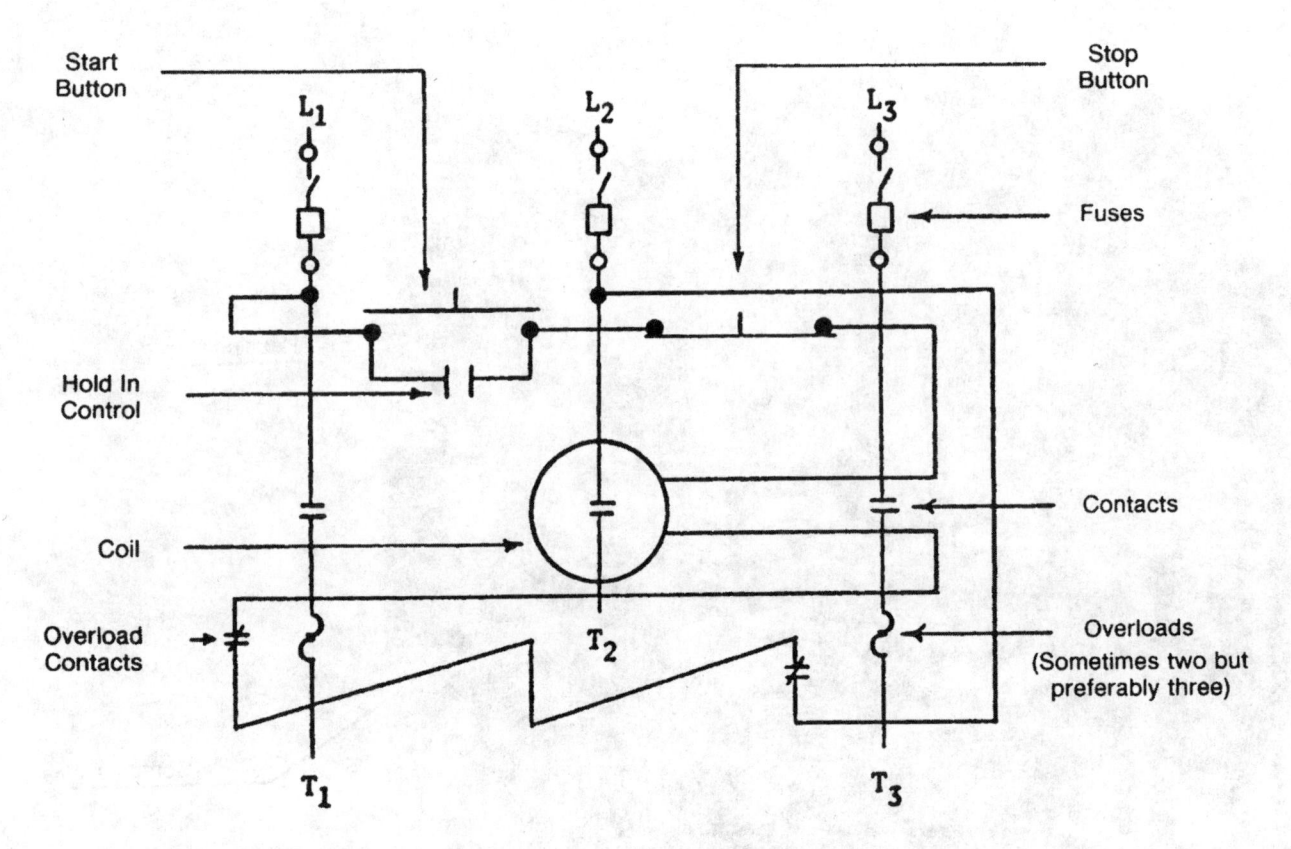

Fig. 18.11 Wiring diagram of three-phase magnetic starter

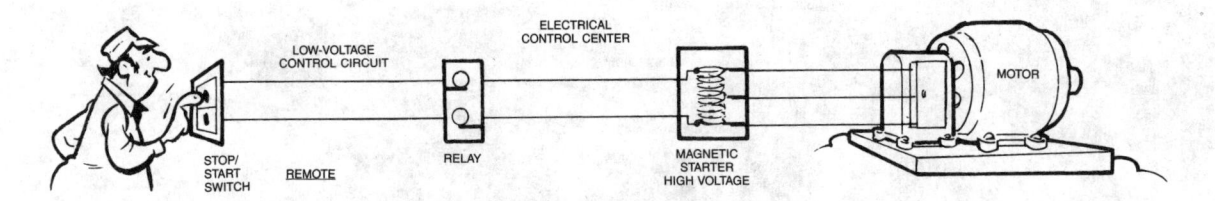

Fig. 18.12 Application of magnetic starter

QUESTIONS

Write your answers in a notebook and then compare your answers with those on pages 379 and 380.

18.13A What are two types of safety devices found in main electrical panels or control units?

18.13B What are fuses used to protect?

18.13C Why must a fuse never be bypassed or jumped?

18.13D How does a circuit breaker work?

18.13E How are motor starters controlled?

18.13F When are magnetic starters used?

18.14 Electric Motors

18.140 Classifications

Electric motors are the machines most commonly used to convert electrical energy into mechanical energy. A motor usually consists of a *STATOR,*[6] *ROTOR,*[7] *END BELLS,*[8] and windings. The rotor has an extended shaft that allows a machine to be coupled to it.

Motors are of many different types (Figure 18.13), such as squirrel cage induction motors, wound rotor motors, synchronous motors, and many others. The most common of these is the squirrel cage induction motor. Some pumping stations use wound rotor induction motors when speed control is needed.

---

[6] *Stator.* That portion of a machine that contains the stationary (nonmoving) parts that surround the moving parts (rotor).
[7] *Rotor.* The rotating part of a machine. The rotor is surrounded by the stationary (nonmoving) parts (stator) of the machine.
[8] *End Bells.* Devices used to hold the rotor and stator of a motor in position.

DRIP PROOF

ITEM NO.	PART NAME
1	Wound Stator w/ Frame
2	Rotor Assembly
3	Rotor Core
4	Shaft
5	Bracket
6	Bearing Cap
7	Bearings
8	Seal, Labyrinth
9	Thru Bolts/Caps
10	Seal, Lead Wire
11	Terminal Box
12	Terminal Box Cover
13	Fan
14	Deflector
15	Lifting Lug

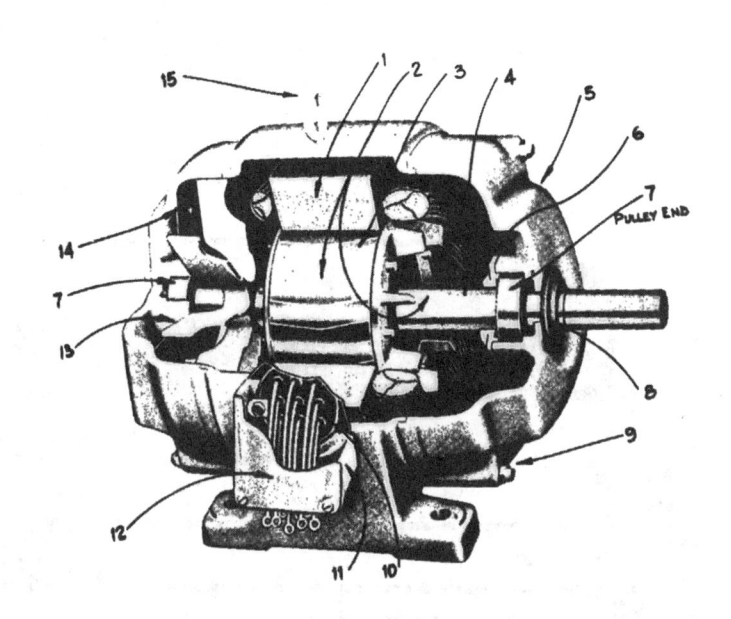

TOTALLY ENCLOSED FAN COOLED

ITEM NO.	PART NAME
1	Wound Stator w/ Frame
2	Rotor Assembly
3	Rotor Core
4	Shaft
5	Brackets
6	Bearings
7	Seal, Labyrinth
8	Thru Bolts/Caps
9	Seal, Lead Wire
10	Terminal Box
11	Terminal Box Cover
12	Fan, Inside
13	Fan, Outside
14	Fan Grill
15	Fan Cover
16	Fan Cover Bolts
17	Lifting Lug

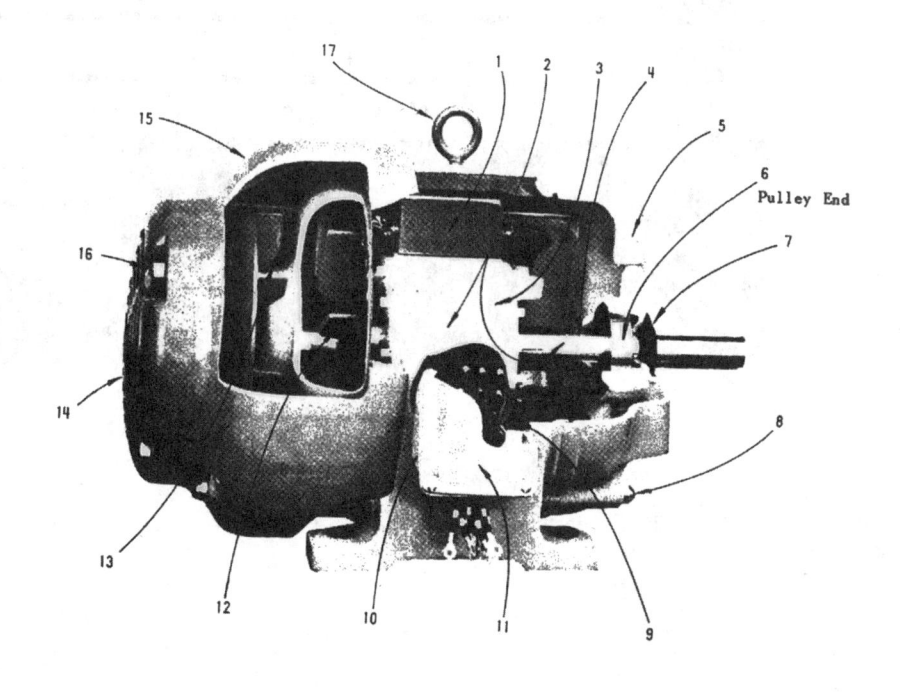

Fig. 18.13   Typical motors
(Courtesy of Sterling Power Systems, Inc.)

Three-phase electric motors are used for operating pumps, compressors, fans, and other machinery. Motors are generally trouble free and, when lubricated properly, cause very few problems. The amperage and voltage readings on motors should be taken periodically to ensure proper operation.

Motors are classified by NEMA (National Electrical Manufacturers Association) with code letters from A through V with "A" having the lowest starting torque and inrush current and "V" having the highest starting torque and inrush current. The most commonly available motors have code letters from "F" through "L," which have inrush currents on start of from 500 to 1,000 percent of full load.

Another important consideration in selecting a motor is the class of insulation. The class of insulation designates how high the temperature of the operating motor surface may rise above ambient temperature. The insulation class is indicated on the motor nameplate (Figure 18.14). Temperature ratings listed below are the allowable temperature increases above 40°C (104°F) ambient temperature. Motor insulation classes are as follows:

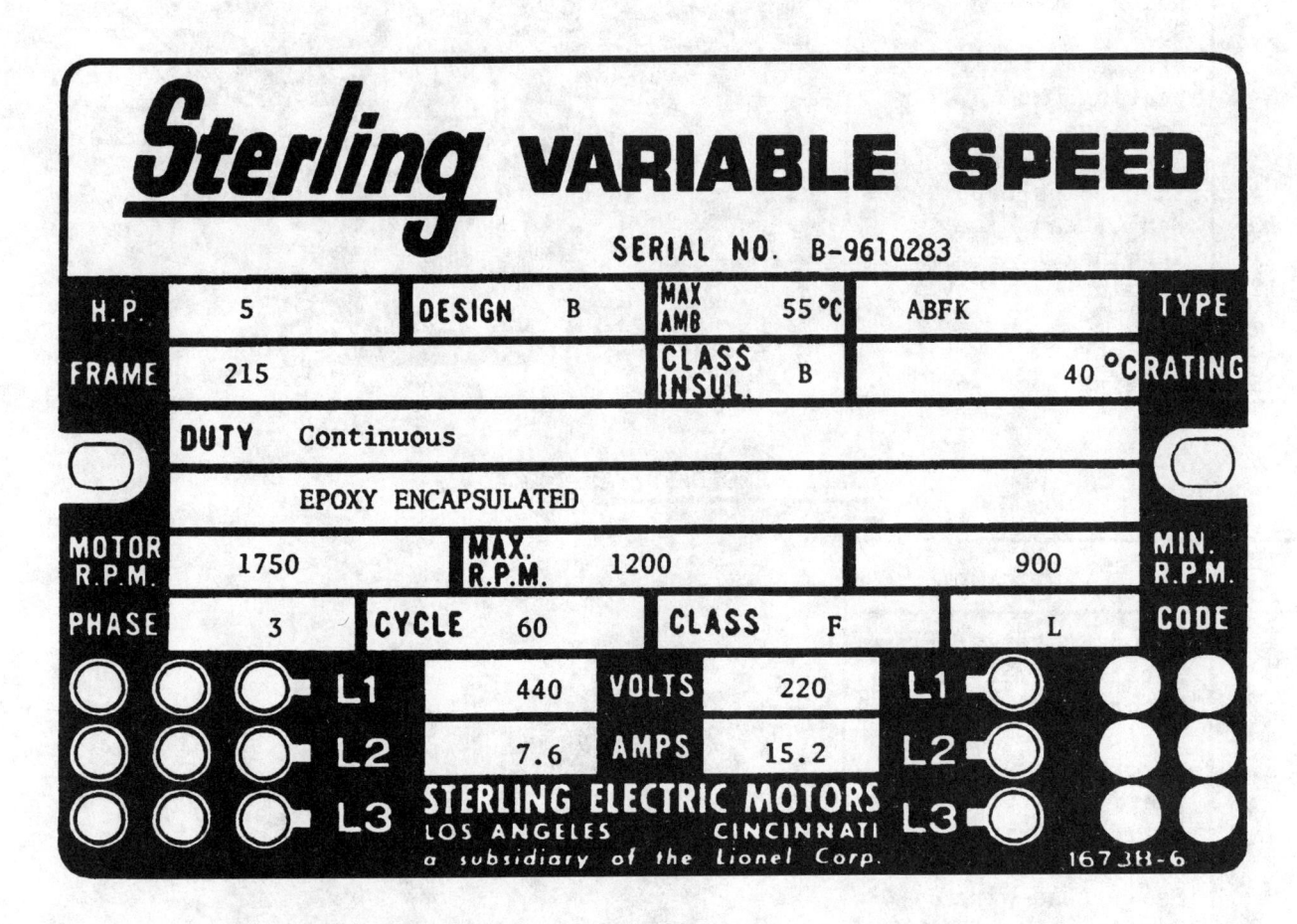

Sterling Variable Speed nameplate

SERIAL NO. B-961Q283

| H.P. | 5 | DESIGN | B | MAX AMB | 55°C | ABFK | | TYPE |
| FRAME | 215 | | | CLASS INSUL. | B | | 40°C | RATING |

DUTY: Continuous

EPOXY ENCAPSULATED

| MOTOR R.P.M. | 1750 | MAX. R.P.M. | 1200 | | 900 | MIN. R.P.M. |
| PHASE | 3 | CYCLE | 60 | CLASS | F | L | CODE |

L1 440 VOLTS 220
L2 7.6 AMPS 15.2
L3

STERLING ELECTRIC MOTORS
LOS ANGELES CINCINNATI
a subsidiary of the Lionel Corp.

16738-6

NOTE:
1. The motor for this unit is rated at 1,750 RPM and the maximum speed for the variable drive unit is 1,200 RPM.
2. The insulation is Class B.

Fig. 18.14 *Typical nameplate*
(Courtesy of Sterling Power Systems, Inc.)

Class	Temperature Rating
A	105°C (221°F)
B	130°C (266°F)
F	155°C (311°F)
H	180°C (356°F)

At present most motors are Class B insulated. Try to keep the actual operating temperature below the temperature rating or limit in order to prolong the life of the insulation.

All of this information can be found on the motor nameplate and should be taken into consideration when evaluating a motor. Most of the trouble encountered with electric motors results from bad bearings, shorted windings due to insulation breakdown, or excessive moisture.

All of the information on the motor nameplate (Figure 18.14) should be recorded and placed in a file for future reference. Many times, the nameplate is painted, corroded, or missing from the unit when the information is needed to repair the motor or replace parts. Also, record the date of installation and service start-up. See Section 18.142, "Recordkeeping," for a typical data sheet for recording the essential information. This information also should be on the manufacturer's data sheet and in the instruction manual. Compare the information for consistency and file in an appropriate location. Be sure you have the correct serial or model numbers.

QUESTIONS

Write your answers in a notebook and then compare your answers with those on page 380.

18.14A How is electrical energy converted into mechanical energy?

18.14B What are the important parts of an electric motor?

18.14C How can motors be kept trouble free?

18.14D What should be done with motor nameplate data?

18.141 Troubleshooting

18.1410 STEP-BY-STEP PROCEDURES

The key to effective troubleshooting is the use of practical, step-by-step procedures combined with a common-sense approach.

"NEVER TAKE ANYTHING FOR GRANTED"

1. Gather preliminary information. The first step in troubleshooting any motor control that has developed trouble is to understand the circuit operation and other related functions. In other words, what is supposed to happen, operate, and so forth when it is working properly? Also, what is it doing now? The qualified maintenance operator should be able to do the following:

   a. *KNOW WHAT SHOULD HAPPEN WHEN A SWITCH IS PUSHED:* When switches are pushed or tripped, what coils go in, contacts close, relays operate, and motors run?

   b. *EXAMINE ALL OTHER FACTORS:* What other unusual things are happening in the pump station (facility) now that this circuit does not work properly? Lights dimmed, other pumps ran faster, lights went out when it broke, everything was flooded, operators were hosing down area, and many other possible factors.

   c. *ANALYZE WHAT YOU KNOW:* What part of it is working correctly? Is switch arm tripped? Everything but this is all right, except pump gets plugged with rags frequently. Is it a mechanical failure or an electrical problem caused by a mechanical failure?

   d. *SELECT SIMPLE PROCEDURES:* To localize the problem, select logical ways that can be simply and quickly accomplished.

   e. *MAKE A VISUAL INSPECTION:* Look for burned wires, loose wires, area full of water, coil burned, contacts loose, or strange smells.

   f. *CONVERGE ON SOURCE OF TROUBLE:* Mechanical or electrical. Motor or control, whatever it might be. Electrical problems result from some type of mechanical failure.

   g. *PINPOINT THE PROBLEM:* Exactly where is the problem and what do you need for repair?

   h. *FIND THE CAUSE:* What caused the problem? Moisture, wear, poor design, voltage, or overloading.

   i. *REPAIR THE PROBLEM AND ELIMINATE THE CAUSE, IF POSSIBLE:* If the problem is inside switch gear or motors, call an electrician. Give the electrician the information you have regarding the equipment. Do not attempt electrical repairs unless you are qualified and authorized, otherwise you could cause excessive damage to yourself and to the equipment.

2. Some of the things to look for when troubleshooting are given in the remainder of this section.

18.1411 TROUBLESHOOTING GUIDE FOR ELECTRIC MOTORS

Symptom	Cause	Result[a]	Remedy
1. Motor does not start (switch is on and not defective)	a. Incorrectly connected	a. Burnout	a. Connect correctly per diagram on motor.
	b. Incorrect power supply	b. Burnout	b. Use only with correctly rated power supply.
	c. Fuse out, loose or open connection	c. Burnout	c. Correct open circuit condition.
	d. Rotating parts of motor may be jammed mechanically	d. Burnout	d. Check and correct: 1. Bent shaft. 2. Broken housing. 3. Damaged bearing. 4. Foreign material in motor.
	e. Driven machine may be jammed	e. Burnout	e. Correct jammed condition.
	f. No power supply	f. None	f. Check for voltage at motor and work back to power supply.
	g. Internal circuitry open	g. Burnout	g. Correct open circuit condition.
2. Motor starts but does not come up to speed	a. Same as 1-a, b, c above	a. Burnout	a. Same as 1-a, b, c above.
	b. Overload	b. Burnout	b. Reduce load to bring current to rated limit. Use proper fuses and overload protection.
	c. One or more phases out on a 3-phase motor	c. Burnout	c. Look for open circuits.
3. Motor noisy (electrically)	a. Same as 1-a, b, c above	a. Burnout	a. Same as 1-a, b, c above.
4. Motor runs hot (exceeds rating)	a. Same as 1-a, b, c above	a. Burnout	a. Same as 1-a, b, c above.
	b. Overload	b. Burnout	b. Reduce load.
	c. Impaired ventilation	c. Burnout	c. Remove obstruction.
	d. Frequent starts or stops	d. Burnout	d. 1. Reduce number of starts or reversals. 2. Secure proper motor for this duty.
	e. Misalignment between rotor and stator laminations	e. Burnout	e. Realign.
5. Motor noisy (mechanically)	a. Misalignment of coupling or sprocket	a. Bearing failure, broken shaft, stator burnout due to motor drag	a. Correct misalignment.
	b. Mechanical unbalance of rotating parts	b. Same as 5-a	b. Find unbalanced part, then balance.
	c. Lack of or improper lubricant	c. Bearing failure	c. Use correct lubricant, replace parts as necessary.

18.1411 TROUBLESHOOTING GUIDE FOR ELECTRIC MOTORS (continued)

Symptom	Cause	Result[a]	Remedy
5. Motor noisy (mechanically) (continued)	d. Foreign material in lubricant	d. Bearing failure	d. Clean out and replace bearings.
	e. Overload	e. Bearing failure	e. Remove overload condition. Replace damaged parts.
	f. Shock loading	f. Bearing failure	f. Correct causes and replace damaged parts.
	g. Mounting acts as amplifier of normal noise	g. Annoying	g. Isolate motor from base.
	h. Rotor dragging due to worn bearings, shaft, or bracket	h. Burnout	h. Replace bearings, shaft, or bracket as needed.
6. Bearing failure	a. Same as 5-a, b, c, d, e	a. Burnout, damaged shaft, damaged housing	a. Replace bearings and follow 5-a, b, c, d, e.
	b. Entry of water or foreign material into bearing housing	b. Burnout, damaged shaft, damaged housing	b. Replace bearings and seals and shield against entry of foreign material (water, dust, etc.). Use proper motor.

Symptom	Caused By	Appearance
1. Shorted motor winding	a. Moisture, chemicals, foreign material in motor, damaged winding	a. Black or burned coil with remainder of winding good.
2. All windings completely burned	a. Overload	a. Burned equally all around winding.
	b. Stalling	b. Burned equally all around winding.
	c. Impaired ventilation	c. Burned equally all around winding.
	d. Frequent reversal or starting	d. Burned equally all around winding.
	e. Incorrect power	e. Burned equally all around winding.
3. Single-phase condition	a. Open circuit in one line. The most common causes are loose connection, one fuse out, loose contact in switch.	a. If 1,800 RPM motor—four equally burned groups at 90° intervals.
		b. If 1,200 RPM motor—six equally burned groups at 60° intervals.
		c. If 3,600 RPM motor—two equally burned groups at 180° intervals.
		NOTE: If wye connected, each burned group will consist of two adjacent phase groups. If delta connected, each burned group will consist of one-phase group.
4. Other	a. Improper connection	a. Irregularly burned groups or spot burns.
	b. Ground	

---

[a] Many of these conditions should trip protective devices rather than burn out motors. Also, many burnouts occur within a short period of time after motor is started up. This does not necessarily indicate that the motor was defective, but usually is due to one or more of the above-mentioned causes. The most common causes of failure shortly after start-up are improper connections, open circuits in one line, incorrect power supply, or overload.

18.1412 TROUBLESHOOTING GUIDE FOR MAGNETIC STARTERS

Trouble	Possible Cause	Remedy
CONTACTS		
Contact chatter	1. Broken *POLE SHADER*[9]	1. Replace.
	2. Poor contact in control circuit	2. Improve contact or use holding circuit interlock.
	3. Low voltage	3. Correct voltage condition. Check momentary voltage drop.
Welding or freezing	1. Abnormal surge of current	1. Use larger contactor and check for grounds, shorts, or excessive motor load current.
	2. Frequent *JOGGING*[10]	2. Install larger device rated for jogging service or caution operators.
	3. Insufficient contact pressure	3. Replace contact spring; check contact carrier for damage.
	4. Contacts not positioning properly	4. Check for voltage drop during start-up.
	5. Foreign matter preventing magnet from seating	5. Clean contacts.
	6. Short circuit	6. Remove short fault and check that fuse and breaker are right.
Short contact life or overheating of tips	1. Contacts poorly aligned, poorly spaced, or damaged	1. Do not file silver-faced contacts. Rough spots or discoloration will not harm contacts. Replace.
	2. Excessively high currents	2. Install larger device. Check for grounds, shorts, or excessive motor currents.
	3. Excessive starting and stopping of motor	3. Caution operators. Check operating controls.
	4. Weak contact pressure	4. Adjust or replace contact springs.
	5. Dirty contacts	5. Clean with approved solvent.
	6. Loose connections	6. Check terminals and tighten.
Coil overheated	1. Starting coil may not kick out	1. Repair coil.
	2. Overload will not let motor reach minimum speed	2. Remove overload.
	3. Overvoltage or high ambient temperature	3. Check application and circuit.
	4. Incorrect coil	4. Check rating; if incorrect, replace with proper coil.
	5. Shorted turns caused by mechanical damage or corrosion	5. Replace coil.

---

[9] *Pole Shader.* A copper bar circling the laminated iron core inside the coil of a magnetic starter.
[10] *Jogging.* The frequent starting and stopping of an electric motor.

18.1412 **TROUBLESHOOTING GUIDE FOR MAGNETIC STARTERS** (continued)

Trouble	Possible Cause	Remedy
CONTACTS (continued)		
Coil overheated (continued)	6. Undervoltage, failure of magnet to seal it	6. Correct system voltage.
	7. Dirt or rust on pole faces increasing air gap	7. Clean pole faces.
Overload relays tripping	1. Sustained overload	1. Check for grounds, shorts, or excessive motor currents. Mechanical overload.
	2. Loose connection on all or any load wires	2. Check, clean, and tighten.
	3. Incorrect heater	3. Replace with correct size heater unit.
	4. Fatigued heater blocks	4. Inspect and replace.
Failure to trip	1. Mechanical binding, dirt, or corrosion	1. Clean or replace.
	2. Wrong heater, or heaters omitted and jumper wires used	2. Check ratings. Apply heaters of proper rating.
	3. Motor and relay in different temperatures	3. Adjust relay rating accordingly, or install temperature compensating relays.
MAGNETIC AND MECHANICAL PARTS		
Noisy magnet (humming)	1. Broken shading coil	1. Replace shading coil.
	2. Magnet faces not mating	2. Replace magnet assembly or realign.
	3. Dirt or rust on magnet faces	3. Clean and realign.
	4. Low voltage	4. Inspect system voltage and voltage dips or drops during start-up.
Failure to pick up and seal	1. Low voltage	1. Inspect system voltage and correct.
	2. Coil open or shorted	2. Replace.
	3. Wrong coil	3. Check coil number and voltage rating.
	4. Mechanical obstruction	4. With power off, check for free movement of contact and armature assembly. Repair.
Failure to drop out	1. Gummy substance on pole	1. Clean with solvent.
	2. Voltage not removed from coil	2. Check coil circuit.
	3. Worn or rusted parts causing binding	3. Replace or clean parts as necessary.
	4. Residual magnetism due to lack of air gap in magnet path	4. Replace worn magnet parts or align, if possible.
	5. Welded contacts	5. Shorted circuit, grounded, overloaded.

18.1413 TROUBLE/REMEDY PROCEDURES FOR INDUCTION MOTORS

1. Motor will not start.

    Overload control tripped. Wait for overload to cool, then try to start again. If motor still does not start, check for the causes outlined below.

    a. Open fuses: test fuses.

    b. Low voltage: check nameplate values against power supply characteristics. Also, check voltage at motor terminals when starting motor under load to check for allowable voltage drop.

    c. Wrong control connections: check connections with control wiring diagram.

    d. Loose terminal-lead connection: turn power off and tighten connections.

    e. Drive machine locked: disconnect motor from load. If motor starts satisfactorily, check driven machine.

    f. Open circuit in stator or rotor winding: check for open circuits.

    g. Short circuit in stator winding: check for short.

    h. Winding grounded: test for grounded wiring.

    i. Bearing stiff: free bearing or replace.

    j. Overload: reduce load.

2. Motor noisy.

    a. Three-phase motor running on single phase: stop motor, then try to start. It will not start on single phase. Check for open circuit in one of the lines.

    b. Electrical load unbalanced: check current balance.

    c. Shaft bumping (sleeve-bearing motor): check alignment and conditions of belt. On pedestal-mounted bearing, check for play and axial centering of rotor.

    d. Vibration: driven machine may be unbalanced. Remove motor from load. If motor is still noisy, rebalance.

    e. Air gap not uniform: center the rotor and, if necessary, replace bearings.

    f. Noisy ball bearing: check lubrication. Replace bearings if noise is excessive and persistent.

    g. Rotor rubbing on stator: center the rotor and replace bearings, if necessary.

    h. Motor loose on foundation: tighten hold-down bolts. Motor may possibly have to be realigned.

    i. Coupling loose: insert feelers at four places in coupling joint before pulling up bolts to check alignment. Tighten coupling bolts securely.

3. Motor at higher than normal temperature or smoking. (Measure temperature with thermometer or thermister and compare with nameplate value.)

    a. Overload: measure motor loading with ammeter. Reduce load.

    b. Electrical load imbalance: check for voltage imbalance or single-phasing.

    c. Restricted ventilation: clean air passage and windings.

    d. Incorrect voltage and frequency: check nameplate values with power supply. Also, check voltage at motor terminals with motor under full load.

    e. Motor stalled by driven tight bearings: remove power from motor. Check machine for cause of stalling.

    f. Stator winding shorted or grounded: test windings by standard method.

    g. Rotor winding with loose connection: tighten, if possible, or replace with another rotor.

    h. Belt too tight: remove excessive pressure on bearings.

    i. Motor used for rapid reversing service: replace with motor designed for this service.

4. Bearings hot.

    a. End shields loose or not replaced properly: make sure end shields fit squarely and are properly tightened.

    b. Excessive belt tension or excessive gear side thrust: reduce belt tension or gear pressure and realign shafts. See that thrust is not being transferred to motor bearing.

    c. Bent shaft: straighten shaft or send to motor repair shop.

5. Sleeve bearings.

    a. Insufficient oil: add oil—if supply is very low, drain, flush, and refill.

    b. Foreign material in oil or poor grade of oil: drain oil, flush, and relubricate using industrial lubricant recommended by a reliable oil manufacturer.

    c. Oil rings rotating slowly or not rotating at all: oil too heavy; drain and replace. If oil ring has worn spot, replace with new ring.

    d. Motor tilted too far: level motor or reduce tilt and realign, if necessary.

    e. Rings bent or otherwise damaged in reassembling: replace rings.

    f. Rings out of slot (oil-ring retaining clip out of place): adjust or replace retaining clip.

    g. Defective bearings or rough shaft: replace bearings. Resurface shaft.

6. Ball bearings.
   a. Too much grease: remove relief plug and let motor run. If excess grease does not come out, flush and relubricate.
   b. Wrong grade of grease: flush bearing and relubricate with correct amount of proper grease.
   c. Insufficient grease: remove relief plug and grease bearing.
   d. Foreign material in grease: flush bearing, relubricate; make sure grease supply is clean (keep can covered when not in use).

18.142 Recordkeeping

Records are a very important part of electrical maintenance. They must be accurate and complete. Whenever something is changed, repaired, or tested, it should be recorded on a material history card of some type. Pages 294 and 295 are examples of typical record sheets.

QUESTIONS

Write your answers in a notebook and then compare your answers with those on page 380.

18.14E   What is the key to effective troubleshooting?

18.14F   What are some of the steps that should be taken when troubleshooting magnetic starters?

18.14G   What kind of information should be recorded regarding electrical equipment?

18.15 Auxiliary Electrical Power

18.150 Safety First

Always remember that a qualified electrician should perform most of the necessary maintenance and repair of electrical equipment. If you do not know the how, why, and when of the job, do not do it. You could endanger your life as well as your fellow operators or damage equipment. Never attempt work that you are not qualified to do or are not authorized to perform.

18.151 Standby Power Generation

There are three ways of providing standby power. One is by providing the treatment plant with an engine-driven generator set. The limit of how much power can be produced is determined only by the size of the generator. The second possibility for standby power is batteries. Batteries should only be considered for low power consumption uses such as emergency lighting, communication, and possibly some control and instrumentation functions. The other possibility for standby power is a connection to an alternate power source, such as a different substation or another power company.

Because the treatment of water is considered a critical service, it is important to be able to provide drinking water even with the loss of commercial power. A power outage of a short duration probably will not have adverse effects on plant operation. The question you must ask yourself is, "Can my plant meet the needs of the public if a brownout or catastrophic event eliminates commercial power for an extended length of time?" If the answer is "no," then perhaps a form of standby power generation should be considered.

The following six conditions must be analyzed to determine the need for and size of standby power generation:

1. Frequency of power outages in the last 10 to 15 years
2. Duration of the power outage in each occurrence
3. Availability of an additional source of power supply from a different substation in the vicinity
4. Method by which raw water reaches the plant (Is flow by gravity or by a raw water pumping station?)
5. Total storage capacity of reservoirs in the distribution system
6. Possibility of obtaining a potable water supply from adjacent cities (Is there a reasonably sized pipe connection between your system and the distribution system of an adjacent city?)

If the frequency of power outages is once or twice a year with a 10- to 30-minute duration, the capacity of a standby power generator can be relatively small. The minimum size of a standby power generator may require sufficient capacity to operate essential equipment such as:

1. Coagulant and chlorine feeders
2. One-third of flocculators
3. Major electric valve operators and plant control system
4. One-third of pumping capacity (if necessary)
5. Minimum lighting

Where do you begin? You have to consider whether you would like to have all of your facility operating or whether just the vital or key equipment would be sufficient. Since the characteristics and operating conditions of every plant are different, it is difficult to make specific suggestions.

For the sake of illustration, let us pose a hypothetical situation. Consider a 10-MGD (38-MLD) capacity plant with an average flow rate of 6 MGD (23 MLD). Prepare a list of needs that must be met to ensure minimal operation:

1. Raw water pumping
2. Clarification
3. Clear water pumping
4. Chlorination
5. Minimal lighting

PUMP RECORD CARD

NAME_____ MAKE_____ MODEL_____
TYPE_____ SIZE_____ SERIAL #_____
ORDER NUMBER_____ SUPPLIER_____ DATE PURCHASED_____
DATE INSTALLED_____ APPLICATION_____ PLANT #_____

Name Plate Data and Pump Info   Stuffing Box Data   Motor Data

GPM _____   Diameter____ Depth____   Name____ Serial #____
TDH _____   Pack. Size____ Type____ H.P.____ Speed____
RPM _____   Length____ No. Rings____ Ambient°____
Gage Press Disc____ ____   Lantern Ring____ Flushed____ RPM____ Frame____
Gage Press Suc ____ ____   Mech. Seal Name____ Size____ Volts____ Amps____
Shut off Press ____ ____   Type____   Phase____ Cycle____
Suction Head ____ ____   Shaft Size____ Key____
Rotation ____ ____   **Pump Materials**
                           Casing_____   Bearing Front____
Impeller Type____   Shaft_____   Rear____
Impeller Dia.____   Wearing Rings Casing____   Code____ Type____
Impeller Clear____   Wearing Rings Impeller____ Amps @ Max. Speed____
Coupl Type & Size____   Shaft Sleeve____   Amps @ Shut Off____
Front Brg #____   Slinger____   Control Data Info
Rear Brg #____   Shims____   Starter____
Lub Interval____   Gaskets____   MENA Size____
Lubricant____   "O" Rings____   Cat. #____
Wearing Rings____   Brg. Seals Front____   Heater Size____
Shaft Sleeve Size____   Rear____   Rated @____
Pump Shaft Size____   Casing Wear Ring Size ID____ Control Voltage____
Pump Keyway____   OD____   Variable Speed
                                                         Type____
_____   Width____   Speed Max____
Other Related Information: Impeller Wear Ring ID____ Speed Min____
                                       OD____
                                       Width____

MOTOR STARTERS   Number_____

Title:_____

Mfg.:_____ Address_____

Style:_____ Class_____ Size_____

Type:_____ _____ _____

_____

O.L. HEATERS   O.L. TRIP UNITS

Style_____ Code_____   Mfg:_____ Style:_____

Amps_____ _____   Type:_____ _____

_____ _____   Amps Range:_____ _____

_____

CIRCUIT BREAKER

Mfg:_____ Address_____

Style:_____ Frame:_____ Volts_____ Amps Setting_____

Cat. No._____ _____ _____

MOTOR   Number_____

TITLE_____

Mfg:_____ Address_____

HP:_____ Volts:_____ Ser. No._____ Duty:_____

Phase:_____ Amps:_____ Frame:_____ Temp:_____

Cycles:_____ RPM:_____ Type_____ Class:_____

Code:_____ S.F.:_____ Model_____ Spec.:_____

SO#_____ S#_____ Style:_____ CSA App:_____

Form_____ Spec._____ Shft. Brg._____ Rear Brg._____

50 Cycle Data_____

Suitable for 208V Network:_____   Connection Diagram

Additional data_____   (6) (5) (4)   (6) (5) (4)

_____   (7) (8) (9)   (7) (8) (9)

_____   (1) (2) (3)   (1) (2) (3)

_____

Calculate the maximum horsepower or total kilowatts necessary to maintain the limited operation:

1. Raw Water Pump—75 horsepower	56.00 kW
2. Clarification—2½ horsepower	2.24 kW
3. Clear Water Pump—40 horsepower	30.00 kW
4. Chlorination—15 horsepower	11.20 kW
5. Lighting	5.00 kW
	104.44 kW

The minimum power required is 104.44 kW. When sizing a generator for emergency power, you have to make sure that the operator will be able to start the needed motors. Since the locked rotor current of the 75-horsepower induction motor on the raw water pump is approximately four times running current, the generator must be able to handle 224 kW at that instant. Size the generator not only by total load, but also for the highest horsepower motor being started. Consider the sequence in which motors will be started. The starting of all the motors simultaneously (without sequence starting) would be nearly impossible. Consult experts in power generation for answers to your specific questions regarding your plant because each plant has different needs. If you are considering standby power, shop around and get ideas from the equipment manufacturers. You may be able to reduce the size of the generator by using reduced-voltage starters on the larger motors.

After you have determined the size of generator needed, you must be able to connect it to your power distribution system. This may require some sophisticated switch gear. Besides the mechanical functions necessary in connecting the emergency power with your normal system, it is important that the two systems cannot be electrically coupled. (Two electrical systems must be in phase with each other before parallel coupling.) For this reason, mechanical interlocks are used to ensure that one circuit is always open. A kirk-key system, where one key is used for two locks, locking one switch open before the other can be closed, is sometimes used. The manufacturers of most packaged motor-generator systems can provide automatic transfer switches that will automatically start the generator when a power failure occurs and connect the generated power into the plant power distribution wiring.

Looking back at the plant described, a generator of 125 kW with intermittent overload capabilities should handle the load. (Note: This is an assumption. Actual calculation may indicate a different size.) An engine-generation system of this size could handle your minimal power needs. If your water distribution system has ample capacity, it may be possible to cut the plant production rate to reduce power requirements to what can be handled with a smaller capacity generator.

If you do not have standby power generation at your facility, talk to others in the water treatment field who do and obtain ideas and information. After due consideration, take the necessary steps to ensure your plant against interrupted power.

Standby power generators should be operated on a regular basis (once a week) to be sure they will operate properly, when needed. Be sure to operate your generator at full load for at least an hour. Commercial power into your plant must be shut off to operate standby power at full load.

18.152 Emergency Lighting

The most practical form of emergency lighting, in most instances, is that provided by battery-powered lighting units. Because they are used primarily for exit lighting, they are more economical than engine-driven power sources. If you have a momentary power outage, the system responds without an engine-generator start-up. All emergency lighting unit equipment is basically the same and consists of a rechargeable battery, a battery charger, low-voltage flood lights, and test monitoring and control accessories. Proper selection of a unit for a particular location requires careful consideration of the following items:

1. Initial cost
2. Types of batteries
3. Maintenance requirements
4. Lighting requirements

The three types of batteries most commonly used are: lead acid, lead calcium, and nickel cadmium. Because poor battery maintenance is quite common in emergency lighting systems, maintenance-free batteries are becoming increasingly popular. These batteries can have a gelatin or acid (wet) *ELECTROLYTE*.[11] The gelatin type is completely spillproof and can be handled safely without the dangers of acid spills. These batteries have a shorter life span than the wet type. Since all batteries undergo evaporation, the gelatin electrolyte will be exhausted before that of a battery containing liquid. Wet-type, maintenance-free batteries require no refilling and, when handled properly, acid spillage is minimal.

In terms of cost, the maintenance-free battery is more expensive; but when you consider the human factor, they may be more reliable and cheaper in the long run. Most systems use a battery charger that monitors the battery voltage. When required, the charger then charges the batteries. In older units, a trickle charger was used. This constant charging resulted in inoperative batteries in a short time because of overcharging.

The lamps used are normally 6- to 12-volt, sealed-beam, 25-watt lamps. The light pattern provided is most effective when illuminating a work area. A rule of thumb is that one lamp will be sufficient for about 1,000 square feet, providing that the full light pattern can be used. Consult emergency light level codes (Table 18.2) for your particular application.

---

[11] *Electrolyte* (ee-LECK-tro-lite). A substance that dissociates (separates) into two or more ions when it is dissolved in water.

TABLE 18.2 IES[a] RECOMMENDED EMERGENCY LIGHT LEVELS[b]

Hazard Requiring Visual Detection	Slight		High	
Normal Activity Level[c]	Low	High	Low	High
Areas	Conference rooms Reception rooms Exterior floodlighting Closets	Lobbies Corridors Concourse Restrooms, washrooms Telephone switchboard rooms Exterior entrance Exterior floodlighting	Elevators (freight) File rooms Mailrooms Offices Stairways Stockrooms Exterior entrance with stairs	Elevators Escalators Computer rooms Drafting rooms Offices Stairways Transformer vaults Engine rooms Electrical, mechanical, plumbing rooms
Footcandles	0.5	1.0	2.0	5.0
Dekalux	0.54	1.1	2.2	2.2

[a] Illuminating Engineering Society.
[b] Reprinted from December, 1978 issue Electrical Construction and Maintenance. Copyright 1978 McGraw-Hill, Inc. All rights reserved. (IES does not have a more current recommendation at the time of this publication.)
[c] Special conditions may require different levels of illumination. In some cases, higher levels may be required, as for example where security is a factor. In some other cases, greatly reduced levels of illumination, including total darkness, may be necessary, specifically in situations involving manufacturing, handling, use, or processing of light-sensitive materials (notably in connection with photographic products). In these situations, alternate methods of ensuring safe operation must be relied upon. Emergency light level codes and standards vary widely throughout the country. Recommended minimum lighting levels of the Illuminating Engineering Society are being considered as a possible standard by ANSI and the Life Safety Code. These are minimum lighting levels recommended for safety of personnel.

When selecting an emergency lighting system, check it very thoroughly to ensure that it will give you the protection needed. If it fails to work when the chips are down and the main power is out, you have wasted your money.

18.153 Batteries

This section will discuss wet storage batteries since they are the most prevalent. Automotive and equipment batteries are usually of the lead-acid type in which the dissimilar plates are made of two types of lead and the electrolyte is sulfuric acid. Wet-type batteries can also be nickel cadmium or nickel iron.

Most batteries are a series of cells enclosed in a common case. Each of these cells develops a potential (voltage) of 2.3 volts per cell when fully charged. Hence, a six-volt battery contains three cells and a 12-volt battery has six cells. The voltage output of a 12-volt battery is 13.8 volts when fully charged. Once a lead-acid battery has been placed in service, the addition of sulfuric acid is not necessary. The water portion of the electrolyte solution evaporates as the battery is charged and discharged. Lost water must be replaced. Deionized or distilled water should be used. Tap water contains impurities that shorten the lifetime of a battery if used to replace lost water. These minute particles become attached to the lead plates and do not allow the battery to rejuvenate itself fully when charged.

When batteries are placed on charge, remove the cell covers to allow the gas (hydrogen) caused by charging to escape and not to build excessive pressure in the battery. A battery on charge is as lethal as a small bomb if you ignite the gas. Do not smoke or cause electrical arcing near the battery. Do not breathe the gas. Make sure that the area where a battery is being charged is well ventilated.

The keys to prolonged life of a battery are to keep the electrolyte level above the cell plates, to keep the battery fully charged, and, above all, to keep the terminals and top clean. When dirt and residue accumulate on the top of a battery, it forms a path for current to flow between the negative and positive posts. Take a multimeter, connect one lead to the proper post (it will cause up-scale deflection), and slowly slide the other lead across the top of the battery toward the other post. If the top is dirty, the meter will deflect more as you proceed across the top.

To clean the battery, use a stiff-bristled brush (not a wire brush) and remove the heavy material. Then, wash with a solution of baking soda and water (four teaspoons of baking soda to one quart of water). This will remove the acid film from the top and neutralize corrosion on the battery terminals. Rinse with fresh water and dry the top with a dry, lintless cloth. Remove cell caps and wipe between them, then replace. At this time, check to be sure that the battery terminals are clean and tight. If a battery is charged, but the terminals are loose, proper voltage and current cannot be delivered.

> ## QUESTIONS
>
> Write your answers in a notebook and then compare your answers with those on page 380.
>
> 18.15A Why should a qualified electrician perform most of the necessary maintenance and repair of electrical equipment?
>
> 18.15B What is the purpose of a kirk-key system?
>
> 18.15C Why are battery-powered backup lighting units considered better than engine-driven power sources?
>
> 18.15D Why should the water lost from a lead-acid battery be replaced with deionized or distilled water?

18.16 High Voltage

18.160 Transmission

In general terms, high voltage is the voltage transmitted to the plant site by the utility company. The voltage level can vary, but 12,000 volts is quite common. After the power reaches the plant, it is transformed down to a usable voltage (460 to 480 volts) either through utility-owned or customer-owned transformers. The NEC (National Electrical Code) denotes high voltages as those over 600 volts.

Why have high voltage? Since current (amperes) varies inversely with voltage, a load of 500 amps on the low-voltage side of the transformer would create a 20-amp load on the high-voltage side of a 12,000-volts/480-volt transformer. Transmission lines would have to be enormous in order to carry the load if a lower voltage were used. Where high-voltage cables terminate at a transformer or switch gear, certain conditions must be adhered to. If outdoor transformers are used that have high-voltage wires exposed, an eight-foot (2.4-m) high fence is required to prevent accessibility by unqualified or unauthorized persons. Signs attached to the fence must indicate "High Voltage." Specifications for clearances, grounding, access, and enclosures vary with installations. Any modification or repair work must be completed by qualified people only.

18.161 Switch Gear

When we see the term "switch gear," it is usually associated with the equipment used in the interruption, transfer, or disconnecting of voltages over 600 volts. The enclosure is designed and manufactured to safely control high-voltage switching. Most distribution systems have a load-interrupting switch that is capable of disconnecting high-voltage lines that are under load. Because of the arc that is caused in breaking the circuit, special "arc shoes" (arc-suppressant devices) are used to ensure that the contact points are not pitted. A keyed lock system is used to prevent opening of the enclosure in the energized state.

Probably the best preventive maintenance that a treatment plant operator can provide for switch gear is to keep the exterior and its surroundings clean. If you encounter difficulties in the course of operating the switches, please obtain qualified help to do the inspection or repairs needed. Check with your particular manufacturer to determine what is needed and when this has to be done to keep your system functioning as designed. If your equipment is in a corrosive atmosphere, it may be necessary to remove it from service and epoxy paint the internal buses. All pivoting points should be lubricated with a lubricant specified by the manufacturer.

18.162 Power Distribution Transformers

If the high-voltage transformers are owned by the utility, the inspection and maintenance is carried out by the utility. Any peculiar changes, smells, or noises should be reported to the utility. When transformers are customer owned, a regular inspection program should be established.

Most transformers use an oil to insulate as well as to cool the windings. As heat is generated in the windings, it is transferred to the oil. The oil is then cooled by air passing the cooling fins of the transformer. The primary requirements of the oil are:

1. High dielectric strength
2. Freedom from inorganic acid, alkali, and sulfur to prevent damage to insulation and conductors
3. Low viscosity to provide good heat transfer
4. Freedom from sludging under normal operation conditions

The principal causes of deterioration of insulating oil are water and oxidation. The oil may be exposed to moisture through condensation of moist air due to breathing of the transformer, especially when the transformer is not continuously in service. The moist air condenses on the surface of the oil and on the inside of the tank. Oxidation causes sludging. The amount of sludge formed in a given oil depends upon the temperature and the time of exposure of the oil to the air. Excessive operating temperatures may cause sludging of any transformer oil. Check with the manufacturer to determine how often the oil should be tested. Oil can be revitalized by a cleaning procedure that is accomplished at the transformer site.

Any symptoms such as unusual noises, high or low oil levels, oil leaks, or high operating temperatures should be investigated at once. If your transformer has a thermometer, it is of the alcohol type and should be replaced with that type only. A mercury-type thermometer could cause insulation failures due to the proximity of a metallic substance, regardless of whether the thermometer is intact or broken.

The tank of every power transformer should be grounded to eliminate the possibility of obtaining static shocks from it or from being injured by accidental grounding of the winding to the case.

If repairs are indicated, use the expertise of a qualified person to ensure that the repairs are made safely as well as correctly. Your life and the lives of others may depend on the use of qualified people.

QUESTIONS

Write your answers in a notebook and then compare your answers with those on page 380.

18.16A Why is electricity transmitted at high voltage?

18.16B What precautions must be taken if outdoor transformers have exposed high-voltage wires?

18.16C What kind of maintenance should a treatment plant operator perform on switch gear?

18.16D What symptoms indicate that a power distribution transformer may be in need of maintenance or repair?

18.17 Electrical Safety Checklist

Throughout this manual and throughout this chapter the need for electrical safety is always being stressed. This section contains an electrical safety checklist, which is provided to help you ensure that you have minimized electrical hazards in your plant. This list is provided to make you aware of potential electrical hazards. You should add to the list additional electrical hazards that could injure someone at your water treatment plant.

1. Are there any conduits rusted to the point where they might have lost their explosion-proof integrity?

2. Are there any electrical conduit hangers that are rusted so bad that they are allowing the conduit to sag?

3. Are there any fasteners on the conduit hangers that are rusted and allowing the conduit to hang by the wires?

4. Do all of the extension cords and power tools meet code requirements for use in wet areas?

5. Does the agency or utility have a policy covering the proper placement of portable ventilation equipment when operators work inside enclosed tanks, vaults, and other confined spaces?

6. Does the agency use proper grounding units (ground-fault circuit interrupters) when working in wet areas?

7. Is the grounding of electrical equipment and systems inspected regularly?

8. Are electrical breakers and controls clearly marked?

9. Is there a formal program for locking out, tagging, and blocking out of electrical devices?

If you can answer these questions properly, you are working in the right direction to minimize electrical hazards in your water treatment plant.

QUESTIONS

Write your answers in a notebook and then compare your answers with those on page 380.

18.17A What is the purpose of an electrical safety checklist?

18.17B Why are rusted conduits of concern to a water treatment plant operator?

18.18 Additional Reading

1. *BASIC ELECTRICITY* by Van Valkenburgh, Nooger & Neville, Inc. Published by Delmar, an imprint of Thomson Learning. Obtain from Thomson Learning, Order Fulfillment, PO Box 6904, Florence, KY 41022-6904. ISBN 0-7906-1041-8. Price, $44.95, plus shipping and handling.

2. "Mechanical Maintenance" by Stan Walton, Chapter 15 in *OPERATION OF WASTEWATER TREATMENT PLANTS*, Volume II. Obtain from the Office of Water Programs, California State University, Sacramento, 6000 J Street, Sacramento, CA 95819. Price, $49.00.

3. "Instrumentation and Control Systems" by Leonard Ainsworth (revised by William H. Hendrix), Chapter 9 in *ADVANCED WASTE TREATMENT*. Obtain from the Office of Water Programs, California State University, Sacramento, 6000 J Street, Sacramento, CA 95819. Price, $49.00.

4. "Control and Automation,"* Chapter XII, in *WATER DISTRIBUTION OPERATOR TRAINING HANDBOOK*, by I. E. Nichols and B. W. Jex. Obtain from American Water Works Association (AWWA), Bookstore, 6666 West Quincy Avenue, Denver, CO 80235. Order No. 20428. ISBN 1-58321-014-8. Price to members, $54.50; nonmembers, $78.50; price includes cost of shipping and handling.

5. *MAINTENANCE ENGINEERING HANDBOOK* by L. Higgins and K. Mobley. Obtain from the McGraw-Hill Companies, Order Services, PO Box 182604, Columbus, OH 43272-3031. ISBN 0-07-028819-4. Price, $150.00, plus nine percent of order total for shipping and handling.

* Depends on edition.

End of Lesson 1 of 5 Lessons on MAINTENANCE

Please answer the discussion and review questions next.

DISCUSSION AND REVIEW QUESTIONS
Chapter 18. MAINTENANCE
(Lesson 1 of 5 Lessons)

At the end of each lesson in this chapter you will find some discussion and review questions. The purpose of these questions is to indicate to you how well you understand the material in the lesson. Write the answers to these questions in your notebook.

1. Why should operators thoroughly read and understand manufacturers' literature before attempting to maintain plant equipment?
2. Why must administrators or supervisors be made aware of the need for an adequate maintenance program?
3. What is the purpose of a maintenance recordkeeping program?
4. What items should be included in a plant library?
5. Why should your plant have an emergency team to repair chlorine leaks?
6. Why should one person never be permitted to repair a chlorine leak alone?
7. Why should inexperienced, unqualified, or unauthorized persons and even qualified and authorized persons be extremely careful around electrical panels, circuits, wiring, and equipment?
8. What protective or safety devices are used to protect operators and equipment from being harmed by electricity?
9. Why must motor nameplate data be recorded and filed?
10. What might be the cause of a pump motor failing to start?
11. Why should a water treatment plant have standby power?
12. How would you determine the capacity of standby generation equipment?

CHAPTER 18. MAINTENANCE

(Lesson 2 of 5 Lessons)

18.2 MECHANICAL EQUIPMENT

Mechanical equipment commonly used in water treatment plants is described and discussed in this section. Equipment used with specific treatment processes such as flocculation and filtration is not discussed. You must be familiar with equipment and understand what it is intended to do before developing a preventive maintenance program and maintaining equipment.

18.20 Repair Shop

Many large plants have fully equipped machine shops staffed with competent mechanics. But for smaller plants, adequate machine shop facilities often can be found in the community. In addition, most pump manufacturers maintain pump repair departments where pumps can be fully reconditioned.

The pump repair shop in a large plant commonly includes such items as welding equipment, lathes, drill press and drills, power hacksaw, flame-cutting equipment, micrometers, calipers, gauges, portable electric tools, grinders, a forcing press, metal-spray equipment, and sandblasting equipment. You must determine what repair work you can and should do and when you need to request assistance from an expert.

Some agencies have their own repair shops or local machine shops rebuild parts rather than buying direct from manufacturers. Many agencies try to select equipment on the basis of the reputations of distributors for supplying repair parts when needed. A parts inventory is essential for key pieces of equipment.

18.21 Pumps

Pumps serve many purposes in water treatment plants. They may be classified by the character of the material handled, such as raw or filtered water. Or, they may relate to the conditions of pumping: high lift, low lift, or high capacity. They may be further classified by principle of operation, such as centrifugal, propeller, reciprocating, and turbine (Figure 18.15).

The type of material to be handled and the function or required performance of the pump vary so widely that the designing engineer must use great care in preparing specifications for the pump and its controls. Similarly, the operator must conduct a maintenance and management program adapted to the peculiar characteristics of the equipment.

18.210 *Centrifugal Pumps*

A centrifugal pump is basically a very simple device: an impeller rotating in a casing called the volute. The impeller is supported on a shaft, which is, in turn, supported by bearings. Liquid coming in at the center (eye) of the impeller (Figure 18.16) is picked up by the vanes and by the rotation of the impeller and then is thrown out by centrifugal force into the discharge.

To help you understand how pumps work and the purpose of the various parts, a section titled "Let's Build a Pump!" has been included on the following pages. This material has been reprinted with the permission of Allis-Chalmers Corporation. Originally, the material was printed in Allis-Chalmers Bulletin No. OBX62568.

18.211 *Let's Build a Pump!*

A student of medicine spends long years learning exactly how the human body is built before attempting to prescribe for its care. Knowledge of pump anatomy is equally basic in caring for centrifugal pumps.

But, whereas the medical student must take a body apart to learn its secrets, it will be far more instructive to us if we put a pump together (on paper, of course). Then we can start at the beginning—adding each new part as we need it in logical sequence.

As we see what each part does, how it does it... we will see how it must be cared for.

Another analogy between medicine and maintenance: there are various types of human bodies, but if you know basic anatomy, you understand them all. The same is true of centrifugal pumps. In building one basic type, we will learn about all types.

Part of this will be elementary to some maintenance people... but they will find it a valuable "refresher" course, and, after all, maintenance just cannot be too good.

So, with a glance at the centrifugal principle on page 304, let us get on with building our pump...

FIRST WE REQUIRE A DEVICE TO SPIN LIQUID AT HIGH SPEED...

That paddle-wheel device is called the "impeller" (Figure 18.16)... and it is the heart of our pump.

302 Water Treatment

Displacement types of pumps

Dynamic types of pumps

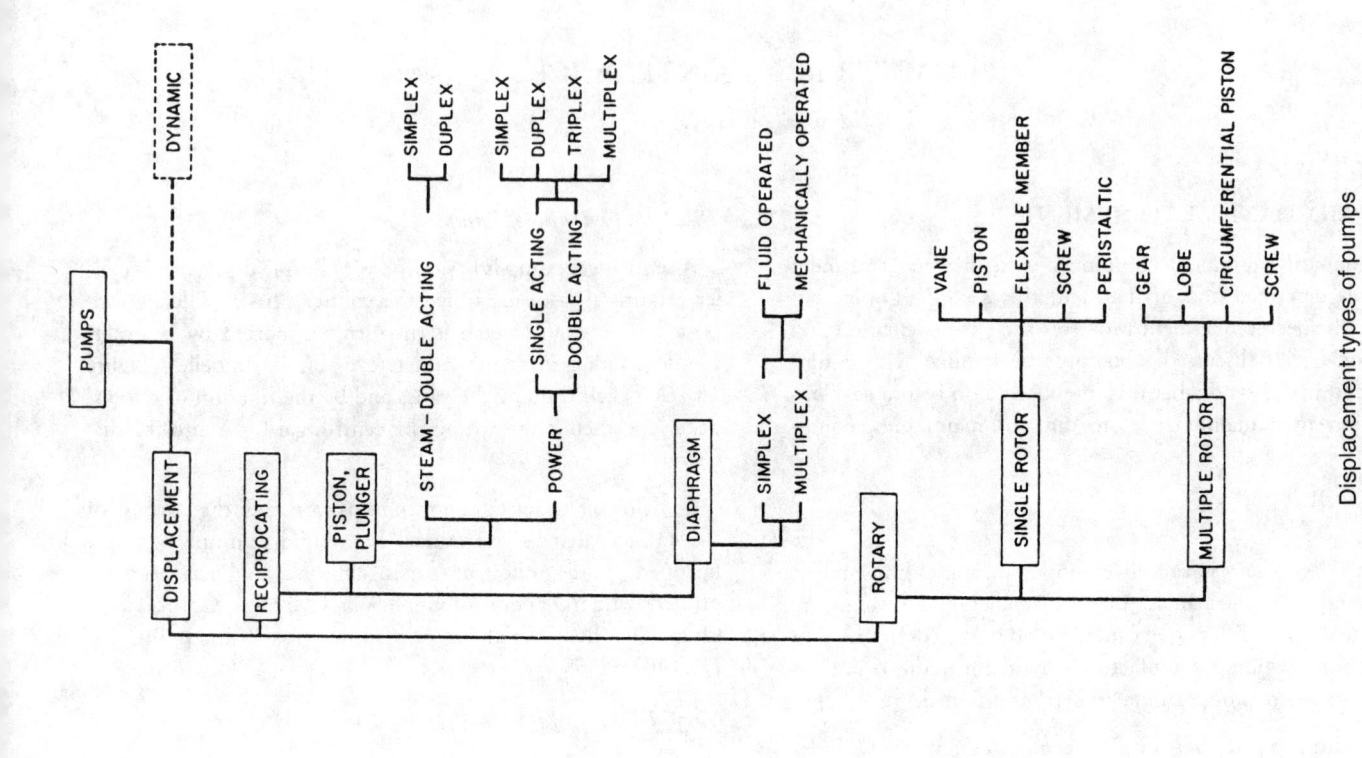

Fig. 18.15 Classification of pumps

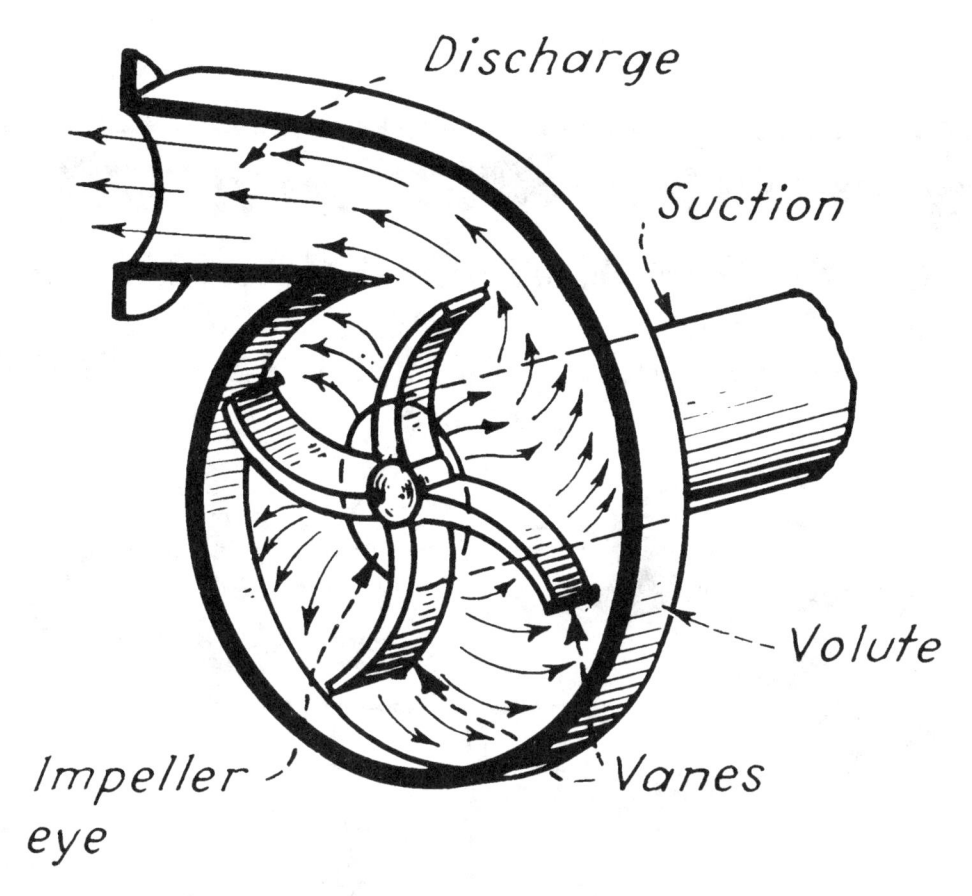

Fig. 18.16 Diagram showing details of centrifugal pump impeller
(Source: *CENTRIFUGAL PUMPS* by Karassik and Carter of Worthington Corporation)

Refer to Figure 18.19, page 311, for location of impeller in pump.

Note that the blades curve out from its hub. As the impeller spins, liquid between the blades is impelled outward by centrifugal force.

Note, too, that our impeller is open at the center—the "eye." As liquid in the impeller moves outward, it will suck more liquid in behind it through this eye ... provided it is not clogged.

Maintenance Rule No. 1: If there is any danger that foreign matter (sticks, refuse, etc.) may be sucked into the pump—clogging or wearing the impeller unduly—provide the intake end of the suction piping with a suitable screen.

NOW WE NEED A SHAFT TO SUPPORT AND TURN THE IMPELLER...

Our shaft looks heavy—and it is. It must maintain the impeller in precisely the right place.

CENTRIFUGAL FORCE IN ACTION—

ALL MOVING BODIES TEND TO TRAVEL IN A STRAIGHT LINE. WHEN FORCED TO TRAVEL IN A CURVE, THEY CONSTANTLY *TRY* TO TRAVEL ON A TANGENT...

... IN AN "AIRPLANE RIDE"

Centrifugal force pushes dummy planes swung in a circle *away* from center of rotation.

... IN A WHIRLPOOL

Centrifugal force tends to push swirling water outward... forming vortex in center.

But that ruggedness does not protect the shaft from the corrosive or abrasive effects of the liquid pumped ... so we must protect it with sleeves slid on from either end.

What these sleeves—and the impeller, too—are made of depends on the nature of the liquid we are to pump. Generally, they are bronze, but various other alloys, ceramics, glass, or even rubber-coating are sometimes required.

Maintenance Rule No. 2: Never pump a liquid for which the pump was not designed.

Whenever a change in pump application is contemplated and there is any doubt as to the pump's ability to resist the different liquid, check with your pump manufacturer.

WE MOUNT THE SHAFT ON SLEEVE, BALL, OR ROLLER BEARINGS...

As we will see later, clearances between moving parts of our pump are quite small.

If bearings supporting the turning shaft and impeller are allowed to wear excessively and lower the turning units within a pump's closely fitted mechanism, the life and efficiency of that pump will be seriously threatened.

Maintenance Rule No. 3: Keep the right amount of the right lubricant in bearings at all times. Follow your pump manufacturer's lubrication instructions to the letter.

Main points to keep in mind are ...

1. Although too much oil will not harm sleeve bearings, too much grease in antifriction type bearings (ball or roller) will promote friction and heat. The main job of grease in antifriction bearings is to protect steel elements against corrosion, not friction.

2. Operating conditions vary so widely that no one rule as to frequency of changing lubricant will fit all pumps. So play it safe: if anything, change lubricant *before* it is too worn or too dirty.

TO CONNECT WITH THE MOTOR, WE ADD A COUPLING FLANGE...

Some pumps are built with pump and motor on one shaft and, of course, offer no alignment problem.

But our pump is to be driven by a separate motor ... and we attach a flange to one end of the shaft through which bolts will connect with the motor flange.

Use a straightedge or dial indicator to ensure shaft alignment (see pages 334 and 335, Figures 18.30 and 18.31).

Maintenance Rule No. 4: See that pump and motor flanges are parallel vertically and axially ... and that they are kept that way.

If shafts are eccentric or meet at an angle, every revolution throws tremendous extra load on bearings of both pump and motor. Flexible couplings will not correct this condition, if excessive.

Checking alignment should be regular procedure in pump maintenance. Foundations can settle unevenly, piping can change pump position, bolts can loosen. Misalignment is a major cause of pump and coupling wear.

NOW WE NEED A "STRAW" THROUGH WHICH LIQUID CAN BE SUCKED...

Notice two things about the suction piping: (1) the horizontal piping slopes upward toward the pump; (2) any reducer that connects between the pipe and pump intake nozzle should be horizontal at the top—(eccentric, not concentric).

This up-sloping prevents air pocketing in the top of the pipe ... where trapped air might be drawn into the pump and cause loss of suction.

Maintenance Rule No. 5: Any down-sloping toward the pump in suction piping (as exaggerated in the previous diagrams) should be corrected.

This rule is *VERY* important. Loss of suction greatly endangers a pump ... as we will see shortly.

WE CONTAIN AND DIRECT THE SPINNING LIQUID WITH A CASING...

We got a little ahead of our story in the previous paragraphs ... because we did not yet have the casing to which the suction piping bolts. And the manner in which it is attached is of great importance.

Maintenance Rule No. 6: See that piping puts absolutely no strain on the pump casing.

When the original installation is made, all piping should be in place and self-supporting before connection. Openings should meet with no force. Otherwise, the casing is apt to be cracked ... or sprung enough to allow closely fitted pump parts to rub.

It is good practice to check the piping supports regularly to see that loosening, or settling of the building, has not put strains on the casing.

NOW OUR PUMP IS ALMOST COMPLETE, BUT IT WOULD LEAK LIKE A SIEVE...

We are far enough along now to trace the flow of water through our pump. It is not easy to show suction piping in the cross-section view above, so imagine it stretching from your eye to the lower center of the pump.

Our pump happens to be a "double suction" pump, which means that water flow is divided inside the pump casing ... reaching the eye of the impeller from either side.

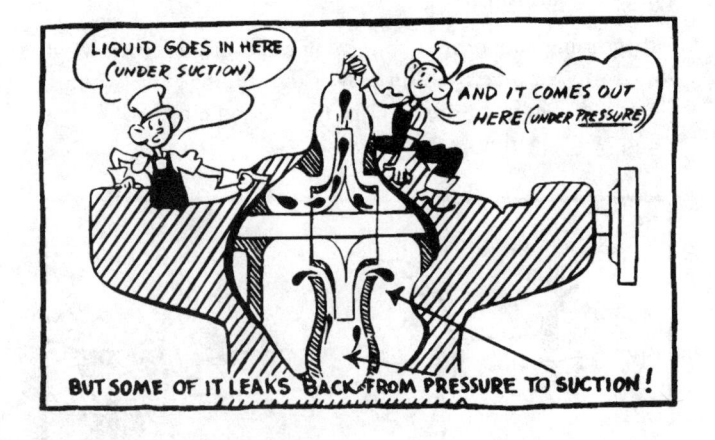

As water is drawn into the spinning impeller, centrifugal force causes it to flow outward ... building up high pressure at the outside of the pump (which will force water out) and creating low pressure at the center of the pump (which will draw water in). This situation is diagrammed in the upper half of the pump, above.

So far so good ... except that water tends to be drawn back from pressure to suction through the space between impeller and casing—as diagrammed in the lower half of the pump, above—and our next step must be to plug this leak, if our pump is to be very efficient.

SO WE ADD WEARING RINGS TO PLUG INTERNAL LIQUID LEAKAGE ...

You might ask why we did not build our parts closer fitting in the first place—instead of narrowing the gap between them by inserting wearing rings (see Figure 18.19 on page 311, item 16).

The answer is that those rings are removable and replaceable ... when wear enlarges the tiny gap between them and the impeller. (Sometimes rings are attached to impeller rather than casing—or rings are attached to both so they face each other.)

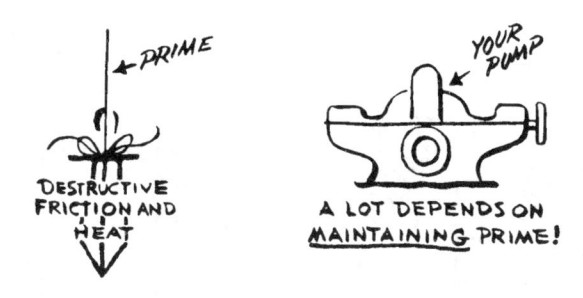

Maintenance Rule No. 7: Never allow a pump to run dry (either through lack of proper priming when starting or through loss of suction when operating). Water is a lubricant between rings and impeller.

Maintenance Rule No. 8: Examine wearing rings at regular intervals. When seriously worn, their replacement will greatly improve pump efficiency.

TO KEEP AIR FROM BEING SUCKED IN, WE USE STUFFING BOXES ...

We have two good reasons for wanting to keep air out of our pump: (1) we want to pump water, not air; (2) air leakage is apt to cause our pump to lose suction.

Each stuffing box we use consists of a casing, rings of packing, and a gland at the outside end. A mechanical seal may be used instead.

Maintenance Rule No. 9: Packing should be replaced periodically—depending on conditions—using the packing recommended by your pump manufacturer. Forcing in a ring or two of new packing instead of replacing worn packing is bad practice. It is apt to displace the seal cage.

Put each ring of packing in separately, seating it firmly before adding the next. Stagger adjacent rings so the points where their ends meet do not coincide. See Figure 18.26, "How to pack a pump," on pages 322 and 323.

Maintenance Rule No. 10: Never tighten a gland more than necessary ... as excessive pressure will wear shaft sleeves unduly.

Maintenance Rule No. 11: If shaft sleeves are badly scored, replace them immediately ... or packing life will be entirely too short.

TO MAKE PACKING MORE AIRTIGHT, WE ADD WATER SEAL PIPING ...

In the center of each stuffing box is a "seal cage." By connecting it with piping to a point near the impeller rim, we bring liquid under pressure to the stuffing box.

308 Water Treatment

This liquid acts both to block out air intake and to lubricate the packing. It makes both packing and shaft sleeves wear longer ... providing it is clean liquid.

WATER IS A LUBRICANT!

Maintenance Rule No. 12: If the liquid being pumped contains grit, a separate source of sealing liquid should be obtained (for example, it may be possible to direct some of the pumped liquid into a container and allow the grit to settle out).

To control liquid flow, draw up the gland just tight enough so a thin stream (approximately one drop per second) flows from the stuffing box during pump operation.

DISCHARGE PIPING COMPLETES THE PUMP INSTALLATION—AND NOW WE CAN ANALYZE THE VARIOUS FORCES WE ARE DEALING WITH ...

SUCTION. At least 75 percent of centrifugal pump troubles trace to the suction side. To minimize them ...

1. Total suction lift (distance between centerline of pump and liquid level when pumping, plus friction losses) generally should not exceed 15 feet (4.5 meters).

2. Piping should be at least a size (diameter) larger than pump suction nozzle.

3. Friction in piping should be minimized ... use as few and as easy bends as possible ... avoid scaled or corroded pipe.

DISCHARGE lift, plus suction lift, plus friction in the piping from the point where liquid enters the suction piping to the end of the discharge piping equals total head.

Pumps should be operated near their rated heads. Otherwise, the pump is apt to operate under unsatisfactory and unstable conditions, which reduce efficiency and operating life of the unit, or "cavitation" could occur. Note the description of cavitation on page 309. Cavitation can seriously damage your pump.

PUMP CAPACITY generally is measured in gallons per minute (liters per second or cu m per second). A new pump is guaranteed to deliver its rating in capacity and head. But whether a pump retains its actual capacity depends to a great extent on its maintenance.

Wearing rings must be replaced when necessary—to keep internal leakage losses down.

Friction must be minimized in bearings and stuffing boxes by proper lubrication ... and misalignment must not be allowed to force scraping between closely fitted pump parts.

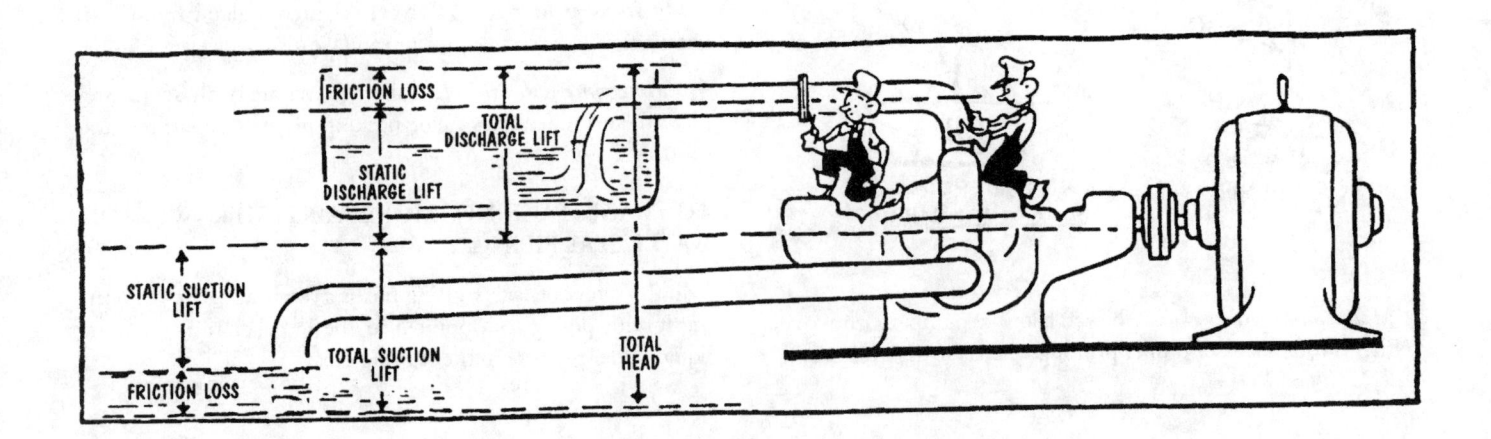

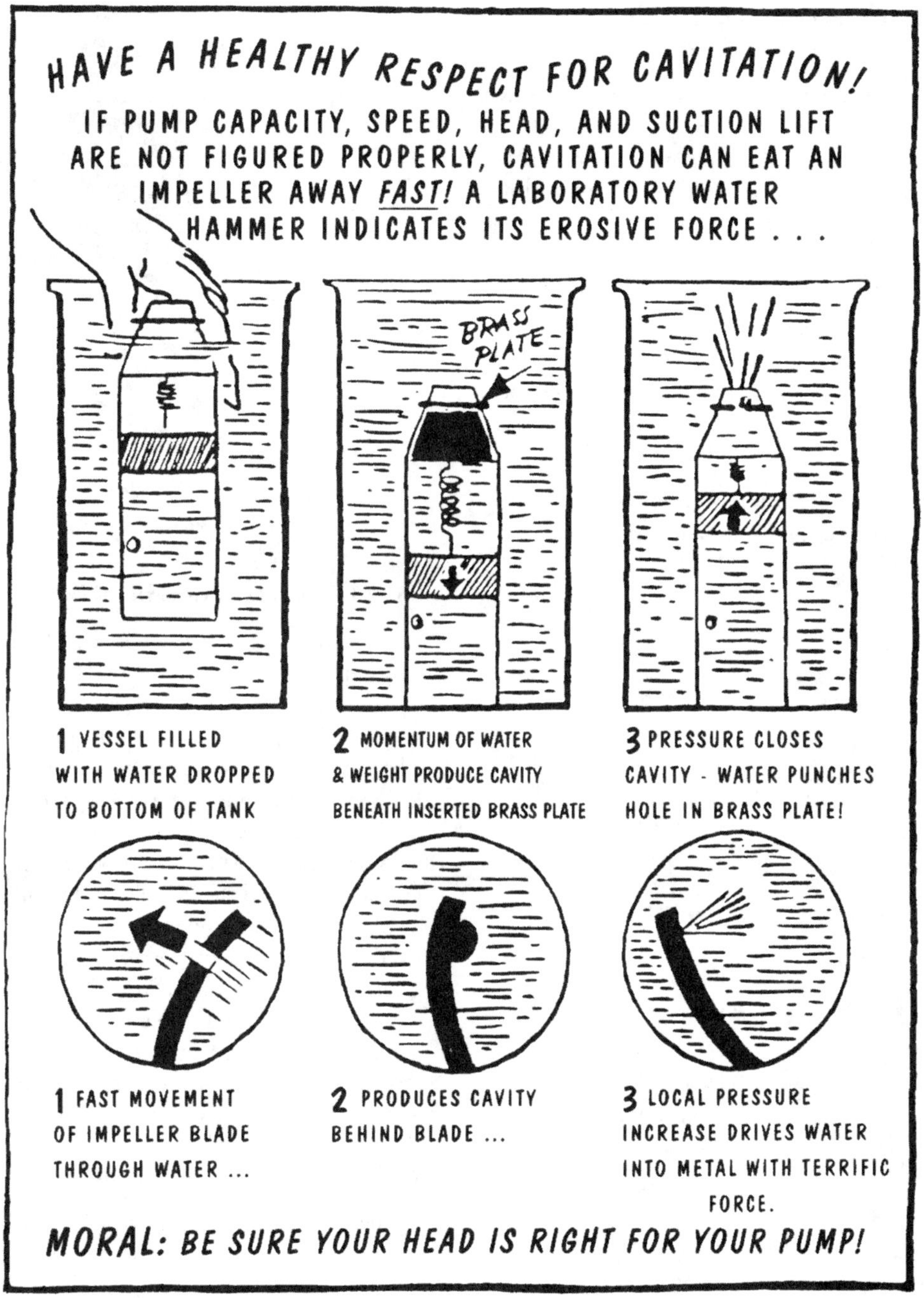

> **HAVE A HEALTHY RESPECT FOR CAVITATION!**
> IF PUMP CAPACITY, SPEED, HEAD, AND SUCTION LIFT ARE NOT FIGURED PROPERLY, CAVITATION CAN EAT AN IMPELLER AWAY *FAST!* A LABORATORY WATER HAMMER INDICATES ITS EROSIVE FORCE...
>
> **1** VESSEL FILLED WITH WATER DROPPED TO BOTTOM OF TANK
>
> **2** MOMENTUM OF WATER & WEIGHT PRODUCE CAVITY BENEATH INSERTED BRASS PLATE
>
> **3** PRESSURE CLOSES CAVITY - WATER PUNCHES HOLE IN BRASS PLATE!
>
> **1** FAST MOVEMENT OF IMPELLER BLADE THROUGH WATER...
>
> **2** PRODUCES CAVITY BEHIND BLADE...
>
> **3** LOCAL PRESSURE INCREASE DRIVES WATER INTO METAL WITH TERRIFIC FORCE.
>
> **MORAL:** BE SURE YOUR HEAD IS RIGHT FOR YOUR PUMP!

Cavitation is a condition that can cause a drop in pump efficiency, vibration, noise, and rapid damage to the impeller of a pump. Cavitation occurs due to unusually low pressures within a pump. These low pressures can develop when pump inlet pressures drop below the design inlet pressures or when the pump is operated at flow rates considerably higher than design flows. When the pressure within the flowing water drops very low, the water starts to boil and gas pockets or bubbles form on the blade of the impeller. The collapse of these gas pockets or bubbles drives water into the impeller with a terrific force that can knock metal particles off the impeller and cause pitting on the impeller surface. This same action can and does occur on pressure reducing valves and partially closed gate and butterfly valves. Cavitation is accompanied by loud noises that sound like someone is pounding on the impeller with a hammer.

POWER of the driving motor, like capacity of the pump, will not remain at a constant level without proper maintenance.

Starting load on motors can be reduced by throttling or closing the pump discharge valve (*NEVER* the suction valve!) ... but the pump must not be operated for long with the discharge valve closed. Power then is converted into friction, overheating the water with serious consequences.

18.212 Vertical Centrifugal Pumps (Figures 18.17 and 18.18)

Another common configuration for centrifugal pumps is the vertical-suction cased centrifugal pump. This is an adaptation of the deep-well turbine pump for booster pump service. They are very flexible in design as the engineers can specify either single- or multistage in a wide variety of sizes and characteristics.

Besides the usual lubrication of the electric motor, the only routine maintenance required is to adjust and repair, as needed, the single packing gland.

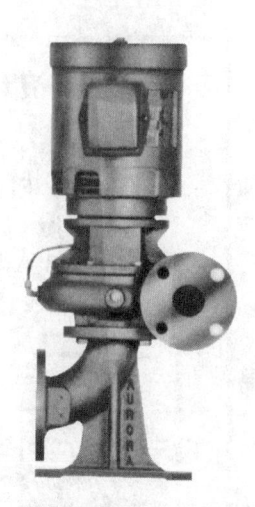

Fig. 18.18 Vertical centrifugal pump (single stage)
(Permission of Aurora Pump Company)

18.213 Horizontal Centrifugal Pumps (Figure 18.19)

Horizontal centrifugal pumps, like the one we just constructed on paper in Section 18.211, are available in a number of configurations. The one we built is best described as a single-stage, horizontal, double-suction, split-case centrifugal pump. The pump is a single-stage pump because it has only one impeller. Some horizontal pumps have two impellers that are working in series to create higher heads than can readily be obtained with only one impeller. Our paper pump was double suction in that water entered the impeller from both sides. The advantage of this design over a single-suction design is that the longitudinal thrust from the water entering the impeller is balanced. This greatly reduces the thrust load that the pump's bearings must carry. The split-case designation indicates that the pump case is made in two halves. Some centrifugal pumps have a single suction in line with the shaft. These are described as single-stage, end-suction centrifugal pumps.

18.214 Reciprocating or Piston Pumps

The word "reciprocating" means moving back and forth, so a reciprocating pump is one that moves a liquid by a piston that moves back and forth. A simple reciprocating pump is shown in Figure 18.20. If the piston is pulled to the left, Check Valve A will be open and the liquid will enter the pump and fill the casing. When the piston reaches the end of its travel to the left and is pushed back to the right, Check Valve A will close, Check Valve B will open, and the liquid will be forced out the exit line.

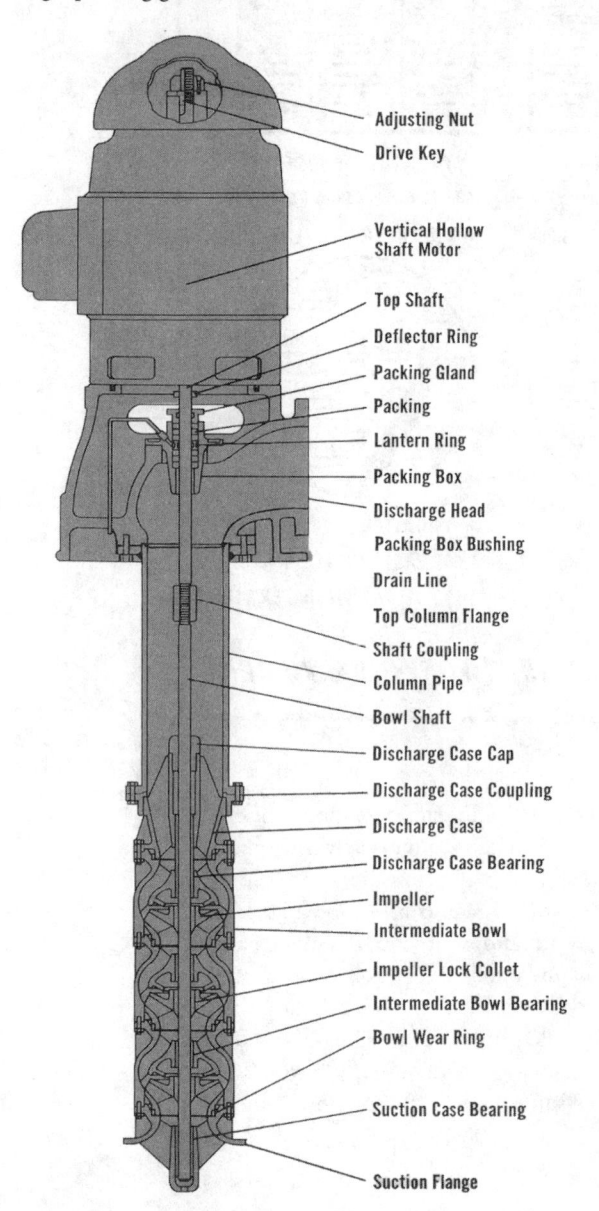

Fig. 18.17 Vertical centrifugal pump (multistage)
(Permission of Aurora Pump Company)

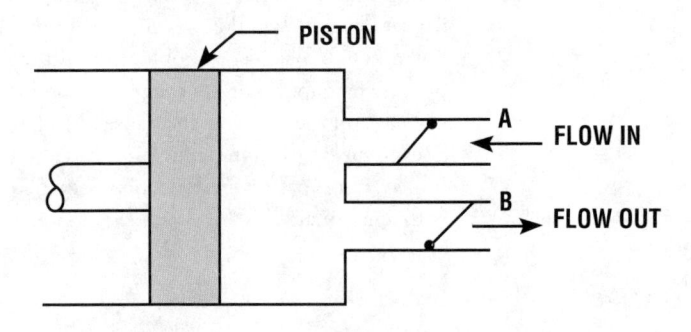

Fig. 18.20 Simple reciprocating pump

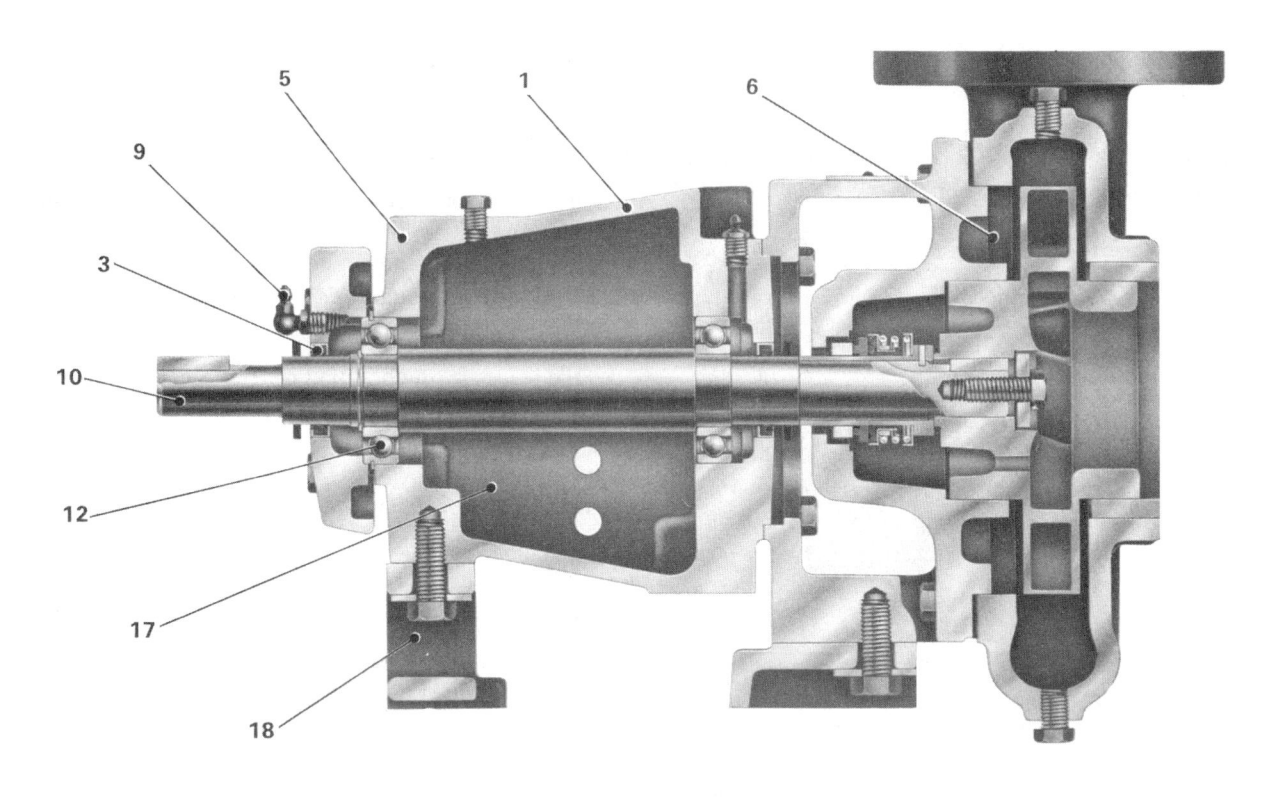

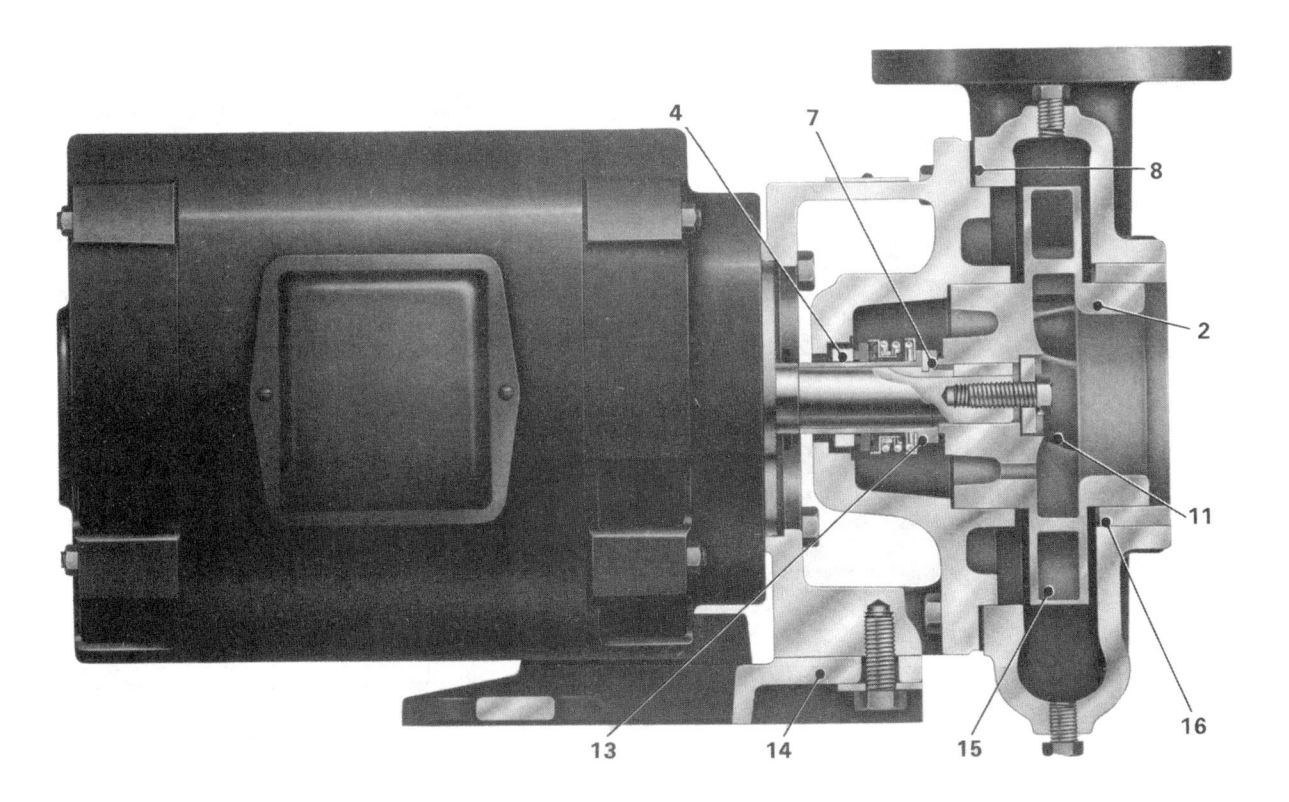

Maintenance 311

1. Motor frame	6. Casing	11. Impeller	16. Wearing ring
2. Impeller	7. Shaft sleeve	12. Bearings	17. Oil reservoir
3. Oil seal	8. Back pull-out	13. Shaft sleeve	18. Rear support foot
4. Mechanical seal	9. Lubrication fittings	14. Close coupled motor support	
5. Frame	10. Shaft	15. Impeller	

Fig. 18.19 Centrifugal pump parts
(Permission of Aurora Pump Company)

A piston pump is a positive displacement pump. Never operate it against a closed discharge valve or the pump, valve, or pipe could be damaged by excessive pressures. Also, the suction valve should be open when the pump is started. Otherwise, an excessive suction or vacuum could develop and cause problems.

18.215 Progressive Cavity (Screw-Flow) Pumps (Figure 18.21)

The progressive cavity pump consists of a screw-shaped rotor snugly enclosed in a nonmoving stator or housing. The threads of the screw-like rotor (commonly manufactured of chromed steel) make contact along the walls of the stator (usually made of synthetic rubber). The gaps between the rotor threads are called "cavities." When water is pumped through an inlet valve, it enters the cavity. As the rotor turns, the material is moved along until it leaves the conveyor (rotor) at the discharge end of the pump. The size of the cavities along the rotor determines the capacity of the pump.

All progressive cavity pumps operate on the basic principle described above. To further increase capacity, some models have a shaped inside surface of the stator (housing) with a similarly shaped rotor. In addition, some models use a rotor that moves up and down inside the stator as well as turning on its axis (Figure 18.22). This allows a further increase in the capacity of the pump.

Progressive cavity pumps are recommended for materials containing higher concentrations of suspended solids. They are commonly used to pump sludges. Progressive cavity pumps should never be operated dry (without liquid in the cavities), and should never be run against a closed discharge valve.

18.216 Chemical Metering Pumps

Many chemical metering pumps are a type of positive displacement pump. For information on chemical metering pumps, see Chapter 13, "Fluoridation," Section 13.30, "Chemical Feeders," and Section 18.4, "Chemical Storage and Feeders," in this chapter.

QUESTIONS

Write your answers in a notebook and then compare your answers with those on pages 380 and 381.

18.20A   List the pieces of equipment and special tools commonly found in a pump repair shop.

18.21A   What is the purpose of a pump impeller?

18.21B   Why should the intake end of suction piping have a suitable screen?

18.21C   Why must suction piping always be up-sloping?

18.21D   What is cavitation?

18.21E   What is an advantage of a double-suction pump over a single-suction pump?

18.22 Lubrication

18.220 Purpose of Lubrication

Lubrication of equipment is probably one of the most important phases of a maintenance operator's job. Without proper lubrication, the tools and equipment used for operating and maintaining water treatment plants would fail. Proper lubrication of tools and equipment is probably one of the maintenance operator's easiest jobs, but often it is the most neglected.

The purpose of lubrication is to reduce friction between two surfaces. Lubrication also removes heat that is caused by friction. Solid friction of two dry surfaces in contact is changed to a fluid friction of a separating layer of liquid or liquid lubricant. Actually, water is a lubricant, although not a good lubricant.

18.221 Properties of Lubricants

A good lubricant must have the following properties:

1. Form a slippery coating on contacting surfaces so they can slide freely past each other

2. Exert sufficient pressure to keep the surfaces apart when running

To be a good lubricant for a particular job, the lubricant used must have the following qualities:

1. Thickness of the lubricant layer must be sufficient to keep the roughness of the metal parts from touching.

2. Lubricity (slipperiness) must be sufficient to allow molecules to slide freely past each other.

3. Viscosity (resistance to flow) must be sufficient to build up a pressure necessary to keep the surfaces apart. If viscosity alone cannot provide enough pressure, an external pressure must be supplied by a pump.

Viscosity in the United States is the number of seconds it takes 60 cubic centimeters (cc) of an oil to flow through the standard orifice of a Saybolt Universal Viscometer at 100, 130, or 210 degrees Fahrenheit. A 300 - SSU[12] @ 130 oil means that it took 300 seconds for 60 cc to flow through a Saybolt Universal Viscometer at 130 degrees Fahrenheit. Viscosity decreases with temperature rise because oil becomes thinner. The specific gravity of an oil is measured by comparing the weight of oil with an equal volume of water, both at 60 degrees Fahrenheit.

Some other important information to know about lubricants is their "pour point," "flash point," and "fire point." Pour point is the temperature at which a lubricant refuses to run. This is important in low-temperature work. Flash point is the temperature at which oil vaporizes enough to ignite momentarily when near a flame. A low flash point means that oil evaporates more readily in service. Fire point is the temperature at which oil vaporizes enough to keep on burning. Oils in service tend to become acid

---

[12] SSU. Standard Saybolt Units.

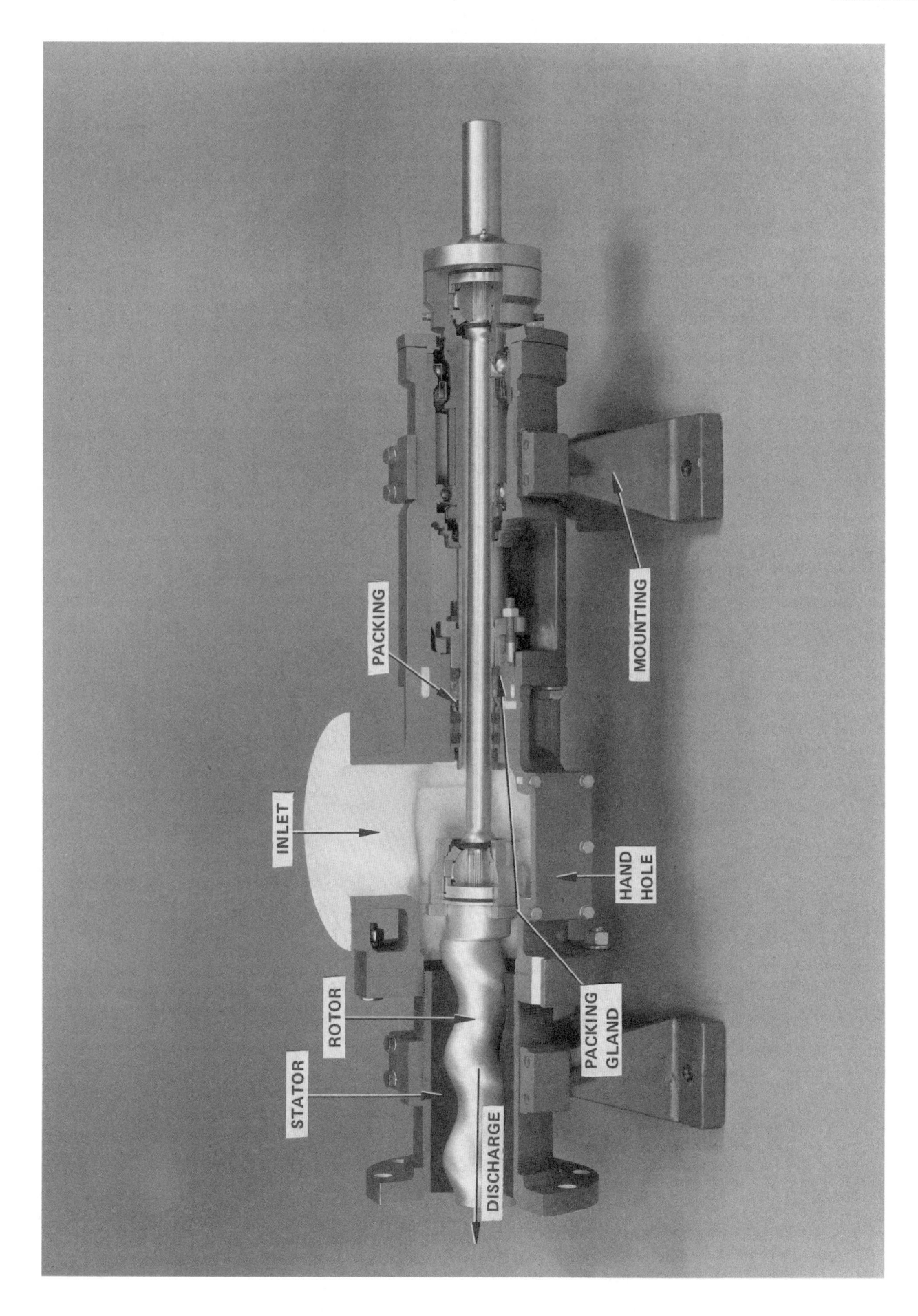

Fig. 18.21 Progressive cavity (screw-flow) pump
(Permission of Moyno Pump Division, Robbins & Meyer, Inc.)

314   Water Treatment

Pumping principle

0°

45°

90°

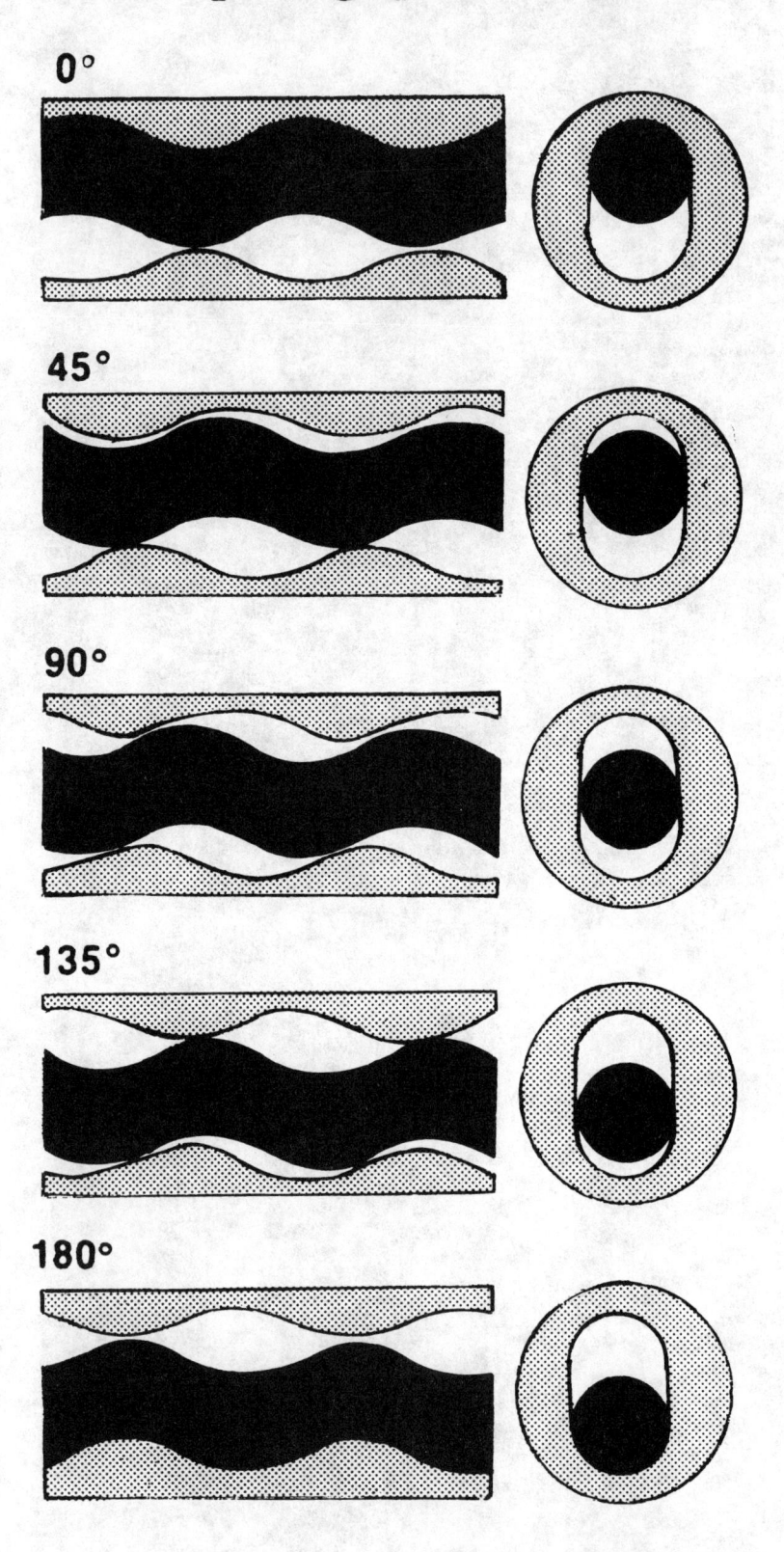

135°

180°

Fig. 18.22   Pumping principle of a progressive cavity pump
(Permission of Allweiler Pumps, Inc.)

and may cause corrosion, deposits, sludging, and other problems. This condition may not be visible when you look at the oil. Therefore, do not extend the time for an oil change because the oil looks clean.

To detect acid conditions in oils, the neutralization number of an oil is used. The neutralization number is the weight in milligrams of potassium hydroxide required to neutralize one gram of oil. This is used by laboratories that test the oil on large engines, turbines, compressors, and other equipment that have large-volume oil reservoirs to determine when oil changes or additives are needed.

Most lubricants in general use are fluid at room temperature. Mostly, these are petroleum based, but others are used. Greases are mixtures of petroleum products with soaps such as lime, soda, aluminum, and metallic. Metallic soaps, forms of calcium, sodium, potassium, and lithium, have good retention in bearings and can withstand high temperatures and pressures. A sodium-base grease has sodium as the soap mixed with the petroleum.

Solid materials such as graphite and finely ground mica are sometimes used as lubricants. Some recently developed silicon compounds (silicones) work very well under heavy loads and widely varying temperatures.

There are many oil additives on the market today and they are worth investigating. Oil additives are chemical compounds added to an oil to improve certain chemical or physical properties such as stability, lubricity, and foaming. They are used to prevent rust or deposits and many other items that could cause problems.

18.222 Lubrication Schedule

To have proper lubrication, you must first set up a lubrication schedule. This can be a simple check-off sheet or card system or an elaborate computer system. The first thing to do is make a list of everything that needs lubrication down to the smallest item, including chains, rollers, and sprockets. After you have listed every item on paper, go through the manufacturer's instruction books to determine the frequency and type of lubrication required. Is the frequency daily, weekly, monthly, semiannually, or annually? The manufacturer's literature usually lists several different name brands of lubricants that are equal. If you need help determining the type of lubricant or cross-referencing it to your particular brand, contact your supplier. Most oil distributors have a service representative who will come to your facility and go over the individual equipment and specify which lubricants you should use. Next, determine the amount of each lubricant required. This is achieved by counting the number of grease fittings. Determine the locations of fill plugs, drain plugs, oil levels, sight glasses, dipsticks, and other important items. To find these locations, physically inspect each piece of equipment thoroughly and look for all lubrication points. Also, the manufacturer's maintenance manual should show the lubrication points for each piece of equipment.

When you have gathered all this information, transfer it to the equipment history cards for future reference. From this information, you can make up a lubrication chart or form.

As stated earlier, use whatever type of lubrication form you prepare, but follow it. Always record each lubrication job when completed and have the operator who did the job initial the record card. Always keep your lubrication schedules up to date. If there are failures due to the wrong or insufficient lubricant, change or increase the lubrication frequency on the schedule. Also, new equipment must be added and discarded equipment removed from the schedule. Someone must be assigned to take care of the lubrication and records. Assign more than one operator or rotate this job so if an individual is off work or leaves the crew, there is a continuity in the lubrication schedule.

18.223 Precautions

When handling or storing oils and greases, some special precautions must be followed. Make sure the storage area does not create a fire hazard. Most lubricants are highly flammable and should not be stored where there is an open flame. "NO SMOKING" signs must be posted outside the building. Be sure to keep any spills wiped up and make sure that all the lids are tight on their containers.

Keep materials and containers clean. Sand, grit, and other substances can contaminate lube supplies and create an equipment failure that lubrication maintenance is intended to prevent. Another good idea is to direct the first shot of grease from a gun into a waste can.

18.224 Pump Lubrication

Pumps, motors, and drives should be oiled and greased in strict accordance with the recommendations of the manufacturer. Cheap lubricants may often be the most expensive in the end. Oil should not be put in the housing while the pump shaft is rotating because the rotary action of the ball bearings will pick up and retain a considerable amount of oil. When the unit comes to rest, an overflow of oil around the shaft or out of the oil cup will result.

Greased bearings should be lubricated as follows:

1. Shut off, lock out, tag, and block the unit if moving parts that might be a safety hazard are close to the grease fitting or drain plugs.

2. Remove the drain plug from the bearing housing.

3. Remove the grease fitting protective cap and wipe off the grease fitting. Be sure that you do not force dirt into the bearing housing along with the clean grease.

4. Pump in clean grease until the grease coming out of the drain hole is clean. Do not pump grease into a bearing with the drain plug in place. This could easily build up enough pressure to blow out the seals.

5. Put the protective cap back on the grease fitting.

6. With the drain plug still removed, put the unit back in service. As the bearing warms up, excess grease will be expelled from the drain hole. After the unit has been running for a few hours, the drain plug may be put back in place. Special drain plugs with spring-loaded check valves are recommended because they will protect against further buildup.

7. Unless you intend to be very careful, we recommend that bearing grease be purchased in cartridge form to minimize the chance of getting dirt into the lubricant.

QUESTIONS

Write your answers in a notebook and then compare your answers with those on page 381.

18.22A What is the purpose of lubrication?

18.22B What happens to oils in service?

18.22C What should be done to ensure proper lubrication of equipment?

18.225 Equipment Lubrication

Different authorities may make conflicting lube recommendations for essentially the same item; however, general reference material is available to help select the correct lubricant for a specific application.

Grease is graded on a number scale, or viscosity index, by the National Lubricating Grease Institute. For example, No. 0 is very soft; No. 6 is quite stiff. A typical grease for most treatment plant applications might be a No. 2 lithium or sodium compound grease, which is used for operating temperatures up to 250°F (120°C).

Generally, the time between flushing and repacking for greased bearings should be divided by 2 for every 25°F (14°C) above 150°F (65°C) operating temperature. Also, generally, the time between lubrications should not be allowed to exceed 48 months, since lube component separation and oxidation can become significant after this period of time, regardless of amount of use.

Another point worth noting is that grease is normally not suitable for moving elements with speeds exceeding 12,000 in/min (5 m/sec). Usually, oil lubricating systems are used for higher speeds. Lighter viscosity oils are recommended for high speeds, and, within the same speed and temperature range, a roller bearing will normally require one grade heavier viscosity than a ball bearing.

A good rule of thumb is to change and flush oil completely at the end of 600 hours of operation, or every 3 months, whichever occurs first. More specific procedures for flushing and changing lubricants are outlined by most equipment manufacturers.

Every operator should be aware of the dangers of overfilling with either grease or oil. Overfilling can result in high pressures and temperatures, and ruined seals or other components. It has been observed that more antifriction bearings are ruined by overgreasing than by neglect.

A thermometer can tell a great deal about the condition of a bearing. Ball bearings are generally in trouble above 180°F (82°C). Grease-packed bearings typically run 10 to 50 degrees above ambient temperature.

For clarifier drive units, which are almost always located outdoors, condensation presents a dangerous problem for the lubrication system. Most units of current design have a condensate bailing system to remove water from the gear housing by displacement. These units should be checked often for proper operation, particularly during seasons of wide air temperature fluctuation.

Pumps incorporate many types of seals and gaskets constructed of combinations of elastomers and metals. As for lubricants, conflicting advice can be obtained. A file containing data on general properties of materials used can help in the choice of lubricant.

QUESTIONS

Write your answers in a notebook and then compare your answers with those on page 381.

18.22D Does a soft grease have a high or low viscosity index as compared with a hard grease?

18.22E Is oil or grease used with higher speeds?

18.22F What problems can result from overfilling with oil or grease?

End of Lesson 2 of 5 Lessons on MAINTENANCE

Please answer the discussion and review questions next.

DISCUSSION AND REVIEW QUESTIONS
Chapter 18.   MAINTENANCE
(Lesson 2 of 5 Lessons)

Write the answers to these questions in your notebook. The question numbering continues from Lesson 1.

13. What is the purpose of a pump shaft?
14. What is the purpose of pump sleeves?
15. Why should a pump never be allowed to run dry?
16. How would you develop a lubrication schedule for a pump?
17. Why is cleanliness important in the storage and use of lubricants?

CHAPTER 18. MAINTENANCE

(Lesson 3 of 5 Lessons)

18.23 Pump Maintenance

18.230 Section Format

The format of this section differs from the other chapters. This format was designed specifically to assist you in planning an effective preventive maintenance program. The paragraphs are numbered for easy reference when you use the Equipment Service Cards and Service Record Cards mentioned in Section 18.00, page 267, and shown in Figure 18.1.

An entire book could be written on the topics covered in this section. Step-by-step details for maintaining equipment are not provided because manufacturers are continually improving their products and these details could soon be out of date. You are assumed to have some familiarity with the equipment being discussed. For details concerning a particular piece of equipment, you should contact the manufacturer. This section indicates to you the kinds of maintenance you should include in your program and how you could schedule your work. Carefully read the manufacturer's instructions and be sure you clearly understand the material before attempting to maintain and repair equipment. If you have any questions or need any help, do not hesitate to contact the manufacturer or your local representative.

A glossary is not provided in this section because of the large number of technical words that require familiarization with the equipment being discussed. The best way to learn the meaning of these new words is from manufacturers' literature or from their representatives. Some new words are described in the lessons, where necessary.

18.231 Preventive Maintenance

The following paragraphs list some general preventive maintenance services and indicate frequency of performance. There are many makes and types of equipment and the wide variation of functions cannot be included; therefore, you will have to use some judgment as to whether the services and frequencies will apply to your equipment. If something goes wrong or breaks in your plant, you may have to disregard your maintenance schedule and fix the problem now.

NOTE: If you need to shut a unit down, make sure it is also locked out and tagged properly. (Figure 18.23)

CODE
D MEANS DAILY; W, WEEKLY; M, MONTHLY;
Q, QUARTERLY; S, SEMIANNUALLY; A, ANNUALLY

Paragraph 1: Pumps, General

This paragraph lists some general preventive maintenance services and indicates frequency of performance. Typical centrifugal pump sections are shown in Figure 18.19 on page 311.

Frequency
of
Service

D   1. CHECK WATER-SEAL PACKING GLANDS FOR LEAKAGE. See that the packing box is protected with a clear water supply from an outside source, make sure that water seal pressure is at least 5 psi (35 kPa or 0.35 kg/sq cm) greater than maximum pump suction pressure. See that there are no *CROSS CONNECTIONS*.[13] Check packing glands for leakage during operation. Allow a slight seal leakage when pumps are running to keep packing cool and in good condition. The proper amount of leakage depends on equipment and operating conditions. Sixty drops of water per minute is a good rule of thumb. If excessive leakage is found, hand tighten gland nuts evenly, but not too tight. After adjusting packing glands, be sure shaft turns freely by hand. If serious leakage continues, renew packing, shaft, or shaft sleeve.

D   2. CHECK GREASE-SEALED PACKING GLANDS. When grease is used as a packing gland seal, maintain constant grease pressure on packing during operation. When a spring-loaded grease cup is used, keep it loaded with grease. Force grease through packing at a rate of about one ounce (30 gm) per day. When water is used, adjust seal pressure to 5 psi (35 kPa or 0.35 kg/sq cm) above maximum pump suction pressure. Never allow the seal to run dry.

---

[13] *Cross Connection.* (1) A connection between drinking (potable) water and an unapproved water supply. (2) A connection between a storm drain system and a sanitary collection system. (3) Less frequently used to mean a connection between two sections of a collection system to handle anticipated overloads of one system.

Fig. 18.23   Typical warning tag
(Source: Industrial Indemnity/Industrial Underwriters/Insurance Companies)

Frequency of Service	
W	3. OPERATE PUMPS ALTERNATELY. If two or more pumps of the same size are installed, alternate their use to equalize wear, keep motor windings dry, and distribute lubricant in bearings.
W	4. INSPECT PUMP CONTROL. Inspect the pump controls to see that the pump responds properly to changes in the controlling variable. This variable may be either a pressure or a water level. This check could be done physically or by analyzing recording gauge records.
D	5. CHECK MOTOR CONDITION. See Paragraph 6: Electric Motors.
W	6. CHECK PACKING GLAND ASSEMBLY. Check packing gland, the unit's most abused and troublesome part. If stuffing box leaks excessively when gland is pulled up with mild pressure, remove packing and examine shaft sleeve carefully. Replace or repair grooved or scored shaft sleeve because packing cannot be held in stuffing box with roughened shaft or shaft sleeve. Replace the packing a strip at a time, tamping each strip thoroughly and staggering joints. (See Figure 18.24.) Position lantern ring (water-seal ring) properly. If grease sealing is used, completely fill lantern ring with grease before putting remaining rings of packing in place. The type of packing used (Figure 18.25) is less important than the manner in which packing is placed. Never use a continuous strip of packing. This type of packing wraps around and scores the shaft sleeve or is thrown out against the outer wall of the stuffing box, allowing water to leak through and score the shaft. The proper size of packing should be available in your plant's equipment files. See Figure 18.26 for illustrated steps on how to pack a pump.

If a bronze shaft sleeve is not too badly scored, the shaft sleeve can be restored to service. The repair procedure consists of turning the sleeve down to a uniform diameter with a rough cut. Then spray the sleeve with stainless steel to a slightly oversized outside diameter followed by machining and polishing to bring the sleeve back to its original diameter. You will probably find that these reworked sleeves will outlast the originals.

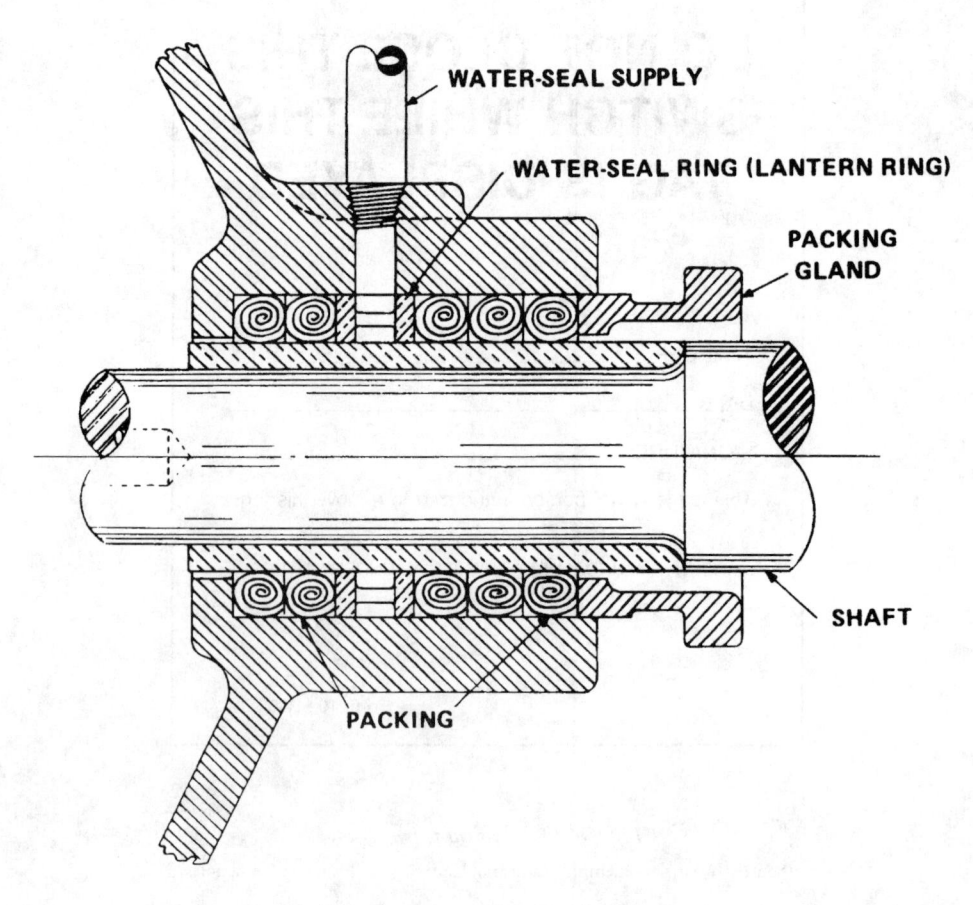

Fig. 18.24 Method of packing shaft
(Source: War Department Technical Manual TM5-666)

Maintenance 321

Teflon Packing

Graphite Packing

Fig. 18.25 Packing
(Courtesy A. W. Chesteron Co.)

Frequency of Service

W   7. CHECK MECHANICAL SEALS. Mechanical seals usually consist of two subassemblies: (1) a rotating ring assembly, and (2) a stationary assembly.

Inspect seal for leakage and excessive heat. If any part of the seal needs replacing, replace the entire seal (both subassemblies) with a new seal that has been provided by the manufacturer. Before installing a new seal, be sure that there are no chips or cracks on the carbide sealing surface. Keep a new mechanical seal clean at all times.

322 Water Treatment

HOW TO PACK A PUMP

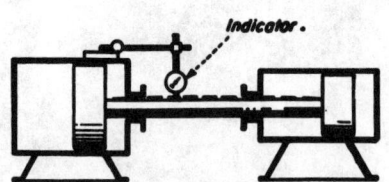

1 Remove *all* old packing. Aim packing hook at bore of the box to keep from scratching the shaft. Clean box thoroughly so the new packing won't hang up

2 Check for bent rod, grooves or shoulders. If the neck bushing clearance in bottom of box is great, use stiffer bottom ring or replace the neck bushing

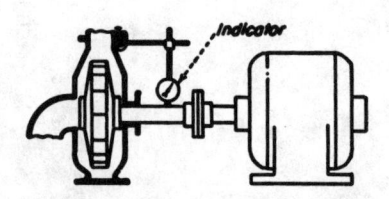

3 Revolve rotary shaft. If the indicator runs out over 0.003-in., straighten shaft, or check bearings, or balance rotor. Gyrating shaft beats out packing

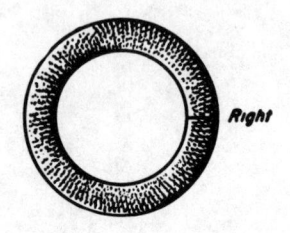

6 Cutting off rings while packing is wrapped around shaft will give you rings with parallel ends. This is very important if packing is to do job

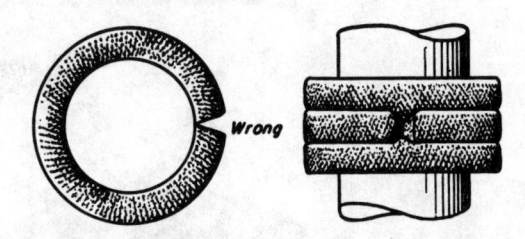

7 If you cut packing while stretched out straight, the ends will be at an angle. With gap at angle, packing on either side squeezes into top of gap and ring, cannot close. This brings up the question about gap for expansion. Most packings need none. Channel-type packing with lead core may need slight gap for expansion

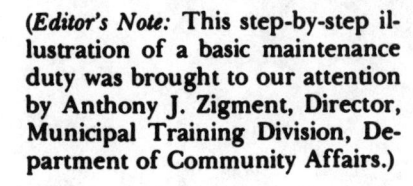

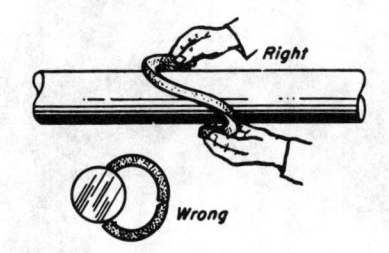

11 Open ring joint sidewise, especially lead-filled and metallic types. This prevents distorting molded circumference—breaking the ring opposite gap

12 Use split wooden bushing. Install first turn of packing, then force into bottom of box by tightening gland against bushing. Seat each turn this way

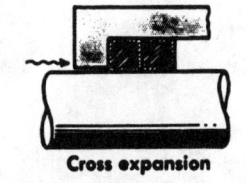

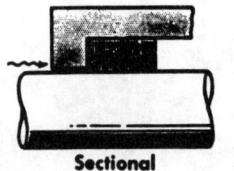

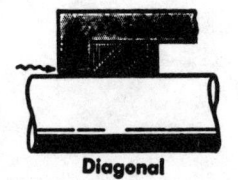

15 Always install cross-expansion packing so plies slope toward the fluid pressure from housing. Place sectional rings so slope between inside and outside ring is toward the pressure. Diagonal rings must also have slope toward the fluid pressure. Watch these details for best results when installing new packing in a box

(*Editor's Note:* This step-by-step illustration of a basic maintenance duty was brought to our attention by Anthony J. Zigment, Director, Municipal Training Division, Department of Community Affairs.)

Fig. 18.26   How to pack a pump

(Source: Water Pollution Control Association of Pennsylvania Magazine, January-February, 1976)

Maintenance 323

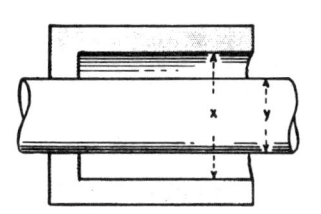

4 To find the right size of packing to install, measure stuffing-box bore and subtract rod diameter, divide by 2. Packing is too critical for guesswork.

5 Wind packing, needed for filling stuffing box, snugly around rod (for same size shaft held in vise) and cut through each turn while coiled, as shown. If the packing is slightly too large, never flatten with a hammer. Place each turn on a clean newspaper and then roll out with pipe as you would with a rolling pin

8 Install foil-wrapped packing so edges on inside will face direction of shaft rotation. This is a must; otherwise, thin edges flake off, reduce packing life

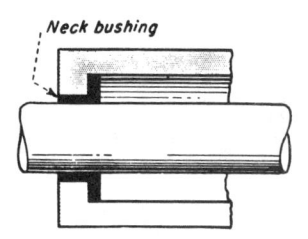

9 Neck bushing slides into stuffing box. Quick way to make it is to pour soft bearing metal into tin can, turn and bore for sliding fit into place

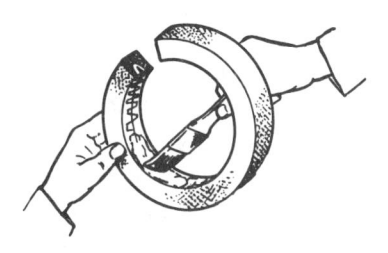

10 Swabbing new metallic packings with lubricant supplied by packing maker is OK. These include foil types, lead-core, etc. If the rod is oily, don't swab it

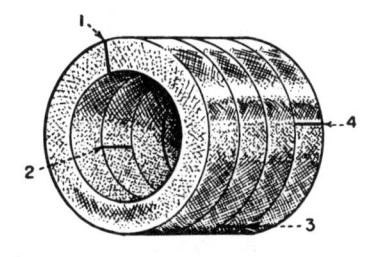

13 Stagger joints 180 degrees if only two rings are in stuffing box. Space at 120 degrees for three rings, or 90 degrees if four rings or more are in set

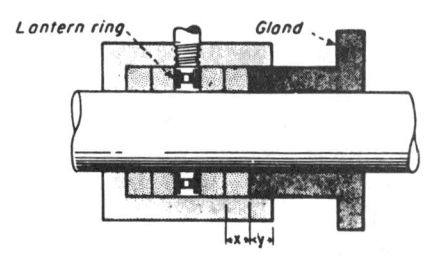

14 Install packing so lantern ring lines up with cooling-liquid opening. Also, remember that this ring moves back into box as packing is compressed. Leave space for gland to enter as shown. Tighten gland with wrench—back off finger-tight. Allow the packing to leak until it seats itself, then allow a slight operating leakage.

Hydraulic-packing pointers

First, clean stuffing box, examine ram or rod. Next, measure stuffing-box depth and packing set—find difference. Place 1/8-in. washers over gland studs as shown. Lubricate ram and packing set (if for water). If you can use them, endless rings give about 17% more wear than cut rings. Place male adapter in bottom, then carefully slide each packing turn home—don't harm lips. Stagger joints for cut rings. Measure from top of packing to top of washers, then compare with gland. Never tighten down new packing set until all air has chance to work out. As packing wears, remove one set of washers, after more wear, remove other washer.

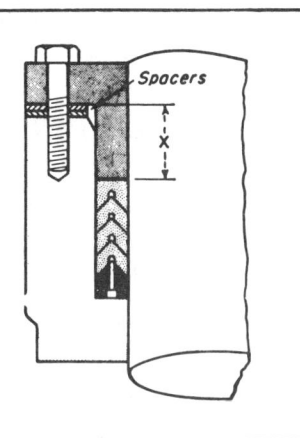

Fig. 18.26 How to pack a pump (continued)

Frequency of Service		

Always be sure that a mechanical seal is surrounded with water before starting and running the pump.

Q 8. INSPECT AND LUBRICATE BEARINGS. Unless otherwise specifically directed for a particular pump model, lubricate according to the procedures covered in Section 18.224, page 315. Check sleeve bearings to see that oil rings turn freely with the shaft. Repair or replace if defective.

Measure sleeve bearings and replace those worn excessively. Generally, allow clearance of 0.002 inch plus 0.001 inch for each inch or fraction of inch of shaft-journal diameter.

Q 9. CHECK OPERATING TEMPERATURE OF BEARINGS. Check bearing temperature with thermometer, not by hand. If antifriction bearings are running hot, check for overlubrication and relieve if necessary. If sleeve bearings run too hot, check for lack of lubricant. If proper lubrication does not correct condition, disassemble and inspect bearings. Check alignment of pump and motor if high temperatures continue.

S 10. CHECK ALIGNMENT OF PUMP AND MOTOR. For method of aligning pump and motor, see Paragraph 10: Couplings. If misalignment recurs frequently, inspect entire piping system. Unbolt piping at suction and discharge nozzles to see if it springs away, indicating strain on casing. Check all piping supports for soundness and effective support of load.

Vertical pumps usually have flexible shafting, which permits slight angular misalignment; however, if solid shafting is used, align exactly. If beams carrying intermediate bearings are too light or are subject to contraction or expansion, replace beams and realign intermediate bearings carefully.

S 11. INSPECT AND SERVICE PUMPS.

a. Remove rotating element of pump and inspect thoroughly for wear. Order replacement parts where necessary. Check impeller clearance between volute.

b. Remove any deposit or scaling. Clean out water-seal piping.

c. Determine pump capacity by pumping into empty tank of known size or by timing the draining of pit or sump.

$$\text{Pump Capacity, GPM} = \frac{\text{Volume, gallons}}{\text{Time, minutes}}$$

or

$$\text{Pump Capacity, } \frac{\text{liters}}{\text{sec}} = \frac{\text{Volume, liters}}{\text{Time, seconds}}$$

See Example 1 (page 325) for procedures on how to calculate pump capacity.

d. Test pump efficiency. Refer to pump manufacturer's instructions on how to collect data and perform calculations. Also see Chapter 3, Section 3.17, "Pump Testing and Evaluation," in SMALL WATER SYSTEM OPERATION AND MAINTENANCE in this series of manuals.

e. Measure total dynamic suction head or lift and discharge head to test pump and pipe condition. Record figures for comparison with later tests.

f. Inspect foot and check valves, paying particular attention to check valves, which can cause water hammer when pump stops. (See Paragraph 13: Check Valves also.) Foot valves, which are a type of check valve, are used when pumping raw water.

g. Examine wearing rings. Replace seriously worn wearing rings to improve efficiency. Check wearing ring clearances, which generally should be no more than 0.003 inch per inch of wearing ring diameter.

CAUTION: To protect rings and casings, never allow pump to run dry through lack of proper priming when starting or loss of suction when operating.

A 12. DRAIN PUMP FOR LONG-TERM SHUTDOWN. When shutting down pump for a long period, open motor disconnect switch to disconnect the motor, and, if so equipped, turn on the electric motor winding heaters. Shut all valves on suction, discharge, water-seal, and priming lines; drain pump completely by removing vent and drain plugs. This procedure protects pump against corrosion, sedimentation, and freezing. Inspect pump and bearings thoroughly and perform all necessary servicing. Drain bearing housings and replenish with fresh oil, purge old grease and replace. When a pump is out of service, run it monthly to warm it up and to distribute lubrication so the packing will not "freeze" to the shaft. Resume periodic checks after pump is put back in service.

FORMULAS

To find the volume of a rectangle in cubic feet, multiply the length times width times depth.

Volume, cu ft = (Length, ft)(Width, ft)(Depth, ft)

To find the volume of a cylinder in cubic feet, multiply 0.785 times the diameter squared times the depth.

Volume, cu ft = (0.785)(Diameter, ft)^2(Depth, ft)

To convert a volume from cubic feet to gallons, multiply the volume in cubic feet times 7.48 gallons per cubic foot.

Volume, gal = (Volume, cu ft)(7.48 gal/cu ft)

To calculate the output or capacity of a pump in gallons per minute, divide the volume pumped in gallons by the pumping time in minutes.

$$\text{Pump Capacity, GPM} = \frac{\text{Volume Pumped, gallons}}{\text{Pumping Time, minutes}}$$

EXAMPLE 1

A pump's capacity is measured by recording the time in minutes for water to rise 3 feet in an 8-foot diameter tank. What is the pumping rate or capacity in gallons per minute when the pumping time is 9 minutes?

Known	Unknown
Diameter, ft = 8 ft	Pump Capacity, GPM
Depth, ft = 3 ft	
Time, min = 9 min	

Calculate the tank volume in cubic feet.

Volume, cu ft = (0.785)(Diameter, ft)^2(Depth, ft)

= (0.785)(8 ft)^2(3 ft)

= 151 cu ft

Convert the tank volume from cubic feet to gallons.

Volume, gal = (Volume, cu ft)(7.48 gal/cu ft)

= (151 cu ft)(7.48 gal/cu ft)

= 1,129 gallons

Calculate the pump capacity in gallons per minute.

$$\text{Pump Capacity, GPM} = \frac{\text{Volume Pumped, gal}}{\text{Pumping Time, min}}$$

$$= \frac{1,129 \text{ gal}}{9 \text{ min}}$$

$$= 125 \text{ GPM}$$

QUESTIONS

Write your answers in a notebook and then compare your answers with those on page 381.

18.23A What is a cross connection?

18.23B Is a slight water-seal leakage desirable when a pump is running? If so, why?

18.23C How would you measure the capacity of a pump?

18.23D Estimate the capacity of a pump (in GPM) if it lowers the water in a 10-foot wide by 15-foot long wet well 1.7 feet in five minutes.

18.23E What should be done to a pump before it is shut down for a long time, and why?

Paragraph 2: Reciprocating Pumps, General

The general procedures in this paragraph apply to all reciprocating pumps described in this section.

Frequency of Service

W 1. CHECK SHEAR PIN ADJUSTMENT. Set eccentric by placing shear pin through proper hole in eccentric flanges to give required stroke. Tighten the two ⅝- or ⅞-inch hexagonal nuts on connecting rods just enough to take spring out of lock washers. (See Paragraph 11: Shear Pins.) When a shear pin fails, eccentric moves toward neutral position, preventing damage to the pump. Remove cause of obstruction and insert new shear pin. Shear pins fail because of one of three common causes:

   a. Solid object lodged under piston

   b. Clogged discharge line

   c. Stuck or wedged valve

D 2. CHECK PACKING ADJUSTMENT. Give special attention to packing adjustment. If packing is too tight, it reduces efficiency and scores piston walls. Keep packing just tight enough to keep sludge from leaking through gland. Before pump is installed or after it has been idle for a time, loosen all nuts on packing gland. Run pump with sludge suction line closed and valve covers open for a few minutes to break in the packing. Turn down gland nuts no more than necessary to prevent sludge from getting past packing. Tighten all packing nuts uniformly.

   When packing gland bolts cannot be taken up farther, remove packing. Remove old packing

Frequency
of
Service

and thoroughly clean cylinder and piston walls. Place new packing into cylinder, staggering packing-ring joints, and tamp each ring into place. Break in and adjust packing as explained above. When chevron type packing is used, tighten gland nuts only finger tight because excessive pressure ruins packing and scores plunger.

Q 3. CHECK BALL VALVES. When valve balls are so worn that diameter is ⅝-inch (1.5-cm) smaller than original size, they may jam into guides in valve chamber. Check size of valve balls and replace if badly worn.

Q 4. CHECK VALVE-CHAMBER GASKETS. Valve-chamber gaskets on most pumps serve as a safety device and blow out under excessive pressure. Check gaskets and replace if necessary. Keep additional gaskets on hand for replacement.

A 5. CHECK ECCENTRIC ADJUSTMENT. To take up babbitt bearing, remove brass shims provided on connecting rod. After removing shims, operate pump for at least one hour and check to see that eccentric does not run hot.

D 6. NOTE UNUSUAL NOISES. Check for noticeable water hammer when pump is operating. This noise is most pronounced when pumping water or very thin sludge; it decreases or disappears when pumping heavy sludge. Eliminate noise by opening the ¼-inch (0.6-cm) petcock on pump body slightly; this draws in a small amount of air, keeping discharge air chamber full at all times.

D 7. CHECK CONTROL VALVE POSITIONS. Because any plunger pump may be damaged if operated against closed valves in the pipeline, especially the discharge line, make all valve setting changes with pump shut down; otherwise, pumps that are installed to pump from two sources or to deliver to separate tanks at different times may be broken if all discharge line valves are closed simultaneously for a few seconds or discharge valve directly above pump is closed.

W 8. GEAR REDUCER. Check oil level by removing plug on the side of the gear case. Unit should not be in operation.

Q 9. CHANGE OIL AND CLEAN MAGNETIC DRAIN PLUG.

W 10. CONNECTING RODS. Set oilers to disperse two drops per minute.

W 11. PLUNGER CROSSHEAD. Fill plunger as required to half cover the wrist pin with oil.

D 12. PLUNGER TROUGH. Keep small quantity of oil in trough to lubricate the plunger.

M 13. MAIN SHAFT BEARING. Grease bearings monthly. Pump should be in operation when lubricating to avoid excessive pressure on seals.

14. CHECK ELECTRIC MOTOR. See Paragraph 6: Electric Motors.

Paragraph 3: Propeller Pumps, General

D 1. CHECK MOTOR CONDITION. See Paragraphs 6.1 and 6.2.

D 2. CHECK PACKING GLAND ASSEMBLY. See Paragraph 1.6.

W 3. INSPECT PUMP ASSEMBLY. See Paragraph 1.4.

W 4. LUBE LINE SHAFT AND DISCHARGE BOWL BEARING. Maintain oil in oiler at all times. Adjust feed rate to approximately four drops per minute.

W 5. LUBE SUCTION BOWL BEARING. Lube through pressure fitting. Usually three or four strokes of gun are enough.

W 6. OPERATE PUMPS ALTERNATELY. See Paragraph 1.3.

A 7. LUBE MOTOR BEARINGS. See Paragraph 6.3.

Paragraph 4: Progressive Cavity Pumps, General (Figure 18.21, page 313).

D 1. CHECK MOTOR CONDITION. See Paragraphs 6.1 and 6.2.

D 2. CHECK PACKING GLAND ASSEMBLY. See Paragraph 1.6.

D 3. CHECK DISCHARGE PRESSURE. A higher than normal discharge pressure may indicate a line blockage or a closed valve downstream. An abnormally low discharge pressure can mean reduced rate of discharge.

S 4. INSPECT AND LUBRICATE BEARINGS—GREASE. If possible, remove bearing cover and visually inspect grease. When greasing, remove relief plug and cautiously add 5 or 6 strokes of the grease gun. Afterward, check bearing temperature with thermometer. If over 220°F (104°C), remove some grease.

Frequency of Service	
S	5. LUBEFLUSH MOTOR BEARINGS. See Paragraph 6.3.
S	6. CHECK PUMP OUTPUT. Check how long it takes to fill a vessel of known volume or quantity; or check performance against a meter, if available. See Paragraph 1.11 c.
A	7. SCOPE MOTOR BEARINGS. See Paragraph 6.4.
A	8. SCOPE PUMP BEARINGS. See Paragraph 6.4.

Paragraph 5: Pump Controls

To ensure the best operation of the pump, a systematic inspection of the controls should be made at least once a week.

W	1. CHECK CONTROLS. Controls respond to the control variable.
W	2. START-UP. The unit starts when the control system makes contact, and the pump stops at the prescribed control setting.
W	3. MOTOR SPEED. The motor comes up to speed quickly and is maintained.
W	4. SPARKING. A brush-type motor does not spark profusely in starting or running.
W	5. INTERFERENCE WITH CONTROLS. Grease and dirt are not interfering with controls.
W	6. ADJUSTMENTS. Any necessary adjustments are properly completed.

QUESTIONS

Write your answers in a notebook and then compare your answers with those on page 381.

18.23F What are some of the common causes of shear pin failure in reciprocating pumps?

18.23G What may happen when water or a thin sludge is being pumped by a reciprocating pump?

18.23H What could be the causes of a higher than normal discharge pressure in a progressive cavity pump?

Paragraph 6: Electric Motors (Figure 18.27)

In order to ensure the proper and continuous function of electric motors, the items listed in this paragraph must be performed at the designated intervals. If operational checks indicate a motor is not functioning properly, these items will have to be checked to locate the problem.

D	1. CHECK MOTOR CONDITIONS.

  a. Keep motors free from dirt, dust, and moisture.

  b. Keep operating space free from articles that may obstruct air circulation.

  c. Check for excessive grease leakage from bearings.

D	2. NOTE ALL UNUSUAL CONDITIONS.

  a. Unusual noises in operation.

  b. Motor failing to start or come to speed normally, sluggish operation.

  c. Motor or bearings that feel or smell hot.

  d. Continuous or excessive sparking at commutator or brushes. Blackened commutator.

  e. Intermittent sparking at brushes.

  f. Fine dust under coupling having rubber buffers or pins.

  g. Smoke, charred insulation, or solder whiskers extending from armature.

  h. Excessive humming.

  I. Regular clicking.

  j. Rapid knocking.

  k. Brush chatter.

  l. Vibration.

  m. Hot commutator.

A	3. LUBRICATE BEARINGS (Figure 18.28). Check grease in ball bearing and relubricate when necessary.

Follow instructions in Section 18.224, "Pump Lubrication," when lubricating greased bearings.

A	4. USING A *STETHOSCOPE*,[14] CHECK BOTH BEARINGS.

  Listen for whines, gratings, or uneven noises. Listen all around the bearing and as near as possible to the bearing. Listen while the motor is being started and shut off. If unusual noises are heard, pinpoint the location.

  5. IF YOU THINK THE MOTOR is running unusually hot, check with a thermometer.

---

[14] *Stethoscope.* An instrument used to magnify sounds and carry them to the ear.

DRIP PROOF

ITEM NO.	PART NAME
1	Wound Stator w/ Frame
2	Rotor Assembly
3	Rotor Core
4	Shaft
5	Bracket
6	Bearing Cap
7	Bearings
8	Seal, Labyrinth
9	Thru Bolts/Caps
10	Seal, Lead Wire
11	Terminal Box
12	Terminal Box Cover
13	Fan
14	Deflector
15	Lifting Lug

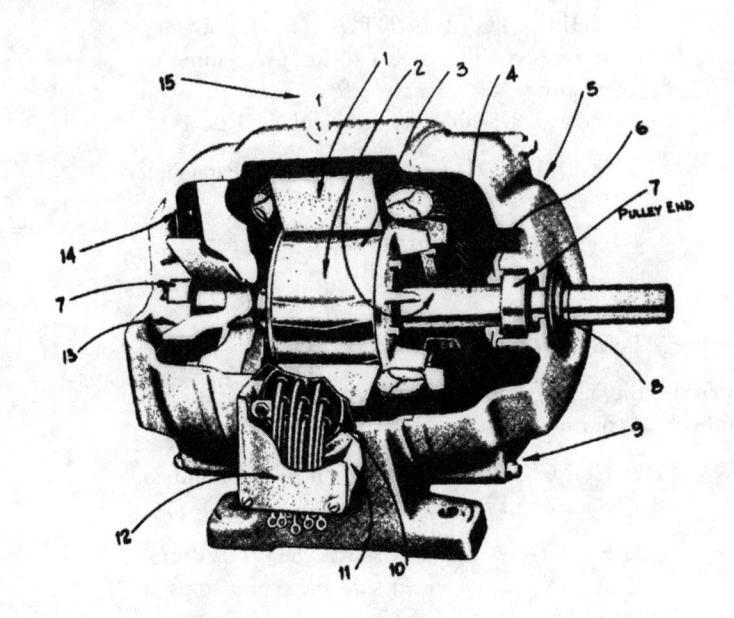

TOTALLY ENCLOSED FAN COOLED

ITEM NO.	PART NAME
1	Wound Stator w/ Frame
2	Rotor Assembly
3	Rotor Core
4	Shaft
5	Brackets
6	Bearings
7	Seal, Labyrinth
8	Thru Bolts/Caps
9	Seal, Lead Wire
10	Terminal Box
11	Terminal Box Cover
12	Fan, Inside
13	Fan, Outside
14	Fan Grill
15	Fan Cover
16	Fan Cover Bolts
17	Lifting Lug

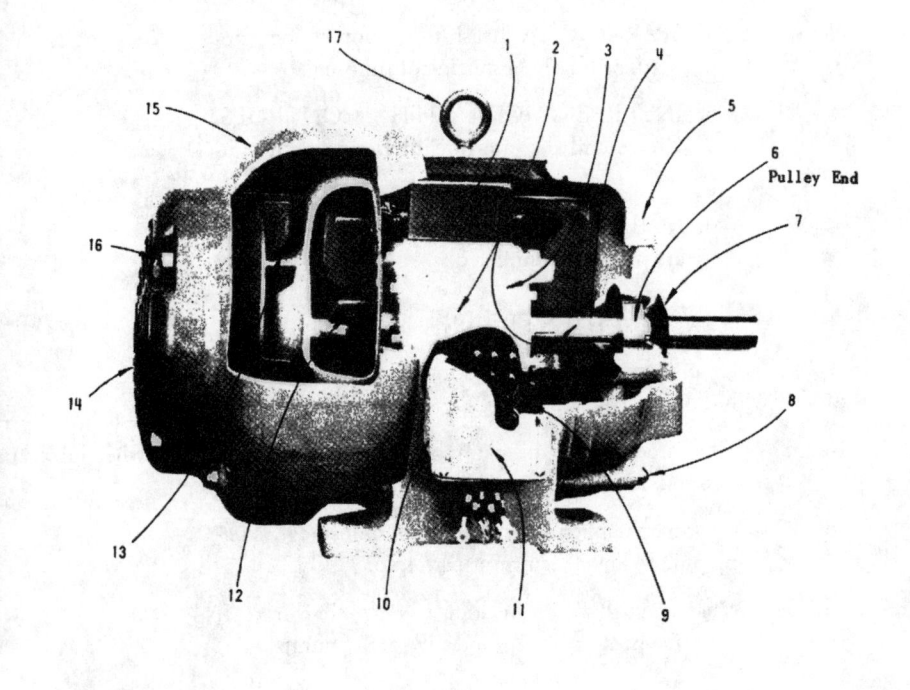

Fig. 18.27 Typical motors
(Courtesy of Sterling Power Systems, Inc.)

Maintenance 329

ELECTRIC MOTOR

MOTOR LUBRICATION

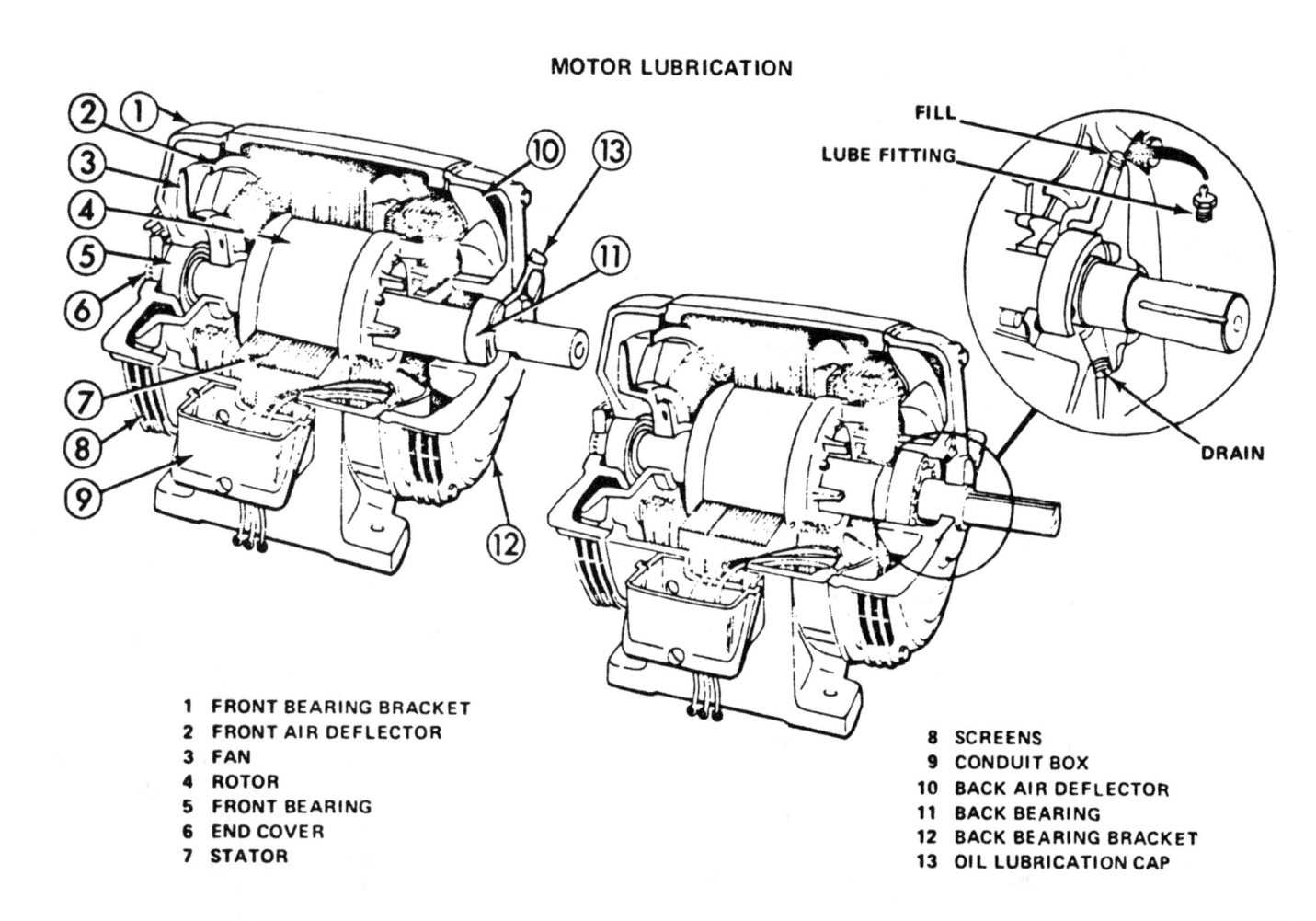

1 FRONT BEARING BRACKET
2 FRONT AIR DEFLECTOR
3 FAN
4 ROTOR
5 FRONT BEARING
6 END COVER
7 STATOR
8 SCREENS
9 CONDUIT BOX
10 BACK AIR DEFLECTOR
11 BACK BEARING
12 BACK BEARING BRACKET
13 OIL LUBRICATION CAP

Fig. 18.28 Electric motor lubrication

Frequency of Service

Place the thermometer on the casing near the bearing, holding it there with putty or clay. Check the current on each leg to determine if the currents are balanced and within the motor nameplate limits.

A   6. *DATEOMETER.*[15] If there is a dateometer on the motor, after changing the oil in the motor, loosen the dateometer screw and set to the corresponding year.

QUESTIONS

Write your answers in a notebook and then compare your answers with those on page 381.

18.23I   What are the major items you would include when checking an electric motor?

18.23J   What is the purpose of a stethoscope?

Paragraph 7: Belt Drives

1. GENERAL. Maintaining a proper tension and alignment of belt drives ensures long life of belts and sheaves. Incorrect alignment causes poor operation and excessive belt wear. Inadequate tension reduces the belt grip and causes high belt loads, snapping, and unusual wear.

    a. Cleaning belts. Keep belts and sheaves clean and free of oil, which causes belts to deteriorate. To remove oil, take belts off sheaves and wipe belts and sheaves with a rag moistened in a nonoil-based solvent. Carbon tetrachloride is *NOT* recommended because exposure to its fumes has many toxic effects on humans. Carbon tetrachloride also is absorbed into the skin on contact and its effects become stronger with each contact.

    b. Installing belts. Before installing belts, replace worn or damaged sheaves, then slack off on adjustments. Do not try to force belts into position. Never use a screwdriver or similar lever to get belts onto sheaves. After belts are installed, adjust tension; recheck tension after eight hours of operation. (See Table 18.3.)

    c. Replacing belts. Replace belts as soon as they become frayed, worn, or cracked.

Never replace only one V-belt on a multiple drive. Replace the complete set with a set of matched belts, which can be obtained from any supplier. All belts in a matched set are machine checked to ensure equal size and tension.

    d. Storing spare belts. Store spare belts in a cool, dark place. Tag all belts in storage to identify them with the equipment on which they can be used.

2. V-BELTS. A properly adjusted V-belt has a slight bow in the slack side when running; when idle, it has an alive springiness when thumped with the hand. An improperly tightened belt feels dead when thumped.

If the slack side of the drive is less than 45° from the horizontal, vertical sag at the center of the span may be adjusted in accordance with Table 18.3 below:

TABLE 18.3 HORIZONTAL BELT TENSION

Span (inches)		10	20	50	100	150	200
Vertical Sag (inches)	From	.01	.03	.20	.80	1.80	3.30
	To	.03	.09	.58	2.30	4.90	8.60
Span (millimeters)		250	500	1,250	2,500	3,750	5,000
Vertical Sag (millimeters)	From	0.25	0.75	5.00	20.0	45.0	82.5
	To	0.75	2.25	14.50	57.5	122.5	215.0

M   a. Check tension. If tightening belt to proper tension does not correct slipping, check for overload, oil on belts, or other possible causes. Never use belt dressing to stop belt slippage. Rubber wearings near the drive are a sign of improper tension, incorrect alignment, or damaged sheaves.

M   b. Check sheave (pulley) alignment. Lay a long straightedge or string across outside faces of pulley, and allow for differences in dimensions from centerlines of grooves to outside faces of the pulleys being aligned. Be very careful in aligning drives with more than one V-belt on a sheave, as misalignment can cause unequal tension.

---

[15] *Dateometer* (day-TOM-uh-ter). A small calendar disk attached to motors and equipment to indicate the year in which the last maintenance service was performed.

Paragraph 8: Chain Drives

Frequency
of
Service

1. GENERAL. Chain drives may be designated for slow, medium, or high speeds.

    a. Slow-speed drives. Because slow-speed drives are usually enclosed, adequate lubrication is difficult. Heavy oil applied to the outside of the chain seldom reaches the working parts; in addition, the oil catches dirt and grit and becomes abrasive. For lubricating and cleaning methods, see 5 and 6 below.

    b. Medium- and high-speed drives. Medium-speed drives should be continuously lubricated with a device similar to a sight-feed oiler. High-speed drives should be completely enclosed in an oil-tight case and the oil maintained at proper level.

D   2. CHECK OPERATION. Check general operating condition during regular tours of duty.

Q   3. CHECK CHAIN SLACK. The correct amount of slack is essential to proper operation of chain drives. Unlike other belts, chain belts should not be tight around the sprocket; when chains are tight, working parts carry a much heavier load than necessary. Too much slack is also harmful; on long centers particularly, too much slack causes vibrations and chain whip, reducing life of both chain and sprocket. A properly installed chain has a slight sag or looseness on the return run.

S   4. CHECK ALIGNMENT. If sprockets are not in line or if shafts are not parallel, excessive sprocket and chain wear and early chain failure result. Wear on inside of chain, side walls, and sides of sprocket teeth are signs of misalignment. To check alignment, remove chain and place a straightedge against sides of sprocket teeth.

S   5. CLEAN. On enclosed types, flush chain and enclosure with a petroleum solvent (kerosene). On exposed types, remove chain and soak and wash it in solvent. Clean sprockets, install chain, and adjust tension.

S   6. CHECK LUBRICATION. Soak exposed-type chains in oil to restore lubricating film. Remove excess lubricant by hanging chains up to drain.

Do not lubricate underwater chains that operate in contact with considerable grit. If water is clean, lubricate by applying waterproof grease with brush while chain is running.

Do not lubricate chains on elevators or on conveyors of feeders that handle dirty or gritty materials. Dust and grit combine with lubricants to form a cutting compound that reduces chain life.

S   7. CHANGE OIL. On enclosed types only, drain oil and refill case to proper level.

S   8. INSPECT. Note and correct abnormal conditions before serious damage results. Do not put a new chain on worn sprockets. Always replace worn sprockets when replacing a chain because out-of-pitch sprockets cause as much chain wear in a few hours as years of normal operation.

9. TROUBLESHOOTING. Some common symptoms of improper chain drive operation and their remedies follow:

    a. Excessive noise. Correct alignment, if misaligned. Adjust centers for proper chain slack. Lubricate in accordance with aforementioned methods. Be sure all bolts are tight. If chain or sprockets are worn, reverse or renew if necessary.

    b. Wear on chain, side walls, and sides of teeth. Remove chain and correct alignment.

    c. Chain climbs sprockets. Check for poorly fitting sprockets and replace if necessary. Make sure tightener is installed on drive chain.

    d. Broken pins and rollers. Check for chain speed that may be too high for the pitch, and substitute chain and sprockets with shorter pitch, if necessary. Breakage also may be caused by shock loads.

    e. Chain clings to sprockets. Check for incorrect or worn sprockets or heavy, tacky lubricants. Replace sprockets or lubricants if necessary.

    f. Chain whip. Check for too-long centers or high, pulsating loads and correct cause.

    g. Chains get stiff. Check for misalignment, improper lubrication, or excessive overloads. Make necessary corrections or adjustments.

Paragraph 9: Variable Speed Belt Drives (Figure 18.29)

Frequency
of
Service

D	1.	CLEAN DISKS. Remove grease, acid, and water from disk faces.
D	2.	CHECK SPEED-CHANGE MECHANISM. Shift drive through entire speed range to make sure shafts and bearings are lubricated and disks move freely in lateral direction on shafts.
W	3.	CHECK V-BELT. Make sure it runs level and true. If one side rides high, a disk is sticking on shaft because of insufficient lubrication or wrong lubricant. In this case, stop the drive at once, remove V-belt, and clean disk hub and shaft thoroughly with petroleum solvent until disk moves freely. Relubricate with soft ball-bearing grease and replace V-belt in opposite direction from that in which it formerly ran.
M		If drive is not operated for 30 days or more, shift unit to minimum speed position, placing spring on variable-speed shaft at minimum tension and relieving belt of excessive pressure.
	4.	LUBRICATE DRIVE. Make sure to apply lubricant at all the six force-feed lubrication fittings (Figure 18.29: A, B, D, E, G, and H) and the one cup-type fitting (C). *NOTE: If the drive is used with a reducer, fitting E is not provided.*
W	a.	Once every ten days to two weeks, use two or three strokes of a grease gun through fittings A and B at ends of shifting screw and variable-speed shaft, respectively, to lubricate bearings of movable disks. Then, with unit running, shift drive from one extreme speed position to the other to ensure thorough distribution of lubricant over disk-hub bearings.
Q	b.	Add two or three shots of grease through fittings D and E to lubricate frame bearing on variable-speed shaft.
Q	c.	Every 90 days, add two or three cupfuls of grease to Cup C, which lubricates thrust bearing on constant-speed shaft.
Q	d.	Every 90 days, use two or three strokes of grease gun through fittings G and H to lubricate motor frame bearings.

CAUTION: Be sure to follow manufacturer's recommendation on type of grease. After lubricating, wipe excessive grease from sheaves and belt.

QUESTIONS

Write your answers in a notebook and then compare your answers with those on page 382.

18.23K How can you tell if a V-belt on belt-drive equipment has proper tension and alignment?

18.23L Why should worn sprockets be replaced when replacing a chain in a chain drive unit?

Paragraph 10: Couplings

1. GENERAL. Unless couplings between the driving and driven elements of a pump or any other piece of equipment are kept in proper alignment, breaking and excessive wear results in either or both the driven machinery and the driver. Burned-out bearings, sprung or broken shaft, and excessively worn or ruined gears are some of the damages caused by misalignment. To prevent outages and the expense of installing replacement parts, check the alignment of all equipment before damage occurs.

   a. Improper original installation of the equipment may not necessarily be the cause of the trouble. Settling of foundations, heavy floor loadings, warping of bases, excessive bearing wear, and many other factors cause misalignment. A rigid base is not always security against misalignment. The base may have been mounted off level, which could cause it to warp.

   b. Flexible couplings permit easy assembly of equipment, but they must be aligned as exactly as flanged couplings if maintenance and repair are to be kept to a minimum. Rubber-bushed types cannot function properly if the bolts cannot move in their bushings.

S 2. CHECK COUPLING ALIGNMENT (straightedge method). Excessive bearing and motor temperatures caused by overload, noticeable vibration, or unusual noises may all be warnings of misalignment. Realign when necessary (Figure 18.30) using a straightedge and thickness gauge or wedge. To ensure satisfactory operation, level up to within 0.005 inch (0.13 mm) as follows:

   a. Remove coupling pins.

   b. Rigidly tighten driven equipment; slightly tighten bolts holding drive.

   c. To correct horizontal and vertical misalignment, shift or shim drive to bring coupling halves into position so no light can be seen under a straightedge laid across them. Place

Maintenance 333

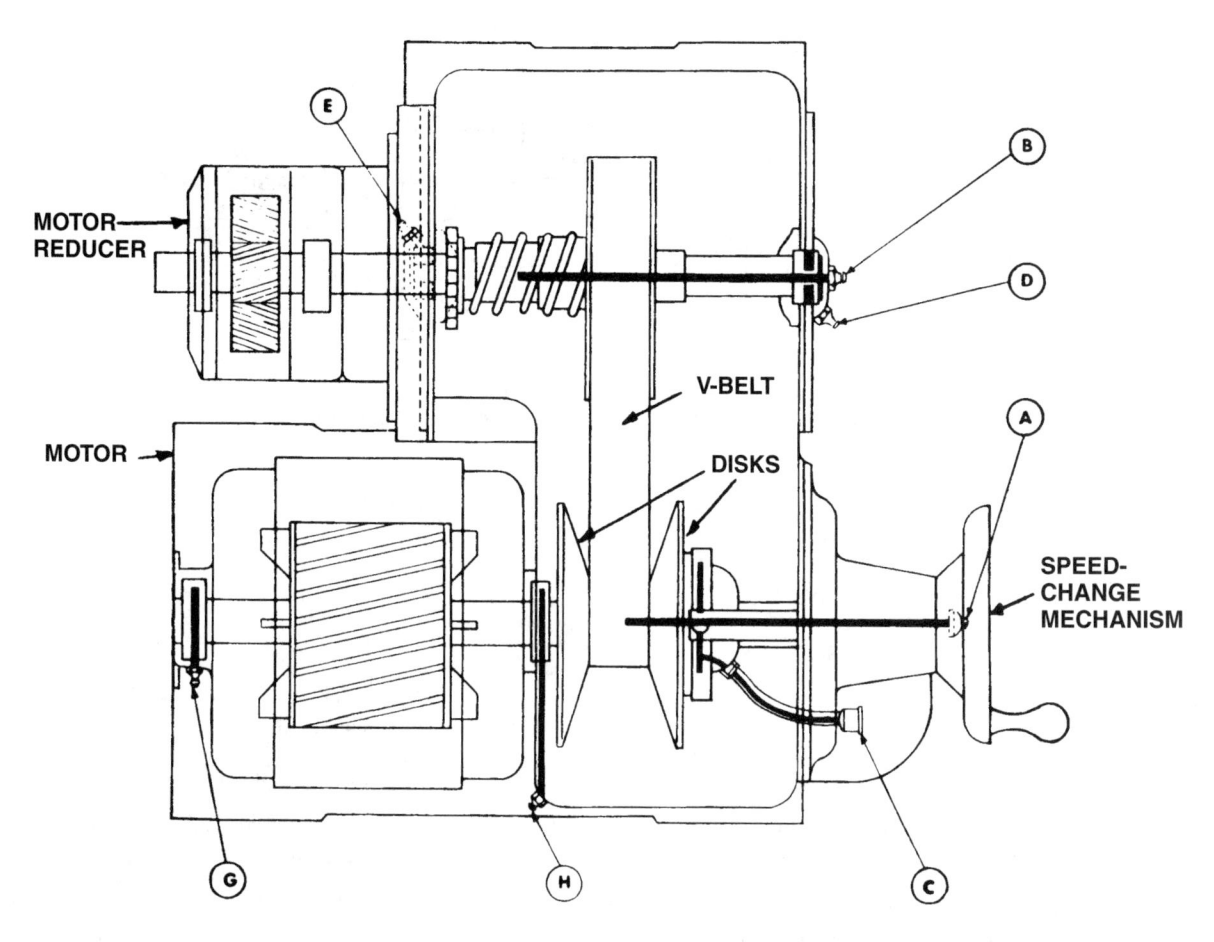

NOTE: A, B, D, E, G, and H are force-feed lubrication fittings. C is a cup-type lubrication fitting.

Fig. 18.29 Reeves varidrive
(Source: War Department Technical Manual TM5-666)

Frequency of Service

straightedge in four positions, holding a light in back of straightedge to help ensure accuracy.

d. Check for angular misalignment with a thickness or feeler gauge inserted at four places to make certain space between coupling halves is equal.

e. If proper alignment has been secured, coupling pins can be put in place easily using only finger pressure. Never hammer pins into place.

f. If equipment is still out of alignment, repeat the procedure.

S   3. CHECK COUPLING ALIGNMENT (dial indicator method). Dial indicators also are used to measure coupling alignment. This method produces better results than the straightedge method. The dial indicates very small movements or distances, which are measured in mils (one mil equals 1/1,000 of an inch). The indicator consists of a dial with a graduated face (with "plus" and "minus" readings), a pedestal, and a rigid indicator bar (or fixture) as shown in Figure 18.31.

The dial indicator is attached to one coupling via the fixture and adjusted to the zero position or reading. When the shaft of the machine is rotated, misalignment will cause the pedestal to compress (a "plus" reading), or extend (a "minus" reading). Literature provided by the manufacturer of machinery usually will

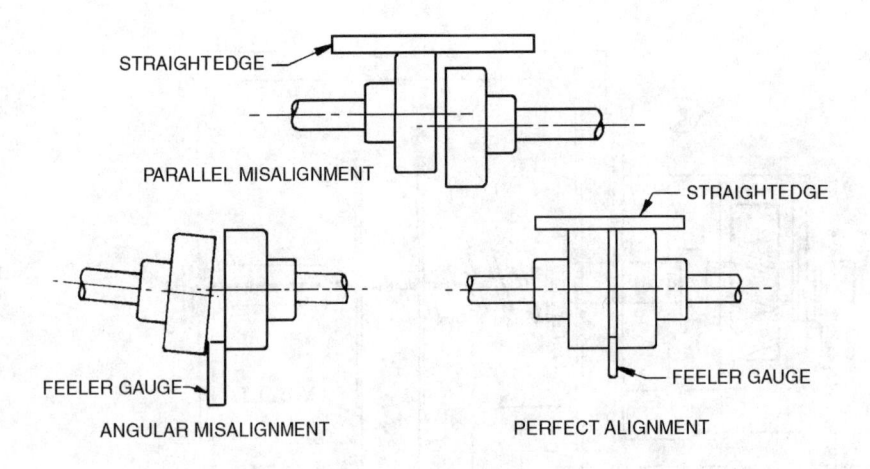

Fig. 18.30 Testing alignment, straightedge

indicate maximum allowable tolerances or movement.

Carefully study the manufacturer's literature provided with your dial indicator before attempting to use the device.

A 4. CHANGE OIL IN FAST COUPLINGS. Drain out old oil and add oil to proper level. Correct quantity is given on instruction card supplied with each coupling.

Paragraph 11: Shear Pins

Some water treatment units use shear pins as protective devices to prevent damage in case of sudden overloads. A shear pin is a straight pin that will fail (break) when a certain load or stress is exceeded. The purpose of the pin is to protect equipment from damage due to excessive loads or stresses. To serve this purpose, these devices must be in operational condition at all times. Under some operating conditions, shearing surfaces of a shear pin device may freeze together so solidly that an overload fails to break them.

Manufacturers' drawings for particular installations usually specify shear pin material and size. If this information is not available, obtain the information from the manufacturer, giving the model, serial number, and load conditions of unit. When necessary to determine shear pin size, select the lowest strength that does not break under the unit's usual loads. When proper size is determined, never use a pin of greater strength, such as a bolt or a nail.

If necked pins are used, be sure the necked-down portion is properly positioned with respect to shearing surfaces. When a shear pin breaks, determine and remedy the cause of failure before inserting new pin and starting drive in operation.

M 1. GREASE SHEARING SURFACES.

Q 2. REMOVE SHEAR PIN. Operate motor for a short time to smooth out any corroded spots.

A 3. CHECK SPARE INVENTORY. Make sure an adequate supply is on hand, properly identified and with record of proper pin size, necked diameter, and longitudinal dimensions.

QUESTIONS

Write your answers in a notebook and then compare your answers with those on page 382.

18.23M What factors could cause couplings to become out of alignment?

18.23N What is the purpose of shear pins?

18.24 Pump Operation

18.240 Starting a New Pump

The initial start-up work described in this paragraph should be done by a competent and trained person, such as a manufacturer's representative, consulting engineer, or an experienced operator. The operator can learn a lot about pumps and motors by accompanying and helping a competent person put new equipment into operation.

Before starting a pump, lubricate it according to manufacturer's recommendations. Quality lubricants should be used. Turn the shaft by hand to see that it rotates freely. Then check to see that the shafts of the pump and motor are aligned and the flexible coupling adjusted. (Refer to Paragraph 10: Couplings, page 332; also see Section 18.23, "Pump Maintenance," page

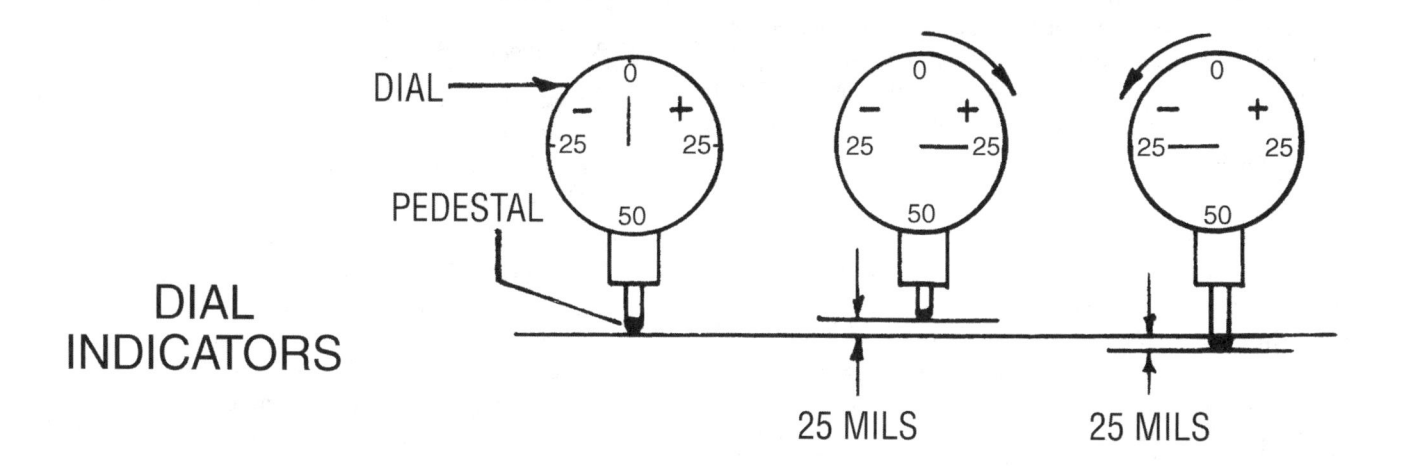

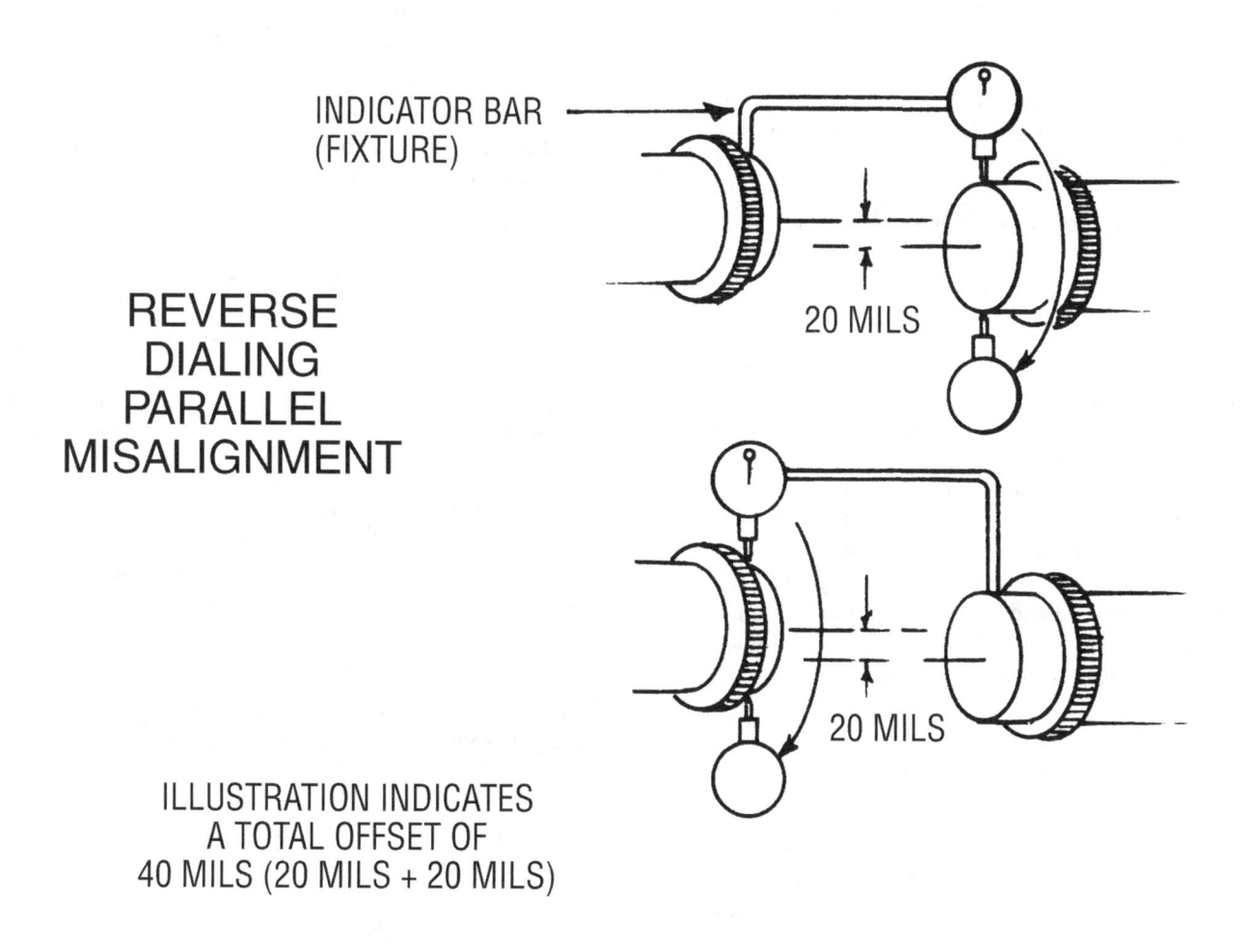

Fig. 18.31 Use of a dial indicator
(Permission of DYMAC, a Division of Spectral Dynamics Corporation)

318.) If the unit is belt driven, sheave (pulley) alignment and belt adjustment should be checked. (Refer to Paragraph 7: Belt Drives.) Check the electric voltage with the motor characteristics and inspect the wiring. See that thermal overload units in the starter are set properly. Turn on the motor just long enough to see that it turns the pump in the direction indicated by the rotational arrows marked on the pump. If separate water seal units or vacuum primer systems are used, these should be started. Finally, make sure lines are open. Sometimes there is an exception (see following paragraph) in the case of the discharge valve.

A pump should not be run without first having been primed. To prime a pump, the pump must be completely filled with water. In some cases, automatic primers are provided. If they are not, it is necessary to vent the casing. Most pumps are provided with a valve to accomplish this. Allow the trapped air to escape until water flows from the vent; then replace the vent cap. In the case of suction lift applications, the pump must be filled with water unless a self-primer is provided. In nearly every case, you may start a pump with the discharge valve open. Exceptions to this, however, are where water hammer or pressure surges might result, or where the motor does not have sufficient margin of safety or power. Sometimes there are no check valves in the discharge line. In this case (with the exception of positive displacement pumps), it is necessary to start the pump and then open the discharge lines. Where there are common discharge headers, it is essential to start the pump and then open the discharge valve. A positive displacement pump (reciprocating or piston types) should never be operated against a closed discharge line.

After starting the pump, again check to see that the direction of rotation is correct. Packing gland boxes (stuffing boxes) should be observed for slight leakage (approximately 60 drops per minute) as described in Paragraph 1: Pumps, General. Check to see that the bearings do not overheat from over- or underlubrication. The flexible coupling should not be noisy; if it is, the noise may be caused by misalignment or improper clearance or adjustment. Check to be sure pump anchorage is tight. Compare delivered pump flows and pressures with pump performance curves. If pump delivery falls below performance curves, look for obstructions in the pipelines and inspect piping for leaks.

18.241 Pump Shutdown

When shutting down a pump for a long period, the motor disconnect switch should be opened, locked out, and tagged with the reason for the tag noted. If the electric motor is equipped with winding heaters, check to be sure they are turned on. This helps to prevent condensation from forming, which can weaken the insulation on the windings. All valves on the suction, discharge, and water-seal lines should be shut tightly. Completely drain the pump by removing the vent and drain plugs.

Inspect the pump and bearings thoroughly so that all necessary servicing may be done during the inactive period. Drain the bearing housing and then add fresh lubricant. Follow any additional manufacturer's recommendations.

18.242 Pump-Driving Equipment

Driving equipment used to operate pumps includes electric motors and internal combustion engines. In rare instances, pumps are driven with steam turbines, steam engines, or air and hydraulic motors.

In all except the large installations, electric motors are used almost exclusively, with synchronous and induction types being the most commonly used. Synchronous motors operate at constant speeds and are used chiefly in large sizes. Three-phase, squirrel-cage induction motors are most often used in water treatment plants. These motors require little attention and, under average operating conditions, the factory lubrication of the bearing will last approximately one year. (Check with the manufacturer for average number of operating hours for bearings.) When lubricating motors, remember that too much grease may cause bearing trouble or damage the winding.

Clean and dry all electrical contacts. Inspect for loose electrical contacts. Make sure that hold-down bolts on motors are secure. Check voltage while the motor is starting and running. Examine bearings and couplings.

18.243 Electrical Controls

A variety of electrical equipment is used to control the operation of pumps or to protect electric motors. If starters, disconnect switches, and cutouts are used, they should be installed in accordance with the local regulations (city or county codes) regarding this equipment. In the case of larger motors, the power company often requires starters that do not overload the power lines.

The electrode-type, bubbler-type, and diaphragm-type water-level control systems are all similar in effect to the floatswitch system. Scum is a problem with most water-level controls that operate pumps, and it must be removed on a regular basis.

QUESTIONS

Write your answers in a notebook and then compare your answers with those on page 382.

18.24A Where would you find out how to lubricate a pump?

18.24B What problems can develop if too much grease is used in lubricating a motor?

18.244 Operating Troubles

The following list of operating troubles includes most of the causes of pump failure or reduced operating efficiency. The remedy or cure is either obvious or may be identified from the description of the cause.

SYMPTOM A—PUMP WILL NOT START

CAUSES:

1. Blown fuses or tripped circuit breakers due to:
   A. Rating of fuses or circuit breakers not correct
   B. Switch (breakers) contacts corroded or shorted
   C. Terminal connections loose or broken somewhere in the circuit
   D. Automatic control mechanism not functioning properly
   E. Motor shorted or burned out
   F. Wiring hookup or service not correct
   G. Switches not set for operation
   H. Contacts of the control relays dirty and arcing
   I. Fuses or thermal units too warm
   J. Wiring short-circuited
   K. Shaft binding or sticking due to rubbing impeller, tight packing glands, or clogging of pump
2. Loose connections, fuse, or thermal unit

SYMPTOM B—REDUCED RATE OF DISCHARGE

CAUSES:

1. Pump not primed
2. Air in the water
3. Speed of motor too low
4. Improper wiring
5. Defective motor
6. Discharge head too high
7. Suction lift greater than anticipated
8. Impeller clogged
9. Discharge line clogged
10. Pump rotating in wrong direction
11. Air leaks in suction line or packing box
12. Inlet to suction line too high, permitting air to enter
13. Valves partially or entirely closed
14. Check valves stuck or clogged
15. Incorrect impeller adjustment
16. Impeller damaged or worn
17. Packing worn or defective
18. Impeller turning on shaft because of broken key
19. Flexible coupling broken
20. Loss of suction during pumping may be caused by leaky suction line or ineffective water or grease seal
21. Belts slipping
22. Worn wearing ring

SYMPTOM C—HIGH POWER REQUIREMENTS

CAUSES:

1. Speed of rotation too high
2. Operating heads lower than rating for which pump was designed, resulting in excess pumping rates
3. Sheaves on belt drive misaligned or maladjusted
4. Pump shaft bent
5. Rotating elements binding
6. Packing too tight
7. Wearing rings worn or binding
8. Impeller rubbing

SYMPTOM D—NOISY PUMP

CAUSES:

1. Pump not completely primed
2. Inlet clogged
3. Inlet not submerged
4. Pump not lubricated properly
5. Worn impellers
6. Strain on pumps caused by unsupported piping fastened to the pump
7. Foundation insecure
8. Mechanical defects in pump
9. Misalignment of motor and pump where connected by flexible shaft
10. Rocks in the impeller
11. Cavitation

QUESTIONS

Write your answers in a notebook and then compare your answers with those on page 382.

18.24C   What items would you check if a pump will not start?

18.24D   How would you attempt to increase the discharge from a pump if the flow rate is lower than expected?

18.245 Starting and Stopping Pumps

The operator must determine what treatment processes will be affected by either starting or stopping a pump. The pump discharge point must be known and valves either opened or closed to direct flows as desired by the operator when a pump is started or stopped.

18.2450 CENTRIFUGAL PUMPS

Basic rules for the operation of centrifugal pumps include the following items.

1. Do not operate the pump when safety guards are not installed over or around moving parts.
2. Do not start a pump that has been locked or tagged out for maintenance or repairs. An operator working on the pump could be injured or the equipment could be damaged.
3. Never run a centrifugal pump when the impeller is dry. Always be sure the pump is primed.
4. Never attempt to start a centrifugal pump whose impeller or shaft is spinning backward.
5. Do not operate a centrifugal pump that is vibrating excessively after start-up. Shut unit down and isolate pump from system by closing the pump suction and discharge valves. Look for a blockage in the suction line and the pump impeller.

There are several situations in which it may be necessary to start a centrifugal pump against a closed discharge valve. Once the pump is primed, running, and indicating a discharge pressure, slowly open the pump discharge valve until the pump is fully on line. This procedure is used with treatment processes or piping systems with vacuums or pressures that cannot be dropped or allowed to fluctuate greatly while an alternate pump is put on the line.

Most centrifugal pumps used in water treatment plants are designed so that they can be easily started even if they have not been primed. This is accomplished with a positive static suction head or a low suction lift. On most of these arrangements, the pump will not require priming as long as the pump and the piping system do not leak. Leaks would allow the water to drain out of the pump volute. When pumps in water systems lose their prime, the cause is often a faulty check valve on the pump discharge line. When the pump stops, the discharge check valve will not seal (close) properly. Water previously pumped then flows back through the check valve and through the pump. The pump is drained and has lost its prime.

About ninety-five percent of the time, the centrifugal pumps in water treatment plants are ready to operate with suction and discharge valves open and seal water turned on. When the automatic start or stop command is received by the pump from the controller, the pump is ready to respond properly.

When the pumping equipment must be serviced, take it off the line by locking and tagging out the pump controls until all service work is completed.

QUESTIONS

Write your answers in a notebook and then compare your answers with those on page 382.

18.24E Why should a pump that has been locked or tagged out for maintenance or repairs not be started?

18.24F Under what conditions might a centrifugal pump be started against a closed discharge valve?

STOPPING PROCEDURES

This section contains a typical sequence of procedures to follow to stop a centrifugal pump. Exact stopping procedures for any pumping system depend upon the condition of the discharge system. The sudden stoppage of a pump could cause severe WATER HAMMER[16] problems in the piping system.

1. Inspect process system affected by pump, start alternate pump if required, and notify supervisor or log action.
2. Before stopping an operating pump, check its operation. This will give an indication of any developing problems, required adjustments, or problem conditions of the unit. This procedure only requires a few minutes. Items to be inspected include:

    a. Pump packing gland.

    (1) Seal water pressure

    (2) Seal leakage (too much, sufficient, or too little leakage)

    (3) Seal leakage drain flowing clear

    (4) Mechanical seal leakage (if equipped)

    b. Pump operating pressures.

    (1) Pump suction (pressure, vacuum)

    A higher vacuum than normal may indicate a partially plugged or restricted suction line. A lower vacuum may indicate a higher suction water level or a worn pump impeller or wearing rings.

    (2) Pump discharge pressure

    System pressure is indicated by the pump discharge pressure. Lower than normal discharge pressures can be caused by:

    (a) Worn impeller or wearing rings in the pump

    (b) A different point of discharge can change discharge pressure conditions

    (c) A broken discharge pipe can change the discharge head

---

[16] *Water Hammer.* The sound like someone hammering on a pipe that occurs when a valve is opened or closed very rapidly. When a valve position is changed quickly, the water pressure in a pipe will increase and decrease back and forth very quickly. This rise and fall in pressures can cause serious damage to the system.

NOTE: To determine the maximum head a centrifugal pump can develop, slowly close the discharge valve at the pump. Read the pressure gauge between the pump and the discharge valve when the valve is fully closed. This is the maximum pressure the pump is capable of developing. Do not operate the pump longer than a few minutes with the discharge valve closed completely because the energy from the pump is converted to heat and water in the pump can become hot enough to damage the pump.

  c. Motor temperature and pump bearing temperature.

  If motor or bearings are too hot to touch, further checking is necessary to determine if a problem has developed or if the temperature is normal. High temperatures may be measured with a thermometer.

  d. Unusual noises, vibrations, or conditions about the equipment.

  If any of the above items indicate a change from the pump's previous operating condition, additional service or maintenance may be required during shutdown.

3. Actuate stop switch for pump motor and lock out switch. If possible, use switch next to equipment so that you may observe the equipment stop. Observe the following items:

  a. Check valve closes and seats.

  Valve should not slam shut, or discharge piping will jump or move in their supports. There should not be any leakage around the check valve shaft. If check valve is operated automatically, it should close smoothly and firmly to the fully closed position.

  NOTE: If the pump is not equipped with a check valve, close discharge valve before stopping pump.

  b. Motor and pump should wind down slowly and not make sudden stops or noises during shutdown.

  c. After equipment has completely stopped, pump shaft and motor should not start backspinning. A pump shaft or motor will spin backward if water being pumped flows back through the pump when the pump is shut off. This will occur if there is a faulty check valve or foot valve in the system. If backspinning is observed in a pump with a check valve or foot valve, close the pump discharge valve slowly. Be extra careful if there is a plug valve on a line with a high head because when the discharge valve is partway closed, the plug valve could slam closed and damage the pump or piping.

4. Go to power control panel containing the pump motor starters just shut down and open motor breaker switch, lock out, and tag.

5. Return to pump and close:

  a. Discharge valve

  b. Suction valve

  c. Seal water supply valve

  d. Pump volute bleed line (if so equipped)

6. If required, close and open appropriate valves along piping system through which pump was discharging.

STARTING PROCEDURES

This section contains a typical sequence of procedures to follow to start a centrifugal pump.

1. Check motor control panel for lock and tags. Examine tags to be sure that no item is preventing start-up of equipment.

2. Inspect equipment.

  a. Be sure stop switch is locked out at equipment location.

  b. Guards over moving parts must be in place.

  c. Cleanout on pump volute and drain plugs should be installed and secure.

  d. Valves should be in closed position.

  e. Pump shaft must rotate freely.

  f. Pump motor should be clean and air vents clear.

  g. Pump, motor, and auxiliary equipment lubricant levels must be at proper elevations.

  h. Determine if any special considerations or precautions are to be taken during start-up.

3. Follow pump discharge piping route. Be sure all valves are in the proper position and that the pump flow will discharge where intended.

4. Return to motor control panel.

  a. Remove tag.

  b. Remove padlock.

  c. Close motor main breaker.

  d. Place selector switch to manual (if you have automatic equipment).

5. Return to pump equipment.

  a. Open seal water supply line to packing gland. Be sure seal water supply pressure is adequate.

  b. Open pump suction valve slowly.

  c. Bleed air out of top of pump volute in order to prime pump. Some pumps are equipped with air relief valves or bleed lines back to the wet well for this purpose.

  d. When pump is primed, slowly open pump discharge valve and recheck prime of pump. Be sure no air is escaping from volute.

  e. Unlock stop switch and actuate start switch. Pump should start.

6. Inspect equipment.

   a. Motor should come up to speed promptly. If ammeter is available, test for excessive draw of power (amps) during start-up and normal operation. Most three-phase induction motors used in water treatment plants will draw 5 to 7 times their normal running current during the brief period when they are coming up to speed.

   b. Unusual noise or vibrations should not be observed during start-up.

   c. Check valve should be open and no chatter or pulsation should be observed.

   d. Pump suction and discharge pressure readings should be within normal operating range for this pump.

   e. Packing gland leakage should be normal.

   f. If a flowmeter is on the pump discharge, record pump output.

7. If the unit is operating properly, return to the motor control panel and place the motor mode of operation selector in the proper operating position (MANUAL/AUTO/OFF).

8. The pump and auxiliary equipment should be inspected routinely after the pump has been placed back into service.

QUESTIONS

Write your answers in a notebook and then compare your answers with those on page 382.

18.24G What should be done before stopping an operating pump?

18.24H What could cause a pump shaft or motor to spin backward?

18.24I Why should the position (open or closed) of all valves be checked before starting a pump?

18.2451 POSITIVE DISPLACEMENT PUMPS

Steps for starting and stopping positive displacement pumps are outlined in this section. There are two basic differences in the operation of positive displacement pumps as compared with centrifugal pumps. Centrifugal pumps (due to their design) will permit an operator error, but a positive displacement pump will not and someone will have to pay for correcting the damages.

Important rules for operating positive displacement pumps include:

1. **Never operate a positive displacement pump against a closed valve, especially a discharge valve.** The pipe, valve, or pump could rupture from excessive pressure. The rupture will damage equipment and possibly seriously injure or kill someone standing nearby.

2. Positive displacement pumps are used to pump solids (sludge). Certain precautions must be taken to prevent injury or damage. If the valves on both ends of a sludge line are closed tightly, the line becomes a closed vessel. Gas from decomposition of the sludge can build up and rupture pipes or valves.

3. Positive displacement pumps also are used to meter and pump chemicals. Care must be exercised to avoid venting chemicals to the atmosphere.

4. Never operate a positive displacement pump when it is dry or empty, especially the progressive-cavity types that use rubber stators. A small amount of liquid is needed for lubrication in the pump cavity between the rotor and the stator.

In addition to never closing a discharge valve on an operating positive displacement pump, the only other difference (when compared with a centrifugal pump) may be that the positive displacement pump system may or may not have a check valve in the discharge piping after the pump. Installation of a check valve depends upon the designer and the material being pumped.

Other than the specific differences mentioned in this section, the starting and stopping procedures for positive displacement pumps are similar to the procedures for centrifugal pumps.

QUESTIONS

Write your answers in a notebook and then compare your answers with those on page 382.

18.24J What is the most important rule regarding the operation of positive displacement pumps?

18.24K What could happen if a positive displacement pump is operated against a closed discharge valve?

18.24L Why should both ends of a sludge line never be closed tight at the same time?

End of Lesson 3 of 5 Lessons on MAINTENANCE

Please answer the discussion and review questions next.

DISCUSSION AND REVIEW QUESTIONS
Chapter 18. MAINTENANCE
(Lesson 3 of 5 Lessons)

Write the answers to these questions in your notebook. The question numbering continues from Lesson 2.

18. When two or more pumps of the same size are installed, why should they be operated alternately?

19. What should be checked if pump bearings are running hot?

20. What happens when the packing is too tight on a reciprocating pump?

21. Why should adjustments in control valves for reciprocating pumps be made when the pump is shut down?

22. Why would you use a stethoscope to check an electric motor?

23. How would you determine if a motor is running unusually hot?

24. How would you clean belts on a belt drive?

25. Why should you never replace only one belt on a multiple-drive unit?

26. What do rubber wearings near a belt drive indicate?

27. How can you determine if a chain in a chain-drive unit has the proper slack?

28. What happens when couplings are not in proper alignment?

29. How can you determine if a new pump will turn in the direction intended?

30. How can you determine if a new pump is delivering design flows and pressures?

31. When shutting down a pump for a long period, what precautions should be taken with the motor disconnect switch?

CHAPTER 18. MAINTENANCE

(Lesson 4 of 5 Lessons)

18.25 Compressors

Compressors (Figure 18.32) are commonly used in the operation and maintenance of water treatment plants. They are used to activate and control pump control systems (bubblers), valve operators, and water pressure systems. They are also used to operate portable pneumatic tools, such as jackhammers, compactors, air drills, sandblasters, tapping machines, and air pumps.

A compressor is a device used to increase the pressure of air or gas. It can be of a very simple diaphragm or bellows type such as are found in aquarium pumps, or an extremely complex rotary, piston, or sliding vane type compressor. A compressor usually has a suction pipe with a filter and a discharge pipe that connects to an air receiver or storage tank. The compressed air or gas is then used from the air receiver.

Due to the complexity of compressors, the water treatment plant operator usually will not be repairing them. You will, however, be required to maintain these compressors. With proper maintenance, a compressor should give years of trouble-free service.

The first step for compressor maintenance, and this pertains to any mechanical equipment, is to get the manufacturer's instruction book and read it completely. Each compressor is different and the particular manufacturer will provide its recommended maintenance schedules and procedures. Some of the maintenance procedures are discussed in the following paragraphs.

1. Inspect the suction filter of the compressor regularly. The frequency of cleaning depends upon the use of the compressor and the atmosphere around it. Under normal operations, the filter should be inspected at least monthly and cleaned or replaced every three to six months. Inspect and replace the filter more frequently in areas with excavation and dust. When breaking up concrete, inspect the filters daily.

   There are several types of filters, such as paper, cloth, wire screen, oil bath, and others. The impregnated paper filters must be replaced when dirty. The cloth-type filters can be washed with soap and water, dried, and reinstalled. If a cloth-type filter is used, it is recommended that a spare be kept so one can be cleaned while the other is being used. The wire mesh and oil-bath type filters can be cleaned with a standard solvent, reoiled or oil bath filled, and used again. Never operate a compressor without the suction filter because dirt and foreign materials will collect on the rotors, pistons, or blades and cause excessive wear.

2. Lubrication. Improper or lack of lubrication is probably the biggest cause of compressor failures. Most compressors require oiling of the bearings. They can have crankcase reservoirs, oil cups, grease fittings, a pressure system, or separate pump. Whatever type, it must be inspected daily. Examine the reservoir dipstick or sight glass. Make sure that drip-feed oilers are dripping at the proper rate, force-feed oilers have the proper pressure, and grease fittings are greased at the proper interval. Compressors use a certain amount of oil in their operation and special attention is needed to keep the reservoirs full. Care also must be used to not overfill the crankcase. On some compressors, it is possible for the oil to get into the compression side and lock up the compressor or damage it. Remember:

   A LIQUID CANNOT BE COMPRESSED.

   When air or gas is compressed, it gives off heat and the compressor becomes very hot. This tends to break down oil faster, so most compressor manufacturers have special oils recommended for their particular compressor. Also, due to the heat and contamination, it is necessary to change oil quite frequently. Compressor oil should be changed at least every three months, unless the manufacturer states differently. If there are filters in the oil system, these also should be changed.

3. Cylinder or casing fins should be cleaned weekly with compressed air or vacuumed off. The fins must be clean to ensure proper cooling of the compressor.

4. Unloader. Many compressors have unloaders that allow the compressor to start under a no-load condition. These can be inspected by observing the compressor. When the compressor starts, it should come up to speed and the unloader will change, starting the compression cycle. This can usually be heard by a change in sound. When it stops, you can hear a small pop and hear the air bleed off the cylinders. If the unloader is not working properly, the compressor will stall when starting, not start, or if belt-driven, burn off the belts.

5. Test the safety valves weekly. The pop-off or safety valves are located on the air receiver or storage tank. They prevent the pressure from building up above a specified pressure by

Fig. 18.32 Two-stage piston compressor
(Courtesy Worthington Corporation)

opening and venting to the atmosphere. In gas compressors, they vent to the suction side of the compressor. Some compressors have high pressure cutoff switches, low oil pressure switches, and high temperature cutoff switches. These switches have preset cutoff settings and must not be changed without proper authorization. If for any reason any of the safety switches are not functioning properly, the problem must be corrected before starting the compressor again. The safety switch settings should be recorded and kept in the equipment file.

6. Drain the condensate (condensed water) from the air receiver daily. Due to temperature changes, the air receiver will fill with condensate. Each day the condensate should be drained from the bottom of the tank. There is usually a small valve at the bottom of the air receiver for this purpose. Some air receivers are equipped with automatic drain valves. These must be inspected periodically to ensure they are operating satisfactorily.

7. Inspect belt tension on compressors. Usually, you should be able to press the belt down with hand pressure approximately three-fourths of an inch. This is done at the center between the two pulleys. Make sure the compressor is locked off before making this test. Do not overtighten belts because it will cause overheating and excessive wear on bearings and motor overloading.

8. Examine operating controls. Make sure the compressor is starting and stopping at the proper settings. If it is a dual installation, make sure they are alternating if so designed; inspect gauge for accuracy. Compare readings with recorded start-up values or other known, accurate readings.

9. Many portable compressors are equipped with tool oilers on the receivers. These are used for mixing a small quantity of oil with the compressed air for lubrication of the tools being used. Tool oilers are located on the discharge side of the air receiver. They have a reservoir that must be filled with rock drill oil.

10. All compressors should be thoroughly cleaned at least monthly. Dirt, oil, grease, and other material must be thoroughly cleaned off the compressor and surrounding area. Compressors have a tendency to lose oil around piping, fittings, and shafts; therefore, constant cleaning is required by the maintenance operator to ensure proper and safe operation.

> **QUESTIONS**
>
> Write your answers in a notebook and then compare your answers with those on page 382.
>
> 18.25A List some of the uses of a compressor in connection with operation and maintenance of a water treatment plant.
>
> 18.25B How often should the suction filter of a compressor be cleaned?
>
> 18.25C How often should compressor oil be changed?
>
> 18.25D How often should the condensate from the air receiver be drained?
>
> 18.25E What must be done before testing belt tension on compressors with your hands?

18.26 Valves[17]

18.260 Uses of Valves

Valves are the controlling devices placed in piping systems to stop, regulate, check, divert, or otherwise modify the flow of liquids or gases. There are specific valves that are more suitable for certain jobs than others. The six most common valves that you will find in a water treatment facility are discussed in this section.

18.261 Gate Valves (Figures 18.33 and 18.34)

The basic parts of a gate valve are: the operator (handle), the shaft packing assembly, the bonnet, the valve body with seats, the stem, and the disc. Gate valves come in a large number of sizes, but the principle of operation is quite similar for all sizes. One could associate the action of a gate valve to that of a guillotine having a screw shaft instead of the rope. The valve disc is raised or lowered by a threaded shaft and is guided on each side to ensure that it will not hang up in the operation. The disc is screwed down until it wedges itself between two machined valve seats. This makes a leakproof seat on both sides of the disc. The discs are replaceable. Some gate valves have discs with wedges inside. As more force is applied to the screwed stem, the wedges force the discs into tighter contact with the valve seats.

Gate valves are either of the rising (Figure 18.33) or nonrising stem (Figure 18.34) type. The rising stem has companion threads in the valve bonnet. As the valve is opened, the stem is threaded out, lifting the wedged disc. In the nonrising type, the stem is held in place in the bonnet by a collar. The stem is threaded with companion threads in the wedged disc. As the valve opens, the disc rises on the stem. Consequently, the handwheel stays on the same plane.

Gate valves are not commonly used to control flows. With the valve partially open, the water velocity is increased through the valve and minute particles transported in the water can cause undue seat wear. However, the V-ported gate valve can be used in controlling flows. As the valve is opened, the V is widened to allow more flow. Because of the valve design, little damage is done to the valve seats in the V-ported type of gate valve.

Suggested gate valve operation and maintenance procedures are listed below:

1. Open valve fully. When at stop, reverse and close valve one-half turn.
2. Operate all large valves at least yearly to ensure proper operation.
3. Inspect valve stem packing for leaks. Tighten, as needed.
4. If the valve has a rising stem, keep stem threads clean and lubricated.
5. Close valves slowly in pressure lines to prevent water hammer.
6. If a valve will not close by using the handwheel, check for the cause. Using a "cheater" (bar-pipe wrench) will only aggravate the problem.

18.262 Maintenance of Gate Valves

Paragraph 12: Gate Valves

The most common maintenance required by gate valves is oiling, tightening, or replacing the stem stuffing box packing.

Frequency of Service

A 1. REPLACE PACKING. Modern gate valves can be repacked without removing them from service. Before repacking, open valve wide. This prevents excessive leakage when the packing or the entire stuffing box is removed. It draws the stem collar tightly against the bonnet on a nonrising stem valve, and tightly against the bonnet bushing on a rising stem valve.

   a. Stuffing box. Remove all old packing from stuffing box with a packing hook or a rattail file with bent end. Clean valve stem of all adhering particles and polish it with fine emery cloth. After polishing, remove the fine grit with a clean cloth to which a few drops of oil have been added.

   b. Insert packing. Insert new split-ring packing in stuffing box and tamp it into place with packing hook. Stagger ring splits. After stuffing box is filled, place a few drops of oil on stem, assemble gland, and tighten it down on packing.

---

[17] For additional information on valves, see *WATER DISTRIBUTION SYSTEM OPERATION AND MAINTENANCE*, Chapter 3, "Distribution System Facilities," Section 3.670, "Valves," in this series of manuals.

Maintenance 345

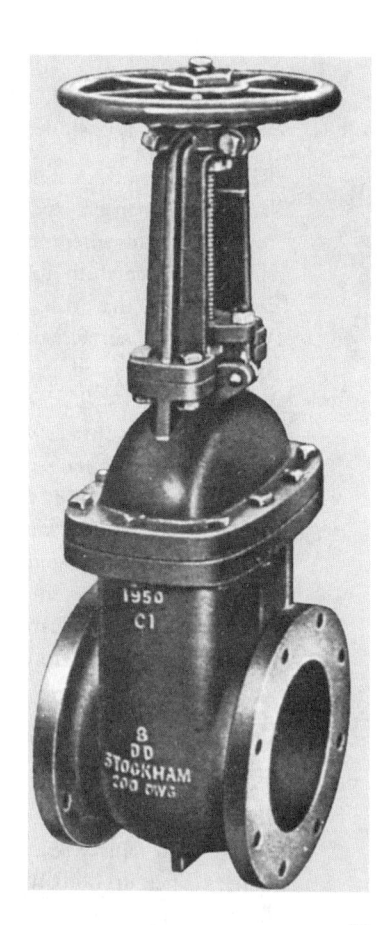

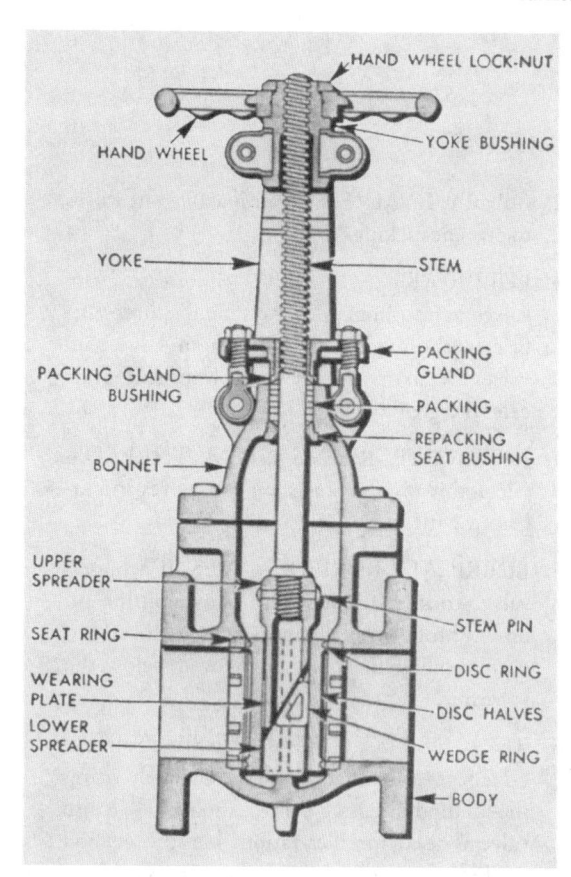

Fig. 18.33 Rising stem gate valve
(Permission of Stockham Valves & Fittings, Copyright, 1976)

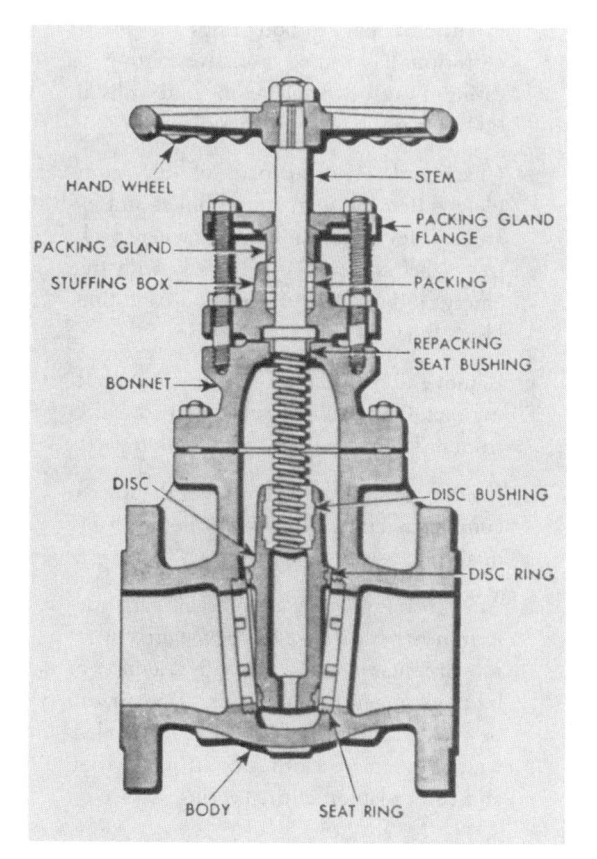

Fig. 18.34 Nonrising stem gate valve
(Permission of Stockham Valves & Fittings, Copyright, 1976)

Frequency of Service	
S	2. OPERATE VALVE. Operate inactive gate valves to prevent sticking.
A	3. LUBRICATE GEARING. Lubricate gate valves as recommended by manufacturer. Lubricate thoroughly any gearing in large gate valves. Wash open gears with solvent and lubricate with grease.
S	4. LUBRICATE RISING STEM THREADS. Clean threads on rising stem gate valves and lubricate with grease.
	5. LUBRICATE BURIED VALVES. If a buried valve works hard, lubricate it by pouring oil down through a pipe that is bent at the end to permit oiling the packing follower below the valve nut.
A	6. REFACE LEAKY GATE VALVE SEATS. If gate valve seats leak, reface them immediately, using the method discussed below. A solid wedge disc valve is used for illustration, but the general method also applies to other types of repairable gate valves. Proceed as follows:

  a. Remove bonnet and clean and examine disc and body thoroughly. Carefully determine extent of damage to body rings and disc. If corrosion has caused excessive pitting or eating away of metal, as in guide ribs in body, repairs may be impractical.

  b. Check and service all parts of valve completely. Remove stem from bonnet and examine it for scoring and pitting where packing makes contact. Polish lightly with fine emery cloth to put stem in good condition. Use soft jaws if stem is put in vise.

  c. Remove all old packing and clean out stuffing box. Clean all dirt, scale, and corrosion from inside of valve bonnet and other parts.

  d. Do not salvage an old gasket. Remove it completely and replace with one of proper quality and size.

  e. After cleaning and examining all parts, determine whether valve can be repaired by removing cuts from disc and body seat faces or by replacement of body seats. If repair can be made, set disc in vise with face leveled, wrap fine emery cloth around a flat tool, and rub or lap off entire bearing surface on both sides to a smooth, even finish. Remove as little metal as possible.

  f. Repair cuts and scratches on body rings, lapping with an emery block small enough to permit convenient rubbing all around rings. Work carefully to avoid removing so much metal that disc will seat too low. When seating surfaces of disc and seat rings are properly lapped in, coat faces of disc with *PRUSSIAN BLUE*[18] and drop disc in body to check contact. When good, continuous contact is obtained, the valve is tight and ready for assembly. Insert stem in bonnet, install new packing, assemble other parts, attach disc to stem, and place assembly in body. Raise stem to prevent contact with seats so bonnet can be properly seated on body before tightening the joint.

  g. Test repaired valve before putting it back in line to ensure that repairs have been properly made.

  h. If leaky gate valve seats cannot be refaced, remove and replace seat rings with a power lathe. Chuck up body with rings vertical to lathe and use a strong steel bar across ring lugs to unscrew them. They can be removed by hand with a diamond point chisel if care is taken to avoid damaging threads. Drive new rings home tightly. Use a wrench on a steel bar across lugs when putting in rings by hand. Always coat threads with a good lubricant before putting threads into the valve body. This helps to make the threads easier to remove the next time the seats have to be replaced. Lap in rings to fit disc perfectly.

18.263 *Globe Valves* (Figure 18.35)

The globe valve seating configuration is quite different from the gate valve. Globe valves use a circular disc to make a flat surface contact with a ground-fitted valve seat. This is similar to placing your thumb over the end of a tube. The parts of the valve are similar in name and function to the gate valve. They can be of the rising or nonrising stem type.

What is unique about the globe valve is its internal design (Figure 18.35). This design enables the valve to be used in a controlling mode. The valve seats are not subject to excessive wear when partially opened like the gate valve. After extended use, the valve may not have a positive shutoff but it will still be effective in throttling flows. Procedures for operating and maintaining globe valves are similar to the procedures outlined for gate valves in Section 18.262.

---

[18] *Prussian Blue.* A blue paste or liquid (often on a paper like carbon paper) used to show a contact area. Used to determine if gate valve seats fit properly.

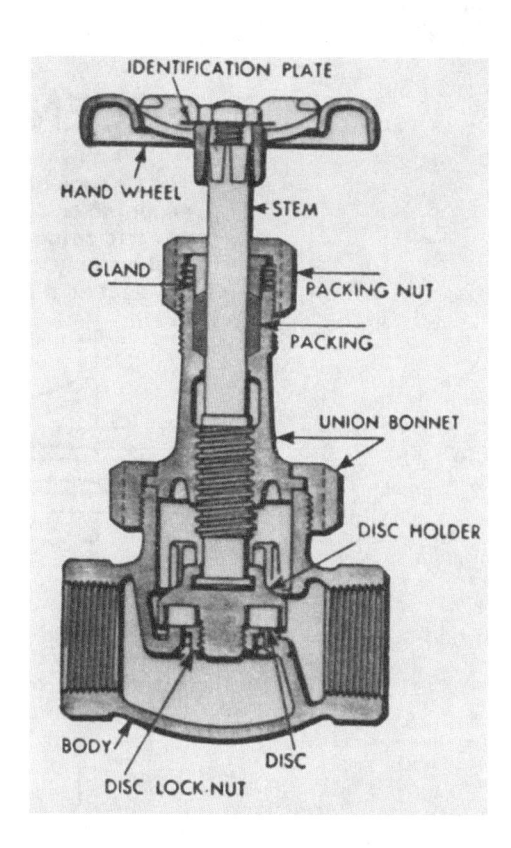

Fig. 18.35 Globe valve
(Permission of Stockham Valves & Fittings)

18.264 Eccentric Valves (Figures 18.36 and 18.37)

The eccentric valve has many desirable features. These features include allowance for high flow capacity, quarter-turn operation, no lubrication, excellent resistance to wear, and good throttling characteristics. The eccentric valve uses a cam-shaped plug to match an eccentric valve seat. As the valve is closed, the plug throttles the flow yet maintains a smooth flow rate. The plug does not come into contact with the valve seat until it is in the closed position.

Because the plug has a resilient coating, it ensures a leak-tight seal at the valve seats. The Buna-N, neoprene, or viton plug coating is a very wear-resistant compound and can function well under a wide temperature range. This valve is excellent for controlling the flows of slurries and sludges found in water treatment facilities.

18.265 Butterfly Valves (Figure 18.38)

The butterfly valve is used primarily as a control valve. The flow characteristics allow the water to move in straight lines with little turbulence in the area of the valve disc (butterfly). Complete flow shutoff can be accomplished but the psi rating is relatively low in comparison to eccentric or gate valves.

The butterfly valve uses a machined disc that can be opened to 90 degrees to allow full flow through the valve. Quarter-turn operation moves the valve from the "closed" to "open" position. The disc is mounted on a shaft eccentric that allows the disc to come into its seat with minimum seating torque and scuffing of the rubber seat. There is no contact between the disc and the seat until the last few degrees of valve closure.

A resilient rubber is used as the seat and is of a continuous form that is not interrupted by a shaft connection. Wear resistance characteristics are good when used in slurry and sludge applications.

When the valve is closed, the disc is forced against the rubber seat. Wedges with jacking screws compress the rubber seat via a jack ring. The rubber seat then conforms to the entire disc circumference. The rubber can be readily replaced, when necessary, without complete valve dismantling. Large valves do not need to be removed from the line for seat replacement.

18.266 Check Valves (Figure 18.39)

The term "check valve" describes its function. A check valve allows water to flow in one direction only. If the water attempts to flow in the opposite direction, an internal mechanism closes the valve and "checks" (prevents) the flow. Three types of check mechanisms may be used—the swing check, the wafer check, or the lift check. In the swing check, a movable disc rests at a right angle to the flow and seats against a ground

348 Water Treatment

ECCENTRIC ACTION

The DeZurik design matches a single-faced eccentric or cam-shaped plug with an eccentric raised body seat. With rotary motion only, the plug advances against the seat as it closes. Here's how it works:

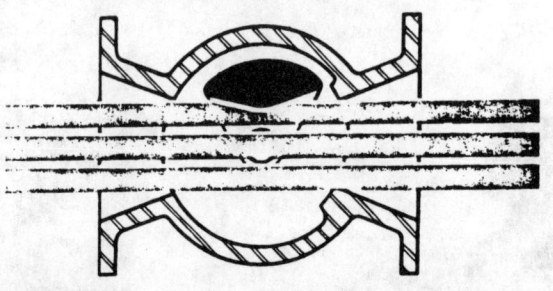

OPEN—The plug is out of the flow path. There is no bonnet or other cavity to fill with slurry material. Flow is straight-through with minimum pressure drop.

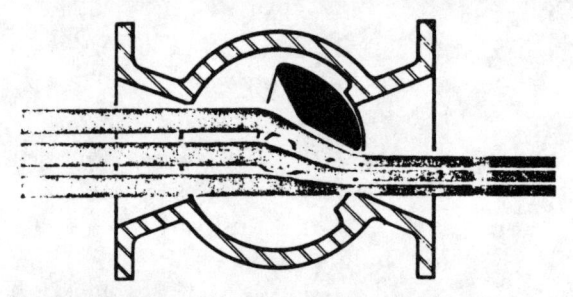

CLOSING—At any position between open and closed, the eccentric plug still has not touched the seat. There is no friction to cause wear or binding. Flow is still smooth and straight. Throttling action is excellent on all types of services from slurries to gas.

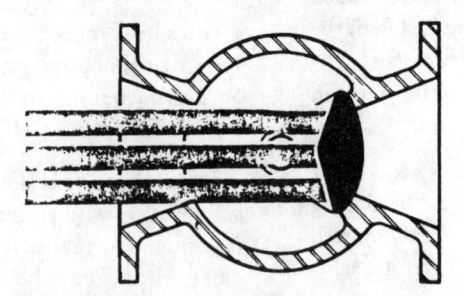

CLOSED—The eccentric plug makes contact with the eccentric seat only in the fully closed position. Action is easy, without binding or scraping. There is no continual seat wear. The plug is moved firmly into the seat to provide a positive, drip-tight, long-lasting seal.

Fig. 18.36   How eccentric valves work
(Permission of DeZurik Corporation, Sartell, MN)

Fig. 18.37   Eccentric valve
(Permission of DeZurik Corporation, Sartell, MN)

Fig. 18.38 Butterfly valve
(Permission of American-Darling Valve, Birmingham, AL)

seat. The movable disc is called the clapper. The clapper can be one of three types: gravity operated, lever and weight operated, or lever and spring operated. In many installations, the water being pumped must be delivered at a desired flow rate and pressure. A clapper with an external means of adjusting the opening in the check valve may be necessary to produce desired flows and pressures. By positioning the weight on the lever or adjusting the spring tension, a check valve can be made to operate either partially or fully open at various pressures and flows. The spring or counterweight also ensures that the check valve closes at no flow. This is very helpful if the valve is not in a position that will enable gravity alone to operate the clapper. The gravity-operated clapper does not have an external adjustment and relies on the weight of the clapper to close the valve at no-flow conditions.

Most swing check valves provide for full opening, that is, the clapper can move up into the bonnet and thus be completely out of the flow. Head loss in swing check valves may be relatively high and this factor must be considered in selecting the device for a particular application. This type of check valve is quite common in pump installations and often has a dampening feature to cushion the closing of the clapper.

The wafer check has a circular disc that hinges in the center (diameter) of the disc. Water passing through collapses the disc and the stoppage of flow allows the disc to return to its circular

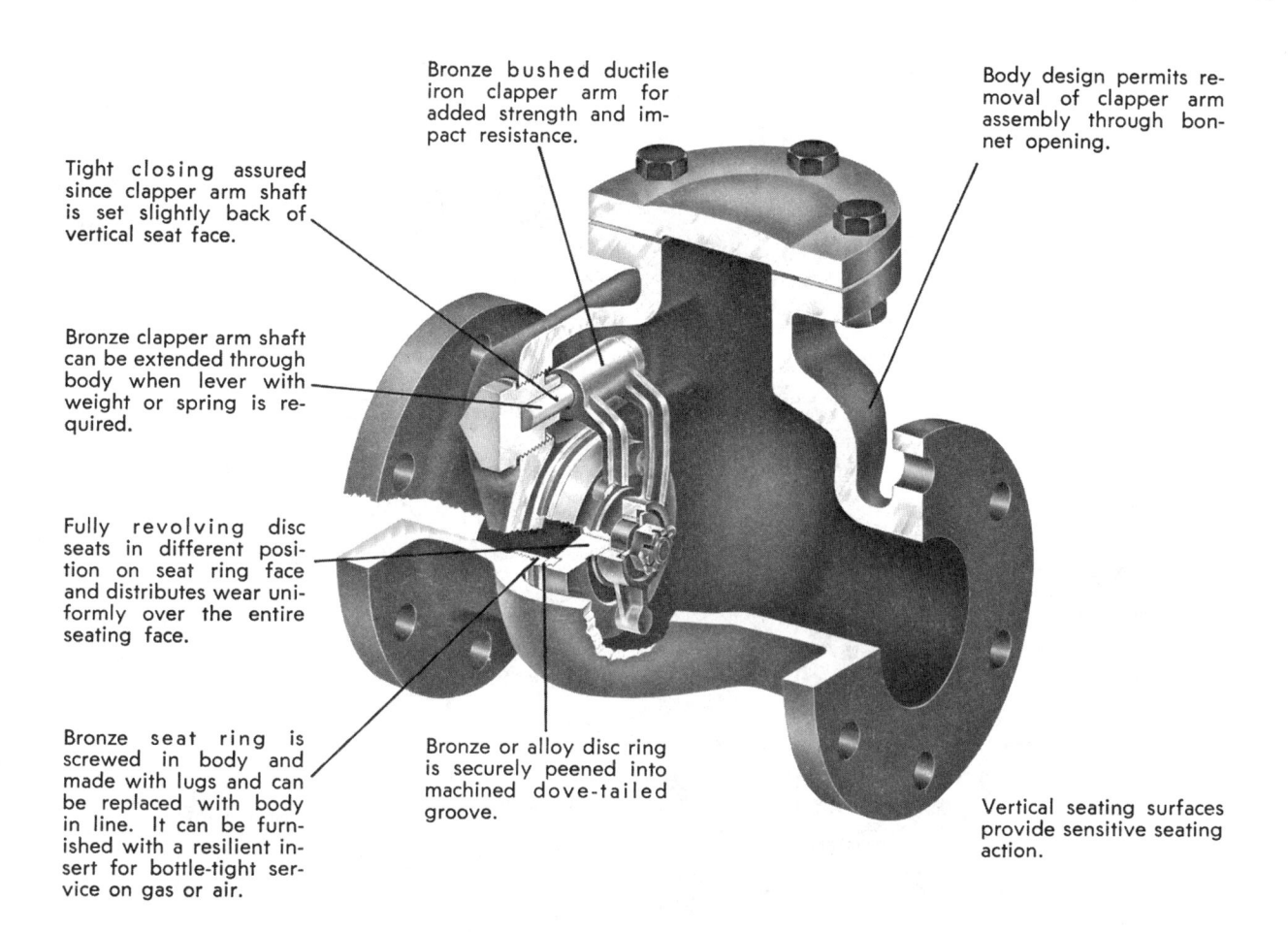

Fig. 18.39 Check valve
(Permission of American-Darling Valve, Birmingham, AL)

form. Because the valve has a tendency to be fouled up by stringy material, it is not commonly used in handling raw water. Wafer check valves are very effective when used with clean water.

The lift check uses a vertical lift disc or ball. When there is flow, the disc or ball is lifted from its ground seat and fluid passes through the valve. As flow stops, the check realigns itself with its seat and checks or prevents water backflow. The movable portion can be a spring or gravity return.

The foot valves used in pump suctions are nearly always of the vertical lift disc design. A check valve of this type is usually applied to handle clean water.

Backflow prevention by check valves is essential in many applications to:

1. Prevent pumps from reversing when power is removed
2. Protect water systems from being cross-connected
3. Aid in pump operation as a dampener
4. Ensure full-pipe operation (pipe is full of water)

Table 18.4 provides a comparison of various types of check valves with features of these valves. Figures 18.40 through 18.46 provide drawings and photographs of the different types of check valves listed in Table 18.4.

18.267 Maintenance of Check Valves

Paragraph 13: Check Valves

Frequency
of
Service

A   1. INSPECT DISC FACING. Open valves to observe condition of facing on swing check valves equipped with leather or rubber seats on disc. If metal seat ring is scarred, dress it with a fine file and lap with fine emery paper wrapped around a flat tool.

A   2. CHECK PIN WEAR. Check pin wear on balanced disc check valve, since disc must be accurately positioned in seat to prevent leakage.

TABLE 18.4 COMPARISON OF FEATURES OF DIFFERENT TYPES OF CHECK VALVES[a]

Globe Silent Check Valve	Wafer Silent Check Valve	Double Door Check Valve	Rubber Flapper Check Valve	Cushion Swing Check Valve	Slanting Disc Check Valve	Automatic Control Check Valve	Feature
		X					1. Lowest initial cost
		X					2. Shortest laying length
	X						3. Highest head loss (see head loss curves)
X	X	X		X			4. Resilient seat (optional)
			X	X			5. For waste and raw sewage
X	X	X			X	X	6. Clean water only
		X	X	X	X		7. Cushion closing
X	X					X	8. Silent closing (positively silent)
			X		X		9. Free open – Free close
				X	X	X	10. Control open or close or both (optional)
X	X					X	11. Vertical installation flow up or down
			X	X		X	12. Can be rubber lined
				X	X	X	13. Disc position indicator
			X				14. Buried service
				X			15. Outside lever
					X	X	16. Surge pressure control
			X	X		X	17. Reverse flow
X	X	X			X	X	18. Up to 600# class
	X						19. Up to 1500# class
					X		20. Lowest head loss (see head loss curves)
	X						21. Up to 2500# class
						X	22. Control open and close standard
						X	23. Shut off valve
						X	24. Throttling valve
		X	X	X	X		25. Vertical installation flow up only
						X	26. Electric motor operated
						X	27. Remote control
						X	28. Control closure upon power failure
			X	X	X	X	29. Resilient seat standard
			X		X	X	30. Velocities in excess of 15 FPS
					X		31. Velocities up to 5 FPS
X	X		X				32. Velocities up to 10 FPS

[a] Permission of APCO/Valve and Primer Corporation.

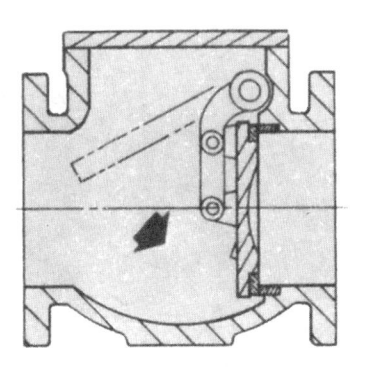

Fig. 18.40   Swing check valves (single disc)
(Permission of APCO/Valve and Primer Corporation)

354 Water Treatment

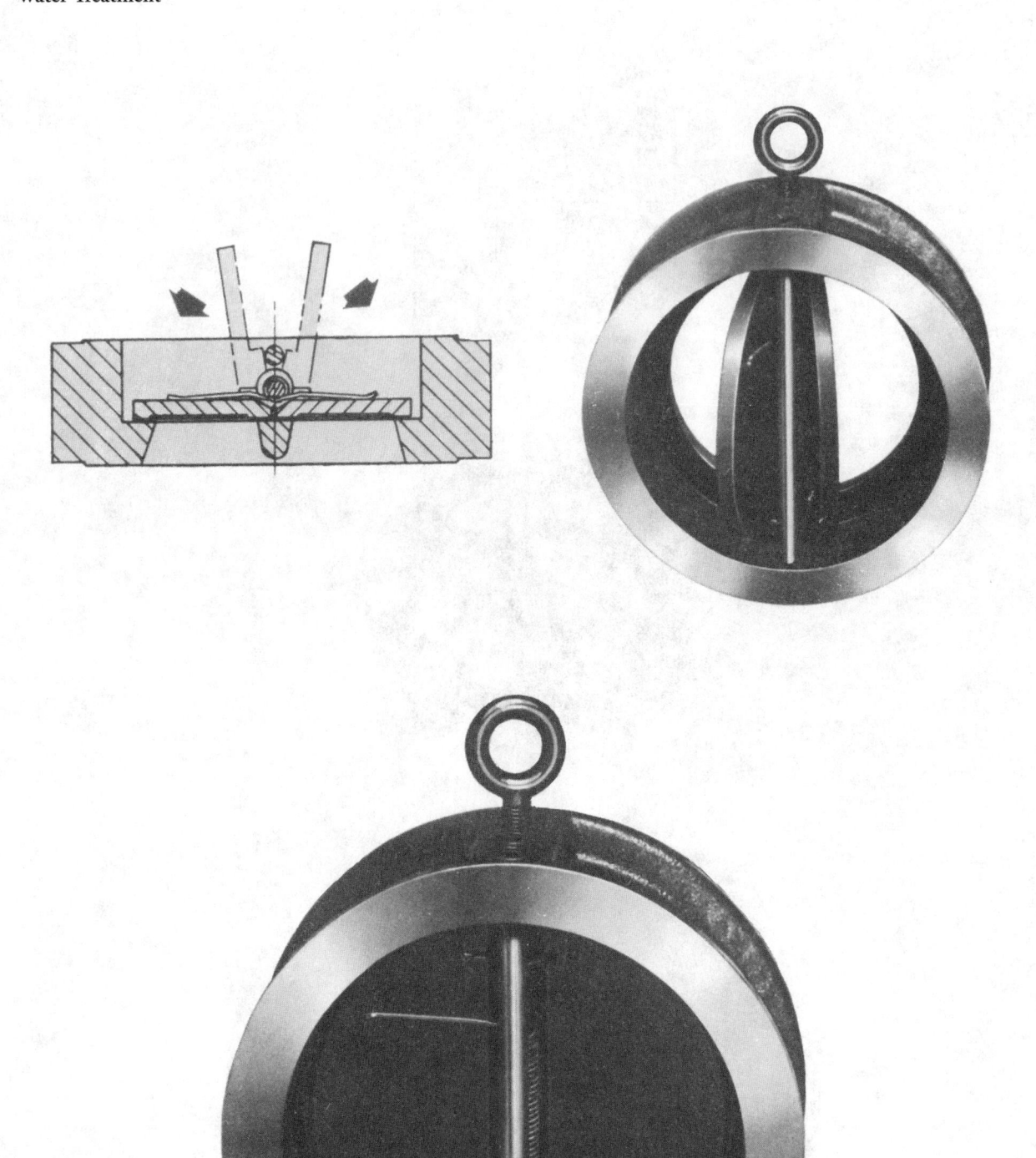

Fig. 18.41 Double disc swing check valves (split swing discs)
(Permission of APCO/Valve and Primer Corporation)

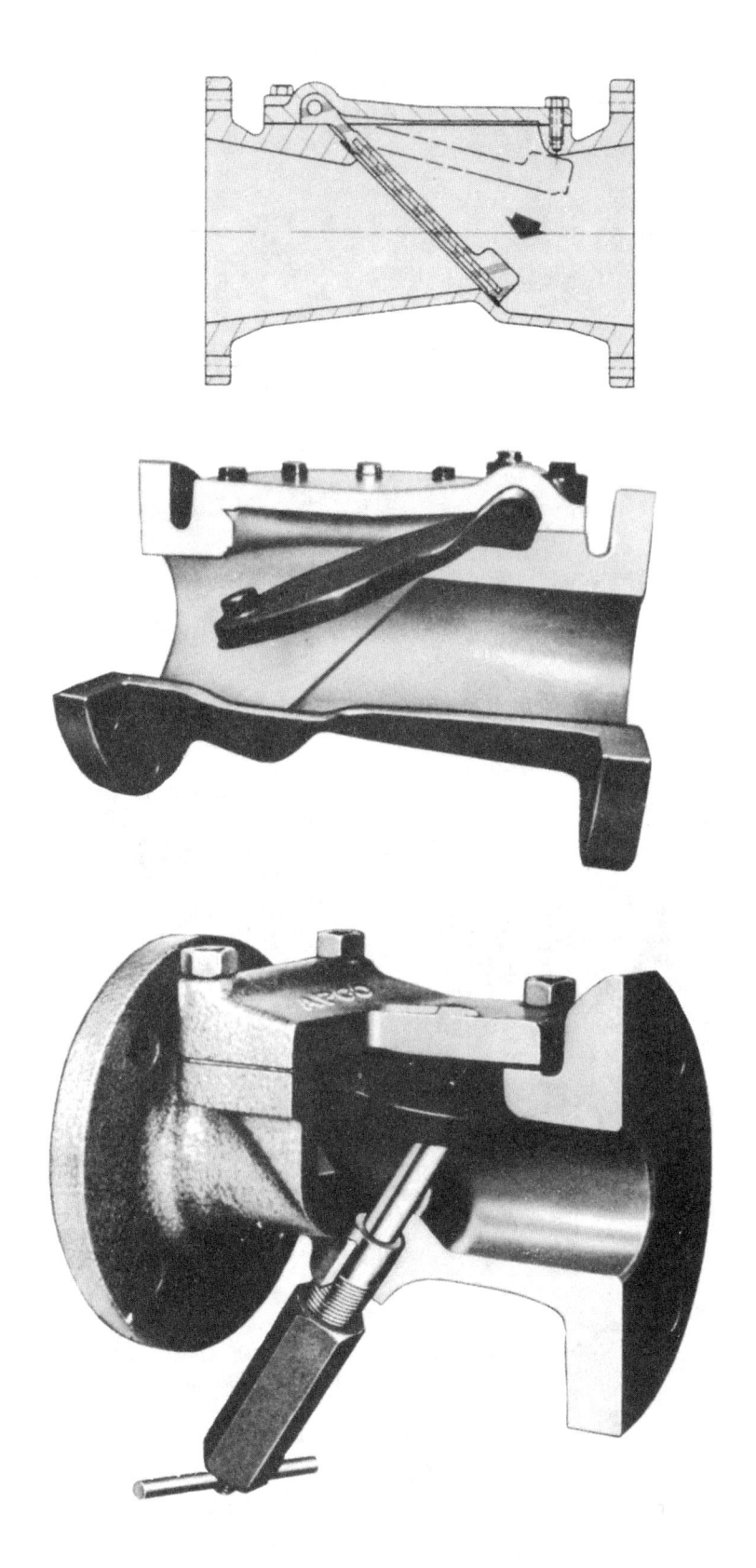

Fig. 18.42 Rubber flapper check valves (angle seating)
(Permission of APCO/Valve and Primer Corporation)

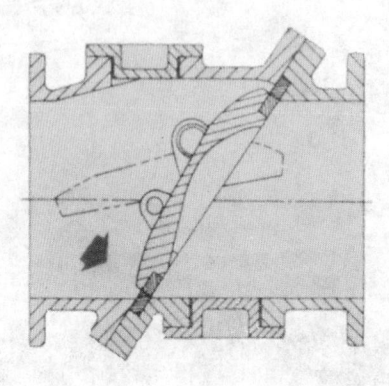

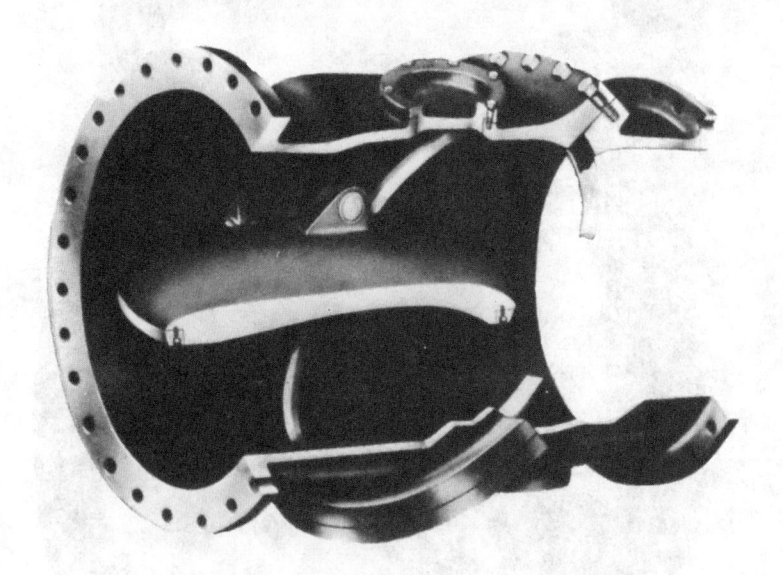

Fig. 18.43 Slanting disc check valves (pivot off center)
(Permission of APCO/Valve and Primer Corporation)

Maintenance 357

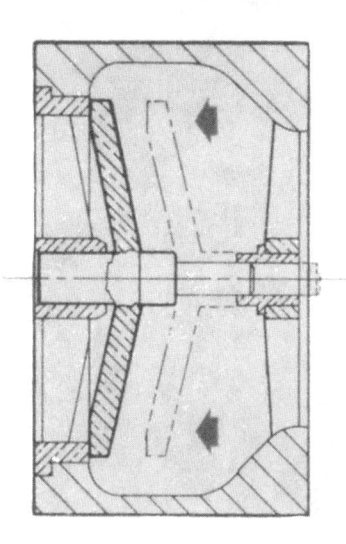

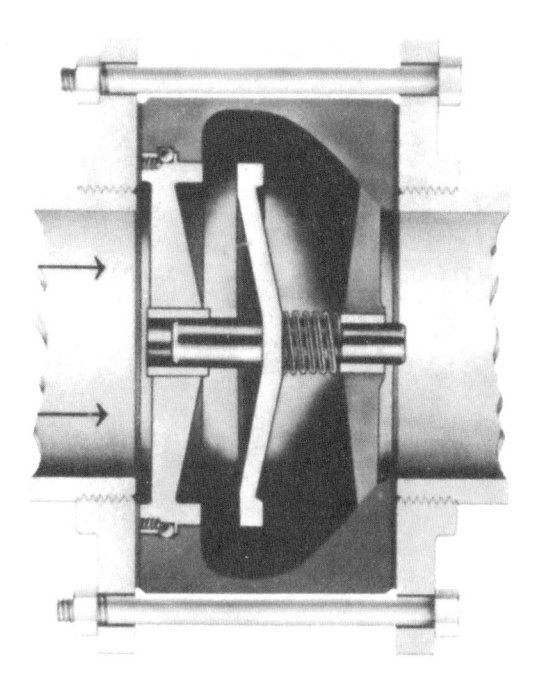

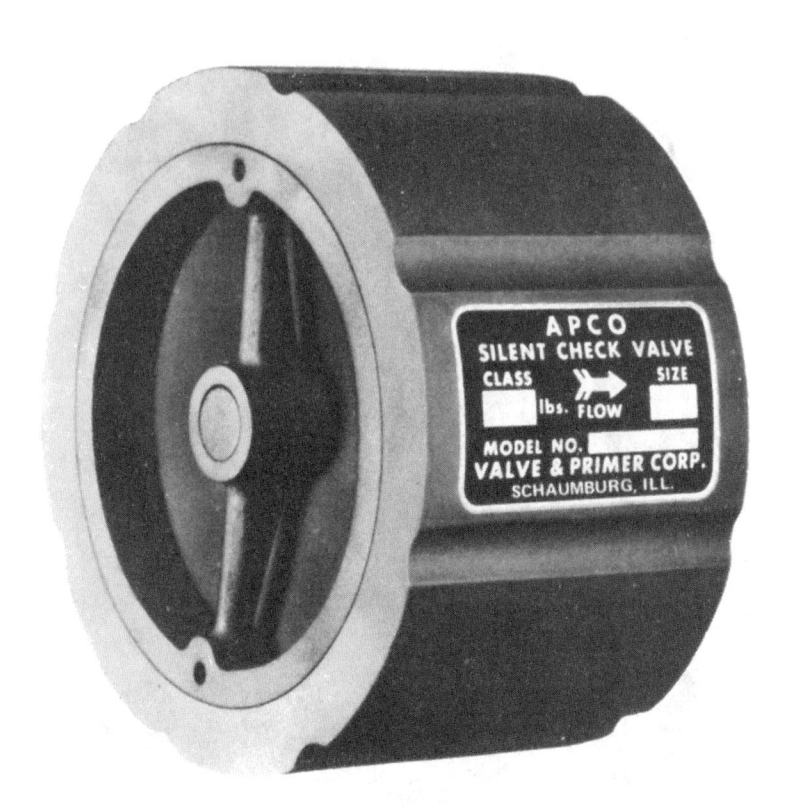

Fig. 18.44   Silent check valves (wafer)
(Permission of APCO/Valve and Primer Corporation)

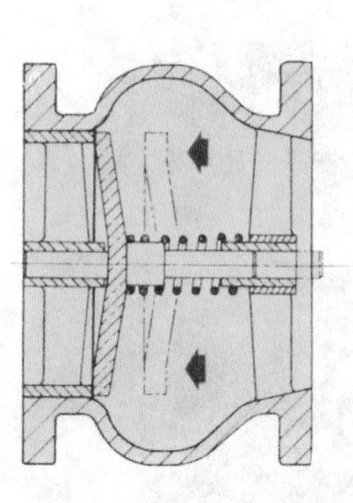

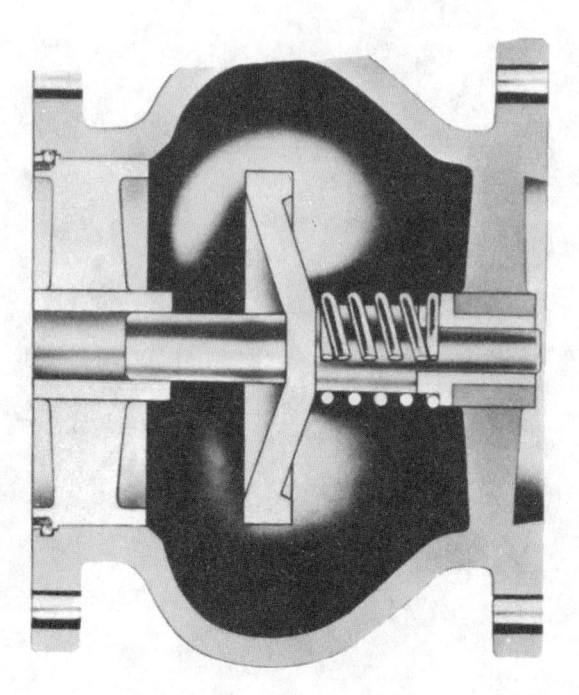

Fig. 18.45 Silent check valves (globe)
(Permission of APCO/Valve and Primer Corporation)

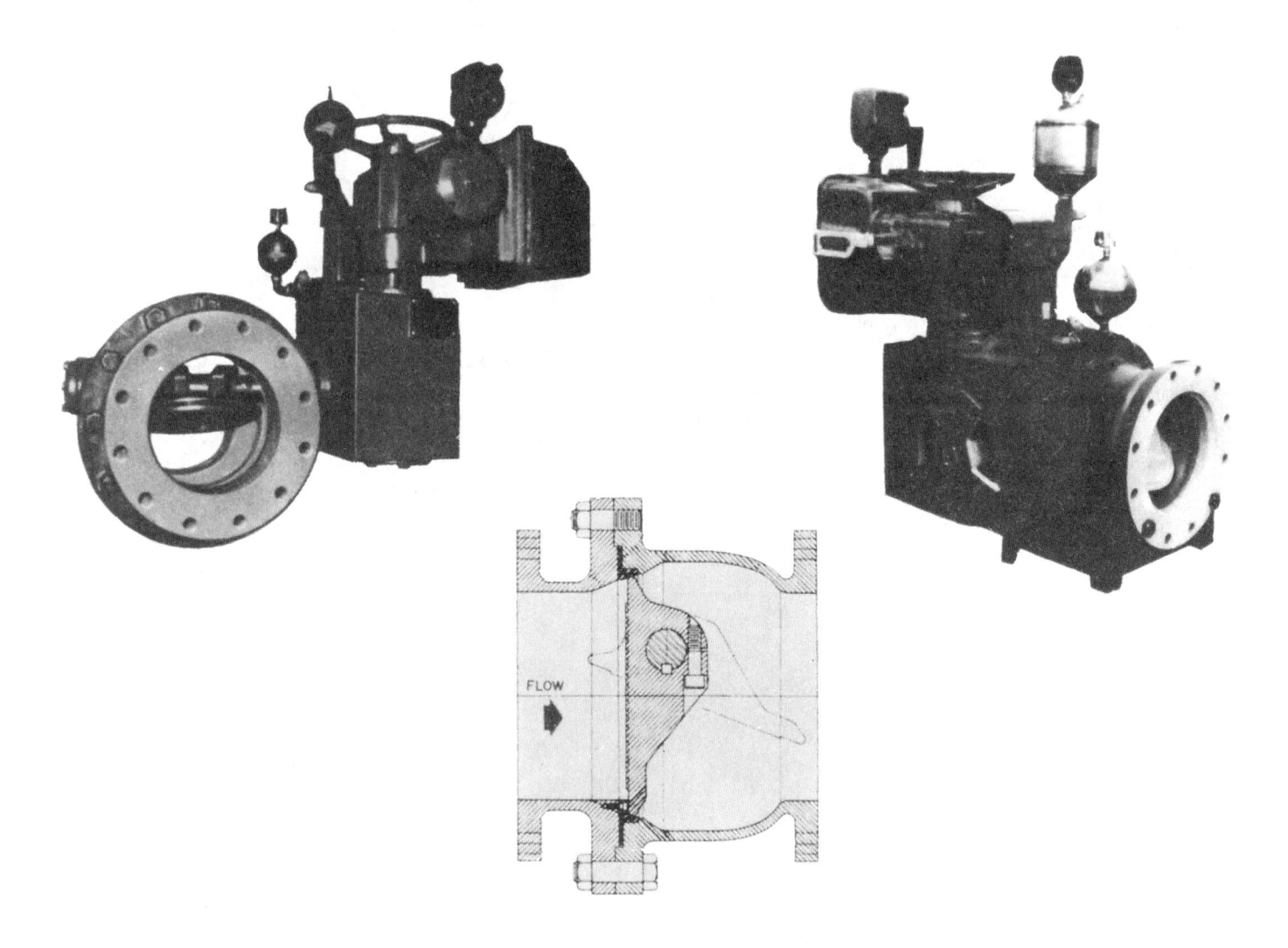

Fig. 18.46 Automatic control check valves
(Permission of APCO/Valve and Primer Corporation)

18.268 Automatic Valves

Water treatment plants usually have a number of automatically operated valves. The simplest type is either open or closed and is not required to operate in an intermediate position. Frequently, these valves are similar to gate valves that have had their threaded stems and handwheels replaced by a smooth shaft and hydraulic piston. Maintenance on these valves is essentially the same as for gate valves.

Other automatically operated valves are used to control flow in water treatment plants and are usually located at some point between tight shut and wide open. These are commonly called modulating valves. A butterfly valve with a hydraulic cylinder operator can be used for this type of service.

The diaphragm-operated globe valve (Figure 18.47) is also used for modulating service. These valves can be equipped with pilot control devices to control pressure, flow, or level either singly or in combination. Maintenance on these valves consists of the following:

1. Periodically clean any strainers in the pilot control system. Scheduling should be adjusted to accommodate the rate at which the strainer collects foreign material.

2. Check the operation of the valve to see that the controls are, in fact, correctly positioning the valve to accomplish the job.

3. If the valve is used in an application where it seldom or never is wide open, it should periodically be exercised manually to cycle from tight shut to wide open. This is to ensure that there is no buildup on the stem that could jam the valve. These valves can be opened wide by drawing all the pressure from the cover chamber. If water does not stop flowing out of the cover chamber when the valve is wide open, this is an indication that the diaphragm is leaking and should be replaced.

QUESTIONS

Write your answers in a notebook and then compare your answers with those on page 383.

18.26A   What is the purpose of valves?

18.26B   List six common types of valves found in water treatment facilities.

18.26C   What maintenance is required by gate valves?

18.26D   What is the purpose of a check valve?

18.26E   Why is backflow prevention by check valves essential in many applications?

End of Lesson 4 of 5 Lessons on MAINTENANCE

Please answer the discussion and review questions next.

DISCUSSION AND REVIEW QUESTIONS
Chapter 18.   MAINTENANCE
(Lesson 4 of 5 Lessons)

Write the answers to these questions in your notebook. The question numbering continues from Lesson 3.

32. What are the uses of a compressor?

33. What items should be maintained on a compressor?

34. What factors can cause wear on gate valve seats?

35. Why should inactive gate valves be operated periodically?

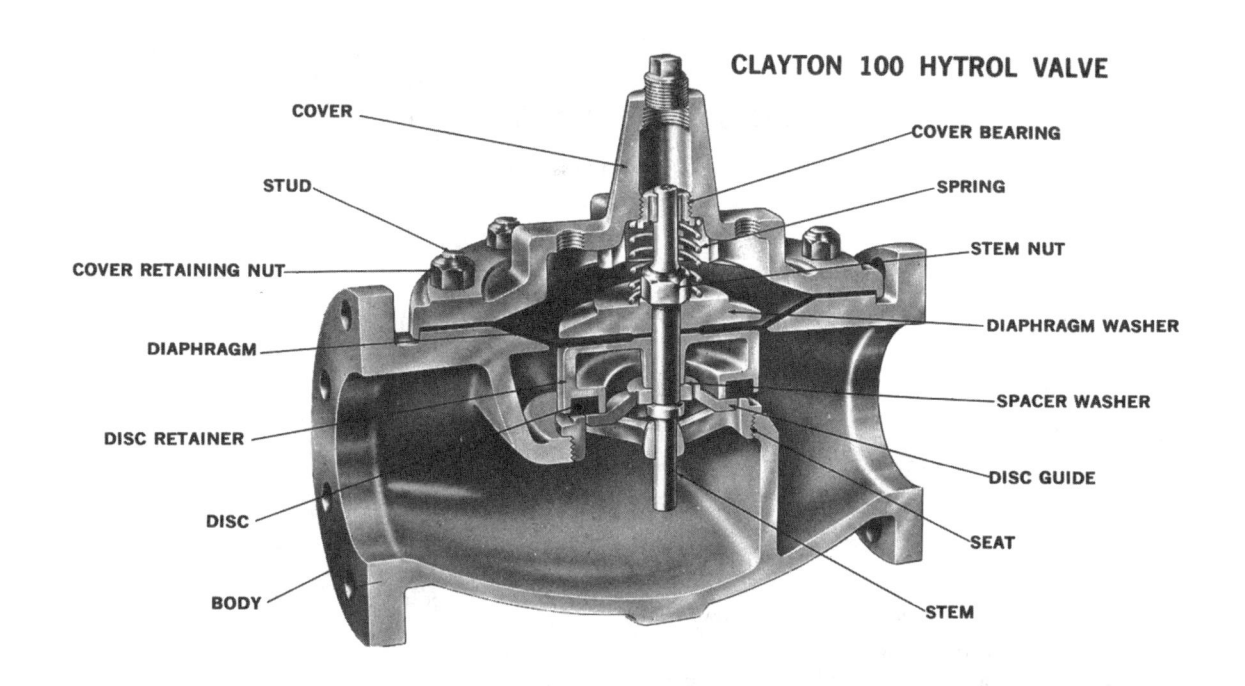

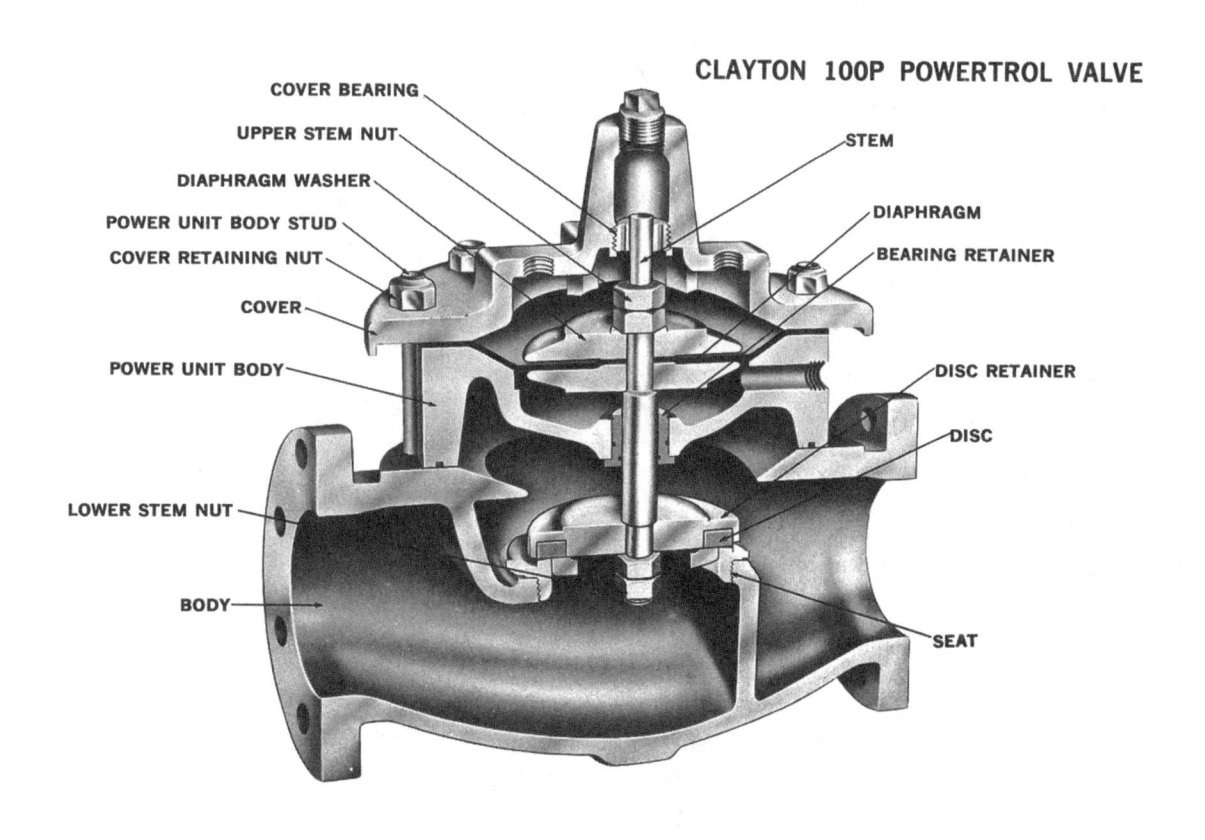

Fig. 18.47 *Diaphragm-operated globe valve*
(Permission of CLA-VAL Co.)

CHAPTER 18. MAINTENANCE

(Lesson 5 of 5 Lessons)

18.3 INTERNAL COMBUSTION ENGINES

18.30 Gasoline Engines

18.300 Need to Maintain Gasoline Engines

In the water treatment departments of all cities, there is occasion to use gasoline-powered engines that drive pumps, generators, tractors, and vehicles. Although we all drive automobiles that are powered by internal combustion engines, are you aware of the fundamentals?

Very few operators actually do the repair of gasoline-powered engines. Although you may not be able to perform the duties of an engine mechanic, there are a number of steps you can take to ensure that your particular engine is well maintained.

At the end of this section, you will have an adequate knowledge of how a gasoline engine operates in order to maintain it so as to provide many hours at optimum performance.

18.301 Maintenance

In order to have an engine that will provide you with many hours of trouble-free operation, it must be well cared for. Please refer to the owner/operator manual for your particular engine. Typical maintenance procedures are as follows:

1. Change engine oil regularly every 25 hours
2. Clean carburetor air filter every 25 hours
3. Blow dust and chaff from louvered engine vanes regularly
4. Clean carburetor fuel filter/screen every 100 hours
5. Lubricate generator or starter motor, as recommended, every 100 hours
6. Lubricate throttle linkage every 100 hours
7. Clean, gap, or replace spark plug every 100 hours
8. Remove carbon deposits from top of piston and valves every 100 to 300 hours

18.302 Starting Problems

Listed below are some items to check if you have problems starting a gasoline engine.

1. No fuel in tank, valve closed
2. Carburetor not choked
3. Water or dirt in fuel lines or carburetor
4. Carburetor flooded
5. Low compression
6. Loose spark plug
7. No spark at plug
    a. Dirty or improperly gapped plug
    b. Broken or wet ignition cables
    c. Breaker points not opening or closing
    d. Magneto grounded

18.303 Running Problems

Check the following items if a gasoline engine does not run properly.

1. Engine misses
    a. Faulty spark plug/gapping
    b. Weak ignition spark
    c. Loose ignition cable
    d. Worn breaker points
    e. Water in fuel
    f. Poor compression
2. Engine surges
    a. Carburetor flooding
    b. Governor spring connected improperly
3. Engine stops
    a. Fuel tank empty
    b. Vapor lock
    c. Tank air vent plugged
4. Engine overheats
    a. Low crankcase oil
    b. Ignition timing wrong
    c. Engine overloaded
    d. Restricted air circulation/high ambient temperature
    e. Poor grade of gasoline

5. Engine knocks
   a. Poor grade of gasoline
   b. Engine under heavy load at low speed
   c. Carbon deposits in cylinder head
   d. Spark advanced too far
   e. Loose connecting rod bearing
   f. Worn or loose piston pin
6. Engine backfires through carburetor
   a. Water or dirt in fuel
   b. Cold engine
   c. Poor grade of gasoline
   d. Sticking inlet valves
   e. Spark plug heat range too hot

QUESTIONS

Write your answers in a notebook and then compare your answers with those on page 383.

18.30A List some possible uses of gasoline engines in water treatment plants.

18.30B What items would you check if you had problems starting a gasoline engine?

18.30C What items could cause a gasoline engine to not run properly?

18.304 How to Start a Gasoline Engine

Because of the wide variety of uses for gasoline-powered engines, no one starting sequence will apply to all engines. In general, gasoline engines can be divided into two groups. In the first group are small engines with magneto ignition and recoil start. Larger engines, with battery-powered ignition and electric start, are in the second group.

18.3040 SMALL ENGINES

The procedure for starting small engines is as follows:

1. Check fuel tank for adequate fuel
2. Ensure fuel shutoff valve from the tank to the carburetor is open
3. Disengage ignition ground (kill switch or mechanism that grounds the spark plug)
4. Check crankcase lubricating oil
5. Set throttle to start position or ¾-full throttle
6. Set choke lever or pull out choke on carburetor
7. Pull recoil starter twice
8. If engine has started, push choke to OFF
9. If engine does not start after two pulls, disengage the choke and try three or four more times

If repeated efforts at starting have been unsuccessful, remove the high-tension voltage wire from the spark. Hold the end of the wire (grasp the insulated portion, not the connector) ⅛ inch (3 mm) from the spark plug. Pull the recoil starter. You should see a small blue spark. This will indicate that the points are opening and closing and providing ignition voltage.

Next, using a ¹³⁄₁₆-inch (20.6-mm) deep socket, remove the spark plug from the cylinder head. Check for a carbon buildup on the electrode. A piece of carbon may have lodged between the center electrode and the side electrode. Also, check to see if the plug is wet with fuel or oil. This could indicate that you have flooded the cylinder with fuel by having the choke on too long. If there is oil residue, it could indicate worn piston rings.

Replace the spark plug with a new one if in doubt. If you must use the one you have, clean it by buffing with a wire brush. Check the gap between the center electrode and side electrode; it should be approximately 0.030 inch (30 thousandths of an inch or 0.76 mm).

Try starting the engine as previously described. If the engine does not sputter or pop, close the fuel shutoff valve, remove fuel sedimentation bowl, and clean it. Open fuel valve. Catch a small amount of fuel in the palm of your hand and examine the fuel for grit or water. If everything looks OK, replace sedimentation bowl and open fuel valve.

Try to start the engine. If you still cannot achieve ignition, you may have other problems that will require further checking by a small-engine mechanic. Do not feel disgruntled; you have checked for the most common problems.

18.3041 LARGE ENGINES

The procedure for starting large engines is as follows:

1. Check fuel tank for fuel
2. Check crankcase for oil
3. Check radiator for coolant (if water-cooled)
4. Set throttle to ½-full position
5. Pull out choke
6. Turn on ignition switch and press start button
7. After four or five engine revolutions, push in the choke
8. Engine should start

After repeated tries, further investigation by a mechanic may be needed.

NOTE: Do not crank engine with the starter motor for more than one minute initially. Wait two minutes and try again for 45 seconds. After three tries, let starter motor cool for 5 minutes before trying again. This will avoid starter motor damage.

Preliminary checks for a large engine that will not start are similar to procedures for small engines. Remove spark plug wires. Test each one by holding it 1/8 inch (3 mm) from the spark plug or ground, and turn engine over with the starter. You should see a small blue spark. If you have no spark, the points are not opening or high-tension voltage is not present from the ignition coil. Check further, as needed. If spark is present, inspect spark plugs. Clean or replace, if needed.

After checking the ignition system, make sure fuel is present at the carburetor. Remove the fuel line at the carburetor and direct it away from you and the engine. Engage starter motor for two revolutions. Fuel should spurt from the line if the fuel pump is working satisfactorily. Replace fuel line and wipe away any fuel that may be present on the engine.

With fuel and ignition voltage present, the engine should start. Repeat start procedure. If you still cannot start the engine, call on your mechanic to look for the problem.

NOTE: Some engines have a low-oil pressure switch that must be manually held in until sufficient oil pressure is present.

Do not use a starting fluid on gasoline engines unless it is a last resort effort to get a critical piece of equipment running. Hard-starting engines should be inspected and repaired by a reliable mechanic.

After an engine has been started, give it an opportunity to warm up before applying the load. Follow manufacturer's recommendations for the starting procedure since there is some variation between different makes of engines.

QUESTIONS

Write your answers in a notebook and then compare your answers with those on page 383.

18.30D If a gasoline engine will not start and the spark plug is wet with fuel or oil, what has happened?

18.30E If a gasoline engine will not start and there is an oil residue on the spark plug, what has happened?

18.30F After an engine has started, what should be done before applying the load?

18.31 Diesel Engines

18.310 How Diesel Engines Work (Figure 18.48)

Diesel engines are similar to gasoline engines and are either two or four cycle. They can be air or water cooled. In general, the diesel engine is of heavier construction to withstand the higher pressures resulting from higher compression ratios.

The diesel does not use spark plugs, but instead relies on heat generated by air compressed in the cylinder (1,000 degrees F (540°C)) to ignite the fuel mixture. The fuel is a petroleum product that is heavier than gasoline and has a higher flash point. Gasoline cannot be used in a diesel because it would start to burn from the heat generated by compression before the piston reached the top of the stroke.

A diesel has no carburetor. The fuel is sprayed (injected) into the cylinder while the cylinder is compressing air. The heat of compression ignites the fuel-air mixture and burns, producing power similar to a gasoline engine. The introduction of fuel into the cylinder must be timed in the same manner as spark to the plug in a gasoline engine. Fuel is pumped by a pumping device that is geared to the crankshaft.

Diesel fuel, unlike gasoline, does not vaporize readily. The fuel must be broken up into fine particles and sprayed into the cylinder. The atomization of fuel is accomplished by forcing the fuel through a nozzle at the top of the combustion chamber. As the fuel combines with the air in the cylinder, it becomes a combustible mixture. Since the diesel engine depends upon the heat of compressed air to ignite the fuel-air mixture, compression pressure must be maintained. Leaking valves or piston rings (causing blowby) cannot be tolerated.

The fuel is also important. The automotive-type diesel is designed to run on a specific type or grade of fuel. Trouble can be expected if an attempt is made to use other than the proper type.

18.311 Operation

In the two-cycle engine, intake and exhaust take place during part of the compression and power strokes; whereas, the four-cycle engine requires four strokes to complete the operating cycle. During one-half of the cycle, the four-stroke acts as an air pump. The two-stroke must have a blower (air pump) to provide the necessary air to expel the exhaust gases and recharge the cylinder with fresh air.

In the two-cycle engine, a series of ports surround the cylinder at a point higher than the lowest position of the piston. These are the intake ports that allow air into the cylinder. The four-cycle engine uses intake valves. The incoming air forces the expended gases out the exhaust valve, leaving the cylinder full of clean air.

As the piston starts its upward stroke, the exhaust valve closes, the intake ports are sealed off by the piston, and the air in the cylinder is compressed. Shortly before the piston reaches the top of the stroke, the required amount of fuel is sprayed into the combustion chamber by the fuel injector. The intense heat of compression ignites the fuel-air mixture with the resulting combustion driving the piston down on its power stroke.

As the piston nears the bottom of the stroke, the exhaust valve opens and the spent gases are released, assisted by the incoming fresh air. The cycle is complete.

18.312 Fuel System (Figure 18.49)

The basic parts of the fuel system are:

1. Primary fuel filter
2. Secondary fuel filter
3. Fuel injection pump
4. Fuel injector

Maintenance 365

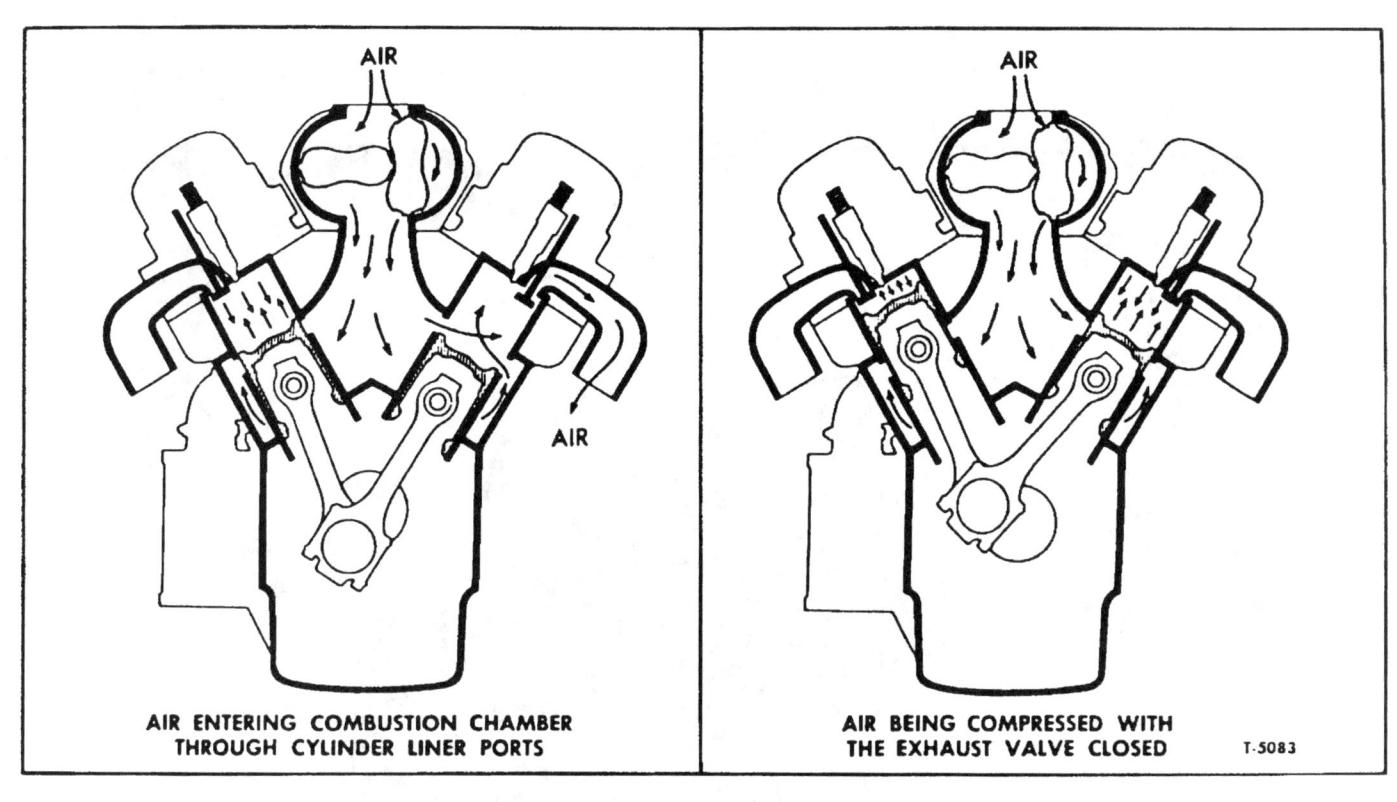

Scavenging and Compression

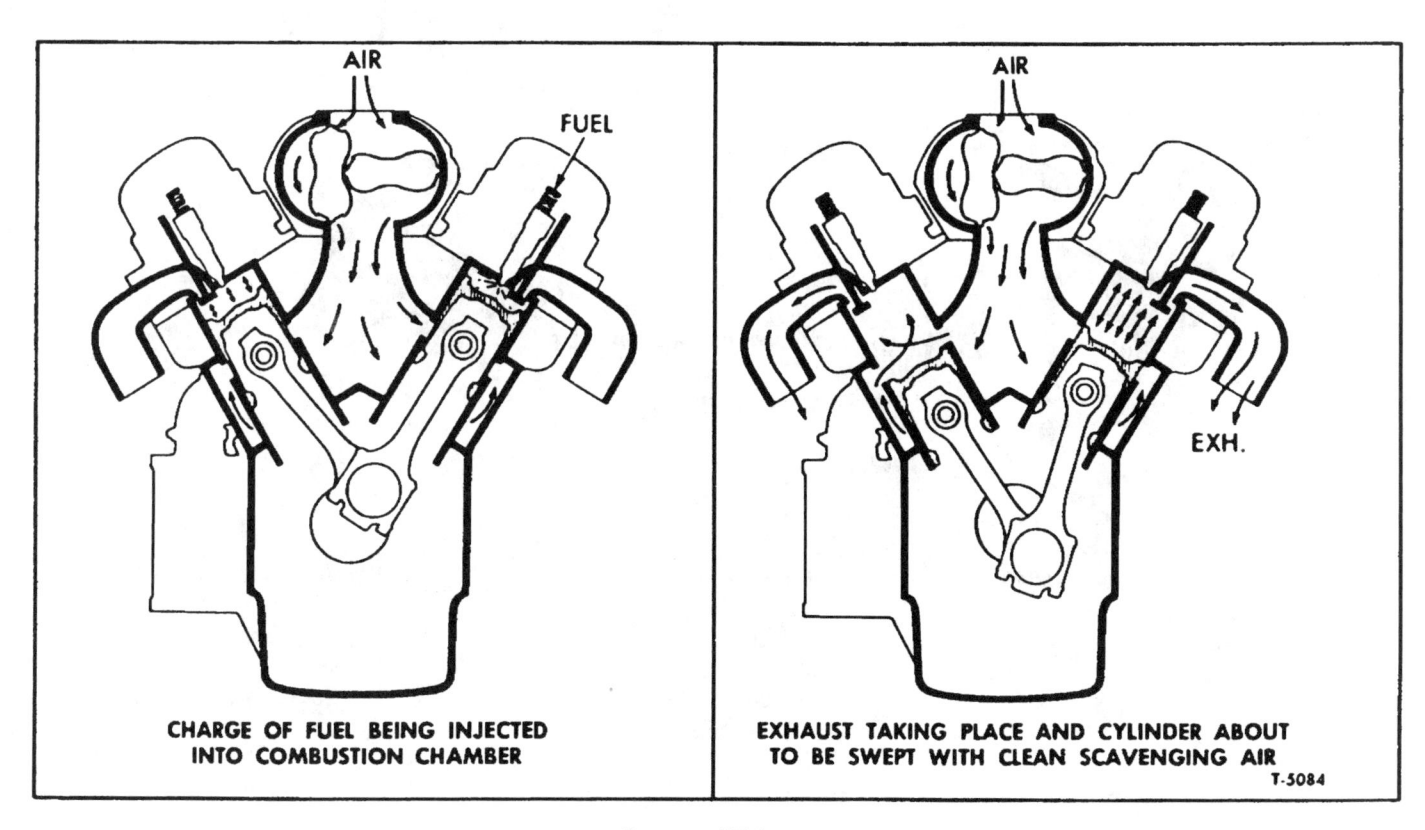

Power and Exhaust

Fig. 18.48 How diesel engines work
(Source: GMC Truck Overhaul Manual Series 53, permission of General Motors Corp.)

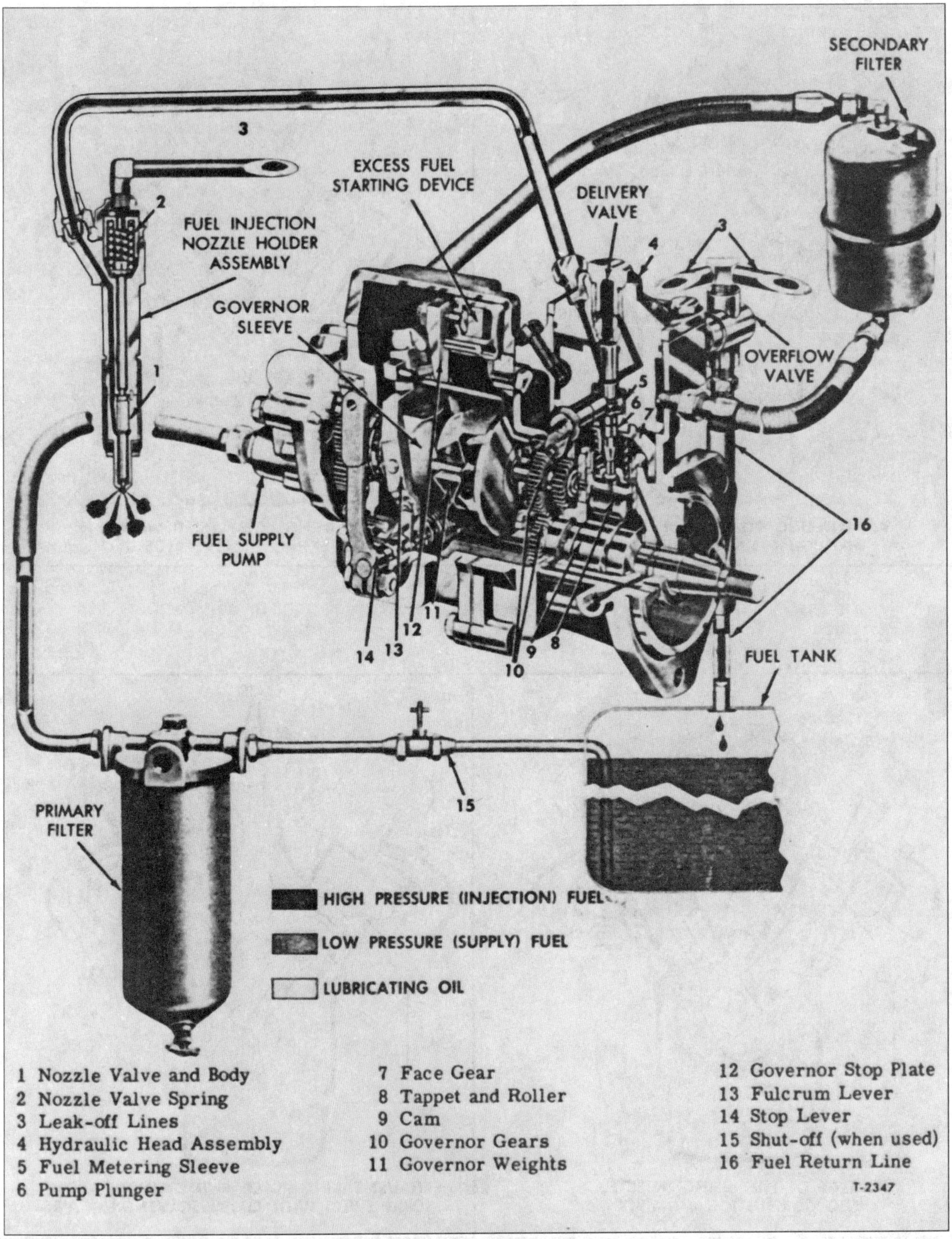

Fig. 18.49 Diesel engine fuel system
(Source: Maintenance Manual, permission of General Motors Corp.)

The primary filter removes all coarse particles from the fuel and the secondary filter removes any minute particles that remain. This ensures a clean fuel that will not clog the injector pump or fuel injectors. The heart of the fuel system is the injection pump (Figure 18.50). This pump is a gear-type positive displacement pump that can deliver fuel to the injector at a very high pressure. Incorporated into the pump is a timing advance mechanism to advance or retard the instant when fuel is injected into the cylinder. At high engine speed, injection would take place sooner in the cycle. The reverse happens for lower speeds.

A governor, which uses centrifugal weights and is driven by the pump shaft, activates a fuel control unit. When engine speed increases, the weights are thrown toward their outer limit. Geared to the assembly, the fuel control valve is opened wider allowing more fuel to flow to the injector.

We now have higher engine speed, advanced timing of injection, and the necessary fuel to sustain the faster operation. When the engine is slowed, the reverse takes place.

Fuel under pressure is fed from the injection pump to the appropriate fuel nozzles. When the pressure reaches approximately 3,000 psi (20,700 kPa or 207 kg/sq cm), the valve in the injector opens allowing fuel to be injected into the combustion chambers. As line pressure drops, the return spring closes the nozzle valve. Fuel left in the line is fed back to the pump through leak-off lines.

18.313 Water-Cooled Diesel Engines

Usually, the larger diesel engines are of the water-cooled type, similar to gasoline engines. In order to deliver a sustained

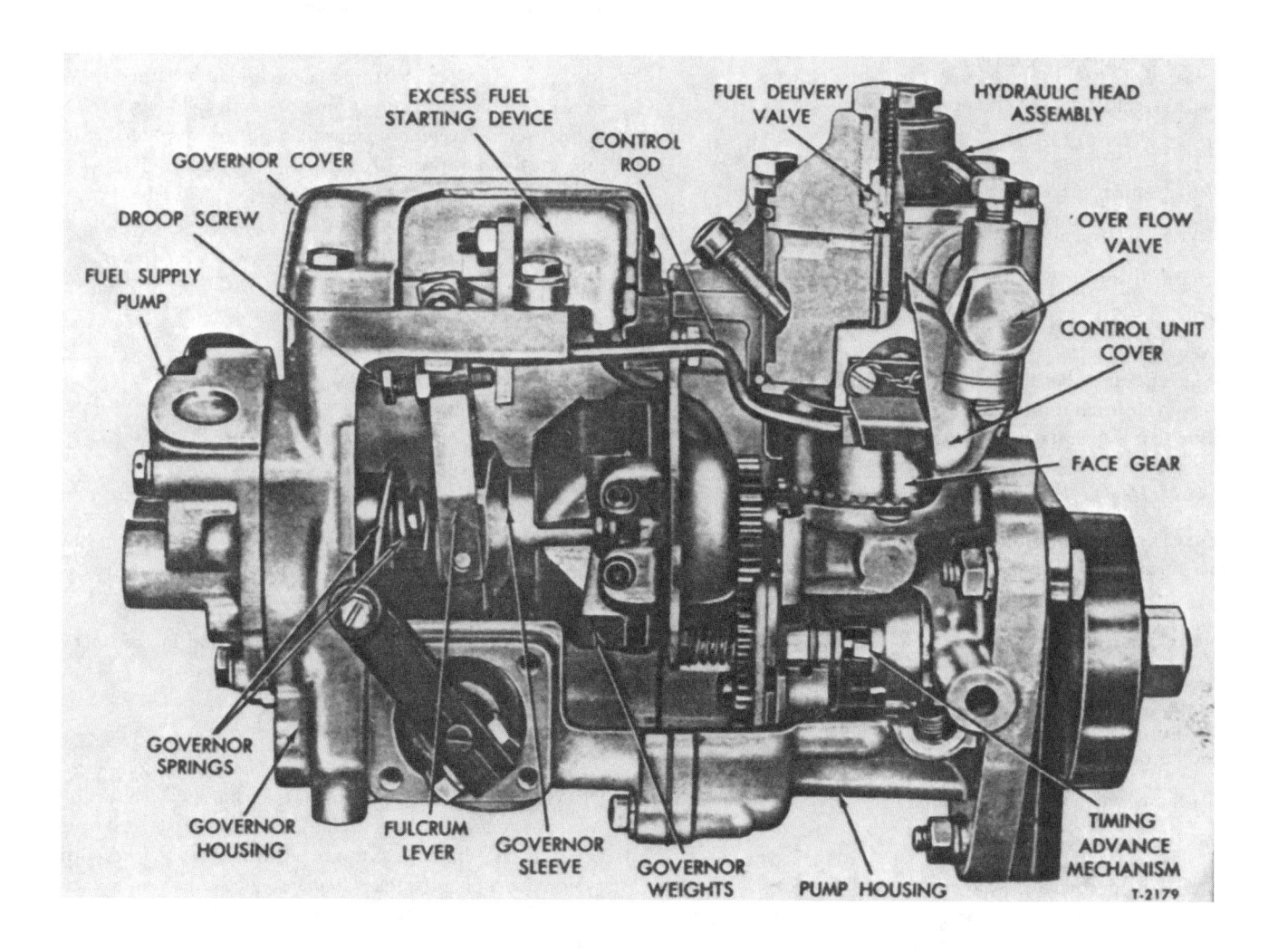

Fig. 18.50 Cutaway view of fuel injection pump for 6-cylinder engine
(Source: Maintenance Manual, permission of General Motors Corp.)

amount of high horsepower, an effective cooling system is necessary to dissipate the extreme heat of combustion. Because of this fact, a water-cooled engine of comparative horsepower to the air-cooled will cost more to manufacture, and, subsequently, to maintain.

18.314 Air-Cooled Diesel Engines

When a lighter weight, lower horsepower, and more compact engine is desired, the air-cooled engine will serve your needs. You get the benefits of a diesel engine in a smaller package.

There are some definite advantages to the diesel engine over the gasoline engine. The initial cost is greater for the diesel; however, the diesel:

1. Requires less maintenance because there are:
   a. No plugs
   b. No contact points to pit
   c. No ignition coils or high-tension wires
   d. Fewer tune-ups
2. Is cheaper to operate because:
   a. Diesel fuel may be cheaper
   b. Fuel efficiency is better

Perhaps the biggest drawbacks against diesel engines are:

1. Initial investment costs
2. Repair costs

The pros and cons must be weighed to provide you with an engine that will meet your particular needs. Whichever engine you select, remember that a well-cared-for engine will be there to serve you when it is needed and will provide trouble-free operation that is essential to most users.

18.315 How to Start Diesel Engines

Diesel engines vary in size and use and have varied starting procedures. Follow the manufacturer's suggested procedures for your particular engine. As with the gasoline engine, check fuel, oil, and coolant.

To start a diesel engine, the procedures are as follows:

1. Push in "stop" control
2. Set throttle to ⅓-full
3. Turn on switch and engage starter
4. Engine should start

Some engines have glow plugs that are energized when the switch is placed in the start position. They preheat the air-fuel mixture in the cylinder to aid in starting. After the engine is started, maintain the lower RPMs on the engine tachometer and allow the engine to warm up. The warm-up period is vital to the diesel engine for efficient engine performance. When operating the engine, maintain adequate engine RPMs as recommended by the manufacturer.

When a diesel engine will not start after repeated tries, a small amount of starting fluid sprayed into the air intake may be needed to start the engine. If you use starting fluid, do not get carried away with its use; a little goes a long way. Use it only as a last resort or as specified by the manufacturer. If your efforts have failed to start the engine, have a mechanic who is familiar with diesel engines determine the cause of the problem.

18.316 Maintenance and Troubleshooting

For detailed maintenance procedures for your diesel engine, see your diesel manufacturer's service manual.

TROUBLESHOOTING

Certain abnormal conditions that sometimes interfere with satisfactory engine operation are listed in this section.

Satisfactory engine operation depends primarily on:

1. An adequate supply of air compressed to a sufficiently high compression pressure
2. The injection of the proper amount of fuel at the right time

Lack of power, uneven running, excessive vibration, stalling at idle speed, and hard starting may be caused by either low compression, faulty injection in one or more cylinders, or lack of sufficient air. Several problems that can reduce engine performance are listed below:

1. Misfiring cylinders
2. Improper compression pressure
3. Engine out of fuel
4. Inadequate fuel flow
5. Excessive crankcase pressure
6. Excessive back pressure
7. Improper air box pressure
8. Restricted air inlet
9. Low oil pressure
10. Improper engine coolant operating temperature

Solutions to these problems can be found in the operation and maintenance instructions for the engines.

18.32 Cooling Systems (Figure 18.51)

In an air-cooled engine, the heat generated by combustion is dissipated by the air circulating past the louvered cylinder block. With a water-cooled system, the same effect is achieved by using water. Each cylinder is surrounded with a water jacket through which the coolant (water) circulates. This is accomplished by a water pump that is belt driven from the crankshaft. The heat transfers from the cylinder wall to the water which, in turn, is pumped back to the radiator where the heat is dissipated. A fan mounted on the same shaft as the water pump ensures that a large volume of air is blown across the radiator coils to facilitate rapid dispersal of heat. The cooled water is then pumped back into the engine.

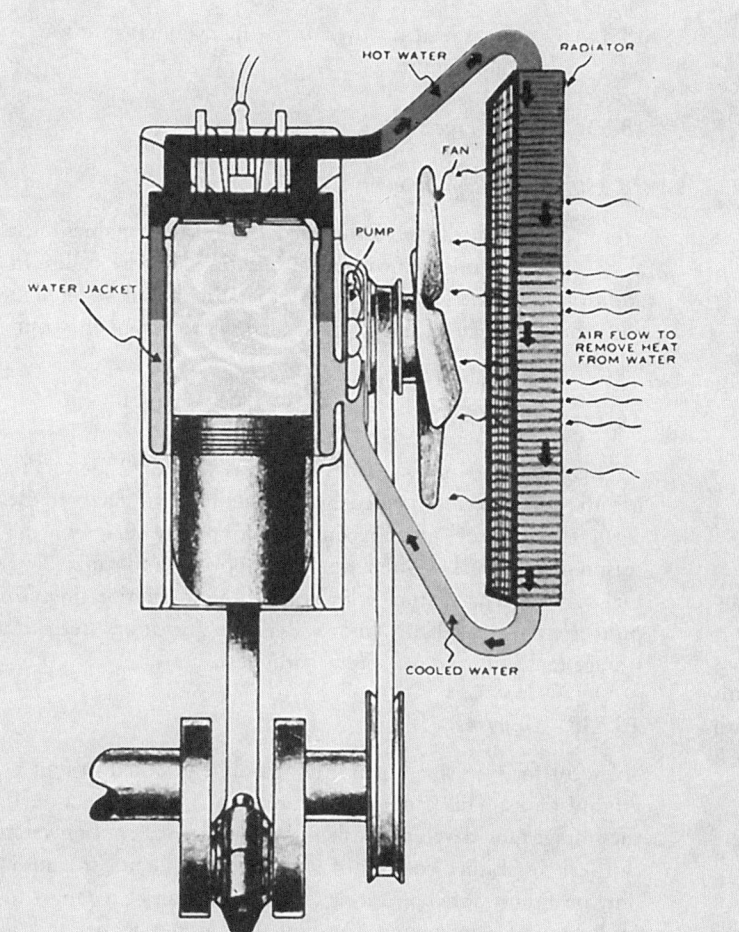

Maintenance 369

Engine temperatures are regulated by transferring excess heat to surrounding air.

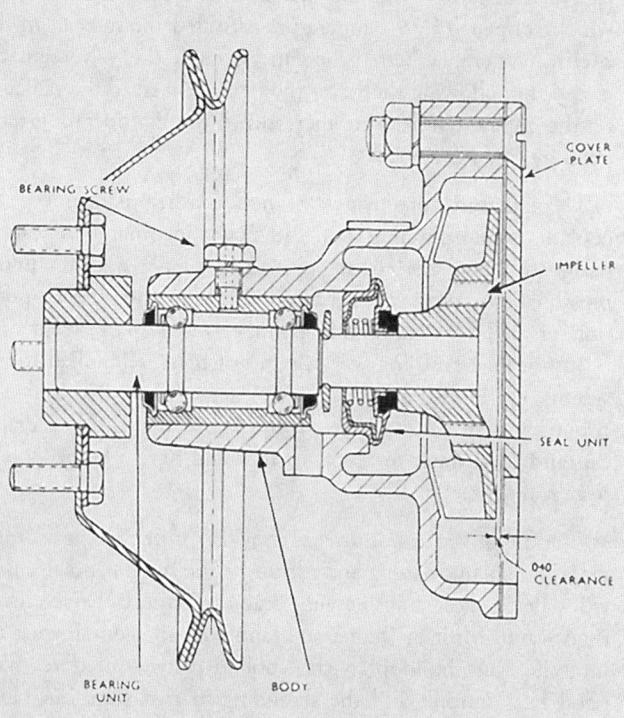

Shaft ball bearings are sealed at each end to keep lubricant in and water out of bearings. A spring-loaded seal (in color) is used to avoid water leakage around pump shaft. Note clearance between impeller and cover plate.

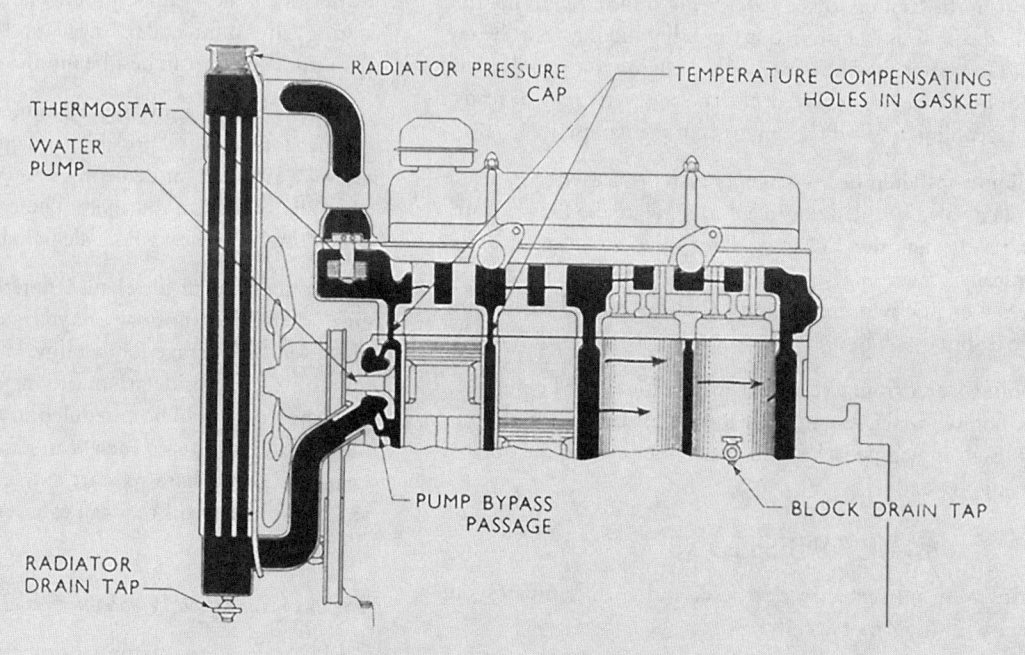

With water jackets entirely around each cylinder and valve, there is a great amount of area exposed to the circulating coolant.

Fig. 18.51 Water cooling system
(Source: Automotive Encyclopedia, permission of the Goodheart-Wilcox Co., Inc.)

Internal combustion engines operate more efficiently when their temperature is maintained within narrow limits. This objective is achieved with the insertion of a thermostatic valve, which is opened by a temperature-activated thermostat, in the cooling system. When the engine is cold, the valve remains closed, not allowing the water to circulate back to the radiator. As the engine temperature increases to normal operating temperature, the valve opens.

The radiator cap provides a function other than preventing coolant from splashing out the filter opening. The cap is designed to seal the cooling system so that it operates under pressure. This improves cooling efficiency and prevents evaporation of coolant. The boiling point of water is 212 degrees F (100°C). However, for every pound of pressure applied to the system, the boiling point rises 3.25 degrees F (1.8°C). If your cooling system had a 15 psi (100 kPa or 1 kg/sq cm) radiator cap and used water for coolant, it would have a boiling point near 260 degrees F (127°C).

The use of coolant/antifreeze provides protection against the radiator coolant freezing and rupturing the system and also provides better heat transfer and heat dissipation characteristics than water. Most of the name-brand coolants contain rust inhibitors. Rust buildup in the cooling system interferes with good heat transfer and the sloughing of rust scale can block narrow passages.

Stationary internal combustion engines, such as those that are used to drive pumps and generators at water treatment plants, are often installed in a building where free circulation of air for radiator cooling may not be possible. For these installations, a liquid-to-liquid heat exchanger often replaces the radiator (which is a liquid-to-air heat exchanger). In this case, instead of the heat in the cooling jacket water being transferred to the surrounding air, it is transferred to another liquid, usually tap water. This water may be wasted if the engine is a standby unit and not operated very much, or the cooling water may be recovered in a cooling tower if the engine is in regular use.

In liquid-to-liquid heat exchanger systems, a thermostatically controlled valve is usually installed to regulate the flow of cooling water through the exchanger. This valve should be checked periodically to see that it: (1) provides sufficient water flow when the engine is running, and (2) closes off tight when the engine is shut down to prevent waste.

Cooling water from a heat exchanger should not be put back into a potable water system. Any leakage in the heat exchanger could result in engine jacket coolant contaminating the potable water supply.

QUESTIONS

Write your answers in a notebook and then compare your answers with those on page 383.

18.31A Why is gasoline not used as a fuel in diesel engines?

18.31B List the four basic parts of a diesel fuel system.

18.31C What is the purpose of the fuel injection pump?

18.32A How is heat removed from the cylinders in a water-cooled engine?

18.33 Fuel Storage

18.330 Code Requirements

Storage and use of fuels for internal combustion engines must always be in accordance with local building and fire codes. In addition, the water treatment plant operator should be familiar with the particular problems associated with each of the commonly used fuels.

18.331 Diesel Fuel

This fuel comes in two grades known as #1 and #2. Be sure to use the grade recommended by the engine manufacturer. Be aware that the fuel grade recommendation may vary with the season. Diesel fuel is often stored in aboveground tanks. The fuel may be kept in storage for years without deteriorating. To protect stored diesel fuel from water contamination, keep the storage tanks full and use special additives.

18.332 Gasoline

Except for very small quantities, gasoline is stored in underground tanks. This can result in problems for the operator. If the storage tank develops a leak, either fuel can leak out or water can leak in. Either condition is undesirable. Fuel loss cannot only be an unwanted operating expense, but can be a danger to underground plant piping. Gasoline deteriorates rubber and, if the piping is put together with rubber gaskets or rings, the deterioration can result in major leaks and broken couplings. The owner of a leaking underground tank can be liable for any resulting environmental damage and also responsible for the cleanup. Fuel loss can best be monitored by careful accounting.

Water leakage into underground fuel tanks can result in engine stoppages and possible damage to the engine. Special devices are available for detecting water in gasoline tanks and such a test should be run routinely. These devices can be obtained by contacting your local wholesale fuel distributor.

Gasoline, unlike diesel fuel, deteriorates in storage. For engines that are in normal everyday use, this is not a problem. However, fuel storage for standby, engine-driven equipment requires further consideration. Engine operation and fuel tank replenishment should be scheduled so that at least half of the gasoline in storage is used each year. Failure to do this can result in engines that are hard to start and in the formation of varnish and gummy deposits that can cause malfunctions in the parts of the fuel system.

18.333 Liquified Petroleum Gas (LPG)

LPG is usually a mixture of propane and butane. The proportion of each is varied according to the weather temperature. The cooler the weather, the greater the proportion of propane.

This fuel is always stored under pressure in aboveground tanks that are located out in the open. LPG does not deteriorate in storage and, therefore, can be kept for many years.

LPG is heavier than air and will collect in low areas if there is any leakage. This poses an extremely dangerous explosive threat that treatment plant operators must constantly guard against.

18.334 Natural Gas

This fuel is usually obtained from the local gas company through a metered connection from their distribution system. There is no on-site storage.

Natural gas, being lighter than air, tends to rise and dissipate from leaks and therefore is less dangerous to handle than LPG. Explosions can occur, however, if the leakage is confined inside a building.

18.34 Standby Engines

Internal combustion engines must be run periodically to ensure that, when needed, they will function properly. An engine that is not in regular service should be started up and test run at least once a week. The test run should be long enough for the engine to come up to its normal operating temperature before the engine is shut down. If at all possible, run the engine under its normal load. Just idling an engine for 20 minutes does not give you much of an indication whether it can handle a load. Check and make note of the engine instruments. Look for changes that may indicate a need for repairs. Lube oil pressure and intake manifold pressure (on spark ignition engines without supercharging or fuel injection) are two key indicators of engine condition.

QUESTIONS

Write your answers in a notebook and then compare your answers with those on page 383.

18.33A The storage and use of fuels for internal combustion engines must be in accordance with what codes?

18.33B List four types of fuels commonly used by internal combustion engines.

18.34A How often should standby internal combustion engines be test run when not in regular service?

18.34B Under what conditions should standby engines be test run?

18.4 CHEMICAL STORAGE AND FEEDERS[19]

18.40 Chemical Storage

Certain dry chemicals such as alum, ferric chloride, and soda ash are *HYGROSCOPIC*.[20] These chemicals require special considerations to protect them from moisture during storage. Dry quicklime should be kept dry because of the tremendous heat that is generated when it comes in contact with water. This heat is sufficient to cause a fire.

Some liquid chemicals such as sodium hydroxide (caustic soda) should not be exposed to air because of the formation of calcium carbonate (a solid) due to the carbon dioxide in the air. Also, some liquid chemicals may "freeze." A 50-percent sodium hydroxide solution becomes crystalized (forms a solid) at temperatures below 55°F (13°C). Therefore, a heater may be required to keep the storage area warm or the solution may have to be diluted down to a 25-percent solution.

Potassium permanganate ($KMnO_4$) can be kept indefinitely if stored in a cool, dry area in closed containers. The drums should be protected from damage that could cause leakage. Potassium permanganate should be stored in fire-resistant buildings, having concrete floors as opposed to wooden floors. It should not be exposed to intense heat, or stored next to heated pipes. Any organic solvent, such as greases and oils in general, should be kept away from stored $KMnO_4$.

Potassium permanganate spills should be swept up and removed immediately. Flushing with water is an effective way to eliminate spillage on floors. Potassium permanganate fires should be extinguished with water.

Carbon should be stored in a clean, dry place, in single or double rows, and with access aisles around every stack for frequent fire inspections. The removal of burning carbon will thus be facilitated. Carbon should never be stored in large stacks.

The storage area should be of fireproof construction, with self-closing fire doors separating the carbon room from other sections. Storage bins for dry bulk carbon should be of fireproof construction equipped for fire control by the installation of carbon dioxide equipment, or should be so arranged that they can be flooded with a fine spray of water.

Carbon storage areas should be protected from contact with flammable materials. (Carbon dust mixed with oily rags or chlorine compounds can ignite in spontaneous combustion.) *SMOKING SHOULD BE PROHIBITED AT ALL TIMES DURING THE HANDLING AND UNLOADING OF CARBON AND IN THE STORAGE AREA.* Carbon should not be stored where a spark from overhead electric equipment could start a fire. If a fire occurs, the carbon monoxide hazard should be taken into account.

Electric equipment should be protected from carbon dust and cleaned frequently or, better, explosion-proof electric wiring and equipment should be used. (The heat from a motor may ignite the accumulated carbon dust; this material, especially when damp, is a good conductor of electricity and could short-circuit the mechanism.)

---

[19] For additional information on chemical feeders, see Chapter 13, "Fluoridation," Section 13.30, "Chemical Feeders."
[20] *Hygroscopic* (hi-grow-SKAWP-ick). Absorbing or attracting moisture from the air.

Polymer solutions will be degraded (lose their strength) by biological contamination. A good cleaning of polymer storage tanks is recommended before a new shipment is delivered to the plant.

Liquid chemical storage tanks should have a berm or earth bank around the tanks to contain any chemicals released if the tank fails due to an earthquake, corrosion, or any other reason.

Some chemicals such as chlorine and fluoride compounds are harmful to the human body when they are released as the result of a leak. Continual surveillance and maintenance of the storage and feeding systems are required.

18.41 Drainage from Chemical Storage and Feeders

Safety regulations prohibit a single drainage pit that can accept and contain both acid and alkali chemicals because of the possibility of an explosion whenever these two types of chemicals come in contact. Also, any organic chemical waste such as a polymer solution should not be allowed to be discharged into a pit or sump that could also receive a waste from oxidizing chemicals such as potassium permanganate ($KMnO_4$) because of the possibility of a fire. Therefore, separate drainage systems or a high dilution of certain chemicals is necessary for a safe drainage system.

18.42 Use of Feeder Manufacturer's Manual

Water treatment plants will have a number of chemical feeders to accurately control the rate at which chemicals are fed into the water as a part of the treatment processes. There are many types of feeders and they work on many different principles. Study the feeder manufacturer's manual that you should find in the treatment plant library for details on maintaining the equipment. Additional information on chemical feeders is contained in specific chapters on treatment processes that require the use of chemical feeders.

18.43 Solid Feeders

Solid feeders usually handle powdered material and usually have many moving mechanical parts that need adjustment, lubrication, and replacement when worn. The chemical supply is usually stored in a hopper. Keep the hopper and feeder clean and dry in order to prevent bridging (a hardened layer that can form an arch and prevent flow) of the chemical in the hopper and clogging in the feeder.

18.44 Liquid Feeders

Liquid feeders handle many types of chemicals, some of which may be corrosive or have a tendency to plug up the mechanism. The key to reliable operations is constant vigilance and cleaning, as needed.

18.45 Gas Feeders

The principal chemical found in gaseous form at water treatment plants is chlorine. Chlorine is quite poisonous to humans and must be handled with great caution.

18.46 Calibration of Chemical Feeders[21]

To ensure the accuracy of chemical feed rates, liquid-chemical metering pumps and dry-chemical feed systems should be tested and calibrated when first installed and at regular intervals thereafter. This section presents general procedures for calibrating several types of liquid- and dry-chemical feeders.

18.460 Large-Volume Metering Pumps

Pumps metering chemicals such as liquid alum deliver a relatively large volume of chemical in a short time period. These pumps can be accurately calibrated with a clear plastic sight tube and a stopwatch (Figure 18.52).

To calibrate the pump, fill the sight tube from the chemical solution tank, then set the valve so the tube is the only source of liquid chemical entering the pump. Run the pump for exactly one minute (use the stopwatch) at each of five or six representative settings of the pump-control scale. Record the amount pumped at each setting as observed in the sight tube. Use this information to develop curves of pump setting versus chemical dose in mg/L or chemical feed in gallons per day for your plant (Figure 18.53).

The graph developed by this process is called a calibration curve. It can be used to determine the pump setting needed to deliver a required chemical feed rate, or the commonly used range of feed rates can be marked in gallons per day directly on the pump control panel.

18.461 Small-Volume Metering Pumps

Pumps metering a chemical such as sodium hexametaphosphate, a lime feed solubility enhancer, feed a very small volume per day. The procedure for calibration of these pumps is similar to the procedure for large-volume units. For very low feed rates, pumping times of longer than one minute may be required to give accurately measurable results.

Once the test data have been recorded, convert the test results to appropriate units and draw a calibration curve to be used as for the larger pumps.

18.462 Dry-Chemical Systems

Dry-chemical feed systems are used for chemicals such as activated carbon, fluoride, and lime. Two types of systems are common, the rocker-dump type and the helix-feed type. The rocker-dump chemical feed system uses a scraper moving back and

---

[21] For additional information on calibration of chemical feeders, see Volume I, Appendix, "How to Solve Water Treatment Plant Arithmetic Problems," Section A.131, "Chemical Doses," pages 616–618.

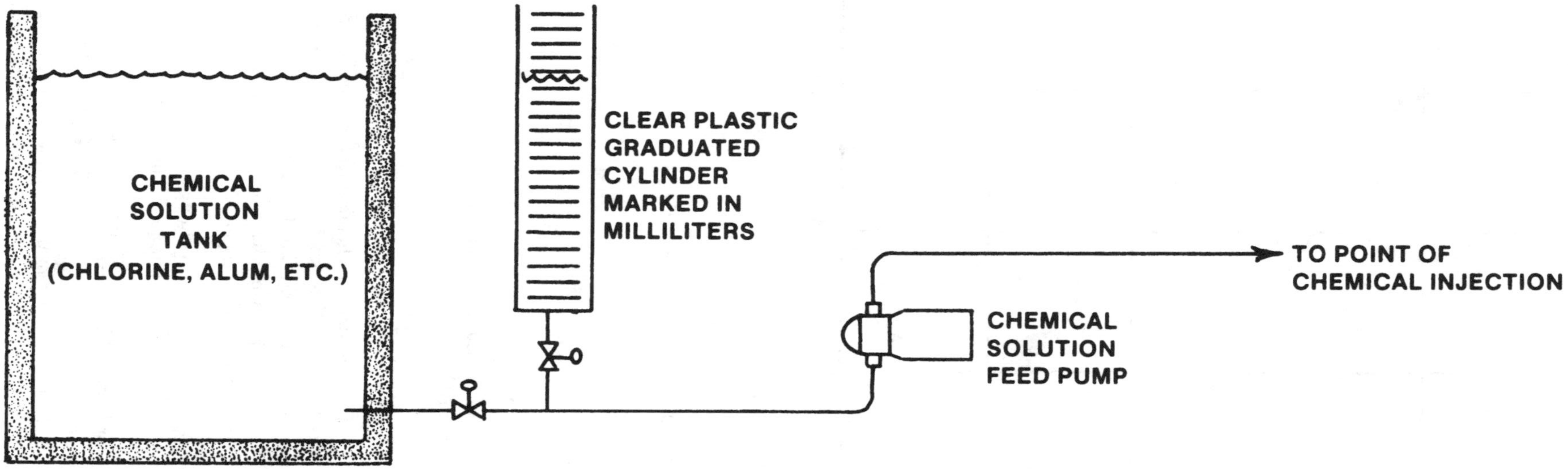

THE FEED RATE OF A CHEMICAL SOLUTION FEED PUMP CAN BE DETERMINED BY MEASURING THE AMOUNT OF SOLUTION WITHDRAWN FROM A GRADUATED CYLINDER IN A GIVEN TIME PERIOD. ALLOW THE CYLINDER TO FILL WITH SOLUTION. THEN CLOSE THE VALVE ON THE LINE FROM THE TANK SO THE FEED PUMP TAKES SUCTION FROM THE CYLINDER ONLY. OBSERVE THE MILLILITERS OF SOLUTION USED IN ONE MINUTE. COMPARE THIS RESULT WITH THE DESIRED FEED RATE AND ADJUST THE FEED PUMP ACCORDINGLY.

Fig. 18.52 Calibration of a chemical feed pump

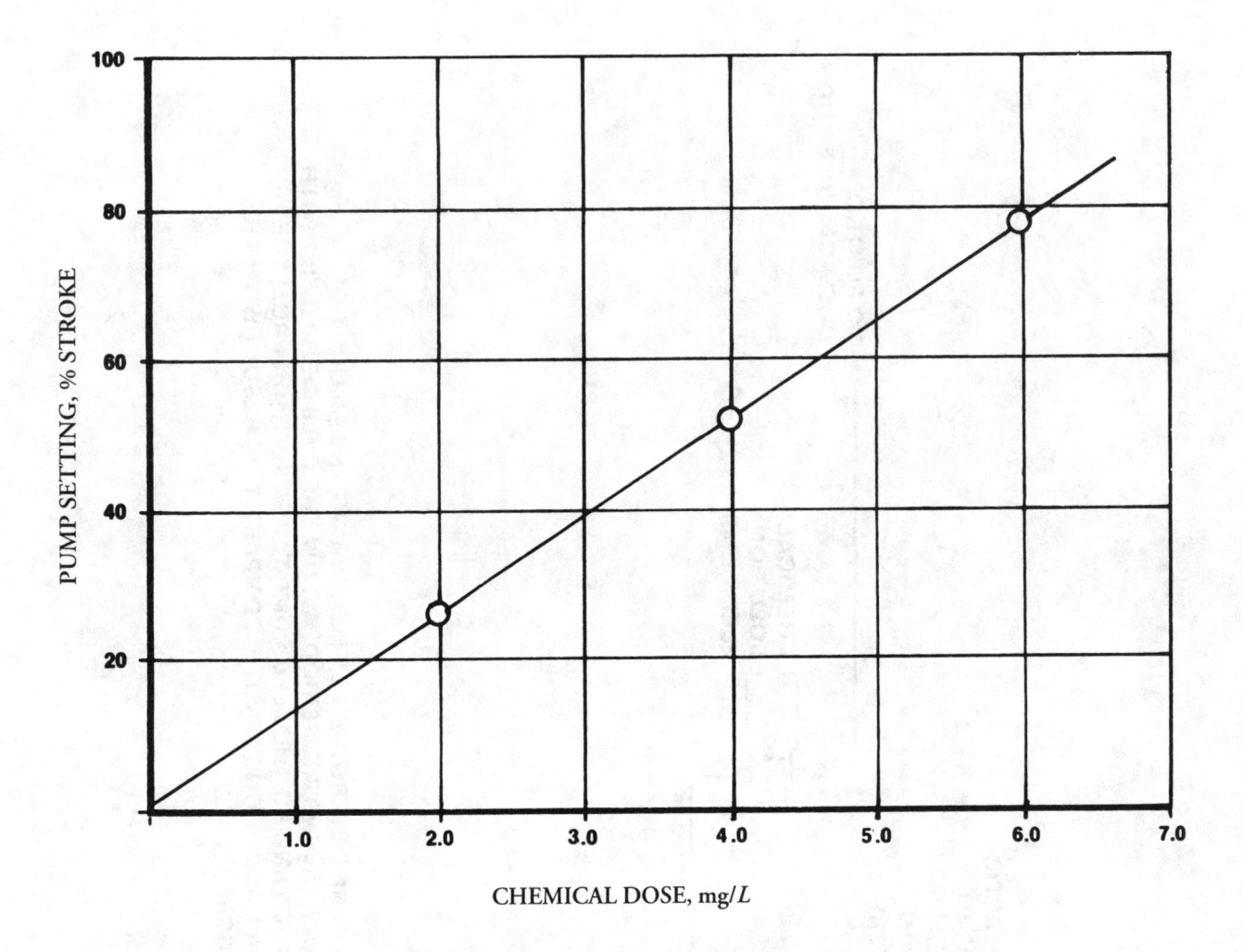

Fig. 18.53 Chemical feed pump settings for various chemical doses

forth on a platform located at the bottom of a hopper filled with dry chemical. The platform may be adjusted up and down to regulate the thickness of the ribbon of chemical, and the length of stroke for the scraper can be adjusted, usually by means of an indicator on an exterior arm.

The helix-type feeder feeds the dry chemical with a rotating screw (helix). The feed rate is adjusted by varying the drive-motor speed. The speed can usually be varied from 0 to 100 percent.

To calibrate either type of feed system, choose five or six representative settings of the arm (rocker-dump) or of the motor speed control (helix type), and at each of the settings catch the amount of chemical fed during a precisely measured time interval. Next, weigh each volume of chemical as accurately as possible and convert the information into pounds per day. Use the data to construct a calibration curve with one axis representing feeder settings and the other representing pounds per day. The curve is used in the same manner as the curves for liquid-feed pumps.

FORMULAS

To determine the chemical feed rate or flow from a chemical feeder, we need to know the amount or volume fed during a known time period. The flow from a chemical feeder can be calculated by knowing the volume pumped from a chemical storage tank and the time period.

$$\text{Flow, GPM} = \frac{\text{Volume Pumped, gal}}{\text{Pumping Time, minutes}}$$

$$\text{or Flow, GPD} = \frac{(\text{Volume Pumped, gal})(24 \text{ hr/day})}{\text{Pumping Time, hour}}$$

Liquid polymer feed rates are often measured in pounds per day. To calculate this feed rate, we need to know the strength of the polymer solution as a percent or as milligrams per liter, the specific gravity of the solution, the volume pumped, and the time period.

$$\text{Polymer Feed, lbs/day} = \frac{(\text{Poly Conc, mg}/L)(\text{Vol Pumped, m}L)(60 \text{ min/hr})(24 \text{ hr/day})}{(\text{Time Pumped, min})(1{,}000 \text{ m}L/L)(1{,}000 \text{ mg/gm})(454 \text{ gm/lb})}$$

To determine the actual feed from a dry chemical feeder, we need to know the pounds of chemical fed and the time period.

$$\text{Chemical Feed, lbs/day} = \frac{(\text{Chemical Fed, lbs})(60 \text{ min/hr})(24 \text{ hr/day})}{\text{Time, minutes}}$$

EXAMPLE 2

A chemical feed pump lowered the chemical solution in a four-foot diameter chemical storage tank two feet during a seven-hour period. Estimate the flow delivered by the pump in gallons per minute and gallons per day.

Known	Unknown
Tank Diameter, ft = 4 ft	1. Flow, GPM
Chemical Drop, ft = 2 ft	2. Flow, GPD
Time, hr = 7 hr	

1. Determine the volume of water pumped in gallons.

$$\begin{aligned}\text{Volume, gal} &= (0.785)(\text{Diameter, ft})^2(\text{Drop, ft})(7.48 \text{ gal/cu ft}) \\ &= (0.785)(4 \text{ ft})^2(2 \text{ ft})(7.48 \text{ gal/cu ft}) \\ &= 188 \text{ gal}\end{aligned}$$

2. Calculate the flow from the chemical feed pump in gallons per minute.

$$\begin{aligned}\text{Flow, GPM} &= \frac{\text{Volume Pumped, gal}}{(\text{Pumping Time, hr})(60 \text{ min/hr})} \\ &= \frac{188 \text{ gal}}{(7 \text{ hr})(60 \text{ min/hr})} \\ &= 0.45 \text{ GPM}\end{aligned}$$

3. Calculate the flow from the chemical feed pump in gallons per day.

$$\begin{aligned}\text{Flow, GPD} &= \frac{\text{Volume Pumped, gal}}{\text{Pumping Time, hr}}\left(\frac{24 \text{ hr}}{\text{day}}\right) \\ &= \frac{188 \text{ gal}}{7 \text{ hr}}\left(\frac{24 \text{ hr}}{\text{day}}\right) \\ &= 645 \text{ GPD}\end{aligned}$$

EXAMPLE 3

Determine the chemical feed in pounds of polymer per day from a chemical feed pump. The polymer solution is 2.0 percent or 20,000 mg polymer per liter. Assume a specific gravity of the polymer solution of 1.0. During a test run, the chemical feed pump delivered 750 mL of polymer solution during six minutes.

Known	Unknown
Polymer Solution, % = 2.0%	Polymer Feed, lbs/day
Polymer Conc, mg/L = 20,000 mg/L	
Polymer Sp Gr = 1.0	
Volume Pumped, mL = 750 mL	
Time Pumped, min = 6 min	

Calculate the polymer feed by the chemical feed pump in pounds of polymer per day.

$$\begin{aligned}\text{Polymer Feed, lbs/day} &= \frac{(\text{Poly Conc, mg}/L)(\text{Vol Pumped, m}L)(60 \text{ min/hr})(24 \text{ hr/day})}{(\text{Time Pumped, min})(1{,}000 \text{ m}L/L)(1{,}000 \text{ mg/gm})(454 \text{ gm/lb})} \\ &= \frac{(20{,}000 \text{ mg}/L)(750 \text{ m}L)(60 \text{ min/hr})(24 \text{ hr/day})}{(6 \text{ min})(1{,}000 \text{ m}L/L)(1{,}000 \text{ mg/gm})(454 \text{ gm/lb})} \\ &= 7.9 \text{ lbs Polymer/day}\end{aligned}$$

EXAMPLE 4

Determine the actual chemical fed in pounds per day from a dry chemical feeder. A pie tin placed under a chemical feeder collected 1,000 grams of chemical in five minutes.

Known	Unknown
Dry Chemical, gm = 1,000 gm	Chemical Feed, lbs/day
Time, min = 5 min	

Determine the chemical feed in pounds of chemical applied per day.

$$\begin{aligned}\text{Chemical Feed, lbs/day} &= \frac{(\text{Chemical Applied, gm})(60 \text{ min/hr})(24 \text{ hr/day})}{(454 \text{ gm/lb})(\text{Time, min})} \\ &= \frac{(1{,}000 \text{ gm})(60 \text{ min/hr})(24 \text{ hr/day})}{(454 \text{ gm/lb})(5 \text{ min})} \\ &= 635 \text{ lbs/day}\end{aligned}$$

18.47 Chlorinators

Chlorine gas leaks around chlorinators or containers of chlorine will cause corrosion of equipment. Check every day for leaks. Large leaks will be detected by odor; small leaks may go unnoticed until damage results. A green or reddish deposit on metal indicates a chlorine leak. Any chlorine gas leakage in the presence of moisture will cause corrosion. Always plug the ends of any open connection to prevent moisture from entering the lines. Never pour water on a chlorine leak because this will only create a bigger problem by enlarging the leak. Chlorine gas reacts with water to form hydrochloric acid.

WARNING

Another important reason for preventing chlorine leaks is that the gas is toxic to humans.

Ammonia water will detect any chlorine leak. A small piece of cloth, soaked with ammonia water[22] and wrapped around the end of a short stick, makes a good leak detector. Wave this stick in the general area of the suspected leak (do not touch the equipment with it). If chlorine gas leakage is occurring, a white cloud of ammonium chloride will form. You should make this test at all gas pipe joints, both inside and outside the chlorinators, at regular intervals. Bottles of ammonia water should be kept tightly capped to avoid loss of strength. All pipe fittings must be kept tight to avoid leaks. New gaskets should be used for each new connection.

Do not use a spray bottle in a room where large amounts of chlorine gas have already leaked into the air. After one squeeze, the entire area may be full of white vapor and you will have trouble locating the leak. Under these conditions, use a cloth soaked in ammonia water to look for leaks.

CAUTION

To locate a chlorine leak, do not spray or swab equipment with ammonia water! Wave an ammonia-soaked rag or paintbrush in the general area and you can detect the presence of many leaks. Some operators prefer to wave a stick with a cloth on the end in front of them when they are looking for chlorine leaks.

The exterior casing of chlorinators should be painted, as required; however, most chlorinators manufactured recently have plastic cases that do not require protective coatings. A clean machine is a better operating machine. Parts of a chlorinator handling chlorine gas must be kept dry to prevent the chlorine and moisture from forming hydrochloric acid. Some parts may be cleaned, when required, first with water to remove water-soluble material, then with wood alcohol, followed by drying. The above chemicals leave no moisture residue. Another method would be to wash them with water and dry them over a pan or heater to remove all traces of moisture.

Water strainers on chlorinators frequently clog and require attention. They may be cleaned by flushing with water or, if badly fouled, they may be cleaned with diluted hydrochloric acid, followed with a water rinse.

The atmosphere vent lines from chlorinators must be open and free. These vent lines evacuate the chlorine to the outside atmosphere when the chlorinator is being shut down. Place a screen over the end of the pipe to keep insects from building a nest in the pipe and clogging it.

When chlorinators are removed from service, as much chlorine gas as possible should be removed from the supply lines and machines. The chlorine valves at the containers are shut off and the chlorinator injector is operated for a period to remove the chlorine gas. In V-notch chlorinators, the rotameter goes to the bottom of the manometer tube when the chlorine gas has been expelled.

All chlorinators will give continuous, trouble-free operation if properly maintained and operated. Each chlorinator manufacturer provides with each machine a maintenance and operations instruction booklet with line diagrams showing the operation of the component parts of the machine. Manufacturer's instructions should be followed for maintenance and lubrication of your particular chlorinator. If you do not have an instruction booklet, you may obtain one by contacting the manufacturer's representative in your area.

QUESTIONS

Write your answers in a notebook and then compare your answers with those on page 383.

18.4A How can an operator locate information on how to operate, control, and maintain chemical feeders?

18.4B List three common types of chemical feeders.

18.4C Why should chlorine leaks be detected and repaired?

18.4D How would you search for chlorine leaks?

18.5 TANKS AND RESERVOIRS[23]

18.50 Scheduling Inspections

Plant tanks should be drained and inspected at regular intervals. If the interior is well protected, five-year intervals between inspections may be sufficient. If the tank is below the surface of the ground, be sure the groundwater level is down far enough (below the bottom of the tank) so the tanks will not float on the groundwater when empty or develop cracks from groundwater pressure.

Schedule inspections of tanks and channels during periods of low plant demand so that plant operation will not be disrupted.

18.51 Steel Tanks

All steel tanks must be protected from rusting. Once metal is lost because of rusting, it cannot be recovered. The exteriors of the tanks are easily inspected—do not forget the roof—and should be repainted, as needed, not only to protect the steel surface but to provide a pleasing appearance. The interiors of steel tanks are exposed to a much harsher environment due either to being constantly submerged or to constant high humidity.

Protective coatings for steel tank interiors must be carefully selected to provide superior protection and, at the same time,

---

[22] Use a concentrated ammonia solution containing 28 to 30 percent ammonia as NH_3 (this is the same as 58 percent ammonium hydroxide, NH_4OH, or commercial 26° Baumé).

[23] Also see *WATER DISTRIBUTION SYSTEM OPERATION AND MAINTENANCE*, Chapter 2, "Storage Facilities," Section 2.4, "Maintenance," in this series of manuals.

impart neither taste nor odors to the water. Proper surface preparation and application is as important as the coating materials in getting interior protection that will last a reasonable period of time.

When interior tank recoating is required, schedule the work when plant demand is low but not during rainy weather when it may be impossible to maintain a dry steel surface warm enough to ensure proper curing of the coating. This type of work is usually done by outside contractors. Constant inspection is a must if the work is going to be completed according to the specifications.

18.52 Cathodic Protection[24]

An alternative to repainting the submerged interior surfaces of a steel water tank is installation of a cathodic protection system. The rusting of steel is accompanied by the flow of small electrical currents. Cathodic protection systems prevent rusting of bare steel surfaces by causing an electrical current to flow from anodes hung in the water to the tank surface. The polarity of this current is opposite to what it would be if rust were forming. The current can be obtained from sacrificial anodes that make the tank into a giant low-voltage battery, or from electronic rectifiers that are powered from the commercial power lines.

Cathodic protection systems provide protection only so long as they are operating and properly adjusted. Systems with rectifiers should be checked weekly. The inspection consists of reading and recording the DC multimeter and ammeter readings. Compare the readings with previous readings and with the readings recommended by the corrosion engineer or technician. Deviations from normal should be investigated without delay.

Once a year, a corrosion specialist should be called in to take potential profile readings on the inside of the tank and to set the rectifier and recommend new normal current settings. Over the years, as more and more of the interior tank coating fails, the bare surface area to be protected by the cathodic protection system will increase. For this reason, it is to be expected that the current required to provide protection will increase in small amounts or increments each year.

18.53 Concrete Tanks

Concrete tanks are not usually coated on the inside and are painted on the exterior for appearance purposes only. This would seem to indicate that maintenance on concrete tanks is minimal, but this may not be true. Concrete tanks are all reinforced with steel. Steel can rust. If too much steel is lost to rust, the structural strength of the tank can be threatened. Periodically inspect the tank for signs of rusting. This is particularly important for prestressed concrete tanks that have a tensioned wire wrap on the exterior. The wires are small in diameter and even a small amount of rust could reduce the size of the wire to the point where it might fail.

QUESTIONS

Write your answers in a notebook and then compare your answers with those on page 384.

18.5A How often should tanks and reservoirs be drained and inspected?

18.5B Why must the groundwater level be below the bottom of a tank before it is drained?

18.5C What is an alternative to applying a protective coating to prevent corrosion of a steel tank?

18.6 BUILDING MAINTENANCE

Building maintenance is another program that should receive attention on a regular schedule. Buildings in a treatment plant are usually built of sturdy materials to last for many years, if they are kept in good repair. In selecting paint for a treatment plant, it is always a good idea to have a painting expert help the operator select the types of paint needed to protect the buildings from deterioration. The expert also will have some good ideas as to color schemes to help blend the plant in with the surrounding area. Consideration should also be given to the quality of paint. A good quality, more expensive material will usually give better service over a longer period of time than the economy-type products.

Building maintenance programs depend on the age, type, and use of a building. New buildings require a thorough check to be certain essential items are available and working properly. Older buildings require careful watching and prompt attention to keep ahead of leaks, breakdowns, replacements when needed, and changing uses of the building. Attention must be given to the maintenance requirements of many items in all plant buildings, such as electrical systems, plumbing, heating, cooling, ventilating, floors, windows, roofs, and drainage around the buildings. Regularly scheduled examinations and necessary maintenance of these items can prevent many costly and time-consuming problems in the future.

In each plant building, periodically check all stairways, ladders, catwalks, and platforms for adequate lighting, head clearance, and sturdy and convenient guardrails. Protective devices should surround all moving equipment. Whenever any repairs, alterations, or additions are built, avoid building accident traps such as pipes laid on top of floors or hung from the ceiling at head height, which could create serious safety hazards.

Organized storage areas should be provided and maintained in an accessible and neat manner.

Keep all buildings clean and orderly. Janitorial work should be done on a regular schedule. All tools and plant equipment should be kept clean and in their proper place. Floors, walls, and windows should be cleaned at regular intervals in order to maintain a neat appearance. A treatment plant kept in a clean, orderly

---

[24] Also see *WATER TREATMENT PLANT OPERATION*, Volume I, Chapter 8, "Corrosion Control," Section 8.45, "Cathodic Protection," in this series of manuals.

condition makes a safe place to work and aids in building good public and employee relations.

QUESTIONS

Write your answers in a notebook and then compare your answers with those on page 384.

18.6A What items should be included in a building maintenance program?

18.6B What factors influence the type of building maintenance program that might be needed for your water treatment plant?

18.7 ARITHMETIC ASSIGNMENT

Turn to the Arithmetic Appendix at the back of this manual. Read and work the problems in Section A.35, "Maintenance." Check the arithmetic in this section using an electronic calculator. You should be able to get the same answers. Section A.55 contains similar problems using metric units.

18.8 ADDITIONAL READING

1. *NEW YORK MANUAL,* Chapter 19,* "Treatment Plant Maintenance and Accident Prevention."

2. *TEXAS MANUAL,* Chapter 13,* "Pumps and Measurement of Pumps."

3. *PUMP HANDBOOK,* Third Edition, edited by Igor Karassik, Joseph Messina, Paul Cooper, and Charles Heald. Obtain from the McGraw-Hill Companies, Order Services, PO Box 182604, Columbus, OH 43272-3031. ISBN 0-07-034032-3. Price, $135.00, plus nine percent of order total for shipping and handling.

* Depends on edition.

18.9 ACKNOWLEDGMENTS

Major portions of this chapter were taken from the following California State University, Sacramento, Operator Manuals:

1. *OPERATION OF WASTEWATER TREATMENT PLANTS,* Volume II, Chapter 15, "Maintenance," by Norman Farnum, Stan Walton, John Brady, Roger Peterson, Malcolm Carpenter, and Rick Arbour.

2. *OPERATION AND MAINTENANCE OF WASTEWATER COLLECTION SYSTEMS,* Volume II, Chapter 9, "Equipment Maintenance," by Lee Doty and Rick Arbour.

3. *INDUSTRIAL WASTE TREATMENT,* Volume II, Chapter 8, "Maintenance," by Norman Farnum, Stan Walton, John Brady, Roger Peterson, and Malcolm Carpenter.

End of Lesson 5 of 5 Lessons on MAINTENANCE

Please answer the discussion and review questions next.

DISCUSSION AND REVIEW QUESTIONS

Chapter 18. MAINTENANCE

(Lesson 5 of 5 Lessons)

Write the answers to these questions in your notebook. The question numbering continues from Lesson 4.

36. What factors could cause gasoline engine starting problems?

37. What is the purpose of the filters in the diesel fuel system?

38. What are the advantages of air-cooled diesel engines as compared with water-cooled types?

39. Why is rust a problem in water-cooled systems?

40. How should large quantities of gasoline be stored?

41. Why is "idling" not a satisfactory method of testing standby engines?

42. Why should solid feeders and hoppers be kept clean and dry?

43. What problems can be caused by chlorine gas leaks around chlorinators or containers of chlorine?

44. Why should a treatment plant be kept in a clean and orderly condition?

SUGGESTED ANSWERS
Chapter 18. MAINTENANCE

ANSWERS TO QUESTIONS IN LESSON 1

Answers to questions on page 271.

18.0A A good maintenance program is essential to maintain successful operation of the treatment plant.

18.0B A successful maintenance program will include everything from mechanical equipment to the care of the plant grounds, buildings, and structures.

18.0C A good recordkeeping system tells when maintenance is due and also provides a record of equipment performance. Poor performance is a good justification for replacement or new equipment. Good records also help keep your warranty in force.

18.0D An equipment service card tells what service or inspection work should be done on a piece of equipment and when, while the service record card is a record of what was done and when.

18.0E Emergency phone numbers for a treatment plant should include the phone numbers for police, fire, hospital or physician, responsible plant officials, local emergency disaster office, emergency team, and CHEMTREC, (800) 424-9300.

18.0F Stored energy must be bled down to prevent its sudden release, which could injure an operator working on the system or equipment.

Answers to questions on page 272.

18.10A Unqualified or inexperienced people must be extremely careful when attempting to troubleshoot or repair electrical equipment because they can be seriously injured and damage costly equipment if a mistake is made.

18.10B When machinery is not shut off, locked out, and tagged properly, the following accidents could occur:

1. Maintenance operator could be cleaning pump and have it start, thus losing an arm, hand, or finger.
2. Electric motors or controls not properly grounded could lead to possible severe shock, paralysis, or death.
3. Improper circuits such as a wrong connection, safety devices jumped, wrong fuses, or improper wiring can cause fires or injuries due to incorrect operation of machinery.

Answers to questions on page 274.

18.11A The proper voltage and allowable current in amps for a piece of equipment can be determined by reading the nameplate information or the instruction manual for the equipment.

18.11B The two types of current are Direct Current (DC) and Alternating Current (AC).

18.11C Amperage is the measurement of current or electron flow and is an indication of work being done or "how hard the electricity is working."

Answers to questions on page 281.

18.12A You test for voltage by using a multimeter.

18.12B A multimeter can be used to test for voltage, open circuits, blown fuses, single phasing of motors, and grounds.

18.12C Before attempting to change fuses, turn off power and check both power lines for voltage. Use a fuse puller.

18.12D If the voltage is unknown and the multimeter has different scales that are manually set, always start with the highest voltage range and work down. Otherwise, the multimeter could be damaged.

18.12E Amp readings different from the nameplate rating could be caused by low voltage, bad bearings, poor connections, or excessive load.

18.12F Motors and wirings should be megged at least once a year, and twice a year, if possible.

18.12G An ohmmeter is used to test the control circuit components such as coils, fuses, relays, resistors, and switches.

Answers to questions on page 284.

18.13A The two types of safety devices in main electrical panels or control units are fuses or circuit breakers.

18.13B Fuses are used to protect operators, wiring, circuits, heaters, motors, and various other electrical equipment.

18.13C A fuse must never be bypassed or jumped because the fuse may be the only protection the circuit has; without it, serious damage to equipment and possible injury to operators can occur.

18.13D A circuit breaker is a switch that is opened automatically when the current or the voltage exceeds or falls below a certain limit. Unlike a fuse that has to be replaced each time it blows, a circuit breaker can be reset after a short delay to allow time for cooling.

18.13E Motor starters can be either manually or automatically controlled.

18.13F Magnetic starters are commonly used to start pumps, compressors, blowers, and anything where automatic or remote control is desired.

Answers to questions on page 287.

18.14A Electrical energy is commonly converted into mechanical energy by electric motors.

18.14B An electric motor usually consists of a stator, rotor, end bells, and windings.

18.14C Motors can be kept trouble free with proper lubrication and maintenance.

18.14D Motor nameplate data should be recorded, compared with manufacturer's data sheet and instruction manual for consistency, and placed in a file for future reference. Many times the nameplate is painted, corroded, or missing from the unit when the information is needed to repair the motor or replace parts.

Answers to questions on page 293.

18.14E The key to effective troubleshooting is practical, step-by-step procedures combined with a common-sense approach.

18.14F When troubleshooting magnetic starters:
1. Gather preliminary information.
2. Inspect:
   a. Contacts
   b. Mechanical parts
   c. Magnetic parts

18.14G Information that should be recorded regarding electrical equipment includes nameplate data and every change, repair, and test.

Answers to questions on page 298.

18.15A A qualified electrician should perform most of the necessary maintenance and repair of electrical equipment to avoid endangering lives and to avoid damage to equipment.

18.15B The purpose of a kirk-key system (one key is used for two locks) is to ensure proper connection of standby power into your power distribution system. The commercial power system must be locked out by the use of switch gear before the standby power is connected to your power distribution system.

18.15C Battery-powered backup lighting units are considered better than engine-driven power sources because they are more economical. If you have a momentary power outage, the system responds without an engine-generator start-up.

18.15D If water lost from a lead-acid battery is replaced with tap water, the impurities in the water will become attached to the lead plates and shorten the life of the battery.

Answers to questions on page 299.

18.16A Electricity is transmitted at high voltage to reduce the size of transmission lines.

18.16B If outdoor transformers have exposed high-voltage wires, the following precautions must be taken:
1. An eight-foot (2.4-m) high fence is required to prevent accessibility by unqualified or unauthorized persons.
2. Signs attached to the fence must indicate "High Voltage."

18.16C The treatment plant operator must keep the exterior and surroundings of the switch gear clean.

18.16D Symptoms that a power distribution transformer may be in need of maintenance or repair include unusual noises, high or low oil levels, oil leaks, or high operating temperatures.

Answers to questions on page 299.

18.17A Electrical safety checklists are used to make operators aware of potential electrical hazards in their water treatment plant.

18.17B Rusted conduits are of concern because they could become the source of a spark that could cause an explosion.

ANSWERS TO QUESTIONS IN LESSON 2

Answers to questions on page 312.

18.20A Pieces of equipment and special tools commonly found in a pump repair shop include welding equipment, lathes, drill press and drills, power hacksaw, flame-cutting equipment, micrometers, calipers, gauges, portable electric tools, grinders, a forcing press, metal-spray equipment, and sandblasting equipment.

18.21A The purpose of a pump impeller is to suck water in the suction piping and to throw water out between the impeller blades.

18.21B A suitable screen should be installed on the intake end of suction piping to prevent foreign matter (sticks, refuse) from being sucked into the pump and clogging or wearing the impeller.

18.21C Suction piping must be up-sloping to prevent air pockets from forming in the top of a pipe where air could be drawn into the pump and cause the loss of suction.

18.21D Cavitation is a condition that occurs when the pressure within the flowing water drops very low, the water starts to boil, and gas pockets or bubbles form on the blade of the impeller. The collapse of these gas pockets or bubbles drives water into the impeller with a terrific force that can knock metal particles off the impeller and cause pitting on the impeller surface. This same action can and does occur on pressure reducing valves and partially closed gate and butterfly valves. Cavitation is accompanied by loud noises that sound like someone is pounding on the impeller with a hammer.

18.21E An advantage of a double-suction pump over a single-suction design is that the longitudinal thrust from the water entering the impeller is balanced.

Answers to questions on page 316.

18.22A The purpose of lubrication is to reduce friction between two surfaces and to remove heat caused by friction.

18.22B Oils in service tend to become acid (contaminated) and may cause corrosion, deposits, sludging, and other problems.

18.22C To ensure proper lubrication of equipment, determine the proper lubrication schedule, lubricant, and amount of lubricant, and prepare a lubrication chart.

Answers to questions on page 316.

18.22D A soft grease has a low viscosity index as compared with a hard grease.

18.22E Oil is used with higher speeds.

18.22F Overfilling with oil or grease can result in high pressures and temperatures, and ruined seals or other components.

ANSWERS TO QUESTIONS IN LESSON 3

Answers to questions on page 325.

18.23A A cross connection is a connection between drinking (potable) water and an unapproved water supply.

18.23B A slight water-seal leakage is desirable when a pump is running to keep the packing cool and in good condition.

18.23C To measure the capacity of a pump, measure the volume pumped during a specific time period.

$$\text{Capacity, GPM} = \frac{\text{Volume, gallons}}{\text{Time, minutes}}$$

or

$$\text{Capacity, } \frac{\text{liters}}{\text{sec}} = \frac{\text{Volume, liters}}{\text{Time, seconds}}$$

18.23D Estimate the capacity of a pump (in GPM) if it lowers the water in a 10-foot wide by 15-foot long wet well 1.7 feet in five minutes.

Known	Unknown
Width, ft = 10 ft	Pump Capacity, GPM
Length, ft = 15 ft	
Depth, ft = 1.7 ft	
Time, min = 5 min	

$$\text{Capacity, GPM} = \frac{\text{Volume, gallons}}{\text{Time, minutes}}$$

$$= \frac{(10 \text{ ft})(15 \text{ ft})(1.7 \text{ ft})(7.48 \text{ gal/cu ft})}{5 \text{ minutes}}$$

$$= 381.5 \text{ GPM}$$

or

$$\text{Capacity, } \frac{\text{liters}}{\text{sec}} = \frac{\text{Volume, liters}}{\text{Time, sec}}$$

$$= \frac{(3 \text{ m})(5 \text{ m})(0.5 \text{ m})(1{,}000 \, L/\text{cu m})}{(5 \text{ minutes})(60 \text{ sec/min})}$$

$$= 25 \text{ liters/sec}$$

18.23E Before a prolonged shutdown, the pump should be drained to prevent damage from corrosion, sedimentation, and freezing. Also, the motor disconnect switch should be opened to disconnect the motor.

Answers to questions on page 327.

18.23F Shear pins commonly fail in reciprocating pumps because of: (1) a solid object lodged under piston, (2) a clogged discharge line, or (3) a stuck or wedged valve.

18.23G A noise may develop when pumping thin sludge due to water hammer, but it will disappear when heavy sludge is pumped.

18.23H Higher than normal discharge pressures in a progressive cavity pump may indicate a line blockage or a closed valve downstream.

Answers to questions on page 330.

18.23I When checking an electric motor, the following items should be checked periodically, as well as when trouble develops:

1. Motor conditions
2. Note all unusual conditions
3. Lubricate bearings
4. Listen to motor
5. Check temperature

18.23J The purpose of a stethoscope is to magnify sounds and carry them to the ear. This instrument is used to detect unusual sounds in electric motors such as whines, gratings, or uneven noises.

Answers to questions on page 332.

18.23K A properly adjusted V-belt has a slight bow in the slack side when running. When idle, it has an alive springiness when thumped with the hand. To check for proper alignment, lay a long straightedge or string across outside faces of the pulley, and allow for differences in dimensions from centerlines of grooves to outside faces of the pulleys being aligned. Be very careful in aligning drives with more than one V-belt on a sheave, as misalignment can cause unequal tension.

18.23L Always replace worn sprockets when replacing a chain because out-of-pitch sprockets cause as much chain wear in a few hours as years of normal operation.

Answers to questions on page 334.

18.23M Improper original installation of equipment, settling of foundations, heavy floor loadings, warping of bases, and excessive bearing wear could cause couplings to become out of alignment.

18.23N A shear pin is a straight pin that will fail (break) when a certain load or stress is exceeded. The purpose of the pin is to protect equipment from damage due to excessive loads or stresses.

Answers to questions on page 336.

18.24A Pumps must be lubricated in accordance with the manufacturer's recommendations. Quality lubricants should be used.

18.24B In lubricating motors, too much grease may cause bearing trouble or damage the winding.

Answers to questions on page 337.

18.24C If a pump will not start, check for blown fuses or tripped circuit breakers and the cause. Also, check for a loose connection, fuse, or thermal unit.

18.24D To increase the rate of discharge from a pump, you should look for something causing the reduced rate of discharge, such as pumping air, motor malfunction, plugged lines or valves, impeller problems, or other factors.

Answers to questions on page 338.

18.24E If a pump that has been locked or tagged out for maintenance or repairs is started, an operator working on the pump could be seriously injured and also equipment could be damaged.

18.24F Normally, a centrifugal pump should be started after the discharge valve is opened. Exceptions are treatment processes or piping systems with vacuums or pressures that cannot be dropped or allowed to fluctuate greatly while an alternate pump is put on the line.

Answers to questions on page 340.

18.24G Before stopping an operating pump:
1. Start another pump (if appropriate).
2. Inspect the operating pump by looking for developing problems, required adjustments, and problem conditions of the unit.

18.24H A pump shaft or motor will spin backward if water being pumped flows back through the pump when the pump is shut off. This will occur if there is a faulty check valve or foot valve in the system.

18.24I The position of all valves should be checked before starting a pump to ensure that the water being pumped will go where intended.

Answers to questions on page 340.

18.24J The most important rule regarding the operation of positive displacement pumps is to never operate the pump against a closed valve, especially a discharge valve.

18.24K If a positive displacement pump is operated against a closed discharge valve, the pipe, valve, or pump could rupture from excessive pressure. The rupture will damage equipment and possibly seriously injure or kill someone standing nearby.

18.24L Both ends of a sludge line should never be closed tight at the same time because gas from decomposition can build up and rupture pipes or valves.

ANSWERS TO QUESTIONS IN LESSON 4

Answers to questions on page 344.

18.25A Compressors are used with water ejectors, pump control systems (bubblers), valve operators, and water pressure systems. Also, they are used to operate portable pneumatic tools, such as jackhammers, compactors, air drills, sandblasters, tapping machines, and air pumps.

18.25B The frequency of cleaning a suction filter on a compressor depends on the use of the compressor and the atmosphere around it. The filter should be inspected at least monthly and cleaned or replaced every three to six months. More frequent inspection, cleaning, and replacement are required under dusty conditions such as near jackhammer operation on a street.

18.25C Compressor oil should be changed at least every three months, unless the manufacturer states differently. If there are filters in the oil system, these also should be changed.

18.25D Drain the condensate from the air receiver daily.

18.25E Before testing belt tension on a compressor with your hands, make sure the compressor is locked off.

Answers to questions on page 360.

18.26A Valves are the controlling devices placed in piping systems to stop, regulate, check, divert, or otherwise modify the flow of liquids or gases.

18.26B Six common types of valves found in water treatment facilities include gate valves, globe valves, eccentric valves, butterfly valves, check valves, and automatic valves.

18.26C The most common maintenance required by gate valves is oiling, tightening, or replacing the stem stuffing box packing.

18.26D The purpose of the check valve is to allow water to flow in one direction only.

18.26E Backflow prevention by check valves is essential in many applications to:

1. Prevent pumps from reversing when power is removed
2. Protect water systems from being cross-connected
3. Aid in pump operation as a dampener
4. Ensure full-pipe operation

ANSWERS TO QUESTIONS IN LESSON 5

Answers to questions on page 363.

18.30A Gasoline engines may be used in water treatment plants to drive pumps, generators, tractors, and vehicles.

18.30B If a gasoline engine will not start, check the following items:

1. No fuel in tank, valve closed
2. Carburetor not choked
3. Water or dirt in fuel lines or carburetor
4. Carburetor flooded
5. Low compression
6. Loose spark plug
7. No spark at plug

18.30C A gasoline engine may not run properly due to:

1. Engine misses
2. Engine surges
3. Engine stops
4. Engine overheats
5. Engine knocks
6. Engine backfires through carburetor

Answers to questions on page 364.

18.30D If a gasoline engine will not start and the spark plug is wet with fuel or oil, this could indicate that the cylinder is flooded with fuel from having the choke on too long.

18.30E If a gasoline engine will not start and there is an oil residue on the spark plug, this could indicate worn piston rings.

18.30F After an engine has started, give it an opportunity to warm up before applying the load.

Answers to questions on page 370.

18.31A Gasoline is not used as a fuel in diesel engines because it would start to burn from the heat generated by compression before the piston reached the top of the stroke.

18.31B The four basic parts of a diesel fuel system are:

1. Primary fuel filter
2. Secondary fuel filter
3. Fuel injection pump
4. Fuel injector

18.31C The purpose of the fuel injection pump is to deliver fuel to the injector at a very high pressure.

18.32A Heat is removed from the cylinders by a water cooling system. Each cylinder is surrounded with a water jacket through which the coolant (water) circulates and pulls heat from the cylinder. This is accomplished by a water pump that is belt driven from the crankshaft.

Answers to questions on page 371.

18.33A The storage and use of fuels for internal combustion engines must be in accordance with local building and fire codes.

18.33B Four types of fuels commonly used by internal combustion engines include: (1) diesel, (2) gasoline, (3) liquified petroleum gas (LPG), and (4) natural gas.

18.34A Standby internal combustion engines not in regular service should be started up and test run at least once a week.

18.34B Standby engines should be test run long enough for the engine to come up to its normal operating temperature. If at all possible, the engine should be run under its normal load.

Answers to questions on page 376.

18.4A Information on how to operate, control, and maintain chemical feeders may be found in the feeder manufacturer's literature.

18.4B The three common types of chemical feeders are: (1) solid feeders, (2) liquid feeders, and (3) gas feeders.

18.4C Chlorine is toxic to humans and will cause corrosion damage to equipment.

18.4D Large chlorine leaks can be detected by smell. Small leaks are detected by soaking a cloth with ammonia water and holding the cloth near areas where leaks might develop. A white cloud will indicate the presence of a leak.

Answers to questions on page 377.

18.5A Tanks and reservoirs should be drained and inspected at least once every five years if the interior is well protected, more often if it is not well protected.

18.5B The groundwater level should be below the bottom of a tank before it is drained so the tank will not float on the groundwater when empty or develop cracks from groundwater pressure.

18.5C Cathodic protection is an alternative to applying a protective coating to prevent corrosion of a steel tank.

Answers to questions on page 378.

18.6A A building maintenance program will keep the building in good shape and includes painting, when necessary. Attention also must be given to electrical systems, plumbing, heating, cooling, ventilating, floors, windows, roofs, and drainage around the buildings. The building should be kept clean, tools should be stored in their proper place, and essential storage should be available.

18.6B Factors that influence the type of building maintenance program needed by a water treatment plant include the age, type, and use of each building.

CHAPTER 19

INSTRUMENTATION AND CONTROL SYSTEMS

by

Leonard Ainsworth

Revised by

William H. Hendrix

TABLE OF CONTENTS

Chapter 19. INSTRUMENTATION AND CONTROL SYSTEMS

			Page
OBJECTIVES			389
WORDS			390
ABBREVIATIONS AND SYMBOLS			395

LESSON 1

19.0	INSTRUMENTATION AND CONTROL SYSTEMS		399
	19.00	Importance and Nature of Instrumentation and Control Systems	399
	19.01	Importance to the Water Treatment Operator	399
	19.02	Nature of the Measurement Process	400
	19.03	Explanation of Control Systems	401
		19.030 Modulating Control Systems	401
		19.031 Motor Control Stations	404
19.1	SAFETY HAZARDS OF INSTRUMENTATION AND CONTROL SYSTEMS		404
	19.10	General Precautions	404
	19.11	Electrical Hazards	404
	19.12	Mechanical and Pneumatic Hazards	408
	19.13	Confined Spaces	409
	19.14	Oxygen Deficiency or Enrichment	412
	19.15	Explosive Gas Mixtures	412
	19.16	Falls and Associated Hazards	413
19.2	MEASURED VARIABLES AND TYPES OF SENSORS		414
	19.20	General Principles of Sensors	414
	19.21	Pressure Measurements	414
	19.22	Level Measurements	414
	19.23	Flow (Rate of Flow and Total Flow)	418
	19.24	Chemical Feed Rate	426
	19.25	Process Instrumentation	426
	19.26	Signal Transmitters/Transducers	428
DISCUSSION AND REVIEW QUESTIONS			428

LESSON 2

19.3 CATEGORIES OF INSTRUMENTATION ... 429
- 19.30 Primary Elements ... 429
- 19.31 Panel Instruments ... 429
  - 19.310 Indicators ... 429
  - 19.311 Recorders ... 430
  - 19.312 Totalizers ... 430
  - 19.313 Alarms ... 433
- 19.32 Automatic Controllers ... 433
- 19.33 Pump Controllers ... 434
- 19.34 Air Supply Systems ... 434
- 19.35 Laboratory Instruments ... 437
- 19.36 Test and Calibration Equipment ... 437
- 19.37 Process Computer Control Systems ... 439
  - 19.370 Computer Control Systems ... 439
  - 19.371 Typical Computer Control System Functions ... 441

19.4 OPERATION AND PREVENTIVE MAINTENANCE ... 443
- 19.40 Proper Care of Instruments ... 443
- 19.41 Indications of Proper Function ... 443
- 19.42 Start-Up/Shutdown Considerations ... 444
- 19.43 Preventive Maintenance ... 444
- 19.44 Operational Checks ... 445

19.5 ADDITIONAL READING ... 446

DISCUSSION AND REVIEW QUESTIONS ... 446

SUGGESTED ANSWERS ... 447

OBJECTIVES

Chapter 19. INSTRUMENTATION AND CONTROL SYSTEMS

Following completion of Chapter 19, you should be able to:

1. Explain the purpose and nature of instrumentation and control systems.

2. Identify, avoid, and correct safety hazards associated with instrumentation work.

3. Recognize various types of sensors and transducers.

4. Read instruments and make proper adjustments in the operation of water treatment facilities.

5. Identify symptoms of measurement and control system problems.

WORDS

Chapter 19. INSTRUMENTATION AND CONTROL SYSTEMS

ACCURACY
How closely an instrument measures the true or actual value of the process variable being measured or sensed.

ALARM CONTACT
A switch that operates when some preset low, high, or abnormal condition exists.

ANALOG
The continuously variable signal type sent to an analog instrument (for example, 4–20 mA).

ANALOG READOUT
The readout of an instrument by a pointer (or other indicating means) against a dial or scale. Also see DIGITAL READOUT.

ANALYZER
A device that conducts a periodic or continuous measurement of turbidity or some factor such as chlorine or fluoride concentration. Analyzers operate by any of several methods including photocells, conductivity, or complex instrumentation.

CALIBRATION
A procedure that checks or adjusts an instrument's accuracy by comparison with a standard or reference.

CONTACTOR
An electric switch, usually magnetically operated.

CONTROL LOOP
The combination of one or more interconnected instrumentation devices that are arranged to measure, display, and control a process variable. Also called a loop.

CONTROL SYSTEM
An instrumentation system that senses and controls its own operation on a close, continuous basis in what is called proportional (or modulating) control.

CONTROLLER

A device that controls the starting, stopping, or operation of a device or piece of equipment.

DANGEROUS AIR CONTAMINATION

An atmosphere presenting a threat of causing death, injury, acute illness, or disablement due to the presence of flammable and/or explosive, toxic, or otherwise injurious or incapacitating substances.

(1) Dangerous air contamination due to the flammability of a gas, vapor, or mist is defined as an atmosphere containing the gas, vapor, or mist at a concentration greater than 10 percent of its lower explosive (lower flammable) limit (LEL).

(2) Dangerous air contamination due to a combustible particulate is defined as a concentration that meets or exceeds the particulate's lower explosive limit (LEL).

(3) Dangerous air contamination due to the toxicity of a substance is defined as the atmospheric concentration that could result in employee exposure in excess of the substance's permissible exposure limit (PEL).

NOTE: A dangerous situation also occurs when the oxygen level is less than 19.5 percent by volume (OXYGEN DEFICIENCY) or more than 23.5 percent by volume (OXYGEN ENRICHMENT).

DESICCANT (DESS-uh-kant)

A drying agent that is capable of removing or absorbing moisture from the atmosphere in a small enclosure.

DIGITAL

The encoding of information that uses binary numbers (ones and zeros) for input, processing, transmission, storage, or display, rather than a continuous spectrum of values (an analog system) or non-numeric symbols such as letters or icons.

DIGITAL READOUT

The readout of an instrument by a direct, numerical reading of the measured value or variable.

DISCRETE CONTROL

ON/OFF control; one of the two output values is equal to zero.

DISCRETE I/O (INPUT/OUTPUT)

A digital signal that senses or sends either ON or OFF signals. For example, a discrete input would sense the position of a switch; a discrete output would turn on a pump or light.

DISTRIBUTED CONTROL SYSTEM (DCS)

A computer control system having multiple microprocessors to distribute the functions performing process control, thereby distributing the risk from component failure. The distributed components (input/output devices, control devices, and operator interface devices) are all connected by communications links and permit the transmission of control, measurement, and operating information to and from many locations.

EFFECTIVE RANGE

That portion of the design range (usually from 10 to 90+ percent) in which an instrument has acceptable accuracy. Also see RANGE and SPAN.

FAIL-SAFE

Design and operation of a process control system whereby failure of the power system or any component does not result in process failure or equipment damage.

FEEDBACK

The circulating action between a sensor measuring a process variable and the controller that controls or adjusts the process variable.

HERTZ (Hz)

The number of complete electromagnetic cycles or waves in one second of an electric or electronic circuit. Also called the frequency of the current.

HUMAN MACHINE INTERFACE (HMI)

The device at which the operator interacts with the control system. This may be an individual instrumentation and control device or the graphic screen of a computer control system. Also called MAN MACHINE INTERFACE (MMI) and OPERATOR INTERFACE.

INTEGRATOR

A device or meter that continuously measures and sums a process rate variable in cumulative fashion over a given time period. For example, total flows displayed in gallons per minute, million gallons per day, cubic feet per second, or some other unit of volume per time period. Also called a TOTALIZER.

INTERLOCK

A physical device, equipment, or software routine that prevents an operation from beginning or changing function until some condition or set of conditions is fulfilled. An example would be a switch that prevents a piece of equipment from operating when a hazard exists.

LAG TIME

The time period between the moment a process change is made and the moment such a change is finally sensed by the associated measuring instrument.

LINEARITY (lin-ee-AIR-it-ee)

How closely an instrument measures actual values of a variable through its effective range.

MAN MACHINE INTERFACE (MMI)

The device at which the operator interacts with the control system. This may be an individual instrumentation and control device or the graphic screen of a computer control system. Also called HUMAN MACHINE INTERFACE (HMI) and OPERATOR INTERFACE.

MEASURED VARIABLE

A factor (flow, temperature) that is sensed and quantified (reduced to a reading of some kind) by a primary element or sensor.

OPERATOR INTERFACE

The device at which the operator interacts with the control system. This may be an individual instrumentation and control device or the graphic screen of a computer control system. Also called HUMAN MACHINE INTERFACE (HMI) and MAN MACHINE INTERFACE (MMI).

ORIFICE (OR-uh-fiss)

An opening (hole) in a plate, wall, or partition. An orifice flange or plate placed in a pipe consists of a slot or a calibrated circular hole smaller than the pipe diameter. The difference in pressure in the pipe above and at the orifice may be used to determine the flow in the pipe. In a trickling filter distributor, the wastewater passes through an orifice to the surface of the filter media.

PRECISION

The ability of an instrument to measure a process variable and repeatedly obtain the same result. The ability of an instrument to reproduce the same results.

PRIMARY ELEMENT

(1) A device that measures (senses) a physical condition or variable of interest. Floats and thermocouples are examples of primary elements. Also called a SENSOR.

(2) The hydraulic structure used to measure flows. In open channels, weirs and flumes are primary elements or devices. Venturi meters and orifice plates are the primary elements in pipes or pressure conduits.

PROCESS VARIABLE

A physical or chemical quantity that is usually measured and controlled in the operation of a water, wastewater, or industrial treatment plant. Common process variables are flow, level, pressure, temperature, turbidity, chlorine, and oxygen levels.

PROGRAMMABLE LOGIC CONTROLLER (PLC)

A microcomputer-based control device containing programmable software; used to control process variables.

RANGE

The spread from minimum to maximum values that an instrument is designed to measure. Also see EFFECTIVE RANGE and SPAN.

READOUT

The reading of the value of a process variable from an indicator or recorder or on a computer screen.

RECEIVER

A device that indicates the result of a measurement, usually using either a fixed scale and movable indicator (pointer), such as a pressure gauge, or a moving chart with a movable pen like those used on a circular flow-recording chart. Also called an INDICATOR.

RECORDER

A device that creates a permanent record, on a paper chart, magnetic tape, or in a computer, of the changes in a measured variable.

REFERENCE

A physical or chemical quantity whose value is known exactly, and thus is used to calibrate instruments or standardize measurements. Also called a STANDARD.

ROTAMETER (ROTE-uh-ME-ter)

A device used to measure the flow rate of gases and liquids. The gas or liquid being measured flows vertically up a tapered, calibrated tube. Inside the tube is a small ball or bullet-shaped float (it may rotate) that rises or falls depending on the flow rate. The flow rate may be read on a scale behind or on the tube by looking at the middle of the ball or at the widest part or top of the float.

SCADA (SKAY-dah) SYSTEM

Supervisory Control And Data Acquisition system. A computer-monitored alarm, response, control, and data acquisition system used to monitor and adjust treatment processes and facilities.

SCALE

(1) A combination of mineral salts and bacterial accumulation that sticks to the inside of a collection pipe under certain conditions. Scale, in extreme growth circumstances, creates additional friction loss to the flow of water. Scale may also accumulate on surfaces other than pipes.

(2) The marked plate against which an indicator or recorder reads, usually the same as the range of the measuring system. See RANGE.

SENSITIVITY

The smallest change in a process variable that an instrument can sense.

SENSOR

A device that measures (senses) a physical condition or variable of interest. Floats and thermocouples are examples of sensors. Also called a PRIMARY ELEMENT.

SET POINT

The position at which the control or controller is set. This is the same as the desired value of the process variable. For example, a thermostat is set to maintain a desired temperature.

SOFTWARE PROGRAM

Computer program; the list of instructions that tell a computer how to perform a given task or tasks. Some software programs are designed and written to monitor and control treatment processes.

SOLENOID (SO-luh-noid)

A magnetically operated mechanical device (electric coil). Solenoids can operate small valves or electric switches.

SPAN

The scale or range of values an instrument is designed to measure. Also see RANGE.

STANDARD

A physical or chemical quantity whose value is known exactly, and thus is used to calibrate instruments or standardize measurements. Also called a REFERENCE.

STANDARDIZE

To compare with a standard.

(1) In wet chemistry, to find out the exact strength of a solution by comparing it with a standard of known strength. This information is used to adjust the strength by adding more water or more of the substance dissolved.

(2) To set up an instrument or device to read a standard. This allows you to adjust the instrument so that it reads accurately, or enables you to apply a correction factor to the readings.

STARTERS (MOTOR)

Devices used to start up large motors gradually to avoid severe mechanical shock to a driven machine and to prevent disturbance to the electrical lines (causing dimming and flickering of lights).

TELEMETRY (tel-LEM-uh-tree)

The electrical link between a field transmitter and the receiver. Telephone lines are commonly used to serve as the electrical line.

THERMOCOUPLE

A heat-sensing device made of two conductors of different metals joined together. An electric current is produced when there is a difference in temperature between the ends.

TIMER

A device for automatically starting or stopping a machine or other device at a given time.

TOTALIZER

A device or meter that continuously measures and sums a process rate variable in cumulative fashion over a given time period. For example, total flows displayed in gallons per minute, million gallons per day, cubic feet per second, or some other unit of volume per time period. Also called an INTEGRATOR.

TRANSDUCER (trans-DUE-sir)

A device that senses some varying condition measured by a primary sensor and converts it to an electrical or other signal for transmission to some other device (a receiver) for processing or decision making.

VARIABLE, MEASURED

A factor (flow, temperature) that is sensed and quantified (reduced to a reading of some kind) by a primary element or sensor.

VARIABLE, PROCESS

A physical or chemical quantity that is usually measured and controlled in the operation of a water, wastewater, or industrial treatment plant.

ABBREVIATIONS AND SYMBOLS
Chapter 19. INSTRUMENTATION AND CONTROL SYSTEMS

Special symbols are used for simplicity and clarity on circuit drawings for instruments. Usually, instrument manufacturers and design engineers provide lists of symbols they use with an explanation of the meaning of each symbol. This section contains a list of typical instrumentation abbreviations and symbols used in this chapter and also used by the waterworks profession.

ABBREVIATIONS

A — **Analyzer**, such as a device used to measure a water quality indicator (pH, temperature).

C — **Controller**, such as a device used to start, operate, or stop a pump.

D — **Differential**, such as a "differential pressure" (DP) cell used with a flowmeter.

E — **Electrical** or **Voltage**.
Element, such as a primary element.

F — **Flow rate** (*NOT* total flow).

H — **Hand** (manual operation).
High as in high-level.

I — **Indicator**, such as the indicator on a flow recording chart.
I = E/R where I is the electric current in amps.

L — **Level**, such as the level of water in a tank.
Low, as in a low-level switch.
Light, as in indicator light.

M — **Motor**.
Middle, as in a mid-level switch.

P — **Pressure** (or vacuum).
Pump.
Program, as in a software program.

Q — **Quantity**, such as a totalized volume (Σ for summation is also used).

R — **Recorder** (or printer), such as a chart recorder.
Receiver.
Relay.

S — **Switch**.
Speed, such as the RPM (revolutions per minute) of a motor.
Starter, such as a motor starter.
Solenoid.

T — **Transmitter**.
Temperature.
Tone.

V — **Valve**.
Voltage.

W — **Weight**.
Watt.

X — **Special** or unclassified variable.

Y — **Computing function**, such as a square root ($\sqrt{}$) extraction.

Z — **Position**, such as a percent valve opening.

TYPICAL PROCESS AND ELECTRICAL SYMBOLS

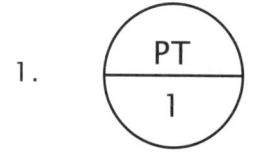

1. (PT/1) — Pressure transmitter #1

2. (LIR/2) — Level indicator/recorder #2

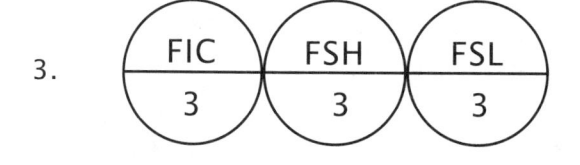

3. (FIC/3) (FSH/3) (FSL/3) — Flow indicator/controller #3 with high-low control switches in the same instrument

396 Water Treatment

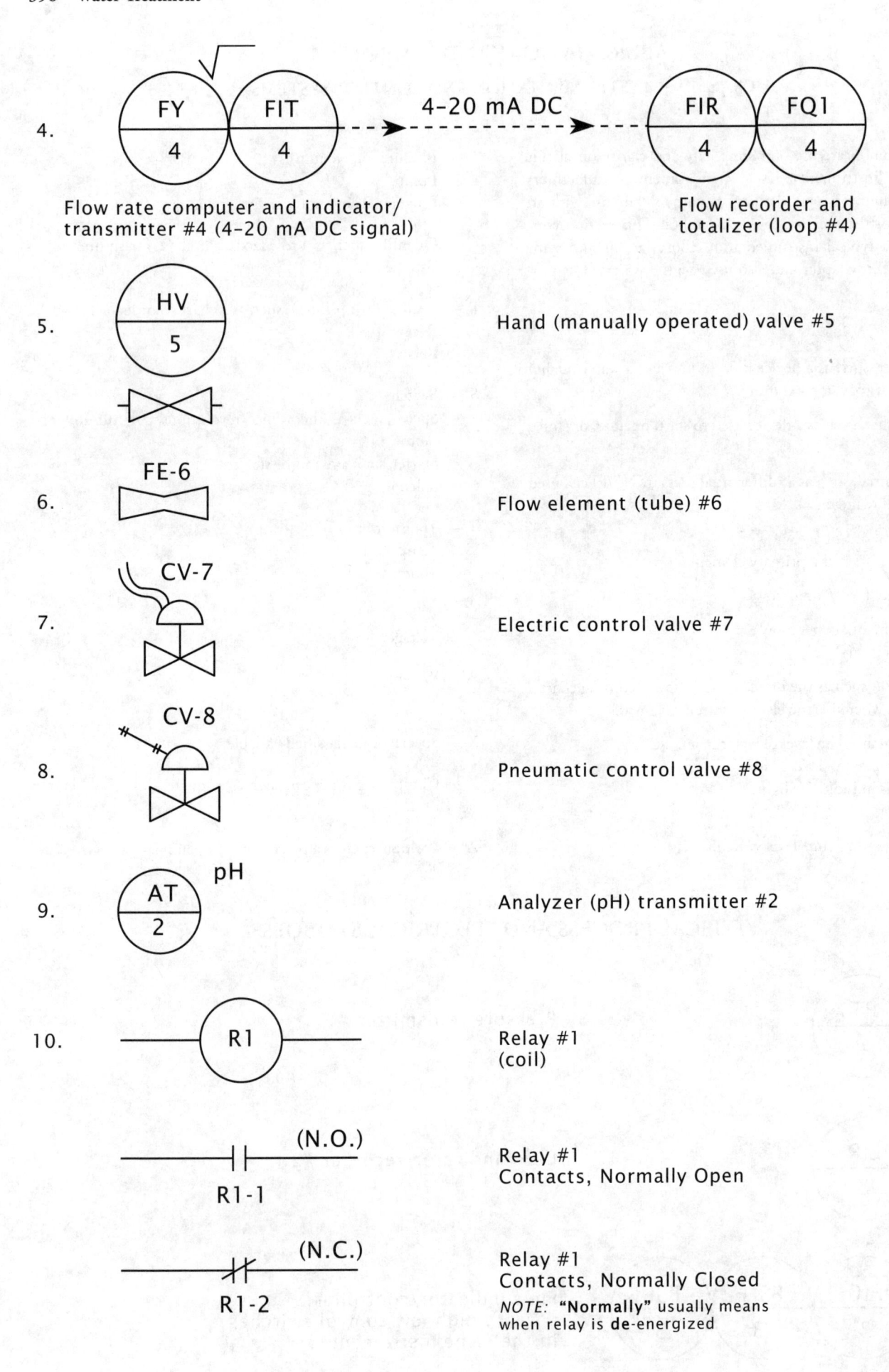

4. (FY/4) (FIT/4) - - - 4-20 mA DC - - -> (FIR/4) (FQ1/4)

Flow rate computer and indicator/
transmitter #4 (4–20 mA DC signal)

Flow recorder and
totalizer (loop #4)

5. HV/5   Hand (manually operated) valve #5

6. FE-6   Flow element (tube) #6

7. CV-7   Electric control valve #7

8. CV-8   Pneumatic control valve #8

9. AT/2 pH   Analyzer (pH) transmitter #2

10. R1   Relay #1 (coil)

R1-1 (N.O.)   Relay #1 Contacts, Normally Open

R1-2 (N.C.)   Relay #1 Contacts, Normally Closed

NOTE: "**Normally**" usually means when relay is **de**-energized

11.	(Hi, Red)	High-level indicator light (red)
12.	100 Ω	Resistor, 100 ohms
13.	480V AC / 110 V AC	Power transformer (step-down)
14.	S-1 OFF/ON	Hand switch #1 (SPST)*
15.	S-2	Hand switch #2 (DPST)*
16.	S-3	Hand switch #3 (SPDT)*
17.	S-4	Hand switch #4 (DPDT)*
18.	PB-1, PB-2, LOS	Manual push button switches: PB #1 push to make PB #2 push to break "Lock-out stop" safety feature

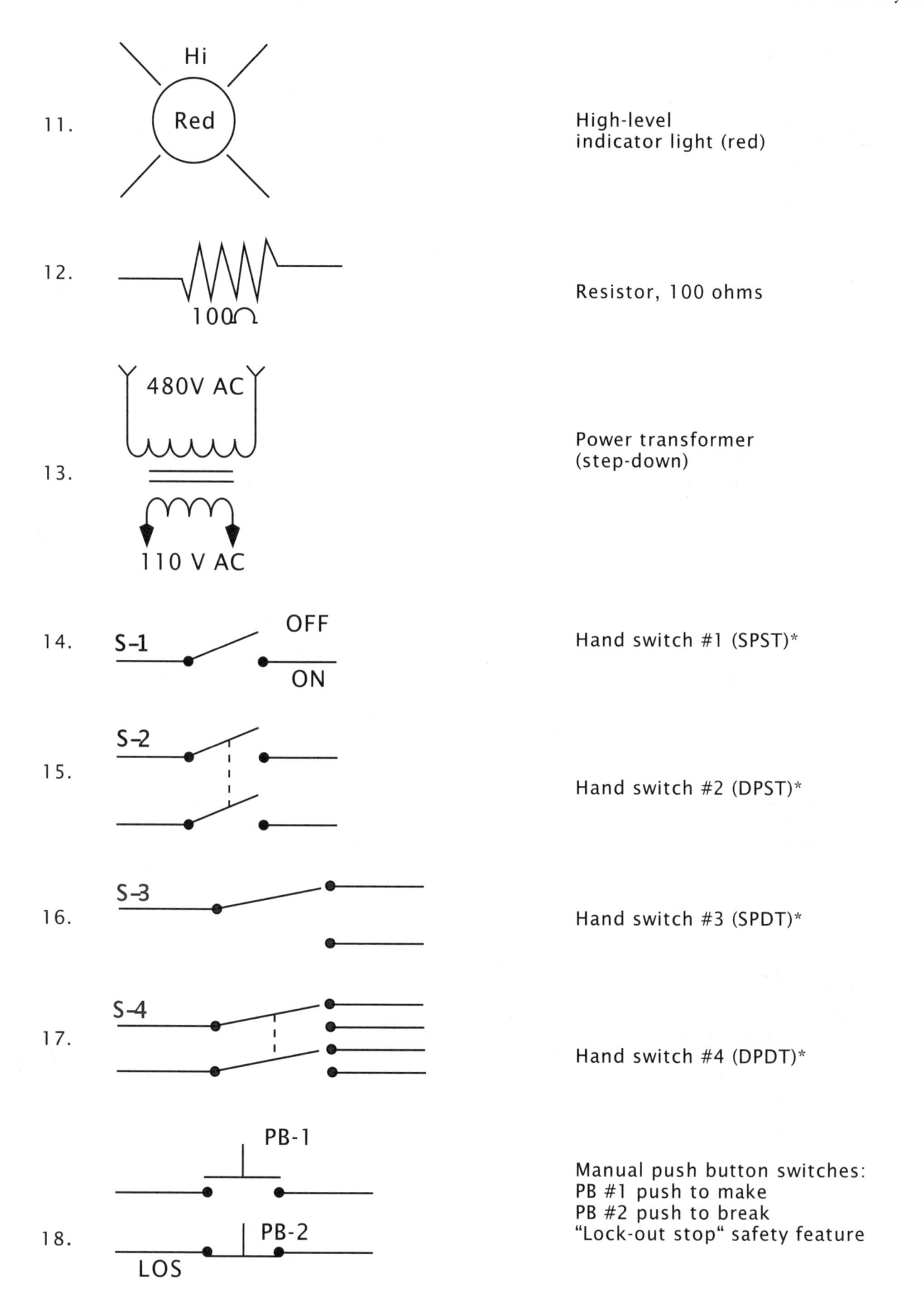

* SPST means single-pole, single-throw; DPST means double-pole, single-throw; SPDT means single-pole, double-throw; DPDT means double-pole, double throw

398 Water Treatment

19. Fuses and Circuit Breakers

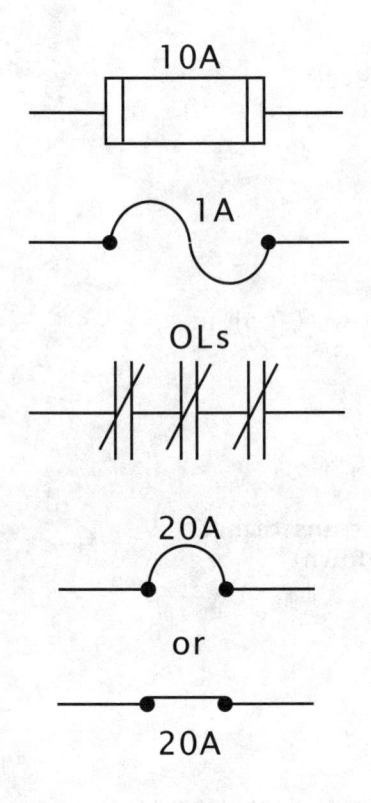

10A — 10-amp cartridge fuse

1A — 1-amp in-line fuse

OLs — Thermal overload contacts (motor)

20A — 20-amp circuit breaker

or

20A

20. L1   L2 (OR N)

110 V AC — Line 1 and Line 2 (neutral) to standard duplex wall plug outlet

21. H - O - A Function Switch

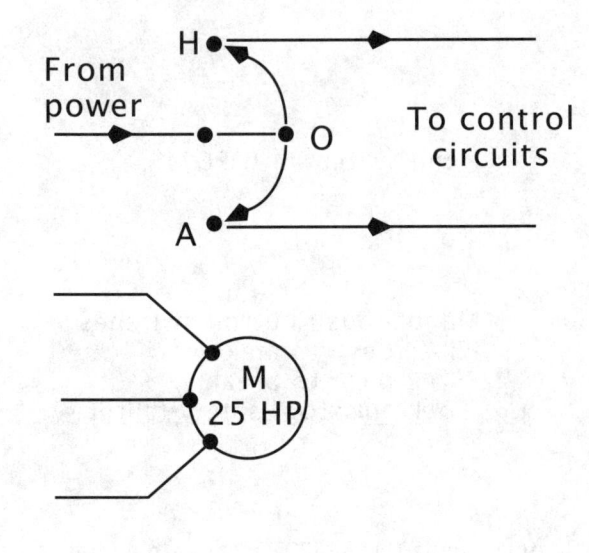

From power → H / O / A → To control circuits — (Hand - Off - Automatic)

22. M 25 HP — Electric motor, 3-phase power, 25 horsepower

CHAPTER 19. INSTRUMENTATION AND CONTROL SYSTEMS

(Lesson 1 of 2 Lessons)

19.0 INSTRUMENTATION AND CONTROL SYSTEMS

19.00 Importance and Nature of Instrumentation and Control Systems

The plant process instrumentation and control system can include manual controls, remote controls, and central computer processing units in any combination. The instrumentation and control system may be computerized or manual: both types provide similar functionality in operating the plant. Today, most facilities are operated through computerized control systems. Classic instrumentation *LOOP*[1] controllers are still used in new facilities for specific purposes and are in general use in older process facilities. The operator's interface with the control system is a small window into the system. In this chapter, you will learn instrumentation and control system concepts and practices that apply to computerized and manual control systems.

Instrumentation is used to sense process variables that can and should be measured. The treatment plant operator must have a good working familiarity with instrumentation and control systems to properly monitor and control the treatment processes. Operators' primary, and sometimes only, links to the process are the instruments and automatic controls in the treatment plant. The range of instrumentation and control system capabilities is extremely large and is beyond what you will cover in this chapter. Here, you will learn general instrumentation functions and control system capabilities. Once the treatment plant is built, the main processes will be permanently in place. However, by using instrumentation and controls, these processes can be manipulated to achieve and maintain maximum process efficiency. The operator's knowledge and use of instrumentation and control will benefit plant operations.

Your knowledge about instrumentation and control systems can only enhance the effective and efficient operation of your facility. Specifically, if you can recognize that instrumentation and control equipment is not operating properly, your process decisions can be based on that knowledge rather than blind faith. This is especially true for computer automation systems, which tend to be trusted implicitly. The automated instrumentation and control is your main link to the process. Some systems provide backup controls to continue critical operation when the computer control system is out of service. Familiarize yourself with these backup controls and how to operate them for your facility. Instrumentation and control systems have self-diagnostic capabilities that will inform the operator of failures and malfunctions. Again, you will want to familiarize yourself with the control systems to understand the capabilities of their diagnostic features.

Basic process variable measurement functions and concepts are the same, whether applied as a single-loop instrumentation and control system or a fully automated computer control system. Even the most modern plants have many instruments that require the operator to perform some minor preventive maintenance. The operator who understands how to recognize failures, troubleshoot, and make adjustments and minor repairs will be more comfortable in operating the plant without the need to call out maintenance personnel. Thus, the more you know about your plant's instrumentation and control systems, the better operator you become.

19.01 Importance to the Water Treatment Operator

Instrumentation and control systems are essential components in treatment facility operation, providing the operator the ability to monitor and manipulate plant processes remotely.

In a very real sense, measurement instruments can be considered extensions of and improvements to your senses of vision, touch, hearing, and even smell. Not only can instruments provide continuous and simultaneous monitoring of the many process variables throughout the plant, they do so in a more precise and consistent manner than the human senses. In addition, instruments often provide a permanent record of measurements taken. The automated control systems, in turn, provide operators with far-reaching and powerful hands to manipulate switches, valves, motors, and pumps in specific ways. In effect, the control and instrumentation system provides you with a staff of obedient and hard-working assistants, always on the job to help you operate your process more easily and efficiently. To appreciate these great advantages of automation, consider what a treatment plant would be like without modern controls. In the not-too-distant past, and even in some treatment operations today, the situation described in the paragraph below was and is the daily *modus operandi* for some operations:

The plant experiences a power failure in the main and backup electrical systems supplying the main automated

---

[1] *Control Loop.* The combination of one or more interconnected instrumentation devices that are arranged to measure, display, and control a process variable. Also called a loop.

control system; other plant power and process equipment continue operating. The equipment failure is such that the operator cannot repair or reset to restore power. As the operator, you must try to keep the plant on line manually by attempting to control all process variables without any of the normal data indications of the control system. That is, flows and levels have to be estimated visually, valves turned manually, and pumps started and stopped on "HAND" to control them. Additionally, you are forced to observe and try to regulate the other process variables using only your eyes, ears, sense of touch, and probably, in time, even your sense of smell.

Impossible, you say? No, it can be, has been, and is done here and there in old or poorly maintained plants. But, even if you could do it (and on some shifts it does seem like this *is* what you have to do), you certainly could not exercise close control or do it for any extended period. Operating a larger or sophisticated plant would be impossible without its process instrumentation and controls, even with a sizeable crew. Your plant emergency systems and operator knowledge and skills are the final line of defense in such a case as described.

The plant instrumentation and control system is designed with safeguards that nearly eliminate such failures. However, infrequent failures do occur, even in the best-designed instrumentation and control system. Operators who develop their knowledge of the benefits and limitations of the instrumentation and control systems will have enhanced skills during such upset conditions of operation.

QUESTIONS

Write your answers in a notebook and then compare your answers with those on page 447.

19.0A Why must a treatment plant operator have a good working familiarity with instrumentation and control systems?

19.0B What kind of information is presented in this chapter with regard to instrumentation?

19.0C Measurement instruments can serve as an extension of and improvement to which of the operator's physical senses?

19.02 Nature of the Measurement Process

Our senses provide us with qualitative (for example, color, sound, odor) and relatively short-term quantitative (for example, brighter, louder, stronger) information; instrumentation measurements provide us with exact quantitative data. That is to say, measurements give us numbers. Without this objective quantification of our environment (that is, the numbers we use to describe it), we could accomplish very little—just as primitive tribes with no real number systems have no technology worthy of the term. Their processing of natural materials is very limited because the information they use consists only of the relative terms "more than" or "less than." Without objective measurements, they have little control over their environment. Modern water treatment plants, on the other hand, depend absolutely upon accurate and reproducible measurements of chemical and physical processes, which is what good instrumentation provides.

A measurement is, by definition, the comparison of a quantity with a standard unit of measure. Thus, a tape measure is marked in feet or meters, standard units of length; clocks are marked in hours, minutes, and seconds as the standard units of time; and a small container may have fluid ounces marked on its side. These units, usually of a convenient magnitude (size), have been agreed upon internationally to serve as the accepted standard units. The primary standard for mass (weight) is the weight of an actual physical object, which is located in Paris, France; a standard kilogram is defined as the weight of this one specific object. The primary standards for time are defined in wavelengths of light, and the primary standard for length is the wavelength of a certain spectrum line of the element cesium. Other primary standards can be set up for certain quantities—for example, solution strengths—but generally, industrial measurements depend on secondary standards that are based on more or less exact comparisons to a primary standard. A machinist's caliper is calibrated against precision measuring blocks, which are such secondary standards.

Of course, all the measurements used in water and wastewater treatment ultimately refer back to a few primary and secondary standards. Measurements of length, and the related calculations for area and volume, are in effect comparisons to the standard foot (meter). Weights of chemicals are traceable to the standard pound (kilogram). Timepieces are designed around the standard second. All the other measurements such as flow rate (volume per unit of time), pressure (weight per unit of area), chemical concentration/dosage (weight of chemical per unit of liquid volume), chemical feed (weight of chemical per unit of time), and others, are derived from the fundamental units of measure.

Some important terms directly relating to the measurement process also need to be defined: a *PROCESS VARIABLE* is a physical or chemical quantity, such as flow rate or pH, that is measured or controlled in a water treatment process. *ACCURACY* refers to how closely an instrument measures the true or actual value of a process variable. Accuracy is usually expressed as within plus or minus a given small percent of the true value, for example, ±2 percent is typical instrument accuracy of measurement. Accuracy depends partly upon the *PRECISION* of an instrument, which amounts to how closely the device can reproduce a given reading (or measurement) time after time (a good example of a *precise* but *inaccurate* instrument is a precision electrical gauge—with a bent needle). SENSITIVITY refers to the smallest change in a process variable that an instrument can sense. It is usually expressed as a percent of the full-scale value of the instrument. Typical values are ±0.1 percent of full scale for a high-sensitivity instrument and ±2 percent of full scale for an average-sensitivity instrument. CALIBRATION is the complete test of an instrument by measuring several (secondary) standards and its complete adjustment (if required) to read the standard values at several points in its range. STANDARDIZATION is an abbreviated calibration where only one standard is measured and the instrument is adjusted to read the proper value, such as standardizing a pH meter with pH 7 buffer. The RANGE of an instrument is the spread between the minimum and maximum values it is designed to measure, most often from zero to a maximum value, as with, say, a 0–100 psi pressure gauge. And the EFFECTIVE RANGE is that portion of an instrument's complete range within which the instrument has acceptable accuracy, commonly between 10 percent and 90+ percent of its design range. SPAN is a closely related term to express that procedure in an instrument's calibration that adjusts it to read both low and high standards accurately; an improperly spanned 0–100 psi pressure gauge might read 10 psi "right on" but then indicate a true 80 psi as 70 psi. LINEARITY is the term to describe how well a device tracks a series of standards throughout its effective range—the preceding poorly spanned gauge is "off" in linearity by –12.5 percent at 80 percent of scale (off 10 psi at 80 psi). It is evident that to be truly accurate, an instrument must have acceptable precision, be in calibration, and be designed with an effective range that is wide enough to measure the range of the treatment plant's process variables.

Finally, two other common terms should be defined. ANALOG displays show a reading as a pointer (or other cursor) against a marked scale, such as simple pressure or level gauges. Numeric DIGITAL[2] displays provide direct, numerical readouts, as with most modern instrumentation. The best example of both is a watch that has both readouts; the hands give a quick analog reference and the digital time display gives a more precise reading. Analog process variable values may be transmitted by a continuously variable 4–20 mA (milliamp) signal or a digital-data signal (not to be confused with a numeric digital *display*). Figure 19.1 illustrates the difference between the connections required for each. Computer readouts are analog input signals that have been encoded as digital data. The OPERATOR INTERFACE[3] on the computer graphic may be programmed to display the value as an analog or digital readout.

An instrument loop encompasses all devices from the process sensing element to the indicators on a control panel or control system computer screen. Typically, loop components are connected together by a 4–20 mA signal or digital-data link (Figure 19.2).

QUESTIONS

Write your answers in a notebook and then compare your answers with those on page 447.

19.0D What is the main difference between measurements by our senses and measurements by instruments?

19.0E What is the definition of a measurement?

19.0F Explain the difference between accuracy and precision.

19.0G What is the difference between analog and digital displays?

19.03 Explanation of Control Systems

Control systems are the means by which such process variables as pressure, level, weight, or flow are controlled. The terms "controller" and "control systems" are used to refer to two different types of instrumentation control systems: (1) MODULATING SYSTEMS, which sense and control their own operation on a close, continuous basis, and (2) MOTOR CONTROL STATIONS, which control only the ON/OFF operation of motors and other devices.

19.030 Modulating Control Systems

The technically proper use of the terms "controller" and "control system" refers to those systems that provide modulating control of process variables. Examples of modulating control systems include: (1) chlorine residual analyzers/controllers; (2) flow-paced (open-loop) chemical feeders; (3) pressure- or flow-regulating valves; (4) continuous level control of process basins; and (5) variable-speed pumping systems for flow/level control.

In order for a process variable, whether pressure, level, weight, or flow, to be closely controlled, it must be measured precisely and continuously. In a modulating control system, the measuring device or primary element sends an electrical or pneumatic signal, proportional to the value of the variable, to the actual system controller.

Within the controller, the signal is compared to the set-point value. A difference between the actual and set-point values results in the controller sending out a command signal to the controlled element, usually a valve, pump, or chemical feeder. Such an error signal produces an adjustment in the system that causes a corresponding change in the original measured variable, making it more closely match the set point. This continuous "cut-and-try" process can result in very fine, ongoing control of variables requiring constant values, such as some flow rates, pressures, levels, or chemical feed rates. The term applied to this circulating action of the variable in such a controller is FEEDBACK. The path through the control system is the CONTROL LOOP. A diagram of such a control loop, measuring the process variable of water flow, is shown in Figure 19.3.

---

[2] *Digital.* The encoding of information that uses binary numbers (ones and zeros) for input, processing, transmission, storage, or display, rather than a continuous spectrum of values (an analog system) or non-numeric symbols such as letters or icons.

[3] *Operator Interface.* The device at which the operator interacts with the control system. This may be an individual instrumentation and control device or the graphic screen of a computer control system. Also called HUMAN MACHINE INTERFACE (HMI) and MAN MACHINE INTERFACE (MMI).

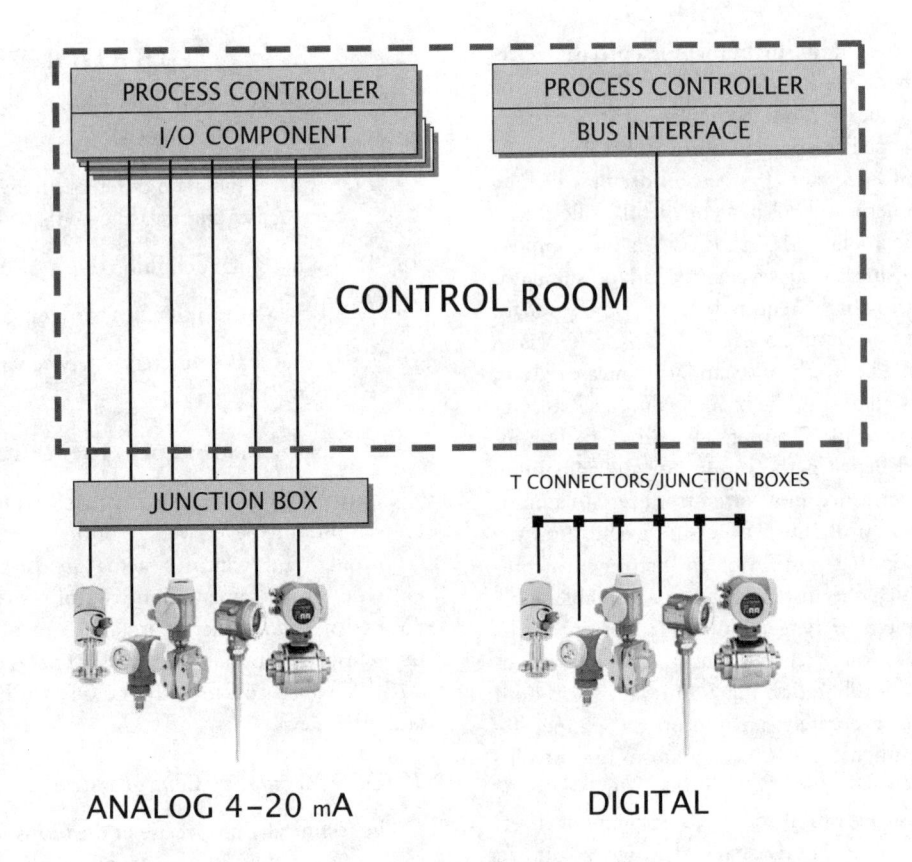

Fig. 19.1 Connections required for analog and digital transmissions
(Adapted from figure provided by courtesy of Endress+Hauser)

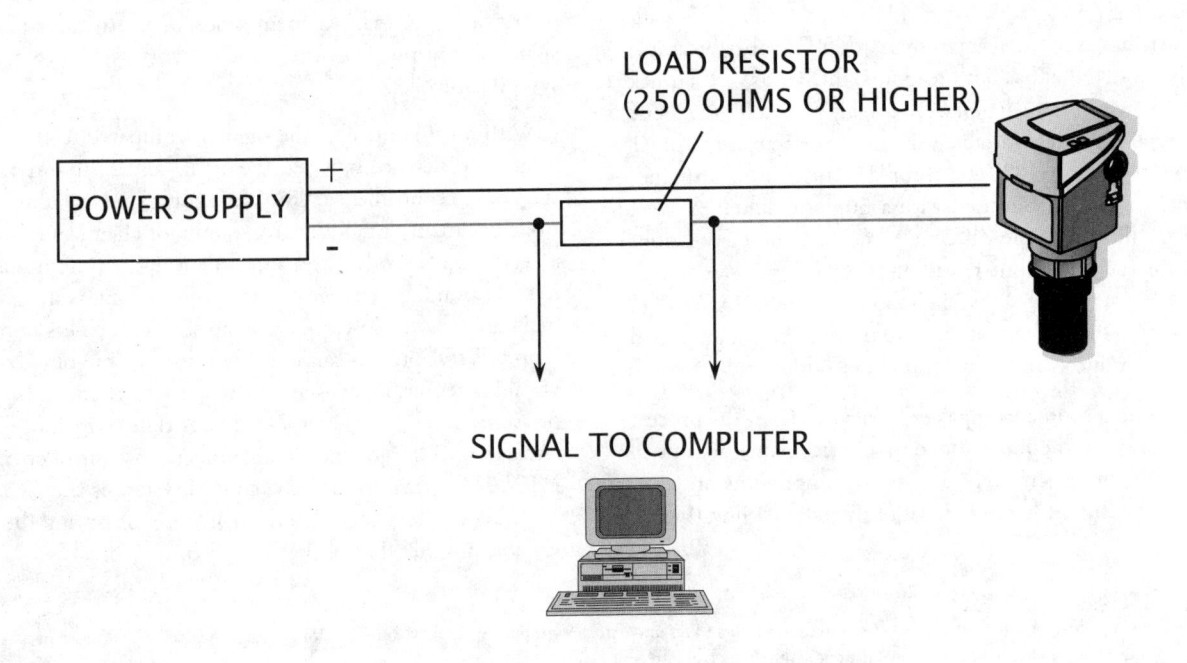

Fig. 19.2 Typical 4–20 mA instrument loop
(Adapted from figure provided by courtesy of Endress+Hauser)

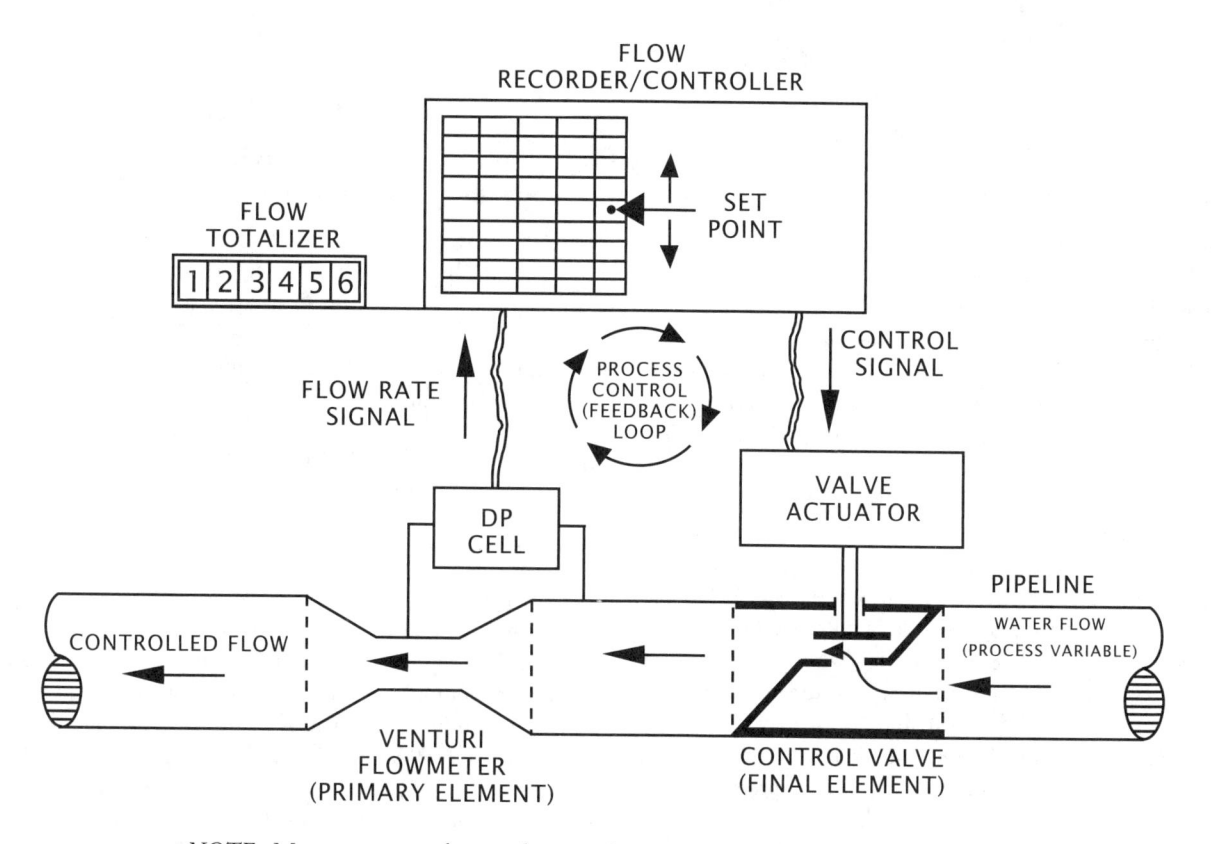

NOTE: Measurement and control system instruments may be electric or pneumatic.

Fig. 19.3 Automatic control system diagram (closed-loop modulating control)

The internal settings of the controller can be quite critical, since close control depends on sensitive adjustments. Thus, you should not try to adjust any such control system unless you know exactly what you are doing. Many plant and system operations have been drastically upset due to such efforts, however well intentioned, by unqualified personnel.

19.031 Motor Control Stations

A motor control station or panel (Figure 19.4) and the related circuitry (Figure 19.5) essentially provide only ON/OFF operation of an electric motor or other device. The electric motor or device, in turn, might power a pump, control valve, or chemical feeder. With this system, control could be manual or automatic, with a switch responding to a preset value of level, pressure, or some other variable. Motor control stations are typically made up of a standard electric power control panel with manual push buttons, overload relays, and Hand-Off-Automatic (H-O-A) or similar On-Off-Remote switch. Additionally, they may include, in good electrical FAIL-SAFE[4] design practice, provisions for power failure or loss-of-phase, and such protective devices as high or low pressure/temperature/level cutoff switches. For this type of panel to be considered a controller (within our secondary meaning of the term), its operation must be directly controlled by the value or values of some variable, not merely by a device such as an ON/OFF timer. In other words, it must be turned on and off as a result of a measurement of a level, pressure, flow, chemical concentration, or other variable as it reaches a predetermined setting or settings. In the automatic mode (A on the H-O-A switch), its operation thus is, in fact, automatic in the sense that the variable is automatically controlled, even though the limits of its value may be quite wide compared to those attainable with a modulating controller as previously described. While a basin level modulating controller may allow only an inch (centimeter) or so of water level change, an ON/OFF system might operate within a few feet (meters) of level difference. In many applications, however, such wide control is of no particular disadvantage, and sometimes is even desirable, as with a tank level where regular exchange of contents is desirable.

Terms used in control practice can now be defined operationally: FEEDBACK and CONTROL LOOP have been mentioned previously, however the term CONTROL LOOP needs some qualification. An OPEN-LOOP control system controls one variable on the basis of another. A good example of this is a chlorinator "paced" by process flow signals (rather than by a chlorine residual analyzer). CLOSED-LOOP control remains as discussed previously, the true control system with measurement and control of the same variable (and feedback). PROPORTIONAL-BAND, RESET, and DERIVATIVE actions are adjustments of a controller that relate to the effectiveness and speed of its control action. OFFSET is the difference between the desired value of the variable (the SET POINT) and the controlled (actual) value. LAG TIME, common to chlorinator control systems, refers to the time period between the moment a process change is made and the moment such a change is finally sensed by the associated measuring instrument. Long lag times can result in unstable processes and poor control.

QUESTIONS

Write your answers in a notebook and then compare your answers with those on page 447.

19.0H What kinds of water treatment process variables do control systems control?

19.0I List three examples of modulating control systems.

19.0J What is the essential purpose of a motor control station or panel?

19.1 SAFETY HAZARDS OF INSTRUMENTATION AND CONTROL SYSTEMS

19.10 General Precautions

The general principles for safe performance on the job, summed up as always avoiding unsafe acts and correcting unsafe conditions, apply as much to instrumentation work as to other plant operations. There are several specific dangers associated with instrument systems that are worth repeating in this section for the sake of safe practice. These include electrical hazards, mechanical and pneumatic hazards, confined spaces, oxygen deficiency and enrichment, explosive gas mixtures, and falls and associated hazards.

19.11 Electrical Hazards

A hidden aspect of energized electrical equipment is that it looks normal; that is, there are no moving parts or other obvious signs that tend to discourage one from touching its components. In fact, there seems to be a peculiar fascination to "see if it is really live" by quickly touching circuit components with a tool, often a screwdriver (usually having an insulated handle, fortunately) but sometimes even with one's finger. Only training (coupled with bad experience, at times) is effective in squelching this morbid curiosity. Even so, most practicing electricians' tools still have the "arc-mark trademark" or two, evidence of the need

---

[4] *Fail-Safe.* Design and operation of a process control system whereby failure of the power system or any component does not result in process failure or equipment damage.

Instrumentation and Control Systems 405

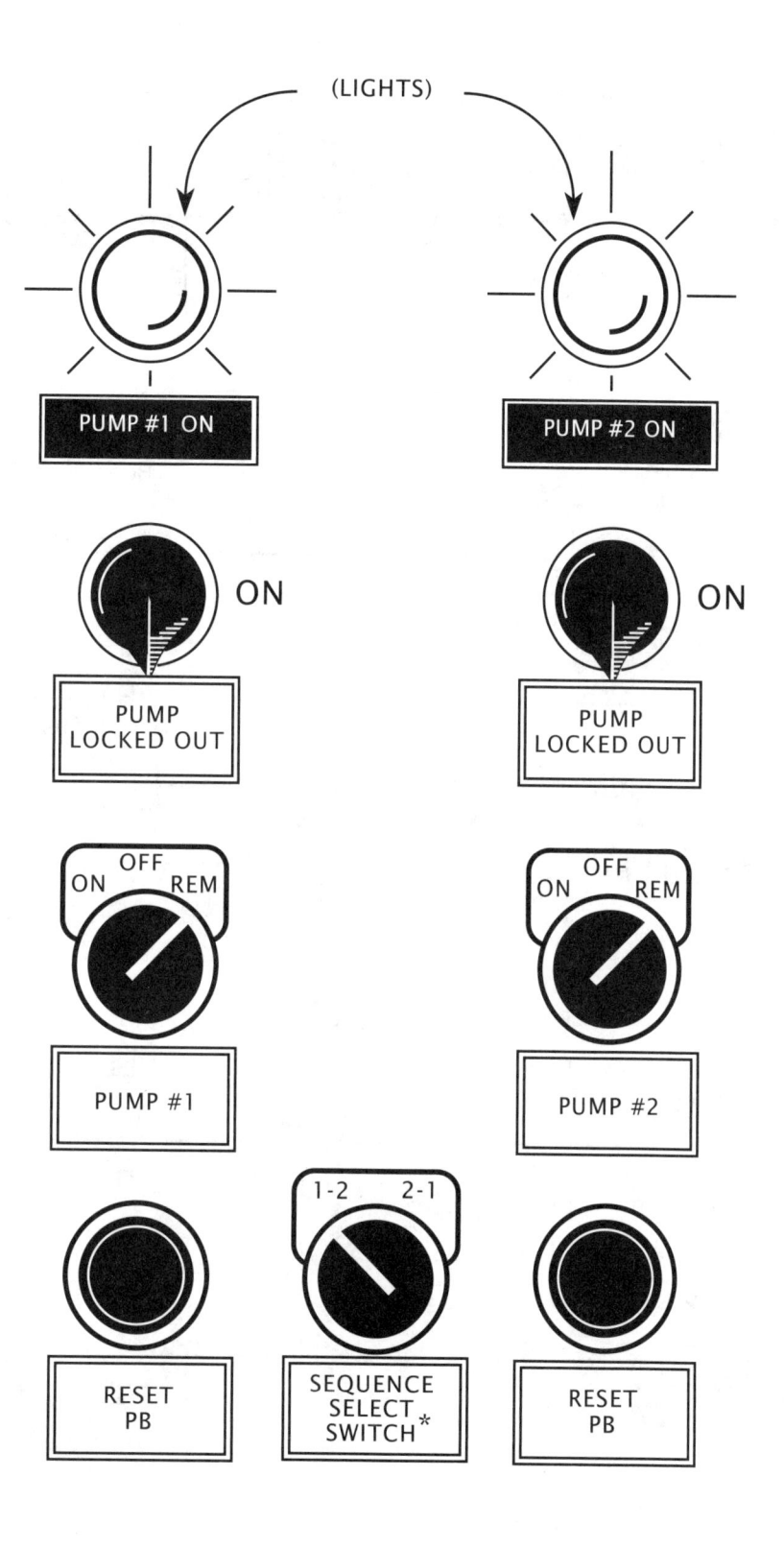

Note: "REM" on switch above means "remote," which is automatic operation in this application

*Determines which is "lead" (first to come on) and "lag" (next to come on) pump in automatic operation

Fig. 19.4   Typical duplex (two-pump) motor control panel

406 Water Treatment

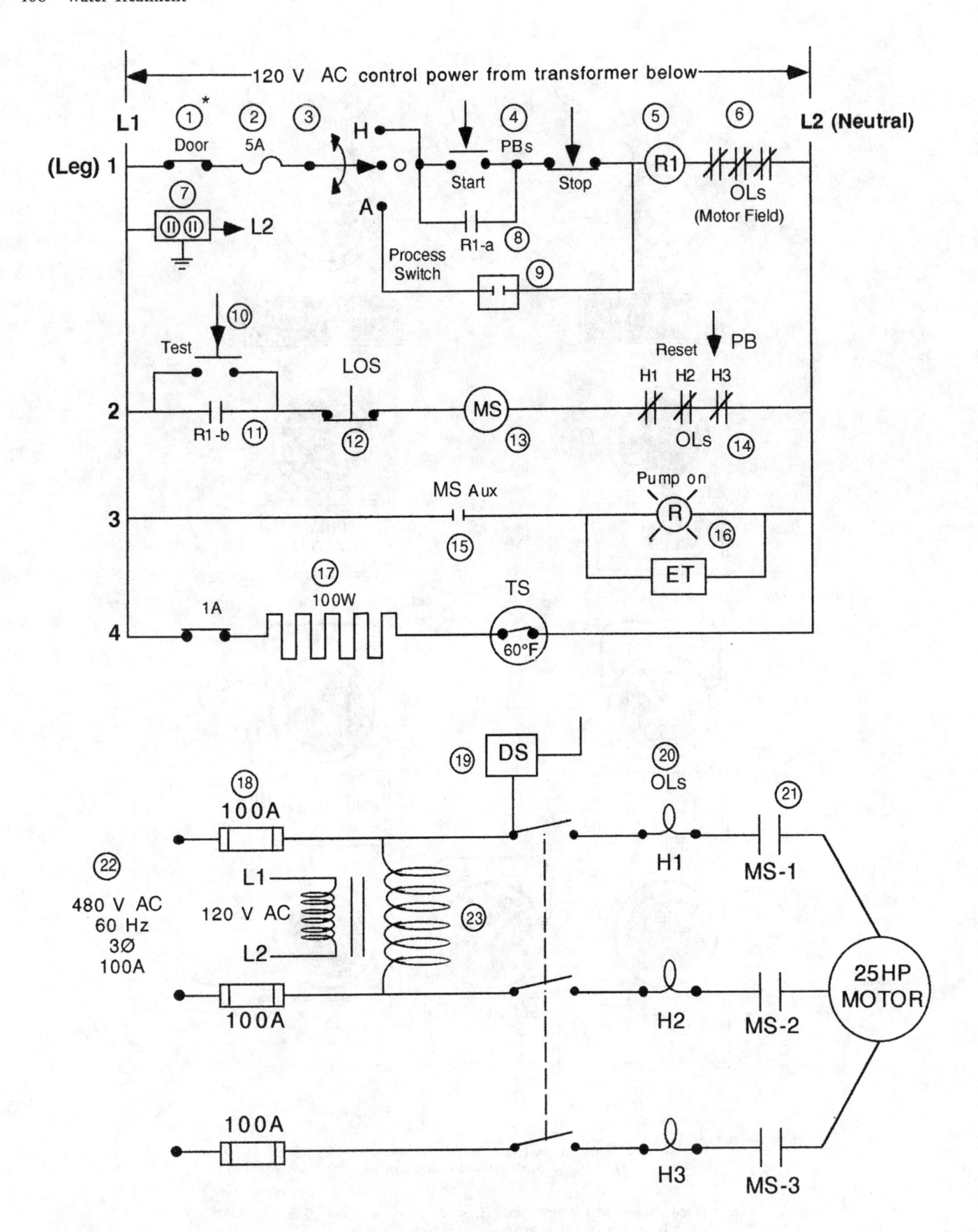

Fig. 19.5 *Typical motor control elementary diagram circuitry (do not use for design)*
(simplified, self-explanatory ladder diagram)
(see page 407 for parts legend)

FIGURE 19.5 PARTS LEGEND

1. Door or "interlock" switch to kill control power if panel door is opened.

2. Control circuit fuse, 5 amp.

3. Hand-Off-Auto (H-O-A) function switch for manual or automatic operation (shown Off).

4. Momentary contact (spring-loaded) push buttons (Start-Stop), "PB station."

5. Control relay coil (other relays, if used, would be shown as R2, R3, etc.).

6. Motor thermal protection elements embedded in 3 phase motor field ("stator") coils.

7. Duplex 120 V AC outlet within panel (for test equipment usage).

8. "Holding contacts" of control relay; start PB energizes relay coil to close contacts, which remain closed until coil is de-energized.

9. Process control contacts (ON/OFF control); for example, level, pressure, or analytical process variable.

10. Push-to-test motor starter relay circuit leg (if MS coil and OLs OK, pump will start); used for troubleshooting purposes.

11. Main control contacts of relay R1, energizes MS coil to start pump.

12. "Lock-Out Stop" switch to prevent motor operation while being serviced, key lockable in open position.

13. Motor Starter ("Mag starter") relay coil.

14. Manually (push button) resettable overload relay contacts, opened by "heaters" (see 20).

15. Auxiliary contacts in MS relay.

16. Pump running light (red) and elapsed time (total hours run) meter.

17. Enclosure heater in panel (keeps components dry) with 1 amp circuit breaker and thermostat switch.

18. 100 amp 480 V AC cartridge-type fuses.

19. Main 480 V AC manual disconnect switch ("Main switch") local at the motor.

20. Motor circuit thermal overload "heaters" (set for maximum load motor current).

21. Motor starter relay contactors (large contacts).

22. Motor power, 480 volts AC, 60 cycle, 3 phase, 100 amp service.

23. Control power transformer (for control circuit above), 120 volts AC.

DESCRIPTION OF TYPICAL MOTOR CONTROL CIRCUIT OPERATION
(Refer to Figure 19.5)

Control circuit is shown Off; the H-O-A function switch is at "O." To start motor manually, place H-O-A in "H" and depress "Start" push button (PB) to energize main control relay R1. With power through coil, "holding contacts" R1-a close around the start PB to keep R1 energized (and motor running) when the operator's finger is removed from the button (such spring-loaded PBs are termed "momentary" contacts).

When R1 is energized, it also "pulls-in" (closes) contacts R1-b to energize motor starter relay MS in turn. With control power to the MS coil, its contactors MS-1, 2, and 3 "make" (close) to supply 480 V AC to the motor—assuming resettable thermal overloads H1, H2, and H3 are all closed (and the main disconnect is closed and 100A fuses are good, of course).

To stop the motor manually, with the H-O-A still in the "H" position, one pushes the "Stop" PB to break the circuit to the R1 coil, allowing contacts R1-a to "fall out" (open) so the control relay is de-energized (even after this momentary open PB is released).

For automatic operation, H-O-A switch is placed in "A" so motor will start and stop with the open and close, respectively, of the process controller contacts. Note that the PBs cannot start or stop the motor with H-O-A switch in the Auto position.

The "Test" PB, in Leg 1 of the schematic, will start stopped pump as long as it is depressed (being a momentary close PB), by bypassing the control contacts R1-b. This feature permits isolation of a no-start problem to Leg 1 circuitry, in that the motor will start if Leg 2 is OK.

In Leg 3, the "motor on" light and "elapsed time" meter are turned on when the MS relay is energized, through its auxiliary contacts (integral to the relay). Leg 4 consists of a small fused and thermostated space heater to keep all electrical panel components dry.

120 V AC control power for the control circuit (L1 and L2) is transformed from the main 480 V AC 3Ø motor power service. A main disconnect switch kills all power to the motor, for servicing, and all three phases are protected with 100-amp cartridge fuses.

This typical motor control circuitry has two types of motor overheating protection:

1. In-line heater elements H1, H2, and H3, which open any or all of their respective contacts above when the motor draws higher currents for a longer time than it is designed for. These interrupt the power to the MS relay (Leg 2).

2. Small, heat-sensitive contacts embedded in all three stator motor windings, which open at a predetermined maximum temperature to interrupt the control circuit in Leg 1, stopping the motor before it gets too hot.

Only the former (overload heaters) are manually resettable using the PB shown; the latter (stator-imbedded) must be allowed to cool before they remake contact.

for continuing self-discipline in this area. Though such mention may conjure up humorous images of the maintenance person's surprise and shock upon such an incident, one only need consider that electric shock can and does regularly cause serious burns and even death (by asphyxiation due to paralysis of the muscles used in breathing) to bring the problem into sober perspective. Also, the expense and effort caused by a needless shorting-out of an electrical device could be significant. The point is, resist the urge to test any electrical device with a hand tool or part of your body.

If there is *any* doubt in your mind that *all* sources of voltage (not merely the local switch) to a device have been switched off or disconnected, then *DO NOT TOUCH,* except possibly with the insulated probes of a test meter. Remember, you cannot see even the highest voltage, and assuming that a circuit is dead can be very hazardous, maybe even deadly.

Do not simulate an electrical action, for example by pressing down a relay armature, within an electrical panel without a positive understanding of the circuitry. Your innocent action may cause an electrical "explosion" to shower you with molten metal, or startle you into a bumped head or elbow, or cause a bad fall. Remember, *WHEN IN DOUBT—DO NOT* when it comes to electricity.

Usually, a plant operator does not have the test equipment or the technical knowledge to correct an electrical malfunction, other than possibly resetting a circuit breaker, regardless of how critical the device's function is to plant operations. In addition to the shock hazards of motor control centers, there also may be a shock hazard within measurement instrument cases. Expensive instrument components can easily be damaged or destroyed when a foolhardy operator uses the tool-touch-test method.

Most panels have an *INTERLOCK* on the door that interrupts all (local) power to a panel or device when the door is opened or the circuit is exposed for service. Never disconnect or disable interlocks that interrupt control power. Warning labels, insulating covers (over "hot" terminals), safety switches, lockouts, and other safety provisions on electrical equipment must remain functional at all times. Your attention to this crucial aspect of your workplace may save a life, and, as the saying goes, it could be your own.

Operators often use hand power tools around electrical equipment. All such power tools (even double-insulated ones) can present a shock hazard, treatment plants being damp places, at times. Power tools can be a mechanical hazard as well. For your own safety, use power tools only when you can have an observer on hand in case of an accident. *NEVER* stand in water with a power tool, even when it is turned off. Brace yourself, if necessary, in such a way that electric current cannot flow from arm to arm in case of a faulty power tool. Shocks through the upper body could involve your heart or your head, whose importance to you is self-evident.

QUESTIONS

Write your answers in a notebook and then compare your answers with those on page 447.

19.1A What are the general principles for safe performance on the job?

19.1B How can electric shock cause death?

19.1C What could happen to you as a result of an electrical "explosion"?

19.1D Why should you brace yourself when operating power equipment so that electric current cannot flow from arm to arm in case of a faulty tool?

19.12 Mechanical and Pneumatic Hazards

There exists a special danger when working around powered mechanical equipment, such as electric motors, valve operators, and chemical feeders that are operated remotely or by an automatic control system. Directly stated, the machinery may suddenly start or move when you are not expecting it. Most devices are powered by motors with enough torque or RPM to severely injure anyone in contact with a moving part. Even when the exposed rotating or meshing elements are fitted with guards in compliance with safety regulations, a danger may exist. A motor started remotely may catch a shirt sleeve, finger, or tool hanging near a loose or poorly fitted shaft or gear train guard.

The sudden, automatic operation of equipment, even when half-expected, may startle a nearby operator into a fall or slip. Signs indicating that "This Equipment May Start At Any Time" tend to be ignored after a while. Accordingly, you must stay alert to the fact that any automatic device may begin to operate at any time. Thus, you must stay well clear of active automatic equipment, especially when it is not operating.

Lockout devices on electric switches must be respected at all times. The electrician who attaches a lockout device to physically prevent the operation of an electric circuit is, in effect, trusting his or her life and health to the device. Once the lockout device is attached to the switch (whether the switch is tagged off or actually locked with lock and key), the electrician will consider the circuit and its connected equipment de-energized and safe and will feel free to work on it. Consider the potential consequences, then, of a careless operator who removes a lockout to place needed equipment back into service, *presuming* the electrician is finished (as might occur after several hours). The point cannot be overstressed:

RESPECT ALL LOCKOUT DEVICES AND ALL TAGGED-OFF EQUIPMENT AS IF A LIFE IS ENTRUSTED TO YOU ... IT MAY WELL BE.

PNEUMATIC HAZARDS

Working with and around pneumatic instrumentation presents an additional hazard associated with high-pressure air. Pneumatic devices are powered by air at supply pressures from about 50 to 100 psi (350 to 700 kPa). Serious injury can be caused by the high air (and trapped particle) velocities that can be produced. A pressure regulator normally reduces the high air pressures to a safer 30 psi (210 kPa) or so for each device, but even these lower pressures can cause injury if directed toward the eye or other delicate tissues.

Before disconnecting any air line, put on your safety goggles (not just glasses, but wraparound goggles). Then valve the line off, from both directions ideally, and crack (open slightly) a fitting on the pipe or tubing to permit the pressurized air to bleed out slowly. If it is necessary to purge a line of moisture after inspection or repair, do so through the filter trap valve (pointing down, if standard) on each of a supply line's connected components. If a very dirty, oily, or moist line must be purged with supply air, set up a temporary purge line to a low area and tape a sock or closed rag to the open end to trap particles and minimize dirt pickup by the exhausting air. *NEVER* direct a pressurized air stream toward any part of your body or anyone else's body. High-velocity air can easily penetrate the body's tissues, even the skin of the hands. The tiny particles that are always present, even in filtered air (picked up from the inside of piping), can enter the eye's delicate tissues and cause an irritation at the least, or corneal damage and infection at worst. Breathing oil-laden air (many compressors seal with oil) from extensive line and filter purging can cause "chemical pneumonia" or even bacterial lung infections in sensitive individuals. Wearing a simple painter's (filter) mask will keep the suspended oil droplets and other particulates out of your respiratory system.

Though it is recognized that plant operators do little, if any, corrective maintenance around instrumentation and control components, it seems there is always some exposure to electrical, mechanical, and pneumatic hazards. Operators often do preventive maintenance tasks, and even minor fixing of many devices. Therefore, always play it smart and safe around all instrumentation.

19.13 Confined Spaces

Many measurement and control systems include remotely installed sensors and control valves. Quite often, these are found in vaults or other closed concrete structures. This section outlines procedures for preventing personal exposure to dangerous air contamination and/or oxygen deficiency/enrichment when working within such spaces. If you enter confined spaces, you must develop and implement written, understandable procedures in compliance with OSHA standards and you must provide training in the use of these procedures for all persons whose duties may involve confined space entry. *The procedures presented here are intended as guidelines. Exact procedures for work in confined spaces may vary with different agencies and geographical locations and must be confirmed with the appropriate regulatory safety agency.*

A confined space may be defined as any space that: (1) is large enough and so configured that an employee can bodily enter and perform assigned work; and (2) has limited or restricted means for entry or exit (for example, manholes, tanks, vessels, silos, storage bins, hoppers, vaults, and pits are spaces that may have limited means of entry); and (3) is not designed for continuous employee occupancy. One easy way to identify a confined space is by whether or not you can enter it by simply walking while standing fully upright. In general, if you must duck, crawl, climb, or squeeze into the space, it is considered a confined space.

A major concern in confined spaces is whether the existing ventilation is capable of removing DANGEROUS AIR CONTAMINATION[5] and/or oxygen deficiency/enrichment that

---

[5] *Dangerous Air Contamination.* An atmosphere presenting a threat of causing death, injury, acute illness, or disablement due to the presence of flammable and/or explosive, toxic, or otherwise injurious or incapacitating substances.
  (1) Dangerous air contamination due to the flammability of a gas, vapor, or mist is defined as an atmosphere containing the gas, vapor, or mist at a concentration greater than 10 percent of its lower explosive (lower flammable) limit (LEL).
  (2) Dangerous air contamination due to a combustible particulate is defined as a concentration that meets or exceeds the particulate's lower explosive limit (LEL).
  (3) Dangerous air contamination due to the toxicity of a substance is defined as the atmospheric concentration that could result in employee exposure in excess of the substance's permissible exposure limit (PEL).
  NOTE: A dangerous situation also occurs when the oxygen level is less than 19.5 percent by volume (OXYGEN DEFICIENCY) or more than 23.5 percent by volume (OXYGEN ENRICHMENT).

may exist or develop. In water treatment, we are concerned primarily with oxygen deficiency (less than 19.5 percent oxygen by volume), oxygen enrichment (greater than 23.5 percent oxygen by volume), methane (explosive), and hydrogen sulfide (toxic).

The potential for buildup of toxic or explosive gas mixtures and/or oxygen deficiency/enrichment exists in all confined spaces. The atmosphere must be checked with reliable, calibrated instruments before every entry. When testing the atmosphere, first test for oxygen deficiency/enrichment, then combustible gases and vapors, and then toxic gases and vapors. The oxygen concentration in normal breathing air is 20.9 percent. The atmosphere in the confined space must not fall below 19.5 percent or exceed 23.5 percent oxygen. Engineering controls are required to prevent low or high oxygen levels. However, personal protective equipment is necessary if engineering controls are not possible. In atmospheres where the oxygen content is less than 19.5 percent, supplied air or self-contained breathing apparatus (SCBA) is required. SCBAs are sometimes referred to as scuba gear because they look and work much like the air tanks used by divers.

Entry into confined spaces is never permitted until the space has been properly ventilated using specially designed forced-air ventilators. These blowers force all the existing air out of the space, replacing it with fresh air from outside. This crucial step must *ALWAYS* be taken even if atmospheric monitoring instruments show the atmosphere to be safe. Because some of the gases likely to be encountered in a confined space are combustible or explosive, the blowers must be specially designed so that the blower itself will not create a source of ignition that could cause an explosion.

There are two general classifications of confined spaces: (1) non-permit confined spaces, and (2) permit-required confined spaces (permit spaces).

A *NON-PERMIT CONFINED SPACE* is a confined space that does not contain or, with respect to atmospheric hazards, have the potential to contain any hazard capable of causing death or serious physical harm. The following steps are recommended *PRIOR* to entry into *ANY* confined space:

1. Ensure that all employees involved in confined space work have been effectively trained.

2. Identify and close off or reroute any lines that may carry harmful substance(s) to, or through, the work area.

3. Empty, flush, or purge the space of any harmful substance(s) to the extent possible.

4. Monitor the atmosphere at the work site and within the space to determine if dangerous air contamination and/or oxygen deficiency/enrichment exists.

5. Record the atmospheric test results and keep them at the site throughout the work period.

6. If the space is interconnected with another space, each space must be tested and the most hazardous conditions found must govern subsequent steps for entry into the space.

7. If an atmospheric hazard is noted, use portable blowers to further ventilate the area; retest the atmosphere after a suitable period of time. Do not place the blowers inside the confined space.

8. If the *ONLY* hazard posed by the space is an actual or potential hazardous atmosphere and the preliminary ventilation has eliminated the atmospheric hazard or continuous forced ventilation *ALONE* can maintain the space safe for entry, entry into the area may proceed.

A *PERMIT-REQUIRED CONFINED SPACE* (permit space) is a confined space that has one or more of the following characteristics:

1. Contains or has the potential to contain a hazardous atmosphere

2. Contains a material that has the potential for engulfing an entrant

3. Has an internal configuration such that an entrant could be trapped or asphyxiated by inwardly converging walls or by a floor that slopes downward and tapers to a smaller cross section

4. Contains any other recognized serious safety or health hazard

OSHA regulations require that a confined space entry permit be completed for each permit-required confined space entry (Figure 19.6). The permit must be renewed each time the space is left and re-entered, even if only for a break or lunch, or to go get a tool. The confined space entry permit is "an authorization and approval in writing that specifies the location and type of work to be done, certifies that all existing hazards have been evaluated by a competent person, and that necessary protective measures have been taken to ensure the safety of each worker." A competent person, in this case, is a person designated in writing as capable, either through education or specialized training, of anticipating, recognizing, and evaluating employee exposure to hazardous substances or other unsafe conditions in a confined space. This person is authorized to specify control procedures and protective actions necessary to ensure worker safety.

The following procedures must be observed before entry into a permit-required confined space:

1. Ensure that personnel are effectively trained.

2. If the confined space has both side and top openings, enter through the side opening if it is within 3½ feet (1.1 meters) of the bottom.

3. Wear appropriate, approved, respiratory protective equipment.

4. Ensure that written operating and rescue procedures are at the entry site.

5. Wear an approved harness with an attached line. The free end of the line must be secured outside the entry point.

6. Test for atmospheric hazards as often as necessary to determine that acceptable entry conditions are being maintained.

Confined Space Pre-Entry Checklist/Confined Space Entry Permit

Date and Time Issued: _____ Date and Time Expires: _____ Job Site/Space I.D.: _____

Job Supervisor: _____ Equipment to be worked on: _____ Work to be performed: _____

Standby personnel: _____ _____ _____

1. Atmospheric Checks: Time _____ Oxygen _____ %   Toxic _____ ppm
                                              Explosive _____ % LEL   Carbon Monoxide _____ ppm

2. Tester's signature: _____

3. Source isolation: (No Entry)   N/A   Yes   No
      Pumps or lines blinded,
         disconnected, or blocked   ( )   ( )   ( )

4. Ventilation Modification: N/A   Yes   No
      Mechanical   ( )   ( )   ( )
      Natural ventilation only   ( )   ( )   ( )

5. Atmospheric check after isolation and ventilation: Time _____
      Oxygen _____ % > 19.5% < 23.5%   Toxic _____ ppm   < 10 ppm H_2S
      Explosive _____ % LEL   < 10%   Carbon Monoxide _____ ppm   < 35 ppm CO

Tester's signature: _____

6. Communication procedures: _____
7. Rescue procedures: _____
   _____

8. Entry, standby, and backup persons   Yes   No
      Successfully completed required training? ( ) ( )
      Is training current?   ( ) ( )

9. Equipment: N/A   Yes   No
      Direct reading gas monitor tested   ( ) ( ) ( )
      Safety harnesses and lifelines for entry and standby persons ( ) ( ) ( )
      Hoisting equipment ( ) ( ) ( )
      Powered communications ( ) ( ) ( )
      SCBAs for entry and standby persons ( ) ( ) ( )
      Protective clothing ( ) ( ) ( )
      All electric equipment listed for Class I, Division I,
         Groups A, B, C, and D, and nonsparking tools ( ) ( ) ( )

10. Periodic atmospheric tests:
       Oxygen: ___% Time ___; ___% Time ___; ___% Time ___; ___% Time ___;
       Explosive: ___% Time ___; ___% Time ___; ___% Time ___; ___% Time ___;
       Toxic: ___ppm Time ___; ___ppm Time ___; ___ppm Time ___; ___ppm Time ___;
       Carbon Monoxide: ___ppm Time ___; ___ppm Time ___; ___ppm Time ___; ___ppm Time ___;

We have reviewed the work authorized by this permit and the information contained herein. Written instructions and safety procedures have been received and are understood. Entry cannot be approved if any brackets ( ) are marked in the "No" column. This permit is not valid unless all appropriate items are completed.

Permit Prepared By: (Supervisor) _____ Approved By: (Unit Supervisor) _____

Reviewed By: (CS Operations Personnel) _____
                            (Entrant)   (Attendant)   (Entry Supervisor)

This permit to be kept at job site. Return job site copy to Safety Office following job completion.

Fig. 19.6 Confined space pre-entry checklist/confined space entry permit

7. Station at least one person to stand by on the outside of the confined space and at least one additional person within sight or call of the standby person.

8. Maintain effective communication between the standby person and the entry person.

9. The standby person, equipped with appropriate respiratory protection, should only enter the confined space in case of emergency.

10. If the entry is made through a top opening, use a hoisting device with a harness that suspends a person in an upright position. A mechanical device must be available to retrieve personnel from vertical spaces more than five feet (1.5 meters) deep.

11. If the space contains, or is likely to develop, flammable or explosive atmospheric conditions, do not use any tools or equipment (including electrical) that may provide a source of ignition.

12. Wear appropriate protective clothing when entering a confined space that contains corrosive substances or other substances harmful to the skin.

13. At least one person trained in first aid and cardiopulmonary resuscitation (CPR) should be immediately available during any confined space job.

Individuals designated to provide first aid or CPR should be included in a Bloodborne Pathogens (BBP) program. These employees may be exposed to contact with blood or other potentially infectious materials from the performance of their duties. The BBP program includes training in exposure potential determination, engineering and work practice controls, personal protective equipment (PPE), and the availability of the hepatitis B vaccination series (29 CFR 1910.1030). If the operator(s) must enter confined spaces to perform rescue services, they must be trained specifically to perform the assigned rescue duty and to use required personal protective equipment (PPE) and rescue equipment. Rescue practice sessions must be held at least once every 12 months.

If you arrange to have a contractor perform work in confined spaces at your facility or within your collection system, you must inform the contractor:

- That the contractor must comply with confined space regulations
- Of hazards that you have identified and your experience with the space(s)
- Of precautions or procedures you have implemented for the protection of employees in or near the space where the contractor's personnel will be working
- That a debriefing must occur at the conclusion of the entry operations regarding the confined space program followed and any hazards encountered or created during the entry operations

To enhance safety, communications, and coordination of confined space activities, the contractor is also required to obtain available information from you and to inform you of the confined space program the contractor will follow. This exchange of information must occur before the confined space is entered by any operator.

Confined space work can present serious hazards if you are uninformed or untrained. The procedures presented are only guidelines and exact requirements for confined space work for your locale may vary. Contact your local regulatory safety agency for specific requirements.

19.14 Oxygen Deficiency or Enrichment

Low oxygen levels may exist in any poorly ventilated, low-lying structure where gases such as hydrogen sulfide, gasoline vapor, carbon dioxide, or chlorine may be produced or may accumulate. Oxygen in a concentration above 23.5 percent (oxygen enrichment) also can be dangerous because it speeds up combustion.

Oxygen deficiency is most likely to occur when structures or channels are installed below grade (ground level). Several gases (including hydrogen sulfide and chlorine) have a tendency to collect in low places because they are heavier than air. The specific gravity of a gas indicates its weight as compared to an equal volume of air. Since air has a specific gravity of exactly 1.0, any gas with a specific gravity greater than 1.0 may sink to low-lying areas and displace the air from that area or structure. (On the other hand, methane may rise out of a manhole because it has a specific gravity of less than 1.0, which means that it is lighter than air.) You should never rely solely on the specific gravity of a gas to tell you where it is. Air movement or temperature differences within a confined space may affect the location of atmospheric hazards. The only effective way of ensuring safe atmospheric conditions prior to entering a confined space is to test the atmosphere with an appropriate monitor(s) at various levels and locations throughout the space.

When oxygen deficiency or enrichment is discovered, the area should be ventilated with fans or blowers and checked again for oxygen deficiency/enrichment before anyone enters the area to work. Ventilation may be provided by fans or blowers. Follow confined space procedures before entering and during occupancy of any suspect area. *ALWAYS* get air into the confined space *BEFORE* you enter to work and maintain the ventilation until you have left the space. Equipment is available to measure oxygen concentration as well as toxic and combustible atmospheric conditions. You must use this equipment whenever you encounter a potential confined space situation. Ask your local safety regulatory agency or water association about sources of this type of equipment in your area.

19.15 Explosive Gas Mixtures

Explosive gas mixtures may develop in many areas of a treatment facility from mixtures of air and methane, natural gas, manufactured fuel gas, hydrogen, or gasoline vapors. The upper explosive limit (UEL) and lower explosive limit (LEL) indicate the range of concentrations at which combustible or explosive gases will ignite when an ignition source is present at ambient temperature. No explosion or ignition occurs when the concentration is outside these ranges. Gas concentrations below the LEL are too lean to ignite; there is not enough flammable gas or vapor to support combustion. Gas concentrations higher than the UEL are too rich to ignite; there is too much flammable gas or vapor and not enough oxygen to support combustion (Figure 19.7).

Explosive ranges can be measured by using a combustible gas detector calibrated for the gas of concern. Do not rely on your nose to detect gases. The sense of smell is absolutely unreliable for evaluating the presence of dangerous gases. Some gases have no smell and hydrogen sulfide can paralyze the sense of smell.

Avoid explosions by eliminating all sources of ignition in areas potentially capable of developing explosive mixtures. Only explosion-proof electrical equipment and fixtures should be used in these areas (inflow/screen rooms, gas compressor areas, battery charging stations). Provide adequate ventilation in all areas that have the potential to develop an explosive atmosphere.

The National Fire Protection Association Standard 820 (NFPA 820), *FIRE PROTECTION IN WASTEWATER TREATMENT AND COLLECTION FACILITIES*, lists requirements for electrical classifications, ventilation, gas detection, and fire control methods in various wastewater treatment and collection system areas. (For ordering information, see Section 19.5, "Additional Reading," item 4.) This information is also applicable to water treatment facilities. Comparing these requirements with your plant's existing design and equipment may indicate deficiencies, which should be remedied to minimize potential hazards.

19.16 Falls and Associated Hazards

Falls from or into tanks, wet wells, catwalks, or conveyors can be disabling or deadly. Most of these can be avoided by the proper use of ladders, hand tools, and safety equipment, and by following established safety procedures. Strains and sprains are probably the most frequent injuries in water treatment facilities. Keep in mind that an electric shock of even minor intensity can result in a serious fall. When working above ground on a ladder, use the proper, nonconductive type of ladder (fiberglass rails) and position it safely. Even an alert, safety-conscious operator could be knocked off balance by a slight electric shock. When required to do preventive maintenance from a ladder, turn off the power to the equipment being serviced, if at all possible. If this is not feasible, take special care to stay out of contact with any component inside the enclosure of an operating mechanism, and well away from terminal strips, unconduited wiring, and electrical black boxes. Though not commonly considered essential, wearing thin rubber or plastic gloves can greatly reduce your chances of electric shock (whether on a ladder or off).

Make provisions for carrying tools or other required objects on an electrician's belt rather than in your hands when climbing up or down ladders. Finally, never leave tools or any object on a step or platform of the ladder when you climb down, even temporarily. You might be the one upon whom they fall if the ladder is moved or even steadied from below. In this regard, it is always a good idea (even if not required) for preventive maintenance personnel to wear a hard hat whenever working on or near equipment, especially when a ladder must be used.

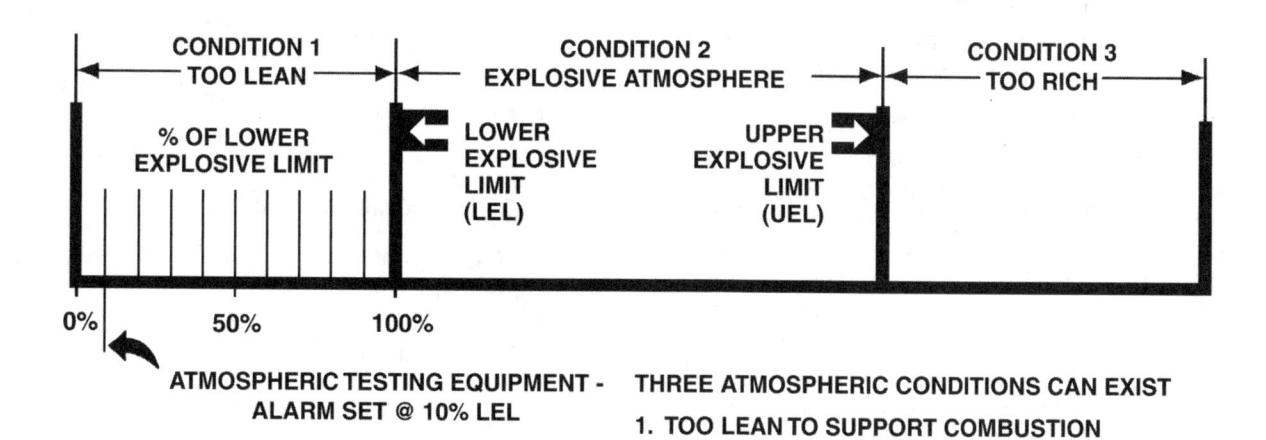

Fig. 19.7 Relationship between the lower explosive limit (LEL) and the upper explosive limit (UEL) of a mixture of air and gas

QUESTIONS

Write your answers in a notebook and then compare your answers with those on page 447.

19.1E Why should operators be especially careful when working around powered mechanical equipment?

19.1F What is the purpose of an electrical lockout device?

19.1G When testing the atmosphere before entry in any confined space, what procedure should be used?

19.1H What do the initials UEL and LEL stand for?

19.1I What kind of specific protective clothing could be worn to protect you from electric shock?

19.2 MEASURED VARIABLES AND TYPES OF SENSORS

19.20 General Principles of Sensors

A measured variable is any factor that is sensed and quantified by the primary element or sensor. In water treatment, the variables of pressure, level, and flow are the most common ones measured; at times, chemical feed rates and some physical or chemical process characteristics are also sensed.

The primary element performs the initial conversion of measurement energy to a signal. An example is a pressure gauge that converts the hydraulic action of the water into a mechanical motion to drive a meter indicator, or produces an electrical signal for a remote readout device. For transmitters not having a local readout, the sensing portion is the primary element. If such a signal is produced, be it electric or pneumatic, the sensor is also then considered a transmitter.

The signal produced is not necessarily a continuous one proportional to the variable (that is, an analog signal), but often merely a switch set to detect when the variable goes above or below preset limits. In this type of ON/OFF control, the predetermined setting is called a *DISCRETE*[6] point. This distinction between continuous and ON/OFF operation relates to the two types of controllers discussed previously.

Process measuring instruments have a common goal: to detect a variable and create an output for use by the control system or the operator. There are two basic types of instrument signals: (1) continuously variable analog, and (2) discrete ON/OFF. Signals to and from a control device are termed as input/output (I/O). The signals can further be classified as analog input (AI), analog output (AO), *DISCRETE INPUT (DI)*, and *DISCRETE OUTPUT (DO)*.[7] Computer control system inputs are converted from the analog and discrete signals and coded into digital-data representations of the inputs. Digital-data output from the computer control system is converted to analog and discrete outputs.

19.21 Pressure Measurements

Since pressure is defined as a force per unit of area (pound per square inch, or kiloPascal), you might expect that sensing pressure would thus entail the small movement of some flexible element subjected to a force. In fact, that is how pressure is measured in practice. There are many classes and brands of sensors, but the most common types contain mechanically deformable components such as the Bourdon tube (Figures 19.8), bellows, or diaphragm arrangements (Figure 19.9). The slight motion each exhibits, directly proportional to the applied force, is then amplified mechanically by levers or gears to position a pointer on a scale or provide an input for an associated transmitter. A blind transmitter for pressure, or any variable, has no local readout.

Some pressure sensors are fitted with surge or overrange protection (snubbers) to limit the effect pressure spikes or water hammer have on the instrument. In most cases, such protection devices function by restricting flow into the sensing element. Surge protection devices thus prevent sudden pressure surges, which can easily damage most pressure sensors.

A snubber (Figure 19.10) consists of a restriction through which the pressure-producing fluid must flow. A more elaborate mechanical snubber responds to surges by moving a piston or plunger that effectively controls the size of an orifice. Some snubbers are subject to clogging or being adjusted so tight as to prevent any response at all to pressure changes. If a pressure sensor is not performing properly, look first for such clogging or adjustment that has become too restrictive.

Another dampening device is an air cushion chamber (Figure 19.11), which is simply constructed yet very effective. The top part of the chamber contains air; water flows into the bottom part. A sudden change in water pressure compresses the air within the chamber, taking the shock. The rate of response can be further dampened by also placing a snubber into the air chamber line.

QUESTIONS

Write your answers in a notebook and then compare your answers with those on page 447.

19.2A What is a primary element or sensor?

19.2B How is pressure measured?

19.2C Why are some pressure sensors fitted with surge or overrange protection?

19.22 Level Measurements

Systems for sensing the level of water or any other liquid, either continuously or at a single point, are very common in municipal and water treatment plants. Pumps are controlled,

---

[6] *Discrete Control.* ON/OFF control; one of the two output values is equal to zero.

[7] *Discrete I/O (Input/Output).* A digital signal that senses or sends either ON or OFF signals. For example, a discrete input would sense the position of a switch; a discrete output would turn on a pump or light.

filters operated, basins and tanks filled and emptied, chemicals fed and ordered, sumps emptied, and many other variables controlled on the basis of liquid levels. Fortunately, level sensors usually are simple devices. A float, for example, can be a very reliable liquid level sensor both for single-point and continuous-level sensing. Other types of liquid level-sensing devices include displacers, electrical probes, direct hydrostatic pressure, pneumatic bubbler tubes, and ultrasonic (sonar) devices.

Single-point detection of level is very common for levels controlled by pumps or valves within fairly wide limits. Single-point ball float levels are very simple devices and have proven reliable in applications where set-point changes are rarely necessary. Single-point detectors mount in a sump or tank to sense a fixed elevation. When the water level rises, the ball floats and rolls onto its side, actuating an internal mercury switch. The switch provides a discrete ON/OFF signal used to control equipment or send level data to the control system.

An alternative to a float for single-point sensing, is the use of a probe to sense liquid level (Figure 19.12). Only single-point determinations can be made this way, though several probes or

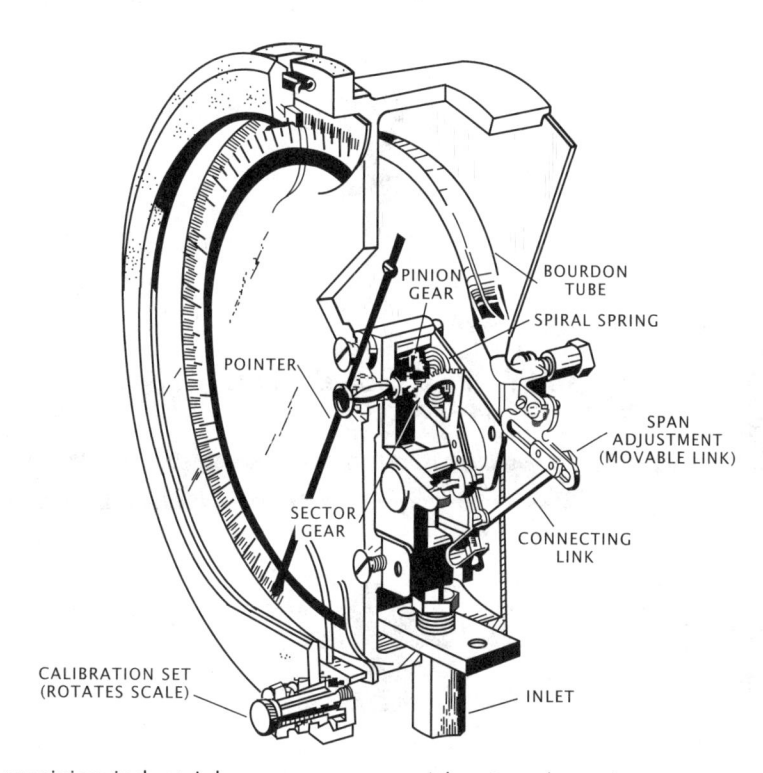

A precision industrial pressure gauge with a Bourdon-tube pressure element.

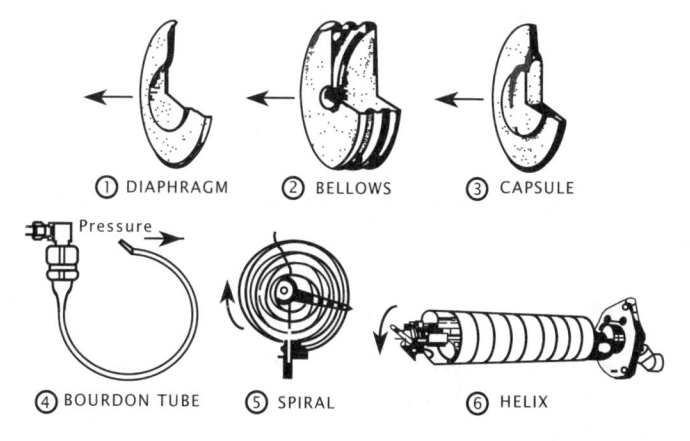

Elastic deformation elements. Pressure tends to expand or unroll elements to indicate pressure as shown by arrows.

Fig. 19.8 Bourdon tube and other pressure-sensing elements

(Permission of Heise Gauge, Dresser Industries)

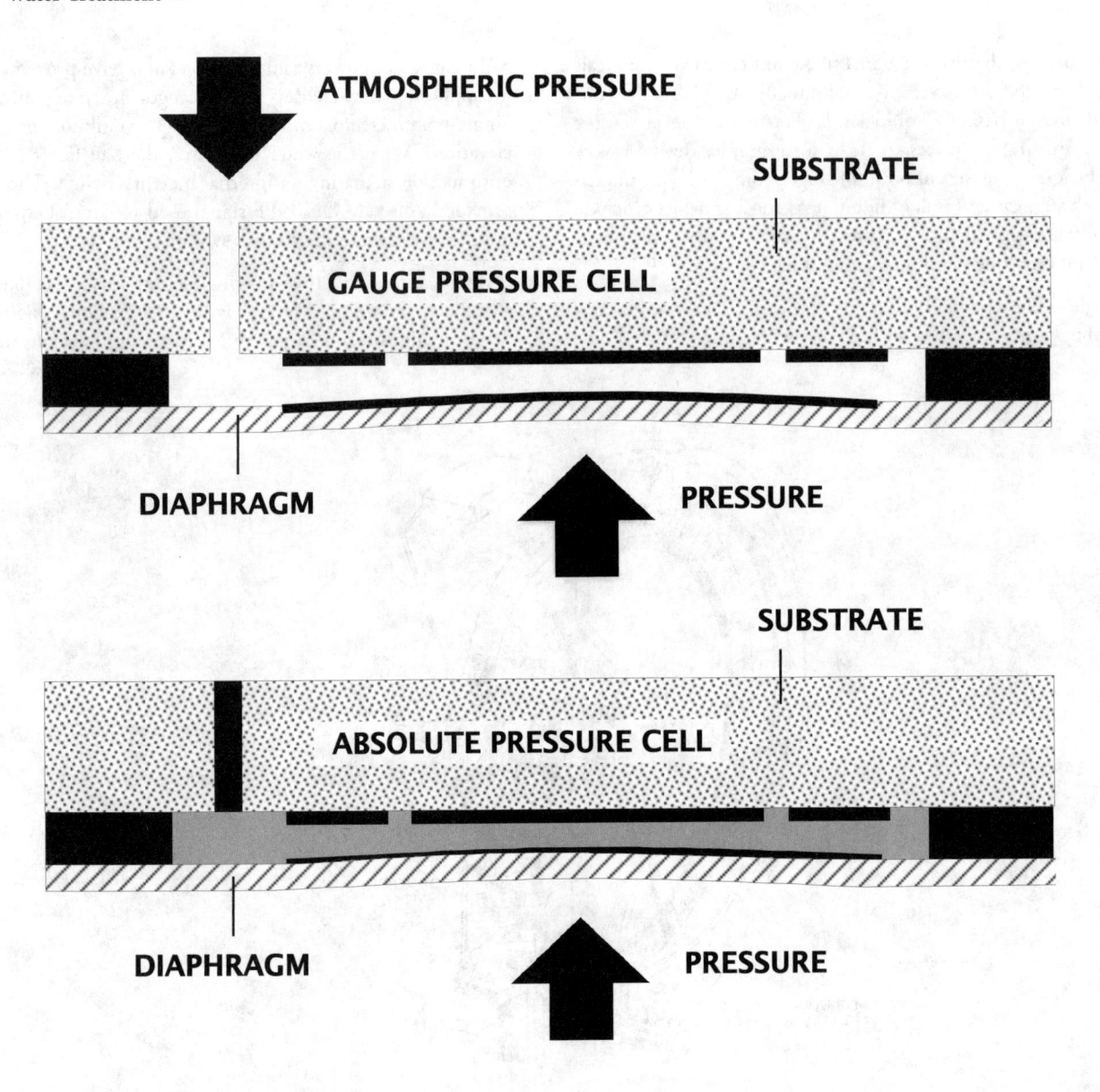

Fig. 19.9 Ceramic diaphragms to measure atmospheric pressure

one with several sensors along its length can be set up to detect several different levels. Level probes are used where a mechanical system is impractical, such as within sealed or pressurized tanks or with chemically active liquids.

The probes can be small-diameter rods of electrically conductive material inserted into a tank through a fitting, usually through the top but at times in the side of a vessel (Figure 19.13). Each rod is cut to length corresponding to a specific liquid level in the case of the top-entering probes; in the side-entering setup, a short rod merely enters the vessel at the appropriate height or depth. One problem encountered with probes is the accumulation of scum or caking (such as grease) on the surface of the rods.

A number of methods are used to detect whether the probe is immersed in liquid. A simple method for conductive liquids is to apply a small voltage to the probe(s) by the system's power supply, with current flowing only when the probe just becomes immersed in the liquid. When immersion is detected, a switch activates a pump/valve control or alarm at as many places as necessary through a discrete output (DO) from the control system. Though at times only a single probe is used, with the metal tank completing the circuit as a ground, usually at least two probes are found. The ground probe extends to within a short distance of the bottom of the tank so as to be in constant contact with the electrically conducting liquid (a liquid ground, as it were). Nonconductive liquids must use other methods to detect immersion of the probe in the liquid.

Instrumentation and Control Systems 417

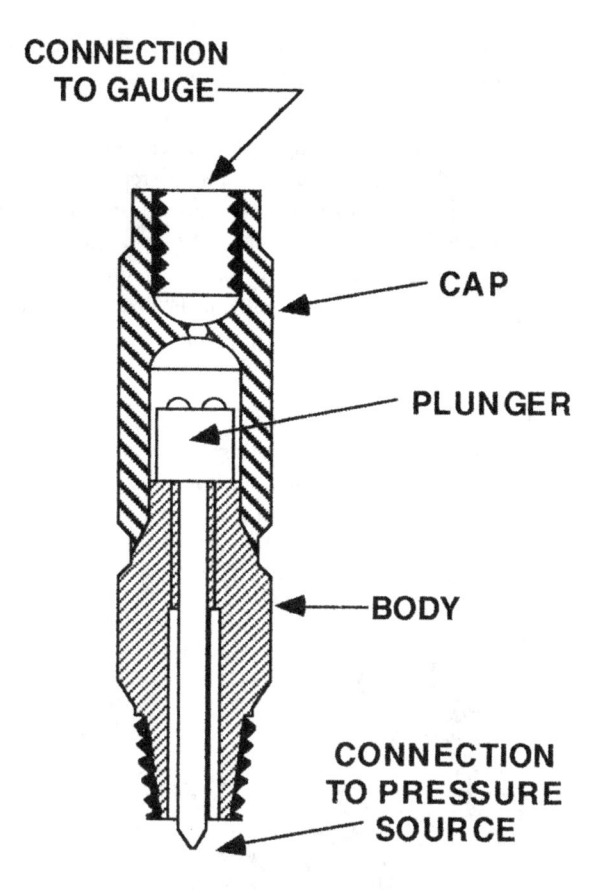

Fig. 19.10 Internal parts of a plunger-type snubber for surge protection

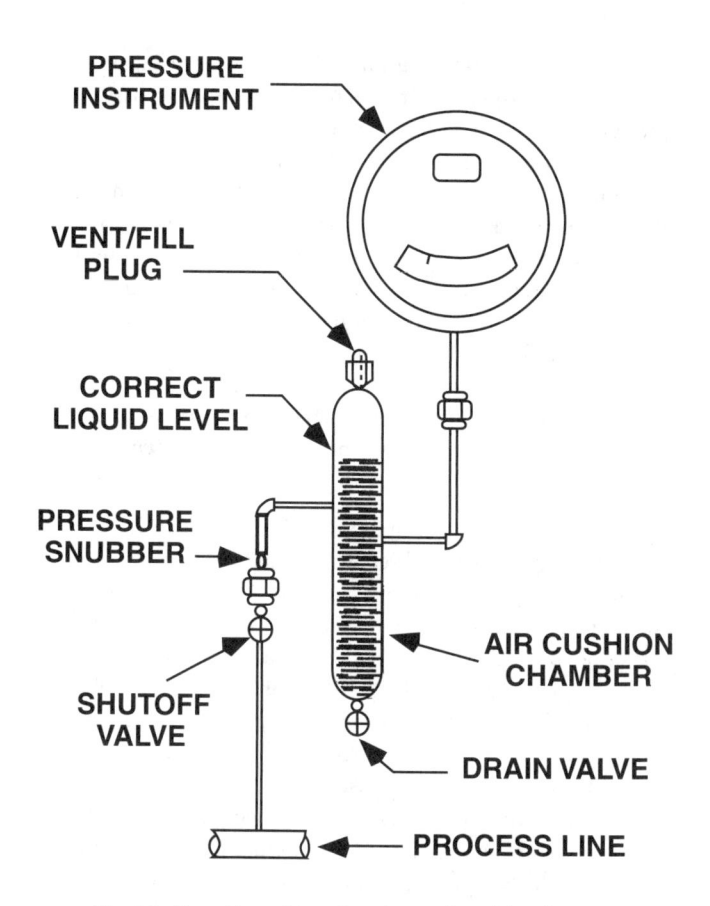

Fig. 19.11 Air cushion chamber and snubber for surge protection of pressure gauge

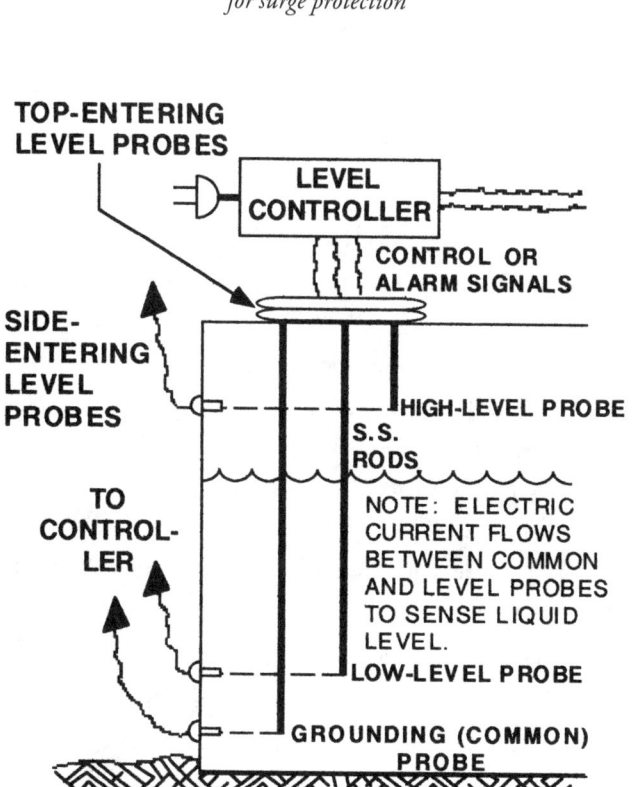

Fig. 19.12 Electrical probes (multi-point)

Fig. 19.13 Single-point level probe
(Courtesy Endress+Hauser)

Levels can be sensed continuously by measurement of liquid hydrostatic pressure near the bottom of a vessel or basin. The pressure elements used for level sensing must be quite sensitive to the low pressures created by liquid level (23 feet of water column equals only 10 psi, or 70 kPa). Therefore, simple off-the-shelf pressure gauges such as are found on pumps are not used to measure water levels. Instead, very sensitive water level sensors are used to measure levels of filter basins or in process tanks where control or monitoring must be close, continuous, and positive. Rather than being calibrated in units of pressure (psi or kPa), these gauges read directly in units of liquid level (feet or meters). One or more single-point control/alarm contacts can be made a part of this, or any, continuous type of level-sensing system.

Ultrasonic, microwave, and TDR (Time Domain Reflectometry) level sensors have proven very reliable for a wide variety of applications. Typically, the instruments in this group detect level by emitting a pulse that is reflected back to the receiver. The sensor periodically generates a pulse of waves that bounce off the liquid surface and reflect back. The sensor detects the reflected pulse. Based on the speed of the pulse, the travel time between sending and receiving is measured and the distance is calculated and converted into a level measurement. The instruments have many display options: digital readout, analog 4–20 mA, discrete ON/OFF, and digital data. Any combination of one or more may be available on a single device. They are very versatile and reliable and can be used to detect levels of liquid and solid materials.

A very precise method of measuring liquid level is the bubbler tube with its associated pneumatic instrumentation (Figure 19.14). The pressure created by the liquid level is sensed, but not directly as with a liquid pressure element. A bubbler measures the level of a liquid by sensing the air pressure necessary to cause bubbles to just flow out the end of a submerged tube. Air pressure is created in a bubbler tube to just match the pressure applied by the liquid above the open end of the tube when it is immersed to a precisely determined depth in a tank or basin. The air pressure in the tube is then measured as proportional to the liquid level *above* the end of the tube. This indirect determination of level using air permits the placement of the instrumentation anywhere above or below the liquid's surface, whereas direct pressure-to-level gauges must be installed at the very point (or very close to) where liquid pressure must be sensed.

Bubbler systems are adjusted so air *just begins* to bubble slowly out of the submerged end of the sensing tube. They automatically compensate for changes in liquid level by providing a small, constant flow of air through the bubbler tube by means of the constant air flow (also called constant differential) regulator. There is no advantage to turning up the amount of air to create more intense bubbling because the back pressure will still mainly depend on the water level. In fact, increasing the air flow may create a sizeable measuring error in the system, so any air flow changes should be left to qualified instrument service personnel.

Bubbler tube systems are common in basin-level controllers that must maintain water levels within a range of a few inches (centimeters) and for measurement of flow in open channels or over weirs. Usually, the level transmitter for basin-level controllers is blind since it only controls liquid level and need not provide a local readout of the level.

QUESTIONS

Write your answers in a notebook and then compare your answers with those on page 448.

19.2D List the different types of liquid level-sensing devices.

19.2E How does a single-point ball float level generate a level signal?

19.2F Under what circumstances are probes used to measure liquid level?

19.2G How does an ultrasonic sensor measure the level of a liquid?

19.23 Flow (Rate of Flow and Total Flow)

The two basic types of flow readings are *RATE OF FLOW*, such as MGD, CFS, and GPM or cu m/sec or liters/sec (volume per unit of time), or *TOTAL FLOW*, in simple units of volume, such as the corresponding million gallons, cubic feet, or gallons (liters or cubic meters). Total flow volumes are usually obtained as a running total, with a comparatively long time period for the flow delivery, such as a day or month. The flow instrument may

Instrumentation and Control Systems 419

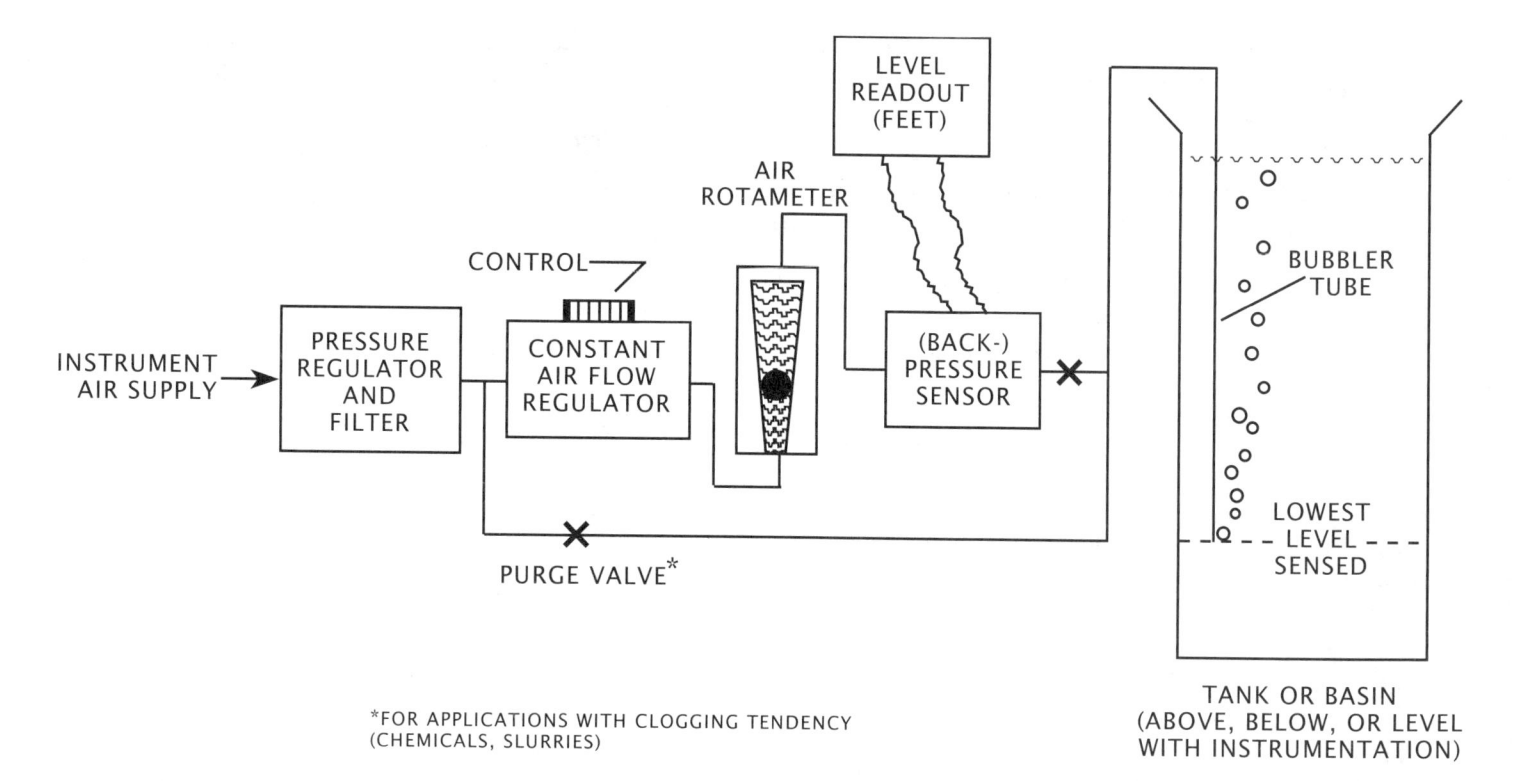

FUNCTIONAL DIAGRAM

*FOR APPLICATIONS WITH CLOGGING TENDENCY (CHEMICALS, SLURRIES)

Constant-flow regulator and rotameter on left, back-pressure sensor (DP cell) to right.

Fig. 19.14   Diagram and photo of a bubbler tube system for measuring liquid level

provide both the rate of flow and total flow values. Flow instruments applied to computer control systems typically only provide flow rate; total flow is calculated by accumulating the flow rate over some period, such as a day, month, or year.

While it is possible, in principle, to measure process flows directly, it is quite impractical. Direct measurement would involve the constant filling and emptying of, say, a gallon container with water flowing from a pipe on a timed basis. This method is obviously not practical. Therefore, sensing of process flows in waterworks practice is done inferentially, that is by inferring what the flow is from the observation of some associated hydraulic action or effect of the water. Figure 19.15 illustrates measuring level in a flume as a process flow instrument, inferring flow from the level measured. The inferential flow-sensing techniques that are used in flow measurement are: (1) velocity, (2) differential pressure, (3) magnetic, and (4) ultrasonic. First, let us look at one device, the rotameter, used in flow sensing for some specialized applications before studying the devices for process stream flows.

ROTAMETERS (Figures 19.16 and 19.17) are transparent (usually) tubes with a tapered bore containing a ball or float. The float rises up within the tube to a point corresponding to a particular rate of flow. The rotameter tube is set against, or has etched upon it, marks calibrated in whatever flow rate unit is appropriate. Rotameters are used to indicate approximate liquid or gas flow. For example, the readout for a gas chlorinator could be a rotameter. Sometimes a simple rotameter is installed merely to indicate a flow or no-flow condition in a pipe, for example, on a chlorinator injector supply line.

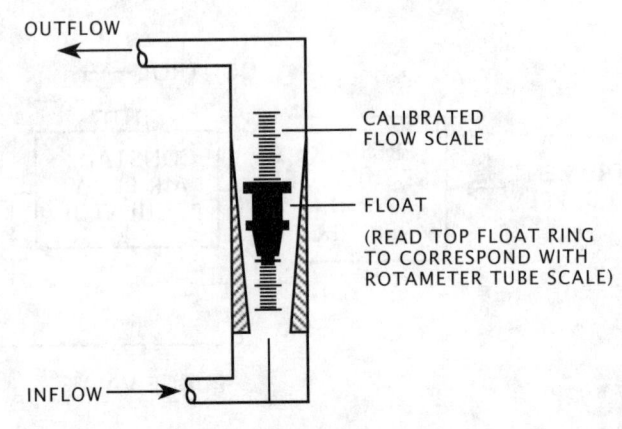

Fig. 19.16   Rotameter

For process flows, as mentioned, *VELOCITY-SENSING* meters measure water speed within a pipeline or channel. One way of doing this is by sensing the rate of rotation of a special impeller (Figure 19.18) placed within the flowing stream; the

Fig. 19.15   Inferring flow ultrasonically in a flume
(Courtesy Endress+Hauser)

Rotameters (gas **rate**-of-flow)

Fig. 19.17 Flow-sensing devices for fluids

rate of flow is directly proportional to impeller RPM (within certain limits). Since normal water velocities in pipes and channels are under 10 feet per second (about 7 MPH or 3 m/sec), the impeller turns rather slowly. This rotary motion drives a train of gears, which indicates *RATE* of flow, in the same way a speedometer indicates the rate of travel for a car. *TOTAL* flow then appears as the cumulative number of revolutions, denoting total volume flowed, like the odometer on your car's speedometer indicates total mileage traveled.

Rotation of the velocity-sensing element is not always transferred by gears, but may be picked up as magnetic or electric pulses by a transducer. Nor is velocity always sensed mechanically; it may also be detected or measured purely electrically (the thermistor type), or hydraulically (the pitot tube). In each case, the principle of equating water velocity with rate of flow (within a constant flow area) is the same. Of course, all such flowmeters are calibrated to read out in an appropriate unit of flow rate, rather than velocity units.

Typically, a velocity-sensing element transmits its reading to a remote site as electric pulses, although other devices can be used in order to convert to any standard electrical or pneumatic signal. Figure 19.19 illustrates an electrical method of sensing velocity to measure flow.

Preventive maintenance of impeller-type flowmeters centers around regular lubrication of rotating parts, at least for the older types. Propeller meters, as they are called, have a long history of reliability and acceptable accuracy in both municipal and industrial applications. When propeller meters become old, they can become susceptible to underregistration (read low) due to bearing wear and gear train friction. Accordingly, annual teardown for inspection is indicated. Overregistration is rare, but a partially full pipeline, wrong gears installed, or a malfunctioning transmitter or receiver can cause high readings.

DIFFERENTIAL PRESSURE-SENSING TUBES (Figures 19.20 and 19.21), also called Venturi or differential meters (or flow tubes), depend for their operation upon a basic principle of hydraulics, the Bernoulli Effect: When a liquid is forced to go faster in a pipe or channel, its internal pressure drops. If a carefully sized restriction is placed within the pipe or flow channel, the flowing water must speed up to get through it. In doing so, its pressure drops a little, and it drops an exact amount for a given flow rate. This small pressure drop, the "pressure differential," is the difference between the water pressure *before* the restriction and *within* the restriction. This difference is proportional in a certain way (but not directly proportional) to the rate of flow. The difference in pressure is measured very precisely by the instrumentation associated with the particular flow tube

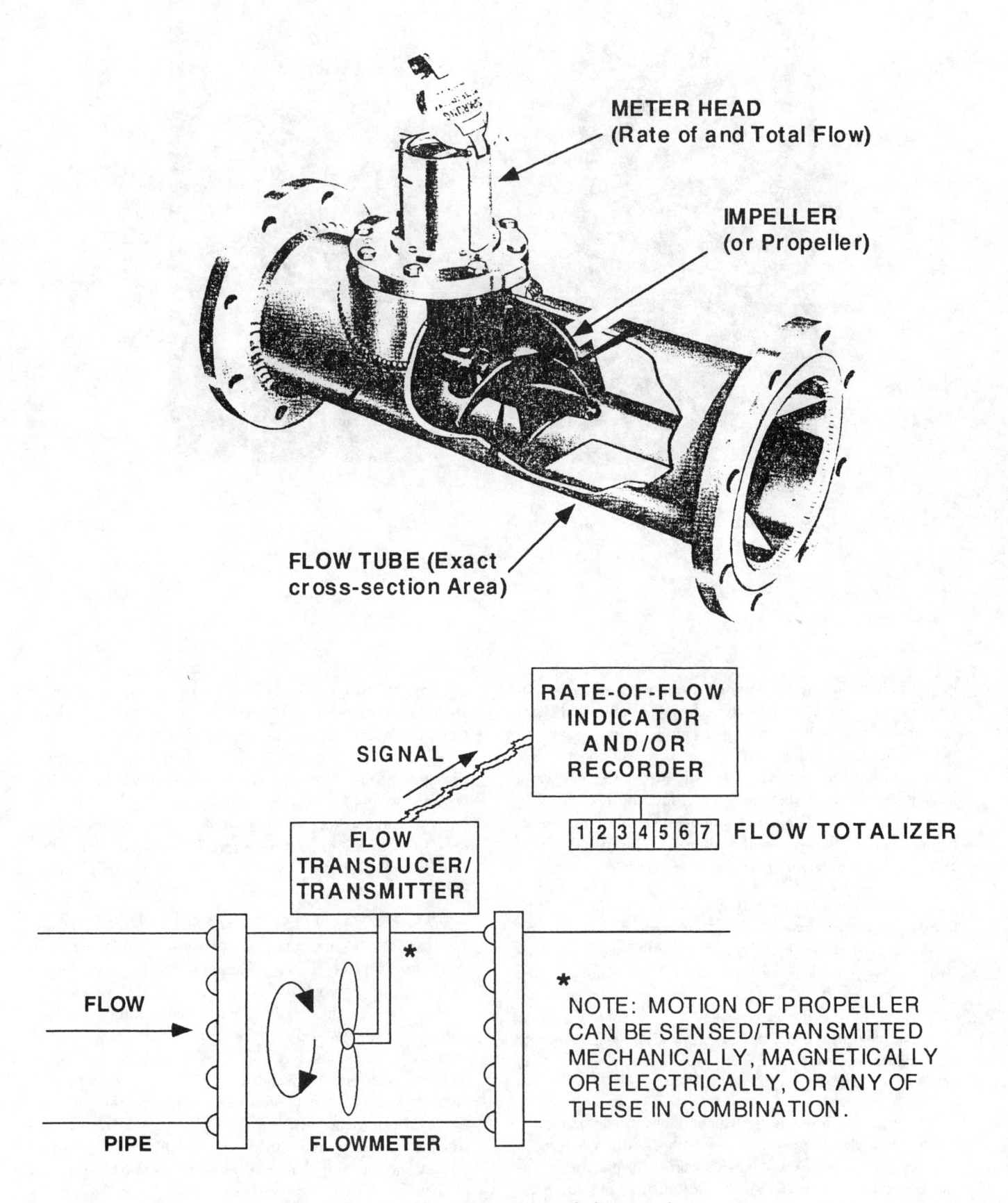

Fig. 19.18 *Propeller meter (a type of velocity meter)*

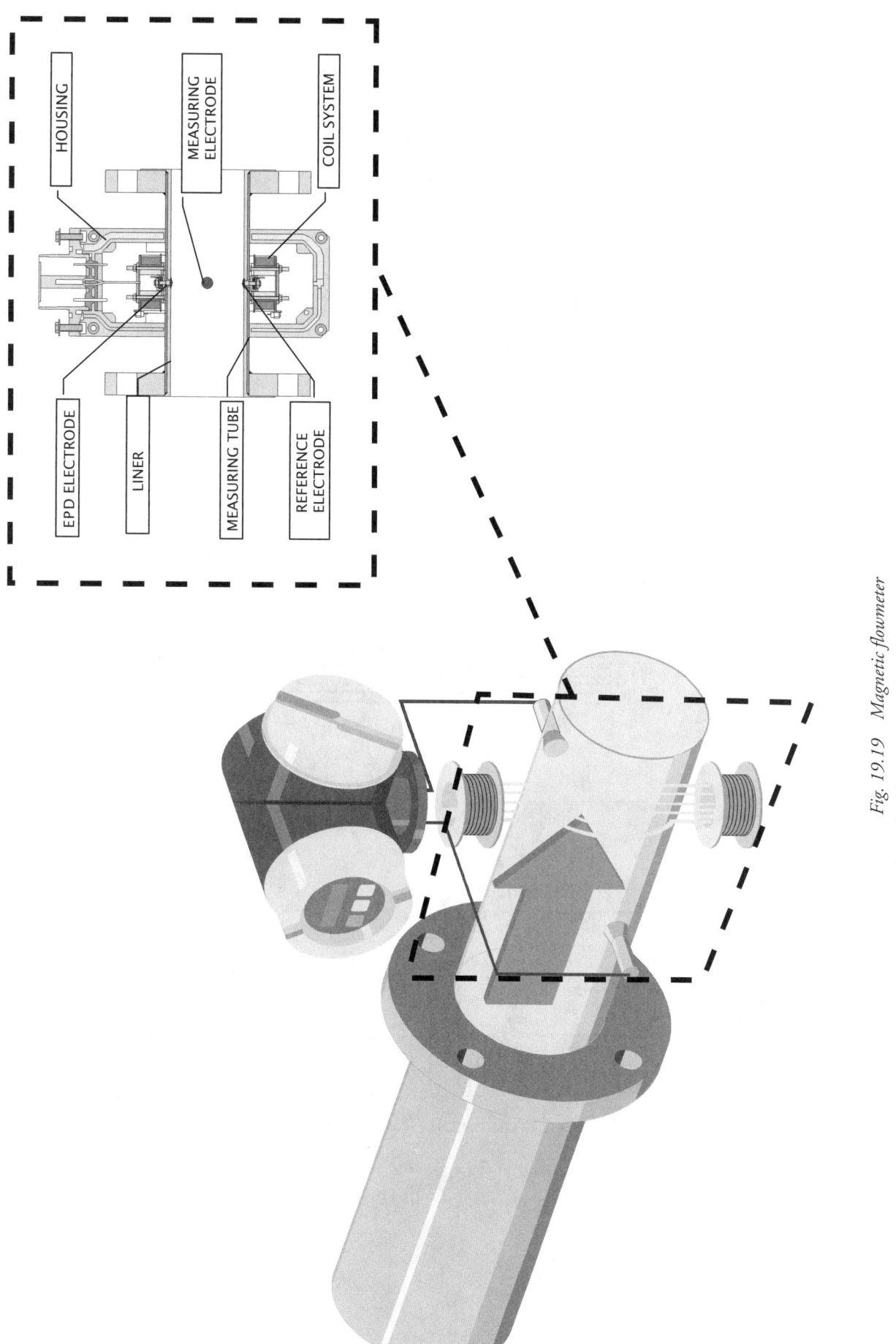

Instrumentation and Control Systems 423

Fig. 19.19 Magnetic flowmeter
(Adapted from figure provided by courtesy of Endress+Hauser)

424 Water Treatment

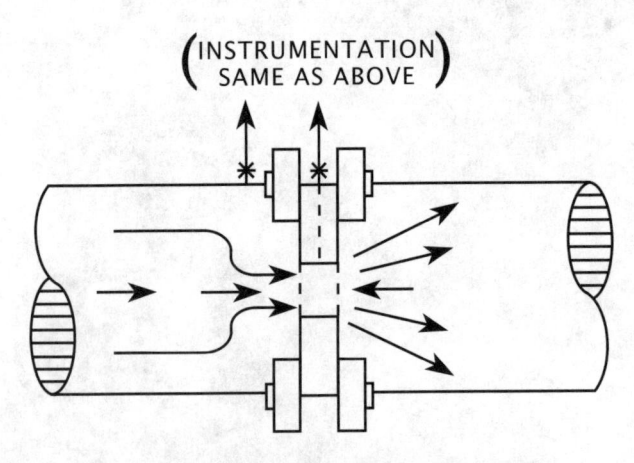

Fig. 19.20 *Schematic diagrams of differential pressure flow measuring devices*

Instrumentation and Control Systems 425

54-inch Venturi tube

24-inch orifice plate

Fig. 19.21   Photos of differential pressure flow tubes

installed. Typically, a difference of only a few psi (kPa) is required. This small value of pressure difference is often described in inches (centimeters) of water (head).

Measuring flow by the differential pressure method removes a little hydraulic energy from the water. However, the modern flow tube, with its carefully tapered form, allows recovery of well over 95 percent of the original pressure throughout its range of flows. Other ways of constricting the flow, such as installation of orifice plates, do not allow such high recoveries of pressure or the accuracy possible with other modern flow tubes.

An orifice plate (Figures 19.20 and 19.21) is a steel plate with a precisely sized hole (orifice) in it. The plate is inserted between flanges in a pipe. The pressure drop is sensed right at the orifice, or immediately downstream, to yield a less accurate flow indication than a Venturi meter. This drop in pressure is not recovered; that is, a permanent pressure loss occurs with orifice plate installations, unlike Venturi flow tubes.

Differential devices require little, if any, preventive maintenance by the operator since there are no moving parts. Occasional flushing of the hydraulic sensing lines is good practice. However, flushing should only be done by a qualified person. When dealing with an instrument sensitive to fractions of a psi or pascals, opening the wrong valve can instantly damage the internal parts severely. Also, if an older DP (differential pressure) cell containing mercury is used, this toxic (and expensive) metal can easily be blown out of the device and into the process pipeline. Thus, all valve manipulations must be understood and done deliberately after careful planning by a qualified person.

In nearly all cases, the signals from larger flow tubes are transmitted to a remote readout station. Local readout is also provided (sometimes inside the case only) for purposes of calibration. Differential pressure transmitters may be electrical or pneumatic types. The signal transmitted is proportional to the square root of the differential pressure.

Venturi meters have been in use for many decades and can produce very close accuracies year after year. Older flow tubes are quite long physically (to yield maximum accuracy and pressure recovery). Newer units are much shorter but have very good pressure recovery and even better accuracy. With no moving parts, the Venturi-type meter is not subject to mechanical failure as is the comparable propeller meter. Flow tubes, however, must be kept internally clean and without obstructions upstream (and even downstream) to provide the designed accuracies.

The piping design for flowmeters must provide for adequate lengths of straight pipe runs upstream and downstream of the meter. Flowmeters in pipes will produce accurate flowmeter readings when the meter is located at least five pipe diameters distance downstream from any pipe bends, elbows, or valves and also at least two pipe diameters distance upstream from any pipe bends, elbows, or valves. Flowmeters also should be calibrated in place to ensure accurate flow measurements.

In summary, all of the flowmeters described in this section provide rate of flow indication. The rate of flow can also be (and usually is) continuously totalized to give a reading of total flow past the measuring point. Total flow is usually indicated at the readout instrument in units of gallons or cubic feet (liters or cubic meters).

QUESTIONS

Write your answers in a notebook and then compare your answers with those on page 448.

19.2H   What are two basic types of flow readings?

19.2I   List the inferential flow-sensing techniques that are used in flow measurement.

19.2J   How do velocity-sensing meters measure process flows?

19.2K   Flows measured with Venturi meters take advantage of what hydraulic principle?

19.24 Chemical Feed Rate

Chemical feed rate indicators are often an integral part of a particular chemical feed system and thus are usually not considered instrumentation as such. For example, a dry feeder for lime may be provided with an indicator for feed rate in units of weight per time, such as pounds/hour or grams/minute. In a fluid (liquid or gas) feeder, the indication of quantity per unit of time, such as gallons/hour or pounds/day (liters/hour or kilograms/day), may be provided by use of a rotameter (Figure 19.16, page 420) or built-in calibrated pump with indicated output settings.

19.25 Process Instrumentation

Process instrumentation, by definition, provides for continuous analysis of physical or chemical indicators of process variables in a municipal or industrial plant. This does not include laboratory "bench" instruments (unless set up to measure sample water continuously), although the operating principles are usually quite similar. The process variables of turbidity and pH are often monitored closely in a water treatment plant (Figure 19.22). Very frequently, chlorine residuals are also continuously measured and controlled. These variables are usually measured at several locations. Additionally, other indicators of process status may be sensed on a continuous basis, such as electrical conductivity (Total Dissolved Solids, TDS), Oxidation-Reduction Potential (ORP), water alkalinity, and temperature. In every case, the instrumentation is specific as to operating principle, standardization procedures, preventive maintenance, and operational checks. The manufacturer's technical manual describes routine procedures to check and operate this sensitive type of equipment.

Operators must realize that most process instrumentation is quite delicate and thus requires careful handling and special training to service. No adjustments should be made without a true understanding of the specific device. Generally speaking, this category of instrumentation must be maintained by the plant's instrument specialist, or the factory representative, rather than by an operator (unless specially instructed).

Instrumentation and Control Systems 427

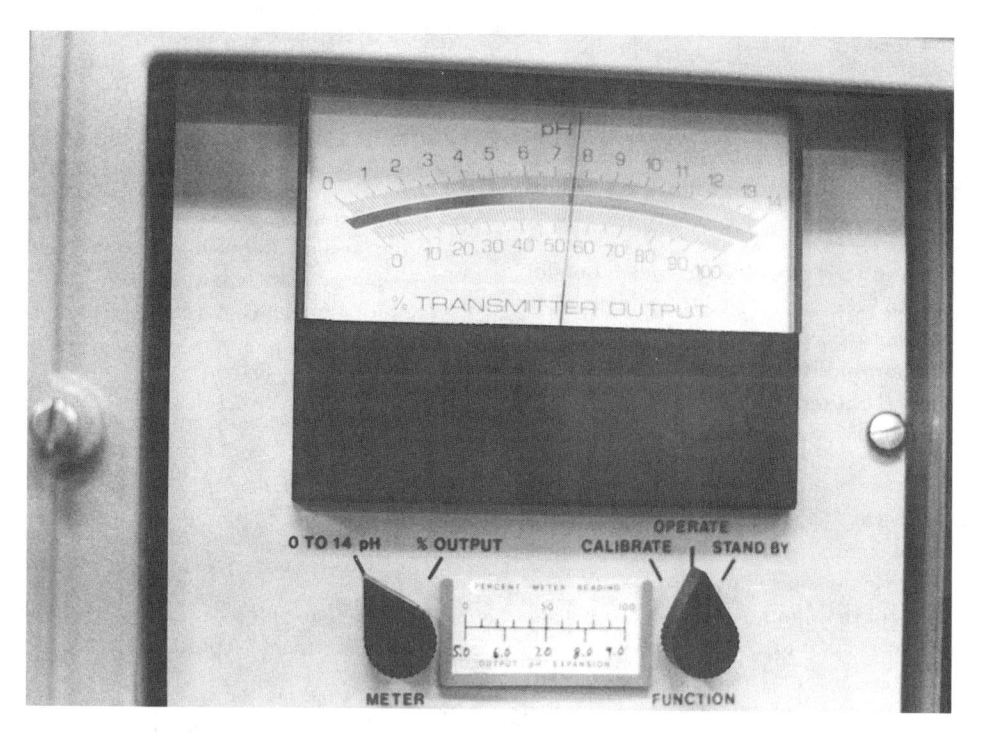

pH meter

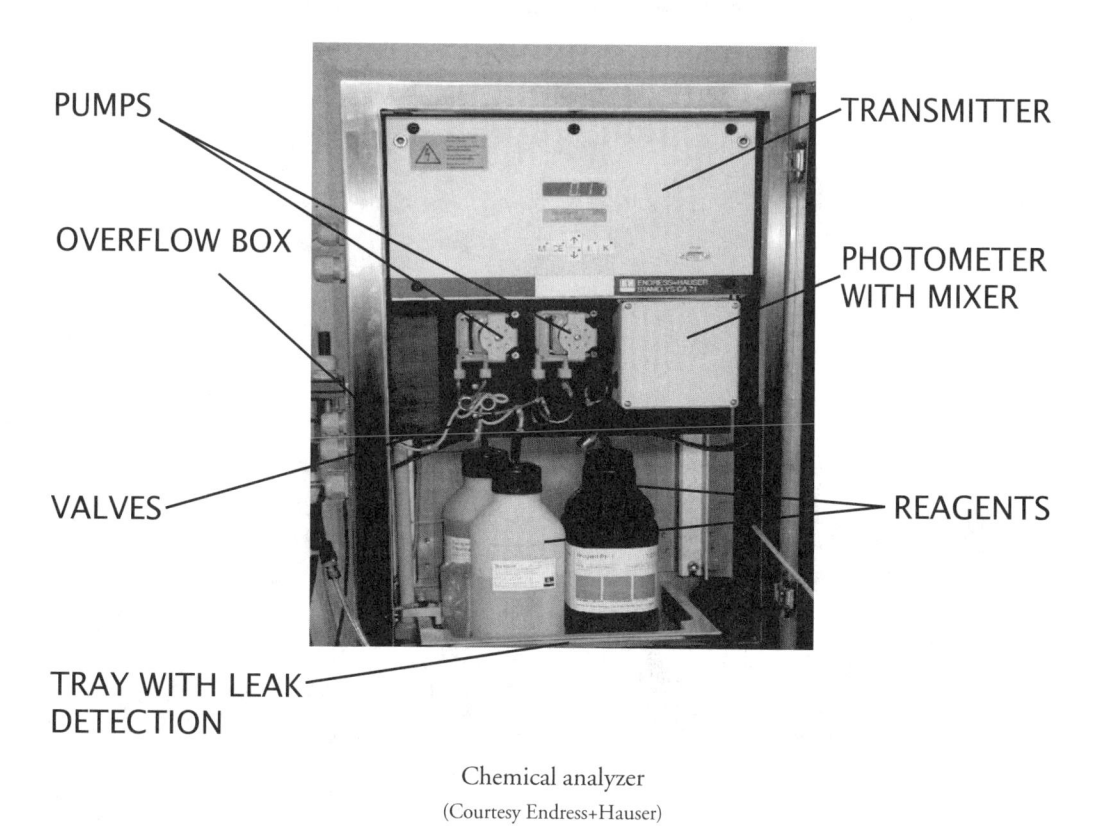

Chemical analyzer
(Courtesy Endress+Hauser)

Fig. 19.22   Water treatment plant analytical process instrumentation

19.26 Signal Transmitters/Transducers

Common system installations measure a variable at one location and provide a readout of the value at a remote location, such as a main control room. Except in the case of a blind transmitter, local indication is provided at the field site as well as being presented at the remote site.

In order to transmit a measured value to a remote location for readout, it is necessary to generate an analog or digital-data signal directly proportional to the value measured. This signal is then transmitted to a remote receiver, which provides a reading based on the signal. Also, a controller may use the signal to control the measured variable and a totalizer to integrate it.

Presently, two general systems for transmission of signals are used in most water treatment facilities: analog or digital data. Analog 4–20 mA transmission has a limited range, typically a few hundred feet or meters, and is being replaced by computer digital-data transmission. For transmission over longer distances, the transmitter output is converted to a digital-data representation of the process variable's measured value. The digital-data transmission simplifies the wired connections that are required for signal transmission.

A power supply to generate the required electrical energy may be located at the analog 4–20 mA transmitter, at the receiver, or at another location in the control instrument loop. The transmitter may be an integral part of the measurement or readout transducer, or it may be housed separately. In either case, the transmitter adjusts the signal to a corresponding value of the measured variable, and the receiver, in turn, converts this signal to a visible indication, the readout. Digital-data transmission distance is limited only by the computer network capability and reliability.

Instrumentation power is required for all components in the instrument loop, for both analog and digital devices. A reliable source of power such as a battery-backed power supply is used to ensure availability during power interruptions. Digital devices have lower power requirements, allowing for smaller and simpler backup power systems.

QUESTIONS

Write your answers in a notebook and then compare your answers with those on page 448.

19.2L What is one way to measure liquid chemical feed rates?

19.2M What process variables are commonly monitored or controlled by process instrumentation?

19.2N What are the two general systems for transmission of measurement signals?

END OF LESSON 1 OF 2 LESSONS

on

INSTRUMENTATION AND CONTROL SYSTEMS

Please answer the discussion and review questions next.

DISCUSSION AND REVIEW QUESTIONS
Chapter 19. INSTRUMENTATION AND CONTROL SYSTEMS
(Lesson 1 of 2 Lessons)

At the end of each lesson in this chapter you will find some discussion and review questions. The purpose of these questions is to indicate to you how well you understand the material in the lesson. Write the answers to these questions in your notebook.

1. Why should operators understand instrumentation and control systems?
2. How can control and instrumentation systems make an operator's job easier?
3. What is the difference between accuracy and precision?
4. Why should a screwdriver not be used to test an electric circuit?
5. What should you do when you discover an area with an oxygen deficiency or enrichment?
6. How can water levels be measured?
7. What problems can develop with propeller meters when they become old or worn?

CHAPTER 19. INSTRUMENTATION AND CONTROL SYSTEMS

(Lesson 2 of 2 Lessons)

19.3 CATEGORIES OF INSTRUMENTATION

19.30 Primary Elements

The first system element that responds quantitatively to the measured variable is the primary element or sensor. Transducers convert the sensor's minute actions to a usable indication or a signal. If remote transmission of the value is required, a transmitter may be part of the transducer. An illustrative example of these three components is the typical Venturi meter (shown in Figure 19.20, page 424): (1) the flow tube is the primary element, (2) the differential pressure-sensing device (DP cell) is the transducer, and (3) the signal-producing component is the transmitter. An understanding of the separate functions of each section of such a flowmeter is important to the proper understanding of equipment problems.

19.31 Panel Instruments

Process variable readouts are provided locally with indicators and repeated in the control panel and the process computer. These particular components are important to the operator and, hence, to plant operation itself, because they display or control the variable directly. The main control panel devices can also produce alarm signals to indicate if a variable is outside its range of expected values. The controllers are often installed on (or behind) the main panel along with the operating buttons, switches, and indicator lights for the plant's equipment. The controller in an instrument loop control system produces the alarm and control signals.

19.310 Indicators

Indicators give a visual representation of a variable's present value, either as an analog or digital display (Figures 19.23 and 19.24). The analog display uses some type of pointer or graphical display against a scale. A digital display is a direct numerical readout. Chart recorders, which by nature also serve as indicators, give a permanent record of how the variable changes with time by way of a moving chart. These historical records are now largely stored electronically as part of the computer control system. They may be displayed on a computer screen or printed out. Indicators out in the plant or field provide operators with local readouts of the process variables.

Indicators with digital readouts may be devices in a 4–20 mA instrument loop or may receive process variable data from a control system communication network. Digital readouts may be read more quickly and precisely from a longer distance, and can respond virtually instantly to variable changes. But analog indicators are cheaper, more rugged, and may not even require electrical power, an advantage during a power failure or in

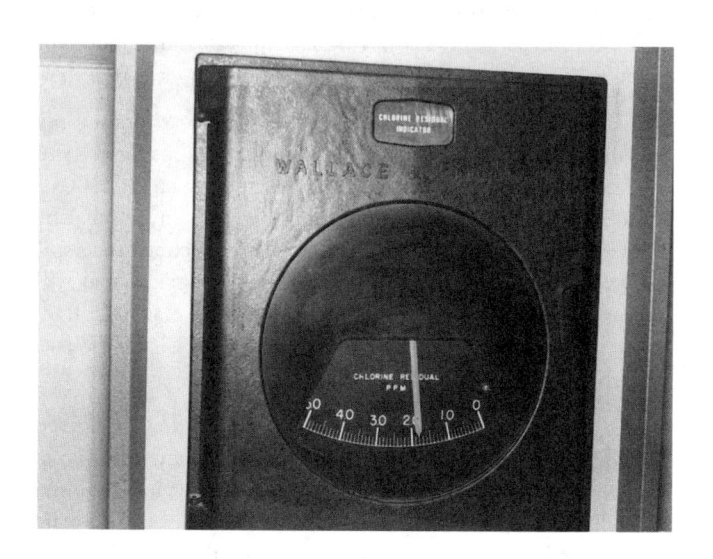

Fig. 19.23 Analog chlorine residual indicator

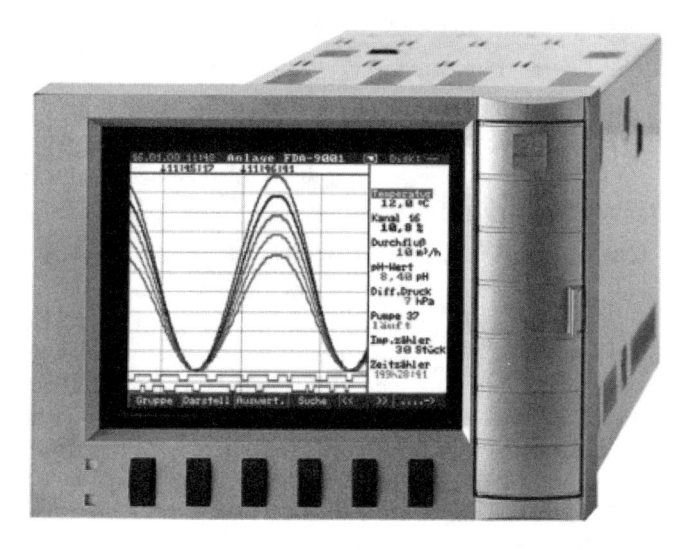

Fig. 19.24 Paperless chart recorder
(Courtesy Endress+Hauser)

hazardous environments. Another advantage of the analog display is that an incorrect readout may be more recognizable than with a digital readout or computer control system, and also is more easily corrected by the operator. For example, the pointer on a flowmeter gauge may merely be stuck, as evidenced by a perfectly constant reading.

Computer control systems provide diagnostic tools to monitor the functionality of instrumentation and troubleshoot process instrumentation loops. Because the capabilities of these systems vary widely, operators must learn the specifics of the system installed. Learning to use the diagnostic tools allows the operator to determine the reliability of the control system.

With all-electronic instrumentation, as advantageous as it may seem from a technical and economic standpoint, the operator has little recourse in case of malfunction of critical instrumentation. Temporary power failures, tripped panel circuit breakers, voltage surges resulting in blown fuses, static electricity, and excessive heat can disrupt the instrument and control systems. Electromechanical or pneumatic instruments may keep operating, or recover operation readily, after such power or heat problems. Computer control systems are typically designed so that the operator can, with training, easily restart the control system and initiate plant operation. Accordingly, the operator should insist upon some input into the design phase of instrument and control systems to ensure that the plant is still operable during power outages, hot weather, and other contingencies. Standby power generators and battery-backed power supplies are used to keep plants operating during commercial power outages, but even they have shortcomings if plant operations depend entirely on electrical power.

QUESTIONS

Write your answers in a notebook and then compare your answers with those on page 448.

19.3A What is the purpose of instrumentation indicators?

19.3B Describe an analog display.

19.3C What is the purpose of chart recorders?

19.3D What factors can disrupt instrument and control systems?

19.311 Recorders

Chart recorder functions have been incorporated into computer control systems (Figure 19.24). However, local recorders are still necessary at some locations throughout a treatment plant to ensure reliable, continuous recording of critical process variables. Recorders are indicators designed to show how the value of the variable has changed with time (Figure 19.25). Usually, this is done by attaching a pen (or stylus) to an indicator's arm, which then marks or scribes the value of the variable onto a continuously moving chart. The chart is marked on a horizontal, vertical, or circular scale in time units. Chart records are also stored on a computer disk or in memory for download from the recorders.

There are two main types of chart recorders: the strip-chart type and the circular-chart type. The strip-chart type carries its chart on a roll or as folded stock, with typically several weeks' supply of chart available. Several hours of charted data are usually visible, or easily available, for the operator to read. On a circular recorder, the chart makes one revolution every day, week, or month, with the advantage that the record of the entire elapsed time period is visible at any time.

Changing of charts is usually the operator's duty. It is easier with circular recorders, though not that difficult with most strip-chart units with some practice.

Chart recorders are typically electrical, powered by the instrumentation and control system electrical supply. Battery-backed power for the recorders may be used, if required, as a backup power source. Recorders are most commonly described by the nominal size of the strip-chart width or circular-chart diameter (for example, a 4-inch (100-mm) strip-chart, or a 10-inch (250-mm) circular-chart recorder). Figure 19.25 shows a combination indicator/recorder and Figure 19.26 presents two models of recorders.

19.312 Totalizers

Rate of flow, as a variable, is a time rate; that is, it involves time directly, such as in gallons per minute, or million gallons per day, or cubic meters per second. Flow rate units become units of volume with the passage of time. For example, flow in gallons per minute accumulates as total gallons during an hour or day. The process of calculating and presenting an ongoing running total of flow volumes passing through a meter is termed "integration" or totalizing.

The totalized flow functions are commonly incorporated into a computer control system, which performs the calculations and stores the historical record of the totals. The historical data may be displayed on the operator interface in trend displays, which are graphical displays similar to strip-chart recorder outputs. The data can be displayed on the computer screen, printed, or sent electronically for analysis by charting and graphing programs.

Large quantities of water (or liquid chemical) are commonly read out in units of hundreds or thousands of gallons (liters). On the face of a totalizer you may find a multiplier such as × 100 or × 1,000. This indicates that the reading is to be multiplied by this factor to yield the full amount of gallons or cubic meters. If the readout uses a large unit, such as mil gal, a decimal will appear between appropriate numbers on the display, or a fractional multiplier (× 0.001, for example) may appear on the face of the totalizer.

Every operator should be able to calculate total flow for a given time period in order to verify that the totalizer is actually producing the correct value. Accuracy to one or two parts in a hundred (1 or 2 percent) is usually acceptable in a totalizer. There are methods to integrate (add up) the area under the flow-rate curve on a recorder chart, to check for long-term accuracy of total flow calculations, but it is cumbersome and rarely necessary.

Fig. 19.25   Chart recorder with digital indicator

432 Water Treatment

Strip-chart recorders

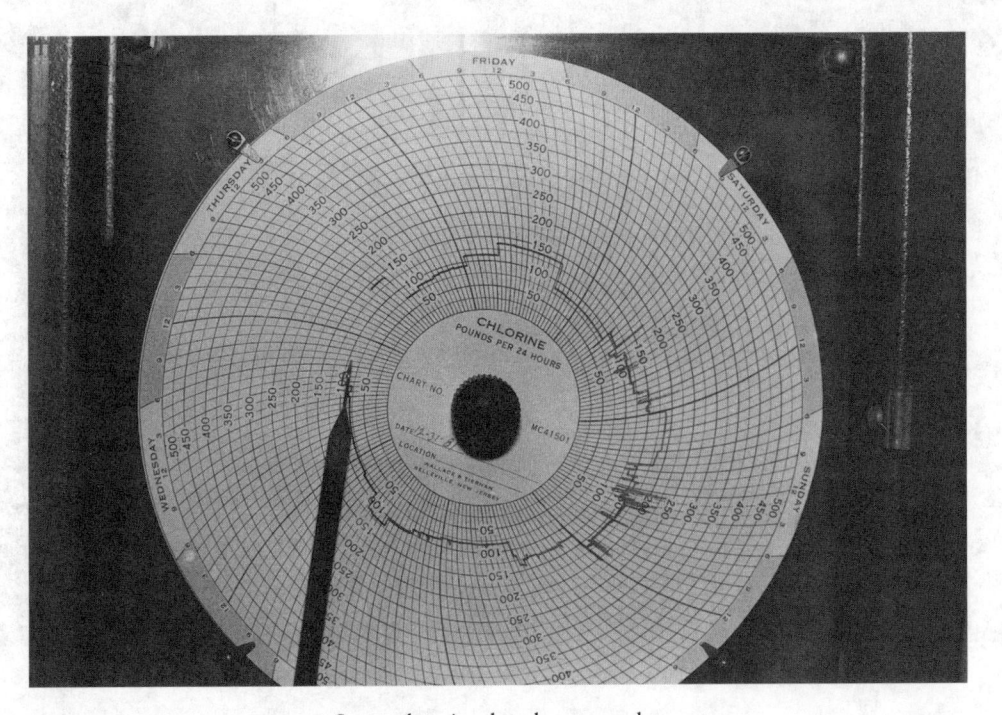

Seven-day circular-chart recorder

Fig. 19.26 Recorders, strip-chart and circular-chart

19.313 Alarms

Alarms are visual or audible signals that a variable is out of bounds, or that a condition exists in the plant requiring the operator's attention. Computer control systems have incorporated the alarm annunciator functions into the human machine interface (HMI). Only conditions that require operator attention need alarm signals. For noncritical conditions, a change in color on the operator interface graphic is sufficient notice. For more important variables or conditions, and especially when no computer system is available, an attention-getting annunciator panel (Figure 19.27) with flashing lights and an unmistakable and penetrating alarm horn is commonly used.

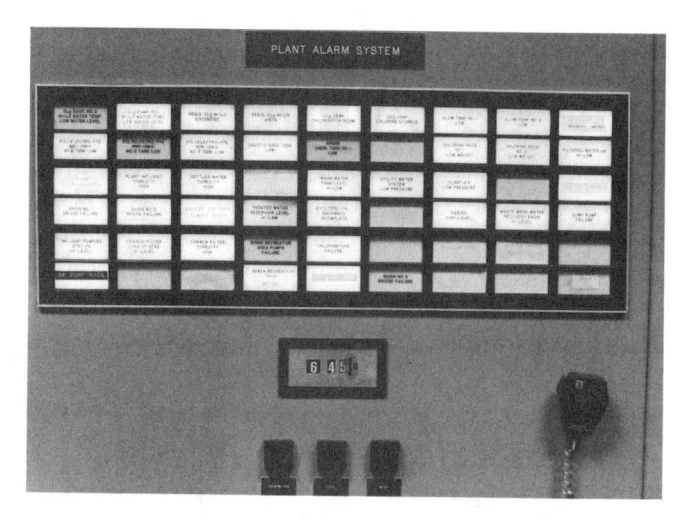

*Fig. 19.27 Annunciator (alarm) panel
(each rectangle represents a monitored location)*

Operator interface and annunciator panels should all have "acknowledge" and "reset" features to allow the operator to squelch the alarm sound (leaving the visible indication alone), and then to reset the system after the alarm condition is corrected. Annunciator panels should also have a "test" button so that an operator can confirm that no alarm lamps have burned out. The alarm contacts that activate annunciator panel alarms are within the field instrumentation devices. Computer control system alarms are activated by the software's alarm set points and may be programmable from the operator interface. The operator may be responsible for setting the alarm set point and must use judgment as to the actual limits of the particular variables that will ensure meeting proper operational goals. Each system is different, so no attempt will be made here to instruct operating personnel in alarm resetting procedures.

Sometimes operators fail to reset alarm limits as conditions and judgments change in the plant. It is not uncommon to see a plant's operator ignoring alarm conditions on an annunciator panel as normal practice. Such practice is not advised because a true alarm condition requiring immediate operator attention may be lost in the resulting general indifference to the alarm system. For some operators, acknowledging an alarm sound to get rid of the noise is second nature, without due attention to each and every activating condition. All alarm contact limits should be reset (or deactivated), as necessary, to ensure that the operator is as attentive to the alarm system as necessary to handle real emergencies. The system design should include provisions for disabling alarms when the associated instrument or device is out of service for long periods of time.

QUESTIONS

Write your answers in a notebook and then compare your answers with those on page 448.

19.3E   What is the purpose of recorders?

19.3F   What are the two main types of chart recorders?

19.3G   What are typical power sources for chart recorders?

19.3H   List the two kinds of warning signals that are produced by alarms.

19.32 Automatic Controllers

Section 19.0 explains the nature of control systems as they are used in water treatment plant operations. Automatic controllers may be individual instrumentation devices or a function programmed into a computer control system. Indications of proper and improper control need to be recognized by operators, but adjustment of the controller is left to a qualified instrument technician. By shifting to the manual mode, the operator can bypass the operation of any controller, whether electric or pneumatic. Learn how to shift all your controllers to manual operation. This will allow you to take over control of a critical system, when necessary, in an emergency, as well as at any other time it suits your purposes. For example, you may be able to quickly correct a cycling or sluggish variable by using manual control rather than waiting for the controller to correct the condition in time (if it ever does).

A controller is limited in its capability. It can only do what it has been programmed to do. You, as the operator, can exercise judgment based on your experience and observations, so do not hesitate to intervene if a controller is not exercising control within sensible limits. Of course, you must be sure of your conclusions and competent to take over control if you decide to operate manually.

To repeat a few of the more important operational control considerations, remember that ON/OFF control is quite different in operation from proportional control. Both methods can exercise close control of a variable; however, proportional control is better suited for close control. Attempting to set up an ON/OFF control system to maintain a variable within too close a tolerance may result in rapid ON/OFF operation of equipment. Such operation can damage both the equipment and the switching devices. Therefore, the basic principle to guide you in setting up the frequency of ON/OFF operation of a piece of equipment is to set the ON/OFF controls to operate or cycle associated equipment on and off no more often than actually necessary for plant operation. A level controller, for instance, should be set to cycle a pump or valve only as often as needed.

In the case of a modulating controller, it too may begin to cycle its final control element (pump or valve) through a wide range if any of the internal settings, namely proportional band or reset, are adjusted so as to attempt closer control of the variable than is reasonable. Accordingly, it may be better to accommodate to a small offset (difference between set point and control point) than risk an upset in control by attempting too close control.

19.33 Pump Controllers

Control of pumping systems can be achieved, as we have seen, by an ON/OFF type of controller starting and stopping pumps according to a level, pressure, or flow measurement.

Usually, an ON/OFF pump control system responds to level changes in a tank of some type. Water level can be sensed directly with a float or by a pressure change at the tank or pump site. The pump is thus turned off or on as the tank level rises above or falls below predetermined level or pressure limits. Control is rather simple in this case.

However, such systems may include several extra electrical control features to ensure fail-safe operation. To prevent the pump from running after a loss of level signal, electrical circuitry should be designed so the pump will turn OFF on an open signal circuit and ON only with a closed circuit. (Ideally, the controller would be able to distinguish between an open or closed remote level/pressure contact and an open or shorted signal line.) Larger pump systems also will often have a low-pressure cutoff switch on the suction side to prevent the pump from running when no water is available, such as with an empty tank or closed suction valve.

Controllers may also protect against overheating a pump (as happens when continuing to pump against a closed discharge valve) by a high-pressure (or low-flow) cutoff switch on the discharge piping. Both the high- and low-pressure switches should shut off a pump through a time delay circuit so that short-term pressure surges (dips and spikes) in the pump's piping can be tolerated. Ideally, the low- or high-pressure switches also key alarms to notify the operator of the condition. For remote stations, a plant's main panel may include indicator lights to show the pumps' operating conditions. Figure 19.28 shows a simplified diagram of pump control circuitry.

Pump control panels (Figure 19.29) may also include automatic or manual alternators (two pumps) or sequencers (more than two pumps). This provision allows the total pump operating time required for the particular system to be distributed equally among all the pumps at a pump station. A manual switch for a two-pump station, for example, may read "1-2" in one position and "2-1" in the other position. In the first position, pump #1 is the lead pump (which runs most of the time) and #2, the lag pump (which runs less). When the operator changes the switch to "2-1," the lead-lag order of pump operation is reversed, as it should be periodically, to keep the running time (as read on the elapsed time meters, or as estimated) of both pumps to about the same number of total hours. In a station with multiple pumping units, an automatic alternator or sequencer regularly changes the order of the pumps' start-up to maintain similar operating times for all pumps. In order to protect a pump's electric motor from overheating, level controls should be set so that the pump starts no more often than six times per hour (average one start every 10 minutes).

> ### QUESTIONS
>
> Write your answers in a notebook and then compare your answers with those on page 448.
>
> 19.3I Under what conditions might an operator decide to bypass the operation of a controller? How could this be done?
>
> 19.3J What basic principle should guide you in setting up the frequency of ON/OFF operation of a piece of equipment?
>
> 19.3K How can pumps be prevented from running after a loss of level signal?
>
> 19.3L What provision allows the total pump operating time required for a particular system to be distributed equally among all the pumps at a pump station?

19.34 Air Supply Systems

Pneumatic instrumentation depends on a constant source of clean, dry, pressurized air for reliable operation. Given a quality air supply, pneumatic devices can operate for long periods without significant problems. Without a quality air supply, operational problems can be frequent. The operator of a plant is usually assigned the task of ensuring that the instrument air is always available and dry, so operators must learn how to accomplish this; it cannot be assumed that clean air is there automatically.

The plant's instrument air supply system consists of a compressor with its own controls, master air pressure regulator, air filter, and air dryer, as well as the individual pressure regulator/filters in the line at each pneumatic plant instrument (Figure 19.30). Only the instrument air is filtered and dried; the plant

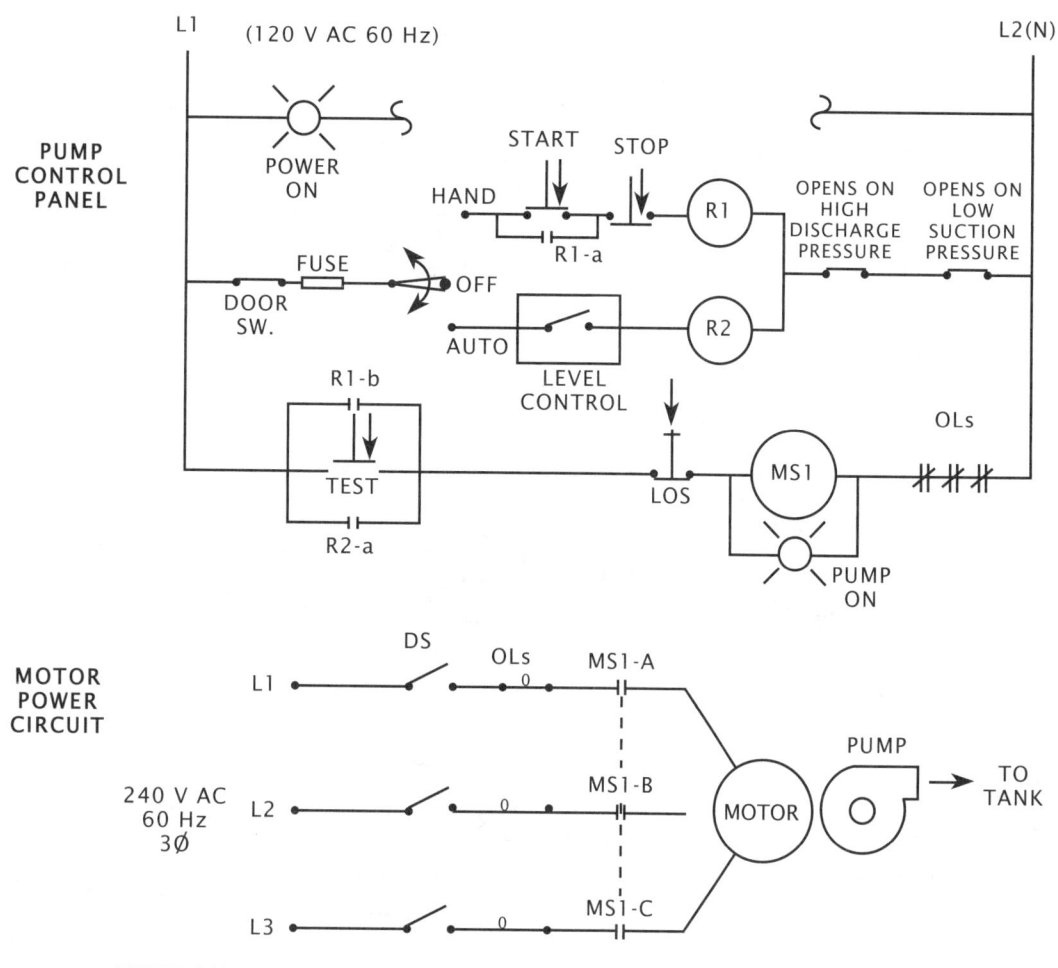

Instrumentation and Control Systems 435

Fig. 19.28 Pump control station ladder diagram (ON/OFF control) (simplified schematic)

Fig. 19.29 Photo of pump control station

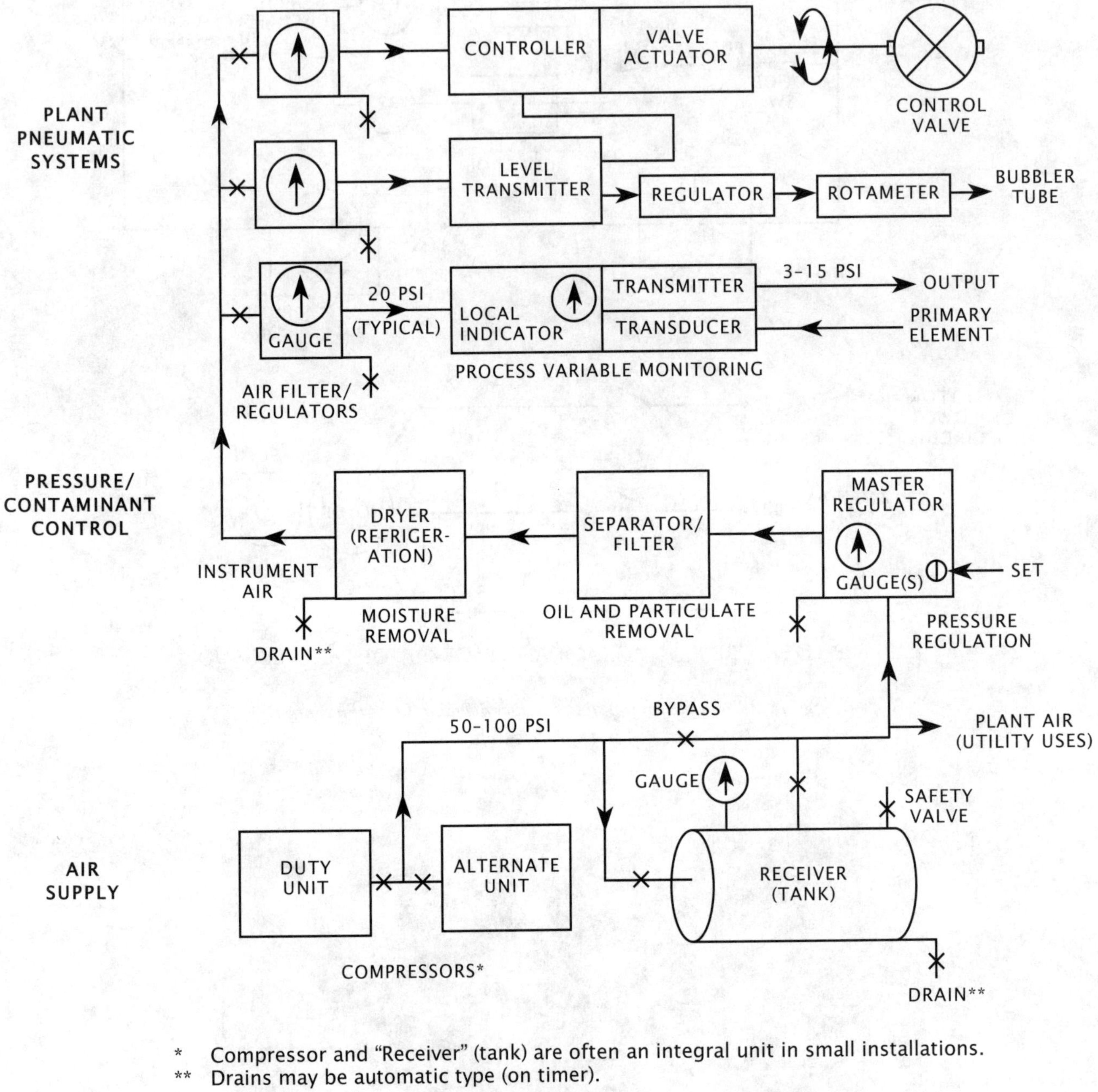

Fig. 19.30 *Typical plant instrument air system functional diagram (simplified, not all valves and piping shown)*

* Compressor and "Receiver" (tank) are often an integral unit in small installations.
** Drains may be automatic type (on timer).

air usually does not require such measures since it is being used only for other purposes.

As air passes through a compressor, it not only can pick up some oil but the air's moisture content is concentrated by the compression process. Special measures must be taken to remove both of the liquids. Oil is removed by filtering the air through special oil-absorbent elements. Water is removed by passing the moisture-laden air through an air-drying system consisting of a moisture separation column and air dryers. Air-drying systems are typically provided in pairs to allow one dryer to regenerate while the other is in service. You must recognize that the capacity of any of these systems of oil or water removal is limited to amounts of liquid encountered under normal conditions.

If the compressor is worn so as to pass more oil than usual, the oil separation process may permit troublesome amounts of oil to pass into the air supply. If the air source contains excessive humidity (due to a rainy day, perhaps, or a wet compressor room), the air-drying system may not handle the excess moisture. Learn enough about the instrument air system to be able to open the drain valves, cycle the air dryer, or even bypass the tank, in order to prevent instrumentation problems due to an oily or moisture-saturated air supply.

Operators should regularly crack the regulator/filter drain valves at each of the plant's process instruments. An unusual quantity of liquid drainage may indicate an overloading or failure in the instrument air filter/drying parts. Also, pneumatic indicators/recorders should be watched for erratic pointer/pen movements, which usually are indicative of air quality problems.

The seriousness of plant power or compressor failures can be lessened if you temporarily turn off all nonessential usages of compressed air in the plant. The air storage tank is usually sized so that there is enough air on hand to last for several hours, if conserved. Knowing this, you may be able to wait out a power failure without undue drastic action by conserving the remaining available pressure of the air supply.

QUESTIONS

Write your answers in a notebook and then compare your answers with those on pages 448 and 449.

19.3M What are the essential qualities of the air supply needed for reliable operation of pneumatic instrumentation?

19.3N How are moisture and oil removed from instrument air?

19.35 Laboratory Instruments

This category of instrumentation includes those analytical units typically found in water treatment plant laboratories (Figure 19.31). Some examples are turbidimeters, colorimeters and comparators, pH, and conductivity (TDS) meters. We have already seen that process models of each of the units monitor these same variables out in the plant. The models used in the laboratory are usually referred to as "bench" models rather than "process" instrumentation.

Operators often are required to make periodic readings from laboratory instruments, and periodic standardizing of particular instruments is often required before determinations are made. Preventive, and certainly corrective, maintenance is handled by the laboratory staff, factory representative, or instrument technician since each unit can be quite complex. Some of these countertop instruments or devices are very delicate and replacement parts, such as glass vessels or pH electrodes, are quite expensive. Moreover, the use of some of these instruments requires the regular handling of laboratory glassware and other breakable items. The operator who, through carelessness, lack of knowledge, or simple hurrying, consistently breaks glassware or "finds the darn meter broken again" does not become popular with the chemist, supervisor, or other operators. The byword in the laboratory is *WORK WITH CAUTION*. Protect valuable and essential instrumentation and supplies.

19.36 Test and Calibration Equipment

Plant measuring systems must be periodically calibrated to ensure accurate measurements. In most larger water treatment facility operations, the plant operating staff has little occasion to use testing and calibration devices on the plant instrumentation systems. A trained technician will usually be responsible for using such equipment. There are, however, some general considerations the operator should understand concerning the testing and calibration of plant measuring and control systems. With this basic knowledge, you may be able to discuss needed repairs or adjustments with an instrument technician and perhaps assist with that work. A better understanding of your plant's instrument systems may also enable you to analyze the effects of instrument problems on continued plant operation, and to handle emergency situations created by instrument failure. Your skills in instrument testing and calibration may even eventually result in a job promotion or pay raise.

The most useful piece of general electrical test equipment is the V-O-M, that is the Volt-Ohm-Milliammeter, commonly referred to as a multimeter (Figure 19.32). To use this instrument, you will need a basic understanding of electricity, but once you learn to use it, the V-O-M has potential for universal usage in instrument and general electrical work. Local colleges and other educational institutions may offer courses in basic electricity, which undoubtedly include practice with a V-O-M. You, as a

438 Water Treatment

Fig. 19.31 Water treatment plant laboratory

SPECIFICATIONS

Voltage Ranges
0–199.9/750 V AC   15 kV AC
0–1.999/19.99/199.9/1,000 V DC 15 kV DC
0–1,999 mV DC

Resistance Ranges
0–199.9/1,999 ohms
0–19.99/199.9/1,999 K ohms

Current Ranges
0–1,999 µA DC
0–19.99/199.9/1,999 mA DC
0–10 Amps DC
0–19.99/199.9/1,999 mA AC
0–10 Amps AC
NOTE: AC accuracy may be affected by outside interference.

Accuracy
DC V: ± 0.5% of rdg ± 2LSD
AC V: ± 1.5% of rdg ± 2LSD
DC Amps: All ranges ± 1.0% of rdg ± 2LSD except 10 Amp range, which is ± 1.5% of rdg ± 3LSD.
AC Amps: All ranges ± 1.5% of rdg ± 2LSD except 10 Amp range, which is ± 2.0% of rdg ± 3LSD.
Ohms: All ranges ± 0.75% of rdg ± 2LSD except 2 megohm range, which is ± 1% of rdg ± 2LSD.
15 kV AC/DC high voltage probe: add up to ± 2% of rdg.

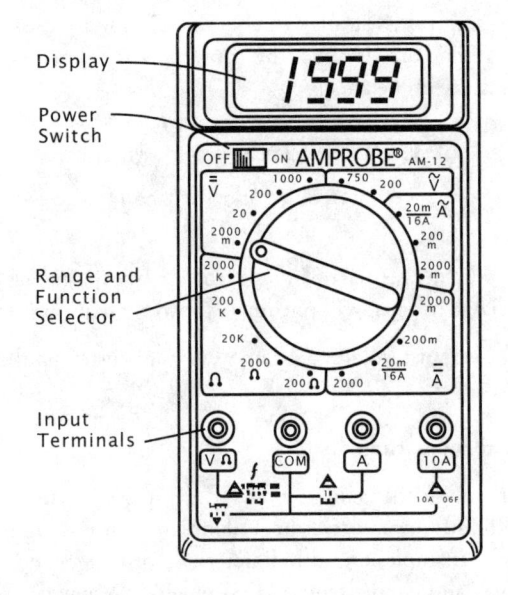

Fig. 19.32 Digital multimeter (V-O-M)

professional plant operator, are unlikely to find technical training of greater practical value than this type of course or program. Your future use of test and calibration equipment, in general, certainly should be preceded by instruction in the fundamentals of electricity.

QUESTIONS

Write your answers in a notebook and then compare your answers with those on page 449.

19.3O Why should an operator be especially careful when working with laboratory instruments?

19.3P Why should an operator become familiar with the testing and calibration of plant measuring and control systems?

19.3Q What is a V-O-M?

19.37 Process Computer Control Systems

19.370 Computer Control Systems

The computer control system is a computer-monitored alarm, response, control, and data acquisition system used by operators to monitor and adjust their treatment processes and facilities. Computer control systems used for process control may be classified by two commonly used terms: the *DISTRIBUTED CONTROL SYSTEM (DCS)*[8] and the *SUPERVISORY CONTROL AND DATA ACQUISITION (SCADA)*[9] system. The DCS and the SCADA system perform the same functions in different settings. DCSs are typically used to control and monitor processes in treatment plants. SCADA systems are most commonly used to control and monitor distribution system facilities that are widely separated geographically. Larger water facilities may have both DCSs and SCADA systems to provide treatment process control and distribution system controls. Smaller utilities often combine all the controls necessary into a SCADA system. In both large and small utilities, the operator interface for both distribution system and plant processes are provided in a single control room.

The computer control system collects, stores, and analyzes information about all aspects of operation and maintenance, transmits alarm signals, when necessary, and allows fingertip control of alarms, equipment, and processes. The computer control system provides the information that operators need to solve minor problems before they become major incidents. As the nerve center at the treatment plant, the system allows operators to enhance the efficiency of their facility by keeping them fully informed and fully in control. Figure 19.33 shows the basic interaction cycle of the operator with the control systems and plant treatment processes.

The five basic components of the computer control system (Figure 19.34) are as follows:

- Process instrumentation and control devices sense process variables in the field and actuate equipment.

- The input/output (I/O) interface sends and receives data with the process instrumentation and control devices.

- The central processing unit (CPU) is the system component that contains the program instructions for the control system. These instructions are programmed to react based on a control strategy. The CPU gathers data from the various interfaces and sends commands to field devices to operate the plant processes.

- The communication interfaces provide the means for the computer control system to send data to and from outside computer systems, business systems, other process control systems, and equipment.

- The *HUMAN MACHINE INTERFACE (HMI)*[10] is commonly a computer workstation that is running the computer control system software that provides the plant data to the operator on the workstation screen.

These components are the means by which the control system gathers and distributes information for the human operator and the process instrumentation and other equipment.

The computer control systems may be used in various capacities, from data collection and storage only, to total data analysis, interpretation, and process control.

Computer control systems monitor levels, pressures, and flows and also operate pumps, valves, and alarms. They monitor

---

[8] *Distributed Control System (DCS).* A computer control system having multiple microprocessors to distribute the functions performing process control, thereby distributing the risk from component failure. The distributed components (input/output devices, control devices, and operator interface devices) are all connected by communications links and permit the transmission of control, measurement, and operating information to and from many locations.

[9] *SCADA (SKAY-dah) System.* Supervisory Control And Data Acquisition system. A computer-monitored alarm, response, control, and data acquisition system used to monitor and adjust treatment processes and facilities.

[10] *Human Machine Interface (HMI).* The device at which the operator interacts with the control system. This may be an individual instrumentation and control device or the graphic screen of a computer control system. Also called MAN MACHINE INTERFACE (MMI) and OPERATOR INTERFACE.

440 Water Treatment

Fig. 19.33 Operator interaction with plant processes

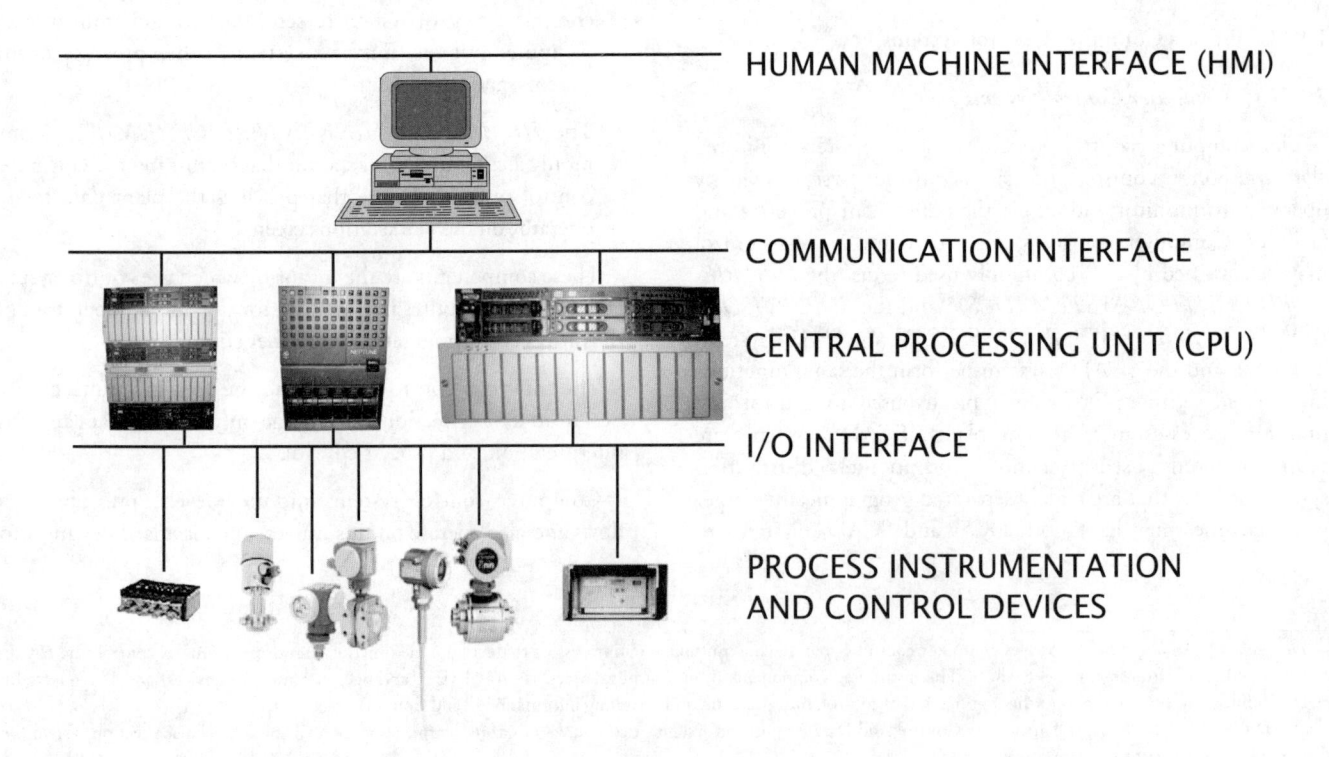

Fig. 19.34 Computer control system components

temperatures, speeds, motor currents, pH, turbidity, and other operating parameters, and provide control, as necessary. Computer control systems provide a log of historical data for events, analog signal trends, and equipment operating time for maintenance purposes. The information collected may be read by an operator on computer screen readouts or analyzed and plotted by the computer as trend charts.

Computer control systems provide a picture of the plant's overall status on a computer screen. In addition, detailed pictures of specific portions of the system can be examined by the operator through the computer workstation. The graphical displays on the computer screens can include current operating information, which the operator can use to determine if the guidelines are within acceptable operating ranges or if any adjustments are necessary.

Computer control systems are capable of analyzing data and providing operating, maintenance, regulatory, and annual reports. Operation and maintenance personnel rely on a computer control system to help them prepare daily, weekly, and monthly maintenance schedules, monitor the spare parts inventory status, order additional spare parts, when necessary, print out work orders, and record completed work assignments.

Computer control systems can also be used to enhance energy conservation programs. For example, operators can develop energy management control strategies that allow for both maximum energy savings and maximum treatment flow prior to peak-flow periods. In this type of system, power meters are used to accurately measure and record power consumption. The information can then be reviewed by operators to watch for changes that may indicate equipment problems.

Emergency response procedures can also be programmed into a computer control system. Operator responses can be provided for different operational scenarios that might be encountered as a result of adverse weather changes, fires, earthquakes, or other emergency situations.

QUESTIONS

Write your answers in a notebook and then compare your answers with those on page 449.

19.3R Computer control systems used for process control are classified by which two commonly used terms?

19.3S What are the five basic components of the computer control system?

19.3T How do computer control systems help operation and maintenance personnel?

19.371 *Typical Computer Control System Functions*

Computer control systems for water treatment plants and distribution systems are usually operated together, with the controls located at the treatment plant. Information that historically was recorded on paper strip-charts is now being recorded and stored (archived) by computers. This information can be retrieved and reviewed easily by the operator, whereas before, years of strip-chart records would need to be examined to find needed information. Therefore, computer control systems are more efficient in providing operators with the information they need to make informed and timely decisions. Figure 19.35 is a typical modern water treatment plant control room.

Computer control systems give the treatment plant operator the tools to optimize plant processes based on current and historical operating information. The treatment plant influent and effluent are monitored continuously for many process variables such as flow, turbidity, pH, ammonium, chlorine, and nitrogen. If these indicators change significantly or exceed predetermined levels, the computer control system alerts the operator or changes the process based on a preprogrammed control strategy defined by the operator.

Historical operating data stored in a computer control system is readily available at any time. The computer control system can be queried to identify, for instance, when peak plant influent flow was greater than 50 MGD during the previous two years. Plant performance under these conditions can be recalled using the computer control system, analyzed by the operators, and the results used to operate the plant accordingly.

Electrical energy consumption can be optimized by the use of computer control systems. Most power companies are eager to help operators save money by structuring their rates to encourage electrical energy consumption when demands for power are low and to discourage consumption when demands for power are high. Computer controls can be programmed with a control strategy to reduce energy costs by automatically operating equipment when demands for power are low.

Computer control systems are being continually improved to help operators do a better job. Operators can create their own display screens, their own graphics, and show whatever operating characteristics they wish to display. The main screen could be a flow diagram from influent to effluent showing the main treatment and auxiliary process areas. Critical operating information could be displayed for the main treatment flow path and process area, with navigation capabilities to easily access detailed screens for each piece of equipment.

Information on the screen should be color coded to indicate if a pump is running, ready, unavailable, or failed, or if a valve is open, closed, moving, unavailable, or failed. A "failed" signal is used by the computer to inform the operator that something is wrong with the information or the signal it is receiving or is being instructed to display. For example, if there is no power to a motor, then the motor cannot be running even though the computer is receiving a signal that indicates it is running. In this case, the computer would send a "failed" signal, indicating that the information it is receiving is not consistent with the rest of the information available.

The operator can request a computer to display a summary of all alarm conditions in a plant, a particular plant area, or a process system. A blinking alarm signal indicates that the alarm condition has not yet been acknowledged by the operator. A steady alarm signal, one that is not blinking, indicates that the alarm

Fig. 19.35 Control room

has been acknowledged but the condition causing it has not yet been fixed. Also, the screen could be set up to automatically designate certain alarm conditions as *PRIORITY* alarms, requiring immediate operator attention.

With proper security implementation, computer control systems allow operators to have remote access to plant controls from anywhere using a laptop or remote workstation. This provides the flexibility for off-duty staff to help on-duty operators solve operational problems. Computer networking systems allow operators at terminals in offices, in plants, and in the field to work together and use the same information or whatever information they need from one central computer database.

A drawback of some computer control systems is that when the system "goes down" due to a power failure, the numbers displayed will be the numbers that were registered immediately before the failure, not the current numbers. The operator may therefore experience a period of time when accurate, current information about the system is not immediately available.

Customer satisfaction with the performance of a water utility can be enhanced by the use of an effective computer control system. Coordination of the treatment facility control and the distribution system control is used to avoid water shortages and low pressures.

When operators decide to initiate or expand a computer control system for their plant process system, the first step is to decide what the computer control system should do to make the operators' jobs easier, more efficient, and safer, and to make their facilities' performance more reliable and cost effective. Cost savings associated with the use of a computer control system frequently include reduced labor costs for operation, maintenance, and monitoring functions that were formerly performed manually. Precise control of chemical feed rates by a computer control system eliminates wasteful overdosing. Preventive maintenance monitoring can save on equipment and repair costs, and, as previously noted, energy savings may result from off-peak (reduced) electrical power rates. Operators should visit facilities with computer control systems and talk to the operators about what they find beneficial, and also detrimental, with regard to computer control systems and how the systems contribute to their performance as operators.

The greatest challenge for operators using computer control systems is to realize that just because a computer says something (a pump is operating as expected), *this does not mean that the computer is always correct.* Also, when the system fails due to a power failure or for any other reason, operators will be required to operate the plant manually and without critical information. Could you do this?

Operators will always be needed to question and analyze the results from computer control systems. They will be needed to *see* if the effluent looks OK, to *listen* to a pump to be sure it sounds right, and to *smell* the process and the equipment to determine if unexpected or unidentified changes are occurring. Treatment plants and distribution systems will always need alert, knowledgeable, and experienced operators who have a "feel" for their plant and their distribution system.

QUESTIONS

Write your answers in a notebook and then compare your answers with those on page 449.

19.3U What do computer control systems do with information that historically was recorded on paper stripcharts?

19.3V Cost savings associated with the use of a computer control system frequently include which items?

19.3W What is the greatest challenge for operators using computer control systems?

19.4 OPERATION AND PREVENTIVE MAINTENANCE

19.40 Proper Care of Instruments

Usually, instrumentation systems are remarkably reliable year after year, assuming proper application, setup, operation, and maintenance. Reliable measurement systems, though outdated by today's standards, are still found in regular service at some plants up to 50 years after installation. To a certain extent, good design and application account for such long service life, but most important is the careful operation and regular maintenance of the instruments' components. The key to proper operation and maintenance (O & M) is the operator's practical understanding of the system. Operators must know how to: (1) recognize malfunctioning instruments so as to prevent prolonged damaging operation, (2) shut down and prepare devices for seasonal or other long-term nonoperation, and (3) perform preventive (and minor corrective) maintenance tasks to ensure proper operation in the long term. A sensitive instrumentation system can be ruined in short order with neglect in any one of these three areas.

Operators should be familiar with the Technical Manual (also called the Instruction Book or Operating Manual) of each piece of equipment and instrument encountered in a plant. Each manual will have a section devoted to the operation of a certain component of a complete measuring or control system (although frequently not for the entire system). Detailed descriptions of maintenance tasks and operating checks will usually be found in the manual. Depending on the general type of instrument (electromechanical, pneumatic, or electronic), the suggested frequency of the operation and maintenance/checking tasks can range from none to monthly. Accordingly, this section of the course only describes those common and general tasks an operator might be expected to perform to operate and maintain instrumentation systems. From an operations standpoint, these tasks include learning, and constant attention to, what constitutes normal function. From a maintenance standpoint, they include ensuring proper and continuing protection and care of each component.

19.41 Indications of Proper Function

The usual pattern of day-to-day operation of every measuring and control system in a plant should become so familiar to operators that they almost unconsciously sense any significant change. This will be especially evident and true for systems with recorders where the pen trace is visible. An operator should thus watch indicators and controllers for their characteristic actions and pay close attention to trend records. Using the trend capability of the computer control system provides a method to analyze the reaction of one process variable to a change in another or other process variables.

Two of the surest indications of a serious electrical problem in instruments or power circuits are, of course, smoke or a burning odor. Such signs of a problem should *never* be ignored. Smoke/odor means heat, and no device can operate long at unduly high temperatures. Any electrical equipment that begins to show signs of excessive heating must be shut down *immediately*, regardless of how critical it is to plant operation. Overheated equipment will very likely fail very soon anyway, with the damage being aggravated by continued usage. Fuses and circuit breakers do not always de-energize circuits before damage occurs, so they cannot be relied upon to do so.

Finally, operators frequently forget to reset an individual alarm, either after an actual occurrence or after a system test. This is especially prevalent when an annunciator panel is allowed to operate day after day with lit-up alarm indicators (contrary to good practice) and one light (or more) is not easily noticeable. Also, when a plant operator must be away from the main duty station, the system may be set so the audible part of the alarm system is temporarily squelched. When the operator returns, the audible system may inadvertently not be reactivated. In both instances (individual or collective loss of audible alarm), the consequences of such inattention can be serious. Therefore, develop the habit of checking your annunciator system often.

QUESTIONS

Write your answers in a notebook and then compare your answers with those on page 449.

19.4A List the three areas of operator responsibility that are key to proper instrument operation and maintenance (O & M).

19.4B What general tasks are expected of an operator of instrumentation systems from an operations standpoint and from a maintenance standpoint?

19.4C What are two of the surest indications of a serious electrical problem in instruments or power circuits?

19.42 Start-Up/Shutdown Considerations

The start-up and periodic or prolonged shutdown of instrumentation equipment usually require very little extra work by the operator. Start-up is limited mainly to undoing or reversing the shutdown measures taken.

When shutting down any pressure-, flow-, or level-measuring system, valve off the liquid to the measuring element. Exercise particular care, as explained previously, regarding the order in which the valves are manipulated for any flow tube installation. Also, the power source of some instruments may be shut off, unless the judgment is made that keeping an instrument case warm (and thus dry) is in order. Constantly moving parts, such as chart drives, should be turned off. With an electrical panel room containing instrumentation, it is good practice to leave some power components on (such as a power transformer) to provide space heat for moisture control. In a known moist environment, sealed instrument cases may be protected for a while with a container of *DESICCANT*[11] (indicating silica gel, which is blue if OK and pink when the moisture-absorbent capacity is exhausted).

Although preventing the access of insects and rodents into any area is difficult, general cleanliness seems to help considerably. Rodenticides are available to control rodents; this is good preventive maintenance practice in any electrical space. Rats and mice will chew off wire and transformer insulation, and may urinate on other insulator material, leading to serious problems for equipment (not to mention the rodent).

Nest-building activities of some birds can also be a problem. Screening some buildings and equipment against entry by birds has become a design practice of necessity. Insects and spiders are not generally known to cause specific functional problems, but start-up and operation of systems invaded by ants, bees, or spiders should await cleanup of each such component of the system. All of these pests can bite or sting, so take care.

With pneumatic instrumentation, it is desirable to purge each device with dry air before shutdown. This measure helps rid the individual parts of residual oil and moisture to minimize internal sticking and corrosion while standing idle. As before, periodic blow-off of air receivers and filters keeps these liquids out of the instruments to a large degree. Before shutdown, however, extra attention should be paid to instrument air quality for purging. Before start-up, each filter/receiver should again be purged.

Finally, pay attention to the pens and chart drives of recorders upon shutdown. Ink containers (capsules) may be removed if deemed necessary, and chart drives turned off. A dry pen bearing against one track (such as zero) of a chart for weeks on end is an invitation to start-up problems. Re-inking and chart replacement at start-up are easy if the proper shutdown procedures were followed.

> ### QUESTIONS
>
> Write your answers in a notebook and then compare your answers with those on page 449.
>
> 19.4D How can moisture be controlled in an electrical panel room containing instrumentation or in sealed instrument cases?
>
> 19.4E Why should pneumatic instrumentation be purged with dry air before shutdown?

19.43 Preventive Maintenance

Preventive maintenance (PM) means that attention is given periodically to equipment in order to prevent future malfunctions. Corrective maintenance involves actual, significant repairs, which are beyond the scope of this chapter and, in most cases, are not the responsibility of the operator. Routine operational checks are part of all PM programs in that a potential problem may be discovered and thereby corrected before it becomes serious.

PM duties for instrumentation should be included in the plant's general PM program. If your plant has no formal, routine PM program, it should have. Such a program must be set up on paper (or on a computer program). That is, the regular duties required are printed on forms or cards (or appear on a computer screen) that the operator (or technician) uses as a reminder, guide, and record of PM tasks performed. Without such explicit measures, experience shows that preventive maintenance will almost surely be put off indefinitely. Eventually, the press of critical corrective maintenance (often due to lack of preventive maintenance) and even equipment replacement projects may well eliminate forever any hope of a regular PM program. The fact that instrumentation is usually very reliable (being of quality design) may keep it running long after pumps and other equipment have failed. Nevertheless, instrumentation does require proper attention periodically to maximize its effective life. PM tasks and checks on modern instrument systems are quite minimal (even virtually nonexistent on some), so there are no valid reasons for ever failing to perform these tasks.

The technical manual for each item of instrumentation in your plant should be available so you can refer to it for O & M purposes. When a manual cannot be located, contact the manufacturer of the unit. Be sure to give all relevant serial/model numbers in your request for the manual. Request two manuals, one to use and one to put in reserve. All equipment manuals should be kept in one protected location and signed out, as needed. Become familiar with the sections of these manuals related to O & M, and follow their procedures and recommendations closely.

A good practice is to have on hand any supplies and spare parts that are or may be necessary for instrument operation (such as charts) or service (such as pens, pen cleaners, and ink). Some technical manuals contain a list of recommended spare

---

[11] *Desiccant* (DESS-uh-kant). A drying agent that is capable of removing or absorbing moisture from the atmosphere in a small enclosure.

parts that you could use as a guide. Try to obtain these supplies/parts for your equipment. A new pen on hand for a critical recorder can be a lifesaver, at times.

Since PM measures can be so diverse for different types, brands, and ages of instrumentation, only the few general considerations applicable to all will be covered in this section.

1. Protect all instrumentation from moisture (except as needed by design), vibration, mechanical shock, vandalism (a very real problem in the field), and unauthorized access.

2. Keep instrument components clean on the outside, and closed/sealed against inside contamination (for example, spider webs and rodent wastes).

3. *DO NOT* presume to lubricate, tweak, fix, calibrate, free up, or modify any component of a system arbitrarily. If you are not qualified to take any of these measures, then do not do it.

4. *DO* keep recorder pens and charts functioning as designed by frequent checking and service, bleed pneumatic systems regularly, as instructed, ensure continuity of power for electrical devices, and do not neglect routine analytical instrument cleanings and standardizing duties as required by your plant's established procedures.

As a final note, it is a good idea to get to know and cooperate fully with your plant's instrument service person. Good communication between this person and the operating staff can only result in better all-around operation. If your agency is too small to staff such a specialist, it may be a good idea to enter into an instrumentation service contract with an established company or possibly even with the manufacturer of the majority of the components. With rare exceptions, general maintenance persons (even journeyman electricians) are *not* qualified to perform extensive maintenance on modern instrumentation. Be sure that someone takes good care of your instruments and they will take good care of you.

QUESTIONS

Write your answers in a notebook and then compare your answers with those on page 450.

19.4F Why should regular preventive maintenance duties be printed on forms or cards?

19.4G How can the technical manual for an instrument be obtained if the only copy in a plant is lost?

19.4H What instrument supplies and spare parts should always be available at your plant?

19.44 Operational Checks

Computer control systems provide the best tools for observing the operating functions of plant process systems. The operator interface (also called the human machine interface (HMI)) provides the ability to view all areas of plant operation. Most systems provide the ability to display trends of multiple process variables on the same graphical display. These trends are tools you will learn to use extensively to monitor and control your facility's processes.

Operational checks are most efficiently made by always observing each system for its continuing signs of normal operation. However, some measuring systems may be cycled within their range of action as a check on the responsiveness of components. For instance, if a pressure-sensing system indicates only one pressure for months on end, and some doubt arises as to whether it is working or not, the operator may bleed off a little pressure at the primary element to produce a small fluctuation. Or, if a flow has appeared constant for an overly long period, the bypass valve in the DP (differential pressure) cell piping may be cracked open briefly to cause a drop in the reading.[12] Be sure you crack the *bypass* valve, not one of the others on the piping. If you open the wrong valve, the pressure may be excessive and be beyond the range of the DP cell, which could cause problems. A float for a level recorder suspected of being stuck (very constant level indication) may be freed by jiggling its cable, or by taking other measures to cause a slight fluctuation in the reading.

Whenever an operator or a technician disturbs normal operation during checking or for any reason, plant process operating personnel must be informed—ideally prior to the disturbance. If a recorder trace is altered from its usual pattern in the process, the person causing the upset should initial the chart appropriately and note the time. Some plants require operators to mark or date each chart at midnight (or noon) of each day for easy reference and filing.

In the case where a pen or pointer is thought to be stuck mechanically, that is, it does not respond at all to simulated or actual change in the measured variable, it is normally permissible to open an instrument's case and try to move the pointer or pen, but only to the minimum extent necessary to free it. Further deflection may well bend or break the device's linkage. A dead pen often is due only to loss of power or air to the readout mechanism. Any hard or repeated striking of an instrument to make it work identifies the striker as ignorant of good operational practice and can ruin the equipment. Insertion of tools into an instrument case in a random fix-it attempt can easily damage an instrument. Generally speaking, any extensive operating check of instrumentation should be performed by the instrument technician during routine PM program activities.

QUESTIONS

Write your answers in a notebook and then compare your answers with those on page 450.

19.4I How are operational checks performed on instrumentation equipment?

19.4J What should be done if a recorder trace is altered from its usual pattern during the process of checking an instrument?

---

[12] There is no similar easy way to check a propeller meter's response.

19.5 ADDITIONAL READING

1. *INSTRUMENTATION—HANDBOOK FOR WATER AND WASTEWATER TREATMENT PLANTS,* Robert G. Skrentner. Obtain from CRC Press LLC, Attn: Order Entry, 6000 Broken Sound Parkway, NW, Suite 300, Boca Raton, FL 33487. ISBN 0873711262. Order No. L126. Price, $189.95, includes shipping and handling.

2. *INSTRUMENTATION IN WASTEWATER TREATMENT FACILITIES* (MOP 21). Obtain from Water Environment Federation (WEF), Publications Order Department, 601 Wythe Street, Alexandria, VA 22314-1994. Order No. MO2021. Price to members, $41.75; nonmembers, $61.75; price includes cost of shipping and handling.

3. *HANDS-ON WATER AND WASTEWATER EQUIPMENT MAINTENANCE.* Obtain from American Public Works Association (APWA) Bookstore, PO Box 802296, Kansas City, MO 64180-2296. Order No. PB.XHOW. Price to members, $110.00; nonmembers, $115.00; plus shipping and handling.

4. National Fire Protection Association Standard 820 (NFPA 820), *FIRE PROTECTION IN WASTEWATER TREATMENT AND COLLECTION FACILITIES.* Obtain from National Fire Protection Association (NFPA), 11 Tracy Drive, Avon, MA 02322. Item No. 82003. Price to members, $31.50; nonmembers, $35.00; plus $7.95 shipping and handling.

5. *INSTRUMENTATION AND CONTROL* (M2). Obtain from American Water Works Association (AWWA), Bookstore, 6666 West Quincy Avenue, Denver, CO 80235. Order No. 30002. Price to members, $87.50; nonmembers, $129.50; price includes cost of shipping and handling.

6. *SUCCESSFUL INSTRUMENTATION AND CONTROL SYSTEMS DESIGN,* Michael D. Whitt. Obtain from ISA, 67 Alexander Drive, Research Triangle Park, NC 27709. ISBN 978-1-55617-844-3. Price to members, $89.00; nonmembers, $99.00.

7. *SCADA SUPERVISORY CONTROL AND DATA ACQUISITION,* Stuart A. Boyer. Obtain from ISA, 67 Alexander Drive, Research Triangle Park, NC 27709. ISBN 978-1-55617-877-1. Price to members, $69.00; nonmembers, $79.00.

END OF LESSON 2 OF 2 LESSONS
on
INSTRUMENTATION AND CONTROL SYSTEMS

Please answer the discussion and review questions next.

DISCUSSION AND REVIEW QUESTIONS
Chapter 19. INSTRUMENTATION AND CONTROL SYSTEMS
(Lesson 2 of 2 Lessons)

Write the answers to these questions in your notebook. The question numbering continues from Lesson 1.

8. What are the advantages and limitations of analog versus digital indicators?

9. Why is it poor practice to ignore the lamps that are lit up (alarm conditions) on an annunciator panel?

10. How should the constantly lit-up lamps (alarm conditions) on an annunciator panel be handled?

11. What electrical control features are available to protect pumps from damage?

12. What problems are created by oil and moisture in instrument air, and how can these contaminants be removed?

13. Why should plant measuring systems be periodically calibrated?

14. Why should rodents be kept out of instrumentation equipment?

15. How could you tell if a float for a level recorder might be stuck, and how would you determine if it was actually stuck?

SUGGESTED ANSWERS
Chapter 19. INSTRUMENTATION AND CONTROL SYSTEMS

ANSWERS TO QUESTIONS IN LESSON 1

Answers to questions on page 400.

19.0A A treatment plant operator must have a good working familiarity with instrumentation and control systems to properly monitor and control the treatment processes.

19.0B This chapter contains information on general instrumentation functions and control system capabilities.

19.0C Measurement instruments can serve as an extension of and improvement to an operator's senses of vision, touch, hearing, and even smell.

Answers to questions on page 401.

19.0D Our senses provide us with qualitative and relatively short-term quantitative information; instrumentation measurements provide us with exact quantitative data.

19.0E A measurement is, by definition, the comparison of a quantity with a standard unit of measure.

19.0F *ACCURACY* refers to how closely an instrument measures the true or actual value of a process variable, while *PRECISION* refers to how closely the device can reproduce a given reading (or measurement) time after time.

19.0G *ANALOG* displays show a reading as a pointer against a marked scale and *DIGITAL* displays provide direct, numerical readouts.

Answers to questions on page 404.

19.0H Control systems are the means by which such process variables as pressure, level, weight, or flow are controlled.

19.0I Examples of modulating control systems include: (1) chlorine residual analyzers/controllers; (2) flow-paced (open-loop) chemical feeders; (3) pressure- or flow-regulating valves; (4) continuous level control of process basins; and (5) variable-speed pumping systems for flow/level control.

19.0J A motor control station or panel and the related circuitry essentially provide ON/OFF operation of an electric motor or other device.

Answers to questions on page 408.

19.1A The general principles for safe performance on the job are to always avoid unsafe acts and correct unsafe conditions immediately.

19.1B Electric shock can cause serious burns and even death (by asphyxiation due to paralysis of the muscles used in breathing).

19.1C An electrical "explosion" could shower you with molten metal, startle you into a bumped head or elbow, or cause a bad fall.

19.1D If electric current flows through your upper body, electric shock could harm your heart or your head.

Answers to questions on page 414.

19.1E Operators should be especially careful when working around powered mechanical equipment because the equipment could start unexpectedly and cause serious injury.

19.1F The purpose of an electrical lockout device is to physically prevent the operation of an electric circuit, or to de-energize the circuit temporarily.

19.1G When testing the atmosphere before entry in any confined space, first test for oxygen deficiency/enrichment, then combustible gases and vapors, and then toxic gases and vapors.

19.1H UEL stands for upper explosive limit and LEL stands for lower explosive limit. UEL and LEL indicate the range of concentrations at which combustible or explosive gases will ignite when an ignition source is present at ambient temperature.

19.1I Wearing thin rubber or plastic gloves can greatly reduce your chances of electric shock.

Answers to questions on page 414.

19.2A A primary element or sensor senses and quantifies measured variables such as pressure, level, and flow; at times, chemical feed rates and some physical or chemical process characteristics are also sensed. The primary element performs the initial conversion of measurement energy to a signal.

19.2B Pressure is measured by the movement of a flexible element or a mechanically deformable device subjected to the force of the pressure being measured.

19.2C Some pressure sensors are fitted with surge or over-range protection to limit the effect of pressure spikes or water hammer on the device.

Answers to questions on page 418.

19.2D Different types of liquid level-sensing devices include floats, displacers, electrical probes, direct hydrostatic pressure, pneumatic bubbler tubes, and ultrasonic (sonar) devices.

19.2E Single-point ball float levels mount in a sump or tank to sense a fixed elevation. When the water level rises, the ball floats and rolls onto its side, actuating an internal mercury switch. The switch provides a discrete ON/OFF signal used to control equipment or send level data to the control system.

19.2F Probes are used to measure liquid levels where mechanical systems are impractical, such as within sealed or pressurized tanks or with chemically active liquids.

19.2G Ultrasonic sensors measure the level of a liquid by emitting a pulse that is reflected back to the receiver. The sensor periodically generates a pulse of waves that bounce off the liquid surface and reflect back. The sensor detects the reflected pulse. Based on the speed of the pulse, the travel time between sending and receiving is measured and the distance is calculated and converted into a level measurement.

Answers to questions on page 426.

19.2H The two basic types of flow readings are rate of flow (volume per unit of time) and total flow (volume units).

19.2I The inferential flow-sensing techniques that are used in flow measurement are: (1) velocity, (2) differential pressure, (3) magnetic, and (4) ultrasonic.

19.2J Velocity-sensing meters measure process flows by measuring water speed within a pipeline or channel. One way of doing this is by sensing the rate of rotation of a special impeller placed within the flowing stream; the rate of flow is directly proportional to impeller RPM (within certain limits).

19.2K Flows measured with Venturi meters take advantage of a basic principle of hydraulics, the Bernoulli Effect: When a liquid is forced to go faster in a pipe or channel, its internal pressure drops.

Answers to questions on page 428.

19.2L In a fluid (liquid or gas) chemical feeder, the indication of quantity per unit of time, or liquid chemical feed rate, may be provided by use of a rotameter or built-in calibrated pump with indicated output settings.

19.2M Process variables commonly monitored or controlled by process instrumentation include turbidity, pH, chlorine residuals, electrical conductivity, ORP, alkalinity, and temperature.

19.2N The two general systems for transmission of measurement signals are analog (4–20 mA) or digital data.

ANSWERS TO QUESTIONS IN LESSON 2

Answers to questions on page 430.

19.3A The purpose of instrumentation indicators is to give a visual representation of a variable's present value, either as an analog or digital display.

19.3B An analog display uses some type of pointer or graphical display against a scale.

19.3C Chart recorders, which by nature also serve as indicators, give a permanent record of how a variable changes with time by way of a moving chart.

19.3D Factors that can disrupt instrument and control systems include temporary power failures, tripped panel circuit breakers, voltage surges resulting in blown fuses, static electricity, and excessive heat.

Answers to questions on page 433.

19.3E Recorders are indicators designed to show how the value of the variable has changed with time.

19.3F The two main types of chart recorders are the strip-chart type and the circular-chart type.

19.3G Chart recorders are typically electrical, powered by the instrumentation and control system electrical supply. Battery-backed power for the recorders may be used, if required, as a backup power source.

19.3H Alarms may produce visual or audible warning signals.

Answers to questions on page 434.

19.3I An operator might bypass the operation of a controller in an emergency or when, in the judgment of the operator, the controller is not exercising control within sensible limits. To bypass a controller, switch to the manual mode of operation.

19.3J The basic principle to guide you in setting up the frequency of ON/OFF operation of a piece of equipment is to set the ON/OFF controls to operate or cycle associated equipment on and off no more often than actually necessary for plant operation.

19.3K Pumps can be prevented from running after a loss of level signal by electrical circuitry designed so the pump will turn *OFF* on an open signal circuit and *ON* only with a closed circuit.

19.3L Pump control panels may include automatic or manual alternators (two pumps) or sequencers (more than two pumps). This provision allows the total pump operating time required for a particular system to be distributed equally among all the pumps at a pump station.

Answers to questions on page 437.

19.3M Pneumatic instrumentation depends on a constant source of clean, dry, pressurized air for reliable operation.

19.3N Oil is removed from instrument air by filtering the air through special oil-absorbent elements. Water is removed from the air by passing the moisture-laden air through an air-drying system consisting of a moisture separation column and air dryers.

Answers to questions on page 439.

19.3O An operator should be especially careful when working with laboratory instruments because some of these instruments are very delicate and replacement parts, such as glass vessels or pH electrodes, are quite expensive. Moreover, the use of some of these instruments requires the regular handling of laboratory glassware and other breakable items.

19.3P Operators should become familiar with the testing and calibration of plant measuring and control systems in order to discuss needed repairs or adjustments with an instrument technician and perhaps assist with that work. This knowledge also enables operators to analyze the effects of instrument problems on continued plant operation and to handle emergency situations created by instrument failure.

19.3Q The most useful piece of general electrical test equipment is the V-O-M, that is the Volt-Ohm-Milliammeter, commonly referred to as a multimeter.

Answers to questions on page 441.

19.3R Computer control systems used for process control are classified by two commonly used terms: the Distributed Control System (DCS) and the Supervisory Control And Data Acquisition (SCADA) system.

19.3S The five basic components of the computer control system are: (1) the process instrumentation and control devices, (2) the input/output (I/O) interface, (3) the central processing unit (CPU), (4) the communication interface, and (5) the human machine interface (HMI). These components are the means by which the control system gathers and distributes information for the human operator and the process instrumentation and equipment.

19.3T Computer control systems help operation and maintenance personnel in preparing daily, weekly, and monthly maintenance schedules, monitoring the spare parts inventory status, and ordering additional spare parts, when necessary, printing out work orders, and recording completed work assignments.

Answers to questions on page 443.

19.3U Computer control systems record and store (archive) information that historically was recorded on paper strip-charts.

19.3V Cost savings associated with the use of a computer control system frequently include reduced labor costs for operation, maintenance, and monitoring functions that were formerly performed manually. Precise control of chemical feed rates by a computer control system eliminates wasteful overdosing. Preventive maintenance monitoring can save on equipment and repair costs, and, as previously noted, energy savings may result from off-peak (reduced) electrical power rates.

19.3W The greatest challenge for operators using computer control systems is to realize that just because a computer says something, this does not mean that the computer is always correct. Operators will always be needed to question and analyze the results from computer control systems.

Answers to questions on page 443.

19.4A The key to proper instrument operation and maintenance (O & M) is the operator's practical understanding of the system. Operators must know how to: (1) recognize malfunctioning instruments so as to prevent prolonged damaging operation, (2) shut down and prepare devices for seasonal or other long-term nonoperation, and (3) perform preventive (and minor corrective) maintenance tasks to ensure proper operation in the long term.

19.4B General tasks expected of operators of instrumentation systems can be summed up, from an operations standpoint, as learning, and constant attention to, what constitutes normal function, and, from a maintenance standpoint, as ensuring proper and continuing protection and care of each component.

19.4C Two of the surest indications of a serious electrical problem in instruments or power circuits are smoke or a burning odor.

Answers to questions on page 444.

19.4D Moisture can be controlled in an electrical panel room containing instrumentation by leaving some power components on (such as a power transformer) to provide space heat. In a known moist environment, sealed instrument cases may be protected for a while with a container of desiccant.

19.4E Pneumatic instrumentation should be purged with dry air before shutdown to rid the individual parts of residual oil and moisture to minimize internal sticking and corrosion while standing idle.

450 Water Treatment

Answers to questions on page 445.

19.4F Regular preventive maintenance duties should be printed on forms or cards (or appear on a computer screen) for use by operators as a reminder, guide, and record of preventive maintenance tasks performed.

19.4G To obtain a technical manual for an instrument, write to the manufacturer. Be sure to provide all relevant serial/model numbers in your request to the manufacturer for a manual.

19.4H Instrument supplies and spare parts that should always be available include charts, pens, pen cleaners, and ink, and any other parts listed in the technical manual as necessary for instrument operation or service.

Answers to questions on page 445.

19.4I Operational checks on instrumentation equipment are performed by always observing each system for its continuing signs of normal operation, and cycling some indicators by certain test methods.

19.4J If a recorder trace is altered from its usual pattern during the process of checking an instrument, the operator causing the upset should initial the chart appropriately and note the time.

CHAPTER 20

SAFETY

by

Joe Monscvitz

TABLE OF CONTENTS
Chapter 20. SAFETY

			Page
OBJECTIVES			456
WORDS			457

LESSON 1

20.0 IMPORTANCE OF SAFETY 459
    20.00   What Is Safety? 459
    20.01   Causes of Accidents 459
    20.02   Steps to Avoid Accidents 460

20.1 CHEMICAL HANDLING 460
    20.10   Safe Handling of Chemicals 460
    20.11   Acids 460
        20.110   Acetic Acid (Glacial) (CH_3COOH) 460
        20.111   Hydrofluosilicic Acid (H_2SiF_6) 461
        20.112   Hydrogen Fluoride (HF) 463
        20.113   Hydrochloric Acid (HCl) 463
        20.114   Nitric Acid (HNO_3) 463
        20.115   Sulfuric Acid (H_2SO_4) 463
    20.12   Bases 464
        20.120   Ammonia (NH_3) 464
        20.121   Calcium Hydroxide ($Ca(OH)_2$) 465
        20.122   Sodium Hydroxide (Caustic Soda) (NaOH) 465
        20.123   Sodium Silicate (Na_2SiO_2) 466
        20.124   Hypochlorite (OCl^-) 466
        20.125   Sodium Carbonate (Na_2CO_3) 466
    20.13   Gases 467
        20.130   Chlorine (Cl_2) 467
        20.131   Carbon Dioxide (CO_2) 469
        20.132   Sulfur Dioxide (SO_2) 469
    20.14   Salts 471
        20.140   Aluminum Sulfate (Alum) ($Al_2(SO_4)_3 \cdot 14H_2O$) 472
        20.141   Ferric Chloride ($FeCl_3$) 472
        20.142   Ferric Sulfate ($Fe_2(SO_4)_3$) 472

	20.143	Ferrous Sulfate (FeSO$_4$·7H$_2$O)	472
	20.144	Sodium Aluminate (Na$_2$O·Al$_2$O$_3$)	473
	20.145	Fluoride Compounds (NaF and H$_3$SiF$_6$)	473
20.15	Powders		473
	20.150	Potassium Permanganate (KMnO$_4$)	473
	20.151	Powdered Activated Carbon	474
	20.152	Other Powders	474
20.16	Labeling of Chemical Containers	475
20.17	Chemical Storage Drains	475

DISCUSSION AND REVIEW QUESTIONS ... 475

LESSON 2

20.2	FIRE PROTECTION	476
20.20	Fire Prevention	476
20.21	Classification of Fires and Extinguishers	476
20.22	Fire Extinguisher Operation and Maintenance	476
20.23	Fire Hoses	478
20.24	Storage of Flammables	479
20.25	Exits	479

20.3	PLANT MAINTENANCE	479
20.30	Maintenance Hazards	479
20.31	Cleaning	479
20.32	Painting	479
20.33	Cranes	480
20.34	Confined Spaces	480
20.35	Manholes	483
20.36	Power Tools	483
20.37	Welding	484
20.38	Safety Valves	484

20.4	VEHICLE MAINTENANCE AND OPERATION	485
20.40	Types of Vehicles	485
20.41	Maintenance	485
20.42	Seat Belts	485
20.43	Accident Prevention	485
20.44	Forklifts	487

DISCUSSION AND REVIEW QUESTIONS ... 487

LESSON 3

20.5 ELECTRICAL EQUIPMENT ... 489
    20.50 Electrical Safety ... 489
    20.51 Current—Voltage ... 489
    20.52 Transformers ... 489
    20.53 Electric Starters ... 489
    20.54 Electric Motors ... 490
    20.55 Instrumentation ... 490
    20.56 Control Panels ... 490
    20.57 Lockout/Tagout Procedure ... 490

20.6 LABORATORY SAFETY ... 492
    20.60 Laboratory Hazards ... 492
    20.61 Glassware ... 492
    20.62 Chemicals ... 492
    20.63 Biological Considerations ... 493
    20.64 Radioactivity ... 493
    20.65 Laboratory Equipment ... 493
        20.650 Hot Plates ... 493
        20.651 Water Stills ... 493
        20.652 Sterilizers ... 493
        20.653 Pipet Washers ... 493

20.7 OPERATOR PROTECTION ... 494
    20.70 Operator Safety ... 494
    20.71 Respiratory Protection ... 494
    20.72 Safety Equipment ... 494
    20.73 Eye Protection ... 495
    20.74 Foot Protection ... 495
    20.75 Hand Protection ... 495
    20.76 Head Protection ... 496
    20.77 Water Safety ... 496

20.8 PREPARATION FOR EMERGENCIES ... 497
20.9 ARITHMETIC ASSIGNMENT ... 497
20.10 ADDITIONAL READING ... 499
DISCUSSION AND REVIEW QUESTIONS ... 499
SUGGESTED ANSWERS ... 500

OBJECTIVES
Chapter 20. SAFETY

Following completion of Chapter 20, you should be able to:

1. List the responsibilities of operators and management for waterworks safety.

2. Identify and safely handle hazardous chemicals.

3. Recognize fire hazards and properly extinguish various types of fires.

4. Safely maintain waterworks equipment and facilities.

5. Properly operate and maintain vehicles.

6. Recognize electrical hazards.

7. Safely perform duties in a laboratory.

8. Protect other operators and yourself while working in and around waterworks facilities.

WORDS
Chapter 20. SAFETY

ACUTE HEALTH EFFECT
An adverse effect on a human or animal body, with symptoms developing rapidly.

ARCH
(1) The curved top of a sewer pipe or conduit.

(2) A bridge or arch of hardened or caked chemical that will prevent the flow of the chemical.

CONFINED SPACE
Confined space means a space that:

(1) Is large enough and so configured that an employee can bodily enter and perform assigned work; and

(2) Has limited or restricted means for entry or exit (for example, manholes, tanks, vessels, silos, storage bins, hoppers, vaults, and pits are spaces that may have limited means of entry); and

(3) Is not designed for continuous employee occupancy.

Also see DANGEROUS AIR CONTAMINATION and OXYGEN DEFICIENCY.

CONFINED SPACE, CLASS A
A confined space that presents a situation that is immediately dangerous to life or health (IDLH). These include but are not limited to oxygen deficiency, explosive or flammable atmospheres, and concentrations of toxic substances.

CONFINED SPACE, CLASS B
A confined space that has the potential for causing injury and illness, if preventive measures are not used, but is not immediately dangerous to life and health.

CONFINED SPACE, CLASS C
A confined space in which the potential hazard would not require any special modification of the work procedure.

CONFINED SPACE, PERMIT-REQUIRED (PERMIT SPACE)
A confined space that has one or more of the following characteristics:

(1) Contains or has a potential to contain a hazardous atmosphere,

(2) Contains a material that has the potential for engulfing an entrant,

(3) Has an internal configuration such that an entrant could be trapped or asphyxiated by inwardly converging walls or by a floor that slopes downward and tapers to a smaller cross section, or

(4) Contains any other recognized serious safety or health hazard.

DECIBEL (DES-uh-bull)
A unit for expressing the relative intensity of sounds on a scale from zero for the average least perceptible sound to about 130 for the average level at which sound causes pain to humans. Abbreviated dB.

MATERIAL SAFETY DATA SHEET (MSDS)

A document that provides pertinent information and a profile of a particular hazardous substance or mixture. An MSDS is normally developed by the manufacturer or formulator of the hazardous substance or mixture. The MSDS is required to be made available to employees and operators or inspectors whenever there is the likelihood of the hazardous substance or mixture being introduced into the workplace. Some manufacturers are preparing MSDSs for products that are not considered to be hazardous to show that the product or substance is not hazardous.

OSHA (O-shuh)

The Williams-Steiger Occupational Safety and Health Act of 1970 (OSHA) is a federal law designed to protect the health and safety of workers, including the operators of water supply and treatment systems and wastewater collection and treatment systems. The Act regulates the design, construction, operation, and maintenance of water and wastewater systems. OSHA regulations require employers to obtain and make available to workers the Material Safety Data Sheets (MSDSs) for chemicals used at industrial facilities and treatment plants. OSHA also refers to the federal and state agencies that administer the OSHA regulations.

OLFACTORY (all-FAK-tore-ee) FATIGUE

A condition in which a person's nose, after exposure to certain odors, is no longer able to detect the odor.

PERMIT-REQUIRED CONFINED SPACE (PERMIT SPACE)

See CONFINED SPACE, PERMIT-REQUIRED (PERMIT SPACE).

TAILGATE SAFETY MEETING

Brief (10 to 20 minutes) safety meetings held every 7 to 10 working days. The term comes from the safety meetings regularly held by the construction industry around the tailgate of a truck.

CHAPTER 20. SAFETY

(Lesson 1 of 3 Lessons)

20.0 IMPORTANCE OF SAFETY

20.00 What Is Safety?

Webster states that safety is the following:

"The condition of being safe; freedom from exposure to danger; exemption from hurt, injury, or loss; to protect against failure, breakage, or other accidents; knowledge of or skill in methods of avoiding accident or disease."

Safety is more than words. Safety is the action of Webster's definition. Safety is using one's knowledge or skill to avoid accidents or to protect oneself and others from accidents. Safety is a form of preventive maintenance that includes equipment and machinery and its proper handling. Who is responsible for safety? Everyone should be responsible, from top management to all employees.

Management has its responsibilities for safety; its main function is to set the tone and provide the training and monies for an effective safety program. (See Chapter 23, "Administration," for a description of the elements of an effective safety program and a detailed discussion of management's responsibilities for safety.) Management should be responsible for the following:

1. Establish a safety policy
2. Assign responsibility for accident prevention, which includes descriptions of the duties and responsibilities of a safety officer, department heads, line supervisors, safety committees, and operators
3. Appoint a safety officer or coordinator
4. Establish realistic goals and periodically revise them to ensure continuous and maximum effort
5. Evaluate the results of the program

While management has its responsibilities, the operators also have their particular responsibilities. Operators should:

1. Perform their jobs in accordance with established safe procedures
2. Recognize the responsibility for their own safety and that of fellow operators
3. Report all injuries
4. Report all observed hazards
5. Actively participate in the safety program

QUESTIONS

Write your answers in a notebook and then compare your answers with those on page 500.

20.0A What are management's responsibilities for safety?

20.0B What are the operators' responsibilities for safety?

20.01 Causes of Accidents

A safety program has one objective: To prevent accidents. Accidents do not happen, they are caused. They may be caused by unsafe acts of operators or result from hazardous conditions or may be a combination of both. An analysis of accident statistics reveals that operator negligence and carelessness are the causes of most accidents. We know how to do our job safely, but we just do not do it safely.

Accidents reduce efficiency and effectiveness. An accident is that occurrence in a sequence of events that usually produces unintended injury, death, or property damage. Accidents affect the lives and morale of operators. They raise the costs not only to management but also to operators.

Safety authorities tell us that nine out of ten injuries are the result of unsafe acts of either the person injured or someone else. Here are some of the principal reasons for unsafe acts.

IGNORANCE. This may be due to lack of experience or training, or to a temporary condition that prevents the recognition of a hazard.

INDIFFERENCE. Some people know better but do not care. They take unnecessary risks and disregard the rules or instructions.

POOR WORK HABITS. Some people either do not learn the right way of doing things, or they develop a wrong way. Supervisors or fellow operators who see things done unsafely must speak up.

LAZINESS. Laziness affects speed and quality of work. Laziness also affects safety because safety requires an effort. In most jobs, you cannot reduce your safety effort and still maintain the same level of safety, even if you slow down or lower your quality standards.

HASTE. When we rush, we work too fast to think about what we are doing, we take dangerous shortcuts, and we are more likely to be injured.

POOR PHYSICAL CONDITION. Some of us just will not take reasonable care of ourselves. We ignore our bodily needs in regard to exercise, rest, and diet, lessening our endurance and alertness.

TEMPER. Impatience and anger cause many accidents. Again, our thinking is interfered with and the way prepared for an accident.

Every one of these reasons for unsafe acts could be considered operator negligence or carelessness.

20.02 Steps to Avoid Accidents

Some specific steps you can take to avoid accidents are the following:

1. Ask about the safety program established by management and then actively participate in safety training sessions and *TAILGATE SAFETY MEETINGS*.[1]

2. Routinely use the safety equipment described in Section 20.7 (for example, lockout/tagout devices and self-contained breathing apparatus) and personal protective gear (hard hat, safety goggles, gloves) provided by your employer.

3. Follow safe procedures at all times. This means taking the time needed to do every job *SAFELY*.

QUESTIONS

Write your answers in a notebook and then compare your answers with those on page 500.

20.0C What is an accident?

20.0D List the principal reasons for unsafe acts.

20.0E What problems are caused by laziness?

20.1 CHEMICAL HANDLING

20.10 Safe Handling of Chemicals

The water treatment plant operator may be required to handle a wide variety of chemicals, depending on the type of plant. In a simple well system, chlorine may be the only chemical used. In more complicated plants, there may be chlorine, other gases, sulfuric acid, lime, alum, powdered activated carbon, and anhydrous ammonia. All of these chemicals fall into groups: acids, hydroxides, gases, salts, organics, and solvents. Each group requires you, the operator, to have a good understanding of all types of chemicals. You must know how to handle the many problems associated with each of these elements or compounds. For example, you must know how to store chemicals, understand the fire problem, be aware of the tendency of chemicals to *ARCH*[2] in a storage bin, and know how to feed dry, how to feed liquid, and how to make up solutions.

All of these situations may cause an unsafe condition. Overheating gas containers, dust problems with powdered carbon, burns caused by acid, reactivity of each chemical under a variety of conditions that may cause fire and explosions are other safety hazards. You will need to know the usable limits of each chemical because of toxicity, the protective equipment required for each chemical, each chemical's antidote, and how to control fires caused by each chemical.

Although you may not regularly handle all of the chemicals listed here, you may come into contact with them from time to time. Try at least to learn all of the characteristics of the chemicals you will use regularly. For example, learn the boiling point, explosive limits, reactivity, flammability, first aid used for each chemical, and other characteristics that may prove helpful in preventing a safety hazard. Study the chemistry of each chemical used in the plant in order to have a safe plant in which to work. The following discussions concentrate on the characteristics of each compound and point out its hazards, reactivity, and the information needed to avoid conditions that may cause a safety problem.

20.11 Acids

Acids are used extensively in water treatment. For example, hydrofluoric acid is used in fluoride addition and hydrochloric acid is often used in cleaning. Table 20.1 lists many of the acids used in water treatment and gives their characteristics. This section can serve as a quick reference and a guide for learning some of each acid's limitations and its reactivity with other compounds.

The antidote to all acids is neutralization. However, one must be careful in how this is performed. Most often, large amounts of water will serve the purpose, but if the acid is ingested (swallowed), then milk of magnesia may be needed. If vapors are inhaled, first aid usually consists of providing fresh air, artificially restoring breathing (CPR), or supplying oxygen. In general, acids are neutralized by a base or alkaline substance. Baking soda is often used to neutralize acids on skin because it is not harmful on contact with your skin. To understand these reactions, you will need to know some acid/base chemistry. The knowledge of acid/base chemistry and fast reactions on your part may reduce the safety hazards involved in handling acids in water treatment.

20.110 *Acetic Acid (Glacial) (CH$_3$COOH)*

This chemical is stable when stored and handled properly. However, it may react violently with certain compounds such as ammonium nitrate, potassium hydroxide, and other alkaline materials. Strong oxidizing glacial acetic acid is a combustible material. Fires involving the acid may be extinguished with water, dry chemical, or carbon dioxide. Under such conditions as adding water, the diluted acid may produce hydrogen gas

---

[1] *Tailgate Safety Meeting.* Brief (10 to 20 minutes) safety meetings held every 7 to 10 working days. The term comes from the safety meetings regularly held by the construction industry around the tailgate of a truck.

[2] *Arch.* (1) The curved top of a sewer pipe or conduit. (2) A bridge or arch of hardened or caked chemical that will prevent the flow of the chemical.

TABLE 20.1 ACIDS USED IN WATER TREATMENT

Name, Formula	Common Name	Available Forms	Specific Gravity	Flammability	Color	Odor	Containers
Acetic Acid, CH_3COOH	Ethanoic Acid	Solution	1.05	Yes	Clear	Sharp, pungent	Carboys, drums
Hydrofluosilicic Acid, H_2SiF_6	Fluosilicic Acid	Solution	1.4634	N/A[a]	Clear	Pungent fumes	Drums, trucks, R.R. tank cars
Hydrogen Fluoride Acid, HF	Hydrofluoric Acid	Liquid	0.987	N/A	Clear	Fumes, toxic	Drums, tank cars
Hydrochloric Acid, HCl	Muriatic Acid	Solution	1.16	N/A	Clear to yellow	Pungent, suffocating	Drums, carboys
Nitric Acid, HNO_3	—	Liquid	1.5027	N/A	Colorless, yellowish	Toxic fumes in presence of light	Drums, carboys, bottles
Sulfuric Acid, H_2SO_4	Oil of Vitriol; Vitriol	Solution	(60–66° Bé[b]) 1.841	N/A	Clear	Odorless	Bottles, carboys, drums, trucks, tank cars

[a] N/A = Not Applicable.
[b] Bé = Baumé scale, a scale for measuring specific gravity of liquids.

---

when it comes in contact with metals. When the chemical is involved in a fire situation, a self-contained breathing apparatus must be used to protect the operator against suffocation and problems caused by corrosive vapors.

Most people find inhalation of acetic acid vapors in concentrations over 50 ppm (parts per million) intolerable, resulting in nose and throat irritations. Repeated exposure to high concentrations may produce congestion of the larynx. Skin contact with concentrated acetic acid can produce deep burns with skin destruction. High vapor concentration may blacken the skin and produce allergic reactions and eye irritation. Possible permanent damage or immediate burns are caused when the acid comes into contact with the eyes. If the acid is ingested, severe intestinal irritation will result as well as burns to the mouth and upper respiratory tract.

Operators should be protected by adequate exhaust facilities to ensure ventilation when working with acetic acid. At a minimum, exhaust hoods should have air velocity of 100 FPM (30 m/min). Wear rubber gloves and an apron to prevent skin contact. Wear splash-proof goggles or a face shield to prevent any eye contact. Gas-tight goggles may also be needed to prevent vapors from irritating your eyes. An eye wash station must be readily available where this chemical is being handled. Also, respiratory equipment should be available for emergency use. Acetic acid can be handled safely by using adequate ventilation and safety equipment to prevent skin and eye contact. Remember also that acetic acid vapors can cause *OLFACTORY* (all-FAK-tore-ee) *FATIGUE*. This is a condition in which a person's nose, after exposure to certain odors, is no longer able to detect the odor. Acetic acid is detectable by your nose at 1 ppm, but documentation has shown operators tolerating up to 200 ppm.

If a leak or spill should occur, notify safety personnel and provide adequate ventilation. When cleaning up large spills, wear a self-contained breathing apparatus and equipment to prevent contact with eyes and skin. To clean up spill areas and remove chemical residue, cover the area with sodium bicarbonate and flush away with large quantities of water.

First aid for acetic acid exposure calls for removal of the victim to fresh air, rinsing the mouth and nasal passages with water, and checking for inhalation problems. If the acid made contact with the eyes, immediately irrigate with water for at least 15 minutes. Obtain medical attention. For skin contact problems, wash with water immediately; if the acid was swallowed, give three glasses of milk or water and obtain medical attention quickly. Acetic acid exposure, like all other acids, must be treated immediately to prevent damage to the victim.

20.111 Hydrofluosilicic Acid (H_2SiF_6)

This chemical is hazardous to handle under any conditions. Be extremely careful using this acid. The acid is colorless, transparent, fuming, corrosive, and is a liquid. A pungent odor is created by the acid and contact causes skin irritation. When the acid vaporizes, it decomposes into hydrofluoric acid and silicon tetrafluoride. Hydrofluoric acid can attack glass.

When handling hydrofluosilicic acid, always use complete protective equipment, rubber gloves, goggles or face shield, rubber apron, rubber boots, and have lime slurry barrels, Epsom salt solution, and safety showers (Figure 20.1) available. Always provide adequate ventilation because its vapor can cause irritation to the respiratory system. Careful maintenance of protective equipment is essential because the fumes of the acid corrode metal or etch glass on the protective equipment.

462　Water Treatment

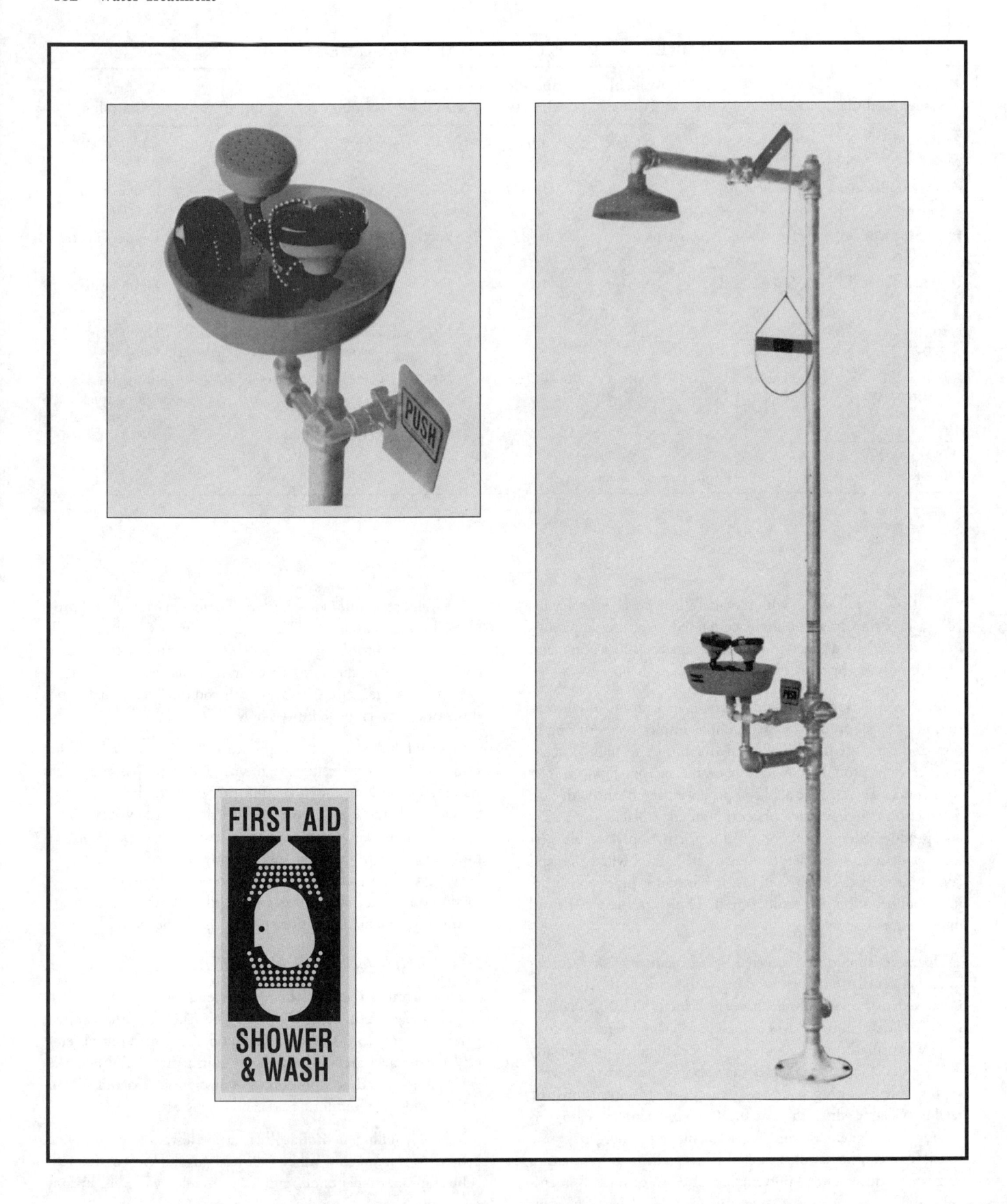

Fig. 20.1 Safety shower with face/eye wash facility
(Permission of Nevada Safety & Supply)

First aid for eye contact is to thoroughly flush with water for 15 minutes and get medical aid as soon as possible. For skin contact, wash the affected areas with water. For gross (large) contact, remove contaminated clothing under a safety shower and thoroughly wash entire body for 15 minutes or longer. In case of inhalation, remove operator to fresh air, restore breathing, if required, and get medical aid.

20.112 Hydrogen Fluoride (HF)

This acid is extremely poisonous, and it produces terrible sores when allowed to come into contact with the skin. The acid is a clear, corrosive liquid that has a pungent odor. All of the precautions discussed for hydrofluosilicic acid apply to this acid also.

20.113 Hydrochloric Acid (HCl)

This acid is used most often for cleaning in and around the treatment plant and is known as muriatic acid. The acid is also used very frequently in the laboratory. Hydrochloric acid is stable when properly contained and handled. This acid is one of the strong mineral acids and, therefore, is highly reactive with metals and these oxides: hydrocarbon, amine, and carbonate compounds. The acid liberates significant levels of hydrogen chloride gas (HCl) because of its vapor pressure at room temperature, and gives off large amounts of gas when heated. In reactions with most metals, hydrochloric acid will produce hydrogen gas.

Inhalation of HCl vapors or mists for long periods can cause damage to teeth and irritation to the nasal passages. Concentrations of 750 ppm or more will cause coughing, choking, and produce severe damage to the mucous membranes of the respiratory tract. In concentrations of 1,300 ppm, HCl is dangerous to life. Ingestion can cause burns of the mouth and digestive tract.

When handling HCl, provide adequate exhaust facilities to ensure ventilation and wear protective clothing and equipment to prevent body contact with the acid. Use rubber gloves, rubber apron, rubber boots, and wear a long-sleeved shirt when handling hydrochloric acid. To protect your eyes against splashing of the acid, you must wear safety goggles or a face shield. There should always be an eye wash station and safety shower located near areas where this acid is to be used.

First aid consists of thoroughly flushing the eyes with running water for 15 minutes and securing medical aid. If hydrochloric acid comes in contact with skin, wash the affected areas with water. For gross contact, remove clothing under the safety shower and continue showering for 15 minutes or longer. Should the acid be ingested, give water and milk of magnesia. Do not induce vomiting; get medical aid. In case of inhalation, remove the victim to fresh air, restore breathing if required, and get medical aid.

Store acid containers closed in a clean, cool, open, and well-ventilated area. Keep out of the sun. Keep the acid away from oxidizing agents or alkaline materials. Provide emergency neutralization materials in use areas.

20.114 Nitric Acid (HNO₃)

Like hydrochloric acid, nitric acid is one of the most commonly used acids in the water treatment laboratory and plant. The acid is a powerful oxidizing agent and attacks most metals. Nitric acid is stable when properly handled and placed into a proper container. The acid is one of the strong mineral acids, and is highly reactive with materials such as metals. When handling nitric acid, use protective clothing and equipment to prevent body contact with the liquid. Such equipment includes rubber gloves, rubber apron, and safety goggles or a face shield for eye protection against splashing of the acid. Nitric acid is a strong, poisonous, and highly corrosive liquid and must be handled carefully. The acid forms toxic fumes in the presence of light; therefore, it should be kept out of the sun.

Like other acids, nitric acid should be stored in clean, cool, well-ventilated areas. The areas should have an acid-resistant floor and adequate drainage. Keep it away from oxidizing agents and alkaline materials. Protect containers from damage or breakage. Avoid contact with skin and provide emergency neutralization materials and safety equipment in use areas.

First aid for skin contact is to flush thoroughly with water for 15 minutes. Get medical aid if needed. This acid will cause burns, but these can be greatly reduced if the contact area is immediately flushed. For gross (large) contact, remove contaminated clothing under a safety shower. For eye contact, flush with water for 15 minutes and get medical aid. If acid is ingested, give water with milk of magnesia; get medical aid. For inhalation, remove to fresh air, restore breathing if required, and get medical aid.

20.115 Sulfuric Acid (H₂SO₄)

This mineral acid is highly corrosive and will attack most metals. Sulfuric acid is also very reactive to the skin and must be handled with extreme care or you will suffer severe burns. Even when the acid is diluted, it is highly corrosive and must be contained in rubber-, glass-, or plastic-lined equipment. The acid will decompose clothing and shoes. Sulfuric acid should not come in contact with potassium permanganate or similar compounds. Sulfuric acid reacts violently with water. *ALWAYS POUR ACID INTO WATER* while stirring to prevent the generation of steam and hot water, which could boil over the container and cause serious acid burns.

As with other mineral acids, this material is stable when properly contained and handled. When you are handling sulfuric acid, you must use protective clothing and equipment to prevent body contact with the acid. Wear rubber gloves and safety goggles or a face shield for eye protection against splashing. Also, wear a rubber apron, rubber boots, and long-sleeved shirt. The eye wash station and safety shower must be located nearby where the acid is being handled. The area should be well ventilated, the acid should be stored in closed containers in a clean, cool, open area. The area should have an acid-resistant floor that drains well. Keep away from oxidizing agents and alkaline materials and protect the containers from damage or breakage.

Because the acid is highly corrosive and causes severe burns, first aid for sulfuric acid exposure must be immediate to avoid substantial damage to human tissue. For eye contact, flush thoroughly with running water for 15 minutes, get medical aid—the first seconds are important. For skin contact, wash affected areas thoroughly with water. For gross (large) contact, remove contaminated clothing under the safety shower with prolonged washing for at least 15 minutes. In cases of ingestion, give water and milk of magnesia; get medical aid. When working with sulfuric acid, avoid skin contact and always provide emergency neutralization materials and a safety shower near the work areas.

QUESTIONS

Write your answers in a notebook and then compare your answers with those on page 500.

20.10A What does an operator need to know about chemicals used in a water treatment plant?

20.11A What should be done if an operator inhales acid vapors?

20.11B Acetic acid will react violently with which compounds?

20.11C Under what conditions can acetic acid be handled safely?

20.11D What protective equipment is necessary for handling hydrofluosilicic acid?

20.11E How can the inhalation of hydrochloric acid (HCl) vapors or mists cause damage to operators?

20.11F How should nitric acid be stored?

20.12 Bases

The bases that are used in water treatment are known as hydroxides. From a functional standpoint, they are used to raise pH. Most common bases are compounds of sodium, calcium, and ammonium, which are strong bases. However, there are other weak bases, such as silicate, carbonate, and hypochlorite. From the standpoint of safety, both weak and strong bases must be given the same consideration when being handled. Some are very toxic and will attack human tissue very rapidly and cause burns. Explosive reactions will occur when bases come in contact with an acid and hazardous decomposition products are created under certain conditions.

Bases must be neutralized with dilute acids. However, the operator must work carefully because, under some conditions, there may be other reactions, such as with hypochlorite compounds. Therefore, you should understand acid/base chemistry before handling any of the basic compounds used in water treatment. Table 20.2 gives some of the common basic compounds used in water treatment. The following sections will discuss some of their characteristics and the precautions the operator must use to safely handle such compounds.

20.120 Ammonia (NH_3)

The operator may use one of two forms of ammonia: anhydrous or hydroxide. The first (anhydrous) is in a gas form and requires one type of consideration. The hydroxide is a liquid and requires another type of consideration. Anhydrous ammonia in the gaseous state is colorless; it is about 0.6 times as heavy as air. In a liquid state, ammonia is also colorless, 0.68 times as heavy as water, and it vaporizes rapidly. Ammonia gas is capable of forming explosive mixtures with air. For your own safety, be

TABLE 20.2 BASES USED IN WATER TREATMENT

Name, Formula	Common Name	Available Forms	Spec. Grav. or lbs/cu ft	Flammability	Color	Odor	Containers
Ammonia, NH_3	Ammonia	Liquid, gas	0.04813 lb/cu ft @ 0°C	Explosive with air; will burn in air	Colorless	Irritating	Cylinders, tanks, trucks
Calcium Hydroxide and Oxide, $Ca(OH)_2$ and CaO	Hydrated Lime and Quicklime	Dry powder, lump	50–70 lbs/cu ft	N/A[a]	White	Dust	Bags, bulk, trucks
Sodium Hydroxide, $NaOH$	Caustic Soda, Lye	Lump, liquid, flake	1.524	May cause flammable condition	Opaque white	Toxic, pungent	Drums, bulk, trucks
Sodium Silicate, Na_2SiO_2	Water Glass	Liquid	1.35–1.42	N/A	Opaque	N/A	Drums, bulk, trucks
Hypochlorite Compounds, $NaOCl$, $Ca(OCl)_2$	HTH	Powder, liquid	—	Explosive with antifreeze	White	Toxic Cl_2, pungent	Cans, drums
Sodium Carbonate, Na_2CO_3	Soda Ash	Powder	23, 35, and 65 lbs/cu ft	N/A	White	Dust	Bags, bulk, trucks

[a] N/A = Not Applicable.

aware of the possibility of suffocation since the gas can displace air, which contains oxygen. Although the vapors are not poisonous, they can and will irritate the mucous membranes of the eyes, nose, throat, and lungs. Irritation will be detected in concentrations of 5.0 ppm and when human tissue comes in contact with the liquid, it will cause severe burns.

When handling ammonia or working in an ammonia environment, respiratory protection is a requirement. For entry into emergency areas, use only a self-contained breathing apparatus. Install a good ventilation system to control vapors in the application room. Use protective clothing, rubber gloves, apron, boots, and face and eye protection if you are going to work with ammonia for long periods.

Care must be used when storing or transporting containers. Always keep cylinder caps in place when cylinders are not in use. Store cylinders in a cool, dry location away from heat, and protect from direct sunlight. Storage near radiators, steam pipes, or other sources of heat may raise the pressure to a dangerous point, whereas dampness may cause excessive corrosion. Do not store in the same room with chlorine. Always use lifting clamps or cradles. Avoid hoisting the cylinders using ropes, cables, or slings, and never drop the containers. Control ammonia leaks. They can be detected by odor or by using a cloth swab soaked with hydrochloric acid. This will form a white cloud of ammonium chloride.

Ammonia gas will burn if it is blended with air in a mixture containing 15 to 28 percent ammonia by volume. Check cylinder valve stems for leaks; tighten the packing gland nut only with a special wrench provided for such purposes. If a serious leak in a cylinder cannot be controlled, place the container in a vat of water. Fifty-three pounds of ammonia will dissolve in 100 pounds of water at 68°F (20°C). *NEVER* neutralize liquid ammonia with an acid. The reaction generates a lot of heat, which may speed up the release of ammonia gas.

First aid for skin contact with ammonia is to flush with large amounts of water for 5 to 10 minutes, and get medical aid. Remove contaminated clothing under a safety shower. For eye contact, flush thoroughly with water for 15 minutes immediately, and get medical aid. In the case of inhalation, remove to fresh air and restore breathing. If required, get medical aid. Throat burns should be rinsed with two percent boric acid solution. Urge the patient to drink large amounts of milk. For irritation of nose and throat because of exposure, see a physician.

Ammonium hydroxide is an aqueous (watery) solution of anhydrous ammonia and is quite volatile (will evaporate) at atmospheric temperatures and pressures. This solution can cause local skin irritations. A strong solution will cause human tissue destruction on contact with eyes, skin, and mucous membranes of the respiratory system, so avoid contact with the compound. The solution will cause severe burns depending upon solution concentration and length of contact time. The solution's vapor causes the same effects as the gas. First aid should be the same as for anhydrous ammonia.

20.121 Calcium Hydroxide ($Ca(OH)_2$)

Hydrated lime (calcium hydroxide) and quicklime (calcium oxide) are two forms of lime commonly used in water treatment plants. Hydrated lime is the least troublesome of the two forms. Hydrated lime is less caustic and is, therefore, less irritating to the skin, but can cause injury to eyes. However, as a dust it is just as hazardous as quicklime. Quicklime is a strong caustic and irritating to personnel exposed to the compound. When quicklime is mixed with water, a great deal of heat is generated, which can cause explosions.

Both quicklime and hydrated lime should be stored in cool, dry areas. Care must be taken to avoid mixtures of alum and quicklime, since quicklime tends to absorb the water away from the alum that forms as alum crystallizes (water of crystallization). In a closed container, this could lead to a violent explosion. Equal care should be taken to avoid mixtures of ferric sulfate and lime.

When handling both forms of lime, the operator should use chemical goggles and a suitable dust mask to protect the eyes and mucous membranes. Also, wear proper clothing to protect the skin because, with long contact, the lime can cause dermatitis (skin irritation) or burns, particularly at perspiration points. Always shower after handling quicklime. All operators should wear a face shield when inspecting lime slakers. Hot lime suspension that splatters on the operator may cause severe burns of the eyes or skin. The hot mist coming from the slakers is also dangerous. The loss of water supply to a lime slaker can create explosive temperatures.

First aid for lime burns, which are like burns from other caustics, consists of alternately washing with water and a mild acetic acid solution. One may also use large amounts of soap and water. For eye contact, wash immediately with large amounts of warm water and rinse with a boric acid solution. Get medical aid. For irritation of nose and throat because of exposure, see a physician.

20.122 Sodium Hydroxide (Caustic Soda) (NaOH)

Sodium hydroxide is available in pellet, liquid, and flake forms. Caustic soda usually comes as a 50-percent solution of sodium hydroxide. This base is a strong caustic alkali and very hazardous to the operator. This compound is extremely reactive. Sodium hydroxide absorbs carbon dioxide from the air, reacting violently or explosively with acid and a number of organic compounds. Caustic soda: (1) dissolves human skin, (2) generates heat when mixed with water, and (3) reacts with amphoteric metals (such as aluminum) generating hydrogen gas, which is flammable and may explode if ignited. Sodium hydroxide can be dissolved in water; the resulting solution is often used for the adjustment of pH because it is a liquid and easy to feed. This base is extensively used in water treatment. Because of its everyday use, you may forget just how hazardous this compound is, and through neglect may injure yourself or another operator. Only trained and protected operators should undertake spill cleanup. The operator must act cautiously, dilute the spill with water, and neutralize with a dilute acid, preferably acetic.

When handling caustic soda, control the mists with good ventilation. Protect your nose and throat with an approved respiratory system. For eye protection, you must wear chemical worker's goggles or a full face shield to protect your eyes. There must be an eye wash station and safety shower at or near the work station for this chemical. Protect your body by being fully clothed, and by using impervious gloves, boots, apron, and face shield.

Special precautions to be taken when handling or storing caustic soda include: (1) prevent eye and skin contact, (2) do not breathe dusts or mists, and (3) avoid storing this chemical next to strong acids. Dissolving sodium hydroxide in water or other substances generates excessive heat and causes splattering and mists. Solutions of sodium hydroxide are viscous (thick) and slippery. Caustic can quickly and permanently cloud vision if not immediately flushed out of eyes. Determine location of safety showers and eye wash stations BEFORE starting to work with caustic soda.

First aid for the eyes consists of irrigating the eyes immediately and continuously with flowing water for at least 30 minutes. Prompt medical attention is essential. For skin burns, immediate and continuous, thorough washing in flowing water for 30 minutes is important to prevent damage to the skin. Consult a physician, if required. In case of inhalation, remove victim to fresh air, call physician, or transport injured person to a medical facility. For ingestion, give large amounts of water or milk and immediately transport injured person to a medical facility; *DO NOT INDUCE VOMITING.*

You may also have occasion to use sodium hydroxide as flakes or pellets. All of the precautions stated for liquid caustic also apply for the flake form.

20.123 Sodium Silicate (Na_2SiO_2)

This chemical is a liquid as used in water treatment. However, it is nontoxic, nonflammable, and nonexplosive, but presents the same hazards to the eyes and skin as any other base compounds. Sodium silicate is a strong alkali and should be handled with care by using goggles or face shield and wearing gloves and protective clothing. The chemical will cause damage to the eyes and skin, but it is less dangerous than other alkaline compounds used in water treatment.

First aid for the eyes is to flush immediately and thoroughly with flowing water for at least 15 minutes. Get medical attention. If sodium silicate makes contact with skin, wash thoroughly with water, particularly if the solution is hot. Then, wash the skin with a 10-percent solution of ammonium chloride or 10 percent acetic acid. For ingestion, give plenty of water and dilute vinegar, lemon, or orange juice. Follow this with milk, white of eggs beaten with water, or olive oil. Call a physician.

20.124 Hypochlorite (OCl^-)

A number of hypochlorite compounds are commercially available for use in water treatment. If you understand the precautions for one such compound, you will know what steps must be taken with other hypochlorite compounds, such as calcium, sodium, or lithium. These chemicals may be used in either a liquid or dry form. There are several grades of hypochlorite compounds, but all are good oxidizers and are used for disinfection. When these compounds come into contact with organic materials, their decomposition releases heat very rapidly and produces oxygen and chlorine. Although hypochlorite compounds are nonflammable, they may cause fires when they come in contact with heat, acids, organic matter, or other oxidizable substances.

All solutions of hypochlorite compounds attack the skin, eyes, or other body tissues with which they come into contact. When handling hypochlorite, liquid or dry, use suitable protective clothing such as rubber gloves, aprons, and goggles or a face shield. Be aware that many times these compounds are stored in containers and give off chlorine gas when opened. Store these compounds in a cool, dry, dark area.

First aid for eyes is to flush with plenty of water for at least 15 minutes and see a physician. If hypochlorite compounds come in contact with the skin, flush thoroughly with water for at least 15 minutes, and get medical attention, as needed. In case of ingestion, wash out mouth thoroughly with water, give plenty of water to drink, and get medical attention. For inhalation, move the victim into fresh air and get medical attention.

Overexposure to any of the hypochlorite compounds may produce severe burns, so avoid contact with these compounds. They are hazardous and can attack skin, eyes, mucous membranes, and clothing.

20.125 Sodium Carbonate (Na_2CO_3)

Soda ash (sodium carbonate) is a mild alkaline compound, but requires safety precautions to minimize hazards when handling the chemical. An adequate ventilation system is needed to control the dust generated by the compound. Wear protective gear, such as chemical safety goggles or a face shield, a well-fitting dust respirator, and protective clothing to avoid skin contact. You should protect yourself by using a suitable cream or petroleum jelly on exposed skin surfaces, such as neck and hands. This compound's dust irritates the mucous membranes and prolonged exposure can cause sores in your nasal passage.

First aid for exposure to eyes (dust or solution) requires irrigation with water immediately for at least 15 minutes. Consult a physician if the exposure has been severe. For skin exposure, wash with large amounts of water; for contaminated clothing, wash before reusing. For inhalation or irritation of the respiratory tract, gargle or spray with warm water, and consult a physician, as needed.

QUESTIONS

Write your answers in a notebook and then compare your answers with those on pages 500 and 501.

20.12A   What are the two forms of ammonia used by operators?

20.12B   How should ammonia be stored?

20.12C   What are the two forms of lime used in water treatment plants?

20.12D   What would you do if someone swallowed sodium hydroxide?

20.12E   What would you do if sodium silicate came in contact with your skin?

20.13 Gases

There are a number of gases used in water treatment (Table 20.3). Most are supplied in steel drum containers; others must be generated on site. Some gases can be seen; others call the operator's attention by odor; and still others cannot be seen or detected by odor, yet are deadly. In this section, we will only discuss those gases that are supplied in containers that the operator must connect, disconnect, handle, or store. (See Section 20.120 for a discussion of ammonia.)

Exposure to the liquid forms of these gases usually will cause damage to human tissue, such as skin burns, but the most important factor to remember is the displacement of oxygen. Most gases are heavier than air and remove air from a room by displacement. Therefore, it is very important to have the right type of ventilation and respiratory protection. Use only the self-contained breathing apparatus when working in emergency areas.

20.130 Chlorine (Cl_2)

Safety is of the utmost importance when handling chlorine. Do not treat chlorine cylinders roughly; never drop them or permit collision of two or more cylinders. Never hoist chlorine cylinders by the neck. Always use lifting clamps or cradles—do not use ropes, cables, or chains. Store the cylinders in such a way that they cannot fall. Do not store chlorine cylinders below ground level and always keep the protective cap on the cylinder when it is not in use. Mark the empty containers and store them apart from full cylinders. Always store containers in an upright position in a clean, dry location free of flammable materials. The storage area must be equipped with forced-exhaust ventilation with starting switches located on the outside of the storage room. Some systems have a microswitch on the door so the ventilation will start automatically whenever the door is opened. Ventilation must provide at least one complete air change per minute. The temperature of the storage room should never be permitted to approach 140°F (60°C). Protect the chlorine cylinder from heat sources and never use an open flame on cylinders or pipes carrying chlorine. If chlorine is heated, the increase in temperature will cause an expansion of the gas, which results in an increase in pressure inside of the cylinders or piping, resulting in rupture of the containers.

When working with chlorine, be equipped to control chlorine leaks, which are most often found in the control valve. Repair kits are available for the 100- and 150-pound (45- and 68-kg) cylinders, as is emergency equipment for the one-ton (909-kg) tanks for controlling leaks. Each operator must be trained in the use of these emergency kits and must practice with the equipment at least once a year. Always check out even the slightest odor of chlorine; it may indicate a leak. Chlorine leaks only get worse. Small leaks can grow very rapidly causing serious problems that could have been easily solved as a small leak. There should always be two operators attending a chlorine leak, one to do the repairs and the other to act as safety observer. Some repairs require two operators to do the job (depends on the leak and the repair kit). Once again, use only the self-contained breathing apparatus when repairing a chlorine leak.

When connecting chlorine cylinders, be very careful with the threaded connections; never use two washers, use only one. If it does not work well, remove the washer and use another one. Do not reuse an old or used washer; always use a new washer. By taking this precaution, by cleaning the threads and washer, and by being careful with the thread setting, many chlorine leaks will be prevented. You *MUST* be aware whether you are using gas or liquid when connecting the container. On the one-ton (909-kg) tank, the top valve is for gas, the bottom valve is for liquid. If liquid chlorine is allowed into a gas feed system it will

TABLE 20.3   GASES USED IN WATER TREATMENT

Name, Formula	Common Name	Available Forms	Spec. Grav. or lbs/cu ft	Flammability	Color	Odor	Containers
Ammonia, NH_3	Ammonia	Liquid, gas	0.04813 lb/cu ft @ 0°C	Explosive with air; will burn in air	Colorless	Irritating	Cylinders, tanks, trucks
Chlorine, Cl_2	Liquid Chlorine	Liquid, gas	1.468 @ 0°C	None	Greenish yellow	Irritating	Cylinders, one-ton units, tank cars
Carbon Dioxide, CO_2	Dry Ice	Liquid, gas	0.914	None	Colorless	Odorless	Bulk liquid under pressure
Sulfur Dioxide, SO_2	Sulfurous Acid Anhydride	Liquid, gas	1.436 @ 0°C	None	Colorless	Suffocating, pungent	Cylinders, one-ton units, tank cars

cause "freezing"[3] and shut down (plug) the system. Similarly, if liquid gets into the gas outlet, it will cause a problem by freezing. You must not panic in this situation. *DO NOT* do anything as foolish as adding heat by open flame or electrical heaters to clear a frozen (plugged) gas line. Get help from someone experienced with chlorine cylinders.

Never make repairs to the valve or chlorine container. Just stop the leak, perhaps by tightening the packing on the valve stem or placing the safety device onto the cylinder. Let the chlorine supplier repair the container. Never use a wrench longer than six inches (15 cm) to open the cylinder valve; make only one complete turn of the valve stem in a counterclockwise direction. The one turn will open the valve sufficiently for the maximum discharge. As a safety consideration, cylinders are equipped with fusible metal plugs that are designed to melt at 158 to 168°F (70 to 76°C). This will allow the cylinder contents to discharge and prevent rupturing of the tank. On 100- to 150-pound (45- to 68-kg) cylinders, the plug is located just below the valve seat. The one-ton (909-kg) tanks have six such plugs, three on each end. Should one of these plugs melt, permitting liquid chlorine to discharge, place the tank in a position with the leak at the top of the tank so that it permits the chlorine gas (rather than the liquid) to discharge. This action will reduce the amount of chlorine being discharged because the liquid will change to a gas to escape. In doing so, it will lower the temperature of the container, reducing the discharge rate.

All employees, maintenance personnel, and operators who handle chlorine must have access to an approved chlorine gas mask (Figure 20.2). They must be instructed in the use and maintenance of this equipment. A monthly program should be conducted to familiarize and train each user of the safety equipment. Those employees who are to use the chlorine emergency equipment should practice with this equipment at least once a year while wearing a self-contained breathing apparatus. The emergency kit consists of clamps, gaskets, drift pins, hammers, wrenches, and other tools needed for repairing leaks. The operator may not be able to practice with all of the tools, but inspection and practice gives the operator an opportunity to do maintenance on the emergency equipment.

All operators working with chlorine should be familiar with methods of detecting chlorine leaks. When testing for leaks, use ammonia water on a small cloth or swab on a stick or use an aspirator containing ammonia water. This will form a white cloud of ammonium chloride. Leaks should be repaired immediately. Do not apply the ammonia swab directly to the equipment surface. Also, do not spray ammonia into a room full of chlorine because a white cloud will form and you will not be able to see anything.

Many plants are equipped with chlorine gas detectors. This equipment must be maintained weekly. If not properly maintained, it may not be operable when you need it. Change the electrolyte regularly, test the alarm, and keep the detectors clean and in good repair.

Someone must be assigned the responsibility for maintenance of the self-contained breathing apparatus. That operator must keep records of the maintenance problems and of monthly drills using the gear. The assigned operator should check the masks for leaks, loose eyepieces, faulty tubing, or other worn or defective spots. If inspection indicates any defective parts, they should be discarded or repaired by a properly trained employee. Remember, in high concentrations of chlorine within a confined space where oxygen can be displaced, *DO NOT USE THE CANISTER TYPE OF MASK. NO CANISTER CAN PROVIDE OXYGEN.* Therefore, use only a self-contained breathing apparatus or a hose-type mask supplied with air (Figure 20.3).

If you are caught in an area containing chlorine, do not panic, but leave immediately. Do not breathe or cough, and keep your head high until you are out of the affected area.

The first safety measure you can take when entering a chlorination room is to make sure the ventilating system is working. The ventilating system for the chlorination room should be working all the time. Doors of chlorination rooms should have panic bars as door openers so that in an emergency you will not have to search for the door opener. All safety equipment should be located outside of the chlorination room, but close enough so you can find the equipment, when needed.

First aid for eyes exposed to liquid chlorine is immediate irrigation with flowing water for at least 30 minutes. Medical attention is essential. If the eyes are exposed to chlorine gas, immediately irrigate with flowing water for a period of 15 minutes. Get medical aid. If skin is exposed to liquid chlorine, it will most likely cause burns. The skin should be washed with flowing water for 30 minutes. If the skin is burned, get medical attention. Chlorine gas can become trapped in the clothing and react with body moisture to form hydrochloric acid, which could burn the skin. Remove the clothing of the victim and wash the body down with water. In case of inhalation, remove the victim to fresh air, administer oxygen, if available, call a physician, or transport the injured person to a medical facility. Ingestion is not a problem because chlorine is a gas at room temperature.

Each operator should have a copy of the Chlorine Institute's *CHLORINE MANUAL*.[4] You should read this manual and review it at least once every year. The manual gives data concerning chlorine as an element and gives you suggestions for safely handling this hazardous liquid or gas.

---

[3] Liquid chlorine becomes a solid around −103 to −105°C. The liquid can plug a chlorine gas line, which operators refer to as a "frozen" line.

[4] *CHLORINE MANUAL*, Sixth Edition. Obtain from the Chlorine Institute, Inc., Bookstore, PO Box 1020, Sewickley, PA 15143-1020. Pamphlet 1. Price to members, $28.00; nonmembers, $70.00; plus $6.95 shipping and handling.

SCOTT PRESUR-PAK IIa

SPECIFICATIONS	BACK-PAK STYLE
AIR SUPPLY	
Rated Duration at moderate exertion (MESA/NIOSH test procedure)*	30 min.
Cylinder Capacity at 2216 psi	45 cu. ft.
USE FACTORS	
Weight, as worn, fully charged (approx.)	32 lbs.
Donning Speed (trained personnel)	under 30 secs.
Facepiece (Scottoramic w/nose cup)	Wide vision, anti-fogging
Cylinder and Valve Connection	Straight thread, gasket seal
Cylinder Change	Hand disconnect, no tools req'd
Harness Webbing (replaceable without tools or rivets)	Polypropylene
Transport and Storage	Custom Molded High Density Polyethylene Case
SHIPPING WEIGHT	48 lbs.

NOTES: * POSITIVE PRESSURE

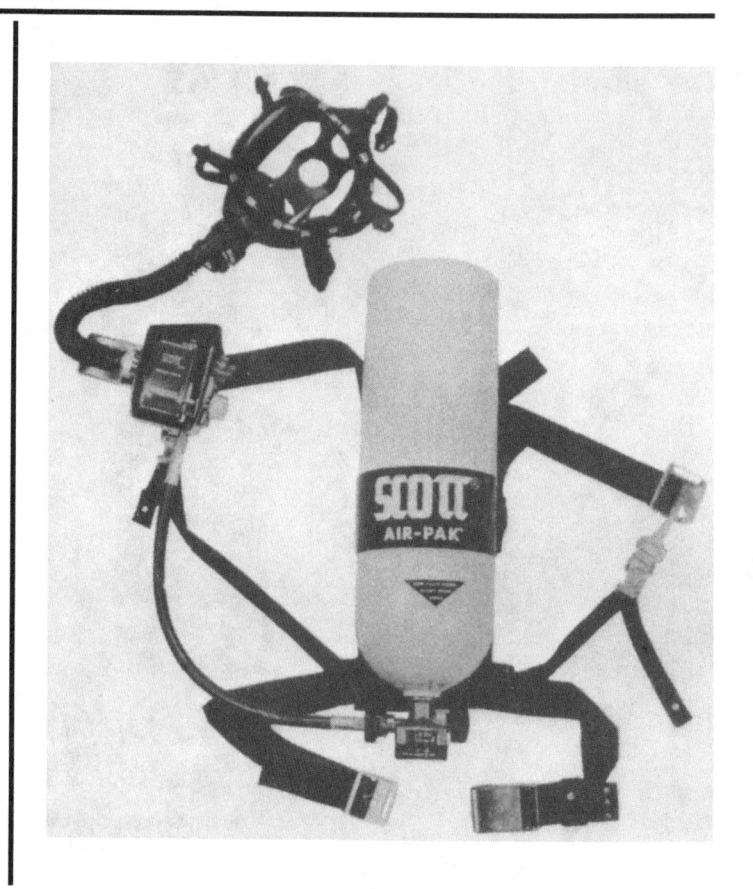

Fig. 20.2 Chlorine gas mask
(Permission of Nevada Safety & Supply)

20.131 Carbon Dioxide (CO_2)

Water treatment plant operators are not often exposed to carbon dioxide because of its limited use, but it is hazardous and can cause suffocation due to the lack of oxygen. Therefore, when using CO_2, keep in mind the carbon dioxide safety considerations. The problem with carbon dioxide is that it is odorless, colorless, and will accumulate at the lowest possible level because it is heavier than air.

Carbon dioxide is obtained in bulk lots, as a liquid (under pressure), which must be vaporized before using. CO_2 is also prepared by generation on site. In either case, good ventilation will reduce the hazards of using CO_2. This will control the accumulating effects of the gas. If you must go into a room filled with CO_2, use a self-contained breathing apparatus, not a canister gas mask. Carbon dioxide displaces oxygen and you may suffocate with the canister type of mask. Exposure to carbon dioxide does not require any protection of the eyes, skin, or other parts of the body, but take precautions when entering rooms, low spots, or manholes that may be filled with carbon dioxide.

First aid involves moving the victim to fresh air, giving resuscitation if the victim has stopped breathing, and getting medical attention.

20.132 Sulfur Dioxide (SO_2)

This gas is about 2.3 times as heavy as air and, therefore, will accumulate in low areas. Sulfur dioxide is colorless in the gaseous form. As a liquid, it is also colorless and, when unconfined, will vaporize rapidly into a gas. The gas is extremely irritating and, like chlorine, will react readily with the respiratory system if inhaled. Sulfur dioxide causes varying degrees of irritation to the mucous membranes—eyes, nose, throat, and lungs. The damage is caused by the formation of sulfurous acid in reaction

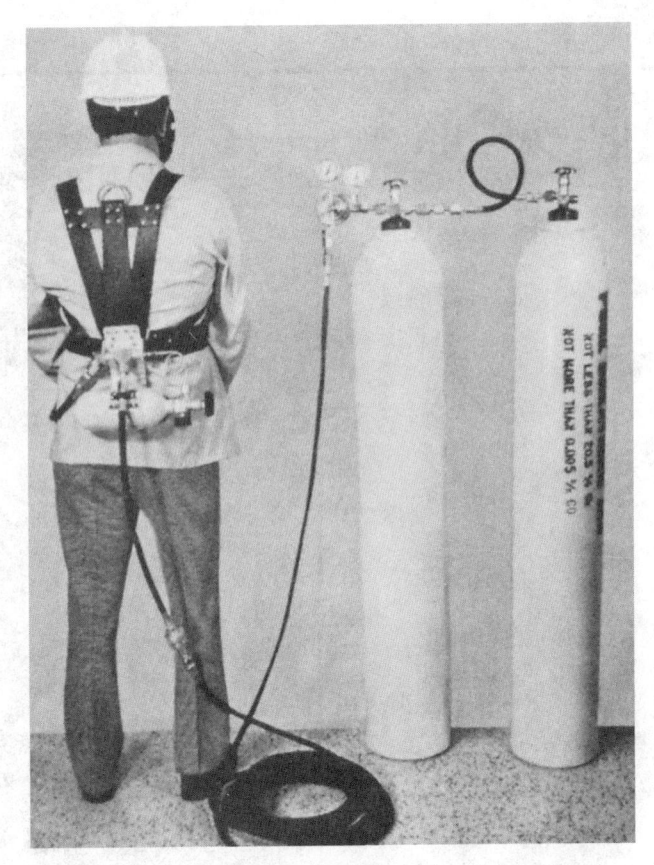

Dual air supply cylinder being used with 900007 series hoseline Air-Pak with Egress.

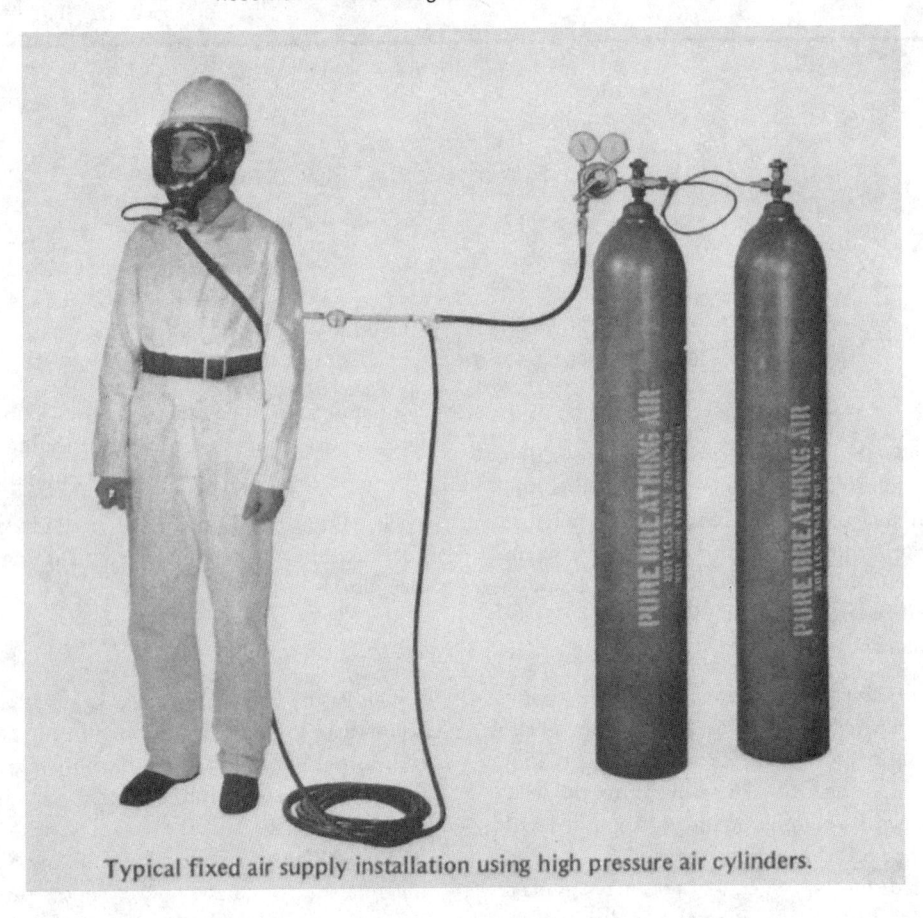

Typical fixed air supply installation using high pressure air cylinders.

Fig. 20.3   Hose-type mask supplied with air
(Permission of Nevada Safety & Supply)

with moisture in these locations. Sulfur dioxide can be readily detected in concentrations of 3 to 5 ppm. In higher concentrations, it is unlikely that you will remain in the area unless you are unconscious or trapped. If the liquid comes into contact with the skin, it may cause local freezing as the liquid evaporates.

Always use a self-contained breathing apparatus around sulfur dioxide and never use the canister type. As with other gases, good ventilation is essential in a room where sulfur dioxide is being used. The fans should be used to dissipate any gas vapors that may occur. There should always be an eye wash fountain close to the work area where sulfur dioxide is used.

First aid for eyes exposed to or splashed with sulfur dioxide is washing immediately with water for at least 15 minutes, then getting medical attention. In case of inhalation, remove victim to fresh air, give resuscitation if needed, consult a physician, or transport the injured person to a medical facility. Sulfur dioxide leaks and injuries should be treated similar to chlorine problems.

QUESTIONS

Write your answers in a notebook and then compare your answers with those on page 501.

20.13A Where are chlorine leaks most often found?

20.13B What is the purpose of the fusible metal plugs on chlorine cylinders?

20.13C How can chlorine leaks be detected?

20.13D Why is carbon dioxide a safety hazard and what safety precautions must be taken when using it?

20.14 Salts

There are many salts (chemicals) used in water treatment. Table 20.4 lists the salts that will be discussed in this section. If you would like additional information about compounds not listed, request a data sheet about specific compounds from the suppliers of the chemicals. Review the chemistry of the compounds in Table 20.4 and become familiar with each chemical's

TABLE 20.4 SALTS USED IN WATER TREATMENT

Name, Formula	Common Name	Available Forms	Density, lbs/cu ft	Flammability	Color	Odor	Containers
Aluminum Sulfate, $Al_2(SO_4)_3 \cdot 14H_2O$	Alum, Filter Alum	Liquids, powder, lump	1.69 (S.G.) 38–67	None	Ivory	N/A[a]	Bags, tank truck, bulk
Ferric Chloride, $FeCl_3$	Ferrichlor, Chloride of Iron	Syrup, liquid, lump	60–90	None	Dark brown, yellow-brown	N/A	Carboys, tank cars
Ferric Sulfate, $Fe_2(SO_4)_3$	Ferrifloc, Ferrisul	Powder, granule	70–72	None	Red-brown	N/A	Bags, drums
Ferrous Sulfate, $FeSO_4 \cdot 7H_2O$	Coppras, Green Vitriol	Crystal, liquid, granule, lump	63–66	None	Green	N/A	Bags, drums, bulk
Sodium Aluminate, $Na_2O \cdot Al_2O_3$	Soda Alum	Dry crystal, liquid	(27° Bé[b])	None	White, green-yellow	N/A	Bags, bulk
Fluoride Compounds, NaF and H_2SiF_6	—	Liquid and powder	50–75	None	Blue, white	Dust	Bags, carboys, tank trucks
Sodium Hexametaphosphate, $(NaPO_3)_6$	Calgon, Glassy Phosphate	Crystal, flake	47	None	White	N/A	Bags, drums
Copper Sulfate, $CuSO_4$	Blue Vitriol, Blue Stone	Crystal, lump, powder	60–90	None	Blue	None	Bags, drums
Sodium Chlorite, NaOCl	Technical Sodium Chlorite	Powder, flake, liquid	70 dry	Oxidizer	Light orange	None	Tank truck, 100-lb drums
Potassium Permanganate, $KMnO_4$	Permanganate	Crystal, powder	90–100	Oxidizer	Purple	None	Drums, bulk

[a] N/A = Not Applicable.
[b] Bé = Baumé scale, a scale for measuring specific gravity of liquids.

characteristics. You should become well trained in how to handle each chemical and know and observe the appropriate safety precautions. For most of these salts, ventilation, respiratory protection, and eye protection will prove adequate. Other problems may involve chemical solutions and dust. The solution may attack skin and clothing. The dust may attack the respiratory system or cause an explosion. Even though these chemicals do not normally react violently, use the following procedures when handling such salts to reduce the hazards and to provide a safe working location.

When handling, storing, or preparing solutions of chemicals, treat them all as being hazardous. All chemicals require careful consideration. They may be sources of an explosion, violent reaction, loss of eyesight, burns, and illness.

Do not store acidic or basic compounds with salts. Keep these chemicals in a clean, dry area. When handling dry, bulk materials, store in a fire-safe area. Keep all lids on containers and follow the instructions on the container. Make sure that the operator who is mixing or dispensing these chemicals is well trained and wears proper clothing to meet all safety requirements, such as chemical goggles, face shield, rubber gloves, rubber boots, rubber apron, and chemical respirator. When working with chemical salts, be aware that fumes, gases, vapors, dusts, or mists may be given off and this represents a hazard to the safety of the operator. *PROTECT YOURSELF!*

20.140 Aluminum Sulfate (Alum) ($Al_2(SO_4)_3 \cdot 14H_2O$)

There are two forms of alum: dry and liquid. Both have to be handled with care. Dry alum is available in lump, ground, or powdered forms and should be stored in a dry location because moisture can cause caking. Liquid alum is acidic and very corrosive. Store liquid alum in corrosion-resistant storage tanks such as:

1. Steel, wood (Douglas Fir), or concrete-lined, all lined with a material corrosion-resistant to alum
2. Steel, lined with 3/16-inch (5-mm) soft rubber
3. Stainless steel
4. Steel, lined with plastic if temperature remains below 150°F (65°C)
5. Glass-reinforced epoxy or polyester plastic

When working with dry alum, use respiratory protection and ensure adequate ventilation of the work area. There should be a good mechanical dust collection system to minimize any dust accumulation. Exposure to alum dust concentrations greater than 15 milligrams per cubic meter of air for more than an 8-hour period is dangerous. Avoid skin exposure to this chemical by using long-sleeved, loose-fitting, dust-proof clothing.

Never use the same conveyor for quicklime and alum. This mixture may explode under proper conditions.

Liquid alum is an acidic solution and should be handled as you would handle a weak acid. Reduce exposures to the skin and eyes. Avoid ingestion. Although the chemical will not cause any lasting internal damage, it will cause discomfort. Use good ventilation for removing any mists. Rubber gloves and protective clothing are recommended.

As a general precaution, avoid prolonged exposure to dry or liquid forms of alum. If used dry, a dust mask and goggles are desirable for the comfort of the operator. Alum dust can be extremely irritating to the eyes. When handling the liquid, normal precautions should be used to prevent splashing of the compound onto the operator, particularly if the liquid is hot. Wear a face shield to protect your eyes and a rubber apron to protect clothing.

First aid for liquid or dry alum is immediate flushing of the eyes for 15 minutes with large amounts of water. Alum should also be washed off the skin with water because prolonged contact will cause irritation.

20.141 Ferric Chloride ($FeCl_3$)

This is a very corrosive compound and should be treated as you would treat any acid. The salt is highly soluble in water, but in the presence of moist air or light, it decomposes to give off hydrochloric acid, which may cause other safety hazards. Avoid prolonged exposure to this liquid (there is a dry form but it is not often used). When handling liquid ferric chloride, normal precautions should be taken to prevent splashing, particularly if the liquid is hot. Use a face shield to protect your eyes and rubber aprons to protect clothing. This compound will not only attack the clothing, but also stain it. First aid for eyes exposed to the liquid is that the eyes must be flushed out immediately for 15 minutes with large amounts of water. Ferric chloride should also be washed off the skin with water because prolonged contact will cause irritation and staining of the skin.

20.142 Ferric Sulfate ($Fe_2(SO_4)_3$)

This compound produces an acidic solution when mixed with water. Because of its acidic nature, operators using this compound should be provided with protection suitable for dry or liquid alum. The hazards associated with the use of dry ferric sulfate are those usually connected with an acid. Use protective clothing, neck cloths, gloves, goggles or face shield, and a respirator. Avoid prolonged exposure to the dry form because of its acidic reaction with moisture on the skin, eyes, and throat. The normal precautions should be used including a dust mask and protective clothing. First aid for exposure to the eyes requires the eyes to be flushed immediately with lots of water. The skin should also be flushed with large amounts of water. Prolonged contact may cause irritation.

20.143 Ferrous Sulfate ($FeSO_4 \cdot 7H_2O$)

This chemical may be obtained in liquid or dry form. The safety hazards are some of those for dry or liquid forms of alum. The operator should be provided with adequate ventilation and respiratory protection. The material should be stored in a clean, dry location. Mechanical dust collecting equipment must be used to minimize the dust. Wear chemical goggles or a face shield, loose-fitting, long-sleeved clothing, and make an effort to minimize all skin exposure.

First aid for ferrous sulfate in the eyes is to flush out immediately with large amounts of water for 15 minutes. The chemical should be washed off the skin to reduce irritations.

20.144 Sodium Aluminate (Na$_2$O•Al$_2$O$_3$)

Sodium aluminate dissolved in water produces a noncorrosive solution. In the dry form, its powder consistency raises the usual dust problems. There are few hazards with this compound, but as with other chemicals, you should use precautions when handling it. Use respiratory protection when handling the dry compound to prevent the inhalation of dust. First aid for eyes that are exposed is to flush with water; keep the skin clean with water.

20.145 Fluoride Compounds (NaF and H$_3$SiF$_6$)

All fluoride compounds should be treated with care when you are handling them because of their long-term accumulative effects. Provide good ventilation; always wear respiratory protection; and be careful not to expose any open cuts, lesions (wounds), or sores to fluoride compounds. Clean up any spills promptly and wash immediately after handling such compounds.

When handling acid compounds of fluoride, always wear a face shield or chemical goggles, rubber gloves, rubber apron, and rubber boots. Your wearing apparel should always be washed after working around fluoride, and the respirator should be kept clean and sanitary. Keep the acid feeder for fluoride in good repair. Use plastic guards to prevent acid spray from glands or other parts of the chemical feeder. This prevents attack upon the equipment and protects operators. All fluoride compounds must be regarded as hazardous chemicals that are toxic to operators. Every means possible must be taken to prevent exposure to these compounds by use of a respirator and protective clothing.

Once it is established that fluoride is the cause of the poisoning, first aid should be started while waiting for medical help. The following are recommended first-aid procedures:

1. Move the person away from any contact with fluoride and keep warm.

2. Give the person three teaspoons (one tablespoon) of table salt in a glass of warm water.

3. If the person is conscious, and has swallowed sodium fluoride, induce vomiting by rubbing the back of the throat with a spoon or your finger; if available, use syrup of ipecac. *DO NOT* induce vomiting if hydrofluosilicic acid is swallowed. Vomiting may cause the acid to go the wrong way into the lungs and cause more serious problems.

4. Give the person a glass of milk.

5. Repeat the salt and vomiting several times.

6. Take the person to the hospital as soon as possible.

First aid for a person with a nosebleed from inhaling a high concentration of fluoride is:

1. Take the person away from the source of the fluoride.

2. Tip the person's head back while placing cotton, cloth, or tissues inside the nostrils (change these often).

3. Take the person to a doctor if you cannot stop the bleeding.

QUESTIONS

Write your answers in a notebook and then compare your answers with those on page 501.

20.14A   What kind of protection does an operator need when handling salts?

20.14B   What is the recommended first aid when either liquid or dry alum comes in contact with your skin or your eyes?

20.14C   What happens when ferric chloride is exposed to moist air or light?

20.15 Powders

20.150 Potassium Permanganate (KMnO$_4$)

Under normal conditions in a water treatment plant, potassium permanganate is considered to be a safe chemical (see Table 20.4, page 471). However, potassium permanganate is a strong oxidizing agent and will react with certain easily oxidizable substances. Keep potassium permanganate away from the possibility of reacting with sulfuric acid, hydrogen peroxide, metallic powders, elemental sulfur, phosphorus, carbon, hydrochloric acid, hydrazine, hydroxylamine, and metal hydrides. When in contact with potassium permanganate, the following compounds may ignite: ethylene glycol (antifreeze), glycerine, sawdust compounds, propylene glycol, and sulfuric oxide.

Potassium permanganate is available as pellets, as a powder, or in crystal form. This chemical can be kept indefinitely if stored in a cool, dry area in closed containers. The drums should be protected from damage that could cause leaks or spills. Potassium permanganate should be stored in fire-resistant buildings having concrete floors instead of wooden floors. The chemical must not be exposed to intense heat, or stored next to heated pipes. Organic solvents, such as greases and oils, should be kept away from stored potassium permanganate.

Potassium permanganate spills should be swept up and removed immediately. Flushing with water is an effective way to eliminate any residue remaining on floors. Potassium permanganate fires should be extinguished with water.

To avoid inhalation of potassium permanganate dust, use an approved mask that is an air-purifying, half-mask respirator with an outblower. Wear safety glasses or a full face shield to protect your eyes. Protective clothing that should be worn includes rubber or plastic gloves and apron, and a long-sleeved shirt for handling both dry and dissolved potassium permanganate.

Mild exposure will cause sneezing and mild irritation of the mucous membranes. Prolonged inhalation of potassium permanganate should be avoided. If potassium permanganate gets on your skin, flood the contacted skin with water. If it gets in your eyes, flush with plenty of water and call a physician immediately.

20.151 Powdered Activated Carbon

Powdered activated carbon is the most dangerous powder that you will be exposed to as a treatment plant operator. If you understand how to handle activated carbon properly, other dust problems or powdered chemicals will not be very difficult for you to handle.

There are two problems when handling activated carbon. One is dust and the second is fire. The two may or may not be related. The dust causes uncomfortable working conditions; fire causes damage to equipment and a hazard to personnel. If the two problems are treated together, it will reduce the hazards to operators. Left unattended they may cause loss of life and property. If you will use the following safety precautions, you can minimize the hazards of handling activated carbon and aid the other operators in handling other powders.

Store activated carbon in a clean, dry, fireproof location. Keep the area free of dust, protect it from flammable materials, and do not permit smoking in the area at any time when handling or unloading activated carbon. Install carbon dioxide fire extinguishers. Store bagged carbon in single rows. Keep access aisles clear to prevent damage to the bags and thus reduce the dust and fire potential.

Electrical equipment in and around activated carbon storage should be explosion-proof and protected from the carbon dust. Keep the equipment clean and dry. Wet or damp carbon is a good conductor of electrical current and can cause short-circuit fires. Heat can also build up from the motors if covered with carbon dust, causing fires. The key to preventing fires with activated carbon is keeping the storage area clean and dust free.

Next to electrical fires, activated carbon gives the operator the most difficult fire to control. The carbon gives off an intense heat; it burns without smoke or visible flame. The fires are difficult to locate and are very hard to control. They cannot readily be detected in a large storage bin or in large stacks of bags.

Bags of activated carbon should be stored in single rows.

You will detect the indications of the fire, such as the smell of charred paper, burned paint, or other odor, before seeing any evidence of flames.

Do not douse a carbon fire with a stream of water. The water may cause burning carbon particles to fly, resulting in a greater fire problem. The carbon fire should be controlled with carbon dioxide (CO_2) extinguishers or hoses equipped with fog nozzles. However, when using CO_2, be aware that there is a potential of carbon monoxide formation and take the precaution of using a self-contained breathing apparatus.

Activated carbon supports fire without atmospheric oxygen because it may have adsorbed sufficient oxygen for combustion. The best means of controlling a carbon fire is to reduce its temperature below the ignition point. This can be done by applying cold water with fog or spray nozzles and soaking the burning carbon, but do not hit the carbon with a stream of water. For a small fire involving just a few bags, move the bags to a safe location and use CO_2 or spray nozzles to extinguish the fire. Blocks of dry ice may help control fires in storage bins or other confined areas, but do not expect this method to be very effective.

A final word about fire and explosions involving activated carbon. Tests performed by carbon manufacturers have not shown that dust mixtures of carbon have explosive tendencies. Activated carbon is a charcoal and performs in a like manner; the carbon burns without smoke or visible flame, burns very hot, and will spread if doused with a large stream of water.

There are no specific first-aid methods for carbon exposure because the carbon will not attack the human body. Carbon does sometimes cause problems with the nasal passages, however, and may be difficult to wash off your hands and body. Therefore, methods here are those of prevention. Provide good dust collection at the point where carbon is being unloaded, in storage bins for liquid preparation, and in dry storage bins. Wear an approved dust mask, loose-fitting and dust-proof clothing. If there is an excess of dust, you should use chemical goggles, close your shirt collar, and tape your trousers to cover your ankles. You will need adequate shower facilities and should use mild liquid soap. Most, if not all, hard soap bars are ineffective in removing activated carbon dust from pores of the human body.

20.152 Other Powders

You may come in contact with other powdered compounds in the water treatment plant, but they do not present the problem in handling that activated carbon does. Bentonite creates no significant hazard other than dust, and this can be controlled by using dust collection systems.

Similarly, calcium carbonate presents a slight dust hazard that can be controlled by a dust collection system. Also, when handling calcium carbonate in bags, use the same methods of control that you would use with bags of carbon.

Of the many organic coagulant aids used in water treatment, only a few are applied in powder form. Most of these compounds are used in the liquid form, which reduces the danger to operators. The dry compounds present a slight hazard in dust irritation to the nasal passages; this can be prevented by use of approved dust masks. The liquid compounds can and will attack the skin, but can be treated by washing with ample amounts of water or soap and water. Organic coagulant aids (polymers) are extremely slippery when wet. Floors and walkways should be clean and dry to prevent slips and falls. Spilled polymers may be cleaned up using inert, absorbent materials, such as sand or earth.

As new compounds are introduced to the waterworks field, ask for training in their use. Such a training program should provide information about detailed safety precautions, the toxicity of the compound, and the appropriate first-aid methods. Supervisors have the responsibility to provide this type of training, either in conjunction with the supplier or sponsored entirely by the utility. The training must be reinforced periodically for those compounds that are not often used by the operator. A review and updating of information will go a long way in preventing accidents.

QUESTIONS

Write your answers in a notebook and then compare your answers with those on page 501.

20.15A   How can potassium permanganate spills be cleaned up?

20.15B   What is the most dangerous powder the water treatment plant operator will be exposed to?

20.15C   How should activated carbon be stored?

20.15D   How can fires caused by activated carbon be prevented?

20.15E   How should an activated carbon fire be extinguished?

20.16 Labeling of Chemical Containers

For information on proper labeling of chemical containers, see Section 4.12, "Hazard Communication (Worker Right-To-Know Laws)," in OPERATION AND MAINTENANCE OF WASTEWATER COLLECTION SYSTEMS, Volume I, in this series of operator training manuals.

20.17 Chemical Storage Drains

Safety regulations prohibit the use of a common drain and sump for acid and alkali chemicals, oxidizing chemicals, and organic chemicals because of the possibility of the release of toxic gases, explosions, and fires. If both an acid and an alkali chemical come in contact, an explosion could occur. If an organic chemical, such as a polymer solution, comes in contact with an oxidizing chemical, such as potassium permanganate, a fire could develop. Be sure that if a leak develops from any chemical container or storage facility, the chemical will not be able to reach, mix, or react with another chemical.

QUESTIONS

Write your answers in a notebook and then compare your answers with those on page 501.

20.17A   Why should drains from chemical storage areas not use common drains and sumps?

20.17B   What could happen if a leak from a polymer storage container comes in contact with potassium permanganate?

END OF LESSON 1 OF 3 LESSONS

on

SAFETY

Please answer the discussion and review questions next.

DISCUSSION AND REVIEW QUESTIONS

Chapter 20. SAFETY

(Lesson 1 of 3 Lessons)

At the end of each lesson in this chapter you will find some discussion and review questions. The purpose of these questions is to indicate to you how well you understand the material in the lesson. Write the answers to these questions in your notebook.

1. What is safety?
2. Who is responsible for safety?
3. What are the causes of accidents?
4. What does an operator need to know about chemicals used in a water treatment plant?
5. How can hydrochloric acid be handled safely?
6. How should hydrochloric acid be stored?
7. How can ammonia leaks be detected?
8. What is the first-aid treatment for lime burns?
9. What special precautions should be taken when handling and storing caustic soda?
10. How should hypochlorite be handled and stored?
11. What first aid is required for a person overcome by carbon dioxide?
12. What safety hazards may be caused by salt dust?
13. What safety precautions should an operator take when handling alum?
14. What are the two major problems encountered when handling activated carbon?
15. How can an operator detect an activated carbon fire?

CHAPTER 20. SAFETY

(Lesson 2 of 3 Lessons)

20.2 FIRE PROTECTION

20.20 Fire Prevention

Fire prevention is the best fire protection the plant operator can afford. Fire protection is just good housekeeping. The word "housekeeping" best describes the action any water plant operator can take to protect from or prevent fires. This means a well-kept, neat, and orderly plant represents a good fire safety policy.

Fire hazards can be easily removed. The prompt disposal of cartons, crates, and other packing materials, a system of wastepaper collection, and the removal of other debris can greatly reduce fire hazards. Provide suitable containers for used wiping cloths and have fire extinguishers conspicuously located in hallways, near work areas, and near potential fire problem areas. All of these housekeeping activities are low-cost measures that also improve the appearance of the plant and create a better work environment.

You can call upon the local fire department for advice on fire prevention in and around the treatment plant. You may also ask the utility insurance underwriter for cooperation in your fire prevention program. All operators should be trained in the proper use and maintenance of fire control equipment. These simple steps can reduce fire losses to a minimum and prevent most fires from happening at very low cost to the utility.

You should make a fire analysis of your plant once a year to determine what new measures should be taken to prevent fires. As activity changes occur, there may be a need to change the location of hoses and extinguishers or it may be necessary to add fire control equipment. Fire and police departments' telephone numbers must be posted in a conspicuous location along with escape routes. Post emergency numbers near all telephones throughout the plant. In hazardous locations, the means of exit should be lighted and all doors should be equipped with panic bars. As indicated above, your best fire protection or prevention is good housekeeping.

20.21 Classification of Fires and Extinguishers

Fire classifications are important for determining the type of fire extinguisher needed to control the fire. Classifications also aid in recordkeeping. Fires are classified as A, B, C, or D fires based on the type of material being consumed: A, ordinary combustibles; B, flammable liquids and vapors; C, energized electrical equipment; and D, combustible metals. Fire extinguishers are also classified as A, B, C, or D to correspond with the class of fire each will extinguish.

Class A fires: ordinary combustibles such as wood, paper, cloth, rubber, many plastics, dried grass, hay, and stubble. Use foam, water, soda-acid, carbon dioxide gas, or almost any type of extinguisher.

Class B fires: flammable and combustible liquids such as gasoline, oil, grease, tar, oil-based paint, lacquer, and solvents, and also flammable gases. Use foam, carbon dioxide, or dry chemical extinguishers.

Class C fires: energized electrical equipment, such as starters, breakers, and motors. Use carbon dioxide or dry chemical extinguishers to smother the fire; both types are nonconductors of electricity.

Class D fires: combustible metals such as magnesium, sodium, zinc, and potassium. Operators rarely encounter this type of fire. Use a Class D extinguisher or use fine, dry soda ash, sand, or graphite to smother the fire. Consult with your local fire department about the best methods to use for specific hazards that exist at your facility.

Multipurpose extinguishers are also available, such as a Class BC carbon dioxide extinguisher that can be used to smother Class B and Class C fires. A multipurpose ABC carbon dioxide extinguisher will handle most laboratory fire situations. (When using carbon dioxide extinguishers, remember that the carbon dioxide can displace oxygen—take appropriate precautions.)

There is no single type of fire extinguisher that is effective for all fires so it is important that you understand the class of fire you are trying to control. You must be trained in the use of the different types of extinguishers, and the proper type should be located near the area where that class of fire may occur.

20.22 Fire Extinguisher Operation and Maintenance

There are many types of hand-held fire extinguishers. Each type of extinguisher has different operation and maintenance requirements.

Water Extinguishers

There are four types of water extinguishers: stored pressure, cartridge operated, water pump tank, and soda-acid. All of these perform well in Class A fires, but they do require maintenance. A preventive maintenance schedule on all water extinguishers should include a monthly check by the operator responsible for the maintenance and completion of appropriate maintenance records. Some agencies make the safety officer responsible for ensuring that an operator checks the fire extinguishers and records each inspection.

1. The method of operation for a stored pressure extinguisher is simply to squeeze the handle or turn a valve. The maintenance is also simple: check air pressure and recharge the extinguisher, as needed.

2. For the cartridge type, the maintenance consists of weighing the gas cartridge and adding water, as required. To operate, turn upside down and pump.

3. To use the water pump tank type of extinguisher, simply operate the pump handle. For maintenance, one has only to discharge the contents and refill with water annually or as needed.

4. The soda-acid type must be turned upside down to operate; it also requires annual recharging.

Foam Extinguishers

The foam type of extinguishers will control Class A and Class B fires well. They, like soda-acid types, operate by turning upside down, and require annual recharging.

The foam and water type extinguishers should not be used for fires involving electrical equipment. However, they can be used in controlling flammable liquids such as gasoline, oil, paints, grease, and other Class B fires.

Carbon Dioxide Extinguishers

The carbon dioxide (CO_2) extinguishers are common (Figures 20.4 and 20.5). They are easy to operate, just pull the pin and squeeze the lever. For maintenance, they must be weighed at least semiannually. Many of these extinguishers will discharge with age. They can be used on a Class C (electrical) fire. All electric circuits should be killed, if possible, before trying to control this type of fire. A carbon dioxide extinguisher is also satisfactory for Class B fires, such as gasoline, oil, and paint, and may be used on surface fires of the Class A type.

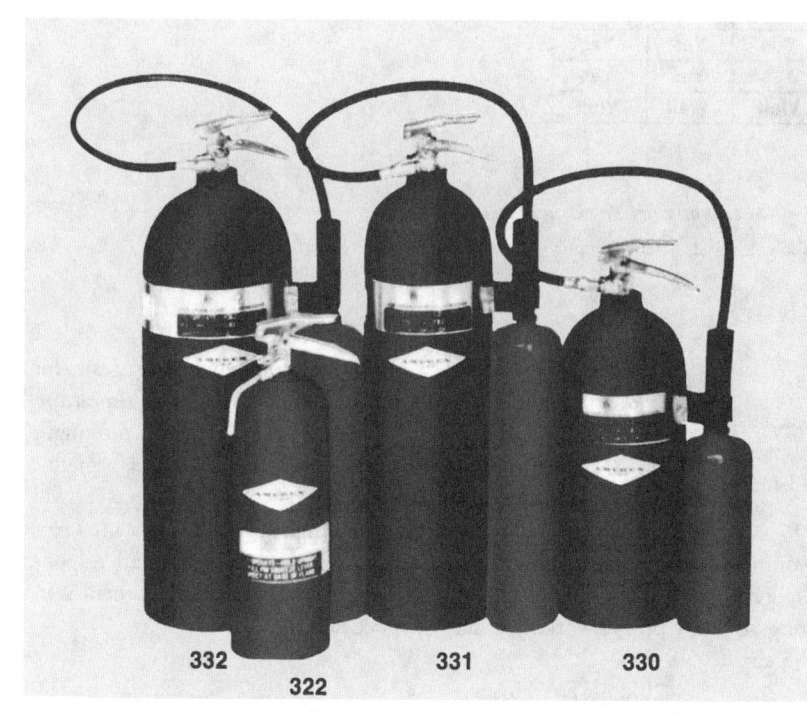

SPECIFICATIONS	CARBON DIOXIDE			
Size/Type	5 Horn	10 Hose	15 Hose	20 Hose
Model Number	322	330	331	332
U/L Rating	5B:C	10B:C	10B:C	10B:C
Capacity (lbs.)	5	10	15	20
Shipping Wt. (lbs.)	15	29½	39½	51½
Height	17¾"	24"	30"	30"
Width	8¼"	12"	12"	13"
Depth (Diam.)	5¼"	7"	7"	8"
Range (Ft.)	3-8	3-8	3-8	3-8
Discharge Time—Seconds	10	10	12.5	19
Coast Guard App.	Yes	Yes	Yes	Yes
FM Approved	Yes	Yes	Yes	Yes
Bracket	Wall	Wall	Wall	Wall

Fig. 20.4 Carbon dioxide extinguishers
(Permission of Nevada Safety & Supply)

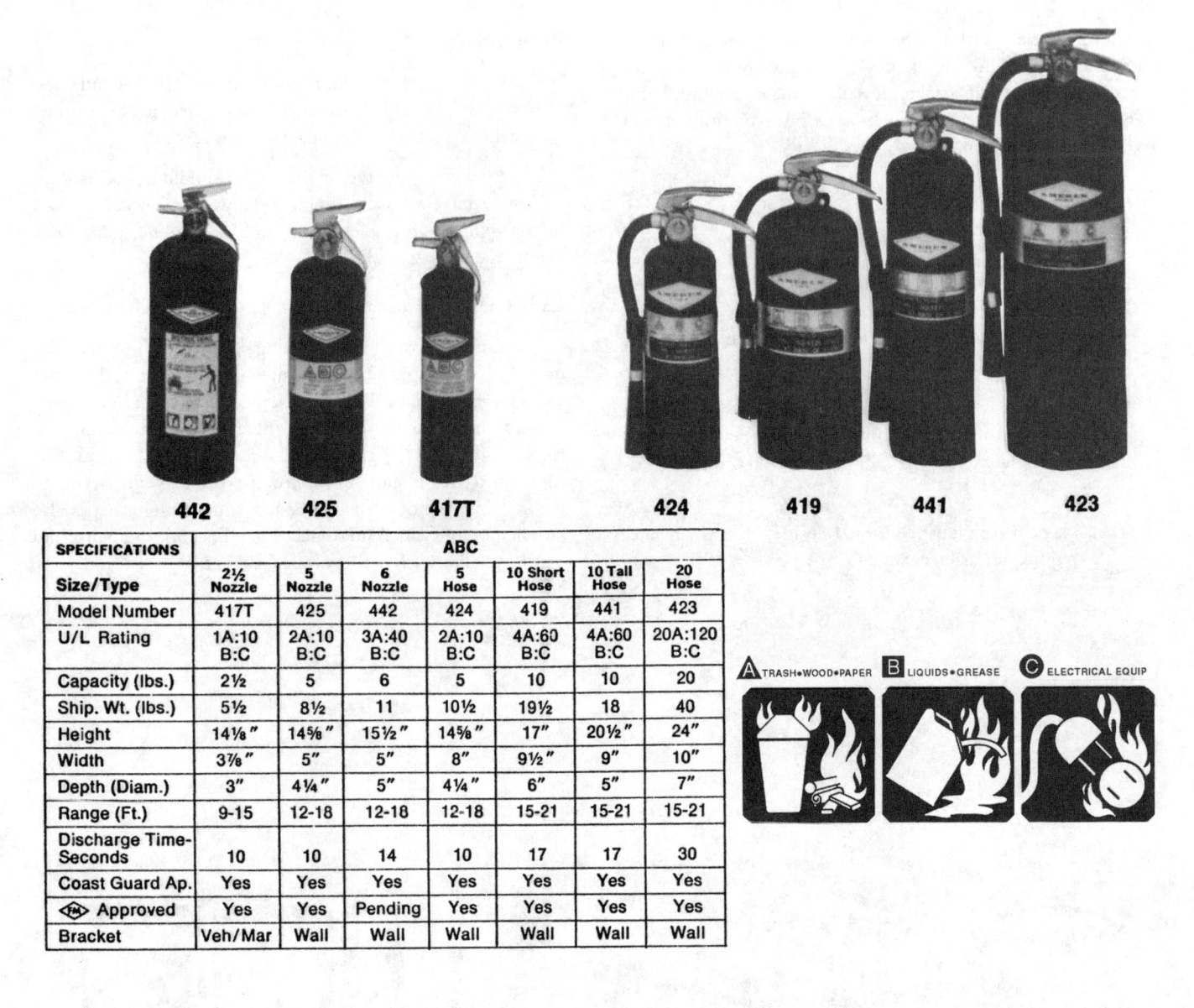

478 Water Treatment

SPECIFICATIONS	ABC						
Size/Type	2½ Nozzle	5 Nozzle	6 Nozzle	5 Hose	10 Short Hose	10 Tall Hose	20 Hose
Model Number	417T	425	442	424	419	441	423
U/L Rating	1A:10 B:C	2A:10 B:C	3A:40 B:C	2A:10 B:C	4A:60 B:C	4A:60 B:C	20A:120 B:C
Capacity (lbs.)	2½	5	6	5	10	10	20
Ship. Wt. (lbs.)	5½	8½	11	10½	19½	18	40
Height	14⅛"	14⅝"	15½"	14⅝"	17"	20½"	24"
Width	3⅞"	5"	5"	8"	9½"	9"	10"
Depth (Diam.)	3"	4¼"	5"	4¼"	6"	5"	7"
Range (Ft.)	9-15	12-18	12-18	12-18	15-21	15-21	15-21
Discharge Time-Seconds	10	10	14	10	17	17	30
Coast Guard Ap.	Yes	Yes	Yes	Yes	Yes	Yes	Yes
Approved	Yes	Yes	Pending	Yes	Yes	Yes	Yes
Bracket	Veh/Mar	Wall	Wall	Wall	Wall	Wall	Wall

A TRASH•WOOD•PAPER   B LIQUIDS•GREASE   C ELECTRICAL EQUIP

Fig. 20.5 Typical carbon dioxide extinguishers
(Permission of Nevada Safety & Supply)

Dry Chemical Extinguishers

There are two types of dry chemical extinguishers. These extinguishers are either: (1) cartridge operated, or (2) stored pressure. These are recommended for Class B and C fires and may work on small surface Class A fires.

1. The cartridge-operated extinguishers only require you to rupture the cartridge, usually by squeezing the lever. The maintenance is a bit more difficult, requiring weighing of the gas cartridge and checking the condition of the dry chemical.

2. For the stored-pressure extinguishers, the operation is the same as the CO_2 extinguisher. Just pull the pin and squeeze the lever. The maintenance requires a check of the pressure gauges and condition of the dry chemical.

As suggested above, a preventive maintenance program for fire extinguishers requires a considerable amount of time from the operator and requires a system of recordkeeping. You might consider hiring a local fire prevention agency to perform this part of your maintenance program. These service agencies will check and maintain the plant's firefighting equipment on a regular basis. This does not relieve the operator of ultimate responsibility for the equipment, but ensures that the equipment is in proper working order when needed.

20.23 Fire Hoses

Fire hoses are usually stationed throughout the treatment and pumping plants. These are the type of firefighting equipment that an operator may see every day, but never give due consideration as to their maintenance. Without proper maintenance, the hoses may develop dry rot and be untrustworthy at the time they are needed. Under some conditions, you may be tempted to use these hoses for cleaning settling basins or filters. The fire hoses should only be used for fighting fires and, after their use, they must be cleaned and stored properly. The hose should be tested periodically and replaced as required, or at regular time intervals. Check with the local fire department for recommendations.

20.24 Storage of Flammables

The storage of flammable material should be isolated, if possible, from other plant structures. Ideally, these storage areas should have explosion-proof lighting. The floor should be grounded and the operator should only use spark-proof tools when working near or handling flammable materials. The room should have an alarm system, be equipped with automatic extinguishers, and have supplementary equipment located outside of the room. In and around the storage area, smoking or welding must be prohibited. The flammable storage areas must be clearly marked with distinctive signs and all entrances should be lighted.

More often than not, however, you will be compelled to use rooms within the plant for storage of flammable material. Here you must make the room fireproof, equip the room with a fire door, automatic extinguishers, and alarms. Keep passageways free from obstructions. Station firefighting equipment at a suitable location, readily accessible and with plainly labeled operating instructions. The room must be equipped with explosion-proof lights, grounded floor, no-smoking signs, and other distinctive signs indicating that the room is a flammable storage area.

20.25 Exits

Access and exit are very important in plant safety. Therefore, all exit signs should be distinctly marked and well lighted. All doors should open outward and, in hazardous areas, there should be panic bars on the doors. To provide positive protection around the filter and sedimentation basins, install handrails or other enclosures for the protection of operating personnel as well as visitors.

In high-fire-hazard occupied areas, there should be at least two means of emergency exit located, if possible, at opposite ends of the room or building. These would include areas containing woodworking and paint-spraying residues that burn rapidly or give off poisonous fumes.

QUESTIONS

Write your answers in a notebook and then compare your answers with those on page 501.

20.2A   Class A fires involve what types of materials?

20.2B   What kinds of fires can be controlled by a foam type of extinguisher?

20.2C   How can an electrical fire be extinguished?

20.3 PLANT MAINTENANCE

20.30 Maintenance Hazards

Plant maintenance, housekeeping, cleaning up, or whatever you wish to call it, is a very important function of the treatment plant and essential for plant equipment. This function requires the use of cleaning materials and hand tools. Maintenance may require you to go into a manhole, repair electric motors, lift boxes, and use power tools. All of these functions may in some way be hazardous and may cause injury, fire, disease, or even death unless proper safety precautions are taken.

20.31 Cleaning

Any effort spent keeping the entire plant clean and sanitary will provide a much nicer place for you to work and will also make visitors feel as if the water being produced is safe. Even if you can just keep all working areas free of tripping hazards, this will add greatly to the safety in the plant.

Cleaning duties should be performed at such times of day or night as to cause a minimum of exposure to other operators. For example, floors become slippery when wet so give some consideration to the time of day and the type of wax to be used. When cleaning floors, there are problems of exposure to others of cleaning equipment, mops, mop and broom handles, other tools, cleaning compounds and, most of all, wet floors. When cleaning, try to keep others out of the area. Warn others about newly waxed floors. Use wax compounds containing nonslip ingredients. Try to do such cleaning and waxing on weekends or at night when foot traffic through the plant is minimal.

As part of your maintenance program, provide trash containers for collecting wastepaper and for separating used, oily rags. Dispose of garbage and flammable refuse on a routine, frequent basis. Hazardous waste, acids, and caustics should be cleaned up immediately. These steps will add to the safety in the plant and to the safety of operators in the plant.

Keeping aisles, doorways, stairs, and work areas free of refuse reduces hazards of tripping and other injuries, as well as reducing the possibility of fires.

Cleaning windows is a hazardous occupation, but the operator who gives due consideration to the task can perform it safely. If windows are high, time of day is important. Cleaning tools may be dropped and fall on pedestrians or vehicles. There may be a need for safety harnesses; check the harness each time it is used. Make sure all parts of the harness are in good working condition. Cleaning compounds that are acidic or alkaline may attack the harness or the safety rope. Also, such compounds may attack human skin; therefore, use rubber gloves, when appropriate.

20.32 Painting

There are a number of considerations when painting in a treatment plant. First, is the paint exposed to the drinking water being treated and are there toxic compounds in the paint? Next, is the paint being applied by brush or spray? In either case, is there sufficient ventilation if the operator is painting indoors or in a closed area?

When working with toxic paints, for example, those containing zinc or organics, be sure to clean your hands before eating or handling food. Also, avoid exposing your skin to solvents and thinners and try not to use compounds such as carbon tetrachloride. When spray painting, always use a respirator to avoid inhaling fumes. Do not allow smoking or open flames of any kind around areas being painted. Also, when painting or cleaning the spraying equipment, avoid closed containers where heat is involved. At a certain temperature, called the flash point, spray or vapors could ignite and burn the operator or start fires. Always clean the spray equipment in an area having sufficient ventilation. If the painting operation is taking place in a paint booth, use only explosion-proof lighting and permit no open flame and no electric switch that may cause a spark.

Other considerations when painting involve the safe use of scaffolding and ladders, proper disposal of solvent-soaked rags, and protection of your personal health. Be very careful when using scaffolding and ladders. The scaffolding must be in good repair and conform to current safety regulations. Ladders must also be in good repair. If they are broken or badly worn, they should be replaced with new ones. Rags are always a problem if they contain oils, paint, or other cleaning compounds; there is always the possibility of fire. The rags should be placed into a closed metal container to reduce the fire hazard. To protect your skin, consider using creams to help reduce direct exposure to paint and solvents. Always use an approved respirator to reduce inhalation of fumes and paint mists.

20.33 Cranes

Overhead traveling cranes require safety considerations. First, only authorized personnel should be allowed to operate them. Inspections should be made to check out the circuit breaker, limit switches, the condition of the hook, the wire rope, and other safety devices. The load limits should be posted on the crane and you should never overload the unit. Always check out each lift for proper balance. Use only a standard set of hand signals, and make sure that each operator involved with the crane knows all of the signals. Personnel in and around an overhead crane should be required to wear hard hats. When making repairs to the crane, lock out the main power switch, physically block any movable parts, and allow only authorized personnel to make repairs.

Never move loads over areas where operators or other people are working. Do not let the load remain over the heads of operators or other workers or allow anyone to work under loaded cranes. If loads must be moved over populated areas, give a warning signal and make sure everyone is in a safe location. Set up monthly safety inspection forms to be filled out and placed into the maintenance file. The plant supervisor should review the forms and authorize any maintenance necessary on the crane in addition to following a good preventive maintenance program.

20.34 Confined Spaces[5]

Special safety rules and regulations have been developed to protect the life and health of persons who must enter confined spaces. A confined space may be defined as a space that has limited or restricted means for entry and exit and that is not designed for continuous occupancy. One easy way to identify a confined space is by whether or not you can enter it by simply walking while standing fully upright. In general, if you must duck, crawl, climb, or squeeze into the space, it is considered a confined space. Typical examples of confined spaces found in water treatment plants and within water systems are pits, tanks, basins, and manholes.

Dangerous air contamination presents a threat of causing death, injury, ACUTE[6] illness, or disablement due to the presence of flammable and/or explosive, toxic, or otherwise injurious or incapacitating substances. Accumulations of gases or vapors, for example, may produce an explosive or flammable atmosphere. Similarly, if particles of chemical dust become airborne when filling hoppers or during operation of chemical feed equipment, the dust concentration in a poorly ventilated space may reach a level where a spark could cause the dust to ignite or explode.

An oxygen-deficient atmosphere is one in which the air with its oxygen content has been reduced by another gas or the oxygen has been removed. Normal air contains approximately 20.9 percent oxygen at sea level with the remaining constituents being primarily nitrogen (about 78.1 percent), argon (about 1 percent), and a few traces of other inert gases. When the oxygen content of air drops below 19.5 percent by volume, a potentially dangerous condition exists. A person breathing air with only 17 percent oxygen will experience shortness of breath. Loss of consciousness occurs rapidly when the oxygen content drops to 6 to 10 percent, and death occurs rapidly (within minutes) at less than 6 percent.

Oxygen deficiency in an enclosed atmosphere can occur for several reasons. Some of the more common causes include bacterial action that uses up the oxygen, displacement by other gases (can be both toxic and inert gases like nitrogen or carbon dioxide), oxidation of metals or other materials that deplete the oxygen level, adsorption of the oxygen onto surfaces, and combustion. Since the oxygen content of an enclosed atmosphere can change quickly, it must be continually monitored to ensure that the oxygen content does not drop below an acceptable level. High levels of oxygen enrichment (above 22.5 percent) are also dangerous because they increase the possibility of an explosion.

---

[5] *Confined Space.* Confined space means a space that: (1) Is large enough and so configured that an employee can bodily enter and perform assigned work; and (2) Has limited or restricted means for entry or exit (for example, manholes, tanks, vessels, silos, storage bins, hoppers, vaults, and pits are spaces that may have limited means of entry); and (3) Is not designed for continuous employee occupancy. Also see DANGEROUS AIR CONTAMINATION and OXYGEN DEFICIENCY.

[6] *Acute Health Effect.* An adverse effect on a human or animal body, with symptoms developing rapidly.

To check the safety of the atmosphere in a confined space, use a gas-detection instrument (Figure 20.6). These devices can detect explosive gases, oxygen deficiency, and toxic conditions. Remember, just because there are no toxic or explosive gases present does not mean that you may not lose your life because of a deficiency of oxygen.

According to OSHA regulations, each entry into a confined space requires a confined space entry permit (Figure 20.7). The purpose of the permit is to ensure that a supervisor or other qualified person has taken the necessary steps to identify any hazards associated with entry into the confined space and has taken appropriate safety measures to protect operators from injury. The permit should be renewed each time the space is left and re-entered, even for a break or lunch, or to go get a tool.

A comprehensive set of written, understandable operating procedures should be developed and provided to all persons whose duties may involve work in *PERMIT-REQUIRED*[7] or non-permit confined spaces. Training in the use of procedures should be provided, along with training in the use of safety equipment, rescue techniques, and cardiopulmonary resuscitation (CPR). Exact procedures for work in confined spaces may vary with different agencies and geographical locations and must be confirmed with the appropriate regulatory safety agency.

The following steps are recommended *PRIOR* to entry into a confined space.

1. Identify and close off or reroute any lines that may convey harmful substance(s) to, or through, the work area.

2. Empty, flush, or purge the space of any harmful substance(s) to the extent possible.

3. Monitor the atmosphere at the work site and within the space to determine if dangerous air contamination or oxygen deficiency exists.

4. Record the atmospheric test results and keep them at the site throughout the work period.

5. If the space is interconnected with another space, each space should be tested and the most hazardous conditions found should govern subsequent steps for entry into the space.

6. If an atmospheric hazard is noted, use portable blowers to further ventilate the area; retest the atmosphere after a suitable period of time.

7. If dangerous air contamination and oxygen deficiency do not exist prior to or following ventilation, entry into the area may proceed.

8. Provide appropriate, approved respiratory protective equipment for the standby person and place it where it will be readily available for immediate use in case of emergency.

Whenever an atmosphere free of dangerous air contamination or oxygen deficiency cannot be ensured through source isolation, ventilation, flushing, or purging, observe the following procedures.

1. If the confined space has both side and top openings, enter through the side opening, whenever possible.

2. Wear appropriate, approved respiratory protective equipment.

3. Wear an approved safety belt with an attached line. The free end of the line should be secured outside the entry point.

4. Station at least one person to stand by on the outside of the confined space and at least one additional person within sight or call of the standby person.

5. Maintain frequent, regular communication between the standby person and the entry person.

6. The standby person, equipped with appropriate respiratory protection, should only enter the confined space in case of emergency.

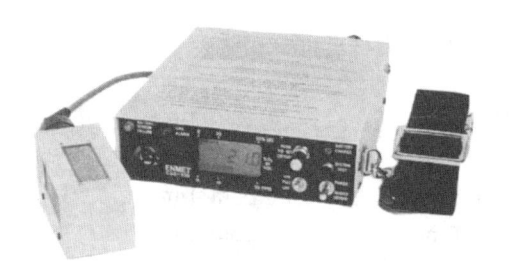

Fig. 20.6 *Gas-detection instruments (toxic gas, combustible gas, and oxygen deficiency)*
(Permission of ENMET Corporation)

---

[7] *Confined Space, Permit-Required (Permit Space).* A confined space that has one or more of the following characteristics: (1) Contains or has a potential to contain a hazardous atmosphere, (2) Contains a material that has the potential for engulfing an entrant, (3) Has an internal configuration such that an entrant could be trapped or asphyxiated by inwardly converging walls or by a floor that slopes downward and tapers to a smaller cross section, or (4) Contains any other recognized serious safety or health hazard.

Confined Space Pre-Entry Checklist/Confined Space Entry Permit

Date and Time Issued: _____ Date and Time Expires: _____ Job Site/Space I.D.: _____

Job Supervisor: _____ Equipment to be worked on: _____ Work to be performed: _____

Standby personnel: _____ _____ _____

1. Atmospheric Checks: Time _____ Oxygen _____ % Toxic _____ ppm
   Explosive _____ % LEL Carbon Monoxide _____ ppm

2. Tester's signature: _____

3. Source isolation: (No Entry)   N/A   Yes   No
   Pumps or lines blinded
      disconnected, or blocked   ( )   ( )   ( )

4. Ventilation Modification: N/A   Yes   No
   Mechanical   ( )   ( )   ( )
   Natural ventilation only   ( )   ( )   ( )

5. Atmospheric check after isolation and ventilation: Time _____
   Oxygen _____ % > 19.5% < 23.5%   Toxic _____ ppm < 10 ppm H_2S
   Explosive _____ % LEL   < 10%   Carbon Monoxide _____ ppm < 35 ppm CO

Tester's signature: _____

6. Communication procedures: _____
7. Rescue procedures: _____

8. Entry, standby, and backup persons   Yes   No
   Successfully completed required training? ( )   ( )
   Is training current?   ( )   ( )

9. Equipment: N/A   Yes   No
   Direct reading gas monitor tested   ( )   ( )   ( )
   Safety harnesses and lifelines for entry and standby persons   ( )   ( )   ( )
   Hoisting equipment   ( )   ( )   ( )
   Powered communications   ( )   ( )   ( )
   SCBAs for entry and standby persons   ( )   ( )   ( )
   Protective clothing   ( )   ( )   ( )
   All electric equipment listed for Class I, Division I,
      Groups A, B, C, and D and nonsparking tools   ( )   ( )   ( )

10. Periodic atmospheric tests:
    Oxygen: ___% Time ___; ___% Time ___; ___% Time ___; ___% Time ___;
    Explosive: ___% Time ___; ___% Time ___; ___% Time ___; ___% Time ___;
    Toxic: ___ppm Time ___; ___ppm Time ___; ___ppm Time ___; ___ppm Time ___;
    Carbon Monoxide: ___ppm Time ___; ___ppm Time ___; ___ppm Time ___; ___ppm Time ___;

We have reviewed the work authorized by this permit and the information contained herein. Written instructions and safety procedures have been received and are understood. Entry cannot be approved if any brackets ( ) are marked in the "No" column. This permit is not valid unless all appropriate items are completed.

Permit Prepared By: (Supervisor) _____ Approved By: (Unit Supervisor) _____

Reviewed By: (CS Operations Personnel) _____
                        (Entrant)   (Attendant)   (Entry Supervisor)

This permit to be kept at job site. Return job site copy to Safety Office following job completion.

Fig. 20.7 Confined space pre-entry checklist/permit

7. If the entry is made through a top opening, use a hoisting device with a harness that suspends a person in an upright position.

8. If the space contains, or is likely to develop, flammable or explosive atmospheric conditions, do not use any tools or equipment (including electrical) that may provide a source of ignition.

9. Wear appropriate protective clothing when entering a confined space that contains corrosive substances, or other substances harmful to the skin.

10. At least one person trained in first aid and cardiopulmonary resuscitation (CPR) should be immediately available during any confined space job.

Confined space work can present serious hazards if you are uninformed or untrained. The procedures presented are only guidelines and exact requirements for confined space work for your locale may vary. Contact your local regulatory safety agency for specific requirements.

20.35 Manholes

There are many hazards involved with manholes and all of them can cause injury to the operator. Just removing the manhole cover can cause the loss of hands or fingers. You should never remove the manhole cover with your hands. Use a manhole hook or special tool such as a pick with a bent point to remove the lid. Be very careful when lifting the lid. Use your legs, not your back, for lifting. This will help prevent back strains. Locate the cover outside the working area to provide adequate working area around the manhole opening.

Next is the problem of traffic around an open manhole. The public, other operators, and vehicles must be protected. Traffic can be warned that operators are working in a manhole by the use of barricades, signs, flags, lights, and other warning devices. Warning devices and procedures must conform to local and state regulations. There also should be a barricade around the manhole to protect the operators. All personnel around manholes should wear hard hats for their safety.

Always inspect and test the ladder rungs in the manhole before using them. They may become loose or corroded and could collapse under your full weight. One should never enter a manhole alone; there should be at least one other person standing by at the top and at least one or more people within hearing distance in case of injury.

Perhaps the greatest threats to operators working in manholes are air contamination or depletion of oxygen. If possible, test the atmosphere in the manhole before removing the lid. Many operators have lost their lives because of leaking gas mains, gases produced by decaying vegetation, or other gases. Never enter a manhole without checking the atmosphere for: (1) sufficient oxygen, (2) presence of toxic gases (hydrogen sulfide), or (3) explosive conditions (methane or natural gas). In any event, always provide adequate ventilation. This will remove any hazardous gases.

Smoking should never be permitted in or within ten feet (three meters) of manholes. Always use a mechanical lifting aid (rope and bucket) for raising or lowering tools and equipment into and out of a manhole. The use of a bucket or basket will keep your hands free when climbing down into or out of the manhole.

To review the hazards of underground structures, remember to use the proper tools for opening and closing the manhole. Keep in mind the need for barricades and lights to warn traffic and to prevent endangering other operators. Be sure that operators are trained in cardiopulmonary resuscitation (CPR) methods and in the way to test the manhole for oxygen deficiency, oxygen enrichment, explosive gases, and toxic gases.

QUESTIONS

Write your answers in a notebook and then compare your answers with those on pages 501 and 502.

20.3A What safety precautions should be taken when waxing floors?

20.3B How should rags containing oils, paint, or other cleaning compounds be stored?

20.3C What safety precautions should be exercised when operating an overhead crane?

20.3D How can traffic be warned that operators are working in a manhole?

20.3E How should tools and equipment be lowered into and removed from manholes?

20.36 Power Tools

The two general classes of portable power tools are: (1) pneumatic, and (2) electrical. Safety precautions for handling these types of tools are much the same for both types. Wear eye and ear protection when operating grinding, chipping, buffing, or pavement-breaking equipment. When using grinding or buffing tools, you will sometimes encounter toxic materials and will need respiratory protection. At other times, there is a need for full face protection because of flying particles; you should use a face shield or at least goggles. In the use of electrical tools, always replace worn extension cords, and never expose cords to oils or chemicals. Extension cords also present a tripping hazard if left in the way. Avoid leaving extension cords in aisles or in work areas. Do not hang extension cords over sharp edges that could cut the cord, and always store the cords in a clean, dry location.

When working in a wet or damp location, consider using rubber mats or insulated platforms. Use only grounded tools. For protection against electric shock, use a ground-fault circuit interrupter (GFCI), especially for outdoor work.

When using pneumatic tools, never use the compressed air to clean off your clothing or parts of your body. Air can enter your tissues or other openings and cause problems. Always check hose clamps. If they are loose or worn, tighten or replace, as needed. Air hoses, like extension cords, are a tripping hazard. Therefore, consider their location when working with pneumatic tools. For the large (¾-inch or 18-mm) hoses, always use an approved safety-type hose connection with a short safety chain or other safety device attached. Air hoses that come apart can cause injuries as they are whipping about. Like electrical cords, keep air hoses away from oils, chemicals, or sharp objects.

Sandblasting with a pneumatic tool requires some special precautions. The operator should protect all skin surfaces with protective clothing, wear eye and face protection, use a respirator, and be very careful of toxic fumes that are discharged from a blasting operation.

The grinding wheel, pneumatic or electric, requires the same safety considerations. Eye and face protection is required. Do not use this tool without safety guards. Be careful of gloves being caught on the grinding wheel. Never operate a wheel with loose nuts on its spindle. When the grinding wheel is badly worn, replace it and use the proper wheel and speed of rotation.

All persons using power tools must be trained in their use and maintenance. Use the manufacturer's operations and maintenance guide for details of proper training. Most injuries by power tools are caused by incorrect setup and operation due to poor training.

Finally, power tools often produce a high level of noise. For example, air drills produce 95 *DECIBELS*[8] and circular saws, 105 dB. Ear protection must be provided when operators are exposed to long periods of high levels of noise. In areas of noise exposure, all operators should be provided with approved ear protective devices.

20.37 Welding

The first safety rule in the operation of gas or electric welding equipment is that the operator must be thoroughly trained in the correct operating procedures. The second rule concerns fire protection. The third rule is personnel protection. None of these rules is first or last—they should *ALL* be followed.

If you are not thoroughly trained in the use of the welding equipment, do not use it. If you absolutely must use the equipment, do so only under the supervision of a trained welder. Whenever such work must be performed in or around a water treatment plant, take time to consider the fire problem. For example, welding can be very dangerous in an oxidizing chemical location, near powdered activated carbon storage, and in storage areas for other bagged chemicals. Avoid welding around oil and grease, when possible; when that is not possible, at least provide for ventilation of fumes. When welding or cutting is done in the vicinity of any combustible material, you must take special precautions to prevent sparks or slag from reaching the combustible material and causing a fire. Some insurance companies require a fire watch to be present during welding operations. The fire watch is responsible for making the welder immediately aware of fire and to assist in putting out any fires that develop.

Regarding the safety of other personnel in the welding area, eye protection comes first. The person using the welding equipment must wear protective clothing, gloves, helmet, and goggles. Others in and around the welding operation should be kept at a safe distance. Always be careful of overhead welding because of falling sparks and slag. If other operators are (or must be) working in the vicinity of the welding operation, they too must be protected from the rays of arc welding; never look at the welding operation without eye protection.

Store welding gas cylinders in the same manner as other gas cylinders. They are stored upright, kept out of radiation of heat and sunlight, and stored with protective covers in place when not in use. Store cylinders away from elevators and stairs, and secure them with a chain or other suitable device.

20.38 Safety Valves

There are quite a number of safety valves in a water treatment plant; operators are not always aware of their locations or functions. For example, most operators know of the safety plugs on chlorine cylinders, but there are also large safety valves in any plant that stores large amounts of chlorine on site. These containers take on truckload lots of 17 tons (15,540 kilograms). The safety valves on such containers should be certified at least every two years or as often as the state requires. Such relief valves must be maintained on a regular basis. Inspect the inside of these tanks at regular time intervals and keep a record of the findings, for example, evidence of deposits and corrosion.

Water heater safety valves should be checked on an annual basis and maintained or replaced, as needed. If the plant has a boiler room, the steam safety valve should be maintained and checked for proper operating pressures. These valves should not discharge in such a manner as to be a hazard to operating personnel.

There also may be surge relief valves on discharge piping (high lift) of the treatment plant. These valves also act as a safety valve to the pumping equipment and must be maintained on some regular time interval. They should be checked for proper pressure setting, with all pilot valves being reconditioned or replaced, as needed.

There may be other safety valves located in the pumping plant's hydraulic system for opening and closing discharge valves that require maintenance. In the maintenance of water treatment plants, you or your supervisor must set up a maintenance system for all equipment. Hand tools, power tools, and other maintenance equipment must also be kept in safe working condition. Operators must be furnished protection for the eyes, the ears, the hands, the head, the feet and, at other times, the body. Work areas should be well ventilated and noise should be reduced, whenever possible. Each operator should always be on the lookout for additional ways of making the treatment plant a safer place to work.

---

[8] *Decibel* (DES-uh-bull). A unit for expressing the relative intensity of sounds on a scale from zero for the average least perceptible sound to about 130 for the average level at which sound causes pain to humans. Abbreviated dB.

> **QUESTIONS**
>
> Write your answers in a notebook and then compare your answers with those on pages 502.
>
> 20.3F   What type of protection do operators need when operating portable power tools?
>
> 20.3G   How can operators be protected from high noise levels when operating air drills and circular saws?
>
> 20.3H   What personal protection should be used when operating welding equipment?

20.4 VEHICLE MAINTENANCE AND OPERATION

20.40 Types of Vehicles

Many types of vehicles are used in the waterworks industry. However, the plant operator may only come into contact with a few. Cars, pickup trucks, forklifts, dump trucks, and some electrically driven cars are the types of vehicles the operator is most likely to be involved with and to need to maintain. In addition to motor vehicle safety, this section will also consider the storage of fuel for these and other engines in the plant.

20.41 Maintenance

To have a safe motor vehicle, there must be a preventive maintenance program. Figure 20.8 gives a checklist for finding potential safety problems and a means of recording the preventive maintenance.

Tire inflation is a good example of proper safety checks. Not only is it unsafe to operate on underinflated tires, but it also causes undue wear on the tire. Therefore, tires should be checked regularly for wear, which may be caused by misalignment or low inflation. If tires are badly worn, they should be replaced. Always maintain the recommended pressure in the tires; when checking, set the hand brake and turn off the motor.

Next, when changing tires, be sure the jack you are using has sure footing. Position the jack at right angles to the direction of the lift. Jacks are a problem in general, and you should make sure that proper jacks are in each vehicle. In other words, select the proper jack for each job and choose only one that is safe and strong enough. If blocking is required, only use safe supports, avoid leaning on the jack, and protect hands. Always stay a safe distance from the jack handle as many injuries are caused by flying jack handles. Also, injuries are caused by overloading jacks. In addition, where needed, use braces or other supports to prevent tipping the vehicle over, another cause of serious injuries.

Fueling motor vehicles also involves some hazards. Always stop the vehicle's engine. Remember to remove the fuel hose immediately after using it. You could start a fire if you carelessly drive off with the hose still attached. Also, make sure the cap is replaced tightly on the tank. Do not permit smoking in the vicinity of gas delivery pumps at any time. Avoid any sparks and skin contact. Use only high-flash-point solvent for cleaning up any gasoline. Some other safety tips when refueling are: do not let the tank overflow, always set the brake, hang nozzle up properly onto the pump, and eliminate any leaks on the hose connections.

Most small water treatment plants do not have hoists or pits for vehicle lubrication. However, most of the following suggestions are applicable. First, keep all walkways, steps, tools, and containers free from grease, oil, and dirt. This will reduce the possibility of accidents caused by these items. As when maintaining any equipment, use the right tools, keep shoes free of all oil or grease, and wear only shoes with nonslip soles.

If the plant is equipped with a hoist, do not permit anyone to remain inside the vehicle when it is on the hoist. When lifting the vehicle, do not permit tools on the hoist or vehicle that may fall onto you or other personnel in the work area. Keep the driveway free of hoses, tools, and cars, and always keep your hand on the operating lever when raising or lowering the vehicle.

Most water treatment plants have assigned specific areas for washing or steam cleaning vehicles. If your plant does not have such an area, you should have one assigned. This area does not have to be elaborate, but should have water hoses and steam-cleaning equipment and be adequately drained. The same safety rules apply to makeshift installations as apply to completely equipped cleaning areas. The most important consideration is the steam cleaner. Keep the nozzle clean, check water level on coils before turning on the flame, and always wear protection for your eyes and face. Make sure that the cleaner is adequately grounded and be careful of cleaning compounds (see Section 20.1, "Chemical Handling," for precautions to prevent burns). Maintain steam hoses, check connections, and never permit horseplay with steam-cleaning equipment. As in other work areas, keep the wash rack free from grease and oil and oily rags. Hoses should be stored on the rack when not in use. Always use scaffolding or platforms when cleaning the tops of vehicles.

20.42 Seat Belts

Many water treatment plant operators have some excuse for not wearing seat belts. The excuses may sound good, but they will not protect you in the event of an accident. Many lives would have been saved if seat belts were used. The water utility should equip all vehicles with seat belts and require every operator to use them.

20.43 Accident Prevention

The best overall means of preventing vehicle accidents is defensive driving. This method requires training and a certain mental outlook on the part of the vehicle operator. Most, if not all, drivers think they are good at what they do and this may be true to some extent. However, if each driver would operate all vehicles as if all other drivers were the world's worst drivers, accidents would be greatly reduced.

Good drivers check out their vehicles each time they use them and have any maintenance performed, when needed. They use proper signals for directional change, always observe traffic regulations, and show courtesy to others. Remember that drivers in

Month _____	Vehicle Number _____		Assignment _____		
			WEEKS		
CHECK	1st	2nd	3rd	4th	5th
1 Oil					
2 Water					
3 Tires					
4 Horn					
5 Headlights, High - Low					
6 Tail Lights					
7 Turn Signals					
8 Stop Lights					
9 Battery Water					
10 Fire Extinguisher					
11 First Aid Kits					
12 Windshield Wipers					
13 Visual Inspection - Wire Rope					
14 " " - Hook					
15 " " - Sheaves					
16 " " - Boom					
17 " " - Hydraulic Level					
18 Operational Test Controls					

SERVICE MILEAGE READINGS				FUEL CONSUMPTION	
Week	Present	Last Service	Difference	Start Mileage	
1st				End Mileage	
2nd				Total Miles	
3rd				Fuel Used	
4th				MPG Average	
5th					

Fig. 20.8 Mobile equipment checklist

an agency vehicle represent the agency. Therefore, good driving skills are good for public relations.

Another way to avoid accidents is not to tailgate. Tailgating is a very unwise practice that is dangerous to the vehicle and hazardous to its operator. A good rule to use when following another vehicle is the old one-car-length for every 10 MPH, and if there is limited visibility, increase that distance. Another rule is the "Three Second Rule," which says you must be at least three seconds behind the car in front of you. Take precautions when backing up. Always set the brake or shift to "Park" when parking the vehicle. Be cautious at intersections. As a defensive driver, always be ready to give the right of way. No right of way is worth injuring oneself.

In some cases, even the most defensive and careful driver has an accident. Because of this, each vehicle should carry flashlights, flares, flags, and a fire extinguisher, along with a first-aid kit. In the event of an accident, the driver should know how to fill out all the forms, a supply of which should be provided in the vehicle.

Remember, when operating a vehicle, an accident can be prevented by defensive driving. The plant supervisor should have each member of the staff take a defensive driver training course. Each driver should develop a defensive driver frame of mind. Developing a good attitude and driving skills are the key to accident prevention when operating a vehicle. All new employees should be given road tests in operating the types of vehicles they will be using.

Here are a few reminders when operating a vehicle. During a storm, roadways and pavement are likely to be slippery. Slow down, pump the brakes when stopping, and remember the minimum distance rules. No driver should be required to operate an unsafe vehicle. Keep copies of a suitable form for reporting mechanical problems in each vehicle and encourage operators to use them.

20.44 Forklifts

Most water treatment plants and pumping stations have a forklift. Most, if not all, plant operators use this vehicle to move chemicals, repair parts, and even use it when making repairs to lift heavy objects. Therefore, every plant operator should be trained in the use of the forklift.

Following are a few points regarding safe operation of the forklift with suggestions for operator safety as well as protection of others who may be in the operating area of the forklift. Keep all aisles free of boxes and other debris. Do not permit anyone to ride on the forklift except the operator. Never overload the forklift. Always be sure the warning signals are operational and never leave the power on when leaving the forklift. Like other vehicles, check out the brakes before operating. Be careful at intersections of aisles and always face the direction of travel. If a loaded forklift is to be placed on an elevator, be sure that the load on the forklift and the weight of the vehicle do not exceed the weight limit of the elevator. Also, make sure the forklift load is stacked properly before lifting or moving. When handling drums, special lifting and retaining devices are needed.

Figure 20.9 is a typical forklift inspection form.

QUESTIONS

Write your answers in a notebook and then compare your answers with those on page 502.

20.4A   What causes tire wear on motor vehicles?

20.4B   Motor vehicles should contain what safety devices?

END OF LESSON 2 OF 3 LESSONS

on

SAFETY

Please answer the discussion and review questions next.

DISCUSSION AND REVIEW QUESTIONS

Chapter 20.   SAFETY

(Lesson 2 of 3 Lessons)

Write the answers to these questions in your notebook. The question numbering continues from Lesson 1.

16. How can an operator prevent fires in a water treatment plant?

17. How would you maintain water-type fire extinguishers?

18. How would you maintain fire hoses in your water treatment plant?

19. What precautions should be taken in areas where flammable material is stored?

20. How can an operator make visitors feel as if the water being produced is safe to drink?

21. How would you remove a manhole cover?

22. What precautions should be taken when operating power tools in a wet or damp location?

23. What fire hazards should be considered before doing any welding?

24. How would you safely refuel a motor vehicle?

25. What safety precautions should be taken when driving during a storm?

SOUTHERN NEVADA WATER SYSTEM
For Truck Operator Inspection

Operator	Brake	Boom Up	Boom Dn.	Tilt F	Tilt R	Back-up Horn	Horn	Wheel	Remarks

FORM #268 — S.N.W.S.

Fig. 20.9 *Forklift inspection form*

CHAPTER 20. SAFETY

(Lesson 3 of 3 Lessons)

20.5 ELECTRICAL EQUIPMENT

20.50 Electrical Safety

As a water treatment plant operator, you are not expected to be an expert in electrical equipment, but you must have a working understanding of electricity. This includes an understanding of the safety precautions needed to operate the electrical equipment. After all, electrical energy is required to power most of the treatment plant operations. The objective of this section is to show you how to operate safely and to become involved, to a limited degree, in the maintenance of electrical equipment. Electricity is unforgiving to the careless treatment plant operator. If you have any doubts about your understanding of this section, you may wish to reread Chapter 18, "Maintenance," Section 18.1, "Electrical Equipment."

20.51 Current—Voltage

Many types of electrical current are used in water treatment plants and the associated pumping plant. Each day the plant operator is exposed to this equipment, giving little thought to the potential hazards of the equipment. Current may come into the plant at a high voltage, for example, 69 KVA, reduced to 4,160 volts or 2,300 volts. This current may power pump motors, blowers, and other equipment at lower voltages of 440, 220, or 120 volts and within starters may be reduced to 24 or 12 volts or changed over into DC voltages. Given all of these various voltages, the operator must be careful not to become careless working with equipment. Therefore, become familiar with the types of current and voltage in the plant. By knowing this, you will avoid the mistake of becoming involved with unsafe electric currents or practices for which you are not trained. This will also enable you to know when to ask for a qualified person to perform any necessary repairs.

20.52 Transformers

Electrical power entering a plant is routed through transformers to reduce the voltage, in most cases. There are many types of transformers although the operator may only think of the larger ones that bring the power into the plant. Sometimes, these are owned by the waterworks agency and, therefore, the maintenance is the responsibility of the operator. There are few, if any, plant operators who are qualified to perform such maintenance. Never attempt to work on a high-voltage transformer without the assistance of qualified personnel. Such personnel can be located at the power company or contact an electrical contractor who specializes in the repair and maintenance of electrical transformers. You will, however, need to keep records of the transformer's operation. This information is helpful to repair personnel and is useful to operators who need to know the status of the transformer.

There are many small transformers within the plant's operating gear and it is these you may have to maintain. Most often these low-voltage transformers become overworked; they overheat and burn out. Any fire in the electrical gear can be hazardous. Be very careful when opening a starter, breaker box, or indicating instrumentation if a fire or overheated transformer is suspected. Operators have been badly burned by not thinking before opening such devices when they smell smoke in pumping stations or treatment plants. When solving problems with hot, overheated, or burning transformers, remember what you learned in Section 20.2 about electrical fires. For the safety of operating personnel and the safety of the plant, regularly inspect, or have someone inspect, both large and small transformers. If you detect any overheating, have a qualified electrician inspect and replace any transformer that is not functioning properly.

There should be a fence around the transformer station with a locked gate and only a limited number of keys issued to plant personnel. The operator may perform routine preventive maintenance such as removing weeds or brush and general cleanup. Replacing fuses and major maintenance or repairs should be made by the power company or qualified electricians. Maintenance must be performed by qualified and well-trained personnel.

20.53 Electric Starters

As a treatment plant operator, your most frequent contact with electrical power will probably be with electric starters on the motor control panels. These devices are used throughout the plant and provide an interface between the operator and the flow of energy. The starter may be located on a switch panel or there may be a switch that is remotely located from the starter. One of the first safety procedures you should take is to use a special insulated mat on the floor at all switchboards. The starter should be provided with adequate lighting and clearly marked Start/Stop buttons. Replace indicating lights as needed without delay. There should always be clear and adequate working space around the starter or switch panels. To reduce the hazards of fire in electric starters, they should be cleaned and maintained on a regular basis. Such maintenance must be performed by trained, qualified personnel. In electrical starting equipment, fires can easily occur because of accumulation of dust and dirt on the contactors. When they become so badly burned that they do

20.54 Electric Motors

The treatment plant operator is exposed to many types and sizes of electric motors. In some plants, the motors are old and require more attention because of exposed parts. The newer electric motors are enclosed and have all parts protected. For the old motors, you should install safety guards or guardrails to prevent accidental contact with live parts of the motors.

Some of the electric motors may have exposed couplings, pulleys, gears, or sprockets that also require consideration. For these and other moving devices, a wire cloth gear guard may be installed. The gear guard can also be made of sheet metal. However, no matter which type of guard is used, it must be securely fastened onto the floor or some other solid support. The safety guards must be constructed and fitted to prevent material being handled by operators from coming into contact with the moving parts driven by the motors.

Another consideration is projections on couplings, pulley shafts, and other revolving parts on the motor or on the device being driven. These projections can be bolts, keys, set screws, or other projections. The projections should be removed, reduced, or protected by one of the above guards.

Check grounding on all electric motors as part of a routine maintenance program. The motor frames themselves must be grounded if the wires to the motor are not enclosed in an armored conduit or other metallic raceway. Check that all joints are mechanically secure to ensure good grounding. In the case of portable electric motors, the simplest way of grounding is an extra conductor in the cord serving the motor. The best way is to install a ground-fault circuit interrupter (GFCI) receptacle. This device will automatically disconnect the tool from the power supply if the ground is not connected and will supply the greatest protection to the operator and to the equipment.

When using portable electric motors, always check the service cord. If the cord or receptacles are in poor condition or showing signs of wear, they should be replaced. A badly worn cord must never be used in a wet location.

20.55 Instrumentation

In this area of water treatment, the operator is not exposed to a great deal of hazard, but must give some consideration to instrumentation devices since they are operated by electric current. This is also true of all other automatic equipment. Although most instruments protect the operator, there is still a degree of hazard when changing charts, calibrating, or performing other maintenance. First, when calibrating an instrument, you are exposed to at least 12 volts DC to 120 volts AC. If you become grounded with the 120 volts AC, you may be killed or severely injured. Also, when maintaining automatic control equipment, adjustment of one instrument may start another device, exposing another operator to a hazard because of an unexpected start. As mentioned above, electronic devices operate on low current, but do not forget that there is still high voltage located somewhere in the instrument.

20.56 Control Panels

Always provide adequate working space in and around control panels. As with electric motors, the panels must be well grounded. At some locations, there may be a need for special insulating mats, such as in wet locations. Adequate lighting must be available inside the control panel as well as outside for those who do the maintenance. Moisture or corrosive gases must be kept away from the control panels. To reduce fire hazard, never store any hazardous material next to switchboards or control panels. Panels carrying greater than 600 volts must be permanently marked warning of the hazards. Areas of high voltage should be screened off and locked with a limited number of keys given to authorized personnel only.

> **CAUTION**
>
> Unless you are a qualified electrician, stay out of the inside of all electrical panels. If you do not know or are not familiar with the equipment, leave it alone!

20.57 Lockout/Tagout Procedure

Any equipment that could suddenly start up or release stored energy is a potential hazard to operators. Stored energy may be in the form of electrical power, hydraulic pressure, pneumatic pressure, or even the tension in a spring. Two hazards due to the lack of a good lockout procedure are: (1) accidentally starting a piece of equipment, exposing a fellow operator to a hazard, and (2) turning electrical power on when someone is still working on the equipment, exposing that person to danger. Every water treatment plant should develop a standard operating procedure (SOP) (Figure 20.10) for lockout of equipment to protect personnel from accidental injury.

A lockout device is a positive means of holding a switch, valve, or other energy isolating device in a safe position to prevent the equipment from becoming energized, either directly or by release of stored energy. Positive control of the energy source is ensured by the use of a keyed lock or a combination lock. Besides physically locking out or isolating the source of energy, it is also important to warn other operators, by means of a tagout device, that the switch or piece of equipment has been shut down. A tagout device is a prominent warning tag that can be securely fastened to the energy isolating device to indicate that both it and the equipment being controlled may not be operated until the tagout device is removed.

Even after equipment is locked out and tagged, it still may not be safe to work on. Stored energy in gas, air, water, steam, and hydraulic systems (such as water pumping systems) must be drained or bled down to release the stored energy. Before bleeding down pressurized systems, think about the pressures involved and what will be discharged to the atmosphere and work area (toxic or explosive gases, corrosive chemicals). Will other

---

not make proper contact, they could overheat and start fires. The key to preventing fires in starting equipment is a good preventive maintenance program.

SOUTHERN NEVADA WATER SYSTEM Standard Operating Procedure	
SUBJECT: REMOVING EQUIPMENT FROM SERVICE	Number: 35
DIVISION: SAFETY	Prepared By: JAR 4/22/06

PURPOSE

This procedure prescribes the method for safely removing a piece of equipment from service prior to performing the required maintenance or repairs.

GENERAL

Equipment may be removed from service for the following reasons:
1. To prevent an unsafe or damaging operating condition.
2. To perform maintenance or repairs that would be unsafe or impractical to perform while equipment is in service.

Equipment should only be removed from service when one of the above conditions exists.

METHOD

RESPONSIBILITY: Operations or Maint. Employee
ACTION:
1. Upon receipt of a work order requiring removal of equipment from service or upon discovering a condition that requires removal of equipment from service, checks the following as applicable:
   – motor(s) is off
   – discharge valves are closed
   – selector switches are off
   – alarm system is bypassed
   – etc.
2. Attaches a "lock-out" device to the piece of equipment, if possible.

NOTE: Once a "lock-out" device is installed, it is not to be removed until repairs are complete, unless removal is authorized by the Director or a Division Head.

3. Prepares a "Do Not Operate" tag, entering the following:
   – date
   – reason for equipment being removed from service
   – signature of employee preparing tag
4. Affixes tag to the piece of equipment.
5. Notifies operations by radio or telephone of the removal from service.
6. If the removal from service was not initiated by a work order, checks to make sure that no work order is already scheduled for the repair.
7. If no active work order is on file for the repair, prepares and processes a work order in accordance with SOP #39, subject: "Work Orders."
8. Removes the "lock-out" device, if possible, or requests that the device be removed by the employee who installed it.

NOTE: If the employee who installed the device is unavailable, requests removal or authorization for removal from the Superintendent.

9. Performs scheduled maintenance or repairs and installs own "lock-out" device as needed until work is completed.
10. Maintenance Technician, upon completion of scheduled maintenance or repairs, removes the tag and the "lock-out" device, if applicable, and returns the piece of equipment to service status.
11. Notifies Operations that the equipment has been returned to service.

Fig. 20.10 Standard operating procedure for locking out electrical equipment

safety precautions have to be taken? What volume of bleed-down material will there be and where will it go? If dealing with chemicals, greases, or lubricants, how can they be cleaned up or contained? Many times, an operator has removed a pump volute cleanout only to find that the discharge check valve was not seated or one of the isolation valves was not fully seated. Once the pump was open, however, the pump intake structure drained into the pump room through the opened pump.

Many types of equipment also require the use of a blockout procedure. Blockout means the insertion of some device that prevents moving parts in equipment from moving. Physically block or secure in place any elevated machine members, flywheels, and springs to prevent movement that could injure an operator working on the equipment.

All rotating mechanical equipment must have guards installed to protect operators from becoming entangled or caught up in belts, pulleys, drive lines, flywheels, and couplings. Keep the guards installed even when no one is working on the equipment.

Various pieces of machinery are equipped with travel limit switches, pressure sensors, pressure reliefs, shear pins, and stall or torque switches to ensure proper and safe operation of the equipment. Never disconnect a device, or install larger shear pins than those specified on the original design, or modify pressure or temperature settings.

In a safe lockout procedure, the switches are locked open and are properly tagged; only the operator who is doing the maintenance should have a key. In fact, all people who perform electrical maintenance should have their own individual lock and key so as to maintain control over the equipment being worked on by each individual. Only the operator who installed the lockout device and warning tag may remove them unless the employer has provided specific procedures and training for removal by others.

The basic elements of a proper lockout/tagout procedure[9] are as follows:

1. Notify all affected employees that a lockout or tagout system is going to be used and the reason why. The authorized employee is responsible for knowing the type and magnitude of energy that the equipment uses and must understand the potential hazards.

2. If the equipment is operating, shut it down by the normal stopping procedure.

3. Operate the switch, valve, or other energy isolating device(s) so that the equipment is isolated from its energy source(s). Stored energy such as that in springs, elevated machine members, rotating flywheels, hydraulic systems, and air, gas, steam, or water pressure must be dissipated or restrained by methods such as repositioning, blocking, or bleeding down.

4. Lock out and tag the energy isolating device with assigned individual lock or tag.

5. After ensuring that no personnel are exposed, and as a check that the energy source is disconnected, operate the push button or other normal operating controls to make certain the equipment will not operate. A common problem when motor control centers (MCCs) contain many breakers is to lock out the wrong equipment (locking pump #3 and thinking it is #2). Always confirm the dead circuit. *CAUTION:* Return operating controls to the neutral or OFF position after the test.

6. The equipment is now locked out or tagged out and work on the equipment may begin.

7. After the work on the equipment is complete, all tools have been removed, guards have been reinstalled, and employees are in the clear, remove all lockout/tagout devices. Operate the energy isolating devices to restore energy to the equipment.

8. Notify affected employees that the lockout or tagout device(s) has been removed before starting the equipment.

QUESTIONS

Write your answers in a notebook and then compare your answers with those on page 502.

20.5A Why should each plant operator become familiar with the type of current and voltage in the water treatment plant?

20.5B List the moving parts on electric motors that require safety guards.

20.5C What are two hazards created by the lack of a good lockout procedure for control panels and switchboards?

20.5D List four examples of stored energy that could be hazardous to operators.

20.6 LABORATORY SAFETY[10]

20.60 Laboratory Hazards

In general, water plant operators do not experience a great deal of exposure to hazardous laboratory conditions. However, you will be in contact with potentially hazardous glassware, toxic chemicals, flammable chemicals, corrosive acids, and alkalies. There may be times when you will be exposed to hazardous bacteriological agents. The seriousness of the hazards depends mainly upon the size of the plant and the operating procedures in the treatment plant. For your own safety, learn the proper procedures for handling laboratory equipment and chemicals.

20.61 Glassware

An important item in laboratory safety is handling of glassware. Almost all tests performed by an operator will require the use of some glassware. The operator's hands, of course, are exposed to the greatest hazard. To reduce accidents when handling glassware, never used chipped, cracked, or broken glassware in any testing procedures. All such glassware should be disposed of in a container marked "For Broken Glass Only." Never put broken glass in wastebaskets. Although it may not be a hazard to the operator, it is a danger to those who clean out the wastebaskets. Clean up any broken glass or spilled chemicals to reduce hazards to others. Never let broken pieces of glass remain in the sink or in sink drains. This may cause cuts to others who unknowingly try to clean the sink.

Washing glassware is always potentially hazardous. The glassware can be broken while being washed, causing cuts, or cuts can be caused by chipped or cracked glassware. Also, the cleaning compounds themselves can be a problem. Sometimes strong acid cleaners are used to remove stains from the glassware. Without protective gloves, your hands could be seriously burned by these acids.

20.62 Chemicals

When handling liquid chemicals, such as acids and bases, always use safety glasses or face shields. If working with ether or chloroform, avoid inhalation of fumes and always do this type of work under the ventilation hood. Be sure to turn the ventilation fan on. Of course, be careful of open flames when using flammables such as ether. As a general rule, do not permit smoking in the laboratory. All chemicals should be stored in proper locations; do not set chemicals on the laboratory benches where they may cause an accident if spilled or if the container is broken.

When handling laboratory gases, give consideration to their location and potential accident hazards. Gas cylinders must be prevented from falling by using safety retaining devices such as

---

[9] Adapted from *WASTEWATER SYSTEM OPERATIONS*, State of Oklahoma Certification Study Guide.
[10] See *FISHER SAFETY CATALOG*. Obtain from Fisher Scientific Company, Safety Division, 4500 Turnberry Drive, Suite A (Customer Service), Hanover Park, IL 60103, or phone (800) 772-6733.

chains. The valve and cylinder regulator should be protected from being struck by stools, ladders, and other objects.

When mixing acid with water, always pour the acid into the water while stirring.

Never add water to acid because the water may remain on top of the acid, thus causing splattering and excess heat generation.

Always use safety goggles, gloves, and protective garments. When cleaning up acid or alkali spills, dilute with lots of water even if you flush them down the sink drain. Baking soda can be used to neutralize acids, and vinegar is used to neutralize bases. Never allow mercury, gasoline, oil, or organic compounds into the laboratory drains. Use only a toxic waste disposal drain system for these items. Pouring such compounds down sink drains can cause an explosion, allow toxic gases and vapors to enter the lab, or destroy the piping.

You must never use your mouth with the pipet for transferring toxic chemicals, acids, or alkalis. Use a suction bulb, aspirator, pump, or vacuum line. If you use your mouth, there is always the danger of getting the toxic solutions into your mouth.

20.63 Biological Considerations

Do not take chances with bacteria. A good policy is to have each operator immunized with antityphoid vaccine and to keep their booster shots current. Always use good sanitary practices, particularly when working with unknown bacteria or known pathogens. Never pipet bacteriological samples by mouth. *ALWAYS USE A PIPET BULB.*

When exposed to any bacteria, you should make it a habit to always wash your hands before eating or smoking. If you have any cuts or broken skin areas, these wounds should not come in contact with bacterial agents. You should wear protective gloves or cover the wound with a bandage when working with any kind of bacteria.

All work areas should be swabbed down with a good bactericidal disinfectant before and after preparing samples. As a general policy, the preparation or serving of food should never be permitted in the laboratory. Also, give some consideration to proper ventilation, because some bacteria may be transmitted through the air system.

20.64 Radioactivity

There are many laboratory and treatment plant instruments that use radioactive isotopes in laboratory tests and research. A plant operator may be exposed to radioactive compounds when calibrating sludge density meters or using research isotopes. From a safety standpoint, only qualified personnel should be involved in the use of radioactive compounds. If radioactive compounds are present in the laboratory, warning signs should be posted. The disposal of all radioactive compounds must be performed strictly by qualified laboratory personnel in accordance with government regulations.

20.65 Laboratory Equipment

20.650 Hot Plates

You will probably use a hot plate in your threshold odor number (TON) tests. You should turn the hot plate off when not in use; never place bare hands on the hot plate to check if it is hot. When using the hot plate, always use the hood and turn on the hood ventilation fan to remove gas or fumes. Never place glassware onto a hot plate if the outside of the glassware has water or moisture on the surface between the glass and the hot plate. Steam will form at this interface and cause the glass to break. When taking hot glassware off the hot plate, always use gloves.

20.651 Water Stills

Most water stills in the laboratory today are the electrical type. To observe good safety practices, check the items described in the electrical safety section of this chapter, such as good grounding. Set up an SOP (Standard Operating Procedure) for proper operation of the still and follow the manufacturer's instructions for proper starting and stopping. The still will require cleaning from time to time. Be very careful when disassembling the still. Parts may be frozen together because of hardness in the water and may require an acid wash to separate. Be sure that the boiler unit is full of water before turning the still on. Never allow cold water into a hot boiler unit because it may cause the unit to break.

20.652 Sterilizers

There are two types of sterilizers: (1) dry electrical sterilizers, and (2) wet sterilizers (autoclaves). In the dry electrical sterilizers, check the cords frequently because the high heat may damage the wiring. Always let the unit cool off before removing its contents. Wet sterilizers (autoclaves) are under pressure by steam and should be opened very slowly. Use gloves and protective clothing when unloading the autoclave. Cover the steam exhaust with an insulated covering to prevent burns. Any leakage around the door should be repaired by replacing door gaskets or even the door if it is worn or bent. Always load the autoclave in accordance with the manufacturer's recommendations. Never allow an operator to work with this equipment without proper instruction in its operation.

20.653 Pipet Washers

Cleaning pipets can be hazardous. Many laboratories have a pipet cleaner that contains an acid compound or other cleaning agents. Always use protective clothing and a face shield when working with the washer unit. Try to avoid dripping or spilling the cleaning compound when transferring the pipets. If the acid comes into contact with your skin, use the remedies recommended in Section 20.11, "Acids."

QUESTIONS

Write your answers in a notebook and then compare your answers with those on page 502.

20.6A Why is washing glassware always a potential hazard?

20.6B What precautions should be taken when handling liquid chemicals, such as acids and bases?

20.6C Why should mercury, gasoline, oil, or organic compounds never be allowed into laboratory drains?

20.6D How can an operator be exposed to radioactive compounds?

20.6E Why should cold water never be allowed into the hot boiler unit of a water still?

20.7 OPERATOR PROTECTION

20.70 Operator Safety

So far in this chapter, we have discussed many means by which you can protect yourself and your equipment. In this section, we wish to discuss your own personal protection. Take a look at the means of protecting the eye, the foot, the head, and, most of all, look at water safety. After all, a plant operator is always in contact with water. The water may be found in raw water reservoirs, settling basins, clear wells, or filters. Operators have lost their lives by falling into the backwash gullet. Operators have lost their lives in the finished water reservoirs. As unlikely as it seems, fatal accidents have happened in the past and will happen again in the future. Therefore, it is the responsibility of each operator to watch for any safety problems in the plant's reservoirs, pumping stations, or filters. You, the operator, are responsible for yourself. You should never expose yourself to unsafe conditions.

A major problem confronting many operators is locating reliable safety vendors and equipment. State, regional, and national professional meetings, such as those sponsored by the American Water Works Association, often have displays or exhibits featuring manufacturers of safety equipment. This is an excellent opportunity to meet the representatives of these companies and discuss with them what equipment they would recommend for your situation. Also, other operators who have had experience with safety equipment of interest to you often attend these meetings and are eager to share their experiences with you.

If you are unable to attend these meetings, the program announcements will often have a short description of the types of safety equipment that will be featured by each vendor exhibiting at the conference. You can obtain the vendor's address by looking in professional journals, buyers' guides, or by writing to the sponsor of the conference.

20.71 Respiratory Protection

There are many respiratory hazards in and around the treatment plant that an operator is exposed to daily including chemical dusts, chemical fumes, and chemical gases, such as chlorine, ammonia, sulfur dioxide, and acid fumes, to name only a few. Whenever working around or handling these and other compounds, you must take adequate precautions.

Two types of conditions for which you should be prepared are: (1) oxygen-deficient atmosphere, and (2) sufficient oxygen, but a contaminated atmosphere containing toxic gases or explosive conditions. In either circumstance you will need an independent oxygen supply. However, an independent oxygen supply will not protect you from an explosion.

Never enter a confined space with an explosive atmosphere.

Call your local gas company and ask their experts to enter the explosive area, if entry is essential. Your independent oxygen supply should be of the positive-pressure type to protect you if there are any leaks in your mask. Good ventilation can reduce explosive conditions.

20.72 Safety Equipment

All waterworks safety equipment, such as life lines, life buoys, fire extinguishers, fencing, guards, and respiratory apparatus, must be kept in good repair. This and other safety equipment is necessary to protect operators or visitors from injury or death. Safety equipment may fall into disrepair because it is only used occasionally and may deteriorate due to heat, time, and other environmental factors. First-aid equipment should also be provided and kept resupplied as it is used. The operating staff should be given regular instructions in the use and maintenance of the safety equipment.

Provide protective clothing for all operators handling chemicals or dangerous materials. Keep the clothing clean and store it in a protected area when not in use.

The water utility is responsible for providing outward opening doors, remote-controlled ventilation, inspection windows, and similar safety devices where appropriate. This equipment should be exercised and kept clean and well maintained so that it will operate when needed. Respiratory (self-contained breathing) apparatus must be stored in unlocked cabinets outside of chlorination, sulfur dioxide, carbon dioxide, ozone, and ammonia rooms. The storage cabinets must have a controlled environment to prevent deterioration of the equipment.

The operator has the responsibility to inspect each apparatus for deterioration and need for repair. Safety equipment is of no use to the operator if it fails when put to use, and it may cost you your life if it is in poor condition. Some self-contained breathing apparatus (air packs) depend on compressed air to supply the oxygen. Under conditions of deficient oxygen supply, the canister type of respirator is useless. You could lose your life by entering a room containing chlorine gas (which is heavier than air) while depending on a poorly maintained respirator. Although you might have protection from the chlorine, you would not have adequate oxygen. Never take a chance with a toxic gas. In water treatment plants, use only the positive-pressure type of self-contained breathing apparatus.

Many newer plants are being constructed with independent air supplies consisting of a helmet, hose, and compressed air. The helmet is connected by a hose to an uncontaminated air source. The key word here is "uncontaminated." Not only should the operator follow a maintenance program for the hose and mouthpieces of the apparatus, but the operator must maintain the air supply. The air is supplied by mechanical equipment that requires maintenance. The air pressure is controlled by a reducer or regulator that must be kept clean and maintained to be available when needed. Set up a preventive maintenance program for this equipment. It should be checked out on a weekly basis, and records should be kept of each inspection. The record should show conditions of the hoses, regulators, air filters, compressors, helmet, and any other apparatus furnished with the system.

The old standby, of course, has been the air packs or self-breathing apparatus. These units are carried by the user, giving the operator an independent source of air (oxygen). The unit can be used in any concentration of contamination of gases, dust, or anywhere the atmosphere is oxygen deficient. There are two types of units. One type of unit depends on compressed air or oxygen, and the second system generates oxygen by use of chemicals in a canister. The oxygen is generated by the moisture exhaled by the user. Because this equipment is not used daily in the water treatment plant, there must be a preventive maintenance program, with records, inspection, and operator checkout. As with any system, self-contained breathing equipment requires maintenance. This is vital because the operator's life will depend on how well this apparatus performs.

Training is another important consideration. Even though you may have used the breathing apparatus many times in the past, you should be checked out each month on the equipment. There should be a maximum allowable time for putting on the apparatus. The apparatus should be checked out under field conditions, such as using ammonia or some other nontoxic gas. Remember, it is too late to learn how fast an operator can put on a breathing apparatus when a room is filled with chlorine.

Be aware that there have been cases where operators have been saved because they knew how to use the breathing apparatus properly. Only repeated practice will enable you to master this survival skill.

20.73 Eye Protection

The water treatment plant operator has only two eyes. You may think that everyone is aware of this fact. However, some operators behave as though they have many eyes and are very careless about protecting the eyes they *DO* have from hazards.

Because some operators fail to see the value of eye protection, it will take a maximum effort on the part of the supervisory staff to enforce an adequate eye protection program. There must be an intense program of education, persuasion, and appeal to guarantee compliance with an eye protection program.

Most conditions in which a plant operator needs eye protection are not too difficult to understand. Eye protection is needed when handling many of the liquid chemicals, acids, and caustics. Some of the tests performed in the laboratory require eye protection. Only a few moments are required to put on a face shield or safety glasses, and remember—the loss of an eye will last a lifetime.

QUESTIONS

Write your answers in a notebook and then compare your answers with those on pages 502 and 503.

20.7A When entering an oxygen-deficient atmosphere, what type of oxygen supply is recommended?

20.7B Where should respiratory apparatus be stored?

20.7C How frequently should independent air supply equipment be checked out and what should be inspected?

20.7D How can compliance with an eye protection program be encouraged?

20.7E Under what conditions does an operator need eye protection?

20.74 Foot Protection

There are few situations under which a water treatment plant operator needs foot protection. In the normal routine of daily operation of the plant, there are not many hazards to the operator's feet. But in some plants the operator also performs plant maintenance. Here the steel-toed safety shoes are useful. The shoe should be able to resist at least a 300-pound (136-kg) impact. An important consideration in any plant under operating conditions is the use of rubber boots. The rubber boots are needed when handling acid or caustic or when the operator is working in wet conditions such as reservoirs, filters, or chemical tanks. Under these circumstances, the agency should have an adequate supply of boots with nonskid soles in various sizes. If these conditions are something that the operator is exposed to daily or weekly, the agency should give the operator a pair of boots for personal use.

20.75 Hand Protection

The treatment plant operator's hands are always exposed to hazards. These include not only minor scratches or cuts but also exposure to chemicals that may not attack immediately. Some compounds, such as alum, attack the skin slowly. Because there is no immediate pain, you may think there is no damage. This is not true; the attack on the skin is slow and may cause an infection at a later date. Therefore, when handling chemicals, always use rubber gloves. As part of a safety package, each operator should be issued a pair of rubber gloves and also a pair of leather gloves. These gloves should be replaced when they no longer provide the necessary protection.

There are other compounds, such as solvents, that will absorb through the skin and can cause long-term effects. For such special problems, there is a need for neoprene gloves. Another problem is that of handling hot materials, such as laboratory flasks and beakers. Here you may need insulated fabric gloves.

When working around machinery that is revolving, wearing gloves or other hand protection can be dangerous. If a glove gets caught in the machinery, you could become injured. Do not let your protective equipment itself become a hazard.

Be sure that the gloves you are wearing are the right type for the job you are doing. The gloves should allow for quick removal and be in good condition. Always check for cracks and holes, flexibility, and grip. Keep them clean and in good condition. There are many types of gloves and the proper type should be worn for each job.

1. *CLOTH GLOVES* protect from general wear, dirt, chafing, abrasions, wood slivers, and low heat.
2. *LEATHER GLOVES* protect from sparks, chips, rough material, and moderate heat.
3. *RUBBER GLOVES* protect against acids and some chemical burns.
4. *NEOPRENE AND CORK-DIPPED GLOVES* give better grip on slippery or oily jobs.
5. *ALUMINIZED GLOVES* are heat-resistant to protect against sparks, flames, and heat.
6. *METAL MESH GLOVES* protect from cuts, rough materials, and blows from edge tools.
7. *PLASTIC GLOVES* protect from chemicals and corrosive substances.
8. *INVISIBLE GLOVES* (barrier cream, such as petroleum jelly) protect from excessive water contact and from substances that dissolve in skin oil.

20.76 Head Protection

In most areas of a water treatment plant, there is really no need for a hard hat. However, there are certain hazardous conditions under which the operators should be required to wear a metal, plastic-impregnated fabric, or fiberglass hat. The hard hat should have a suspended crown with an adjustable head band, provide good ventilation, and be water resistant. Operators should be required to wear the hard hats when work is being performed overhead or in any location where there is danger of tools or other materials falling, for example, working in filters, settling basins, manholes, or trenches. There has been a long history showing the value of hard hats in reducing injuries and death.

20.77 Water Safety

Every operator in a treatment plant is exposed to situations in which the operator's life can be lost due, either directly or indirectly, to water. Although during your daily activity you may never think in terms of drowning, this hazard is always present in the treatment plant. If you are working at a reservoir or a lake in a boat, you may think of water safety, but still never pay real attention to the danger.

Starting at the treatment plant, you can take simple measures that will reduce hazards. To reduce the hazard of slipping on catwalks or when working around clarifiers or settling basins, use nonskid surfaces on ladders and walkways going into and out of clarifiers or sedimentation basins. Be very cautious during cold or wet weather. Water freezes into ice, which is slippery.

Keep all handrails or other guards in good repair; replace any that become unsafe. Many older plants do not have protective handrails; install rails or chain off the unsafe area to all employees and mark off with warning signs. The unsafe areas can be guarded with ⅜-inch (9-mm) Manila rope, chains, or cables that you may have around the plant.

Filters are an important area of safety consideration because there is always activity in or around each filter, such as washing or maintenance. Make repairs to handrails immediately, when needed. Station emergency gear around the filter areas; equipment such as life jackets are good, but buoys, ⅜-inch (9-mm) Manila line, or a long wooden pole are much more useful. These types of devices can be used to rescue someone who has fallen into the filter. An operator should never work in the filter when it is being backwashed. There is always the danger of falling into the wash-water gullet and being unable to get out before drowning.

Sedimentation basins, flocculation basins, or clarifiers present many of the same problems as filters. Maintain handrails, place warning signs, or put up guard ropes or chains. Also, keep life rings and Manila or nylon lines in good repair. A lift ring, life pole, and lines should be stationed at each basin. A good idea is to shelter the safety gear from the weather, but do not cause the gear to become inaccessible.

In reservoir operation and maintenance, you will encounter two types of water: (1) raw water, and (2) treated water. In a raw water reservoir or lake, you have to worry only about personnel safety. In a treated water reservoir, you must also be concerned about the safety of the water going to the customer. If you are working out of a boat, make sure that everyone in the boat is wearing a life jacket. Also, take on board both a safety line and buoys. Cold-weather conditions are an added problem. Even though you may be a good or excellent swimmer, the thermal shock of cold water may quickly paralyze you, making you unable to save yourself. Under such conditions, if a second operator goes into the water to save you, there may be two lives lost.

Of course, when taking a boat out on the water, it should first be checked out for safety. Check the bilge pump, ventilation in the compartments, safety cushions, fire extinguishers, battery, and engine. Also, check for safety equipment, oars, life jackets, lights, mooring lines, and fuel. If you are applying copper sulfate powder or solution, other safety equipment will be needed, such as respiratory and eye protection equipment. Prepare a detailed equipment checklist to use each time the boat goes out onto the lake or reservoir. The boat should never be taken out on choppy waters or when the wind is high.

On some occasions, there is a need for underwater examination of valves, intake, or other underwater equipment or apparatus. Such work should only be performed by employees who are trained in underwater diving. If there are no qualified divers on your staff, you should hire such personnel to do the diving and underwater inspections. There are organizations with people who do this type of work and they are well qualified in underwater examinations. If an operator on staff is to do the diving, the operator should be certified by a local diving school or other certifying agency. The operator's certificate should always be kept current and the operator should be required to perform the number of dives necessary to keep this certification current.

In closing, all plant operators should know how to swim. If they do not, they should take a Red Cross class and learn the minimum fundamentals to save their own lives. Any operator working over open water should be required to wear a buoyant vest. All basins should have approved safety vests, buoys, and life lines stationed at outside edges.

QUESTIONS

Write your answers in a notebook and then compare your answers with those on page 503.

20.7F Under what operating conditions should an operator wear rubber boots?

20.7G Under what specific conditions should an operator be very careful wearing gloves?

20.7H Why should operators never work in a filter when it is being backwashed?

20.7I What items should be checked before an operator takes a boat out on the water?

20.8 PREPARATION FOR EMERGENCIES[11]

Emergencies are very difficult to plan and prepare for because you never know what will happen and when it will occur. Catastrophic events could include floods, tornados, hurricanes, fires, and earthquakes. Serious injury to anyone on the plant grounds is an emergency.

Conduct periodic tours of your facilities with the local fire, police, and emergency response organizations to familiarize them with the site, potentially hazardous locations, and location of fire hydrants. Their familiarity with the plant layout will be very helpful if an emergency ever occurs. Emphasize to these people that, if a disaster occurs, it is important for your plant to be a top priority for assistance because the entire community relies on you for its drinking water.

You should know the names and phone numbers of your local and state civil preparedness coordinators.

If a chemical emergency occurs, such as a chemical spill, leak, fire, exposure, or accident, phone *CHEMTREC*, (800) 424-9300. *CHEMTREC* (Chemical Transportation Emergency Center) provides immediate advice for those at the scene of a chemical emergency, and then quickly and promptly alerts experts from the manufacturers whose products are involved for more detailed assistance and appropriate followup. Also, notify local and state civil preparedness coordinators.

Prepare a procedure for quick and efficient handling of all accidents or injuries occurring in your treatment facilities or involving your outside crews. All personnel must be familiar with these procedures and must be prepared to carry them out with a minimum amount of delay or confusion.

A copy of these procedures must be posted in all working areas accessible to a phone and in all vehicles containing work crews. Names, addresses, and phone numbers of operators in each working area should be listed in that area and also those immediately available (day or night) by telephone.

Everyone must study these procedures carefully and be able to respond properly and quickly. Your health and life may depend on these procedures.

Figure 20.11 is an example of a typical safety procedure and Figure 20.12 is a checklist of what must be done if someone is seriously injured.

QUESTIONS

Write your answers in a notebook and then compare your answers with those on page 503.

20.8A What types of emergencies should operators be prepared to handle?

20.8B Who should be contacted if a serious chemical emergency occurs, such as a chemical spill, leak, fire, exposure, or accident?

20.8C What is *CHEMTREC?*

20.9 ARITHMETIC ASSIGNMENT

Turn to the Arithmetic Appendix at the back of this manual. Read and work the problems in Section A.37, "Administration, Safety." Check the arithmetic in this section using an electronic calculator. You should be able to get the same answers.

---

[11] Some of the information in this section was provided by Mr. Richard R. Metcalf, Training Officer, County of Onondaga, NY.

498 Water Treatment

EMERGENCY SAFETY PROCEDURE

1. *DO NOT MOVE THE INJURED PERSON* except when conditions would cause additional injury, such as a gas leak or a fire.
2. *ADMINISTER ONLY SUCH AID AS NECESSARY TO PRESERVE LIFE—TREAT FOR SHOCK*
   - (a) clear throat and restore breathing
   - (b) stop bleeding
   - (c) closed heart massage
3. *DO NOT ATTEMPT MEDICAL TREATMENT* such as
   - (a) do not apply splints or attempt to set broken bones
   - (b) do not remove foreign objects from the body
   - (c) do not administer liquids or oxygen
4. *NOTIFY YOUR SUPERVISOR*

IF AN AMBULANCE IS REQUIRED:
1. CALL AMBULANCE—phone _____
2. Give this information carefully and accurately:
   - (a) Location of the injured—be specific
      1. Street location and number, town or city
         **Remember some streets have north or south or east or west designation—use the full street name. Also, many streets in different towns have the same name—specify the town.
      2. Location within the plant area
      (b) Phone number from which you are calling
      (c) Number of persons injured and nature of the injury
      (d) Post an operator to direct the ambulance to the victim
3. Upon arrival of the ambulance:
   - (a) give name, address, and phone number of the injured person to the ambulance crew
   - (b) notify relatives of injury and hospital to which person is being taken
     (Medical treatment cannot be given without the permission of the injured or a relative, if a minor)
4. Call your supervisor

IF AN AMBULANCE IS NOT REQUIRED:
1. Take injured: (see map) (phone _____), ask for Emergency Room)
   Emergency Room
   St. Joseph's Hospital
   301 Prospect Avenue
2. If possible, call ahead. Give the names of injured and nature of injury
3. Notify relatives of injury and address of hospital
4. Call your supervisor

Fig. 20.11 *Emergency safety procedure*

INJURED PERSON CHECKLIST

1. *CALL AMBULANCE SERVICE*, phone _____

   LOCATION OF INJURED _____ STREET _____

   _____ TOWN _____

   _____ BUILDING LOCATION _____

   NUMBER OF PERSONS INJURED _____ PHONE NUMBER _____

   NATURE OF INJURY _____

2. *POST OPERATOR TO DIRECT THE AMBULANCE*
   NAME, ADDRESS, PHONE OF INJURED PERSON

   NAME _____

   ADDRESS _____

   PHONE _____

3. *GIVE ABOVE INFORMATION TO AMBULANCE CREW*
   NAME AND LOCATION OF HOSPITAL

   HOSPITAL NAME _____

   ADDRESS _____

   PHONE _____

4. *NOTIFY RELATIVES*
5. *NOTIFY SUPERVISORS*
6. *MAKE OUT ACCIDENT REPORT AS SOON AS POSSIBLE*

Fig. 20.12 *Injured person checklist*

20.10 ADDITIONAL READING

This chapter cannot provide you with all of the information you need to respond to the many safety problems that occur daily. You should have other sources of ready references. Because this chapter does not give all of the detail you may need, the following bibliography is given. You should have copies of these documents for your reference or have your employer obtain them.

1. *CHLORINE MANUAL,* Sixth Edition. Obtain from the Chlorine Institute, Inc., Bookstore, PO Box 1020, Sewickley, PA 15143-1020. Pamphlet 1. Price to members, $28.00; nonmembers, $70.00; plus $6.95 shipping and handling.

2. *FISHER SAFETY CATALOG.* Obtain from Fisher Scientific Company, Safety Division, 4500 Turnberry Drive, Suite A (Customer Service), Hanover Park, IL 60103, or phone (800) 772-6733.

3. The American Red Cross is another source of up-to-date information on safety and first aid. Contact your local Area Chapter for a catalog of materials.

4. *SAFETY PRACTICES FOR WATER UTILITIES* (M3). Obtain from American Water Works Association (AWWA), Bookstore, 6666 West Quincy Avenue, Denver, CO 80235. Order No. 30003. ISBN 1-58321-190-X. Price to members, $72.50; nonmembers, $102.50; price includes cost of shipping and handling.

5. *LET'S TALK SAFETY—SAFETY TALKS 2006.* A series of 52 lectures on common water utility safety practices. Obtain from American Water Works Association (AWWA), Bookstore, 6666 West Quincy Avenue, Denver, CO 80235. Order No. 10123. ISBN 1-58321-402-X. Price to members, $45.50; nonmembers, $66.50; price includes cost of shipping and handling.

END OF LESSON 3 OF 3 LESSONS

on

SAFETY

Please answer the discussion and review questions next.

DISCUSSION AND REVIEW QUESTIONS

Chapter 20. SAFETY

(Lesson 3 of 3 Lessons)

Write the answers to these questions in your notebook. The question numbering continues from Lesson 2.

26. How can the hazards of fire be reduced in electric starters?

27. How should gas cylinders be stored in laboratories?

28. What safety precautions should be taken with regard to bacteria while working in the laboratory?

29. What is the most common reason safety equipment falls into disrepair?

30. How can an operator reduce the hazard of slipping when working around clarifiers or settling basins?

SUGGESTED ANSWERS
Chapter 20. SAFETY

ANSWERS TO QUESTIONS IN LESSON 1

Answers to questions on page 459.

20.0A Management's responsibilities for safety include:
1. Establishing a safety policy
2. Assigning responsibility for accident prevention
3. Appointing a safety officer or coordinator
4. Establishing realistic goals and periodically revising them
5. Evaluating the results of the program

20.0B Operators' responsibilities for safety include:
1. Performing their jobs in accordance with established safe procedures
2. Recognizing the responsibility for their own safety and that of fellow operators
3. Reporting all injuries
4. Reporting all observed hazards
5. Actively participating in the safety program

Answers to questions on page 460.

20.0C An accident is that occurrence in a sequence of events that usually produces unintended injury, death, or property damage.

20.0D The principal reasons for unsafe acts include ignorance, indifference, poor work habits, laziness, haste, poor physical condition, and temper.

20.0E Laziness affects the speed and quality of work. Laziness also affects safety because safety requires an effort. In most jobs, you cannot reduce your safety effort and still maintain the same level of safety, even if you slow down or lower your quality standards.

Answers to questions on page 464.

20.10A An operator needs to know how to handle the problems associated with the chemicals used in a water treatment plant. The operator needs to know how to store chemicals, understand the fire problem, be aware of the tendency of chemicals to "arch" in a storage bin, and know how to feed dry, how to feed liquid, and how to make up solutions. Overheating gas containers, dust problems with powdered carbon, burns caused by acid, reactivity of each chemical under a variety of conditions that may cause fire and explosions are other safety hazards that an operator needs to know about and know how to control. Also, the operator needs to know the usable limits of each chemical because of toxicity, the protective equipment required for each chemical, each chemical's antidote, and how to control fires caused by each chemical.

20.11A To give first aid when acid vapors are inhaled, remove the victim to fresh air, restore breathing, or give oxygen, when necessary.

20.11B Acetic acid will react violently with ammonium nitrate, potassium hydroxide, and other alkaline materials.

20.11C Acetic acid can be handled safely if the operator uses adequate ventilation and prevents skin and eye contact.

20.11D When handling hydrofluosilicic acid, always use complete protective equipment including rubber gloves, goggles or face shield, rubber apron, rubber boots, and have lime slurry barrels, Epsom salt solution, and safety showers available. Always provide adequate ventilation.

20.11E Inhalation of hydrochloric acid (HCl) vapors or mists can cause damage to teeth and irritation to the nasal passages. Concentrations of 750 ppm or more will cause coughing, choking, and produce severe damage to the mucous membranes of the respiratory tract. In concentrations of 1,300 ppm, HCl is dangerous to life.

20.11F Like other acids, nitric acid should be stored in clean, cool, well-ventilated areas. The area should have an acid-resistant floor and adequate drainage. Keep away from oxidizing agents and alkaline materials. Protect containers from damage or breakage. Avoid contact with skin and provide emergency neutralization materials and safety equipment in use areas.

Answers to questions on pages 466 and 467.

20.12A Operators use two forms of ammonia: the gaseous form (anhydrous) and the liquid form (hydroxide).

20.12B Care must be used when storing or transporting ammonia containers. Always keep cylinder caps in place when cylinders are not in use. Store cylinders in a cool, dry location away from heat, and protect from direct sunlight. Do not store in the same room with chlorine.

20.12C The two forms of lime used in water treatment plants are: (1) hydrated lime (calcium hydroxide), and (2) quicklime (calcium oxide).

20.12D If someone swallowed sodium hydroxide, give large amounts of water or milk and immediately transport to a medical facility; *DO NOT INDUCE VOMITING*.

20.12E If sodium silicate comes in contact with your skin, wash thoroughly with water, followed by washing with a 10-percent solution of ammonium chloride or 10 percent acetic acid.

Answers to questions on page 471.

20.13A Chlorine leaks are most often found in the control valve.

20.13B The purpose of the fusible metal plugs is to melt at 158 to 168°F. If a cylinder becomes overheated, the plugs will melt and let the gas escape rather than the cylinder bursting.

20.13C Chlorine leaks can be detected by the odor, by the use of ammonia water on a small cloth or swab on a stick, or by the use of an aspirator containing ammonia water. (Remember not to spray ammonia into a room full of chlorine because a white cloud will form and you will not be able to see anything.) Also, a chlorine gas detector may be used.

20.13D Carbon dioxide is a safety hazard because it is odorless, colorless, and will accumulate at the lowest possible level. Carbon dioxide will displace oxygen so you must use a self-contained breathing apparatus, not a canister gas mask.

Answers to questions on page 473.

20.14A For handling most salts, ventilation, respiratory protection, and eye protection will prove adequate.

20.14B First aid when liquid or dry alum gets into the eyes consists of flushing them immediately for 15 minutes with large amounts of water. Alum should be washed off the skin with water because prolonged contact will cause skin irritation.

20.14C When exposed to moist air or light, ferric chloride decomposes and gives off hydrochloric acid.

Answers to questions on page 475.

20.15A Potassium permanganate spills can be swept up. Flushing with water is an effective way to eliminate any residue remaining on floors.

20.15B Powdered activated carbon is the most dangerous powder the water treatment plant operator will be exposed to.

20.15C Activated carbon should be stored in a clean, dry, fireproof location. Keep the area free of dust, protect it from flammable materials, and do not permit smoking in the area at any time when handling or unloading activated carbon.

20.15D The key to preventing activated carbon fires is keeping the storage area clean and free of dust.

20.15E Carbon fires should be controlled by carbon dioxide (CO_2) extinguishers or hoses equipped with fog nozzles. An activated carbon fire should not be doused with a stream of water. The water may cause burning carbon particles to fly, resulting in a greater fire problem.

Answers to questions on page 475.

20.17A Safety regulations prohibit the use of common drains and sumps from chemical storage areas to avoid the possibility of chemicals reacting and producing toxic gases, explosions, and fires.

20.17B If a polymer solution comes in contact with potassium permanganate, a fire could develop.

ANSWERS TO QUESTIONS IN LESSON 2

Answers to questions on page 479.

20.2A Class A fires involve ordinary combustible materials. These include wood, paper, cloth, rubber, many plastics, dried grass, hay, and stubble.

20.2B Foam extinguishers can control Class A and Class B fires. Class A fires are those that consume ordinary combustibles such as wood, paper, cloth, rubber, many plastics, dried grass, hay, and stubble. Class B fires are those that consume flammable and combustible liquids such as gasoline, oil, grease, tar, oil-based paint, lacquer, and solvents, and also flammable gases.

20.2C An electrical fire can be extinguished by the use of carbon dioxide (CO_2) extinguishers or with a dry chemical extinguisher.

Answers to questions on page 483.

20.3A When waxing floors, use compounds containing nonslip ingredients. Warn others about newly waxed floors. Try to do cleaning and waxing during weekends or at night when foot traffic is light.

20.3B Rags are always a problem if they contain oils, paint, or other cleaning compounds; there is always the possibility of fire. The rags should be placed into a closed metal container to reduce the fire hazard.

20.3C When operating an overhead crane, the following safety precautions must be exercised:

1. Allow only trained and authorized personnel to operate the overhead crane.
2. Inspect the circuit breaker, limit switches, the condition of the hook, the wire rope, and other safety devices.
3. Post load limits on the crane and never overload the crane.
4. Check each lift for proper balance.
5. Use a standard set of hand signals.
6. Be sure everyone in the vicinity wears a hard hat.
7. Allow only authorized personnel to make repairs.
8. Lock out the main power switch before repairs begin.
9. Try to avoid moving loads over populated areas.
10. Set up monthly safety inspection forms to be filled out and placed into the maintenance file.

20.3D Traffic can be warned that operators are working in a manhole by the use of barricades, signs, flags, lights, and other warning devices. Warning devices and procedures must conform to local and state regulations.

20.3E Operators should always use a mechanical lifting aid (rope and bucket) for raising or lowering tools and equipment into and out of a manhole. The use of a bucket or basket will keep your hands free when climbing down into or out of the manhole.

Answers to questions on page 485.

20.3F Operators should wear eye and ear protection when operating grinding, chipping, buffing, or pavement-breaking equipment. Sometimes, when using grinding or buffing tools, operators encounter toxic dusts or fumes and therefore need respiratory protection. At other times, there is a need for full face protection because of flying particles.

20.3G Operators can be protected from high noise levels by wearing approved ear protective devices.

20.3H When operating welding equipment, the operator should wear protective clothing, gloves, helmet, and goggles.

Answers to questions on page 487.

20.4A Tire wear is caused by misalignment and low inflation.

20.4B Motor vehicles should have flashlights, flares, flags, fire extinguisher, and first-aid kit.

ANSWERS TO QUESTIONS IN LESSON 3

Answers to questions on page 492.

20.5A Each operator should become familiar with the type of current and voltage in the plant in order to avoid any mistake of becoming involved with unsafe electric currents or practices for which they are not trained. This will also enable operators to know when to ask for a qualified person to perform any necessary repairs.

20.5B Moving parts on electric motors that require safety guards include exposed couplings, pulleys, gears, and sprockets, as well as projections such as bolts, keys, or set screws.

20.5C Two hazards due to the lack of a good lockout procedure for control panels and switchboards are: (1) accidentally starting a piece of equipment, exposing a fellow operator to a hazard, and (2) turning electrical power on when someone is still working on the equipment, exposing that person to danger.

20.5D Examples of stored energy that could be hazardous to operators include electrical power; air, gas, or steam pressure; water (hydraulic) pressure; spring tension; rotational forces in flywheels; and the energy in elevated machine parts if they suddenly swing free.

Answers to questions on page 494.

20.6A Washing glassware is always a potential hazard because the glassware can be broken while being washed, causing cuts, or cuts can be caused by chipped or cracked glassware. If your hands come in contact with strong acid cleaners, the acids may cause serious burns.

20.6B When handling liquid chemicals, such as acids and bases, always use safety glasses or face shields.

20.6C Never allow mercury, gasoline, oil, or organic compounds into the laboratory drains. Use only a toxic waste disposal drain system for these items. Flushing such compounds down sink drains can cause an explosion, allow toxic gases and vapors to enter the lab, or destroy the piping.

20.6D The plant operator may be exposed to radioactive compounds when calibrating sludge density meters or using research isotopes.

20.6E Never allow cold water into the hot boiler unit of a water still because it may cause the unit to break.

Answers to questions on page 495.

20.7A When entering an oxygen-deficient atmosphere, you should have an independent oxygen supply of the positive-pressure type to protect you if there are any leaks in your mask.

20.7B Respiratory apparatus must be stored outside of chlorination, sulfur dioxide, carbon dioxide, ozone, and ammonia rooms in an unlocked cabinet. The storage cabinets must have a controlled environment to prevent deterioration of the equipment.

20.7C Independent air supply equipment should be checked out on a weekly basis, and records kept of each inspection. The record should show conditions of the hoses, regulators, air filters, compressors, helmet, and any other apparatus furnished with the system.

20.7D To obtain compliance with an eye protection program, supervisors should undertake an intense program of education, persuasion, and appeal.

20.7E An operator needs eye protection when handling many of the liquid chemicals, acids, and caustics. Many of the tests performed in the laboratory also require eye protection.

Answers to questions on page 497.

20.7F Rubber boots are needed when handling acid or caustic, or when the operator is working in wet conditions such as reservoirs, filters, or chemical tanks.

20.7G An operator should be very careful wearing gloves when working around machinery that is revolving.

20.7H Operators should never work in a filter when it is being backwashed because there is always the danger of falling into the wash-water gullet and being unable to get out before drowning.

20.7I Before taking a boat out on the water, it should be checked out for safety. The operator should check the bilge pump, ventilation in the compartments, the safety cushions, fire extinguishers, battery, and the engine. Also, check for safety equipment, oars, life jackets, lights, mooring lines, and fuel.

Answers to questions on page 497.

20.8A Operators should be prepared for catastrophic events such as floods, tornados, hurricanes, fires, and earthquakes. Serious injury to anyone on the plant grounds is an emergency.

20.8B If a serious chemical emergency occurs, such as a chemical spill, leak, fire, exposure, or accident, phone *CHEMTREC,* (800) 424-9300. Also, local and state civil preparedness coordinators should be notified.

20.8C *CHEMTREC* is an emergency response service that provides immediate advice (by telephone) about what to do in a chemical emergency, such as a spill, leak, fire, exposure, or accident.

CHAPTER 21

ADVANCED LABORATORY PROCEDURES

by

Jim Sequeira

TABLE OF CONTENTS

Chapter 21. ADVANCED LABORATORY PROCEDURES

	Page
OBJECTIVES	508

LESSON 1

21.0	USE OF A SPECTROPHOTOMETER	509
21.1	TEST PROCEDURES[1]	510
	1. Algae Counts	510
	2. Calcium	511
	3. Chloride	512
	4. Color	514
	5. Dissolved Oxygen	515
	6. Fluoride	519

DISCUSSION AND REVIEW QUESTIONS ... 522

LESSON 2

7. Iron (Total)	523
8. Manganese	526
9. Marble Test (Calcium Carbonate Saturation Test)	528
10. Metals	530
11. Nitrate	530
12. pH	534
by Jack Rossum	
13. Specific Conductance (Conductivity)	535
14. Sulfate	535
15. Taste and Odor	538
16. Trihalomethanes	542
17. Total Dissolved Solids	543

DISCUSSION AND REVIEW QUESTIONS ... 545
SUGGESTED ANSWERS ... 546

---

[1] Test Procedures in Chapter 11 (Volume I) include alkalinity, chlorine residual, chlorine demand, coliform bacteria, hardness, jar test, pH, temperature, and turbidity.

OBJECTIVES

Chapter 21. ADVANCED LABORATORY PROCEDURES

Following completion of Chapter 21, you should be able to:

1. Explain how a spectrophotometer analyzes samples of water.

2. Perform the following field or laboratory tests—algae counts, calcium, chloride, color, dissolved oxygen, fluoride, iron, manganese, marble test, metals, nitrate, pH, specific conductance, sulfate, taste and odor, trihalomethanes, and total dissolved solids.

CHAPTER 21. ADVANCED LABORATORY PROCEDURES
(Lesson 1 of 2 Lessons)

21.0 USE OF A SPECTROPHOTOMETER

In the field of water analysis, many determinations such as iron, manganese, and phosphorus are based on the color intensity formed when a specific color developing reagent is added to the sample being tested. Measuring the intensity of the color enables the concentration of the substance to be determined. The simplest means of accomplishing this is through either Nessler tubes or a pocket comparator. The color developed in a sample is compared by the operator to a series of known standards, to each of which has been added the same color developing reagents. For the analysis of phosphorus present in a water sample, for example, ammonium molybdate reagent is added as the color developing reagent. If phosphorus is present, a blue color develops. The more phosphorus there is, the deeper and darker the blue color.

The human eye can detect some differences in color intensity; however, for very precise measurements an instrument called a spectrophotometer (SPEK-tro-fo-TOM-uh-ter) is used.

THE SPECTROPHOTOMETER. A spectrophotometer is an instrument generally used to measure the color intensity of a chemical solution. A spectrophotometer in its simplest form consists of a light source that is focused on a prism or other suitable light dispersion device to separate the light into its separate bands of energy. Each different wavelength or color may be selectively focused through a narrow slit. This beam of light then passes through the sample to be measured. The sample is usually contained in a glass tube called a cuvette (kyoo-VET). Most cuvettes are standardized to have a 1.0-cm light path length; however, many other sizes are available.

After the selected beam of light has passed through the sample, it emerges and strikes a photoelectric cell. If the solution in the sample cell has absorbed any of the light, the total energy content will be reduced. If the solution in the sample cell does not absorb the light, then there will be no change in energy. When the transmitted light beam strikes the photoelectric tube, it generates an electric current that is proportional to the intensity of light energy striking it. By connecting the photoelectric tube to a galvanometer (a device for measuring electric current) with a graduated scale, a means of measuring the intensity of the transmitted beam is achieved.

The diagram below illustrates the working parts of a spectrophotometer.

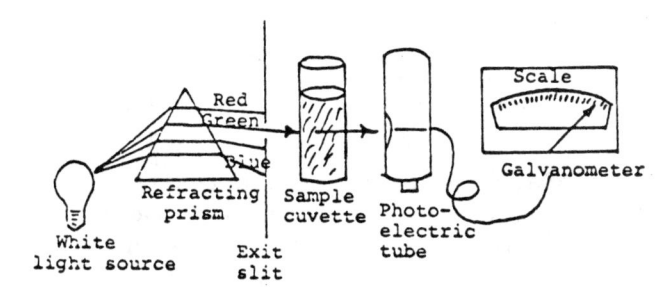

The operator should always follow the working instructions provided with the instrument.

UNITS OF SPECTROSCOPIC MEASUREMENT. The scale[2] on spectrophotometers is generally graduated in two ways:

1. In units of percent transmittance (%T), an arithmetic scale with units graded from 0 to 100%

2. In units of absorbance (A), a logarithmic scale of nonequal divisions graduated from 0.0 to 2.0

Both the units, percent transmittance and absorbance, are associated with color intensity. That is, a sample that has a low color intensity will have a high percent transmittance but a low absorbance.

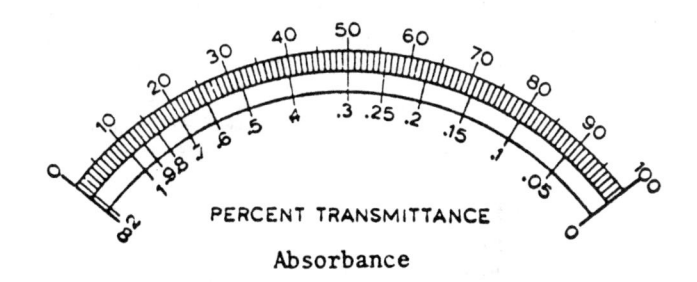

As illustrated above, the absorbance scale is ordinarily calibrated on the same scale as percent transmittance on spectrophotometers. The chief usefulness of absorbance lies in the fact that it is a logarithmic function rather than linear (arithmetic) and a law known as Beer's Law states that the concentration of a light-absorbing colored solution is directly proportional to absorbance

---

[2] Some specialized spectrophotometers also contain a scale that directly reads the concentration of one chemical constituent in milligrams per liter.

(Algae Counts)

over a given range of concentrations. If one were to plot a graph showing percent transmittance (%T) versus concentration on straight graph or line paper and another showing absorbance versus concentration on the same paper, the following curves (graphs) would result:

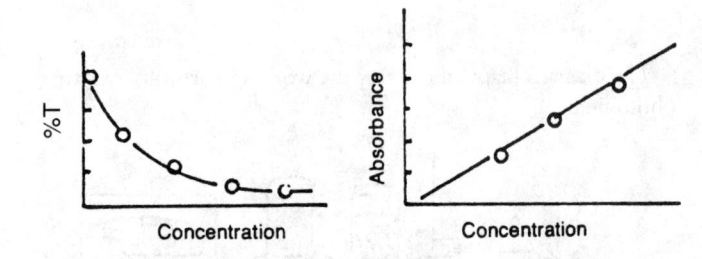

CALIBRATION CURVES: The calibration curve is used to determine the concentration of the water quality indicator (iron or manganese) contained in a sample. Three steps must be completed in order to prepare a calibration graph.

First, a series of standards must be prepared. A standard is a solution that contains a known amount of the same chemical constituent that is being determined in the sample.

Second, these standard solutions and a sample containing none of the constituent being tested for (usually distilled water and generally referred to as a blank) must be treated with the developing reagent in the same manner as the sample would be treated.

Third, using a spectrophotometer, the absorbance or transmittance at the specified wavelength of the standards and blank must be determined. From the values obtained, a calibration curve of absorbance (or %T) versus concentration can be plotted. Once you have plotted these several points, you can then extend the plotted points by connecting the known points with a straight line. For example, with the data given below you could construct the following calibration curve.

Absorbance	Concentration, mg/L
0.0	0.0
0.30	0.25
0.55	0.50
0.80	0.75

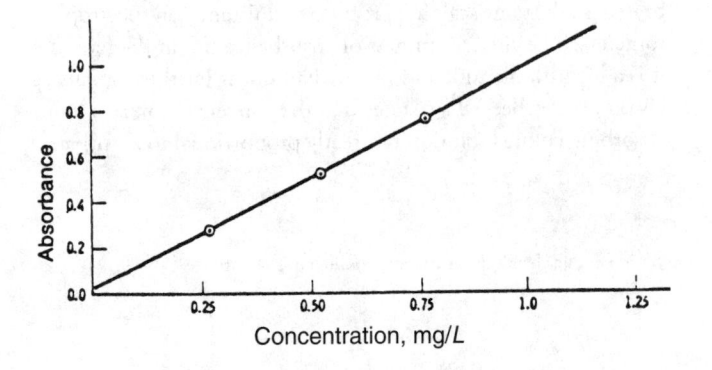

Once you have established a calibration curve for the water quality indicator in question, you can easily determine the amount of that substance contained in a solution of unknown concentration. You merely take an absorbance reading on the color developed by the unknown and locate it on the vertical axis. Draw a straight line to the right on the graph until it intersects with the experimental standard curve. Then drop a line to the horizontal axis; this value identifies the concentration of your unknown water quality indicator.

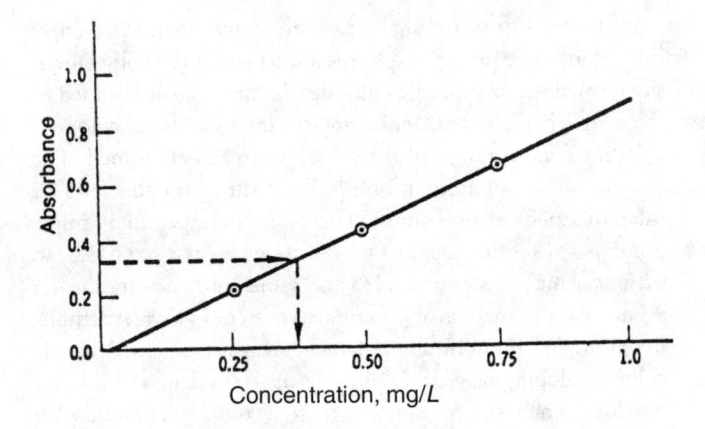

In this example, an absorbance reading of 0.32 was read on the unknown solution or sample, which indicates a concentration of about 0.37 mg/L.

QUESTIONS

Write your answers in a notebook and then compare your answers with those on page 546.

21.0A When measuring phosphorus concentrations, the intensity of what color is measured?

21.0B What are the units of measurements for spectrophotometers?

21.0C Using the previous absorbance versus concentration calibration graph, if an absorbance of 0.60 was read on an unknown solution or sample, what would be the concentration of the unknown?

21.1 TEST PROCEDURES

1. Algae Counts

A. Discussion

The quality of water in any lake, reservoir, or stream has a very direct effect on the abundance and types of aquatic organisms found. By knowing the nature and numbers of these aquatic organisms, one can obtain a good idea of the water quality. A biological method used for measuring water quality is the collection, counting, and identification of algae contained in a particular body of water. Information from algae counts can serve one or more of the following purposes:

1. Help explain the cause of color and turbidity and the presence of tastes and odors in the water

(Calcium)

2. Help explain the clogging of screens or filters

3. Help document variability in the water quality

Algae counting and identification may be done very simply or it may be developed into a highly technical operation. The beginner should use great caution applying the results of algae identifications until considerable experience has been gained.

Some operators perform algae counts on both the raw water and treated water. Taking algae counts on treated water is a means of studying the effectiveness of coagulation and the performance of filters. If the filters are performing properly, there should not be any countable algae in the treated water.

B. Materials and Procedures

See page 10-2, STANDARD METHODS, 21st Edition.[3]

2. Calcium

A. Discussion

In most natural waters, calcium is the principal cation. The element is widely distributed in the common minerals of rocks and of soil. Calcium in the form of lime or calcium hydroxide may be used to soften water or to control corrosion through pH adjustment.

B. What is Tested?

Sample	Common Range, mg/L
Raw and Treated Surface Water	5 to 50
Well Water	10 to 100

C. Apparatus Required

Buret, 25 mL
Buret support
Graduated cylinder, 100 mL
Beaker, 100 mL
Magnetic stirrer
Magnetic stir-bar

D. Reagents

NOTE: Standardized solutions are commercially available for most reagents. Refer to STANDARD METHODS if you wish to prepare your own reagents.[4]

1. Sodium hydroxide (NaOH), 1 N.

2. Eriochrome Blue Black R indicator.

3. Standard EDTA titrant, 0.01 M. Standardize against Standard Calcium Solution and store in plastic polyethylene bottle.

4. Standard Calcium Solution. Store in polyethylene plastic bottle. 1 mL of this solution = 1 mg calcium hardness as $CaCO_3$ or 400.8 micrograms (µg) Ca.

5. 1 + 1 HCl: Carefully add 50 mL concentrated HCl to 50 mL distilled water.

6. Ammonium hydroxide, 3 N.

7. Methyl Orange indicator solution.

E. Procedure

1. Take a clean beaker and add 50 mL of sample.

2. Add 2.0 mL NaOH solution.

3. Add 0.1 to 0.2 gm indicator mixture.

4. Titrate immediately with EDTA titrant until last reddish-purple tinge disappears. Mix with magnetic stirrer during titration.

5. Calculate calcium concentration.

F. Example

Results from calcium testing of a treated water sample were as follows:

Sample Size $\quad = 50$ mL

mL EDTA Titrant Used, A = 7.3 mL

G. Calculation

$$\text{mg Ca}/L = \frac{A \times 400.8^*}{mL \text{ of Sample}}$$

$$= \frac{7.3 \text{ mL} \times 400.8}{50 \text{ mL}}$$

$$= 59 \text{ mg}/L$$

---

* 400.8 is a constant for this calculation. It contains the conversion units to produce mg/L.

H. Precautions

1. Titrate immediately after adding NaOH solution.

2. Use 50 mL or a smaller portion of sample diluted to 50 mL with distilled water so that the calcium content is about 5 to 10 mg.

3. For hard waters with alkalinity greater than 300 mg $CaCO_3/L$, use a smaller portion or neutralize alkalinity with acid, boiling for one minute, and cooling before beginning the titration.

I. Reference

See page 3-65, STANDARD METHODS, 21st Edition.

---

[3] STANDARD METHODS FOR THE EXAMINATION OF WATER AND WASTEWATER, 21st Edition. Obtain from American Water Works Association (AWWA), Bookstore, 6666 West Quincy Avenue, Denver, CO 80235. Order No. 10084. ISBN 0-87553-047-8. Price to members, $198.50; nonmembers, $266.00; price includes cost of shipping and handling.

[4] See "Prepared vs. Do-It-Yourself Reagents," by Josephine W. Boyd, OPFLOW, Vol. 9, No. 10, October 1983.

(Chloride)

OUTLINE OF PROCEDURE FOR CALCIUM

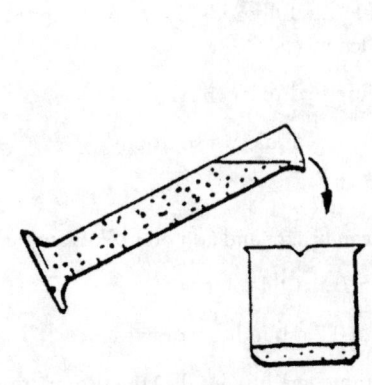

1. Add 50 mL to a clean beaker.
2. Add 2 mL NaOH and 0.2 gm indicator mixture.
3. Titrate with EDTA. Mix with magnetic stirrer.

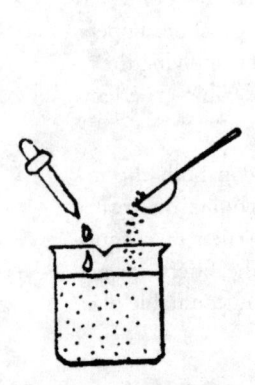

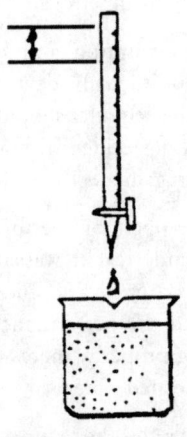

3. Chloride

A. Discussion

Chloride occurs in all natural waters, usually as a metallic salt. In most cases, the chloride content increases as mineral content increases. Mountain water supplies usually are quite low in chloride while groundwaters and valley rivers often contain a considerable amount. The maximum allowable chloride concentration of 250 mg/L in drinking water has been established for reasons of taste rather than as a safeguard against a physical or a health hazard. At concentrations above 250 mg/L, chloride may give a salty taste to the water that is objectionable to many people.

B. What is Tested?

Sample	Common Range, mg/L
Surface or Groundwater	2 to 100

C. Apparatus Required

Graduated cylinder, 100 mL
Buret, 50 mL
Erlenmeyer flask, 250 mL
Pipet, 10 mL
Magnetic stirring apparatus

D. Reagents

NOTE: Standardized solutions may be purchased from chemical suppliers.

1. Chloride-free water—distilled or deionized water.
2. Potassium chromate (K_2CrO_4) indicator solution.
3. Standard Silver Nitrate Titrant, 0.0141 N.
4. Standard Sodium Chloride, 0.0141 N.

E. Procedure

1. Place 100 mL or a suitable portion of sample diluted to 100 mL in a 250-mL Erlenmeyer flask.
2. Add 1.0 mL K_2CrO_4 indicator solution.
3. Titrate with standard silver nitrate to a pinkish yellow end point. Be consistent in end point recognition. Compare with known standards of various chloride concentrations.

F. Calculation

$$\text{Chloride (as Cl), mg}/L = \frac{(A - B) \times N \times 35{,}450^*}{mL \text{ of Sample}}$$

A = mL $AgNO_3$ used for titration of sample
B = mL $AgNO_3$ used for blank
N = normality of $AgNO_3$

---

* 35,450 is a constant for this calculation. It contains the conversion units to produce mg/L.

G. Example

Sample Size = 100 mL

A = mL $AgNO_3$ used for titration of sample = 10.0 mL
B = mL $AgNO_3$ used for blank = 0.4 mL
N = normality of $AgNO_3$ = 0.0141 N

$$\text{Chloride, mg}/L = \frac{(10.0 \text{ mL} - 0.4 \text{ mL}) \times (0.141\ N) \times 35{,}450}{100 \text{ mL}}$$

$$= 48 \text{ mg}/L$$

(Chloride)

OUTLINE OF PROCEDURE FOR CHLORIDE

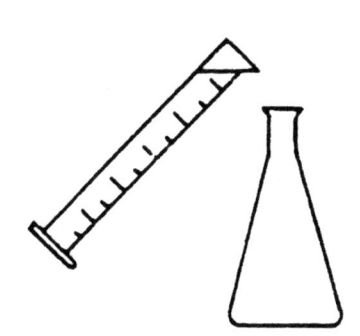

1. Place 100 mL or other measured sample in flask.

2. Add 1 mL chromate indicator.

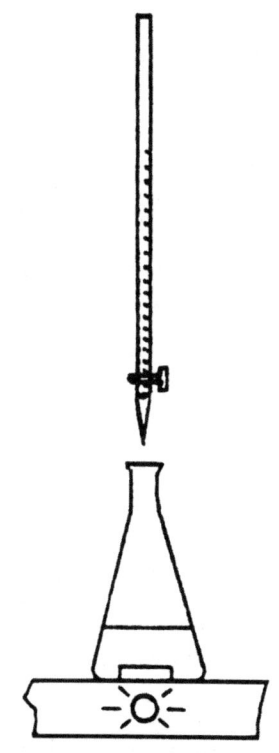

3. Place flask on magnetic stirrer and titrate with standard silver nitrate.

H. Special Notes

1. Sulfide, thiosulfate, and sulfite ions interfere, but can be removed by treatment with 1 mL of 30 percent hydrogen peroxide (H_2O_2).

2. Highly colored samples must be treated with an aluminum hydroxide suspension and then filtered.

3. Orthophosphate in excess of 25 mg/L and iron in excess of 10 mg/L also interfere.

4. If the pH of the sample is not between 7 and 10, adjust with 1 N sulfuric acid or 1 N sodium hydroxide.

5. Procedure for standardization of $AgNO_3$:

   a. Add 10 mL (1 mg Cl) standard sodium chloride solution to a clean 250-mL Erlenmeyer flask.

   b. Add 90 mL distilled water.

   c. Titrate as in Section E above.

(Color)

d. Calculation

$$\text{Normality, } N, \text{ AgNO}_3 = \frac{mL \text{ CaCl Standard} \times 0.0141}{mL \text{ AgNO}_3 \text{ Used in Titration}}$$

e. Example

10.0 mL NaCl standard used

10.0 mL AgNO$_3$ used in titration

0.0141 N = normality of NaCl standard

$$\text{Normality, } N, \text{ AgNO}_3 = \frac{10.0 \text{ mL} \times 0.0141}{10 \text{ mL}}$$

$$= 0.0141$$

I. Reference

See page 4-70, *STANDARD METHODS*, 21st Edition.

QUESTIONS

Write your answers in a notebook and then compare your answers with those on page 546.

21.1A Does the quality of water in any lake, reservoir, or stream affect the abundance and types of aquatic organisms found in the water? Yes or No?

21.1B How are calcium compounds used to treat water?

21.1C How soon should a sample be titrated for calcium after the sodium hydroxide (NaOH) solution has been added?

21.1D Why are concentrations of chloride above 250 mg/L objectionable to many people?

4. Color

A. Discussion

Color in water supplies may result from the presence of metallic ions (iron, manganese, and copper), organic matter of vegetable or soil origin, and industrial wastes. The most common colors that occur in raw water are yellow and brown. There are two general types of color found in water. True color results from the presence of dissolved organic substances or from certain minerals such as copper sulfate dissolved in the water. Suspended materials (including colloidal substances) can add what is called apparent color. True color is normally removed or at least reduced by coagulation and chlorination or ozonation. The method given below is suitable only for the measurement of color in clear treated water supplies having a turbidity of less than one unit of turbidity. When greater amounts of turbidity are present in the sample, some form of pretreatment for turbidity removal must be used before measuring the color.

B. What is Tested?

Sample	Common Range, units
Treated Surface Water	1 to 10
Groundwater	0 to 5

C. Apparatus Required

Nessler tubes, matched, 50 mL tall form
Pipet, 1.0 mL

D. Reagents

1. Color Standard. Use a stock standard with a color of 500 units.

2. Prepare color standards by adding the following increments of stock color standard to a Nessler tube and diluting to 50 mL.

Color Unit Standard	mL of Stock Color Standard
1	0.1
2	0.2
3	0.3
4	0.4
5	0.5

Protect these standards against evaporation and contamination when not in use.

E. Procedure

1. Fill a clean matched Nessler tube to the 50-mL mark with sample.

2. Compare the sample with the various color standards by looking downward vertically through the tubes toward a white surface.

3. Match as closely as possible sample color with a color standard.

F. Other Procedures

Color may also be measured by the use of:

1. Color comparator kits

2. Spectrophotometers

G. Notes

1. If the color exceeds 70 units, dilute sample with distilled water in known proportions until the color is within range of the standards. Calculate color units by the following equation:

$$\text{Color Units} = \frac{A \times 50}{B}$$

(Color)

OUTLINE OF PROCEDURE FOR COLOR

1. Fill Nessler tube to 50 mL.

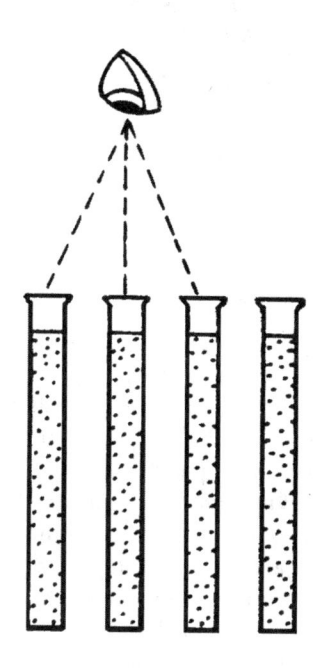

2. Compare the sample to color standards.

where:

A = estimated color of diluted sample

B = mL of sample taken from dilution

2. If turbidity is greater than one unit, consult *STANDARD METHODS* for pretreatment for turbidity removal.

H. Reference

See page 2-1, *STANDARD METHODS*, 21st Edition.

QUESTIONS

Write your answers in a notebook and then compare your answers with those on page 546.

21.1E What are the most common colors that occur in raw water?

21.1F How can true color be removed from water?

21.1G When not in use, stock color standards should be protected against what?

5. Dissolved Oxygen

A. Discussion

Dissolved oxygen (DO) is important to the water treatment plant operator for a number of reasons. In surface waters, dissolved oxygen must be present in order for fish and smaller aquatic organisms to survive. The taste of water is improved by dissolved oxygen. However, the presence of dissolved oxygen in water can contribute to corrosion of piping systems. Low or zero dissolved oxygen levels at the bottom of lakes or reservoirs often cause taste and odor problems in drinking water.

B. What is Tested?

Sample	Common Range, mg/L
Surface Water	5 to 11*
Groundwaters	0 to 2

* Some reservoirs and lakes may have zero DO near the bottom.

(Dissolved Oxygen)

C. Apparatus Required

METHOD A (SODIUM AZIDE MODIFICATION OF THE WINKLER METHOD)

Buret, graduated to 0.1 mL
Buret support
BOD bottle, 300 mL
Magnetic stirrer
Magnetic stir-bar
Pipets, 10 mL

METHOD B (MEMBRANE ELECTRODE METHOD)

Follow manufacturer's instructions. To ensure that the DO probe reading is accurate, the probe must be calibrated frequently. Take a sample that does not contain substances that interfere with either the probe reading or the Modified Winkler procedure. Split the sample. Measure the DO in one portion of the sample using the Modified Winkler procedure and compare this result with the DO probe reading on the other portion of the sample. Adjust the probe reading to agree with the results from the Modified Winkler procedure. To obtain good results when using a probe, you should be aware of the following precautions:

1. Periodically check the calibration of the probe.
2. Keep the membrane in the tip of the probe from drying out.
3. Dissolved inorganic salts, such as found in seawater, can influence the readings from a probe.
4. Reactive compounds, such as reactive gases and sulfur compounds, can interfere with the output of a probe.
5. Do not place the probe directly over a diffuser because you want to measure the dissolved oxygen in the water being treated, not the oxygen in the air supply to the aerator.

D. Reagents

   NOTE: Standardized solutions may be purchased from chemical suppliers.

   1. Manganous Sulfate solution.
   2. Alkaline Iodide-Sodium Azide solution.
   3. Sulfuric Acid: Use concentrated reagent-grade acid (H_2SO_4). Handle carefully, since this material will burn hands and clothes. Rinse affected parts with tap water to prevent injury.

      CAUTION: When working with alkaline azide and sulfuric acid, keep a nearby water faucet running for frequent hand rinsing.

   4. 0.025 N Phenylarsine Oxide (PAO) solution.
   5. 0.025 N Sodium Thiosulfate solution. For preservation, add 0.4 gm or 1 pellet of sodium hydroxide (NaOH). Solutions of "thio" should be used within two weeks to avoid loss of accuracy due to decomposition of solution.
   6. Starch solution.

E. Procedure

SODIUM AZIDE MODIFICATION OF THE WINKLER METHOD

NOTE: The sodium azide destroys nitrate, which would otherwise interfere with this test.

The reagents are to be added in the quantities, order, and methods as follows:

1. Collect a sample to be tested in 300-mL BOD bottle taking special care to avoid aeration of the liquid being collected. Fill bottle completely and stopper.
2. Remove stopper and add 1 mL of manganous sulfate solution below surface of the liquid.
3. Add 1 mL of alkaline iodide-sodium azide solution below the surface of the liquid.
4. Replace the stopper, avoid trapping air bubbles, and mix well by inverting the bottle several times. Repeat this mixing after the floc has settled halfway. Allow the floc to settle halfway a second time.
5. Acidify with 1 mL of concentrated sulfuric acid by allowing the acid to run down the neck of the bottle above the surface of the liquid.
6. Restopper and mix well until the precipitate has dissolved. The solution will then be ready to titrate. Handle the bottle carefully to avoid acid burns.
7. Pour 201 mL from bottle into an Erlenmeyer flask.
8. If the solution is brown in color, titrate with 0.025 N PAO until the solution is a pale yellow color. Add a small quantity of starch indicator and proceed with Step 10. (Note: Either PAO or 0.025 N sodium thiosulfate can be used.)
9. If the solution has no brown color, or is only slightly colored, add a small quantity of starch indicator. If no blue color develops, there is zero dissolved oxygen. If a blue color does develop, proceed to Step 10.
10. Titrate to the first disappearance of the blue color. Record the number of mL of PAO used.
11. The amount of oxygen dissolved in the original solution will be equal to the number of mL of PAO used in the titration provided significant interfering substances are not present.

mg DO/L = mL PAO

Laboratory Procedures 517

(Dissolved Oxygen)

OUTLINE OF PROCEDURE FOR DO

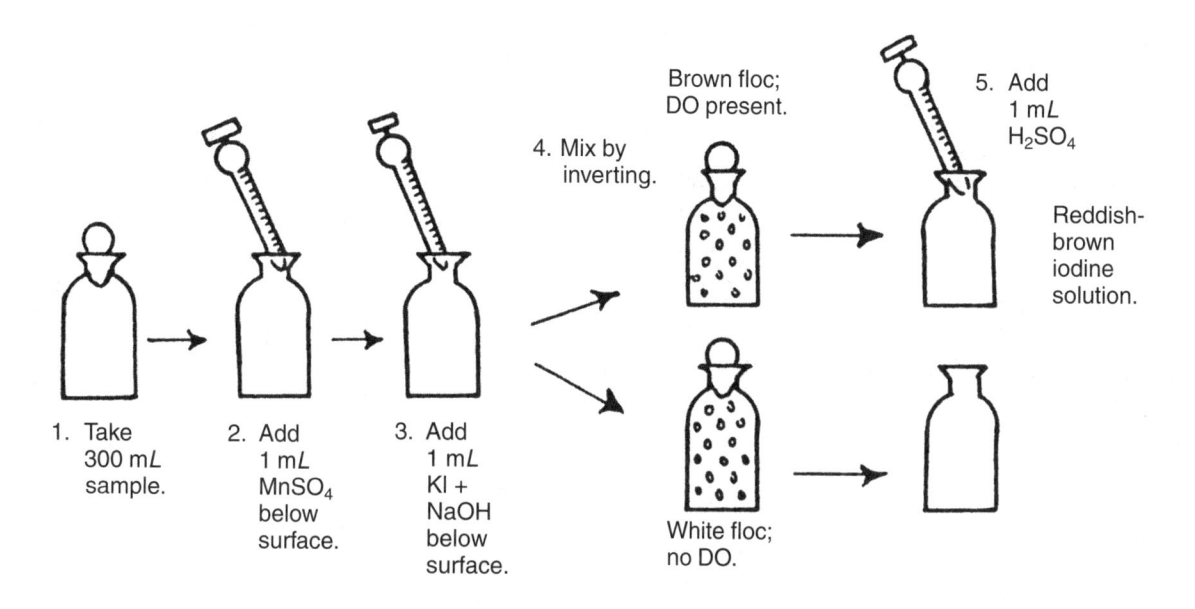

1. Take 300 mL sample.
2. Add 1 mL MnSO$_4$ below surface.
3. Add 1 mL KI + NaOH below surface.
4. Mix by inverting.

Brown floc; DO present.

White floc; no DO.

5. Add 1 mL H$_2$SO$_4$

Reddish-brown iodine solution.

Titration of Iodine Solution:

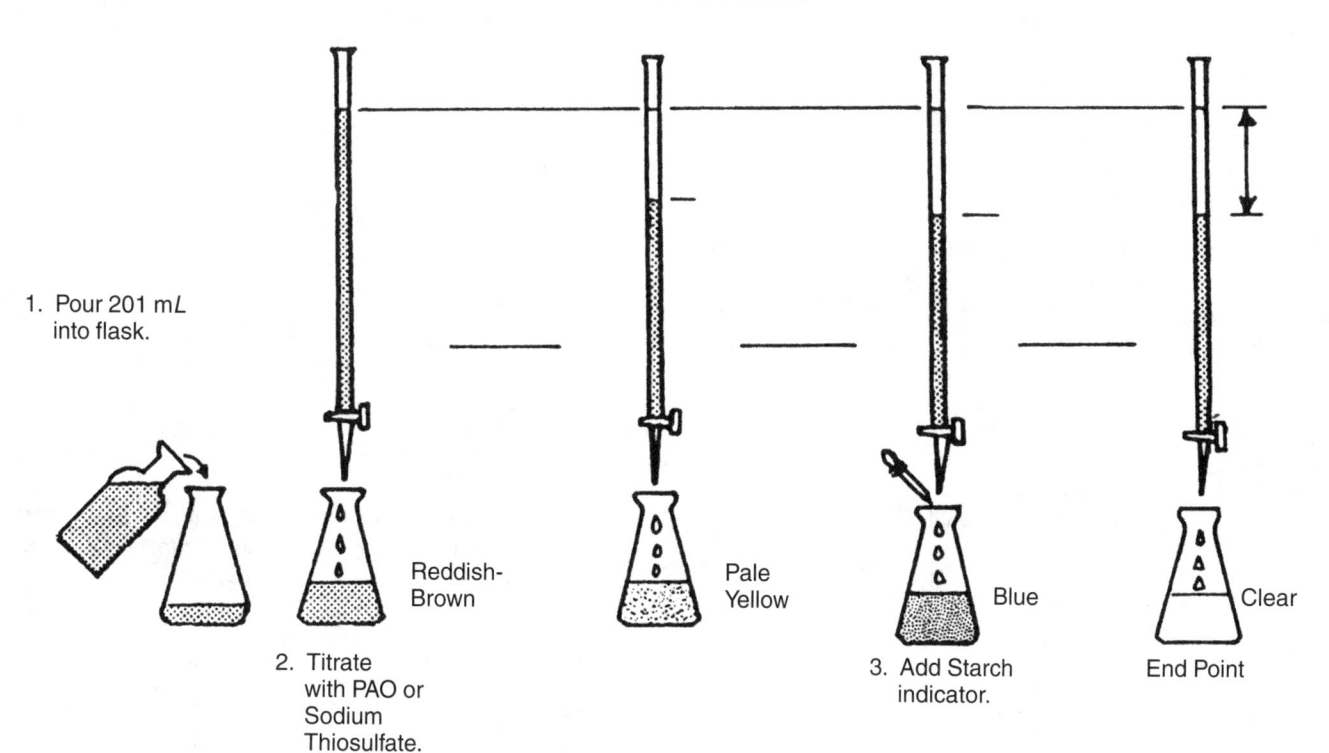

1. Pour 201 mL into flask.
2. Titrate with PAO or Sodium Thiosulfate.

Reddish-Brown

Pale Yellow

3. Add Starch indicator.

Blue

Clear

End Point

(Dissolved Oxygen)

F. Example

A sample is collected from just upstream of a river intake to a water treatment facility. The water temperature is 18°C. The sample is tested for DO and the operator uses 9.1 mL of 0.025 N PAO titrant.

G. Calculation

The DO titration of the 201-mL sample required 9.1 mL of 0.025 N PAO. Therefore, the dissolved oxygen (DO) concentration in the sample is 9.1 mg/L.

The percent saturation of DO in the river can be calculated using the dissolved oxygen saturation values given in Table 21.1. Note that as the temperature of water increases, the DO saturation value (Saturation column) decreases. Table 21.1 gives 100 percent DO saturation values for temperatures in °C and °F.

$$\text{DO Saturation, \%} = \frac{\text{DO of Sample, mg}/L \times 100\%}{\text{DO at 100\% Saturation, mg}/L}$$

For example, given

$$\text{DO Saturation, \%} = \frac{9.1 \text{ mg}/L}{9.5 \text{ mg}/L} \times 100\%$$

$$= 0.96 \times 100\%$$

$$= 96\%$$

where

9.1 mg/L = DO of Sample

9.5 mg/L = DO at 100% Saturation at 18°C (river temperature)

H. Precautions

1. Samples for dissolved oxygen measurements should be collected very carefully. Do not let sample remain in contact with air or be agitated. Collect samples in a 300-mL BOD bottle. Avoid entraining (trapping) or dissolving atmospheric oxygen.

2. When sampling from a water line under pressure, attach a tube to the tap and extend tube to bottom of bottle. Let bottle overflow two or three times its volume and replace glass stopper so no air bubbles are entrapped.

3. Use suitable sampler for streams, reservoirs, or tanks of moderate depth such as that shown in Figure 21.1. Use a Kemmerer-type sampler for samples collected from depths greater than 6½ feet (2 m).

4. Always record temperature of water at time of sampling.

5. Use the proper bottle with matched stopper.

TABLE 21.1
EFFECT OF TEMPERATURE ON DISSOLVED OXYGEN SATURATION FOR A CHLORIDE CONCENTRATION OF ZERO mg/L

°C	°F	DO at 100% Saturation, mg/L
0	32.0	14.6
1	33.8	14.2
2	35.6	13.8
3	37.4	13.5
4	39.2	13.1
5	41.0	12.8
6	42.8	12.5
7	44.6	12.2
8	46.4	11.9
9	48.2	11.6
10	50.0	11.3
11	51.8	11.1
12	53.6	10.8
13	55.4	10.6
14	57.2	10.4
15	60.0	10.2
16	61.8	10.0
17	63.6	9.7
18	65.4	9.5
19	67.2	9.4
20[a]	68.0	9.2[a]
21	69.8	9.0
22	71.6	8.8
23	73.4	8.7
24	75.2	8.5
25	77.0	8.4

[a] The standard temperature for a BOD test is 20°C (68°F).

6. When working with a lake or reservoir, examine the temperature and DO profile (measure temperature and DO at surface and at various depths all the way down to the bottom).

7. Measure the DO in the sample as soon as possible.

I. Reference

See page 4-138, *STANDARD METHODS*, 21st Edition.

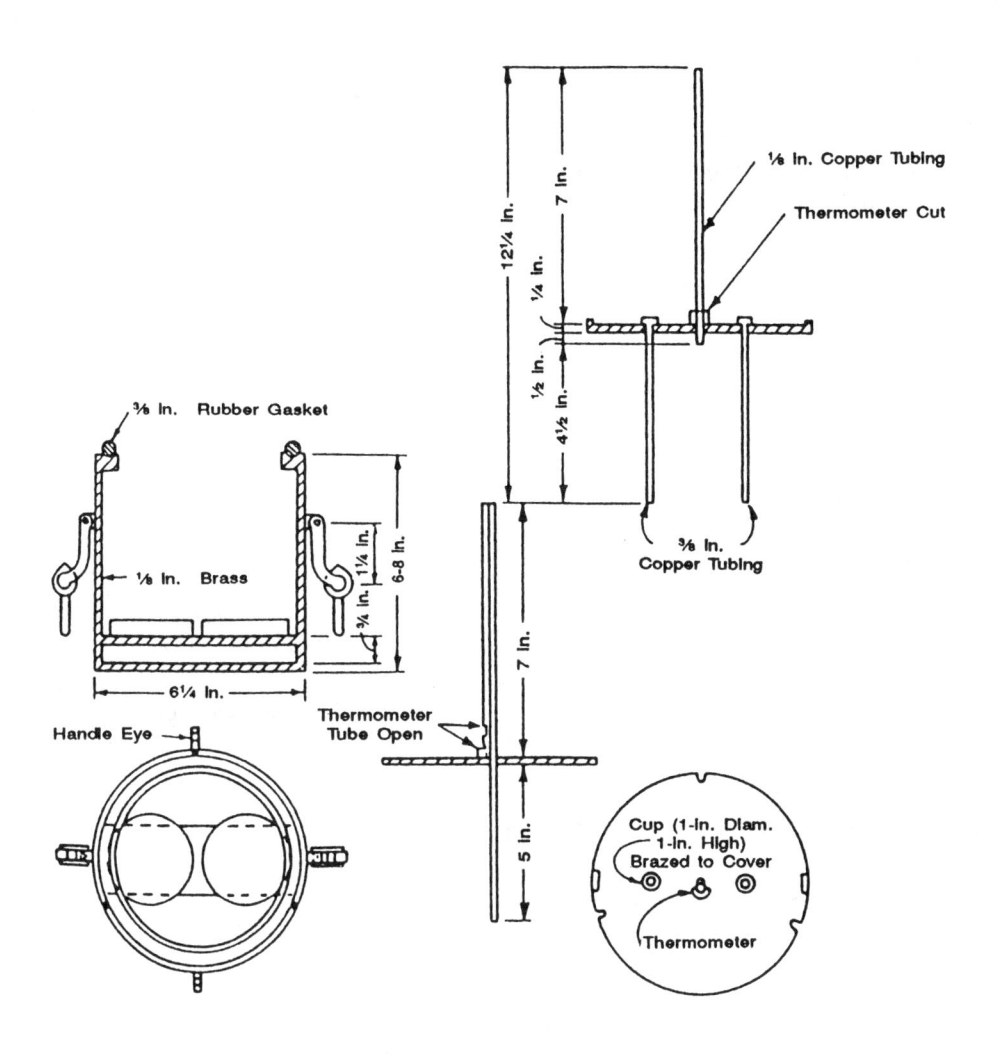

Fig. 21.1 DO Sampler
(Reprinted from *STANDARD METHODS*, 15th Edition, by permission. Copyright 1980, the American Public Health Association)

6. Fluoride

A. Discussion

Fluoride may occur naturally or it may be added in controlled amounts. The concentration of fluoride in most natural waters is less than 1.0 mg/L. There are, however, several areas in the United States that have natural fluoride concentrations as high as 30 mg/L. The importance of fluoride in forming human teeth and the role of fluoride intake from drinking water in controlling the characteristics of tooth structure have been accurately documented. Studies have shown that a fluoride concentration of approximately 1.0 mg/L reduces dental caries of young people without harmful effects on health.

B. What is Tested?

Sample	Common Range, mg/L
Fluoridated Water	0.8 to 1.2

C. Apparatus Required

Spectrophotometer for use at 570 nanometers (nm) wavelength
Pipets, 5 mL
Erlenmeyer flasks, 125 mL

(Fluoride)

D. Reagents

1. Stock fluoride solution. 1.0 mL = 0.100 mg F.

2. Standard fluoride solution: Dilute 100 mL stock fluoride solution to 1,000 mL with distilled water; 1.0 mL = 0.010 mg F.

3. SPADNS solution. This solution is stable indefinitely if protected from direct sunlight.

4. Zirconyl-acid reagent.

5. Acid zirconyl-SPADNS reagent: Mix equal volumes of SPADNS solution and zirconyl-acid reagent. The combined reagent is stable for at least 2 years.

6. Reference solution: Add 10 mL SPADNS solution to 100 mL distilled water. Dilute 7 mL concentrated HCl to 10 mL and add to the diluted SPADNS solution. The resulting solution, used for setting the instrument reference point (zero), is stable and may be reused indefinitely. Alternatively, use a prepared standard as a reference.

7. Sodium arsenite solution. (*CAUTION*: Toxic—avoid ingestion.)

E. Procedure

1. Measure 50 mL of sample and add to a clean 125-mL Erlenmeyer flask. (If sample contains residual chlorine, add one drop $NaAsO_2$ solution per 0.1 mg chlorine residual and mix.)

2. Add 5.0 mL each of SPADNS solution and zirconyl-acid reagent, or 10.0 mL acid zirconyl-SPADNS reagent. Mix.

3. Set spectrophotometer to 0.730 absorbance with reference solution containing zero mg/L of fluoride (see G. Example).

4. Read absorbance at 570 nm with spectrophotometer and determine the amount of fluoride from standard curve.

NOTE: A colorimeter may also be used to measure fluoride.

F. Construction of Standard Calibration Curve

1. Using the standard fluoride solution, prepare the following standards in 100-mL volumetric flasks.

mL of Standard Fluoride Solution Placed in 100-mL Volumetric Flask	Fluoride Concentration, mg/L
5.0	0.50
7.5	0.75
10.0	1.00
12.5	1.25

OUTLINE OF PROCEDURE FOR FLUORIDE

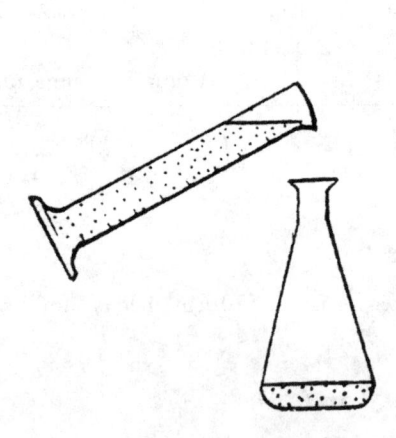

1. Measure 50 mL of sample into flask. Dechlorinate, if necessary.

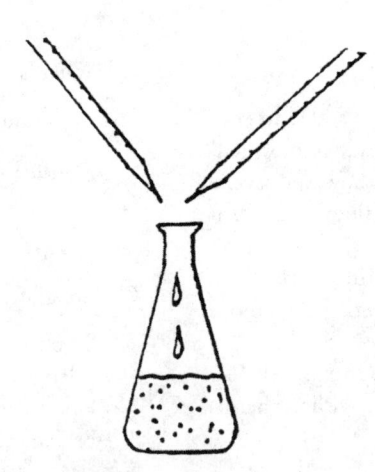

2. Add 5 mL each of SPADNS solution and zirconyl-acid reagent.

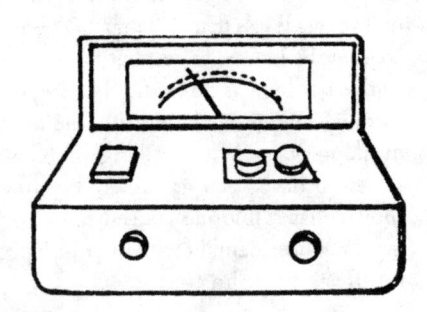

3. Measure absorbance at 570 nm with spectrophotometer.

2. Dilute flasks to 100 mL.
3. Transfer 50 mL to 125-mL Erlenmeyer flask.
4. Determine amount of fluoride as outlined previously.
5. Prepare a standard curve by plotting the absorbance values of standards versus the corresponding fluoride concentrations.

G. Example

Results from a series of tests for fluoride were as follows:

Flask No.	Sample	Volume, mg/L	Absorbance
1	Distilled Water	50	0.730
2	C Street Well	50	0.470
3	Plant Effluent	50	0.520
4	0.5 mg/L F Standard	50	0.625
5	0.75 mg/L F Standard	50	0.560
6	1.0 mg/L F Standard	50	0.500
7	1.25 mg/L F Standard	50	0.444

H. Calculation

1. Prepare a standard curve by using data from prepared standards. From above example:

Fluoride Concentration, mg/L	Absorbance
0.0	0.730
0.5	0.625
0.75	0.560
1.0	0.500
1.25	0.444

The graph below is a result of plotting concentration of fluoride standards versus their corresponding absorbance.

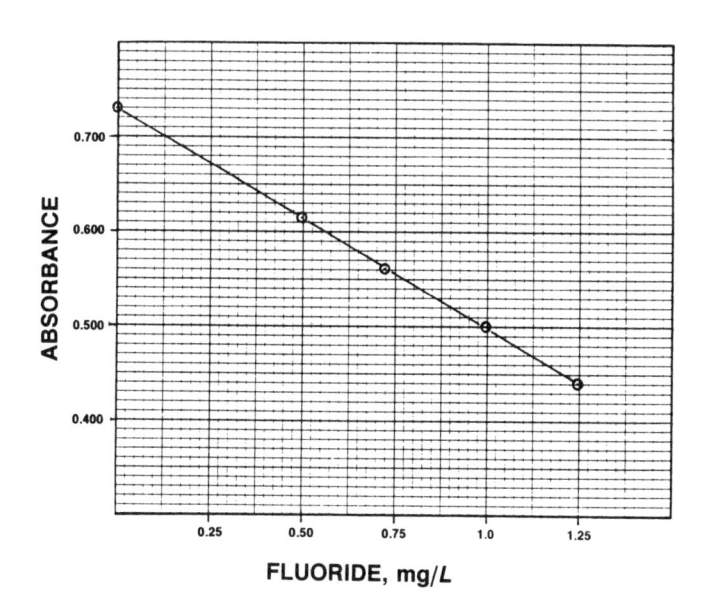

2. Obtain concentration of unknown samples from curve.

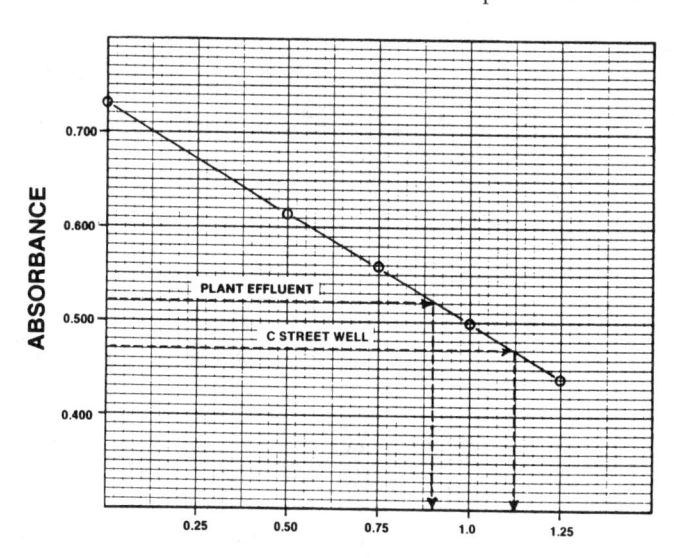

PLANT EFFLUENT = 0.90 mg/L
C STREET WELL = 1.12 mg/L

I. Precautions

1. Whenever any of the following substances are present in the listed quantities, the sample must be distilled prior to analysis.

Substance	Concentration, mg/L
Alkalinity	5,000
Aluminum	0.1
Chloride	7,000
Iron	10
Hexametaphosphate	1.0
Phosphate	16
Sulfate	200

2. Samples and standards should be at the same temperature throughout color development.

J. Reference

See page 4-85, *STANDARD METHODS,* 21st Edition.

> ### QUESTIONS
>
> Write your answers in a notebook and then compare your answers with those on page 546.
>
> 21.1H Why is the presence of dissolved oxygen (DO) in water in piping systems of concern to operators?
>
> 21.1I What is the common range of fluoride in fluoridated drinking water?

END OF LESSON 1 OF 2 LESSONS

on

ADVANCED LABORATORY PROCEDURES

Please answer the discussion and review questions next.

DISCUSSION AND REVIEW QUESTIONS
Chapter 21. ADVANCED LABORATORY PROCEDURES
(Lesson 1 of 2 Lessons)

At the end of each lesson in this chapter you will find some discussion and review questions. The purpose of these questions is to indicate to you how well you understand the material in the lesson. Write the answers to these questions in your notebook.

1. What is the purpose of spectrophotometer calibration curves?

2. How would you prepare a spectrophotometer calibration graph?

3. Why are algae counts in raw waters important to operators?

4. The maximum allowable chloride concentration in drinking water has been established on what basis?

5. What are the two general types of color found in water and what is the cause of each type?

6. Why is dissolved oxygen (DO) in water important to the treatment plant operator?

7. What precautions would you take when collecting a lake sample for a dissolved oxygen measurement?

8. How does fluoride get into drinking waters?

CHAPTER 21. ADVANCED LABORATORY PROCEDURES

(Lesson 2 of 2 Lessons)

7. Iron (Total)

A. Discussion

Iron is an abundant and widespread constituent of rocks and soils. The most common form of iron in solution in groundwater and in water under anaerobic conditions (bottom of a lake or reservoir) is the ferrous ion, Fe^{2+}. Ferric iron can occur in soils, in aerated water, and in acid solutions as Fe^{3+}, ferric hydroxide, and polymeric forms depending upon pH. Above pH of 4.8, however, the solubility of the ferric species is less than 0.1 mg/L. Colloidal ferric hydroxide is commonly present in surface water and small quantities may persist even in water that appears clear.

Iron in a domestic water supply can stain laundry, concrete, and porcelain. A bitter, astringent taste can be detected by some people at levels above 0.3 mg/L. When iron reacts with oxygen, a red precipitate (rust) is formed.

B. What is Tested?

Source	Common Range, mg/L
Untreated Surface Water	0.10 to 1.0
Treated Surface Water	<0.01 to 0.20
Groundwater	<0.01 to 10

C. Apparatus Required

Spectrophotometer for use at 510 nm
Acid-washed glassware. Wash all glassware with concentrated HCl and rinse with distilled water to remove deposits of iron oxide, which could give false results.
Erlenmeyer flasks, 125 mL
Pipets, 5 and 10 mL
Volumetric flasks, 100 mL
Hot plate

D. Reagents

Use reagents low in iron. Use iron-free distilled water. Store reagents in glass-stoppered bottles. The hydrochloric acid and ammonium acetate solutions are stable indefinitely if tightly stoppered. The hydroxylamine, phenanthroline, and stock iron solutions are stable for several months. The standard iron solutions are not stable; prepare daily, as needed, by diluting the stock solution. Visual standards in Nessler tubes are stable for several months if sealed and protected from light.

1. Hydrochloric acid, HCl.

2. Hydroxylamine solution.

3. Ammonium acetate buffer solution. Because even a good grade of $NH_4C_2H_3O_2$ contains a significant amount of iron, prepare new reference standards with each buffer preparation.

4. Sodium acetate solution.

5. Phenanthroline solution. (*NOTE:* One milliliter of this reagent is sufficient for no more than 100 µg Fe.)

6. Stock iron solution: 1.00 mL = 0.200 mg Fe.

7. Standard iron solutions. Prepare daily for use. Pipet 50.00 mL stock solution into a one-liter volumetric flask and dilute to mark with iron-free distilled water; 1.00 mL = 0.010 mg Fe.

E. Procedure

TOTAL IRON

1. Measure 50 mL of thoroughly mixed sample into a 125-mL Erlenmeyer flask.

2. Add 2 mL concentrated HCl and 1 mL hydroxylamine solution.

3. Heat to boiling. Boil sample until volume is reduced to 20 mL. Cool to room temperature.

4. Transfer to 100-mL volumetric flask.

5. Add 10 mL acetate buffer solution and 2 mL phenanthroline solution. Dilute to 100-mL mark with iron-free distilled water. Mix thoroughly.

6. After 15 minutes, measure the absorbance at 510 nm and determine the amount of iron from the standard curve.

524 Water Treatment

(Iron)

OUTLINE OF PROCEDURE FOR IRON

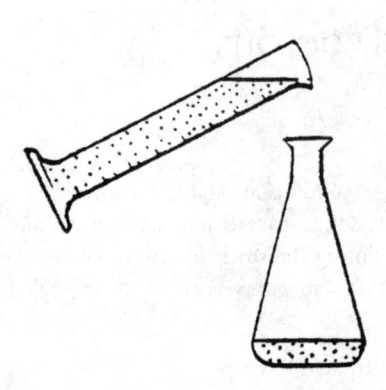

1. Measure 50 mL into flask.

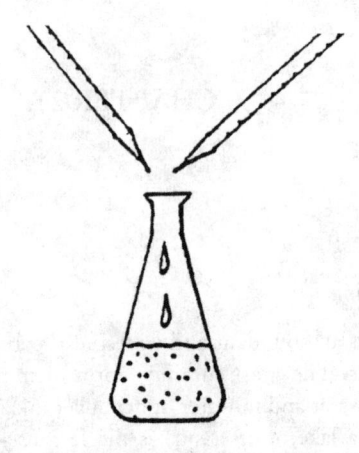

2. Add 2 mL conc. HCl and 1 mL hydroxylamine solution.

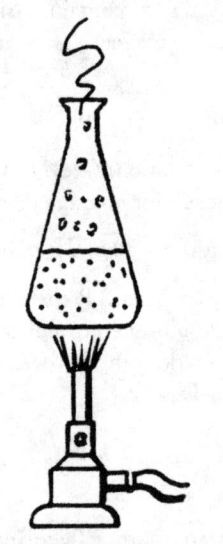

3. Heat to boiling. Reduce volume to 20 mL. Cool.

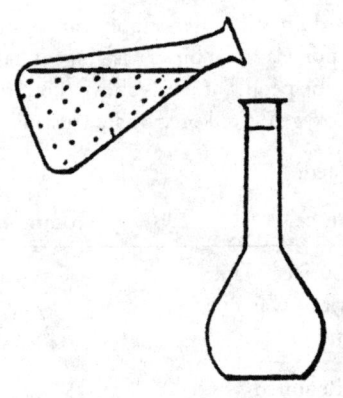

4. Transfer to 100-mL volumetric flask.

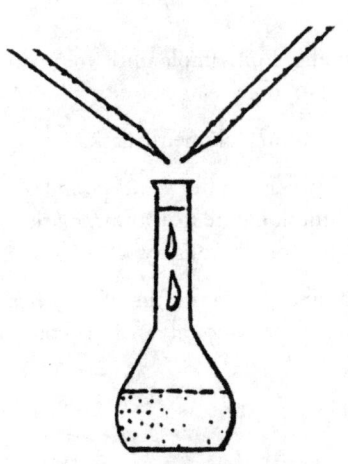

5. Add 10 mL acetate buffer and 2 mL phenanthroline solution. Dilute to 100 mL.

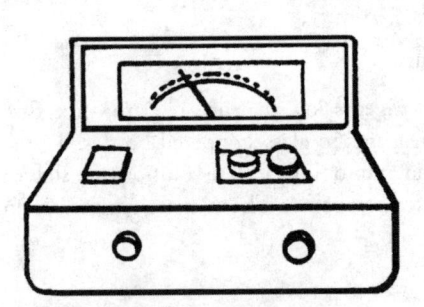

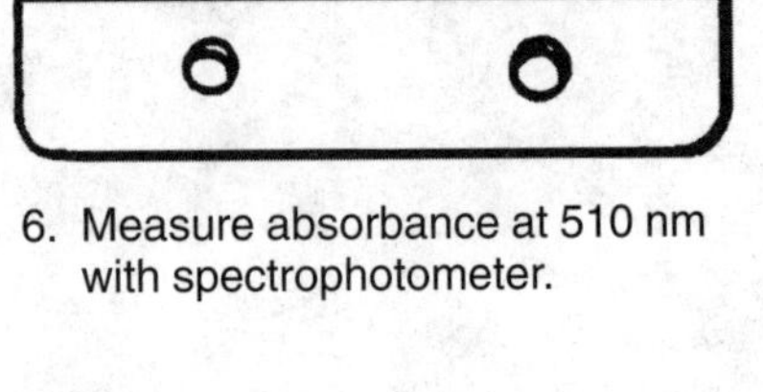

6. Measure absorbance at 510 nm with spectrophotometer.

F. Construction of Standard Calibration Curve

1. Using the standard solution, prepare the following standards in 100-mL volumetric flasks.

mL of Standard Iron Solution Placed in 100-mL Volumetric Flask	Iron Concentration, mg/L
0	0
1.0	0.10
2.5	0.25
5.0	0.50
7.5	0.75
10.0	1.00

2. Dilute flasks to 100 mL.

3. Transfer 50 mL to 100-mL volumetric flask.

4. Add 1.0 mL hydroxylamine solution and 1 mL acetate solution to each flask.

5. Dilute to about 75 mL, add 10 mL phenanthroline solution, dilute to 100-mL mark. Mix thoroughly.

6. Measure absorbance at 510 nm against the reference blank.

7. Prepare a standard curve by plotting the absorbance values of standards versus the corresponding iron concentrations.

G. Example

A series of tests for total iron produced these results:

Flask No.	Sample	Absorbance
1	Distilled Water	0.000
2	Plant Clear Well	0.100
3	River Sample	0.420
4	0.10 mg/L Fe Standard	0.066
5	0.25 mg/L Fe Standard	0.161
6	0.50 mg/L Fe Standard	0.328
7	0.75 mg/L Fe Standard	0.495
8	1.00 mg/L Fe Standard	0.658

H. Calculation

1. Prepare a standard curve by using data from prepared standards. From the above example:

Concentration Iron, mg/L	Absorbance
0.0	0.000
0.10	0.066
0.25	0.161
0.50	0.328
0.75	0.495
1.00	0.658

(Iron)

The graph below is a result of plotting concentration of standards versus their corresponding absorbance.

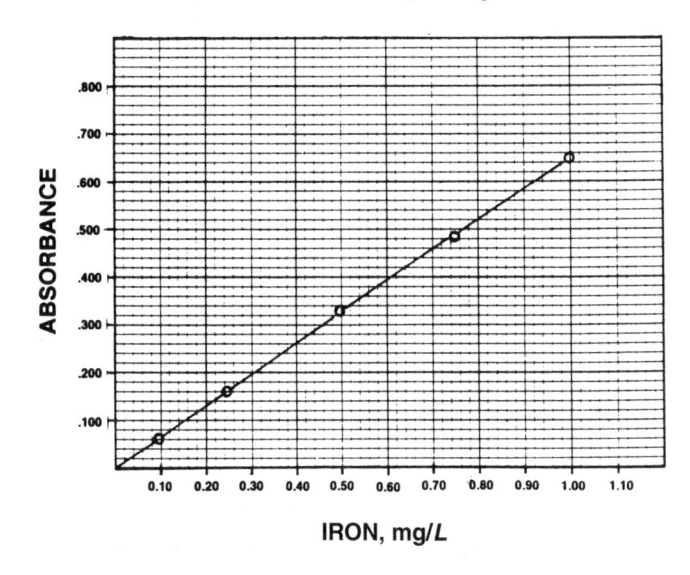

2. Obtain concentration of unknown clear well and river samples from curve.

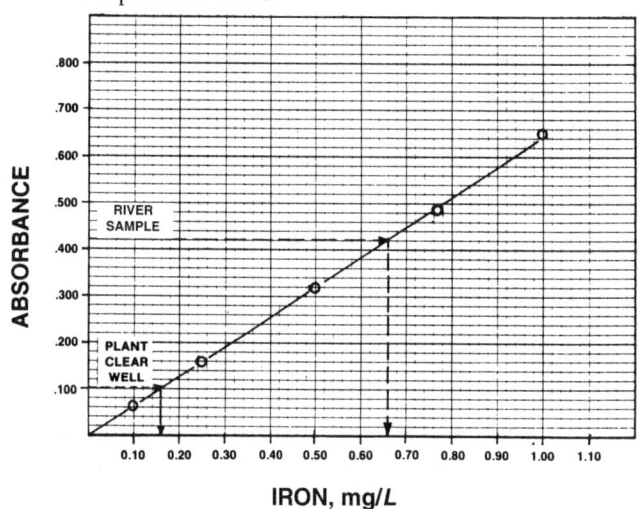

PLANT CLEAR WELL = 0.16 mg/L Fe
RIVER SAMPLE = 0.66 mg/L Fe

I. Notes

1. Iron in well water or tap samples may vary in concentration and form with the duration and degree of flushing before and during sampling.

2. For precise determination of total iron, use a separate container for sample collection. Treat with acid at time of collection to place iron in solution and prevent deposition on walls of sample container.

3. Exercise caution when handling sulfuric acid.

J. Reference

See page 3-76, *STANDARD METHODS*, 21st Edition.

(Manganese)

8. Manganese

A. Discussion

Although manganese is much less abundant than iron in the earth's crust, it is one of the most common elements and widely distributed in rocks and soils. Some groundwaters that contain objectionable amounts of iron also contain considerable amounts of manganese, but groundwaters that contain more manganese than iron are rather unusual. Manganese in surface waters occurs both in suspension and as a soluble complex. Although rarely present in excess of 1 mg/L, manganese imparts objectionable stains to laundry and plumbing fixtures. Manganese will also cause stains on the walls of tanks and driveways in treatment plants.

B. What is Tested?

Source	Common Range, mg/L
Treated and Untreated Surface Water	<0.01 to 0.10
Groundwater	<0.01 to 1.0

C. Apparatus Required

Spectrophotometer for use at 525 nm
Hot plate
Erlenmeyer flask, 250 mL
Pipets, 5 and 10 mL
Volumetric flasks, 100 and 500 mL

D. Reagents

1. Special reagent.
2. Ammonium persulfate, $(NH_4)_2S_2O_8$, solid.
3. Standard manganese solution. 1 mL = 0.01 mg Mn. Prepare dilute solution daily.
4. 1% HCl: Add 10 mL concentrated HCl carefully to 990 mL distilled water.
5. Hydrogen peroxide, H_2O_2, 30 percent.

E. Procedure

1. Measure 100 mL of thoroughly mixed sample into a 250-mL Erlenmeyer flask that has been marked with a line at the 90-mL level.
2. Add 5 mL special reagent and 1 drop H_2O_2.
3. Concentrate to 90 mL by boiling. Add 1 gram ammonium persulfate. Cool immediately under water tap.
4. Dilute to 100 mL.
5. Measure the absorbance at 525 nm with a spectrophotometer and determine the amount of manganese from the standard curve.

F. Construction of Calibration Curve

1. Using the standard manganese solution, prepare the following standards in 100-mL volumetric flasks.

mL of Standard Manganese Solution Placed in 100-mL Volumetric Flask	Manganese Concentration, mg/L
0	0
1.0	0.10
2.0	0.20
3.0	0.30
4.0	0.40

2. Dilute flasks to 100 mL.
3. Transfer to 250-mL Erlenmeyer flask.
4. Determine amount of manganese as outlined previously.
5. Prepare a standard curve by plotting the absorbance values of standards versus the corresponding manganese concentrations.

G. Example

Results from a series of tests for manganese were as follows:

Flask No.	Sample	Absorbance
1	Distilled Water	0.000
2	Plant Effluent	0.000
3	Jones St. Well	0.030
4	0.05 mg/L Mn Standard	0.009
5	0.10 mg/L Mn Standard	0.018
6	0.20 mg/L Mn Standard	0.036
7	0.30 mg/L Mn Standard	0.053
8	0.40 mg/L Mn Standard	0.071

H. Calculation

1. Prepare a standard curve by using data from prepared standards. From the above example:

Concentration Manganese, mg/L		Absorbance
0.0	(distilled water)	0.000
0.05		0.009
0.10		0.018
0.20		0.036
0.30		0.053
0.40		0.071

(Manganese)

OUTLINE OF PROCEDURE FOR MANGANESE

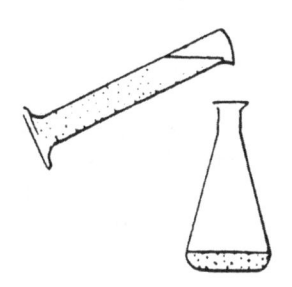

1. Measure 100 mL into flask.

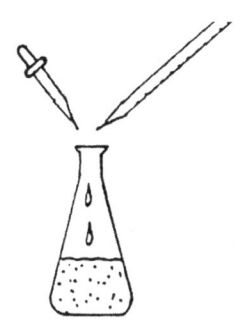

2. Add 5 mL special reagent and 1 drop H_2O_2.

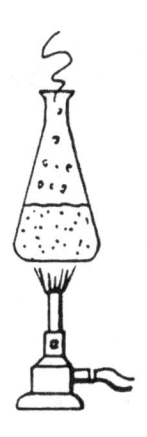

3. Concentrate to 90 mL, then add 1 gm ammonium persulfate. Dilute to 100 mL after cooling.

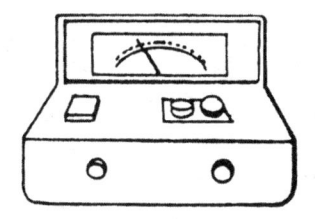

4. Measure absorbance at 525 nm with spectrophotometer.

(Manganese)

The graph below is the result of plotting concentration of standards versus their corresponding absorbance.

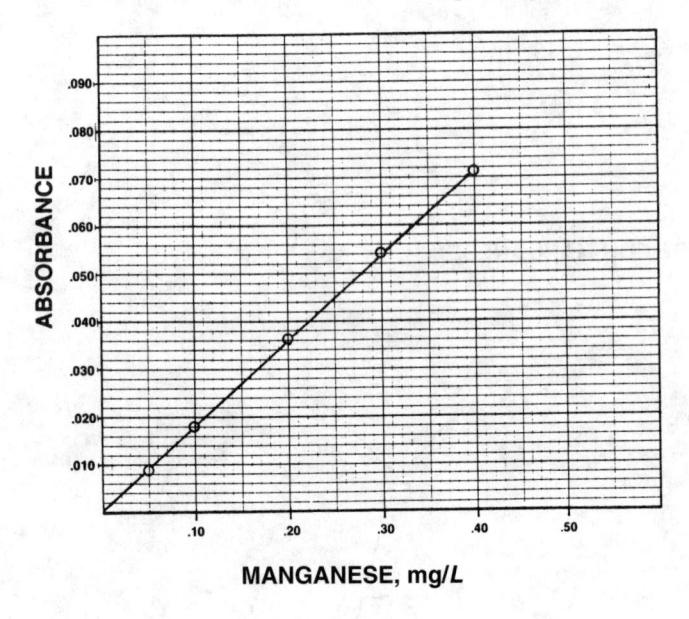

MANGANESE, mg/L

2. Obtain concentration of unknown plant effluent and well sample from curve.

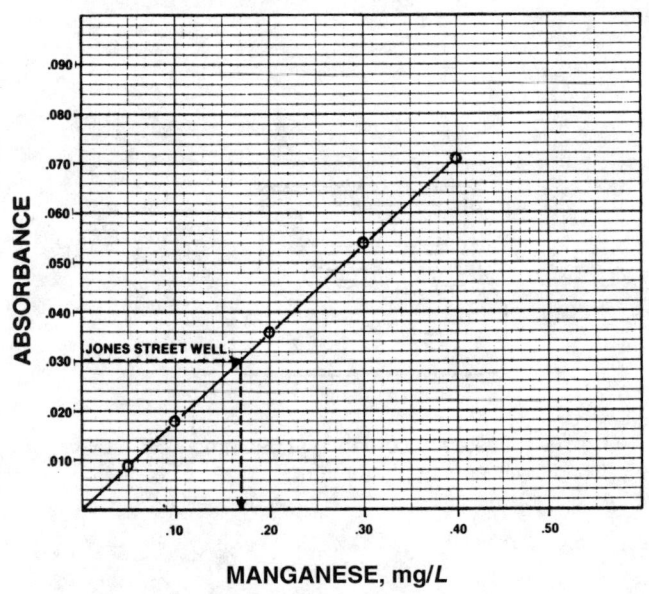

MANGANESE, mg/L

PLANT EFFLUENT = <0.01 mg/L Mn
JONES ST. WELL = 0.17 mg/L Mn

I. Notes

1. If turbidity or interfering color is present, use the following "bleaching" method: as soon as the spectrophotometer reading has been made, add 0.05 mL hydrogen peroxide solution directly to the optical cell. Mix and read again as soon as the color has faded. Deduct the absorbance of the bleached solution from the initial absorbance to obtain the absorbance due to manganese.

2. Determine manganese as soon as possible after sample collection. If this is not possible, acidify the sample with nitric acid to pH less than 2.

J. Reference

See page 3-84, *STANDARD METHODS*, 21st Edition.

QUESTIONS

Write your answers in a notebook and then compare your answers with those on page 546.

21.1J Iron in a domestic water supply may cause what problems?

21.1K Why must all glassware be acid washed when analyzing samples for iron?

21.1L In what forms does manganese occur in surface waters?

21.1M If the manganese concentration in a sample cannot be measured immediately, what would you do?

9. Marble Test (Calcium Carbonate Saturation Test)

A. Discussion

The Marble Test is used to determine the degree to which a sample of water is saturated with calcium carbonate. Marble Test results are used in the lime–soda softening process and to control corrosion. Water in intimate contact with powdered calcium carbonate (calcite) will approach saturation. The water being tested should not be exposed to atmospheric carbon dioxide. The Marble Test must be conducted at the specific temperature because the solubility of calcium carbonate varies with temperature. However, equipment that will maintain a constant temperature (either lower or higher than room temperature) while mixing the solution is not commonly available in water treatment plants. The only other way to keep a reasonably uniform temperature is to run the test as rapidly as possible.

B. What is Tested?

Source	Common Range*
Untreated Surface Water	−1 to +1
Treated Surface Water	−0.2 to +0.2
Well Water	−0.1 to +1

* Initial pH − Final pH.

C. Apparatus Required

BOD bottle, 300 mL
Magnetic stirrer
Stir-bar
Thermometer
Glass funnel, 125 mm
Filter paper, Whatman #50 (18.5 inch)
Equipment for determining pH and hardness

(Marble Test)

D. Reagents

1. Calcium carbonate, reagent grade.

2. Reagents for determining pH and hardness.

E. Procedure

1. Measure the temperature of the water to be tested.

2. Measure the pH, hardness, and, if desired, the alkalinity of the sample being tested.

3. Insert the stirring bar in the BOD bottle and fill with the water being tested. Adjust the water temperature to within 1°C of the initial temperature. Add approximately one (1) gram of calcium carbonate and stir for five minutes at a rate high enough to keep the calcium carbonate in suspension and the sample vigorously agitated.

4. Recheck the temperature. If the temperature has changed more than one degree Celsius, repeat the stirring with a fresh sample whose temperature has been adjusted so that the final temperature will be within one degree Celsius of the initial temperature.

5. Immediately measure the final pH.

6. Filter the remaining sample. Determine the hardness and, if desired, the final alkalinity on the filtrate (water that passed through the filter).

OUTLINE OF PROCEDURE FOR MARBLE TEST

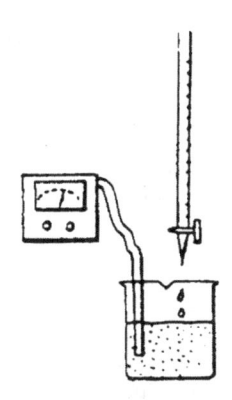

1. Measure temperature, pH, hardness, and alkalinity of sample being tested.

2. Transfer to BOD bottle and add 1 gm calcium carbonate. Mix.

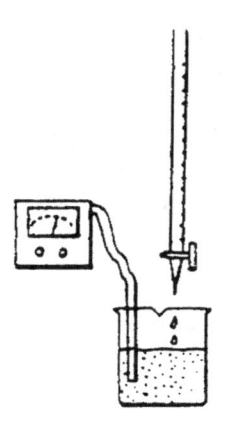

3. Measure final pH and temperature.

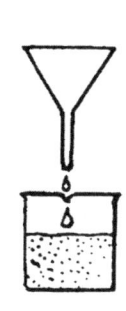

4. Filter.

5. Determine hardness and alkalinity of filtrate.

(Marble Test)

F. Example

Results from a series of tests for the calcium carbonate precipitation potential were as follows:

Filtered Water Sample

Initial Temperature	14°C
Final Temperature	14°C
Initial pH	8.7
Final pH	9.1
Initial Hardness	34 mg/L
Final Hardness	38 mg/L
Initial Alkalinity	24 mg/L
Final Alkalinity	27 mg/L

G. Calculation

$$\text{Calcium Carbonate Precipitation Potential, mg}/L = \text{Initial Hardness, mg}/L - \text{Final Hardness, mg}/L$$

The Langelier Index[5] is approximately equal to the initial pH – final pH. If the value of this index is less than 0.2, this value will indicate that the water is very near the saturation level. In any event, the sign of this value will be the same as the sign of the Langelier Index. That is to say, both the Langelier Index and the calcium carbonate precipitation potential will be negative if the water is undersaturated, and positive if the water is supersaturated.

From the example above:

$$\text{Calcium Carbonate Precipitation Potential, mg}/L = \text{Initial Hardness, mg}/L - \text{Final Hardness, mg}/L$$

$$= 34 \text{ mg}/L - 38 \text{ mg}/L$$

$$= -4 \text{ mg}/L$$

Langelier Index $\cong$ Initial pH – Final pH

$$\cong 8.7 - 9.1$$

$$\cong -0.4$$

This water is undersaturated (and therefore corrosive) with respect to calcium carbonate.

Also see page 2-30, *STANDARD METHODS*, 21st Edition.

10. Metals

A. Discussion

The presence of certain metals in drinking water can be a matter of serious concern because of the toxic properties of these materials. The analyses of these metals are generally done by using atomic absorption spectroscopy or colorimetric methods. The term "metals" includes the following elements:

Aluminum	Cobalt	Potassium
Antimony	Copper	Selenium
Arsenic	Iron	Silver
Barium	Lead	Sodium
Beryllium	Magnesium	Thallium
Cadmium	Manganese	Tin
Calcium	Mercury	Titanium
Chromium	Molybdenum	Vanadium
	Nickel	Zinc

B. Reference

For materials and procedures see page 3-1, *STANDARD METHODS*, 21st Edition.

QUESTIONS

Write your answers in a notebook and then compare your answers with those on page 546.

21.1N Why is temperature important when running the Marble Test?

21.1O The results from the Marble Test produce an initial pH of 8.9 and a final pH of 8.6. Would this water be considered corrosive?

21.1P How are the concentrations of most metals in water measured?

11. Nitrate

A. Discussion

Nitrate represents the most completely oxidized form of nitrogen found in water. High levels of nitrate in raw water samples indicate biological wastes in the final stage of stabilization or runoff from fertilized areas. High nitrate levels degrade water quality by stimulating excessive algal growth. Drinking water that contains excessive amounts of nitrate can cause infant methemoglobinemia (blue babies). For this

---

[5] *Langelier Index (LI).* An index reflecting the equilibrium pH of a water with respect to calcium and alkalinity. This index is used in stabilizing water to control both corrosion and the deposition of scale.

Langelier Index = $pH - pH_s$

where pH = actual pH of the water

pH_s = pH at which water having the same alkalinity and calcium content is just saturated with calcium carbonate

(Nitrate)

reason, a level of 10 mg/L (as nitrogen) has been established as a maximum level. The procedure given below measures the amount of both nitrate and nitrite nitrogen present in a sample by reducing all nitrate to nitrite through the use of a copper-cadmium column. The total nitrate (any nitrite present originally plus the reduced nitrate) is then measured colorimetrically.

B. What is Tested?

Sample	Common Range, mg/L
Treated Surface Water	<0.1 to 5
Groundwater	0.5 to 10

C. Apparatus

Reduction column. The column in Figure 21.2 was constructed from a 100-mL volumetric pipet by removing the top portion. This column may also be constructed from two pieces of tubing joined end to end. A 10-cm length of 3-cm I.D. tubing is joined to a 25-cm length of 3.5-mm I.D. tubing.
Spectrophotometer for use at 540 nm, providing a light path of 1 cm or longer
Beakers, 125 mL
Glass wool
Glass-fiber filter or 0.45-micron membrane filter
Filter holder assembly
Filter flask
pH meter
Separatory funnel, 250 mL
Volumetric pipets, 1, 2, 5, and 10 mL

D. Reagents

1. Granulated cadmium: 40 to 60 mesh.

2. Copper-Cadmium: The cadmium granules (new or used) are cleaned with 6 N HCl and copperized with a 2-percent solution of copper sulfate in the following manner:

   a. Wash the cadmium with 6 N HCl and rinse with distilled water. The color of the cadmium should be silver.

   b. Swirl 25 gm cadmium in 100-mg/L portions of a 2-percent solution of copper sulfate for 5 minutes or until the blue color partially fades, decant, and repeat with fresh copper until a brown precipitate forms.

   c. Wash the copper-cadmium with distilled water at least 10 times to remove all the precipitated copper. The color of the cadmium should now be black.

3. Preparation of reaction column: Insert a glass wool plug into the bottom of the reduction column and fill with distilled water. Add sufficient copper-cadmium granules to produce a column 18.5 cm in length. Maintain a level of distilled water above the copper-cadmium granules to eliminate entrapment of air. Wash the column with 200 mL of dilute ammonium chloride-EDTA solution (reagent 5). The column is then activated by passing through the column 100 mL of solution composed of 25 mL of a 1.0-mg/L NO_2-N standard and 75 mL of concentrated ammonium chloride-EDTA solution. Use a flow rate of 7 to 10 mL per minute. Collect the reduced standard until the level of solution is 0.5 cm above the top of the granules. Close the screw clamp to stop flow. Discard the reduced standard.

4. Measure about 40 mL of concentrated ammonium chloride-EDTA and pass through column at 7 to 10 mL per minute to wash nitrate standard off column. Always leave at least 0.5 cm of liquid above top of granules. The column is now ready for use.

5. Dilute ammonium chloride-EDTA solution. Dilute 300 mL of concentrated ammonium chloride-EDTA solution (reagent 4) to 500 mL with distilled water.

6. Color reagent.

7. Zinc sulfate solution.

8. Sodium hydroxide, 6 N.

9. Ammonium hydroxide, concentrated.

10. Hydrochloric acid, 6 N. Dilute 50 mL concentrated HCl to 100 mL with distilled water.

11. Copper sulfate solution, 2 percent.

12. Nitrate stock solution. 1.0 mL = 1.00 mg NO_3-N. Preserve with 2 mL of chloroform per liter. This solution is stable for at least six months.

13. Nitrate standard solution. 1.0 mL = 0.01 mg NO_3-N. Dilute 10.0 mL of nitrate stock solution (reagent 12) to 1,000 mL with distilled water.

14. Chloroform.

E. Procedure

REMOVAL OF INTERFERENCES (IF NECESSARY)

1. Turbidity removal. Use one of the following methods to remove suspended matter that can clog the reduction column.

   a. Filter sample through a glass-fiber filter or a 0.45-micron pore size filter as long as the pH is less than 8.

   b. Add 1 mL zinc solution (reagent 7) to 100 mL of sample and mix thoroughly. Add enough (usually 8 to 10 drops) sodium hydroxide solution (reagent 8) to obtain a pH of 10.5. Let treated sample stand a few minutes to allow the heavy flocculant precipitate to settle. Clarify by filtering through a glass-fiber filter.

(Nitrate)

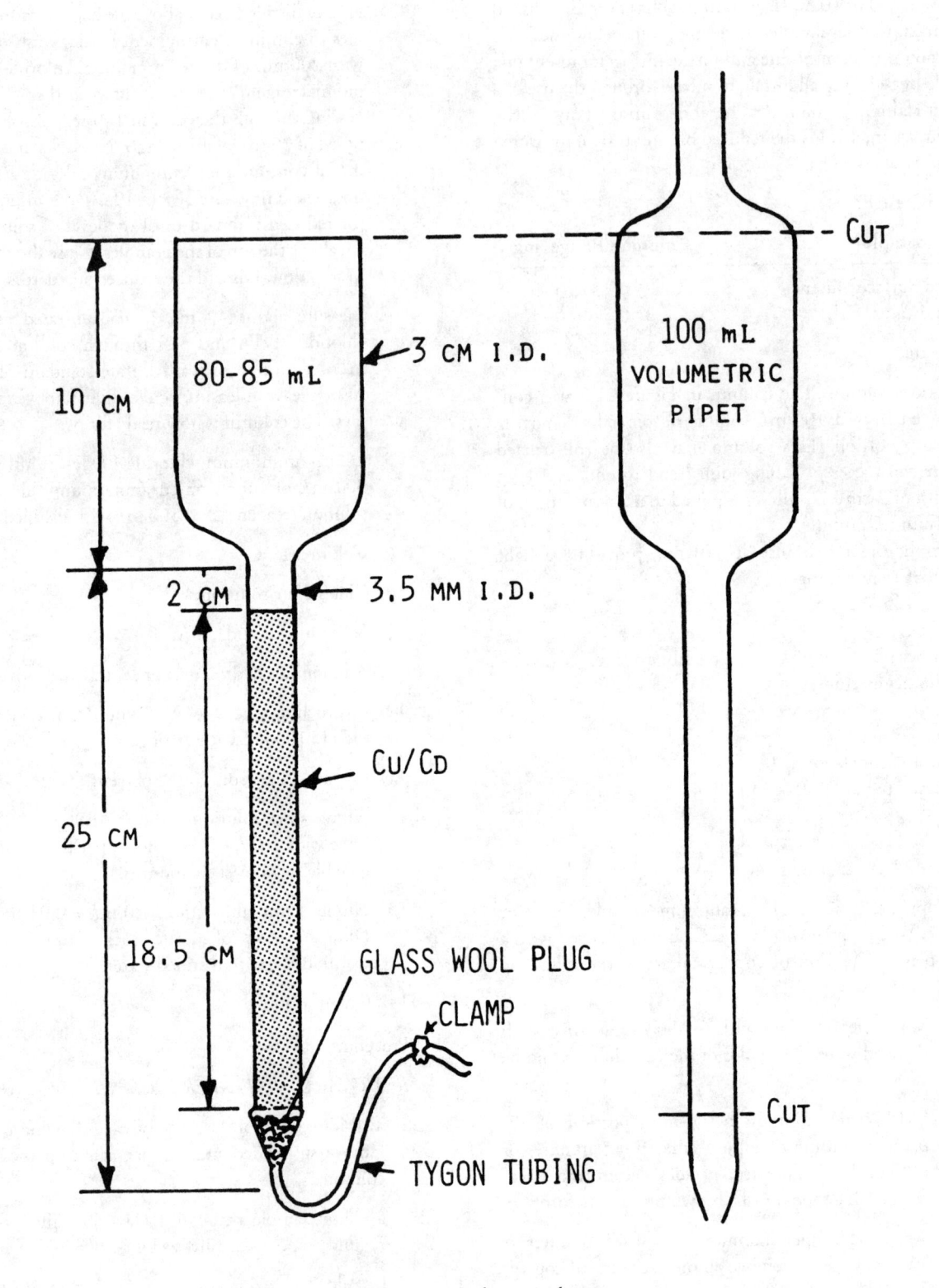

Fig. 21.2   Reduction column

(Nitrate)

REDUCTION OF NITRATE TO NITRITE

1. Using a pH meter, adjust the pH of sample (or standard) to between 5 and 9 either with concentrated HCl or concentrated NH_4OH.

2. To 25 mL of sample (or standard) or aliquot diluted to 25 mL, add 75 mL of concentrated ammonium chloride-EDTA solution and mix.

3. Pour sample into column and collect reduced sample at a rate of 7 to 10 mL per minute.

4. Discard the first 25 mL. Collect the rest of the sample (approximately 70 mL) in the original sample flask. Reduced samples should not be allowed to stand longer than 15 minutes before addition of color reagent.

5. Add 2.0 mL of color reagent to 50 mL of sample. Allow 10 minutes for color development. Within two hours, measure the absorbance at 540 nm against a reagent blank (50 mL distilled water to which 2.0 mL color reagent has been added).

F. Construction of Standard Calibration Graph

1. Prepare working standards by pipeting the following volumes of nitrate standard solution into each of five 100-mL volumetric flasks.

Add This Volume of Nitrate Standard Solution to 100-mL Flask	Concentration of NO_3-N in mg/L
0.0	0.00
1.0	0.10
2.0	0.20
5.0	0.50
10.0	1.00

Dilute each to 100 mL with distilled water and mix.

2. Determine the amount of nitrate-nitrite as outlined above in the procedure for reduction of nitrate to nitrite.

3. Plot on a sheet of graph paper the absorbance versus concentration.

G. Example

Results from the analyses of samples and working standards for nitrate-nitrite were as follows:

Flask No.	Sample	Volume	Absorbance
1	Jones St. Well	25 mL	0.440
2	Blank (distilled water)	25 mL	0.00
3	0.10 mg/L NO_3-N Standard	25 mL	0.075
4	0.20 mg/L NO_3-N Standard	25 mL	0.142
5	0.50 mg/L NO_3-N Standard	25 mL	0.355
6	1.00 mg/L NO_3-N Standard	25 mL	0.700

H. Calculation

1. Using graph paper, plot the absorbance values of working standards versus their known concentrations. For example, from the above data the following graph can be constructed.

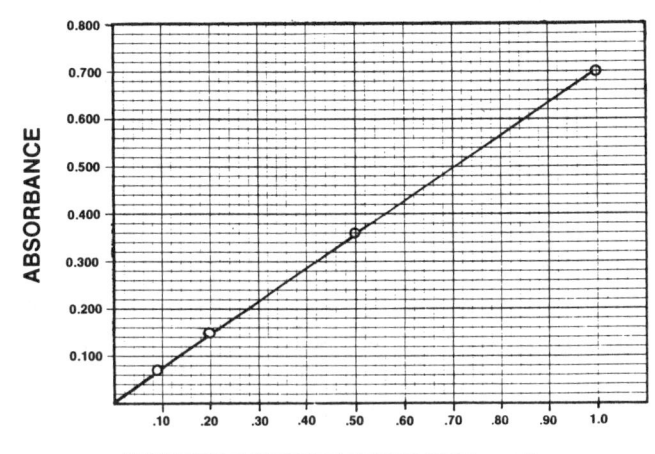

NITRATE & NITRITE-NITROGEN, mg/L

2. Read concentration of $NO^-_3 + NO^-_2$ nitrogen from graph shown below.

mg/L Nitrate + Nitrite Nitrogen in Sample = 0.62 mg/L

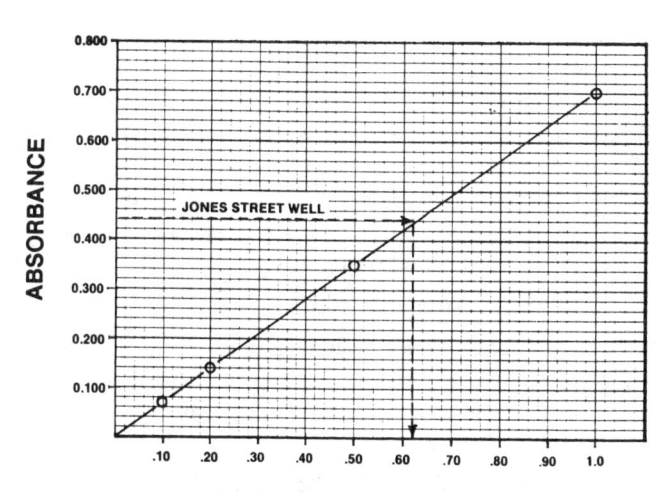

NITRATE & NITRITE-NITROGEN, mg/L at the Jones Street Well

3. Determine concentration of nitrite-nitrogen (NO_2-N) in sample using nitrite procedure.

4. Subtract nitrite from $NO^-_2 + NO^-_3$ nitrogen concentration. The result is the amount of nitrate nitrogen in sample.

5. For example, if the sample of Jones St. Well used in the above example contained no nitrite nitrogen then the nitrate nitrogen (NO_3-N) would be 0.62 mg/L.

(pH)

I. Notes

1. If concentration of nitrate in the sample is greater than 1 mg/L, then the sample must be diluted.

2. Cadmium metal is highly toxic, thus caution must be exercised in its use. Rubber gloves should be used whenever it is handled.

J. Reference

See page 4-123, *STANDARD METHODS*, 21st Edition.

12. pH
by Jack Rossum

DISCUSSION

This discussion is presented to give you a better understanding of what a pH value actually represents. Procedures for measuring pH are given in Volume I, Chapter 11, "Laboratory Procedures."

Pure water dissociates according to the following reaction:

$H_2O = H^+ + OH^-$

At 25°C and a pH of 7, the activity of the hydrogen ion is equal to the activity of the hydroxyl ion at .000 000 1 moles/liter. "Activity" is a term used by chemists to allow real atoms, molecules, and ions to behave as if they were perfect particles (having zero size). Activity is obtained by multiplying the concentration by an activity coefficient. The value of the activity coefficient depends on the electrical charge on the particle, the temperature, and the other substances dissolved in the water. For hydrogen ion, the activity coefficient at 25°C varies from 0.996 in pure water to 0.900 in water containing 400 mg/L of dissolved solids. Activities are expressed in moles per liter, which is assumed to be the number of grams per liter since the molecular weight of hydrogen ion is 1.008 (or near 1.0).

When the activities of the hydrogen and hydroxyl ions are equal, the solution is neutral. If hydrogen ions are in excess, the solution is acid and if hydroxyl ions are in excess, the solution is basic. An important property of water is that for any temperature, the product of the activities of these ions is a constant. At 25°C, this constant is .000 000 000 000 01.

In a strong solution of hydrochloric acid, the hydrogen ion activity may be as high as 1 mole per liter, while in a strong solution of lye, the hydroxyl ion concentration may be as high as 1 mole per liter. To avoid the inconvenience of writing these very small numbers, hydrogen ion activities are expressed in terms of pH, with

$$pH = \log \frac{1}{\{H^+\}}$$

The relation between pH, H^+, and OH^- at 25°C is shown in Table 21.2.

TABLE 21.2 RELATION BETWEEN pH, H^+, AND OH^- AT 25°C

pH	Activity of H^+, moles/L	Activity of OH^-, moles/L
0	1.0	0.000 000 000 000 01
1	0.1	0.000 000 000 000 1
2	0.01	0.000 000 000 001
3	0.001	0.000 000 000 01
4	0.000 1	0.000 000 000 1
5	0.000 01	0.000 000 001
6	0.000 001	0.000 000 01
7	0.000 000 1	0.000 000 1
8	0.000 000 01	0.000 001
9	0.000 000 001	0.000 01
10	0.000 000 000 1	0.000 1
11	0.000 000 000 01	0.001
12	0.000 000 000 001	0.01
13	0.000 000 000 000 1	0.1
14	0.000 000 000 000 01	1.0

Most natural waters have pH values between 6.5 and 8.5. The pH of natural waters is controlled by the relative amounts of carbon dioxide, bicarbonate, and carbonate ions. Rainwater usually has a pH of slightly less than 7 because carbon dioxide from the air dissolves to form carbonic acid.

Human blood has a pH of 7.4 and the gastric juices in your stomach have a pH of approximately 0.9 to aid in the digestion of food.

Alum coagulates most effectively at pH values near 6.8.

REFERENCE

See page 4-90, *STANDARD METHODS*, 21st Edition.

QUESTIONS

Write your answers in a notebook and then compare your answers with those on pages 546 and 547.

21.1Q How is nitrate measured in the nitrate test?

21.1R If turbidity is interfering with a nitrate analysis, how can turbidity be removed?

21.1S The pH of natural waters is usually controlled by the relative amounts of what ions?

(Sulfate)

13. Specific Conductance (Conductivity)

A. Discussion

Specific conductance or conductivity is a numerical expression (expressed in micromhos per centimeter) of the ability of a water to conduct an electric current. This number depends on the total concentration of the minerals dissolved in the sample (TDS) and the temperature. Changes in conductivity from normal levels may indicate changes in mineral composition of the water, seasonal variations in lakes and reservoirs, or intrusion of pollutants. The custom of reporting conductivity values in micromhos/cm at 25°C requires the accurate determination of each sample's temperature at the time of conductivity measurement.

Specific conductance is measured by the use of a conductivity meter.

B. What is Tested?

Sample	Common Range, micromhos/cm
Raw and Treated Surface Waters	30 to 500
Groundwater	100 to 1,000

C. Materials and Procedure

Follow instrument manufacturer's instructions. Also see page 2-44, *STANDARD METHODS*, 21st Edition.

14. Sulfate

A. Discussion

The sulfate ion is one of the major anions occurring in natural waters. Sulfate ions are of importance in water supplies because of the tendency of appreciable amounts to form hard scales in boilers and heat exchangers. EPA is evaluating the need to regulate sulfate in drinking water.

B. What is Tested?

Sample	Common Range, mg/L
Raw or Treated Water Supply	5 to 100

C. Apparatus Required

Turbidimeter or spectrophotometer
Stopwatch or timer
Measuring spoon, 0.3 mL
Magnetic stirrer
Magnetic stir-bar
Pipet, 10 mL
Erlenmeyer flasks, 250 mL
Volumetric flasks, 100 mL

D. Reagents

NOTE: Standardized solutions are commercially available.

1. Conditioning reagent.

2. Barium chloride, $BaCl_2$, crystals: Sized for turbidimetric work (Baker No. 0974 or equivalent). To ensure uniformity of results, construct a standard curve for each batch of $BaCl_2$ crystals.

3. Standard sulfate solution: Prepare a standard sulfate solution as described in a. or b. below; 1.00 mL = 0.10 mg SO_4.

   a. Dilute 10.41 mL standard 0.0200 N H_2SO_4 titrant specified in Alkalinity Test (Chapter 11, Volume I, of this manual) to 100 mL with distilled water.

   b. Dissolve 147.9 mg anhydrous Na_2SO_4 in distilled water and dilute to 1,000 mL.

E. Procedure

1. Place 100 mL of sample or a suitable portion diluted to 100 mL into a clean 250-mL Erlenmeyer flask.

2. Add 5.0 mL of conditioning reagent and mix.

3. While stirring, add a spoonful (0.3 mL) of barium chloride crystals. Stir for exactly 1 minute.

4. Measure turbidity at 30-second intervals for 4 minutes. Consider turbidity to be the maximum reading obtained in the 4-minute interval.

F. Construction of Standard Calibration Curve

1. Using the standard solution prepare the following standards in 100-mL volumetric flasks.

mL of Standard Sulfate Solution Placed in 100-mL Volumetric Flask	Sulfate Concentration, mg/L
5.0	5.0
10.0	10.0
15.0	15.0
20.0	20.0
25.0	25.0

2. Dilute flasks to 100 mL.

3. Transfer to 250-mL Erlenmeyer flask.

4. Determine amount of sulfate as outlined previously.

5. Prepare a standard curve by plotting turbidity values of standards versus the corresponding sulfate concentrations. Set nephelometer (or spectrophotometer) at zero sulfate concentration using distilled water as a control.

(Sulfate)

OUTLINE OF PROCEDURE FOR SULFATE

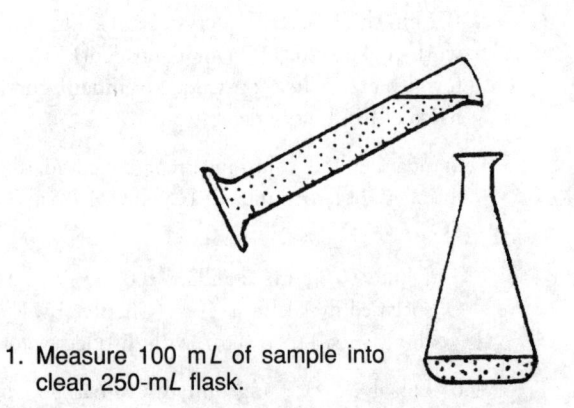

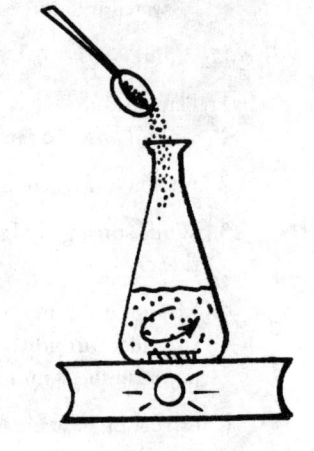

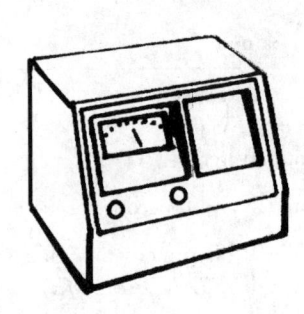

1. Measure 100 mL of sample into clean 250-mL flask.
2. Add 5.0 mL conditioning reagent. Mix.
3. Add barium chloride and stir for 1 minute.
4. Measure turbidity.

G. Example

Results from a series of tests for sulfate were as follows:

Flask No.	Sample	Volume	Turbidity
1	Distilled Water	100 mL	0
2	Plant Effluent	100 mL	35
3	Jones St. Well	50 mL	45
4	5.0 mg/L SO$_4$ Standard	100 mL	11
5	10.0 mg/L SO$_4$ Standard	100 mL	29
6	15.0 mg/L SO$_4$ Standard	100 mL	40
7	20.0 mg/L SO$_4$ Standard	100 mL	53

H. Calculation

1. Prepare a standard curve by using data from prepared standards. From the above example:

Concentration Sulfate, mg/L	Turbidity, TU
0.0	0
5.0	11
10.0	29
15.0	40
20.0	53

(Sulfate)

The graph below is the result of plotting concentration of standards versus their corresponding turbidity.

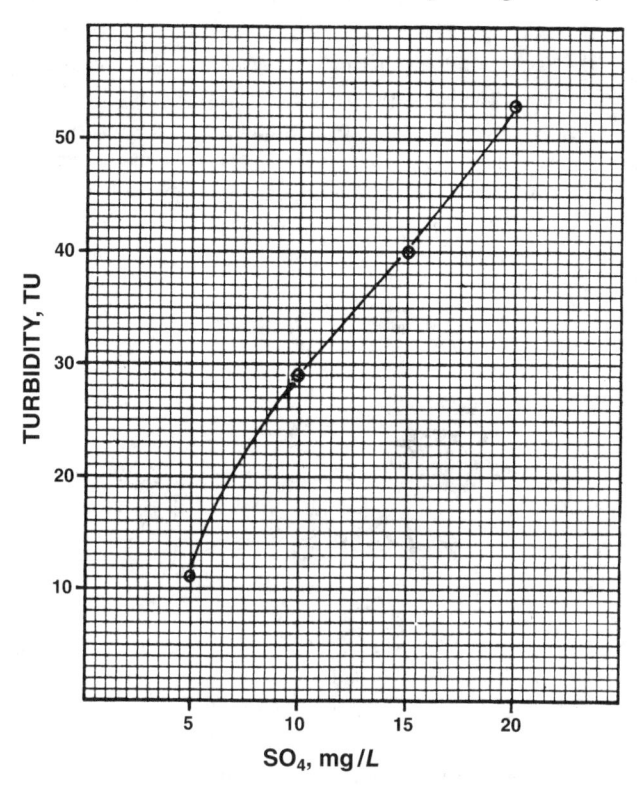

2. Obtain concentration of unknown plant and well samples from curve.

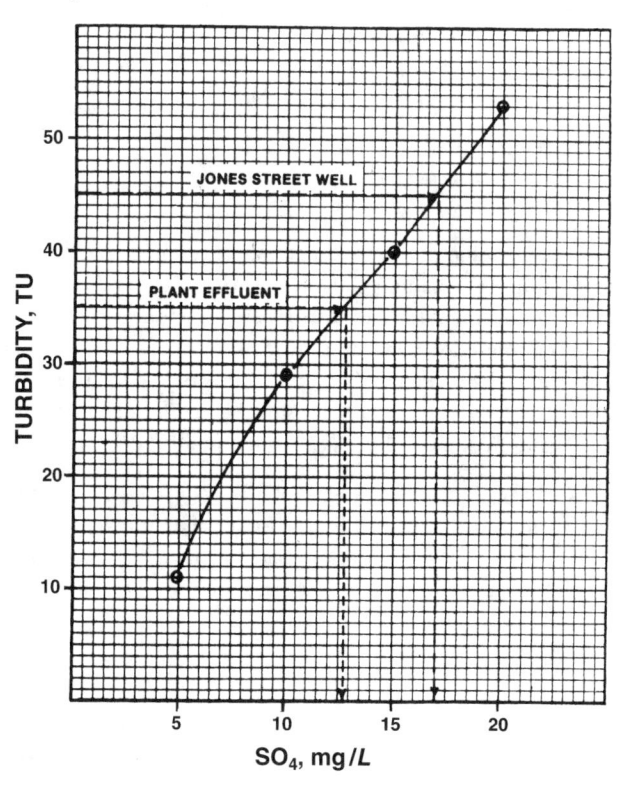

3. Correct (if necessary) for samples of less than 100 mL by using the following formula:

$$\text{Sulfate (SO}_4\text{), mg}/L = \frac{(\text{Graph Sulfate, mg}/L)(100 \text{ m}L)}{\text{Sample Size, m}L}$$

Using data from example:

	Sample Volume	Turbidity	Concentration from Graph
Plant Effluent	100 mL	35 TU	13 mg/L

Sulfate, mg/L = 13 mg/L

	Sample Volume	Turbidity	Concentration from Graph
Jones St. Well	50 mL	45 TU	17 mg/L

$$\text{Sulfate, mg}/L = \frac{(\text{Sulfate, mg}/L)(100 \text{ m}L)}{\text{Sample Size, m}L}$$

$$= \frac{(17 \text{ mg}/L)(100 \text{ m}L)}{50 \text{ m}L}$$

$$= 34 \text{ mg}/L \text{ SO}_4$$

I. Notes

1. A spectrophotometer can be used to measure absorbance of barium sulfate suspension. Use at 420 nanometers (nm) wavelength.

2. Color or suspended matter will interfere when present in large amounts. Correct for these items by testing blanks from which barium chloride is withheld.

3. Analyze samples and standards with their temperatures in the range of 20 to 25°C.

J. Reference

See page 4-188, *STANDARD METHODS*, 21st Edition.

QUESTIONS

Write your answers in a notebook and then compare your answers with those on page 547.

21.1T What is the meaning of specific conductance or conductivity?

21.1U Sulfate ions are of concern in drinking water for what reason?

21.1V A 50-mL sample from a well produced a turbidity reading of 40 TU using a nephelometer (turbidimeter). What was the sulfate concentration in mg/L?

(Taste and Odor)

15. Taste and Odor

A. Discussion

Taste and odor are sensory clues that provide the first warning of potential hazards in the environment. Water, in its pure form, cannot produce odor or taste sensations. However, algae, *Actinomycetes,* bacteria, decaying vegetation, metals, and pollutants can cause tastes and odors in drinking water. Corrective measures designed to reduce unpleasant tastes and odors include aeration or the addition of chlorine, chlorine dioxide, potassium permanganate, or activated carbon. (For additional details, see Volume I, Chapter 9, "Taste and Odor Control.")

Odor is considered a quality factor affecting acceptability of drinking water (and foods prepared with it), tainting of fish and other aquatic organisms, and aesthetics of recreational waters. Most organic and some inorganic chemicals contribute to taste and odor. These chemicals may originate from municipal and industrial waste discharges, from natural sources (such as decomposition of vegetable matter), or from associated microbial activity.

Some substances, such as certain inorganic salts, produce taste without odor. Many taste sensations actually are odor sensations, even though the sensation is not noticed until the water is in the mouth.

Taste, like odor, is one of the chemical senses. Taste and odor are different in that odors are sensed high up in our nose and tastes are sensed on our tongue. Taste is simpler than odor because there may be only four true taste sensations: sour, sweet, salty, and bitter. Dissolved inorganic salts of copper, iron, manganese, potassium, sodium, and zinc can be detected by taste.

Operators must remember that a tasteless water is not the most desirable water. Distilled water is considered less pleasant to drink than a high-quality water. The taste test must determine the taste intensity by the threshold test and also evaluate the quality of the drinking water on the basis of desirability for consumers.

B. Apparatus Required

Sample bottles, glass-stoppered or with TFE-lined closures
Constant temperature bath
Odor flasks (500-mL glass-stoppered Erlenmeyer flasks)
Transfer and volumetric pipets or graduated cylinders (200, 100, 50, and 25 mL)
Measuring pipets (10 mL, graduated in 0.1-mL increments)
Thermometer (0 to 110°C)

C. Precautions

Use preliminary tests to select the persons to make taste or odor tests. Use only persons who want to participate in the test. Avoid distracting odors such as those caused by smoking, foods, soaps, perfumes, and shaving lotions. The testers should not have colds or allergies that affect odor response. Do not have the testers perform too many tests and allow frequent rests so the testers will not become tired and lose their sensitivity. Keep the room in which the tests are conducted free from distractions, drafts, and odors.

A panel of five or more testers is recommended for precise work. Do not allow the testers to prepare the samples or to know the dilution concentrations being evaluated. Familiarize testers with the procedure before they participate in a panel test. Present the most dilute sample first to avoid tiring the senses with a concentrated sample. Keep temperature of sample during test within 1°C of the specified temperature. Use opaque or darkly colored flasks to avoid biasing the results due to turbid or colored waters being tested.

D. Procedure

ODOR

1. Determine the approximate range of the threshold odor number by adding 200 mL, 50 mL, 12 mL, and 2.8 mL of sample to 500-mL glass-stoppered Erlenmeyer flasks containing odor-free water[6] to make a total volume of 200 mL. Use a separate flask containing only odor-free water as a reference for comparison. Heat dilutions and reference to desired test temperature (usually 60°C or 140°F).

2. Shake flask containing odor-free water, remove stopper, and sniff vapors. Test sample containing the least amount of odor-bearing water in the same way. If an odor can be detected in this dilution, prepare more dilute samples.

---

[6] See *STANDARD METHODS,* 21st Edition, page 2-11, for directions on how to prepare odor-free water.

(Taste and Odor)

To prepare more dilute samples, prepare an intermediate dilution consisting of 20-mL sample diluted to 200 mL with odor-free water. Use this dilution for the threshold determination. Multiply the threshold odor number (TON) obtained by 10 to correct for the intermediate dilution.

If an odor cannot be detected in the first dilution, repeat the above procedure using sample containing the next higher concentration of odor-bearing water and continue this process until odor is detected clearly.

3. Based on the results obtained in the preliminary test, prepare a set of dilutions using Table 21.3 as a guide. Prepare the five dilutions shown in the appropriate column and the three next most concentrated in the next column to the right in Table 21.3. For example, if odor was first noted in the flask containing the 50 mL sample in the preliminary test, prepare flasks containing 50, 35, 25, 17, 12, 8.3, 5.7, and 4.0 mL sample, each diluted to 200 mL with odor-free water. This procedure is necessary to challenge the range of sensitivities of the entire panel of testers.

TABLE 21.3 DILUTIONS FOR VARIOUS ODOR INTENSITIES

PRELIMINARY TEST

Sample Volume in Which Odor First Noted

| 200 mL | 50 mL | 12 mL | 3.8 mL |

FINAL TEST

Volume in mL of Sample to be Diluted to 200 mL

200	50	12	(Intermediate dilution)
140	35	8.3	
100	25	5.7	
70	17	4.0	
50	12	2.8	

Insert two or more blanks near the expected threshold, but avoid any repeated patterns. Do not let the testers know which dilutions are odorous and which are blanks. Instruct each tester to smell each flask in sequence, beginning with the least concentrated sample, until odor is detected with certainty.

4. Record observations by indicating whether odor is noted in each flask. For example,

mL Sample Diluted to 200 mL	12	0	17	25	0	35	50
Response	−	−	−	+	−	+	+

5. Calculate the threshold odor number (TON) as shown in E. Calculations.

TASTE THRESHOLD TEST

1. The taste threshold test is used when the purpose is quantitative measurement of detectable taste. When odor is the predominant sensation, as in the case of chlorophenols, the threshold odor test takes priority.

2. Use the dilution and random blank system described for odor tests when preparing taste samples.

3. Present each dilution and blank to the tester in a clean 50-mL plastic container filled to the 30-mL level. Use high-quality clear plastic containers. Discard the plastic containers when finished. Do not use glass containers because the soap used to clean the glass could leave a residue that may affect the results.

4. *STANDARD METHODS* recommends maintaining the sample presentation at 40 ± 1°C (104 ± 2°F).

   NOTE: Some operators use normal water temperatures for taste tests or a temperature of 15°C (59°F).

5. Present the series of samples to each tester. Pair each sample with a known blank.

6. Have each tester taste the sample by taking into the mouth whatever volume is comfortable, holding it in the mouth for several seconds, and discharging the sample without swallowing the water.

7. Have the tester compare the sample with the blank and record whether a taste or aftertaste is detectable in the sample.

8. Submit samples in an increasing order of concentration until the tester's taste threshold has been passed.

9. Calculate individual threshold and threshold of the panel as shown in E. Calculations.

TASTE RATING TEST

1. When the purpose of the test is to estimate the taste acceptability, use the taste rating test procedure described below.

2. Samples for this test usually represent treated water ready for human consumption. If experimentally treated water is tested, *BE CERTAIN THAT THE WATER IS SAFE TO DRINK* (no pathogens and no toxic chemicals present).

(Taste and Odor)

3. Give testers thorough instructions and trial or orientation sessions followed by questions and discussions of procedures.

4. Select panel members on the basis of performance in trial sessions.

5. When testing samples, testers work alone.

6. Present samples at a temperature that testers find pleasant for drinking water. Maintain this temperature by the use of a water bath apparatus. A temperature of 15°C (59°F) is recommended. Do not let the test temperature exceed tap water temperatures that are customary at the time of the test. Specify the test temperature in reporting results.

7. Present each dilution and blank to the tester in a clean 50-mL plastic container filled to the 30-mL level. Use high-quality clear plastic containers. Discard the plastic containers when finished. Do not use glass containers because the soap used to clean the glass could leave a residue that may affect the results.

8. Each tester is presented with a list of nine statements about the water, ranging on a scale from very favorable to very unfavorable (Table 21.4). The tester's task is to select the statement that best expresses the tester's opinion. The scored rating is the scale number of the statement selected. The panel rating is the arithmetic mean (average) of the scale numbers of all testers.

TABLE 21.4 ACTION TENDENCY RATING SCALE FOR TASTE RATING TEST

1. I would be very happy to drink this water as my everyday drinking water.
2. I would be happy to accept this water as my everyday drinking water.
3. I am sure that I could accept this water as my everyday drinking water.
4. I could accept this water as my everyday drinking water.
5. Maybe I could accept this water as my everyday drinking water.
6. I do not think I could accept this water as my everyday drinking water.
7. I could not accept this water as my everyday drinking water.
8. I could never drink this water.
9. I cannot stand this water in my mouth and I could never drink it.

9. Rating involves the following steps:

   a. Initial tasting of about half the sample by taking water into the mouth, holding it for several seconds, and discharging it without swallowing

   b. Forming an initial judgment on the rating scale

   c. A second tasting is made in the same manner as the first

   d. A final rating is made for the sample and the result is recorded on the appropriate data form

   e. Rinse mouth with taste- and odor-free water

   f. Rest one minute before repeating steps (a) through (e) on the next sample

10. Independently randomize sample order for each tester. Allow at least 30 minutes rest between repeated rating sessions. Testers should not know the composition or source of the samples.

E. Calculations

FORMULAS

1. ODOR

The threshold odor number (TON) for an individual tester is calculated using the following formula:

$$TON = \frac{A + B}{A}$$

where:

A = mL sample

B = mL odor-free water

The threshold odor number for a group is presented as the geometric mean of the individual tester thresholds.

Geometric Mean = $(X_1 \times X_2 \times X_3 \ldots X_n)^{1/n}$

where:

X_1 = threshold odor number for tester number 1

X_2 = threshold odor number for tester number 2

X_n = threshold odor number for the nth tester

n = total number of testers

2. TASTE THRESHOLD

Calculate the individual tester's threshold taste number and the threshold taste number for a panel using the same formulas that are used for the threshold odor tests.

(Taste and Odor)

3. TASTE RATING

Determine the taste rating for a water by calculating the arithmetic mean and *STANDARD DEVIATION*[7] of all ratings given for each sample.

$$\text{Arithmetic Mean, } \overline{X} = \frac{X_1 + X_2 + X_3 + \ldots X_n}{n}$$

where: X_1 = taste rating for tester number 1

X_2 = taste rating for tester number 2

X_n = taste rating for the nth tester

n = total number of testers

$$\text{Standard Deviation} = \left[\frac{(X_1 - \overline{X})^2 + (X_2 - \overline{X})^2 + \ldots (X_n - \overline{X})^2}{n - 1}\right]^{0.5}$$

$$\text{or} = \left[\frac{(X_1^2 + X_2^2 + \ldots X_n^2) - (X_1 + X_2 + \ldots X_n)^2/n}{n - 1}\right]^{0.5}$$

EXAMPLE 1

Calculate the threshold odor number (TON) for a sample when the first detectable odor occurred when the 25-mL sample was diluted to 200 mL (175 mL of odor-free water was added to the 25-mL sample.)

Known	Unknown
A or Sample Size, mL = 25 mL	TON
B or Odor-Free Water, mL = 175 mL	

Calculate the threshold odor number (TON).

$$\text{TON} = \frac{A + B}{A}$$

$$= \frac{25 \text{ mL} + 175 \text{ mL}}{25 \text{ mL}}$$

$$= 8$$

EXAMPLE 2

Determine the geometric mean threshold odor number for a panel of five testers given the following results.

Known	Unknown
Tester 1, X_1 = 8	Geometric Mean Threshold Odor Number
Tester 2, X_2 = 6	
Tester 3, X_3 = 12	
Tester 4, X_4 = 8	
Tester 5, X_5 = 4	

Calculate the geometric mean.

$$\text{Geometric Mean TON} = (X_1 \times X_2 \times X_3 \times X_4 \times X_5)^{1/n}$$

$$= (8 \times 6 \times 12 \times 8 \times 4)^{1/5}$$

$$= (18{,}432)^{0.2}$$

$$= 7.1$$

EXAMPLE 3

Calculate the threshold taste number for a sample when the first detectable taste occurred when the 50-mL sample was diluted to 200 mL (150 mL of taste-free water was added to the 50-mL sample).

Known	Unknown
A or Sample Size, mL = 50 mL	Threshold Taste Number
B or Taste-Free Water, mL = 150 mL	

Calculate the threshold taste number.

$$\text{Threshold Taste Number} = \frac{A + B}{A}$$

$$= \frac{50 \text{ mL} + 150 \text{ mL}}{50 \text{ mL}}$$

$$= 4$$

EXAMPLE 4

Determine the taste rating for a water by calculating the arithmetic mean and standard deviation for the panel ratings given below.

Known	Unknown
Tester 1, X_1 = 2	1. Arithmetic Mean, $\overline{X}$
Tester 2, X_2 = 5	2. Standard Deviation, S
Tester 3, X_3 = 3	
Tester 4, X_4 = 6	
Tester 5, X_5 = 2	
Tester 6, X_6 = 6	

1. Calculate the arithmetic mean, $\overline{X}$, taste rating.

$$\text{Arithmetic Mean, } \overline{X} \text{ Taste Rating} = \frac{X_1 + X_2 + X_3 + X_4 + X_5 + X_6}{n}$$

$$= \frac{2 + 5 + 3 + 6 + 2 + 6}{6}$$

$$= \frac{24}{6}$$

$$= 4$$

---

[7] *Standard Deviation.* A measure of the spread or dispersion of data.

(Taste and Odor)

2. Calculate the standard deviation, S, of the taste rating.

$$\text{Standard Deviation, S} = \left[\frac{(X_1 - \bar{X})^2 + (X_2 - \bar{X})^2 + (X_3 - \bar{X})^2 + (X_4 - \bar{X})^2 + (X_5 - \bar{X})^2 + (X_6 - \bar{X})^2}{n-1}\right]^{0.5}$$

$$= \left[\frac{(2-4)^2 + (5-4)^2 + (3-4)^2 + (6-4)^2 + (2-4)^2 + (6-4)^2}{6-1}\right]^{0.5}$$

$$= \left[\frac{(-2)^2 + (1)^2 + (-1)^2 + (2)^2 + (-2)^2 + (2)^2}{5}\right]^{0.5}$$

$$= \left[\frac{4+1+1+4+4+4}{5}\right]^{0.5}$$

$$= \left[\frac{18}{5}\right]^{0.5}$$

$$= (3.6)^{0.5}$$

$$= 1.9$$

or

$$\text{Standard Deviation, S} = \left[\frac{(X_1^2 + X_2^2 + X_3^2 + X_4^2 + X_5^2 + X_6^2) - (X_1 + X_2 + X_3 + X_4 + X_5 + X_6)^2/n}{n-1}\right]^{0.5}$$

$$= \left[\frac{(2^2 + 5^2 + 3^2 + 6^2 + 2^2 + 6^2) - (2+5+3+6+2+6)^2/6}{6-1}\right]^{0.5}$$

$$= \left[\frac{(4+25+9+36+4+36) - (24)^2/6}{5}\right]^{0.5}$$

$$= \left[\frac{114 - 96}{5}\right]^{0.5}$$

$$= \left[\frac{18}{5}\right]^{0.5}$$

$$= (3.6)^{0.5}$$

$$= 1.9 \text{ (same answer as before)}$$

F. Reference

Odor: See page 2-11, *STANDARD METHODS*, 21st Edition.

Taste: See page 2-16, *STANDARD METHODS*, 21st Edition.

QUESTIONS

Write your answers in a notebook and then compare your answers with those on page 547.

21.1W List the items that can cause tastes and odors in drinking water.

21.1X Calculate the threshold odor number (TON) for a sample when the first detectable odor occurred when the 12-mL sample was diluted to 200 mL (188 mL of odor-free water was added to the 12-mL sample).

16. Trihalomethanes

A. Discussion

The trihalomethanes (THMs) are members of the family of organohalogen compounds, which are named as derivatives of methane. Current analytical chemistry applied to drinking water has thus far detected chloroform, bromodichloromethane, dibromochloromethane, bromoform, and dichloroiodomethane.

The principal source of chloroform and other trihalomethanes in drinking water is the chemical interaction of chlorine added for disinfection and other purposes with the commonly present natural humic substances and other precursors produced either by normal organic decomposition or by the metabolism of aquatic organisms. Since these natural organic precursors are more commonly found in surface water, water taken from a surface source is more likely to produce high THM levels than most groundwaters.

(Total Dissolved Solids)

Generally, the THM-producing reaction is:

Chlorine + Precursors = Chloroform + Other THMs

Chloroform is the most common THM found in drinking water and it is usually present in the highest concentration. The presence in drinking water of chloroform and other THMs and synthetic organic chemicals may have an adverse effect on the health of consumers; therefore, human exposure to these chemicals should be reduced.

B. Reference

For materials and procedures, see pages 5-26 and 5-62, STANDARD METHODS, 21st Edition.

NOTE: A gas chromatography analyzer is required for this analysis.

17. **Total Dissolved Solids**

A. Discussion

"Total dissolved solids" (TDS) refers to material that passes through a standard glass-fiber filter disk and remains after evaporation at 180°C. The amount of dissolved solids present in water is a consideration in its suitability for domestic use. In general, waters with a TDS content of less than 50 mg/L are most desirable for such purposes. The higher the TDS concentration, the greater the likelihood of tastes and odors and also scaling problems. As TDS increase, the number of times the water can be recycled and reclaimed before requiring demineralization decreases. In potable waters, TDS consist mainly of inorganic salts, small amounts of organic matter, and dissolved gases.[8]

B. What is Tested?

Sample	Common Range, mg/L
Raw and Treated Surface Waters	20 to 700
Groundwater	100 to 1,000

C. Apparatus Required

Glass-fiber filter disks (Millipore AP40; or Gelman Type A/E)
Suction flask, 500 mL
Filter holder or Gooch crucible adapter
Gooch crucibles (25 mL if 2.2-cm filter used)
Evaporating dishes, 100 mL (high-silica glass)
Drying oven, 180°C
Steam bath
Vacuum source
Desiccator
Analytical balance
Muffle furnace, 550°C

D. Procedure

PREPARATION OF DISH

1. Ignite a clean evaporating dish at 550±50°C for one hour in muffle furnace.

2. Cool in desiccator, then weigh and record weight. Store in desiccator until needed.

PREPARATION OF GLASS-FIBER FILTER DISK

1. Place the disk on the filter apparatus or insert into the bottom of a suitable Gooch crucible. While vacuum is applied, wash the filter disk with three successive 20-mL volumes of distilled water. Continue the suction to remove all traces of water from the disk and discard the washings.

SAMPLE ANALYSIS

1. Shake the sample vigorously and transfer 100 to 150 mL to the funnel or Gooch crucible by means of a 150-mL graduated cylinder.

2. Filter the sample through the glass-fiber filter and continue to apply vacuum for about three minutes after filtration is complete to remove as much water as possible.

3. Transfer 100 mL of the filtrate to the weighed evaporating dish and evaporate to dryness on a steam bath.

4. Dry the evaporated sample for at least one hour at 180°C. Cool in desiccator and weigh. Repeat drying cycle until constant weight is obtained or until weight loss is less than 0.5 mg.

E. Example

Results from weighings were:

Clean Dish	= 47.0028 grams (47,002.8 mg)
Dissolved Residue + Dish	= 47.0453 grams (47,045.3 mg)
Sample Volume	= 100 mL

---

[8] Reference. *CHEMISTRY FOR ENVIRONMENTAL ENGINEERS*, Fifth Edition, 2003, by Clair N. Sawyer, Perry L. McCarty, and Gene F. Parkin. Obtain from the McGraw-Hill Companies, Order Services, PO Box 182604, Columbus, OH 43272-3031. ISBN 0-07-248066-1. Price, $133.75, plus nine percent of order total for shipping and handling.

544 Water Treatment

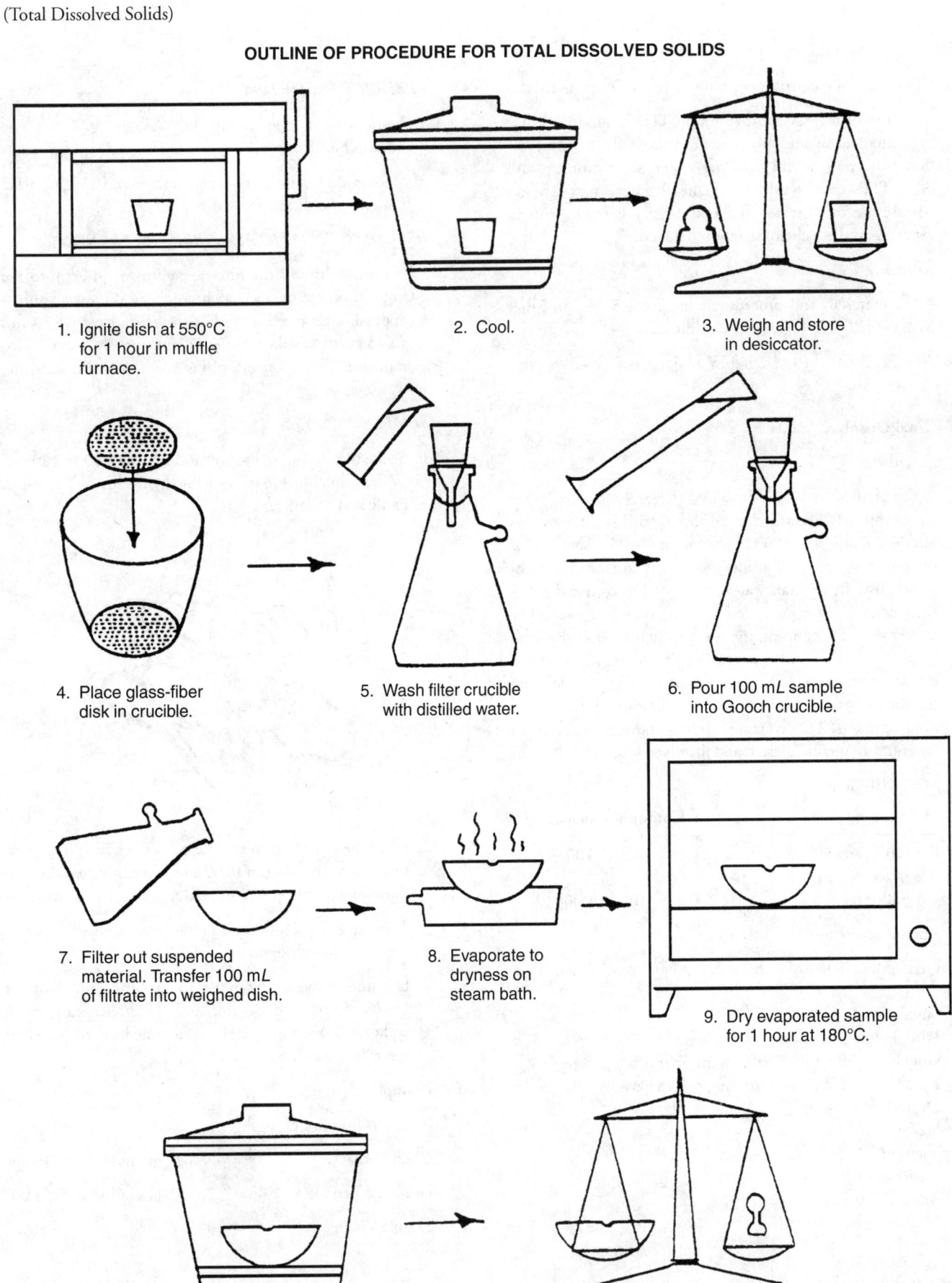

(Total Dissolved Solids)

OUTLINE OF PROCEDURE FOR TOTAL DISSOLVED SOLIDS

1. Ignite dish at 550°C for 1 hour in muffle furnace.
2. Cool.
3. Weigh and store in desiccator.
4. Place glass-fiber disk in crucible.
5. Wash filter crucible with distilled water.
6. Pour 100 mL sample into Gooch crucible.
7. Filter out suspended material. Transfer 100 mL of filtrate into weighed dish.
8. Evaporate to dryness on steam bath.
9. Dry evaporated sample for 1 hour at 180°C.
10. Cool in desiccator.
11. Weigh.

F. Calculations

1. Total Dissolved Solids, $\text{mg}/L = \dfrac{(A-B) \times 1{,}000}{mL \text{ Sample Volume}}$

   where: A = weight of dish and dissolved material in milligrams (mg)

   B = weight of clean dish in milligrams (mg)

2. From example:

$$\begin{aligned}\text{Total Dissolved Solids, mg}/L &= \dfrac{(A-B) \times 1{,}000}{mL \text{ Sample Volume}} \\ &= \dfrac{(47{,}045.3 \text{ mg} - 47{,}002.8 \text{ mg})(1{,}000 \text{ m}L/L)}{100 \text{ m}L} \\ &= 425 \text{ mg}/L\end{aligned}$$

G. Comments

Because excessive residue in the evaporating dish may form a water-entrapping crust, use a sample that yields no more than 200 mg of residue.

H. Reference

See page 2-57, *STANDARD METHODS*, 21st Edition.

QUESTIONS

Write your answers in a notebook and then compare your answers with those on page 547.

21.1Y How are trihalomethanes produced?

21.1Z What are total dissolved solids (TDS)?

END OF LESSON 2 OF 2 LESSONS

on

ADVANCED LABORATORY PROCEDURES

Please answer the discussion and review questions next.

DISCUSSION AND REVIEW QUESTIONS

Chapter 21. ADVANCED LABORATORY PROCEDURES

(Lesson 2 of 2 Lessons)

Write the answers to these questions in your notebook. The question numbering continues from Lesson 1.

9. Why is iron undesirable in a domestic water supply?

10. What precautions must be exercised when collecting samples to be analyzed for iron?

11. How would you obtain the manganese concentration in a sample by using a spectrophotometer if turbidity or color was interfering with the results?

12. What is the purpose of the Marble Test?

13. Why is the presence of certain metals in drinking water of serious concern?

14. How would you interpret the results of lab tests that indicate high levels of nitrate in a raw water sample?

15. When performing the nitrate determination, why should caution be exercised when using cadmium and what precautions should be used?

16. How would you interpret changes (away from normal) in conductivity in water?

17. Why are sulfate ions of concern in water supplies?

18. How would you attempt to reduce unpleasant tastes and odors in drinking water?

19. Why should exposure to trihalomethanes (THMs) be reduced?

20. Why is the amount of dissolved solids present in water a consideration in its suitability for domestic use?

SUGGESTED ANSWERS

Chapter 21. ADVANCED LABORATORY PROCEDURES

ANSWERS TO QUESTIONS IN LESSON 1

Answers to questions on page 510.

21.0A The intensity of a blue color is measured when measuring the concentration of phosphorus in water.

21.0B The scale on spectrophotometers is usually graduated in two ways:

1. In units of percent transmittance (%T), an arithmetic scale with units graded from 0 to 100%
2. In units of absorbance (A), a logarithmic scale of nonequal divisions graduated from 0.0 to 2.0

21.0C If an absorbance reading was 0.60, the unknown concentration would be 0.70 mg/L.

Answers to questions on page 514.

21.1A Yes, the quality of water in any lake, reservoir, or stream has a very direct effect on the abundance and types of aquatic organisms found.

21.1B Calcium in the form of lime or calcium hydroxide may be used to soften water or to control corrosion through pH adjustment.

21.1C Titrate a sample for calcium immediately after adding sodium hydroxide (NaOH) solution.

21.1D Chloride concentrations above 250 mg/L are objectionable to many people due to a salty taste.

Answers to questions on page 515.

21.1E The most common colors that occur in raw water are yellow and brown.

21.1F True color is normally removed or at least decreased by coagulation and chlorination or ozonation.

21.1G Stock color standards should be protected against evaporation and contamination when not in use.

Answers to questions on page 521.

21.1H The presence of dissolved oxygen (DO) in water can contribute to corrosion of piping systems.

21.1I The common range of fluoride in fluoridated drinking water is 0.8 to 1.2 mg/L.

ANSWERS TO QUESTIONS IN LESSON 2

Answers to questions on page 528.

21.1J Problems that may be caused by iron in a domestic water supply include staining of laundry, concrete, and porcelain. A bitter, astringent taste can be detected by some people at levels above 0.3 mg/L.

21.1K All glassware must be acid washed when analyzing samples for iron to remove deposits of iron oxide, which could give false results.

21.1L Manganese occurs both in suspension and as a soluble complex in surface waters.

21.1M If the manganese concentration cannot be determined immediately, acidify the sample with nitric acid to pH less than 2.

Answers to questions on page 530.

21.1N Temperature is important in the Marble Test because the solubility of calcium carbonate varies with temperature. Therefore, the test must be performed immediately after the sample is collected and as rapidly as possible.

21.1O Langelier Index $\cong$ Initial pH – Final pH

$\cong$ 8.9 – 8.6

$\cong$ 0.3

Since the Langelier Index is positive, the water is supersaturated with calcium carbonate and not considered corrosive.

21.1P The concentrations of most metals in water are determined by using atomic absorption spectroscopy or colorimetric methods.

Answers to questions on page 534.

21.1Q In the nitrate test, all nitrate is reduced to nitrite and then measured colorimetrically.

21.1R Removal of turbidity interfering with nitrate analyses can be accomplished by one of the following methods to remove suspended matter that can clog the reduction column.

1. Filter sample through a glass-fiber filter or a 0.45-micron pore size filter as long as the pH is less than 8.
2. Add 1 mL zinc solution to 100 mL of sample and mix thoroughly. Add enough sodium hydroxide solution to obtain a pH of 10.5. Let treated sample stand a few minutes to allow the heavy flocculant precipitate to settle. Clarify by filtering through a glass-fiber filter.

21.1S The pH of natural waters is controlled by the relative amounts of carbon dioxide, bicarbonate, and carbonate ions.

Answers to questions on page 537.

21.1T Specific conductance or conductivity is a numerical expression (expressed in micromhos per centimeter) of the ability of a water to conduct an electric current. This number depends on the total concentration of the minerals dissolved in the sample (TDS) and the temperature.

21.1U Sulfate ions are of importance in water supplies because of the tendency of appreciable amounts to form hard scales in boilers and heat exchangers.

21.1V A 50-mL sample from a well produced a turbidity reading of 40 TU using a nephelometer. What was the sulfate concentration in mg/L?

Known	Unknown
Sample Size, mL = 50 mL	Sulfate, mg/L
Turbidity, TU = 40 TU	

1. Determine the sulfate concentration from the graph.

   Sulfate Concentration, mg/L = 15 mg/L

2. Calculate the sulfate concentration in mg/L.

$$\text{Sulfate, mg/}L = \frac{(\text{Graph Sulfate, mg/}L)(100 \text{ mL})}{\text{Sample Size, mL}}$$

$$= \frac{(15 \text{ mg/}L)(100 \text{ mL})}{50 \text{ mL}}$$

$$= 30 \text{ mg/}L$$

Answers to questions on page 542.

21.1W Tastes and odors can be caused in drinking water by algae, *Actinomycetes*, bacteria, decaying vegetation, metals, and pollutants (most organic chemicals and some inorganic chemicals). Dissolved inorganic salts of copper, iron, manganese, potassium, sodium, and zinc can be detected by taste.

21.1X Calculate the threshold odor number (TON) for a sample when the first detectable odor occurred when the 12-mL sample was diluted to 200 mL (188 mL of odor-free water was added to the 12-mL sample).

Known	Unknown
A or Sample Size, mL = 12 mL	TON
B or Odor-Free Water, mL = 188 mL	

Calculate the threshold odor number (TON).

$$\text{TON} = \frac{A + B}{A}$$

$$= \frac{12 \text{ mL} + 188 \text{ mL}}{12 \text{ mL}}$$

$$= 17$$

Answers to questions on page 545.

21.1Y The principal source of chloroform and other trihalomethanes in drinking water is the chemical interaction of chlorine added for disinfection and other purposes with the commonly present natural humic substances and other precursors produced either by normal organic decomposition or by the metabolism of aquatic organisms.

21.1Z Total dissolved solids (TDS) refers to material that passes through a standard glass-fiber filter disk and remains after evaporation at 180°C.

CHAPTER 22

DRINKING WATER REGULATIONS

by

Tim Gannon

Revised by

Jim Sequeira

and

Ken Kerri

(1988, 1990, 1991, 1993, 1995, 1997, 2000, 2001, 2004, 2006)

NOTICE

EPA rules and regulations are continually changing. State rules and regulations may be stricter. Keep in contact with your state drinking water agency to obtain the rules and regulations that currently apply to your water utility. For current or additional information, phone EPA's toll-free Safe Drinking Water Hotline at (800) 426-4791.

Also see the Safe Drinking Water Act poster provided with this manual.

TABLE OF CONTENTS
Chapter 22. DRINKING WATER REGULATIONS

	Page
OBJECTIVES	555
WORDS	556

LESSON 1

- **22.0** HISTORY OF DRINKING WATER LAWS AND STANDARDS ... 559
- **22.1** HOW EPA DEVELOPS DRINKING WATER STANDARDS ... 560
  - 22.10 Types of Contaminants ... 560
  - 22.11 Identifying Contaminants to Be Regulated ... 561
  - 22.12 Unregulated Contaminants ... 561
  - 22.13 Newer and Proposed Regulations ... 561
    - 22.130 Arsenic Rule ... 562
    - 22.131 Lead and Copper Rule ... 562
    - 22.132 Total Coliform Rule (TCR) ... 562
    - 22.133 Surface Water Treatment Rules ... 562
    - 22.134 Filter Backwash Recycling Rule (FBRR) ... 562
    - 22.135 Disinfectants and Disinfection By-Products (D/DBP) Rules ... 563
    - 22.136 Ground Water Rule (GWR) ... 563
    - 22.137 Radionuclides Rule ... 564
    - 22.138 Standardized Monitoring Framework (SMF) ... 564
    - 22.139 Consumer Confidence Report (CCR) Rule ... 564
  - 22.14 Setting Standards ... 564
  - 22.15 Types of Water Systems ... 565
- **22.2** PRIMARY DRINKING WATER STANDARDS ... 565
  - 22.20 Inorganic Chemical Standards ... 566
    - 22.200 Antimony ... 566
    - 22.201 Arsenic ... 566
    - 22.202 Asbestos ... 566
    - 22.203 Barium ... 566
    - 22.204 Beryllium ... 566
    - 22.205 Bromate ... 566

	22.206	Cadmium	566
	22.207	Chlorite	569
	22.208	Chromium	569
	22.209	Copper	569
	22.2010	Cyanide	569
	22.2011	Fluoride	569
	22.2012	Lead and Copper	569
	22.2013	Mercury	571
	22.2014	Nitrate and Nitrite	571
	22.2015	Selenium	572
	22.2016	Thallium	572
22.21	Organic Chemical Standards	572
	22.210	Trichloroethylene (TCE)	572
	22.211	1,1-Dichloroethylene	572
	22.212	Vinyl Chloride	572
	22.213	1,1,1-Trichloroethane	572
	22.214	1,2-Dichloroethane	573
	22.215	Carbon Tetrachloride	573
	22.216	Benzene	573
	22.217	Para-Dichlorobenzene (p-Dichlorobenzene)	573
22.22	Microbial Standards	573
	22.220	Total Coliform Rule	573
		22.2200 Sanitary Survey	573
		22.2201 Sampling Plan	574
		22.2202 Laboratory Procedures	574
		22.2203 Monitoring Frequency	575
		22.2204 Determining Compliance	576
		22.2205 Reporting and Notification Requirements	576
	22.221	Surface Water Treatment Rules	578
		22.2210 Criteria for Avoiding Filtration	578
		22.2211 Requirements for Filtered Water Systems	579
		22.2212 Chlorine Residual Substitution	579
		22.2213 Turbidity Requirements	579
		22.2214 Monitoring Requirements	580
		22.2215 CT Values	580
		22.2216 Interim Enhanced Surface Water Treatment Rule (IESWTR)	581
		22.2217 Long Term 1 Enhanced Surface Water Treatment Rule (LT1ESWTR)	581

		22.2218	Long Term 2 Enhanced Surface Water Treatment Rule (LT2ESWTR)	582
		22.2219	Additional Reading	582
	22.23	Disinfectants and Disinfection By-Products (D/DBP)		583
	22.24	Radiological Standards		584

DISCUSSION AND REVIEW QUESTIONS 585

LESSON 2

22.3	SECONDARY DRINKING WATER STANDARDS		586
	22.30	Enforcement of Regulations	586
	22.31	Secondary Maximum Contaminant Levels	586
	22.32	Monitoring	587
	22.33	Secondary Contaminants	587

		22.330	Aluminum	587
		22.331	Chloride	587
		22.332	Color	587
		22.333	Copper	587
		22.334	Corrosivity	587
		22.335	Fluoride	588
		22.336	Foaming Agents	588
		22.337	Iron and Manganese	588
		22.338	Iron	588
		22.339	Manganese	589
		22.3310	Odor	589
		22.3311	pH	589
		22.3312	Silver	590
		22.3313	Sulfate	590
		22.3314	Total Dissolved Solids (TDS)	590
		22.3315	Zinc	590

22.4	SAMPLING PROCEDURES		591
	22.40	Safe Drinking Water Act Regulations	591
	22.41	Initial Sampling	591
	22.42	Routine Sampling	591
	22.43	Check and Repeat Sampling	591
	22.44	Sampling Points	591
	22.45	Sampling Point Selection	591
	22.46	Sampling Schedule	591
	22.47	Sampling Route	592
	22.48	Sample Collection	592

22.5	REPORTING PROCEDURES	592
22.6	NOTIFICATION REQUIREMENTS	592
22.7	RECORDKEEPING	602
22.8	CONSUMER CONFIDENCE REPORTS (CCRs)	602
DISCUSSION AND REVIEW QUESTIONS		604
SUGGESTED ANSWERS		605

OBJECTIVES

Chapter 22. DRINKING WATER REGULATIONS

Following completion of Chapter 22, you should be able to:

1. Identify the two basic types of water systems.
2. List the types of primary contaminants.
3. Explain the Total Coliform Rule.
4. Explain the Surface Water Treatment Rules.
5. Describe the Primary Drinking Water Standards.
6. List the secondary contaminants.
7. Develop and conduct a sampling program.
8. Record and report results.
9. Comply with notification requirements.
10. Prepare a Consumer Confidence Report (CCR).

WORDS
Chapter 22. DRINKING WATER REGULATIONS

ACUTE HEALTH EFFECT

An adverse effect on a human or animal body, with symptoms developing rapidly.

CT VALUE

Residual concentration of a given disinfectant in mg/L times the disinfectant's contact time in minutes.

CHECK SAMPLING

Whenever an initial or routine sample analysis indicates that a Maximum Contaminant Level (MCL) has been exceeded, check sampling is required to confirm the routine sampling results. Check sampling is in addition to the routine sampling program.

CHLORAMINATION (KLOR-uh-min-NAY-shun)

The application of chlorine and ammonia to water to form chloramines for the purpose of disinfection.

CHRONIC HEALTH EFFECT

An adverse effect on a human or animal body with symptoms that develop slowly over a long period of time or that recur frequently.

CONSUMER CONFIDENCE REPORT (CCR)

An annual report prepared by a water utility to communicate with its consumers. The report provides consumers with information on the source and quality of their drinking water and is an opportunity for positive communication with consumers.

ENTERIC

Of intestinal origin, especially applied to wastes or bacteria.

HETEROTROPHIC (HET-er-o-TROF-ick)

Describes organisms that use organic matter for energy and growth. Animals, fungi, and most bacteria are heterotrophs.

INFORMATION COLLECTION RULE (ICR)

The Information Collection Rule (ICR) was promulgated on May 14, 1996 and approved by the Director of the Federal Register on June 18, 1996. It was to remain effective until December 31, 2000. The rule specified requirements for monitoring microbial contaminants and disinfection by-products (DBPs) by large public water systems (PWSs). It required large PWSs to conduct either bench- or pilot-scale testing of advanced treatment techniques. The data reported under the ICR were used by EPA to learn more about the occurrence of microbial contamination and disinfection by-products, the health risks posed, appropriate analytical methods, and effective forms of treatment. The ICR data form the scientific basis for EPA's development of the Enhanced Surface Water Treatment Rule and the Disinfectants and Disinfection By-Products (D/DBP) Rule.

INITIAL SAMPLING

The very first sampling conducted under the Safe Drinking Water Act (SDWA) for each of the applicable contaminant categories.

MBAS

Methylene Blue Active Substance. Another name for surfactants or surface active agents. The determination of surfactants is accomplished by measuring the color change in a standard solution of methylene blue dye.

MCL

Maximum Contaminant Level. The largest allowable amount. MCLs for various water quality indicators are specified in the National Primary Drinking Water Regulations (NPDWR).

MCLG

Maximum Contaminant Level Goal. MCLGs are health goals based entirely on health effects. They are a preliminary standard set but not enforced by EPA. MCLs consider health effects, but also take into consideration the feasibility and cost of analysis and treatment of the regulated MCL. Although often less stringent than the corresponding MCLG, the MCL is set to protect health.

MRDL

Maximum Residual Disinfectant Level. The highest level of a disinfectant allowed in drinking water without causing an unacceptable possibility of adverse health effects.

NTU

Nephelometric Turbidity Units. See TURBIDITY UNITS.

pCi/L

picoCurie per liter. A picoCurie is a measure of radioactivity. One picoCurie of radioactivity is equivalent to 0.037 nuclear disintegrations per second.

ROUTINE SAMPLING

Sampling repeated on a regular basis.

SANITARY SURVEY

A detailed evaluation or inspection of a source of water supply and all conveyances, storage, treatment, and distribution facilities to ensure protection of the water supply from all pollution sources.

SURFACTANT (sir-FAC-tent)

Abbreviation for surface-active agent. The active agent in detergents that possesses a high cleaning ability.

THM

See TRIHALOMETHANES.

THRESHOLD ODOR NUMBER (TON)

The greatest dilution of a sample with odor-free water that still yields a just-detectable odor.

TRIHALOMETHANES (THMs) (tri-HAL-o-METH-hanes)

Derivatives of methane, CH_4, in which three halogen atoms (chlorine or bromine) are substituted for three of the hydrogen atoms. Often formed during chlorination by reactions with natural organic materials in the water. The resulting compounds (THMs) are suspected of causing cancer.

TURBIDITY (ter-BID-it-tee) UNITS (TU)

Turbidity units are a measure of the cloudiness of water. If measured by a nephelometric (deflected light) instrumental procedure, turbidity units are expressed in nephelometric turbidity units (NTU) or simply TU. Those turbidity units obtained by visual methods are expressed in Jackson turbidity units (JTU), which are a measure of the cloudiness of water; they are used to indicate the clarity of water. There is no real connection between NTUs and JTUs. The Jackson turbidimeter is a visual method and the nephelometer is an instrumental method based on deflected light.

CHAPTER 22. DRINKING WATER REGULATIONS

(Lesson 1 of 2 Lessons)

All water treatment plant operators need to be thoroughly familiar with the state and federal laws and standards that apply to domestic water supply systems. These regulations are the goals and guideposts for the water supply industry. Their purpose is to ensure the uniform delivery of safe and aesthetically pleasing drinking water to the public.

This chapter will introduce the major drinking water regulations and explain the monitoring and reporting requirements. For more detailed information, you will need to refer to current copies of your state's regulations and the most recent federal standards. These publications should be made readily available to all operators since operators will only know whether their system is in compliance by comparing monitoring test data with the actual current regulations. For the most current and accurate information, visit the EPA's website at www.epa.gov.

22.0 HISTORY OF DRINKING WATER LAWS AND STANDARDS

Up until shortly after 1900, there were no standards for drinking water. The first standards, established in 1914, were designed in large part to control waterborne bacteria and viruses that cause diseases such as cholera, typhoid, and dysentery. These new standards were overwhelmingly successful in curbing the spread of such diseases. However, with time and technology, other types of contaminants, this time chemicals, again stirred public concern. In 1962, the US Public Health Service (the forerunner of the US Environmental Protection Agency) revised the national drinking water standards to include limits on selected organic chemicals.

In 1972, a series of reports detailing organic contamination in the drinking water supplied to the residents of New Orleans from the Mississippi River triggered profound changes in drinking water regulations. A study by the Environmental Defense Fund found that people drinking treated Mississippi River water in New Orleans had a greater chance of developing certain cancers than those in neighboring areas whose drinking water came from groundwater sources. Heightened public awareness and concern regarding cancer became major factors behind the push for legislative action on the issue of drinking water contamination. The finding of suspected carcinogens in drinking water established a widespread sense of urgency that led to the passage and signing into law of the Safe Drinking Water Act in December, 1974.

The Safe Drinking Water Act (SDWA) gave the federal government, through the US Environmental Protection Agency (EPA), the authority to:

- Set national standards regulating the levels of contaminants in drinking water

- Require public water systems to monitor and report the levels of their identified contaminants

- Establish uniform guidelines specifying the acceptable treatment technologies for removing unsafe levels of pollutants from drinking water

While the SDWA gave EPA responsibility for promulgating (passing into law) drinking water regulations, it gave state regulatory agencies the opportunity to assume primary responsibility for enforcing those regulations.

Implementation of the SDWA has greatly improved basic drinking water purity across the nation. However, EPA surveys of surface water and groundwater indicate the presence of synthetic organic chemicals in 20 percent of the nation's water sources, with a small percentage at levels of concern. In addition, research studies suggest that some naturally occurring contaminants may pose even greater risks to human health than the synthetic contaminants. Further, there is growing concern about microbial and radon contamination.

In the years following passage of the SDWA, Congress felt that EPA was slow to regulate contaminants and states were lax in enforcing the law. Consequently, in 1986 Congress enacted amendments designed to strengthen the 1974 SDWA. These amendments set deadlines for the establishment of maximum contaminant levels, placed greater emphasis on enforcement, authorized penalties for tampering with drinking water supplies, mandated the complete elimination of lead from drinking water, and placed considerable emphasis on the protection of underground drinking water sources.

The 1986 SDWA amendments set up a timetable under which EPA was required to develop primary standards for 83 contaminants. Other major provisions of the 1986 SDWA amendments required EPA to:

1. Define an approved treatment technique for each regulated contaminant
2. Specify criteria for filtration of surface water supplies
3. Specify criteria for disinfection of surface and groundwater supplies
4. Prohibit the use of lead products in materials used to convey drinking water

To comply with the provisions of the 1986 SDWA amendments, the EPA, the states, and the water supply industry undertook significant new programs to clean up the country's water supplies.

On August 6, 1996, the President signed new Safe Drinking Water Act (SDWA) amendments into law as Public Law (PL) 104–182. These amendments made sweeping changes to the existing SDWA, created several new programs, and included a total authorization of more than $12 billion in federal funds for various drinking water programs and activities from fiscal year (FY) 1997 through FY 2003.

Topics covered in the amendments include arsenic research, assistance for water infrastructure and watersheds, assistance to colonias (low-income communities located along the US-Mexico border), backwash water recycling, bottled water, capacity development (technical, financial, and managerial), consumer awareness, contaminant selection and standard-setting authority, definitions of public water system, community water system and noncommunity water system, disinfectants and disinfection by-products (D/DBP), drinking water studies and research, effective date of regulations, enforcement, environmental finance centers and capacity clearinghouse, estrogenic substances screening program, conditions that could qualify a water system for an exemption, groundwater disinfection, groundwater protection programs, lead plumbing and pipes, monitoring and information gathering, monitoring relief, monitoring for unregulated contaminants, occurrence of contaminants in drinking water database, operator certification, primacy, public notification, drinking water regulations for radon, review of National Primary Drinking Water Regulations (NPDWRs), risk assessment application to establishing NPDWRs, small systems (technical assistance, treatment technology, variances), source water quality assessment and petition programs, state revolving loan fund, authorization to promulgate an NPDWR for sulfate, Surface Water Treatment Rule (SWTR) compliance, variance treatment technologies, water conservation, and waterborne disease study and training. For additional information and details, see "Overview of the Safe Drinking Water Act Amendments of 1996," by Frederick W. Pontius, *JOURNAL AMERICAN WATER WORKS ASSOCIATION,* October 1996, pages 22–33.

On December 16, 1998, two major regulations were signed into law: the Disinfectants/Disinfection By-Products (D/DBP) Rule (see Section 22.23 for details), and the Interim Enhanced Surface Water Treatment Rule (IESWTR) (see Section 22.2216 for details). The D/DBP Rule was developed to protect the public from harmful concentrations of disinfectants and from trihalomethanes, which could form when disinfection by-products combine with organic matter in drinking water. The goal of the IESWTR is to reduce the occurrence of *Cryptosporidium* and other pathogens in drinking water. This rule was developed to ensure that protection against microbial contaminants is not lowered as water systems comply with the D/DBP Rule.

Operators are urged to develop close working relationships with their local regulatory agencies to keep themselves informed of the frequent changes in regulations and requirements.

QUESTIONS

Write your answers in a notebook and then compare your answers with those on page 605.

22.0A What were the first drinking water standards designed to control?

22.0B Why is it important for operators to establish good working relationships with local regulatory agencies?

22.1 HOW EPA DEVELOPS DRINKING WATER STANDARDS

The process by which EPA sets drinking water standards is both long and complicated. It involves deciding which contaminants may endanger public health, conducting studies of the effects of these contaminants, defining the maximum safe level of each contaminant, estimating the costs and benefits of regulation, proposing a standard or limit, listening to and evaluating public reactions to the proposed standard, revising the standard (if needed), proposing the standard in final form, seeking and evaluating additional public input, and finally, publishing the standard. The entire process by which a standard is proposed and promulgated is governed by strict procedural guidelines. It often takes three or more years to produce a standard. The remainder of this section describes the general types of contaminants that are considered public health threats and how EPA decides which contaminants to regulate first, how it approaches the task of regulating contaminants, and how it defines a standard.

22.10 Types of Contaminants

Five types of primary contaminants are considered to be of public health importance:

1. *INORGANIC CONTAMINANTS,* such as lead and mercury
2. *ORGANIC CONTAMINANTS,* which include pesticides, herbicides, trihalomethanes, solvents, and other synthetic organic compounds
3. *TURBIDITY,* such as small particles suspended in water that interfere with light penetration and disinfection
4. *MICROBIAL CONTAMINANTS,* such as bacteria, viruses, and protozoa
5. *RADIOLOGICAL CONTAMINANTS,* which include natural and manmade sources of radiation

22.11 Identifying Contaminants to Be Regulated

The 1986 SDWA required EPA to establish a priority list of contaminants that may have adverse health effects and may require regulation. The 1996 SDWA amendments retained the list of 83 contaminants developed earlier but revised the process by which contaminants are selected for regulation. Before selecting a contaminant for regulation, the EPA must now consult with the scientific community, solicit public comments, and demonstrate that the contaminant actually occurs in public water systems. For this latter purpose, the 1996 amendments required EPA to establish an occurrence database by August 1999. The database contains information on regulated and unregulated contaminants and the information is available to the public.

To regulate a substance, EPA must demonstrate that the contaminant meets three criteria:

1. The contaminant has an adverse effect on human health

2. It occurs, or is likely to occur, in public water systems at a frequency and concentration of significance to public health

3. Regulation of the contaminant offers a meaningful opportunity to reduce health risks for people served by public water systems

In addition to meeting these three criteria, EPA must also weigh the relative health risks of various contaminants being considered for regulation, the risk reduction that regulation would accomplish, and the costs of implementing the regulations.

Working on a five-year cycle, the EPA is required to select at least five contaminants from its Drinking Water Contaminant Candidate List (DWCCL) and decide whether there is sufficient reason to regulate them. The first DWCCL included 50 chemicals and 10 microbial contaminants. In June 2002, the EPA announced a preliminary finding that their research revealed no need to regulate ten of the listed contaminants. Based on public comments and continued research and evaluation, EPA will make a final determination about regulating any of these contaminants and will begin to develop proposed regulations as necessary.

22.12 Unregulated Contaminants

EPA uses the Unregulated Contaminant Monitoring (UCM) program to collect data for contaminants suspected to be present in drinking water, but that do not have health-based standards set under the Safe Drinking Water Act (SDWA). Every five years, EPA reviews the list of contaminants, largely based on the Contaminant Candidate List. The SDWA Amendments of 1996 provide for:

- Monitoring no more than 30 contaminants per 5-year cycle

- Monitoring only a representative sample of public water systems serving fewer than 10,000 people

- Storing analytical results in a National Contaminant Occurrence Database (NCOD)

The UCM program progressed in several stages. Currently, EPA manages the program directly as specified in the Unregulated Contaminant Monitoring Regulation (UCMR).

Under the UCMR, community water systems and nontransient noncommunity water systems serving more than 10,000 people are required to monitor for unregulated contaminants; smaller systems may also have to conduct monitoring if required by the state. Transient water systems are not affected by the UCMR.

Always keep in mind that EPA's regulatory process is basically just that—a process. No list is permanent; regulations change frequently as new information becomes available. Water treatment plant operators must continually seek opportunities to learn about the regulations that affect the water industry. Also, operators should feel free to express their concerns regarding the regulations and the regulatory process to EPA. For additional information on EPA's Unregulated Contaminant Monitoring (UCM) program, visit EPA's website at www.epa.gov/safewater/ucmr/index.html.

QUESTIONS

Write your answers in a notebook and then compare your answers with those on page 605.

22.1A List the five types of primary contaminants considered to be of public health importance.

22.1B What criteria must be met before EPA selects a contaminant for regulation?

22.1C Why are water systems required to monitor some unregulated contaminants?

22.13 Newer and Proposed Regulations

To help operators and managers quickly learn the basics of newer and proposed drinking water regulations, the US EPA has developed a "Compilation of Quick Reference Guides." These guides provide information such as: (1) an overview of the rules, (2) public health-related benefits, (3) critical deadlines and requirements, (4) compliance determination, and (5) monitoring requirements.

The purpose of this section is to provide you with the important information and requirements of these newer and proposed regulations. Additional information and details are provided in sections in the remainder of this chapter.

If you would like to know how these rules and regulations apply specifically to your water supply system, contact your local drinking water regulator or visit the US EPA website at www.epa.gov/safewater/standards.html. Also, a phone call to EPA's Safe Drinking Water Hotline at (800) 426-4791 can be very helpful.

The information in this section is based on EPA's "Compilation of Quick Reference Guides" at the time this manual was revised. Each portion of this section also refers to more details in other sections of this chapter.

22.130 Arsenic Rule

The purpose of the Arsenic Rule is to improve public health by reducing exposure to arsenic in drinking water. The Arsenic Rule reduces the arsenic Maximum Contaminant Level (MCL) to 10 µg/L. The rule requires new systems and new drinking water sources to demonstrate compliance as specified by the state. The public health benefits from the Arsenic Rule include the avoidance of bladder and lung cancers and a reduction in the frequency of noncarcinogenic diseases.

All samples must be collected at each entry point to the distribution system, unless otherwise specified by the state. Compliance with the MCLs for IOCs (Inorganic Chemicals), SOCs (Synthetic Organic Chemicals), and VOCs (Volatile Organic Chemicals) is based on a running annual average at each sampling point.

Monitoring requirements for total arsenic include one sample after the effective date of the MCL (January 23, 2006). Surface water systems must take annual samples. Groundwater systems must take one sample during the 2005–2007 compliance period. If the monitoring result is less than the MCL, groundwater systems must collect one sample every three years and surface water systems must continue to collect annual samples. Also see Section 22.201, "Arsenic."

22.131 Lead and Copper Rule

The purpose of the Lead and Copper Rule is to protect the public health by minimizing lead (Pb) and copper (Cu) levels in drinking water, primarily by reducing water corrosivity. Lead and copper enter drinking water mainly from corrosion of lead and copper in plumbing materials.

The Lead and Copper Rule establishes an action level (AL) of 0.015 mg/L for lead and 1.3 mg/L for copper, based on the 90th percentile level of tap-water samples. An action level exceedance is not a violation of the rule, but can trigger other requirements that include water quality parameter (WQP) monitoring, corrosion control treatment (CCT), source water monitoring/treatment, public education, and lead service line replacement (LSLR).

The public health benefits from the Lead and Copper Rule include a reduced risk of exposure to lead, which can cause damage to the brain, red blood cells, and kidneys, especially for young children and pregnant women. Benefits also include a reduction in the risk of exposure to copper, which can cause stomach and intestinal distress, liver and kidney damage, and complications of Wilson's disease in genetically predisposed people.

Treatment technique and sampling requirements if the action level (AL) is exceeded include:

1. Water quality parameter (WQP) monitoring
2. Public education (PE)
3. Source water monitoring and treatment (SOWT)
4. Corrosion control treatment (CCT)

If the system continues to exceed the action level after installing CCT or SOWT, then requirements include:

5. Lead service line (LSL) monitoring
6. Lead service line replacement (LSLR)

Also see Section 22.209, "Copper," and Section 22.2012, "Lead and Copper."

22.132 Total Coliform Rule (TCR)

The purpose of the Total Coliform Rule (TCR) is to improve public health protection by reducing fecal pathogens to minimal levels through control of total coliform bacteria, including fecal coliforms and *Escherichia coli (E. coli)*. The Total Coliform Rule establishes a maximum contaminant level (MCL) based on the presence or absence of total coliforms or *E. coli*, requires use of a sample siting plan, and also requires sanitary surveys for systems collecting fewer than five samples per month.

The public health benefits from the Total Coliform Rule include a reduction in the risk of illness from disease-causing organisms associated with wastewater or animal wastes. Disease symptoms may include diarrhea, cramps, nausea, and possible jaundice, and associated headaches and fatigue.

Also see Section 22.220, "Total Coliform Rule."

22.133 Surface Water Treatment Rules

On January 14, 2002, the EPA promulgated the Long Term 1 Enhanced Surface Water Treatment Rule (LT1ESWTR). The purpose of this rule is to improve public health protection through the control of microbial contaminants, particularly *Cryptosporidium* and to prevent significant increases in microbial risk that might otherwise occur when systems implement the Stage 1 Disinfectants and Disinfection By-Products Rule.

On December 15, 2005, the EPA promulgated the Long Term 2 Enhanced Surface Water Treatment Rule (LT2), which increases the monitoring and treatment requirements for water systems that are prone to outbreaks of microorganisms such as *Cryptosporidium*. A purpose of this rule is to reduce the risk of disease-causing microorganisms from entering drinking water systems. This rule also contains provisions to reduce risks from uncovered finished water reservoirs and to ensure that systems maintain microbial protection when they take steps to decrease the formation of disinfection by-products that results from chemical water treatment.

The major provisions of these rules include the control of *Cryptosporidium* and combined filter effluent (CFE) turbidity performance standards. Turbidity monitoring requirements are also included in these rules.

For additional information, see Section 22.221, "Surface Water Treatment Rules."

22.134 Filter Backwash Recycling Rule (FBRR)

The Filter Backwash Recycling Rule was promulgated by the EPA on June 8, 2001. The purpose of this rule is to improve public health protection by assessing and changing, where needed, recycle practices for improved contaminant control, particularly microbial contaminants. The FBRR requires systems that recycle certain flows to return specific recycle flows through all processes of the system's existing conventional or direct filtration system or to recycle to an alternate location approved by the regulatory agency.

The FBRR applies to public water systems that use surface water or groundwater under the direct influence of surface water, practice conventional or direct filtration, and recycle spent filter backwash, thickener supernatant, or liquids from dewatering processes. Conventional filtration is a series of processes including coagulation, flocculation, sedimentation, and filtration resulting in substantial particulate removal. Direct filtration is a series of processes including coagulation and filtration, but excluding sedimentation, and resulting in substantial particulate removal. Typically, direct filtration can be used only with high-quality raw water that has low levels of turbidity and suspended solids.

Recycle flows include spent filter backwash water, which is a stream containing particles that are dislodged from filter media when water is forced back through a filter (backwashed) to clean the filter. Thickener supernatant is a stream containing the decant from a sedimentation basin, clarifier, or other treatment unit that is used to treat water, solids, or semisolids from the primary treatment processes. Liquids from dewater processes include streams containing liquids generated from a treatment unit used to concentrate solids for disposal.

The public health benefits from the Filter Backwash Recycling Rule include a reduction in the risk of illness from microbial pathogens in drinking water, particularly *Cryptosporidium*.

For additional information, see Section 22.2211, "Requirements for Filtered Water Systems."

22.135 Disinfectants and Disinfection By-Products (D/DBP) Rules

EPA promulgated the Stage 1 Disinfectants and Disinfection By-Products (Stage 1 D/DBP) Rule on December 16, 1998. The purpose of this rule is to improve public health protection by reducing exposure to disinfection by-products (DBPs). Some disinfectants and disinfection by-products have been shown to cause cancer and also cause reproductive effects in laboratory animals, which suggests the potential to cause bladder cancer and reproductive effects in humans. The Stage 1 D/DBP Rule is the first of a staged set of rules that will reduce the allowable levels of disinfection by-products (DBPs) in drinking water. The rule establishes seven new standards and a treatment technique of enhanced coagulation or enhanced softening to further reduce DBP exposure. The rule is designed to limit capital investments and avoid major shifts in disinfection technologies until additional information is available on the occurrence and health effects of DBPs. For additional information, see Section 4.8, "Enhanced Coagulation," in *WATER TREATMENT PLANT OPERATION*, Volume I.

The public health benefits from the Stage 1 D/DBP Rule include a reduction in trihalomethane (THM) levels in drinking water, and a reduction in exposure to the major disinfection by-products (DBPs) from the use of ozone (bromate) and chlorine dioxide (chlorite).

On December 15, 2005, EPA promulgated the Stage 2 Disinfection By-Products (Stage 2 D/DBP) Rule, which establishes standards for controlling the harmful by-products of drinking water disinfection measures. A purpose of this rule is to limit the amount of potentially harmful disinfection by-products that end up in drinking water systems. This rule strengthens public health protection for consumers by tightening compliance monitoring requirements for two groups of DBPs: total trihalomethanes (TTHMs) and haloacetic acids (HAA5). The rule targets systems with the greatest risk. It will reduce potential health risks related to DBP exposure and provide more equitable public health protection.

Also see Section 22.23, "Disinfectants and Disinfection By-Products (D/DBP)."

22.136 Ground Water Rule (GWR)

On October 11, 2006, the final Ground Water Rule (GWR) was signed by the USEPA Administrator to protect the public from pathogen contamination in water systems that use groundwater. Pathogens are microorganisms, such as bacteria *(E. Coli)*, viruses, and protozoa, such as *Cryptosporidium* and *Giardia lamblia*. Essentially, the GWR aims to identify operating deficiencies in water systems and requires them to do more frequent monitoring.

The GWR applies to all systems using groundwater and it takes a targeted, risk-based approach. There are no mandatory disinfection requirements, but the rule does build on existing state programs and provides flexibility in defining significant deficiencies.

A significant deficiency refers to a problem that could have immediate potential to affect public health. Some possible examples include:

- Maximum contaminant level violations
- Not enough chlorine contact time
- Insufficient water for normal demand
- Not enough disinfectant residual to meet the minimum requirement
- No licensed operator for the water system

Under the GWR, primacy agencies must complete a sanitary survey of treatment plants that identifies any problems that could cause contamination. A hydrogeology sensitivity assessment must also be completed to determine the groundwater's vulnerability to contamination. For those water systems that use groundwater and that do not disinfect, the rule requires continuous monitoring for susceptible systems. For systems that disinfect, a 4-log (99.9 percent) inactivation of viruses should be demonstrated.

22.137 Radionuclides Rule

EPA promulgated the Radionuclides Rule on December 7, 2000. The purpose of the Radionuclides Rule is to reduce the exposure to radionuclides in drinking water, which will reduce the risk of cancer. This rule will also improve public health protection by reducing exposure to all radionuclides. The rule applies to all community water systems. The public health benefits from the rule include reduced exposure to uranium, which can cause toxic kidney effects and cancer. For additional information, see Section 22.24, "Radiological Standards."

22.138 Standardized Monitoring Framework (SMF)

The Standardized Monitoring Framework (SMF) was promulgated by EPA in the Phase II Rule on January 30, 1991. The purpose of the Standardized Monitoring Framework is to standardize, simplify, and consolidate monitoring requirements across contaminant groups. The SMF increases public health protection by simplifying monitoring plans and synchronizing monitoring schedules leading to increased compliance with monitoring requirements. The SMF reduces the variability within monitoring requirements for chemical and radiological contaminants across system sizes and types.

The benefits of the Standardized Monitoring Framework (SMF) include:

- Increased public health protection through monitoring consistency
- A reduction in the complexity of water quality monitoring from a technical and managerial perspective for both primacy agencies and water systems
- Equalizing of resource expenditures for monitoring and vulnerability assessments
- Increased water system compliance with monitoring requirements

The contaminants regulated by the Standardized Monitoring Framework (SMF) include:

- Inorganic Contaminants (IOCs)
- Synthetic Organic Contaminants (SOCs)
- Volatile Organic Contaminants (VOCs)
- Radionuclides

22.139 Consumer Confidence Report (CCR) Rule

The purpose of the Consumer Confidence Report (CCR) Rule is to improve public health protection by providing educational material to allow consumers to make educated decisions regarding any potential health risks pertaining to the quality, treatment, and management of their drinking water supply. The Consumer Confidence Report (CCR) Rule requires all community water systems to prepare and distribute a brief annual water quality report summarizing information regarding source, any detected contaminants, compliance, and educational information.

The public health-related benefits from the CCR Rule include:

- Increased consumer knowledge of drinking water quality, sources, susceptibility, treatment, and drinking water supply management
- Increased awareness of consumers to potential health risks, so they may make informed decisions to reduce those risks, including taking steps toward protecting their water supply
- Increased dialog with drinking water utilities and increased understanding of consumers to take steps toward active participation in decisions that affect public health

Information that should be included in the Consumer Confidence Report includes:

- Water system information
- Source of water
- Definitions of water terms
- Detected contaminants
- Compliance with drinking water regulations
- Required educational information

For additional information, see Section 22.8, "Consumer Confidence Reports (CCRs)."

QUESTIONS

Write your answers in a notebook and then compare your answers with those on page 605.

22.1D   What are the public health benefits from the Arsenic Rule?

22.1E   What is the purpose of the Total Coliform Rule (TCR)?

22.1F   The Filter Backwash Recycling Rule (FBRR) refers to which recycle flows?

22.1G   What is the purpose of the Consumer Confidence Report (CCR) Rule?

22.14 Setting Standards

A standard is usually the maximum level of a substance that EPA has deemed acceptable in drinking water. The first step in the setting of a standard is to study the human and animal health effects of a given chemical. These studies are normally performed using rats or mice. Based on these studies, EPA establishes a "no observed adverse effect level" (NOAEL). A safety factor is added to the NOAEL and the result is an acceptable daily intake limit of the chemical in question. The limit is adjusted to take into account the average weight and water consumption of the consumer, and the resulting figure is called a maximum contaminant level goal, or MCLG. By policy, the EPA sets MCLGs at zero for known or probable human carcinogens. MCLGs for noncarcinogens are set at a level where no adverse health effects would occur with a margin of safety.

The maximum contaminant level goal represents what EPA believes to be a safe level of consumption based solely on its studies of health effects. It is, however, a goal rather than an immediately achievable constituent limit. To develop more realistic, enforceable limits, EPA further revises the MCLG to take into account existing laboratory detection technology, costs, and reasonableness. After adjusting for these factors, EPA sets the maximum contaminant level (MCL) as close to the MCLG as is realistically feasible. The important difference between the two levels is that the MCLG is a nonenforceable goal and the MCL is an enforceable standard.

EPA has established standards (maximum contaminant levels) for chemicals, pesticides, bacteria and viruses, radioactivity, turbidity, and trihalomethanes. (Refer to the poster included with this manual for a complete listing of MCLs.) Most of these substances occur naturally in our environment and in the foods we eat. However, the MCLs apply whether the contaminant is from naturally occurring sources or from manmade pollution. The national drinking water standards set by EPA reflect the levels we can safely consume in our water, taking into account the amounts we are exposed to from these other sources.

Drinking water regulations are sometimes called "interim" regulations because research continues on drinking water contaminants. The existing standards may be strengthened and new standards may be established for other substances based on studies being conducted by the National Academy of Sciences, EPA, and others. The 1996 SDWA amendments gave EPA the authority to issue interim regulations for contaminants that do not appear on the Drinking Water Contaminant Candidate List if the agency determines that the contaminant presents an urgent threat to public health. In making this determination, EPA must consult with the Department of Health and Human Services.

22.15 Types of Water Systems

All of the drinking water regulations apply to all public water systems. It makes no difference whether the water system is publicly or privately owned. A PUBLIC WATER SYSTEM (PWS) is defined as any system that:

1. Has at least 15 service connections, or
2. Regularly serves an average of at least 25 individuals daily at least 60 days out of the year.

Any water system that provides services for fewer connections or persons than this is not covered by the SDWA. Certain other individuals and residences also are excluded, such as those whose water is supplied by an irrigation, mining, or industrial water system. However, regardless of size, all operators must strive to provide consumers with a potable drinking water.

Drinking water regulations also take into account the type of population served by the system and classify water systems as community or noncommunity systems. Therefore, in order to understand what requirements apply to any specific system, it is first necessary to determine whether the system is considered a community system or a noncommunity system. A COMMUNITY WATER SYSTEM is defined as a public water system that:

1. Has at least 15 service connections used by all-year residents, or
2. Regularly serves at least 25 all-year residents.

Any public water system that is not a community water system is classified as a NONCOMMUNITY WATER SYSTEM. Restaurants, campgrounds, and hotels could be considered noncommunity systems for purposes of drinking water regulations.

In addition to distinguishing between community and noncommunity water systems, EPA identifies some small systems as NONTRANSIENT NONCOMMUNITY systems if they regularly serve at least 25 of the same persons over 6 months per year. This classification applies to water systems for facilities such as schools or factories where the consumers served are nearly the same every day but do not actually live at the facility. In general, nontransient noncommunity systems must meet the same requirements as community systems.

A TRANSIENT NONCOMMUNITY water system is a system that does not regularly serve drinking water to at least 25 of the same persons over 6 months per year. This classification is used by EPA only in regulating nitrate levels and total coliform. Examples of a transient noncommunity system might be campgrounds or service stations if those facilities do not meet the definition of a community, noncommunity, or nontransient noncommunity system.

QUESTIONS

Write your answers in a notebook and then compare your answers with those on page 605.

22.1H What is the MCLG for known or probable human carcinogens?

22.1I What is the difference between a maximum contaminant level (MCL) and a maximum contaminant level goal (MCLG)?

22.1J Define a community water system.

22.2 PRIMARY DRINKING WATER STANDARDS

Primary standards or MCLs are set for substances that are thought to pose a threat to health when present in drinking water at certain levels. Because these substances are of health concern, primary standards are enforceable by law. (In contrast, secondary standards relate to cosmetic factors and are not federally enforceable.) A primary standard is usually expressed as a maximum contaminant level (MCL). Some contaminants, such as pathogenic organisms, are very difficult or expensive to measure so EPA requires the use of specific treatment techniques (disinfection or filtration, for example) that are known to be effective in reducing the health risks of these contaminants. Treatment technique requirements, therefore, are also referred to as primary standards. Table 22.1 lists the primary standards and health concerns associated with the contaminants.

22.20 Inorganic Chemical Standards

Inorganic chemicals are metals, salts, and other chemical compounds that do not contain carbon. The health concerns about inorganic chemicals are not centered on cancer, but rather on their suspected links to several different human disorders. For example, lead is suspected of contributing to mental retardation in children. The following paragraphs briefly discuss each of the inorganic contaminants regulated by the national drinking water standards. Waters exceeding the MCL for these elements for short periods of time will pose no immediate threat to health. However, studies show that these substances must be controlled because consumption of drinking water that exceeds these standards over long periods of time may prove harmful.

22.200 Antimony

Antimony is used in the production of ceramics, glassware, and pigments. Most exposures to antimony occur in industrial settings where workers may inhale dust containing particles of the metal. Ingestion of antimony can alter cholesterol and glucose levels and may cause chromosome damage in humans. The MCL for antimony is 0.006 mg/L.

22.201 Arsenic

This element occurs naturally in the environment, especially in the western United States, and it is also used in insecticides. Arsenic is found in foods, tobacco, shellfish, drinking water, and in the air in some locations. The national standard for arsenic is 0.010 mg/L (or 10 ppb). Anyone who drinks water that continuously exceeds the national standard by a substantial amount over a lifetime may experience fatigue and loss of energy. Extremely high levels can cause poisoning.

The regulation of arsenic has been vigorously debated. In 1997, EPA released a health effects study plan. Under this plan, EPA worked with the American Water Works Association (AWWA) Research Foundation and the Association of California Water Agencies to assess the health risks from exposure to low levels of arsenic. Based on the results of these studies, EPA proposed an MCL for arsenic in June 2000, and a final rule was issued January 22, 2001. Technologies available for removing arsenic include ion exchange, iron hydroxide coagulation followed by microfiltration, and activated alumina.

22.202 Asbestos

The general term "asbestos" refers to a family of fibrous silicate minerals that have been widely used in the manufacture of commercial products. Asbestos fibers are flexible enough to be woven into fabric that does not conduct electricity, is heat resistant and is chemically inert. Some examples of products containing asbestos include floor and ceiling tiles, paper products, paint and caulking, plastics, brake linings, insulation cement, and filters. Asbestos-cement pipe is a potential source of contamination of drinking water. Deterioration of this pipe material with age and with exposure to corrosive water is thought to release asbestos particles. Mining for asbestos minerals has also contributed to the contamination of some drinking water sources in Canada and the United States.

Scientific evidence on the harmful effects of asbestos consumed in drinking water is less conclusive than the evidence relating to inhaled asbestos. Nonetheless, the MCL of 7 million fibers/liter is presently thought to provide an adequate margin of safety.

22.203 Barium

Although not as widespread as arsenic, this element also occurs naturally in the environment in some areas. Barium can also enter water supplies through industrial waste discharges. Small doses of barium are not harmful. However, it is quite dangerous when consumed in large quantities and will bring on increased blood pressure, nerve damage, and even death. The maximum amount of barium allowed in drinking water by the national standard is 2 milligrams per liter of water.

22.204 Beryllium

The major source of beryllium in the environment is the combustion of coal and oil. Airborne particulates containing beryllium can be inhaled or washed by precipitation into drinking water supplies. Animal studies show that when large doses of beryllium are ingested, it is carried by the bloodstream from the stomach to the liver but it is gradually transferred to the animal's bones. The maximum contaminant level for beryllium is 0.004 mg/L.

22.205 Bromate

Bromate is formed as a by-product of ozone disinfection of drinking water. Ozone reacts with naturally occurring bromide in the water to form bromate. Bromate has been shown to cause cancer in rats. EPA has set a drinking water standard at 0.010 mg/L to limit exposure to bromate.

22.206 Cadmium

Only extremely small amounts of this element are found in natural waters in the United States. Waste discharges from the electroplating, photography, insecticide, and metallurgy industries can increase cadmium levels. The most common source of cadmium in our drinking water is from galvanized pipes and fixtures. The maximum amount of cadmium allowed in drinking water by the national standard is 0.005 milligram per liter of water.

TABLE 22.1 NATIONAL PRIMARY DRINKING WATER STANDARDS

Contaminant	MCL	Health Effects
Inorganics		
Antimony	0.006 mg/L	decreases longevity, alters cholesterol and glucose levels
Arsenic	0.01 mg/L	skin/nervous system toxicity; possible cancer
Asbestos	7 million fibers/L	possible cancer
Barium	2.0 mg/L	circulatory system effects
Beryllium	0.004 mg/L	bone/lung damage; possible cancer
Bromate	0.010 mg/L	possible cancer
Cadmium	0.005 mg/L	kidney effects
Chlorite	1.0 mg/L	blood; developing nervous system
Chromium	0.1 mg/L	skin sensitization; liver/kidney/circulatory/nervous system disorders and respiratory problems
Copper	1.3 mg/L[a]	nervous system/kidney/gastrointestinal effects; toxicity
Cyanide	0.2 mg/L	spleen/liver/brain effects
Fluoride	4.0 mg/L	brown staining or pitting of teeth
Lead	0.015 mg/L[a] (at tap)	interference with red blood cell chemistry; developmental delays in children; blood pressure effects in some adults; toxic to infants and pregnant women
Mercury	0.002 mg/L	kidney disorders
Nitrate (as N) and	10.0 mg/L	Methemoglobinemia ("blue baby"
Nitrite (as N)[b]	1.0 mg/L	syndrome)
Selenium	0.05 mg/L	nervous system damage
Thallium	0.002 mg/L	liver/kidney/intestines/brain effects
Organics		
VOLATILE ORGANICS		
Benzene	0.005 mg/L	possible cancer
Carbon tetrachloride	0.005 mg/L	possible cancer
o-Dichlorobenzene	0.6 mg/L	kidney and liver effects; blood cell damage
p-Dichlorobenzene	0.075 mg/L	kidney and liver effects
1,2-Dichloroethane	0.005 mg/L	possible cancer
1,1-Dichloroethylene	0.007 mg/L	kidney and liver damage
cis-1,2-Dichloroethylene	0.07 mg/L	nervous system/liver/circulatory system damage
trans-1,2-Dichloroethylene	0.1 mg/L	nervous system/liver/circulatory system damage
Dichloromethane	0.005 mg/L	possible cancer
1,2-Dichloropropane	0.005 mg/L	possible cancer
Ethylbenzene	0.7 mg/L	kidney/liver/nervous system damage
Monochlorobenzene	0.1 mg/L	kidney/liver/nervous system damage
Styrene	0.1 mg/L	nervous system/liver damage
Tetrachloroethylene	0.005 mg/L	possible cancer
Toluene	1.0 mg/L	nervous system/kidney/liver damage
1,2,4-Trichlorobenzene	0.07 mg/L	adrenal gland and internal organ damage
1,1,1-Trichloroethane	0.2 mg/L	liver/nervous system/circulatory system damage
1,1,2-Trichloroethane	0.005 mg/L	liver/kidney effects
Trichloroethylene (TCE)	0.005 mg/L	possible cancer
Vinyl chloride	0.002 mg/L	possible cancer
Xylenes (total)	10.0 mg/L	liver/kidney/nervous system damage
PESTICIDES AND SYNTHETIC ORGANICS		
Acrylamide	treatment technique	possible cancer; nervous system effects
Alachlor	0.002 mg/L	possible cancer
Atrazine	0.003 mg/L	cardiac/reproductive system damage
Benzo[a]pyrene	0.0002 mg/L	possible cancer
Carbofuran	0.04 mg/L	nervous system/reproductive system damage
Chlordane	0.002 mg/L	possible cancer
Dalapon	0.2 mg/L	kidney/liver damage
Dibromochloropropane (DBCP)	0.0002 mg/L	possible cancer

TABLE 22.1 NATIONAL PRIMARY DRINKING WATER STANDARDS (continued)

Contaminant	MCL	Health Effects
PESTICIDES AND SYNTHETIC ORGANICS (continued)		
Di(2-ethylhexyl)adipate	0.4 mg/L	liver/reproductive system effects
Di(2-ethylhexyl)phthalate	0.006 mg/L	possible cancer
Dinoseb	0.007 mg/L	thyroid/reproductive system damage
Diquat	0.02 mg/L	kidney/gastrointestinal damage; cataract risk
Endothall	0.1 mg/L	liver/kidney/gastrointestinal and reproductive system effects
Endrin	0.002 mg/L	liver/kidney/heart damage
Epichlorohydrin	treatment technique	possible cancer
Ethylene dibromide (EDB)	0.00005 mg/L	possible cancer
Glyphosate	0.7 mg/L	liver/kidney damage
Heptachlor	0.0004 mg/L	possible cancer
Heptachlor epoxide	0.0002 mg/L	possible cancer
Hexachlorobenzene	0.001 mg/L	possible cancer
Hexachlorocyclopentadiene	0.05 mg/L	stomach/kidney damage
Lindane	0.0002 mg/L	nervous system/immune system/liver/kidney effects
Methoxychlor	0.04 mg/L	nervous system/reproductive system/kidney/liver effects
Oxamyl (Vydate)	0.2 mg/L	kidney damage
PCBs	0.0005 mg/L	possible cancer
Pentachlorophenol	0.001 mg/L	liver/kidney damage
Picloram	0.5 mg/L	liver/kidney damage
Simazine	0.004 mg/L	possible cancer
Toxaphene	0.003 mg/L	possible cancer
2,4-D	0.07 mg/L	liver/kidney/nervous system damage
2,3,7,8-TCDD (Dioxin)	0.00000003 mg/L	possible cancer
2,4,5-TP (Silvex)	0.05 mg/L	liver/kidney damage
Microbial		
Total Coliform	1 per 100 mL	indicators of disease-causing organisms
	<40 samples/mo—no more than 1 positive	
	>40 samples/mo—no more than 5% positive	
Giardia lamblia	3-log (99.9%) removal[c]	Giardiasis; gastrointestinal effects
Legionella	treatment technique	Legionnaire's Disease; affects respiratory system
Enteric viruses	4-log (99.9%) removal[c]	gastrointestinal and other viral infections
Heterotrophic bacteria	treatment technique[c]	gastrointestinal infections
Physical		
Turbidity[c]	0.5 to 5 NTU	interferes with disinfection
Turbidity[d]	0.3 to 1 NTU	interferes with disinfection
Radionuclides		
Gross alpha particles	15 pCi/L	cancer risk
Gross beta particles[e]	4 mrem/yr	cancer risk
Radium 226 & 228	5 pCi/L	bone cancer
Uranium	30 µg/L	kidney damage; possible cancer
Disinfection By-Products		
TTHMs[f]	0.080 mg/L	cancer risk
Haloacetic Acids (HAA5)[f]	0.060 mg/L	nervous system/liver effects

a Action level for treatment.
b Applies to community, nontransient noncommunity, and transient noncommunity water systems.
c Applies to systems using surface water or groundwater under the influence of surface water.
d Applies to conventional and direct filtration systems.
e Applies to surface water systems serving more than 100,000 persons and any system determined by the state to be vulnerable.
f Applies to systems serving more than 10,000 persons and also to all surface water systems that meet the criteria for avoiding filtration.

22.207 Chlorite

The main source of chlorite in drinking water is chlorine dioxide used for disinfection. Chlorite forms as the chlorine dioxide breaks down during the disinfection process. Chlorite in drinking water has been shown to affect blood and the developing nervous system; therefore, EPA has set a drinking water standard at 1.0 mg/L to protect against these effects.

22.208 Chromium

This metal is found in cigarettes, some of our foods, and the air. Some studies suggest that in very small amounts, chromium may be essential to human beings, but this has not been proven. The national standard for chromium is 0.1 milligram per liter of water.

22.209 Copper

Copper in drinking water usually results from the reaction of aggressive water on copper plumbing. Treatment of surface water in storage reservoirs to control algae may also cause high levels of copper. Copper is an essential nutrient; adults require 2 mg daily and children of preschool age require about 0.1 mg daily for normal growth. Excess copper intake and exposure over a long period of time can cause liver damage.

The presence of copper in drinking water is also undesirable because of its aesthetic effects. At low levels (0.5 mg/L in soft waters) blue or blue-green staining of porcelain occurs. At higher levels (4 mg/L), it will stain clothing and blond hair. Concentrations greater than 1 mg/L can produce insoluble green curds when reacting with soap.

The primary standard for copper is a treatment technique that requires action to be taken when more than 10 percent of the taps sampled exceed the action level of 1.3 mg/L. Treatment technique requirements are contained in the Lead and Copper Rule, which is described in Section 22.2012, "Lead and Copper."

22.2010 Cyanide

Cyanide is a common ingredient of rat poisons, silver and metal polishes, and photo processing chemicals. Ingestion of even very small amounts of sodium cyanide or potassium cyanide (both readily available to the general public) can cause death within a few minutes to a few hours. Typical symptoms of cyanide poisoning include nausea without vomiting, anxiety, vertigo, lower jaw stiffness, convulsions, paralysis, irregular heartbeat, and coma. The primary standard for cyanide is 0.2 mg/L.

22.2011 Fluoride

This is a natural mineral and many drinking waters contain some fluoride. Fluoride produces two effects, depending on its concentration, and EPA has set both primary and secondary limits to regulate it. At levels of 6 to 8 mg/L, fluoride may cause skeletal fluorosis (the bones become brittle) and stiffening of the joints. On the basis of this health hazard, fluoride has been added to the list of primary standards.

At levels of 2 mg/L and greater, fluoride may cause dental fluorosis, which is discoloration and mottling of the teeth, especially in children. EPA has reclassified dental fluorosis as a cosmetic effect, raised the primary drinking water standard from 1.4–2 mg/L to 4 mg/L, and established a secondary standard of 2 mg/L for fluoride.

22.2012 Lead and Copper

Lead is found in the air and in our food. This metal comes from galvanized pipes, solder used with copper pipes, auto exhausts, and other sources. Excessive amounts may cause nervous system disorders or brain or kidney damage and can retard the physical and mental development of children.

Copper is an essential nutrient but excessive amounts can cause liver damage. Excess intake of copper and the inability to metabolize copper is called Wilson's Disease. Corrosion of copper water pipes releases this metal into drinking water. Since corrosive waters are a primary cause of both lead and copper contamination, the EPA has focused its regulatory approach on reducing drinking water corrosivity and removing pipes or solder containing lead from drinking water distribution systems.

EPA promulgated the Lead and Copper Rule on June 7, 1991. Revisions of the rule were published January 12, 2000. The rule applies to all community water systems and to non-transient noncommunity systems. Each system was required to begin an initial lead and copper sampling program by July 1993. The Lead and Copper Rule sets action levels for lead and copper and specifies treatment techniques that must be implemented when the action levels are exceeded. The lead level that triggers enforcement of treatment technique requirements is 0.015 mg/L and the level for copper is 1.3 mg/L. If a water system exceeds the action levels in more than 10 percent of its samples, steps must be taken to control corrosion in the distribution system, educate the public about the harmful effects of lead, and advise the public what steps should be taken to reduce their exposure. Continued failure to meet the action levels after installation of optimum corrosion control measures will usually mean the water system must conduct source water quality studies and implement treatment to reduce source water lead levels. If corrosion control and source water treatment measures fail to bring the contaminant levels below the action levels, water systems will be required to begin a program of replacing lead service lines.

All water systems are required to optimize corrosion control; that is to say, they must do the best they can to control corrosion. Small and medium-sized systems are considered to have optimized corrosion control if they meet the lead and copper action levels during each of two consecutive six-month sampling periods. Large systems have optimized corrosion control if the difference between the source water lead level and the 90th percentile tap-water lead level is less than 0.005 mg/L, and the 90th percentile copper concentration is less than or equal to half the copper action level (0.65 mg/L) for two consecutive six-month periods. Any system that is achieving optimal corrosion control (and therefore meeting the action limits) is permitted to reduce monitoring and reporting frequencies.

Large water systems that do not meet the guidelines for optimal corrosion control are required to conduct studies of various types of corrosion control treatment and then submit to the state a plan that would adequately control corrosion in their distribution system. Small and medium-sized systems exceeding the action levels must make a written recommendation to the state for a proposed treatment method, but may also be required to conduct similar studies before the state makes a final determination. Once the state directs a water supplier to install treatment, the agency has 24 months to install the specified treatment and 12 months to collect follow-up samples to determine if the system is working. The state will also determine maximum acceptable levels for pH, alkalinity, calcium (if carbonate stabilization is used), orthophosphate (if an inhibitor with a phosphate compound is used), and silica (if an inhibitor with a silicate compound is used). The system must then continue to operate within these water quality guidelines.

If corrosion control methods fail to reduce lead and copper concentrations below the action levels, the water supplier must collect and analyze source water samples. Data from this sampling is sent to the state along with the water agency's recommended treatment plan for reducing contaminant levels. The state may accept the agency's recommended plan or may require some alternative treatment such as ion exchange, reverse osmosis, lime softening, or coagulation-filtration. When required to install source water treatment, the agency is allowed 24 months for installation and 12 months to collect follow-up samples of specific water quality indicators. Based on the results of the follow-up sampling, the state determines the maximum allowable source water lead and copper concentrations and the water system is required to meet these state limits on an ongoing basis.

Water systems that continue to exceed the lead action level even after installing optimal corrosion control treatment and source water treatment will have to begin a program to replace lead service lines within a maximum of 15 years.

SAMPLING GUIDELINES

The sampling required by the Lead and Copper Rule includes tap-water sampling at the consumer's faucet and distribution system sampling. Tap-water samples for initial and routine lead and copper monitoring must be collected at high-risk locations, which are defined as homes with lead solder installed after 1982, homes with lead pipes, and homes with lead service lines. Tap-water samples for water quality indicators (when required) should be taken from representative taps throughout the distribution system. The representative taps can be the same as total coliform sampling sites. These samples will be analyzed for water quality indicator levels to identify optimal treatment and monitor compliance with the rule.

The rule calls for first-draw or first-flush samples, which are water samples taken after the water stands motionless in the plumbing pipes for at least six hours. This usually means taking a sample early in the day before water is used in the kitchen or bathroom. The rule permits water utilities to instruct consumers in sampling techniques so that they may collect the samples. One-liter samples are needed.

Distribution system sampling conducted in connection with corrosion control studies requires samples from representative taps throughout the system as well as samples from entry points to the distribution system.

INITIAL AND BASE MONITORING SAMPLING FREQUENCIES

Compliance with the Lead and Copper Rule depends on meeting the action levels during two consecutive six-month monitoring periods. All public water systems are required to collect one sample for lead and copper analysis from the following number of sites during each six-month monitoring period.

System Size (Population)	No. of Sampling Sites (Initial Base Monitoring)	No. of Sampling Sites (Reduced Monitoring)
>100,000	100	50
10,001 to 100,000	60	30
3,301 to 10,000	40	20
501 to 3,300	20	10
101 to 500	10	5
≤100	5	5

If a system meets the lead and copper action levels or maintains optimal corrosion control treatment for two consecutive six-month periods, tap-water sampling for lead and copper can be reduced to once per year and half the number of sites (see chart above). After three consecutive years of acceptable performance, tap-water sampling can be reduced to once every three years at the reduced number of sites listed above.

WATER QUALITY INDICATOR SAMPLING FREQUENCIES

All large systems (more than 50,000 persons) and any small or medium-sized system that exceeds the lead or copper action level will be required to monitor for the following water quality indicators: pH, alkalinity, calcium, conductivity, orthophosphate, silica, and water temperature. Samples must be taken both from representative taps throughout the system and at entry points to the distribution system. Two tap-water samples must be collected for each applicable water quality indicator from the following number of sites during each six-month monitoring period.

System Size (Population)	No. of Sampling Sites (Initial Base Monitoring)	No. of Sampling Sites (Reduced Monitoring)
>100,000	25	10
10,001 to 100,000	10	7
3,301 to 10,000	3	3
501 to 3,300	2	2
101 to 500	1	1
≤100	1	1

In addition to the sampling from representative taps, one sample must be collected every two weeks for each applicable water quality indicator at each entry point to the distribution system.

All large water systems and any small and medium-sized system that exceed the lead or copper action levels after installing optimal corrosion control treatment must continue to collect:

- Two samples for each applicable water quality indicator at each of the sampling sites (taps) every six months, and
- One sample for each applicable water quality indicator at each entry point to the distribution system every two weeks.
- After meeting water quality guidelines for two consecutive six-month monitoring periods, systems may reduce the number of tap samples collected (as indicated in the chart above), and after three consecutive years of acceptable performance they will be permitted to collect annual samples at the reduced number of sampling points.

CALCULATION OF 90TH PERCENTILE

The action levels that trigger implementation of treatment techniques to reduce lead and copper concentrations are based on the principle that no more than 10 percent of the samples should exceed the limit. To determine whether a system meets this test, it is necessary to figure out the lead or copper concentration on the sample that falls at the 90th percentile.

To calculate the 90th percentile value, write down the results from a set of samples arranging them in order of increasing or decreasing concentration. Number each sample beginning with the number 1 for the smallest concentration. Multiply the total number of samples (which will be the same as the number assigned to the highest concentration) by 0.9. Then, use the value from this calculation to identify the sample with the same number. The concentration of the sample whose number matches your calculated value is the 90th percentile and is used in determining compliance with the action level. (For systems serving fewer than 100 people, the 90th percentile is the average of the highest concentration and the second highest concentration.)

PUBLIC EDUCATION

A public education program developed by the EPA must be delivered to water system customers within 60 days whenever the system exceeds the lead and/or copper action levels. The information describes the harmful effects of lead in drinking water and tells customers what they can do to reduce their exposure. The education program should continue as long as the system fails to meet the lead and/or copper action levels.

In conjunction with the public education program, the water system must offer to sample a customer's water if asked to do so. However, the water system is not required to pay for the sampling and analysis and is not required to collect and analyze the sample itself.

LEAD SERVICE LINE REPLACEMENT

Any water system that continues to exceed the lead and copper action levels even after optimizing corrosion control and installing source water treatment will be required to replace its lead distribution pipes within a maximum of 15 years. A lead pipeline qualifies for replacement if it contributes more than 15 parts per billion to the total tap-water lead levels. Seven percent of the lines meeting this standard must be replaced annually.

22.2013 Mercury

Mercury is found naturally throughout the environment. Large increases in mercury levels in water can be caused by industrial and agricultural use. The health risk from mercury is greater from mercury in fish than simply from waterborne mercury. Mercury poisoning may have an *ACUTE HEALTH EFFECT*[1] in large doses, or a *CHRONIC HEALTH EFFECT*[2] in lower doses taken over an extended time period.

22.2014 Nitrate and Nitrite

Nitrate in drinking water above the national standard of 10 mg/L (as N) poses an immediate threat to children under three months of age. In some infants, excessive levels of nitrate have been known to react with intestinal bacteria that change nitrate to nitrite. Nitrite reacts with hemoglobin in the blood. This reaction will reduce the oxygen-carrying ability of the blood and produce an anemic condition commonly known as "blue baby."

---

[1] *Acute Health Effect.* An adverse effect on a human or animal body, with symptoms developing rapidly.
[2] *Chronic Health Effect.* An adverse effect on a human or animal body with symptoms that develop slowly over a long period of time or that recur frequently.

Noncommunity systems *MAY* be allowed to serve water containing up to 20 mg/L (as N) nitrate if:

1. The water is not available to infants six months of age and younger;
2. Posting of the potential health hazard is maintained;
3. State and local health authorities are notified annually of nitrate levels that exceed 10 mg/L; and
4. No threat to health will result.

22.2015 Selenium

This mineral occurs naturally in soil and plants, especially in western states. Selenium is found in meat and other foods. Although it is believed to be essential in the diet, there are indications that excessive amounts of selenium may be toxic. Studies are underway to determine the amount required for good nutrition and the amount that may be harmful.

The national standard for selenium is 0.05 milligram per liter of water. If a person's intake of selenium came only from drinking water, it would take an amount many times greater than the standard to produce any ill effects.

22.2016 Thallium

Thallium is an extremely toxic metal that is used in refining iron, cadmium, and zinc. It is also used for production of optical lenses, jewelry, semiconductors, dyes, and pigments. The most common source of thallium poisoning is ingestion of rat poisons and insecticides. Ingestion of thallium can injure a person's liver, kidneys, and nervous system and may also cause deafness and loss of vision. The primary standard (MCL) for thallium is 0.002 mg/L.

QUESTIONS

Write your answers in a notebook and then compare your answers with those on pages 605 and 606.

22.2A Why are primary standards sometimes expressed as treatment techniques rather than MCLs?

22.2B What are inorganic chemicals?

22.2C Why is arsenic listed as a primary contaminant?

22.2D What are two harmful effects of excessive levels of fluoride?

22.2E What are the sources of lead in drinking water?

22.2F Why is nitrate in drinking water above the national standard of 10 mg/L considered an immediate threat to public health?

22.21 Organic Chemical Standards

Organic chemicals are either natural or synthetic chemical compounds that contain carbon. Synthetic organic chemicals (SOCs) are manmade compounds and are used throughout the world as pesticides, paints, dyes, solvents, plastics, and food additives. Volatile organic chemicals (VOCs) are a subcategory of organic chemicals. These chemicals are termed "volatile" because they evaporate easily. The most commonly found VOCs are trihalomethanes (THMs), trichloroethylene (TCE), tetrachloroethylene, and 1,1-dichloroethylene. THMs were the first regulated VOCs when EPA finalized regulations in 1979.

The most common sources of organic contamination of drinking water are pesticides and herbicides, industrial solvents, and disinfection by-products (trihalomethanes and haloacetic acids). Millions of pounds of pesticides are used on croplands, forests, lawns, and gardens in the United States each year. They drain off into surface waters or seep into underground water supplies. Spills, poor storage, improper application, and haphazard disposal of organic chemicals have resulted in widespread groundwater contamination. This is a critical problem because once groundwater is contaminated, it may remain that way for a long time.

Many organic chemicals pose health problems if they get into drinking water and the water is not properly treated. The maximum limits for pesticides in drinking water are shown in Table 22.1 on pages 567 and 568. The next several paragraphs describe the sources and maximum contaminant levels of some of the most commonly used organic chemical contaminants.

22.210 Trichloroethylene (TCE)

Although the use of trichloroethylene is declining because of stringent regulations, it was, for many years, a common ingredient in household products (spot removers, rug cleaners, air fresheners), dry cleaning agents, industrial metal cleaners and polishes, refrigerants, and even anesthetics. Its wide range of use is perhaps why TCE is the organic contaminant most frequently encountered in groundwater. The MCL for TCE is 0.005 mg/L (5 µg/L).

22.211 1,1-Dichloroethylene

This solvent is used in manufacturing plastics and, more recently, in the production of 1,1,1-trichloroethane. At high doses, 1,1-dichloroethylene causes liver and kidney damage in animals. It also affects the central nervous system and heart of humans. The MCL for this chemical is 0.007 mg/L (7 µg/L).

22.212 Vinyl Chloride

Billions of pounds of this solvent are used annually in the United States to produce polyvinyl chloride (PVC), the most widely used ingredient for manufacturing plastics throughout the world. There is also evidence that vinyl chloride may be a biodegradation end-product of tri- and tetrachloroethylene under certain environmental conditions. There is strong laboratory evidence that vinyl chloride causes cancer in humans. The MCL for vinyl chloride is 0.002 mg/L (2 µg/L).

22.213 1,1,1-Trichloroethane

This chemical has replaced TCE in many industrial and household products. It is the principal solvent in septic tank degreasers, cutting oils, inks, shoe polishes, and many other products.

Among the VOCs found in groundwaters, 1,1,1-trichloroethane and TCE are encountered most frequently and in the highest concentrations. It causes depression of the central nervous system and changes in the cardiovascular system and liver in humans and animals but as yet there is not enough evidence to conclude that it causes cancer in humans. The MCL for this contaminant is 0.2 mg/L (200 µg/L).

22.214 1,2-Dichloroethane

1,2-dichloroethane is used as a solvent for fats, oils, waxes, gums, and resins. This chemical is a probable cause of cancer in humans. The MCL for 1,2-dichloroethane is 0.005 mg/L (5 µg/L).

22.215 Carbon Tetrachloride

Carbon tetrachloride was once a popular household solvent, a frequently used dry cleaning agent, and a charging agent for fire extinguishers. Since 1970, however, carbon tetrachloride has been banned from all use in consumer goods in the United States and in 1978 it was banned as an aerosol propellant. Currently, its principal use is in the manufacture of fluorocarbons, which are used as refrigerants. Carbon tetrachloride is a probable human carcinogen. The MCL for carbon tetrachloride is 0.005 mg/L (5 µg/L).

22.216 Benzene

Benzene is used primarily in the synthesis of styrene (for plastics), phenol (for resins), and cyclohexane (for nylon). Other uses include the production of detergents, drugs, dyes, and insecticides. Benzene is still being used as a solvent and as a component of gasoline. It is considered a probable human carcinogen. The MCL for benzene is 0.005 mg/L (5 µg/L).

22.217 Para-Dichlorobenzene (p-Dichlorobenzene)

The principal uses of this chemical are in moth control (balls, powders) and as lavatory deodorants. Para-dichlorobenzene causes liver damage and is suspected of being an animal carcinogen. The MCL is 0.075 mg/L (75 µg/L).

QUESTIONS

Write your answers in a notebook and then compare your answers with those on page 606.

22.2G What are organic chemicals?

22.2H What have been the common uses of trichloroethylene (TCE)?

22.22 Microbial Standards

Bacteria, viruses, and other organisms have long been recognized as serious contaminants of drinking water. Organisms such as *Giardia*, *Cryptosporidium*, and *E. coli* (a type of coliform bacteria) cause almost immediate gastrointestinal illness when people consume them in water. Waterborne diseases such as typhoid, cholera, infectious hepatitis, and dysentery have been traced to improperly disinfected drinking water.

Many types of coliform bacteria from human and animal wastes may be found in drinking water if the water is not properly treated. Often, the bacteria themselves do not cause diseases transmitted by water, although certain coliforms have been identified as the cause of "traveler's" diarrhea. In general, however, the presence of coliform bacteria indicates that other harmful organisms may be present in the water.

Throughout the early 1990s, chemical contaminants in drinking water were the focus of EPA's regulatory efforts. With passage of the 1996 SDWA amendments, Congress directed EPA to focus on microbial contaminants and to promulgate a series of new regulations to improve treatment effectiveness. Significant new regulations of the microbial contaminant *Cryptosporidium* were promulgated in December 1998. The new regulations consist of mandatory treatment techniques similar to the current regulations for *Giardia* and coliform bacteria. Filtration and disinfection requirements for surface water systems are currently in place and the final Ground Water Rule was signed on October 11, 2006. These new regulations will significantly affect the water treatment processes of most water suppliers.

For additional information about the new regulations, see Section 22.2216, "Interim Enhanced Surface Water Treatment Rule (IESWTR)," and Section 22.23, "Disinfectants and Disinfection By-Products (D/DBP)." For more information about *Cryptosporidium*, see Chapter 23, Section 23.1026, "*Cryptosporidium.*"

22.220 Total Coliform Rule

The primary standards (MCLs) for coliform bacteria have been established to indicate the likely presence of disease-causing bacteria. The Total Coliform Rule uses a presence-absence approach (rather than an estimation of how many coliforms are present) to determine compliance with the standards. The MCLG for coliform is zero.

Compliance under the Total Coliform Rule is determined on a monthly basis. In general, coliforms must be absent in at least 95 percent of the samples for those larger systems that collect more than 40 samples per month. Smaller systems that collect fewer than 40 samples cannot have coliform-positive results in more than one sample per month.

Whenever a routine coliform sample is coliform-positive, the regulation calls for determination of the presence of fecal coliform or *E. coli* and for repeat sampling. Whenever fecal coliforms or *E. coli* are present in the routine sample and the repeat samples are coliform-positive, there is a violation of the MCL, additional repeat sampling is required, and notification of both the state and the public is required.

22.2200 SANITARY SURVEY

Sanitary surveys are conducted to identify possible health risks that may not be discovered by routine coliform sampling. The Total Coliform Rule required community water systems serving fewer than 4,100 persons to complete an initial sanitary survey by June 1994 and to conduct subsequent surveys every five years. Noncommunity water systems were required to complete an initial sanitary survey by June 1999 and subsequent surveys every five years with the exception of those systems with protected and disinfected groundwater sources, which are allowed ten years for subsequent surveys.

A sanitary survey requires detailed planning, a thorough system survey, and reporting of the results. The planning portion involves a review of water quality records for compliance with applicable microbial, inorganic chemical, organic chemical, and radiological contaminant MCLs as well as the records of compliance with the monitoring requirements for those contaminants. The actual field survey is a detailed evaluation and inspection of the source of the water supply and all conveyances, storage, treatment, and distribution facilities to ensure protection from all pollution sources. The final report includes the date(s) of the survey, who was present during the survey, the survey findings, the recommended improvements to correct identified problems, and the target dates of completion for any improvements.

22.2201 SAMPLING PLAN

The Total Coliform Rule requires each water supply system to develop and follow a written sampling plan. Each plan must specifically identify sampling points throughout the distribution system. Sampling plans must be approved by the state regulatory agency and it will be necessary to check with your state agency to determine the details of their review process and what documents need to be submitted.

22.2202 LABORATORY PROCEDURES

With regard to coliform testing methodology, EPA will accept any one of five analytical methods for determination of total coliforms:

- Multiple-tube fermentation technique (MTF)
- Membrane filter technique (MF)
- Presence-absence coliform test (P-A)
- Colilert™ system (ONPG-MUG test)
- Colisure test

Regardless of the method used, the standard sample volume for total coliform testing is now 100 mL. This is an increase over the past testing method using 50 mL for the MTF technique.

The multiple-tube fermentation method of testing for coliforms determines the presence or absence of coliforms by the multiple-tube dilution method. This is a process whereby equal portions of a sample (100 mL) are added to 10 tubes containing a culture medium and an inverted vial. If gas accumulates in the inverted vial, it indicates presumptive evidence of coliform organisms and the sample is considered coliform-positive. If no gas forms in any of the vials, the sample is coliform-negative.

The membrane filter method provides for filtering a 100-mL water sample through a thin, porous, cellulose membrane filter under a partial vacuum. The filter is placed in a sterile container and incubated in contact with a special liquid called a "culture medium," which the bacteria use as a food source. Colonies of bacteria then grow on the media. The coliform colonies are visually identified, counted, and recorded as the number of coliform colonies per 100 mL of sample.

The presence-absence (P-A) test for the coliform group is a simple modification of the multiple-tube procedure. Even though this simplified test uses only one large test portion (100 mL) in a single culture bottle, the results still indicate the presence or absence of coliform bacteria and thus meet the testing requirements.

The Colilert™ system is a privately developed version of the ONPG-MUG test procedure that meets the EPA's laboratory methods requirements. The ONPG-MUG test indicates the presence or absence of coliform bacteria within 24 hours.

Although the Colilert™ method can yield presence-absence results within 24 hours and is easier to perform than the Membrane Filtration (MF) method, operators should be aware of the limitations of these tests in evaluating samples for regulatory purposes. The results for the Colilert and MF tests are not always comparable, which may be due to:

- Interferences in the sample that may suppress or mask bacterial growth
- Greater sensitivity of the Colilert media
- Added stress to organisms related to filtering
- The fact that different media may obtain better growth for some bacteria

Both test methods are approved by the EPA for reporting under the Safe Drinking Water Program. When these tests produce conflicting results, however, the safest course of action is to increase monitoring and treatment efforts until the results for *both* tests are negative.

The Colisure test is a presence-absence test for coliform bacteria. A sample of water is added to a dehydrated medium and, after 24 to 48 hours, the medium is examined for the presence of coliforms. (The Colisure test is available from IDEXX Laboratories, 1 IDEXX Drive, Westbrook, ME 04092.)

Water supply systems may contain various other types of bacteria besides coliforms. These are nonpathogenic bacteria and are sometimes referred to as *HETEROTROPHIC*[3] bacteria. During the analyses of water samples for coliforms, heterotrophic bacteria sometimes interfere with the test causing one of the following reactions:

- MTF technique: a turbid culture with no gas production
- P-A test: a turbid culture in the absence of an acid reaction
- MF technique: confluent growth or a coliform colony number too numerous to count

---

[3] *Heterotrophic* (HET-er-o-TROF-ick).   Describes organisms that use organic matter for energy and growth. Animals, fungi, and most bacteria are heterotrophs.

The sample being tested can be considered invalid if any of these situations occurs *and* total coliforms are not detected. A replacement sample must be taken from the same location within 24 hours of receiving the laboratory results. If the replacement sample also shows heterotrophic interference but tests positive for total coliforms, the sample is considered valid.

22.2203 MONITORING FREQUENCY

The routine monitoring frequency for community water systems is based on the population served, as shown in Table 22.2. Routine monitoring frequencies for noncommunity systems are based on the source of supply and, in some cases, the population served. See Table 22.2, Footnote a, for the specific details about monitoring by noncommunity water systems.

Whenever a routine sample tests positive for total coliform, two provisions of the Total Coliform Rule take effect: (1) repeat samples (formerly referred to as check samples) must be taken, and (2) the original coliform-positive sample must be tested for the presence of fecal coliforms or *E. coli* to determine whether there is an actual or potential violation of the coliform MCL. The second provision above will be discussed in the next section, "Determining Compliance."

The reason repeat samples are required is to investigate whether the original coliform-positive sample was caused by a contamination problem that exists throughout the distribution system or if it is a localized problem that exists only at that one sampling point. With this information, appropriate corrective action can be taken to eliminate the problem as quickly as possible.

The Total Coliform Rule specifies how many repeat samples must be taken, when, and from what locations. It also directs the water utility to collect a specific number of samples the

TABLE 22.2 TOTAL COLIFORM MONITORING FREQUENCY FOR COMMUNITY WATER SYSTEMS

Population Served	Minimum Number of Routine Samples per Month[a]	Population Served	Minimum Number of Routine Samples per Month
25 to 1,000[b]	1[c]	59,001 to 70,000	70
1,001 to 2,500	2	70,001 to 83,000	80
2,501 to 3,300	3	83,001 to 96,000	90
3,301 to 4,100	4	96,001 to 130,000	100
4,101 to 4,900	5	130,001 to 220,000	120
4,901 to 5,800	6	220,001 to 320,000	150
5,801 to 6,700	7	320,001 to 450,000	180
6,701 to 7,600	8	450,001 to 600,000	210
7,601 to 8,500	9	600,001 to 780,000	240
8,501 to 12,900	10	780,001 to 970,000	270
12,901 to 17,200	15	970,001 to 1,230,000	300
17,201 to 21,500	20	1,230,001 to 1,520,000	330
21,501 to 25,000	25	1,520,001 to 1,850,000	360
25,001 to 33,000	30	1,850,001 to 2,270,000	390
33,001 to 41,000	40	2,270,001 to 3,020,000	420
41,001 to 50,000	50	3,020,001 to 3,960,000	450
50,001 to 59,000	60	3,960,001 or more	480

[a] A noncommunity water system using groundwater and serving 1,000 persons or fewer may monitor at a lesser frequency specified by the state until a sanitary survey is conducted and the state reviews the results. Thereafter, noncommunity water systems using groundwater and serving 1,000 persons or fewer must monitor in each calendar quarter during which the system provides water to the public, unless the state determines that some other frequency is more appropriate and notifies the system (in writing). In all cases, noncommunity water systems using groundwater and serving 1,000 persons or fewer must monitor at least once/year.
A noncommunity water system using surface water, or groundwater under the direct influence of surface water, regardless of the number of persons served, must monitor at the same frequency as a like-sized community public system. A noncommunity water system using groundwater and serving more than 1,000 persons during any month must monitor at the same frequency as a like-sized community water system, except that the state may reduce the monitoring frequency for any month the system serves 1,000 persons or fewer.
[b] Includes public water systems that have at least 15 service connections, but serve fewer than 25 persons.
[c] For a community water system serving 25–1,000 persons, the state may reduce this sampling frequency if a sanitary survey conducted in the last five years indicates that the water system is supplied solely by a protected groundwater source and is free of sanitary defects. However, in no case may the state reduce the sampling frequency to less than once/quarter.
NOTE: These sampling frequencies took effect December 31, 1990.

following month, based on the number of routine samples ordinarily collected. Table 22.3 lists the repeat sampling frequencies triggered by a coliform-positive routine sample.

TABLE 22.3 COLIFORM MONITORING—ROUTINE AND REPEAT SAMPLING FREQUENCIES

Number of Routine Samples per Month	Minimum Number of Repeat Samples[a]	Samples per Month Next Month[b]
1 or fewer	4	5
2	3	5
3	3	5
4	3	5
5 or more	3	See Table 22.2

[a] Number of repeat samples in the same month for each total coliform-positive routine sample.
[b] Except where the state has invalidated the original routine sample, substitutes an on-site evaluation of the problem, or waives the requirement on a case-by-case basis.

Noncommunity systems serving more than 1,000 people have the same requirements as community water systems (Table 22.2). Noncommunity systems serving fewer than 1,000 people are required to collect one routine sample per quarter. When the routine sample tests positive, four repeat samples are required in the same quarter and five samples are required the following quarter.

As Table 22.3 indicates, whenever a routine or repeat sample tests positive for total coliform, the water agency must collect a set of either three or four repeat samples within 24 hours of receiving the laboratory results. At least one of the repeat samples must be taken from the same tap as the original coliform-positive sample; the remaining repeat samples in the set must be collected at nearby taps (within five service connections of the original sampling point), both upstream and downstream of the original. Repeat samples must be taken until no coliforms are detected or until the MCL is exceeded and the state is notified.

For additional information about sampling and compliance procedures, see "Rule Changes Way Systems Will Look at Coliform," *OPFLOW,* Vol. 16, No. 12, December 1990, pages 1, 4, and 5, and "Corrections," *OPFLOW,* Vol. 17, No. 2, February 1991, page 2.

22.2204 DETERMINING COMPLIANCE

As discussed earlier, the MCLG for total coliforms (including fecal coliforms and *E. coli*) is zero. The MCL, based on the presence-absence concept, is as follows:

- For water systems analyzing at least 40 samples per month, no more than 5.0 percent of the samples (including routine and repeat samples) may be positive for total coliforms;

- For water systems analyzing fewer than 40 samples per month, no more than one sample per month may be positive for total coliforms.

The Total Coliform Rule makes another significant change in the way compliance is calculated. That is, all valid coliform-positive samples—both routine *and* repeat samples—must be counted when calculating compliance with the monthly MCL. Under previous regulations, check or repeat samples were not included in the monthly MCL calculation.

On a case-by-case basis, the state may declare a sample invalid for one of several reasons, including interference by heterotrophic bacteria during laboratory analysis, as previously discussed. Total-coliform-positive samples may be invalidated by the state under any of the following conditions:

- The analytical laboratory acknowledges that improper sample analysis caused the positive result;

- The state determines that the contamination is a local plumbing problem; and/or

- The state has substantial grounds to believe that the positive result was unrelated to the quality of drinking water in the distribution system.

A simplified decision tree for determining compliance with the Total Coliform Rule is shown in Table 22.4. It is strongly recommended that operators work closely with their state regulatory agencies in implementing this rule.

22.2205 REPORTING AND NOTIFICATION REQUIREMENTS

Reporting frequencies for coliform test results increase in step with the urgency of the problem. A water agency must report the results of monthly coliform testing to the state regulatory agency within the first 10 days of the following month.

Any time a water agency fails to collect a sample as required, the state must be notified within 10 days after the system learns of the violation. An invalid sample result is considered a failure to monitor and must be reported.

If the MCL is exceeded, the state must be notified no later than the end of the next business day and the public within 14 days. This could occur when the water agency exceeds its monthly coliform-positive limit or when test results show the presence of fecal coliforms or *E. coli* in any sample. (See Section 22.6 for further details about public notification procedures.)

The most critical situation exists when either of two situations occurs:

1. A routine sample tests positive for total coliforms and for fecal coliforms or *E. coli,* and any repeat sample tests positive for total coliforms.

2. A routine sample tests positive for total coliforms and negative for fecal coliforms or *E. coli,* and any repeat sample tests positive for fecal coliforms or *E. coli.*

Either of these situations is considered an "acute risk to health." This is a Tier 1 violation, which requires that the state and the public be notified within 24 hours, as described in Section 22.6.

TABLE 22.4 SIMPLIFIED DECISION TREE FOR DETERMINING COMPLIANCE WITH THE TOTAL COLIFORM RULE[a]

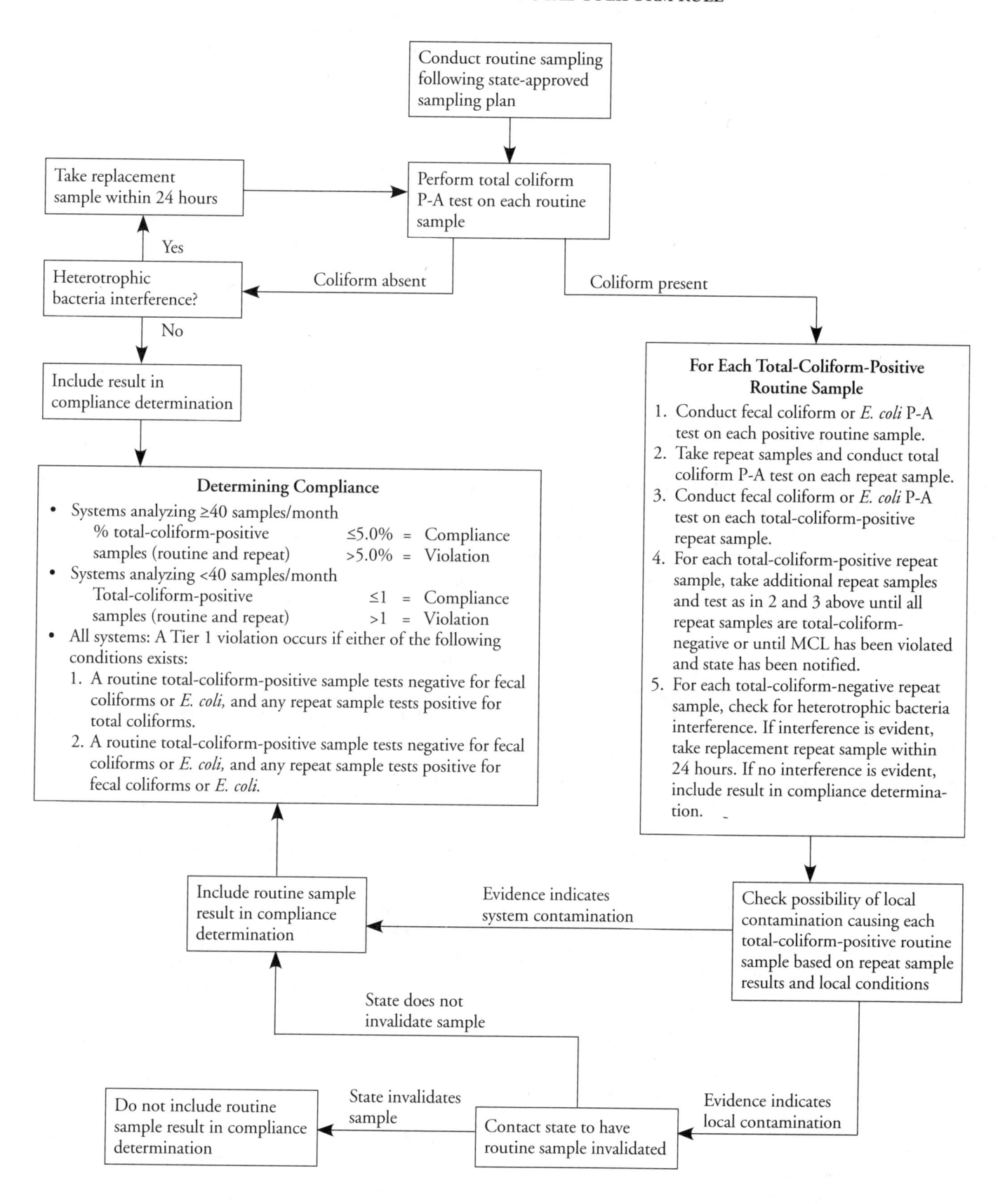

```
                        ┌─────────────────────────┐
                        │ Conduct routine sampling│
                        │ following state-approved│
                        │ sampling plan           │
                        └───────────┬─────────────┘
                                    ▼
┌──────────────────┐     ┌─────────────────────────┐
│ Take replacement │     │ Perform total coliform  │
│ sample within    │     │ P-A test on each routine│
│ 24 hours         │     │ sample                  │
└────────▲─────────┘     └──┬───────────────────┬──┘
         │ Yes               │ Coliform absent   │ Coliform present
┌────────┴─────────┐         │                   │
│ Heterotrophic    │◄────────┘                   ▼
│ bacteria         │                  
│ interference?    │
└────────┬─────────┘
         │ No
         ▼
┌──────────────────┐
│ Include result in│
│ compliance       │
│ determination    │
└────────┬─────────┘
         ▼
```

For Each Total-Coliform-Positive Routine Sample

1. Conduct fecal coliform or *E. coli* P-A test on each positive routine sample.
2. Take repeat samples and conduct total coliform P-A test on each repeat sample.
3. Conduct fecal coliform or *E. coli* P-A test on each total-coliform-positive repeat sample.
4. For each total-coliform-positive repeat sample, take additional repeat samples and test as in 2 and 3 above until all repeat samples are total-coliform-negative or until MCL has been violated and state has been notified.
5. For each total-coliform-negative repeat sample, check for heterotrophic bacteria interference. If interference is evident, take replacement repeat sample within 24 hours. If no interference is evident, include result in compliance determination.

Determining Compliance

- Systems analyzing ≥40 samples/month
  - % total-coliform-positive samples (routine and repeat)   ≤5.0% = Compliance
  -      >5.0% = Violation
- Systems analyzing <40 samples/month
  - Total-coliform-positive samples (routine and repeat)   ≤1 = Compliance
  -      >1 = Violation
- All systems: A Tier 1 violation occurs if either of the following conditions exists:
  1. A routine total-coliform-positive sample tests negative for fecal coliforms or *E. coli*, and any repeat sample tests positive for total coliforms.
  2. A routine total-coliform-positive sample tests negative for fecal coliforms or *E. coli*, and any repeat sample tests positive for fecal coliforms or *E. coli*.

Check possibility of local contamination causing each total-coliform-positive routine sample based on repeat sample results and local conditions

- Evidence indicates system contamination → Include routine sample result in compliance determination
- Evidence indicates local contamination → Contact state to have routine sample invalidated
  - State does not invalidate sample → Include routine sample result in compliance determination
  - State invalidates sample → Do not include routine sample result in compliance determination

---

[a] Reprinted from *OPFLOW*, Vol. 16, No. 12, December 1990, by permission. Copyright © 1990, American Water Works Association.

> **QUESTIONS**
>
> Write your answers in a notebook and then compare your answers with those on page 606.
>
> 22.2I Why are microbial contaminants a public health concern?
>
> 22.2J What is the basic approach used to determine compliance with the Total Coliform Rule?
>
> 22.2K List the five analytical methods EPA will accept for coliform testing.
>
> 22.2L What are reasons for collecting repeat coliform samples?
>
> 22.2M What sampling locations are used for repeat coliform samples?

22.221 Surface Water Treatment Rules

On June 29, 1989, the EPA promulgated the Surface Water Treatment Rule (SWTR). The 1996 SDWA amendments imposed deadlines by which EPA must complete work on several major regulations, including promulgation of an Enhanced Surface Water Treatment Rule (ESWTR). On December 16, 1998, EPA promulgated the Interim Enhanced Surface Water Treatment Rule (IESWTR). On January 14, 2002 and December 15, 2002, respectively, the EPA promulgated the Long Term 1 Enhanced Surface Water Treatment Rule (LT1ESWTR) and the Long Term 2 Enhanced Surface Water Treatment Rule (LT2ESWTR). These regulations set forth primary drinking water regulations requiring treatment of surface water supplies or groundwater supplies under the direct influence of surface water. These regulations require surface water systems to use a specific water treatment technique (filtration and/or disinfection) rather than meeting maximum contaminant levels (MCLs) for *Cryptosporidium, Giardia lamblia*, viruses, *Legionella*, and heterotrophic bacteria. Also, the regulations require that all systems must be operated by qualified operators.

The protozoan *Giardia lamblia* is presently the organism most implicated in waterborne disease outbreaks in the United States. These microscopic creatures are found mainly in mountain streams. Once inside the body, they cause a painful and disabling illness. The infection caused by *Giardia* is called Giardiasis. The symptoms of Giardiasis are usually severe diarrhea, gas, cramps, nausea, vomiting, and fatigue.

Giardia and viruses have been added to the traditional coliform and turbidity indicators of microbiological quality. In this case, the Maximum Contaminant Level Goals (MCLGs) are zero because the organisms are pathogens, or indicators of pathogens, and should not be present in drinking water.

The regulations define surface water as "all water open to the atmosphere and subject to surface runoff." This would include rivers, lakes, streams, and reservoirs. This surface water definition also includes groundwaters that are directly influenced by surface water. The determination as to whether a supply uses surface water is left up to the state. Generally, if a groundwater source has significant and relatively rapid shifts in water quality such as turbidity, temperature, conductivity, or pH that closely correlate to climatological or nearby surface water conditions, that source can be considered surface water.

At a minimum, the treatment required for surface water would include disinfection systems for very clean and protected source waters. These systems would only be required to disinfect to achieve removal of coliforms to meet the requirements.

Turbidity measurements are a key element in determining compliance with both the filtration and the disinfection requirements of the Surface Water Treatment Rules. The effectiveness of a filtration system in removing particles and organisms is easily obtained by measuring the water's turbidity before and after filtration. High turbidity also reduces the effectiveness of disinfection processes and, therefore, must be controlled and closely monitored.

Water supply systems that are required to filter can use a variety of treatment technologies, including disinfection, to meet the expected performance levels. These technologies include conventional treatment, direct filtration, slow sand filtration, and diatomaceous earth filtration.

The two-pronged treatment technique of the regulations aims to ensure that harmful microorganisms are either removed by filtration processes or inactivated (and thus made harmless) by disinfection processes. Success, therefore, depends heavily on consistently well-operated filtration and disinfection systems. The general performance standards to be met by all surface water systems and any systems using groundwater under the influence of surface water are:

1. At least 99.9 percent (also called 3 log for the 3 nines) removal and/or inactivation of *Giardia lamblia* cysts, and

2. At least 99.99 percent (4 log) removal and/or inactivation of enteric (intestinal) viruses.

In general, compliance by the surface water purveyor could be through one of the following alternatives:

1. Meeting the criteria for which filtration is not required and providing disinfection according to the specific requirements in the regulations, or

2. Providing filtration and meeting disinfection criteria required for those supplies that are filtered.

Operators are urged to work closely with their state regulatory agencies to stay informed about changes in water regulations as they occur in the future.

22.2210 CRITERIA FOR AVOIDING FILTRATION

To avoid mandatory filtration, a water utility must meet the following requirements to ensure adequate disinfection:

1. The fecal coliform concentration in the water prior to disinfection must not exceed 20/100 mL, or total coliform concentration must not exceed 100/100 mL in more than 10 percent of the samples analyzed in the previous six months;

2. The system must currently provide at least 99.9 percent (3 log) *Cryptosporidium* inactivation;

3. The turbidity of the water prior to disinfection cannot exceed 5 *NTUs*[4] based on grab samples collected every four hours or by continuous monitoring; and

4. Certain site-specific conditions:

   (a) has disinfection that achieves 99.9 percent inactivation of *Giardia* and 99.99 percent inactivation of viruses (CT requirement—see Section 22.2215);

   (b) watershed control or sanitary surveys that satisfy regulatory requirements;

   (c) no history of waterborne disease outbreak without making treatment corrections;

   (d) compliance with long-term coliform maximum contaminant level (MCL);

   (e) compliance with total trihalomethanes MCL, if the system serves more than 10,000 people;

   (f) disinfection system must have an alternative power supply and automatic alarm and start-up to ensure continuous disinfection; and

   (g) ensure that disinfectant residual at entrance to distribution system does not drop below 0.2 mg/L for more than four hours.

If a system cannot meet the source water quality criteria and site-specific conditions listed above, then the system must install and operate appropriate filtration facilities.

22.2211 REQUIREMENTS FOR FILTERED WATER SYSTEMS

For systems that filter, the primary concern is adequate disinfection and filtration performance. The requirements are:

1. For conventional or direct filtration systems, the filtered water turbidity must be less than or equal to 0.3 NTU for at least 95 percent of each month's measurements. For slow sand or diatomaceous earth filtration, the filtered water turbidity must be less than 1 NTU in at least 95 percent of the measurements taken each month.

2. Filtered water turbidity must never exceed 5 NTUs.

3. A disinfectant residual at the entrance to the distribution system of at least 0.2 mg/L.

4. Filtered water supplies must achieve the same disinfection as required for unfiltered systems (that is, 99.9 percent and 99.99 percent removal/inactivation of *Giardia lamblia* and viruses) through a combination of filtration and application of a disinfectant.

22.2212 CHLORINE RESIDUAL SUBSTITUTION

It is important to remember that this is a two-pronged approach and both elements are considered when determining compliance by systems that are required to install filtration. Filtration alone accomplishes part of the job of removing harmful organisms; disinfection does the rest of the job.

At the discretion of the state and based on a review of the water system, chlorine residual testing may be substituted for some of the bacteriological testing. Chlorine residual testing could give the operator a quicker indication of the condition of the system. However, the following requirements must be met:

1. Samples must be taken at points that are representative of conditions within the distribution system.

2. Chlorine residual testing can replace only up to 75 percent of the bacteriological testing.

3. At least four chlorine residual tests must be taken to substitute for one bacteriological sample.

4. A free chlorine residual of at least 0.2 mg/L must be maintained throughout the distribution system.

5. If free chlorine residual falls below 0.2 mg/L, check samples must be taken for bacteriological testing and a report must be submitted to the state within 48 hours.

6. Chlorine residual must be determined daily.

22.2213 TURBIDITY REQUIREMENTS

Turbidity is undesirable in a finished or treated water because it causes cloudiness resulting in an unattractive water. However, the major reason that turbidity is undesirable is that it causes a health hazard by:

1. Interfering with disinfection by reducing the ability of the disinfectant to inactivate or kill disease-causing organisms

2. Exerting a chlorine demand that makes it difficult to maintain a residual throughout the distribution system

3. Interfering with the bacteriological examination of the water

4. Preventing satisfactory reduction of tastes and odors and asbestos fibers

The MCLs and monitoring requirements for turbidity (which apply to surface water only) are as follows:

- For unfiltered surface water supplies, the turbidity of the raw water prior to disinfection cannot exceed 5 NTU based on grab samples collected every four hours or by continuous turbidity monitoring;

- For surface water supplies using conventional or direct filtration, the process must achieve a turbidity level of less than 1

---

[4] *Turbidity* (ter-BID-it-tee) *Units (TU).* Turbidity units are a measure of the cloudiness of water. If measured by a nephelometric (deflected light) instrumental procedure, turbidity units are expressed in nephelometric turbidity units (NTU) or simply TU. Those turbidity units obtained by visual methods are expressed in Jackson turbidity units (JTU), which are a measure of the cloudiness of water; they are used to indicate the clarity of water. There is no real connection between NTUs and JTUs. The Jackson turbidimeter is a visual method and the nephelometer is an instrumental method based on deflected light.

NTU at all times and not more than 0.3 NTU in more than 5 percent of the samples analyzed during a given month. Turbidity must be sampled every four hours or by continuous monitoring; and

- Those systems using slow sand or diatomaceous earth filtration must achieve a turbidity level of less than 5 NTU at all times and not more than 1 NTU in more than 5 percent of the samples collected each month. Systems using slow sand filtration must sample at least once a day; diatomaceous earth filtration systems must monitor turbidity every four hours or by continuous sampling.

Other filtration technologies may be used if they meet, as a minimum, the turbidity requirements for slow sand filters, the disinfection requirements, and are approved by the state.

Unfiltered systems are required to begin with a clean source water and have a watershed that is protected from human activities that might otherwise have an adverse impact on water quality. Unfiltered systems would have very little, if any, virus contamination. For these systems, the major concern is *Giardia* contamination from animal activities that cannot be prevented by watershed protection. The purpose of the turbidity limit for unfiltered water is to ensure a high probability that turbidity does not interfere with disinfection of *Giardia* cysts. The turbidity limit of 5 NTU serves this purpose.

For filtered water systems, the major burden for *Giardia* removal rests with filtration. With conventional treatment and direct filtration, low turbidity levels (<0.5 NTU) are needed to ensure effective *Giardia* cyst removals. Disinfection of either *Giardia* or viruses will not be hampered at these turbidity levels.

For slow sand filtration and diatomaceous earth filtration, effective *Giardia* removal does not necessarily correlate with low treated water turbidities. However, to ensure effective virus inactivation, a low filtered water turbidity is needed. Viruses are much smaller than *Giardia,* and thus a lower turbidity limit of 1 NTU is needed compared with the turbidity level of 5 NTU for unfiltered supplies to ensure effective disinfection.

22.2214 MONITORING REQUIREMENTS

Unfiltered surface water systems must:

1. Monitor raw water supply for fecal and total coliform. Samples must be collected and analyzed at a minimum frequency based on system size as shown below.

Number of People Served	Samples/Week
<501	1
501–3,300	2
3,301–10,000	3
10,001–25,000	4
>25,000	5

If the fecal coliform concentration in the water prior to disinfection exceeds 20 coliforms/100 mL, or the total coliform concentration exceeds 100 coliforms/100 mL in more than 10 percent of the samples analyzed in a six-month period, the system no longer meets the criteria to avoid filtration.

2. Monitor turbidity every four hours or by continuous measurement.

3. Monitor and control activities on the watershed in order to minimize adverse impacts on water quality. Annual on-site inspections are also required.

4. Continuously monitor the disinfectant residual to ensure a minimum of 0.2 mg/L at the entrance to the distribution system. Systems serving fewer than 3,300 persons may take grab samples at the frequencies shown below.

Number of People Served	Samples/Week
<501	1
501–1,000	2
1,001–2,500	3
2,501–3,300	4

In addition, disinfectant residuals in the distribution system must be monitored at the same frequency and location as total coliform samples.

5. Monitor daily to demonstrate that the level of disinfection achieved is 99.9 percent inactivation of *Giardia* and 99.99 percent inactivation/removal of enteric viruses.

Filtered systems must meet the following requirements:

1. Perform turbidity measurements of representative water every 4 hours (which can be continuous monitoring). For any system using slow sand filtration or any other filtration treatment other than conventional treatment, such as direct filtration or diatomaceous earth filtration, the state may reduce this monitoring to once per day.

2. Continuously monitor the disinfectant residual entering the distribution system to ensure a minimum of 0.2 mg/L at the entrance to the distribution system. Systems serving fewer than 3,300 people may take grab samples at the frequencies shown previously. Disinfectant residuals in the distribution system must be monitored at the same frequency and location as total coliform samples.

3. Monitor daily to demonstrate that a combination of filtration and application of disinfectant achieves 99.9 percent and 99.99 percent removal or inactivation for *Giardia* and viruses, respectively.

4. Measurements for heterotrophic plate count (HPC) bacteria may be substituted for disinfectant residual measurements. If the HPC is less than 500 colonies per mL, then the sample is considered equivalent to a detectable disinfectant residual. For systems serving fewer than 500 people, the regulatory agency may determine the adequacy of the disinfectant residual in place of monitoring.

22.2215 CT VALUES

The purpose of the SWTR is to ensure that pathogenic organisms are removed or inactivated by the treatment process. To meet this goal, all systems are required to disinfect their water supplies. For some water systems using very clean source water and meeting the other criteria to avoid filtration, disinfection alone can achieve the 3-log (99.9 percent) *Giardia* and 4-log (99.99 percent) virus inactivation levels required by the Surface Water Treatment Rule.

Several methods of disinfection are in common use, including free chlorination, chloramination, use of chlorine dioxide, and application of ozone. The concentration of chemical needed and the length of contact time needed to ensure disinfection are different for each disinfectant. Therefore, the effectiveness of the disinfectant is measured by the time (T) in minutes of the disinfectant's contact in the water and the concentration (C) of the disinfectant residual in mg/L measured at the end of the contact time. The product of these two factors (C × T) provides a measure of the degree of pathogenic inactivation. The required CT value to achieve inactivation is dependent upon the organism in question, type of disinfectant, pH, and temperature of the water supply.

Time or "T" is measured from point of application to the point where "C" is determined. "T" must be based on peak hour flow rate conditions. In pipelines, "T" is calculated by dividing the volume of the pipeline in gallons by the flow rate in gallons per minute (GPM). In reservoirs and basins, dye tracer tests must be used to determine "T." In this case, "T" is the time it takes for 10 percent of the tracer to pass the measuring point.

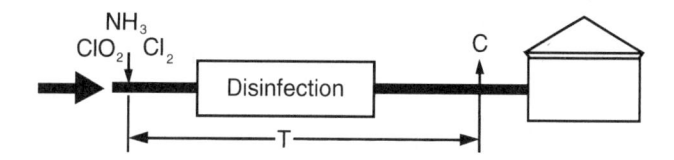

A properly operated filtration system can achieve limited removal or inactivation of microorganisms. Because of this, systems that are required to filter their water are permitted to apply a factor that represents the microorganism removal value of filtration when calculating CT values to meet the disinfection requirements. The factor (removal credit) varies with the type of filtration system. Its purpose is to take into account the combined effect of both disinfection and filtration in meeting the SWTR microbial standards.

Please refer to the Arithmetic Appendix at the end of this manual (Section A.6, "Calculation of CT Values") for instructions on how to perform these calculations for a water treatment plant.

22.2216 INTERIM ENHANCED SURFACE WATER TREATMENT RULE (IESWTR)

The Interim Enhanced Surface Water Treatment Rule (IESWTR) was promulgated by EPA on December 16, 1998 (*FEDERAL REGISTER* 63, No. 241). Most provisions of the IESWTR apply to all public water systems using surface water or groundwater under the direct influence of surface water and serving 10,000 or more people.

The principal health goal of the IESWTR is to reduce the risk from *Cryptosporidium*. The Maximum Contaminant Level Goal (MCLG) was set at zero for *Cryptosporidium*. *Cryptosporidium* was included in filtration avoidance criteria and groundwater under the direct influence of surface water determinations. These include tightening of combined filter effluent turbidities to <0.3 NTU 95 percent of the time and <1 NTU all the time for conventional and direct filtration plants. Monitoring of individual filter effluent turbidities is also required, although exceedances are not violations, but only trigger filter studies. Systems must prepare disinfection profiles and benchmarks to be used in discussions with their primacy agencies. The IESWTR also prohibits the construction of any new uncovered finished water facilities after February 1999.

One provision of the IESWTR applies to all systems using surface water or groundwater under the influence of surface water, regardless of size: the state must conduct a periodic sanitary survey that covers the entire system from sources through treatment plant, distribution, and storage. Significant deficiencies must be addressed in a timely manner.

Large systems (serving more than 10,000 persons) that use surface water or groundwater under the direct influence of surface water were required to comply with the IESWTR by January 2002. Small systems (serving fewer than 10,000) that use surface water or groundwater under the direct influence of surface water must comply by January 1, 2004. The full text of the rule is available at the EPA home page (www.epa.gov).

22.2217 LONG TERM 1 ENHANCED SURFACE WATER TREATMENT RULE (LT1ESWTR)

The Long Term 1 Enhanced Surface Water Treatment Rule (LT1ESWTR) was promulgated by EPA on January 14, 2002. The provisions of the LT1ESWTR apply to public water systems that use surface water or groundwater under the direct influence of surface water (GWUDI) and serve fewer than 10,000 people.

The major provisions of the LT1ESWTR include control of *Cryptosporidium* and combined filter effluent (CFE) turbidity performance standards. Provisions for the control of *Cryptosporidium* include the maximum contaminant level goal (MCLG) of zero for *Cryptosporidium*. Filtered water systems must physically remove 99 percent (2-log removal) of the *Cryptosporidium*. Unfiltered systems must update their watershed control programs to minimize the potential for contamination of *Cryptosporidium* oocysts. Also, *Cryptosporidium* is included as an indicator that the groundwater is under the direct influence of surface water (GWUDI).

The specific combined filter effluent (CFE) turbidity requirements depend on the type of filtration used by the system. For conventional and direct filtration systems, the turbidity must be less than or equal to 0.3 nephelometric turbidity units (NTU) in at least 95 percent of the measurements taken each month and the maximum level of turbidity is 1 NTU. Slow sand and diatomaceous earth (DE) filtration systems must meet the CFE turbidity limits specified in the SWTR of less than 1 NTU in at least 95 percent of the measurements taken each month and a maximum level of turbidity of 5 NTU. Alternative technologies (other than conventional, direct, slow sand, or DE) must meet turbidity levels established by the drinking water authority based on filter demonstration data submitted by the system, but these limits must not exceed 1 NTU (in at least 95 percent of the measurements) or 5 NTU (maximum).

Turbidity monitoring requirements for combined filter effluent (CFE) include monitoring performed at least every four hours to ensure compliance with CFE turbidity performance standards. Since the combined filter effluent (CFE) may meet regulatory requirements even though one filter is producing high-turbidity water, the individual filter effluent (IFE) is measured to assist conventional and direct filtration treatment plant operators in understanding and assessing individual filter performance. The individual filter effluent (IFE) monitoring requirements include monitoring continuously (recorded at least every 15 minutes). Systems with two or fewer filters may conduct continuous monitoring of CFE turbidity in place of individual filter effluent monitoring. Follow-up actions (i.e., additional reporting, filter self-assessment, or comprehensive performance evaluations (CPEs)) are required if the IFE turbidity (or CFE for systems with two filters) exceeds 1.0 NTU in two consecutive readings or more.

22.2218 LONG TERM 2 ENHANCED SURFACE WATER TREATMENT RULE (LT2ESWTR)

On December 15, 2005, the EPA promulgated the Long Term 2 Enhanced Surface Water Treatment Rule (LT2ESWTR or LT2). The purpose of this rule is to reduce illness linked with the contaminant *Cryptosporidium* and other pathogenic microorganisms in drinking water. The LT2ESWTR *supplements existing regulations* by targeting additional *Cryptosporidium* treatment requirements to higher-risk systems. This rule also contains provisions to reduce risks from uncovered finished water reservoirs and provisions to ensure that systems maintain microbial protection when they take steps to decrease the formation of disinfection by-products that result from chemical water treatment.

Current regulations require filtered water systems to reduce source water *Cryptosporidium* levels by 2-log (99 percent). Data on *Cryptosporidium* infectivity and occurrence indicate that this treatment requirement is sufficient for most systems, but additional treatment is necessary for certain higher-risk systems. These higher-risk systems include filtered water systems with high levels of *Cryptosporidium* in their water sources and all unfiltered water systems that do not treat for *Cryptosporidium*.

The LT2ESWTR was promulgated simultaneously with the Stage 2 Disinfection By-Products Rule to address concerns about risk tradeoffs between pathogens and DBPs.

Monitoring. Under the LT2ESWTR, systems will monitor their water sources to determine treatment requirements. This monitoring includes an initial two years of monthly sampling for *Cryptosporidium*. To reduce monitoring costs, small filtered water systems will first monitor for *E. coli* (a bacterium that is less expensive to analyze than *Cryptosporidium*) and will monitor for *Cryptosporidium* only if their *E. coli* results exceed specified concentration levels.

Monitoring dates are staggered by system size, with smaller systems beginning monitoring after larger systems. Systems must conduct a second round of monitoring six years after completing the initial round to determine if source water conditions have changed significantly. Systems may use (grandfather) previously collected data in lieu of conducting new monitoring, and systems are not required to monitor if they provide the maximum level of treatment required under the rule.

Cryptosporidium **Treatment.** Filtered water systems will be classified in one of four treatment categories (bins) based on their monitoring results. The majority of systems will be classified in the lowest treatment bin, which carries no additional treatment requirements. Systems classified in higher treatment bins must provide 90 to 99.7 percent (1.0- to 2.5-log) additional treatment for *Cryptosporidium*. Systems will select from a wide range of treatment and management strategies in the "microbial toolbox" to meet their additional treatment requirements. All unfiltered water systems must provide at least 99 or 99.9 percent (2- or 3-log) inactivation of *Cryptosporidium* depending on the results of their monitoring. These *Cryptosporidium* treatment requirements reflect consensus recommendations of the Stage 2 Microbial and Disinfection By-Products Federal Advisory Committee.

Other Requirements. Systems that store treated water in open reservoirs must either cover the reservoir or treat the reservoir discharge to inactivate 4-log virus, 3-log *Giardia lamblia*, and 2-log *Cryptosporidium*. These requirements are necessary to protect against the contamination of water that occurs in open reservoirs. In addition, systems must review their current level of microbial treatment before making a significant change in their disinfection practice. This review will assist systems to reduce the formation of disinfection by-products under the Stage 2 Disinfection By-Products Rule, which EPA has finalized along with the LT2ESWTR.

22.2219 ADDITIONAL READING

For more detailed information about the requirements and application of the Surface Water Treatment Rules, you may wish to order a copy of the publication, *GUIDANCE MANUAL FOR COMPLIANCE WITH THE FILTRATION AND DISINFECTION REQUIREMENTS FOR PUBLIC WATER SYSTEMS USING SURFACE WATER SOURCES*. It is available from the American Water Works Association (AWWA), 6666 West Quincy Avenue, Denver, CO 80235. Order No. 20271. ISBN 0-89867-558-8. Price to members, $57.50; nonmembers, $84.50; price includes cost of shipping and handling.

QUESTIONS

Write your answers in a notebook and then compare your answers with those on page 606.

22.2N What do the Surface Water Treatment Rules specifically require?

22.2O How do the Surface Water Treatment Rules define surface water?

22.2P How can a water utility avoid mandatory filtration?
22.2Q What are the MCLs for turbidity for surface waters?
22.2R How is the effectiveness of a disinfectant measured?

22.23 Disinfectants and Disinfection By-Products (D/DBP)

Disinfection of drinking water by the addition of chlorine has long been considered a highly effective yet relatively low-cost method of preventing widespread outbreaks of waterborne diseases. In addition to reacting with disease-causing organisms in water, however, chlorine also reacts with many other types of organic materials. There is growing scientific evidence that the by-products of these chemical reactions can produce adverse health effects in humans.

The highest priority health risk concern in the regulation of drinking water is the potential risk-risk tradeoff between the control of microbiological contamination (bacteria, viruses, and protozoa) on the one hand and disinfection by-products on the other hand. This risk-risk tradeoff arises because typically the least expensive way for a public water system to increase microbial control is to increase disinfection (which generally increases by-product formation) and the easiest way to reduce by-products is to decrease disinfection (which generally increases microbial risk). Microbiological contamination often causes flu-like symptoms, but can also cause serious diseases such as hepatitis, giardiasis, cryptosporidiosis, and Legionnaire's Disease. Disinfection by-products may pose the risk of cancer and developmental effects.

Trihalomethanes (THMs) are an example of a compound formed by the reaction of chlorine with organic matter in water (also see Chapter 15, "Specialized Treatment Processes"). THMs are suspected of being carcinogenic and have been regulated by EPA in the 1996 SDWA amendments. The maximum contaminant level (MCL) for total trihalomethanes (TTHMs) is 0.080 milligram per liter or 80 micrograms per liter.

In May 1996, EPA published the Information Collection Rule (ICR). This rule required large public water systems to undertake extensive monitoring of microbial contaminants and disinfection by-products in their water systems. Also, some water systems conducted studies on the use of granular activated carbon and membrane processes. The data reported under the ICR were used by EPA to learn more about the occurrence of microbial contamination and disinfection by-products, the health risks posed, appropriate analytical methods, and effective forms of treatment. The ICR data form the scientific basis for EPA's development of the Enhanced Surface Water Treatment Rule and the Disinfectants and Disinfection By-Products (D/DBP) Rule.

The Stage 1 Disinfectants and Disinfection By-Products (Stage 1 D/DBP) Rule was promulgated by EPA on December 16, 1998 (*FEDERAL REGISTER* 63, No. 241). This rule set new Maximum Contaminant Level Goals (MCLGs) and Maximum Contaminant Levels (MCLs) for total trihalomethanes (TTHMs) and five haloacetic acids (HAA5), bromate, and chlorite. Maximum Residual Disinfectant Level Goals (MRDLGs) and Maximum Residual Disinfectant Levels (MRDLs) have also been set for chlorine, chloramine, and chlorine dioxide, as follows:

Disinfection By-Products	MCLG (mg/L)	MCL (mg/L)
TTHM		0.080
Bromoform	0	
Chloroform	0	
Bromodichloromethane	0	
Dibromochloromethane	0.06	
HAA5		0.060
Dichloroacetic acid	0	
Trichloroacetic acid	0.3	
Bromate	0	0.010
Chlorite	0.8	1.0
Disinfectants	**MRDLG (mg/L)**	**MRDL (mg/L)**
Chlorine	4 (as Cl_2)	4.0 (as Cl_2)
Chloramine	4 (as Cl_2)	4.0 (as Cl_2)
Chlorine dioxide	0.8 (as ClO_2)	0.8 (as ClO_2)

The D/DBP Rule attempts to further reduce potential formation of harmful disinfection by-products by requiring the removal of THM precursors. A treatment technique of enhanced coagulation, enhanced softening, or use of granular activated carbon (GAC) applies to conventional filtration systems. In most cases, systems must reduce total organic carbon (TOC) levels based on specific source water quality factors.

For large systems (serving more than 10,000 persons) that use surface water or groundwater under the direct influence of surface water, the compliance date for the Stage 1 D/DBP Rule was January 1, 2002. Small systems (serving fewer than 10,000) that use surface water or groundwater under the direct influence of surface water and all groundwater systems must comply by January 1, 2004. The full text of the D/DBP Rule is available by visiting EPA's website at www.epa.gov/safewater/standards.html.

The Stage 1 Disinfectants and Disinfection By-Products Rule has very specific laboratory and monitoring requirements. The routine monitoring requirements include the following regulated contaminants/disinfectants:

- TTHM/HAA5 (total trihalomethanes and sum of five haloacetic acids)
- Bromate
- Chlorite
- Chlorine/chloramines
- Chlorine dioxide
- DBP (disinfection by-product) precursors (TOC/alkalinity/SUVA (specific UV absorbance))

Also, the D/DBP Rule specifies the monitoring coverage in terms of surface waters, groundwaters, and groundwater under direct influence (GWUDI), population served, and also the type of filtration system and disinfection system. Monitoring frequency depends on the type of source water, population served, and type of treatment and disinfection system.

The routine monitoring requirements are based on the regulated contaminants/disinfectants and also include the MCL, MRDL, analytical method, preservation/quenching agent, holding time for sample/extract, and sample container size and type.

On December 15, 2005, EPA promulgated the Stage 2 Disinfection By-Products (Stage 2 D/DBP) Rule. This rule will reduce potential cancer and reproductive and developmental health risks from disinfection by-products (DBPs) in drinking water, which form when disinfectants are used to control microbial pathogens.

This final rule strengthens public health protection for consumers by tightening compliance monitoring requirements for two groups of DBPs: total trihalomethanes (TTHMs) and haloacetic acids (HAA5). The rule targets systems with the greatest risk and builds incrementally on existing rules. This regulation will reduce DBP exposure and related potential health risks and provide more equitable public health protection.

The Stage 2 D/DBP Rule was promulgated simultaneously with the Long Term 2 Enhanced Surface Water Treatment Rule to address concerns about risk tradeoffs between pathogens and DBPs.

Under the Stage 2 D/DBP Rule, systems will conduct an evaluation of their distribution systems, known as an Initial Distribution System Evaluation (IDSE), to identify the locations with high disinfection by-product concentrations. These locations will then be used by the systems as the sampling sites for Stage 2 D/DBP Rule compliance monitoring.

Compliance with the maximum contaminant levels for two groups of disinfection by-products (TTHMs and HAA5) will be calculated for each monitoring location in the distribution system. This approach, referred to as the locational running annual average (LRAA), differs from previous requirements, which determine compliance by calculating the running annual average of samples from all monitoring locations across the system.

The Stage 2 D/DBP Rule also requires each system to determine if they have exceeded an operational evaluation level, which is identified using their compliance monitoring results. The operational evaluation level provides an early warning of possible future MCL violations, which allows the system to take proactive steps to remain in compliance. A system that exceeds an operational evaluation level is required to review their operational practices and submit a report to their state that identifies actions that may be taken to mitigate future high DBP levels, particularly those that may jeopardize their compliance with the DBP MCLs.

22.24 Radiological Standards

Radioactivity is the only contaminant that has been shown to cause cancer for which standards have been set. Three radioactive elements, radon, radium, and uranium, occur naturally in the ground and dissolve into groundwater supplies. However, the possible exposure to radiation in drinking water is only a fraction of the exposure from all natural sources. The main source of radioactive material in surface water is fallout from nuclear testing. Other sources could be nuclear power plants, nuclear fuel processing plants, and uranium mines. Those sources are monitored constantly and there is no great risk of contamination, barring accidents.

Alpha and radium radioactivity occur naturally in groundwater in parts of the West, Midwest, and Northeast. Standards for those types of radioactivity and for manmade, or beta radiation, have been set at levels of safety comparable to other contaminants.

The MCLs for radiological contaminants are divided into two categories: (1) natural radioactivity that results from well water passing through deposits of naturally occurring radioactive materials, and (2) manmade radioactivity such as might result from industrial wastes, hospitals, or research laboratories. Table 22.5 summarizes the MCLs for radioactivity.

TABLE 22.5 MCLs FOR RADIOACTIVITY

Constituent	Maximum Contaminant Level
Combined Radium 226 and Radium 228	5 pCi/L[a]
Gross Alpha Activity (including Radium 226 but excluding Radon and Uranium)	15 pCi/L[a]
Beta/Photon Emitters[b]	4 mrem/yr
Uranium	30 µg/L

a pCi/L. picoCurie per Liter. A picoCurie is a measure of radioactivity. One picoCurie of radioactivity is equivalent to 0.037 nuclear disintegrations per second.
b For surface water systems serving more than 100,000 persons and systems determined by the state to be vulnerable.

Monitoring for natural radioactivity contamination is required every four years for both surface water and groundwater community systems. Routine monitoring procedures to follow are:

1. Test for gross alpha activity; if gross alpha exceeds 5 pCi/L, then
2. Test for radium 226; if radium 226 exceeds 3 pCi/L, then
3. Test for radium 228.

The MCL for gross beta particle activity is 4 mrem/yr for water systems using surface water and serving more than 100,000 people, and for any other community system determined by the state to be vulnerable to this type of contamination. After the initial monitoring period, beta radiation must be monitored every four years unless the MCL is exceeded or the state determines that more frequent monitoring is appropriate.

Regulation of radon continues to be a source of considerable debate. EPA proposed a maximum contaminant level in 1991 and then withdrew the proposed MCL in 1997 in response to the 1996 SDWA amendments. EPA published a proposed regulation for radon in 1999, but the final MCL has not been published.

The final Radionuclides Rule was published on December 7, 2000. The rule retains the current standards for radium 226 and 228, gross alpha activity, and gross beta particle activity (see Table 22.5), but establishes a new MCL 30 µg/L for uranium. The EPA has indicated that it will continue to review the standard for beta particle activity and may revise the MCL in the future. EPA must base its new rules on a scientific risk assessment and an analysis of both the amount of health risk reduction that will be achieved and the cost of implementing the new rules.

QUESTIONS

Write your answers in a notebook and then compare your answers with those on page 607.

22.2S What are trihalomethanes (THMs)?

22.2T How does drinking water become contaminated with radioactive elements?

22.2U The MCLs for radiological contaminants are divided into what two categories?

END OF LESSON 1 OF 2 LESSONS
on
DRINKING WATER REGULATIONS

Please answer the discussion and review questions next.

DISCUSSION AND REVIEW QUESTIONS
Chapter 22. DRINKING WATER REGULATIONS
(Lesson 1 of 2 Lessons)

At the end of each lesson in this chapter you will find some discussion and review questions. The purpose of these questions is to indicate to you how well you understand the material in the lesson. Write the answers to these questions in your notebook.

1. What criteria must be met before EPA selects a contaminant for regulation?
2. What are Maximum Contaminant Levels (MCLs)?
3. What is the definition of a public water system?
4. What is a community water system?
5. Inorganic chemicals pose what type of health concerns?
6. What steps must be taken if a water system exceeds the lead and copper action levels in more than 10 percent of its samples?
7. What are the most common sources of organic contamination of drinking water?
8. Why are coliform bacteria found in drinking water and what does their presence indicate?
9. What is the purpose of the Surface Water Treatment Rules?
10. Why is turbidity undesirable in a finished or treated water?
11. Why are trihalomethanes (THMs) regulated?
12. What were the requirements and purposes of the Information Collection Rule?
13. What are the two categories of MCLs for radiological contaminants?

CHAPTER 22. DRINKING WATER REGULATIONS

(Lesson 2 of 2 Lessons)

22.3 SECONDARY DRINKING WATER STANDARDS

22.30 Enforcement of Regulations

The National Secondary Drinking Water Regulations control contaminants in drinking water that primarily affect the aesthetic qualities relating to the public acceptance of drinking water. At considerably higher concentrations of these contaminants, health implications may also exist as well as aesthetic degradation. These regulations are not federally enforceable; however, some states have passed laws requiring the state health agency to enforce the regulations.

22.31 Secondary Maximum Contaminant Levels

Secondary Maximum Contaminant Levels (SMCLs) apply to public water systems and, in the judgment of the EPA Administrator, are necessary to protect the public welfare or for public acceptance of the drinking water. The SMCL means the maximum permissible level of a contaminant that is delivered to the free-flowing outlet of the ultimate user of a public water system. Contaminants added to the water under circumstances controlled by the user, except those resulting from corrosion of piping and plumbing caused by water quality, are excluded by definition. Currently there are 15 secondary standards (see Table 22.6).

States may establish higher or lower levels depending on local conditions, providing that public health and welfare are adequately protected.

Aesthetic qualities are important factors in public acceptance and confidence in a public water system. States are encouraged to implement SMCLs so that the public will not be driven to obtain drinking water from potentially lower-quality, higher-risk sources. Many states have chosen to enforce both Primary and Secondary MCLs to ensure that the consumer is provided with the best quality water available.

TABLE 22.6 SECONDARY DRINKING WATER STANDARDS

Contaminant	SMCL	Effect
Aluminum	0.05 to 0.2 mg/L	precipitates cause cloudy looking water
Chloride	250 mg/L	taste and corrosion of water pipes
Color	15 Color Units	aesthetic
Copper	1.0 mg/L	staining
Corrosivity	Noncorrosive	corrosion control
Fluoride	2 mg/L	dental fluorosis (a brownish discoloration of the teeth)
Foaming Agents	0.5 mg/L	aesthetic and taste
Iron	0.3 mg/L	taste and staining
Manganese	0.05 mg/L	taste and staining
Odor	3 Threshold Odor Number[a]	aesthetic and taste
pH	6.5 to 8.5	corrosion control and taste
Silver	0.10 mg/L	blue-gray discoloration of the skin, mucous membranes, and eyes
Sulfate	250 mg/L	taste and laxative effects
Total Dissolved Solids (TDS)	500 mg/L	taste
Zinc	5 mg/L	taste

[a] Threshold Odor Number (TON). The greatest dilution of a sample with odor-free water that still yields a just-detectable odor.

22.32 Monitoring

Collect samples for secondary contaminants at a free-flowing outlet of water being delivered to the consumer. Monitor contaminants in these regulations at least as often as the monitoring performed for inorganic chemical contaminants listed in the Primary Drinking Water Regulations as applicable to community water systems. For surface water systems, this means yearly monitoring is required; for groundwater systems, monitor at least once every three years. Collect monthly distribution system physical water quality monitoring samples for color and odors. More frequent monitoring would be appropriate for specific contaminants such as pH, color, odor, or others, under certain circumstances as directed by the state.

QUESTIONS

Write your answers in a notebook and then compare your answers with those on page 607.

22.3A Under what conditions are secondary drinking water regulations enforceable?

22.3B List the secondary drinking water contaminants.

22.3C How frequently should the contaminants in the secondary regulations be monitored?

22.33 Secondary Contaminants

22.330 *Aluminum*

Until relatively recently, most environmental aluminum was found in forms that did not greatly affect humans and other animals. Acid rain, however, has caused a significant increase in aluminum exposure. The average daily intake of aluminum in the general population is about 20 milligrams per day; the SMCL is 0.05 to 0.2 mg/L. Although aluminum is regulated as a secondary contaminant because precipitates of the metal can cause cloudy-looking water, aluminum compounds may also interfere with absorption of fluoride in the gastrointestinal tract and may decrease the absorption of calcium, iron, and cholesterol. Many over-the-counter antacids contain aluminum and use of these products frequently causes constipation, which is thought to be the result of the gastrointestinal effects described above.

22.331 *Chloride*

The SMCL for chloride is 250 mg/L.

UNDESIRABLE EFFECTS

1. Objectionable salty taste in water
2. Corrosion of the pipes in hot water systems and other pipelines

STUDIES ON THE MINERALIZATION OF WATER INDICATE THE FOLLOWING

1. Major taste effects are produced by anions (where TDS was studied).
2. Chloride produces a taste effect somewhere between the milder sulfate and the stronger carbonate.
3. Laxative effects are caused by highly mineralized waters. Mineralized waters often contain chloride as well as high levels of sodium and magnesium sulfate.

CORROSION EFFECT

1. Studies indicate that corrosion depends on concentration of TDS (TDS may contain 50 percent chloride ions).
2. Domestic plumbing, water heaters, and municipal waterworks equipment will deteriorate when high concentrations of chloride ions are present.

EXAMPLE

Where the TDS = 200 mg/L (Chloride = 100 mg/L), water heater life will range from 10 to 13 years. Water heater life declines uniformly as a function of TDS—1 year shortened life per 200 mg/L additional TDS.

22.332 *Color*

The SMCL for color is 15 color units. The level of this water quality indicator is not known to be a measure of the safety of water. However, high color content may indicate:

1. High organic chemical contamination
2. Inadequate treatment
3. High disinfectant demand and the potential for production of excess amounts of disinfection by-products

Color may be caused by:

1. Natural color-causing solids such as aromatic, polyhydroxy, methoxy, and carboxylic acids
2. Fulvic and humic acid fractions
3. Presence of metals such as copper, iron, and manganese

Rapid changes in color levels may provoke more citizen complaints than relatively high, constant color levels.

22.333 *Copper*

The SMCL for copper is 1.0 mg/L. Soft water containing a low level (0.5 mg/L) of copper may cause blue or blue-green staining of porcelain. Higher levels (4 mg/L) of copper will stain clothing and blond hair. When soap is used with water having a copper concentration greater than 1 mg/L, insoluble green curds will form.

22.334 *Corrosivity*

The corrosivity of water indicates the rate at which water causes the gradual decomposition or destruction of a material (such as a metal or cement lining). The severity and type of corrosivity are dependent on the chemical and physical characteristics of the water and the material.

Corrosivity causes materials to deteriorate and go into solution (be carried by the water). Corrosion of toxic metal pipe materials such as lead can create a serious health hazard. Corrosion of iron may produce a flood of unpleasant telephone calls from consumers complaining about rusty water, stained laundry, and bad tastes.

Corrosivity can cause the reduction of the carrying capacity of a water main. This reduced carrying capacity can cause an increase in pump energy costs and may reduce distribution system pressures. Leaks in water mains due to corrosivity may eventually require replacement of a water main.

22.335 Fluoride

Fluoride produces two effects, depending on its concentration. At levels of 6 to 8 mg/L, fluoride may cause skeletal fluorosis (the bones become brittle) and stiffening of the joints. For this reason, fluoride has been added to the list of primary standards (those that have health effects).

At levels of 2 mg/L and greater, fluoride may cause dental fluorosis, which is discoloration and mottling of the teeth, especially in children. EPA has recently reclassified dental fluorosis as a cosmetic effect, raised the primary drinking water standard from 1.4 to 2 mg/L to 4 mg/L, and established a secondary standard of 2 mg/L for fluoride.

22.336 Foaming Agents

The SMCL for foaming agents is 0.5 mg/L.

UNDESIRABLE EFFECTS

1. Causes frothing and foaming, which are associated with contamination (greater than 1.0 mg/L)
2. Imparts an unpleasant taste (oily, fishy, perfume-like) (less than 1.0 mg/L)

INFORMATION ITEMS

1. Because no convenient foamability test exists and because SURFACTANTS[5] are one major class of substances that cause foaming, this property is determined indirectly by measuring the anionic surfactant (MBAS[6]) concentration in the water.
2. Surfactants are synthetic organic chemicals and are the principal ingredient of household detergents.
3. The requirement for biodegradability led to the widespread use of Linear Alkylbenzene Sulfonate (LAS), an anionic surfactant.
4. Concentrations of anionic surfactants found in drinking waters range from 0 to 2.6 mg/L in well supplies and 0 to 5 mg/L in surface water supplies.
5. LAS is essentially odorless. The odor and taste characteristics are likely to arise from the degradation of waste products rather than the detergents.
6. If water contains an average concentration of 10 mg/L surfactants, the water is likely to be entirely of wastewater origin.
7. From a toxicological standpoint, an MCL of 0.5 mg/L, assuming a daily adult human water intake of 2 liters, would give a safety factor of 15,000.

22.337 Iron and Manganese

1. Iron and manganese are frequently found together in natural waters and produce similar adverse environmental effects and color problems. Excessive amounts of iron and manganese are usually found in groundwater and in surface water contaminated by industrial waste discharges.
2. Prior to 1962, both were covered by a single recommended limit.
3. In 1962, the US Public Health Service recommended separate limits for both iron and manganese to reflect more accurately the levels at which adverse effects occur for each.
4. Both are highly objectionable in large amounts in water supplies for either domestic or industrial use.
5. Both impart color to laundered goods and plumbing fixtures.
6. Taste thresholds in drinking water are considerably higher than the levels that produce staining effects.
7. Both are part of our daily nutritional requirements, but these requirements are not met by the consumption of drinking water.

22.338 Iron

The SMCL for iron is 0.3 mg/L.

UNDESIRABLE EFFECTS

1. At levels greater than 0.05 mg/L some color may develop, staining of fixtures may occur, and precipitates may form.
2. The magnitude of the staining effect is directly proportional to the concentration.
3. Depending on the sensitivity of taste perception, a bitter, astringent taste can be detected from 0.1 mg/L to 1.0 mg/L.
4. Precipitates that are formed create not only color problems but also lead to bacterial growth of slimes and of the iron-loving bacteria, *Crenothrix*, in wells and distribution piping.

---

[5] *Surfactant* (sir-FAC-tent). Abbreviation for surface-active agent. The active agent in detergents that possesses a high cleaning ability.
[6] MBAS. Methylene Blue Active Substance. Another name for surfactants or surface active agents. The determination of surfactants is accomplished by measuring the color change in a standard solution of methylene blue dye.

NUTRITIONAL REQUIREMENTS

1. Daily requirement is 1 to 2 mg, but intake of larger quantities is required as a result of poor absorption.

2. The limited amount of iron permitted in water (because of objectionable taste or staining effects) constitutes only a small fraction of the amount normally consumed and does not have toxicologic (poisonous) significance.

22.339 Manganese

The SMCL for manganese is 0.05 mg/L.

UNDESIRABLE EFFECTS

1. A concentration of more than 0.02 mg/L may cause buildup of coatings in distribution piping.

2. If these coatings slough off, they can cause brown blotches in laundry items and black precipitates.

3. Manganese imparts a taste to water above 0.15 mg/L.

4. The application of chlorine, even at low levels, increases the likelihood of precipitation of manganese at low levels.

5. Unless the precipitate is removed, precipitates reaching pipelines will promote bacterial growth.

TOXIC EFFECTS

1. Toxic effects are reported as a result of inhalation of manganese dust or fumes.

2. Liver cirrhosis has arisen in controlled feeding of rats.

3. Neurological effects have been suggested; however, these effects have not been scientifically confirmed.

NUTRITIONAL REQUIREMENTS

1. Daily intake of manganese from a normal diet is about 10 mg.

2. Manganese is essential for proper nutrition.

3. Diets deficient in manganese will interfere with growth, blood and bone formation, and reproduction.

QUESTIONS

Write your answers in a notebook and then compare your answers with those on page 607.

22.3D Why is chloride a secondary contaminant?

22.3E What are surfactants?

22.3F What is the impact of chlorine on manganese?

22.3310 Odor

The SMCL for odor is a *THRESHOLD ODOR NUMBER (TON[7])* of 3. Important facts to remember when dealing with odors include:

1. Taste and odor go hand-in-hand.

2. Absence of taste and odor helps to maintain the consumers' confidence in the quality of their water, even though it does not guarantee that the water is safe.

3. Research indicates that there are only four true taste sensations: sour, sweet, salty, and bitter.

4. All other sensations ascribed to the sense of taste are actually odors even though the sensation is not noticed until the material is taken into the mouth.

5. Odor tests are less fatiguing to people than taste tests when testing for tastes and odors.

6. Taste and odor tests are useful:

   a. As a check on the quality of raw[8] and treated water

   b. To help control odor throughout the plant

7. Odor is a useful test:

   a. For determining the effectiveness of different kinds of treatment

   b. As a means for tracing the source of contaminants

8. Hydrogen sulfide is included under the odor SMCL.

22.3311 pH

The SMCL for pH is defined as pH values beyond the acceptable range from 6.5 to 8.5. A wide range of pH values in drinking water can be tolerated by consumers.

UNDESIRABLE EFFECTS

1. When the pH increases, the disinfection activity of chlorine falls significantly.

2. High pH may cause increased halogen reactions, which produce chloroform and other trihalomethanes during chlorination.

3. Both excessively high and low pHs may cause increased corrosivity, which can, in turn, create taste problems, staining problems, and significant health hazards.

4. Metallic piping in contact with low-pH water will impart a metallic taste.

5. If the piping is iron or copper, high pH will cause oxide and carbonate compounds to be deposited, leaving red or green stains on fixtures and laundry.

---

[7] *Threshold Odor Number (TON).* The greatest dilution of a sample with odor-free water that still yields a just-detectable odor.
[8] When testing raw water, be sure there are no pathogens or toxic chemicals present.

6. At a high pH, drinking water acquires a bitter taste.

7. The high degree of mineralization often associated with basic waters results in encrustation of water pipes and water-using appliances.

22.3312 Silver

Silver is a nonessential element, providing no beneficial effects from its ingestion in trace amounts. Chronic toxicity causes an unsightly blue-gray discoloration of the skin, mucous membranes, and eyes. Apparently, beside cosmetic changes, there are no physiologic effects. Ingestion of trace amounts of silver or silver salts results in its accumulation in the body, particularly the skin and eyes. There is some evidence that changes to the kidneys, liver, and spleen can occur. The SMCL for silver is 0.10 mg/L.

22.3313 Sulfate

The SMCL for sulfate is 250 mg/L.

UNDESIRABLE EFFECTS AT HIGH LEVELS

1. Tends to form hard scales in boilers and heat exchangers.

2. Causes taste effects.

3. Causes a laxative effect. This effect is commonly noted by newcomers or casual or intermittent users of water high in sulfate. Water containing more than 750 mg/L of sulfate usually produces the laxative effect while water with less than 600 mg/L sulfate usually does not. An individual can adjust to sulfate in drinking water.

4. Sodium sulfate and magnesium sulfate are more active as laxatives, whereas calcium sulfate is less active.

5. When the magnesium sulfate content is 200 mg/L, the most sensitive person will feel the laxative effect; however, magnesium sulfate levels between 500 mg/L and 1,000 mg/L will induce diarrhea in most individuals.

6. Tastes may sometimes be detected at 200 mg/L of sulfate, but generally are detected in the range of 300 to 400 mg/L.

22.3314 Total Dissolved Solids (TDS)

The SMCL for total dissolved solids is 500 mg/L.

UNDESIRABLE EFFECTS

1. TDS imparts adverse taste effects at greater than 500 mg/L.

2. Highly mineralized water influences the deterioration of distribution systems as well as domestic plumbing and appliances (the life of a water heater will decrease one year with each additional 200 mg/L of TDS above a typical 200 mg/L value).

3. Mineralization can also cause precipitates to form in boilers and other heating units, sludge in freezing processes, rings on utensils, and precipitates in food being cooked.

4. There may be a great difference between a detectable concentration and an objectionable concentration of the neutral salts. Many people can become accustomed to high levels.

5. Studies show that the temperature of mineralized waters influences their acceptability to the public.

22.3315 Zinc

The SMCL for zinc is 5 mg/L.

UNDESIRABLE EFFECTS

1. High concentrations of zinc produce adverse physiological effects.

2. Zinc imparts a bitter, astringent taste that is distinguishable at 4 mg/L. Also, at 4 mg/L a metallic taste will exist.

3. Zinc will cause a milky appearance in water at 30 mg/L.

4. Zinc may increase lead and cadmium concentrations.

5. The activity of several enzymes is dependent on zinc.

6. Cadmium and lead are common contaminants of zinc used in galvanizing steel pipe.

PHYSIOLOGICAL EFFECTS

1. A concentration of 30 mg/L can cause nausea and fainting.

2. Zinc salts act as gastrointestinal irritants. This symptom of illness is acute and transitory.

3. The vomiting concentration range is 675 to 2,280 mg/L.

4. A wide margin of safety exists between normal food intake and concentrations in water high enough to cause oral toxicity.

DIETARY REQUIREMENTS

1. The daily requirement for preschool children is 0.3 mg Zn/kg of weight.

2. Total zinc in an adult human body averages two grams.

3. Zinc most likely concentrates in the retina of the eye and in the prostate.

4. Zinc deficiency in animals leads to growth retardation.

QUESTIONS

Write your answers in a notebook and then compare your answers with those on page 607.

22.3G   What are the undesirable effects of abnormal pH values?

22.3H   Why are high levels of sulfate undesirable in drinking water?

22.3I   Why are high levels of zinc undesirable in drinking water?

22.4 SAMPLING PROCEDURES

22.40 Safe Drinking Water Act Regulations

The Safe Drinking Water Act and accompanying regulations require that you must take the following actions to comply:

1. Sampling
2. Testing
3. Recordkeeping
4. Reporting

Understanding and implementing each of these steps will help ensure the success of your operation.

The poster provided with this manual summarizes the essential elements of drinking water regulations, including the MCLs and monitoring requirements.

Operators are urged to stay in close contact with their state regulatory agencies and become thoroughly familiar with their state requirements. Although the poster is revised frequently and every effort is made to ensure accuracy at the time of publication, it is extremely difficult to keep pace with changing regulations. An excellent source of up-to-the-minute information about drinking water regulations is EPA's toll-free Safe Drinking Water Hotline at (800) 426-4791.

22.41 Initial Sampling

"Initial sampling" refers to the very first sampling conducted under the SDWA for each of the applicable contaminant categories. The timing of this sampling depends on:

1. The type of contaminant being monitored
2. Whether the system is a community or noncommunity water system
3. Whether the water source is a surface water or groundwater supply

22.42 Routine Sampling

Routine sampling refers to sampling repeated on a regular basis. Coliform sampling must be conducted in accordance with a written sampling plan that has been approved by the state regulatory agency. The poster provided with this manual summarizes the routine sampling requirements for each contaminant category.

22.43 Check and Repeat Sampling

Whenever an initial or routine sample analysis indicates that an MCL has been exceeded, check sampling is required to confirm the routine sampling results. Check sampling is in addition to the routine sampling program. Although check sampling cannot be scheduled in advance, there are specific check sampling procedures to follow. The number of samples, sampling points, and frequency of sampling vary according to the particular contaminant. For example, the regulations specify that wherever coliform bacteria repeat samples are required, at least one of the repeat samples must be taken from the same sampling point as the original and two of the repeat samples must be collected from nearby points upstream and downstream of the original.

22.44 Sampling Points

Some of the samples required to determine compliance with the primary regulations can be taken from the routine sampling points. By coordinating the present sampling points with the sampling program required by the regulations, additional monitoring costs can be minimized. The poster summarizes what, where, and how often you need to sample for all types of water systems (defined in Section 22.15, "Types of Water Systems"). The number of sampling points required will depend on the specific size of the population served and layout of each water system.

Samples for organics, inorganics, and turbidity must be taken at the points where water enters the distribution system, and samples collected for coliform bacteria must be taken at representative points within the distribution system.

At the very minimum, a small system (with a population of 25 to 1,000) must sample for turbidity and coliform bacteria and also must have two sampling points:

1. One where the water enters the distribution system
2. One at a consumer faucet at a representative point in the distribution system

QUESTIONS

Write your answers in a notebook and then compare your answers with those on page 607.

22.4A What do the words "Initial Sampling" mean?

22.4B What is routine sampling?

22.4C What is check (repeat) sampling?

22.4D What are the minimum sampling requirements for a small system with a population of 100 people?

22.45 Sampling Point Selection

The two major considerations in determining the number and location of sampling points are that they should be:

1. Representative of each different surface water source entering the system
2. Representative of conditions within the system such as dead ends, loops, storage facilities, and pressure zones

22.46 Sampling Schedule

A sampling schedule should be prepared that indicates all of the samples that will be collected during a yearly period. The schedule should include the following information:

1. Sampling frequency
2. Sampling point designation

3. Location (address)
4. Type of test
5. Sample volume
6. Special handling instructions

This schedule should be reviewed with your state regulatory agency to determine adequacy to meet the SDWA regulations. A written sampling plan for coliform testing must be approved by the state regulatory agency.

22.47 Sampling Route

After selection of the sampling points and preparation of the sampling schedule, the next step is to select a route. Arrange your route so that samples that must be analyzed immediately are not delayed while other sampling is done. Field data forms must be completed by the person doing the sampling and submitted to the laboratory with the samples.

22.48 Sample Collection

Good sampling techniques are the key to a meaningful and useful sampling program. The following eight steps are necessary to the collection of an acceptable sample:

1. Obtain a sample that is truly representative of the existing conditions.
2. Flush the line before sample collection (except when a first-draw sample is required, such as in testing for lead).
3. Fill the sample bottle without leaving any air pocket.
4. Analyze residual chlorine when the sample is taken.
5. Handle and store the sample so that it does not become contaminated before it reaches the laboratory.
6. Use preservation techniques. These preservation methods are generally:
   a. pH control
   b. Refrigeration
7. Label the sample immediately and keep accurate records of every sample collected including:
   a. Date and time sampled
   b. Location sampled
   c. Name of sample collected
   d. Bottle number
   e. Type of sample
   f. Name of person collecting sample. Any person collecting samples should be required to complete a form providing the above information at the time of sample collection. This form should be supplied by the laboratory.
8. Keep the time between the collection of the sample and analysis as short as possible.

QUESTIONS

Write your answers in a notebook and then compare your answers with those on page 608.

22.4E List the information that should be included in a sampling schedule.

22.4F List the elements necessary to the collection of an acceptable sample.

22.5 REPORTING PROCEDURES

The primary purpose of the SDWA is to protect the public's health. Two general categories of reporting are called for by the act:

1. Reporting to the public (public notification)
2. Reporting to the state

There are four types of reports that must be sent to the state:

1. Initial reports
2. Routine sampling reports
3. Check or repeat sampling reports
4. Violation reports

Tables 22.7 through 22.15 outline reporting procedures for various contaminants.

22.6 NOTIFICATION REQUIREMENTS

Public notification regulations were developed to make sure that consumers will be informed whenever there is a serious problem with their drinking water. Operators are urged to contact their state regulatory agency to determine how the public notification regulations affect their water system. Also, contact the EPA to request the guidance documents it has developed for implementation of the public notification regulations.

The regulations define three levels (tiers) of violations and specify which groups of consumers must be notified in each instance, what type of notice (electronic broadcast, print) must be provided, and what information the notice must contain.

All notices should be written in language that is free of technical jargon, printed in a type size that is easy to read, and published or posted in a conspicuous location. In addition, copies of all public notices must be sent to the state regulatory agency within 10 days after they are issued.

Tier 1 Violations

Tier 1 violations are any violations or situations in which even a brief violation or short-term exposure to a contaminant has the potential to cause a serious, negative effect on human health. Some examples of Tier 1 violations are:

- Exceeding the total coliform MCL when fecal coliforms or *E. coli* are present in the water distribution system *or* failure to test for fecal coliforms or *E. coli* after any repeat sample tests positive for coliforms.

**TABLE 22.7 REPORTING PROCEDURES—
INORGANIC CHEMICALS (EXCEPT NITRATE) AND ORGANIC CHEMICALS**

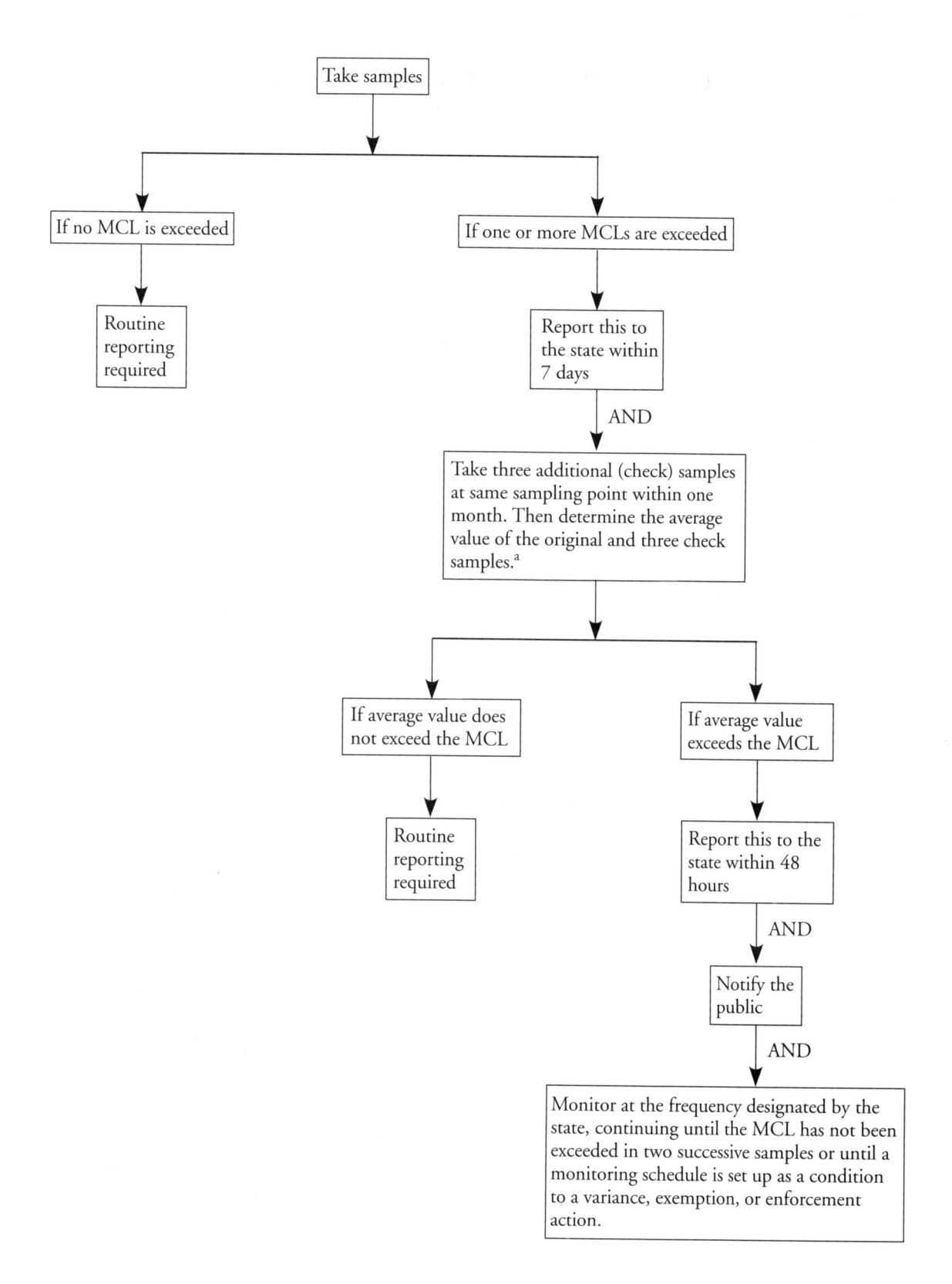

```
                          ┌─────────────┐
                          │ Take samples│
                          └──────┬──────┘
                   ┌─────────────┴─────────────┐
                   ▼                           ▼
         ┌──────────────────┐       ┌──────────────────────────┐
         │ If no MCL is     │       │ If one or more MCLs are  │
         │ exceeded         │       │ exceeded                 │
         └────────┬─────────┘       └────────────┬─────────────┘
                  ▼                              ▼
         ┌──────────────┐              ┌──────────────────┐
         │ Routine      │              │ Report this to   │
         │ reporting    │              │ the state within │
         │ required     │              │ 7 days           │
         └──────────────┘              └────────┬─────────┘
                                                │ AND
                                                ▼
                                  ┌──────────────────────────────┐
                                  │ Take three additional (check)│
                                  │ samples at same sampling     │
                                  │ point within one month. Then │
                                  │ determine the average value  │
                                  │ of the original and three    │
                                  │ check samples.[a]            │
                                  └──────────────┬───────────────┘
                              ┌──────────────────┴──────────────────┐
                              ▼                                     ▼
                    ┌──────────────────┐                  ┌──────────────────┐
                    │ If average value │                  │ If average value │
                    │ does not exceed  │                  │ exceeds the MCL  │
                    │ the MCL          │                  │                  │
                    └────────┬─────────┘                  └────────┬─────────┘
                             ▼                                     ▼
                    ┌──────────────┐                     ┌──────────────────┐
                    │ Routine      │                     │ Report this to   │
                    │ reporting    │                     │ the state within │
                    │ required     │                     │ 48 hours         │
                    └──────────────┘                     └────────┬─────────┘
                                                                  │ AND
                                                                  ▼
                                                         ┌──────────────┐
                                                         │ Notify the   │
                                                         │ public       │
                                                         └──────┬───────┘
                                                                │ AND
                                                                ▼
                                              ┌────────────────────────────────┐
                                              │ Monitor at the frequency       │
                                              │ designated by the state,       │
                                              │ continuing until the MCL has   │
                                              │ not been exceeded in two       │
                                              │ successive samples or until a  │
                                              │ monitoring schedule is set up  │
                                              │ as a condition to a variance,  │
                                              │ exemption, or enforcement      │
                                              │ action.                        │
                                              └────────────────────────────────┘
```

[a] Average Value = $\dfrac{\text{Total of Original Sample} + 3 \text{ Check Samples}}{4}$

TABLE 22.8 REPORTING PROCEDURES—NITRATE

```
                         ┌─────────────┐
                         │ Take samples│
                         └──────┬──────┘
                    ┌───────────┴───────────┐
                    ▼                       ▼
        ┌───────────────────────┐  ┌───────────────────────┐
        │ If the MCL is not     │  │ If the MCL is exceeded│
        │ exceeded              │  │                       │
        └───────────┬───────────┘  └───────────┬───────────┘
                    ▼                          ▼
           ┌─────────────────┐       ┌──────────────────────────┐
           │ Routine         │       │ An additional (check)    │
           │ reporting       │       │ sample must be taken     │
           │ required        │       │ within 24 hours          │
           └─────────────────┘       └────────────┬─────────────┘
                                ┌─────────────────┴─────────────────┐
                                ▼                                   ▼
                   ┌──────────────────────────┐      ┌──────────────────────────┐
                   │ If the average (mean)    │      │ If the average (mean)    │
                   │ of original and check    │      │ of original and check    │
                   │ sample does not          │      │ sample does exceed       │
                   │ exceed the MCL           │      │ the MCL                  │
                   └────────────┬─────────────┘      └────────────┬─────────────┘
                                ▼                                 ▼
                       ┌─────────────────┐              ┌──────────────────────┐
                       │ Routine         │              │ Report this to the   │
                       │ reporting       │              │ state within 24      │
                       │ required        │              │ hours                │
                       └─────────────────┘              └──────────┬───────────┘
                                                                   │ AND
                                                                   ▼
                                                          ┌────────────────┐
                                                          │ Notify the     │
                                                          │ public         │
                                                          └────────┬───────┘
                                                                   │ AND
                                                                   ▼
                                           ┌──────────────────────────────────────────┐
                                           │ Monitor at the frequency designated by   │
                                           │ the state, continuing until the MCL has  │
                                           │ not been exceeded in two successive      │
                                           │ samples or until a monitoring schedule   │
                                           │ is set up as a condition to a variance,  │
                                           │ exemption, or enforcement action.        │
                                           └──────────────────────────────────────────┘
```

TABLE 22.9 REPORTING PROCEDURES—TURBIDITY MONITORING
(Surface Water Using Conventional or Direct Filtration)

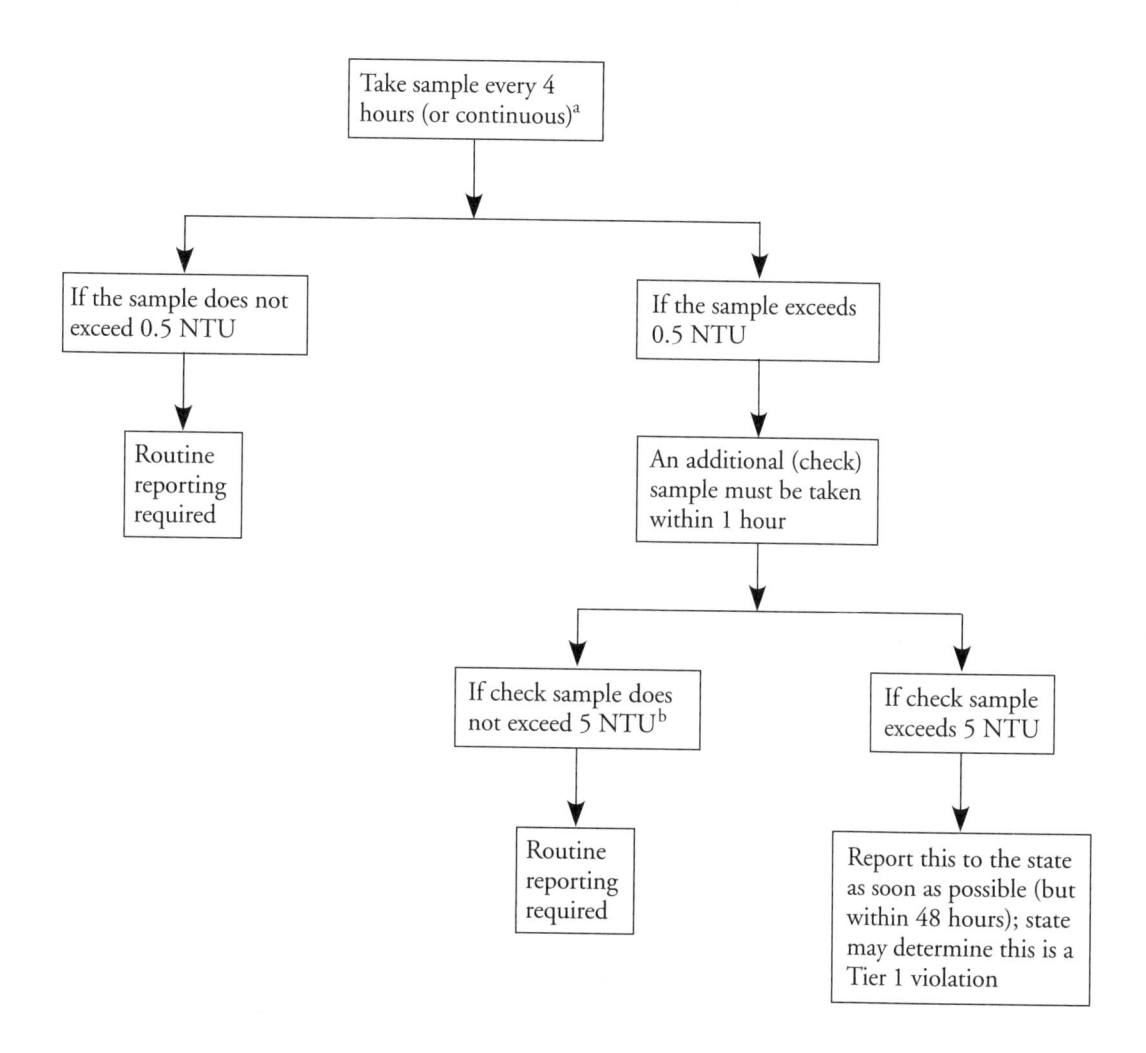

```
                    ┌─────────────────────┐
                    │ Take sample every 4 │
                    │ hours (or continuous)[a]│
                    └──────────┬──────────┘
                  ┌────────────┴────────────┐
                  ▼                         ▼
        ┌──────────────────┐      ┌──────────────────┐
        │ If the sample does│      │ If the sample    │
        │ not exceed 0.5 NTU│      │ exceeds 0.5 NTU  │
        └─────────┬─────────┘      └─────────┬────────┘
                  ▼                          ▼
         ┌───────────────┐        ┌─────────────────────┐
         │ Routine       │        │ An additional (check)│
         │ reporting     │        │ sample must be taken │
         │ required      │        │ within 1 hour        │
         └───────────────┘        └──────────┬──────────┘
                              ┌──────────────┴──────────────┐
                              ▼                             ▼
                  ┌────────────────────┐      ┌──────────────────┐
                  │ If check sample does│      │ If check sample  │
                  │ not exceed 5 NTU[b] │      │ exceeds 5 NTU    │
                  └──────────┬─────────┘      └─────────┬────────┘
                             ▼                          ▼
                    ┌───────────────┐         ┌────────────────────┐
                    │ Routine       │         │ Report this to the │
                    │ reporting     │         │ state as soon as   │
                    │ required      │         │ possible (but      │
                    └───────────────┘         │ within 48 hours);  │
                                              │ state may determine│
                                              │ this is a Tier 1   │
                                              │ violation          │
                                              └────────────────────┘
```

[a] Less frequent sampling may be established at state option.
[b] Filter effluent turbidity must be 1 NTU in 95% of samples.

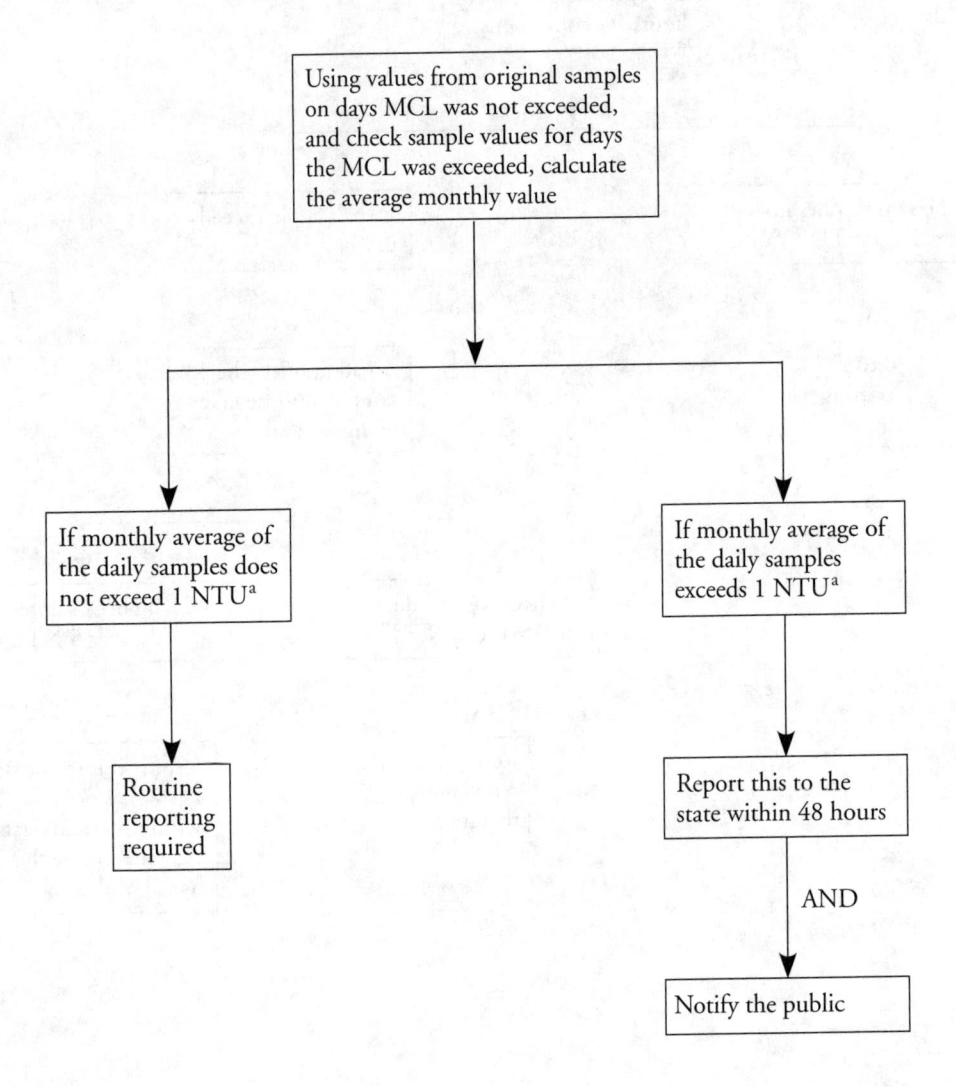

**TABLE 22.10   REPORTING PROCEDURES—
WHEN CALCULATING MONTHLY AVERAGE TURBIDITY VALUE**

```
┌─────────────────────────────────────┐
│ Using values from original samples  │
│ on days MCL was not exceeded,       │
│ and check sample values for days    │
│ the MCL was exceeded, calculate     │
│ the average monthly value           │
└─────────────────────────────────────┘
                 │
      ┌──────────┴──────────┐
      ▼                     ▼
┌──────────────┐     ┌──────────────┐
│ If monthly   │     │ If monthly   │
│ average of   │     │ average of   │
│ the daily    │     │ the daily    │
│ samples does │     │ samples      │
│ not exceed   │     │ exceeds      │
│ 1 NTU[a]     │     │ 1 NTU[a]     │
└──────────────┘     └──────────────┘
      │                     │
      ▼                     ▼
┌──────────┐         ┌────────────────┐
│ Routine  │         │ Report this to │
│ reporting│         │ the state      │
│ required │         │ within 48 hours│
└──────────┘         └────────────────┘
                             │ AND
                             ▼
                     ┌────────────────┐
                     │ Notify the     │
                     │ public         │
                     └────────────────┘
```

[a] MCL of 5 NTU may be established at state option.

Drinking Water Regulations

TABLE 22.11 REPORTING PROCEDURES—MICROBIAL CONTAMINANTS—TOTAL COLIFORM RULE

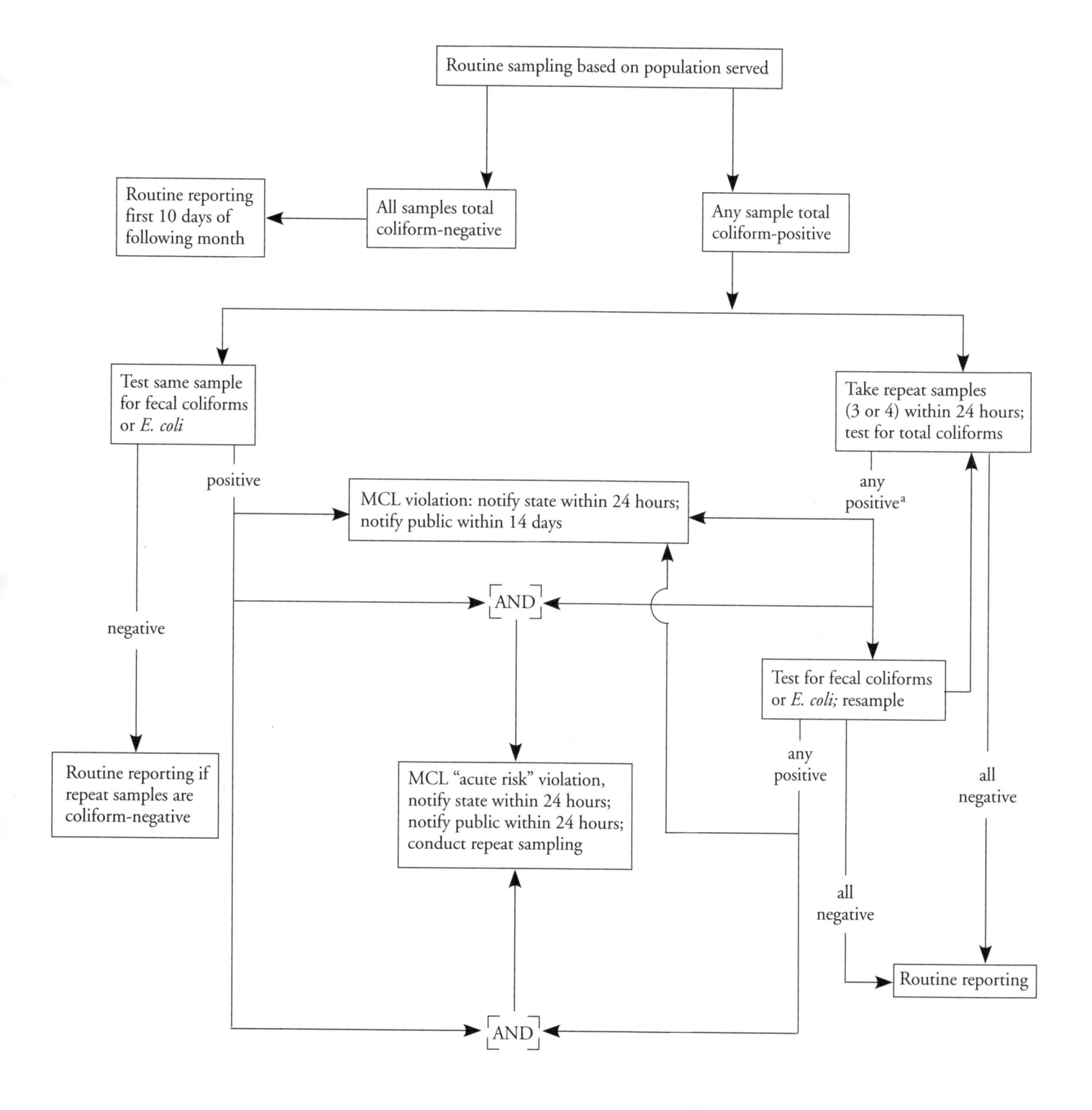

```
                    Routine sampling based on population served
                           │                        │
                           ▼                        ▼
  Routine reporting ◄── All samples total     Any sample total
  first 10 days of       coliform-negative    coliform-positive
  following month
                    ┌──────────────────────────────┬──────────────┐
                    ▼                                              ▼
          Test same sample                              Take repeat samples
          for fecal coliforms                           (3 or 4) within 24 hours;
          or E. coli                                    test for total coliforms
              │                                                    │
         positive                                            any positive[a]
              ├──► MCL violation: notify state within 24 hours; ◄──┤
              │    notify public within 14 days
              │
              ├────────────► [AND] ◄────────────────────────────────┤
         negative                                                    │
              │                                              Test for fecal coliforms
              ▼                                              or E. coli; resample
       Routine reporting if                                         │
       repeat samples are              MCL "acute risk" violation,   ├── any positive
       coliform-negative               notify state within 24 hours; │
                                       notify public within 24 hours;├── all negative
                                       conduct repeat sampling       │
                                                                     ▼
                                                             Routine reporting
                                       
                              [AND]
```

[a] For systems collecting more than 40 samples per month, MCL is violated when more than 5% are total coliform-positive. For systems collecting fewer than 40 samples per month, MCL is violated when more than one sample is total coliform-positive. Monthly calculations must include repeat samples.

TABLE 22.12 REPORTING PROCEDURES—MICROBIAL CONTAMINANTS—CHLORINE RESIDUAL

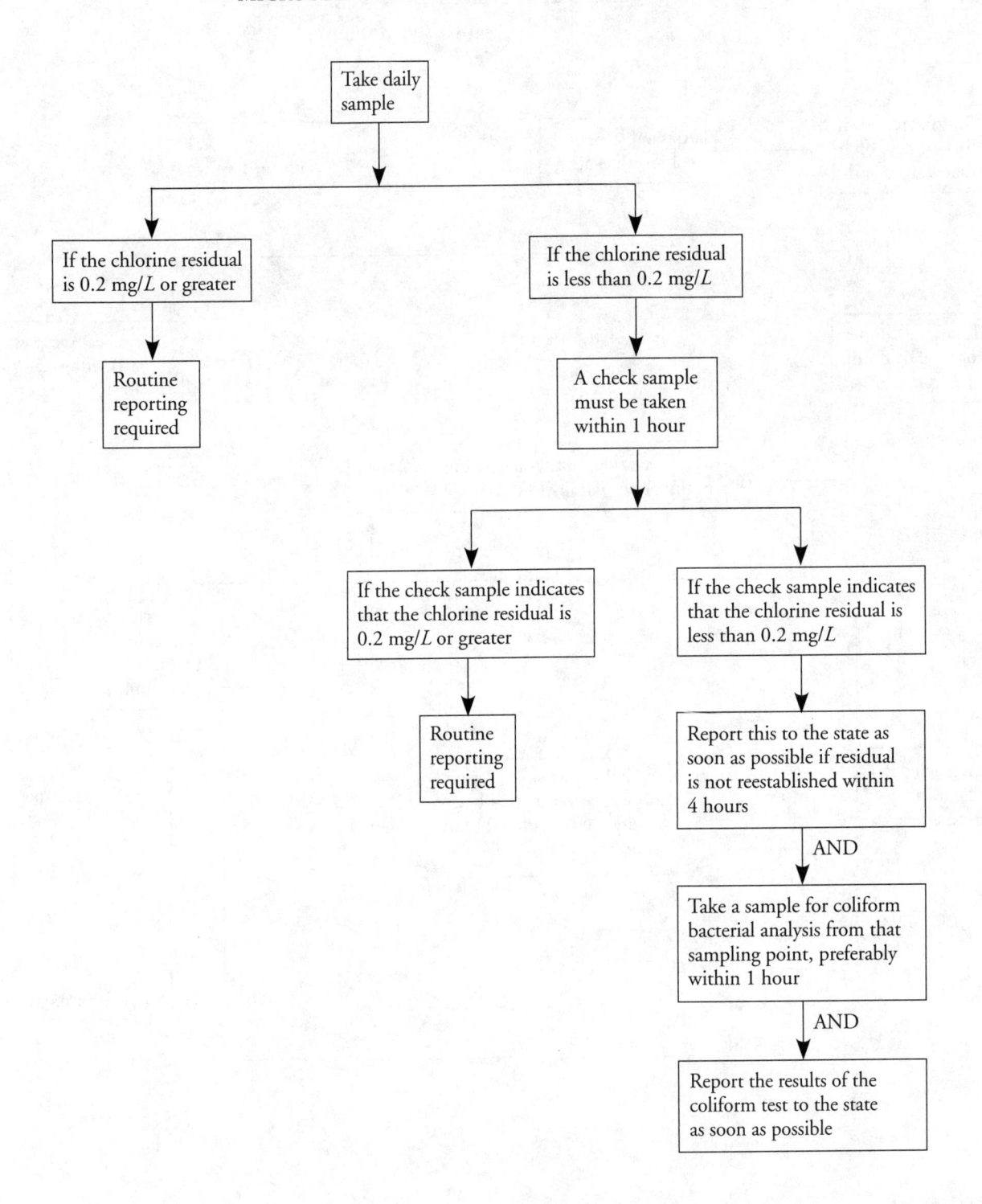

```
                    Take daily
                     sample
                        │
           ┌────────────┴────────────┐
           ▼                         ▼
  If the chlorine residual   If the chlorine residual
  is 0.2 mg/L or greater     is less than 0.2 mg/L
           │                         │
           ▼                         ▼
      Routine                  A check sample
      reporting                must be taken
      required                 within 1 hour
                                     │
                         ┌───────────┴───────────┐
                         ▼                       ▼
               If the check sample indicates   If the check sample indicates
               that the chlorine residual is   that the chlorine residual is
               0.2 mg/L or greater             less than 0.2 mg/L
                         │                       │
                         ▼                       ▼
                    Routine              Report this to the state as
                    reporting            soon as possible if residual
                    required             is not reestablished within
                                         4 hours
                                                 │
                                                AND
                                                 ▼
                                         Take a sample for coliform
                                         bacterial analysis from that
                                         sampling point, preferably
                                         within 1 hour
                                                 │
                                                AND
                                                 ▼
                                         Report the results of the
                                         coliform test to the state
                                         as soon as possible
```

Drinking Water Regulations 599

TABLE 22.13 REPORTING PROCEDURES—RADIOLOGICAL CONTAMINANTS—NATURAL
(Test for Gross Alpha Activity)

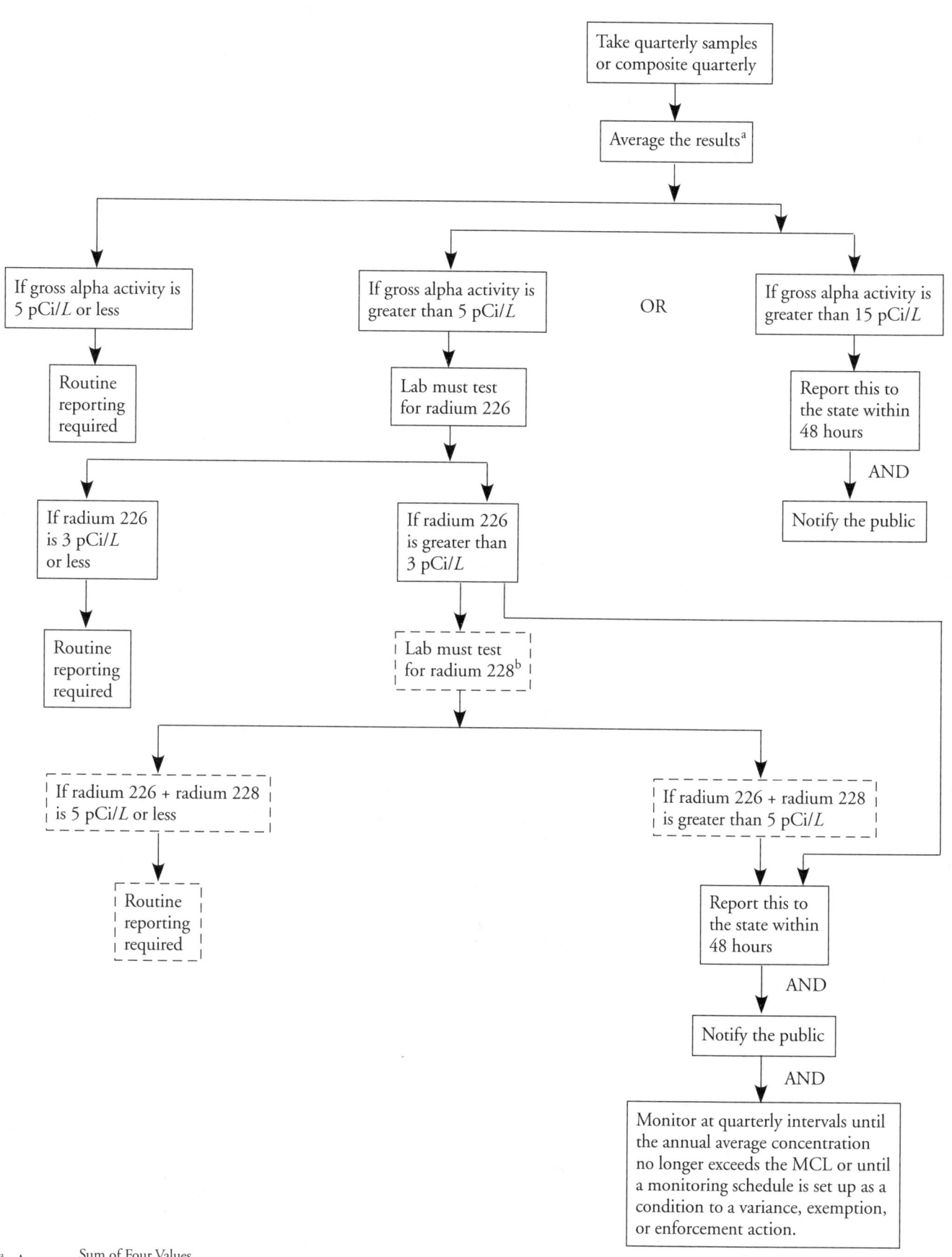

```
                    ┌─────────────────────┐
                    │ Take quarterly      │
                    │ samples or          │
                    │ composite quarterly │
                    └──────────┬──────────┘
                               ▼
                    ┌─────────────────────┐
                    │ Average the results[a] │
                    └──────────┬──────────┘
          ┌────────────────────┼────────────────────┐
          ▼                    ▼                    ▼
  ┌───────────────┐    ┌───────────────┐     ┌───────────────┐
  │ If gross alpha│    │ If gross alpha│ OR  │ If gross alpha│
  │ activity is   │    │ activity is   │     │ activity is   │
  │ 5 pCi/L or    │    │ greater than  │     │ greater than  │
  │ less          │    │ 5 pCi/L       │     │ 15 pCi/L      │
  └───────┬───────┘    └───────┬───────┘     └───────┬───────┘
          ▼                    ▼                     ▼
  ┌───────────────┐    ┌───────────────┐     ┌───────────────┐
  │ Routine       │    │ Lab must test │     │ Report this to│
  │ reporting     │    │ for radium 226│     │ the state     │
  │ required      │    │               │     │ within 48 hrs │
  └───────────────┘    └───────┬───────┘     └───────┬───────┘
                   ┌───────────┴─────────┐          AND
                   ▼                     ▼           ▼
           ┌───────────────┐    ┌───────────────┐  ┌──────────┐
           │ If radium 226 │    │ If radium 226 │  │ Notify   │
           │ is 3 pCi/L or │    │ is greater    │  │ the      │
           │ less          │    │ than 3 pCi/L  │  │ public   │
           └───────┬───────┘    └───────┬───────┘  └──────────┘
                   ▼                    ▼
           ┌───────────────┐    ┌ ─ ─ ─ ─ ─ ─ ─ ┐
           │ Routine       │      Lab must test
           │ reporting     │    │ for radium 228[b]│
           │ required      │    └ ─ ─ ─ ┬ ─ ─ ─ ┘
           └───────────────┘            │
                         ┌──────────────┴──────────────┐
                         ▼                             ▼
             ┌ ─ ─ ─ ─ ─ ─ ─ ─ ─ ─ ─ ┐   ┌ ─ ─ ─ ─ ─ ─ ─ ─ ─ ─ ─ ┐
               If radium 226 + 228        If radium 226 + 228
             │ is 5 pCi/L or less    │   │ is greater than 5 pCi/L│
             └ ─ ─ ─ ─ ┬ ─ ─ ─ ─ ─ ─ ┘   └ ─ ─ ─ ─ ─ ┬ ─ ─ ─ ─ ─ ┘
                       ▼                             ▼
             ┌ ─ ─ ─ ─ ─ ─ ─ ─ ┐           ┌───────────────┐
               Routine                     │ Report this to│
             │ reporting       │           │ the state     │
               required                    │ within 48 hrs │
             └ ─ ─ ─ ─ ─ ─ ─ ─ ┘           └───────┬───────┘
                                                  AND
                                                   ▼
                                          ┌───────────────┐
                                          │ Notify the    │
                                          │ public        │
                                          └───────┬───────┘
                                                  AND
                                                   ▼
                                          ┌───────────────────────┐
                                          │ Monitor at quarterly  │
                                          │ intervals until the   │
                                          │ annual average        │
                                          │ concentration no      │
                                          │ longer exceeds the MCL│
                                          │ or until a monitoring │
                                          │ schedule is set up as │
                                          │ a condition to a      │
                                          │ variance, exemption,  │
                                          │ or enforcement action.│
                                          └───────────────────────┘
```

[a] Average = $\dfrac{\text{Sum of Four Values}}{4}$

No averaging is required if the quarterly samples were composited. In that case, use the results of the single sample.

[b] This step is required only for the initial monitoring period and not for routine monitoring, *except as required by the state*.

600 Water Treatment

TABLE 22.14 REPORTING PROCEDURES—RADIOLOGICAL CONTAMINANTS—MANMADE[a]

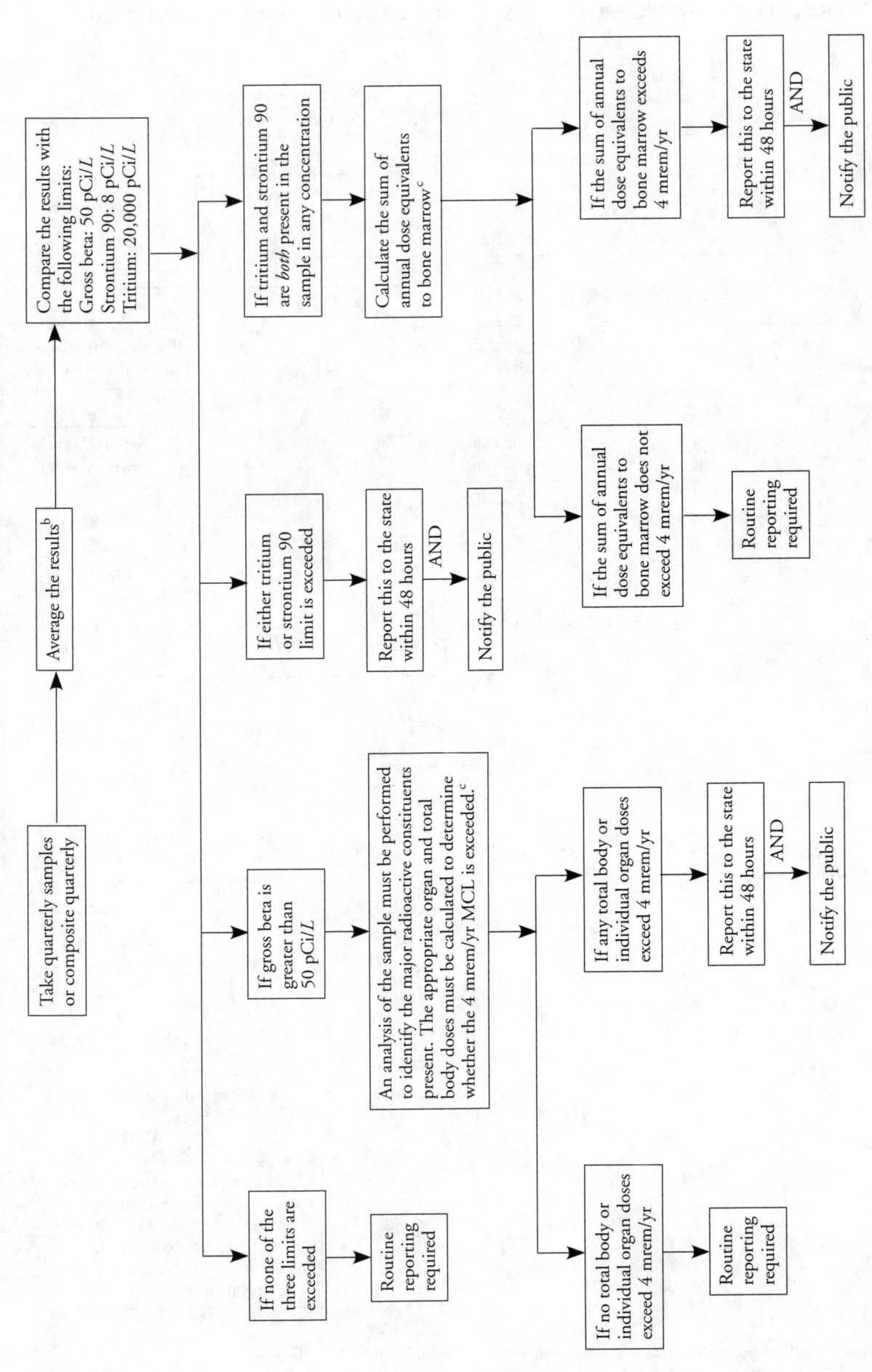

```
Take quarterly samples
or composite quarterly
          │
          ▼
   Average the results[b]
          │
          ▼
Compare the results with
the following limits:
  Gross beta: 50 pCi/L
  Strontium 90: 8 pCi/L
  Tritium: 20,000 pCi/L
```

Branch 1: If tritium and strontium 90 are *both* present in the sample in any concentration → Calculate the sum of annual dose equivalents to bone marrow[c]
 - If the sum of annual dose equivalents to bone marrow exceeds 4 mrem/yr → Report this to the state within 48 hours AND Notify the public
 - If the sum of annual dose equivalents to bone marrow does not exceed 4 mrem/yr → Routine reporting required

Branch 2: If either tritium or strontium 90 limit is exceeded → Report this to the state within 48 hours AND Notify the public

Branch 3: If gross beta is greater than 50 pCi/L → An analysis of the sample must be performed to identify the major radioactive constituents present. The appropriate organ and total body doses must be calculated to determine whether the 4 mrem/yr MCL is exceeded.[c]
 - If any total body or individual organ doses exceed 4 mrem/yr → Report this to the state within 48 hours AND Notify the public
 - If no total body or individual organ doses exceed 4 mrem/yr → Routine reporting required

Branch 4: If none of the three limits are exceeded → Routine reporting required

[a] Applies to surface water systems serving populations of 100,000 or more and systems determined by the state to be vulnerable.

[b] Average = $\dfrac{\text{Sum of Four Values}}{4}$

No averaging is required if the quarterly samples were composited. In that case, use the results of the single sample.

[c] It is likely that the laboratory will not make these calculations. You will probably have to get help from state water supply personnel in making these calculations.

Drinking Water Regulations 601

TABLE 22.15 REPORTING PROCEDURES—TOTAL TRIHALOMETHANES[a]

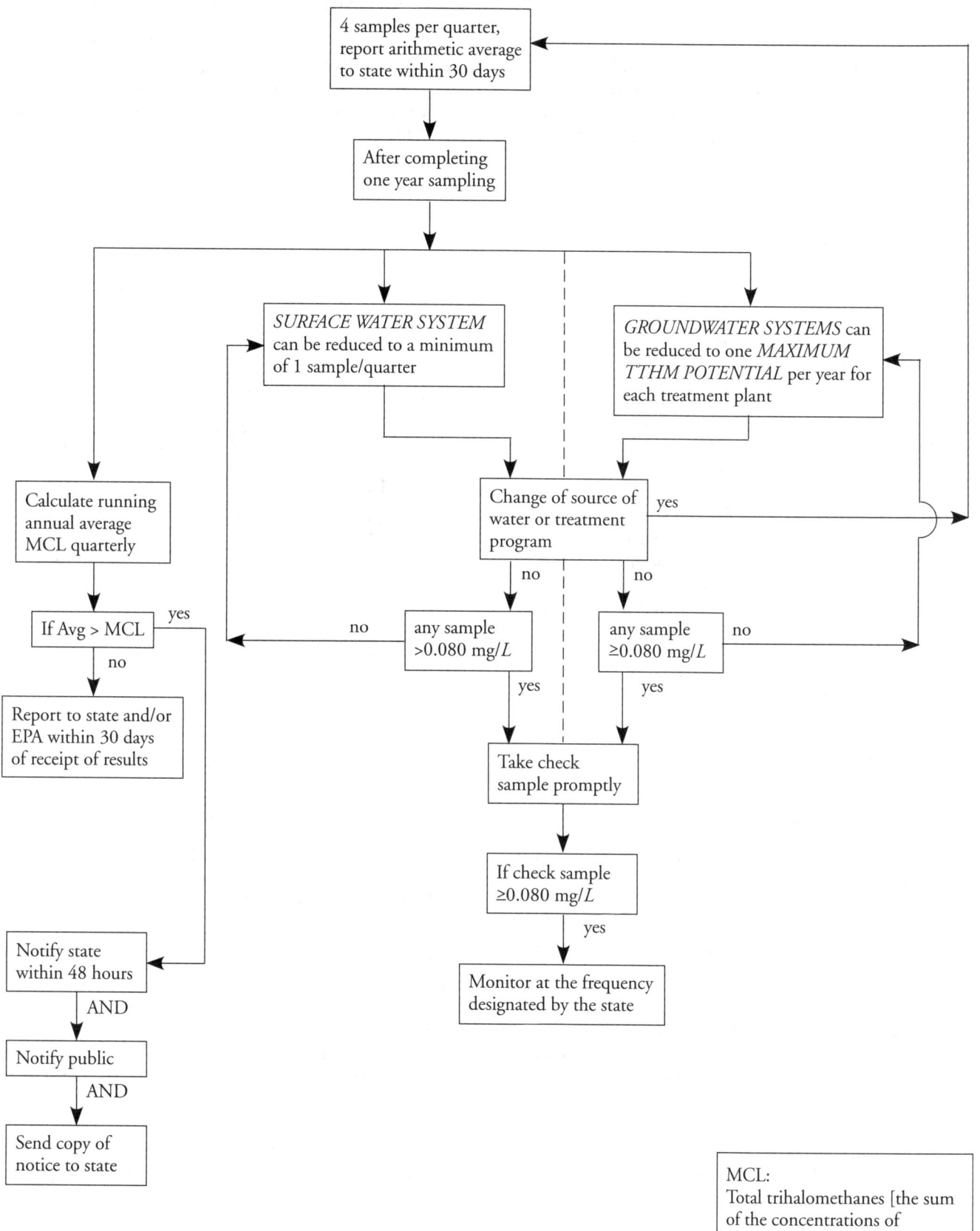

```
                    ┌─────────────────────────┐
                    │ 4 samples per quarter,  │◄──────────────────┐
                    │ report arithmetic average│                   │
                    │ to state within 30 days │                   │
                    └───────────┬─────────────┘                   │
                                ▼                                 │
                    ┌─────────────────────────┐                   │
                    │ After completing        │                   │
                    │ one year sampling       │                   │
                    └───────────┬─────────────┘                   │
                ┌───────────────┴───────────────┐                 │
                ▼                               ▼                 │
   ┌──────────────────────┐         ┌──────────────────────┐     │
   │ SURFACE WATER SYSTEM │         │ GROUNDWATER SYSTEMS  │     │
   │ can be reduced to a  │         │ can be reduced to one│     │
   │ minimum of 1         │         │ MAXIMUM TTHM         │     │
   │ sample/quarter       │         │ POTENTIAL per year   │     │
   │                      │         │ for each treatment   │     │
   │                      │         │ plant                │     │
   └──────────┬───────────┘         └──────────┬───────────┘     │
              ▼                                ▼                 │
              ┌────────────────────────────────┐                 │
              │ Change of source of water or   │── yes ──────────┤
              │ treatment program              │                 │
              └────┬───────────────────────┬───┘                 │
                   no                      no                    │
                   ▼                       ▼                     │
          ┌─────────────────┐     ┌──────────────────┐           │
  ◄─ no ──│ any sample      │     │ any sample       │── no ─────┤
          │ >0.080 mg/L     │     │ ≥0.080 mg/L      │           │
          └────────┬────────┘     └────────┬─────────┘           │
                   yes                     yes                   │
                   └───────────┬───────────┘                     │
                               ▼                                 │
                    ┌──────────────────────┐                     │
                    │ Take check sample    │                     │
                    │ promptly             │                     │
                    └──────────┬───────────┘                     │
                               ▼                                 │
                    ┌──────────────────────┐                     │
                    │ If check sample      │                     │
                    │ ≥0.080 mg/L          │                     │
                    └──────────┬───────────┘                     │
                               yes                               │
                               ▼                                 │
                    ┌──────────────────────┐                     │
                    │ Monitor at the       │                     │
                    │ frequency designated │                     │
                    │ by the state         │                     │
                    └──────────────────────┘                     │

   ┌─────────────────────┐
   │ Calculate running   │
   │ annual average      │
   │ MCL quarterly       │
   └──────────┬──────────┘
              ▼
   ┌─────────────────────┐
   │ If Avg > MCL        │── yes ──┐
   └──────────┬──────────┘         │
              no                   ▼
                         ┌─────────────────────┐
                         │ Report to state     │
                         │ and/or EPA within   │
                         │ 30 days of receipt  │
                         │ of results          │
                         └──────────┬──────────┘
                                    ▼
                         ┌─────────────────────┐
                         │ Notify state        │
                         │ within 48 hours     │
                         └──────────┬──────────┘
                                   AND
                                    ▼
                         ┌─────────────────────┐
                         │ Notify public       │
                         └──────────┬──────────┘
                                   AND
                                    ▼
                         ┌─────────────────────┐
                         │ Send copy of        │
                         │ notice to state     │
                         └─────────────────────┘
```

MCL:
Total trihalomethanes [the sum of the concentrations of bromodichloromethane, dibromochloromethane, tribromomethane (bromoform), and trichloromethane (chloroform)] 0.080 mg/L

[a] Applies to community surface water systems serving populations of 10,000 or more and all surface water systems that meet the SWTR criteria for avoiding filtration.

- Exceeding the MCLs for nitrate, nitrite, or total nitrate and nitrite *or* failure to take a confirmation sample within 24 hours of learning that the MCL has been exceeded.

- Exceeding the maximum residual disinfectant level (MRDL) for chlorine dioxide in a distribution system sample the day after a violation of the MRDL has occurred at the entrance to the distribution system. Failure to collect the required samples in the distribution system is also a Tier 1 violation.

- Violations of the turbidity MCL or treatment technique requirements may be considered Tier 1 violations at the discretion of the primacy agency.

- Occurrence of a waterborne disease outbreak or emergency.

- Any other violations or conditions that the state determines could pose a significant potential to cause adverse health effects upon short-term exposure.

Public notification of Tier 1 violations must take place within 24 hours after the occurrence of the violation. Bill-paying customers of the water system must be notified as well as other people in the service area, even though they may not actually be customers of the water supplier. The precise wording of the notices, called "mandatory health effects language," is provided in the Public Notification Rule. The method of delivery should include at least one of the following three methods: (1) appropriate broadcast media (radio, television), (2) newspapers, (3) posting of notices in conspicuous locations, or (4) hand delivery of notices.

Tier 2 Violations

Tier 2 violations include any other situations or violations not included in the Tier 1 notice category that have the potential to cause negative effects on human health. Some examples of Tier 2 violations include:

- Violation of the MCL, MRDL, or treatment technique requirements except when Tier 1 notice is required

- Violation of monitoring and testing requirements, if directed by the primacy agency

- Failure to comply with the terms of a variance or exemption

- Any other violation that the primacy agency decides is serious enough to warrant Tier 2 notification

Tier 2 notices should be distributed as soon as practical within 30 days and efforts should be made to reach everyone using the water, not just bill-paying customers. Community systems should mail or deliver the notices to their customers and other people in the area; noncommunity systems may mail or deliver the notices to their customers or post notices in conspicuous public places. If notices are posted in public places, they should remain posted for a minimum of at least seven days. After the first seven days, the notices may be removed when the situation has been corrected, but if the situation remains uncorrected, new notices should be posted every 30 days.

It is the responsibility of the public water system to consult with their primacy agency regarding any violation of the turbidity MCL, treatment techniques, or both to allow the primacy agency to determine if the incident is serious enough to require Tier 1 notification procedures.

Tier 3 Violations

Tier 3 violations consist mainly of monitoring and testing procedure violations. The water system must notify its customers and other people in the area within one year of learning of the violation and repeat the notice annually for as long as the violation continues. As with Tier 2 notices, community systems may mail or deliver the notices to their customers and others in the area; noncommunity systems may mail or deliver the notices to their customers or post notices in conspicuous public places.

22.7 RECORDKEEPING

Water suppliers are required to keep certain information on file, as follows:

- Bacteriological results: five years

- Chemical results: ten years

- Actions taken to correct violations: three years after the action was taken

- Sanitary survey reports: ten years

- Variance or Exemption records: five years

22.8 CONSUMER CONFIDENCE REPORTS (CCRs)

EPA has developed regulations requiring every community water system to prepare and distribute to its customers an annual Consumer Confidence Report (CCR). The reports are an opportunity for positive communication with the utility's customers and an effective means for a water utility to inform consumers that their water is safe to drink. CCRs also provide an opportunity to convince consumers of the importance of paying for good-quality water and the need for sufficient funds to properly operate and maintain the water supply, treatment, and distribution systems. If higher rates are necessary to fund a capital improvement program, the CCR can explain the importance of having sufficient water with adequate pressure and high quality.

Items that should be covered in Consumer Confidence Reports include:

1. The name(s), location(s), and type(s) of source water.

2. EPA's definitions of Maximum Contaminant Level Goal (MCLG) and Maximum Contaminant Level (MCL).

3. EPA's definitions of "treatment technique" and "action level" (only required if the CCR contains data on a contaminant for which EPA has specified use of a treatment technique or action level).

4. EPA's specific descriptions of health effects for any regulated contaminant for which there was a violation of the MCL during the year covered by the report. The rule lists specific instructions about how the data should be reported and explained for turbidity, lead and copper, total and fecal coliform, *Cryptosporidium,* radon, arsenic, nitrate, and any other contaminants. The CCR must also include EPA's specific warnings to people who may be more vulnerable to contaminants than the general population.

5. The most likely source(s) of detected contaminants. Information from sanitary surveys or source water assessments should be used, if available and appropriate. If this type of site-specific information is not available, the operator should include one or more of the typical sources for that contaminant listed in the CCR Rule.

6. Any violation(s) of the National Primary Drinking Water Regulations that occurred during the year covered by the CCR must be identified in the report. For certain violations, such as a surface water system's failure to install filtration or disinfection, failure to comply with lead and copper requirements, or violation of treatment technique requirements for acrylamide and epichlorohydrin, the EPA-prescribed wording must be used to describe the violation.

7. If the system is operating under a variance or exemption, the CCR must include EPA's definition of "variances and exemptions," the reasons for the variance or exemption, the date it was granted, and a status report on the steps the system is taking to comply with the terms and schedules of the variance or exemption.

8. Notification that the presence of contaminants in drinking water does not necessarily mean the drinking water poses a health risk. Consumers should be advised that they can call the Safe Drinking Water Hotline (800) 426-4791 for more information about contaminants and their potential health effects.

9. Telephone number of the owner, operator, or other authorized person customers can call for additional information about the CCR.

10. If the state primacy agency determines there is a large proportion of non-English-speaking residents in an area, the CCR must use appropriate language(s) to inform consumers about the importance of the report and tell them how to obtain a translated copy or how to get assistance in the appropriate language.

11. Information about how the public may participate in decisions affecting water quality (for example, time and place of regularly scheduled board meetings).

12. Any other information the water utility wishes to communicate to its customers regarding issues covered in the CCR.

Some utility agencies try to have an article published in the local newspaper explaining the report before it is made available to consumers. This advance information helps the consumers understand the report. Reports should be a short, concise letter report of one or two pages that can be mailed directly to consumers or mailed in the envelope with the utility bill. All utility staff should be familiar with the contents of the report because consumers who know utility personnel frequently will ask personnel questions about the report. Water utilities should emphasize the good job they are doing as guardians of health in these Consumer Confidence Reports.

The CCR regulations not only require water utilities to send annual reports to their direct customers, but also to make a reasonable attempt to reach other people who regularly use water from the system, such as renters and people who work in the area but live elsewhere. A wide variety of methods are available for reaching these individuals, including the following: posting the report on the Internet, mailing the report to postal patrons in metropolitan areas, announcing availability of the report in the news media, publication in a local newspaper, posting in public places, delivery of multiple copies to apartment buildings or large private employers, and delivery of copies to community organizations.

Large water systems (100,000 or more) must post their current year's report on the Internet, and all systems must make their reports available to the public upon request. The utility should send a copy to the state primacy agency at the same time it distributes copies to its customers and should keep a copy on file for five years. The deadline for state primacy agencies to adopt CCR regulations was August 19, 2000.

The CCR regulations give state governors authority to modify some of the notification requirements for small water systems serving fewer than 10,000 persons. Instead of mailing the CCR directly to customers, these systems may be permitted to advise their customers that the CCR will be published in one or more local newspapers and that a copy of the report is available from the utility upon request.

Very small community water systems (500 persons or fewer) that are not required to meet the direct mailing requirements still must prepare a CCR annually but may make it available to customers on request. In such cases, the water utility still must advise its customers (by mail, door-to-door notices, posted signs, or other means authorized by the regulations) that the report is available.

Contact your local or state drinking water supply agency for suggestions and details on how to prepare a Consumer Confidence Report. EPA, AWWA, and Rural Water all have examples of typical Consumer Confidence Reports that can serve as a guide for the preparation of a Consumer Confidence Report for your drinking water system.

QUESTIONS

Write your answers in a notebook and then compare your answers with those on page 608.

22.5A What are the two general categories of reporting called for by the SDWA?

22.5B What are the four types of reports that must be sent to the state?

22.6A If a Tier 1 violation occurs, by what means is the public notified?

22.8A Why should all utility staff be familiar with the contents of the consumer confidence report?

END OF LESSON 2 OF 2 LESSONS

on

DRINKING WATER REGULATIONS

Please answer the discussion and review questions next.

DISCUSSION AND REVIEW QUESTIONS

Chapter 22. DRINKING WATER REGULATIONS

(Lesson 2 of 2 Lessons)

Write the answers to these questions in your notebook. The question numbering continues from Lesson 1.

14. What do secondary drinking water regulations control?
15. Why are secondary drinking water regulations important?
16. How may color be caused in water?
17. Why are iron and manganese undesirable in drinking water?
18. Why is the absence of tastes and odors in drinking water important?
19. Why is hydrogen sulfide not listed under the Secondary Drinking Water Standards?
20. Why are high levels of total dissolved solids undesirable in drinking water?
21. How is the number of sampling points determined?
22. What are the major considerations in determining the number and location of sampling points?
23. How is a sampling route selected?
24. How can samples be preserved?
25. What communication opportunities are provided by Consumer Confidence Reports?

SUGGESTED ANSWERS
Chapter 22. DRINKING WATER REGULATIONS

ANSWERS TO QUESTIONS IN LESSON 1

Answers to questions on page 560.

22.0A The first drinking water standards were designed to control waterborne bacteria and viruses that cause diseases such as cholera, typhoid, and dysentery.

22.0B Operators should develop good working relationships with their local regulatory agencies in order to keep up to date on the latest changes in regulations that apply to their water treatment system.

Answers to questions on page 561.

22.1A The five types of primary contaminants considered to be of public health importance are:

1. Inorganic contaminants
2. Organic contaminants
3. Turbidity
4. Microbial contaminants
5. Radiological contaminants

22.1B To regulate a substance, EPA must demonstrate that the contaminant meets three criteria:

1. The contaminant has an adverse effect on human health
2. It occurs, or is likely to occur, in public water systems at a frequency and concentration of significance to public health
3. Regulation of the contaminant offers a meaningful opportunity to reduce health risks for people served by public water systems

22.1C EPA requires water systems to monitor some unregulated contaminants to assist the agency in determining the extent of contamination by these substances.

Answers to questions on page 564.

22.1D The public health benefits from the Arsenic Rule include the avoidance of bladder and lung cancers and a reduction in the frequency of noncarcinogenic diseases.

22.1E The purpose of the Total Coliform Rule (TCR) is to improve public health protection by reducing fecal pathogens to minimal levels through control of total coliform bacteria, including fecal coliforms and *Escherichia coli (E. coli)*.

22.1F Recycle flows referred to in the Filter Backwash Recycling Rule (FBRR) include spent filter backwash water, thickener supernatant, and liquids from dewater processes.

22.1G The purpose of the Consumer Confidence Report (CCR) Rule is to improve public health protection by providing educational material to allow consumers to make educated decisions regarding any potential health risks pertaining to the quality, treatment, and management of their drinking water supply.

Answers to questions on page 565.

22.1H EPA sets MCLGs at zero for known or probable human carcinogens as a matter of policy.

22.1I A maximum contaminant level goal (MCLG) represents what EPA believes to be a safe level of consumption of a contaminant based solely on health effects, but this is not a legally enforceable standard. The maximum contaminant level (MCL) is a legally enforceable standard that protects public health but takes into account factors such as existing laboratory detection technology, costs, and reasonableness.

22.1J A community water system is defined as a public water system that:

1. Has at least 15 service connections used by all-year residents, or
2. Regularly serves at least 25 all-year residents.

Answers to questions on page 572.

22.2A Primary standards for some contaminants are specified as treatment techniques because the contaminants are too difficult or expensive to detect and measure. The specified treatment technique is known to adequately reduce the public health threat of the contaminant.

22.2B Inorganic chemicals are metals, salts, and other chemical compounds that do not contain carbon.

22.2C Arsenic is listed as a primary contaminant because anyone who drinks water that continuously exceeds the national standard by a substantial amount over a lifetime may experience fatigue and loss of energy. Extremely high levels can cause poisoning.

22.2D At levels of 6 to 8 mg/L, fluoride may cause skeletal fluorosis (the bones become brittle) and stiffening of the joints. At levels of 2 mg/L and greater, fluoride may cause dental fluorosis (discoloration and mottling of the teeth), especially in children.

22.2E Lead may enter drinking water from galvanized pipes, solder used with copper pipes, auto exhausts, and other sources.

22.2F Nitrate above 10 mg/L in drinking water is considered an immediate threat to public health because it causes a serious disorder called "blue baby" in infants under three months of age. Intestinal bacteria change nitrate to nitrite and the nitrite reduces the oxygen-carrying ability of the blood. This produces a life-threatening anemic condition.

Answers to questions on page 573.

22.2G Organic chemicals are either natural or synthetic chemical compounds that contain carbon. Synthetic organic chemicals (SOCs) are manmade compounds that are widely used as pesticides, paints, dyes, solvents, plastics, and food additives.

22.2H Trichloroethylene (TCE) has been widely used as an ingredient in many household products (spot removers, rug cleaners, air fresheners), dry cleaning agents, industrial metal cleaners and polishes, refrigerants, and even anesthetics.

Answers to questions on page 578.

22.2I Microbial contaminants are a public health concern because they cause almost immediate gastrointestinal illness when people consume them in water. Waterborne diseases such as typhoid, cholera, infectious hepatitis, and dysentery have been traced to improperly disinfected drinking water.

22.2J The Total Coliform Rule uses a presence-absence approach to determine compliance rather than estimating the number of coliform organisms found in a sample.

22.2K The five analytical methods EPA will accept for coliform testing include:

1. Multiple-tube fermentation technique (MTF)
2. Membrane filter technique (MF)
3. Presence-absence coliform test (P-A)
4. Colilert™ system (ONPG-MUG test)
5. Colisure test

22.2L Repeat coliform samples are collected to confirm the presence of coliform bacteria and to help determine whether the contamination is a localized problem or a system-wide problem.

22.2M When repeat samples must be taken, one must be taken from the same tap as the original coliform-positive sample and at least two must be taken from nearby points (within five service connections) upstream and downstream.

Answers to questions on pages 582 and 583.

22.2N The Surface Water Treatment Rules set forth primary drinking water regulations requiring treatment of surface water supplies or groundwater supplies under the direct influence of surface water. These regulations require a specific water treatment technique (filtration and/or disinfection) in place of the establishment of maximum contaminant levels (MCLs) for *Cryptosporidium, Giardia lamblia,* viruses, *Legionella,* and bacteria. Also, the regulations require that all systems be operated by qualified operators.

22.2O The Surface Water Treatment Rules define surface water as "all water open to the atmosphere and subject to surface runoff." This would include rivers, lakes, streams, reservoirs, and groundwaters directly influenced by surface water.

22.2P A water utility can avoid mandatory filtration by meeting the following requirements:

1. The fecal coliform concentration in the water prior to disinfection must not exceed 20/100 mL, or total coliform concentration must not exceed 100/100 mL in more than 10 percent of the samples analyzed in the previous six months;
2. The turbidity of the water prior to disinfection cannot exceed 5 NTUs based on grab samples collected every four hours or by continuous monitoring; and
3. Certain site-specific conditions.

22.2Q The MCLs for turbidity for surface waters are:

1. For unfiltered surface water supplies, the turbidity of the raw water prior to disinfection cannot exceed 5 NTU based on grab samples collected every four hours or by continuous turbidity monitoring;
2. For filtered surface water supplies, the filtration process must achieve a turbidity level of less than 1 NTU at all times and not more than 0.3 NTU in more than 5 percent of the samples analyzed during a given month; and
3. Those systems using slow sand or diatomaceous earth filtration must achieve a turbidity level of less than 5 NTU at all times and not more than 1 NTU in more than 5 percent of the samples collected each month.

22.2R The effectiveness of a disinfectant is measured by the time (T) in minutes of the disinfectant's contact in the water and the concentration (C) of the disinfectant residual in mg/L measured at the end of the contact time. The product of these two factors (C × T) provides a measure of the degree of pathogenic inactivation.

Answers to questions on page 585.

22.2S Trihalomethanes (THMs) are the product of chlorine combining with organic material in the water; they are suspected carcinogens.

22.2T Drinking water can become contaminated with radioactive elements when these naturally occurring elements dissolve and are carried to groundwater basins as water passes through the soil.

22.2U The MCLs for radiological contaminants are divided into two categories: (1) natural radioactivity that results from well water passing through deposits of naturally occurring radioactive materials, and (2) manmade radioactivity such as might result from industrial wastes, hospitals, or research laboratories.

ANSWERS TO QUESTIONS IN LESSON 2

Answers to questions on page 587.

22.3A Secondary drinking water regulations are enforceable after a state has passed a law requiring the state health agency to enforce the regulations.

22.3B The secondary drinking water contaminants include:

1. Aluminum
2. Chloride
3. Color
4. Copper
5. Corrosivity
6. Fluoride
7. Foaming Agents
8. Iron
9. Manganese
10. Odor
11. pH
12. Silver
13. Sulfate
14. Total Dissolved Solids (TDS)
15. Zinc

22.3C Contaminants in the secondary regulations should be monitored at least as often as the monitoring performed for inorganic chemical contaminants listed in the Primary Drinking Water Regulations as applicable to community water systems. More frequent monitoring would be appropriate for specific contaminants such as pH, color, odor, or others, under certain circumstances as directed by the state.

Answers to questions on page 589.

22.3D Chloride is a secondary contaminant because it affects the aesthetic quality of water by imparting an objectionable salty taste in water and because it causes corrosion of the pipes in hot water systems and other pipelines.

22.3E Surfactants are synthetic organic chemicals and are the principal ingredient of household detergents.

22.3F The application of chlorine to waters containing manganese increases the likelihood of precipitation of manganese at low levels. Unless the precipitate is removed, precipitates reaching pipelines will promote bacterial growth.

Answers to questions on page 590.

22.3G The undesirable effects of abnormal pH values include:

1. When the pH increases, the disinfection activity of chlorine falls significantly.
2. High pH may cause increased halogen reactions, which produce chloroform and other trihalomethanes during chlorination.
3. Both excessively high and low pHs may cause increased corrosivity, which can, in turn, create taste problems, staining problems, and significant health hazards.
4. Metallic piping in contact with low pH water will impart a metallic taste.
5. If piping is iron or copper, high pH will cause oxide and carbonate compounds to be deposited, leaving red or green stains on fixtures and laundry.
6. At a high pH, drinking water acquires a bitter taste.
7. The high degree of mineralization often associated with basic waters results in encrustation of water pipes and water-using appliances.

22.3H High levels of sulfate are undesirable in drinking water because they:

1. Tend to form hard scales in boilers and heat exchangers
2. Cause taste effects
3. Cause a laxative effect

22.3I High levels of zinc are undesirable in drinking water because they:

1. Produce adverse physiological effects
2. Impart undesirable tastes
3. Cause a milky appearance in the water
4. May increase lead and cadmium concentrations

Answers to questions on page 591.

22.4A "Initial Sampling" refers to the very first sampling conducted under the Safe Drinking Water Act for each of the applicable contaminant categories.

22.4B Routine sampling refers to sampling repeated on a regular basis.

22.4C Whenever an initial or routine sample analysis indicates that an MCL has been exceeded, check (repeat) sampling is required to confirm the routine sampling results. Check sampling is in addition to the routine sampling program.

22.4D At the very minimum, a small system (with a population of 25 to 1,000 people) must sample for turbidity and coliform bacteria and also must have two sampling points:

1. One where the water enters the distribution system
2. One at a consumer faucet at a point representative of the distribution system

Answers to questions on page 592.

22.4E Information that should be specified in a sampling schedule includes:

1. Sampling frequency
2. Sampling point designation
3. Location
4. Type of test
5. Sample volume
6. Special handling instructions

22.4F The following elements are necessary to the collection of an acceptable sample:

1. Obtain a sample that is truly representative of the existing conditions.
2. Flush the line before sample collection (except for when a first-draw sample is required).
3. Fill the sample bottle without leaving any air pocket.
4. Analyze residual chlorine when the sample is taken.
5. Maintain the sample so that it does not become contaminated before it reaches the laboratory.
6. Use preservation techniques (pH control and refrigeration).
7. Keep accurate records of every sample collected (date, time, location, name of sample, bottle number, type of sample, and name of person collecting sample).
8. Keep the time between the collection of the sample and analysis as short as possible.

Answers to questions on page 604.

22.5A The two general categories of reporting called for by the SDWA are:

1. Reporting to the public (public notification)
2. Reporting to the state

22.5B The four types of reports that must be sent to the state are:

1. Initial reports
2. Routine sampling reports
3. Check or repeat sampling reports
4. Violation reports

22.6A When a Tier 1 violation occurs, the public must be notified by: (1) appropriate broadcast media (radio, television), (2) newspapers, (3) posting of notices in conspicuous locations, or (4) hand delivery of notices.

22.8A All utility staff should be familiar with the contents of the report because consumers who know utility personnel frequently will ask personnel questions about the report.

CHAPTER 23

ADMINISTRATION

by

Lorene Lindsay

With Portions by

Tim Gannon

and

Jim Sequeira

TABLE OF CONTENTS
Chapter 23. ADMINISTRATION

			Page
OBJECTIVES			615
WORDS			616

LESSON 1

23.0	NEED FOR UTILITY MANAGEMENT		619
23.1	FUNCTIONS OF A MANAGER		619
23.2	PLANNING		621
23.3	ORGANIZING		621
23.4	STAFFING		624
	23.40	The Utility Manager's Responsibilities	624
	23.41	How Many Employees Are Needed?	625
	23.42	Qualifications Profile	625
	23.43	Applications and the Selection Process	626
		23.430 Advertising the Position	626
		23.431 Paper Screening	626
		23.432 Interviewing Applicants	626
		23.433 Selecting the Most Qualified Candidate	628
	23.44	New Employee Orientation	629
	23.45	Employment Policies and Procedures	629
		23.450 Probationary Period	629
		23.451 Compensation	629
		23.452 Training and Certification	630
		23.453 Performance Evaluation	631
		23.454 Dealing with Disciplinary Problems	631
		23.455 Example Policy: Harassment	635
		23.456 Labor Laws Governing Employer/Employee Relations	638
		23.457 Personnel Records	639
	23.46	Unions	639
DISCUSSION AND REVIEW QUESTIONS			640

LESSON 2

23.5 COMMUNICATION 641
    23.50   Oral Communication 641
    23.51   Written Communication 641

23.6 CONDUCTING MEETINGS 643

23.7 PUBLIC RELATIONS 644
    23.70   Establish Objectives 644
    23.71   Utility Operations 644
    23.72   The Mass Media 644
    23.73   Being Interviewed 644
    23.74   Public Speaking 645
    23.75   Telephone Contacts 645
    23.76   Consumer Inquiries 645
    23.77   Plant Tours 646

23.8 FINANCIAL MANAGEMENT 646
    23.80   Financial Stability 646
    23.81   Budgeting 647
    23.82   Equipment Repair/Replacement Fund 648
    23.83   Water Rates 648
    23.84   Capital Improvements and Funding in the Future 649
    23.85   Financial Assistance 650

DISCUSSION AND REVIEW QUESTIONS 651

LESSON 3

23.9 OPERATIONS AND MAINTENANCE 652
    23.90   The Manager's Responsibilities 652
    23.91   Purpose of O & M Programs 652
    23.92   Types of Maintenance 652
    23.93   Benefits of Managing Maintenance 653
    23.94   Computer Control Systems 653
            23.940   Description of SCADA Systems 653
            23.941   Typical Water Treatment and Distribution SCADA Systems 654
    23.95   Cross Connection Control Program 655
            23.950   Importance of Cross Connection Control 655
            23.951   Program Responsibilities 655
            23.952   Water Supplier Program 659
            23.953   Types of Backflow Prevention Devices 659
            23.954   Devices Required for Various Types of Situations 659
    23.96   Geographic Information System (GIS) 661

Administration

23.10 EMERGENCY RESPONSE 661
    23.100   Planning 661
    23.101   Homeland Defense 663
    23.102   Handling the Threat of Contaminated Water Supplies 664
        23.1020   Importance 664
        23.1021   Toxicity 664
        23.1022   Emergency Contaminant Limits 664
        23.1023   Protective Measures 668
        23.1024   Emergency Countermeasures 668
        23.1025   In Case of Contamination 668
        23.1026   *Cryptosporidium* 669

23.11 SAFETY PROGRAM 670
    23.110   Responsibilities 670
        23.1100   Everyone Is Responsible for Safety 670
        23.1101   Regulatory Agencies 671
        23.1102   Managers 671
        23.1103   Supervisors 672
        23.1104   Operators 673
    23.111   First Aid 673
    23.112   Hazard Communication Program and Worker Right-To-Know (RTK) Laws 673
    23.113   Confined Space Entry Procedures 674
    23.114   Reporting 678
    23.115   Training 678
    23.116   Measuring 682
    23.117   Human Factors 683

23.12 RECORDKEEPING 684
    23.120   Purpose of Records 684
    23.121   Types of Records 684
    23.122   Types of Plant Operations Data 684
    23.123   Maintenance Records 685
    23.124   Procurement Records 685
    23.125   Inventory Records 685
    23.126   Equipment Records 687
    23.127   Computer Recordkeeping Systems 687
    23.128   Disposition of Plant Records 687

23.13 WATER AND ENERGY CONSERVATION 688
    23.130   Need for Conservation 688
    23.131   What Is Water Conservation? 688

	23.132	Elements of Conservation Programs	688
		23.1320 Residential Water Surveys	688
		23.1321 Residential Plumbing Retrofits	689
		23.1322 System Water Audits, Leak Detection, and Repair	689
		23.1323 Metering with Commodity Rates	689
		23.1324 Large Landscape Conservation Programs	690
		23.1325 High-Efficiency Clothes Washers	690
		23.1326 Public Information Programs	690
		23.1327 School Education Programs	690
		23.1328 Conservation Programs for Commercial, Industrial, and Institutional (CII) Sectors	691
		23.1329 Wholesale Agency Assistance Programs	691
		23.13210 Conservation Pricing	692
		23.13211 Conservation Coordinator	692
		23.13212 Water Waste Prohibition	692
		23.13213 Residential ULFT Replacement Programs	692
		23.13214 Potential Best Management Practices	693
	23.133	EPA's WaterSense: Efficiency Made Easy	693
	23.134	Acknowledgment	693
	23.135	Additional Reading	693
23.14	**ACKNOWLEDGMENTS**		693
23.15	**ADDITIONAL READING**		693
DISCUSSION AND REVIEW QUESTIONS			694
SUGGESTED ANSWERS			695

OBJECTIVES

Chapter 23. ADMINISTRATION

Following completion of Chapter 23, you should be able to:

1. Identify the functions of a manager.

2. Describe the benefits of short-term, long-term, and emergency planning.

3. Define the following terms:
   a. Authority
   b. Responsibility
   c. Delegation
   d. Accountability
   e. Unity of command

4. Read and construct an organizational chart identifying lines of authority and responsibility.

5. Write a job description for a specific position within the utility.

6. Write good interview questions.

7. Conduct employee evaluations.

8. Describe the steps necessary to provide equal and fair treatment to all employees.

9. Prepare a written or oral report on the utility's operations.

10. Communicate effectively within the organization, with media representatives, and with the community.

11. Describe the financial strength of your utility.

12. Calculate your utility's operating ratio, coverage ratio, and simple payback.

13. Prepare a contingency plan for emergencies.

14. Prepare a plan to strengthen the security of your utility's facilities.

15. Set up a safety program for your utility.

16. Collect, organize, file, retrieve, use, and dispose of plant records.

17. Describe the basic elements of water and energy conservation and associated best management practice (BMP) strategies.

WORDS
Chapter 23. ADMINISTRATION

ACCOUNTABILITY
When a manager gives power/responsibility to an employee, the employee ensures that the manager is informed of results or events.

AUTHORITY
The power and resources to do a specific job or to get that job done.

BACK PRESSURE
A pressure that can cause water to backflow into the water supply when a user's water system is at a higher pressure than the public water system.

BACKFLOW
A reverse flow condition, created by a difference in water pressures, that causes water to flow back into the distribution pipes of a potable water supply from any source or sources other than an intended source. Also see BACKSIPHONAGE.

BACKSIPHONAGE
A form of backflow caused by a negative or below atmospheric pressure within a water system. Also see BACKFLOW.

BOND
(1) A written promise to pay a specified sum of money (called the face value) at a fixed time in the future (called the date of maturity). A bond also carries interest at a fixed rate, payable periodically. The difference between a note and a bond is that a bond usually runs for a longer period of time and requires greater formality. Utility agencies use bonds as a means of obtaining large amounts of money for capital improvements.

(2) A warranty by an underwriting organization, such as an insurance company, guaranteeing honesty, performance, or payment by a contractor.

CALL DATE
First date a bond can be paid off.

CERTIFICATION EXAMINATION
An examination administered by a state agency or professional association that operators take to indicate a level of professional competence. In the United States, certification of operators of water treatment plants, wastewater treatment plants, water distribution systems, and small water supply systems is mandatory. In many states, certification of wastewater collection system operators, industrial wastewater treatment plant operators, pretreatment facility inspectors, and small wastewater system operators is voluntary; however, current trends indicate that more states, provinces, and employers will require these operators to be certified in the future. Operator certification is mandatory in the United States for the Chief Operators of water treatment plants, water distribution systems, and wastewater treatment plants.

CODE OF FEDERAL REGULATIONS (CFR)
A publication of the US government that contains all of the proposed and finalized federal regulations, including safety and environmental regulations.

CONFINED SPACE

Confined space means a space that:

(1) Is large enough and so configured that an employee can bodily enter and perform assigned work; and

(2) Has limited or restricted means for entry or exit (for example, manholes, tanks, vessels, silos, storage bins, hoppers, vaults, and pits are spaces that may have limited means of entry); and

(3) Is not designed for continuous employee occupancy.

Also see DANGEROUS AIR CONTAMINATION and OXYGEN DEFICIENCY.

CONFINED SPACE, PERMIT-REQUIRED (PERMIT SPACE)

A confined space that has one or more of the following characteristics:

(1) Contains or has a potential to contain a hazardous atmosphere,

(2) Contains a material that has the potential for engulfing an entrant,

(3) Has an internal configuration such that an entrant could be trapped or asphyxiated by inwardly converging walls or by a floor that slopes downward and tapers to a smaller cross section, or

(4) Contains any other recognized serious safety or health hazard.

COVERAGE RATIO

The coverage ratio is a measure of the ability of the utility to pay the principal and interest on loans and bonds (this is known as debt service) in addition to any unexpected expenses.

DEBT SERVICE

The amount of money required annually to pay the (1) interest on outstanding debts, or (2) funds due on a maturing bonded debt or the redemption of bonds.

DELEGATION

The act in which power is given to another person in the organization to accomplish a specific job.

MATERIAL SAFETY DATA SHEET (MSDS)

A document that provides pertinent information and a profile of a particular hazardous substance or mixture. An MSDS is normally developed by the manufacturer or formulator of the hazardous substance or mixture. The MSDS is required to be made available to employees and operators or inspectors whenever there is the likelihood of the hazardous substance or mixture being introduced into the workplace. Some manufacturers are preparing MSDSs for products that are not considered to be hazardous to show that the product or substance is not hazardous.

OSHA (O-shuh)

The Williams-Steiger Occupational Safety and Health Act of 1970 (OSHA) is a federal law designed to protect the health and safety of workers, including the operators of water supply and treatment systems and wastewater collection and treatment systems. The Act regulates the design, construction, operation, and maintenance of water and wastewater systems. OSHA regulations require employers to obtain and make available to workers the Material Safety Data Sheets (MSDSs) for chemicals used at industrial facilities and treatment plants. OSHA also refers to the federal and state agencies that administer the OSHA regulations.

OPERATING RATIO

The operating ratio is a measure of the total revenues divided by the total operating expenses.

ORGANIZING

Deciding who does what work and delegating authority to the appropriate persons.

OUCH PRINCIPLE

This principle says that as a manager when you delegate job tasks you must be **O**bjective, **U**niform in your treatment of employees, and the tasks must be **C**onsistent with utility policies, and **H**ave job relatedness.

PLANNING

Management of utilities to build the resources and financial capability to provide for future needs.

RESPONSIBILITY

Answering to those above in the chain of command to explain how and why you have used your authority.

SCADA (SKAY-dah) SYSTEM

Supervisory Control And Data Acquisition system. A computer-monitored alarm, response, control, and data acquisition system used to monitor and adjust treatment processes and facilities.

TAILGATE SAFETY MEETING

Brief (10 to 20 minutes) safety meetings held every 7 to 10 working days. The term comes from the safety meetings regularly held by the construction industry around the tailgate of a truck.

CHAPTER 23. ADMINISTRATION

(Lesson 1 of 3 Lessons)

23.0 NEED FOR UTILITY MANAGEMENT

The management of a public or private utility, large or small, is a complex and challenging job. Communities are concerned about their drinking water and their wastewater. They are aware of past environmental disasters and they want to protect their communities, but they want this protection with a minimum investment of money. In addition to the local community demands, the utility manager must also keep up with increasingly stringent regulations and monitoring from regulatory agencies. While meeting these external (outside the utility) concerns, the manager faces the normal challenges from within the organization: personnel, resources, equipment, and preparing for the future. For the successful manager, all of these responsibilities combine to create an exciting and rewarding job.

A brief quiz is given in Table 23.1 that asks some basic management questions. This quiz can be used as a guide to management areas that may need some attention in your utility. You should be able to answer yes to most of the questions; however, all utilities have areas that can be improved.

In the environmental field, as well as other fields, the workforce itself is changing. Minorities, women, and people with disabilities provide new opportunities for growth in the utility. For the employee, however, overcoming employment barriers can be difficult, especially when the workload is demanding and physically challenging. The utility manager must provide adequate support services for these workers and learn to deal with organized worker groups.

Changes in the environmental workplace also are created by advances in technology. The environmental field has exploded with new technologies, such as computer-controlled water treatment processes and distribution systems. The utility manager must keep up with these changes and provide the leadership to keep everyone at the utility up to speed on new ways of doing things. In addition, the utility manager must provide a safer, cleaner work environment while constantly training and retraining employees to understand new technologies.

QUESTIONS

Write your answers in a notebook and then compare your answers with those on page 695.

23.0A What are the local community demands on a utility manager?

23.0B What has created changes in the environmental workplace?

23.1 FUNCTIONS OF A MANAGER

The functions of a utility manager are the same as for the CEO (Chief Executive Officer) of any big company: planning, organizing, staffing, directing, and controlling. In small communities, the utility manager may be the only one who has these responsibilities and the community depends on the manager to handle everything.

Planning (see Section 23.2) consists of determining the goals, policies, procedures, and other elements to achieve the goals and objectives of the agency. Planning requires the manager to collect and analyze data, consider alternatives, and then make decisions. Planning must be done before the other managing functions. Planning may be the most difficult in smaller communities, where the future may involve a decline in population instead of growth.

Organizing (see Section 23.3) means that the manager decides who does what work and delegates authority to the appropriate operators. The organizational function in some utilities may be fairly loose while some communities are very tightly controlled.

Staffing (see Section 23.4) is the recruiting of new operators and staff and determining if there are enough qualified operators and staff to fill available positions. The utility manager's staffing responsibilities include selecting and training employees, evaluating their performance, and providing opportunities for advancement for operators and staff in the agency.

Directing includes guiding, teaching, motivating, and supervising operators and utility staff members. Direction includes issuing orders and instructions so that activities at the facilities or in the field are performed safely and are properly completed.

TABLE 23.1 HOW WELL DOES YOUR SYSTEM MANAGE?

The following self-test is designed for water treatment facilities to provide a guide for identifying areas of concern and for improving system management.

1. Is the treatment system budget separate from other accounts so that the true cost of treatment can be determined?
2. Are the funds adequate to cover operating costs, debt service, and future capital improvements?
3. Do operational personnel have input into the budget process?
4. Is there a monthly or quarterly review of the actual operating costs compared to the budgeted costs?
5. Does the user charge system adequately reflect the cost of treatment?
6. Are all users properly metered and does the unaccounted for water not exceed 20 percent of the total flow?
7. Are finished water quality tests representative of plant performance?
8. Are operational control decisions based on process control testing within the plant?
9. Are provisions made for continued training for plant personnel?
10. Are qualified personnel available to fill job vacancies and is job turnover relatively low?
11. Are the energy costs for the system not more than 20 to 30 percent of the total operating costs?
12. Is the ratio of corrective (reactive) maintenance to preventive (proactive) maintenance remaining stable and is it less than 1.0?
13. Are maintenance records available for review?
14. Is the spare parts inventory adequate to prevent long delays in equipment repairs?
15. Are old or outdated pieces of equipment replaced as necessary to prevent excessive equipment downtime, inefficient process performance, or unreliability?
16. Are technical resources and tools available for repairing, maintaining, and installing equipment?
17. Is the utility's pump station equipment providing the expected design performance?
18. Are standby units for key equipment available to maintain process performance during breakdowns or during preventive maintenance activities?
19. Are the plant processes adequate to meet the demand for treatment?
20. Does the facility have an adequate emergency response plan including an alternate water source?

Controlling involves taking the steps necessary to ensure that essential activities are performed so that objectives will be achieved as planned. Controlling means being sure that progress is being made toward objectives and taking corrective action as necessary. The utility manager is directly involved in controlling the treatment process to ensure that water is being properly treated and to make sure that the utility is meeting its short- and long-term goals.

QUESTIONS

Write your answers in a notebook and then compare your answers with those on page 695.

23.1A What are the functions of a utility manager?

23.1B In small communities, what does the community depend on the utility manager to do?

23.2 PLANNING[1]

A very large portion of any manager's typical work day will be spent on activities that can be described as planning activities since nearly every area of a manager's responsibilities require some type of planning.

Planning is one of the most important functions of utility management and one of the most difficult. Communities must have good, safe drinking water and the management of water utilities must include building the resources and financial capability to provide for future needs. The utility must plan for future growth, including industrial development, and be ready to provide the water that will be needed as the community grows. The most difficult problem for some small communities is recognizing and planning for a decline in population. The utility manager must develop reliable information to plan for growth or decline. Decisions must be made about goals, both short- and long-term. The manager must prepare plans for the next two years and the next 10 to 20 years. Remember that utility planning should include operational personnel, local officials (decision makers), and the public. Everyone must understand the importance of planning and be willing to contribute to the process.

Operation and maintenance of a utility also involves planning by the utility manager. A preventive maintenance program should be established to keep the system performing as intended and to protect the community's investment in water supply and distribution facilities. (Section 23.9 describes the various types of maintenance and the benefits of establishing maintenance programs.)

The utility also must have an emergency response plan to deal with natural or human disasters. Without adequate planning your utility will be facing system failures, inability to meet compliance regulations, and inadequate service capacity to meet community needs. Plan today and avoid disaster tomorrow.

(Section 23.10, "Emergency Response," describes the basic elements of an emergency operations plan.)

23.3 ORGANIZING[2]

A utility should have a written organizational plan and written policies. In some communities, the organizational plan and policies are part of the overall community plan. In either case, the utility manager and all plant personnel should have a copy of the organizational plan and written policies of the utility.

The purpose of the organizational plan is to show who reports to whom and to identify the lines of authority. The organizational plan should show each person or job position in the organization with a direct line showing to whom each person reports in the organization. Remember, an employee can serve only one supervisor (unity of command) and each supervisor should ideally manage only six or seven employees. The organizational plan should include a job description for each of the positions on the organizational chart. When the organizational plan is in place, employees know who is their immediate boss and confusion about job tasks is eliminated. A sample organizational plan for a water/wastewater utility is shown in Figure 23.1. The basic job duties for some typical utility positions are described in Table 23.2.

To understand organization and its role in management, we need to understand some other terms including authority, responsibility, delegation, and accountability. *AUTHORITY* means the power and resources to do a specific job or to get that job done. Authority may be given to an employee due to their position in the organization (this is formal authority) or authority may be given to the employee informally by their co-workers when the employee has earned their respect. *RESPONSIBILITY* may be described as answering to those above in the chain of command to explain how and why you have used your authority. *DELEGATION* is the act in which power is given to another person in the organization to accomplish a specific job. Finally, when a manager gives power/responsibility to an employee, then the employee is *ACCOUNTABLE*[3] for the results.

Organization and effective delegation are very important to keep any utility operating efficiently. Effective delegation is uncomfortable for many managers since it requires giving up power and responsibility. Many managers believe that they can do the job better than others, they believe that other employees are not well trained or experienced, and they are afraid of mistakes. The utility manager retains some responsibility even after delegating to another employee and, therefore, the manager is often reluctant to delegate or may delegate the responsibility but not the authority to get the job done. For the utility manager, good organization means that employees are ready to accept responsibility and have the power and resources to make sure that the job gets done.

---

[1] *Planning.* Management of utilities to build the resources and financial capability to provide for future needs.
[2] *Organizing.* Deciding who does what work and delegating authority to the appropriate persons.
[3] *Accountability.* When a manager gives power/responsibility to an employee, the employee ensures that the manager is informed of results or events.

622 Water Treatment

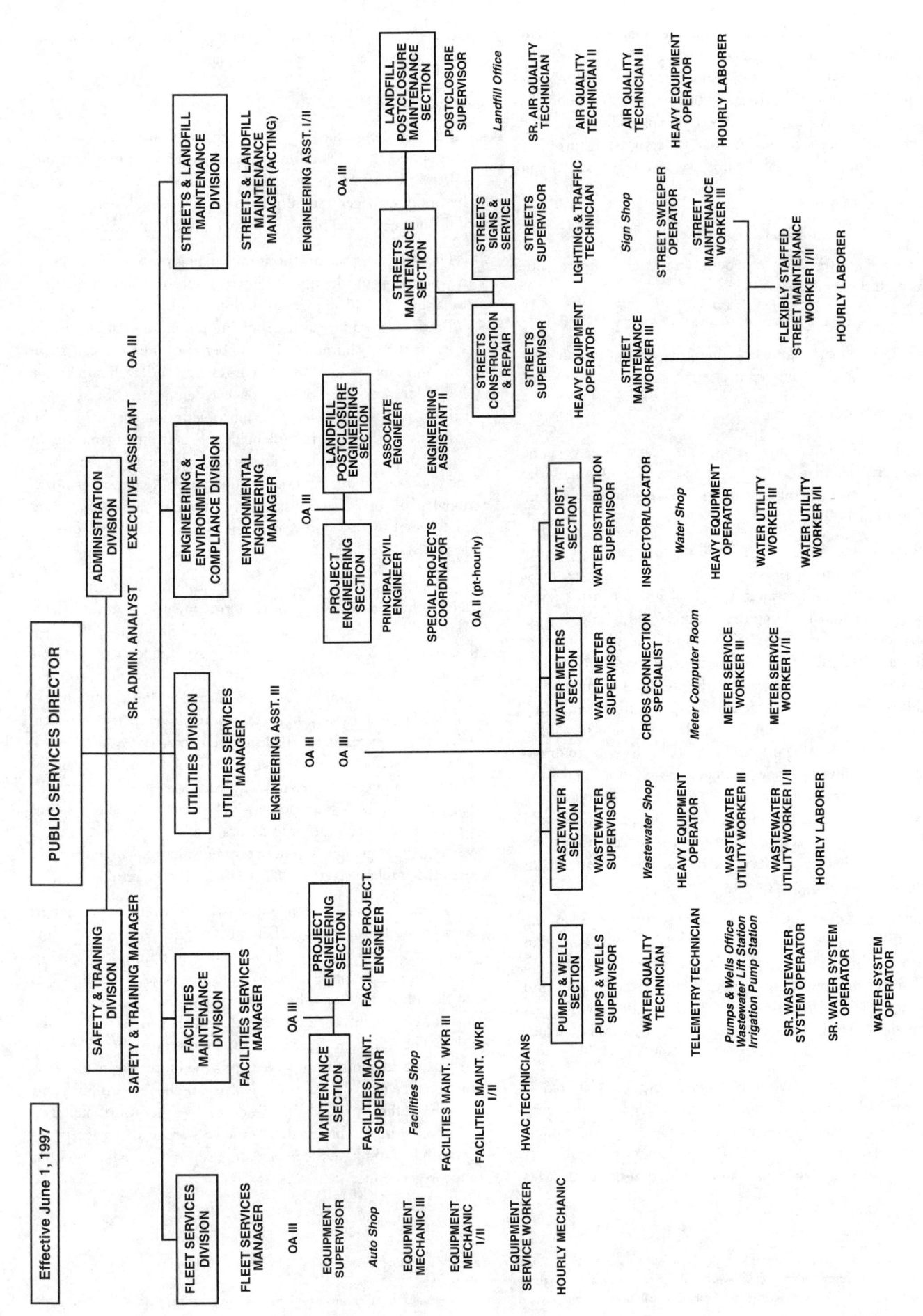

Fig. 23.1 Organizational chart for medium-sized utility
(Courtesy of City of Mountain View, California)

TABLE 23.2 JOB DUTIES FOR STAFF OF A MEDIUM-SIZED UTILITY

Job Title	Job Duties
Superintendent	Responsible for administration, operation, and maintenance of entire facility. Exercises direct authority over all plant functions and personnel.
Assistant Superintendent	Assists Superintendent in review of operation and maintenance function, plans special operation and maintenance tasks.
Clerk/Typist	Performs all clerical duties.
Operations Supervisor	Coordinates activities of plant operators and other personnel. Prepares work schedules, inspects plant, and makes note of operational and maintenance requirements.
Lead Utility Worker	Supervises operations and manages all operators.
Utility Worker II (Journey Level)	Controls treatment processes. Collects samples and delivers them to the lab for analysis. Makes operational decisions.
Utility Worker I	Performs assigned job duties.
Maintenance Supervisor	Supervises all maintenance for plant. Plans and schedules all maintenance work. Responsible for all maintenance records.
Maintenance Foreman	Supervises mechanical maintenance crew. Performs inspections and determines repair methods. Schedules all maintenance including preventive maintenance.
Maintenance Mechanic II	Selects proper tools and assigns specific job tasks. Reports any special considerations to Foreman.
Maintenance Mechanic I	Performs assigned job duties.
Electrician II	Schedules and coordinates electrical maintenance with other planned maintenance. Plans and selects specific work methods.
Electrician I	Performs assigned job duties.
Chemist	Directs all laboratory activities and makes operational recommendations to Operations Supervisor. Reports and maintains all required laboratory records. Oversees laboratory quality control.
Laboratory Technician	Performs laboratory tests. Manages day-to-day laboratory operations.

Employees should not be asked to accept responsibilities for job tasks that are beyond their level of authority in the organizational structure. For example, an operator or lead utility worker should not be asked to accept responsibility for additional lab testing. The responsibility for additional lab testing must be delegated to the lab supervisor. Authority and responsibility must be delegated properly to be effective. When these three components—proper job assignments, authority, and responsibility—are all present, the supervisor has successfully delegated. The success of delegation is dependent upon all three components.

An important and often overlooked part of delegation is *follow-up* by the supervisor. A good manager will delegate and follow up on progress to make sure that the employee has the necessary resources to get the job done. Well-organized managers can delegate effectively and do not try to do all the work themselves, but are responsible for getting good results. The Management Muddle No. 1 that follows describes what can happen when delegation is improperly conducted, and illustrates how an organizational plan can prevent disaster.

Management Muddle No. 1

The City Manager of Pleasantville calls the Director of Public Works and asks for a report on the need for and cost of a new utility truck to be presented at the September 13 meeting of the City Council. The Director of Public Works calls the Plant Manager and asks for a report on the need and cost for a new utility truck with a deadline of September 12. The Plant Manager calls the Lead Utility Worker, an operator, who has been asking for a new utility truck and has been looking into the details. The Plant Manager requests that the Lead Utility Worker provide a report on September 12 about the purchase of the truck. The Lead Utility Worker gathers all the notes and hand writes a report identifying the need for the truck, the features required, and the cost. The Lead Utility Worker takes the report to City Hall to be typed and leaves it with a secretary on September 12. On September 13, the City Manager is preparing for the City Council meeting and does not have the report. Who is responsible? Who is accountable? How could this situation have been avoided?

Responsibility: The Lead Utility Worker's responsibility has been carried out with the authority and resources made available. Both the Director of Public Works and the Plant Manager failed to follow up on the report on September 12. No one informed the Lead Utility Worker that the report must be presented to the City Council on September 13, nor was the Lead Utility Worker supplied with the resources for getting the report in final form. However, the City Manager is ultimately responsible for reporting to the City Council.

Accountability: Starting with the Lead Utility Worker and working upward, each employee is accountable to his or her supervisor and should have communicated the status of the report.

How to avoid this situation: Good communication and follow-up by each of these supervisors could have prevented this situation completely. The City Manager should have asked to see the report on September 12; the Director of Public Works should have asked the Plant Manager to deliver the report no later than September 11; and the Plant Manager should have asked the Lead Utility Worker to submit the typed report (to the Plant Manager) no later than September 10. When delegating this task, the Plant Manager should have arranged for a secretary or clerk to assist the Lead Utility Worker in typing the report. Providing clerical support enables the Lead Utility Manager to complete the assigned task in a timely manner.

At each step in this chain of delegation, setting an early deadline gives the individual receiving the report an opportunity to review the document and make revisions, if necessary, before forwarding it up the chain of authority and ensures that the report reaches the City Manager no later than September 12.

QUESTIONS

Write your answers in a notebook and then compare your answers with those on page 695.

23.2A Who must be included in utility planning?

23.3A What is the purpose of an organizational plan?

23.3B Why is it sometimes difficult or uncomfortable for supervisors or managers to delegate effectively?

23.3C What is an important and often overlooked part of delegation?

NOTICE

The information provided in this section on staffing should *not* be viewed as *legal advice.* The purpose of this section is simply to identify and describe in general terms the major components of a utility manager's responsibilities in the area of staffing. One issue, harassment, is discussed in somewhat greater detail to illustrate the broad scope of a manager's responsibilities within a single policy area. Personnel administration is affected by many federal and state regulations. Legal requirements of legislation such as the Americans with Disabilities Act (ADA), Equal Employment Opportunity (EEO) Act, Family and Medical Leave Act (FMLA), and wages and hours laws are complex and beyond the scope of this manual. If your utility does not have established personnel policies and procedures, consider getting help from a labor law attorney to develop appropriate policies. At the very least, you should get help from a recruitment specialist to develop and document hiring procedures that meet the federal guidelines for Equal Employment Opportunity.

23.4 STAFFING

23.40 The Utility Manager's Responsibilities

The utility manager is also responsible for staffing, which includes hiring new employees, training employees, and evaluating job performance. The utility should have established procedures for job hiring that include requirements for advertising the position, application procedures, and the procedures for conducting interviews.

In the area of staffing, more than any other area of responsibility, a manager must be extremely cautious and consider the consequences before taking action. Personnel management practices are changing dramatically and continue to be redefined almost daily by the courts. A manager who violates an employee's or job applicant's rights can be held both personally and professionally liable in court. Throughout this section on staffing you will repeatedly find references to two terms: **job-related** and **documentation**. These are key concepts in personnel management today. Any personnel action you take must be job-related, from the questions you ask during interviews to disciplinary actions or promotions. And while almost no one wants more paperwork, documentation of personnel actions detailing what you did, when you did it, and why you did it (the reasons will be job-related, of course) is absolutely essential. There is no way to predict when you might be called upon to defend your actions in court. Good records not only serve to refresh your memory about past events but can also be used to demonstrate your pattern of lawful behavior over time.

23.41 How Many Employees Are Needed?

There is a common tendency for organizations to add personnel in response to changing conditions without first examining how the existing workforce might be reorganized to achieve greater efficiency and meet the new work demands. In water supply and distribution utilities, aging of the system, changes in use, and expansion of the system often mean changes in the operation and maintenance tasks being performed. The manager of a utility should periodically review the agency's work requirements and staffing to ensure that the utility is operating as efficiently as possible. A good time to conduct such a review is during the annual budgeting process or when you are considering hiring a new employee because the workload seems to be greater than the current staff can adequately handle.

The staffing analysis procedure outlined in this section illustrates how to conduct a comprehensive analysis of the type needed for a complete reorganization of the agency. In practice, however, a complete reorganization may not be desirable or even possible. Frequent organizational changes can make employees anxious about their jobs and may interfere with their work performance. Some employees show strong resistance to any change in job responsibilities. Nonetheless, by thoroughly examining the functions and staffing of the utility on a periodic basis, the manager may spot trends (such as an increase in the amount of time spent maintaining certain equipment or portions of the system) or discover inefficiencies that could be corrected over a period of time.

The first step in analyzing the utility's staffing needs is to prepare a detailed list of all the tasks to be performed to operate and maintain the utility. Next, estimate the number of staff hours per year required to perform each task. Be sure to include the time required for supervision and training.

When you have completed the task analysis, prepare a list of the utility's current employees. Assign tasks to each employee based on the person's skills and abilities. To the extent possible, try to minimize the number of different work activities assigned to each person but also keep in mind the need to provide opportunities for career advancement. One full-time staff year equals 260 days, including vacation and holiday time: (52 wks/yr)(5 days/wk) = 260 days/yr.

You can expect to find that this ideal staffing arrangement does not exactly match up with your current employees' job assignments. Most likely, you will also find that the number of staff hours required does not exactly equal the number of staff hours available. Your responsibility as a manager is to create the best possible fit between the work to be done and the personnel/skills available to do it. In addition to shifting work assignments between employees, other options you might consider are contracting out some types of work, hiring part-time or seasonal staff, or setting up a second shift (to make fuller use of existing equipment). Of course, you may find that it is time to hire another full- or part-time operator.

23.42 Qualifications Profile

Hiring new employees requires careful planning before the personal interview process. In an effort to limit discriminatory hiring practices, the law and administrative policy have carefully defined the hiring methods and guidelines employers may use. The selection method and examination process used to evaluate applicants must be limited to the applicant's knowledge, skills, and abilities to perform relevant job-related activities. In all but rare cases, factors such as age and level of education may not be used to screen candidates in place of performance testing. A description of the duties and qualifications for the job must be clearly defined in writing. The job description may be used to develop a qualifications profile. This qualifications profile clearly and precisely identifies the required job qualifications. All job qualifications must be relevant to the actual job duties that will be performed in that position. The following list of typical job qualifications may be used to help you develop your own qualifications profiles with advice from a recruitment specialist.

1. General Requirements:

    a. Knowledge of methods, tools, equipment, and materials used in water utilities

    b. Knowledge of work hazards and applicable safety precautions

    c. Ability to establish and maintain effective working relations with employees and the general public

    d. Possession of a valid state driver's license for the class of equipment the employee is expected to drive

2. General Educational Development:

    a. Reasoning: Apply common-sense understanding to carry out instructions furnished in oral, written, or diagrammatical form.

    b. Mathematical: Use a pocket calculator to make arithmetic calculations relevant to the utility's operation and maintenance processes.

    c. Language: Communicate with fellow employees and train subordinates in work methods. Fill out maintenance report forms.

3. Specific Vocational Preparation: Three years of experience in water utility operation and maintenance.

4. Interests: May or may not be relevant to knowledge, skills, and ability; for example, an interest in activities concerned with objects and machines, ecology, or business management.

5. Temperament: Must adjust to a variety of tasks requiring frequent change and must routinely use established standards and procedures.

6. Physical Demands: Medium to heavy work involving lifting, climbing, kneeling, crouching, crawling, reaching, hearing, and seeing. Must be able to lift and carry_____number of pounds for a distance of_____feet.

7. Working Conditions: The work involves wet conditions, cramped and awkward spaces, noise, risks of bodily injury, and exposure to weather.

QUESTIONS

Write your answers in a notebook and then compare your answers with those on page 695.

23.4A What do staffing responsibilities include?

23.4B What are two key personnel management concepts a manager should always keep in mind?

23.4C List the steps involved in a staffing analysis.

23.4D What is a qualifications profile?

23.43 Applications and the Selection Process

23.430 Advertising the Position

To advertise a job opening, first prepare a written description of the required job qualifications, compensation, job duties, selection process, and application procedures (with a closing date). The utility should have established procedures about how to advertise the position and conduct the application process. The application procedure may require that the job be posted first within the utility to allow existing personnel first chance at the job opportunity.

23.431 Paper Screening

The next step in the selection process is known as paper screening. The personnel department and the utility manager review each application and eliminate those who are not qualified. The qualified applicants may be given examinations to verify their qualifications. Usually, the top three to twelve applicants are selected for an interview, depending on the agency's preference.

23.432 Interviewing Applicants

The purpose of the job interview is to gain additional information about the applicants so that the most qualified person can be selected. The utility manager should prepare for the interview in advance. Review the background information on each applicant. Draw up a list of job-related questions that will be asked of each applicant. During the interviews, briefly note the answers each applicant gives.

It used to be thought that the best way to learn about applicants was to give them plenty of time to talk about themselves because the content and type of information applicants volunteer might provide a deeper insight into the person and what type of employee they will become. Be very careful about open-ended, unstructured conversations with job applicants, even the friendly remarks you make initially to put the applicant at ease during the interview. If the applicant begins to volunteer information that you could not otherwise legally ask for (such as marital status, number of children, religious affiliation, or age), be polite but firm in promptly redirecting the conversation. Even if this information was provided to you voluntarily, an applicant who did not get the job could later allege that you discriminated against them based on age or religion.

The only type of information you may legally request is information about the applicant's job skills, abilities, and experience relating directly to the job for which the person is applying. You must always be sensitive to the civil rights of the applicant and the affirmative action policies of the utility, which is another good reason to prepare a list of questions before the interview process begins. Structure the questions so that you avoid simple yes-and-no answers. Table 23.3 summarizes acceptable and unacceptable pre-employment inquiries to guide you in developing a good list of questions.

If other utility staff members are participating in the interviews, their participation should be confined to the pre-selected questions. Under no circumstances should front line employees conduct interviews in the absence of the manager or another person knowledgeable about personnel policies and practices.

The interview should be conducted in a quiet room without interruptions. Most applicants will be nervous, so start the interview on a positive note with introductions and some general remarks to put the applicant at ease. Explain the details of the job, working conditions, wages, benefits, and potential for advancement. Allow the applicant a chance to ask questions about the job. Ask each applicant the questions you have prepared and jot down brief notes on their responses. Taking notes while interviewing is awkward for some people but it becomes easier

TABLE 23.3   ACCEPTABLE AND UNACCEPTABLE PRE-EMPLOYMENT INQUIRIES[a]

Acceptable Pre-Employment Inquiries	Subject	Unacceptable Pre-Employment Inquiries
"Have you ever worked for this agency under a different name?"	NAME	Former name of applicant whose name has been changed by court order or otherwise.
Applicant's place of residence. How long applicant has been resident of this state or city.	ADDRESS OR DURATION OF RESIDENCE	
"If hired, can you submit a birth certificate or other proof of US citizenship or age?"	BIRTHPLACE	Birthplace of applicant. Birthplace of applicant's parents, spouse, or other relatives. Requirement that applicant submit a birth certificate, naturalization, or baptismal record.
"If hired, can you furnish proof of age?" /or/ Statement that hire is subject to verification that applicant's age meets legal requirements.	AGE	Questions that tend to identify applicants 40 to 64 years of age.
Statement by employer of regular days, hours, or shift to be worked.	RELIGION	Applicant's religious denomination or affiliation, church, parish, pastor, or religious holidays observed. "Do you attend religious services /or/ a house of worship?" Applicant may not be told, "This is a Catholic/Protestant/Jewish/atheist organization."
	RACE OR COLOR	Complexion, color of skin, or other questions directly or indirectly indicating race or color.
Statement that photograph may be required after employment.	PHOTOGRAPH	Requirement that applicant affix a photograph to his/her application form. Request applicant, at his/her option, to submit photograph. Requirement of photograph after interview but before hiring.
Statement by employer that, if hired, applicant may be required to submit proof of eligibility to work in the United States.	CITIZENSHIP	"Are you a United States citizen?" Whether applicant or applicant's parents or spouse are naturalized or native-born US citizens. Date when applicant or parents or spouse acquired US citizenship. Requirement that applicant produce naturalization papers or first papers. Whether applicant's parents or spouse are citizens of the US
Applicant's work experience.	EXPERIENCE	"Are you currently employed?"
Applicant's military experience in armed forces of United States, in a state militia (US), or in a particular branch of the US armed forces.		Applicant's military experience (general). Type of military discharge.
Applicant's academic, vocational, or professional education; schools attended.	EDUCATION	Date last attended high school.

TABLE 23.3 ACCEPTABLE AND UNACCEPTABLE PRE-EMPLOYMENT INQUIRIES (continued)

Acceptable Pre-Employment Inquiries	Subject	Unacceptable Pre-Employment Inquiries
Language applicant reads, speaks, or writes fluently.	NATIONAL ORIGIN OR ANCESTRY	Applicant's nationality, lineage, ancestry, national origin, descent, or parentage.
		Date of arrival in United States or port of entry; how long a resident.
		Nationality of applicant's parents or spouse; maiden name of applicant's wife or mother.
		Language commonly used by applicant. "What is your mother tongue?"
		How applicant acquired ability to read, write, or speak a foreign language.
	CHARACTER	"Have you ever been arrested?"
Names of applicant's relatives already employed by the agency.	RELATIVES	Marital status or number of dependents.
		Name or address of relative, spouse, or children of adult applicant.
		"With whom do you reside?"
		"Do you live with your parents?"
Organizations, clubs, professional societies, or other associations of which applicant is a member, excluding any names the character of which indicate the race, religious creed, color, national origin, or ancestry of its members.	ORGANIZATIONS	"List all organizations, clubs, societies, and lodges to which you belong."
"By whom were you referred for a position here?"	REFERENCES	Requirement of submission of a religious reference.
"Do you have any physical condition that may limit your ability to perform the job applied for?"	PHYSICAL CONDITION	"Do you have any physical disabilities?"
		Questions on general medical condition.
Statement by employer that offer may be made contingent on passing a physical examination.		Inquiries as to receipt of Workers' Compensation.
Notice to applicant that any misstatements or omissions of material facts in his/her application may be cause for dismissal.	MISCELLANEOUS	Any inquiry that is not job-related or necessary for determining an applicant's eligibility for employment.

[a] Courtesy of Marion B. McCamey, Affirmative Action Officer, California State University, Sacramento, CA.

with practice. Notes are important because after interviewing several candidates you may not be able to remember what each one said. Also, as mentioned earlier, notes taken at the time of an interview can be valuable evidence in court if an unsuccessful applicant files a lawsuit for unfair hiring practices. At the end of the interview, tell the applicant when a decision will be made and how the applicant will be informed of the decision.

If an applicant's responses during the interview indicate that the person clearly is not qualified for this job but may be qualified for another job, briefly describe the other opportunity and how the person can apply for that position. The applicant may ask to be interviewed immediately for the second position. However, do not violate the utility's hiring procedures for the convenience of a job applicant. The same sequence of hiring procedures should be followed each time a position is filled. Tell this applicant that it will still be necessary to apply for the other position and that another interview may or may not follow, depending on the qualifications of the other applicants for the position.

23.433 Selecting the Most Qualified Candidate

Once the interviews are over, the job of evaluating and selecting the successful candidate begins. Review your interview notes and check the candidates' references. Checking references will verify the job experience of the applicant and may provide insight into the applicant's work habits. Questions you might ask previous employers include: Was the employee reliable and punctual? How well did the employee relate to co-workers? Did the employee consistently practice safe work procedures? Would you rehire this employee?

The rights of certain protected groups in the workforce today, such as minorities, women, disabled persons, persons over 40 years of age, and union members, are protected by law. A manager's responsibilities regarding protected groups begins with the hiring process and continues for as long as the employer/employee relationship lasts. The best principle to deal with protected groups (and all other employees, for that matter) is the "OUCH" principle. The OUCH principle says that when you hire new employees or delegate job tasks to current employees, you must be **O**bjective, **U**niform in your treatment of applicants or employees, **C**onsistent with utility policies, and **H**ave job relatedness. If you do not manage with all of these characteristics, you may find yourself in a "hurting" position with regard to protected workers.

Objectivity is the first hurdle. Often the physical characteristics of a person, such as large or small size, may make a person seem more or less job capable. However, many utility agency jobs are done with power tools or other technology that allows all persons, regardless of size, to manage most tasks. Try to remain objective but reasonable in assessing job applicants and making job assignments.

Uniform treatment of job applicants and employees is necessary to protect yourself and other employees. Nothing will destroy morale more quickly than unequal treatment of employees. Your role as a manager is to consistently apply the policies and procedures that have been adopted by the utility. Often, policies and procedures exist that are not popular and may not even be appropriate. However, the job of the utility manager is to consistently uphold and apply the policies of the utility.

The last part of the OUCH principle is having job relatedness. Any hiring decision must be based on the applicant's qualifications to meet the specific job requirements and any job assignment given to an employee must be related to that employee's job description. Extra assignments, such as buying personal gifts for the boss's family or washing the boss's car, are not appropriate. These types of job assignments will eventually catch up to the manager and can be particularly embarrassing if the public gets involved. So to protect yourself and your utility, remember the OUCH principle as you hire and manage your employees.

Once you have made your selection, the applicant is usually required to pass a medical examination. When this has been successfully completed and the applicant has accepted the position, notify the other applicants that the position has been filled.

23.44 New Employee Orientation

During the first day of work, a new employee should be given all the information available in written and verbal form on the policies and practices of the utility including compensation, benefits, attendance expectations, alcohol and drug testing (if the utility does this), and employer/employee relations. Answer any questions from the new employee at this time and try to explain the overall structure of the utility as well as identify who can answer employee questions when they arise. Introduce the new employee to co-workers and tour the work area. Every utility should have a safety training session for all new employees and specific safety training for some job categories. Provide safety training (see Section 23.11, "Safety Program") for new employees on the first day of employment or as soon thereafter as possible. Establishing safe work practices is a very important function of management.

QUESTIONS

Write your answers in a notebook and then compare your answers with those on page 695.

23.4E   What is the purpose of a job interview?

23.4F   List four "protected groups."

23.4G   What does the "OUCH" principle stand for?

23.4H   When should a new employee's safety training begin?

23.45 Employment Policies and Procedures

23.450 Probationary Period

Many employers now use a probationary period for all new employees. The probationary period is typically three to six months but may be as long as a year. This period begins on the first day of work. Management may reserve the right to terminate employment of the person with or without cause during this probationary period. The employee must be informed of this probationary period and must understand that successful completion of the probationary period is required in order to move into regular employment status.

The probationary period provides a time during which both the employer and employee can assess the fit between the job and the person. Normally, a performance evaluation is completed near the end of the probationary period. A satisfactory performance evaluation is the mechanism used to move an employee from probationary status into regular employment.

23.451 Compensation

The compensation an employee receives for the work performed includes satisfaction, recognition, security, appropriate pay, and benefits. All are important to keep good employees satisfied. Salaries should be a function of supply and demand. Pay should be high enough to attract and retain qualified employees. Salaries are usually determined by the governing body of the utility in negotiation with employee groups, when appropriate. The salary structure should meet all state and federal regulations and accurately reflect the level of service given by the employee. A survey of salaries from other utilities in the area may provide valuable information in the development of a salary structure.

The benefits supplied by the employer are an important part of the compensation package. Benefits generally include the following: retirement, health insurance, life insurance, employer's portion of social security, holiday and vacation pay, sick leave, personal leave, parental leave, worker's compensation, and protective clothing. Many employers now provide dental and vision insurance, long-term disability insurance, educational bonus or costs and leave, bereavement leave, and release time for jury duty. Some employers also include in their benefit package cash bonus programs and longevity pay. The value of an employee's entire benefit package is often computed and printed on the pay stub as a reminder that salary alone is not the only compensation being provided.

23.452 Training and Certification[4]

Training has become an ongoing process in the workplace. The utility manager must provide new employee training as well as ongoing training for all employees. Safety training is particularly important for all utility operators and staff members and is discussed in detail later in Section 23.11. Certified water supply system and treatment plant operators earn their certificates by knowing how to do their jobs safely. Preparing for certification examinations is one means by which operators learn to identify safety hazards and to follow safe procedures at all times under all circumstances.

Although it is extremely important, safety is not the sole benefit to be derived from a certification program. Other benefits include protection of the public's investment in water supply and treatment facilities and employee pride and recognition.

Vast sums of public funds have been invested in the construction of water supply and treatment facilities. Certification of operators assures utilities that these facilities will be operated and maintained by qualified operators who possess a certain level of competence. These operators should have the knowledge and skills not only to prevent unnecessary deterioration and failure of the facilities, but also to improve operation and maintenance techniques.

Achievement of a level of certification is a public acknowledgment of a water supply system or treatment plant operator's skills and knowledge. Presentation of certificates at an official meeting of the governing body will place the operators in a position to receive recognition for their efforts and may even get press coverage and public opinion that is favorable. An improved public image will give the certified operator more credibility in discussions with property owners.

Recognition for their personal efforts raises the self-esteem of all certified operators. Certification also gives water supply system and treatment plant operators an upgraded image. If properly publicized, certification ceremonies give the public a more accurate image of the many dedicated, well-qualified operators working for them.

All states and most Canadian provinces now require that water treatment plant operators be certified. To maintain current certification, these operators must complete additional training classes every one to five years. In the environmental field, new technologies and regulations require operators to attend training to keep up with their field. The utility manager has the responsibility to provide employees with high-quality training opportunities. Many types of training are available to meet the different training needs of utility operators, for example, in-house training, training conducted by training centers, professional organizations, engineering firms, or regulatory agencies, and also online courses and correspondence courses such as this one by the Office of Water Programs.

ABC stands for the Association of Boards of Certification for Operating Personnel in Water Utilities and Pollution Control Systems. If you wish to find out how operators can become certified in your state or province, contact:

Executive Director, ABC
208 Fifth Street
Ames, IA 50010-6259
Phone: (515) 232-3623

ABC will provide you with the name and address of the appropriate contact person.

One area of training that is frequently overlooked is training for supervisors. Managing people requires a different set of skills than performing the day-to-day work of operating and maintaining a water treatment facility. Supervisors need to know how to communicate effectively and how to motivate others, as well as how to delegate responsibility and hold people accountable for their performance. Supervisors share management's responsibility for fair and equitable treatment of all workers and are required to act in accordance with applicable state and federal personnel regulations. Making the transition from operator to supervisor also requires a change in attitude. A supervisor is part of the management team and is therefore obliged to promote the best interests of the utility at all times. When the interests of the utility conflict with the desires of one or more employees, the supervisor must support management's decisions and policies regardless of the supervisor's own personal opinion about the issue. It is the responsibility of the utility manager to ensure that supervisors receive appropriate training in all of these areas.

Training on how to motivate people, deal with co-workers, and supervise or manage people working for you has become a very highly specialized field of training. These are complex topics that are beyond the scope of this manual. If you have a need for or wish to learn more about how to deal with people,

---

[4] *Certification Examination.* An examination administered by a state agency or professional association that operators take to indicate a level of professional competence. In the United States, certification of operators of water treatment plants, wastewater treatment plants, water distribution systems, and small water supply systems is mandatory. In many states, certification of wastewater collection system operators, industrial wastewater treatment plant operators, pretreatment facility inspectors, and small wastewater system operators is voluntary; however, current trends indicate that more states, provinces, and employers will require these operators to be certified in the future. Operator certification is mandatory in the United States for the Chief Operators of water treatment plants, water distribution systems, and wastewater treatment plants.

consider enrolling in courses or reading books on supervision or personnel management. An excellent book is *MANAGE FOR SUCCESS* in this series of operator training manuals from the Office of Water Programs.

QUESTIONS

Write your answers in a notebook and then compare your answers with those on pages 695 and 696.

23.4I What is the purpose of a probationary period for new employees?

23.4J What kind of compensation does an employee receive for work performed?

23.4K Why should utility managers provide training opportunities for employees?

23.453 Performance Evaluation

Most organizations conduct some type of performance evaluation, usually on an annual basis. The evaluation may be written or oral; however, a written evaluation is strongly recommended because it will provide a record of the employee's performance. Documentation of this type may be needed in the future to support taking disciplinary action if the employee's performance consistently fails to meet expectations. The evaluation of employee performance can be a challenging task, especially when performance has not been acceptable. However, evaluations are also an opportunity to provide employees with positive feedback and let them know their contributions to the organization have been noticed and appreciated.

A formal performance evaluation typically begins with an employee's immediate supervisor filling out the performance evaluation form (a sample evaluation form is shown in Figure 23.2). Complete the entire form and be specific about the employee's achievements as well as areas needing improvement. Next, schedule a private meeting with the employee to discuss the evaluation. Give the employee frequent opportunities to be heard and listen carefully. If some of the employee's accomplishments were overlooked, note them on the evaluation form and consider whether this new information changes your overall rating of performance in one or more categories.

After reviewing the employee's performance for the past year, set performance goals for the next year. Be sure to document the goals you have agreed upon. Setting performance goals is particularly important if an employee's performance has been poor and improvement is needed. If appropriate, develop a written performance improvement plan that includes specific dates when you will again review the employee's progress in meeting the performance goals. Some supervisors find it helpful to schedule an informal mid-year meeting with each employee to review their progress and to avoid surprises during the next performance evaluation.

Many employees and managers dread even the thought of a performance evaluation and see it as an ordeal to be endured. When properly conducted, however, a performance review can strengthen the lines of communication and increase trust between the employee and the manager. Use this opportunity to acknowledge the employee's unique contributions and to seek solutions to any problems the employee may be having in completing work assignments. Ask the employee how you can be of assistance in removing any obstacles to getting the job done. If necessary, provide coaching to help the employee understand both how and why certain tasks are performed. Be generous (but sincere) with praise for the good work the employee does well every day and try to keep the employee's shortcomings in perspective. If the person is doing a good job 95 percent of the time, do not let the entire discussion consist of criticism about the remaining 5 percent of the person's job assignments.

At the end of the meeting, ask the employee to sign the evaluation form to acknowledge having seen and discussed it. Give the employee a copy of the evaluation. If the employee disagrees with any part of the evaluation, invite the person to submit a written statement describing the reasons for their disagreement. The written statement should be filed with the completed performance evaluation form.

23.454 Dealing with Disciplinary Problems

Handling employee discipline problems is difficult, even for an experienced manager. But remember, **no discipline problem ever solves itself and the sooner you deal with the problem, the better the outcome will be.** If problem behavior is not corrected, then other employees will become dissatisfied and the problems will increase.

Every utility, no matter how small, should have written employment policies enabling the manager to deal effectively with employee problems. It should also provide a formal complaint or grievance procedure by which employees can have their complaint heard and resolved without fear of retaliation by the supervisor.

Dealing with employee discipline requires tact and skill. You will have to find your own style and then try to stay flexible, calm, and open-minded when the situation gets really tough. If you repeatedly find yourself unable to deal successfully with disciplinary problems, consult with the utility's personnel office (if available) or consider enrolling in a management training course designed specifically around strategies and techniques for disciplining employees.

A commonly accepted method for dealing with job-related employee problems is to first discuss the problem with the employee in private. Most employers will give a person two or three verbal warnings; then the warnings should be written with copies given to the employee. Finally, if the written warnings do not produce positive results, the employee may have to be suspended or dismissed. Your job is to make sure that all warnings are documented with specific descriptions of unsatisfactory behaviors and to make sure that all employees are treated fairly.

Start the disciplinary discussion with a positive comment about the employee. Then, identify the problem but keep emotion and blame out of the discussion. The best approach is to

EMPLOYEE EVALUATION FORM

Employee Name _____ Date _____

Job Title _____ Department _____

Evaluate the employee on the job now being performed. Circle the number which most nearly expresses your overall judgment. In the space for comments, consider the employee's performance since their last evaluation and make notes about the progress or specific concerns in that area. The care and accuracy of this appraisal will determine its value to you, the employee, and your employer.

JOB KNOWLEDGE: (Consider knowledge of the job gained through experience, education, and special training)

5. Well informed on all phases of work
4. Knowledge thorough enough to perform well without assistance
3. Adequate grasp of essentials, some assistance required
2. Requires considerable assistance to perform
1. Inadequate knowledge

Comments: _____
_____
_____

QUALITY OF WORK: (Consider accuracy and dependability of the results)

5. Exceptionally accurate, practically no mistakes
4. Usually accurate, seldom necessary to check results
3. Acceptable, occasional errors
2. Often unacceptable, frequent errors, needs supervision
1. Unacceptable, too many errors

Comments: _____
_____
_____

INITIATIVE: (Consider the speed with which the employee grasps new job skills)

5. Excellent, grasps new ideas and suggests improvements, is a leader with others
4. Very resourceful, can work unsupervised, manages time well, is reliable
3. Shows initiative on occasion, is reliable
2. Lacks initiative, must be reminded to complete tasks
1. Needs constant prodding to complete job tasks, is unreliable

Comments: _____
_____
_____

Fig. 23.2   Employee evaluation form

COOPERATION AND RELATIONSHIPS: (Consider manner of handling relationships with co-workers, superiors, and the public)

5. Excellent cooperation and communication with co-workers, supervisors, and others, takes and gives instructions well
4. Gets along well with co-workers
3. Acceptable, usually gets along well, occasionally complains
2. Shows a reluctance to cooperate, complains
1. Very poor cooperation, does not follow instruction, dislikes fellow employees

Comments: _____
_____
_____

ATTENDANCE: (Consider frequency of absences, reasons for absences or tardiness, and promptness in giving notice about absences)

5. Excellent, absent only for emergencies, illness, civic duties, always on time, gives notice when absent
4. Rarely absent or late, always gives notice and good reason
3. Occasionally absent, less important reasons, usually gives notice, but not always in time
2. Often absent, lack of adequate notice or reasons for absenteeism
1. Unexcusable absenteeism, does not give notice, reasons are unacceptable, cannot be depended upon

Comments: _____
_____
_____

OVERALL EVALUATION: Superior _____ Good _____ Satisfactory _____ Unsatisfactory _____

Comments: _____
_____
_____

I hereby certify that this appraisal is my best judgment of the service value of this employee and is based on personal observation and knowledge of the employee's work.

Supervisor's Signature _____ Date _____

I hereby certify that I have personally reviewed this report.

Employee's Signature _____ Date _____

Fig. 23.2 Employee evaluation form (continued)

state the problem and then ask the employee to suggest a solution. If they respond inappropriately, you must restate the problem and explain that you are trying to find a positive solution that is acceptable to everyone.

Try to keep the discussion focused on solving the problems and do not permit the employee to heap on general complaints, report on what other employees do, or wander from the topic. The following is an example of how you might start the discussion for an employee who is tardy every day. "Joe, you have done a good job in keeping that north side pump station running. You are an asset to this operation. Your tardiness every morning, however, is causing problems. Is there some reason for you to be tardy? We need to find a solution to this problem because your being late creates a bad situation for the night shift. What do you suggest?"

Always remain calm and do not allow yourself to become angry when dealing with an employee about performance issues. If you begin to feel angry and are about to lose control, or if the employee becomes combative or abusive, suggest that the meeting is not producing positive results and schedule an alternative meeting time. Do not let the emotions of the moment carry you into a rage in front of employees. If either you or one of your employees expresses extreme emotions, then the discussion should be postponed until everyone cools down. The following steps may serve as a guide to dealing with confrontation; they apply equally to the employee and the supervisor.

- Maintain an adult approach—positive criticism should be taken/given to improve job skills.
- Create a private environment—job performance issues should be discussed in private between the employee and supervisor.
- Listen very carefully—be sure that both you and the employee understand the situation in the same way. If not, you need to keep talking until both parties are in agreement about the problem and the solution.
- Keep your language appropriate—anger and use of bad language will cause the situation to escalate. Keep your cool and hold your tongue.
- Stay focused in the present—let go of all the past slights, misunderstandings, and dissatisfaction. Problems must be solved one at a time.
- Aim for a permanent solution—changes in job performance need to be permanent to be effective.

Reports of violence in the workplace appear regularly in newspapers and on television. Managers and supervisors should be alert for signs that an employee might become violent and should take any threat of violence seriously.

Violence in the workplace may take the form of physical harm, psychological harm, or property damage. Common warning signs include abusive language, threatening or confrontational behavior, and brandishing a weapon. Some examples of physical harm include pushing, hitting, shoving, or any other form of physical assault. Threats and harassment are forms of psychological harm, and property damage can range from theft or destruction of equipment to sabotage of the employer's computer systems.

The utility's safe workplace policy should be a zero tolerance policy. Any employee who is the target of violent behavior or who witnesses such behavior should be encouraged to immediately report the behavior to a supervisor and the incident should be investigated promptly. If necessary for the immediate safety of other employees, the offending employee should be placed on administrative leave, escorted from the work environment, and permitted to return to work only after the investigation has been completed.

Management Muddle No. 2

Sue has been working for five years as a laboratory technician and was recently passed over for promotion. A lab director who has more college experience than Sue was hired from another plant. Since then, Sue's work has not been very good, she has come to work late, and she does not always get all of the lab tests done during her workday. What should be done about Sue? If you were the supervisor, how would you handle this problem with Sue?

Actions: As the supervisor, you should ask Sue to come by your office. In private, you should discuss with Sue why the new lab director was hired. Discuss with her the good work record she has maintained over the past five years, and explain the changes you have seen in her work recently. At this point you might ask her to evaluate her own performance or what she would do if she were in your situation. She might need to express her resentment about the new lab director. If she does, let her ramble and rave just for a few minutes, then stop her. You might say, "OK, you are unhappy, but what are we going to do to change this situation? How can I help you to regain your motivation and improve your work habits?" If possible you might help her figure out a way to continue her college education, go to additional training classes, or reorganize the lab so that her job duties change somewhat. There are many other possibilities for helping Sue to become motivated again but she should be part of the process. It is important to communicate clearly that her job performance must improve.

QUESTIONS

Write your answers in a notebook and then compare your answers with those on page 696.

23.4L   Who is the appropriate person to conduct an employee's performance evaluation?

23.4M   What should be the attitude of a supervisor or manager when dealing with disciplinary problems?

23.4N   What are some common warning signs that an employee could become violent?

23.455 Example Policy: Harassment

Harassment is any behavior that is offensive, annoying, or humiliating to an individual and that interferes with a person's ability to do a job. This behavior is uninvited, often repeated, and creates an uncomfortable or even hostile environment in the workplace. Harassment is not limited to physical behavior but also may be verbal or involve the display of offensive pictures or other images. Sexual harassment is legally defined as unwanted sexual advances, or visual, verbal, or physical conduct of a sexual nature. Any type of harassment is inappropriate in the workplace. A manager's responsibilities with regard to harassment include:

- Establish a written policy (such as the one shown in Figure 23.3) that clearly defines and prohibits harassment of any type.
- Distribute copies of the harassment policy to all employees and take whatever steps are necessary (small group discussions, general staff meetings, training sessions) to ensure that all employees understand the policy.
- Encourage employees to report incidents of harassment to their immediate supervisor, a manager, or the personnel department.
- Investigate every reported case of harassment.
- Document all aspects of the complaint investigation, including the procedures followed, statements by witnesses, the complainant, and the accused person, the conclusions reached in the case, and the actions taken (if any).

How do you know when offensive behavior could be considered sexual harassment? Unwelcome sexual advances or other verbal or physical conduct of a sexual nature could be interpreted by an employee as sexual harassment under the following conditions:

- A person is required, or feels they are required, to accept unwelcome sexual conduct in order to get a job or keep a job.
- Decisions about an employee's job or work status are made based upon either the employee's acceptance or rejection of unwelcome sexual conduct.
- The conduct interferes with the employee's work performance or creates an intimidating, hostile, or offensive working environment.

The following is a list of examples of the kind of behavior that is unacceptable and illegal. It is only a partial list to give you an idea of the scope of the requirements.

- Unwanted hugging, patting, kissing, brushing up against someone's body, or other inappropriate sexual touching
- Subtle or open pressure for sexual activity
- Persistent sexually explicit or sexist statements, jokes, or stories
- Repeated leering or staring at a person's body
- Suggestive or obscene notes or phone calls
- Display of sexually explicit pictures or cartoons

The best way to prevent harassment is to set an example by your own behavior and to keep communication open between employees. In most cases, an open discussion with employees about harassment can help everyone understand that innuendo and slurs about a person's race, religion, sex, appearance, or any other personal belief or characteristic are humiliating. The most productive way to control such behavior is by enlisting the help of all employees to feel that they have the right and the responsibility to stop harassment. When you get employees to think about their behavior and how their behavior makes others feel, they will usually realize they should speak up to prevent harassment.

Here is an example of how employees handled a problem of harassment. A group of operators often had coffee in the office of the utility during their morning break. One of the operators often used foul language, which was embarrassing and offensive to one of the operator's co-workers. The manager sent a memo to all employees that mentioned respect for fellow workers and included a reminder about inappropriate language. The next day all of the other operators had taped their copy of the memo to the mailbox of the one operator who was most vocal. These operators found a way to send their message loud and clear—no one wants to work in an environment that is unpleasant to others. As a manager, you must establish an atmosphere that is open, congenial, and harmonious for all employees.

Occasional flirting, innuendo, or jokes may not meet the legal definition of sexual harassment. Nevertheless, they may be offensive or intimidating to others. Every employee has a right to a workplace free of discrimination and harassment, and every employee has a responsibility to respect the rights of others.

Managers need to *be aware of* and *take action to prevent* any type of harassment in the workplace. It is not enough to simply distribute copies of the utility's harassment policy. If legal action is taken against the utility due to harassment or the existence of a hostile work environment, the utility manager may face both personal and professional liability if it can be shown that the manager *should have known* harassment was occurring or that the manager permitted a hostile environment to continue to exist.

Management Muddle No. 3

The maintenance crew is a group of five men who have been with the utility for many years and are well respected for their work habits. However, in the maintenance shed the walls are covered by calendars of scantily clad women and the language used out in the shed is sometimes pretty rough. One of your operators is a woman; she is well respected by her co-workers and is a very good operator. She comes to your office to complain about the situation in the maintenance shed and demands that

SUBJECT: HARASSMENT POLICY AND COMPLAINT PROCEDURE	**NO:** _____

PURPOSE:

To establish a strong commitment to prohibit harassment in employment, to define discrimination harassment and to set forth a procedure for investigating and resolving internal complaints of harassment.

POLICY:

Harassment of an applicant or employee by a supervisor, management, employee, or co-worker on the basis of race, religion, color, national origin, ancestry, handicap, disability, medical condition, marital status, familial status, sex, sexual orientation, or age will not be tolerated. This policy applies to all terms and conditions of employment, including, but not limited to, hiring, placement, promotion, disciplinary action, layoff, recall, transfer, leave of absence, compensation, and training.

Disciplinary action up to and including termination will be instituted for behavior described in the definition of harassment set forth below:

- Any retaliation against a person for filing a harassment charge or making a harassment complaint is prohibited. Employees found to be retaliating against another employee shall be subject to disciplinary action up to and including termination.

DEFINITION:

Harassment includes, but is not limited to:

A. <u>Verbal Harassment</u>—For example, epithets, derogatory comments or slurs on the basis of race, religious creed, color, national origin, ancestry, handicap, disability, medical condition, marital status, familial status, sex, sexual orientation, or age. This might include inappropriate sex-oriented comments on appearance, including dress or physical features or race-oriented stories.

B. <u>Physical Harassment</u>—For example, assault, impeding or blocking movement, with a physical interference with normal work or movement when directed at an individual on the basis of race, religion, color, national origin, ancestry, handicap, disability, medical condition, marital status, familial status, age, sex, or sexual orientation. This could be conduct in the form of pinching, grabbing, patting, propositioning, leering, or making explicit or implied job threats or promises in return for submission to physical acts.

C. <u>Visual Forms of Harassment</u>—For example, derogatory posters, notices, bulletins, cartoons, or drawings on the basis of race, religious creed, color, national origin, ancestry, handicap, disability, medical conditions, marital status, familial status, sex, sexual orientation, or age.

D. <u>Sexual Favors</u>—Unwelcome sexual advances, requests for sexual favors, and other verbal or physical conduct of a sexual nature which is conditioned upon an employment benefit, unreasonably interferes with an individual's work performance, or creates an offensive work environment.

Fig. 23.3 Harassment policy
(Courtesy of City of Mountain View, California)

SUBJECT: HARASSMENT POLICY AND COMPLAINT PROCEDURE (continued)

COMPLAINT PROCEDURE:

A. <u>Filing</u>:

An employee who believes he or she has been harassed may make a complaint orally or in writing with any of the following:

1. Immediate supervisor.
2. Any supervisor or manager within or outside of the department.
3. Department head.
4. Employee Services Director (or his/her designee).

Any supervisor or department head who receives a harassment complaint should notify the Employee Services Director immediately.

B. Upon notification of the harassment complaint, the Employee Services Director shall:

1. Authorize the investigation of the complaint and supervise and/or investigate the complaint. The investigation will include interviews with:

    (a) The complainant;
    (b) The accused harasser; and
    (c) Any other persons the Employee Services Director has reasons to believe has relevant knowledge concerning the complaint. This may include victims of similar conduct.

2. Review factual information gathered through the investigation to determine whether the alleged conduct constitutes harassment; giving consideration to all factual information, the totality of the circumstances, including the nature of the verbal, physical, visual, or sexual conduct and the context in which the alleged incidents occurred.

3. Report the results of the investigation and the determination as to whether harassment occurred to appropriate persons, including to the complainant, the alleged harasser, the supervisor, and the department head. If discipline is imposed, the discipline may or may not be communicated to the complainant.

4. If harassment occurred, take and/or recommend to the appropriate department head or other appropriate authority prompt and effective remedial action against the harasser. The action will be commensurate with the severity of the offense.

5. Take reasonable steps to protect the victim and other potential victims from further harassment.

6. Take reasonable steps to protect the victim from any retaliation as a result of communicating the complaint.

DISSEMINATION OF POLICY:

All employees, supervisors, and managers shall be sent copies of this policy.

Effective Date: May, 2003
Revision Date: March 1, 2003

_____
City Manager

Fig. 23.3 Harassment policy (continued)
(Courtesy of City of Mountain View, California)

you remove the pictures from the walls in the shed. She goes on to report that when she went out to the shed and requested assistance to check on a pump, which was noisy and running hot, she was told "Kiss my_____, toots!" As the manager, what should you do? Should you immediately go to the maintenance shed and rip down the pictures? What should you say to the female operator in your office?

Actions: Your first response should be to reassure your operator that you understand her anger and frustration. Let her know that you will investigate the matter immediately and take action to correct any problems you find. Ask her to write out a complete statement of the facts, including her concerns about the pump, when and how her request for assistance was made, to whom the request was made, who made the offensive remark, the names of any witnesses, and what responses she has gotten from the maintenance crew in the past. Try to establish if this is a one-time response or if this problem has been going on for some time.

Begin your investigation with a trip to the maintenance shed to observe and evaluate what is hanging on the walls. Discuss the situation with the crew. Try to make this an open discussion so that everybody understands how the pictures affect the atmosphere and the image of professionalism of the utility. The best solution is to get the crew to understand how this type of behavior looks to persons outside their own small group and then let them take down the pictures. (Be sure to follow up later to confirm that the calendars or other offensive material has not reappeared.)

Next, set up private interviews with the person accused of making the offensive remark and each of the witnesses. Ask each person to describe the encounter in the maintenance shed and make detailed notes of their responses. (Depending on the complexity of the situation, it may sometimes be appropriate to have each person involved submit a written statement describing what occurred.)

After you have thoroughly investigated the incident, discuss with the crew the use of acceptable language in response to other employees. An open discussion and increased awareness of sexual harassment should be all that is necessary to change this situation. If it is not, arrange for a training program in sexual harassment awareness, if one is available. Be sure to establish a policy on the consequences of inappropriate behavior and be prepared to enforce the policy when needed.

Retaliation against an employee for filing a complaint about harassment or a hostile work environment is also illegal. Some examples of retaliation are demotion, suspension, failure to hire or consider for hire, failure to make impartial employment recommendations or decisions, adversely changing working conditions, spreading rumors, or denying any employment benefit. Retaliation could be the basis for a lawsuit involving not only the person who is accused of retaliation, but also the immediate supervisor, the manager, and the utility.

Most areas of personnel management, including harassment and retaliation issues, are complex and have significant legal consequences for everyone involved. The discussion in this section is *not* a complete explanation of harassment or retaliation. If your utility is large enough to employ a personnel specialist or a labor law attorney, ask them to review your staffing policies and procedures and consult with them whenever you have questions about personnel matters. If you manage a small utility and have no in-house sources for technical or legal advice, enroll in appropriate training courses or consider working with an attorney on a contract basis.

23.456 *Labor Laws Governing Employer/Employee Relations*

Many employers take pride in advertising that they are an equal opportunity employer, and one often sees this claim in newspaper help wanted ads and other forms of job postings. It means an employer's staffing policies and procedures do not discriminate against anyone based on race, religion, national origin, color, citizenship, marital status, gender, age, Vietnam-era or disabled veteran status, or the presence of a physical, mental, or sensory disability. An employer must meet specific requirements of the federal Equal Employment Opportunity Act to be eligible to advertise as an equal opportunity employer. These requirements include adoption of nondiscriminatory personnel policies and procedures and periodic submission of reports of personnel actions for review by the Equal Employment Opportunity Commission.

The Family and Medical Leave Act of 1993 (FMLA) is a federal law that requires all public agencies as well as companies with 50 or more employees to permit eligible employees to take up to 12 weeks of time off in a 12-month period for the following purposes: (1) the employee's own serious health condition, (2) to care for a child following birth or placement for adoption or foster care, or (3) to care for the employee's spouse, child, or parent with a serious health condition. To be eligible to receive this benefit, an employee must have been employed for at least one year prior to the leave. The employer is not required to pay the employee's salary during the time off work, but many employers permit (or require) employees to use accrued sick leave and vacation time during the period of unpaid FMLA leave.

The Americans with Disabilities Act of 1990 (ADA) prohibits employment discrimination based on a person's mental or physical disability. The law applies to employers engaged in an industry affecting commerce who have 15 or more employees.

In general, the ADA defines disability as a physical or mental impairment that substantially limits one or more of the major life activities of an individual. The exact meaning of this definition is evolving as the courts settle lawsuits in which individuals allege they were discriminated against because of a physical or mental disability. The original ADA legislation listed more than 40 specific types of impairments and the courts continue to expand the list.

Under the ADA, employers must make reasonable accommodations to enable a disabled person to function successfully in the work environment, for example, installing a ramp to make facilities accessible to someone in a wheelchair, or restructuring an individual's job or work schedule, or providing an interpreter. The requirements of each situation are unique. In each case, the nature and extent of the disability and the reasonableness (including cost factors) of the requested accommodation by the employer must be weighed. Employers are not automatically required to do everything possible to accommodate disabled persons, but rather to take whatever reasonable steps they can to do so.

All of these personnel laws are very complex and managers of any utility that may be covered by them are strongly urged to seek the assistance of an experienced labor law attorney or personnel specialist.

23.457 Personnel Records

A personnel file should be maintained for each utility employee. This file should contain all documents related to the employee's hiring, performance reviews, promotions, disciplinary actions, and any other records of employment-related matters. Since these records often contain sensitive, confidential information, access to personnel records should be closely controlled. (Also see Section 23.12, "Recordkeeping," for more information about what records should be kept and how long they should be kept.)

QUESTIONS

Write your answers in a notebook and then compare your answers with those on page 696.

23.4O What is harassment?

23.4P List three types of behavior that could be considered sexual harassment.

23.4Q What is the best way to prevent harassment?

23.4R What is the meaning of "disability" under the Americans With Disabilities Act?

23.46 Unions

Whether your utility operators belong to a union now or may join one in the future, a good employee-management relationship is crucial to keeping an agency functioning properly. Managers, supervisors, crew leaders, and operators all have to work together to develop this relationship.

Most of a manager's union contacts are with a shop steward. The shop steward is elected by the union employees and is their official representative to management and the local union. The steward is in an awkward position because the steward is an employee who is expected to do a full-time job like other employees, while also representing all of the employees. The steward must create an effective link between the utility manager or supervisors and the employees.

During contract negotiations between management and the employees' union representatives, management should be in constant consultation with the supervisors. Many employee demands regarding working conditions originate from the supervisor's daily dealings with the employees. An effective supervisor can minimize unreasonable demands. Also, any demands that are agreed upon must be implemented and carried out by a supervisor. A supervisor can help both sides reach an acceptable contractual agreement.

Once a contract has been agreed upon by both the union and the utility, the utility manager and the other supervisors must manage the organization within the framework of the contract. Do not attempt to ignore or get around the contract even if you disagree with some aspects of it. If you do not understand certain contract provisions, ask for clarification before you begin implementation of those provisions.

Contracts do not change the supervisor's delegated authority or responsibility. Operators must carry out the supervisor's orders and get the work done properly, safely, and within a reasonable amount of time. As a supervisor, you have the right and even the duty to make decisions. However, a contract gives a union the right to protest or challenge your decision. When an operator requires discipline, disciplinary action is a management responsibility.

Handling employee grievances within the framework of a union contract can be a very time-consuming job for the supervisor and the steward. Union contracts usually spell out in great detail the steps and procedures the steward and the supervisor must follow to settle differences. Grievances can develop over disciplinary action, distribution of overtime, transfers, promotions, demotions, and interpretation of labor contracts. The shop steward must communicate complaints and grievances from operators to the supervisor. Then, the supervisor and the steward must work together to settle complaints and adjust grievances. When a shop steward and supervisor can work together, the steward can help the supervisor to be an effective manager.

An effective manager is available to discuss problems. *Dealing with grievances as quickly as possible* often prevents small problems from growing into large problems. When a shop steward presents a grievance to you, listen carefully and sympathetically to the steward. Discuss the problem with the employee directly with the help of the shop steward. Try to identify the facts and cause of the problem and keep a written record of your findings. Focus on the problem and do not get caught up in irrelevant issues. Make every effort to settle the grievance quickly and to everyone's satisfaction.

The consequences of any solution to a grievance must be considered and solutions must be consistent and fair to other operators. The solution or settlement should be clear and understandable to everyone involved. Once a solution has been agreed upon, prepare a written summary of the agreement. Review this final report with the shop steward to be sure the intent of the solution is understood and properly documented. The entire grievance procedure must be documented and properly filed, from initial presentation to final solution and settlement.

Union activities are governed by the National Labor Relations Act. When a union attempts to organize the utility's employees, your rights and actions as a manager are also governed by this Act. Be sure to seek competent legal assistance if your experience in dealing with a union is limited or if you have no such experience.

QUESTIONS

Write your answers in a notebook and then compare your answers with those on page 696.

23.4S What is the role of a shop steward?

23.4T How does a union contract affect a supervisor's authority?

END OF LESSON 1 OF 3 LESSONS

on

ADMINISTRATION

Please answer the discussion and review questions next.

DISCUSSION AND REVIEW QUESTIONS
Chapter 23. ADMINISTRATION
(Lesson 1 of 3 Lessons)

At the end of each lesson in this chapter you will find some discussion and review questions. The purpose of these questions is to indicate to you how well you understand the material in the lesson. Write the answers to these questions in your notebook.

1. What are the different types of demands on a utility manager?

2. List the basic functions of a manager.

3. What can happen without adequate utility planning?

4. What is the purpose of an organizational plan?

5. Define the following terms:
   1. Authority
   2. Responsibility
   3. Delegation
   4. Accountability

6. When has a supervisor successfully delegated?

7. What two concepts should a manager keep in mind to avoid violating the rights of an employee or job applicant?

8. Why should a manager thoroughly examine the functions and staffing of the utility on a periodic basis?

9. When hiring new employees, the selection method and examination process used to evaluate applicants must be based on what criteria?

10. Why should you make notes of applicants' responses during job interviews?

11. What information should be provided to a new employee during orientation?

12. What type of training should be provided for supervisors?

13. Why is it important to formally document each employee's performance on a regular basis?

14. How should discipline problems be solved?

15. What steps can you take to help reach a successful resolution to a confrontation with an employee?

16. What is a manager's responsibility for preventing harassment in the workplace?

17. How are employee grievances usually handled under a union contract?

CHAPTER 23. ADMINISTRATION

(Lesson 2 of 3 Lessons)

23.5 COMMUNICATION

Good communication is an essential part of good management skills. Both written and oral communication skills are needed to effectively organize and direct the operation of a treatment facility. Remember that communication is a two-part process; information must be given and it must be understood. Good listening skills are as important in communication as the information you need to communicate. As the manager of a water utility, you will need to communicate with employees, with your governing body, and with the public. Your communication style will be slightly different with each of these groups but you should be able to adjust easily to your audience.

23.50 Oral Communication

Oral communication may be informal, such as talking with employees, or it may be formal, such as giving a technical presentation. In both cases, your words should be appropriate to the audience, for example, avoid technical jargon when talking with nontechnical audiences. As you talk, you should be observing your audience to be sure that what you are saying is getting across. If you are talking with an employee, it is a good idea to ask for feedback from the employee, especially if you are giving instructions. When the employee is talking, watch and listen carefully and clarify areas that seem unclear. Likewise, in a more formal presentation, watching your audience will give you feedback about how well your message is being received. Some tips for preparing a formal speech are given in Table 23.4.

23.51 Written Communication

Written communication is more demanding than oral communication and requires more careful preparation. Again, keep your audience in mind and use language that will be understood. Written communication requires more organization since you cannot clarify and explain ideas in response to your audience. Before you begin, you should have a clear idea of exactly what you wish to communicate, then keep your language as concise as possible. Extra words and phrases tend to confuse and clutter your message. Good writing skills develop slowly, but you should be able to find good writing classes in your community if you need help improving your skills. In addition, many publications and computer software programs are available that will assist you in writing the most commonly needed documents such as memos, letters, press releases, résumés, monitoring reports, and the annual report.

Before you can write a report you must first organize your thoughts. Ask yourself, what is the objective of this report? Am I trying to persuade someone of something? What information is important to communicate in this report? For whom is the report being written? How can I make it interesting? What does the reader want to learn from this report?

After you have answered the above questions, the next step is to prepare a general outline of how you intend to proceed with the preparation of the report. List not only key topics, but try to list all of the related topics. Then, arrange the key topics in sequence so there is a workable, smooth flow from one topic to the next. Do not attempt to make your outline perfect. It is just a guide. It should be flexible. As you write, you will find that you need to remove nonessential points and expand on more important points.

You might, for example, outline the following points in preparation for writing a report on a polymer testing program.

- A problem condition of high turbidity was discovered
- Polymer testing offered the best means of reducing turbidity
- Funds, equipment, and material were acquired
- Operators were trained
- Tests were conducted
- Results and conclusions were reached
- Corrective actions were planned and taken
- Conclusion, the tests did or did not produce the anticipated results or correct the problem

Once you are fairly sure you have included all the major topics you will want to discuss, go through the outline and write down facts you want to include on each topic. As you work through it, you may decide to move material from one topic to another. The outline will help you organize your ideas and facts.

TABLE 23.4   TIPS FOR GIVING AN ORAL PRESENTATION

1. Arrive early. Give yourself plenty of time to become familiar with the room, practice using your audiovisuals, and make any necessary changes in room setup.

2. Be ready for mistakes. Number the pages in your presentation and your audiovisuals. Check the order of the pages before the meeting begins.

3. Pace yourself. Do not speak too quickly, speak slowly and carefully. Keep a careful eye on audience reaction to be sure that you are speaking at a pace that can be understood.

4. Project yourself. Speak loudly and look at the audience. Do not talk with your back to the audience. Check that those in the back can hear you.

5. Be natural. Try not to read your presentation. Practice ahead of time so that you can speak normally and keep eye contact with your audience.

6. Connect with the audience. Try to smile and make eye contact with the audience.

7. Involve the audience. Allow for audience questions and invite their comments.

8. Repeat audience questions. Always repeat the question so everyone can hear and to be sure that you hear the question correctly.

9. Know when to stop. Keep your remarks within the time allocated for your presentation and be aware that long, rambling speeches create a negative impression on the audience.

10. Use audiovisuals that can be heard and read. Audiovisuals should enhance and reinforce your words. Be sure that all members of the audience can hear, see, and read your audiovisuals. Normal typewritten text is not readable on overheads; use large type so everyone can see. Use no more than five to seven key ideas per overhead.

11. Organize your presentation. Prepare an introduction, body of the speech, and conclusion. "Tell them what you are going to say, say it, and then tell them what you said." The presentation should have three to five main points presented in some logical order, for example, chronologically or from simple to complex.

When your outline is complete, you will have the essentials of your report. Now you need to tailor it to the audience that will be reading it. Take a few minutes to think about your audience. What information do they want? What aspects of the topics will they be most interested in reading? Each of the following groups may be interested in specific topics in the report. Consider these interests as you write.

1. Management

    Management will have specific interests that relate to the cost effectiveness of the program. A report to management should include a summary that presents the essential information, procedures used, an analysis of the data (including trends), and conclusions. Be sure to include complete cost information. Did the benefits warrant the costs? As a result of the tests, can future expenses be reduced? Backup information and field data can be included in an appendix for those who want more information.

2. Other Utilities

    Other utilities will be interested in costs but will also want more detailed information about how the program was performed. They will also be interested in the results and benefits of the program. Explain how the tests were done, the procedures, size of the crew, equipment used, source and availability of materials, and difficulties encountered and how they were overcome.

3. Citizen Groups

    Citizens' interest will be more general. What is a polymer test and why is it needed? Is the polymer harmless? Will it injure fish or birds? How does the polymer test work? Who pays for the test? How much will it cost to implement results and will they be effective?

Your report may be written to include all of these groups. Adjust the outline to include the topics of interest to each group and identify the topics so readers can find the information most interesting to them. Keep the following information in mind as you write your report:

- Drafts. Good reports are not perfect the first time. Re-read and improve your report several times.

- Facts. Confine your writing to the facts and events that occurred. Include figures and statistics only when they make the report more effective. Include only the relevant facts. Large amounts of data should be put in the appendix. Do not clutter the report with unimportant data.

- Continuity. To be interesting and understood, a report must have continuity. It must make sense to the reader and be organized logically. In the report on the polymer testing, the report should be organized to show you had a problem, you had to find a way to identify where the problem existed, you did the testing, and you identified the problems and the corrective actions.

- Effective. To be effective, a report should achieve the objective for which it was written. In this example, we wanted to justify the costs for the program to management, help other utilities in conducting a similar program, and help citizens to understand what we were doing and why.

- Candid. A good report should be frank and straightforward. Keep the language appropriate for the audience. Do not try to impress your readers with technical terms they do not understand. Your purpose is to communicate information. Keep the information accurate and easy to understand.

The annual report is an important part of the management of the utility. It is one of the most involved writing projects that the utility must put together. The annual report should be a review of what and how the utility operated during the past year and it should also include the goals for the next year. In many small communities, the annual report may be presented orally to the city council rather than written. If this report is well written, it can be used to highlight accomplishments and provide support for future planning.

The first step to organizing the report is to make a list of three or four major accomplishments of the last year, then make a list of the top three goals for next year. These accomplishments and goals should be the focus of the report. The annual report should be a summary of the expenses, treatment services provided, and revenues generated over the last year. As you organize this information, keep those accomplishments in mind and let the data tell the story of how the utility accomplished last year's tasks. The data by itself may seem boring but as you organize the data it becomes a meaningful description of the year's accomplishments. Conclude with projections for next year. The facts and figures should tell the audience how you plan to accomplish your goals for the next year. The annual report may be simple or complex depending on your community needs. A sample Table of Contents for a medium-sized utility is given in Table 23.5. When you are finished, the annual report will be a valuable planning tool for the utility and can be used to build support for new projects.

TABLE 23.5 EXAMPLE OF TABLE OF CONTENTS FOR THE ANNUAL REPORT OF A UTILITY

Table of Contents
Executive Summary
Summary of the Treatment Process Including Flows and Costs
Review of Goals and Objectives for the Year
Special Projects Completed
Professional Awards or Recognition for the Utility or Its Staff
General Operating Conditions Including Regulatory Requirements
Expectations for the Next Year—Goals and Objectives
Recommended Changes for the Utility in Organizing, Staffing, Equipment, or Resources Summary

Appendixes: Operating Data
              Budget
              Information on Special Projects

QUESTIONS

Write your answers in a notebook and then compare your answers with those on page 696.

23.5A What kinds of communication skills are needed by a manager?

23.5B What are the most common written documents that a utility manager must write?

23.5C What should be included in the annual report?

23.6 CONDUCTING MEETINGS

As a utility manager, you will be asked to conduct meetings. These meetings may be with employees, your governing board, the public, or with other professionals in your field. Many new managers fail to prepare for these meetings and the meetings end up as a terrible waste of time. As a manager, you need to learn to conduct meetings in a way that is productive and guides the participants into an active role. The following steps should be taken to conduct a productive meeting.

Before the meeting:

- Prepare an agenda and distribute it to all participants.
- Find an adequate meeting room.
- Set a beginning and ending time for the meeting.

During the meeting:

- Start the meeting on time.
- Clearly state the purpose and objectives of the meeting.
- Involve all the participants.
- Do not let one or two individuals dominate the meeting.
- Keep the discussion on track and on time with the agenda.

- When the group makes a decision or reaches consensus, restate your understanding of the results.
- Make clear assignments for participants and review them with everyone during the meeting.

After the meeting:

- Send out minutes of the meeting.
- Send out reminders, when appropriate, about any assignments made for participants, and the next meeting time.

QUESTIONS

Write your answers in a notebook and then compare your answers with those on page 696.

23.6A With whom may a utility manager be asked to conduct meetings?

23.6B What should be done before a meeting?

23.7 PUBLIC RELATIONS

23.70 Establish Objectives

The first step in organizing an effective public relations campaign is to establish objectives. The only way to know whether your program is a success is to have a clear idea of what you expect to achieve—for example, better customer relations, greater water conservation, and enhanced organizational credibility. Each objective must be specific, achievable, and measurable. It is also important to know your audience and tailor various elements of your public relations effort to specific groups you wish to reach, such as community leaders, school children, or the average customer. Your objective may be the same in each case, but what you say and how you say it will depend upon your target audience.

23.71 Utility Operations

Good public relations begin at home. Any time you or a member of your utility comes in contact with the public, you will have an impact on the quality of your public image. Dedicated, service-oriented employees provide for better public relations than paid advertising or complicated public relations campaigns. For most people, contact with an agency employee establishes their first impression of the competence of the organization, and those initial opinions are difficult to change.

In addition to ensuring that employees are adequately trained to do their jobs and knowledgeable about the utility's operations, management has the responsibility to keep employees informed about the organization's plans, practices, and goals. Newsletters, bulletin boards, and regular, open communication between supervisors and subordinates will help build understanding and contribute to a team spirit.

Despite the old adage to the contrary, the customer is not always right. Management should try to instill among its employees the attitude that while the customer may be confused or unclear about the situation, everyone is entitled to courteous treatment and a factual explanation. Whenever possible, employees should phrase responses as positively, or neutrally, as possible, avoiding negative language. For example, "Your complaint" is better stated as "Your question." "You should have …" is likely to make the customer defensive, while "Will you please …" is courteous and respectful. "You made a mistake" emphasizes the negative, "What we will do …" is a positive, problem-solving approach.

23.72 The Mass Media

We live in the age of communications, and one of the most effective and least expensive ways to reach people is through the mass media—radio, television, newspapers, and the Internet. Each medium has different needs and deadlines, and obtaining coverage for your issue or event is easier if you are aware of these constraints. Television must have strong visuals, for example. When scheduling a press conference, provide an interesting setting and be prepared to suggest good shots to the reporter. Radio's main advantage over television and newspapers is immediacy, so have a spokesperson available and prepared to give the interview over the telephone, if necessary. Newspapers give more thorough, in-depth coverage to stories than do the broadcast media, so be prepared to spend extra time with print reporters and provide written backup information and additional contacts.

It is not difficult to get press coverage for your event or press conference if a few simple guidelines are followed:

1. Demonstrate that your story is newsworthy, that it involves something unusual or interesting.
2. Make sure your story will fit the targeted format (television, radio, newspaper, or the Internet).
3. Provide a spokesperson who is interesting, articulate, and well prepared.

23.73 Being Interviewed

Whether you are preparing for a scheduled interview or are simply contacted by the press on a breaking news story, here are some key hints to keep in mind when being interviewed.

1. Speak in personal terms, free of institutional jargon.
2. Do not argue or show anger if the reporter appears to be rude or overly aggressive.
3. If you do not know an answer, say so and offer to find out. Do not bluff.
4. If you say you will call back by a certain time, do so. Reporters face tight deadlines.
5. State your key points early in the interview, concisely and clearly. If the reporter wants more information, he or she will ask for it.
6. If a question contains language or concepts with which you disagree, do not repeat them, even to deny them.

7. Know your facts.

8. Never ask to see a story before it is printed or broadcast. Doing so indicates that you doubt the reporter's ability and professionalism.

23.74 Public Speaking

Direct contact with people in your community is another effective tool in promoting your utility. Though the audiences tend to be small, a personal, face-to-face presentation generally leaves a strong and long-lasting impact on the listener.

Depending upon the size of the organization, your utility may wish to establish a speaker's bureau and send a list of topics to service clubs in the area. Visits to high schools and college campuses can also be beneficial, and educators are often looking for new and interesting topics to supplement their curriculum.

Effective public speaking takes practice. It is important to be well prepared while retaining a personal, informal style. Find out how long your talk is expected to be, and do not exceed that time frame. Have a definite beginning, middle, and end to your presentation. Visual aids such as charts, slides, or models can assist in conveying your message. The use of humor and anecdotes can help to warm up the audience and build rapport between the speaker and the listener. Just be sure the humor is natural, not forced, and that the point of your story is accessible to the particular audience. Try to keep in mind that audiences only expect you to do your best. They are interested in learning about their water supply and will appreciate that you are making a sincere effort to inform them about an important subject.

23.75 Telephone Contacts

First impressions are extremely important, and frequently a person's first contact with your water utility is over the telephone. A person who answers the phone in a courteous, pleasant, and helpful manner goes a long way toward establishing a friendly, cooperative atmosphere. Be sure anyone answering telephone inquiries receives appropriate training and conveys a positive image for the utility.

Following a few simple guidelines will help to start your utility off on the right note with your customers:

1. *Answer calls promptly.* Your conversation will get off to a better start if the phone is answered by the third or fourth ring.

2. *Identify yourself.* This adds a personal note and lets the caller know whom he or she is talking to.

3. *Pay attention.* Do not conduct side conversations. Minimize distractions so you can give the caller your full attention, avoiding repetitions of names, addresses, and other pertinent facts.

4. *Minimize transfers.* Nobody likes to get the run-around. Few things are more frustrating to a caller than being transferred from office to office, repeating the situation, problem, or concern over and over again. Transfer only those calls that must be transferred, and make certain you are referring the caller to the right person. Then, explain why you are transferring the call. This lets the caller know you are referring him or her to a co-worker for a reason and reassures the customer that the problem or question will be dealt with. In some cases, it may be better to take a message and have someone return the call than to keep transferring the customer's call.

23.76 Consumer Inquiries

No single set of rules can possibly apply to all types of consumer questions or complaints about water quality and service. There are, however, basic principles to follow in responding to inquiries and concerns.

1. *Be prepared.* Your employees should be familiar enough with your utility's organization, services, and policies to either respond to the question or complaint or locate the person who can.

2. *Listen.* Ask the customer to describe the problem and listen carefully to the explanation. Take written notes of the facts and addresses.

3. *Do not argue.* Callers often express a great deal of pent-up frustration in their contacts with a utility. Give the caller your full attention. Once you have heard them out, most people will calm down and state their problems in more reasonable terms.

4. *Avoid jargon.* The average consumer lacks the technical knowledge to understand the complexities of water quality. Use plain, nontechnical language and avoid telling the consumer more than he or she wants to know.

5. *Summarize the problem.* Repeat your understanding of the situation back to the caller. This will assure the customer that you understand the problem and offer the opportunity to clear up any confusion or missed communication.

6. *Promise specific action.* Make an effort to give the customer an immediate, clear, and accurate answer to the problem. Be as specific as possible without promising something you cannot deliver.

In some cases, you may wish to have a representative of the utility visit the customer and observe the problem first hand. If the complaint involves water quality, take samples, if necessary, and report back to the customer to be sure the problem has been resolved.

Complaints can be a valuable asset in determining consumer acceptance and pinpointing water quality problems. Customer calls are frequently your first indication that something may be wrong. Responding to complaints and inquiries promptly can save the utility money and staff resources, and minimize the number of customers who are inconvenienced. Still, education can greatly reduce complaints about water quality. Information brochures, utility bill inserts, and other educational tools help to inform customers and avoid future complaints.

23.77 Plant Tours

Tours of water treatment plants can be an excellent way to inform the public about your utility's efforts to provide a safe, high-quality water supply. Political leaders, such as the City Council and members of the Board of Supervisors, should be invited and encouraged to tour the facilities, as should school groups and service clubs.

A brochure describing your utility's goals, accomplishments, operations, and processes can be a good supplement to the tour and should be handed out at the end of the visit. The more visually interesting the brochure is, the more likely that it will be read, and the use of color, photographs, graphics, or other design features is encouraged. If you have access to the necessary equipment, production of a videotape program about the utility can also add interest to the facility tour.

The tour itself should be conducted by an employee who is very familiar with plant operations and can answer the types of questions that are likely to arise. Consider including:

1. A description of the sources of water supply
2. History of the plant, the years of operation, modifications and innovations over the years
3. Major plant design features, including plant capacity and safety features
4. Observation of the treatment processes, including filtration, sedimentation, flocculation, and disinfection
5. A visit to the laboratory, including information on the quality of water distributed to consumers
6. Anticipated improvements, expansions, and long-range plans for meeting future service needs

Plant tours can contribute to a water utility's overall program to gain financing for capital improvements. If the City Council or other governing board has seen the treatment process first hand, it is more likely to understand the need for enhancement and support future funding.

As beneficial as plant tours may be for promoting public interest and confidence in the water utility, security precautions should be carefully considered when planning for visitors to the plant. For more information, see Section 23.101, "Homeland Defense."

QUESTIONS

Write your answers in a notebook and then compare your answers with those on page 696.

23.7A What is the first step in organizing a public relations campaign?

23.7B How can employees be kept informed of the utility's plans, practices, and goals?

23.7C Which news medium is more likely to give a story thorough, in-depth coverage?

23.7D What is the key to effective public speaking?

23.7E How do customer complaints help a utility?

23.8 FINANCIAL MANAGEMENT

Financial management for a utility should include providing financial stability for the utility, careful budgeting, and providing capital improvement funds for future utility expansion. These three areas must be examined on a routine basis to ensure the continued operation of the utility. They may be formally reviewed on an annual basis or more frequently when the utility is changing rapidly. The utility manager should understand what is required for each of the three areas and be able to develop record systems that keep the utility on track and financially prepared for the future.

23.80 Financial Stability

How do you measure financial stability for a utility? Two very simple calculations can be used to help you determine how healthy and stable the finances are for the utility. These two calculations are the *OPERATING RATIO* and the *COVERAGE RATIO*. The operating ratio is a measure of the total revenues divided by the total operating expenses. The coverage ratio is a measure of the ability of the utility to pay the principal and interest on loans and bonds (this is known as *DEBT SERVICE*[5])

---

[5] *Debt Service.* The amount of money required annually to pay the (1) interest on outstanding debts, or (2) funds due on a maturing bonded debt or the redemption of bonds.

in addition to any unexpected expenses. A utility that is in good financial shape will have an operating ratio and coverage ratio above 1.0. In fact, most bonds and loans require the utility to have a coverage ratio of at least 1.25. As state and federal funds for utility improvements have become much more difficult to obtain, these financial indicators have become more important for utilities. Being able to show and document the financial stability of the utility is an important part of getting funding for more capital improvements.

The operating ratio is perhaps the simplest measure of a utility's financial stability. In essence, the utility must be generating enough revenue to pay its operating expenses. The actual ratio is usually computed on a yearly basis, since many utilities may have monthly variations that do not reflect the overall performance. The total revenue is calculated by adding up all revenue generated by user fees, hook-up charges, taxes or assessments, interest income, and special income. Next, determine the total operating expenses by adding up the expenses of the utility, including administrative costs, salaries, benefits, energy costs, chemicals, supplies, fuel, equipment costs, equipment replacement fund, principal and interest payments, and other miscellaneous expenses.

EXAMPLE 1

The total revenues for a utility are $1,686,000 and the operating expenses for the utility are $1,278,899. The debt service expenses are $560,000. What is the operating ratio? What is the coverage ratio?

Known		Unknown
Total Revenue, $ = $1,686,000		Operating Ratio
Operating Expenses, $ = $1,278,899		Coverage Ratio
Debt Service Expenses, $ = $560,000		

1. Calculate the operating ratio.

$$\text{Operating Ratio} = \frac{\text{Total Revenue, \$}}{\text{Operating Expenses, \$}}$$

$$= \frac{\$1,686,000}{\$1,278,899}$$

$$= 1.32$$

2. Calculate nondebt expenses.

$$\text{Nondebt Expenses, \$} = \text{Operating Exp, \$} - \text{Debt Service Exp, \$}$$

$$= \$1,278,899 - \$560,000$$

$$= \$718,899$$

3. Calculate coverage ratio.

$$\text{Coverage Ratio} = \frac{\text{Total Revenue, \$} - \text{Nondebt Expenses, \$}}{\text{Debt Service Expenses, \$}}$$

$$= \frac{\$1,686,000 - \$718,899}{\$560,000}$$

$$= 1.73$$

These calculations provide a good starting point for looking at the financial strength of the utility. Both of these calculations use the total revenue for the utility, which is an important component for any utility budgeting. As managers, we often focus on the expense side and forget to look carefully at the revenue side of utility management. The fees collected by the utility, including hook-up fees and user fees, must accurately reflect the cost of providing service. These fees must be reviewed annually and they must be increased as expenses rise to maintain financial stability. Some other areas to examine on the revenue side include how often and how well user fees are collected, the number of delinquent accounts, and the accuracy of meters. Some small communities have found they can cut their administrative costs significantly by switching to a quarterly billing cycle. The utility must have the support of the community to determine and collect user fees, and the utility must keep track of revenue generation as carefully as resource spending.

23.81 Budgeting

Budgeting for the utility is perhaps the most challenging task of the year for many managers. The list of needs usually is much larger than the possible revenue for the utility. The only way for the manager to prepare a good budget is to have good records from the year before. A system of recording or filing purchase orders (see Section 23.124, "Procurement Records") or a requisition records system must be in place to keep track of expenses and prevent spending money that is not in the budget.

To budget effectively, a manager needs to understand how the money has been spent over the last year, the needs of the utility, and how the needs should be prioritized. The manager also must take into account cost increases that cannot be controlled while trying to minimize the expenses as much as possible. The following problem is an example of the types of decisions a manager must make to keep the budget in line while also improving service from the utility.

EXAMPLE 2

A pump that has been in operation for 25 years pumps a constant 600 GPM through 47 feet of dynamic head. The pump uses 6,071 kilowatt-hours of electricity per month, at a cost of $0.085 per kilowatt-hr. The old pump efficiency has dropped to 63 percent. Assuming a new pump that operates at 86 percent efficiency is available for $9,730.00, how long would it take to pay for replacing the old pump?

Known		Unknown
Electricity, kW-hr/mo	= 6,071 kW-hr/mo	New Pump Payback Time, yr
Electricity Cost, $/kW-hr	= $0.085/kW-hr	
Old Pump Efficiency, %	= 63%	
New Pump Efficiency, %	= 86%	
New Pump Cost, $	= $9,730	

1. Calculate old pump operating costs in dollars per month.

$$\text{Old Pump Operating Costs, \$/mo} = (\text{Electricity, kW-hr/mo})(\text{Electricity Cost, \$/kW-hr})$$
$$= (6{,}071 \text{ kW-hr/mo})(\$0.085/\text{kW-hr})$$
$$= \$516.04/\text{mo}$$

2. Calculate new pump operating electricity requirements.

$$\text{New Pump Electricity, kW-hr/mo} = (\text{Old Pump Electricity, kW-hr/mo}) \frac{(\text{Old Pump Eff, \%})}{(\text{New Pump Eff, \%})}$$
$$= (6{,}071 \text{ kW-hr/mo}) \frac{(63\%)}{(86\%)}$$
$$= 4{,}447 \text{ kW-hr/mo}$$

3. Calculate new pump operating costs in dollars per month.

$$\text{New Pump Operating Costs, \$/mo} = (\text{Electricity, kW-hr/mo})(\text{Electricity Cost, \$/kW-hr})$$
$$= (4{,}447 \text{ kW-hr/mo})(\$0.085/\text{kW-hr})$$
$$= \$378.03/\text{mo}$$

4. Calculate annual cost savings of new pump.

$$\text{Cost Savings, \$/yr} = (\text{Old Costs, \$/mo} - \text{New Costs, \$/mo})(12 \text{ mo/yr})$$
$$= (\$516.04/\text{mo} - \$378.03/\text{mo})(12 \text{ mo/yr})$$
$$= \$1{,}656.12/\text{yr}$$

5. Calculate the new pump payback time in years.

$$\text{Payback Time, yr} = \frac{\text{Initial Cost, \$}}{\text{Savings, \$/yr}}$$
$$= \frac{\$9{,}730.00}{\$1{,}656.12/\text{yr}}$$
$$= 5.9 \text{ years}$$

In this example, a payback time of 5.9 years is acceptable and would probably justify the expense for a new pump. This calculation was a simple payback calculation that did not take into account the maintenance on each pump, depreciation, and inflation. Many excellent references are available from EPA to help utility managers make more complex decisions about purchasing new equipment.

The annual report should be used to help develop the budget so that long-term planning will have its place in the budgeting process. The utility manager must track revenue generation and expenses with adequate records to budget effectively. The manager must also get input from other personnel in the utility as well as community leaders as the budgeting process proceeds. This input from others is invaluable to gain support for the budget and to keep the budget on track once adopted.

23.82 Equipment Repair/Replacement Fund

To adequately plan for the future, every utility must have a repair/replacement fund. The purpose of this fund is to generate additional revenue to pay for the repair and replacement of capital equipment as the equipment wears out. To prepare adequately for this repair/replacement, the manager should make a list of all capital equipment (this is called an asset inventory) and estimate the replacement cost for each item. The expected life span of the equipment must be used to determine how much money should be collected over time. When a treatment plant is new, the balance in repair/replacement fund should be increasing each year. As the plant gets older, the funds will have to be used and the balance may get dangerously low as equipment breakdowns occur. Perhaps the hardest job for the utility manager is to maintain a positive balance in this account with the understanding that this account is not meant to generate a profit for the utility but rather to plan for future equipment needs. In water treatment facilities construction, providing an adequate repair/replacement fund is very important, but if this repair/replacement fund has not been reviewed annually, it must be updated.

To set up a repair/replacement fund for your utility, you should first put together a list of the equipment required for each process in your utility. Once you have this list, you need to estimate the life expectancy of the equipment and the replacement cost. From this list you can predict the amount of money you should set aside each year so that when each piece of equipment wears out, you will have enough money to replace that piece of equipment. The publications listed in Section 23.15, "Additional Reading," at the end of this chapter are excellent references for utility planning.

23.83 Water Rates

The process of determining the cost of water and establishing a water rate schedule for customers is a subject of much controversy. There is no single set of rules for determining water rates. The establishment of a rate schedule involves many factors including the form of ownership (investor or publicly owned), differences in regulatory control over the water utility (state commission or local authority), and differences in individual viewpoints and preferences concerning the appropriate philosophy to be followed to meet local conditions and requirements.

Generally, the development of water rate schedules involves the following procedures:

- A determination of the total revenue requirements for the period that the rates are to be effective (usually one year).

- A determination of all the cost components of system operations. That is, how much does it cost to treat the water? How much does it cost to distribute? How much does it cost to install a water service to a customer? How much are administrative costs?

- Distribution of the various component costs to the various customer classes in accordance with their requirements for service.

- The design of water rates that will recover from each class of customers, within practical limits, the cost to serve that class of customers.

Sales of water to customers may be metered or unmetered. In the case of metered sales, the charge to the customer is based on a rate schedule applied to the amount of water used through each water meter. If meters are not used, the charge per customer is based on a flat rate per period of time per fixture, foot of frontage, number of rooms, or other measurable unit. Although the flat-rate basis still is fairly common, meter-based rates are more widely used.

See *SMALL WATER SYSTEM OPERATION AND MAINTENANCE*, Chapter 8, "Setting Water Rates for Small Water Utilities," for an explanation and examples of how to determine and set water rates. This publication is available from the Office of Water Programs, California State University, Sacramento, 6000 J Street, Sacramento, CA 95819-6025. Price, $49.00.

23.84 Capital Improvements and Funding in the Future

A capital improvements fund must be a part of the utility budget and included in the operating ratio. Your responsibility as the utility manager is to be sure that everyone, your governing body and the public, understands the capital improvement fund is not a profit for the utility but a replacement fund to keep the utility operating in the future.

Capital planning starts with a look at changes in the community. Where are the areas of growth in the community, where are the areas of decline, and what are the anticipated changes in industry within the community? After identifying the changing needs in the community, you should examine the existing utility structure. Identify your weak spots (in the distribution system or with in-plant processes). Make a list of the areas that will be experiencing growth, weak spots in the system, and anticipated new regulatory requirements. The list should include expected capital improvements that will need to be made over the next year, two years, five years, and ten years. You can use the information in your annual reports and other operational logs to help compile the list.

Once you have compiled this information, prioritize the list and make a timetable for improving each of the areas. Starting at the top of the priority list, estimate the costs for improvements and incorporate these costs into your capital improvement budget. The calculations you have made previously, including corrective to preventive maintenance ratios, operating ratio, coverage ratio, and payback time will all be useful in prioritizing and streamlining your list of needs.

You may find that some of your capital improvement needs could be met in more than one way. How do you decide which of several options is most cost-effective? How do you compare fundamentally different solutions? For example, assume your community's population is growing rapidly and you will need to increase treated water production by 20 percent by the end of the next ten years. Your existing wells cannot provide that amount of additional water and your filtration equipment is operating at 90 percent of design capacity. Possible solutions might include a combination of the following options, some of which might be implemented immediately while others might be brought on line in five or ten years:

- Rehabilitate some declining wells.

- Drill additional new wells.

- Develop an available surface water source.

- Install another filtration unit.

- Install additional distribution system storage reservoirs.

To compare alternative plans, you will need to calculate the present value (or present worth) of each plan; that is, the cost of each plan in today's dollars. This is done by adding up all of the costs and benefits of the plan over the entire service life of the facility or equipment. Costs should include not only the initial purchase price or construction costs, but also financing costs over the life of the loans or bonds and all operation and maintenance costs. Benefits include all of the revenue that would be produced by this facility or equipment, including connection and use fees. With the help of an experienced accountant, apply standard inflation, depreciation, and other factors to calculate the present value of each plan. This will give you the cost of each plan in the equivalent of today's dollars.

Remember to involve all of your local officials and the public in development of this capital improvement budget so they understand what will be needed.

Long-term capital improvements such as a new plant or a new treatment process are usually anticipated in your 10-year or 20-year projection. These long-term capital improvements usually require some additional financing. The basic ways for a utility to finance capital improvements are through general obligation bonds, revenue bonds, or loan funding programs.

General obligation bonds or *ad valorem* (based on value) taxes are assessed based on property taxes. These bonds usually have a lower interest rate and longer payback time, but the total bond limit is determined for the entire community. This means that the water utility will have available only a portion of the total bond capacity of the community. These bonds are not often used for funding water utility improvements today.

The second type of bond, the revenue bond, is commonly used to fund utility improvements. This bond has no limit on the amount of funds available and the user charges provide repayment on the bond. To qualify for these bonds, the utility must show sound financial management and the ability to repay the bond. As the utility manager, you should be aware of the provisions of the bond. Be sure the bond has a call date, which is the first date when you can pay off the bond. The common practice is for a 20-year bond to have a 10-year call date and for a 15-year bond to have an 8-year call date. The bond will also have a call premium, which is the amount of extra funds needed to pay off the debt on the call date. You should try to get your bonds a call premium of no more than 102 percent par. This means that for a debt of $200,000 on the call date, the total payoff would be $204,000, which includes the extra two percent for the call premium. You will need to get help from a financial advisor to prepare for and issue the bonds. These advisors will help you negotiate the best bond structure for your community.

Special assessment bonds may be used to extend services into specific areas. The direct users pay the capital costs and the assessment is usually based on frontage or area of real estate. These special assessments carry a greater risk to investors but may be the best way to extend service to some areas.

The most common way to finance water supply system improvements in the past has been federal and state grant programs. Block grants from HUD are still available for some projects and Rural Utilities Service (RUS) loans may also be used as a funding source. In addition, state revolving fund (SRF) programs provide loans (but not direct grants) for improvements. The SRF program has already been implemented with wastewater improvements and the Safe Drinking Water Regulations also include an SRF program for funding water treatment improvements. These SRF programs will be very competitive and utilities must provide evidence of sound financial management to qualify for these loans. You should contact your state regulatory agency to find out more about the SRF program in your state.

23.85 Financial Assistance

Many small water treatment and distribution systems need additional funds to repair and upgrade their systems. Potential funding sources include loans and grants from federal and state agencies, banks, foundations, and other sources. Some of the federal funding programs for small public utility systems include:

- Appalachian Regional Commission (ARC)
- Department of Housing and Urban Development (HUD) (provides Community Development Block Grants)
- Economic Development Administration (EDA)
- Indian Health Service (IHS)
- Rural Utilities Service (RUS) (formerly Farmer's Home Administration (FmHA) and Rural Development Administration (RDA))

For additional information, see "Financing Assistance Available for Small Public Water Systems," by Susan Campbell, Benjamin W. Lykins, Jr., and James A. Goodrich, *Journal American Water Works Association*, June 1993, pages 47–53.

Another valuable contact is the Environmental Financing Information Network (EFIN), which provides information on financing alternatives for state and local environmental programs and projects in the form of abstracts of publications, case studies, and contacts. Contact Environmental Financing Information Network, US Environmental Protection Agency (EPA), EFIN (mail code 2731R), Ariel Rios Building, 1200 Pennsylvania Avenue, NW, Washington, DC 20460. Phone (202) 564-4994 and FAX (202) 565-2587.

Also, many states have one or more special financing mechanisms for small public utility systems. These funds may be in the form of grants, loans, bonds, or revolving loan funds. Contact your state drinking water agency for more information.

QUESTIONS

Write your answers in a notebook and then compare your answers with those on pages 696 and 697.

23.8A List the three main areas of financial management for a utility.

23.8B How is a utility's operating ratio calculated?

23.8C Why is it important for a manager to consult with other utility personnel and with community leaders during the budget process?

23.8D How can long-term capital improvements be financed?

23.8E What is a revenue bond?

END OF LESSON 2 OF 3 LESSONS

on

ADMINISTRATION

Please answer the discussion and review questions next.

DISCUSSION AND REVIEW QUESTIONS
Chapter 23. ADMINISTRATION
(Lesson 2 of 3 Lessons)

Write the answers to these questions in your notebook. The question numbering continues from Lesson 1.

18. With whom do managers need to communicate?

19. What information should be included in the utility's annual report?

20. List four steps that can be taken during a meeting to make sure it is a productive meeting.

21. What happens any time you or a member of your utility comes in contact with the public?

22. What attitude should management try to develop among its employees regarding the customer?

23. What is the value of customer complaints?

24. How do you measure financial stability for a utility?

25. How can a manager prepare a good budget?

CHAPTER 23. ADMINISTRATION

(Lesson 3 of 3 Lessons)

23.9 OPERATIONS AND MAINTENANCE

23.90 The Manager's Responsibilities

A utility manager's specific operation and maintenance (O & M) responsibilities vary depending on the size of the utility. At a small utility, the manager may oversee all utility operations while also serving as chief operator and supervising a small staff of operations and maintenance personnel. In larger utility agencies, the manager may have no direct, day-to-day responsibility for operations and maintenance but is ultimately responsible for efficient, cost-effective operation of the entire utility. Whether large or small, every utility needs an effective operations and maintenance program.

23.91 Purpose of O & M Programs

The purpose of O & M programs is to maintain design functionality (capacity) or to restore the system components to their original condition and thus functionality. Stated another way, does the system perform as designed and intended? The ability to effectively operate and maintain a water supply utility so it performs as intended depends greatly on proper design (including selection of appropriate materials and equipment), construction and inspection, acceptance, and system start-up. Permanent system deficiencies that affect O & M of the system are frequently the result of these phases. O & M staff should be involved at the beginning of each project, including planning, design, construction, acceptance, and start-up. When a utility system is designed with future O & M considerations in mind, the result is a more effective O & M program in terms of O & M cost and performance.

Effective O & M programs are based on knowing what components make up the system, where they are located, and the condition of the components. With that information, proactive maintenance can be planned and scheduled, rehabilitation needs identified, and long-term Capital Improvement Programs (CIP) planned and budgeted. High-performing agencies have all developed performance measurements of their O & M program and track the information necessary to evaluate performance.

23.92 Types of Maintenance

Water supply and distribution system maintenance can be either a proactive or a reactive activity. Commonly accepted types of maintenance include three classifications: corrective maintenance, preventive maintenance, and predictive maintenance.

Corrective maintenance, including emergency maintenance, is reactive. For example, a piece of equipment or a system is allowed to operate until it fails, with little or no scheduled maintenance occurring prior to the failure. Only when the equipment or system fails is maintenance performed. Reliance on reactive maintenance will always result in poor system performance, especially as the system ages. Utility agencies taking a corrective maintenance approach are characterized by:

- The inability to plan and schedule work
- The inability to budget adequately
- Poor use of resources
- A high incidence of equipment and system failures

Emergency maintenance involves two types of emergencies: normal emergencies and extraordinary emergencies. Public utilities are faced with normal emergencies such as water main breaks on a daily basis. Normal emergencies can be reduced by an effective maintenance program. Extraordinary emergencies, such as high-intensity rainstorms, hurricanes, floods, and earthquakes, will always be unpredictable occurrences. However, the effects of extraordinary emergencies on the utility's performance can be minimized by implementation of a planned maintenance program and development of a comprehensive emergency response plan (see Section 23.10).

Preventive maintenance is proactive and is defined as a programmed, systematic approach to maintenance activities. This type of maintenance will always result in improved system performance except in the case where major chronic problems are the result of design or construction flaws that cannot be corrected by O & M activities. Proactive maintenance is performed on a periodic (preventive) basis or an as-needed (predictive) basis. Preventive maintenance can be scheduled on the basis of specific criteria such as equipment operating time since the last maintenance was performed, or passage of a certain amount of time (calendar period). Lubrication of motors, for example, is frequently based on running time.

The major elements of a good preventive maintenance program include the following:

1. Planning and scheduling
2. Records management
3. Spare parts management
4. Cost and budget control

5. Emergency repair procedures
6. Training program

Some benefits of taking a preventive maintenance approach are:

1. Maintenance can be planned and scheduled
2. Work backlog can be identified
3. Adequate resources necessary to support the maintenance program can be budgeted
4. Capital Improvement Program (CIP) items can be identified and budgeted for
5. Human and material resources can be used effectively

Predictive maintenance, which is also proactive, is a method of establishing baseline performance data, monitoring performance criteria over a period of time, and observing changes in performance so that failure can be predicted and maintenance can be performed on a planned, scheduled basis. Knowing the condition of the system makes it possible to plan and schedule maintenance as required and thus avoid unnecessary maintenance.

In reality, every agency operates their system with corrective and emergency maintenance, preventive maintenance, and predictive maintenance methods. The goal, however, is to reduce the corrective and emergency maintenance efforts by performing preventive maintenance, which will minimize system failures that result in stoppages and overflows.

System performance is frequently a reliable indicator of how the system is operated and maintained. Agencies that rely primarily on corrective maintenance as their method of operating and maintaining the system are never able to focus on preventive and predictive maintenance. With most of their resources directed at corrective maintenance activities, it is difficult to free up these resources to begin developing preventive maintenance programs. For an agency to develop an effective proactive maintenance program, they must add initial resources over and above those currently existing.

23.93 Benefits of Managing Maintenance

The goal of managing maintenance is to minimize investments of labor, materials, money, and equipment. In other words, we want to manage our human and material resources as effectively as possible, while delivering a high level of service to our customers. The benefits of an effective operation and maintenance program are as follows:

- Ensuring the availability of facilities and equipment as intended.
- Maintaining the reliability of the equipment and facilities as designed. Utility systems are required to operate 24 hours per day, 7 days per week, 365 days per year. Reliability is a critical component of the operation and maintenance program. If equipment and facilities are not reliable, then the ability of the system to perform as designed is impaired.
- Maintaining the value of the investment. Water supply and distribution systems represent major capital investments for communities and are major capital assets of the community. If maintenance of the system is not managed, equipment and facilities will deteriorate through normal use and age. Maintaining the value of the capital asset is one of the utility manager's major responsibilities. Accomplishing this goal requires ongoing investment to maintain existing facilities and equipment, extend the life of the system, and establish a comprehensive O & M program.
- Obtaining full use of the system throughout its design life.
- Collecting accurate information and data on which to base the operation and maintenance of the system and justify requests for the financial resources necessary to support it.

QUESTIONS

Write your answers in a notebook and then compare your answers with those on page 697.

23.9A What is the purpose of an operation and maintenance (O & M) program?

23.9B What are the three common types of maintenance?

23.9C List the major elements of a good preventive maintenance program.

23.94 Computer Control Systems

Computer control systems are used by operators and administrators to monitor and adjust their treatment processes and facilities. One type of computer control system is the SCADA system. SCADA stands for Supervisory Control And Data Acquisition system.

23.940 Description of SCADA Systems

A SCADA system collects, stores, and analyzes information about all aspects of operation and maintenance, transmits alarm signals, when necessary, and allows fingertip control of alarms, equipment, and processes. SCADA provides the information that operators need to solve minor problems before they become major incidents. As the nerve center of a water utility, the system allows operators to enhance the efficiency of their water facility by keeping them fully informed and fully in control.

A typical SCADA system is made up basically of five groups of components:

1. Process instrumentation and control devices
2. Input/output (I/O) interface
3. Central processing unit (CPU)
4. Communications interface
5. Human machine interface (HMI)

Applications for SCADA systems include water distribution system control and monitoring, water treatment plant control monitoring, and other related applications. SCADA systems can vary from merely data collection and storage to total data analysis, interpretation, and process control.

In water applications, SCADA systems monitor levels, pressures, and flows and also operate pumps, valves, and alarms. They monitor temperatures, speeds, motor currents, pH, chlorine residual, dissolved oxygen levels, and other operating guidelines, and provide control as necessary. SCADA also logs event and analog signal trends and monitors equipment operating time for maintenance purposes.

A SCADA system might include liquid level, pressure, and flow sensors. The measured (sensed) information could be transmitted to a computer system, which stores, analyzes, and presents the information. The information may be read by an operator on dials or as digital readouts or analyzed and plotted by the computer as trend charts.

Most SCADA systems present a graphical picture of the overall system on a computer screen. In addition, detailed pictures of specific portions of the system can be examined by the operator following a request and instructions to the computer. The graphical displays on the computer screen can include current operating information. The operator can observe this information, analyze it for trends, or determine if it is within acceptable operating ranges, and then decide if any adjustments or changes are necessary.

SCADA systems are capable of analyzing data and providing operating, maintenance, regulatory, and annual reports. In some plants, operators rely on a SCADA system to help them prepare daily, weekly, and monthly maintenance schedules; monitor the spare parts inventory status; order additional spare parts when necessary; print out work orders; and record completed work assignments. SCADA systems can also be used to enhance energy conservation programs and emergency response procedures.

QUESTIONS

Write your answers in a notebook and then compare your answers with those on page 697.

23.9D What does SCADA stand for?

23.9E What does a SCADA system do?

23.941 *Typical Water Treatment and Distribution SCADA Systems*

Water treatment and water distribution SCADA systems are usually linked together with the controls located at the water treatment plant. Information that historically was recorded on paper strip-charts is now being recorded and stored (archived) by computers. This information can be easily retrieved and reviewed by a SCADA system, rather than examining years of strip-chart records to find needed information. Therefore, SCADA systems quickly provide operators with the information they need to make informed decisions.

SCADA systems are also used by operators at water treatment plants to:

- Monitor filtration system influent and effluent turbidity levels, head losses through the filters, filter flows, and filter valve settings
- Control filter backwashing
- Study clear well water levels and fluctuations, as well as expected consumer demands

Distribution system service storage tank levels, system demands, and system pressures are all recorded and plotted by SCADA systems. Operators study this information and decide if adjustments are necessary in booster pump operation or target levels in service storage tanks. SCADA systems can plot system pressures over time against flow demands. A steady pressure in spite of fluctuating flows shows that the operators are in control of the system. If a customer phones and complains about a loss of water pressure, a review of system pressures can indicate if a problem exists or if the problem might be in the customer's home.

Hand-held computers are being used to assign work orders to field crews. Crews could be instructed to perform routine preventive maintenance on hydrants and valves or to investigate a drinking water taste or odor complaint. The field crew uses the computer to record standard tasks performed and also to record special comments. The standardized comments allow for field information to be retrieved and analyzed (queried).

Electrical energy consumption can be optimized by the use of SCADA systems. Time-power management describes procedures used to minimize power costs and meet consumer water and pressure demands. Most power companies are eager to work with operators to try to increase water treatment and distribution power consumption during periods when electrical system power demands are low and also to decrease water power consumption during periods of peak demands on electrical power supplies. SCADA systems also monitor power consumption and conduct a diagnostic performance of pumps. Operators can review this information and then identify potential pump problems before they become serious.

With proper security implementation, SCADA systems allow operators to have remote access to plant controls from anywhere using a laptop or remote workstation. This provides the flexibility for off-duty staff to help on-duty operators solve operational problems. Computer networking systems allow operators at terminals in offices, in plants, and in the field to work together and use the same information or whatever information they need from one central computer database.

The greatest challenge for operators using SCADA systems is to realize that just because a computer says something (a pump is operating as expected), *this does not mean that the computer is always correct.* Also, when the system fails due to a power failure or for any other reason (natural disaster), operators will be required to operate the plant manually and without critical information. Could you do this?

Operators will always be needed to question and analyze the results from SCADA systems. They will be needed to *see* if the floc looks OK, to *listen* to a pump to be sure it sounds proper, and to *smell* the water and the equipment to determine if unexpected or unidentified changes are occurring. Water treatment plants and distribution systems will always need alert, knowledgeable, and experienced operators who have a feel for their plant and their distribution system.

For more information on SCADA systems, see Chapter 19, "Instrumentation and Control Systems," Section 19.37, "Process Computer Control Systems."

QUESTIONS

Write your answers in a notebook and then compare your answers with those on page 697.

23.9F List four items that operators at water treatment plants could use a SCADA system to monitor.

23.9G What can an operator do when a customer phones and complains about a loss of water pressure?

23.9H What are the greatest challenges for operators using SCADA systems?

23.95 Cross Connection Control Program

23.950 Importance of Cross Connection Control

BACKFLOW[6] of contaminated water through cross connections into community water systems is not just a theoretical problem. Contamination through cross connections has consistently caused more waterborne disease outbreaks in the United States than any other reported factor. Inspections have often disclosed numerous unprotected cross connections between public water systems and other piped systems on consumers' premises, which might contain wastewater, stormwater, processed waters (containing a wide variety of chemicals), and untreated supplies from private wells, streams, and ocean waters. Therefore, an effective cross connection control program is essential.

Backflow results from either *BACK PRESSURE*[7] or *BACKSIPHONAGE*[8] situations in the distribution system. Back pressure occurs when the user's water supply is at a higher pressure than the public water supply system (Figure 23.4). Typical locations where back pressure problems could develop include services to premises where wastewater or toxic chemicals are handled under pressure or where there are unapproved auxiliary water supplies such as a private well or the use of surface water or seawater for firefighting. Backsiphonage is caused by the development of negative or below atmospheric pressures in the water supply piping (Figure 23.5). This condition can occur when there are extremely high water demands (firefighting), water main breaks, or the use of on-line booster pumps.

The best way to prevent backflow is to permanently eliminate the hazard. Back pressure hazards can be eliminated by severing (eliminating) any direct connection at the pump causing the back pressure with the domestic water supply system. Another solution to the problem is to require an air gap separation device (Figure 23.6) where the water supply service line connects to the private system under pressure from the pump. To eliminate or minimize backsiphonage problems, proper enforcement of plumbing codes and improved water distribution and storage facilities will be helpful. As an additional safety factor for certain selected conditions, a double check valve (Figure 23.7) may be required at the meter.

23.951 Program Responsibilities

Responsibilities in the implementation of cross connection control programs are shared by water suppliers, water users (businesses and industries), health agencies, and plumbing officials.

The water supplier is responsible for preventing contamination of the public water system by backflow. This responsibility begins at the source, includes the entire distribution system, and ends at the user's connection. To meet this responsibility, the water supplier must issue (promulgate) and enforce needed laws, rules, regulations, and policies. Water service should not be provided to premises where the strong possibility of an unprotected cross connection exists. The essential elements of a water supplier cross connection control program are discussed in the next section.

The water user is responsible for keeping contaminants out of the potable water system on the user's premises. When backflow prevention devices are required by the health agency or water supplier, the water user must pay for the installation, testing, and maintenance of the approved devices. The user is also responsible for preventing the creation of cross connections through modifications of the plumbing system on the premises. The health agency or water supplier may, when necessary, require a water user to designate a water supervisor or foreman to be responsible for the cross connection control program within the water user's premises.

---

[6] *Backflow.* A reverse flow condition, created by a difference in water pressures, that causes water to flow back into the distribution pipes of a potable water supply from any source or sources other than an intended source. Also see BACKSIPHONAGE.

[7] *Back Pressure.* A pressure that can cause water to backflow into the water supply when a user's water system is at a higher pressure than the public water system.

[8] *Backsiphonage.* A form of backflow caused by a negative or below atmospheric pressure within a water system. Also see BACKFLOW.

656 Water Treatment

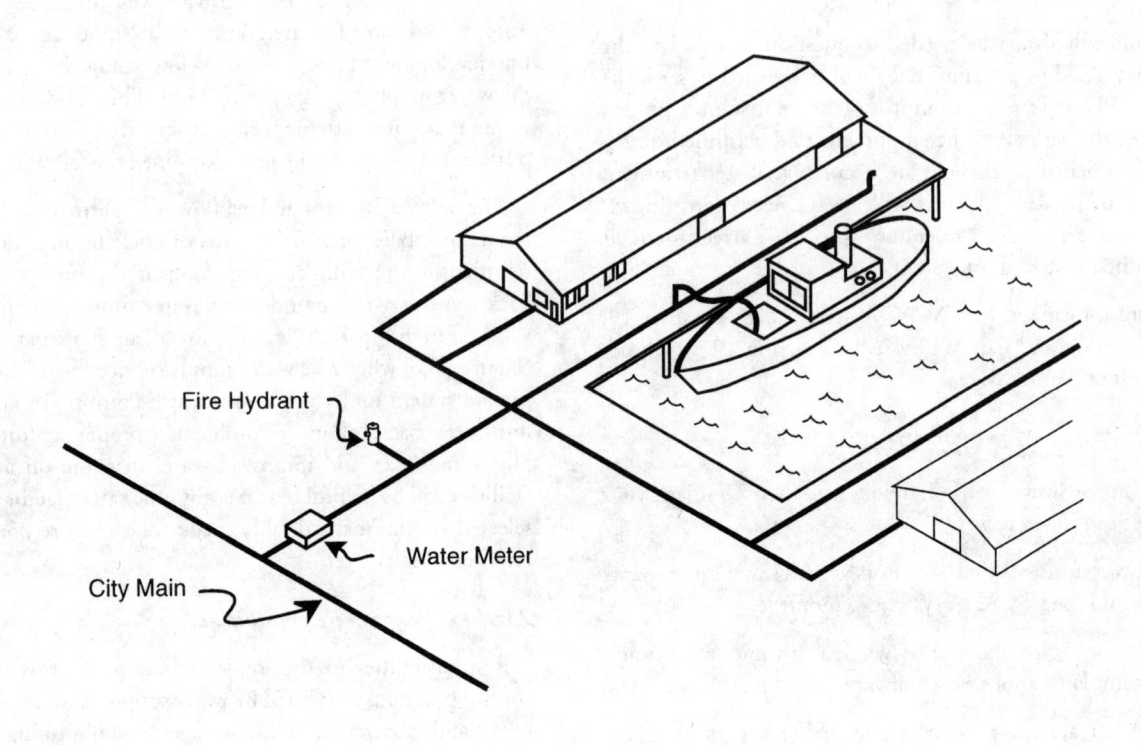

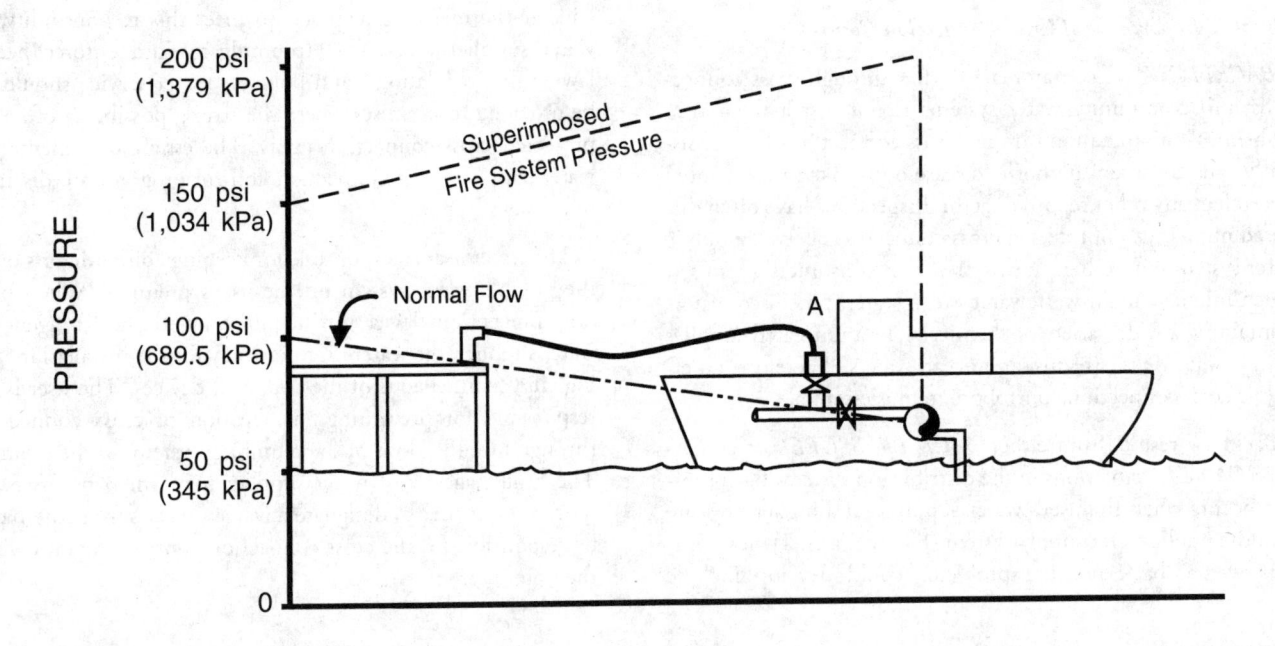

Fig. 23.4 Backflow due to back pressure
(Source: *MANUAL OF CROSS-CONNECTION CONTROL PRACTICES AND PROCEDURES*,
Sanitary Engineering Branch, California Department of Health Services, Berkeley, CA)

Administration 657

DISTRIBUTION SYSTEM

HYDRAULIC GRADIENT

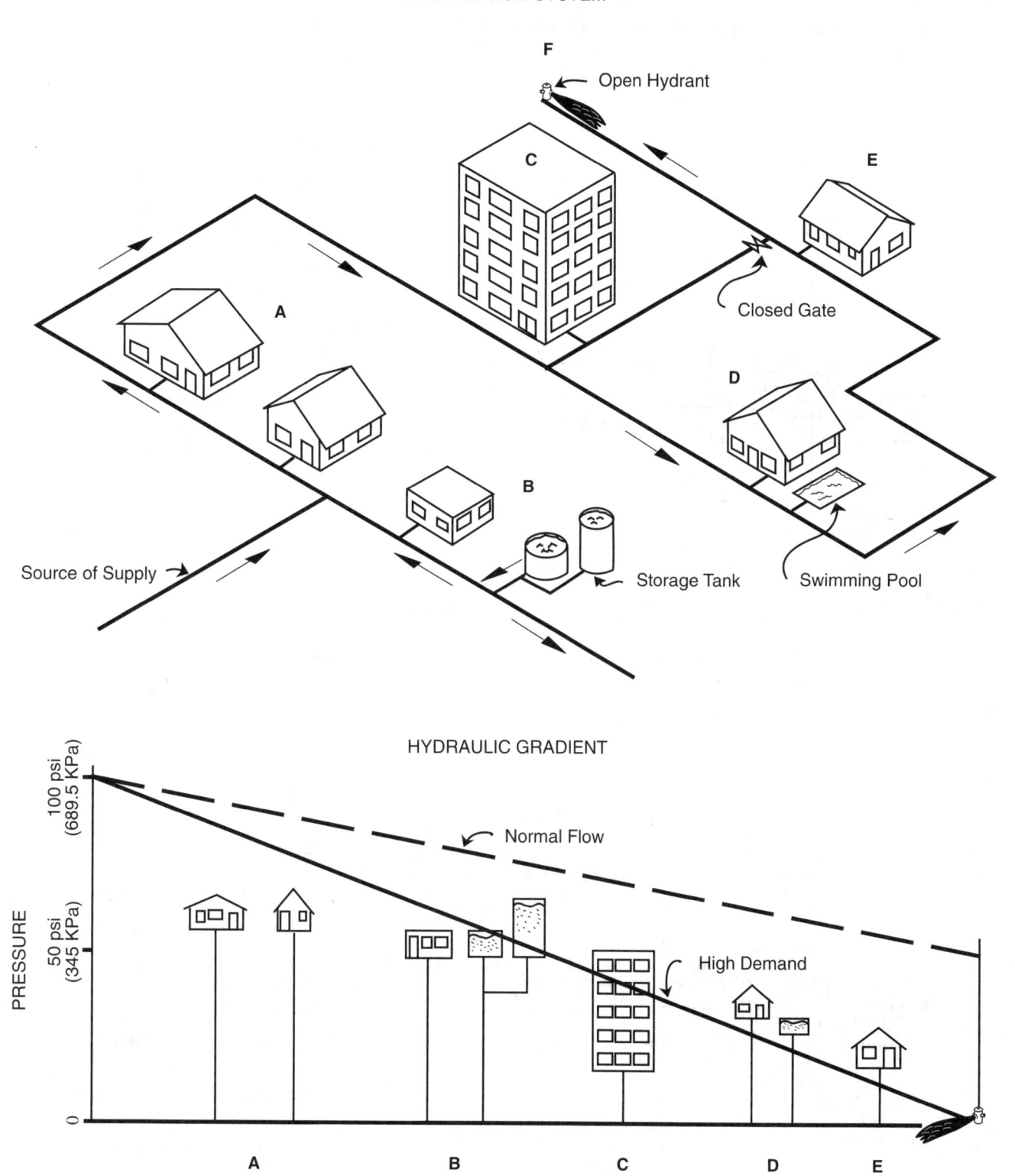

Fig. 23.5 Backsiphonage due to extremely high water demand
(Source: *MANUAL OF CROSS-CONNECTION CONTROL PRACTICES AND PROCEDURES*,
Sanitary Engineering Branch, California Department of Health Services, Berkeley, CA)

TANK SHOULD BE OF SUBSTANTIAL CONSTRUCTION AND OF A KIND AND SIZE TO SUIT CONSUMER'S NEEDS. TANK MAY BE SITUATED AT GROUND LEVEL (WITH A PUMP TO PROVIDE ADEQUATE PRESSURE HEAD) OR BE ELEVATED ABOVE THE GROUND.

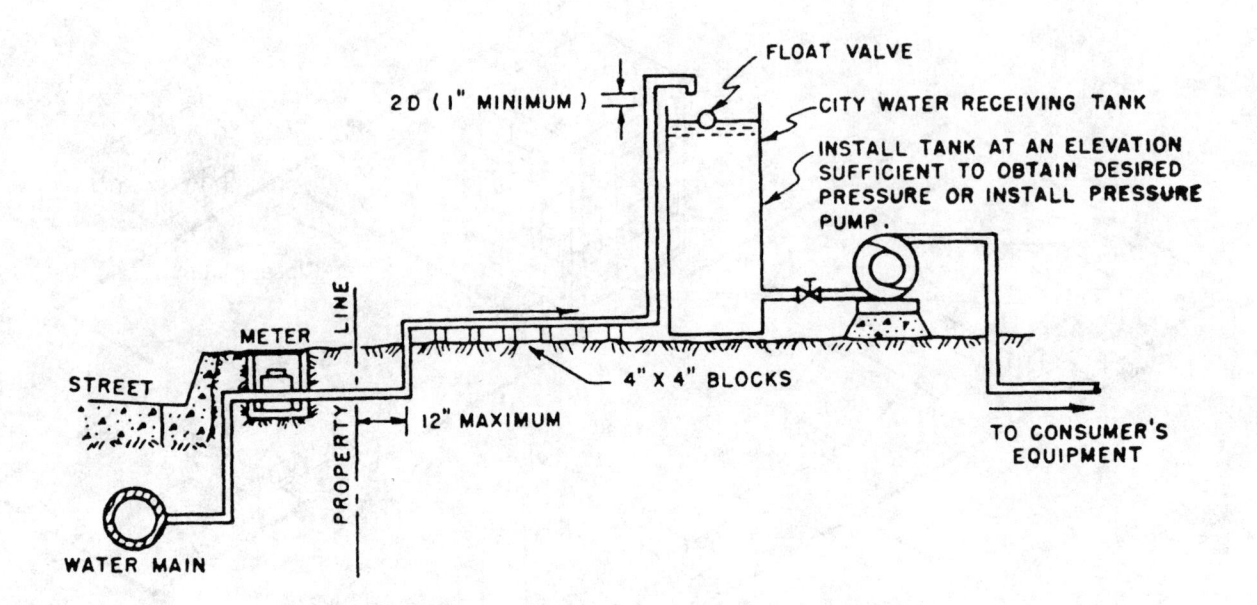

Fig. 23.6 Typical air gap separation
(From *MANUAL OF CROSS-CONNECTION CONTROL PRACTICES AND PROCEDURES*, Sanitary Engineering Branch, California Department of Health Services, Berkeley, CA)

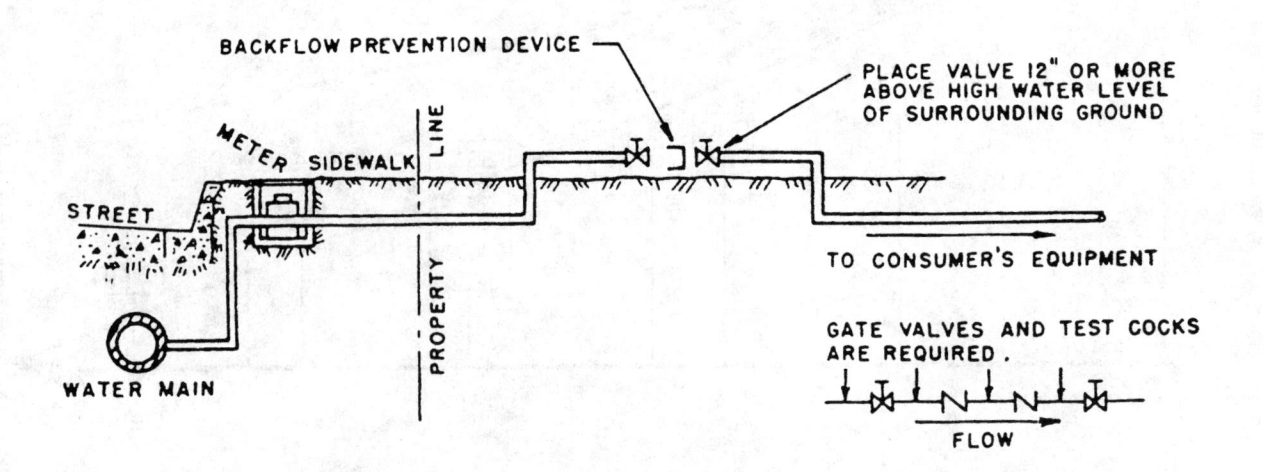

Fig. 23.7 Typical double check valve backflow prevention device
(From *MANUAL OF CROSS-CONNECTION CONTROL PRACTICES AND PROCEDURES*, Sanitary Engineering Branch, California Department of Health Services, Berkeley, CA)

The local or state health agency is responsible for issuing and enforcing laws, rules, regulations, and policies needed to control cross connections. Also, this agency must have a program that ensures maintenance of an adequate cross connection control program. Protection of the system on the user's premises is provided, where needed, by the water utilities.

The plumbing agency (building inspector) is responsible for the enforcement of building regulations relating to prevention of cross connections on the user's premises.

QUESTIONS

Write your answers in a notebook and then compare your answers with those on page 697.

23.9I   What has caused more waterborne disease outbreaks in the United States than any other reported factor?

23.9J   Who is usually responsible for the implementation of cross connection control programs?

23.9K   What is the water user's cross connection control responsibility?

23.952 Water Supplier Program

The following elements should be included in each water supplier's cross connection control program:

1. Enactment of an ordinance providing enforcement authority if the supplier is a government agency, or enactment of appropriate rules of service if the system is investor-owned.[9]

2. Training of personnel on the causes of and hazards from cross connections and procedures to follow for effective cross connection control.

3. Listing and inspection or reinspection on a priority basis of all existing facilities where cross connections are of concern.

4. Review and screening of all applications for new services or modification of existing services for cross connection hazards to determine if backflow protection is needed.

5. Obtaining a list of approved backflow prevention devices and a list of certified testers, if available.

6. Acceptable installation of the proper type of device needed for the specific hazard on the premises.

7. Routine testing of installed backflow prevention devices as required by the health agency or the water supplier. Contact the health agency for approved procedures.

8. Maintenance of adequate records for each backflow prevention device installed, including records of inspection and testing.

9. Notification of each water user when a backflow prevention device has to be tested. This should be done after installation or repair of the device and at least once a year.

10. Maintenance of adequate pressures throughout the distribution system at all times to minimize the hazards from any undetected cross connections that may exist.

All field personnel should be constantly alert for situations where cross connections are likely to exist, whether protection has been installed or not. An example is a contractor using a fire hose from a hydrant to fill a tank truck for dust control or the jetting (for compaction) of pipe trenches. Operators should especially be on the lookout for illegal bypassing of installed backflow prevention devices.

23.953 Types of Backflow Prevention Devices

Different types of backflow prevention devices are available. The particular type of device most suitable for a given situation depends on the degree of health hazard, the probability of backflow occurring, the complexity of the piping on the premises, and the probability of the piping being modified. The higher the assessed risk due to these factors, the more reliable and positive the type of device needed. The types of devices normally approved are listed below according to the degree of assessed risk, with the type of device providing the greatest protection listed first. Only the first three devices are approved for use at service connections.

1. Air gap separation

2. Reduced pressure principle (RPP) device

3. Double check valve

4. Pressure vacuum breaker (only used for internal protection on the premises)

5. Atmospheric (nonpressure) vacuum breaker

Figure 23.6 shows a typical air gap separation device and its recommended location. Figure 23.7 shows the installation of a typical double check valve backflow prevention device. These devices are normally installed on the water user's side of the connection to the utility's system and as close to the connection as practical. Figure 23.8 shows typical installations of atmospheric and pressure vacuum breakers.

Only backflow prevention devices that have passed both laboratory and field evaluations by a recognized testing agency and that have been accepted by the health agency and the water supplier should be used.

23.954 Devices Required for Various Types of Situations

The state or local health agency should be contacted to determine the actual types of devices acceptable for various situations

---

[9] A typical ordinance is available in *CROSS-CONNECTION CONTROL MANUAL,* EPA No. 570-9-89-007, available from National Technical Information Service (NTIS), 5285 Port Royal Road, Springfield, VA 22161. Order No. PB91-145490. Price, $33.50, plus $5.00 shipping and handling per order.

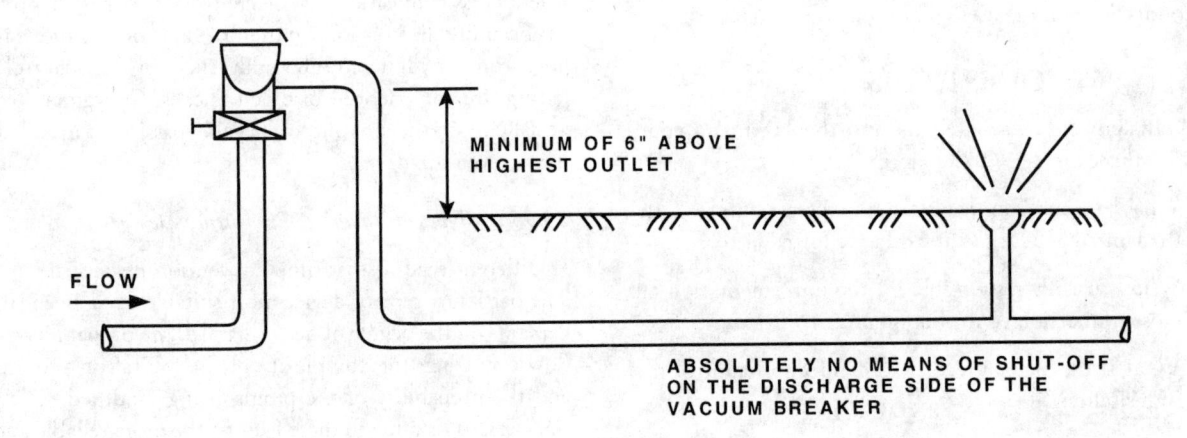

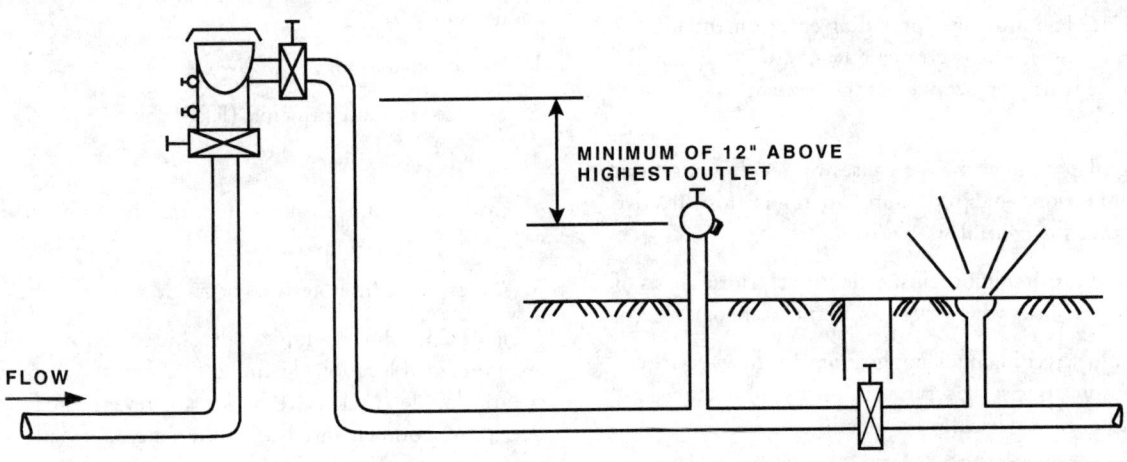

Fig. 23.8 Typical installations of atmospheric (top) and pressure (bottom) vacuum breakers
(From *MANUAL OF CROSS-CONNECTION CONTROL PRACTICES AND PROCEDURES*,
Sanitary Engineering Branch, California Department of Health Services, Berkeley, CA)

inside the consumer's premises. However, the types of devices generally acceptable for particular situations can be mentioned.

An air gap separation or a reduced pressure principle (RPP) device is normally required at services to wastewater treatment plants, wastewater pumping stations, reclaimed water reuse areas, areas where toxic substances in toxic concentrations are handled under pressure, and premises having an auxiliary water supply that is or may be contaminated. The ultimate degree of protection is also needed in cases where fertilizer, herbicides, or pesticides are injected into a sprinkler system.

A double check valve device should be required when a moderate hazard exists on the premises or where an auxiliary supply exists, but adequate protection on the premises is provided.

Atmospheric and pressure vacuum breakers are usually required for irrigation systems; however, they are not adequate in situations where they may be subject to back pressure. If there is a possibility of back pressure, a reduced pressure principle device is needed.

23.96 Geographic Information System (GIS)

The geographic information system (GIS) is a computer program that combines mapping with detailed information about the physical structures within geographic areas. To create the database of information, entities within a mapped area, such as streets, fire hydrants, distribution system line segments, storage facilities, and wells, are given "attributes." Attributes are simply the pieces of information about a particular feature or structure that are stored in a database. The attributes can be as basic as an address, tank storage capacity, or line segment length, or they may be as specific as diameter, tank dimensions, and quadrant (coordinate) location. Attributes of a main line segment might include engineering information, maintenance information, and inspection information. Thus, an inventory of entities and their properties is created. The system allows the operator to periodically update the map entities and their corresponding attributes.

The power of a GIS is that information can be retrieved geographically. An operator can choose an area to look at by pointing to a specific place on the map or outlining (windowing) an area of the map. The system will display the requested section on the screen and show the attributes of the entities located on the map. A printed copy may also be requested. In most cases, CMMS (Computer-based Maintenance Management Systems) software has the ability to communicate with geographic information systems so that attribute information from the distribution system can be copied into the GIS.

A GIS can generate work orders in the form of a map with the work to be performed outlined on the map. This minimizes paper work and gives the work crew precise information about where the work is to be performed. Completion of the work is recorded in the GIS to keep the work history for the area and entity up to date. Reports and other inquiries can be requested, as needed, for example, a listing of all line segments in a specific area could be generated for a report.

In many areas, GISs are being developed on an area-wide basis with many agencies, utilities, counties, cities, and state agencies participating. Usually, a county-wide base map is developed and then all participants provide attributes for their particular systems. For example, information on the sanitary sewer collection system might be one map layer, the second map layer might be the water distribution system, and the third layer might be the electric utility distribution system. In addition to sharing databases with CMMSs, GISs have the ability to operate smoothly with computer-aided design (CAD) systems.

QUESTIONS

Write your answers in a notebook and then compare your answers with those on page 697.

23.9L What factors should be considered when selecting a suitable backflow prevention device for a given situation?

23.9M Air gap separation or reduced pressure principle backflow prevention devices are installed under what conditions?

23.9N What makes a geographic information system (GIS) a potentially powerful tool for a water treatment or distribution system operator?

23.10 EMERGENCY RESPONSE

23.100 Planning

Contingency planning is an essential facet of utility management and one that is often overlooked. Although utilities in various locations will be vulnerable to somewhat different kinds of natural disasters, the effects of these disasters often will be similar. As a first step toward an effective contingency plan, each utility should make an assessment of its own vulnerability and then develop and implement a comprehensive plan of action.

All utilities suffer from common problems such as equipment breakdowns and leaking pipes. During the past few years, there has also been an increasing amount of vandalism, civil disorder, toxic spills, and employee strikes, which have threatened to disrupt utility operations. In observing today's international tension and the potential for nuclear war or the effects of terrorist-induced chemical or biological warfare, water utilities must seriously consider how to respond. Natural disasters such as floods, earthquakes, hurricanes, forest fires, avalanches, and blizzards are a more or less routine occurrence for some utilities. When such catastrophic emergencies occur, the utility must be prepared to minimize the effects of the event and have a plan for rapid recovery. Such preparation should be a specific obligation of every utility manager.

Start by assessing the vulnerability of the utility during various types of emergency situations. If the extent of damage can be estimated for a series of most probable events, the weak elements can be studied and protection and recovery operations can center on these elements. Although all elements are important for the utility to function, experience with disasters points out elements that are most subject to disruption. These elements are:

1. The absence of trained personnel to make critical decisions and carry out orders
2. The loss of power to the utility's facilities
3. An inadequate amount of supplies and materials
4. Inadequate communication equipment

The following steps should be taken in assessing the vulnerability of a system:

1. Identify and describe the system components.
2. Assign assumed disaster characteristics.
3. Estimate disaster effects on system components.
4. Estimate customer demand for service following a potential disaster.
5. Identify key system components that would be primarily responsible for system failure.

If the assessment shows a system is unable to meet estimated requirements because of the failure of one or more critical components, the vulnerable elements have been identified. Repeating this procedure using several typical disasters will usually point out system weaknesses. Frequently, the same vulnerable element appears for a variety of assumed disaster events.

You might consider, for example, the case of the addition of toxic pollutants to water supplies. The list of toxic agents that may have a harmful effect on humans is almost endless. However, it is recognized that there is a relationship between the quantity of toxic agents added to the treatment provided for the supply. Adequate chlorination is effective against most biological agents. Other considerations are the amount of dilution water and the solubility of the chemical agents. There is the possibility that during normal detention times many of the biological agents will die off with adequate chlorination.

Although the drafting of an emergency plan for a water system may be a difficult job, the existence of such a plan can be of critical importance during an emergency situation.

An emergency operations plan need not be too detailed, since all types of emergencies cannot be anticipated and a complex response program can be more confusing than helpful. Supervisory personnel must have a detailed description of their responsibilities during emergencies. They will need information, supplies, equipment, and the assistance of trained personnel. All these can be provided through a properly constructed emergency operations plan that is not extremely detailed.

The following outline can be used as the basis for developing an emergency operations plan:

1. Make a vulnerability assessment.
2. Inventory organizational personnel.
3. Provide for a recovery operation (plan).
4. Provide training programs for operators in carrying out the plan.
5. Coordinate with local and regional agencies such as the health, police, and fire departments to develop procedures for carrying out the plan.
6. Establish a communications procedure.
7. Provide protection for personnel, plant equipment, records, and maps.

By following these steps, an emergency plan can be developed and maintained even though changes in personnel may occur. "Emergency simulation" training sessions, including the use of standby power, equipment, and field test equipment will ensure that equipment and personnel are ready at times of emergency.

A list of phone numbers for operators to call in an emergency should be prepared and posted by a phone for emergency use. The list should include:

1. Plant supervisor
2. Director of public works or head of utility agency
3. Police
4. Fire
5. Doctor (2 or more)
6. Ambulance (2 or more)
7. Hospital (2 or more)

If appropriate for your utility, also include the following phone numbers on the emergency list:

8. Chlorine supplier and manufacturer
9. *CHEMTREC,* (800) 424-9300, for the hazardous chemical spills; sponsored by the Manufacturing Chemists Association
10. US Coast Guard's National Response Center, (800) 424-8802
11. Local and state poison control centers
12. Local hazardous materials spill response team

You should prepare a list for your plant *now,* if you have not already done so, and update the numbers annually.

For additional information on emergencies, see Chapter 7, "Disinfection," Section 7.52, "Chlorine Leaks"; Chapter 10, "Plant Operation," Section 10.9, "Emergency Conditions and Procedures"; and Chapter 18, "Maintenance," Section 18.02, "Emergencies."

QUESTIONS

Write your answers in a notebook and then compare your answers with those on pages 697 and 698.

23.10A What is the first step toward an effective contingency plan for emergencies?

23.10B Why is too detailed an emergency operations plan not needed or even desirable?

23.10C An emergency operations plan should include what specific information?

23.101 Homeland Defense

World events in recent years have heightened concern in the United States over the security of one of America's most valuable resources, the critical drinking water treatment, storage, and distribution infrastructure. Water distribution pipelines form an extensive network that runs near or beneath key buildings and roads and is physically close to many communication and transportation networks. Significant damage to the nation's water supply or distribution facilities could result in: loss of life, contamination of drinking water supplies, long-term public health impacts, catastrophic environmental damage to rivers, lakes, and wetlands, and disruption to commerce, the economy, and our normal way of life.

Water treatment and distribution facilities have been identified as a target for international and domestic terrorism. This knowledge, coupled with the responsibility of the facility to provide a safe and healthful workplace, requires that management establish rules to protect the workers as well as the facilities. Emergency action and fire prevention plans must identify what steps need to be taken when the threat analysis indicates a potential for attack. These plans must be in writing and be practiced periodically so that all workers know what actions to take.

Some actions that should be taken at all times to reduce the possibility of a terrorist attack are:

- Ensure that all visitors sign in and out of the facilities with a positive ID check.
- Reduce the number of visitors to a minimum.
- Discourage parking by the public near critical buildings to eliminate the chances of car bombs.
- Be cautious with suspicious packages that arrive.
- Be aware of the hazardous chemicals used and how to defend against spills.
- Keep emergency numbers posted near telephones and radios.
- Patrol the facilities frequently, looking for suspicious activity or behavior.
- Maintain, inspect, and use your personal protective equipment (PPE) (hard hats, respirators).

The following recommendations by the EPA[10] include many straightforward, common-sense actions a utility can take to increase security and reduce threats from terrorism.

Guarding Against Unplanned Physical Intrusion

- Lock all doors and set alarms at your office, booster pump stations, treatment plants, and vaults, and make it a rule that doors are locked and alarms are set.
- Limit access to facilities and control access to booster pump stations, chemical and fuel storage areas, giving close scrutiny to visitors and contractors.
- Post guards at treatment plants and post "Employee Only" signs in restricted areas.
- Increase lighting in parking lots, treatment bays, and other areas with limited staffing.
- Control access to computer networks and control systems and change the passwords frequently.
- Do not leave keys in equipment or vehicles at any time.

Making Security a Priority for Employees

- Conduct background security checks on employees at hiring and periodically thereafter.
- Develop a security program with written plans and train employees frequently.
- Ensure all employees are aware of established procedures for communicating with law enforcement, public health, environmental protection, and emergency response organizations.
- Ensure that employees are fully aware of the importance of vigilance and the seriousness of breaches in security.
- Make note of unaccompanied strangers on the site and immediately notify designated security officers or local law enforcement agencies.
- If possible, consider varying the timing of operational procedures so that, to anyone watching for patterns, the pattern changes.

---

[10] Adapted from "What Wastewater Utilities Can Do Now to Guard Against Terrorist and Security Threats," US Environmental Protection Agency, Office of Wastewater Management, October 2001.

- Upon the dismissal of an employee, change pass codes and make sure keys and access cards are returned.
- Provide customer service staff with training and checklists of how to handle a threat if it is called in.

Coordinating Actions for Effective Emergency Response

- Review existing emergency response plans and ensure that they are current and relevant.
- Make sure employees have the necessary training in emergency operating procedures.
- Develop clear procedures and chains of command for reporting and responding to threats and for coordinating with emergency management agencies, law enforcement personnel, environmental and public health officials, consumers, and the media. Practice the emergency procedures regularly.
- Ensure that key utility personnel (both on and off duty) have access to critical telephone numbers and contact information at all times. Keep the call list up to date.
- Develop close relationships with local law enforcement agencies and make sure they know where critical assets are located. Ask them to add your facilities to their routine rounds.
- Report to county or state health officials any illness among the employees that might be associated with water contamination.
- Immediately report criminal threats, suspicious behavior, or attacks on water utilities to law enforcement officials and the nearest field office of the Federal Bureau of Investigation.

Investing in Security and Infrastructure Improvements

- Assess the vulnerability of the distribution system, water storage facilities, major pumping stations, water treatment plants, chemical and fuel storage areas, and other key infrastructure elements.
- Improve computer system and remote operational security.
- Use local citizen watches.
- Seek financing for more expensive and comprehensive system improvements.

The US Terrorism Alert System (Figure 23.9) is a color-coded system that identifies the potential for terrorist activity and suggests specific actions to be taken. Your safety plan should identify the actions that your facility will take when the threat level changes. Tables 23.6 and 23.7 show examples of security measures that should be taken to improve safety at a water treatment or distribution facility when the threat level is YELLOW and when it is ORANGE. (The utility's safety plan should include similar lists of actions for the RED, BLUE, and GREEN levels as well.)

23.102 Handling the Threat of Contaminated Water Supplies[11]

23.1020 Importance

More than 50 water utilities in southern Louisiana were threatened with cyanide poisoning in their water supplies in one year. Such threats can occur anywhere, and every water utility should be prepared to handle this type of emergency.

23.1021 Toxicity

The term "toxicity" is often used when discussing contamination of a water supply with the intention of creating a serious health hazard. Toxicity is the ability of a contaminant (chemical or biological) to cause injury when introduced into the body. The degree of toxicity varies with the concentration of a contaminant required to cause injury, the speed with which the injury takes place, and the severity of the injury.

The effect of a toxic contaminant, once added to a water supply, depends on several things. First, the amount of contaminant added can vary, as can the size of the water supply. In general, it takes larger quantities of a contaminant to be toxic in a larger water supply. Second, the solubility of the contaminant can vary. The more soluble the substance is in water, the more likely it is to cause problems. Finally, the detention time of the contaminant in the water can vary. For example, many biological agents will die before they can cause a problem in the water supply.

Generally, the terms acute and chronic are used to describe toxic agents and their effects. An acute toxic agent causes injury quickly. When the contaminant causes illness in seconds, minutes, or hours after a single exposure or a single dose, it is considered an acute toxic agent. A chronic toxic agent causes injury to occur over an extended period of exposure. Generally, the contaminant is ingested in repeated doses over a period of days, months, or years.

23.1022 Emergency Contaminant Limits (See Table 23.8)

When determining the emergency limits of a contaminant, the following factors should be considered:

1. Quantity or concentration of the contaminant
2. Duration of exposure to the contaminant
3. Physical form of the contaminant (size of particle; physical state—solid, liquid, gas)
4. Attraction of the contaminant to the organism being contaminated
5. Solubility of the contaminant in the organism
6. Sensitivity of the organism to the contaminant

---

[11] This section was reprinted from *OPFLOW*, Vol. 9, No. 3, March 1983, by permission. Copyright 1983, the American Water Works Association.

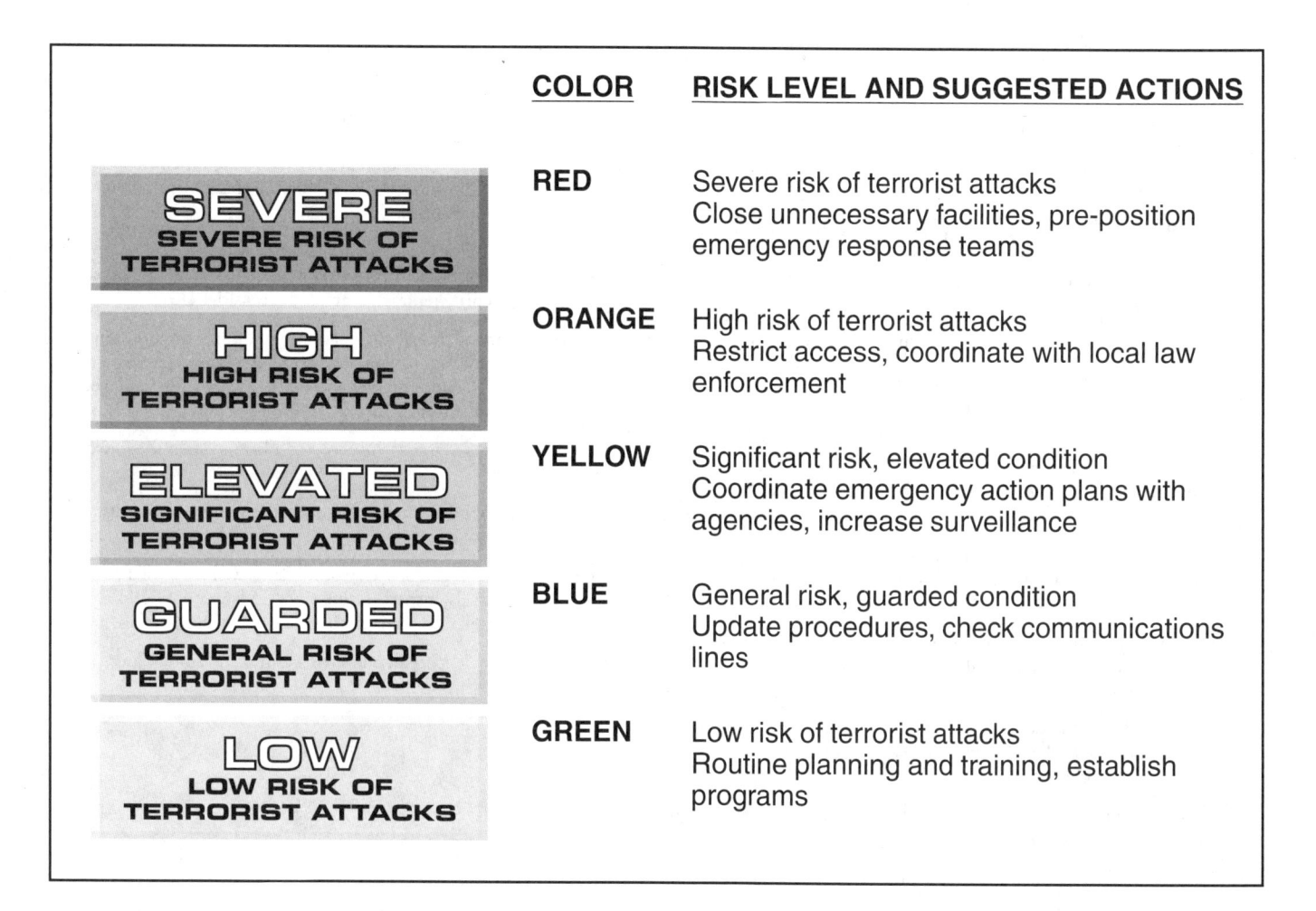

	COLOR	RISK LEVEL AND SUGGESTED ACTIONS
SEVERE — SEVERE RISK OF TERRORIST ATTACKS	RED	Severe risk of terrorist attacks Close unnecessary facilities, pre-position emergency response teams
HIGH — HIGH RISK OF TERRORIST ATTACKS	ORANGE	High risk of terrorist attacks Restrict access, coordinate with local law enforcement
ELEVATED — SIGNIFICANT RISK OF TERRORIST ATTACKS	YELLOW	Significant risk, elevated condition Coordinate emergency action plans with agencies, increase surveillance
GUARDED — GENERAL RISK OF TERRORIST ATTACKS	BLUE	General risk, guarded condition Update procedures, check communications lines
LOW — LOW RISK OF TERRORIST ATTACKS	GREEN	Low risk of terrorist attacks Routine planning and training, establish programs

Fig. 23.9   Threat level categories established by the US Department of Homeland Defense

TABLE 23.6 SECURITY MEASURES FOR THREAT LEVEL YELLOW (CONDITION ELEVATED)

Continue to introduce all measures listed in BLUE: Condition Guarded.	
Detection	**Prevention**
• To the extent possible, increase the frequency and extent of monitoring the flow coming into and leaving the treatment facility and review results against baseline quantities. Increase review of operational and analytical data (including customer complaints) with an eye toward detecting unusual variability (as an indicator of unexpected changes in the system). Variations due to natural or routine operational variability should be considered first. • Increase surveillance activities in water supply, treatment, and distribution facilities.	• Carefully review all facility tour requests before approving. If allowed, implement security measures to include list of names prior to tour, request identification of each attendee prior to tour, prohibit backpacks, duffle bags, and cameras, and identify parking restrictions. • On a daily basis, inspect the interior and exterior of buildings in regular use for suspicious activity or packages, signs of tampering, or indications of unauthorized entry. • Implement mail room security procedures. Follow guidance provided by the United States Postal Service.
Preparedness	**Protection**
• Continue to review, update, and test emergency response procedures and communication protocols. • Establish unannounced security spot checks (such as verification of personal identification and door security) at access control points for critical facilities. • Increase frequency for posting employee reminders of the threat situation and about events that constitute security violations. • Ensure employees understand notification procedures in the event of a security breach. • Conduct security audit of physical security assets, such as fencing and lights, and repair or replace missing/broken assets. Remove debris from along fence lines that could be stacked to facilitate scaling. • Maximize physical control of all equipment and vehicles; make them inoperable when not in use (for example, lock steering wheels, secure keys, chain, and padlock on front-end loaders). • Review draft communications on potential incidents; brief media relations personnel of potential for press contact or issuance of press releases. • Ensure that list of sensitive customers (such as government agencies and medical facilities) within the service area is accurate and shared with appropriate public health officials. • Contact neighboring water utilities to review coordinated response plans and mutual aid during emergencies. • Review whether critical replacement parts are available and accessible. • Identify any work/project taking place in proximity to events where large attendance is anticipated. Consult with the event organizers and local law enforcement regarding contingency plans, security awareness, and site accessibility and control.	• Verify the identity of all persons entering the water utility. Mandate visible use of identification badges. Randomly check identification badges and cards of those on the premises. • At the discretion of the facility manager or security director, remove all vehicles and objects (such as trash containers) located near mission critical facility security perimeters and other sensitive areas. • Verify the security of critical information systems (for example, Supervisory Control And Data Acquisition (SCADA), Internet, e-mail) and review safe computer and Internet access procedures with employees to prevent cyber intrusion. • Consider steps needed to control access to all areas under the jurisdiction of the utility. • Implement critical infrastructure facility surveillance and security plans. • At the beginning and end of each work shift, as well as at other regular and frequent intervals, inspect the interior and exterior of buildings in regular use for suspicious packages, persons, and circumstances. • Lock and regularly inspect all buildings, rooms, and storage areas not in regular use.

TABLE 23.7 SECURITY MEASURES FOR THREAT LEVEL ORANGE (CONDITION HIGH)

Continue to introduce all measures listed in YELLOW: Condition Elevated.	
Detection	Prevention
<ul><li>Increase the frequency and extent of monitoring activities. Review results against baseline measurements.</li><li>Confirm that county and state health officials are on high alert and will inform the utility of any potential waterborne illnesses.</li><li>If a neighborhood watch-type program is in place, notify the community and request increased awareness.</li></ul>	<ul><li>Discontinue tours and prohibit public access to all operational facilities.</li><li>Consider requesting increased law enforcement surveillance, particularly of critical assets and otherwise unprotected areas.</li><li>Limit access to computer facilities. No outside visitors.</li><li>Increase monitoring of computer and network intrusion detection systems and security monitoring systems.</li></ul>
Preparedness	Protection
<ul><li>Confirm that emergency response and laboratory analytical support network are ready for deployment 24 hours per day, 7 days a week.</li><li>Reaffirm liaison with local police, intelligence, and security agencies to determine likelihood of an attack on the water supply, treatment, or distribution utility personnel and facilities and consider appropriate protective measures (such as road closing and extra surveillance).</li><li>Practice communications procedures with local authorities and others cited in the facility's emergency response plan.</li><li>Post frequent reminders for staff and contractors of the threat level, along with a reminder of what events constitute security violations.</li><li>Ensure employees are fully aware of emergency response communication procedures and have access to contact information for relevant law enforcement, public health, environmental protection, and emergency response organizations.</li><li>Have alternative water supply plan ready to implement (for example, bottled water delivery for employees and other critical business uses).</li><li>Place all emergency management and specialized response teams on full alert status.</li><li>Ensure personal protective equipment (PPE) and specialized response equipment is checked, issued, and readily available for deployment.</li><li>Review all plans, procedures, guidelines, personnel details, and logistical requirements related to the introduction of a higher threat condition level.</li></ul>	<ul><li>Evaluate the need to staff the water treatment facility at all times.</li><li>Increase security patrol activity to the maximum level sustainable and ensure tight security in the vicinity of mission critical facilities. Vary the timing of security patrols.</li><li>Request employees change their passwords on critical information management systems.</li><li>Limit building access points to the absolute minimum, strictly enforce entry control procedures. Identify and protect all designated vulnerable points. Give special attention to vulnerable points outside of the critical facility.</li><li>Lock all exterior doors except the main facility entrance(s). Check all visitors' purpose, intent, and identification. Ensure that contractors have valid work orders. Require visitors to sign in upon arrival; verify and record their identifying information. Escort visitors at all times when they are in the facility.</li></ul>

TABLE 23.8 EMERGENCY LIMITS OF SOME CHEMICAL POLLUTANTS IN DRINKING WATER[a]

Chemical	Concentration Limits, mg/L Emergency Short Term (Three Days)	Long Term
Cyanide (CN)	5.0	0.01
Aldrin	0.05	0.032
Chlordane	0.06	0.003
DDT	1.4	0.042
Dieldrin	0.05	0.017
Endrin	0.01	0.001
Heptachlor	0.1	0.018
Heptachlor epoxide	0.05	0.018
Lindane	2.0	0.056
Methoxychlor	2.8	0.035
Toxaphene	1.4	0.005
Beryllium	0.1	0.000
Boron	25.0	1.000
2,4-D	2.0	0.1
Ethylene chlorohydrin	2.0	
Organophosphorus and carbamate pesticides	2.0	0.100
Trinitrotoluene $(NO_2)(C_6H_2CH_3)$	0.75	0.005

[a] These limits, based on current knowledge and informed judgment, have been recommended by knowledgeable persons in the field of toxicology. They are subject to change should new information indicate the need. Additional information on some of the chemicals listed can be found in "Report of the Secretary's Commission on Pesticides and Their Relationship to Environmental Health," Parts I and II, USDHEW, Washington, DC, Dec. 1969.

Concentration of a contaminant can be expressed in two ways. The maximum allowable concentration (MAC) is the maximum concentration of the contaminant allowed in drinking water. Table 23.8 lists several contaminants and their MACs, specifically for short-term emergencies ranging up to three days. The MACs should not be confused with concentration required to have an acute effect on the population. Lethal dose 50 (LD 50) is used to express the concentration of a contaminant that will produce 50 percent fatalities from an average exposure.

23.1023 Protective Measures

A utility can take three approaches to protect its water supply from contamination. First, the utility can isolate those reservoirs that offer easy access to the general public. These reservoirs can be fenced off and patrolled, or they can be covered. If access to on-line reservoirs is limited, persons attempting to contaminate the water supply will generally be forced to look to larger bodies of water. Contamination of these large water bodies requires larger quantities of contaminant, increases the detention time of the contaminant, and increases the likelihood of its detection.

As a second means of protection, the water utility can develop an extensive detection and monitoring program. Detecting any contaminant that might be added to a water supply is difficult and expensive. However, because most contaminants cause secondary effects in a water supply, such as taste, color, odor, or chlorine demand, detection is easier.

Because utility operators know their water supply (they know its characteristics), any subtle changes in taste, odor, color, and chlorine demand are instantly recognized. Once it has been determined that the water supply may be contaminated, water samples can be tested. Tests can either be done at the utility's laboratory, if it is a large utility, or the samples can be sent to the state health department.

Finally, the utility can maintain a high chlorine residual. Generally, chlorine residuals of one mg/L or higher effectively oxidize or destroy most contaminants. For example, infectious hepatitis virus will not survive a free residual chlorine level of 0.7 mg/L.

23.1024 Emergency Countermeasures

Following is a list of emergency countermeasures that, when used over a short time period, can increase protection of a water supply:

1. Maintaining a high chlorine residual in the system
2. Having engineers, chemists, and medical personnel on 24-hour alert
3. Continuously monitoring key points in the distribution system (monitoring chlorine residual is mandatory)
4. Increasing security around exposed on-line reservoirs
5. Sealing off access to manholes within a three- to six-block radius of highly populated areas
6. Setting up emergency crews who can isolate sections of the distribution system
7. Staffing the treatment facility on a 24-hour basis

23.1025 In Case of Contamination

If contamination of the water supply is discovered, the immediate concern must be the safety of the public. If the contaminated water has already entered the distribution system, immediate public notification is the highest priority. The local police chief, sheriff, or other responsible governmental authority will help to spread the word. Alternative sources of water may need to be provided.

If the contaminated water has not entered the distribution system, it may be possible to isolate the contaminated source and continue to supply water from other, unaffected sources. If the contaminated water is the only source for the community, treatment measures may be available that will remove the contaminant or reduce its toxicity.

Table 23.9 lists a series of emergency treatment steps that can be taken when identified chemicals are added to the system.

TABLE 23.9 EMERGENCY TREATMENT FOR REDUCING CONCENTRATION OF SPECIFIC CHEMICALS IN COMMUNITY WATER SUPPLIES[a]

Concentration	Treatment
Arsenicals	
Unknown organic and inorganic arsenicals in groundwater at concentrations of 100 mg/L	Precipitation with ferric sulfate and liming to pH 6.8, followed by sedimentation and filtration.
Cyanides	
Hydrogen cyanide	Prechlorination to free residual with pH 7, followed by coagulation, sedimentation, and filtration. *Caution: housed facilities must be adequately ventilated.*
	Precipitation with ferrous or ferric salts to form Prussian blue (iron ferric cyanide) followed by coagulation, sedimentation, and clarification. As long as an excess of iron coagulant is applied, the filtered water should be nontoxic even though it is blue.
Acetone cyanohydrin	Same as for hydrogen cyanide.
Cyanogen chloride	Same as for hydrogen cyanide.
Hydrocarbons	
Kerosene, peak concentrations of 140 mg/L	Preapplications of bleaching clay and activated carbon, plus some increase in normal dosages of alum, chlorine dioxide, lime, and carbon, to provide treatment enabling continued production of water.
Miscellaneous Organic Chemicals	
LSD (lysergic acid derivative)	Chlorination in alkaline water, or water made alkaline by addition of lime or soda ash, to provide a free chlorine residual. Two parts free chlorine are required to react with each part LSD.
Nerve Agents	
(Organophosphorus compounds)	Superchlorination at pH 7 to provide at least 40 mg/L residual after 30-min chlorine contact time, followed by dechlorination and conventional clarification processes.
Pesticides	
2,4-DCP (2,4-Dichlorophenol), an impurity in commercial 2,4-D herbicides	Adsorption on activated carbon followed by coagulation, sedimentation, and filtration.
DDT (dichlorodiphenyltrichloroethane), concentrations of 10 gm/L	Chemical coagulation, sedimentation, and filtration.
Dieldrin, concentrations of 10 gm/L	Chemical coagulation, sedimentation, and filtration. Supplemental treatment with activated carbon may be necessary.
Endrin, concentrations of 10 gm/L	Chemical coagulation, sedimentation, and filtration. Supplemental treatment with activated carbon may be necessary.
Lindane, concentrations of 10 gm/L	Application of activated carbon followed by chemical coagulation, sedimentation, and filtration.
Parathion, concentrations of 10 gm/L	Chemical coagulation, sedimentation, and filtration. Supplemental treatment with activated carbon may be necessary. Omit prechlorination because chlorine reacts with parathion to form paraoxon, which is more toxic than parathion.

[a] Source: Graham Walton, Chief, Technical Services, National Water Supply Research Laboratory, USSR Program, Oct. 24, 1968.

These emergency treatment methods are effective only if the contaminant has been identified.

QUESTIONS

Write your answers in a notebook and then compare your answers with those on page 698.

23.10D What are the color codes and risk levels for the US Terrorism Alert System?

23.10E What does the word "toxicity" mean?

23.10F The degree of toxicity varies with what factors?

23.10G List possible secondary effects in a water supply that may allow detection of a contaminant without specific testing.

23.1026 Cryptosporidium

Cryptosporidium is a parasite that has become a significant public health concern for drinking water utilities. Even when drinking water meets or exceeds all current state and federal standards, it may contain sufficient *Cryptosporidia* to cause serious illness in sensitive individuals. This section contains suggestions that operators should consider to protect consumers from *"crypto."*

Operators need to develop a strategy for protecting their source water supplies and optimizing the operation of their water treatment plant to protect the public health. *Cryptosporidium* oocyst contamination is widespread in the water environment. Potential sources of the parasite in the watersheds of your water supply need to be identified and controlled. The most serious threat to a public water supply occurs during periods of heavy rains or snow melts that flush areas that are sources of high concentrations of oocysts into waters upstream of a plant intake. In the watershed, these potentially contaminated areas include wastewater treatment plants, cattle feedlots, and pastures where livestock graze.

Conventional water treatment plants using coagulation, flocculation, sedimentation, filtration, and disinfection can provide effective treatment to protect drinking water from *Cryptosporidium*. Be especially alert during periods when the intake water has high turbidity levels resulting from stormwater runoff, snow melt runoff, or lake overturns. During these periods, all treatment processes must operate effectively. Try to achieve turbidity levels of 0.1 NTU or lower at all times. Run frequent jar tests to determine or simply to verify that you are using the optimum coagulant doses as intake turbidity and other water quality indicators change (pH, temperature, alkalinity). Also, monitor your filtration processes continuously to avoid any increases in turbidity in the treated water. If the water used for backwashing filters is routinely returned to the headworks for conservation purposes, avoid recycling the backwash water during periods of high intake water turbidity. Recycling backwash water may tend to concentrate oocysts in the filter media. Instead, consider wasting the backwash water until the high turbidity levels drop back to normal levels.

Chemical inactivation of *Cryptosporidium* oocysts by disinfection is influenced by several factors. The effectiveness of chemical disinfectants such as chlorine, chlorine dioxide, and ozone is reduced by the presence of high levels of total organic carbon (TOC) (caused by color and turbidity), lower water temperatures, and shorter disinfection contact times. Therefore, effective disinfection can be difficult during periods of high turbidity caused by high stormwater or snow melt runoff flows.

The best approach to evaluating the potential threat of the drinking water to the public is the analysis for *Cryptosporidium* oocysts in the treated drinking water. However, current sampling and analytical methods make it difficult for operators to use the detection of *Cryptosporidium* oocysts for determining the efficiency and effectiveness of filtration and chemical inactivation treatment processes. Today, operators are using turbidity and particle counting measurements as a means of determining the treatment processes' ability to remove or inactivate oocysts. Test methods for identifying the presence or potential presence of *Cryptosporidium* are evolving and should be used to analyze both source and treated (finished) waters for *"crypto."*

Keeping the public informed is another good strategy for preventing outbreaks of disease due to *Cryptosporidium*. Operators need to educate the media and consumers regarding the sources of the parasite, possible health risks, monitoring efforts, and treatment processes. Sensitive populations must be informed that even properly treated municipal drinking water, bottled water, and water treated by a home water treatment device still may not be free of *Cryptosporidia*. The Centers for Disease Control and Prevention suggests, "Immunocompromised persons who wish to take independent action to reduce the risk of waterborne *Cryptosporidium* may choose to take precautions similar to those recommended during outbreaks (such as boiling tap water for one minute). Such decisions should be made in conjunction with their health care provider."

There are alternatives to boiling water that may be effective when used with proper precautions. Point-of-use filters that remove particles one micrometer (1 μm) in diameter or smaller are effective. One-micrometer filters for cyst removal and reverse osmosis units are also acceptable. Bottled water from protected springs and wells may be a safe drinking water, especially if treated by reverse osmosis or distillation to remove *Cryptosporidium* before bottling.

Operators must stay current with the efforts of our profession to prevent outbreaks of waterborne *Cryptosporidium*.

QUESTIONS

Write your answers in a notebook and then compare your answers with those on page 698.

23.10H   Why is *Cryptosporidium* a significant public health concern for drinking water utilities?

23.10I   When does the most serious threat of *Cryptosporidium* occur to a public water supply?

23.10J   How can a water treatment plant be operated to protect the public from *Cryptosporidium*?

23.11 SAFETY PROGRAM

23.110 Responsibilities

23.1100 Everyone Is Responsible for Safety

Waterworks utilities, regardless of size, must have a safety program if they are to realize a low frequency of accident occurrence. A safety program also provides a means of comparing frequency, disability, and severity of injuries with other utilities. The utility should identify the causes of accidents and injuries, provide safety training, implement an accident reporting system, and hold supervisors responsible for implementing the safety program. Each utility should have a safety officer or supervisor evaluate every accident, offer recommendations, and keep and apply statistics.

The effectiveness of any safety program will depend upon how the utility holds its supervisors responsible. If the utility holds only the safety officer or the employees responsible, the program will fail. The supervisors are key in any organization. They should be responsible for the implementation of a safety program. If they disregard safety measures, essential parts of the program will not work. The results will be an overall poor safety record. After all, the first line supervisor is where the work is being performed, and some may take advantage of an unsafe situation in order to get the job completed. The organization must discipline such supervisors and make them aware of their responsibility for their own and their operators' safety.

Safety is good business, both for the operator and the agency. For a good safety record to be accomplished, all individuals must be educated and must believe in the program. All individuals involved must have the conviction that accidents can be prevented. The operations should be studied to determine the safe way of performing each job. Safety pays, both in monetary savings and in the health and well-being of the operating staff.

23.1101 Regulatory Agencies

Many state and federal agencies are involved in ensuring safe working conditions. The one law that has had the greatest impact has been the *OCCUPATIONAL SAFETY AND HEALTH ACT OF 1970 (OSHA),*[12] Public Law 91-596, which took effect on December 29, 1970. This legislation affects more than 75,000,000 employees and has been the basis for most of the current state laws covering employees. Also, many state regulatory agencies enforce the OSHA requirements.

The OSHA regulations provide for safety inspections, penalties, recordkeeping, and variances. Managers and supervisors must understand the OSHA regulations and must furnish each operator with the rules of conduct in order to comply with occupational safety and health standards. The intent of the regulations is to create a place of employment that is free from recognized hazards that could cause serious physical harm or death to an operator.

Civil and criminal penalties are allowed under the OSHA law, depending upon the size of the business and the seriousness of the violation. A routine violation could cost an employer or supervisor up to $1,000 for each violation. A serious, willful, or repeated violation could cause the employer or supervisor to be assessed a penalty of not more than $10,000 for each violation. Penalties are assessed against the supervisor responsible for the injured operator. Operators should become familiar with the OSHA regulations as they apply to their organizations. Managers and supervisors must correct violations and prevent others from occurring.

23.1102 Managers

The utility manager is responsible for the safety of the agency's personnel and the public exposed to the water utility's operations. Therefore, the manager must develop and administer an effective safety program and must provide new employee safety training as well as ongoing training for all employees. The basic elements of a safety program include a safety policy statement, safety training and promotion, and accident investigation and reporting.

A safety policy statement should be prepared by the top management of the utility. The purpose of the statement is to let employees know that the safety program has the full support of the agency and its management. The statement should:

1. Define the goals and objectives of the program.

2. Identify the persons responsible for each element of the program.

3. Affirm management's intent to enforce safety regulations.

4. Describe the disciplinary actions that will be taken to enforce safe work practices.

Give a copy of the safety policy statement to every current employee and each new employee during orientation. Figure 23.10 is an example of a safety policy statement for a water supply utility.

The following list of responsibilities for safety is from the *PLANT MANAGER'S HANDBOOK.*[13] These responsibilities represent a typical list but may be incomplete if your agency is subject to stricter local, state, or federal regulations than what is shown here. Check with your safety professional.

Management has the responsibility to:

1. Formulate a written safety policy.

2. Provide a safe workplace.

3. Set achievable safety goals.

4. Provide adequate training.

5. Delegate authority to ensure that the program is properly implemented.

---

[12] *OSHA* (O-shuh). The Williams-Steiger Occupational Safety and Health Act of 1970 (OSHA) is a federal law designed to protect the health and safety of workers, including the operators of water supply and treatment systems and wastewater collection and treatment systems. The Act regulates the design, construction, operation, and maintenance of water and wastewater systems. OSHA regulations require employers to obtain and make available to workers the Material Safety Data Sheets (MSDSs) for chemicals used at industrial facilities and treatment plants. OSHA also refers to the federal and state agencies that administer the OSHA regulations.

[13] *PLANT MANAGER'S HANDBOOK* (MOP SM-4), Water Environment Federation (WEF), no longer in print.

> **SAFETY POLICY STATEMENT**
>
> It is the policy of the Las Vegas Valley Water District that every employee shall have a safe and healthy place to work. It is the District's responsibility; its greatest asset, the employees and their safety.
>
> When a person enters the employ of the District, he or she has a right to expect to be provided a proper work environment, as well as proper equipment and tools, so that they will be able to devote their energies to the work without undue danger. Only under such circumstances can the association between employer and employee be mutually profitable and harmonious. It is the District's desire and intention to provide a safe workplace, safe equipment, proper materials, and to establish and insist on safe work methods and practices at all times. It is a basic responsibility of all District employees to make the SAFETY of human beings a matter for their daily and hourly concern. This responsibility must be accepted by everyone who works at the District, regardless of whether he or she functions in a management, supervisory, staff, or the operative capacity. Employees must use the SAFETY equipment provided; Rules of Conduct and SAFETY shall be observed; and, SAFETY equipment must not be destroyed or abused. Further, it is the policy of the Water District to be concerned with the safety of the general public. Accordingly, District employees have the responsibility of performing their duties in such a manner that the public's safety will not be jeopardized.
>
> The joint cooperation of employees and management in the implementation and continuing observance of this policy will provide safe working conditions and relatively accident-free performance to the mutual benefit of all involved. The Water District considers the SAFETY of its personnel to be of primary importance, and asks each employee's full cooperation in making this policy effective.

Fig. 23.10 Safety policy statement
(Permission of Las Vegas Valley Water District)

The manager is the key to any safety program. Implementation and enforcement of the program is the responsibility of the manager. The manager also has the responsibility to:

1. Ensure that all employees are trained and periodically retrained in proper safe work practices.

2. Ensure that proper safety practices are implemented and continued as long as the policy is in effect.

3. Investigate all accidents and injuries to determine their cause.

4. Institute corrective measures where unsafe conditions or work methods exist.

5. Ensure that equipment, tools, and the work are maintained to comply with established safety standards.

QUESTIONS

Write your answers in a notebook and then compare your answers with those on page 698.

23.11A What should be the duties of a safety officer?

23.11B Who should be responsible for the implementation of a safety program?

23.11C Who enforces the OSHA requirements?

23.11D What are the utility manager's responsibilities with regard to safety?

23.11E What should be included in a utility's policy statement on safety?

23.1103 Supervisors

The success of any safety program will depend upon how the supervisors of the utility view their responsibility. The supervisor who has the responsibility for directing work activities must be safety conscious. This supervisor controls the operators' general environment and work habits and influences whether or not the operators comply with safety regulations. The supervisor is in the best position to counsel, instruct, and review the operators' working methods and thereby effectively ensure compliance with all aspects of the utility's safety program.

The problem, however, is one of the supervisor accepting this responsibility. The supervisor who wishes to complete the job and go on to the next one without taking time to be concerned about working conditions, the welfare of operators, or considering any aspects of safety is a poor supervisor. Only after an accident occurs will a careless supervisor question the need for a work program based on safety. At this point, however, it is too late, and the supervisor may be tempted to simply cover up past mistakes. As sometimes happens, the supervisor may even be partially or fully responsible for the accident by causing unsafe acts to take place, by requiring work to be performed in haste, by disregarding an unsafe work environment, or by overlooking or failing to consider any number of safety hazards. This negligent supervisor could be fined, sentenced to a jail term, or even be barred from working in the profession.

All utilities should make their supervisors bear the greatest responsibility for safety and hold them accountable for planning, implementing, and controlling the safety program. If most accidents are caused and do not just happen, then it is the supervisor who can help prevent most accidents.

Equally important are the officials above the supervisor. These officials include commissioners, managers, public works directors, chief engineers, superintendents, and chief operators. The person in responsible charge for the entire agency or operation must believe in the safety program. This person must budget, promote, support, and enforce the safety program by vocal and visible examples and actions. The top person's support is absolutely essential for an effective safety program.

23.1104 Operators

Each operator also shares in the responsibility for an effective safety program. After all, operators have the most to gain since they are the most likely victims of accidents. A review of accident causes shows that the accident victim often has not acted responsibly. In some way, the victim has not complied with the safety regulations, has not been fully aware of the working conditions, has not been concerned about fellow employees, or just has not accepted any responsibility for the utility's safety program.

Each operator must accept, at least in part, responsibility for fellow operators, for the utility's equipment, for the operator's own welfare, and even for seeing that the supervisor complies with established safety regulations. As pointed out above, the operator has the most to gain. If the operator accepts and uses unsafe equipment, it is the operator who is in danger if something goes wrong. If the operator fails to protect the other operators, it is the operator who must make up the work lost because of injury. If operators fail to consider their own welfare, it is they who suffer the pain of any injury, the loss of income, and maybe even the loss of life.

The operator must accept responsibility for an active role in the safety program by becoming aware of the utility's safety policy and conforming to established regulations. The operator should always call to the supervisor's attention unsafe conditions, environment, equipment, or other concerns operators may have about the work they are performing. Safety should be an essential part of the operator's responsibility.

23.111 First Aid

By definition, first aid means emergency treatment for injury or sudden illness, before regular medical treatment is available. Everyone in an organization should be able to give some degree of prompt treatment and attention to an injury.

First-aid training in the basic principles and practices of lifesaving steps that can be taken in the early stages of an injury are available through the local Red Cross, Heart Association, local fire departments, and other organizations. Such training should periodically be reinforced so that the operator has a complete understanding of water safety, cardiopulmonary resuscitation (CPR), and other life-saving techniques. All operators need training in first aid, but it is especially important for those who regularly work with electrical equipment or must handle chlorine and other dangerous chemicals. (Chapter 20, "Safety," lists specific first-aid procedures for exposure to a variety of water treatment chemicals.)

First aid has little to do with preventing accidents, but it has an important bearing upon the survival of the injured patient. A well-equipped first-aid chest or kit is essential for proper treatment. The kit should be inspected regularly by the safety officer to ensure that supplies are available when needed. First-aid kits should be prominently displayed throughout the treatment plant and in company vehicles. Special consideration must be given to the most hazardous areas of the plant such as shops, laboratories, and chemical handling facilities.

Regardless of size, each utility should establish standard operating procedures (SOPs) for first-aid treatment of injured personnel. All new operators should be instructed in the utility's first-aid program.

QUESTIONS

Write your answers in a notebook and then compare your answers with those on page 698.

23.11F How could a supervisor be responsible for an accident?

23.11G What types of safety-related responsibilities must each operator accept?

23.11H What is first aid?

23.11I First-aid training is most important for operators involved in what types of activities?

23.112 Hazard Communication Program and Worker Right-To-Know (RTK) Laws

In the past few years, there has been an increased emphasis nationally on hazardous materials and wastes. Much of this attention has focused on hazardous and toxic waste dumps and the efforts to clean them up after the long-term effects on human health were recognized. Each year thousands of new chemical compounds are produced for industrial, commercial, and household use. Frequently, the long-term effects of these chemicals are unknown. As a result, federal and state laws have been enacted to control all aspects of hazardous materials handling and use. These laws are more commonly known as Worker Right-To-Know (RTK) laws, which are enforced by OSHA.

As noted earlier, the intent of the OSHA regulations is to create a place of employment that is free from recognized hazards that could cause serious physical harm or death to an operator (or other employee). In many cases, the individual states have the authority under the OSHA Standard to develop their own state RTK laws and most states have adopted their own laws. The Federal OSHA Standard 29 *CFR*[14] 1910.1200—Hazard Communication forms the basis of most of these state RTK laws. Unfortunately, state laws vary significantly from state to state. The state laws that have been passed are at least as stringent as the federal standard and, in most cases, are even more stringent. State laws are also under continuous revision and, because a strong emphasis is being placed on hazardous materials and worker exposure, state laws can be expected to be amended in the future to apply to virtually everybody in the workplace. Managers should become familiar with both the state and federal OSHA regulations that apply to their organizations.

The basic elements of a hazard communication program are described in the following paragraphs.

1. Identify Hazardous Materials—While there are thousands of chemical compounds that could be considered hazardous, focus on the ones to which operators and other personnel in your utility are most likely to be exposed.

2. Obtain Chemical Information and Define Hazardous Conditions—Once the inventory of hazardous chemicals is complete, the next step is to obtain specific information on each of the chemicals. This information is generally incorporated into a standard format form called the *MATERIAL SAFETY DATA SHEET (MSDS)*.[15] This information is commonly available from manufacturers. Many agencies request an MSDS when the purchase order is generated and will refuse to accept delivery of the shipment if the MSDS is not included. Figure 23.11 shows OSHA's standard MSDS form, but other forms are also acceptable provided they contain all of the required information.

The purpose of the MSDS is to have a readily available reference document that includes complete information on common names, safe exposure level, effects of exposure, symptoms of exposure, flammability rating, type of first-aid procedures, and other information about each hazardous substance. Operators should be trained to read and understand the MSDS forms. The forms themselves should be stored in a convenient location where they are readily available for reference.

3. Properly Label Hazards—Once the physical, chemical, and health hazards have been identified and listed, a labeling and training program must be implemented. To meet labeling requirements on hazardous materials, specialized labeling is available from a number of sources, including commercial label manufacturers. Exemptions to labeling requirements do exist, so consult your local safety regulatory agency for specific details.

4. Train Operators—The last element in the hazard communication program is training and making information available to utility personnel. A common-sense approach eliminates the confusing issue of which of the thousands of substances operators should be trained for, and concentrates on those that they will be exposed to or use in everyday maintenance routines.

The Hazard Communication Standard and the individual state requirements are obviously a very complex set of regulations. Remember, however, the ultimate goal of these regulations is to provide additional operator protection. These standards and regulations, once the intent is understood, are relatively easy to implement.

23.113 Confined Space Entry Procedures

CONFINED SPACES[16,17] pose significant risks for a large number of workers, including many utility operators. OSHA has therefore defined very specific procedures to protect the health and safety of operators whose jobs require them to enter or work in a confined space. The regulations (which can be found in the Code of Federal Regulations at 29 CFR 1910.146) require conditions in the confined space to be tested and evaluated before anyone enters the space. If conditions exceed OSHA's limits for safe exposure, additional safety precautions must be taken and a confined space entry permit (Figure 23.12) must be approved by the appropriate authorities prior to anyone entering the space.

The managers of water utilities may or may not be involved in the day-to-day details of enforcing the agency's confined space policy and procedures. However, every utility manager should be aware of the current OSHA requirements and should ensure that the utility's policies not only comply with current regulations, but that the agency's policies are vigorously enforced for the safety of all operators.

---

[14] *Code of Federal Regulations (CFR).* A publication of the US government that contains all of the proposed and finalized federal regulations, including safety and environmental regulations.

[15] *Material Safety Data Sheet (MSDS).* A document that provides pertinent information and a profile of a particular hazardous substance or mixture. An MSDS is normally developed by the manufacturer or formulator of the hazardous substance or mixture. The MSDS is required to be made available to employees and operators or inspectors whenever there is the likelihood of the hazardous substance or mixture being introduced into the workplace. Some manufacturers are preparing MSDSs for products that are not considered to be hazardous to show that the product or substance is not hazardous.

[16] *Confined Space.* Confined space means a space that: (1) Is large enough and so configured that an employee can bodily enter and perform assigned work; and (2) Has limited or restricted means for entry or exit (for example, manholes, tanks, vessels, silos, storage bins, hoppers, vaults, and pits are spaces that may have limited means of entry); and (3) Is not designed for continuous employee occupancy. Also see DANGEROUS AIR CONTAMINATION and OXYGEN DEFICIENCY.

[17] *Confined Space, Permit-Required (Permit Space).* A confined space that has one or more of the following characteristics: (1) Contains or has a potential to contain a hazardous atmosphere, (2) Contains a material that has the potential for engulfing an entrant, (3) Has an internal configuration such that an entrant could be trapped or asphyxiated by inwardly converging walls or by a floor that slopes downward and tapers to a smaller cross section, or (4) Contains any other recognized serious safety or health hazard.

Material Safety Data Sheet

May be used to comply with
OSHA's Hazard Communication Standard,
29 CFR 1910.1200 Standard must be
consulted for specific requirements.

U.S. Department of Labor
Occupational Safety and Health Administration
(Non-Mandatory Form)
Form Approved
OMB No. 1218-0072

IDENTITY *(As Used on Label and List)*

Note: Blank spaces are not permitted. If any item is not applicable, or no information is available, the space must be marked to indicate that.

Section I

Manufacturer's Name	Emergency Telephone Number
Address *(Number, Street, City, State, and ZIP Code)*	Telephone Number for Information
	Date Prepared
	Signature of Preparer *(optional)*

Section II — Hazardous Ingredients/Identity Information

Hazardous Components (Specific Chemical Identity; Common Name(s))	OSHA PEL	ACGIH TLV	Other Limits Recommended	%(optional)

Section III — Physical/Chemical Characteristics

Boiling Point		Specific Gravity ($H_2O = 1$)	
Vapor Pressure (mm Hg)		Melting Point	
Vapor Density (AIR = 1)		Evaporation Rate (Butyl Acetate = 1)	
Solubility in Water			
Appearance and Odor			

Section IV — Fire and Explosion Hazard Data

Flash Point (Method Used)	Flammable Limits	LEL	UEL
Extinguishing Media			
Special Fire Fighting Procedures			
Unusual Fire and Explosion Hazards			

(Reproduce locally)

OSHA 174, Sept. 1985

Fig. 23.11 Material Safety Data Sheet

676 Water Treatment

Section V — Reactivity Data

Stability	Unstable		Conditions to Avoid
	Stable		

Incompatibility *(Materials to Avoid)*

Hazardous Decomposition or Byproducts

Hazardous Polymerization	May Occur		Condition to Avoid
	Will Not Occur		

Section VI — Health Hazard Data

Route(s) of Entry:	Inhalation?	Skin?	Ingestion?

Health Hazards *(Acute and Chronic)*

Carcinogenicity:	NTP?	IARC Monographs?	OSHA Regulated?

Signs and Symptoms of Exposure

Medical Conditions Generally Aggravated by Exposure

Emergency and First Aid Procedures

Section VII — Precautions for Safe Handling and Use

Steps to Be Taken in Case Material is Released or Spilled

Waste Disposal Method

Precautions to Be Taken in Handling and Storing

Other Precautions

Section VIII — Control Measures

Respiratory Protection *(Specify Type)*

Ventilation	Local Exhaust		Special
	Mechanical *(General)*		Other

Protective Gloves		Eye Protection

Other Protective Clothing or Equipment

Work/Hygienic Practices

Page 2 * USGPO:1986-491-529/45775

Fig. 23.11 Material Safety Data Sheet (continued)

Confined Space Pre-Entry Checklist/Confined Space Entry Permit

Date and Time Issued: _____ Date and Time Expires: _____ Job Site/Space I.D.: _____

Job Supervisor: _____ Equipment to be worked on: _____ Work to be performed: _____

Standby personnel: _____ _____ _____

1. Atmospheric Checks: Time _____ Oxygen _____ %   Toxic _____ ppm
                                        Explosive _____ % LEL Carbon Monoxide _____ ppm

2. Tester's signature: _____

3. Source isolation: (No Entry)   N/A   Yes   No
   Pumps or lines blinded
   disconnected, or blocked   ( )   ( )   ( )

4. Ventilation Modification: N/A   Yes   No
   Mechanical   ( )   ( )   ( )
   Natural ventilation only   ( )   ( )   ( )

5. Atmospheric check after isolation and ventilation: Time _____
   Oxygen _____ % > 19.5% < 23.5%   Toxic _____ ppm   < 10 ppm H_2S
   Explosive _____ % LEL   < 10%   Carbon Monoxide _____ ppm   < 35 ppm CO

Tester's signature: _____

6. Communication procedures: _____

7. Rescue procedures: _____
   _____

8. Entry, standby, and backup persons   Yes   No
   Successfully completed required training?   ( )   ( )
   Is training current?   ( )   ( )

9. Equipment: N/A   Yes   No
   Direct reading gas monitor tested   ( )   ( )   ( )
   Safety harnesses and lifelines for entry and standby persons ( )   ( )   ( )
   Hoisting equipment   ( )   ( )   ( )
   Powered communications   ( )   ( )   ( )
   SCBAs for entry and standby persons   ( )   ( )   ( )
   Protective clothing   ( )   ( )   ( )
   All electric equipment listed for Class I, Division I,
   Groups A, B, C, and D and nonsparking tools   ( )   ( )   ( )

10. Periodic atmospheric tests:
    Oxygen: ____% Time ____; ____% Time ____; ____% Time ____; ____% Time ____;
    Explosive: ____% Time ____; ____% Time ____; ____% Time ____; ____% Time ____;
    Toxic: ____ppm Time ____; ____ppm Time ____; ____ppm Time ____; ____ppm Time ____;
    Carbon Monoxide: ____ppm Time ____; ____ppm Time ____; ____ppm Time ____; ____ppm Time ____;

We have reviewed the work authorized by this permit and the information contained herein. Written instructions and safety procedures have been received and are understood. Entry cannot be approved if any brackets ( ) are marked in the "No" column. This permit is not valid unless all appropriate items are completed.

Permit Prepared By: (Supervisor) _____ Approved By: (Unit Supervisor) _____

Reviewed By: (CS Operations Personnel) _____
                                  (Entrant)   (Attendant)   (Entry Supervisor)

This permit to be kept at job site. Return job site copy to Safety Office following job completion.

Fig. 23.12   Confined space pre-entry checklist/permit

QUESTIONS

Write your answers in a notebook and then compare your answers with those on page 698.

23.11J List the basic elements of a hazard communication program.

23.11K What are a manager's responsibilities for ensuring the safety of operators entering or working in confined spaces?

23.114 Reporting

The mainstay of a safety program is the method of reporting and keeping of statistics. These records are needed regardless of the size of the utility because they provide a means of identifying accident frequencies and causes as well as the personnel involved. The records can be looked upon as the operator's safety report card. Therefore, it becomes the responsibility of each injured operator to fill out the utility's accident report.

All injuries should be reported, even if they are minor in nature, so as to establish a record in case the injury develops into a serious injury. It may be difficult at a later date to prove the accident did occur on the job and have the utility accept the responsibility for costs. The responsibility for reporting accidents affects several levels of personnel. First, of course, is the injured person. Next, it is the responsibility of the supervisor, and finally, the

Responsibility of Management to review the causes and take steps to prevent such accidents from happening in the future.

Accident report forms may be very simple. However, they must record all details required by law and all data needed for statistical purposes. The forms shown in Figures 23.13 and 23.14 are examples for you to consider for use in your plant. The report must show the name of the injured, employee number, division, time of accident, nature of injury, cause of accident, first aid administered, and remarks for items not covered elsewhere. There should be a review process by foreman, supervisor, safety officer, and management. Recommendations are needed as well as a follow-up review to be sure that proper action has been taken to prevent recurrence.

In addition to reports needed by the utility, other reports may be required by state or federal agencies. For example, vehicle accident reports must be submitted to local police departments. If a member of the public is injured, additional forms are needed because of possible subsequent claims for damages. If the accident is one of occupational injury, causing lost time, other reports may be required. Follow-up investigations to identify causes and responsibility may require the development of other specific types of record forms.

In the preparation of accident reports, it is the operator's responsibility to correctly fill out each form, giving complete details. The supervisor must be sure no information is overlooked that may be helpful in preventing recurrence.

The Safety Officer must review the reports and determine corrective actions and make recommendations.

In day-to-day actions, operators, supervisors, and management often overlook opportunities to counsel individual operators in safety matters. Then, when an accident occurs, they are not inclined to look too closely at accident reports. First, the accident is a series of embarrassments, to the injured person, to the supervisor, and to management. Therefore, there is a reluctance to give detailed consideration to accident reports. However, if a safety program is to function well, it will require a thorough effort on the part of the operator, supervisor, and management in accepting their responsibility for the accident and making a greater effort through good reporting to prevent future similar accidents. Accident reports must be analyzed, discussed, and the real cause of the accident identified and corrected.

Emphasis on the prevention of future accidents cannot be overstressed. We must identify the causes of accidents and implement whatever measures are necessary to protect operators from becoming injured.

23.115 Training

If a safety program is to work well, management will have to accept responsibility for the following three components of training:

1. Safety education of all employees
2. Reinforced education in safety
3. Safety education in the use of tools and equipment

Or to put it another way, the three most important controlling factors in safety are education, education, and more education.

Date _____

Name of injured employee _____ Employee # _____ Area _____
Date of accident _____ Time _____ Employee's Occupation _____
Location of accident _____ Nature of injury _____
First aid administered _____
Name of doctor _____ Address _____
Name of hospital _____ Address _____
Witnesses (name & address) _____

PHYSICAL CAUSES

Indicate below by an "X" whether, in your opinion, the accident was caused by:

_____	Improper guarding	_____	No mechanical cause
_____	Defective substances or equipment	_____	Working methods
_____	Hazardous arrangement	_____	Lack of knowledge or skill
_____	Improper illumination	_____	Wrong attitude
_____	Improper dress or apparel	_____	Physical defect

_____ Not listed (describe briefly) _____

UNSAFE ACTS

Sometimes the injured person is not directly associated with the causes of an accident. Using an "X" to represent the injured worker and an "O" to represent any other person involved, indicate whether, in your opinion, the accident was caused by:

_____	Operating without authority	_____	Unsafe loading, placement & etc.
_____	Failure to secure or warn	_____	Took unsafe position
_____	Working at unsafe speed	_____	Worked on moving equipment
_____	Made safety device inoperative	_____	Teased, abused, distracted & etc.
_____	Unsafe equipment or hands instead of equipment	_____	Did not use safe clothing or personal protective equipment
_____	No unsafe act		

_____ Not listed (describe briefly) _____

What job was the employee doing? _____
What specific action caused the accident? _____
What steps will be taken to prevent recurrence? _____

Date of Report _____ Immediate Supervisor _____

REVIEWING AUTHORITY

Comments: | Comments:

Safety Officer _____ Date _____ | Department Director _____ Date _____

Fig. 23.13 Supervisor's accident report

INJURED: COMPLETE THIS SECTION

Name _____ Age _____ Sex _____

Address _____

Title _____ Dept. Assigned _____

Place of Accident _____

Street or Intersection _____

Date _____ Hour _____ A.M. _____ P.M. _____

Type of Job You Were Doing When Injured

_____

Object Which Directly Injured You   Part of Body Injured

_____

How Did Accident Happen? (Be specific and give details; use back of sheet if necessary.)

_____

First Aid Administered

_____

Did You Report Accident or Exposure at Once? (Explain "No")	Yes ☐	No ☐
Did You Report Accident or Exposure to Supervisor? Give Name	Yes ☐	No ☐
Were There Witnesses to Accident or Exposure? Give Names	Yes ☐	No ☐
Did You See a Doctor? (If Yes, Give Name)	Yes ☐	No ☐
Are You Going to See a Doctor? (Give Name)	Yes ☐	No ☐

_____   _____
              Date   Signature

SUPERVISOR: COMPLETE THIS SECTION — (Return to Personnel as Soon as Possible)

Was an Investigation of Unsafe Conditions and/or Unsafe Acts Made? If Yes, Please Submit Copy.	Yes ☐	No ☐
Was Injured Intoxicated or Behaving Inappropriately at Time of Accident? (Explain "Yes")	Yes ☐	No ☐

Date Disability   Last Day   Date Back
Commenced _____   Wages Earned _____   on Job _____

Date Report Completed _____ 20 ____ Signed By _____

                                                   Title _____

Distribution: Canary - Department Head, Pink - Supervisor, White - Personnel

Fig. 23.14   Accident report

Responsibility for overall training must be that of upper management. A program that will educate operators and then reinforce this education in safety must be planned systematically and promoted on a continuous basis. There are many avenues to achieving this goal.

The safety education program should start with the new operator. Even before employment, verify the operator's past record and qualifications and review the pre-employment physical examination. In the new operator's orientation, include instruction in the importance of safety at your utility or plant. Also, discuss the matter of proper reporting of accidents as well as the organization's policies and practices. Give new operators copies of all safety SOPs and direct their attention to parts that directly involve them. Ask the safety officer to give a talk about utility policy, safety reports, and past accidents, and to orient the new operator toward the importance of safety to operators and to the organization.

The next consideration must be one of training the new operator in how to perform assigned work. Most supervisors think in terms of On-the-Job Training (OJT). However, OJT is not a good way of preventing accidents with an inexperienced operator. The idea is all right if the operator comes to the organization trained in how to perform the work, such as a treatment operator from another plant. Then you only need to explain your safety program and how your policies affect the new operator. For a new operator who is inexperienced in water treatment or in utility operation, the supervisor must give detailed consideration to the operator's welfare. In this instance, the training should include not only a safety talk, but the foreman (supervisor) must train the inexperienced operator in all aspects of treatment plant safety. This training includes instruction in the handling of chemicals, the dangers of electrical apparatus, fire hazards, and proper maintenance of equipment to prevent accidents. Special instructions will also be needed for specific work environments such as manholes, gases (chlorine and hydrogen sulfide (H_2S)), water safety, and any specific hazards that are unique to your facility. The new operator must be checked out on any equipment personnel may operate such as vehicles, forklifts, valve operators, and radios. All new operators should be required to participate in a safety orientation program during the first few days of their employment, and an overall training program in the first few months.

The next step in safety education is reinforcement. Even if the operator is well trained, mistakes can occur; therefore, safety education must be continual. Many organizations use the *TAILGATE*[18] method as a means of maintaining the operator's interest in safety. The program should be conducted by the first line supervisor. Schedule the informal tailgate meeting for a suitable location, keep it short, avoid distractions, and be sure that everyone can hear. Hand out literature, if available. Tailgate talks should communicate to the operator specific considerations, new problems, and accident information. These topics should be published. One resource for such meetings can be those operators who have been involved in an accident. Although it is sometimes embarrassing to the injured, the victim is now the expert on how the accident occurred, what could have been done to prevent it, and how it felt to have the injury. Encourage all operators, new and old, to participate in tailgate safety sessions.

Use safety posters to reinforce safety training and to make operators aware of the location of dangerous areas or show the importance of good work habits. Such posters are available through the National Safety Council's catalog.[19]

Awards for good safety records are another means of keeping operators aware of the importance of safety. The awards could be given to individuals in recognition of a good safety record. Publicity about the awards may provide an incentive to the operators and demonstrates the organization's determination to maintain a good safety record. The awards may include safety lapel pins, certificates, or plaques showing number of years without an accident. Consider publishing a utility newsletter on safety tips or giving details concerning accidents that may be helpful to other operators in the organization. Awards may be given to the organization in recognition of its effort in preventing accidents or for its overall safety program. A suggestion program concerning safety will promote and reinforce the program and give recognition to the best suggestions. The goal of all these efforts is to reinforce concerns for the safety of all operators. If safety, as an idea, is present, then accidents can be prevented.

Education of the operator in the use of tools and equipment is necessary. As pointed out above, OJT is not the answer to a good safety record. A good safety record will be achieved only with good work habits and safe equipment. If the operator is trained in the proper use of equipment (hand tools or vehicles), the operator is less likely to misuse them. However, if the supervisor finds an operator misusing tools or equipment, then it is the supervisor's responsibility to reprimand the operator as a means of reinforcing utility policies. The careless operator who misuses equipment is a hazard to other operators. Careless operators will also be the cause of a poor safety record in the operator's division or department.

An important part of every job should be the consideration of its safety aspects by the supervisor. The supervisor should instruct the foreman or operators about any dangers involved in job assignments. If a job is particularly dangerous, then the supervisor must bring that fact to everyone's attention and clarify utility policy in regard to unsafe acts and conditions.

If the operator is unsure of how to perform a job, then it is the operator's responsibility to ask for the training needed. Each operator must think, act, and promote safety if the organization is to achieve a good safety record. Training is the key to achieving this objective and training is everyone's responsibility—management, the supervisors, foremen, and operators.

---

[18] *Tailgate Safety Meeting.* Brief (10 to 20 minutes) safety meetings held every 7 to 10 working days. The term comes from the safety meetings regularly held by the construction industry around the tailgate of a truck.
[19] Write or call your local safety council or the National Safety Council, 1121 Spring Lake Drive, Itasca, IL 60143-3201.

QUESTIONS

Write your answers in a notebook and then compare your answers with those on page 699.

23.11L What is the mainstay of a safety program?

23.11M Why should you report even a minor injury?

23.11N Why should a safety officer review an accident report form?

23.11O A new, inexperienced operator must receive instruction on what aspects of treatment plant safety?

23.11P What should an operator do if unsure of how to perform a job?

23.116 Measuring

To be complete, a safety program must also include some means of identifying, measuring, and analyzing the effects of the program. The systematic classification of accidents, injuries, and lost time is the responsibility of the safety officer. This person should use an analytical method that would refer to types and classes of accidents. Reports should be prepared using statistics showing lost time, costs, type of injuries, and other data, based on a specific time interval.

Such data call attention to the effectiveness of the program and promote awareness of the types of accidents that are happening. Management can use this information to decide where the emphasis should be placed to avoid accidents. However, statistical data are of little value if a report is prepared and then set on the bookshelf or placed in a supervisor's desk drawer. The data must be distributed and read by all operating and maintenance personnel.

As an example, injuries can be classified as fractures, burns, bites, eye injuries, cuts, and bruises. Causes can be referred to as heat, machinery, falls, handling objects, chemicals, unsafe acts, and miscellaneous. Cost can be considered as lost time, lost dollars, lost production, contaminated water, or any other means of showing the effects of the accidents.

Good analytical reporting will provide a great deal of detail without a lot of paper to read and comprehend. Keep the method of reporting simple and easy to understand by all operators, so they can identify with the causes and be aware of how to prevent the accident happening to themselves or other operators. Table 23.10 gives one method of summarizing the causes of various types of injuries.

There are many other methods of compiling data. Table 23.10 could reflect cost in dollars or in work hours lost. Not all accidents mean time lost, but there can be other cost factors. The data analysis should also indicate if the accidents involve vehicles, company personnel, the public, company equipment, loss of chemical, or other factors. Results also should show direct cost and indirect cost to the agency, operator, and the public.

TABLE 23.10 SUMMARY OF TYPES AND CAUSES OF INJURIES

	\multicolumn{11}{c}{Primary Cause of Injury}
	Unsafe Act	Chemical	Falls	Handling Objects	Heat	Machinery	Falling Objects	Stepping	Striking	Miscellaneous	TOTAL
Type of Injury											
Fractures											
Sprains											
Eye Injuries											
Bites											
Cuts											
Bruises											
Burns											
Miscellaneous											

Once the statistical data have been compiled, someone must be responsible for reviewing it in order to take preventive actions. Frequently, such responsibility rests with the safety committee. In fact, safety committees may operate at several levels, for example management committee, working committee, or an accident review board. In any event, the committee must be active, be serious, and be fully supported by management.

Another means of measuring safety is by calculating the injury frequency rate for an indication of the effectiveness of your safety program. Multiply the number of disabling injuries by one million and divide by the total number of employee-hours worked. The number of injuries per year is multiplied by one million in order to obtain injury frequency rate values or numbers that are easy to use. In our example problems, we obtained numbers between one and one thousand.

$$\text{Injury Frequency Rate} = \frac{(\text{Number of Disabling Injuries/yr})(1{,}000{,}000)}{\text{Number of Hours Worked/yr}}$$

These calculations indicate a frequency rate per year, which is the usual means of showing such data. Note that this calculation accounts only for disabling injuries. You may wish to show all injuries, but the calculations are much the same.

EXAMPLE 3

A rural water company employs 36 operators who work in many small towns throughout a three-state area. The operators suffered four injuries in one year while working 74,880 hours. Calculate the injury frequency rate.

Known		Unknown
Number of Operators	= 36	Injury Frequency Rate
Number of Injuries/yr	= 4 Injuries/yr	
Number of Hours Worked/yr	= 74,880 Hr/yr	

Calculate the injury frequency rate.

$$\text{Injury Frequency Rate} = \frac{(\text{Number of Injuries/yr})(1{,}000{,}000)}{\text{Number of Hours Worked/yr}}$$

$$= \frac{(4 \text{ Injuries/yr})(1{,}000{,}000)}{74{,}880 \text{ Hr/yr}}$$

$$= 53.4$$

EXAMPLE 4

Of the four injuries suffered by the operators in Example 3, one was a disabling injury. Calculate the injury frequency rate for the disabling injuries.

Known	Unknown
Number of Disabling Injuries/yr = 1 Injury/yr	Injury Frequency Rate
Number of Hours Worked/yr = 74,880 Hr/yr	

Calculate the injury frequency rate.

$$\text{Injury Frequency Rate (Disabling Injuries)} = \frac{(\text{Number of Disabling Injuries/yr})(1{,}000{,}000)}{\text{Number of Hours Worked/yr}}$$

$$= \frac{(1 \text{ Injury/yr})(1{,}000{,}000)}{74{,}880 \text{ Hr/yr}}$$

$$= 13.4$$

Yet another consideration may be lost-time accidents. The safety officer's analysis may take into account many other considerations, but in any event, the method given here will provide a means of recording and measuring injuries in the treatment plant. In measuring lost-time injuries, a severity rate can be considered.

A severity rate is based on one lost hour for every million operator hours worked. The rate is found by multiplying the number of hours lost by one million and dividing by the total of operator-hours worked.

$$\text{Injury Severity Rate} = \frac{(\text{Number of Hours Lost/yr})(1{,}000{,}000)}{\text{Number of Hours Worked/yr}}$$

EXAMPLE 5

The water company described in Examples 3 and 4 experienced 40 operator-hours lost due to injuries while the operators worked 74,880 hours. Calculate the injury severity rate.

Known	Unknown
Number of Hours Lost/yr = 40 Hr/yr	Injury Severity Rate
Number of Hours Worked/yr = 74,880 Hr/yr	

Calculate the injury severity rate.

$$\text{Injury Severity Rate} = \frac{(\text{Number of Hours Lost/yr})(1{,}000{,}000)}{\text{Number of Hours Worked/yr}}$$

$$= \frac{(40 \text{ hr/yr})(1{,}000{,}000)}{74{,}880 \text{ Hr/yr}}$$

$$= 534$$

Notice that all these data points are based on a one-year time interval, which makes them suitable for use by the safety officer in preparing an annual report.

23.117 Human Factors

First, you may ask, what is a human factor? Well, it is not too often that a safety text considers human factors as part of the safety program. However, if these factors are understood and emphasis is given to their practical application, then many accidents can be prevented. Human factors engineering is the specialized study of technology relating to the design of operator-machine interfaces. That is to say, it examines ways in which machinery might be designed or altered to make it easier to use, safer, and more efficient for the operator. We hear a lot about making computers more user friendly, but human factors engineering is just as important to everyday operation of other machinery in the everyday plant.

Many accidents occur because the operator forgets the human factors. The ultimate responsibility for accidents due to human factors belongs to the management group. However, this does not relieve the operator of the responsibility to point out the human factors as they relate to safety. After all, it is the operator using the equipment who can best tell if it meets all the needs for an interrelationship between an operator and a machine.

The first step in the prevention of accidents takes place in the plant design. Even with excellent designs, accidents can and do happen. However, every step possible must be taken during design to ensure a maximum effort of providing a safe plant environment. Most often, the operator has little to do with design, and, therefore, needs to understand human factors engineering so as to be able to evaluate these factors as the plant is being operated. As newer plants become automated, this type of understanding may even be more important.

Other contributing human factors are the operator's mental and physical characteristics. The operator's decision-making abilities and general behavior (response time, sense of alarm, and perception of problems and danger) are all important factors. Ideally, tools and machines should function as intuitive extensions of the operator's natural senses and actions. Any factors disrupting this flow of action can cause an accident. Therefore, be on the lookout for such factors. When you find a system that cannot be operated in a smooth, logical sequence of steps, change it. You may prevent an accident. If the everyday behavior of an operator is inappropriate with regard to a specific job, reconsider the assignment to prevent an accident.

The human factor in safety is the responsibility of design engineers, supervisors, and operators. However, the operator who is doing the work will have a greater understanding of the operator-machine interface. For this reason, the operator is the appropriate person to evaluate the means of reducing the human factor's contribution to the cause of accidents, thereby improving the plant's safety record.

> **QUESTIONS**
>
> Write your answers in a notebook and then compare your answers with those on page 699.
>
> 23.11Q Statistical accident reports should contain what types of accident data?
>
> 23.11R How can injuries be classified?
>
> 23.11S How can causes of injuries be classified?
>
> 23.11T How can costs of accidents be classified?

23.12 RECORDKEEPING

23.120 Purpose of Records

Accurate records are a very important part of effective operation of a water treatment plant and distribution system facilities. Records are a valuable source of information. They can save time when trouble develops and provide proof that problems were identified and solved. Pertinent and complete records should be used as a basis for plant operation, interpreting results of water treatment, preparing preventive maintenance programs, and preparing budget requests. When accurately kept, records provide an essential basis for design of future changes or expansions of the treatment plant, and also can be used to aid in the design of other water treatment plants where similar water may be treated and similar problems may develop.

If legal questions or problems occur in connection with the treatment of the water or the operation of the plant, accurate and complete records will provide evidence of what actually occurred and what procedures were followed.

Records are essential for effective management of water treatment facilities and to satisfy legal requirements. Some of the important uses of records include:

1. Aiding operators in solving treatment and water quality problems
2. Providing a method of alerting operators to changes in source water quality
3. Showing that the treated water is acceptable to the consumer
4. Documenting that the final product meets plant performance standards, as well as the standards of the regulatory agencies
5. Determining performance of treatment processes, equipment, and the plant
6. Satisfying legal requirements
7. Aiding in answering complaints
8. Anticipating routine maintenance
9. Providing data for cost analysis and preparation of budgets
10. Providing data for future engineering designs
11. Providing information for monthly and annual reports

23.121 Types of Records

There are many different types of records that are required for effective management and operation of water supply, treatment, and distribution system facilities. Below is a listing of some essential records:

1. Source of supply
2. Operation
3. Laboratory
4. Maintenance
5. Chemical inventory and usage
6. Purchases
7. Chlorination station
8. Main disinfection
9. Cross connection control
10. Personnel
11. Accidents
12. Customer complaints

23.122 Types of Plant Operations Data[20]

Plant operations logs can be as different as the plants and water systems whose information they record. The differences in amount, nature, and format of data are so significant that any attempt to prepare a typical log would be very difficult. This section will outline the kinds of data that are usually required to help you develop a useful log for your facilities.

Treatment plant data such as total flows, chemical use, chemical doses, filter performance, reservoir levels, quality control tests, and rainfall and runoff information represent the bulk of the data required for proper plant operation. Frequently, however, source and distribution system data such as reservoir storage and water quality data are included because of the impact of this information on plant operation and operator responsibilities. Typical plant operations data include:

1. Plant title, agency, and location
2. Date
3. Names of operators and supervisors on duty
4. Source of supply:
   a. Reservoir elevation and volume of storage
   b. Reservoir inflow and outflow
   c. Evaporation and precipitation

---

[20] Also see Volume I, Chapter 10, "Plant Operation," Section 10.6, "Operating Records and Reports," for additional details and recordkeeping forms.

d. Apparent runoff, seepage loss, or infiltration gain

   e. Production figures from wells

5. Water treatment plant:

   a. Plant inflow

   b. Treated water flow

   c. Plant operating water (backwash)

   d. Clear well level

6. Distribution system:

   a. Flows to system (system demand)

   b. Distribution system reservoir levels and changes

   c. Comparison of production with deliveries (unaccounted-for water)

7. Chemical inventory and usage:

   a. Chemical inventory/storage (measured use and deliveries)

   b. Metered or estimated plant usages

   c. Calculated usage of chemicals (compare with actual use)

8. Quality control tests:

   a. Turbidity

   b. Chlorine residual

   c. Coliforms

   d. Odor

   e. Color

   f. Other

9. Filter performance:

   a. Operation:

      (1) Total hours, all units

      (2) Filtered water turbidities

      (3) Head losses

      (4) Levels

      (5) Flow rates

   b. Backwash:

      (1) Total hours

      (2) Head losses

      (3) Total wash water used

      (4) Duration and rate of back/surface wash

10. Meteorologic:

    a. Rainfall, evaporation, and temperature of both water and air

    b. Weather (clear, cloudy, windy)

11. Remarks:

    Space should be provided to describe or explain unusual data or events. Extensive notes should be entered on a daily worksheet or diary.

23.123 Maintenance Records

A good plant maintenance effort depends heavily upon good recordkeeping. There are several areas where proper records and documentation can definitely improve overall plant performance.

23.124 Procurement Records

Ordering repair parts and supplies usually is done when the on-hand quantity of a stocked part or chemical falls below the reorder point, a new item is added to stock, or an item has been requested that is not stocked. Most organizations require employees to submit a requisition (similar to the one shown in Figure 23.15) when they need to purchase equipment or supplies. When the requisition has been approved by the authorized person (a supervisor or purchasing agent, in most cases), the items are ordered using a form called a purchase order. A purchase order contains a number of important items. These items include: (1) the date, (2) a complete description of each item and quantity needed, (3) prices, (4) the name of the vendor, and (5) a purchase order number.

A copy of the purchase order should be retained in a suspense file or on a clipboard until the ordered items arrive. This procedure helps keep track of the items that have been ordered but have not yet been received.

All supplies should be processed through the storeroom immediately upon arrival. When an item is received, it should be so recorded on an inventory card. The inventory card will keep track of the numbers of an item in stock, when last ordered, cost, and other information. Furthermore, by always logging in supplies immediately upon receipt, you are in a position to reject defective or damaged shipments and control shortages or errors in billing. Some utilities use personal computers to keep track of orders and deliveries.

23.125 Inventory Records

An inventory consists of the supplies the treatment plant needs to keep on hand to operate the facility. These maintenance supplies may include repair parts, spare valves, electrical supplies, tools, and lubricants. The purpose of maintaining an inventory is to provide needed parts and supplies quickly, thereby reducing equipment downtime and work delays.

In deciding what supplies to stock, keep in mind the economics involved in buying and stocking an item as opposed to depending upon outside availability to provide needed supplies. Is the item critical to continued plant or process operation? Should certain frequently used repair parts be kept on hand? Does the item have a shelf life?

Inventory costs can be held to a minimum by keeping on hand only those parts and supplies for which a definite need

P-1

CITY OF SACRAMENTO
REQUISITION

1 BID NO.

DUE DATE

2 PURCHASE ORDER NO.

DATE

3 DELIVER TO

4 FUND GP.	5 FUND	6 ORGAN.	7 REV. COST CTR	8 OBJECT

9 REQUISITION NO

DATE

10 REFER QUESTIONS TO:

11 ITEM NO.	12 COMMODITY CODE NO.	13 DESCRIPTION	14 QUANTITY UNIT	15 PRICE UNIT	VENDORS: USE THESE COLS. ONLY 16 BRAND OFFERED	17 LOT PRICE FOR EACH ITEM

18 DEPT.

19 ORIG BY

TEL. NO.

THIS REQUISITION WILL BE REPRODUCED TO CREATE A PURCHASE ORDER QUICKLY AND PROVIDE FASTER DELIVERY OF THESE ITEMS TO YOUR DEPARTMENT. PLEASE FOLLOW INSTRUCTIONS CAREFULLY.

SUB TOTAL

SALES TAX

INVOICE AMOUNT

20 EST. OF COST

22 SPECIAL REQUIREMENTS OR INSTRUCTIONS - PREVIOUS P.O. NO. OR BID NO., OR RECOMMENDED VENDORS

21

NUMBER OF REQUISITION ATTACHMENT SHEETS

23 CERTIFICATION IS HEREBY MADE THAT THE ABOVE IS A LEGAL CHARGE AGAINST THE APPROPRIATION INDICATED

DEPT. HEAD ________ BY ________

____
OFFICIAL TITLE

24

VENDOR ________ (A)

ADDRESS ________ (B)

CITY, STATE AND ZIP ________ (C)

TERMS ________ TEL. NO. ________ (D)

DELIVERY ________ FOB DESTINATION ________ (E)

BY ________ TITLE ________

Fig. 23.15 Requisition/purchase order form

exists or that would take too long to obtain from an outside vendor. A definite need for an item is usually demonstrated by a history of regular use. Some items may be infrequently used but may be vital in the event of an emergency; these items should also be stocked. Take care to exclude any parts and supplies that may become obsolete, and do not stock parts for equipment scheduled for replacement.

Tools should be inventoried. Tools that are used by operators on a daily basis should be permanently signed out to them. More expensive tools and tools that are only occasionally used, however, should be kept in a storeroom. These tools should be signed out only when needed and signed back in immediately after use.

23.126 Equipment Records

You will need to keep accurate records to monitor the operation and maintenance of plant equipment. Equipment control cards and work orders can be used to:

- Record important equipment data such as make, model, serial number, and date purchased.
- Record maintenance and repair work performed to date.
- Anticipate preventive maintenance needs.
- Schedule future maintenance work.

See Chapter 18, Section 18.00, "Preventive Maintenance Records," for additional information.

23.127 Computer Recordkeeping Systems

Until fairly recently, water supply system recordkeeping has been done manually. The current availability of low-cost personal computer systems puts automation of many manual bookkeeping functions within the means of all water utilities.

To automate your recordkeeping functions as they relate to customer billing, you will need to develop a simple database management system that will create tables similar to those illustrated in this chapter. This can be readily accomplished by use of standard spreadsheet software programs, which are available in the marketplace at a cost of $300 to $400. Hardware including a personal computer, data storage system, and a printer can be purchased for under $5,000.

Excellent computer software packages are being developed and offered to assist utility managers. SURF (Small Utility Rates and Finances) has been developed by the American Water Works Association (AWWA). SURF is a self-guided, interactive financial spreadsheet application (Excel 97 or later) designed to assist small drinking water systems in developing budgets, setting user rates, and tracking expenses. SURF requires very little computer or software knowledge and can be used by system operators, bookkeepers, and managers to improve the financial management practices of their utilities. SURF can print out three separate modules: (1) system budget, (2) user rate(s), and (3) system expenses. SURF hardware and software requirements are modest.

The SURF software and user's guide are available free from the American Water Works Association. They can be obtained by calling the AWWA Small Systems Program at (800) 366-0107 or by downloading the program from the AWWA website (www.AWWA.org/science/sun/).

23.128 Disposition of Plant Records

Good recordkeeping is very important because records indicate potential problems, adequate operation, and are a good waterworks practice. Usually, the only record required by the health agency is the summary of the daily turbidity of the treated surface water as it enters the distribution system. Chlorine residual and bacterial counts are often required. Other records that may also be required include:

1. Total trihalomethane (TTHM) data (frequency of this report is based on the number of people served).

2. The daily log and records of the analyses to control the treatment process may be required when there are chronic treatment problems.

3. Chlorination, constituent removal, and sequestering records may be required from small systems (especially those demonstrating little understanding of the processes).

4. Records showing the quantity of water from each source in use may be required from systems with sources producing water not meeting state or local health department water quality standards.

An important question is how long records should be kept. Records should be kept as long as they may be useful. Some information will become useless after a short time, while other data may be valuable for many years. Data that might be used for future design or expansion should be kept indefinitely. Laboratory data will always be useful and should be kept indefinitely. Regulatory agencies may require you to keep certain water quality analyses (bacteriological test results) and customer complaint records on file for specified time periods (10 years for chemical analyses and bacteriological tests).

Even if old records are not consulted every day, this does not lessen their potential value. For orderly records handling and storage, set up a schedule to periodically review old records and to dispose of those records that are no longer needed. A decision can be made when a record is established regarding the time period for which it must be retained.

QUESTIONS

Write your answers in a notebook and then compare your answers with those on page 699.

23.12A List some of the important uses of records.

23.12B What is "unaccounted-for water"?

23.12C What chemical inventory and usage records should be kept?

23.12D List the important items usually contained on a purchase order.

23.13 WATER AND ENERGY CONSERVATION

23.130 Need for Conservation

An effective water and energy conservation program is an excellent means of saving money. Water conservation can reduce energy costs. This effort can reduce the flows to wastewater treatment plants, thus saving energy and possibly postponing the need for capacity expansions. Reducing water and energy consumption results in conservation of natural resources.

This section emphasizes possible approaches to water conservation programs. However, many of the concepts used to conserve water also can be applied to an energy conservation program.

23.131 What Is Water Conservation?

When developing a water conservation program, the water utility manager must realize that consumers have different perspectives regarding water conservation. A preservation perspective emphasizes aesthetic and naturalistic goals over all else, including economic development, growth, and efficiency. An ecological perspective emphasizes avoiding the tragedy that occurs when a common resource (water) is overconsumed. A hydrological perspective emphasizes the water cycle and maintaining water quality through sufficient levels of water supply. A traditional economics perspective emphasizes allocating and using water efficiently. A resource economics perspective goes a step further by emphasizing the need to control rates of water use to ensure a sustainable future.

One simple definition equates water conservation with reduced water use. This definition is too simple for practical purposes. Some practices that decrease water use can be undesirable, particularly if the water-use reductions unnecessarily impair consumer lifestyles or come at the expense of other valued resources. The concept of water conservation should take into account the impact of conservation on the mix of all resources.

Developing conservation programs, including conservation-oriented rate structures, requires community involvement through both public participation and consumer education. Community involvement and consumer education can improve the effectiveness of conservation efforts and help mitigate many of the problems associated with implementing conservation programs and rate structures.

The key practical long-term benefit of water conservation is the postponement or deferral of additional treatment and source development capacity. Avoided costs associated with new capacity or sources include not only the capital and operating costs associated with facilities, but also the procedural costs associated with obtaining regulatory approvals, the political costs associated with securing support for projects and necessary revenues, and the environmental costs associated with developing water supplies. By inducing consumers to control their usage, conservation practices and rate structures can help extend the useful service life of existing capacity.

Reductions in water usage can lead to reductions in wastewater flows. These reduced wastewater flows can result in the postponement or deferral of expanded wastewater treatment capacity.

The most important effect of water conservation pricing is improved economic behavior on the part of consumers and improved economic performance on the part of water-supply agencies. Cost-based conservation rates send important economic signals about the value of water. The resulting efficiency benefits water consumers, the water agency, the environment, and society.

23.132 Elements of Conservation Programs

Water utility agencies should develop a water conservation strategy. The basic elements of this strategy consist of water conservation best management practice (BMP) strategies. A BMP is described as a policy, program, practice, rule, regulation, or ordinance or the use of devices, equipment, or facilities that meet either of the following criteria:

1. An established and generally accepted practice among water suppliers that results in more efficient use or conservation of water.

2. A practice for which sufficient data are available from existing water conservation projects to indicate that significant conservation or conservation-related benefits can be achieved; that the practice is technically and economically reasonable and not environmentally or socially unacceptable; and that the practice is not otherwise unreasonable for most water suppliers to carry out.

The BMPs described in this section were developed by the California Urban Water Conservation Council (CUWCC). Water utility managers need to evaluate the BMPs outlined in terms of effort to develop and implement the BMP, including costs, with the anticipated results and benefits.

23.1320 Residential Water Surveys

Residential water survey BMP includes indoor and outdoor audits of residential water use and distribution of water-saving devices. The surveys should include the following elements:

1. Indoor

    a. Check for leaks, including toilets, faucets, and meter check.

    b. Check shower head flow rates, check aerator flow rates, and offer to replace or recommend replacement, as necessary.

    c. Check toilet flow rates and offer to install or recommend installation of displacement device or direct customer to ULFT (ultra-low-flush [flow] toilet) replacement program, as necessary; replace leaking toilet flapper, as necessary.

2. Outdoor

    a. Check irrigation system and timers.

    b. Review or develop customer irrigation schedule.

3. Recommended

    a. Measure currently landscaped area.

    b. Measure total irrigable area.

Following completion of a residential water survey, the customer should be provided with an evaluation of the results and water-saving recommendations. The customer should be provided a water conservation information packet.

23.1321 Residential Plumbing Retrofits

The residential plumbing retrofit BMP includes distribution or installation of water-saving devices in pre-1992 residences. The retrofit program should include at least the following actions:

1. Identify single-family and multifamily residences constructed prior to 1992. Develop a targeting and marketing strategy to distribute or directly install high-quality, low-flow shower heads (rated 2.5 GPM [10.5 liters per minute] or less), toilet displacement devices (as needed), toilet flappers (as needed), and faucet aerators (rated 2.2 GPM [8.3 liters per minute] or less) as practical to residences requiring them.

2. Maintain distribution or direct installation programs until most single-family residences and multifamily units are fitted with high-quality, low-flow shower heads.

3. Track the type and number of retrofits completed, devices distributed, and program costs.

23.1322 System Water Audits, Leak Detection, and Repair

The system water audits BMP consists of annually calculating unaccounted-for water and performing distribution system water audits, as required. Elements of this BMP include:

1. Annually complete a prescreening system audit to determine the need for a full-scale system audit. The prescreening system audit should be calculated as follows:

    a. Determine metered sales.

    b. Determine other system verifiable uses.

    c. Determine total supply into the system.

    d. Divide metered sales plus other verifiable uses by total supply into the system. If this quantity is less than 0.9, a full-scale system audit is indicated.

2. When indicated, water agencies should complete water audits of their distribution systems using methodology consistent with that described in AWWA's guidebook, *WATER AUDITS AND LEAK DETECTION*.[21]

3. Water agencies should advise customers whenever it appears that leaks exist on the customer's side of the meter; perform distribution system leak detection when warranted and cost-effective; and repair leaks when found.

23.1323 Metering with Commodity Rates

This BMP requires the metering of all new connections and retrofitting of existing connections to ensure metering of all consumptions and billing on the basis of volume of water consumed. Elements of this BMP include:

1. Requiring meters for all new connections and billing by volume of water used.

2. Establishing a program for retrofitting existing unmetered connections and billing by volume of water consumed.

3. Identifying intra- and interagency disincentives or barriers to retrofitting mixed-use commercial accounts with dedicated landscape meters, and conducting a feasibility study to assess the merits of a program to provide incentives to switch mixed-use accounts to dedicated landscape meters.

---

[21] Available from American Water Works Association (AWWA), Bookstore, 6666 West Quincy Avenue, Denver, CO 80235. Order No. 30036. ISBN 1-58321-018-0. Price to members, $67.50; nonmembers, $97.50; price includes cost of shipping and handling.

23.1324 Large Landscape Conservation Programs

Large landscape conservation programs and incentives can be a very effective BMP. A major element of this program is evapotranspiration-based (ETo-based)[22] water-use budgets for large landscape irrigators. This BMP includes provisions for customer support, education, and assistance. Water agencies should provide nonresidential customers with support and incentives to improve their landscape water-use efficiency. This support should include the following elements:

1. Accounts with Dedicated Irrigation Meters

   a. Identify accounts with dedicated irrigation meters and assign ETo-based water-use budgets.

   b. Provide notices with each billing cycle to accounts with water-use budgets showing the relationship between the budget and actual consumption; the water agency may choose not to notify customers whose use is less than their water-use budget.

2. Commercial/Industrial/Institutional (CII) Accounts with Mixed-Use Meters or Not Metered

   a. Develop and implement a strategy targeting and marketing large landscape water-use surveys to commercial/industrial/institutional accounts with mixed-use meters.

   b. Unmetered service areas will actively market landscape surveys to existing accounts with large landscapes, or accounts with landscapes that have been determined by the water purveyor not to be water efficient.

   c. Offer the following measures when cost-effective:

      (1) Landscape water-use analysis/surveys

      (2) Voluntary water-use budgets

      (3) Installation of dedicated landscape meters

      (4) Training (multilingual, where appropriate) in landscape maintenance, irrigation system maintenance, and irrigation system design

      (5) Financial incentives to improve irrigation system efficiency such as loans, rebates, and grants for the purchase or installation of water-efficient irrigation systems

      (6) Follow-up water-use analyses/surveys consisting of a letter, phone call, or site visit, where appropriate

   d. Survey elements will include: measurement of landscape area; measurement of total irrigable area; irrigation system check and distribution uniformity analysis; review or development of irrigation schedules, as appropriate; provision of a customer survey report and information packet.

   e. Track survey offers, acceptance, findings, devices installed, savings potential, and survey cost.

3. New or Change-of-Service Accounts

   Provide information on climate-appropriate landscape design and efficient irrigation equipment/management to new customers and change-of-service customer accounts.

4. Recommended

   a. Install climate-appropriate, water-efficient landscaping at water agency facilities, and dual metering, where appropriate.

   b. Provide customer notices prior to the start of the irrigation season alerting them to check their irrigation systems and to make repairs, as necessary. Provide customer notices at the end of the irrigation season advising them to adjust their irrigation system timers and irrigation schedules.

23.1325 High-Efficiency Clothes Washers

This BMP provides rebates for high-efficiency washing machines. Important elements of this BMP include the water agency determining that significant water savings are available from high-efficiency washing machines and that there is widespread product availability. Also, financial incentive programs offered by energy service providers may create an effective partnership with the water utility.

23.1326 Public Information Programs

Public information programs are an effective means of promoting water conservation. Important elements of a water agency's public information programs should include:

1. Information to promote water conservation and water conservation-related benefits.

2. The program should include providing speakers to employees, community groups, and media; using paid and public service advertising; using billing inserts; providing information on customers' bills showing use in gallons per day (liters per day) for the last billing period compared to the same period the year before; providing public information to promote water conservation practices; and coordinating with other government agencies, industry groups, public interest groups, and the media.

23.1327 School Education Programs

School education programs include the provision of educational materials and services to schools. Educating children of

---

[22] ETo is an estimate of the depth of water evaporated and transpired from a reference crop if water is not a limiting factor. The reference crop for landscape is tall fescue grass that is actively growing, completely shading the soil, and cut 4- to 6-inches (9- to 15-cm) high.

the need to conserve water is also a way for the water agency to communicate to adults by information children bring home from school. A school education water conservation program should include the following elements:

1. Information to promote water conservation and water conservation-related benefits.

2. Programs should include working with school districts and private schools in the water supplier's service area to provide instructional assistance, educational materials, and classroom presentations that identify urban, agricultural, and environmental issues and conditions in the local watershed. Educational materials should meet the local education framework requirements, and grade-appropriate materials should be distributed to grade levels K–3, 4–6, 7–8, and high school.

23.1328 *Conservation Programs for Commercial, Industrial, and Institutional (CII) Sectors*

The purpose of this BMP is to increase water-use efficiency in the commercial, industrial, and institutional (CII) sectors. Activities for this BMP include:

1. Identify and rank commercial, industrial, and institutional (CII) accounts according to water use. For the purposes of this BMP, CII accounts are defined as follows:

    a. Commercial Accounts: Any water use that provides or distributes a product or service, such as hotels, restaurants, office buildings, commercial businesses, or other places of commerce. These do not include multifamily residences, agricultural users, or customers that fall within the industrial or institutional classifications.

    b. Industrial Accounts: Any water users that are primarily manufacturers or processors of materials as defined by the Standard Industrial Classifications (SIC) code numbers 2000 through 3999 (or by the North American Industry Classification System (NAICS) code numbers 311611 through 339999).

    c. Institutional Accounts: Any water-using establishment dedicated to public service. This includes schools, courts, hospitals, and government facilities. All facilities serving these functions are to be considered institutions regardless of ownership.

2. A program to accelerate replacement of existing high-water-using toilets with ultra-low-flush (1.6 gallons [6 liters] or less) toilets in commercial, industrial, and institutional facilities.

3. Implementation of a CII water-use survey and customer incentives program.

4. Implementation of programs to achieve annual water-use savings.

23.1329 *Wholesale Agency Assistance Programs*

The purpose of this BMP is for the wholesaler to provide support for the water conservation programs of retail water suppliers. Important elements of this program include:

1. Financial Support

    a. Wholesale water supplier should provide financial incentives or equivalent resources, as appropriate, beneficial, and mutually agreeable, to the retail water agency customers to advance water conservation efforts and effectiveness.

    b. All BMPs implemented by retail water agency customers that can be shown to be cost-effective in terms of avoided cost of water from the wholesaler's perspective, using cost-effectiveness analysis procedures, should be supported.

2. Technical Support

    Wholesale water agencies should provide conservation-related technical support and information to all retail agencies for whom they serve as a wholesale supplier. At a minimum, this requires:

    a. Conducting, funding, or promoting workshops addressing the following topics:

        (1) Procedures for calculating program savings, costs, and cost-effectiveness.

        (2) Retail agencies' BMP implementation reporting requirements.

        (3) The technical, programmatic, strategic, or other pertinent issues and developments associated with water-conservation activities in each of the following areas: ultra-low-flush toilet (ULFT) replacement; residential retrofits; commercial, industrial, and institutional surveys; residential and large turf irrigation; and conservation-related rates and pricing.

    b. Having the necessary staff or equivalent resources available to respond to retail agencies' technical and programmatic questions involving the BMPs and their associated reporting requirements.

3. Program Management

    Wholesale and retail agencies should retain maximum local flexibility in designing and implementing locally cost-effective BMP conservation programs. Cooperatively designed regional programs should be encouraged. When mutually agreeable and beneficial, the wholesaler may operate all or any part of the conservation-related activities that a given retail supplier is obligated to implement under the BMP's cost-effectiveness test.

4. Water Shortage Allocations

    Wholesale agencies should work in cooperation with their customers to identify and remove potential disincentives to long-term conservation created by water shortage allocation policies and to identify opportunities to encourage and reward cost-effective investments in long-term conservation shown to advance regional water supply reliability and sufficiency.

23.13210 Conservation Pricing

Conservation pricing includes uniform or increasing block rate structure, volume-related sewer charges, and service cost recovery. Implementation methods should be at least as effective as eliminating nonconserving pricing and adopting conserving pricing. For agencies supplying both water and sewer service, this BMP applies to pricing of both water and sewer service. Agencies that supply water but not sewer service should make good-faith efforts to work with sewer agencies so that those sewer agencies adopt conservation pricing for sewer service.

1. Nonconserving pricing provides no incentives to customers to reduce use. Such pricing is characterized by one or more of the following components: rates in which the unit price decreases as the quantity used increases (declining block rates); rates that involve charging customers a fixed amount per billing cycle regardless of the quantity used; pricing in which the typical bill is determined by high fixed charges and low commodity (water-use) charges.

2. Conservation pricing provides incentives to customers to reduce average or peak use, or both. Such pricing includes rates designed to cover the cost of providing service and billing for water and sewer service based on metered water use. Conservation pricing is also characterized by one or more of the following components: rates in which the unit rate is constant regardless of the quantity used (uniform rates) or increases as the quantity used increases (increasing block rates); seasonal rates or excess-use surcharges to reduce peak demands during summer months; or rates based upon the long-run marginal cost or the cost of adding the next unit of capacity to the system.

3. Adoption of lifeline rates for low-income customers should neither qualify nor disqualify a rate structure as meeting the requirements of this BMP.

23.13211 Conservation Coordinator

This BMP consists of the designation of a staff coordinator for the agency conservation programs. This BMP includes the following actions:

1. Designation of a water conservation coordinator and support staff (if necessary), whose duties should include the following:

   a. Coordination and oversight of conservation programs and BMP implementation.

   b. Preparation and submittal of the BMP implementation report.

   c. Communication and promotion of water conservation issues to agency senior management; coordination of agency conservation programs with operations and planning staff; preparation of annual conservation budget; and preparation of the conservation elements of the agency's urban water management plan.

2. Agencies jointly operating regional conservation programs are not expected to staff duplicative and redundant conservation coordinator positions.

23.13212 Water Waste Prohibition

The purpose of this BMP is to enforce prohibition of wasteful use of water. Elements of this BMP include the enacting and enforcing of measures prohibiting gutter flooding, single-pass cooling systems in new connections, nonrecirculating systems in all new conveyer car wash and commercial laundry systems, and nonrecycling decorative water fountains.

Water agencies should support efforts to develop laws regarding exchange-type water softeners that would: (1) allow the sale of only more efficient, demand-initiated regenerating (DIR) models; (2) develop minimum appliance efficiency standards that: (a) increase the regeneration efficiency standard to at least 3,350 grains of hardness removed per pound (477 grams/kg) of common salt used; and (b) implement an identified maximum number of gallons discharged per gallon of soft water produced; and (3) allow local agencies, including municipalities and special districts, to set more stringent standards or to ban on-site regeneration of water softeners if it is demonstrated and found by the agency governing board that there is an adverse effect on the reclaimed water or groundwater supply.

Water agencies should also include water softener checks in home water audit programs and include information about DIR and exchange-type water softeners in their educational efforts to encourage replacement of less efficient timer models.

23.13213 Residential ULFT Replacement Programs

This BMP encourages programs that promote the replacement of high-water-using toilets with ULFTs (ultra-low-flow toilets). Important elements of this BMP include:

1. Implementation of programs for replacing existing high-water-using toilets with ultra-low-flush (1.6 gallons [6 liters] or less) toilets in single-family and multifamily residences.

2. Programs should be at least as effective as those programs requiring toilet replacement at time of resale with program effectiveness determined using a methodology for calculating water savings.

23.13214 Potential Best Management Practices

This section lists other potential best management practices that a water utility might wish to consider when developing and implementing a water conservation program.

- Graywater use
- Efficiency standards for water-using appliances and irrigation devices
- Retrofit of existing car washes
- Distribution system pressure regulation
- Swimming pool and spa conservation including covers to reduce evaporation
- Point-of-use water heaters, recirculating hot water systems, and hot water pipe insulation
- Efficiency standards for new industrial and commercial processes

23.133 EPA's WaterSense: Efficiency Made Easy

The US Environmental Protection Agency has implemented a water efficiency program to educate consumers about making smart water choices that save money and maintain high environmental standards. Manufacturers can use the EPA WaterSense label on products and services that perform at least 20 percent more efficiently than less efficient counterparts. To learn more about EPA's WaterSense program, visit EPA's website at www.epa.gov.

23.134 Acknowledgment

Information presented in this section on water and energy conservation was obtained from the many excellent publications of the California Urban Water Conservation Council (CUWCC). Permission to use this well-documented information is greatly appreciated.

23.135 Additional Reading

1. *CONSUMER'S GUIDE TO WATER CONSERVATION.* Obtain from American Public Works Association, 2345 Grand Boulevard, Suite 500, Kansas City, MO 64108-2641, phone (800) 848-2792 ((800) 848-APWA) or visit their website at www.apwa.net/bookstore. Order No. PB.XDSW. Members, $8.00; nonmembers, $13.00.

2. *WATER CONSERVATION PROGRAMS—A PLANNING MANUAL* (M52). Obtain from American Water Works Association (AWWA), Bookstore, 6666 West Quincy Avenue, Denver, CO 80235. Order No. 30052. ISBN 1-58321-391-0. Price to members, $72.50; nonmembers, $102.50; price includes cost of shipping and handling.

3. *WATER AUDITS AND LEAK DETECTION* (M36). Obtain from American Water Works Association (AWWA), Bookstore, 6666 West Quincy Avenue, Denver, CO 80235. Order No. 30036. ISBN 1-58321-018-0. Price to members, $67.50; nonmembers, $97.50; price includes cost of shipping and handling.

Many valuable publications are available on water conservation. For additional reading on this subject, contact:

- California Urban Water Conservation Council (CUWCC), 455 Capital Mall, Suite 703, Sacramento, CA 95814; phone: (916) 552-5885; website: www.cuwcc.org.
- American Water Works Association (AWWA), 6666 West Quincy Avenue, Denver, CO 80235; phone: (800) 926-7337; website: www.awwa.org.

QUESTIONS

Write your answers in a notebook and then compare your answers with those on page 699.

23.13A What are the different perspectives that consumers have regarding water conservation?

23.13B What are the key elements of a residential water survey BMP?

23.13C What is a major element of a large landscape water conservation program BMP?

23.13D What are the important elements of wholesale agency assistance programs?

23.14 ACKNOWLEDGMENTS

During the writing of this material, Lynne Scarpa, Phil Scott, Chris Smith, and Rich von Langen, all members of California Water Environment Association (CWEA), provided many excellent materials and suggestions for improvement. Their generous contributions are greatly appreciated.

23.15 ADDITIONAL READING

1. *MANAGE FOR SUCCESS,* especially Chapter 11, "Emergency Planning." Obtain from the Office of Water Programs, California State University, Sacramento, 6000 J Street, Sacramento, CA 95819-6025. Price, $49.00.

2. *WATER UTILITY MANAGEMENT* (M5). Obtain from American Water Works Association (AWWA), Bookstore, 6666 West Quincy Avenue, Denver, CO 80235. Order No. 30005. ISBN 0-89867-063-2. Price to members, $72.50; nonmembers, $102.50; price includes cost of shipping and handling.

3. "A Water and Wastewater Manager's Guide for Staying Financially Healthy," US Environmental Protection Agency. EPA No. 430-9-89-004. Obtain from National Technical Information Service (NTIS), 5285 Port Royal Road, Springfield, VA 22161. Order No. PB90-114455. Price, $33.50, plus $5.00 shipping and handling per order.

4. "Wastewater Utility Recordkeeping, Reporting and Management Information Systems," US Environmental Protection Agency. EPA No. 430-9-82-006. Obtain from National Technical Information Service (NTIS), 5285 Port Royal Road, Springfield, VA 22161. Order No. PB83-109348. Price, $39.50, plus $5.00 shipping and handling per order.

5. *SUPERVISION: CONCEPTS AND PRACTICES OF MANAGEMENT*, 10th Edition, 2007, Hilgert, Raymond L., and Edwin Leonard, Jr. Obtain from Thomson Learning, Customer Service, 10650 Toebben Drive, Independence, KY 41051. ISBN 0-324-31624-0. Price, $121.95, plus shipping and handling.

6. *TEXAS MANUAL*, Chapter 18,* "Effective Public Relations in Water Works Operations," and Chapter 19,* "Planning and Financing."

7. *PRINCIPLES OF WATER RATES, FEES, AND CHARGES* (M1). Obtain from American Water Works Association (AWWA), Bookstore, 6666 West Quincy Avenue, Denver, CO 80235. Order No. 30001. ISBN 1-58321-069-5. Price to members, $87.50; nonmembers, $129.50; price includes cost of shipping and handling.

8. *EMERGENCY PLANNING FOR WATER UTILITIES* (M19). Obtain from American Water Works Association (AWWA), Bookstore, 6666 West Quincy Avenue, Denver, CO 80235. Order No. 30019. ISBN 0-89867-135-7. Price to members, $72.50; nonmembers, $102.50; price includes cost of shipping and handling.

9. *GIS IMPLEMENTATION FOR WATER AND WASTEWATER TREATMENT FACILITIES* (MOP 26). Obtain from Water Environment Federation (WEF), 601 Wythe Street, Alexandria, VA 22314-1994. Order No. WPM401. ISBN 0-07-145305-9. Price to members, $76.75; nonmembers, $92.75; price includes cost of shipping and handling.

10. To learn more about homeland defense, visit EPA's website at www.epa.gov/safewater/watersecurity/index.html.

* Depends on edition.

END OF LESSON 3 OF 3 LESSONS
on
ADMINISTRATION

Please answer the discussion and review questions next.

DISCUSSION AND REVIEW QUESTIONS
Chapter 23. ADMINISTRATION
(Lesson 3 of 3 Lessons)

Write the answers to these questions in your notebook. The question numbering continues from Lesson 2.

26. What can happen when agencies rely primarily on corrective maintenance to keep the system running?

27. What does a SCADA system do?

28. What items should be included in a water supplier's cross connection control program?

29. What factors should be considered when selecting a backflow prevention device for a particular situation?

30. How would you assess the vulnerability of a water supply system?

31. How can a utility protect its water supply from contamination?

32. What would you do if you discovered that contaminated water has entered your distribution system?

33. Why must waterworks utilities have a safety program?

34. How can a good safety record be accomplished?

35. What is the intent of the OSHA regulations?

36. What are the four main elements of a hazard communication program?

37. What topics should be included in a safety officer's talk to new operators?

38. Why are records important?

39. Why do water utilities need an effective water and energy conservation program?

40. What is the most important effect of water conservation pricing?

SUGGESTED ANSWERS
Chapter 23. ADMINISTRATION

ANSWERS TO QUESTIONS IN LESSON 1

Answers to questions on page 619.

23.0A Local community demands on a utility manager include protection from environmental disasters with a minimum investment of money.

23.0B Changes in the environmental workplace are created by changes in the workforce and advances in technology.

Answers to questions on page 621.

23.1A The functions of a utility manager include planning, organizing, staffing, directing, and controlling.

23.1B In small communities, the community depends on the manager to handle everything.

Answers to questions on page 624.

23.2A Utility planning must include operational personnel, local officials (decision makers), and the public.

23.3A The purpose of an organizational plan is to show who reports to whom and to identify the lines of authority.

23.3B Effective delegation is uncomfortable for many managers since it requires giving up power and responsibility. Many managers believe that they can do the job better than others, they believe that other employees are not well trained or experienced, and they are afraid of mistakes. The utility manager retains some responsibility even after delegating to another employee and, therefore, the manager is often reluctant to delegate or may delegate the responsibility but not the authority to get the job done.

23.3C An important and often overlooked part of delegation is follow-up by the supervisor.

Answers to questions on page 626.

23.4A Staffing responsibilities include hiring new employees, training employees, and evaluating job performance.

23.4B The two personnel management concepts a manager should always keep in mind are "job-related" and "documentation."

23.4C The steps involved in a staffing analysis include:

1. List the tasks to be performed.
2. Estimate the number of staff hours per year required to perform each task.
3. List the utility's current employees.
4. Assign tasks based on each employee's skills and abilities.
5. Adjust the work assignments as necessary to achieve the best possible fit between the work to be done and the personnel/skills available to do it.

23.4D A qualifications profile is a clear statement of the knowledge, skills, and abilities a person must possess to perform the essential job duties of a particular position.

Answers to questions on page 629.

23.4E The purpose of a job interview is to gain additional information about the applicants so that the most qualified person can be selected to fill a job opening.

23.4F Protected groups include minorities, women, disabled persons, persons over 40 years of age, and union members.

23.4G The "OUCH" principle stands for:

Objectivity
Uniform treatment of employees
Consistency
Having job relatedness

23.4H A new employee's safety training should begin on the first day of employment or as soon thereafter as possible.

Answers to questions on page 631.

23.4I The purpose of a probationary period for new employees is to provide a time during which both the employer and employee can assess the "fit" between the job and the person.

23.4J The compensation an employee receives for the work performed includes satisfaction, recognition, security, appropriate pay, and benefits.

23.4K   Utility managers should provide training opportunities for employees so they can keep informed of new technologies and regulations. Training for supervisors is also important to ensure that supervisors have the knowledge, skills, and attitude that will enable them to be effective supervisors.

Answers to questions on page 634.

23.4L   An employee's immediate supervisor should conduct the employee's performance evaluation.

23.4M   Dealing with employee discipline requires tact and skill. The manager or supervisor should stay flexible, calm, and open-minded.

23.4N   Common warning signs of potential violence include abusive language, threatening or confrontational behavior, and brandishing a weapon.

Answers to questions on page 639.

23.4O   Harassment is any behavior that is offensive, annoying, or humiliating to an individual and that interferes with a person's ability to do a job. This behavior is uninvited, often repeated, and creates an uncomfortable or even hostile environment in the workplace.

23.4P   Types of behavior that could be considered sexual harassment include:

- Unwanted hugging, patting, kissing, brushing up against someone's body, or other inappropriate sexual touching
- Subtle or open pressure for sexual activity
- Persistent sexually explicit or sexist statements, jokes, or stories
- Repeated leering or staring at a person's body
- Suggestive or obscene notes or phone calls
- Display of sexually explicit pictures or cartoons

23.4Q   The best way to prevent harassment is to set an example by your own behavior and to keep communication open between employees. A manager must also be aware of and take action to prevent any type of harassment in the workplace.

23.4R   In general, the ADA defines disability as a physical or mental impairment that substantially limits one or more of the major life activities of an individual.

Answers to questions on page 640.

23.4S   The shop steward is elected by the union employees and is their official representative to management and the local union.

23.4T   Union contracts do not change the supervisor's delegated authority or responsibility. Operators must carry out the supervisor's orders and get the work done properly, safely, and within a reasonable amount of time. However, a contract gives a union the right to protest or challenge a supervisor's decision.

ANSWERS TO QUESTIONS IN LESSON 2

Answers to questions on page 643.

23.5A   A manager needs both written and oral communication skills.

23.5B   The most common written documents that a utility manager must write include memos, business letters, press releases, résumés, monitoring reports, and the annual report.

23.5C   The annual report should be a review of what and how the utility operated during the past year and also the goals for the next year.

Answers to questions on page 644.

23.6A   A utility manager may be asked to conduct meetings with employees, the governing board, the public, and with other professionals in your field.

23.6B   Before a meeting: (1) prepare an agenda and distribute, (2) find an adequate meeting room, and (3) set a beginning and ending time.

Answers to questions on page 646.

23.7A   The first step in organizing a public relations campaign is to establish objectives so you will have a clear idea of what you expect to achieve.

23.7B   Employees can be informed about the utility's plans, practices, and goals through newsletters, bulletin boards, and regular, open communication between supervisors and subordinates.

23.7C   Newspapers give more thorough, in-depth coverage to stories than do the broadcast media.

23.7D   Practice is the key to effective public speaking.

23.7E   Complaints can be a valuable asset in determining consumer acceptance and pinpointing water quality problems. Customer calls are frequently the first indication that something may be wrong. Responding to complaints and inquiries promptly can save the utility money and staff resources, and minimize customer inconvenience.

Answers to questions on page 650.

23.8A   The three main areas of financial management for a utility include providing financial stability for the utility, careful budgeting, and providing capital improvement funds for future utility expansion.

23.8B   The operating ratio for a utility is calculated by dividing total revenues by total operating expenses.

23.8C   It is important for a manager to get input from other personnel in the utility as well as community leaders as the budgeting process proceeds in order to gain support for the budget and to keep the budget on track once adopted.

23.8D The basic ways for a utility to finance capital improvements are through general obligation bonds, revenue bonds, or loan funding programs.

23.8E A revenue bond is commonly used to fund utility improvements. This bond has no limit on the amount of funds available and the user charges provide repayment on the bond. To qualify for these bonds, the utility must show sound financial management and the ability to repay the bond.

ANSWERS TO QUESTIONS IN LESSON 3

Answers to questions on page 653.

23.9A The purpose of an O & M program is to maintain design functionality or to restore the system components to their original condition and thus functionality, that is, to ensure that the system performs as designed and intended.

23.9B Commonly accepted types of maintenance include corrective maintenance, preventive maintenance, and predictive maintenance.

23.9C The major elements of a good preventive maintenance program include the following:

1. Planning and scheduling
2. Records management
3. Spare parts management
4. Cost and budget control
5. Emergency repair procedures
6. Training program

Answers to questions on page 654.

23.9D SCADA stands for Supervisory Control And Data Acquisition system.

23.9E A SCADA system collects, stores, and analyzes information about all aspects of operation and maintenance, transmits alarm signals, when necessary, and allows fingertip control of alarms, equipment, and processes.

Answers to questions on page 655.

23.9F SCADA systems are used by operators at water treatment plants to monitor filtration system influent and effluent turbidity levels, head losses through the filters, filter flows, and filter valve settings.

23.9G When a customer phones and complains about a loss of water pressure, the operator can determine the recorded system pressures near the customer's residence. If there is a loss of water pressure, the operator can try to determine the cause (if the water pressure was below an acceptable minimum (30 psi), the SCADA system should have alerted the operators immediately). If there is not a system loss of water pressure, the operator can suggest how the customer can try to locate the cause of the problem on the customer's property.

23.9H The greatest challenges for operators using SCADA systems are to realize that computers may not be correct and to have the ability to operate when the SCADA system fails.

Answers to questions on page 659.

23.9I Contamination through cross connections has consistently caused more waterborne disease outbreaks in the United States than any other reported factor.

23.9J Responsibilities for the implementation of cross connection control programs are shared by water suppliers, water users, health agencies, and plumbing officials.

23.9K The water user's cross connection control responsibility is to keep contaminants out of the potable water system on the user's premises. The user also has the responsibility to prevent the creation of cross connections through modifications of the plumbing system on the premises.

Answers to questions on page 661.

23.9L The particular type of backflow prevention device most suitable for a given situation depends on the degree of health hazard, the probability of backflow occurring, the complexity of the piping on the premises, and the probability of the piping being modified. The higher the assessed risk due to these factors, the more reliable and positive the type of device needed.

23.9M Air gap separation or reduced pressure principle backflow prevention devices are installed at wastewater treatment plants, wastewater pumping stations, reclaimed water reuse areas, areas where toxic substances in toxic concentrations are handled under pressure, premises having an auxiliary water supply that is or may be contaminated, and in locations where fertilizer, herbicides, or pesticides are injected into a sprinkler system.

23.9N The power of a geographic information system (GIS) is that information can be retrieved geographically. An operator can easily look at or print out a specific area of a map. The map will contain an inventory of the distribution system pipes and structures within the selected area and it will provide detailed information about each of the pipes and structures (or entities).

Answers to questions on page 663.

23.10A The first step toward an effective contingency plan for emergencies is to make an assessment of vulnerability. Then, a comprehensive plan of action can be developed and implemented.

23.10B An emergency operations plan need not be too detailed, since all types of emergencies cannot be anticipated and a complex response program can be more confusing than helpful.

23.10C The following outline can be used as the basis for developing an emergency operations plan:

1. Make a vulnerability assessment.
2. Inventory organizational personnel.
3. Provide for a recovery operation (plan).
4. Provide training programs for operators in carrying out the plan.
5. Coordinate with local and regional agencies such as the health, police, and fire departments to develop procedures for carrying out the plan.
6. Establish a communications procedure.
7. Provide protection for personnel, plant equipment, records, and maps.

Answers to questions on page 669.

23.10D The color codes and risk levels for the US Terrorism Alert System include: RED, severe risk of terrorist attacks; ORANGE, high risk of terrorist attacks; YELLOW, significant risk, elevated condition; BLUE, general risk, guarded condition; and GREEN, low risk of terrorist attacks.

23.10E Toxicity is the ability of a contaminant (chemical or biological) to cause injury when introduced into the body.

23.10F The degree of toxicity varies with the concentration of a contaminant required to cause injury, the speed with which the injury takes place, and the severity of the injury.

23.10G Possible secondary effects in a water supply that may allow detection of a contaminant without specific testing include taste, color, odor, and chlorine demand.

Answers to questions on page 670.

23.10H *Cryptosporidium* is a significant public health concern because sensitive populations may become seriously ill even when the drinking water meets or exceeds all current state and federal standards.

23.10I The most serious threat of *Cryptosporidium* to a public water supply occurs during periods of heavy rains or snow melts that flush areas that are sources of high concentrations of oocysts into waters upstream of a plant intake.

23.10J A water treatment plant can be operated to protect the public from *Cryptosporidium* by optimizing the treatment processes. The operator must continually verify optimum coagulant doses, be sure filtered water turbidity levels are low, and provide adequate disinfection chemicals and contact times.

Answers to questions on page 672.

23.11A A safety officer should evaluate every accident, offer recommendations, and keep and apply statistics.

23.11B The supervisors should be responsible for the implementation of a safety program.

23.11C Both state and federal regulatory agencies enforce the OSHA requirements.

23.11D The utility manager is responsible for the safety of the agency's personnel and the public exposed to the water utility's operations. Therefore, the manager must develop and administer an effective safety program and must provide new employee safety training as well as ongoing training for all employees.

23.11E A safety policy statement should:

1. Define the goals and objectives of the program.
2. Identify the persons responsible for each element of the program.
3. Affirm management's intent to enforce safety regulations.
4. Describe the disciplinary actions that will be taken to enforce safe work practices.

Answers to questions on page 673.

23.11F A supervisor may be responsible, in part or completely, for an accident by causing unsafe acts to take place, by requiring that work be performed in haste, by disregarding an unsafe environment of the workplace, or by failing to consider any number of safety hazards.

23.11G Each operator must accept, at least in part, responsibility for fellow operators, for the utility's equipment, for the operator's own welfare, and even for seeing that the supervisor complies with established safety regulations.

23.11H First aid means emergency treatment for injury or sudden illness, before regular medical treatment is available.

23.11I First-aid training is most important for operators who regularly work with electrical equipment and those who must handle chlorine and other dangerous chemicals.

Answers to questions on page 678.

23.11J The basic elements of a hazard communication program include the following:

1. Identify hazardous materials.
2. Obtain chemical information and define hazardous conditions.
3. Properly label hazards.
4. Train operators.

23.11K A utility manager may or may not be involved in the day-to-day details of enforcing the agency's confined space policy and procedures. However, every utility manager should be aware of the current OSHA requirements and should ensure that the utility's policies not only comply with current regulations, but that the agency's policies are vigorously enforced for the safety of all operators.

Answers to questions on page 682.

23.11L The mainstay of a safety program is the method of reporting and keeping statistics.

23.11M All injuries should be reported, even if they are minor in nature, so as to establish a record in case the injury develops into a serious injury. It may be difficult at a later date to prove whether the accident occurred on or off the job and this information may determine who is responsible for the costs.

23.11N A safety officer should review an accident report form to make recommendations as well as to follow up to be sure that proper action has been taken to prevent recurrence.

23.11O A new, inexperienced operator must receive instruction on all aspects of plant safety. This training includes instruction in the handling of chemicals, the dangers of electrical apparatus, fire hazards, and proper maintenance of equipment to prevent accidents. Special instructions are required for specific work environments such as manholes, gases (chlorine and hydrogen sulfide (H_2S)), water safety, and any specific hazards that are unique to your facility. All new operators should be required to participate in a safety orientation program during the first few days of employment, and an overall training program in the first few months.

23.11P If an operator is unsure of how to perform a job, then it is the operator's responsibility to ask for the training needed.

Answers to questions on page 684.

23.11Q Statistical accident reports should contain accident statistics showing lost time, costs, type of injuries, and other data, based on some time interval.

23.11R Injuries can be classified as fractures, burns, bites, eye injuries, cuts, and bruises.

23.11S Causes of injuries can be classified as heat, machinery, falls, handling objects, chemicals, unsafe acts, and miscellaneous.

23.11T Costs of accidents can be classified as lost time, lost dollars, lost production, contaminated water, or any other means of showing the effects of the accidents.

Answers to questions on page 688.

23.12A Some of the important uses of records include:

1. Aiding operators in solving treatment and water quality problems
2. Providing a method of alerting operators to changes in source water quality
3. Showing that the treated water is acceptable to the consumer
4. Documenting that the final product meets plant performance standards, as well as the standards of the regulatory agencies
5. Determining performance of treatment processes, equipment, and the plant
6. Satisfying legal requirements
7. Aiding in answering complaints
8. Anticipating routine maintenance
9. Providing data for cost analysis and preparation of budgets
10. Providing data for future engineering designs
11. Providing information for monthly and annual reports

23.12B "Unaccounted-for water" is the difference between the amount of treated water that enters the distribution system and water delivered to consumers.

23.12C Chemical inventory and usage records that should be kept include:

1. Chemical inventory/storage (measured use and deliveries)
2. Metered or estimated plant usages
3. Calculated usage of chemicals (compare with actual use)

23.12D Important items usually contained on a purchase order include: (1) the date, (2) a complete description of each item and quantity needed, (3) prices, (4) the name of the vendor, and (5) a purchase order number.

Answers to questions on page 693.

23.13A The different perspectives that consumers have regarding water conservation include preservation, ecological, hydrological, traditional economics, and resource economics.

23.13B The key elements of a residential water survey BMP include indoor and outdoor audits of residential water use and distribution of water-saving devices.

23.13C A major element of a large landscape water conservation program BMP is evapotranspiration-based (ETo-based) water-use budgets for large landscape irrigators.

23.13D The important elements of wholesale agency assistance programs include financial support, technical support, program management, and water shortage allocations.

APPENDIX

WATER TREATMENT PLANT OPERATION
(VOLUME II)

Comprehensive Review Questions and
Suggested Answers

How To Solve Water Treatment Plant
Arithmetic Problems

Abbreviations

Water Words

Subject Index

COMPREHENSIVE REVIEW QUESTIONS
VOLUME II

This section was prepared to help you review the material in Volume II. The questions are divided into five types:

1. True-False
2. Best Answer
3. Multiple Choice
4. Short Answer
5. Problems

To work this section:

1. Write the answer to each question in your notebook.

2. After you have worked a group of questions (you decide how many), check your answers with the suggested answers at the end of this section.

3. If you missed a question and do not understand why, reread the material in the manual.

You may wish to use this section for review purposes when preparing for civil service and certification examinations.

Since you have already completed this course, please *DO NOT SEND* your answers to California State University, Sacramento.

True-False

1. There are direct adverse health effects from drinking water containing either iron or manganese.
   1. True
   2. False

2. In some plants, the injection of the potassium permanganate solution in the volute of the pump will produce complete mixing of potassium permanganate.
   1. True
   2. False

3. One way to rid a distribution system of iron bacteria is to develop a flushing program.
   1. True
   2. False

4. Never allow a backsiphon condition to develop in a distribution system.
   1. True
   2. False

5. Fluoride doses must never be metered against a negative or suction head.
   1. True
   2. False

6. Once the chemical feed equipment is in operation and the major bugs are worked out, the feeder will need to be fine-tuned.
   1. True
   2. False

7. Additional benefits result from the overfeeding of fluoride.
   1. True
   2. False

8. The levels of carbon dioxide, bicarbonate ion, and carbonate ion in waters are very sensitive to pH.
   1. True
   2. False

9. Alkalinity is determined by the amount of base required to reach a titration end point (specific color change).
   1. True
   2. False

10. Hardness is completely removed by the chemical precipitation methods used in water treatment plants.
    1. True
    2. False

11. In the lime–soda softening process, extra lime will be required as the coagulant alum or iron feed rate goes down and, therefore, less lime will be required as the alum or iron feed rate is increased.
    1. True
    2. False

12. When selecting a target hardness level for your plant, consider the uses of the softened water and the cost of softening.
    1. True
    2. False

13. Automated operational control systems are maintenance free.
    1. True
    2. False

14. Methane is involved in the THM reaction.

    1. True
    2. False

15. THM samples should be sent to the laboratory immediately after they are collected.

    1. True
    2. False

16. Chloramines are weaker disinfectants than free chlorine, ozone, or chlorine dioxide, but the residuals remain much longer than free chlorine.

    1. True
    2. False

17. Spent activated alumina media may be regenerated or discarded.

    1. True
    2. False

18. Operators must be sure to use only NSF certified media and chemicals.

    1. True
    2. False

19. The constituents in the water being filtered may cause fouling.

    1. True
    2. False

20. Membrane filtration SCADA systems monitor alarm conditions and follow protocols for alarm notification and plant shutdown.

    1. True
    2. False

21. All available water supplies can be classified according to their mineral quality.

    1. True
    2. False

22. The water flux increases as the mineral content of the feed increases because the osmotic pressure contribution increases with increasing mineral content.

    1. True
    2. False

23. In reverse osmosis operation, as the temperature of the feedwater increases, flux decreases.

    1. True
    2. False

24. Water is transmitted through the membrane at a much more rapid rate than minerals.

    1. True
    2. False

25. Sludge from a water treatment plant is considered an industrial waste that requires compliance with the provisions of the Water Pollution Control Act.

    1. True
    2. False

26. During the draining stages of a sedimentation tank, the walls and all the equipment should be completely hosed down and inspected for damage.

    1. True
    2. False

27. Brine from ion exchange units may be discharged into wastewater collection systems.

    1. True
    2. False

28. Operators should thoroughly read manufacturers' literature on plant equipment and understand the procedures.

    1. True
    2. False

29. Everyone must realize that if the equipment cannot work, no one can work.

    1. True
    2. False

30. The most important requirements in the reliable operation of electrical equipment are cleanliness and the elimination of moisture into the insulation.

    1. True
    2. False

31. Misalignment is a major cause of pump and coupling wear.

    1. True
    2. False

32. Improper or lack of lubrication is probably the biggest cause of compressor failures.

    1. True
    2. False

33. Diesel engines use spark plugs.

    1. True
    2. False

34. The more you know about your plant's instrumentation, the better operator you become.

    1. True
    2. False

35. Digital displays show a reading as a pointer against a marked scale.

    1. True
    2. False

36. Differential pressure flow measuring devices require significant preventive maintenance because of all the moving parts.
    1. True
    2. False

37. The computer control system can alert the operator or change a plant process based on a preprogrammed control strategy defined by the operator.
    1. True
    2. False

38. Safety is using one's knowledge or skill to avoid accidents.
    1. True
    2. False

39. If you are caught in an area containing chlorine, keep your head high until you are out of the affected area.
    1. True
    2. False

40. All fluoride compounds must be regarded as hazardous chemicals that are toxic to operators.
    1. True
    2. False

41. Good driving skills are good for utility public relations.
    1. True
    2. False

42. Never pipet bacteriological samples by mouth, always use a pipet bulb.
    1. True
    2. False

43. Both of the spectroscopic measurement units of percent transmittance and absorbance are associated with color intensity.
    1. True
    2. False

44. If the filters are performing properly, there should be countable algae in the treated water.
    1. True
    2. False

45. Alum coagulates most effectively at pH values near 6.8.
    1. True
    2. False

46. Transient water systems are affected by the Unregulated Contaminants Monitoring Rule (UCMR1).
    1. True
    2. False

47. Primary standards or MCLs are set for substances that are thought to pose a threat to health when present in drinking water at certain levels.
    1. True
    2. False

48. Waters exceeding the MCLs for metals and salts for short periods of time will pose an immediate threat to health.
    1. True
    2. False

49. Whenever a routine coliform sample is coliform-positive, the Total Coliform Rule calls for determination of the presence of fecal coliform or *E. coli* and for repeat sampling.
    1. True
    2. False

50. Iron and manganese are frequently found together in natural waters and produce similar adverse environmental effects and color problems.
    1. True
    2. False

51. Planning is one of the most important functions of utility management and one of the most difficult.
    1. True
    2. False

52. Every utility should have a safety training session for all new employees.
    1. True
    2. False

53. Good communication is an essential part of good management skills.
    1. True
    2. False

54. Large amounts of data, including unimportant data, should be included in a report.
    1. True
    2. False

55. The purpose of the equipment repair/replacement fund account is to generate a profit for the utility.
    1. True
    2. False

56. Safety is good business for both the operator and the agency.
    1. True
    2. False

57. By always logging in supplies immediately upon receipt, you are in a position to reject defective or damaged shipments and control shortages or errors in billing.
    1. True
    2. False

Best Answer (Select only the closest or best answer.)

1. When iron bacteria form thick slimes on the walls of distribution system mains, the slimes are what color from the iron?
   1. Black
   2. Blue-green
   3. Orange
   4. Rust

2. When analyzing for iron and manganese, where should samples be collected?
   1. As close to the well or other source as possible
   2. At the entrance to the distribution system
   3. At the far end of the distribution system
   4. From a sampling tap in the laboratory

3. How does potassium permanganate remove iron and manganese from water?
   1. By exchanging iron and manganese for other ions
   2. By filtering out the iron and manganese
   3. By forming insoluble oxides
   4. By forming soluble oxides

4. Why should fluoride saturators be stirred every day?
   1. To ensure that the stirrer is properly maintained
   2. To prevent fluoride solids from building up on the bottom
   3. To prevent the development of safety hazards
   4. To produce a uniform fluoride strength

5. How can fluoride chemical overloading be prevented?
   1. By proper calculations of chemical dosages
   2. By proper examination of consumers' teeth
   3. By proper maintenance of fluoride chemical feed equipment
   4. By proper operation and continuous monitoring of the product water

6. What is carbonate hardness? That hardness in water caused by
   1. Calcium ions
   2. Magnesium ions
   3. That portion of the total hardness in excess of the alkalinity
   4. The alkalinity present in water up to the total hardness

7. Why must storage areas for bagged lime be covered?
   1. To allow the bags to be stacked in multiple layers
   2. To control lime dust from leaving the area
   3. To prevent rain from wetting the bags
   4. To prevent sunlight from causing lime deterioration

8. What happens when quicklime reacts with water in the slaking process?
   1. It becomes acid as a result of chemical reactions
   2. It becomes clearer as a result of coagulation
   3. It gets hot enough to cause serious burns
   4. It gets more corrosive

9. What happens during the brine stage of an ion exchange softener?
   1. A clear rinse is applied from the top of the unit to remove the waste products and excess brine solution from the softener
   2. The actual softening of the water occurs
   3. The sodium ion concentration of the resin is recharged by pumping a concentrated brine solution onto the resin
   4. The unit is taken out of service and the flow pattern through the unit is reversed

10. What are THM precursors?
    1. Disinfection by-products
    2. Natural organic compounds found after preliminary treatment
    3. Natural organic compounds found in all surface and groundwaters
    4. Organic compounds found in drinking water after chlorination

11. Why do some THM sample bottles contain a small amount of a chemical reducing agent (usually sodium thiosulfate or sodium sulfite)?
    1. To facilitate the chemical reaction that occurs between chlorine and the THM precursors
    2. To provide the chlorine essential for the formation of THMs
    3. To provide the natural organics essential for the formation of THMs
    4. To stop the chemical reaction that occurs between chlorine and the THM precursors

12. What is the operator's responsibility regarding the water treatment plant's approved operating plan?
    1. To ensure that the plan is followed at all times
    2. To modify the plan routinely, as needed, to reflect real-world conditions
    3. To prepare and keep appropriate, up-to-date record-keeping forms
    4. To train all other operators in the effective use of the plan

13. What is the purpose of arsenic sampling/monitoring for process control?
    1. To calculate conformance with the MCL
    2. To troubleshoot or investigate problems
    3. To verify proper operation of the arsenic removal treatment plant
    4. To verify settings on process controls

14. How may the flow through a membrane be recovered after fouling?
    1. By backwashing and cleaning
    2. By removing and reinserting
    3. By repairing and replacing
    4. By scraping and reversing

15. What is the first level of maintenance performed at a water treatment plant?
    1. Breakdown maintenance
    2. Membrane maintenance
    3. Preventive maintenance
    4. Routine maintenance

16. What is the term "flux" used to describe?
    1. The feasibility of the membrane to remove salinity
    2. The flexibility of the semipermeable membrane
    3. The fluctuations in flow through the semipermeable membrane
    4. The rate of water flow through the semipermeable membrane

17. Why is acid added as a pretreatment step before demineralization using cellulose acetate membranes?
    1. To control corrosivity
    2. To increase hydrolysis
    3. To reduce compaction
    4. To slow hydrolysis

18. Which scale inhibitor is most frequently used to prevent the coating of the reverse osmosis membrane?
    1. Nonionic polymer
    2. Soda ash
    3. Sodium hexametaphosphate
    4. Sulfuric acid

19. How does water temperature influence the volume of sludge produced?
    1. As the temperature increases, the biological activity of the bacteria will increase
    2. As the temperature increases, the flocculation factor of the solids will increase
    3. As the temperature increases, the sedimentation tank bottom attracts the solids
    4. As the temperature increases, the settling rate of the solids will increase

20. What is sludge thickening?
    1. Pretreatment of sludge to facilitate removal of water in subsequent treatment processes
    2. Treatment by centrifugal or rotational forces to separate the solids from the liquids
    3. Treatment that removes or separates a portion of the water present in a sludge or slurry
    4. Treatment to remove water from the sludge mass to reduce the volume that must be handled

21. Why is a good plant maintenance program important?
    1. To maintain a good relationship with regulatory agencies
    2. To maintain communication with equipment vendors
    3. To maintain successful operation of the plant
    4. To maintain support from the community

22. Why should the electrical panel door be closed when testing circuits?
    1. Because closing the panel door will conserve energy
    2. Because greater accuracy with meter readings can be achieved with the door closed
    3. Because hot copper sparks could seriously injure you when the circuit is energized and the voltage is high
    4. Because the manufacturers recommend closing the doors to avoid interference from outside light

23. What causes cavitation to occur in a pump?
    1. Due to coupling misalignment
    2. Due to excessive water hammer
    3. Due to the location of the pump in a confined space
    4. Due to unusually low pressures within the pump

24. How can coupling alignment be checked? By the use of a
    1. Computer
    2. Dial indicator
    3. Level
    4. Ruler

25. When shutting down a pump for a long period, why should the electric motor winding heaters be turned on?
    1. To heat the motor control center (MCC) during cold weather
    2. To keep the motor warm and ready to start
    3. To prevent condensation from forming, which can weaken the insulation on the windings
    4. To warm the pump to prevent priming from freezing

26. How can an operator ensure the accuracy of chemical feed rates?
    1. By assisting and working with operators of other facilities with chemical feeders
    2. By contacting and meeting with the manufacturers' representatives of the chemical feeders
    3. By reading and reviewing the specifications for the chemical feeders
    4. By testing and calibrating the chemical feeders

27. Which instrumentation and control system tasks should operators be able to perform in operating a treatment plant?
    1. Recognize failures, troubleshoot, and make adjustments and minor repairs
    2. Repair and replace defective instruments
    3. Specify and prepare plans for new instrumentation systems
    4. Write computer programs for instrumentation controls

28. What is a process variable?

    1. A physical or chemical quantity that is measured or controlled in a water treatment process
    2. How closely a device can reproduce a given reading (or measurement) time after time
    3. How closely an instrument measures the true or actual value of a process variable
    4. The smallest change in a process variable that an instrument can sense

29. Why are some pressure sensors fitted with surge or over-range protection (snubbers)?

    1. To identify the source of the surge and take preventive action
    2. To limit the effect pressure spikes or water hammer have on the instrument
    3. To protect the pipe from bursting under excessive pressures
    4. To record instantaneous peak pressure and alert the operator to inspect for leaks

30. Why should operators learn how to shift all controllers to manual operation?

    1. In order to learn the value of controllers
    2. In order to observe changes in downstream processes
    3. In order to prepare operator training programs
    4. In order to take over control of a critical system, when necessary, in an emergency

31. Who is responsible for safety?

    1. Everyone should be responsible, from top management to all employees
    2. The budget administrator
    3. The employees who are safety conscious
    4. The safety trainers

32. Why should carbon fires not be doused with a stream of water?

    1. Water may cause a slippery surface
    2. Water may cause an acidic, corrosive mixture
    3. Water may cause burning carbon particles to fly
    4. Water may react chemically with oxygen in the air

33. How does the electrical power entering a water treatment plant have the voltage reduced? By routing the power through

    1. A multimeter
    2. A resister
    3. A transformer
    4. An insulator

34. What is a lockout device?

    1. A device that locks out electrical energy
    2. A means of keeping intruders out of a facility
    3. A positive means of holding a switch in a safe position
    4. A secure lock on a gate

35. What kind of program is necessary to guarantee compliance with an eye protection program? An intense program of

    1. Education, persuasion, and appeal
    2. Losses resulting from eye accidents
    3. Visits to eye hospitals
    4. Visual graphics of eye injuries

36. How does a spectrophotometer work?

    1. The instrument automatically compares the unknown concentration with concentrations of known samples
    2. The instrument records the rate of decay of the sample and converts the result to the concentration of the chemical
    3. When the sample transmits light to the spectrophotometer, the instrument records the unknown concentration
    4. When the transmitted light beam strikes the photoelectric tube, it generates an electric current that is proportional to the intensity of light energy striking it

37. What are the most common colors that occur in raw water?

    1. Blue and blue-green
    2. Green and blue-green
    3. Red and pink
    4. Yellow and brown

38. What is the principal source of trihalomethanes in drinking water?

    1. Chemical interaction of chlorine with inorganic nitrogen compounds
    2. Chemical interaction of chlorine with natural humic substances
    3. Chemical interaction of chlorine with toxic metals
    4. Chemical interaction of chlorine with weak salts

39. What is the purpose of drinking water regulations?

    1. To ensure the uniform delivery of safe and aesthetically pleasing drinking water to the public
    2. To force all operators to become appropriately certified
    3. To inform the public of the need for adequate funding for water utilities
    4. To require every water treatment plant to be in compliance with the regulations

40. What are the public health benefits from the Filter Backwash Recycling Rule (FBRR)?

    1. A reduction in loss of filter media containing pathogens as a result of improved backwashing processes
    2. A reduction in the risk of illness from microbial pathogens in drinking water, particularly *Cryptosporidium*
    3. A reduction in treated water turbidity and more effective disinfection processes
    4. A reduction in water demands as a result of more efficient backwash recycling

41. What are volatile organic chemicals (VOCs)?

    1. Organic chemicals that evaporate easily
    2. Organic chemicals that freeze easily
    3. Organic chemicals that liquify easily
    4. Organic chemicals that solidify easily

42. Why do the Surface Water Treatment Rules allow chlorine residual testing to be substituted for some of the bacteriological testing?

    1. To allow operators more time to safely operate and maintain the system
    2. To ensure the bacteriological quality of the consumers' drinking water
    3. To give the operator a quicker indication of the condition of the system
    4. To provide operators with a more accurate evaluation of the system

43. What is organizing?

    1. Deciding who does what work and delegating authority to the appropriate operators
    2. Determining the goals, policies, procedures, and other elements to achieve the goals and objectives of the agency
    3. Guiding, teaching, motivating, and supervising operators and utility staff members
    4. Taking the steps necessary to ensure that essential activities are performed so that objectives will be achieved as planned

44. What does delegation mean?

    1. Answering to those above in the chain of command to explain how and why you have used your authority
    2. The act in which power is given to another person in the organization to accomplish a specific job
    3. The power and resources to do a specific job or to get that job done
    4. The responsibility for results

45. How can a manager prepare a good budget?

    1. By estimating the revenue for the budget year
    2. By having good records from the year before
    3. By reviewing the budget from a successful, nearby utility
    4. By using the same budget as last year

46. What is toxicity?

    1. The ability of a contaminant (chemical or biological) to cause injury when introduced into the human body
    2. The concentration of a contaminant required to cause injury
    3. The severity of an injury caused by a contaminant
    4. The speed with which a contaminant causes injury to take place

47. Following completion of a residential water survey, what should be provided to the customer?

    1. A list of water conservation best management practice (BMP) strategies
    2. A warning to control excess water usage
    3. A water-use rate schedule that will emphasize water conservation
    4. An evaluation of the results and water-saving recommendations

Multiple Choice (Select all correct answers.)

1. Iron can be oxidized by aerating the water to form insoluble ferric hydroxide. The rates of iron oxidation can be lowered by which conditions?

    1. Decrease in flows
    2. Decrease in pH
    3. Increase in hardness
    4. Increase in organic substances
    5. Reduction in temperature

2. Pretreatment may be required before a continuous regeneration (CR) manganese greensand plant to oxidize and remove which items?

    1. Alkalinity
    2. Coliforms
    3. Hardness
    4. Iron
    5. Sulfide

3. When troubleshooting manganese greensand systems, operators should watch for what indications of trouble?

    1. Excessive pressure drop across the bed immediately after backwashing
    2. Filter effluent clear, but manganese higher than raw water
    3. Filter effluent turbid with yellow to brownish color
    4. Iron and manganese low
    5. Water quality good on some units of a multiple unit installation, but bad on others

4. What types of feeders are used to add fluoride ions to water?

    1. Electronic pumps
    2. Gravimetric feeders
    3. Peristaltic pumps
    4. Positive displacement diaphragm pumps
    5. Volumetric feeders

5. When reviewing fluoride feeding system designs, the operator should check which items?

    1. Any switches that throw the equipment from automatic into a hand or manual mode should be equipped with a color-coded red warning light to indicate that the equipment is on "hand" or "manual"
    2. Consider the effect of changing head conditions (both feeder suction and discharge head conditions) on the chemical feeder output
    3. Determine the amount of maintenance required
    4. Determine whether locations for monitoring readouts and dosage controls are convenient to the operation center and easy to read and record
    5. The location where fluoride is added to the water should be where there will be the least possible removal of fluoride by other chemicals added to the water

6. Which items should be checked before start-up of a chemical feeder system?

    1. Be sure that the type of fluoride to be fed is available and in the hopper or feeder
    2. Confirm that the manufacturer's lubrication and start-up procedures are being followed
    3. Determine the proper position of all valves
    4. Examine all fittings, inspection plates, and drains to ensure that they will not leak when placed in service
    5. Inspect all equipment for binding or rubbing

7. Alkalinity is caused by a water's content of which items?

    1. Bicarbonate
    2. Carbonate
    3. Hydroxide
    4. Nitrate
    5. Phosphate

8. What happens to the lime and lime–soda softening process sludge removed from the bottom of settling basins?

    1. May be discharged to groundwaters
    2. May be discharged to surface waters
    3. May be disposed of in a landfill after dewatering
    4. May be incinerated
    5. May be recirculated

9. The length of the service stage of an ion exchange softener is dependent on which factors?

    1. Amount of hardness leakage
    2. Brine disposal procedures
    3. Exchange capacity of the resin
    4. Size of the softener
    5. Source water hardness

10. Which existing water treatment processes can be used to control THMs?

    1. Aeration
    2. Chlorine dioxide oxidation
    3. Coagulation/sedimentation/filtration
    4. Permanganate oxidation
    5. Powdered and granular activated carbon

11. What are the general treatment categories used to classify the treatment methods used to remove arsenic from drinking water sources?

    1. Adsorption
    2. Biological
    3. Coprecipitation
    4. pH adjustment
    5. Physical separation

12. Which factors influence the procedures used to operate an ion exchange unit and the efficiency of the arsenic removal process?

    1. Brine concentration and contact time
    2. Characteristics of the ion exchange resin
    3. Quality of the source water
    4. Rate of flow applied to the ion exchange unit
    5. Salt dosage during regeneration

13. Which items should be included on a water treatment plant's daily inspection form?

    1. Laboratory results
    2. Meter readings
    3. Notations of valve operation
    4. Overview for management
    5. Tank levels

14. Reverse osmosis (RO) and nanofiltration (NF) membranes are used to remove which constituents from water?

    1. Arsenic
    2. Dissolved organic matter (DOM)
    3. Nitrate
    4. Oils and greases
    5. Pesticides

15. How can membrane fouling be described?

    1. By the frequency of fouling (hourly or daily)
    2. By the material causing the fouling (biological, organic, particulate, or dissolved)
    3. By the means of fouling (cake formation or membrane pore blockage)
    4. By the thickness of the fouling (millimeters or centimeters)
    5. By whether the cause of the fouling can be removed (reversible or irreversible)

16. The mineral (salt) flux (mineral passage) through a membrane depends on which factors?

    1. Concentration gradient across the membrane
    2. Mineral permeability constant
    3. Osmotic pressure differential across the membrane
    4. Pressure differential applied across the membrane
    5. Water permeability constant

17. Where should sample valves be located on a reverse osmosis unit? Sample valves should be located on the

    1. Concentrate line
    2. Feed line
    3. Permeate line
    4. Permeate line of each pressure vessel
    5. Workbench of each laboratory technician

18. What are the common symptoms of reverse osmosis membrane damage?

    1. High differential pressure (ΔP)
    2. Higher product water flow rate
    3. Higher salt rejection
    4. Lower product water flow rate
    5. Lower salt rejection

19. Electrodialysis pretreatment is required to remove which materials?

    1. Chloride
    2. Chlorine residual
    3. Iron
    4. Manganese
    5. Sodium

20. The amount of sludge accumulated at a water treatment plant depends on which factors?

    1. Amount of suspended matter in the source water
    2. Level of coagulant dosage
    3. Type of coagulant used
    4. Type of consumers served
    5. Type of suspended matter in the source water

21. Which safety precautions must be exercised by operators when working in any closed tank (confined space)?

    1. Do not operate gasoline engines in the tank
    2. Provide a source of water to clean off boots and tools where the operators come out of the tank
    3. Provide adequate ventilation of clean air at all times
    4. Test the atmosphere in the tank for oxygen deficiency/enrichment, flammable or explosive gases, and toxic gases
    5. Use the buddy system

22. Why should water separated or drained from old sludge not be recycled through a water treatment plant?

    1. The water can cause serious problems if recycled
    2. The water could cause taste and odor problems
    3. The water could contain millions of bacteria and microorganisms
    4. The water may be ignored by the operator
    5. The water may cause a cross connection

23. A successful maintenance program requires operators to take care of which items?

    1. Buildings
    2. Mechanical equipment
    3. Operation
    4. Plant grounds
    5. Structures

24. A multimeter can be used for which purposes?

    1. Check power
    2. Test for blown fuses
    3. Test for grounds
    4. Test for open circuits
    5. Test for single phasing of motors

25. What problems can be caused by pump cavitation?

    1. Drop in pump efficiency
    2. Increase in discharge head
    3. Noise
    4. Rapid damage to the impeller
    5. Vibration

26. Which factors can cause misalignment of couplings?

    1. Excessive bearing wear
    2. Heavy floor loadings
    3. Improper original installation of equipment
    4. Settling of foundations
    5. Warping of bases

27. What could be the causes of high power requirements for a pump?

    1. Impeller is rubbing
    2. Pump shaft is bent
    3. Rotating elements are binding
    4. Speed of rotation is too high
    5. Wearing rings are worn or binding

28. Which items should be checked if you are having problems starting a gasoline engine?

    1. Carburetor flooded
    2. Low compression
    3. No fuel in the tank, valve closed
    4. No spark at the plug
    5. Water or dirt in the fuel lines or carburetor

29. Which minor corrective maintenance or preventive maintenance instrumentation tasks should operators be able to perform?

    1. Make adjustments
    2. Make minor repairs
    3. Properly tap an errant gauge
    4. Recognize failures
    5. Troubleshoot

30. What are the most common variables that are measured at water treatment plants?

    1. Flow
    2. Level
    3. Number of dischargers
    4. Pressure
    5. Weather

31. Which factors could cause high readings with a propeller meter?

    1. Bearing wear and gear train friction
    2. Malfunctioning receiver
    3. Malfunctioning transmitter
    4. Partially full pipeline
    5. Wrong gears installed

32. ON/OFF types of controllers are used to start and stop pumps according to which measurements?

    1. Dissolved oxygen
    2. Electrical power supply
    3. Flow
    4. Level
    5. Pressure

33. Management should be responsible for which safety items?

    1. Appoint a safety officer or coordinator
    2. Assign responsibility for accident prevention
    3. Establish a safety policy
    4. Establish realistic goals
    5. Report all injuries

34. Which safety precautions must be practiced when working with chlorine cylinders?

    1. Always keep the protective cap on the cylinder when it is not in use
    2. Always store chlorine containers in an upright position in a clean, dry location free of flammable materials
    3. Never drop chlorine cylinders or permit the collision of two or more cylinders
    4. Never hoist chlorine cylinders by the neck
    5. Store chlorine cylinders in such a way that they cannot fall

35. Which items are proper clothing that should be worn by operators when mixing and dispensing chemicals in a water treatment plant?

    1. Chemical goggles
    2. Chemical respirators
    3. Face shields
    4. Rubber boots
    5. Rubber gloves

36. The operator should always wear the proper type of glove for each job. What types of gloves are available for operators to wear?

    1. Aluminized
    2. Cloth
    3. Leather
    4. Plastic
    5. Rubber

37. Information from algae counts can serve which purposes?

    1. Help to document the change in trihalomethanes in water
    2. Help to document variability in the water quality
    3. Help to explain the cause of color and turbidity in water
    4. Help to explain the clogging of screens or filters
    5. Help to explain the presence of tastes and odors

38. Operators must be aware of which precautions when using a probe to measure dissolved oxygen?

    1. Dissolved inorganic salts, such as found in seawater, can influence the readings from a probe
    2. Do not place the probe directly over a diffuser because you want to measure the dissolved oxygen in the water being treated, not the oxygen in the air supply to the aerator
    3. Keep the membrane in the tip of the probe from drying out
    4. Periodically check the calibration of the probe
    5. Reactive compounds, such as reactive gases and sulfur compounds, can interfere with the output of a probe

39. Why are high nitrate levels of concern in water?

    1. In drinking water they can cause enteric diseases
    2. In drinking water they can cause infant methemoglobinemia (blue babies)
    3. They can create an excessive chlorine demand
    4. They can degrade water quality by stimulating excessive algal growth
    5. They can stimulate pathogens

40. Which items are part of the process by which EPA sets drinking water standards?

    1. Conducting studies of the effects of contaminants that may endanger public health
    2. Deciding which contaminants may endanger public health
    3. Defining the maximum safe level of each contaminant
    4. Estimating the costs and benefits of regulation
    5. Listening to and evaluating public reactions to the proposed standard

41. What information can operators and managers obtain from EPA's "Compilation of Quick Reference Guides"?

    1. An overview of the rule
    2. Compliance determination
    3. Critical deadlines and requirements
    4. Monitoring requirements
    5. Public health-related benefits

42. The actual field sanitary survey is a detailed evaluation and inspection of which items?

    1. The source of the water supply
    2. The water supply conveyances
    3. The water supply distribution facilities
    4. The water supply storage facilities
    5. The water supply treatment facilities

43. Which site-specific conditions must a drinking water system meet to avoid filtration?

    1. Compliance with the long-term coliform maximum contaminant level (MCL)
    2. Disinfection system must have an alternative power supply, automatic alarm and start-up to ensure continuous disinfection
    3. Ensure that the disinfectant residual at the entrance to the disinfection system does not drop below 0.2 mg/L for more than four hours
    4. No history of waterborne disease outbreak without making treatment corrections
    5. Watershed control or sanitary surveys that satisfy regulatory requirements

44. What are the undesirable effects of high levels of chloride in drinking water?

    1. Corrosion of the pipes in hot water systems
    2. Formation of disinfection by-products
    3. Objectionable color in water
    4. Objectionable medicinal odor in water
    5. Objectionable salty taste in water

45. What information should be included in a drinking water sampling schedule?

    1. Location (address)
    2. Sample volume
    3. Sampling frequency
    4. Sampling point designation
    5. Type of test

46. Which items are responsibilities of utility managers?

    1. Keeping up with increasingly stringent regulations and monitoring
    2. Meeting personnel, resources, and equipment challenges
    3. Preparing for the future
    4. Protecting against environmental disasters with a minimum investment of money
    5. Responding to the community's concern about their drinking water and wastewater

47. Under the Family and Medical Leave Act of 1993 (FMLA), eligible employees may take time off for what purposes?

    1. The employee's own serious health conditions
    2. To care for a child following birth or placement for adoption or foster care
    3. To care for the employee's spouse, child, or parent with a serious health condition
    4. To return to college to complete a degree
    5. To take time off to study for a certification examination

48. Which guidelines should be followed to get press coverage for an event or press conference?

    1. Demonstrate that your story is newsworthy, that it involves something unusual or interesting
    2. Inform the media that their competitors have agreed to participate in the event
    3. Make sure your story will fit the targeted format
    4. Phone the media and inform them you have a breaking news story
    5. Provide a spokesperson who is interesting, articulate, and well prepared

49. Which items are the major elements of a good preventive maintenance program?

    1. Cost and budget control
    2. Planning and scheduling
    3. Records management
    4. Spare parts management
    5. Training program

50. Which tasks are performed in water applications by SCADA systems?

    1. Monitor levels, pressures, and flows
    2. Monitor operating guidelines such as pH and chlorine residual
    3. Monitor operating guidelines such as temperatures, speeds, and motor currents
    4. Operate pumps, valves, and alarms
    5. Provide control of operating guidelines as necessary

51. Which items can be used as the basis for developing an emergency operations plan?

    1. Establish a communications procedure
    2. Inventory organizational personnel
    3. Make a vulnerability assessment
    4. Provide for a recovery operation (plan)
    5. Provide protection for personnel, plant equipment, records, and maps

52. Which approaches can a water utility take to protect its water supply from contamination?

    1. Develop an extensive detection and monitoring program
    2. Install extensive screening facilities
    3. Isolate those reservoirs that offer easy access to the general public
    4. Maintain a high chlorine residual
    5. Use sensitive tasters who can continuously taste and observe the water

53. Which items are water conservation best management practice (BMP) strategies?

    1. Conservation pricing
    2. Conservation programs for commercial, industrial, and institutional (CII) sectors
    3. Public information programs
    4. School education programs
    5. Wholesale agency assistance programs

Short Answer

1. Why should iron and manganese be controlled in drinking water?
2. What happens when an ion exchange resin becomes fouled with iron rust or insoluble manganese dioxide?
3. How does an operator decide when to backwash a manganese greensand filter?
4. Why are drinking waters fluoridated?
5. How can water be softened prior to use with fluoridation equipment?
6. How would you protect yourself from the dust of dry fluoride compounds?
7. Why should water be softened?
8. How can operators protect themselves from lime?
9. Why is sludge sometimes recirculated back into the primary mix area of conventional plants?
10. How would you ensure that large amounts of resin are not being lost during the backwash stage?
11. What happens if an ion exchange softener removes iron in the ferrous (soluble) or ferric (solid) form?
12. How would you determine if iron has fouled the resin of an ion exchange softener?
13. How are trihalomethanes formed in drinking water?
14. What are the advantages of removing precursors before free chlorine is added?
15. What items must be considered before an alternative disinfectant is applied to any system?
16. What should a water utility do before considering treatment of a water source to reduce or remove arsenic?
17. What are the required types of arsenic sampling/monitoring?
18. How can membrane fouling be described?
19. What usually happens to water flux with time? Explain.
20. What is the most common and serious problem resulting from concentration polarization?
21. Why does water to be demineralized require pretreatment?
22. What will happen in a reverse osmosis plant if the brine flow valves are accidentally closed during operation?
23. An excessive concentration of any specific ion in the feedwater to an electrodialysis unit can cause what problem?
24. The amount of sludge accumulation at a typical water treatment plant depends on what factors?
25. What precaution must be taken before draining any in-ground tank?
26. Why should the back of the sludge truck face downhill when releasing wet sludge at one point?
27. Why should operators thoroughly read and understand manufacturers' literature before attempting to maintain plant equipment?
28. Why should your plant have an emergency team to repair chlorine leaks?

29. Why must motor nameplate data be recorded and filed?
30. Why should a pump never be allowed to run dry?
31. When two or more pumps of the same size are installed, why should they be operated alternately?
32. How can you determine if a new pump will turn in the direction intended?
33. What factors can cause wear on gate valve seats?
34. Why is rust a problem in water-cooled systems?
35. Why should solid feeders and hoppers be kept clean and dry?
36. Why should operators understand instrumentation and control systems?
37. What is the difference between accuracy and precision?
38. Why is it poor practice to ignore the lamps that are lit up (alarm conditions) on an annunciator panel?
39. Why should plant measuring systems be periodically calibrated?
40. Who is responsible for safety?
41. What special precautions should be taken when handling and storing caustic soda?
42. What safety precautions should an operator take when handling alum?
43. How can an operator prevent fires in a water treatment plant?
44. How would you remove a manhold cover?
45. How should gas cylinders be stored in laboratories?
46. What is the most common reason safety equipment falls into disrepair?
47. What is the purpose of spectrophotometer calibration curves?
48. Why are algae counts in raw water important to operators?
49. Why is iron undesirable in a domestic water supply?
50. How would you interpret changes (away from normal) in conductivity in water?
51. How would you attempt to reduce unpleasant tastes and odors in drinking water?
52. What criteria must be met before EPA selects a contaminant for regulation?
53. Why are coliform bacteria found in drinking water and what does their presence indicate?
54. What do secondary drinking water regulations control?
55. Why is the absence of tastes and odors in drinking water important?
56. How is a sampling route selected?
57. What can happen without adequate utility planning?
58. When has a supervisor successfully delegated?
59. When hiring new employees, the selection method and examination process used to evaluate applicants must be based on what criteria?
60. How should discipline problems be solved?
61. With whom do managers need to communicate?
62. What attitude should management try to develop among its employees regarding the customer?
63. What does a SCADA system do?
64. How would you assess the vulnerability of a water supply system?
65. What topics should be included in a safety officer's talk to new operators?
66. Why do water utilities need an effective water and energy conservation program?

Problems

1. What is the setting on a potassium permanganate chemical feeder in pounds per day if the required chemical dose is 2.4 mg/L and the flow is 0.18 MGD?

2. A flow of 200 GPM is to be treated with a 2.4-percent (0.2 pound per gallon) solution of sodium fluoride (NaF). The water to be treated contains 0.5 mg/L of fluoride ion and the desired fluoride ion concentration is 1.4 mg/L. What is the sodium feed rate in gallons per day? Assume the sodium fluoride has a fluoride purity of 43.4 percent.

3. How many gallons of water with a hardness of 10 grains per gallon may be treated with an ion exchange softener with an exchange capacity of 8,000 kilograins?

4. Calculating the mineral rejection as a percent, what is an estimate of the ability of a reverse osmosis plant to reject minerals? The feedwater contains 1,600 mg/L TDS and the product water TDS is 130 mg/L.

5. What is the feed rate of a dry chemical feeder in pounds per day if 3.0 pounds of chemical are caught in a weighing tin during 15 minutes?

SUGGESTED ANSWERS TO COMPREHENSIVE REVIEW QUESTIONS
VOLUME II

True-False

1. False — There are NO direct adverse health effects from drinking water containing either iron or manganese.

2. True — In some plants, the injection of the potassium permanganate solution in the volute of the pump will produce complete mixing of potassium permanganate.

3. True — One way to rid a distribution system of iron bacteria is to develop a flushing program.

4. True — Never allow a backsiphon condition to develop in a distribution system.

5. True — Fluoride doses must never be metered against a negative or suction head.

6. True — Once the chemical feed equipment is in operation and the major bugs are worked out, the feeder will need to be fine-tuned.

7. False — Additional benefits do NOT result from the overfeeding of fluoride.

8. True — The levels of carbon dioxide, bicarbonate ion, and carbonate ion in waters are very sensitive to pH.

9. False — Alkalinity is determined by the amount of acid (NOT base) required to reach a titration end point (specific color change).

10. False — Hardness is NOT completely removed by the chemical precipitation methods used in water treatment plants.

11. False — In the lime–soda softening process, extra lime will be required as the coagulant alum or iron feed rate goes up (NOT down) and, therefore, less lime will be required as the alum or iron feed rate is reduced (NOT increased).

12. True — When selecting a target hardness level for your plant, consider the uses of the softened water and the cost of softening.

13. False — Automated operational control systems are NOT maintenance free.

14. False — Methane is NOT involved in the THM reaction.

15. True — THM samples should be sent to the laboratory immediately after they are collected.

16. True — Chloramines are weaker disinfectants than free chlorine, ozone, or chlorine dioxide, but the residuals remain much longer than free chlorine.

17. True — Spent activated alumina media may be regenerated or discarded.

18. True — Operators must be sure to use only NSF certified media and chemicals.

19. True — The constituents in the water being filtered may cause fouling.

20. True — Membrane filtration SCADA systems monitor alarm conditions and follow protocols for alarm notification and plant shutdown.

21. True — All available water supplies can be classified according to their mineral quality.

22. False — The water flux decreases (NOT increases) as the mineral content of the feed increases because the osmotic pressure contribution increases with increasing mineral content.

23. False — In reverse osmosis operation, as the temperature of the feedwater increases, flux increases (NOT decreases).

24. True — Water is transmitted through the membrane at a much more rapid rate than minerals.

25. True — Sludge from a water treatment plant is considered an industrial waste that requires compliance with the provisions of the Water Pollution Control Act.

26. True — During the draining stages of a sedimentation tank, the walls and all the equipment should be completely hosed down and inspected for damage.

27. True — Brine from ion exchange units may be discharged into wastewater collection systems.

28. True — Operators should thoroughly read manufacturers' literature on plant equipment and understand the procedures.

718 Water Treatment

29. True — Everyone must realize that if the equipment cannot work, no one can work.

30. True — The most important requirements in the reliable operation of electrical equipment are cleanliness and the elimination of moisture into the insulation.

31. True — Misalignment is a major cause of pump and coupling wear.

32. True — Improper or lack of lubrication is probably the biggest cause of compressor failures.

33. False — Diesel engines do NOT use spark plugs.

34. True — The more you know about your plant's instrumentation, the better operator you become.

35. False — Analog (NOT digital) displays show a reading as a pointer against a marked scale.

36. False — Differential pressure flow measuring devices require little (NOT significant) preventive maintenance because there are NO moving parts.

37. True — The computer control system can alert the operator or change a plant process based on a preprogrammed control strategy defined by the operator.

38. True — Safety is using one's knowledge or skill to avoid accidents.

39. True — If you are caught in an area containing chlorine, keep your head high until you are out of the affected area.

40. True — All fluoride compounds must be regarded as hazardous chemicals that are toxic to operators.

41. True — Good driving skills are good for utility public relations.

42. True — Never pipet bacteriological samples by mouth, always use a pipet bulb.

43. True — Both of the spectroscopic measurement units of percent transmittance and absorbance are associated with color intensity.

44. False — If the filters are performing properly, there should NOT be any countable algae in the treated water.

45. True — Alum coagulates most effectively at pH values near 6.8.

46. False — Transient water systems are NOT affected by the Unregulated Contaminants Monitoring Rule (UCMR1).

47. True — Primary standards or MCLs are set for substances that are thought to pose a threat to health when present in drinking water at certain levels.

48. False — Waters exceeding the MCLs for metals and salts for short periods of time will pose NO immediate threat to health.

49. True — Whenever a routine coliform sample is coliform-positive, the Total Coliform Rule calls for determination of the presence of fecal coliform or *E. coli* and for repeat sampling.

50. True — Iron and manganese are frequently found together in natural waters and produce similar adverse environmental effects and color problems.

51. True — Planning is one of the most important functions of utility management and one of the most difficult.

52. True — Every utility should have a safety training session for all new employees.

53. True — Good communication is an essential part of good management skills.

54. False — Large amounts of data, including unimportant data, should NOT be included in a report.

55. False — The purpose of the equipment repair/replacement fund account is NOT to generate a profit for the utility.

56. True — Safety is good business for both the operator and the agency.

57. True — By always logging in supplies immediately upon receipt, you are in a position to reject defective or damaged shipments and control shortages or errors in billing.

Best Answer

1. 4 — When iron bacteria form thick slimes on the walls of distribution system mains, the slimes are rust colored from the iron.

2. 1 — When analyzing for iron and manganese, collect samples as close to the well or other source as possible.

3. 3 — Potassium permanganate removes iron and manganese from water by forming insoluble oxides.

4. 2 — Fluoride saturators should be stirred every day to prevent fluoride solids from building up on the bottom.

5. 4 — Fluoride chemical overloading can be prevented by proper operation and continuous monitoring of the product water.

6. 4 — Carbonate hardness is that hardness in water caused by the alkalinity present in water up to the total hardness.

7. 3 — Storage areas for bagged lime must be covered to prevent rain from wetting the bags.

8. 3 — When quicklime reacts with water in the slaking process, it gets hot enough to cause serious burns.

9. 3 — During the brine stage of an ion exchange softener, the sodium ion concentration of the resin is recharged by pumping a concentrated brine solution onto the resin.

10. 3 THM precursors are natural organic compounds found in all surface and groundwaters.

11. 4 Some THM sample bottles contain a small amount of chemical reducing agent to stop the chemical reaction that occurs between chlorine and the THM precursors.

12. 2 The operator's responsibility regarding the water treatment plant's approved operating plan is to modify the plan routinely, as needed, to reflect real-world conditions.

13. 3 The purpose of arsenic sampling/monitoring for process control is to verify proper operation of the arsenic removal treatment plant.

14. 1 The flow through a membrane may be recovered after fouling by backwashing and cleaning.

15. 4 The first level of maintenance performed at a water treatment plant is routine maintenance.

16. 4 Flux is used to describe the rate of water flow through the semipermeable membrane.

17. 4 Acid is added as a pretreatment step before demineralization using cellulose acetate membranes to slow hydrolysis.

18. 3 Sodium hexametaphosphate is the scale inhibitor most frequently used to prevent the coating of the reverse osmosis membrane.

19. 4 Water temperature influences the volume of sludge produced because as the temperature increases, the settling rate of the solids will increase.

20. 4 Sludge thickening is treatment to remove water from the sludge mass to reduce the volume that must be handled.

21. 3 A good plant maintenance program is important to maintain successful operation of the plant.

22. 3 The electrical panel door should be closed when testing circuits to protect you from hot copper sparks when the circuit is energized and the voltage is high.

23. 4 Cavitation occurs in a pump due to unusually low pressures within the pump.

24. 2 Coupling alignment can be checked by the use of a dial indicator.

25. 3 When shutting down a pump for a long period, electric motor winding heaters should be turned on to prevent condensation from forming, which can weaken the insulation on the windings.

26. 4 An operator can ensure the accuracy of chemical feed rates by testing and calibrating the chemical feeders.

27. 1 Instrumentation and control system tasks operators should be able to perform include recognizing failures, troubleshooting, and making adjustments and minor repairs.

28. 1 A process variable is a physical or chemical quantity that is measured or controlled in a water treatment process.

29. 2 Some pressure sensors are fitted with surge or overrange protection to limit the effect pressure spikes or water hammer have on the instrument.

30. 4 Operators should learn how to shift all controllers to manual operation in order to take over control of a critical system, when necessary, in an emergency.

31. 1 Everyone should be responsible for safety, from top management to all employees.

32. 3 Carbon fires should not be doused with a stream of water because water may cause burning carbon particles to fly.

33. 3 The electrical power entering a water treatment plant has the voltage reduced by routing the power through a transformer.

34. 3 A lockout device is a positive means of holding a switch in a safe position.

35. 1 To guarantee compliance with an eye protection program there must be an intense program of education, persuasion, and appeal.

36. 4 When the transmitted light beam strikes the photoelectric tube, it generates an electric current that is proportional to the intensity of light energy striking it.

37. 4 Yellow and brown are the most common colors that occur in raw water.

38. 2 The principal source of trihalomethanes in drinking water is the chemical interaction of chlorine with natural humic substances.

39. 1 The purpose of drinking water regulations is to ensure the uniform delivery of safe and aesthetically pleasing drinking water to the public.

40. 2 The public health benefits from the Filter Backwash Recycling Rule (FBRR) include a reduction in the risk of illness from microbial pathogens in drinking water, particularly *Cryptosporidium*.

41. 1 Volatile organic chemicals (VOCs) are organic chemicals that evaporate easily.

42. 3 The Surface Water Treatment Rules allow chlorine residual testing to be substituted for some of the bacteriological testing to give the operator a quicker indication of the condition of the system.

43. 1 Organizing is deciding who does what work and delegating authority to the appropriate operators.

44. 2 Delegation means the act in which power is given to another person in the organization to accomplish a specific job.

45. 2 A manager can prepare a good budget by having good records from the year before.

46. 1 Toxicity is the ability of a contaminant (chemical or biological) to cause injury when introduced into the human body.

47. 4 Following completion of a residential water survey, the customer should be provided with an evaluation of the results and water-saving recommendations.

Multiple Choice

1. 2, 4, 5   The rates of iron oxidation can be lowered by a decrease in pH, an increase in organic substances, or a reduction in temperature.

2. 4, 5   Pretreatment may be required before a continuous regeneration (CR) manganese greensand plant to oxidize and remove iron and sulfide.

3. 1, 2, 3, 5   When troubleshooting manganese greensand systems, operators should watch for trouble indicators that include excessive pressure drop across the bed immediately after backwashing; filter effluent clear, but manganese higher than raw water; filter effluent turbid with yellow to brownish color; and water quality good on some units of a multiple unit installation, but bad on others.

4. 1, 2, 3, 4, 5   The types of feeders used to add fluoride ions to water include electronic pumps, gravimetric feeders, peristaltic pumps, positive displacement diaphragm pumps, and volumetric feeders.

5. 1, 2, 3, 4, 5   When reviewing fluoride feeding system designs, the operator should check the following items: (1) any switches that throw the equipment from automatic into a hand or manual mode should be equipped with a color-coded red warning light to indicate that the equipment is on "hand" or "manual"; (2) consider the effect of changing head conditions (both feeder suction and discharge head conditions) on the chemical feeder output; (3) determine the amount of maintenance required; (4) determine whether locations for monitoring readouts and dosage controls are convenient to the operation center and easy to read and record; and (5) the location where fluoride is added to the water should be where there will be the least possible removal of fluoride by other chemicals added to the water.

6. 1, 2, 3, 4, 5   Before start-up of a chemical feeder system, check the following items: (1) be sure that the type of fluoride to be fed is available and in the hopper or feeder; (2) confirm that the manufacturer's lubrication and start-up procedures are being followed; (3) determine the proper position of all valves; (4) examine all fittings, inspection plates, and drains to ensure that they will not leak when placed in service; and (5) inspect all equipment for binding or rubbing.

7. 1, 2, 3, 5   Alkalinity is caused by a water's content of bicarbonate, carbonate, hydroxide, or phosphate.

8. 3, 5   Lime and lime–soda softening process sludge removed from the bottom of settling basins may be disposed of in a landfill after dewatering or may be recirculated.

9. 1, 3, 4, 5   The length of the service stage of an ion exchange softener is dependent on the amount of hardness leakage, the exchange capacity of the resin, the size of the softener, and source water hardness.

10. 1, 2, 3, 4, 5   Existing water treatment processes that can be used to control THMs include the following: (1) aeration, (2) chlorine dioxide oxidation, (3) coagulation/sedimentation/filtration, (4) permanganate oxidation, and (5) powdered and granular activated carbon.

11. 1, 3, 5   The general treatment categories used to classify the treatment methods used to remove arsenic from drinking water sources include adsorption, coprecipitation, and physical separation.

12. 1, 2, 3, 4, 5   Factors that influence the procedures used to operate an ion exchange unit and the efficiency of the arsenic removal process include the following: (1) brine concentration and contact time, (2) characteristics of the ion exchange resin, (3) quality of the source water, (4) rate of flow applied to the ion exchange unit, and (5) salt dosage during regeneration.

13. 1, 2, 3, 5   Items that should be included on a water treatment plant's daily inspection form include laboratory results, meter readings, notations of valve operation, and tank levels.

14. 1, 2, 3, 5 — Reverse osmosis (RO) and nanofiltration (NF) membranes are used to remove arsenic, dissolved organic matter (DOM), nitrate, and pesticides from water.

15. 2, 3, 5 — Membrane fouling can be described by the material causing the fouling (biological, organic, particulate, or dissolved); by the means of fouling (cake formation or membrane pore blockage); or by whether the cause of the fouling can be removed (reversible or irreversible).

16. 1, 2 — The mineral (salt) flux (mineral passage) through a membrane depends on the concentration gradient across the membrane and the mineral permeability constant.

17. 1, 2, 3, 4 — Sample valves should be located on the concentrate line, the feed line, the permeate line, and the permeate line of each pressure vessel of a reverse osmosis unit.

18. 1, 2, 3, 4, 5 — Common symptoms of reverse osmosis membrane damage include the following: (1) high differential pressure (ΔP), (2) higher product water flow rate, (3) higher salt rejection, (4) lower product water flow rate, and (5) lower salt rejection.

19. 2, 3, 4 — Electrodialysis pretreatment is required to remove chlorine residual, iron, and manganese.

20. 1, 2, 3, 5 — The amount of sludge accumulated at a water treatment plant depends on the amount of suspended matter in the source water, the level of coagulant dosage, the type of coagulant used, and the type of suspended matter in the source water.

21. 1, 2, 3, 4, 5 — Safety precautions that must be exercised by operators when working in any closed tank (confined space) include the following: (1) not operating gasoline engines in the tank; (2) providing a source of water to clean off boots and tools where the operators come out of the tank; (3) providing adequate ventilation of clean air at all times; (4) testing the atmosphere in the tank for oxygen deficiency/enrichment, flammable or explosive gases, and toxic gases; and (5) using the buddy system.

22. 1, 2, 3 — Water separated or drained from old sludge should not be recycled through a water treatment plant for the following reasons: (1) the water can cause serious problems if recycled, (2) the water could cause taste and odor problems, or (3) the water could contain millions of bacteria and microorganisms.

23. 1, 2, 4, 5 — A successful maintenance program requires operators to take care of buildings, mechanical equipment, plant grounds, and structures.

24. 1, 2, 3, 4, 5 — A multimeter can be used to check power and to test for blown fuses, grounds, open circuits, and single phasing of motors.

25. 1, 3, 4, 5 — Problems that can be caused by pump cavitation include a drop in pump efficiency, noise, rapid damage to the impeller, and vibration.

26. 1, 2, 3, 4, 5 — Factors that can cause misalignment of couplings include excessive bearing wear, heavy floor loadings, improper original installation of equipment, settling of foundations, and warping of bases.

27. 1, 2, 3, 4, 5 — High power requirements for a pump could result if the impeller is rubbing, if the pump shaft is bent, if the rotating elements are binding, if the speed of rotation is too high, or if the wearing rings are worn or binding.

28. 1, 2, 3, 4, 5 — If you are having problems starting a gasoline engine, check the following items: (1) carburetor flooded; (2) low compression; (3) no fuel in the tank, valve closed; (4) no spark at the plug; or (5) water or dirt in the fuel lines or carburetor.

29. 1, 2, 4, 5 — Minor corrective maintenance or preventive maintenance instrumentation tasks that operators should be able to perform include making adjustments, making minor repairs, recognizing failures, and troubleshooting.

30. 1, 2, 4 — The most common variables that are measured at water treatment plants are flow, level, and pressure.

31. 2, 3, 4, 5 — Factors that could cause high readings with a propeller meter include a malfunctioning receiver, a malfunctioning transmitter, a partially full pipeline, and wrong gears installed.

32. 3, 4, 5 — ON/OFF types of controllers are used to start and stop pumps according to flow, level, and pressure measurements.

33. 1, 2, 3, 4 — Safety items for which management should be responsible include appointing a safety officer or coordinator, assigning responsibility for accident prevention, establishing a safety policy, and establishing realistic goals.

34. 1, 2, 3, 4, 5 — Safety precautions that must be practiced when working with chlorine cylinders include: (1) always keeping the protective cap on the cylinder when it is not in use; (2) always storing chlorine containers in an upright position in a clean, dry location free of flammable materials; (3) never dropping chlorine cylinders or permitting the collision of two or more cylinders; (4) never hoisting chlorine cylinders by the neck; and (5) storing chlorine cylinders in such a way that they cannot fall.

35. 1, 2, 3, 4, 5 — When mixing and dispensing chemicals in a water treatment plant, operators should wear the following items: (1) chemical goggles, (2) chemical respirators, (3) face shields, (4) rubber boots, and (5) rubber gloves.

36. 1, 2, 3, 4, 5 — The types of gloves available for operators to wear include aluminized, cloth, leather, plastic, and rubber.

37. 2, 3, 4, 5 — Information from algae counts can help to document variability in the water quality, to explain the cause of color and turbidity in water, to explain the clogging of screens or filters, and to explain the presence of tastes and odors.

38. 1, 2, 3, 4, 5 — When using a DO probe to measure dissolved oxygen, operators must be aware of the following precautions: (1) dissolved inorganic salts, such as found in seawater, can influence the readings from a probe; (2) do not place the probe directly over a diffuser because you want to measure the dissolved oxygen in the water being treated, not the oxygen in the air supply to the aerator; (3) keep the membrane in the tip of the probe from drying out; (4) periodically check the calibration of the probe; and (5) reactive compounds, such as reactive gases and sulfur compounds, can interfere with the output of a probe.

39. 2, 4 — High nitrate levels are of concern in drinking water because they can cause infant methemoglobinemia (blue babies) and because they can degrade water quality by stimulating excessive algal growth.

40. 1, 2, 3, 4, 5 — EPA's process for setting drinking water standards includes the following: (1) conducting studies of the effects of contaminants that may endanger public health; (2) deciding which contaminants may endanger public health; (3) defining the maximum safe level of each contaminant; (4) estimating the costs and benefits of regulation; and (5) listening to and evaluating public reactions to the proposed standard.

41. 1, 2, 3, 4, 5 — Information that operators and managers obtain from EPA's "Compilation of Quick Reference Guides" includes an overview of the rule, compliance determination, critical deadlines and requirements, monitoring requirements, and public health-related benefits.

42. 1, 2, 3, 4, 5 — The actual field sanitary survey is a detailed evaluation and inspection of the source of the water supply and of the water supply conveyances, distribution facilities, storage facilities, and treatment facilities.

43. 1, 2, 3, 4, 5 — Site-specific conditions that a drinking water system must meet to avoid filtration include the following: (1) compliance with the long-term coliform maximum contaminant level (MCL); (2) a disinfection system that has an alternative power supply, automatic alarm and start-up to ensure continuous disinfection; (3) ensuring that the disinfectant residual at the entrance to the disinfection system does not drop below 0.2 mg/L for more than four hours; (4) no history of waterborne disease outbreak without making treatment corrections; and (5) watershed control or sanitary surveys that satisfy regulatory requirements.

44. 1, 5 — The undesirable effects of high levels of chloride in drinking water include the corrosion of the pipes in hot water systems and an objectionable salty taste in water.

45. 1, 2, 3, 4, 5 — Information that should be included in a drinking water sampling schedule includes the location (address), the sample volume, the sampling frequency, the sampling point designation, and the type of test.

46. 1, 2, 3, 4, 5 — Utility managers are responsible for the following: (1) keeping up with increasingly stringent regulations and monitoring; (2) meeting personnel, resources, and equipment challenges; (3) preparing for the future; (4) protecting against environmental disasters with a minimum investment of money; and (5) responding to the community's concern about their drinking water and wastewater.

47. 1, 2, 3 — Under the FMLA, eligible employees may take time off to take care of the employee's own serious health conditions, to care for a child following birth or placement for adoption or foster care, and to care for the employee's spouse, child, or parent with a serious health condition.

48. 1, 3, 5 — Guidelines that should be followed to get press coverage for an event or press conference include demonstrating that your story is newsworthy, that it involves something unusual or interesting; making sure your story will fit the targeted format; and providing a spokesperson who is interesting, articulate, and well prepared.

49. 1, 2, 3, 4, 5 — The major elements of a good preventive maintenance program include cost and budget control, planning and scheduling, records management, spare parts management, and a training program.

50. 1, 2, 3, 4, 5 — Tasks performed in water applications by SCADA systems include the following: (1) monitoring levels, pressures, and flows; (2) monitoring operating guidelines such as pH and chlorine residual; (3) monitoring operating guidelines such as temperatures, speeds, and motor currents; (4) operating pumps, valves, and alarms; and (5) providing control of operating guidelines as necessary.

51. 1, 2, 3, 4, 5 — Items that can be used as the basis for developing an emergency operations plan include establishing a communications procedure; making an inventory of organizational personnel; making a vulnerability assessment; providing for a recovery operation (plan); and providing protection for personnel, plant equipment, records, and maps.

52. 1, 3, 4 — Approaches that a water utility can take to protect its water supply from contamination include developing an extensive detection and monitoring program, isolating those reservoirs that offer easy access to the general public, and maintaining a high chlorine residual.

53. 1, 2, 3, 4, 5 — Water conservation best management practice (BMP) strategies include the following: (1) conservation pricing; (2) conservation programs for commercial, industrial, and institutional (CII) sectors; (3) public information programs; (4) school education programs; and (5) wholesale agency assistance programs.

Short Answer

1. Iron and manganese should be controlled in drinking water because when clothes are laundered in water containing iron and manganese, they come out stained. Also, waters containing iron and manganese promote the growth of iron bacteria. These bacteria form thick slimes on the walls of the distribution system mains. The slimes are rust colored from iron and black from manganese. Variations in flow cause these slimes to slough, which results in dirty water. Furthermore, these slimes will cause foul tastes and odors in the water.

2. If an ion exchange resin becomes fouled with iron rust or insoluble manganese dioxide, the resin can be cleaned but this is expensive and the resin capacity is reduced.

3. Typically, a manganese greensand filter should be backwashed when head loss reaches 10 psi or after treating a predetermined number of gallons of water.

4. Drinking waters are fluoridated to reduce the incidence (number) of dental caries (tooth decay) in children.

5. Water can be softened prior to use with fluoridation equipment by the use of zeolite ion exchangers.

6. To protect yourself from the dust of dry fluoride compounds, be sure the dust collector system works properly. Even with the use of dust collector systems, dust will circulate in the air. Always use approved respirators equipped with cartridges for organic dusts and vapors, protective coveralls, and gloves when emptying sacks or cleaning up equipment and plant surfaces.

7. Hard waters cause difficulties in doing the laundry and in dishwashing in the household. There is also a coating that forms inside the hot water heater. Many industrial processes are adversely affected by hard water.

8. Operators can protect themselves from lime by wearing protective clothing. Protective clothing includes long-sleeved shirt with sleeves and collar buttoned, trousers with legs down over tops of shoes or boots, head protection, and gloves. A protective cream should be applied to exposed parts of the body. Operators should also wear a lightweight filter mask and tight-fitting safety glasses with side shields for protection from lime dust.

9. Sludge may be recirculated back into the primary mix area of conventional plants to help seed the process. The advantages are: (1) recirculation speeds up the precipitation process, and (2) some reduction of chemical requirements may result.

10. To ensure that large amounts of resin are not being lost during the backwash stage, a glass beaker can be used to catch a sample of the effluent while the unit is backwashing. A trace amount of resin should cause no alarm, but a steady loss of resin could indicate a problem in the unit and the cause should be located and corrected as soon as possible.

11. If iron is removed in the ferrous (soluble) form, the media can become iron coated and the efficiency of the softener will be greatly reduced. If iron is removed in the ferric (solid) form, the bed could become plugged and force the water to channel or short-circuit through the bed. The result is incomplete contact between the water and media, thus creating hardness leakage and loss of softening efficiency.

12. To determine if iron is fouling the resin of an ion exchange softener, look at the color of the resin. If iron has fouled the resin, the bed will be an orange, rusty color, while the backwash effluent will appear a light orange at the end of the backwash stage. Also, the head loss on the unit will run higher than normal as the bed becomes plugged with iron.

13. Trihalomethanes are formed in drinking water when free chlorine comes in contact with naturally occurring organic compounds (THM precursors). Trihalomethanes are a class of organic compounds in which there has been a replacement of the three hydrogen atoms in the methane molecule with three halogen atoms (chlorine or bromine). The production of THMs can generally be shown as:

Free Chlorine + Natural Organics (precursor) + Bromide → THMs + Other Products

14. Removing precursors before free chlorine is added allows the continued use of free chlorine as a disinfectant and the formation of fewer "other products" whose health significance is not known.

15. Before an alternative disinfectant is applied to any system, the source of the water supply, water quality, and treatment effectiveness for bacteriological control must be evaluated.

16. Before considering treatment of a water source to reduce or remove arsenic, a water utility should investigate a new source as an alternative water supply.

17. Types of required arsenic sampling/monitoring include compliance, process control, calibration/verification, and special sampling for troubleshooting.

18. Membrane fouling can be described by whether the cause of fouling can be removed (reversible or irreversible), by the material causing the fouling (biological, organic, particulate, or dissolved), and by the means of fouling (cake formation or membrane pore blockage).

19. Even under ideal conditions (pure feedwater and no fouling of the membrane surface), there is a decline in water flux with time. This decrease in flux is due to membrane compaction. Flux decline also results from foulants and bacterial growth.

20. The most common and serious problem resulting from concentration polarization is the increasing tendency for precipitation of sparingly soluble salts and the deposition of particulate matter on the membrane surface.

21. Water to be demineralized always contains impurities that should be removed by pretreatment to protect the membrane and to ensure maximum efficiency of the reverse osmosis process.

22. If the brine flow valves are accidentally closed during operation, 100 percent recovery will result in almost certain damage to the membranes due to the precipitation of inorganic salts ($CaSO_4$).

23. An excessive concentration of any specific ion in the feedwater to an electrodialysis unit could lead to chemical fouling due to scaling.

24. The amount of sludge accumulation at a typical water treatment plant depends on the type and amount of suspended matter in the source water being treated as well as on the level of dosage and the type of coagulant used. Also, by using a source water stabilizing reservoir, a plant will accumulate less sludge because the turbidity level of the water to be treated will be reduced, providing a more constant quality of water that requires less alum.

25. Before draining any in-ground tank, always determine the level of the water table. If the water table is high, an empty tank could float like a cork on the water surface and cause considerable damage to the tank and piping.

26. When releasing wet sludge at one point, the back of the sludge truck should face downhill so:

   1. The tank will drain faster
   2. The tank will empty completely
   3. The truck will not become stuck in the sludge

27. The operator should thoroughly read and understand manufacturers' literature before attempting to maintain plant equipment so that the job required to keep the equipment operating can be done properly.

28. An emergency team to repair chlorine leaks is important because chlorine is a hazardous chemical. An emergency team is specially trained and qualified to control and repair emergency chlorine leaks. Unqualified and untrained personnel may injure themselves and create hazards for others.

29. Motor nameplate data must be recorded and filed so the information is available when needed to repair the motor or to obtain replacement parts.

30. A pump should never be allowed to run dry because water acts as a lubricant between the rings and the impeller. Running the pump without this lubrication could cause severe pump damage.

31. When two or more pumps of the same size are installed, they should be operated alternately to equalize wear, keep motor windings dry, and distribute lubricant in the bearings.

32. To determine if a new pump will turn in the direction intended, momentarily start the motor by a quick electrical contact and check to be sure the motor will turn the pump in the direction indicated by the rotational arrows on the pump.

33. Wear on gate valve seats can be caused by: (1) minute particles transported in the water, and (2) operation of the gate valve partially opened.

34. Rust is a problem in water-cooled systems because: (1) rust does not allow good heat transfer, and (2) the sloughing of rust scale can block narrow passages.

35. Solid feeders and hoppers should be kept clean and dry in order to prevent "bridging" of the chemical in the hopper and clogging in the feeder.

36. Operators must have a good working familiarity with instrumentation and control systems to properly monitor and control the treatment processes.

37. *ACCURACY* refers to how closely an instrument measures the true or actual value of a process variable and *PRECISION* refers to how closely the device (instrument) can reproduce a given reading (or measurement) time after time.

38. It is poor practice to ignore the lamps that are lit up (alarm conditions) on an annunciator panel because a true alarm condition requiring immediate operator attention may be lost in the resulting general indifference to the alarm system.

39. Plant measuring systems must be periodically calibrated to ensure accurate measurements.

40. Safety is the responsibility of everyone, from top management to all employees.

41. Special precautions to be taken when handling and storing caustic soda include: (1) prevent eye and skin contact, (2) do not breathe dusts or mists, and (3) avoid storing this chemical next to strong acids.

42. For the dry type of alum, the operator should avoid breathing the powder and skin contact. For the liquid type of alum, the operator should avoid exposures to the skin and eyes, as well as ingestion.

43. Fires can be prevented by good housekeeping. This means a well-kept, neat, and orderly plant.

44. A manhole cover should be lifted with a special tool such as a pick with a bent point or a manhole hook. Use your legs, not your back, for lifting. Locate the cover outside the working area to provide adequate working area around the manhole opening.

45. Gas cylinders must be kept in place by safety retaining devices such as chains around the cylinders to keep them from falling. The valve and cylinder regulator should be protected from being struck by stools, ladders, and other objects.

46. Safety equipment may fall into disrepair because it is only used occasionally and may deteriorate due to heat, time, and other environmental factors.

47. Spectrophotometer calibration curves are used to determine the concentrations of water quality indicators by the use of a spectrophotometer.

48. Algae counts in raw waters are important to operators to help:
    1. Explain the cause of color and turbidity and the presence of tastes and odors in the water
    2. Explain the clogging of screens or filters
    3. Document variability in the water quality

49. Iron in a domestic water supply can stain laundry, concrete, and porcelain. A bitter, astringent taste can be detected by some people at levels above 0.3 mg/L.

50. Changes in conductivity from normal levels may indicate changes in mineral composition of the water, seasonal variations in lakes and reservoirs, or intrusion of pollutants.

51. Corrective measures designed to reduce unpleasant tastes and odors include aeration, or the addition of chlorine, chlorine dioxide, potassium permanganate, or activated carbon. (For additional details, see Volume I, Chapter 9, "Taste and Odor Control.")

52. To regulate a substance, EPA must demonstrate that the contaminant meets three criteria:
    1. The contaminant has an adverse effect on human health
    2. It occurs, or is likely to occur, in public water systems at a frequency and concentration of significance to public health
    3. Regulation of the contaminant offers a meaningful opportunity to reduce health risks for people served by public water systems

53. Many types of coliform bacteria from human and animal wastes may be found in drinking water if the water is not properly treated. Often the bacteria themselves do not cause diseases transmitted by water, although certain coliforms have been identified as the cause of "traveler's" diarrhea. In general, however, the presence of coliform bacteria indicates that other harmful organisms may be present in the water.

54. Secondary drinking water regulations control contaminants that primarily affect the aesthetic qualities relating to the public acceptance of drinking water. At considerably higher concentrations of these contaminants, health implications may also exist as well as aesthetic degradation.

55. The absence of tastes and odors helps to maintain the consumers' confidence in the quality of their water, even though it does not guarantee that the water is safe.

56. The sampling route should be arranged so that samples that must be analyzed immediately are not delayed while other sampling is done.

57. Without adequate planning, your utility will be facing system failures, inability to meet compliance regulations, and inadequate service capacity to meet community needs.

58. A supervisor has successfully delegated when proper job assignments, authority, and responsibility are all present.

59. When hiring new employees, the selection method and examination process used to evaluate applicants must be based on the applicant's knowledge, skills, and abilities to perform relevant job-related activities.

60. Discipline problems should be solved quickly. The sooner you deal with the problem, the better the outcome will be.

61. Managers need to communicate with employees, the governing body, and the public.

62. Management should try to develop among its employees the attitude that even though the customer is not always right, every customer is always entitled to courteous treatment and a proper explanation of anything the customer does not understand.

63. A SCADA system collects, stores, and analyzes information about all aspects of operation and maintenance, transmits alarm signals, when necessary, and allows fingertip control of alarms, equipment, and processes.

64. The following steps should be taken in assessing the vulnerability of a system:
    1. Identify and describe the system components.
    2. Assign assumed disaster characteristics.
    3. Estimate disaster effects on system components.
    4. Estimate customer demand for service following a potential disaster.
    5. Identify key system components that would be primarily responsible for system failure.

65. The safety officer should tell new employees about utility policy, safety reports, past accidents, and orient the new operator toward the importance of safety to operators and to the organization.

66. An effective water and energy conservation program is an excellent means of saving money. Water conservation can reduce energy costs. Reducing water and energy consumption results in conservation of natural resources.

Problems

1. What is the setting on a potassium permanganate chemical feeder in pounds per day if the required chemical dose is 2.4 mg/L and the flow is 0.18 MGD?

Known	Unknown
Flow, MGD = 0.18 MGD	Chemical Feeder, lbs/day
Dose, mg/L = 2.4 mg/L	

   Determine the chemical feeder setting in pounds per day.

   $$\text{Chemical Feeder, lbs/day} = (\text{Flow, MGD})(\text{Dose, mg}/L)(8.34 \text{ lbs/gal})$$
   $$= (0.18 \text{ MGD})(2.4 \text{ mg}/L)(8.34 \text{ lbs/gal})$$
   $$= 3.6 \text{ lbs/day}$$

2. A flow of 200 GPM is to be treated with a 2.4-percent (0.2 pound per gallon) solution of sodium fluoride (NaF). The water to be treated contains 0.5 mg/L of fluoride ion and the desired fluoride ion concentration is 1.4 mg/L. What is the sodium feed rate in gallons per day? Assume the sodium fluoride has a fluoride purity of 43.4 percent.

Known	Unknown
Flow, GPM = 200 GPM	Feed Rate, gal/day
NaF Solution, % = 2.4%	
NaF Solution, lbs/gal = 0.2 lb/gal	
Desired F, mg/L = 1.4 mg/L	
Actual F, mg/L = 0.5 mg/L	
Purity, % = 43.4%	

   1. Convert the flow from gallons per minute to million gallons per day.

   $$\text{Flow, MGD} = \frac{(\text{Flow, gal/min})(60 \text{ min/hr})(24 \text{ hr/day})(1 \text{ million})}{1{,}000{,}000}$$
   $$= \frac{(200 \text{ gal/min})(60 \text{ min/hr})(24 \text{ hr/day})(1 \text{ million})}{1{,}000{,}000}$$
   $$= 0.29 \text{ MGD}$$

   2. Determine the fluoride feed dose in milligrams per liter.

   $$\text{Feed Dose, mg}/L = \text{Desired Dose, mg}/L - \text{Actual F, mg}/L$$
   $$= 1.4 \text{ mg}/L - 0.5 \text{ mg}/L$$
   $$= 0.9 \text{ mg}/L$$

3. Calculate the feed rate in pounds of fluoride per day.

$$\text{Feed Rate, lbs/day} = (\text{Flow, MGD})(\text{Feed Dose, mg}/L)(8.34 \text{ lbs/gal})$$

$$= (0.29 \text{ MGD})(0.9 \text{ mg}/L)(8.34 \text{ lbs/gal})$$

$$= 2.2 \text{ lbs F/day}$$

4. Convert the feed rate from pounds of fluoride per day to gallons of sodium fluoride solution per day.

$$\text{Feed Rate, gal/day} = \frac{(\text{Feed Rate, lbs F/day})(100\%)}{(\text{NaF Solution, lbs F/gal})(\text{Purity, \%})}$$

$$= \frac{(2.2 \text{ lbs F/day})(100\%)}{(0.2 \text{ lb/gal})(43.4\%)}$$

$$= 25.3 \text{ gal/day or } 25 \text{ gal/day}$$

3. How many gallons of water with a hardness of 10 grains per gallon may be treated with an ion exchange softener with an exchange capacity of 8,000 kilograins?

Known	Unknown
Hardness, grains/gal = 10 grains/gal	Water Treated, gallons
Exchange Capacity, grains = 8,000,000 grains	

Calculate the gallons of water that may be treated.

$$\text{Water Treated, gal} = \frac{\text{Exchange Capacity, grains}}{\text{Hardness, grains/gal}}$$

$$= \frac{8,000,000 \text{ grains}}{10 \text{ grains/gal}}$$

$$= 800,000 \text{ gal}$$

Or $= 0.80 \text{ M gal}$

4. Calculating the mineral rejection as a percent, what is an estimate of the ability of a reverse osmosis plant to reject minerals? The feedwater contains 1,600 mg/L TDS and the product water TDS is 130 mg/L.

Known	Unknown
Feedwater TDS, mg/L = 1,600 mg/L	Mineral Rejection, %
Product Water TDS, mg/L = 130 mg/L	

Calculate the mineral rejection as a percent.

$$\text{Mineral Rejection, \%} = \left(1 - \frac{\text{Product TDS, mg}/L}{\text{Feed TDS, mg}/L}\right)(100\%)$$

$$= \left(1 - \frac{130 \text{ mg}/L}{1,600 \text{ mg}/L}\right)(100\%)$$

$$= (1 - 0.08)(100\%)$$

$$= 92\%$$

5. What is the feed rate of a dry chemical feeder in pounds per day if 3.0 pounds of chemical are caught in a weighing tin during 15 minutes?

Known	Unknown
Chemical, lbs = 3.0 lbs	Chemical Feed, lbs/day
Time, min = 15 min	

Calculate the chemical feed rate in pounds of chemical per day.

$$\text{Chemical Feed, lbs/day} = \frac{(\text{Chemical, lbs})(60 \text{ min/hr})(24 \text{ hr/day})}{\text{Time, min}}$$

$$= \frac{(3.0 \text{ lbs})(60 \text{ min/hr})(24 \text{ hr/day})}{15 \text{ min}}$$

$$= 288 \text{ lbs/day}$$

APPENDIX

HOW TO SOLVE WATER TREATMENT PLANT ARITHMETIC PROBLEMS

(VOLUME II)

by

Ken Kerri

TABLE OF CONTENTS
HOW TO SOLVE WATER TREATMENT PLANT ARITHMETIC PROBLEMS

		Page
A.1	BASIC CONVERSION FACTORS (ENGLISH SYSTEM)	733
A.2	BASIC FORMULAS	733
A.3	TYPICAL WATER TREATMENT PLANT PROBLEMS (ENGLISH SYSTEM)	737
	A.30 Iron and Manganese Control	737
	A.31 Fluoridation	738
	A.32 Softening	739
	A.33 Specialized Treatment Processes	743
	A.34 Membrane Treatment Processes	744
	A.35 Maintenance	745
	A.36 Advanced Laboratory Procedures	746
	A.37 Regulations	748
	A.38 Administration, Safety	748
A.4	BASIC CONVERSION FACTORS (METRIC SYSTEM)	748
A.5	TYPICAL WATER TREATMENT PLANT PROBLEMS (METRIC SYSTEM)	749
	A.50 Iron and Manganese Control	749
	A.51 Fluoridation	750
	A.52 Softening	752
	A.53 Specialized Treatment Processes	755
	A.54 Membrane Treatment Processes	756
	A.55 Maintenance	757
	A.56 Advanced Laboratory Procedures	758
	A.57 Regulations	760
	A.58 Administration, Safety	760
A.6	CALCULATION OF CT VALUES	760
A.7	CALCULATION OF LOG REMOVALS	763

HOW TO SOLVE WATER TREATMENT PLANT ARITHMETIC PROBLEMS
(VOLUME II)

A.1 BASIC CONVERSION FACTORS (ENGLISH SYSTEM)

UNITS

1,000,000	= 1 Million	1,000,000/1 Million

LENGTH

12 in	= 1 ft	12 in/ft
3 ft	= 1 yd	3 ft/yd
5,280 ft	= 1 mi	5,280 ft/mi

AREA

144 sq in	= 1 sq ft	144 sq in/sq ft
43,560 sq ft	= 1 acre	43,560 sq ft/ac

VOLUME

7.48 gal	= 1 cu ft	7.48 gal/cu ft
1,000 mL	= 1 liter	1,000 mL/L
3.785 L	= 1 gal	3.785 L/gal
231 cu in	= 1 gal	231 cu in/gal

WEIGHT

1,000 mg	= 1 gm	1,000 mg/gm
1,000 gm	= 1 kg	1,000 gm/kg
454 gm	= 1 lb	454 gm/lb
2.2 lbs	= 1 kg	2.2 lbs/kg

POWER

0.746 kW	= 1 HP	0.746 kW/HP

DENSITY

8.34 lbs	= 1 gal	8.34 lbs/gal
62.4 lbs	= 1 cu ft	62.4 lbs/cu ft

DOSAGE

17.1 mg/L	= 1 grain/gal	17.1 mg/L/gpg
64.7 mg	= 1 grain	64.7 mg/grain

PRESSURE

2.31 ft Water	= 1 psi	2.31 ft Water/psi
0.433 psi	= 1 ft Water	0.433 psi/ft Water
1.133 ft Water	= 1 in Mercury	1.133 ft Water/in Mercury

FLOW

694 GPM	= 1 MGD	694 GPM/MGD
1.55 CFS	= 1 MGD	1.55 CFS/MGD

TIME

60 sec	= 1 min	60 sec/min
60 min	= 1 hr	60 min/hr
24 hr	= 1 day	24 hr/day*

---

* This may be written either as 24 hr/day or 1 day/24 hours depending on which units we wish to convert to obtain our desired results.

A.2 BASIC FORMULAS

IRON AND MANGANESE CONTROL

1a. Stock Solution, mg/mL $= \dfrac{(\text{Polyphosphate, grams})(1{,}000 \text{ mg/gm})}{(\text{Solution, liter})(1{,}000 \text{ m}L/L)}$

1b. Dose, mg/L $= \dfrac{(\text{Stock Solution, mg/m}L)(\text{Volume Added, m}L)}{\text{Sample Volume, }L}$

1c. Dose, lbs/MG $= \dfrac{(\text{Dose, mg}/L)(3.785\ L/\text{gal})(1{,}000{,}000/\text{Million})}{(1{,}000 \text{ mg/gm})(454 \text{ gm/lb})}$

2. Chemical Feeder, lbs/day = (Flow, MGD)(Dose, mg/L)(8.34 lbs/gal)

3. Detention Time, min = $\dfrac{\text{(Basin Vol, gal)(24 hr/day)(60 min/hr)}}{\text{Flow, gal/day}}$

4. $KMnO_4$ Dose, mg/L = 0.2(Iron, mg/L) + 2.0(Manganese, mg$L$)

FLUORIDATION

5. Feed Rate, gal/day = $\dfrac{\text{(Feed Rate, lbs F/day)(100\%)}}{\text{(NaF Solution, lbs F/gal)(Purity, \%)}}$

6. Feed Solution, gal/day = $\dfrac{\text{(Flow, gal/day)(Feed Dose, mg/}L\text{)}}{\text{Feed Solution, mg/}L}$

7. Fluoride Ion Purity, % = $\dfrac{\text{(Molecular Weight of Fluoride)(100\%)}}{\text{Molecular Weight of Chemical}}$

8a. Feed Dose, mg/L = Desired Dose, mg/L – Actual Conc, mg/L

8b. Feed Rate, lbs/day = $\dfrac{\text{Feed Rate, lbs F/day}}{\text{lbs F/lb Commercial Na}_2\text{SiF}_6}$

9. Feed Solution, gal = $\dfrac{\text{(Flow Vol, gal)(Feed Dose, mg/}L\text{)}}{\text{Feed Solution, mg/}L}$

10. Mixture Strength, % = $\dfrac{\text{(Tank, gal)(Tank, \%) + (Vendor, gal)(Vendor, \%)}}{\text{Tank, gal + Vendor, gal}}$

SOFTENING

11. Total Hardness, mg/L as $CaCO_3$ = Calcium Hardness, mg/L as $CaCO_3$ + Magnesium Hardness, mg/L as $CaCO_3$

12. If alkalinity is greater than total hardness,

    Carbonate Hardness, mg/L as $CaCO_3$ = Total Hardness, mg/L as $CaCO_3$

    and

    Noncarbonate Hardness, mg/L as $CaCO_3$ = 0

13. If alkalinity is less than total hardness,

    Carbonate Hardness, mg/L as $CaCO_3$ = Alkalinity, mg/L as $CaCO_3$

    and

    Noncarbonate Hardness, mg/L as $CaCO_3$ = Total Hardness, mg/L as $CaCO_3$ – Alkalinity, mg/L as $CaCO_3$

14a. Phenolphthalein Alkalinity, mg/L as $CaCO_3$ = $\dfrac{A \times N \times 50{,}000}{\text{m}L \text{ of sample}}$

14b. Total Alkalinity, mg/L as $CaCO_3$ = $\dfrac{B \times N \times 50{,}000}{\text{m}L \text{ of sample}}$

15a. Hydrated Lime ($Ca(OH)_2$) Feed, mg/L = $\dfrac{(A + B + C + D)1.15}{\text{Purity of Lime as a decimal}}$

15b. Soda Ash (Na_2CO_3) Feed, mg/L = (Noncarbonate Hardness, mg/L as $CaCO_3$)(106/100)

15c. Total CO_2 Feed, mg/L = ($Ca(OH)_2$ excess, mg/L)(44/74) + (Mg^{2+} residual, mg/L)(44/24.3)

16. Feeder Setting, lbs/day = (Flow, MGD)(Conc, mg/L)(8.34 lbs/gal)

17. Feed Rate, lbs/min $= \dfrac{\text{Feeder Setting, lbs/day}}{(60 \text{ min/hr})(24 \text{ hr/day})}$

18. Hardness, mg/L $= \dfrac{(\text{Hardness, grains/gal})(17.1 \text{ mg/}L)}{1 \text{ grain/gal}}$

19. Exchange Capacity, grains $= (\text{Resin Vol, cu ft})(\text{Removal Capacity, grains/cu ft})$

20. Water Treated, gal $= \dfrac{\text{Exchange Capacity, grains}}{\text{Hardness, grains/gal}}$

21. Operating Time, hr $= \dfrac{\text{Water Treated, gal}}{(\text{Avg Daily Flow, gal/min})(60 \text{ min/hr})}$

22. Salt Needed, lbs $= (\text{Salt Required, lbs/1,000 gr})(\text{Hardness Removed, gr})$

23. Bypass Flow, GPD $= \dfrac{(\text{Total Flow, GPD})(\text{Plant Effl Hardness, gpg})}{\text{Raw Water Hardness, gpg}}$

SPECIALIZED TREATMENT PROCESSES

24. Avg TTHM, µg/L $= \dfrac{\text{Sum of Measurements, µg/}L}{\text{Number of Measurements}}$

25. Annual Running TTHM Average, µg/L $= \dfrac{\text{Sum of Average TTHM for Four Quarters}}{\text{Number of Quarters}}$

26. Blended Concentration, mg/L $= \dfrac{(\text{Flow 1, GPM})(\text{Conc 1, mg/}L) + (\text{Flow 2, GPM})(\text{Conc 2, mg/}L)}{\text{Flow 1, GPM} + \text{Flow 2, GPM}}$

MEMBRANE TREATMENT PROCESSES

27. Flow, GPD/sq ft $= \dfrac{(\text{Flux, gm/sq cm-sec})(2.54 \text{ cm/in})^2 (12 \text{ in/ft})^2 (60 \text{ sec/min})(60 \text{ min/hr})(24 \text{ hr/day})}{(1,000 \text{ gm/}L)(3.785 \text{ }L/\text{gal})}$

28. Mineral Rejection, % $= \left(1 - \dfrac{\text{Product TDS, mg/}L}{\text{Feed TDS, mg/}L}\right)(100\%)$

29. Recovery, % $= \dfrac{(\text{Product Flow, MGD})(100\%)}{\text{Feed Flow, MGD}}$

MAINTENANCE

30. Pump Capacity, GPM $= \dfrac{\text{Tank Volume, gal}}{\text{Pumping Time, min}}$

31. Flow, GPD $= \dfrac{(\text{Volume Pumped, gal})(24 \text{ hr/day})}{\text{Time, hr}}$

32. Polymer Feed, lbs/day $= \dfrac{(\text{Poly Conc, mg/}L)(\text{Vol Pumped, m}L)(60 \text{ min/hr})(24 \text{ hr/day})}{(\text{Time Pumped, min})(1,000 \text{ m}L/L)(1,000 \text{ mg/gm})(454 \text{ gm/lb})}$

33. Chemical Feed, lbs/day $= \dfrac{(\text{Chemical, gm})(60 \text{ min/hr})(24 \text{ hr/day})}{(454 \text{ gm/lb})(\text{Time, min})}$

ADVANCED LABORATORY PROCEDURES

34. Threshold Odor Number (TON) $= \dfrac{\text{Size of Sample, m}L + \text{Odor-Free Water, m}L}{\text{Size of Sample, m}L}$

35. Geometric Mean $= (X_1 \times X_2 \times X_3 \times \ldots X_n)^{1/n}$

36. Threshold Taste Number $= \dfrac{\text{Size of Sample, m}L + \text{Taste-Free Water, m}L}{\text{Size of Sample, m}L}$

37a. Arithmetic Mean, $\bar{X}$, Taste Rating $= \dfrac{X_1 + X_2 + X_3 + \ldots X_n}{n}$

37b. Standard Deviation, S, Taste Rating $= \left[\dfrac{(X_1 - \bar{X})^2 + (X_2 - \bar{X})^2 + \ldots (X_n - \bar{X})^2}{n - 1}\right]^{0.5}$

$\text{or} = \left[\dfrac{(X_1^2 + X_2^2 + \ldots X_n^2) - (X_1 + X_2 + \ldots X_n)^2/n}{n - 1}\right]^{0.5}$

REGULATIONS

38. Portions Positive, %/mo $= \dfrac{(\text{Number Positive/mo})(100\%)}{\text{Total Portions Tested}}$

ADMINISTRATION, SAFETY

39. Injury Frequency Rate $= \dfrac{(\text{Number of Injuries/yr})(1{,}000{,}000)}{\text{Number of Hours Worked/yr}}$

40. Injury Severity Rate $= \dfrac{(\text{Number of Hours Lost/yr})(1{,}000{,}000)}{\text{Number of Hours Worked/yr}}$

LOG REMOVALS

46a. Log Removal $= \text{Log Influent, particles/m}L - \text{Log Effluent, particles/m}L$

46b. Log Removal $= \text{Log}\left[\dfrac{\text{Influent, particles/m}L}{\text{Effluent, particles/m}L}\right]$

A.3 TYPICAL WATER TREATMENT PLANT PROBLEMS (ENGLISH SYSTEM)

A.30 Iron and Manganese Control

EXAMPLE 1

A standard polyphosphate solution is prepared by mixing and dissolving 1.0 gram of polyphosphate in a container and adding distilled water to the one-liter mark. Determine the concentration of the stock solution in milligrams per milliliter. If 6.0 milliliters of the stock solution are added to a one-liter sample, what is the polyphosphate dose in milligrams per liter and pounds per million gallons?

Known	Unknown
Polyphosphate, gm = 1.0 gm	1. Stock Solution, mg/mL
Solution, L = 1.0 L	2. Dose, mg/L
Stock Solution, mL = 6 mL	3. Dose, lbs/MG
Sample, L = 1 L	

1. Calculate the concentration of the stock solution in milligrams per milliliter.

$$\text{Stock Solution, mg/mL} = \frac{(\text{Polyphosphate, gm})(1{,}000 \text{ mg/gm})}{(\text{Solution, } L)(1{,}000 \text{ m}L/L)}$$

$$= \frac{(1.0 \text{ gm})(1{,}000 \text{ mg/gm})}{(1 \, L)(1{,}000 \text{ m}L/L)}$$

$$= 1.0 \text{ mg/m}L$$

2. Determine the polyphosphate dose in the sample in milligrams per liter.

$$\text{Dose, mg/}L = \frac{(\text{Stock Solution, mg/m}L)(\text{Vol Added, m}L)}{\text{Sample Volume, }L}$$

$$= \frac{(1.0 \text{ mg/m}L)(6 \text{ m}L)}{1 \, L}$$

$$= 6.0 \text{ mg/}L$$

3. Determine the polyphosphate dose in the sample in pounds of phosphate per million gallons of water.

$$\text{Dose, lbs/MG} = \frac{(\text{Dose, mg/}L)(3.785 \, L/\text{gal})(1{,}000{,}000/\text{Mil})}{(1{,}000 \text{ mg/gm})(454 \text{ gm/lb})}$$

$$= \frac{(6.0 \text{ mg/}L)(3.785 \, L/\text{gal})(1{,}000{,}000/\text{Mil})}{(1{,}000 \text{ mg/gm})(454 \text{ gm/lb})}$$

$$= 50 \text{ lbs/MG}$$

EXAMPLE 2

Determine the chemical feeder setting in pounds of polyphosphate per day if 0.62 MGD is treated with a dose of 6 mg/L.

Known	Unknown
Flow, MGD = 0.62 MGD	Chemical Feeder, lbs/day
Dose, mg/L = 6 mg/L	

Determine the chemical feeder setting in pounds per day.

$$\text{Chemical Feeder, lbs/day} = (\text{Flow, MGD})(\text{Dose, mg/}L)(8.34 \text{ lbs/gal})$$

$$= (0.62 \text{ MGD})(6 \text{ mg/}L)(8.34 \text{ lbs/gal})$$

$$= 31 \text{ lbs/day}$$

EXAMPLE 3

A reaction basin 14 feet in diameter and 4 feet deep treats a flow of 240,000 gallons per day. What is the average detention time in minutes?

Known	Unknown
Diameter, ft = 14 ft	Detention Time, min
Depth, ft = 4 ft	
Flow, GPD = 240,000 GPD	

1. Calculate the basin volume in cubic feet.

$$\text{Basin Vol, cu ft} = (0.785)(\text{Diameter, ft})^2(\text{Depth, ft})$$

$$= (0.785)(14 \text{ ft})^2(4 \text{ ft})$$

$$= 615 \text{ cu ft}$$

2. Convert the basin volume from cubic feet to gallons.

$$\text{Basin Vol, gal} = (\text{Basin Vol, cu ft})(7.48 \text{ gal/cu ft})$$

$$= (615 \text{ cu ft})(7.48 \text{ gal/cu ft})$$

$$= 4{,}600 \text{ gal}$$

3. Determine the average detention time in minutes for the reaction basin.

$$\text{Detention Time, min} = \frac{(\text{Basin Vol, gal})(24 \text{ hr/day})(60 \text{ min/hr})}{\text{Flow, gal/day}}$$

$$= \frac{(4{,}600 \text{ gal})(24 \text{ hr/day})(60 \text{ min/hr})}{240{,}000 \text{ gal/day}}$$

$$= 28 \text{ minutes}$$

EXAMPLE 4

Calculate the potassium permanganate dose in milligrams per liter for a well water with 2.4 mg/L iron before aeration and 0.3 mg/L after aeration. The manganese concentration is 0.8 mg/L both before and after aeration.

Known		Unknown
Iron, mg/L = 0.3 mg/L		KMnO$_4$ Dose, mg/L
Manganese, mg/L = 0.8 mg/L		

Calculate the potassium permanganate dose in milligrams per liter.

$$\text{KMnO}_4 \text{ Dose, mg}/L = 0.2(\text{Iron, mg}/L) + 2.0(\text{Manganese, mg}/L)$$

$$= 0.2(0.3 \text{ mg}/L) + 2.0(0.8 \text{ mg}/L)$$

$$= 1.66 \text{ mg}/L$$

NOTE: If there are any oxidizable compounds (organic color, bacteria, or hydrogen sulfide) in the water, the dose will have to be increased.

A.31 Fluoridation

EXAMPLE 5

Determine the setting for a chemical feed pump in gallons per day when the desired fluoride dose is 1.8 pounds of fluoride per day. The sodium fluoride solution contains 0.2 pound of fluoride per gallon and the fluoride purity is 43.4 percent.

Known		Unknown
Feed Rate, lbs F/day	= 1.8 lbs F/day	Feed Rate, gal/day
NaF Solution, lbs F/gal	= 0.2 lb F/gal	
Purity, %	= 43.4%	

Determine the setting on the chemical feed pump in gallons per day.

$$\text{Feed Rate, gal/day} = \frac{(\text{Feed Rate, lbs F/day})(100\%)}{(\text{NaF Solution, lbs F/gal})(\text{Purity, \%})}$$

$$= \frac{(1.8 \text{ lbs F/day})(100\%)}{(0.2 \text{ lb F/gal})(43.4\%)}$$

$$= 20.7 \text{ gal/day}$$

$$\text{or} = 21 \text{ gal/day}$$

EXAMPLE 6

Determine the setting on a chemical feed pump in gallons per day if 500,000 gallons per day of water must be treated with 0.9 mg/L of fluoride. The fluoride feed solution contains 18,000 mg/L of fluoride.

Known		Unknown
Flow, gal/day	= 500,000 gal/day	Feed Pump, gal/day
Fluoride, mg/L	= 0.9 mg/L	
Feed Solution, mg/L	= 18,000 mg/L	

Determine the setting on the chemical feed pump in gallons per day.

$$\text{Feed Pump, gal/day} = \frac{(\text{Flow, gal/day})(\text{Feed Dose, mg}/L)}{\text{Feed Solution, mg}/L}$$

$$= \frac{(500,000 \text{ gal/day})(0.9 \text{ mg}/L)}{18,000 \text{ mg}/L}$$

$$= 25 \text{ gal/day}$$

EXAMPLE 7

Determine the fluoride ion purity of Na$_2$SiF$_6$ as a percent.

Known	Unknown
Fluoride Chemical, Na$_2$SiF$_6$	Fluoride Ion Purity, %

Determine the molecular weight of fluoride and Na$_2$SiF$_6$.

Symbol	(No. Atoms)	(Atomic Wt)	=	Molecular Wt
Na$_2$	(2)	(22.99)	=	45.98
Si	(1)	(28.09)	=	28.09
F$_6$	(6)	(19.00)	=	114.00
Molecular Weight of Chemical			=	188.07

Calculate the fluoride ion as a percent.

$$\text{Fluoride Ion Purity, \%} = \frac{(\text{Molecular Weight of Fluoride})(100\%)}{\text{Molecular Weight of Chemical}}$$

$$= \frac{(114.00)(100\%)}{188.07}$$

$$= 60.62\%$$

EXAMPLE 8

A flow of 1.7 MGD is treated with sodium silicofluoride. The raw water contains 0.2 mg/L of fluoride ion and the desired fluoride concentration is 1.1 mg/L. What should be the chemical feed rate in pounds per day? Assume each pound of commercial sodium silicofluoride (Na$_2$SiF$_6$) contains 0.6 pound of fluoride ion.

Known		Unknown
Flow, MGD	= 1.7 MGD	Feed Rate, lbs/day
Raw Water F, mg/L	= 0.2 mg/L	
Desired F, mg/L	= 1.1 mg/L	
Chemical, lbs F/lb	= 0.6 lb F/lb	

1. Determine the fluoride feed dose in milligrams per liter.

$$\text{Feed Dose, mg}/L = \text{Desired Dose, mg}/L - \text{Actual Conc, mg}/L$$

$$= 1.1 \text{ mg}/L - 0.2 \text{ mg}/L$$

$$= 0.9 \text{ mg}/L$$

2. Calculate the fluoride feed rate in pounds per day.

$$\text{Feed Rate, lbs F/day} = (\text{Flow, MGD})(\text{Feed Dose, mg}/L)(8.34 \text{ lbs/gal})$$

$$= (1.7 \text{ MGD})(0.9 \text{ mg}/L)(8.34 \text{ lbs/gal})$$

$$= 12.8 \text{ lbs F/day}$$

3. Determine the chemical feed rate in pounds of commercial sodium silicofluoride per day.

$$\text{Feed Rate, lbs/day} = \frac{\text{Feed Rate, lbs F/day}}{\text{lbs F/lb Commercial Na}_2\text{SiF}_6}$$

$$= \frac{12.8 \text{ lbs F/day}}{0.6 \text{ lb F/lb Commercial Na}_2\text{SiF}_6}$$

$$= 21.3 \text{ lbs/day Commercial Na}_2\text{SiF}_6$$

EXAMPLE 9

The feed solution from a saturator containing 1.8 percent fluoride ion is used to treat a total flow of 250,000 gallons of water. The raw water has a fluoride ion content of 0.2 mg/L and the desired fluoride level in the treated water is 0.9 mg/L. How many gallons of feed solution are needed?

Known		Unknown
Flow Vol, gal	= 250,000 gal	Feed Solution, gal
Raw Water F, mg/L	= 0.2 mg/L	
Desired F, mg/L	= 0.9 mg/L	
Feed Solution, % F	= 1.8% F	

1. Convert the feed solution from a percentage fluoride ion to milligrams fluoride ion per liter of water.

$$1.0\% \text{ F} = 10,000 \text{ mg F}/L$$

$$\text{Feed Solution, mg}/L = \frac{(\text{Feed Solution, \%})(10,000 \text{ mg}/L)}{1.0\%}$$

$$= \frac{(1.8\% \text{ F})(10,000 \text{ mg}/L)}{1.0\%}$$

$$= 18,000 \text{ mg}/L$$

2. Determine the fluoride feed dose in milligrams per liter.

$$\text{Feed Dose, mg}/L = \text{Desired Dose, mg}/L - \text{Raw Water F, mg}/L$$

$$= 0.9 \text{ mg}/L - 0.2 \text{ mg}/L$$

$$= 0.7 \text{ mg}/L$$

3. Calculate the gallons of feed solution needed.

$$\text{Feed Solution, gal} = \frac{(\text{Flow Vol, gal})(\text{Feed Dose, mg}/L)}{\text{Feed Solution, mg}/L}$$

$$= \frac{(250,000 \text{ gal})(0.7 \text{ mg}/L)}{18,000 \text{ mg}/L}$$

$$= 9.7 \text{ gallons}$$

EXAMPLE 10

A hydrofluosilicic acid (H_2SiF_6) tank contains 350 gallons of acid with a strength of 19.3 percent. A commercial vendor delivers 2,500 gallons of acid with a strength of 18.1 percent to the tank. What is the resulting strength of the mixture as a percentage?

Known		Unknown
Tank Contents, gal	= 350 gal	Mixture Strength, %
Tank Strength, %	= 19.3%	
Vendor, gal	= 2,500 gal	
Vendor Strength, %	= 18.1%	

Calculate the strength of the mixture as a percentage.

$$\text{Mixture Strength, \%} = \frac{(\text{Tank, gal})(\text{Tank, \%}) + (\text{Vendor, gal})(\text{Vendor, \%})}{\text{Tank, gal} + \text{Vendor, gal}}$$

$$= \frac{(350 \text{ gal})(19.3\%) + (2,500 \text{ gal})(18.1\%)}{350 \text{ gal} + 2,500 \text{ gal}}$$

$$= \frac{6,755 + 45,250}{2,850}$$

$$= 18.2\%$$

A.32 Softening

EXAMPLE 11

Determine the total hardness as $CaCO_3$ for a sample of water with a calcium content of 33 mg/L and a magnesium content of 6 mg/L.

Known		Unknown
Calcium, mg/L	= 33 mg/L	Total Hardness, mg/L as $CaCO_3$
Magnesium, mg/L	= 6 mg/L	

Calculate the total hardness as milligrams per liter of calcium carbonate equivalent.

$$\text{Total Hardness, mg}/L \text{ as } CaCO_3 = \text{Calcium Hardness, mg}/L \text{ as } CaCO_3 + \text{Magnesium Hardness, mg}/L \text{ as } CaCO_3$$

$$= 2.5(\text{Ca, mg}/L) + 4.12(\text{Mg, mg}/L)$$

$$= 2.5(33 \text{ mg}/L) + 4.12(6 \text{ mg}/L)$$

$$= 82 \text{ mg}/L + 25 \text{ mg}/L$$

$$= 107 \text{ mg}/L \text{ as } CaCO_3$$

EXAMPLE 12

The alkalinity of a water is 120 mg/L as $CaCO_3$ and the total hardness is 105 mg/L as $CaCO_3$. What is the carbonate and noncarbonate hardness in mg/L as $CaCO_3$?

Known	Unknown
Alkalinity, mg/L = 120 mg/L as $CaCO_3$	1. Carbonate Hardness, mg/L as $CaCO_3$
Total Hardness, mg/L = 105 mg/L as $CaCO_3$	2. Noncarbonate Hardness, mg/L as $CaCO_3$

1. Determine the carbonate hardness in mg/L as $CaCO_3$.

   Since the alkalinity is greater than the total hardness (120 mg/L > 105 mg/L),

   $$\text{Carbonate Hardness, mg/}L \text{ as } CaCO_3 = \text{Total Hardness, mg/}L \text{ as } CaCO_3$$
   $$= 105 \text{ mg/}L \text{ as } CaCO_3$$

2. Determine the noncarbonate hardness in mg/L as $CaCO_3$.

   Since the alkalinity is greater than the total hardness,

   Noncarbonate Hardness, mg/L as $CaCO_3$ = 0

   In other words, all of the hardness is in the carbonate form.

EXAMPLE 13

The alkalinity of a water is 92 mg/L as $CaCO_3$ and the total hardness is 105 mg/L. What is the carbonate and noncarbonate hardness in mg/L as $CaCO_3$?

Known	Unknown
Alkalinity, mg/L = 92 mg/L as $CaCO_3$	1. Carbonate Hardness, mg/L as $CaCO_3$
Total Hardness, mg/L = 105 mg/L as $CaCO_3$	2. Noncarbonate Hardness, mg/L as $CaCO_3$

1. Determine the carbonate hardness in mg/L as $CaCO_3$.

   Since the alkalinity is less than the total hardness (92 mg/L < 105 mg/L),

   $$\text{Carbonate Hardness, mg/}L \text{ as } CaCO_3 = \text{Alkalinity, mg/}L \text{ as } CaCO_3$$
   $$= 92 \text{ mg/}L \text{ as } CaCO_3$$

2. Determine the noncarbonate hardness in mg/L as $CaCO_3$.

   Since the alkalinity is less than the total hardness (92 mg/L < 105 mg/L),

   $$\text{Noncarbonate Hardness, mg/}L \text{ as } CaCO_3 = \text{Total Hardness, mg/}L \text{ as } CaCO_3 - \text{Alkalinity, mg/}L \text{ as } CaCO_3$$
   $$= 105 \text{ mg/}L - 92 \text{ mg/}L$$
   $$= 13 \text{ mg/}L \text{ as } CaCO_3$$

EXAMPLE 14

Results from alkalinity titrations on a water sample were as follows:

Known

Sample Size, mL	= 100 mL
mL Titrant Used to pH 8.3, A	= 1.1 mL
Total mL of Titrant Used, B	= 12.4 mL
Acid Normality, N	= 0.02 N H_2SO_4

Unknown

1. Total Alkalinity, mg/L as $CaCO_3$
2. Bicarbonate Alkalinity, mg/L as $CaCO_3$
3. Carbonate Alkalinity, mg/L as $CaCO_3$
4. Hydroxide Alkalinity, mg/L as $CaCO_3$

See Table 14.4, page 81, for alkalinity relationships among constituents.

1. Calculate the phenolphthalein alkalinity in mg/L as $CaCO_3$.

   $$\text{Phenolphthalein Alkalinity, mg/}L \text{ as } CaCO_3 = \frac{A \times N \times 50{,}000}{\text{m}L \text{ of Sample}}$$
   $$= \frac{(1.1 \text{ m}L)(0.02 \ N)(50{,}000)}{100 \text{ m}L}$$
   $$= 11 \text{ mg/}L \text{ as } CaCO_3$$

2. Calculate the total alkalinity in mg/L as $CaCO_3$.

   $$\text{Total Alkalinity, mg/}L \text{ as } CaCO_3 = \frac{B \times N \times 50{,}000}{\text{m}L \text{ of Sample}}$$
   $$= \frac{(12.4 \text{ m}L)(0.02 \ N)(50{,}000)}{100 \text{ m}L}$$
   $$= 124 \text{ mg/}L \text{ as } CaCO_3$$

3. Refer to Table 14.4 for alkalinity constituents. The second row indicates that since P is less than ½T (11 mg/L < ½(124 mg/L)), bicarbonate alkalinity is T − 2P and carbonate alkalinity is 2P.

$$\text{Bicarbonate Alkalinity, mg}/L \text{ as } CaCO_3 = T - 2P$$

$$= 124 \text{ mg}/L - 2(11 \text{ mg}/L)$$

$$= 102 \text{ mg}/L \text{ as } CaCO_3$$

$$\text{Carbonate Alkalinity, mg}/L \text{ as } CaCO_3 = 2P$$

$$= 2(11 \text{ mg}/L)$$

$$= 22 \text{ mg}/L \text{ as } CaCO_3$$

$$\text{Hydroxide Alkalinity, mg}/L \text{ as } CaCO_3 = 0 \text{ mg}/L \text{ as } CaCO_3$$

EXAMPLE 15

Calculate the hydrated lime ($Ca(OH)_2$) with 90 percent purity, soda ash, and carbon dioxide requirements in milligrams per liter for the water shown below.

Known

Constituents	Source Water	Softened Water After Recarbonation and Filtration
CO_2, mg/L	= 7 mg/L	= 0 mg/L
Total Alkalinity, mg/L	= 125 mg/L as $CaCO_3$	= 22 mg/L as $CaCO_3$
Total Hardness, mg/L	= 240 mg/L as $CaCO_3$	= 35 mg/L as $CaCO_3$
Mg^{2+}, mg/L	= 38 mg/L	= 8 mg/L
pH	= 7.6	= 8.8
Lime Purity, %	= 90%	

Unknown

1. Hydrated Lime, mg/L
2. Soda Ash, mg/L
3. Carbon Dioxide, mg/L

1. Calculate the hydrated lime ($Ca(OH)_2$) required in milligrams per liter.

$$A = (CO_2, \text{mg}/L)(74/44)$$

$$= (7 \text{ mg}/L)(74/44)$$

$$= 12 \text{ mg}/L$$

$$B = (\text{Alkalinity, mg}/L)(74/100)$$

$$= (125 \text{ mg}/L - 22 \text{ mg}/L)(74/100)$$

$$= 76 \text{ mg}/L$$

$$C = 0 \quad \text{Hydroxide Alkalinity} = 0$$

$$D = (Mg^{2+}, \text{mg}/L)(74/24.3)$$

$$= (38 \text{ mg}/L - 8 \text{ mg}/L)(74/24.3)$$

$$= 91 \text{ mg}/L$$

$$\text{Hydrated Lime } (Ca(OH)_2) \text{ Feed, mg}/L = \frac{(A + B + C + D)1.15}{\text{Purity of Lime, as a decimal}}$$

$$= \frac{(12 \text{ mg}/L + 76 \text{ mg}/L + 0 + 91 \text{ mg}/L)1.15}{0.90}$$

$$= \frac{(179 \text{ mg}/L)(1.15)}{0.90}$$

$$= 229 \text{ mg}/L$$

2. Calculate the soda ash required in milligrams per liter.

$$\text{Total Hardness Removed, mg}/L \text{ as } CaCO_3 = \text{Total Hardness, mg}/L \text{ as } CaCO_3 - \text{Total Hardness Remaining, mg}/L \text{ as } CaCO_3$$

$$= 240 \text{ mg}/L - 35 \text{ mg}/L$$

$$= 205 \text{ mg}/L \text{ as } CaCO_3$$

$$\text{Noncarbonate Hardness, mg}/L \text{ as } CaCO_3 = \text{Total Hardness Removed, mg}/L \text{ as } CaCO_3 - \left(\text{Carbonate Hardness, mg}/L \text{ as } CaCO_3 - \text{Carbonate Hardness Remaining, mg}/L \text{ as } CaCO_3 \right)$$

$$= 205 \text{ mg}/L - (125 \text{ mg}/L - 22 \text{ mg}/L)$$

$$= 102 \text{ mg}/L \text{ as } CaCO_3$$

$$\text{Soda Ash } (Na_2CO_3) \text{ Feed, mg}/L = \left(\text{Noncarbonate Hardness, mg}/L \text{ as } CaCO_3 \right)(106/100)$$

$$= (102 \text{ mg}/L)(106/100)$$

$$= 108 \text{ mg}/L$$

3. Calculate the dosage of carbon dioxide required for recarbonation.

$$\text{Excess Lime, mg}/L = (A + B + C + D)(0.15)$$

$$= (12 \text{ mg}/L + 93 \text{ mg}/L + 0 + 116 \text{ mg}/L)(0.15)$$

$$= (221 \text{ mg}/L)(0.15)$$

$$= 33 \text{ mg}/L$$

$$\text{Total } CO_2 \text{ Feed, mg}/L = (Ca(OH)_2 \text{ Excess, mg}/L)(44/74) + (Mg^{2+} \text{ Residual, mg}/L)(44/24.3)$$

$$= (33 \text{ mg}/L)(44/74) + (8 \text{ mg}/L)(44/24.3)$$

$$= 20 \text{ mg}/L + 15 \text{ mg}/L$$

$$= 35 \text{ mg}/L$$

EXAMPLE 16

The optimum lime dosage from the jar tests is 180 mg/L. If the flow to be treated is 1.7 MGD, what is the feeder setting in pounds per day and the feed rate in pounds per minute?

Known	Unknown
Lime Dose, mg/L = 180 mg/L	1. Feeder Setting, lbs/day
Flow, MGD = 1.7 MGD	2. Feed Rate, lbs/min

1. Calculate the feeder setting in pounds per day.

$$\text{Feeder Setting, lbs/day} = (\text{Flow, MGD})(\text{Lime, mg/}L)(8.34 \text{ lbs/gal})$$
$$= (1.7 \text{ MGD})(180 \text{ mg/}L)(8.34 \text{ lbs/gal})$$
$$= 2{,}550 \text{ lbs/day}$$

2. Calculate the feed rate in pounds per minute.

$$\text{Feed Rate, lbs/min} = \frac{\text{Feeder Setting, lbs/day}}{(60 \text{ min/hr})(24 \text{ hr/day})}$$
$$= \frac{2{,}550 \text{ lbs/day}}{(60 \text{ min/hr})(24 \text{ hr/day})}$$
$$= 1.8 \text{ lbs/min}$$

EXAMPLE 17

How much soda ash is required (pounds per day and pounds per minute) to remove 40 mg/L noncarbonate hardness as $CaCO_3$ from a flow of 1.7 MGD?

Known	Unknown
Noncarbonate Hardness Removed, mg/L as $CaCO_3$ = 40 mg/L	1. Feeder Setting, lbs/day
Flow, MGD = 1.7 MGD	2. Feed Rate, lbs/min

1. Calculate the soda ash dose in milligrams per liter. See Section 14.316, "Calculation of Chemical Dosages," page 85, for the following formula.

$$\text{Soda Ash, mg/}L = \left(\frac{\text{Noncarbonate Hardness,}}{\text{mg/}L \text{ as CaCO}_3}\right)(106/100)$$
$$= (40 \text{ mg/}L)(106/100)$$
$$= 42.4 \text{ mg/}L$$

2. Determine the feeder setting in pounds per day.

$$\text{Feeder Setting, lbs/day} = (\text{Flow, MGD})(\text{Soda Ash, mg/}L)(8.34 \text{ lbs/gal})$$
$$= (1.7 \text{ MGD})(42.4 \text{ mg/}L)(8.34 \text{ lbs/gal})$$
$$= 601 \text{ lbs/day}$$

3. Calculate the soda ash feed rate in pounds per minute.

$$\text{Feed Rate, lbs/min} = \frac{\text{Feeder Setting, lbs/day}}{(60 \text{ min/hr})(24 \text{ hr/day})}$$
$$= \frac{601 \text{ lbs/day}}{(60 \text{ min/hr})(24 \text{ hr/day})}$$
$$= 0.42 \text{ lb/min}$$

EXAMPLE 18

What is the hardness in milligrams per liter for a water with a hardness of 12 grains per gallon?

Known	Unknown
Hardness, gpg = 12 grains/gallon	Hardness, mg/L

Calculate the hardness in milligrams per liter.

$$\text{Hardness, mg/}L = \frac{(\text{Hardness, grains/gal})(17.1 \text{ mg/}L)}{1 \text{ grain/gal}}$$
$$= \frac{(12 \text{ grains/gal})(17.1 \text{ mg/}L)}{1 \text{ grain/gal}}$$
$$= 205 \text{ mg/}L$$

EXAMPLE 19

Estimate the exchange capacity in grains of hardness for an ion exchange unit that contains 600 cubic feet of resin with a removal capacity of 25,000 grains per cubic foot.

Known	Unknown
Resin Volume, cu ft = 600 cu ft	Exchange Capacity, grains
Removal Capacity, gr/cu ft = 25,000 gr/cu ft	

Estimate the exchange capacity in grains of hardness.

$$\text{Exchange Capacity, grains} = (\text{Resin Volume, cu ft})(\text{Removal Capacity, grains/cu ft})$$
$$= (600 \text{ cu ft})(25{,}000 \text{ gr/cu ft})$$
$$= 15{,}000{,}000 \text{ grains of hardness}$$

EXAMPLE 20

How many gallons of water with a hardness of 12 grains per gallon may be treated by an ion exchange softener with an exchange capacity of 15,000,000 grains?

Known	Unknown
Hardness, grains/gal = 12 grains/gal	Water Treated, gallons
Exchange Capacity, grains = 15,000,000 grains	

Calculate the gallons of water that may be treated.

$$\text{Water Treated, gal} = \frac{\text{Exchange Capacity, grains}}{\text{Hardness, grains/gal}}$$
$$= \frac{15{,}000{,}000 \text{ grains}}{12 \text{ grains/gal}}$$
$$= 1{,}250{,}000 \text{ gal}$$

EXAMPLE 21

How many hours will an ion exchange softening unit operate when treating an average daily flow of 750 GPM? The unit is capable of softening 1,250,000 gallons of water before requiring regeneration.

Known	Unknown
Avg Daily Flow, GPM = 750 GPM	Operating Time, hr
Water Treated, gal = 1,250,000 gal	

Estimate how many hours the softening unit can operate before requiring regeneration.

$$\text{Operating Time, hr} = \frac{\text{Water Treated, gal}}{(\text{Avg Daily Flow, gal/min})(60 \text{ min/hr})}$$

$$= \frac{1,250,000 \text{ gal}}{(750 \text{ gal/min})(60 \text{ min/hr})}$$

$$= 27.8 \text{ hours}$$

EXAMPLE 22

Determine the pounds of salt needed to regenerate an ion exchange softening unit capable of removing 15,000,000 grains of hardness if 0.25 pound of salt is required for every 1,000 grains of hardness removed.

Known	Unknown
Hardness Removed, gr = 15,000,000 grains	Salt Needed, lbs
Salt Required, lbs/1,000 gr = 0.25 lb Salt/1,000 gr	

Calculate the pounds of salt needed to regenerate the ion exchange softening unit.

$$\text{Salt Needed, lbs} = (\text{Salt Required, lbs/1,000 gr})(\text{Hardness Removed, gr})$$

$$= \left(\frac{0.25 \text{ lb Salt}}{1,000 \text{ grains}}\right)(15,000,000 \text{ grains})$$

$$= 3,750 \text{ lbs of Salt}$$

EXAMPLE 23

Estimate the bypass flow in gallons per day around an ion exchange softener if the plant treats 250,000 gallons per day with a source water hardness of 20 grains per gallon if the desired product water hardness is 5 grains per gallon.

Known	Unknown
Total Flow, GPD = 250,000 GPD	Bypass Flow, GPD
Source Water Hardness, gpg = 20 grains/gal	
Plant Effl Hardness, gpg = 5 grains/gal	

Estimate the bypass flow in gallons per day.

$$\text{Bypass Flow, GPD} = \frac{(\text{Total Flow, GPD})(\text{Plant Effl Hardness, gpg})}{\text{Source Water Hardness, gpg}}$$

$$= \frac{(250,000 \text{ GPD})(5 \text{ gpg})}{20 \text{ gpg}}$$

$$= 62,500 \text{ GPD}$$

A.33 Specialized Treatment Processes

EXAMPLE 24

A water utility collected and analyzed eight samples from a water distribution system on the same day for TTHMs. The results are shown below.

Sample No.	1	2	3	4	5	6	7	8
TTHM, µg/L	80	90	100	90	110	100	100	90

What was the average TTHM for the day?

Known	Unknown
Results from analyses of 8 TTHM samples	Average TTHM level for the day

Calculate the average TTHM level in micrograms per liter.

$$\text{Avg TTHM, µg/}L = \frac{\text{Sum of Measurements, µg/}L}{\text{Number of Measurements}}$$

$$= \frac{80 \text{ µg/}L + 90 \text{ µg/}L + 100 \text{ µg/}L + 90 \text{ µg/}L + 110 \text{ µg/}L + 100 \text{ µg/}L + 100 \text{ µg/}L + 90 \text{ µg/}L}{8}$$

$$= \frac{760 \text{ µg/}L}{8}$$

$$= 95 \text{ µg/}L$$

EXAMPLE 25

The results of the quarterly average TTHM measurements for two years are given below. Calculate the running annual average of the four quarterly measurements in micrograms per liter.

Quarter	1	2	3	4	1	2	3	4
Avg Quarterly TTHM, µg/L	77	88	112	95	83	87	109	89

Known	Unknown
Results from analyses of two years of TTHM sampling	Running Annual Average of quarterly TTHM measurements

Calculate the running annual average of the quarterly TTHM measurements.

$$\text{Annual Running TTHM Average, µg/}L = \frac{\text{Sum of Average TTHM for Four Quarters}}{\text{Number of Quarters}}$$

QUARTERS 1, 2, 3, AND 4

$$\text{Annual Running TTHM Average, µg/}L = \frac{77 \text{ µg/}L + 88 \text{ µg/}L + 112 \text{ µg/}L + 95 \text{ µg/}L}{4}$$

$$= \frac{372 \text{ µg/}L}{4}$$

$$= 93 \text{ µg/}L$$

QUARTERS 2, 3, 4, AND 1

$$\text{Annual Running TTHM Average, µg/}L = \frac{88 \text{ µg/}L + 112 \text{ µg/}L + 95 \text{ µg/}L + 83 \text{ µg/}L}{4}$$

$$= \frac{378 \text{ µg/}L}{4}$$

$$= 95 \text{ µg/}L$$

QUARTERS 3, 4, 1, AND 2

$$\text{Annual Running TTHM Average, µg/}L = \frac{112 \text{ µg/}L + 95 \text{ µg/}L + 83 \text{ µg/}L + 87 \text{ µg/}L}{4}$$

$$= \frac{377 \text{ µg/}L}{4}$$

$$= 94 \text{ µg/}L$$

QUARTERS 4, 1, 2, AND 3

$$\text{Annual Running TTHM Average, µg/}L = \frac{95 \text{ µg/}L + 83 \text{ µg/}L + 87 \text{ µg/}L + 109 \text{ µg/}L}{4}$$

$$= \frac{374 \text{ µg/}L}{4}$$

$$= 94 \text{ µg/}L$$

QUARTERS 1, 2, 3, AND 4

$$\text{Annual Running TTHM Average, µg/}L = \frac{83 \text{ µg/}L + 87 \text{ µg/}L + 109 \text{ µg/}L + 89 \text{ µg/}L}{4}$$

$$= \frac{368 \text{ µg/}L}{4}$$

$$= 92 \text{ µg/}L$$

SUMMARY OF RESULTS

Quarter	1	2	3	4	1	2	3	4
Avg Quarterly TTHM, µg/L	77	88	112	95	83	87	109	89
Annual Running TTHM Avg, µg/L				93	95	94	94	92

EXAMPLE 26

Well number 1 delivers a flow of 50 GPM with an arsenic concentration of 0.024 mg/L. Well number 2 delivers a flow of 500 GPM with an arsenic concentration of 0.002 mg/L. What would be the blended arsenic concentration in milligrams per liter if the flows from these two wells were blended together?

Known	Unknown
Well 1 Q, GPM = 50 GPM	Blended Conc, mg/L
Well 2 Q, GPM = 500 GPM	
Well 1 Conc, mg/L = 0.024 mg/L	
Well 2 Conc, mg/L = 0.002 mg/L	

Calculate the blended arsenic concentration in milligrams per liter.

$$\text{Blended Conc, mg/}L = \frac{(\text{Flow 1, GPM})(\text{Conc 1, mg/}L) + (\text{Flow 2, GPM})(\text{Conc 2, mg/}L)}{\text{Flow 1, GPM} + \text{Flow 2, GPM}}$$

$$= \frac{(50 \text{ GPM})(0.024 \text{ mg/}L) + (500 \text{ GPM})(0.002 \text{ mg/}L)}{50 \text{ GPM} + 500 \text{ GPM}}$$

$$= \frac{1.2 + 1}{550}$$

$$= \frac{2.2}{550}$$

$$= 0.004 \text{ mg/}L$$

A.34 Membrane Treatment Processes

EXAMPLE 27

Convert a water flux of 12×10^{-4} gm/sq cm-sec to gallon per day per square foot.

Known	Unknown
Water Flux, gm/sq cm-sec = 12×10^{-4} gm/sq cm-sec	Flow, GPD/sq ft

Convert the water flux from gm/sq cm-sec to flow in GPD/sq ft.

$$\text{Flow, GPD/sq ft} = \frac{\left(\text{Flux, gm/sq cm-sec}\right)(2.54 \text{ in/in})^2(12 \text{ in/ft})^2(60 \text{ sec/min})(60 \text{ min/hr})(24 \text{ hr/day})}{(1{,}000 \text{ gm/}L)(3.785 \text{ }L/\text{gal})}$$

$$= \frac{\left(0.0012 \text{ gm/sq cm-sec}\right)(2.54 \text{ cm/in})^2(12 \text{ in/ft})^2(60 \text{ sec/min})(60 \text{ min/hr})(24 \text{ hr/day})}{(1{,}000 \text{ gm/}L)(3.785 \text{ }L/\text{gal})}$$

$$= 25.5 \text{ GPD/sq ft}$$

EXAMPLE 28

Estimate the ability of a reverse osmosis plant to reject minerals by calculating the mineral rejection as a percent. The feedwater contains 1,800 mg/L TDS and the product water TDS is 120 mg/L.

Known	Unknown
Feedwater TDS, mg/L = 1,800 mg/L	Mineral Rejection, %
Product Water TDS, mg/L = 120 mg/L	

Calculate the mineral rejection as a percent.

$$\text{Mineral Rejection, \%} = \left(1 - \frac{\text{Product TDS, mg/}L}{\text{Feed TDS, mg/}L}\right)(100\%)$$

$$= \left(1 - \frac{120 \text{ mg/}L}{1,800 \text{ mg/}L}\right)(100\%)$$

$$= (1 - 0.067)(100\%)$$

$$= 93.3\%$$

EXAMPLE 29

Estimate the percent recovery of a reverse osmosis unit with a 4-2-1 arrangement if the feed flow is 2.0 MGD and the product flow is 1.75 MGD.

Known	Unknown
Product Flow, MGD = 1.75 MGD	Recovery, %
Feed Flow, MGD = 2.0 MGD	

Calculate the recovery as a percent.

$$\text{Recovery, \%} = \frac{(\text{Product Flow, MGD})(100\%)}{\text{Feed Flow, MGD}}$$

$$= \frac{(1.75 \text{ MGD})(100\%)}{2.0 \text{ MGD}}$$

$$= 87.5\%$$

A.35 Maintenance

EXAMPLE 30

Calculate the pumping capacity of a pump in gallons per minute when 12 minutes are required for the water to rise 3 feet in an 8-foot by 6-foot rectangular tank.

Known	Unknown
Length, ft = 8 ft	Pump Capacity, GPM
Width, ft = 6 ft	
Depth, ft = 3 ft	
Time, min = 12 min	

1. Calculate the volume pumped in cubic feet.

$$\text{Volume Pumped, cu ft} = (\text{Length, ft})(\text{Width, ft})(\text{Depth, ft})$$

$$= (8 \text{ ft})(6 \text{ ft})(3 \text{ ft})$$

$$= 144 \text{ cu ft}$$

2. Convert the volume pumped from cubic feet to gallons.

$$\text{Volume Pumped, gal} = (\text{Volume Pumped, cu ft})(7.48 \text{ gal/cu ft})$$

$$= (144 \text{ cu ft})(7.48 \text{ gal/cu ft})$$

$$= 1,077 \text{ gal}$$

3. Calculate the pump capacity in gallons per minute.

$$\text{Pump Capacity, GPM} = \frac{\text{Volume Pumped, gal}}{\text{Pumping Time, min}}$$

$$= \frac{1,077 \text{ gal}}{12 \text{ min}}$$

$$= 90 \text{ GPM}$$

EXAMPLE 31

A small chemical feed pump lowered the chemical solution in a 2.5-foot diameter tank 2.25 feet during seven hours. Estimate the flow delivered by the pump in gallons per minute and gallons per day.

Known	Unknown
Tank Diameter, ft = 2.5 ft	Flow, GPM
Chemical Drop, ft = 2.25 ft	Flow, GPD
Time, hr = 7.0 hr	

1. Determine the gallons of chemical solution pumped.

$$\text{Volume, gal} = (0.785)(\text{Diameter, ft})^2(\text{Drop, ft})(7.48 \text{ gal/cu ft})$$

$$= (0.785)(2.5 \text{ ft})^2(2.25 \text{ ft})(7.48 \text{ gal/cu ft})$$

$$= 83 \text{ gallons}$$

2. Estimate the flow delivered by the pump in gallons per minute and gallons per day.

$$\text{Flow, GPM} = \frac{\text{Volume Pumped, gal}}{(\text{Time, hr})(60 \text{ min/hr})}$$

$$= \frac{83 \text{ gallons}}{(7 \text{ hr})(60 \text{ min/hr})}$$

$$= 0.2 \text{ GPM}$$

or

$$\text{Flow, GPD} = \frac{(\text{Volume Pumped, gal})(24 \text{ hr/day})}{\text{Time, hr}}$$

$$= \frac{(83 \text{ gallons})(24 \text{ hr/day})}{7 \text{ hr}}$$

$$= 285 \text{ GPD}$$

EXAMPLE 32

Determine the chemical feed in pounds of polymer per day from a chemical feed pump. The polymer solution is 1.8 percent or 18,000 mg polymer per liter. Assume a specific gravity of the polymer solution of 1.0. During a test run, the chemical feed pump delivered 650 mL of polymer solution in 4.5 minutes.

Known	Unknown
Polymer Solution, % = 1.8%	Polymer Feed, lbs/day
Polymer Conc, mg/L = 18,000 mg/L	
Polymer Sp Gr = 1.0	
Volume Pumped, mL = 650 mL	
Time Pumped, min = 4.5 min	

Calculate the polymer fed by the chemical feed pump in pounds of polymer per day.

$$\text{Polymer Feed, lbs/day} = \frac{(\text{Poly Conc, mg}/L)(\text{Vol Pumped, mL})(60 \text{ min/hr})(24 \text{ hr/day})}{(\text{Time Pumped, min})(1{,}000 \text{ mL}/L)(1{,}000 \text{ mg/gm})(454 \text{ gm/lb})}$$

$$= \frac{(18{,}000 \text{ mg}/L)(650 \text{ mL})(60 \text{ min/hr})(24 \text{ hr/day})}{(4.5 \text{ min})(1{,}000 \text{ mL}/L)(1{,}000 \text{ mg/gm})(454 \text{ gm/lb})}$$

$$= 8.2 \text{ lbs/day}$$

EXAMPLE 33

Determine the actual chemical feed in pounds per day from a dry chemical feeder. A pie tin placed under the chemical feeder caught 824 grams of chemical during five minutes.

Known	Unknown
Chemical, gm = 824 gm	Chemical Feed, lbs/day
Time, min = 5 min	

Determine the chemical feed in pounds per day.

$$\text{Chemical Feed, lbs/day} = \frac{(\text{Chemical, gm})(60 \text{ min/hr})(24 \text{ hr/day})}{(454 \text{ gm/lb})(\text{Time, min})}$$

$$= \frac{(824 \text{ gm})(60 \text{ min/hr})(24 \text{ hr/day})}{(454 \text{ gm/lb})(5 \text{ min})}$$

$$= 523 \text{ lbs/day}$$

A.36 Advanced Laboratory Procedures

EXAMPLE 34

Calculate the threshold odor number (TON) for a sample when the first detectable odor occurred when the 70-mL sample was diluted to 200 mL (130 mL of odor-free water was added to the 70-mL sample).

Known	Unknown
Size of Sample, mL = 70 mL	TON
Odor-Free Water, mL = 130 mL	

Calculate the threshold odor number (TON).

$$\text{TON} = \frac{\text{Size of Sample, mL} + \text{Odor-Free Water, mL}}{\text{Size of Sample, mL}}$$

$$= \frac{70 \text{ mL} + 130 \text{ mL}}{70 \text{ mL}}$$

$$= 3$$

EXAMPLE 35

Determine the geometric mean threshold odor number for a panel of six testers given the results shown below.

Known	Unknown
Tester 1, X_1 = 2	Geometric Mean Threshold Odor Number
Tester 2, X_2 = 4	
Tester 3, X_3 = 3	
Tester 4, X_4 = 8	
Tester 5, X_5 = 6	
Tester 6, X_6 = 2	

Calculate the geometric mean.

$$\text{Geometric Mean TON} = (X_1 \times X_2 \times X_3 \times X_4 \times X_5 \times X_6)^{1/n}$$

$$= (2 \times 4 \times 3 \times 8 \times 6 \times 2)^{1/6}$$

$$= (2{,}304)^{0.167}$$

$$= 3.6$$

EXAMPLE 36

Calculate the threshold taste number for a sample when the first detectable taste occurred when the 8.3-mL sample was diluted to 200 mL (191.7 mL of taste-free water was added to the 8.3-mL sample).

Known	Unknown
Size of Sample, mL = 8.3 mL	Threshold Taste Number
Taste-Free Water, mL = 191.7 mL	

Calculate the threshold taste number.

$$\text{Threshold Taste Number} = \frac{\text{Size of Sample, mL} + \text{Taste-Free Water, mL}}{\text{Size of Sample, mL}}$$

$$= \frac{8.3 \text{ mL} + 191.7 \text{ mL}}{8.3 \text{ mL}}$$

$$= 24$$

EXAMPLE 37

Determine the taste rating for a water by calculating the arithmetic mean and standard deviation for the panel ratings given below.

Known	Unknown
Tester 1, $X_1 = 2$	1. Arithmetic Mean, $\overline{X}$
Tester 2, $X_2 = 5$	2. Standard Deviation, S
Tester 3, $X_3 = 3$	
Tester 4, $X_4 = 6$	
Tester 5, $X_5 = 2$	
Tester 6, $X_6 = 6$	

1. Calculate the arithmetic mean, $\overline{X}$, taste rating.

$$\text{Arithmetic Mean, } \overline{X}, \text{ Taste Rating} = \frac{X_1 + X_2 + X_3 + X_4 + X_5 + X_6}{n}$$

$$= \frac{2 + 5 + 3 + 6 + 2 + 6}{6}$$

$$= \frac{24}{6}$$

$$= 4$$

2. Calculate the standard deviation, S, of the taste rating.

$$\text{Standard Deviation, S} = \left[\frac{(X_1 - \overline{X})^2 + (X_2 - \overline{X})^2 + (X_3 - \overline{X})^2 + (X_4 - \overline{X})^2 + (X_5 - \overline{X})^2 + (X_6 - \overline{X})^2}{n - 1}\right]^{0.5}$$

$$= \left[\frac{(2-4)^2 + (5-4)^2 + (3-4)^2 + (6-4)^2 + (2-4)^2 + (6-4)^2}{6-1}\right]^{0.5}$$

$$= \left[\frac{(-2)^2 + (1)^2 + (-1)^2 + (2)^2 + (-2)^2 + (2)^2}{5}\right]^{0.5}$$

$$= \left[\frac{4 + 1 + 1 + 4 + 4 + 4}{5}\right]^{0.5}$$

$$= \left[\frac{18}{5}\right]^{0.5}$$

$$= (3.6)^{0.5}$$

$$= 1.9$$

or

$$\text{Standard Deviation, S} = \left[\frac{(X_1^2 + X_2^2 + X_3^2 + X_4^2 + X_5^2 + X_6^2) - (X_1 + X_2 + X_3 + X_4 + X_5 + X_6)^2/n}{n - 1}\right]^{0.5}$$

$$= \left[\frac{(2^2 + 5^2 + 3^2 + 6^2 + 2^2 + 6^2) - (2 + 5 + 3 + 6 + 2 + 6)^2/6}{6-1}\right]^{0.5}$$

$$= \left[\frac{(4 + 25 + 9 + 36 + 4 + 36) - (24)^2/6}{5}\right]^{0.5}$$

$$= \left[\frac{114 - 96}{5}\right]^{0.5}$$

$$= \left[\frac{18}{5}\right]^{0.5}$$

$$= (3.6)^{0.5}$$

$$= 1.9$$

A.37 Regulations

EXAMPLE 38

A small water system collected 14 samples during one month. After each sample was collected, 10 mL of each sample was placed in each of 5 fermentation tubes. At the end of the month, the results indicated that 2 out of a total of 70 fermentation tubes were positive. What percent of the portions tested during the month were positive?

Known	Unknown
Number Positive/mo = 2 Positive/mo	Portions Positive, %/mo
Total Portions Tested = 70 Portions	

Calculate the percent of the portions tested during the month that were positive.

$$\text{Portions Positive, \%/mo} = \frac{(\text{Number Positive/mo})(100\%)}{\text{Total Portions Tested}}$$

$$= \frac{(2 \text{ Positive/mo})(100\%)}{70 \text{ Portions}}$$

$$= 3\%/\text{mo}$$

A.38 Administration, Safety

EXAMPLE 39

Calculate the injury frequency rate for a water utility where there were four injuries in one year and the operators worked 97,120 hours.

Known	Unknown
Number of Injuries/yr = 4 Injuries/yr	Injury Frequency Rate
Number of Hours Worked/yr = 97,120 Hr/yr	

Calculate the injury frequency rate.

$$\text{Injury Frequency Rate} = \frac{(\text{Number of Injuries/yr})(1,000,000)}{\text{Number of Hours Worked/yr}}$$

$$= \frac{(4 \text{ Injuries/yr})(1,000,000)}{97,120 \text{ Hr/yr}}$$

$$= 41.2$$

EXAMPLE 40

Calculate the injury severity rate for a water company that experienced 57 operator-hours lost due to injuries while the operators worked 97,120 hours during the year.

Known	Unknown
Number of Hours Lost/yr = 57 Hr/yr	Injury Severity Rate
Number of Hours Worked/yr = 97,120 Hr/yr	

Calculate the injury severity rate.

$$\text{Injury Severity Rate} = \frac{(\text{Number of Hours Lost/yr})(1,000,000)}{\text{Number of Hours Worked/yr}}$$

$$= \frac{(57 \text{ Hr/yr})(1,000,000)}{97,120 \text{ Hr/yr}}$$

$$= 587$$

A.4 BASIC CONVERSION FACTORS (METRIC SYSTEM)

LENGTH

100 cm	= 1 m	100 cm/m
3.281 ft	= 1 m	3.281 ft/m

AREA

2.4711 ac	= 1 ha*	2.4711 ac/ha
10,000 sq m	= 1 ha	10,000 sq m/ha

VOLUME

1,000 mL	= 1 liter	1,000 mL/L
1,000 L	= 1 cu m	1,000 L/cu m
3.785 L	= 1 gal	3.785 L/gal

WEIGHT

1,000 mg	= 1 gm	1,000 mg/gm
1,000 gm	= 1 kg	1,000 gm/kg

DENSITY

1 kg	= 1 liter	1 kg/L

PRESSURE

10.015 m	= 1 kg/sq cm	10.015 m/kg/sq cm
1 Pascal	= 1 N/sq m	1 Pa/N/sq m
1 psi	= 6,895 Pa	1 psi/6,895 Pa

FLOW

3,785 cu m/day	= 1 MGD	3,785 cu m/day/MGD
3.785 ML/day	= 1 MGD	3.785 ML/day/MGD

* hectare

A.5 TYPICAL WATER TREATMENT PLANT PROBLEMS (METRIC SYSTEM)

A.50 Iron and Manganese Control

EXAMPLE 1

A standard polyphosphate solution is prepared by mixing and dissolving 1.0 gram of polyphosphate in a container and adding distilled water to the one-liter mark. Determine the concentration of the stock solution in milligrams per milliliter. If 6.0 milliliters of the stock solution are added to a one-liter sample, what is the polyphosphate dose in milligrams per liter and milligrams per kilogram?

Known			Unknown
Polyphosphate, gm	= 1.0 gm	1.	Stock Solution, mg/mL
Solution, L	= 1.0 L	2.	Dose, mg/L
Stock Solution, mL	= 6 mL	3.	Dose, mg/kg
Sample, L	= 1 L		

1. Calculate the concentration of the stock solution in milligrams per milliliter.

$$\text{Stock Solution, mg/m}L = \frac{(\text{Polyphosphate, gm})(1{,}000 \text{ mg/gm})}{(\text{Solution, }L)(1{,}000 \text{ m}L/L)}$$

$$= \frac{(1.0 \text{ gm})(1{,}000 \text{ mg/gm})}{(1 \text{ }L)(1{,}000 \text{ m}L/L)}$$

$$= 1.0 \text{ mg/m}L$$

2. Determine the polyphosphate dose in the sample in milligrams per liter.

$$\text{Dose, mg/}L = \frac{(\text{Stock Solution, mg/m}L)(\text{Vol Added, m}L)}{\text{Sample Volume, }L}$$

$$= \frac{(1.0 \text{ mg/m}L)(6 \text{ m}L)}{1 \text{ }L}$$

$$= 6.0 \text{ mg/}L$$

3. Determine the polyphosphate dose in the sample in milligrams of phosphate per kilogram of water.

$$\text{Dose, mg/kg} = \frac{(\text{Stock Solution, mg/}L)(\text{Vol Added, m}L)}{(\text{Sample Volume, }L)(1 \text{ kg/}L)}$$

$$= \frac{(1.0 \text{ mg/m}L)(6 \text{ m}L)}{(1 \text{ }L)(1 \text{ kg/}L)}$$

$$= 6.0 \text{ mg/kg}$$

EXAMPLE 2

Determine the chemical feeder setting in grams per second and kilograms per day if 2.4 MLD (mega or million liters per day) are treated with a dose of 5 mg/L.

Known	Unknown
Flow, MLD = 2.4 MLD	1. Chemical Feeder, gm/sec
Dose, mg/L = 5 mg/L	2. Chemical Feeder, kg/day

1. Determine the chemical feeder setting in grams per second.

$$\text{Chemical Feeder, gm/sec} = \frac{(\text{Flow, M}L\text{D})(\text{Dose, mg/}L)(1{,}000{,}000/\text{M})}{(24 \text{ hr/day})(60 \text{ min/hr})(60 \text{ sec/min})(1{,}000 \text{ mg/gm})}$$

$$= \frac{(2.4 \text{ M}L\text{D})(5 \text{ mg/}L)(1{,}000{,}000/\text{M})}{(24 \text{ hr/day})(60 \text{ min/hr})(60 \text{ sec/min})(1{,}000 \text{ mg/gm})}$$

$$= 0.139 \text{ gm/sec}$$

or = 139 mg/sec

2. Determine the chemical feeder setting in kilograms per day.

$$\text{Chemical Feeder, kg/day} = \frac{(\text{Flow, M}L\text{D})(\text{Dose, mg/}L)(1{,}000{,}000/\text{M})}{(1{,}000 \text{ mg/gm})(1{,}000 \text{ gm/kg})}$$

$$= \frac{(2.4 \text{ M}L\text{D})(5 \text{ mg/}L)(1{,}000{,}000/\text{M})}{(1{,}000 \text{ mg/gm})(1{,}000 \text{ gm/kg})}$$

$$= 12 \text{ kg/day}$$

EXAMPLE 3

A reaction basin 4 meters in diameter and 1.2 meters deep treats a flow of 0.9 MLD. What is the average detention time in minutes?

Known	Unknown
Diameter, m = 4 m	Detention Time, min
Depth, m = 1.2 m	
Flow, MLD = 0.9 MLD	

1. Calculate the basin volume in cubic meters.

$$\text{Basin Vol, cu m} = (0.785)(\text{Diameter, m})^2(\text{Depth, m})$$

$$= (0.785)(4 \text{ m})^2(1.2 \text{ m})$$

$$= 15.1 \text{ cu m}$$

2. Determine the average detention time in minutes for the reaction basin.

$$\text{Detention Time, min} = \frac{(\text{Basin Vol, cu m})(24 \text{ hr/day})(60 \text{ min/hr})(1{,}000 \text{ }L/\text{cu m})}{(\text{Flow, M}L\text{D})(1{,}000{,}000/\text{M})}$$

$$= \frac{(15.1 \text{ cu m})(24 \text{ hr/day})(60 \text{ min/hr})(1{,}000 \text{ }L/\text{cu m})}{(0.9 \text{ M}L\text{D})(1{,}000{,}000/\text{M})}$$

$$= 24 \text{ minutes}$$

EXAMPLE 4

Calculate the potassium permanganate dose in milligrams per liter for a well water with 2.4 mg/L iron before aeration and 0.3 mg/L after aeration. The manganese concentration is 0.8 mg/L both before and after aeration.

Known	Unknown
Iron, mg/L = 0.3 mg/L	KMnO$_4$ Dose, mg/L
Manganese, mg/L = 0.8 mg/L	

Calculate the potassium permanganate dose in milligrams per liter.

$$\text{KMnO}_4 \text{ Dose, mg}/L = 0.2(\text{Iron, mg}/L) + 2.0(\text{Manganese, mg}/L)$$

$$= 0.2(0.3 \text{ mg}/L) + 2.0(0.8 \text{ mg}/L)$$

$$= 1.66 \text{ mg}/L$$

NOTE: If there are any oxidizable compounds (organic color, bacteria, or hydrogen sulfide) in the water, the dose will have to be increased.

A.51 Fluoridation

EXAMPLE 5

Determine the setting for a chemical feed pump in liters per day and milliliters per second when the desired fluoride dose is 0.9 kilogram of fluoride per day. The sodium fluoride solution contains 0.025 kilogram of fluoride per liter and fluoride purity is 43.4 percent.

Known	Unknown
Feed Rate, kg F/day = 0.9 kg F/day	1. Feed Rate, liters/day
NaF Solution, kg F/L = 0.025 kg F/L	
Purity, % = 43.4%	2. Feed Rate, mL/sec

1. Determine the setting on the chemical feed pump in liters per day.

$$\text{Feed Rate, }L/\text{day} = \frac{(\text{Feed Rate, kg F/day})(100\%)}{(\text{NaF Solution, kg F}/L)(\text{Purity, \%})}$$

$$= \frac{(0.9 \text{ kg F/day})(100\%)}{(0.025 \text{ kg F}/L)(43.4\%)}$$

$$= 83 \text{ liters/day}$$

2. Convert the feed rate from kilograms per day to grams per second.

$$\text{Feed Rate, gm/sec} = \frac{(\text{Feed Rate, kg F/day})(1{,}000 \text{ gm/kg})}{(24 \text{ hr/day})(60 \text{ min/hr})(60 \text{ sec/min})}$$

$$= \frac{(0.9 \text{ kg F/day})(1{,}000 \text{ gm/kg})}{(24 \text{ hr/day})(60 \text{ min/hr})(60 \text{ sec/min})}$$

$$= 0.010 \text{ gm/sec}$$

$$\text{or} = 10 \text{ mg/sec}$$

3. Determine the setting on the chemical feed pump in milliliters per second.

$$\text{Feed Rate, m}L/\text{sec} = \frac{(\text{Feed Rate, gm/sec})(100\%)(1{,}000 \text{ m}L/L)}{(\text{NaF Solution, kg F}/L)(\text{Purity, \%})(1{,}000 \text{ gm/kg})}$$

$$= \frac{(0.010 \text{ gm/sec})(100\%)(1{,}000 \text{ m}L/L)}{(0.025 \text{ kg F}/L)(43.4\%)(1{,}000 \text{ gm/kg})}$$

$$= 0.92 \text{ m}L/\text{sec}$$

EXAMPLE 6

Determine the setting on a chemical feed pump in liters per day and milliliters per second if 2 megaliters per day of water must be treated with 0.9 mg/L of fluoride. The fluoride feed solution contains 18,000 mg/L of fluoride.

Known	Unknown
Flow, MLD = 2 MLD	1. Feed Pump, liters/day
Fluoride, mg/L = 0.9 mg/L	
Feed Solution, mg/L = 18,000 mg/L	2. Feed Pump, mL/sec

1. Determine the setting on the chemical feed pump in liters per day.

$$\text{Feed Pump, liters/day} = \frac{(\text{Flow, M}LD)(\text{Feed Dose, mg}/L)(1{,}000{,}000/\text{M})}{\text{Feed Solution, mg}/L}$$

$$= \frac{(2.0 \text{ M}LD)(0.9 \text{ mg}/L)(1{,}000{,}000/\text{M})}{18{,}000 \text{ mg}/L}$$

$$= 100 \text{ liters/day}$$

2. Determine the setting on the chemical feed pump in milliliters per second.

$$\text{Feed Pump, m}L/\text{sec} = \frac{(\text{Flow, M}LD)(\text{Feed Dose, mg}/L)(1{,}000{,}000/\text{M})}{(\text{Feed Solution, mg}/L)(24 \text{ hr/day})(60 \text{ min/hr})(60 \text{ sec/min})}$$

$$= \frac{(2.0 \text{ M}LD)(0.9 \text{ mg}/L)(1{,}000{,}000/\text{M})}{(18{,}000 \text{ mg}/L)(24 \text{ hr/day})(60 \text{ min/hr})(60 \text{ sec/min})}$$

$$= 1.16 \text{ m}L/\text{sec}$$

EXAMPLE 7

Determine the fluoride ion purity of Na$_2$SiF$_6$ as a percent.

Known	Unknown
Fluoride Chemical, Na$_2$SiF$_6$	Fluoride Ion Purity, %

Determine the molecular weight of fluoride and Na$_2$SiF$_6$.

Symbol	(No. Atoms)	(Atomic Wt)	=	Molecular Wt
Na$_2$	(2)	(22.99)	=	45.98
Si	(1)	(28.09)	=	28.09
F$_6$	(6)	(19.00)	=	114.00
	Molecular Weight of Chemical		=	188.07

Calculate the fluoride ion purity as a percent.

$$\text{Fluoride Ion Purity, \%} = \frac{(\text{Molecular Weight of Fluoride})(100\%)}{\text{Molecular Weight of Chemical}}$$

$$= \frac{(114.00)(100\%)}{188.07}$$

$$= 60.62\%$$

EXAMPLE 8

A flow of 6.5 MLD is treated with sodium silicofluoride. The raw water contains 0.2 mg/L of fluoride ion and the desired fluoride concentration is 1.1 mg/L. What should be the chemical feed rate in kilograms per day and milligrams per second? Assume each gram of commercial sodium silicofluoride (Na$_2$SiF$_6$) contains 0.6 gram of fluoride ion.

Known			Unknown
Flow, MLD	= 6.5 MLD	1.	Feed Rate, kg/day
Raw Water F, mg/L	= 0.2 mg/L		
Desired F, mg/L	= 1.1 mg/L	2.	Feed Rate, mg/sec
Chemical, gm F/gm	= 0.6 gm F/gm		

1. Determine the fluoride feed dose in milligrams per liter.

$$\text{Feed Dose, mg/}L = \text{Desired Dose, mg/}L - \text{Actual Conc, mg/}L$$

$$= 1.1 \text{ mg/}L - 0.2 \text{ mg/}L$$

$$= 0.9 \text{ mg/}L$$

2. Calculate the chemical feed rate in kilograms per day.

$$\text{Feed Rate, kg/day} = \frac{(\text{Flow, M}LD)(\text{Feed Dose, mg/}L)(1{,}000{,}000/\text{M})}{(\text{Purity, gm F/gm Chemical})(1{,}000 \text{ mg/gm})(1{,}000 \text{ gm/kg})}$$

$$= \frac{(6.5 \text{ M}LD)(0.9 \text{ mg/}L)(1{,}000{,}000/\text{M})}{(0.6 \text{ gm F/gm Chemical})(1{,}000 \text{ mg/gm})(1{,}000 \text{ gm/kg})}$$

$$= 9.75 \text{ kg/day}$$

3. Calculate the chemical feed rate in milligrams per second.

$$\text{Feed Rate, mg/sec} = \frac{(\text{Flow, M}LD)(\text{Feed Dose, mg/}L)(1{,}000{,}000/\text{M})}{(\text{Purity, gm F/gm Chemical})(24 \text{ hr/day})(60 \text{ min/hr})(60 \text{ sec/min})}$$

$$= \frac{(6.5 \text{ M}LD)(0.9 \text{ mg/}L)(1{,}000{,}000/\text{M})}{(0.6 \text{ gm F/gm Chemical})(24 \text{ hr/day})(60 \text{ min/hr})(60 \text{ sec/min})}$$

$$= 113 \text{ mg/sec}$$

EXAMPLE 9

The feed solution from a saturator containing 1.8 percent fluoride ion is used to treat a total flow of 0.95 megaliter (ML) of water. The raw water has a fluoride ion content of 0.2 mg/L and the desired fluoride level in the treated water is 0.9 mg/L. How many liters of feed solution are needed?

Known		Unknown
Flow Vol, ML	= 0.95 ML	Feed Solution, liters
Raw Water F, mg/L	= 0.2 mg/L	
Desired F, mg/L	= 0.9 mg/L	
Feed Solution, % F	= 1.8% F	

1. Convert the feed solution from a percentage fluoride ion to milligrams fluoride ion per liter of water.

$$1.0\% \text{ F} = 10{,}000 \text{ mg F/}L$$

$$\text{Feed Solution, mg/}L = \frac{(\text{Feed Solution, \%})(10{,}000 \text{ mg/}L)}{1.0\%}$$

$$= \frac{(1.8\% \text{ F})(10{,}000 \text{ mg/}L)}{1.0\%}$$

$$= 18{,}000 \text{ mg/}L$$

2. Determine the fluoride feed dose in milligrams per liter.

$$\text{Feed Dose, mg/}L = \text{Desired Dose, mg/}L - \text{Raw Water F, mg/}L$$

$$= 0.9 \text{ mg/}L - 0.2 \text{ mg/}L$$

$$= 0.7 \text{ mg/}L$$

3. Calculate the liters of feed solution needed.

$$\text{Feed Solution, }L = \frac{(\text{Flow Vol, M}L)(\text{Feed Dose, mg/}L)(1{,}000{,}000/\text{M})}{\text{Feed Solution, mg/}L}$$

$$= \frac{(0.95 \text{ M}L)(0.7 \text{ mg/}L)(1{,}000{,}000/\text{M})}{18{,}000 \text{ mg/}L}$$

$$= 37 \text{ liters}$$

EXAMPLE 10

A hydrofluosilicic acid (H$_2$SiF$_6$) tank contains 1,300 liters of acid with a strength of 19.3 percent. A commercial vendor delivers 10,000 liters of acid with a strength of 18.1 percent to the tank. What is the resulting strength of the mixture as a percentage?

Known		Unknown
Tank Contents, liters	= 1,300 L	Mixture Strength, %
Tank Strength, %	= 19.3%	
Vendor, L	= 10,000 L	
Vendor Strength, %	= 18.1%	

Calculate the strength of the mixture as a percentage.

$$\text{Mixture Strength, \%} = \frac{(\text{Tank, }L)(\text{Tank, \%}) + (\text{Vendor, }L)(\text{Vendor, \%})}{(\text{Tank, }L + \text{Vendor, }L)}$$

$$= \frac{(1{,}300 \text{ }L)(19.3\%) + (10{,}000 \text{ }L)(18.1\%)}{(1{,}300 \text{ }L + 10{,}000 \text{ }L)}$$

$$= \frac{25{,}090 + 181{,}000}{11{,}300}$$

$$= 18.2\%$$

A.52 Softening

EXAMPLE 11

Determine the total hardness as $CaCO_3$ for a sample of water with a calcium content of 33 mg/L and a magnesium content of 6 mg/L.

Known	Unknown
Calcium, mg/L = 33 mg/L	Total Hardness, mg/L as $CaCO_3$
Magnesium, mg/L = 6 mg/L	

Calculate the total hardness as milligrams per liter of calcium carbonate equivalent.

$$\text{Total Hardness, mg}/L \text{ as } CaCO_3 = \text{Calcium Hardness, mg}/L \text{ as } CaCO_3 + \text{Magnesium Hardness, mg}/L \text{ as } CaCO_3$$

$$= 2.5(\text{Ca, mg}/L) + 4.12(\text{Mg, mg}/L)$$

$$= 2.5(33 \text{ mg}/L) + 4.12(6 \text{ mg}/L)$$

$$= 82 \text{ mg}/L + 25 \text{ mg}/L$$

$$= 107 \text{ mg}/L \text{ as } CaCO_3$$

EXAMPLE 12

The alkalinity of a water is 120 mg/L as $CaCO_3$ and the total hardness is 105 mg/L as $CaCO_3$. What is the carbonate and noncarbonate hardness in mg/L as $CaCO_3$?

Known	Unknown
Alkalinity, mg/L = 120 mg/L as $CaCO_3$	1. Carbonate Hardness, mg/L as $CaCO_3$
Total Hardness, mg/L = 105 mg/L as $CaCO_3$	2. Noncarbonate Hardness, mg/L as $CaCO_3$

1. Determine the carbonate hardness in mg/L as $CaCO_3$.

    Since the alkalinity is greater than the total hardness (120 mg/L >105 mg/L),

    $$\text{Carbonate Hardness, mg}/L \text{ as } CaCO_3 = \text{Total Hardness, mg}/L \text{ as } CaCO_3$$

    $$= 105 \text{ mg}/L \text{ as } CaCO_3$$

2. Determine the noncarbonate hardness in mg/L as $CaCO_3$.

    Since the alkalinity is greater than the total hardness,

    Noncarbonate Hardness, mg/L as $CaCO_3$ = 0

    In other words, all of the hardness is in the carbonate form.

EXAMPLE 13

The alkalinity of a water is 92 mg/L as $CaCO_3$ and the total hardness is 105 mg/L. What is the carbonate and noncarbonate hardness in mg/L as $CaCO_3$?

Known	Unknown
Alkalinity, mg/L = 92 mg/L as $CaCO_3$	1. Carbonate Hardness, mg/L as $CaCO_3$
Total Hardness, mg/L = 105 mg/L as $CaCO_3$	2. Noncarbonate Hardness, mg/L as $CaCO_3$

1. Determine the carbonate hardness in mg/L as $CaCO_3$.

    Since the alkalinity is less than the total hardness (92 mg/L <105 mg/L),

    $$\text{Carbonate Hardness, mg}/L \text{ as } CaCO_3 = \text{Alkalinity, mg}/L \text{ as } CaCO_3$$

    $$= 92 \text{ mg}/L \text{ as } CaCO_3$$

2. Determine the noncarbonate hardness in mg/L as $CaCO_3$.

    Since the alkalinity is less than the total hardness (92 mg/L <105 mg/L),

    $$\text{Noncarbonate Hardness, mg}/L \text{ as } CaCO_3 = \text{Total Hardness, mg}/L \text{ as } CaCO_3 - \text{Alkalinity, mg}/L \text{ as } CaCO_3$$

    $$= 105 \text{ mg}/L - 92 \text{ mg}/L$$

    $$= 13 \text{ mg}/L \text{ as } CaCO_3$$

EXAMPLE 14

Results from alkalinity titrations on a water sample were as follows:

Known

Sample Size, mL	= 100 mL
mL Titrant Used to pH 8.3, A	= 1.1 mL
Total mL of Titrant Used, B	= 12.4 mL
Acid Normality, N	= 0.02 N H_2SO_4

Unknown

1. Total Alkalinity, mg/L as $CaCO_3$
2. Bicarbonate Alkalinity, mg/L as $CaCO_3$
3. Carbonate Alkalinity, mg/L as $CaCO_3$
4. Hydroxide Alkalinity, mg/L as $CaCO_3$

See Table 14.4, page 81, for alkalinity relationships among constituents.

1. Calculate the phenolphthalein alkalinity in mg/L as $CaCO_3$.

    $$\text{Phenolphthalein Alkalinity, mg}/L \text{ as } CaCO_3 = \frac{A \times N \times 50{,}000}{\text{m}L \text{ of Sample}}$$

    $$= \frac{(1.1 \text{ m}L)(0.02 \text{ } N)(50{,}000)}{100 \text{ m}L}$$

    $$= 11 \text{ mg}/L \text{ as } CaCO_3$$

2. Calculate the total alkalinity in mg/L as $CaCO_3$.

$$\text{Total Alkalinity, mg}/L \text{ as } CaCO_3 = \frac{B \times N \times 50{,}000}{mL \text{ of Sample}}$$

$$= \frac{(12.4 \text{ m}L)(0.02 \text{ }N)(50{,}000)}{100 \text{ m}L}$$

$$= 124 \text{ mg}/L \text{ as } CaCO_3$$

3. Refer to Table 14.4 for alkalinity constituents. The second row indicates that since P is less than ½T (11 mg/L < ½(124 mg/L)), bicarbonate alkalinity is T − 2P and carbonate alkalinity is 2P.

$$\text{Bicarbonate Alkalinity, mg}/L \text{ as } CaCO_3 = T - 2P$$

$$= 124 \text{ mg}/L - 2(11 \text{ mg}/L)$$

$$= 102 \text{ mg}/L \text{ as } CaCO_3$$

$$\text{Carbonate Alkalinity, mg}/L \text{ as } CaCO_3 = 2P$$

$$= 2(11 \text{ mg}/L)$$

$$= 22 \text{ mg}/L \text{ as } CaCO_3$$

$$\text{Hydroxide Alkalinity, mg}/L \text{ as } CaCO_3 = 0 \text{ mg}/L \text{ as } CaCO_3$$

EXAMPLE 15

Calculate the hydrated lime ($Ca(OH)_2$) with 90 percent purity, soda ash, and carbon dioxide requirements in milligrams per liter for the water shown below.

Known

Constituents	Source Water	Softened Water After Recarbonation and Filtration
CO_2, mg/L	= 7 mg/L	= 0 mg/L
Total Alkalinity, mg/L	= 125 mg/L as $CaCO_3$	= 22 mg/L as $CaCO_3$
Total Hardness, mg/L	= 240 mg/L as $CaCO_3$	= 35 mg/L as $CaCO_3$
Mg^{2+}, mg/L	= 38 mg/L	= 8 mg/L
pH	= 7.6	= 8.8
Lime Purity, %	= 90%	

Unknown

1. Hydrated Lime, mg/L
2. Soda Ash, mg/L
3. Carbon Dioxide, mg/L

1. Calculate the hydrated lime ($Ca(OH)_2$) required in milligrams per liter.

$$A = (CO_2, \text{mg}/L)(74/44)$$

$$= (7 \text{ mg}/L)(74/44)$$

$$= 12 \text{ mg}/L$$

$$B = (\text{Alkalinity, mg}/L)(74/100)$$

$$= (125 \text{ mg}/L - 22 \text{ mg}/L)(74/100)$$

$$= 76 \text{ mg}/L$$

$$C = 0 \qquad \text{Hydroxide Alkalinity} = 0$$

$$D = (Mg^{2+}, \text{mg}/L)(74/24.3)$$

$$= (38 \text{ mg}/L - 8 \text{ mg}/L)(74/24.3)$$

$$= 91 \text{ mg}/L$$

$$\text{Hydrated Lime } (Ca(OH)_2) \text{ Feed, mg}/L = \frac{(A + B + C + D)1.15}{\text{Purity of Lime, as a decimal}}$$

$$= \frac{(12 \text{ mg}/L + 76 \text{ mg}/L + 0 + 91 \text{ mg}/L)1.15}{0.90}$$

$$= \frac{(179 \text{ mg}/L)(1.15)}{0.90}$$

$$= 229 \text{ mg}/L$$

2. Calculate the soda ash required in milligrams per liter.

$$\text{Total Hardness Removed, mg}/L \text{ as } CaCO_3 = \text{Total Hardness, mg}/L \text{ as } CaCO_3 - \text{Total Hardness Remaining, mg}/L \text{ as } CaCO_3$$

$$= 240 \text{ mg}/L - 35 \text{ mg}/L$$

$$= 205 \text{ mg}/L \text{ as } CaCO_3$$

$$\text{Noncarbonate Hardness, mg}/L \text{ as } CaCO_3 = \text{Total Hardness Removed, mg}/L \text{ as } CaCO_3 - \left(\text{Carbonate Hardness, mg}/L \text{ as } CaCO_3 - \text{Carbonate Hardness Remaining, mg}/L \text{ as } CaCO_3 \right)$$

$$= 205 \text{ mg}/L - (125 \text{ mg}/L - 22 \text{ mg}/L)$$

$$= 102 \text{ mg}/L \text{ as } CaCO_3$$

$$\text{Soda Ash } (Na_2CO_3) \text{ Feed, mg}/L = \left(\text{Noncarbonate Hardness, mg}/L \text{ as } CaCO_3 \right)(106/100)$$

$$= (102 \text{ mg}/L)(106/100)$$

$$= 108 \text{ mg}/L$$

3. Calculate the dosage of carbon dioxide required for recarbonation.

$$\text{Excess Lime, mg}/L = (A + B + C + D)(0.15)$$

$$= (12 \text{ mg}/L + 93 \text{ mg}/L + 0 + 116 \text{ mg}/L)(0.15)$$

$$= (221 \text{ mg}/L)(0.15)$$

$$= 33 \text{ mg}/L$$

$$\text{Total } CO_2 \text{ Feed, mg}/L = (Ca(OH)_2 \text{ Excess, mg}/L)(44/74) + (Mg^{2+} \text{ Residual, mg}/L)(44/24.3)$$

$$= (33 \text{ mg}/L)(44/74) + (8 \text{ mg}/L)(44/24.3)$$

$$= 20 \text{ mg}/L + 15 \text{ mg}/L$$

$$= 35 \text{ mg}/L$$

EXAMPLE 16

The optimum lime dosage from the jar tests is 180 mg/L. If the flow to be treated is 6.5 MLD, what is the feeder setting in kilograms per day and the feed rate in grams per second?

Known	Unknown
Lime Dose, mg/L = 180 mg/L	1. Feeder Setting, kg/day
Flow, MLD = 6.5 MLD	2. Feed Rate, gm/sec

1. Calculate the feeder setting in kilograms per day.

$$\text{Feeder Setting, kg/day} = \frac{(\text{Flow, } MLD)(\text{Lime, mg/}L)(1,000,000/M)}{(1,000 \text{ mg/gm})(1,000 \text{ gm/kg})}$$

$$= \frac{(6.5 \text{ } MLD)(180 \text{ mg/}L)(1,000,000/M)}{(1,000 \text{ mg/gm})(1,000 \text{ gm/kg})}$$

$$= 1{,}170 \text{ kg/day}$$

2. Calculate the feed rate in grams per second.

$$\text{Feed Rate, gm/sec} = \frac{(\text{Flow, } MLD)(\text{Lime, mg/}L)(1,000,000/M)}{(1,000 \text{ mg/gm})(24 \text{ hr/day})(60 \text{ min/hr})(60 \text{ sec/min})}$$

$$= \frac{(6.5 \text{ } MLD)(180 \text{ mg/}L)(1,000,000/M)}{(1,000 \text{ mg/gm})(24 \text{ hr/day})(60 \text{ min/hr})(60 \text{ sec/min})}$$

$$= 13.5 \text{ gm/sec}$$

EXAMPLE 17

How much soda ash is required (kilograms per day and grams per second) to remove 40 mg/L noncarbonate hardness as $CaCO_3$ from a flow of 6.5 MLD?

Known	Unknown
Noncarbonate Hardness Removed, mg/L as $CaCO_3$ = 40 mg/L	1. Feeder Setting, kg/day
Flow, MLD = 6.5 MLD	2. Feed Rate, gm/sec

1. Calculate the soda ash dose in milligrams per liter. See Section 14.316, "Calculation of Chemical Dosages," page 85, for the following formula.

$$\text{Soda Ash, mg/}L = \left(\begin{array}{c}\text{Noncarbonate Hardness,} \\ \text{mg/}L \text{ as } CaCO_3\end{array}\right)(106/100)$$

$$= (40 \text{ mg/}L)(106/100)$$

$$= 42.4 \text{ mg/}L$$

2. Determine the feeder setting in kilograms per day.

$$\text{Feeder Setting, kg/day} = \frac{(\text{Flow, } MLD)(\text{Soda Ash, mg/}L)(1,000,000/M)}{(1,000 \text{ mg/gm})(1,000 \text{ gm/kg})}$$

$$= \frac{(6.5 \text{ } MLD)(42.4 \text{ mg/}L)(1,000,000/M)}{(1,000 \text{ mg/gm})(1,000 \text{ gm/kg})}$$

$$= 276 \text{ kg/day}$$

3. Calculate the soda ash feed rate in grams per second.

$$\text{Feed Rate, gm/sec} = \frac{(\text{Flow, } MLD)(\text{Soda Ash, mg/}L)(1,000,000/M)}{(1,000 \text{ mg/gm})(24 \text{ hr/day})(60 \text{ min/hr})(60 \text{ sec/min})}$$

$$= \frac{(6.5 \text{ } MLD)(42.4 \text{ mg/}L)(1,000,000/M)}{(1,000 \text{ mg/gm})(24 \text{ hr/day})(60 \text{ min/hr})(60 \text{ sec/min})}$$

$$= 3.2 \text{ gm/sec}$$

EXAMPLE 18

What is the hardness in grains per gallon for a water with a hardness of 200 mg/L?

Known	Unknown
Hardness, mg/L = 200 mg/L	Hardness, grains/gal

Calculate the hardness in grains per gallon.

$$\text{Hardness, grains/gal} = \frac{(\text{Hardness, mg/}L)(1 \text{ grain/gal})}{17.1 \text{ mg/}L}$$

$$= \frac{(200 \text{ mg/}L)(1 \text{ grain/gal})}{17.1 \text{ mg/}L}$$

$$= 11.7 \text{ grains/gal}$$

EXAMPLE 19

Estimate the exchange capacity in milligrams of hardness for an ion exchange unit that contains 20 cubic meters of resin with a removal capacity of 14,000 milligrams per cubic meter.

Known	Unknown
Resin Volume, cu m = 20 cu m	Exchange Capacity, milligrams
Removal Capacity, mg/cu m = 14,000 mg/cu m	

Estimate the exchange capacity in milligrams of hardness.

$$\text{Exchange Capacity, mg} = (\text{Resin Volume, cu m})(\text{Removal Capacity, mg/cu m})$$

$$= (20 \text{ cu m})(14,000 \text{ mg/cu m})$$

$$= 280{,}000 \text{ mg of hardness}$$

EXAMPLE 20

How many liters of water with a hardness of 200 mg/L may be treated by an ion exchange softener with an exchange capacity of 280,000 milligrams?

Known	Unknown
Hardness, mg/L = 200 mg/L	Water Treated, liters
Exchange Capacity, mg = 280,000 mg	

Calculate the liters of water that may be treated.

$$\text{Water Treated, liters} = \frac{\text{Exchange Capacity, mg}}{\text{Hardness, mg/}L}$$

$$= \frac{280{,}000 \text{ mg}}{200 \text{ mg/}L}$$

$$= 1{,}400 \text{ liters}$$

EXAMPLE 21

How many hours will an ion exchange softening unit operate when treating an average daily flow of 50 liters per second? The unit is capable of softening 4,500,000 liters of water before requiring regeneration.

Known	Unknown
Avg Daily Flow, L/sec = 50 L/sec	Operating Time, hr
Water Treated, L = 4,500,000 L	

Estimate how many hours the softening unit can operate before requiring regeneration.

$$\text{Operating Time, hr} = \frac{\text{Water Treated, } L}{(\text{Avg Daily Flow, } L/\text{sec})(60 \text{ sec/min})(60 \text{ min/hr})}$$

$$= \frac{4{,}500{,}000 \; L}{(50 \; L/\text{sec})(60 \text{ sec/min})(60 \text{ min/hr})}$$

$$= 25 \text{ hours}$$

EXAMPLE 22

Determine the kilograms of salt needed to regenerate an ion exchange softening unit capable of removing 225,000 milligrams of hardness if 7 kilograms of salt are required for every 1,000 milligrams of hardness removed.

Known	Unknown
Hardness Removed, mg = 225,000 mg	Salt Needed, kg
Salt Required, kg/1,000 mg = 7 kg Salt/1,000 mg	

Calculate the kilograms of salt needed to regenerate the ion exchange softening unit.

$$\text{Salt Needed, kg} = (\text{Salt Required, kg/1,000 mg})(\text{Hardness Removed, mg})$$

$$= \frac{(7 \text{ kg Salt})(225{,}000 \text{ mg})}{1{,}000 \text{ mg}}$$

$$= 1{,}575 \text{ kilograms of Salt}$$

EXAMPLE 23

Estimate the bypass flow in cubic meters per day and megaliters per day around an ion exchange softener in a plant that treats 1,000 cubic meters per day with a source water hardness of 350 mg/L if the desired product water hardness is 80 mg/L.

Known	Unknown
Total Flow, cu m/day = 1,000 cu m/day	1. Bypass Flow, cu m/day
Source Water Hardness, mg/L = 350 mg/L	2. Bypass Flow, MLD
Plant Effl Hardness, mg/L = 80 mg/L	

1. Estimate the bypass flow in cubic meters per day.

$$\text{Bypass Flow, cu m/day} = \frac{(\text{Total Flow, cu m/day})(\text{Plant Effl Hardness, mg/}L)}{\text{Source Water Hardness, mg/}L}$$

$$= \frac{(1{,}000 \text{ cu m/day})(80 \text{ mg/}L)}{350 \text{ mg/}L}$$

$$= 229 \text{ cu m/day}$$

2. Estimate the bypass flow in megaliters per day.

$$\text{Bypass Flow, M}LD = \frac{(\text{Total Flow, cu m/day})(\text{Plant Effl Hardness, mg/}L)(1{,}000 \; L/\text{cu m})}{(\text{Source Water Hardness, mg/}L)(1{,}000{,}000/\text{M})}$$

$$= \frac{(1{,}000 \text{ cu m/day})(80 \text{ mg/}L)(1{,}000 \; L/\text{cu m})}{(350 \text{ mg/}L)(1{,}000{,}000/\text{M})}$$

$$= 0.229 \; \text{M}LD$$

A.53 Specialized Treatment Processes

EXAMPLE 24

A water utility collected and analyzed eight samples from a water distribution system on the same day for TTHMs. The results are shown below.

Sample No.	1	2	3	4	5	6	7	8
TTHM, µg/L	80	90	100	90	110	100	100	90

What was the average TTHM for the day?

Known	Unknown
Results from analyses of 8 TTHM samples	Average TTHM level for the day

Calculate the average TTHM level in micrograms per liter.

$$\text{Avg TTHM, µg/}L = \frac{\text{Sum of Measurements, µg/}L}{\text{Number of Measurements}}$$

$$= \frac{80 \text{ µg/}L + 90 \text{ µg/}L + 100 \text{ µg/}L + 90 \text{ µg/}L + 110 \text{ µg/}L + 100 \text{ µg/}L + 100 \text{ µg/}L + 90 \text{ µg/}L}{8}$$

$$= \frac{760 \text{ µg/}L}{8}$$

$$= 95 \text{ µg/}L$$

756 Water Treatment

EXAMPLE 25

The results of the quarterly average TTHM measurements for two years are given below. Calculate the running annual average of the four quarterly measurements in micrograms per liter.

Quarter	1	2	3	4	1	2	3	4
Avg Quarterly TTHM, µg/L	77	88	112	95	83	87	109	89

Known	Unknown
Results from analyses of two years of TTHM sampling	Running Annual Average of quarterly TTHM measurements

Calculate the running annual average of the quarterly TTHM measurements.

$$\text{Annual Running TTHM Average, µg}/L = \frac{\text{Sum of Average TTHM for Four Quarters}}{\text{Number of Quarters}}$$

QUARTERS 1, 2, 3, AND 4

$$\text{Annual Running TTHM Average, µg}/L = \frac{77 \text{ µg}/L + 88 \text{ µg}/L + 112 \text{ µg}/L + 95 \text{ µg}/L}{4}$$

$$= \frac{372 \text{ µg}/L}{4}$$

$$= 93 \text{ µg}/L$$

QUARTERS 2, 3, 4, AND 1

$$\text{Annual Running TTHM Average, µg}/L = \frac{88 \text{ µg}/L + 112 \text{ µg}/L + 95 \text{ µg}/L + 83 \text{ µg}/L}{4}$$

$$= \frac{378 \text{ µg}/L}{4}$$

$$= 95 \text{ µg}/L$$

QUARTERS 3, 4, 1, AND 2

$$\text{Annual Running TTHM Average, µg}/L = \frac{112 \text{ µg}/L + 95 \text{ µg}/L + 83 \text{ µg}/L + 87 \text{ µg}/L}{4}$$

$$= \frac{377 \text{ µg}/L}{4}$$

$$= 94 \text{ µg}/L$$

QUARTERS 4, 1, 2, AND 3

$$\text{Annual Running TTHM Average, µg}/L = \frac{95 \text{ µg}/L + 83 \text{ µg}/L + 87 \text{ µg}/L + 109 \text{ µg}/L}{4}$$

$$= \frac{374 \text{ µg}/L}{4}$$

$$= 94 \text{ µg}/L$$

QUARTERS 1, 2, 3, AND 4

$$\text{Annual Running TTHM Average, µg}/L = \frac{83 \text{ µg}/L + 87 \text{ µg}/L + 109 \text{ µg}/L + 89 \text{ µg}/L}{4}$$

$$= \frac{368 \text{ µg}/L}{4}$$

$$= 92 \text{ µg}/L$$

SUMMARY OF RESULTS

Quarter	1	2	3	4	1	2	3	4
Avg Quarterly TTHM, µg/L	77	88	112	95	83	87	109	89
Annual Running TTHM Avg, µg/L				93	95	94	94	92

EXAMPLE 26

Well number 1 delivers a flow of 0.25 MLD with an arsenic concentration of 0.024 mg/L. Well number 2 delivers a flow of 2.50 MLD with an arsenic concentration of 0.002 mg/L. What would be the blended arsenic concentration in milligrams per liter if the flows from these two wells were blended together?

Known	Unknown
Well 1 Q, MLD = 0.25 MLD	Blended Conc, mg/L
Well 2 Q, MLD = 2.50 MLD	
Well 1 Conc, mg/L = 0.024 mg/L	
Well 2 Conc, mg/L = 0.002 mg/L	

Calculate the blended arsenic concentration in milligrams per liter.

$$\text{Blended Conc, mg}/L = \frac{(\text{Flow 1, M}LD)(\text{Conc 1, mg}/L) + (\text{Flow 2, M}LD)(\text{Conc 2, mg}/L)}{\text{Flow 1, M}LD + \text{Flow 2, M}LD}$$

$$= \frac{(0.25 \text{ M}LD)(0.024 \text{ mg}/L) + (2.50 \text{ M}LD)(0.002 \text{ mg}/L)}{0.25 \text{ M}LD + 2.50 \text{ M}LD}$$

$$= \frac{0.006 + 0.005}{2.75}$$

$$= \frac{0.011}{2.75}$$

$$= 0.004 \text{ mg}/L$$

A.54 Membrane Treatment Processes

EXAMPLE 27

Convert a water flux of 12×10^{-4} gm/sq cm-sec to liters per second per square centimeter and liters per day per square centimeter.

Known	Unknown
Water Flux, gm/sq cm-sec = 12×10^{-4} gm/sq cm-sec	1. Flow, liters per sq cm-sec
	2. Flow, liters per sq cm-day

1. Convert the water flux from gm/sq cm-sec to flow in liters per second per square centimeter.

$$\text{Flow, }L/\text{sq cm-sec} = \frac{\text{Flux, gm/sq cm-sec}}{1{,}000 \text{ gm}/L}$$

$$= \frac{12 \times 10^{-4} \text{ gm/sq cm-sec}}{1{,}000 \text{ m}L}$$

$$= 12 \times 10^{-7} \text{ }L/\text{sq cm-sec}$$

2. Convert the water flux from gm/sq cm-sec to flow in liters per day per square centimeter.

$$\text{Flow, } L/\text{sq cm-day} = \frac{(\text{Flux, gm/sq cm-sec})(60 \text{ sec/min})(60 \text{ min/hr})(24 \text{ hr/day})}{1{,}000 \text{ gm}/L}$$

$$= \frac{(0.0012 \text{ gm/sq cm-sec})(60 \text{ sec/min})(60 \text{ min/hr})(24 \text{ hr/day})}{1{,}000 \text{ gm}/L}$$

$$= 0.10 \ L/\text{sq cm-day}$$

EXAMPLE 28

Estimate the ability of a reverse osmosis plant to reject minerals by calculating the mineral rejection as a percent. The feedwater contains 1,800 mg/L TDS and the product water TDS is 120 mg/L.

Known	Unknown
Feedwater TDS, mg/L = 1,800 mg/L	Mineral Rejection, %
Product Water TDS, mg/L = 120 mg/L	

Calculate the mineral rejection as a percent.

$$\text{Mineral Rejection, \%} = \left(1 - \frac{\text{Product TDS, mg}/L}{\text{Feed TDS, mg}/L}\right)(100\%)$$

$$= \left(1 - \frac{120 \text{ mg}/L}{1{,}800 \text{ mg}/L}\right)(100\%)$$

$$= (1 - 0.067)(100\%)$$

$$= 93.3\%$$

EXAMPLE 29

Estimate the percent recovery of a reverse osmosis unit with a 4-2-1 arrangement if the feed flow is 8.0 MLD and the product flow is 7.0 MLD.

Known	Unknown
Product Flow, MLD = 7.0 MLD	Recovery, %
Feed Flow, MLD = 8.0 MLD	

Calculate the recovery as a percent.

$$\text{Recovery, \%} = \frac{(\text{Product Flow, M}LD)(100\%)}{\text{Feed Flow, M}LD}$$

$$= \frac{(7.0 \text{ M}LD)(100\%)}{8.0 \text{ M}LD}$$

$$= 87.5\%$$

A.55 Maintenance

EXAMPLE 30

Calculate the pumping capacity of a pump in liters per second when 12 minutes are required for the water to rise 1.0 meter in a 2.5-meter by 2.0-meter rectangular tank.

Known	Unknown
Length, m = 2.5 m	Pump Capacity, L/sec
Width, m = 2.0 m	
Depth, m = 1.0 m	
Time, min = 12 min	

1. Calculate the volume pumped in cubic meters.

$$\text{Volume Pumped, cu m} = (\text{Length, m})(\text{Width, m})(\text{Depth, m})$$

$$= (2.5 \text{ m})(2.0 \text{ m})(1.0 \text{ m})$$

$$= 5.0 \text{ cu m}$$

2. Calculate the pump capacity in liters per second.

$$\text{Pump Capacity, liters/sec} = \frac{(\text{Volume Pumped, cu m})(1{,}000 \ L/\text{cu m})}{(\text{Pumping Time, min})(60 \text{ sec/min})}$$

$$= \frac{(5.0 \text{ cu m})(1{,}000 \ L/\text{cu m})}{(12 \text{ min})(60 \text{ sec/min})}$$

$$= 417 \ L/\text{sec}$$

EXAMPLE 31

A small chemical feed pump lowered the chemical solution in a 0.8-meter diameter tank 0.7 meter during 7.0 hours. Estimate the flow delivered by the pump in liters per second and milliliters per second.

Known	Unknown
Tank Diameter, m = 0.8 m	1. Flow, L/sec
Chemical Drop, m = 0.7 m	2. Flow, mL/sec
Time, hr = 7.0 hr	

1. Determine the liters of chemical solution pumped.

$$\text{Volume, liters} = (0.785)(\text{Diameter, m})^2(\text{Drop, m})(1{,}000 \ L/\text{cu m})$$

$$= (0.785)(0.8 \text{ m})^2(0.7 \text{ m})(1{,}000 \ L/\text{cu m})$$

$$= 352 \text{ liters}$$

2. Estimate the flow delivered by the pump in liters per second.

$$\text{Flow, } L/\text{sec} = \frac{\text{Volume Pumped, } L}{(\text{Pumping Time, hr})(60 \text{ min/hr})(60 \text{ sec/min})}$$

$$= \frac{352 \ L}{(7.0 \text{ hr})(60 \text{ min/hr})(60 \text{ sec/min})}$$

$$= 0.014 \ L/\text{sec}$$

3. Estimate the flow delivered by the pump in milliliters per second.

$$\text{Flow, m}L/\text{sec} = \frac{(\text{Volume Pumped, } L)(1{,}000 \text{ m}L/L)}{(\text{Pumping Time, hr})(60 \text{ sec/min})}$$

$$= \frac{(352 \ L)(1{,}000 \text{ m}L/L)}{(7.0 \text{ hr})(60 \text{ min/hr})(60 \text{ sec/min})}$$

$$= 14 \text{ m}L/\text{sec}$$

EXAMPLE 32

Determine the chemical feed in kilograms of polymer per day and grams per second from a chemical feed pump. The polymer solution is 1.8 percent or 18,000 mg polymer per liter. Assume a specific gravity of the polymer solution of 1.0. During a test run, the chemical feed pump delivered 650 mL of polymer solution in 4.5 minutes.

Known	Unknown
Polymer Solution, % = 1.8%	1. Polymer Feed, kg/day
Polymer Conc, mg/L = 18,000 mg/L	2. Polymer Feed, gm/sec
Polymer Sp Gr = 1.0	
Volume Pumped, mL = 650 mL	
Time Pumped, min = 4.5 min	

1. Calculate the polymer fed by the chemical feed pump in kilograms of polymer per day.

$$\frac{\text{Polymer Feed,}}{\text{kg/day}} = \frac{(\text{Vol Pumped, m}L)(\text{Poly Conc, mg}/L)(60 \text{ min/hr})(24 \text{ hr/day})}{(\text{Time Pumped, min})(1{,}000 \text{ m}L/L)(1{,}000 \text{ mg/gm})(1{,}000 \text{ gm/kg})}$$

$$= \frac{(650 \text{ m}L)(18{,}000 \text{ mg}/L)(60 \text{ min/hr})(24 \text{ hr/day})}{(4.5 \text{ min})(1{,}000 \text{ m}L/L)(1{,}000 \text{ mg/gm})(1{,}000 \text{ gm/kg})}$$

$$= 3.7 \text{ kg/day}$$

2. Calculate the polymer fed by the chemical feed pump in grams of polymer per second.

$$\frac{\text{Polymer Feed,}}{\text{gm/sec}} = \frac{(\text{Vol Pumped, m}L)(\text{Poly Conc, mg}/L)}{(\text{Time Pumped, min})(1{,}000 \text{ m}L/L)(60 \text{ sec/min})(1{,}000 \text{ mg/gm})}$$

$$= \frac{(650 \text{ m}L)(18{,}000 \text{ mg}/L)}{(4.5 \text{ min})(1{,}000 \text{ m}L/L)(60 \text{ sec/min})(1{,}000 \text{ mg/gm})}$$

$$= 0.043 \text{ gm/sec}$$

or = 43 mg/sec

EXAMPLE 33

Determine the actual chemical feed in kilograms per day and grams per second from a dry chemical feeder. A pie tin placed under the chemical feeder caught 824 grams of chemical during five minutes.

Known	Unknown
Chemical, gm = 824 gm	1. Chemical Feed, kg/day
Time, min = 5 min	2. Chemical Feed, gm/sec

1. Determine the chemical feed in kilograms per day.

$$\frac{\text{Chemical Feed,}}{\text{kg/day}} = \frac{(\text{Chemical, gm})(60 \text{ min/hr})(24 \text{ hr/day})}{(\text{Time, min})(1{,}000 \text{ gm/kg})}$$

$$= \frac{(824 \text{ gm})(60 \text{ min/hr})(24 \text{ hr/day})}{(5 \text{ min})(1{,}000 \text{ gm/kg})}$$

$$= 237 \text{ kg/day}$$

2. Determine the chemical feed in grams per second.

$$\frac{\text{Chemical Feed,}}{\text{gm/sec}} = \frac{\text{Chemical, gm}}{(\text{Time, min})(60 \text{ sec/min})}$$

$$= \frac{824 \text{ gm}}{(5 \text{ min})(60 \text{ sec/min})}$$

$$= 2.75 \text{ gm/sec}$$

A.56 Advanced Laboratory Procedures

EXAMPLE 34

Calculate the threshold odor number (TON) for a sample when the first detectable odor occurred when the 70-mL sample was diluted to 200 mL (130 mL of odor-free water was added to the 70-mL sample).

Known	Unknown
Size of Sample, mL = 70 mL	TON
Odor-Free Water, mL = 130 mL	

Calculate the threshold odor number (TON).

$$\text{TON} = \frac{\text{Size of Sample, m}L + \text{Odor-Free Water, m}L}{\text{Size of Sample, m}L}$$

$$= \frac{70 \text{ m}L + 130 \text{ m}L}{70 \text{ m}L}$$

$$= 3$$

EXAMPLE 35

Determine the geometric mean threshold odor number for a panel of six testers given the results shown below.

Known	Unknown
Tester 1, X_1 = 2	Geometric Mean Threshold Odor Number
Tester 2, X_2 = 4	
Tester 3, X_3 = 3	
Tester 4, X_4 = 8	
Tester 5, X_5 = 6	
Tester 6, X_6 = 2	

Calculate the geometric mean.

$$\text{Geometric Mean TON} = (X_1 \times X_2 \times X_3 \times X_4 \times X_5 \times X_6)^{1/n}$$

$$= (2 \times 4 \times 3 \times 8 \times 6 \times 2)^{1/6}$$

$$= (2{,}304)^{0.167}$$

$$= 3.6$$

EXAMPLE 36

Calculate the threshold taste number for a sample when the first detectable taste occurred when the 8.3-mL sample was diluted to 200 mL (191.7 mL of taste-free water was added to the 8.3-mL sample).

Known	Unknown
Size of Sample, mL = 8.3 mL	Threshold Taste Number
Taste-Free Water, mL = 191.7 mL	

Calculate the threshold taste number.

$$\frac{\text{Threshold Taste}}{\text{Number}} = \frac{\text{Size of Sample, m}L + \text{Taste-Free Water, m}L}{\text{Size of Sample, m}L}$$

$$= \frac{8.3 \text{ m}L + 191.7 \text{ m}L}{8.3 \text{ m}L}$$

$$= 24$$

EXAMPLE 37

Determine the taste rating for a water by calculating the arithmetic mean and standard deviation for the panel ratings given below.

Known	Unknown
Tester 1, $X_1 = 2$	1. Arithmetic Mean, $\overline{X}$
Tester 2, $X_2 = 5$	2. Standard Deviation, S
Tester 3, $X_3 = 3$	
Tester 4, $X_4 = 6$	
Tester 5, $X_5 = 2$	
Tester 6, $X_6 = 6$	

1. Calculate the arithmetic mean, $\overline{X}$, taste rating.

$$\text{Arithmetic Mean, } \overline{X}, \text{ Taste Rating} = \frac{X_1 + X_2 + X_3 + X_4 + X_5 + X_6}{n}$$

$$= \frac{2 + 5 + 3 + 6 + 2 + 6}{6}$$

$$= \frac{24}{6}$$

$$= 4$$

2. Calculate the standard deviation, S, of the taste rating.

$$\text{Standard Deviation, S} = \left[\frac{(X_1 - \overline{X})^2 + (X_2 - \overline{X})^2 + (X_3 - \overline{X})^2 + (X_4 - \overline{X})^2 + (X_5 - \overline{X})^2 + (X_6 - \overline{X})^2}{n - 1}\right]^{0.5}$$

$$= \left[\frac{(2-4)^2 + (5-4)^2 + (3-4)^2 + (6-4)^2 + (2-4)^2 + (6-4)^2}{6-1}\right]^{0.5}$$

$$= \left[\frac{(-2)^2 + (1)^2 + (-1)^2 + (2)^2 + (-2)^2 + (2)^2}{5}\right]^{0.5}$$

$$= \left[\frac{4 + 1 + 1 + 4 + 4 + 4}{5}\right]^{0.5}$$

$$= \left[\frac{18}{5}\right]^{0.5}$$

$$= (3.6)^{0.5}$$

$$= 1.9$$

or

$$\text{Standard Deviation, S} = \left[\frac{(X_1^2 + X_2^2 + X_3^2 + X_4^2 + X_5^2 + X_6^2) - (X_1 + X_2 + X_3 + X_4 + X_5 + X_6)^2/n}{n-1}\right]^{0.5}$$

$$= \left[\frac{(2^2 + 5^2 + 3^2 + 6^2 + 2^2 + 6^2) - (2 + 5 + 3 + 6 + 2 + 6)^2/6}{6-1}\right]^{0.5}$$

$$= \left[\frac{(4 + 25 + 9 + 36 + 4 + 36) - (24)^2/6}{5}\right]^{0.5}$$

$$= \left[\frac{114 - 96}{5}\right]^{0.5}$$

$$= \left[\frac{18}{5}\right]^{0.5}$$

$$= (3.6)^{0.5}$$

$$= 1.9$$

A.57 Regulations

EXAMPLE 38

A small water system collected 14 samples during one month. After each sample was collected, 10 mL of each sample was placed in each of 5 fermentation tubes. At the end of the month, the results indicated that 2 out of a total of 70 fermentation tubes were positive. What percent of the portions tested during the month were positive?

Known	Unknown
Number Positive/mo = 2 Positive/mo	Portions Positive, %/mo
Total Portions Tested = 70 Portions	

Calculate the percent of the portions tested during the month that were positive.

$$\text{Portions Positive, \%/mo} = \frac{(\text{Number Positive/mo})(100\%)}{\text{Total Portions Tested}}$$

$$= \frac{(2 \text{ Positive/mo})(100\%)}{70 \text{ Portions}}$$

$$= 3\%/\text{mo}$$

A.58 Administration, Safety

EXAMPLE 39

Calculate the injury frequency rate for a water utility where there were four injuries in one year and the operators worked 97,120 hours.

Known	Unknown
Number of Injuries/yr = 4 Injuries/yr	Injury Frequency Rate
Number of Hours Worked/yr = 97,120 Hr/yr	

Calculate the injury frequency rate.

$$\text{Injury Frequency Rate} = \frac{(\text{Number of Injuries/yr})(1,000,000)}{\text{Number of Hours Worked/yr}}$$

$$= \frac{(4 \text{ Injuries/yr})(1,000,000)}{97,120 \text{ Hr/yr}}$$

$$= 41.2$$

EXAMPLE 40

Calculate the injury severity rate for a water company that experienced 57 operator-hours lost due to injuries while the operators worked 97,120 hours during the year.

Known	Unknown
Number of Hours Lost/yr = 57 Hr/yr	Injury Severity Rate
Number of Hours Worked/yr = 97,120 Hr/yr	

Calculate the injury severity rate.

$$\text{Injury Severity Rate} = \frac{(\text{Number of Hours Lost/yr})(1,000,000)}{\text{Number of Hours Worked/yr}}$$

$$= \frac{(57 \text{ Hr/yr})(1,000,000)}{97,120 \text{ Hr/yr}}$$

$$= 587$$

A.6 CALCULATION OF CT VALUES

The Surface Water Treatment Rule requires all surface water systems and all systems using ground water under the influence of surface water to achieve 3-log removal of *Giardia* and 4-log removal or inactivation of viruses. This level of treatment can be accomplished by disinfection alone if the source water is relatively free of turbidity and the system meets other specific conditions. Or, the treatment requirements can be met by a combination of filtration and disinfection.

When both filtration and disinfection are used, it is first necessary to know the effectiveness of filtration in removing *Giardia* and viruses before it will be possible to determine the level of disinfection needed to reach the 3-log and 4-log treatment levels. Research studies of conventional, direct, slow sand, and diatomaceous earth filtration have measured the ability of each of these systems to remove *Giardia* cysts. For example, a well-operated conventional filtration system should be able to achieve 2.5-log removal of *Giardia* and 2.0-log removal/inactivation of viruses. Table A.1 lists the treatment removal efficiencies of each of these four types of filtration systems. Using the example above, if filtration achieves 2.5-log removal, disinfection processes will need to achieve at least 0.5-log *Giardia* removal to meet the SWTR requirement of 3-log *Giardia* removal.[1]

---

[1] Please refer to Chapter 22, "Drinking Water Regulations," for additional information. Also, drinking water supply and distribution utilities and operators are strongly urged to stay in close contact with their state and federal regulatory agencies to keep informed of regulations such as the Surface Water Treatment Rules, the Disinfection/Disinfection By-Products (D/DBP) Rules, and the Filter Backwash Rule.

TABLE A.1 LOG REMOVAL EFFICIENCY

Type of Filtration	Treatment Removal Credit (Log Removals) Giardia	Viruses	Required Disinfection (Log Inactivations[a]) Giardia	Viruses
Conventional	2.5	2.0	0.5	2.0
Direct	2.0	1.0	1.0	3.0
Slow Sand	2.0	2.0	1.0	2.0
Diatomaceous Earth	2.0	1.0	1.0	3.0

[a] Assumed 3-log *Giardia* and 4-log virus removal or inactivation (can be increased based on sanitary hazards in watershed).

Not all disinfectants are equally effective at inactivating *Giardia* and viruses. Research has shown that a free chlorination system that achieves 3-log *Giardia* inactivation also achieves 4-log virus inactivation. On the other hand, a system using chloramination must disinfect to the 4-log level for virus inactivation before the water will be considered safe and free of harmful organisms.

The effectiveness of disinfection by free chlorination also depends on pH and temperature. The required CT value for any given free chlorine residual increases with increasing pH and decreasing temperature. Tables A.2 through A.7 (pages 764–769) provide the CT values for inactivation of *Giardia* cysts by free chlorine, and the remainder of this section provides examples of how to calculate disinfection CT (concentration × contact time) values using free chlorination.

EXAMPLE 41

A 10-MGD direct filtration plant applies free chlorine as a disinfectant. The disinfectant has a contact time of 26 minutes under peak flow conditions. The pH of the water is 7.5 and the temperature is 10°C. The free chlorine residual is 2.0 mg/L. Table A.1 above indicates that direct filtration is capable of achieving 2-log (99-percent) removal of *Giardia* cysts. Therefore, 1-log (90-percent or CT-90) *Giardia* inactivation must be achieved by disinfection.

SOLUTION

1. Using Table A.4 (page 766) with a pH of 7.5, a temperature of 10°C, 1.0-log inactivation, and a free chlorine residual of 2.0 mg/L, find the required CT value of 50 mg/L-min.

2. Calculate the CT provided.

    CT Provided, mg/L-min = (C, mg/L)(T, min)

    = (2 mg/L)(26 min)

    = 52 mg/L-min

    Since 52 mg/L-min is greater than the required 50 mg/L-min, the free chlorine residual is adequate for the peak flow conditions.

3. Using the above procedure, assume appropriate free chlorine residuals for lower flows and determine if the CT values provided are greater than the required CT-90 (1-log) values.

Flow, MGD	Contact Time, min	Free Chlorine Residual (assumed), mg/L	CT Provided, mg/L-min	CT-90 Required,* mg/L-min
10	26	2	52	50
7.5	39	1.4	55	47
5	52	1	52	45
2.5	104	0.5	52	43

* From Table A.4 (CT-90 = 1-log inactivation).

Changing the disinfectant dose to meet the required CT values as the flow changes is not a practical operational procedure. The suggested operational procedure is to adjust the disinfectant dose to meet the highest flow requirement with various flow ranges.

Flow Range, MGD	Free Chlorine Residual, mg/L
7.5–10	2.0
5.0–7.5	1.4
2.5–5.0	1.0

EXAMPLE 42

The 10-MGD plant in Example 41 is a conventional filtration plant and only needs to achieve 0.5-log *Giardia* inactivation (CT-70). Rework Example 41 for 0.5-log *Giardia* inactivation assuming a free chlorine residual of 1.0 mg/L for peak flow conditions.

Flow, MGD	Contact Time, min	Free Chlorine Residual (assumed), mg/L	CT Provided, mg/L-min	CT-70 Required, mg/L-min
10	26	1.0	26	22
7.5	39	0.7	27	22
5	52	0.5	26	21
2.5	104	0.4	42	21

EXAMPLE 43

A conventional water treatment plant consists of flocculation, sedimentation, filtration, and a clear well. Prechlorination is applied before flocculation and postchlorination after the clear well. A dye tracer study has been done to determine the chlorine contact times (T_{10}, time for 10 percent of the tracer to pass through a tank). Assume a water temperature of 15°C and a pH of 7.5. The plant needs to achieve 0.5-log inactivation of *Giardia* cysts. Determine the fraction of the CT value (CTF) the plant provides the first customer in the distribution system given the following information:

Process	Cl$_2$ Free Residual, mg/L	T_{10}, min
Flocculation	0.8	15
Sedimentation	0.4	75
Filtration	0.2	15
Clear Well	0.6	120
Pipe to Customer	0.4	20

SOLUTION

Calculate the actual CT value for each process, determine the CT required from Table A.5 (page 767), calculate the CT actual/CT required ratio and sum the ratios. The sum of the ratios is the CTF value and should be greater than 1.0.

Process	Cl Conc, mg/L	T_{10}, min	CT Actual	CT Required*	CT Ratio
Floc.	0.8 mg/L	15 min	12	15	0.80
Sed.	0.4 mg/L	75 min	30	14	2.14
Filt.	0.2 mg/L	15 min	3	14	0.21
CW	0.6 mg/L	120 min	72	14	5.14
Pipe	0.4 mg/L	20 min	8	14	0.57
					8.86

* From Table A.5.

EXAMPLE 44

If the sum of the CT actual/CT required ratios for the flocculation, sedimentation, and filtration processes for Example 43 was 0.72, determine the free chlorine residual from the clear well necessary to achieve a total CTF (CT fraction) value greater than 1.0 for the plant only (neglect pipe CT). Assume the T_{10} value (actual detention time for the clear well) is 30 minutes.

SOLUTION

Determine the minimum CT ratio for the clear well that will produce a total CTF value greater than 1.0 for the plant.

$$\text{Min Clear Well CT Ratio Needed} = \text{Min CTF} - \text{CTF Provided}$$

$$= 1.0 - 0.72$$

$$= 0.28$$

Determine the CT ratios (CT actual/CT required) for various chlorine residuals in the clear well. Start with the minimum free chlorine residual for the clear well.

Cl$_2$ Free Residual, mg/L	T_{10}, min	CT Actual	CT Required	CT Ratio
0.2 mg/L	30 min	6	14	0.43
0.4 mg/L	30 min	12	14	0.86
0.6 mg/L	30 min	18	14	1.29
0.8 mg/L	30 min	24	15	1.60

Since the free chlorine residual of 0.2 mg/L produces a CT ratio of 0.43, which is greater than 0.28, the free chlorine residual of 0.2 mg/L will be adequate according to these calculations.

EXAMPLE 45

If prechlorination were not practiced in Example 44, what free chlorine residual would be necessary for the clear well alone to provide adequate disinfection?

SOLUTION

The required CT value is 14 mg/L-min.

$$\text{Free Chlorine Residual, mg/}L = \frac{\text{CT, mg/}L\text{-min}}{T_{10}, \text{min}}$$

$$= \frac{14 \text{ mg/}L\text{-min}}{30 \text{ min}}$$

$$= 0.5 \text{ mg/}L \text{ or greater}$$

A.7 CALCULATION OF LOG REMOVALS

Regulations may require the calculation of log removals for inactivation of *Giardia* cysts, viruses, or particle counts. How are log removals calculated? The following example illustrates two methods of calculating log removals.

EXAMPLE 46

Calculate the log removal of 5- to 15-micron particles per milliliter if the influent particle count to a water treatment filter reported 2,100 particles in the 5- to 15-microns range per milliliter of water and the filter effluent reported 30 particles in the 5- to 15-microns range per milliliter of filtered water.

Known	Unknown
Influent, particles/mL = 2,100 particles/mL	Log Removal
Effluent, particles/mL = 30 particles/mL	

Calculate the log removal of particles in the 5- to 15- microns range per mL by the filter.

PROCEDURE 1

$$\text{Log Removal} = \text{Log Influent, particles/mL} - \text{Log Effluent, particles/mL}$$

$$= \text{Log } 2{,}100 \text{ particles/mL} - \text{Log } 30 \text{ particles/mL}$$

$$= 3.30 - 1.5$$

$$= 1.8$$

PROCEDURE 2

$$\text{Log Removal} = \text{Log}\left[\frac{\text{Influent, particles/mL}}{\text{Effluent, particles/mL}}\right]$$

$$= \text{Log}\left[\frac{2{,}100 \text{ particles/mL}}{30 \text{ particles/mL}}\right]$$

$$= \text{Log } 70$$

$$= 1.8$$

TABLE A.2 CT VALUES FOR INACTIVATION OF *GIARDIA* CYSTS BY FREE CHLORINE AT 0.5°C OR LOWER (1)

CHLORINE CONCENTRATION (mg/L)	pH ≤ 6 Log Inactivations 0.5	1.0	1.5	2.0	2.5	3.0	pH = 6.5 Log Inactivations 0.5	1.0	1.5	2.0	2.5	3.0	pH = 7.0 Log Inactivations 0.5	1.0	1.5	2.0	2.5	3.0	pH = 7.5 Log Inactivations 0.5	1.0	1.5	2.0	2.5	3.0
≤ 0.4	23	46	69	91	114	137	27	54	82	109	136	163	33	65	98	130	163	195	40	79	119	158	198	237
0.6	24	47	71	94	118	141	28	56	84	112	140	168	33	67	100	133	167	200	40	80	120	159	199	239
0.8	24	48	73	97	121	145	29	57	86	115	143	172	34	68	103	137	171	205	41	82	123	164	205	246
1	25	49	74	99	123	148	29	59	88	117	147	176	35	70	105	140	175	210	42	84	127	169	211	253
1.2	25	51	76	101	127	152	30	60	90	120	150	180	36	72	108	143	179	215	43	86	130	173	216	259
1.4	26	52	78	103	129	155	31	61	92	123	153	184	37	74	111	147	184	221	44	89	133	177	222	266
1.6	26	52	79	105	131	157	32	63	95	126	158	189	38	75	113	151	188	226	46	91	137	182	228	273
1.8	27	54	81	108	135	162	32	64	97	129	161	193	38	77	116	154	193	231	47	93	140	186	233	279
2	28	55	83	110	138	165	33	66	99	131	164	197	39	79	118	157	197	236	48	95	143	191	238	286
2.2	28	56	85	113	141	169	34	67	101	134	168	201	40	81	121	161	202	242	50	99	149	198	248	297
2.4	29	57	86	115	143	172	34	68	103	137	171	205	41	82	124	165	206	247	50	99	149	199	248	298
2.6	29	58	88	117	146	175	35	70	105	139	174	209	42	84	126	168	210	252	51	101	152	203	253	304
2.8	30	59	89	119	148	178	36	71	107	142	178	213	43	86	129	171	214	257	52	103	155	207	258	310
3	30	60	91	121	151	181	36	72	109	145	181	217	44	87	131	174	218	261	53	105	158	211	263	316

CHLORINE CONCENTRATION (mg/L)	pH = 8.0 Log Inactivations 0.5	1.0	1.5	2.0	2.5	3.0	pH = 8.5 Log Inactivations 0.5	1.0	1.5	2.0	2.5	3.0	pH ≤ 9.0 Log Inactivations 0.5	1.0	1.5	2.0	2.5	3.0
≤ 0.4	46	92	139	185	231	277	55	110	165	219	274	329	65	130	195	260	325	390
0.6	48	95	143	191	238	286	57	114	171	228	285	342	68	136	204	271	339	407
0.8	49	98	148	197	246	295	59	118	177	236	295	354	70	141	211	281	352	422
1	51	101	152	203	253	304	61	122	183	243	304	365	73	146	219	291	364	437
1.2	52	104	157	209	261	313	63	125	188	251	313	376	75	150	226	301	376	451
1.4	54	107	161	214	268	321	65	129	194	258	323	387	77	155	232	309	387	464
1.6	55	110	165	219	274	329	66	132	199	265	331	397	80	159	239	318	398	477
1.8	56	113	169	225	282	338	68	136	204	271	339	407	82	163	245	326	408	489
2	58	115	173	231	288	346	70	139	209	278	348	417	83	167	250	333	417	500
2.2	59	118	177	235	294	353	71	142	213	284	355	426	85	170	256	341	426	511
2.4	60	120	181	241	301	361	73	145	218	290	363	435	87	174	261	348	435	522
2.6	61	123	184	245	307	368	74	148	222	296	370	444	89	178	267	355	444	533
2.8	63	125	188	250	313	375	75	151	226	301	377	452	91	181	272	362	453	543
3	64	127	191	255	318	382	77	153	230	307	383	460	92	184	276	368	460	552

Notes:
(1) CT = CT for 3-log inactivation
99.9

Reprinted from *GUIDANCE MANUAL FOR COMPLIANCE WITH THE FILTRATION AND DISINFECTION REQUIREMENTS FOR PUBLIC WATER SYSTEMS USING SURFACE WATER SOURCES*, a publication of the American Water Works Association. Reproduced with permission. Copyright © 1991 by American Water Works Association.

TABLE A.3 CT VALUES FOR INACTIVATION OF *GIARDIA* CYSTS BY FREE CHLORINE AT 5°C (1)

CHLORINE CONCENTRATION (mg/L)	pH ≤ 6 Log Inactivations						pH = 6.5 Log Inactivations						pH = 7.0 Log Inactivations						pH = 7.5 Log Inactivations					
	0.5	1.0	1.5	2.0	2.5	3.0	0.5	1.0	1.5	2.0	2.5	3.0	0.5	1.0	1.5	2.0	2.5	3.0	0.5	1.0	1.5	2.0	2.5	3.0
≤ 0.4	16	32	49	65	81	97	20	39	59	78	98	117	23	46	70	93	116	139	28	55	83	111	138	166
0.6	17	33	50	67	83	100	20	40	60	80	100	120	24	48	72	95	119	143	29	57	86	114	143	171
0.8	17	34	52	69	86	103	20	41	61	81	102	122	24	49	73	97	122	146	29	58	88	117	146	175
1	18	35	53	70	88	105	21	42	63	83	104	125	25	50	75	99	124	149	30	60	90	119	149	179
1.2	18	36	54	71	89	107	21	42	64	85	106	127	25	51	76	101	127	152	31	61	92	122	153	183
1.4	18	36	55	73	91	109	22	43	65	87	108	130	26	52	78	103	129	155	31	62	94	125	156	187
1.6	19	37	56	74	93	111	22	44	66	88	110	132	26	53	79	105	132	158	32	64	96	128	160	192
1.8	19	38	57	76	95	114	23	45	68	90	113	135	27	54	81	108	135	162	33	65	98	131	163	196
2	19	39	58	77	97	116	23	46	69	92	115	138	28	55	83	110	138	165	33	67	100	133	167	200
2.2	20	39	59	79	98	118	23	47	70	93	117	140	28	56	85	113	141	169	34	68	102	136	170	204
2.4	20	40	60	80	100	120	24	48	72	95	119	143	29	57	86	115	143	172	35	70	105	139	174	209
2.6	20	41	61	81	102	122	24	49	73	97	122	146	29	58	88	117	146	175	36	71	107	142	178	213
2.8	21	41	62	83	103	124	25	49	74	99	123	148	30	59	89	119	148	178	36	72	109	145	181	217
3	21	42	63	84	105	126	25	50	76	101	126	151	30	61	91	121	152	182	37	74	111	147	184	221

CHLORINE CONCENTRATION (mg/L)	pH = 8.0 Log Inactivations						pH = 8.5 Log Inactivations						pH ≤ 9.0 Log Inactivations					
	0.5	1.0	1.5	2.0	2.5	3.0	0.5	1.0	1.5	2.0	2.5	3.0	0.5	1.0	1.5	2.0	2.5	3.0
≤ 0.4	33	66	99	132	165	198	39	79	118	157	197	236	47	93	140	186	233	279
0.6	34	68	102	136	170	204	41	81	122	163	203	244	49	97	146	194	243	291
0.8	35	70	105	140	175	210	42	84	126	168	210	252	50	100	151	201	251	301
1	36	72	108	144	180	216	43	87	130	173	217	260	52	104	156	208	260	312
1.2	37	74	111	147	184	221	45	89	134	178	223	267	53	107	160	213	267	320
1.4	38	76	114	151	189	227	46	91	137	183	228	274	55	110	165	219	274	329
1.6	39	77	116	155	193	232	47	94	141	187	234	281	56	112	169	225	281	337
1.8	40	79	119	159	198	238	48	96	144	191	239	287	58	115	173	230	288	345
2	41	81	122	162	203	243	49	98	147	196	245	294	59	118	177	235	294	353
2.2	41	83	124	165	207	248	50	100	150	200	250	300	60	120	181	241	301	361
2.4	42	84	127	169	211	253	51	102	153	204	255	306	61	123	184	245	307	368
2.6	43	86	129	172	215	258	52	104	156	208	260	312	63	125	188	250	313	375
2.8	44	88	132	175	219	263	53	106	159	212	265	318	64	127	191	255	318	382
3	45	89	134	179	223	268	54	108	162	216	270	324	65	130	195	259	324	389

Notes:

(1) CT$_{99.9}$ = CT for 3-log inactivation

Reprinted from *GUIDANCE MANUAL FOR COMPLIANCE WITH THE FILTRATION AND DISINFECTION REQUIREMENTS FOR PUBLIC WATER SYSTEMS USING SURFACE WATER SOURCES*, a publication of the American Water Works Association. Reproduced with permission. Copyright © 1991 by American Water Works Association.

766 Water Treatment

TABLE A.4 CT VALUES FOR INACTIVATION OF *GIARDIA* CYSTS BY FREE CHLORINE AT 10°C (1)

CHLORINE CONCENTRATION (mg/L)	pH <=6 Log Inactivations						pH =6.5 Log Inactivations						pH =7.0 Log Inactivations						pH =7.5 Log Inactivations					
	0.5	1.0	1.5	2.0	2.5	3.0	0.5	1.0	1.5	2.0	2.5	3.0	0.5	1.0	1.5	2.0	2.5	3.0	0.5	1.0	1.5	2.0	2.5	3.0
<=0.4	12	24	37	49	61	73	15	29	44	59	73	88	17	35	52	69	87	104	21	42	63	83	104	125
0.6	13	25	38	50	63	75	15	30	45	60	75	90	18	36	54	71	89	107	21	43	64	85	107	128
0.8	13	26	39	52	65	78	15	31	46	61	77	92	18	37	55	73	92	110	22	44	66	87	109	131
1	13	26	40	53	66	79	16	31	47	63	78	94	19	37	56	75	93	112	22	45	67	89	112	134
1.2	13	27	40	53	67	80	16	32	48	63	79	95	19	38	57	76	95	114	23	46	69	91	114	137
1.4	14	27	41	55	68	82	16	33	49	65	82	98	19	39	58	77	97	116	23	47	70	93	117	140
1.6	14	28	42	55	69	83	17	33	50	66	83	99	20	40	60	79	99	119	24	48	72	96	120	144
1.8	14	29	43	57	72	86	17	34	51	67	84	101	20	41	61	81	102	122	24	49	74	98	123	147
2	15	29	44	58	73	87	17	35	52	69	87	104	21	41	62	83	103	124	25	50	75	100	125	150
2.2	15	30	45	59	74	89	18	35	53	70	88	105	21	42	64	85	106	127	26	51	77	102	128	153
2.4	15	30	45	60	75	90	18	36	54	71	89	107	22	43	65	86	108	129	26	52	79	105	131	157
2.6	15	31	46	61	77	92	18	37	55	73	92	110	22	44	66	87	109	131	27	53	80	107	133	160
2.8	16	31	47	62	78	93	19	37	56	74	93	111	22	45	67	89	112	134	27	54	82	109	136	163
3	16	32	48	63	79	95	19	38	57	75	94	113	23	46	69	91	114	137	28	55	83	111	138	166

CHLORINE CONCENTRATION (mg/L)	pH =8.0 Log Inactivations						pH =8.5 Log Inactivations						pH <=9.0 Log Inactivations					
	0.5	1.0	1.5	2.0	2.5	3.0	0.5	1.0	1.5	2.0	2.5	3.0	0.5	1.0	1.5	2.0	2.5	3.0
<=0.4	25	50	75	99	124	149	30	59	89	118	148	177	35	70	105	139	174	209
0.6	26	51	77	102	128	153	31	61	92	122	153	183	36	73	109	145	182	218
0.8	26	53	79	105	132	158	32	63	95	126	158	189	38	75	113	151	188	226
1	27	54	81	108	135	162	33	65	98	130	163	195	39	78	117	156	195	234
1.2	28	55	83	111	138	166	33	67	100	133	167	200	40	80	120	160	200	240
1.4	28	57	85	113	142	170	34	69	103	137	172	206	41	82	124	165	206	247
1.6	29	58	87	116	145	174	35	70	106	141	176	211	42	84	127	169	211	253
1.8	30	60	90	119	149	179	36	72	108	143	179	215	43	86	130	173	216	259
2	30	61	91	121	152	182	37	74	111	147	184	221	44	88	133	177	221	265
2.2	31	62	93	124	155	186	38	75	113	150	188	225	45	90	136	181	226	271
2.4	32	63	95	127	158	190	38	77	115	153	192	230	46	92	138	184	230	276
2.6	32	65	97	129	162	194	39	78	117	156	195	234	47	94	141	187	234	281
2.8	33	66	99	131	164	197	40	80	120	159	199	239	48	96	144	191	239	287
3	34	67	101	134	168	201	41	81	122	162	203	243	49	97	146	195	243	292

Notes:
(1) CT = CT for 3-log inactivation
99.9

Reprinted from *GUIDANCE MANUAL FOR COMPLIANCE WITH THE FILTRATION AND DISINFECTION REQUIREMENTS FOR PUBLIC WATER SYSTEMS USING SURFACE WATER SOURCES*, a publication of the American Water Works Association. Reproduced with permission. Copyright © 1991 by American Water Works Association.

TABLE A.5 CT VALUES FOR INACTIVATION OF *GIARDIA* CYSTS BY FREE CHLORINE AT 15°C (1)

CHLORINE CONCENTRATION (mg/L)	pH <= 6 Log Inactivations						pH = 6.5 Log Inactivations						pH = 7.0 Log Inactivations						pH = 7.5 Log Inactivations					
	0.5	1.0	1.5	2.0	2.5	3.0	0.5	1.0	1.5	2.0	2.5	3.0	0.5	1.0	1.5	2.0	2.5	3.0	0.5	1.0	1.5	2.0	2.5	3.0
<=0.4	8	16	25	33	41	49	10	20	30	39	49	59	12	23	35	47	58	70	14	28	42	55	69	83
0.6	8	17	25	33	42	50	10	20	30	40	50	60	12	24	36	48	60	72	14	29	43	57	72	86
0.8	9	17	26	35	43	52	10	20	31	41	51	61	12	24	37	49	61	73	15	29	44	59	73	88
1	9	18	27	35	44	53	11	21	32	42	53	63	13	25	38	50	63	75	15	30	45	60	75	90
1.2	9	18	27	36	45	54	11	21	32	43	53	64	13	25	38	51	63	76	15	31	46	61	77	92
1.4	9	18	28	37	46	55	11	22	33	43	54	65	13	26	39	52	65	78	16	31	47	63	78	94
1.6	9	19	28	37	47	56	11	22	33	44	55	66	13	26	40	53	66	79	16	32	48	64	80	96
1.8	10	19	29	38	48	57	11	23	34	45	57	68	14	27	41	54	68	81	16	33	49	65	82	98
2	10	19	29	39	48	58	12	23	35	46	58	69	14	28	42	55	69	83	17	33	50	67	83	100
2.2	10	20	30	39	49	59	12	23	35	47	58	70	14	28	43	57	71	85	17	34	51	68	85	102
2.4	10	20	30	40	50	60	12	24	36	48	60	72	14	29	43	57	72	86	18	35	53	70	88	105
2.6	10	20	31	41	51	61	12	24	37	49	61	73	15	29	44	59	73	88	18	36	54	71	89	107
2.8	10	21	31	41	52	62	12	25	37	49	62	74	15	30	45	59	74	89	18	36	55	73	91	109
3	11	21	32	42	53	63	13	25	38	51	63	76	15	30	46	61	76	91	19	37	56	74	93	111

CHLORINE CONCENTRATION (mg/L)	pH = 8.0 Log Inactivations						pH = 8.5 Log Inactivations						pH <= 9.0 Log Inactivations					
	0.5	1.0	1.5	2.0	2.5	3.0	0.5	1.0	1.5	2.0	2.5	3.0	0.5	1.0	1.5	2.0	2.5	3.0
<=0.4	17	33	50	66	83	99	20	39	59	79	98	118	23	47	70	93	117	140
0.6	17	34	51	68	85	102	20	41	61	81	102	122	24	49	73	97	122	146
0.8	18	35	53	70	88	105	21	42	63	84	105	126	25	50	76	101	126	151
1	18	36	54	72	90	108	22	43	65	87	108	130	26	52	78	104	130	156
1.2	19	37	56	74	93	111	22	45	67	89	112	134	27	53	80	107	133	160
1.4	19	38	57	76	95	114	23	46	69	91	114	137	28	55	83	110	138	165
1.6	19	39	58	77	97	116	23	47	71	94	118	141	28	56	85	113	141	169
1.8	20	40	60	79	99	119	24	48	72	96	120	144	29	58	87	115	144	173
2	20	41	61	81	102	122	25	49	74	98	123	147	30	59	89	118	148	177
2.2	21	41	62	83	103	124	25	50	75	100	125	150	30	60	91	121	151	181
2.4	21	42	64	85	106	127	26	51	77	102	128	153	31	61	92	123	153	184
2.6	22	43	65	86	108	129		52	78	104	130	156	31	63	94	125	157	188
2.8	22	44	66	88	110	132	27	53	80	106	133	159	32	64	96	127	159	191
3	22	45	67	89	112	134	27	54	81	108	135	162	33	65	98	130	163	195

Notes:

(1) CT$_{99.9}$ = CT for 3-log inactivation

Reprinted from *GUIDANCE MANUAL FOR COMPLIANCE WITH THE FILTRATION AND DISINFECTION REQUIREMENTS FOR PUBLIC WATER SYSTEMS USING SURFACE WATER SOURCES*, a publication of the American Water Works Association. Reproduced with permission. Copyright © 1991 by American Water Works Association.

768 Water Treatment

TABLE A.6 CT VALUES FOR INACTIVATION OF *GIARDIA* CYSTS BY FREE CHLORINE AT 20°C (1)

CHLORINE CONCENTRATION (mg/L)	pH ≤ 6 Log Inactivations						pH = 6.5 Log Inactivations						pH = 7.0 Log Inactivations						pH = 7.5 Log Inactivations					
	0.5	1.0	1.5	2.0	2.5	3.0	0.5	1.0	1.5	2.0	2.5	3.0	0.5	1.0	1.5	2.0	2.5	3.0	0.5	1.0	1.5	2.0	2.5	3.0
≤ 0.4	6	12	18	24	30	36	7	15	22	29	37	44	9	17	26	35	43	52	10	21	31	41	52	62
0.6	6	13	19	25	32	38	8	15	23	30	38	45	9	18	27	36	45	54	11	21	32	43	53	64
0.8	7	13	20	26	33	39	8	15	23	31	38	46	9	18	28	37	46	55	11	22	33	44	55	66
1	7	13	20	26	33	39	8	16	24	31	39	47	9	19	28	37	47	56	11	22	34	45	56	67
1.2	7	13	20	27	33	40	8	16	24	32	40	48	10	19	29	38	48	57	12	23	35	46	58	69
1.4	7	14	21	27	34	41	8	16	25	33	41	49	10	19	29	39	48	58	12	23	35	47	58	70
1.6	7	14	21	28	35	42	8	17	25	33	42	50	10	20	30	39	49	59	12	24	36	48	60	72
1.8	7	14	22	29	36	43	9	17	26	34	43	51	10	20	31	41	51	61	12	25	37	49	62	74
2	7	15	22	29	37	44	9	17	26	35	43	52	10	21	31	41	52	62	13	25	38	50	63	75
2.2	7	15	22	29	37	44	9	18	27	35	44	53	11	21	32	42	53	63	13	26	39	51	64	77
2.4	8	15	23	30	38	45	9	18	27	36	45	54	11	22	33	43	54	65	13	26	39	52	65	78
2.6	8	15	23	31	38	46	9	18	28	37	46	55	11	22	33	44	55	66	13	27	40	53	67	80
2.8	8	16	24	31	39	47	9	19	28	37	47	56	11	22	34	45	56	67	14	27	41	54	68	81
3	8	16	24	31	39	47	10	19	29	38	48	57	11	23	34	45	57	68	14	28	42	55	69	83

CHLORINE CONCENTRATION (mg/L)	pH = 8.0 Log Inactivations						pH = 8.5 Log Inactivations						pH ≤ 9.0 Log Inactivations					
	0.5	1.0	1.5	2.0	2.5	3.0	0.5	1.0	1.5	2.0	2.5	3.0	0.5	1.0	1.5	2.0	2.5	3.0
≤ 0.4	12	25	37	49	62	74	15	30	45	59	74	89	18	35	53	70	88	105
0.6	13	26	39	51	64	77	15	31	46	61	77	92	18	36	55	73	91	109
0.8	13	26	40	53	66	79	16	32	48	63	79	95	19	38	57	75	94	113
1	14	27	41	54	68	81	16	33	49	65	82	98	20	39	59	78	98	117
1.2	14	28	42	55	69	83	17	33	50	67	83	100	20	40	60	80	100	120
1.4	14	28	43	57	71	85	17	34	52	69	86	103	21	41	62	82	103	123
1.6	15	29	44	58	73	87	18	35	53	70	88	105	21	42	63	84	105	126
1.8	15	30	45	59	74	89	18	36	54	72	90	108	22	43	65	86	108	129
2	15	30	46	61	76	91	18	37	55	73	92	110	22	44	66	88	110	132
2.2	16	31	47	62	78	93	19	38	57	75	94	113	23	45	68	90	113	135
2.4	16	32	48	63	79	95	19	38	58	77	96	115	23	46	69	92	115	138
2.6	16	32	49	65	81	97	19	39	59	78	98	117	24	47	71	94	118	141
2.8	17	33	50	66	83	99	20	40	60	79	99	119	24	48	72	95	119	143
3	17	34	51	67	84	101	20	41	61	81	102	122	24	49	73	97	122	146

Notes:
(1) $CT_{99.9}$ = CT for 3-log inactivation

Reprinted from *GUIDANCE MANUAL FOR COMPLIANCE WITH THE FILTRATION AND DISINFECTION REQUIREMENTS FOR PUBLIC WATER SYSTEMS USING SURFACE WATER SOURCES*, a publication of the American Water Works Association. Reproduced with permission. Copyright © 1991 by American Water Works Association.

Arithmetic 769

TABLE A.7 CT VALUES FOR INACTIVATION OF *GIARDIA* CYSTS BY FREE CHLORINE AT 25°C (1)

CHLORINE CONCENTRATION (mg/L)	pH ≤ 6 Log Inactivations						pH = 6.5 Log Inactivations						pH = 7.0 Log Inactivations						pH = 7.5 Log Inactivations					
	0.5	1.0	1.5	2.0	2.5	3.0	0.5	1.0	1.5	2.0	2.5	3.0	0.5	1.0	1.5	2.0	2.5	3.0	0.5	1.0	1.5	2.0	2.5	3.0
≤ 0.4	4	8	12	16	20	24	5	10	15	19	24	29	6	12	18	23	29	35	7	14	21	28	35	42
0.6	4	8	13	17	21	25	5	10	15	20	25	30	6	12	18	24	30	36	7	14	22	29	36	43
0.8	4	9	13	17	22	26	5	10	16	21	26	31	6	12	19	25	31	37	7	15	22	29	37	44
1	4	9	13	17	22	26	5	10	16	21	26	31	6	12	19	25	31	37	8	15	23	30	38	45
1.2	5	9	14	18	23	27	5	11	16	21	27	32	6	13	19	25	32	38	8	15	23	31	38	46
1.4	5	9	14	18	23	27	5	11	16	22	27	33	7	13	20	26	33	39	8	16	24	31	39	47
1.6	5	9	14	19	23	28	6	11	17	22	28	33	7	13	20	27	33	40	8	16	24	32	40	48
1.8	5	10	15	19	24	29	6	11	17	23	28	34	7	14	21	27	34	41	8	16	25	33	41	49
2	5	10	15	19	24	29	6	12	18	23	29	35	7	14	21	27	34	41	8	17	25	33	42	50
2.2	5	10	15	20	25	30	6	12	18	23	29	35	7	14	21	28	35	42	9	17	26	34	43	51
2.4	5	10	15	20	25	30	6	12	18	24	30	36	7	14	22	29	36	43	9	17	26	35	43	52
2.6	5	10	16	21	26	31	6	12	19	25	31	37	7	15	22	29	37	44	9	18	27	35	44	53
2.8	5	10	16	21	26	31	6	12	19	25	31	37	8	15	23	30	38	45	9	18	27	36	45	54
3	5	11	16	21	27	32	6	13	19	25	32	38	8	15	23	31	38	46	9	18	28	37	46	55

CHLORINE CONCENTRATION (mg/L)	pH = 8.0 Log Inactivations						pH = 8.5 Log Inactivations						pH ≤ 9.0 Log Inactivations					
	0.5	1.0	1.5	2.0	2.5	3.0	0.5	1.0	1.5	2.0	2.5	3.0	0.5	1.0	1.5	2.0	2.5	3.0
≤ 0.4	8	17	25	33	42	50	10	20	30	39	49	59	12	23	35	47	58	70
0.6	9	17	26	34	43	51	10	20	31	41	51	61	12	24	37	49	61	73
0.8	9	18	27	35	44	53	11	21	32	42	53	63	13	25	38	50	63	75
1	9	18	27	36	45	54	11	22	33	43	54	65	13	26	39	52	65	78
1.2	9	18	28	37	46	55	11	22	34	45	56	67	13	27	40	53	67	80
1.4	10	19	29	38	48	57	12	23	35	46	58	69	14	27	41	55	68	82
1.6	10	19	29	39	48	58	12	23	35	47	58	70	14	28	42	56	70	84
1.8	10	20	30	40	50	60	12	24	36	48	60	72	14	29	43	57	72	86
2	10	20	31	41	51	61	12	25	37	49	62	74	15	29	44	59	73	88
2.2	10	21	31	41	52	62	13	25	38	50	63	75	15	30	45	60	75	90
2.4	11	21	32	42	53	63	13	26	39	51	64	77	15	31	46	61	77	92
2.6	11	22	33	43	54	65	13	26	39	52	65	78	16	31	47	63	78	94
2.8	11	22	33	44	55	66	13	27	40	53	67	80	16	32	48	64	80	96
3	11	22	34	45	56	67	14	27	41	54	68	81	16	32	49	65	81	97

Notes:

(1) $CT_{99.9}$ = CT for 3-log inactivation

Reprinted from *GUIDANCE MANUAL FOR COMPLIANCE WITH THE FILTRATION AND DISINFECTION REQUIREMENTS FOR PUBLIC WATER SYSTEMS USING SURFACE WATER SOURCES*, a publication of the American Water Works Association. Reproduced with permission. Copyright © 1991 by American Water Works Association.

ABBREVIATIONS

°C	degrees Celsius	km	kilometers
°F	degrees Fahrenheit	kN	kilonewtons
μ	micron	kPa	kiloPascals
μg	microgram	kW	kilowatts
μm	micrometer	kWh	kilowatt-hours
ac	acres	L	liters
ac-ft	acre-feet	lb	pounds
amp	amperes	lbs/sq in	pounds per square inch
atm	atmosphere	M	mega
CFM	cubic feet per minute	M	million
CFS	cubic feet per second	M	molar (or molarity)
Ci	Curie	m	meters
cm	centimeters	mA	milliampere
cu ft	cubic feet	meq	milliequivalent
cu in	cubic inches	mg	milligrams
cu m	cubic meters	MGD	million gallons per day
cu yd	cubic yards	mg/L	milligrams per liter
		min	minutes
D	Dalton	mL	milliliters
dB	decibel	mm	millimeters
ft	feet or foot	N	Newton
ft-lb/min	foot-pounds per minute	N	normal (or normality)
		nm	nanometer
g	gravity		
gal	gallons	ohm	ohm
gal/day	gallons per day	Pa	Pascal
GFD	gallons of flux per square foot per day	pCi	picoCurie
gm	grams	pCi/L	picoCuries per liter
GPCD	gallons per capita per day	ppb	parts per billion
GPD	gallons per day	ppm	parts per million
gpg	grains per gallon	psf	pounds per square foot
GPM	gallons per minute	psi	pounds per square inch
GPY	gallons per year	psig	pounds per square inch gauge
gr	grains		
ha	hectares	RPM	revolutions per minute
HP	horsepower	SCFM	standard cubic feet per minute
hr	hours	sec	seconds
Hz	hertz	SI	Le Système International d'Unités
in	inches	sq ft	square feet
		sq in	square inches
J	joules	W	watt
k	kilos	yd	yards
kg	kilograms		

WATER WORDS

A Summary of the Words Defined

in

WATER TREATMENT PLANT OPERATION,

WATER DISTRIBUTION SYSTEM
OPERATION AND MAINTENANCE,

and

SMALL WATER SYSTEM
OPERATION AND MAINTENANCE

PROJECT PRONUNCIATION KEY

by Warren L. Prentice

The Project Pronunciation Key is designed to aid you in the pronunciation of new words. While this key is based primarily on familiar sounds, it does not attempt to follow any particular pronunciation guide. This key is designed solely to aid operators in this program.

You may find it helpful to refer to other available sources for pronunciation help. Each current standard dictionary contains a guide to its own pronunciation key. Each key will be different from each other and from this key. Examples of the difference between the key used in this program and the *WEBSTER'S NEW WORLD COLLEGE DICTIONARY*[1] "Key" are shown below.

In using this key, you should accent (say louder) the syllable that appears in capital letters. The following chart is presented to give examples of how to pronounce words using the Project Key.

WORD	SYLLABLE				
	1st	2nd	3rd	4th	5th
acid	AS	id			
coliform	KOAL	i	form		
biological	BUY	o	LODGE	ik	cull

The first word, *ACID*, has its first syllable accented. The second word, *COLIFORM*, has its first syllable accented. The third word, *BIOLOGICAL*, has its first and third syllables accented.

We hope you will find the key useful in unlocking the pronunciation of any new word.

Term	Project Key	Webster Key
acid	AS-id	aś id
coliform	KOAL-i-form	kō′ lə fôrm
biological	BUY-o-LODGE-ik-cull	bī ə läj′ i kəl

---

[1] The *WEBSTER'S NEW WORLD COLLEGE DICTIONARY*, Fourth Edition, 1999, was chosen rather than an unabridged dictionary because of its availability to the operator.

WATER WORDS

>GREATER THAN
DO >5 mg/*L* would be read as DO GREATER THAN 5 mg/*L*.

<LESS THAN
DO <5 mg/*L* would be read as DO LESS THAN 5 mg/*L*.

A

ABC
See ASSOCIATION OF BOARDS OF CERTIFICATION.

ACEOPS
See ALLIANCE OF CERTIFIED OPERATORS, LAB ANALYSTS, INSPECTORS, AND SPECIALISTS.

atm
The abbreviation for atmosphere. One atmosphere is equal to 14.7 psi or 100 kPa.

AWWA
See AMERICAN WATER WORKS ASSOCIATION.

ABSORPTION (ab-SORP-shun)
The taking in or soaking up of one substance into the body of another by molecular or chemical action (as tree roots absorb dissolved nutrients in the soil).

ACCOUNTABILITY
When a manager gives power/responsibility to an employee, the employee ensures that the manager is informed of results or events.

ACCURACY
How closely an instrument measures the true or actual value of the process variable being measured or sensed.

ACID RAIN
Precipitation that has been rendered (made) acidic by airborne pollutants.

ACIDIC (uh-SID-ick)
The condition of water or soil that contains a sufficient amount of acid substances to lower the pH below 7.0.

ACIDIFICATION (uh-SID-uh-fuh-KAY-shun)
The addition of an acid (usually nitric or sulfuric) to a sample to lower the pH below 2.0. The purpose of acidification is to fix a sample so it will not change until it is analyzed.

ACRE-FOOT
A volume of water that covers one acre to a depth of one foot, or 43,560 cubic feet (1,233.5 cubic meters).

ACTIVATED ALUMINA

A charged form of aluminum, used with a synthetic, porous media in an ion exchange adsorption process to remove charged contaminants.

ACTIVATED CARBON

Adsorptive particles or granules of carbon usually obtained by heating carbon (such as wood). These particles or granules have a high capacity to selectively remove certain trace and soluble materials from water.

ACUTE HEALTH EFFECT

An adverse effect on a human or animal body, with symptoms developing rapidly.

ADSORBATE (add-SORE-bait)

The material being removed by the adsorption process.

ADSORBENT (add-SORE-bent)

The material (activated carbon) that is responsible for removing the undesirable substance in the adsorption process.

ADSORPTION (add-SORP-shun)

The gathering of a gas, liquid, or dissolved substance on the surface or interface zone of another material.

AERATION (air-A-shun)

The process of adding air to water. Air can be added to water by either passing air through water or passing water through air.

AEROBIC (air-O-bick)

A condition in which atmospheric or dissolved oxygen is present in the aquatic (water) environment.

AESTHETIC (es-THET-ick)

Attractive or appealing.

AGE TANK

A tank used to store a chemical solution of known concentration for feed to a chemical feeder. It usually stores sufficient chemical solution to properly treat the water being treated for at least one day. Also called a DAY TANK.

AIR BINDING

The clogging of a filter, pipe, or pump due to the presence of air released from water. Air entering the filter media is harmful to both the filtration and backwash processes. Air can prevent the passage of water during the filtration process and can cause the loss of filter media during the backwash process.

AIR GAP

An open, vertical drop, or vertical empty space, between a drinking (potable) water supply and potentially contaminated water. This gap prevents the contamination of drinking water by backsiphonage because there is no way potentially contaminated water can reach the drinking water supply.

AIR PADDING

Pumping dry air (dew point −40°F (−40°C)) into a container to assist with the withdrawal of a liquid or to force a liquified gas such as chlorine or sulfur dioxide out of a container.

AIR STRIPPING

A physical treatment process used to remove volatile substances from water or wastestreams. The process uses large volumes of air to transfer volatile pollutants from a high concentration in the water or wastestream into a lower concentration in an air stream.

ALARM CONTACT

A switch that operates when some preset low, high, or abnormal condition exists.

ALGAE (AL-jee)

Microscopic plants containing chlorophyll that live floating or suspended in water. They also may be attached to structures, rocks, or other submerged surfaces. Excess algal growths can impart tastes and odors to potable water. Algae produce oxygen during sunlight hours and use oxygen during the night hours. Their biological activities appreciably affect the pH, alkalinity, and dissolved oxygen of the water.

ALGAL (AL-gull) BLOOM

Sudden, massive growths of microscopic and macroscopic plant life, such as green or blue-green algae, which can, under the proper conditions, develop in lakes, reservoirs, and ponds.

ALGICIDE (AL-juh-side)

Any substance or chemical specifically formulated to kill or control algae.

ALIPHATIC (al-uh-FAT-ick) HYDROXY ACIDS

Organic acids with carbon atoms arranged in branched or unbranched open chains rather than in rings.

ALIQUOT (AL-uh-kwot)

Representative portion of a sample. Often, an equally divided portion of a sample.

ALKALI (AL-kuh-lie)

Any of certain soluble salts, principally of sodium, potassium, magnesium, and calcium, that have the property of combining with acids to form neutral salts and may be used in chemical processes such as water or wastewater treatment.

ALKALINE (AL-kuh-line)

The condition of water or soil that contains a sufficient amount of alkali substances to raise the pH above 7.0.

ALKALINITY (AL-kuh-LIN-it-tee)

The capacity of water or wastewater to neutralize acids. This capacity is caused by the water's content of carbonate, bicarbonate, hydroxide, and occasionally borate, silicate, and phosphate. Alkalinity is expressed in milligrams per liter of equivalent calcium carbonate. Alkalinity is not the same as pH because water does not have to be strongly basic (high pH) to have a high alkalinity. Alkalinity is a measure of how much acid must be added to a liquid to lower the pH to 4.5.

ALLIANCE OF CERTIFIED OPERATORS, LAB ANALYSTS, INSPECTORS, AND SPECIALISTS (ACEOPS)

A professional organization for operators, lab analysts, inspectors, and specialists dedicated to improving professionalism; expanding training, certification, and job opportunities; increasing information exchange; and advocating the importance of certified operators, lab analysts, inspectors, and specialists. For information on membership, contact ACEOPS, PO Box 934, Dakota City, NE 68731-0934, phone (402) 698-2330, or e-mail: Info@aceops.org.

ALLUVIAL (uh-LOO-vee-ul)

Relating to mud or sand deposited by flowing water. Alluvial deposits may occur after a heavy rainstorm.

ALTERNATING CURRENT (AC)

An electric current that reverses its direction (positive/negative values) at regular intervals.

ALTITUDE VALVE

A valve that automatically shuts off the flow into an elevated tank when the water level in the tank reaches a predetermined level. The valve automatically opens when the pressure in the distribution system drops below the pressure in the tank.

AMBIENT (AM-bee-ent) TEMPERATURE

Temperature of the surroundings.

AMERICAN WATER WORKS ASSOCIATION (AWWA)

A professional organization for all persons working in the water utility field. This organization develops and recommends goals, procedures, and standards for water utility agencies to help them improve their performance and effectiveness. For information on AWWA membership and publications, contact AWWA, 6666 West Quincy Avenue, Denver, CO 80235. Phone (303) 794-7711.

AMPERAGE (AM-purr-age)

The strength of an electric current measured in amperes. The amount of electric current flow, similar to the flow of water in gallons per minute.

AMPERE (AM-peer)

The unit used to measure current strength. The current produced by an electromotive force of one volt acting through a resistance of one ohm.

AMPEROMETRIC (am-purr-o-MET-rick)

A method of measurement that records electric current flowing or generated, rather than recording voltage. Amperometric titration is a means of measuring concentrations of certain substances in water.

AMPEROMETRIC (am-purr-o-MET-rick) TITRATION

A means of measuring concentrations of certain substances in water (such as strong oxidizers) based on the electric current that flows during a chemical reaction. Also see TITRATE.

AMPLITUDE

The maximum strength of an alternating current during its cycle, as distinguished from the mean or effective strength.

ANAEROBIC (AN-air-O-bick)

A condition in which atmospheric or dissolved oxygen (DO) is *NOT* present in the aquatic (water) environment.

ANALOG

The continuously variable signal type sent to an analog instrument (for example, 4–20 mA).

ANALOG READOUT

The readout of an instrument by a pointer (or other indicating means) against a dial or scale. Also see DIGITAL READOUT.

ANALYZER

A device that conducts a periodic or continuous measurement of turbidity or some factor such as chlorine or fluoride concentration. Analyzers operate by any of several methods including photocells, conductivity, or complex instrumentation.

ANGSTROM (ANG-strem)

A unit of length equal to one-tenth of a nanometer or one-tenbillionth of a meter (1 Angstrom = 0.000 000 000 1 meter). One Angstrom is the approximate diameter of an atom.

ANION (AN-EYE-en)

A negatively charged ion in an electrolyte solution, attracted to the anode under the influence of a difference in electrical potential. Chloride ion (Cl⁻) is an anion.

ANIONIC (AN-eye-ON-ick) POLYMER

A polymer having negatively charged groups of ions; often used as a filter aid and for dewatering sludges.

ANNULAR (AN-yoo-ler) SPACE

A ring-shaped space located between two circular objects. For example, the space between the outside of a pipe liner and the inside of a pipe.

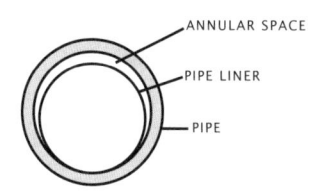

ANODE (AN-ode)

The positive pole or electrode of an electrolytic system, such as a battery. The anode attracts negatively charged particles or ions (anions).

APPARENT COLOR

Color of the water that includes not only the color due to substances in the water but suspended matter as well.

APPROPRIATIVE RIGHTS

Water rights to or ownership of a water supply that is acquired for the beneficial use of water by following a specific legal procedure.

APPURTENANCE (uh-PURR-ten-nans)

Machinery, appliances, structures, and other parts of the main structure necessary to allow it to operate as intended, but not considered part of the main structure.

AQUEOUS (ACK-wee-us)

Something made up of, similar to, or containing water; watery.

AQUIFER (ACK-wi-fer)

A natural, underground layer of porous, water-bearing materials (sand, gravel) usually capable of yielding a large amount or supply of water.

ARCH

(1) The curved top of a sewer pipe or conduit.
(2) A bridge or arch of hardened or caked chemical that will prevent the flow of the chemical.

ARTESIAN (are-TEE-zhun)

Pertaining to groundwater, a well, or underground basin where the water is under a pressure greater than atmospheric and will rise above the level of its upper confining surface if given an opportunity to do so.

ASEPTIC (a-SEP-tick)

Free from the living germs of disease, fermentation, or putrefaction. Sterile.

ASSOCIATION OF BOARDS OF CERTIFICATION (ABC)

An international organization representing over 150 boards that certify the operators of waterworks and wastewater facilities. For information on ABC publications regarding the preparation of and how to study for operator certification examinations, contact ABC, 208 Fifth Street, Ames, IA 50010-6259. Phone (515) 232-3623.

ASYMMETRIC (A-sim-MET-rick)

Not similar in size, shape, form, or arrangement of parts on opposite sides of a line, point, or plane.

ATOM

The smallest unit of a chemical element; composed of protons, neutrons, and electrons.

AUDIT, WATER

A thorough examination of the accuracy of water agency records or accounts (volumes of water) and system control equipment. Water managers can use audits to determine their water distribution system efficiency. The overall goal is to identify and verify water and revenue losses in a water system.

AUTHORITY

The power and resources to do a specific job or to get that job done.

AVAILABLE CHLORINE

A measure of the amount of chlorine available in chlorinated lime, hypochlorite compounds, and other materials that are used as a source of chlorine when compared with that of elemental (liquid or gaseous) chlorine.

AVAILABLE EXPANSION

The vertical distance from the sand surface to the underside of a trough in a sand filter. This distance is also called FREEBOARD.

AVERAGE

A number obtained by adding quantities or measurements and dividing the sum or total by the number of quantities or measurements. Also called the arithmetic mean.

$$\text{Average} = \frac{\text{Sum of Measurements}}{\text{Number of Measurements}}$$

AVERAGE DEMAND

The total demand for water during a period of time divided by the number of days in that time period. This is also called the average daily demand.

AXIAL TO IMPELLER

The direction in which material being pumped flows around the impeller or flows parallel to the impeller shaft.

AXIS OF IMPELLER

An imaginary line running along the center of a shaft (such as an impeller shaft).

B

BOD (pronounce as separate letters)

Biochemical Oxygen Demand. The rate at which organisms use the oxygen in water or wastewater while stabilizing decomposable organic matter under aerobic conditions. In decomposition, organic matter serves as food for the bacteria and energy results from its oxidation. BOD measurements are used as a surrogate measure of the organic strength of wastes in water.

BACK PRESSURE

A pressure that can cause water to backflow into the water supply when a user's water system is at a higher pressure than the public water system.

BACKFLOW

A reverse flow condition, created by a difference in water pressures, that causes water to flow back into the distribution pipes of a potable water supply from any source or sources other than an intended source. Also see BACKSIPHONAGE.

BACKSIPHONAGE

A form of backflow caused by a negative or below atmospheric pressure within a water system. Also see BACKFLOW.

BACKWASHING

The process of reversing the flow of water back through the filter media to remove the entrapped solids.

BACKWATER GATE

A gate installed at the end of a drain or outlet pipe to prevent the backward flow of water or wastewater. Generally used on storm sewer outlets into streams to prevent backward flow during times of flood or high tide. Also called a TIDE GATE. Also see CHECK VALVE and FLAP GATE.

BACTERIA (back-TEER-e-uh)

Bacteria are living organisms, microscopic in size, that usually consist of a single cell. Most bacteria use organic matter for their food and produce waste products as a result of their life processes.

BAFFLE

A flat board or plate, deflector, guide, or similar device constructed or placed in flowing water, wastewater, or slurry systems to cause more uniform flow velocities, to absorb energy, and to divert, guide, or agitate liquids (water, chemical solutions, slurry).

BAILER (BAY-ler)

A length of pipe equipped with a valve at the lower end used to remove slurry from the bottom or the side of a well as it is being drilled.

BASE-EXTRA CAPACITY METHOD

A cost allocation method used by utilities to determine rates for various groups. This method considers base costs (O & M expenses and capital costs), extra capacity costs (additional costs for maximum day and maximum hour demands), customer costs (meter maintenance and reading, billing, collection, accounting), and fire protection costs.

BASE METAL

A metal (such as iron) that reacts with dilute hydrochloric acid to form hydrogen. Also see NOBLE METAL.

BATCH PROCESS

A treatment process in which a tank or reactor is filled, the water (or wastewater or other solution) is treated or a chemical solution is prepared, and the tank is emptied. The tank may then be filled and the process repeated. Batch processes are also used to cleanse, stabilize, or condition chemical solutions for use in industrial manufacturing and treatment processes.

BENCH-SCALE ANALYSIS (TEST)

A method of studying different ways or chemical doses for treating water or wastewater and solids on a small scale in a laboratory. Also see JAR TEST.

BIOCHEMICAL OXYGEN DEMAND (BOD)

See BOD.

BIOLOGICAL GROWTH

The activity and growth of any and all living organisms.

BLANK

A bottle containing only dilution water or distilled water; the sample being tested is not added. Tests are frequently run on a sample and a blank and the differences are compared. The procedure helps to eliminate or reduce test result errors that could be caused when the dilution water or distilled water used is contaminated.

BOND

(1) A written promise to pay a specified sum of money (called the face value) at a fixed time in the future (called the date of maturity). A bond also carries interest at a fixed rate, payable periodically. The difference between a note and a bond is that a bond usually runs for a longer period of time and requires greater formality. Utility agencies use bonds as a means of obtaining large amounts of money for capital improvements.

(2) A warranty by an underwriting organization, such as an insurance company, guaranteeing honesty, performance, or payment by a contractor.

BONNET (BON-it)

The cover on a gate valve.

BOWL, PUMP

The submerged pumping unit in a well, including the shaft, impellers, and housing.

BRAKE HORSEPOWER

(1) The horsepower required at the top or end of a pump shaft (input to a pump).

(2) The energy provided by a motor or other power source.

BREAKPOINT CHLORINATION

Addition of chlorine to water or wastewater until the chlorine demand has been satisfied. At this point, further additions of chlorine will result in a free chlorine residual that is directly proportional to the amount of chlorine added beyond the breakpoint.

BREAKTHROUGH

A crack or break in a filter bed allowing the passage of floc or particulate matter through a filter. This will cause an increase in filter effluent turbidity. A breakthrough can occur (1) when a filter is first placed in service, (2) when the effluent valve suddenly opens or closes, and (3) during periods of excessive head loss through the filter (including when the filter is exposed to negative heads).

BRINELLING (bruh-NEL-ing)

Tiny indentations (dents) high on the shoulder of the bearing race or bearing. A type of bearing failure.

BUFFER

A solution or liquid whose chemical makeup neutralizes acids or bases without a great change in pH.

BUFFER CAPACITY

A measure of the capacity of a solution or liquid to neutralize acids or bases. This is a measure of the capacity of water or wastewater for offering a resistance to changes in pH.

C

C FACTOR

A factor or value used to indicate the smoothness of the interior of a pipe. The higher the C Factor, the smoother the pipe, the greater the carrying capacity, and the smaller the friction or energy losses from water flowing in the pipe. To calculate the C Factor, measure the flow, pipe diameter, distance between two pressure gauges, and the friction or energy loss of the water between the gauges.

$$\text{C Factor} = \frac{\text{Flow, GPM}}{193.75(\text{Diameter, ft})^{2.63}(\text{Slope})^{0.54}}$$

or

$$\text{C Factor} = \frac{\text{Flow, cu m/sec}}{0.278(\text{Diameter, m})^{2.63}(\text{Slope})^{0.54}}$$

CHEMTREC (KEM-trek)

Chemical Transportation Emergency Center. A public service of the American Chemistry Council dedicated to assisting emergency responders deal with incidents involving hazardous materials. Their toll-free 24-hour emergency phone number is (800) 424-9300.

CT VALUE

Residual concentration of a given disinfectant in mg/L times the disinfectant's contact time in minutes.

CAISSON (KAY-sawn)

A structure or chamber that is usually sunk or lowered by digging from the inside. Used to gain access to the bottom of a stream or other body of water.

CALCIUM CARBONATE EQUILIBRIUM

A water is considered stable when it is just saturated with calcium carbonate. In this condition, the water will neither dissolve nor deposit calcium carbonate. Thus, in this water the calcium carbonate is in equilibrium with the hydrogen ion concentration.

CALCIUM CARBONATE ($CaCO_3$) EQUIVALENT

An expression of the concentration of specified constituents in water in terms of their equivalent value to calcium carbonate. For example, the hardness in water that is caused by calcium, magnesium, and other ions is usually described as calcium carbonate equivalent. Alkalinity test results are usually reported as mg/L $CaCO_3$ equivalents. To convert chloride to $CaCO_3$ equivalents, multiply the concentration of chloride ions in mg/L by 1.41, and for sulfate, multiply by 1.04.

CALIBRATION

A procedure that checks or adjusts an instrument's accuracy by comparison with a standard or reference.

CALL DATE

First date a bond can be paid off.

CAPILLARY (KAP-uh-larry) ACTION

The movement of water through very small spaces due to molecular forces.

CAPILLARY (KAP-uh-larry) FORCES

The molecular forces that cause the movement of water through very small spaces.

CAPILLARY (KAP-uh-larry) FRINGE

The porous material just above the water table that may hold water by capillarity (a property of surface tension that draws water upward) in the smaller void spaces.

CARCINOGEN (kar-SIN-o-jen)

Any substance that tends to produce cancer in an organism.

CATALYST (KAT-uh-list)

A substance that changes the speed or yield of a chemical reaction without being consumed or chemically changed by the chemical reaction.

CATALYZE (KAT-uh-lize)

To act as a catalyst. Or, to speed up a chemical reaction.

CATALYZED (KAT-uh-lized)

To be acted upon by a catalyst.

CATHODE (KATH-ode)

The negative pole or electrode of an electrolytic cell or system. The cathode attracts positively charged particles or ions (cations).

CATHODIC (kath-ODD-ick) PROTECTION

An electrical system for prevention of rust, corrosion, and pitting of metal surfaces that are in contact with water, wastewater, or soil. A low-voltage current is made to flow through a liquid (water) or a soil in contact with the metal in such a manner that the external electromotive force renders the metal structure cathodic. This concentrates corrosion on auxiliary anodic parts, which are deliberately allowed to corrode instead of letting the structure corrode.

CATION (KAT-EYE-en)

A positively charged ion in an electrolyte solution, attracted to the cathode under the influence of a difference in electrical potential. Sodium ion (Na^+) is a cation.

CATIONIC (KAT-eye-ON-ic) POLYMER

A polymer having positively charged groups of ions; often used as a coagulant aid.

CAUTION

This word warns against potential hazards or cautions against unsafe practices. Also see DANGER, NOTICE, and WARNING.

CAVITATION (kav-uh-TAY-shun)

The formation and collapse of a gas pocket or bubble on the blade of an impeller or the gate of a valve. The collapse of this gas pocket or bubble drives water into the impeller or gate with a terrific force that can knock metal particles off and cause pitting on the impeller or gate surface. Cavitation is accompanied by loud noises that sound like someone is pounding on the impeller or gate with a hammer.

CENTRATE

The water leaving a centrifuge after most of the solids have been removed.

CENTRIFUGAL (sen-TRIF-uh-gull) PUMP

A pump consisting of an impeller fixed on a rotating shaft that is enclosed in a casing, and having an inlet and discharge connection. As the rotating impeller whirls the liquid around, centrifugal force builds up enough pressure to force the water through the discharge outlet.

CENTRIFUGE

A mechanical device that uses centrifugal or rotational forces to separate solids from liquids.

CERTIFICATION EXAMINATION

An examination administered by a state agency or professional association that operators take to indicate a level of professional competence. In the United States, certification of operators of water treatment plants, wastewater treatment plants, water distribution systems, and small water supply systems is mandatory. In many states, certification of wastewater collection system operators, industrial wastewater treatment plant operators, pretreatment facility inspectors, and small wastewater system operators is voluntary; however, current trends indicate that more states, provinces, and employers will require these operators to be certified in the future. Operator certification is mandatory in the United States for the Chief Operators of water treatment plants, water distribution systems, and wastewater treatment plants.

CERTIFIED OPERATOR

A person who has the education and experience required to operate a specific class of treatment facility as indicated by possessing a certificate of professional competence given by a state agency or professional association.

CHARGE CHEMISTRY

A branch of chemistry in which the destabilization and neutralization reactions occur between stable, negatively charged and stable, positively charged particles.

CHECK SAMPLING

Whenever an initial or routine sample analysis indicates that a Maximum Contaminant Level (MCL) has been exceeded, check sampling is required to confirm the routine sampling results. Check sampling is in addition to the routine sampling program.

CHECK VALVE

A special valve with a hinged disk or flap that opens in the direction of normal flow and is forced shut when flows go in the reverse or opposite direction of normal flows. Also see FLAP GATE and TIDE GATE.

CHELATING (KEY-LAY-ting) AGENT

A chemical used to prevent the precipitation of metals (such as copper).

CHELATION (key-LAY-shun)

A chemical complexing (forming or joining together) of metallic cations (such as copper) with certain organic compounds, such as EDTA (ethylene diamine tetracetic acid). Chelation is used to prevent the precipitation of metals (copper). Also see SEQUESTRATION.

CHLORAMINATION (KLOR-uh-min-NAY-shun)

The application of chlorine and ammonia to water to form chloramines for the purpose of disinfection.

CHLORAMINES (KLOR-uh-means)

Compounds formed by the reaction of hypochlorous acid (or aqueous chlorine) with ammonia.

CHLORINATION (klor-uh-NAY-shun)

The application of chlorine to water or wastewater, generally for the purpose of disinfection, but frequently for accomplishing other biological or chemical results—aiding coagulation and controlling tastes and odors in drinking water, or controlling odors or sludge bulking in wastewater.

CHLORINATOR (KLOR-uh-nay-ter)

A metering device that is used to add chlorine to water.

CHLORINE DEMAND

Chlorine demand is the difference between the amount of chlorine added to water or wastewater and the amount of residual chlorine remaining after a given contact time. Chlorine demand may change with dosage, time, temperature, pH, and nature and amount of the impurities in the water.

$$\text{Chlorine Demand, mg}/L = \text{Chlorine Applied, mg}/L - \text{Chlorine Residual, mg}/L$$

CHLORINE REQUIREMENT

The amount of chlorine that is needed for a particular purpose. Some reasons for adding chlorine are reducing the MPN (Most Probable Number) of coliform bacteria, obtaining a particular chlorine residual, or oxidizing some substance in the water. In each case, a definite dosage of chlorine will be necessary. This dosage is the chlorine requirement.

CHLORINE RESIDUAL

The concentration of chlorine present in water after the chlorine demand has been satisfied. The concentration is expressed in terms of the total chlorine residual, which includes both the free and combined or chemically bound chlorine residuals. Also called RESIDUAL CHLORINE.

CHLOROPHENOLIC (KLOR-o-fee-NO-lick)

Chlorophenolic compounds are phenolic compounds (carbolic acid) combined with chlorine.

CHLOROPHENOXY (KLOR-o-fuh-NOX-ee)

A class of herbicides that may be found in domestic water supplies and cause adverse health effects. Two widely used chlorophenoxy herbicides are 2,4-D (2,4-Dichlorophenoxy acetic acid) and 2,4,5-TP (2,4,5-Trichlorophenoxy propionic acid [silvex]).

CHLORORGANIC (klor-or-GAN-ick)

Organic compounds combined with chlorine. These compounds generally originate from, or are associated with, living or dead organic materials, such as algae in water.

CHRONIC HEALTH EFFECT

An adverse effect on a human or animal body with symptoms that develop slowly over a long period of time or that recur frequently.

CIRCLE OF INFLUENCE

The circular outer edge of a depression produced in the water table by the pumping of water from a well. Also see CONE OF DEPRESSION.

[SEE DRAWING ON PAGE 785]

CIRCUIT

The complete path of an electric current, including the generating apparatus or other source; or, a specific segment or section of the complete path.

CIRCUIT BREAKER

A safety device in an electric circuit that automatically shuts off the circuit when it becomes overloaded. The device can be manually reset.

CISTERN (SIS-turn)

A small tank (usually covered) or a storage facility used to store water for a home or farm. Often used to store rainwater.

CLARIFIER (KLAIR-uh-fire)

A tank or basin in which water or wastewater is held for a period of time during which the heavier solids settle to the bottom and the lighter materials float to the surface. Also called settling tank or SEDIMENTATION BASIN.

CLASS, PIPE AND FITTINGS

The working pressure rating, including allowances for surges, of a specific pipe for use in water distribution systems. The term is used for cast iron, ductile iron, asbestos cement, and some plastic pipe.

CLEAR WELL

A reservoir for the storage of filtered water of sufficient capacity to prevent the need to vary the filtration rate with variations in demand. Also used to provide chlorine contact time for disinfection.

COAGULANT (ko-AGG-yoo-lent)

A chemical that causes very fine particles to clump (floc) together into larger particles. This makes it easier to separate the solids from the liquids by settling, skimming, draining, or filtering.

COAGULANT (ko-AGG-yoo-lent) AID

Any chemical or substance used to assist or modify coagulation.

COAGULATION (ko-agg-yoo-LAY-shun)

The clumping together of very fine particles into larger particles (floc) caused by the use of chemicals (coagulants). The chemicals neutralize the electrical charges of the fine particles, allowing them to come closer and form larger clumps.

CODE OF FEDERAL REGULATIONS (CFR)

A publication of the US government that contains all of the proposed and finalized federal regulations, including safety and environmental regulations.

TOP or PLAN VIEW

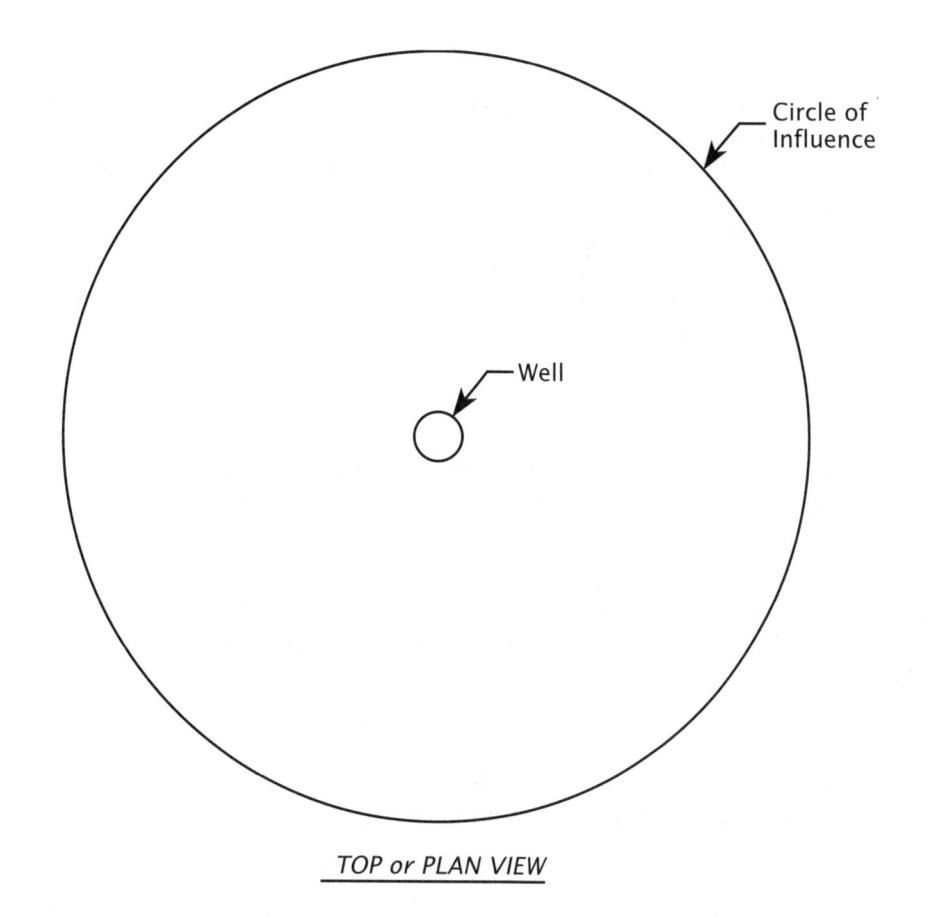

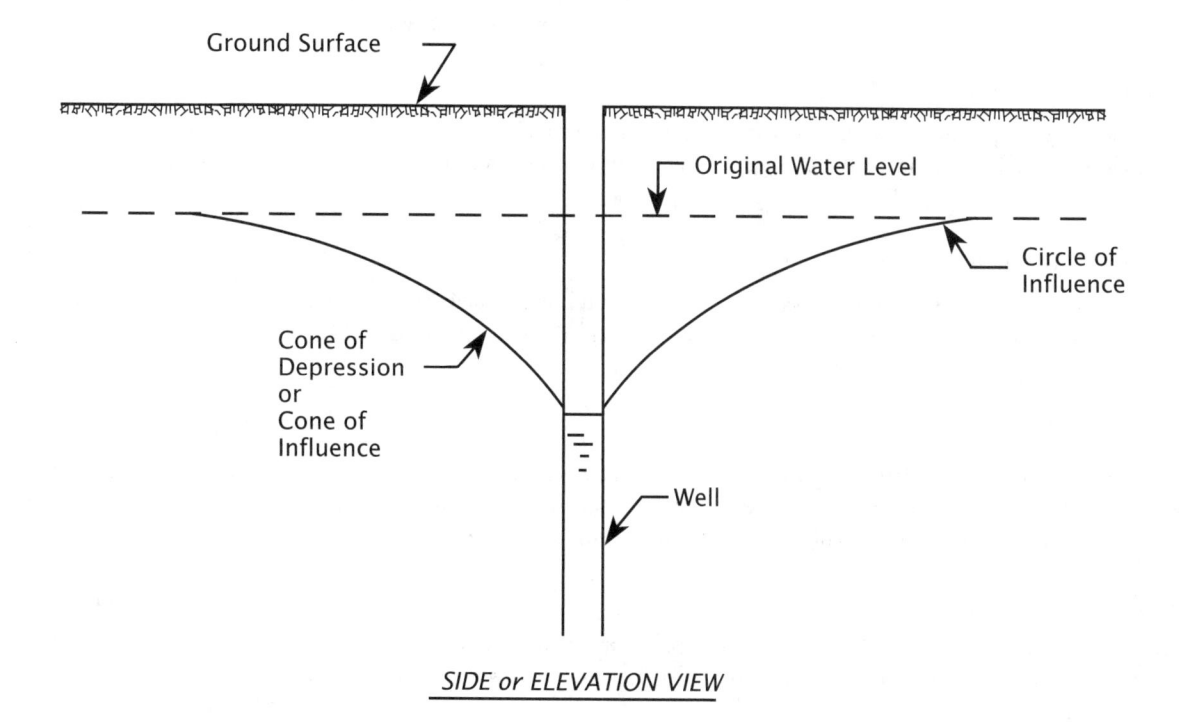

SIDE or ELEVATION VIEW

Circle of influence and cone of depression/cone of influence

COLIFORM (KOAL-i-form)

A group of bacteria found in the intestines of warm-blooded animals (including humans) and also in plants, soil, air, and water. The presence of coliform bacteria is an indication that the water is polluted and may contain pathogenic (disease-causing) organisms. Fecal coliforms are those coliforms found in the feces of various warm-blooded animals, whereas the term "coliform" also includes other environmental sources.

COLLOIDS (KALL-loids)

Very small, finely divided solids (particles that do not dissolve) that remain dispersed in a liquid for a long time due to their small size and electrical charge. When most of the particles in water have a negative electrical charge, they tend to repel each other. This repulsion prevents the particles from clumping together, becoming heavier, and settling out.

COLOR

The substances in water that impart a yellowish-brown color to the water. These substances are the result of iron and manganese ions, humus and peat materials, plankton, aquatic weeds, and industrial waste present in the water. Also see TRUE COLOR.

COLORIMETRIC MEASUREMENT

A means of measuring unknown chemical concentrations in water by measuring a sample's color intensity. The specific color of the sample, developed by addition of chemical reagents, is measured with a photoelectric colorimeter or is compared with color standards using, or corresponding with, known concentrations of the chemical.

COMBINED AVAILABLE CHLORINE

The total chlorine, present as chloramine or other derivatives, that is present in a water and is still available for disinfection and for oxidation of organic matter. The combined chlorine compounds are more stable than free chlorine forms, but they are somewhat slower in disinfection action.

COMBINED AVAILABLE CHLORINE RESIDUAL

The concentration of residual chlorine that is combined with ammonia, organic nitrogen, or both in water as a chloramine (or other chloro derivative) and yet is still available to oxidize organic matter and help kill bacteria.

COMBINED CHLORINE

The sum of the chlorine species composed of free chlorine and ammonia, including monochloramine, dichloramine, and trichloramine (nitrogen trichloride). Dichloramine is the strongest disinfectant of these chlorine species, but it has less oxidative capacity than free chlorine.

COMBINED RESIDUAL CHLORINATION

The application of chlorine to water or wastewater to produce a combined available chlorine residual. The residual may consist of chlorine compounds formed by the reaction of chlorine with natural or added ammonia (NH_3) or with certain organic nitrogen compounds.

COMMODITY-DEMAND METHOD

A cost allocation method used by water utilities to determine water rates for the various water user groups. This method considers the commodity costs (water, chemicals, power, amount of water use), demand costs (treatment, storage, distribution), customer costs (meter maintenance and reading, billing, collection, accounting), and fire protection costs.

COMPETENT PERSON

A competent person is defined by OSHA as a person capable of identifying existing and predictable hazards in the surroundings, or working conditions that are unsanitary, hazardous, or dangerous to employees, and who has authorization to take prompt corrective measures to eliminate the hazards.

COMPLETE TREATMENT

A method of treating water that consists of the addition of coagulant chemicals, flash mixing, coagulation-flocculation, sedimentation, and filtration. Also called CONVENTIONAL FILTRATION. Also see DIRECT FILTRATION and IN-LINE FILTRATION.

COMPOSITE (PROPORTIONAL) SAMPLE

A composite sample is a collection of individual samples obtained at regular intervals, usually every one or two hours during a 24-hour time span. Each individual sample is combined with the others in proportion to the rate of flow when the sample was collected. Equal volume individual samples also may be collected at intervals after a specific volume of flow passes the sampling point or after equal time intervals and still be referred to as a composite sample. The resulting mixture (composite sample) forms a representative sample and is analyzed to determine the average conditions during the sampling period.

COMPOUND

A pure substance composed of two or more elements whose composition is constant. For example, table salt (sodium chloride, NaCl) is a compound.

CONCENTRATION POLARIZATION

(1) A buildup of retained particles on the membrane surface due to dewatering of the feed closest to the membrane. The thickness of the concentration polarization layer is controlled by the flow velocity across the membrane.

(2) Used in corrosion studies to indicate a depletion of ions near an electrode.

(3) The basis for chemical analysis by a polarograph.

CONDITIONING

Pretreatment of sludge to facilitate removal of water in subsequent treatment processes.

CONDUCTANCE

A rapid method of estimating the dissolved solids content of a water supply. The measurement indicates the capacity of a sample of water to carry an electric current, which is related to the concentration of ionized substances in the water. Also called SPECIFIC CONDUCTANCE.

CONDUCTIVITY

A measure of the ability of a solution (water) to carry an electric current.

CONDUCTOR

(1) A pipe that carries a liquid load from one point to another point. In a wastewater collection system, a conductor is often a large pipe with no service connections. Also called a CONDUIT. Also see INTERCEPTOR (INTERCEPTING) SEWER or INTERCONNECTOR.

(2) In plumbing, a pipe installed to drain water from the roof gutters or roof catchment to the storm drain or other means of disposal. Also called a DOWNSPOUT.

(3) In electricity, a substance, body, device, or wire that readily conducts or carries electric current. Also called a CONDUIT.

CONDUCTOR CASING

The outer casing of a well. The purpose of this casing is to prevent contaminants from surface waters or shallow groundwaters from entering a well.

CONDUIT

Any artificial or natural duct, either open or closed, for carrying fluids from one point to another. An electrical conduit carries electricity.

CONE OF DEPRESSION

The depression, roughly conical in shape, produced in the water table by the pumping of water from a well. Also called the cone of influence. Also see CIRCLE OF INFLUENCE.

[SEE DRAWING ON PAGE 785]

CONE OF INFLUENCE

See CONE OF DEPRESSION.

CONFINED SPACE

Confined space means a space that:

(1) Is large enough and so configured that an employee can bodily enter and perform assigned work; and

(2) Has limited or restricted means for entry or exit (for example, manholes, tanks, vessels, silos, storage bins, hoppers, vaults, and pits are spaces that may have limited means of entry); and

(3) Is not designed for continuous employee occupancy.

Also see DANGEROUS AIR CONTAMINATION and OXYGEN DEFICIENCY.

CONFINED SPACE, CLASS A

A confined space that presents a situation that is immediately dangerous to life or health (IDLH). These include but are not limited to oxygen deficiency, explosive or flammable atmospheres, and concentrations of toxic substances.

CONFINED SPACE, CLASS B

A confined space that has the potential for causing injury and illness, if preventive measures are not used, but is not immediately dangerous to life and health.

CONFINED SPACE, CLASS C

A confined space in which the potential hazard would not require any special modification of the work procedure.

CONFINED SPACE, NON-PERMIT

A non-permit confined space is a confined space that does not contain or, with respect to atmospheric hazards, have the potential to contain any hazard capable of causing death or serious physical harm.

CONFINED SPACE, PERMIT-REQUIRED (PERMIT SPACE)

A confined space that has one or more of the following characteristics:

(1) Contains or has a potential to contain a hazardous atmosphere,

(2) Contains a material that has the potential for engulfing an entrant,

(3) Has an internal configuration such that an entrant could be trapped or asphyxiated by inwardly converging walls or by a floor that slopes downward and tapers to a smaller cross section, or

(4) Contains any other recognized serious safety or health hazard.

CONFINING UNIT

A layer of rock or soil of very low hydraulic conductivity that hampers the movement of groundwater in and out of an aquifer.

CONSOLIDATED FORMATION

A geologic material whose particles are stratified (layered), cemented, or firmly packed together (hard rock); usually occurring at a depth below the ground surface. Also see UNCONSOLIDATED FORMATION.

CONSUMER CONFIDENCE REPORT (CCR)

An annual report prepared by a water utility to communicate with its consumers. The report provides consumers with information on the source and quality of their drinking water and is an opportunity for positive communication with consumers.

CONTACTOR

An electric switch, usually magnetically operated.

CONTAMINATION

The introduction into water of microorganisms, chemicals, toxic substances, wastes, or wastewater in a concentration that makes the water unfit for its next intended use.

CONTINUOUS SAMPLE

A flow of water from a particular place in a plant to the location where samples are collected for testing. This continuous stream may be used to obtain grab or composite samples. Frequently, several taps (faucets) will flow continuously in the laboratory to provide test samples from various places in a water treatment plant.

CONTROL LOOP

The combination of one or more interconnected instrumentation devices that are arranged to measure, display, and control a process variable. Also called a loop.

CONTROL SYSTEM

An instrumentation system that senses and controls its own operation on a close, continuous basis in what is called proportional (or modulating) control.

CONTROLLER

A device that controls the starting, stopping, or operation of a device or piece of equipment.

CONVENTIONAL FILTRATION

A method of treating water that consists of the addition of coagulant chemicals, flash mixing, coagulation-flocculation, sedimentation, and filtration. Also called COMPLETE TREATMENT. Also see DIRECT FILTRATION and IN-LINE FILTRATION.

CONVENTIONAL TREATMENT

(1) The common wastewater treatment processes such as preliminary treatment, sedimentation, flotation, trickling filter, rotating biological contactor, activated sludge, and chlorination wastewater treatment processes used by POTWs.

(2) The hydroxide precipitation of metals processes used by pretreatment facilities.

COPRECIPITATION

A treatment process that occurs when ferrous iron is added to water or metallic wastestreams and subsequently oxidized in an aerator. The oxidized iron, which is insoluble, precipitates along with other metallic contaminants present in the water or wastestream, thereby enhancing metals removal.

CORPORATION STOP

A water service shutoff valve located at a street water main. This valve cannot be operated from the ground surface because it is buried and there is no valve box. Also called a corporation cock.

CORROSION

The gradual decomposition or destruction of a material by chemical action, often due to an electrochemical reaction. Corrosion may be caused by (1) stray current electrolysis, (2) galvanic corrosion caused by dissimilar metals, or (3) differential-concentration cells. Corrosion starts at the surface of a material and moves inward.

CORROSION INHIBITORS

Substances that slow the rate of corrosion.

CORROSIVE GASES

Gases that, when dissolved in water, can oxidize materials of construction such as steel and concrete.

CORROSIVITY

An indication of the corrosiveness of a water. The corrosiveness of a water is described by the water's pH, alkalinity, hardness, temperature, total dissolved solids, dissolved oxygen concentration, and the Langelier Index.

COULOMB (KOO-lahm)

A measurement of the amount of electrical charge carried by an electric current of one ampere in one second. One coulomb equals about 6.25×10^{18} electrons (6,250,000,000,000,000,000 electrons).

COUPON

A steel specimen inserted into water or wastewater to measure corrosiveness. The rate of corrosion is measured as the loss of weight of the coupon or change in its physical characteristics. Measure the weight loss (in milligrams) per surface area (in square decimeters) exposed to the water or wastewater per day. 1 meter = 10 decimeters = 100 centimeters.

COVERAGE RATIO

The coverage ratio is a measure of the ability of the utility to pay the principal and interest on loans and bonds (this is known as debt service) in addition to any unexpected expenses.

CROSS CONNECTION

(1) A connection between drinking (potable) water and an unapproved water supply.

(2) A connection between a storm drain system and a sanitary collection system.

(3) Less frequently used to mean a connection between two sections of a collection system to handle anticipated overloads of one system.

CROSS-FLOW FILTRATION

A type of membrane filtration where the water being filtered flows across the surface of the membrane to keep the particle buildup and fouling to a minimum. The flow that is not filtered becomes concentrated and flows out the end of the membrane fiber as a wastestream.

CRYPTOSPORIDIUM (KRIP-toe-spo-RID-ee-um)

A waterborne intestinal parasite that causes a disease called cryptosporidiosis (KRIP-toe-spo-rid-ee-O-sis) in infected humans. Symptoms of the disease include diarrhea, cramps, and weight loss. *Cryptosporidium* contamination is found in most surface waters and some groundwaters. Commonly referred to as "crypto."

CURB STOP

A water service shutoff valve located in a water service pipe near the curb and between the water main and the building. This valve is usually operated by a wrench or valve key and is used to start or stop flows in the water service line to a building. Also called a curb cock.

CURIE (KYOOR-ee)

A measure of radioactivity. One Curie of radioactivity is equivalent to 3.7×10^{10} or 37,000,000,000 nuclear disintegrations per second.

CURRENT

A movement or flow of electricity. Electric current is measured by the number of coulombs per second flowing past a certain point in a conductor. A coulomb is equal to about 6.25×10^{18} electrons (6,250,000,000,000,000,000 electrons). A flow of one coulomb per second is called one ampere, the unit of the rate of flow of current.

CYCLE

A complete alternation of voltage or current in an alternating current (AC) circuit.

D

DBP

See DISINFECTION BY-PRODUCT.

DOC

See DISSOLVED ORGANIC CARBON.

DOM

See DISSOLVED ORGANIC MATTER.

DPD METHOD

A method of measuring the chlorine residual in water. The residual may be determined by either titrating or comparing a developed color with color standards. DPD stands for N,N-diethyl-p-phenylenediamine.

DANGER

The word *DANGER* is used where an immediate hazard presents a threat of death or serious injury to employees. Also see CAUTION, NOTICE, and WARNING.

DANGEROUS AIR CONTAMINATION

An atmosphere presenting a threat of causing death, injury, acute illness, or disablement due to the presence of flammable and/or explosive, toxic, or otherwise injurious or incapacitating substances.

(1) Dangerous air contamination due to the flammability of a gas, vapor, or mist is defined as an atmosphere containing the gas, vapor, or mist at a concentration greater than 10 percent of its lower explosive (lower flammable) limit (LEL).

(2) Dangerous air contamination due to a combustible particulate is defined as a concentration that meets or exceeds the particulate's lower explosive limit (LEL).

(3) Dangerous air contamination due to the toxicity of a substance is defined as the atmospheric concentration that could result in employee exposure in excess of the substance's permissible exposure limit (PEL).

NOTE: A dangerous situation also occurs when the oxygen level is less than 19.5 percent by volume (OXYGEN DEFICIENCY) or more than 23.5 percent by volume (OXYGEN ENRICHMENT).

DATEOMETER (day-TOM-uh-ter)

A small calendar disk attached to motors and equipment to indicate the year in which the last maintenance service was performed.

DATUM LINE

A line from which heights and depths are calculated or measured. Also called a datum plane or a datum level.

DAY TANK

A tank used to store a chemical solution of known concentration for feed to a chemical feeder. A day tank usually stores sufficient chemical solution to properly treat the water being treated for at least one day. Also called an AGE TANK.

DEAD END

The end of a water main that is not connected to other parts of the distribution system by means of a connecting loop of pipe.

DEAD-END FILTRATION

A type of membrane filtration where the water being filtered flows through the membrane, but there is no wastestream from the system. All solids accumulate on the membrane during filtration and are removed during backwash.

DEBT SERVICE

The amount of money required annually to pay the (1) interest on outstanding debts, or (2) funds due on a maturing bonded debt or the redemption of bonds.

DECANT (de-KANT)

To draw off the upper layer of liquid (water) after the heavier material (a solid or another liquid) has settled.

DECANT (de-KANT) WATER

Water that has separated from sludge and is removed from the layer of water above the sludge.

DECHLORINATION (DEE-klor-uh-NAY-shun)

The deliberate removal of chlorine from water. The partial or complete reduction of residual chlorine by any chemical or physical process.

DECIBEL (DES-uh-bull)

A unit for expressing the relative intensity of sounds on a scale from zero for the average least perceptible sound to about 130 for the average level at which sound causes pain to humans. Abbreviated dB.

DECOMPOSITION or DECAY

The conversion of chemically unstable materials to more stable forms by chemical or biological action.

DEFLUORIDATION (DEE-floor-uh-DAY-shun)

The removal of excess fluoride in drinking water to prevent the mottling (brown stains) of teeth.

DEGASIFICATION (DEE-gas-if-uh-KAY-shun)

A water treatment process that removes dissolved gases from the water. The gases may be removed by either mechanical or chemical treatment methods or a combination of both.

DELEGATION

The act in which power is given to another person in the organization to accomplish a specific job.

DEMINERALIZATION (DEE-min-er-al-uh-ZAY-shun)

A treatment process that removes dissolved minerals (salts) from water.

DENSITY

A measure of how heavy a substance (solid, liquid, or gas) is for its size. Density is expressed in terms of weight per unit volume, that is, grams per cubic centimeter or pounds per cubic foot. The density of water (at 4°C or 39°F) is 1.0 gram per cubic centimeter or about 62.4 pounds per cubic foot.

DEPOLARIZATION

The removal or depletion of ions in the thin boundary layer adjacent to a membrane or pipe wall.

DEPRECIATION

The gradual loss in service value of a facility or piece of equipment due to all the factors causing the ultimate retirement of the facility or equipment. This loss can be caused by sudden physical damage, wearing out due to age, obsolescence, inadequacy, or availability of a newer, more efficient facility or equipment. The value cannot be restored by maintenance.

DESALINIZATION (DEE-SAY-lin-uh-ZAY-shun)

The removal of dissolved salts (such as sodium chloride, NaCl) from water by natural means (leaching) or by specific water treatment processes.

DESICCANT (DESS-uh-kant)

A drying agent that is capable of removing or absorbing moisture from the atmosphere in a small enclosure.

DESICCATION (dess-uh-KAY-shun)

A process used to thoroughly dry air; to remove virtually all moisture from air.

DESICCATOR (DESS-uh-kay-tor)

A closed container into which heated weighing or drying dishes are placed to cool in a dry environment in preparation for weighing. The dishes may be empty or they may contain a sample. Desiccators contain a substance (DESICCANT), such as anhydrous calcium chloride, that absorbs moisture and keeps the relative humidity near zero so that the dish or sample will not gain weight from absorbed moisture.

DESTRATIFICATION (DEE-strat-uh-fuh-KAY-shun)

The development of vertical mixing within a lake or reservoir to eliminate (either totally or partially) separate layers of temperature, plant life, or animal life. This vertical mixing can be caused by mechanical means (pumps) or through the use of forced air diffusers that release air into the lower layers of the reservoir.

DETENTION TIME

(1) The time required to fill a tank at a given flow.

(2) The theoretical (calculated) time required for water to pass through a tank at a given rate of flow.

(3) The actual time in hours, minutes, or seconds that a small amount of water is in a settling basin, flocculating basin, or rapid-mix chamber. In septic tanks, detention time will decrease as the volumes of sludge and scum increase. In storage reservoirs, detention time is the length of time entering water will be held before being drafted for use (several weeks to years, several months being typical).

$$\text{Detention Time, hr} = \frac{(\text{Basin Volume, gal})(24 \text{ hr/day})}{\text{Flow, gal/day}}$$

or

$$\text{Detention Time, hr} = \frac{(\text{Basin Volume, cu m})(24 \text{ hr/day})}{\text{Flow, cu m/day}}$$

DEW POINT

The temperature to which air with a given quantity of water vapor must be cooled to cause condensation of the vapor in the air.

DEWATER

(1) To remove or separate a portion of the water present in a sludge or slurry. To dry sludge so it can be handled and disposed of.

(2) To remove or drain the water from a tank or a trench. A structure may be dewatered so that it can be inspected or repaired.

DIATOMACEOUS (DYE-uh-toe-MAY-shus) EARTH

A fine, siliceous (made of silica) earth composed mainly of the skeletal remains of diatoms.

DIATOMS (DYE-uh-toms)

Unicellular (single cell), microscopic algae with a rigid, box-like internal structure consisting mainly of silica.

DIELECTRIC (DYE-ee-LECK-trick)

Does not conduct an electric current. An insulator or nonconducting substance.

DIGITAL

The encoding of information that uses binary numbers (ones and zeros) for input, processing, transmission, storage, or display, rather than a continuous spectrum of values (an analog system) or non-numeric symbols such as letters or icons.

DIGITAL READOUT

The readout of an instrument by a direct, numerical reading of the measured value or variable.

DILUTE SOLUTION

A solution that has been made weaker, usually by the addition of water.

DIMICTIC (dye-MICK-tick)

Lakes and reservoirs that freeze over and normally go through two stratification and two mixing cycles within a year.

DIRECT CURRENT (DC)

Electric current flowing in one direction only and essentially free from pulsation. Also see ALTERNATING CURRENT.

DIRECT FILTRATION

A method of treating water that consists of the addition of coagulant chemicals, flash mixing, coagulation, minimal flocculation, and filtration. The flocculation facilities may be omitted, but the physical-chemical reactions will occur to some extent. The sedimentation process is omitted. Also see CONVENTIONAL FILTRATION and IN-LINE FILTRATION.

DIRECT RUNOFF

Water that flows over the ground surface directly into streams, rivers, or lakes. Also called STORM RUNOFF.

DISCHARGE HEAD

The pressure (in pounds per square inch (psi) or kilopascals (kPa)) measured at the centerline of a pump discharge and very close to the discharge flange, converted into feet or meters. The pressure is measured from the centerline of the pump to the hydraulic grade line of the water in the discharge pipe.

Discharge Head, ft = (Discharge Pressure, psi)(2.31 ft/psi)

or

Discharge Head, m = (Discharge Pressure, kPa)(1 m/9.8 kPa)

DISCRETE CONTROL

ON/OFF control; one of the two output values is equal to zero.

DISCRETE I/O (INPUT/OUTPUT)

A digital signal that senses or sends either ON or OFF signals. For example, a discrete input would sense the position of a switch; a discrete output would turn on a pump or light.

DISINFECTION (dis-in-FECT-shun)

The process designed to kill or inactivate most microorganisms in water or wastewater, including essentially all pathogenic (disease-causing) bacteria. There are several ways to disinfect, with chlorination being the most frequently used in water and wastewater treatment plants. Compare with STERILIZATION.

DISINFECTION BY-PRODUCT (DBP)

A contaminant formed by the reaction of disinfection chemicals (such as chlorine) with other substances in the water being disinfected.

DISSOLVED ORGANIC CARBON (DOC)

That portion of the organic carbon in water that passes through a 0.45 μm pore-diameter filter.

DISSOLVED ORGANIC MATTER (DOM)

That portion of the organic matter in water that passes through a 0.45 μm pore-diameter filter.

DISTILLATE (DIS-tuh-late)

In the distillation of a sample, a portion is collected by evaporation and recondensation; the part that is recondensed is the distillate.

DISTRIBUTED CONTROL SYSTEM (DCS)

A computer control system having multiple microprocessors to distribute the functions performing process control, thereby distributing the risk from component failure. The distributed components (input/output devices, control devices, and operator interface devices) are all connected by communications links and permit the transmission of control, measurement, and operating information to and from many locations.

DIVALENT (dye-VAY-lent)

Having a valence of two, such as the ferrous ion, Fe^{2+}. Also called bivalent.

DIVERSION

Use of part of a stream flow as a water supply.

DOWNSPOUT

In plumbing, a pipe installed to drain water from the roof gutters or roof catchment to the storm drain or other means of disposal. Also called a CONDUCTOR, ROOF LEADER, or roof drain.

DRAFT

(1) The act of drawing or removing water from a tank or reservoir.
(2) The water that is drawn or removed from a tank or reservoir.

DRAWDOWN

(1) The drop in the water table or level of water in the ground when water is being pumped from a well.

(2) The amount of water used from a tank or reservoir.

(3) The drop in the water level of a tank or reservoir.

DRIFT

The difference between the actual value and the desired value (or set point); characteristic of proportional controllers that do not incorporate reset action. Also called OFFSET.

DYNAMIC HEAD

When a pump is operating, the vertical distance (in feet or meters) from a point to the energy grade line. Also see ENERGY GRADE LINE, STATIC HEAD, and TOTAL DYNAMIC HEAD.

DYNAMIC PRESSURE

When a pump is operating, pressure resulting from the dynamic head.

Dynamic Pressure, psi = (Dynamic Head, ft)(0.433 psi/ft)

or

Dynamic Pressure, kPa = (Dynamic Head, m)(9.8 kPa/m)

E

EGL

See ENERGY GRADE LINE.

EPA

United States Environmental Protection Agency. A regulatory agency established by the US Congress to administer the nation's environmental laws. Also called the US EPA.

EDUCTOR (e-DUCK-ter)

A hydraulic device used to create a negative pressure (suction) by forcing a liquid through a restriction, such as a Venturi. An eductor or aspirator (the hydraulic device) may be used in the laboratory in place of a vacuum pump. As an injector, it is used to produce vacuum for chlorinators. Sometimes used instead of a suction pump.

EFFECTIVE RANGE

That portion of the design range (usually from 10 to 90+ percent) in which an instrument has acceptable accuracy. Also see RANGE and SPAN.

EFFECTIVE SIZE (ES)

The diameter of the particles in a granular sample (filter media) for which 10 percent of the total grains are smaller and 90 percent larger on a weight basis. Effective size is obtained by passing granular material through sieves with varying dimensions of mesh and weighing the material retained by each sieve. The effective size is also approximately the average size of the grains.

EFFLUENT (EF-loo-ent)

Water or other liquid—raw (untreated), partially treated, or completely treated—flowing *FROM* a reservoir, basin, treatment process, or treatment plant.

EJECTOR

A device used to disperse a chemical solution into water being treated.

ELECTRICAL LOGGING

A procedure used to search for water-bearing formations (aquifers) by determining the porosity (spaces or voids) of geologic materials. Electrical probes are lowered into wells, an electric current is induced at various depths, and the resistance measured indicates the porosity of the formation.

ELECTROCHEMICAL REACTION

Chemical changes produced by electricity (electrolysis) or the production of electricity by chemical changes (galvanic action). In corrosion, a chemical reaction is accompanied by the flow of electrons through a metallic path. The electron flow may come from an external source and cause the reaction, such as electrolysis caused by a DC (direct current) electric railway, or the electron flow may be caused by a chemical reaction, as in the galvanic action of a flashlight dry cell.

ELECTROCHEMICAL SERIES

A list of metals with the standard electrode potentials given in volts. The size and sign of the electrode potential indicates how easily these elements will take on or give up electrons, or corrode. Hydrogen is conventionally assigned a value of zero.

ELECTROLYSIS (ee-leck-TRAWL-uh-sis)

The decomposition of material by an outside electric current.

ELECTROLYTE (ee-LECK-tro-lite)

A substance that dissociates (separates) into two or more ions when it is dissolved in water.

ELECTROLYTIC (ee-LECK-tro-LIT-ick) CELL

A device in which the chemical decomposition of material causes an electric current to flow. Also, a device in which a chemical reaction occurs as a result of the flow of electric current. Chlorine and caustic (NaOH) are made from salt (NaCl) in electrolytic cells.

ELECTROMOTIVE FORCE (EMF)

The electrical pressure available to cause a flow of current (amperage) when an electric circuit is closed. Also called VOLTAGE.

ELECTROMOTIVE SERIES

A list of metals and alloys presented in the order of their tendency to corrode (or go into solution). Also called the GALVANIC SERIES. This is a practical application of the theoretical ELECTROCHEMICAL SERIES.

ELECTRON

(1) A very small, negatively charged particle that is practically weightless. According to the electron theory, all electrical and electronic effects are caused either by the movement of electrons from place to place or because there is an excess or lack of electrons at a particular place.
(2) The part of an atom that determines its chemical properties.

ELEMENT

A substance that cannot be separated into its constituent parts and still retain its chemical identity. For example, sodium (Na) is an element.

EMPTY BED CONTACT TIME (EBCT)

A measure of the time during which a water to be treated is in contact with the treatment medium in a contact vessel, assuming that all liquid passes through the vessel at the same velocity. EBCT is equal to the volume of the empty bed divided by the flow rate.

ENCLOSED SPACE

See CONFINED SPACE.

END BELLS

Devices used to hold the rotor and stator of a motor in position.

END POINT

The completion of a desired chemical reaction. Samples of water or wastewater are titrated to the end point. This means that a chemical is added, drop by drop, to a sample until a certain color change (blue to clear, for example) occurs. This is called the end point of the titration. In addition to a color change, an end point may be reached by the formation of a precipitate or the reaching of a specified pH. An end point may be detected by the use of an electronic device, such as a pH meter.

ENDEMIC (en-DEM-ick)

Something peculiar to a particular people or locality, such as a disease that is always present in the population.

ENDOCRINE (EN-doe-krin) EFFECTS

An altering of the organs in the human body responsible for secreting hormones into the bloodstream. Endocrine glands include the thyroid gland, the pancreas, and the adrenal glands.

ENDRIN (EN-drin)

A pesticide toxic to freshwater and marine aquatic life that produces adverse health effects in domestic water supplies.

ENERGY GRADE LINE (EGL)

A line that represents the elevation of energy head (in feet or meters) of water flowing in a pipe, conduit, or channel. The line is drawn above the hydraulic grade line (gradient) a distance equal to the velocity head ($V^2/2g$) of the water flowing at each section or point along the pipe or channel. Also see HYDRAULIC GRADE LINE.

[SEE DRAWING ON PAGE 798]

ENTERIC

Of intestinal origin, especially applied to wastes or bacteria.

ENTRAIN

To trap bubbles in water either mechanically through turbulence or chemically through a reaction.

ENVIRONMENTAL PROTECTION AGENCY

See EPA.

ENZYMES (EN-zimes)

Organic or biochemical substances that cause or speed up chemical reactions.

EPIDEMIC (EP-uh-DEM-ick)

A disease that occurs in a large number of people in a locality at the same time and spreads from person to person.

EPIDEMIOLOGY (EP-uh-DE-me-ALL-o-jee)

A branch of medicine that studies epidemics (diseases that affect significant numbers of people during the same time period in the same locality). The objective of epidemiology is to determine the factors that cause epidemic diseases and how to prevent them.

EPILIMNION (EP-uh-LIM-nee-on)

The upper layer of water in a thermally stratified lake or reservoir. This layer consists of the warmest water and has a fairly uniform (constant) temperature. The layer is readily mixed by wind action.

EQUILIBRIUM, CALCIUM CARBONATE

A water is considered stable when it is just saturated with calcium carbonate. In this condition, the water will neither dissolve nor deposit calcium carbonate. Thus, in this water the calcium carbonate is in equilibrium with the hydrogen ion concentration.

EQUITY

The value of an investment in a facility.

PIPE

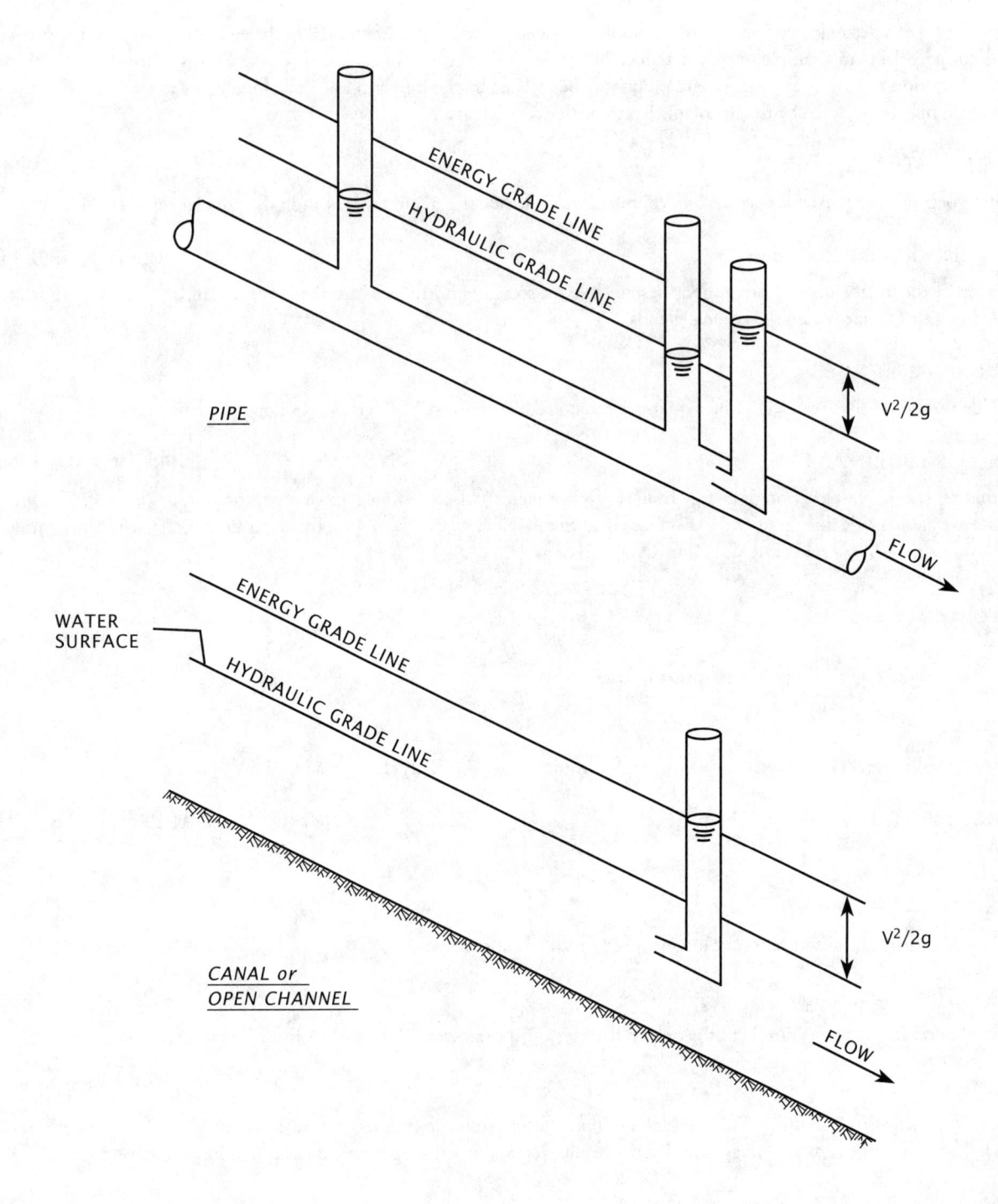

CANAL or OPEN CHANNEL

$V^2/2g$ = Velocity Head

Energy grade line and hydraulic grade line

EQUIVALENT WEIGHT

That weight that will react with, displace, or is equivalent to one gram atom of hydrogen.

ESTER

A compound formed by the reaction between an acid and an alcohol with the elimination of a molecule of water.

EUTROPHIC (yoo-TRO-fick)

Reservoirs and lakes that are rich in nutrients and very productive in terms of aquatic animal and plant life.

EUTROPHICATION (YOO-tro-fi-KAY-shun)

The increase in the nutrient levels of a lake or other body of water; this usually causes an increase in the growth of aquatic animal and plant life.

EVAPORATION

The process by which water or other liquid becomes a gas (water vapor or ammonia vapor).

EVAPOTRANSPIRATION (ee-VAP-o-TRANS-purr-A-shun)

(1) The process by which water vapor is released to the atmosphere from living plants. Also called TRANSPIRATION.

(2) The total water removed from an area by transpiration (plants) and by evaporation from soil, snow, and water surfaces.

F

FACULTATIVE (FACK-ul-tay-tive) BACTERIA

Facultative bacteria can use either dissolved oxygen or oxygen obtained from food materials such as sulfate or nitrate ions. In other words, facultative bacteria can live under aerobic, anoxic, or anaerobic conditions.

FAIL-SAFE

Design and operation of a process control system whereby failure of the power system or any component does not result in process failure or equipment damage.

FEEDBACK

The circulating action between a sensor measuring a process variable and the controller that controls or adjusts the process variable.

FEEDWATER

The water that is fed to a treatment process; the water that is going to be treated.

FINISHED WATER

Water that has passed through a water treatment plant. All the treatment processes are completed or finished. This water is the product from the water treatment plant and is ready to be delivered to consumers. Also called PRODUCT WATER.

FIXED COSTS

Regularly recurring costs that a utility must cover or pay (as rent, insurance, depreciation, interest, etc.) that do not vary as a result of production or demand for service. Also see VARIABLE COSTS.

FIXED SAMPLE

A sample is fixed in the field by adding chemicals that prevent the water quality indicators of interest in the sample from changing before final measurements are performed later in the laboratory.

FLAGELLATES (FLAJ-el-lates)

Microorganisms that move by the action of tail-like projections.

FLAME POLISHED

Melted by a flame to smooth out irregularities. Sharp or broken edges of glass (such as the end of a glass tube) are rotated in a flame until the edge melts slightly and becomes smooth.

FLAP GATE

A hinged gate that is mounted at the top of a pipe or channel to allow flow in only one direction. Flow in the wrong direction closes the gate. Also see CHECK VALVE and TIDE GATE.

FLOAT ON THE SYSTEM

A method of operating a water storage facility. Daily flow into the facility is approximately equal to the average daily demand for water. When consumer demands for water are low, the storage facility will be filling. During periods of high demand, the facility will be emptying.

FLOC

Clumps of bacteria and particles, or coagulants and impurities, that have come together and formed a cluster. Found in flocculation tanks, sedimentation basins, aeration tanks, secondary clarifiers, and chemical precipitation processes.

FLOCCULATION (flock-yoo-LAY-shun)

The gathering together of fine particles after coagulation to form larger particles by a process of gentle mixing. This clumping together makes it easier to separate the solids from the water by settling, skimming, draining, or filtering.

FLOW LINE

(1) The top of the wetted line, the water surface, or the hydraulic grade line of water flowing in an open channel or partially full conduit.

(2) The lowest point of the channel inside a pipe or manhole. See INVERT. *NOTE:* (2) is an improper definition, although used by some contractors.

FLUIDIZED (FLOO-id-i-zd)

A mass of solid particles that is made to flow like a liquid by injection of water or gas is said to have been fluidized. In water and wastewater treatment, a bed of filter media is fluidized by backwashing water through the filter.

FLUORIDATION (floor-uh-DAY-shun)

The addition of a chemical to increase the concentration of fluoride ions in drinking water to a predetermined optimum limit to reduce the incidence (number) of dental caries (tooth decay) in children. Defluoridation is the removal of excess fluoride in drinking water to prevent the mottling (brown stains) of teeth.

FLUSHING

The removal of deposits of material that have lodged in water distribution lines or sewers because of inadequate velocity of flows. Water is discharged into the lines at such rates that the larger flow and higher velocities are sufficient to remove the material.

FLUX

A flowing or flow.

FOOT VALVE

A special type of check valve located at the bottom end of the suction pipe on a pump. This valve opens when the pump operates to allow water to enter the suction pipe but closes when the pump shuts off to prevent water from flowing out of the suction pipe.

FREE AVAILABLE CHLORINE RESIDUAL

That portion of the total available chlorine residual composed of dissolved chlorine gas (Cl_2), hypochlorous acid (HOCl), and/or hypochlorite ion (OCl^-) remaining in water after chlorination. This does not include chlorine that has combined with ammonia, nitrogen, or other compounds. Also called free available residual chlorine.

FREEBOARD

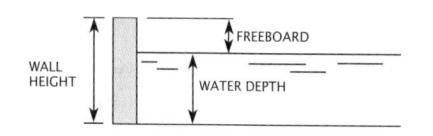

(1) The vertical distance from the normal water surface to the top of the confining wall.

(2) The vertical distance from the sand surface to the underside of a trough in a sand filter. This distance is also called AVAILABLE EXPANSION.

FRICTION LOSS

The head, pressure, or energy (they are the same) lost by water flowing in a pipe or channel as a result of turbulence caused by the velocity of the flowing water and the roughness of the pipe, channel walls, or restrictions caused by fittings. Water flowing in a pipe loses head, pressure, or energy as a result of friction losses. Also called HEAD LOSS.

FULVIC ACID

A complex organic compound that can be derived either from soil or water. Aquatic fulvic acids are major precursors of disinfection by-products.

FUNGI (FUN-ji)

Mushrooms, molds, mildews, rusts, and smuts that are small non-chlorophyll-bearing plants lacking roots, stems, and leaves. They occur in natural waters and grow best in the absence of light. Their decomposition may cause objectionable tastes and odors in water.

FUSE

A protective device having a strip or wire of fusible metal that, when placed in a circuit, will melt and break the electric circuit if heated too much. High temperatures will develop in the fuse when a current flows through the fuse in excess of that which the circuit will carry safely.

G

GIS

See GEOGRAPHIC INFORMATION SYSTEM.

GALVANIC CELL

An electrolytic cell capable of producing electric energy by electrochemical action. The decomposition of materials in the cell causes an electric (electron) current to flow from cathode to anode.

GALVANIC SERIES

A list of metals and alloys presented in the order of their tendency to corrode (or go into solution). Also called the ELECTROMOTIVE SERIES. This is a practical application of the theoretical ELECTROCHEMICAL SERIES.

GALVANIZE

To coat a metal (especially iron or steel) with zinc. Galvanization is the process of coating a metal with zinc.

GARNET

A group of hard, reddish, glassy, mineral sands made up of silicates of base metals (calcium, magnesium, iron, and manganese). Garnet has a higher density than sand.

GAUGE PRESSURE

The pressure within a closed container or pipe as measured with a gauge. In contrast, absolute pressure is the sum of atmospheric pressure (14.7 lbs/sq in (1.0 atm.)) *PLUS* pressure within a vessel (as measured with a gauge). Most pressure gauges read in "gauge pressure" or psig (pounds per square inch gauge pressure or kilopascals).

GEOGRAPHIC INFORMATION SYSTEM (GIS)

A computer program that combines mapping with detailed information about the physical locations of structures, such as pipes, valves, and manholes, within geographic areas. The system is used to help operators and maintenance personnel locate utility system features or structures and to assist with the scheduling and performance of maintenance activities.

GEOLOGICAL LOG

A detailed description of all underground features discovered during the drilling of a well (depth, thickness, and type of formations).

GEOPHYSICAL LOG

A record of the structure and composition of the earth encountered when drilling a well or similar type of test hole or boring.

GERMICIDE (JERM-uh-side)

A substance formulated to kill germs or microorganisms. The germicidal properties of chlorine make it an effective disinfectant.

GIARDIA (jee-ARE-dee-ah)

A waterborne intestinal parasite that causes a disease called giardiasis in infected humans. Symptoms of the disease include diarrhea, cramps, and weight loss. *Giardia* contamination is found in most surface waters and some groundwaters.

GIARDIASIS (jee-are-DYE-uh-sis)

Intestinal disease caused by an infestation of *Giardia* flagellates.

GRAB SAMPLE

A single sample of water collected at a particular time and place that represents the composition of the water only at that time and place.

GRADE

(1) The elevation of the invert (or bottom) of a pipeline, canal, culvert, sewer, or similar conduit.

(2) The inclination or slope of a pipeline, conduit, stream channel, or natural ground surface; usually expressed in terms of the ratio or percentage of number of units of vertical rise or fall per unit of horizontal distance. A 0.5 percent grade would be a drop of one-half foot per hundred feet (one-half meter per hundred meters) of pipe.

GRAVIMETRIC

A means of measuring unknown concentrations of water quality indicators in a sample by weighing a precipitate or residue of the sample.

GRAVIMETRIC FEEDER

A dry chemical feeder that delivers a measured weight of chemical during a specific time period.

GREENSAND

A mineral (glauconite) material that looks like ordinary filter sand except that it is green in color. Greensand is a natural ion exchange material that is capable of softening water. Greensand that has been treated with potassium permanganate ($KMnO_4$) is called manganese greensand; this product is used to remove iron, manganese, and hydrogen sulfide from groundwaters.

GROUND

An expression representing an electrical connection to earth or a large conductor that is at the earth's potential or neutral voltage.

H

HGL

See HYDRAULIC GRADE LINE.

HTH

High Test Hypochlorite. Calcium hypochlorite or $Ca(OCl)_2$.

HALOACETIC (HAL-o-uh-SEE-tick) ACID (HAA)

A class of disinfection by-products, formed mainly during the chlorination of water, containing natural organic matter. HAA5 is the sum of the concentrations, in milligrams per liter, of five haloacetic acid compounds.

HARD WATER

Water having a high concentration of calcium and magnesium ions. A water may be considered hard if it has a hardness greater than the typical hardness of water from the region. Some textbooks define hard water as a water with a hardness of more than 100 mg/L as calcium carbonate.

HARDNESS, WATER

A characteristic of water caused mainly by the salts of calcium and magnesium, such as bicarbonate, carbonate, sulfate, chloride, and nitrate. Excessive hardness in water is undesirable because it causes the formation of soap curds, increased use of soap, deposition of scale in boilers, damage in some industrial processes, and sometimes causes objectionable tastes in drinking water.

HARMFUL PHYSICAL AGENT or TOXIC SUBSTANCE

Any chemical substance, biological agent (bacteria, virus, or fungus), or physical stress (noise, heat, cold, vibration, repetitive motion, ionizing and non-ionizing radiation, hypo- or hyperbaric pressure) that:

(1) Is regulated by any state or federal law or rule due to a hazard to health

(2) Is listed in the latest printed edition of the National Institute of Occupational Safety and Health (NIOSH) Registry of Toxic Effects of Chemical Substances (RTECS)

(3) Has yielded positive evidence of an acute or chronic health hazard in human, animal, or other biological testing conducted by, or known to, the employer

(4) Is described by a Material Safety Data Sheet (MSDS) available to the employer that indicates that the material may pose a hazard to human health

Also see ACUTE HEALTH EFFECT and CHRONIC HEALTH EFFECT.

HEAD

The vertical distance, height, or energy of water above a reference point. A head of water may be measured in either height (feet or meters) or pressure (pounds per square inch or kilograms per square centimeter). Also see DISCHARGE HEAD, DYNAMIC HEAD, STATIC HEAD, SUCTION HEAD, SUCTION LIFT, and VELOCITY HEAD.

HEAD LOSS

The head, pressure, or energy (they are the same) lost by water flowing in a pipe or channel as a result of turbulence caused by the velocity of the flowing water and the roughness of the pipe, channel walls, or restrictions caused by fittings. Water flowing in a pipe loses head, pressure, or energy as a result of friction losses. The head loss through a filter is due to friction losses caused by material building up on the surface or in the top part of a filter. Also called FRICTION LOSS.

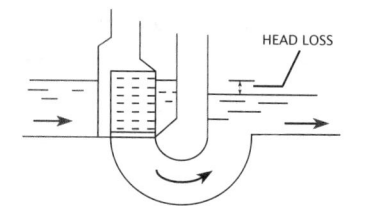

HEADER

A large pipe to which the ends of a series of smaller pipes are connected. Also called a MANIFOLD.

HEAT SENSOR

A device that opens and closes a switch in response to changes in the temperature. This device might be a metal contact, or a thermocouple that generates a minute electric current proportional to the difference in heat, or a variable resistor whose value changes in response to changes in temperature. Also called a TEMPERATURE SENSOR.

HECTARE (HECK-ter)

A measure of area in the metric system similar to an acre. One hectare is equal to 10,000 square meters and 2.4711 acres.

HEPATITIS (HEP-uh-TIE-tis)

Hepatitis is an inflammation of the liver caused by an acute viral infection. Yellow jaundice is one symptom of hepatitis.

HERBICIDE (HERB-uh-side)

A compound, usually a manmade organic chemical, used to kill or control plant growth.

HERTZ (Hz)

The number of complete electromagnetic cycles or waves in one second of an electric or electronic circuit. Also called the frequency of the current.

HETEROTROPHIC (HET-er-o-TROF-ick)

Describes organisms that use organic matter for energy and growth. Animals, fungi, and most bacteria are heterotrophs.

HIGH-LINE JUMPERS

Pipes or hoses connected to fire hydrants and laid on top of the ground to provide emergency water service for an isolated portion of a distribution system.

HOSE BIB

Faucet. A location in a water line where a hose is connected.

HUMAN MACHINE INTERFACE (HMI)

The device at which the operator interacts with the control system. This may be an individual instrumentation and control device or the graphic screen of a computer control system. Also called MAN MACHINE INTERFACE (MMI) and OPERATOR INTERFACE.

HUMIC ACID

A polymeric constituent of soils, lignite, and peat. Aquatic humic acids are major precursors of disinfection by-products.

HUMIC SUBSTANCES

Natural organic matter resulting from partial decomposition of plant or animal matter and forming the organic portion of soil.

HYDRATED LIME

Limestone that has been burned and treated with water under controlled conditions until the calcium oxide portion has been converted to calcium hydroxide ($Ca(OH)_2$). Hydrated lime is quicklime combined with water. $CaO + H_2O \rightarrow Ca(OH)_2$. Also called slaked lime. Also see QUICKLIME.

HYDRAULIC CONDUCTIVITY (K)

A coefficient describing the relative ease with which groundwater can move through a permeable layer of rock or soil. Typical units of hydraulic conductivity are feet per day, gallons per day per square foot, or meters per day (depending on the unit chosen for the total discharge and the cross-sectional area).

HYDRAULIC GRADE LINE (HGL)

The surface or profile of water flowing in an open channel or a pipe flowing partially full. If a pipe is under pressure, the hydraulic grade line is that level water would rise to in a small, vertical tube connected to the pipe. Also see ENERGY GRADE LINE.

[SEE DRAWING ON PAGE 798]

HYDRAULIC GRADIENT

The slope of the hydraulic grade line. This is the slope of the water surface in an open channel, the slope of the water surface of the groundwater table, or the slope of the water pressure for pipes under pressure.

HYDROGEOLOGIST (HI-dro-jee-ALL-uh-jist)

A person who studies and works with groundwater.

HYDROLOGIC (HI-dro-LOJ-ick) CYCLE

The process of evaporation of water into the air and its return to earth by precipitation (rain or snow). This process also includes transpiration from plants, groundwater movement, and runoff into rivers, streams, and the ocean. Also called the WATER CYCLE.

HYDROLYSIS (hi-DROLL-uh-sis)

(1) A chemical reaction in which a compound is converted into another compound by taking up water.

(2) Usually a chemical degradation of organic matter.

HYDROPHILIC (hi-dro-FILL-ick)

Having a strong affinity (liking) for water. The opposite of HYDROPHOBIC.

HYDROPHOBIC (hi-dro-FOE-bick)

Having a strong aversion (dislike) for water. The opposite of HYDROPHILIC.

HYDROPNEUMATIC (hi-dro-new-MAT-ick)

A water system, usually small, in which a water pump is automatically controlled (started and stopped) by the air pressure in a compressed-air tank.

HYDROSTATIC (hi-dro-STAT-ick) PRESSURE

(1) The pressure at a specific elevation exerted by a body of water at rest.

(2) In the case of groundwater, the pressure at a specific elevation due to the weight of water at higher levels in the same zone of saturation.

HYGROSCOPIC (hi-grow-SKAWP-ick)

Absorbing or attracting moisture from the air.

HYPOCHLORINATION (HI-poe-klor-uh-NAY-shun)

The application of hypochlorite compounds to water or wastewater for the purpose of disinfection.

HYPOCHLORINATORS (HI-poe-KLOR-uh-nay-tors)

Chlorine pumps, chemical feed pumps, or devices used to dispense chlorine solutions made from hypochlorites, such as bleach (sodium hypochlorite) or calcium hypochlorite into the water being treated.

HYPOCHLORITE (HI-poe-KLOR-ite)

Chemical compounds containing available chlorine; used for disinfection. They are available as liquids (bleach) or solids (powder, granules, and pellets) in barrels, drums, and cans. Salts of hypochlorous acid.

HYPOLIMNION (HI-poe-LIM-nee-on)

The lowest layer in a thermally stratified lake or reservoir. This layer consists of colder, denser water, has a constant temperature, and no mixing occurs.

I

ICR

See INFORMATION COLLECTION RULE.

IDLH

Immediately Dangerous to Life or Health. The atmospheric concentration of any toxic, corrosive, or asphyxiant substance that poses an immediate threat to life or would cause irreversible or delayed adverse health effects or would interfere with an individual's ability to escape from a dangerous atmosphere.

IMHOFF CONE

A clear, cone-shaped container marked with graduations. The cone is used to measure the volume of settleable solids in a specific volume (usually one liter) of water or wastewater.

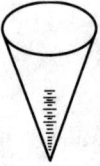

IMPELLER

A rotating set of vanes in a pump or compressor designed to pump or move water or air.

IMPERMEABLE (im-PURR-me-uh-bull)

Not easily penetrated. The property of a material or soil that does not allow, or allows only with great difficulty, the movement or passage of water.

INDICATOR

(1) (Chemical indicator) A substance that gives a visible change, usually of color, at a desired point in a chemical reaction, generally at a specified end point.

(2) (Instrument indicator) A device that indicates the result of a measurement, usually using either a fixed scale and movable indicator (pointer), such as a pressure gauge, or a moving chart with a movable pen like those used on a circular flow-recording chart. Also called a RECEIVER.

INFILTRATION (in-fill-TRAY-shun)

The seepage of groundwater into a sewer system, including service connections. Seepage frequently occurs through defective or cracked pipes, pipe joints and connections, interceptor access risers and covers, or manhole walls.

INFLUENT

Water or other liquid—raw (untreated) or partially treated—flowing *INTO* a reservoir, basin, treatment process, or treatment plant.

INFORMATION COLLECTION RULE (ICR)

The Information Collection Rule (ICR) was promulgated on May 14, 1996 and approved by the Director of the Federal Register on June 18, 1996. It was to remain effective until December 31, 2000. The rule specified requirements for monitoring microbial contaminants and disinfection by-products (DBPs) by large public water systems (PWSs). It required large PWSs to conduct either bench- or pilot-scale testing of advanced treatment techniques. The data reported under the ICR were used by EPA to learn more about the occurrence of microbial contamination and disinfection by-products, the health risks posed, appropriate analytical methods, and effective forms of treatment. The ICR data form the scientific basis for EPA's development of the Enhanced Surface Water Treatment Rule and the Disinfectants and Disinfection By-Products (D/DBP) Rule.

INITIAL SAMPLING

The very first sampling conducted under the Safe Drinking Water Act (SDWA) for each of the applicable contaminant categories.

INJECTOR WATER

Service water in which chlorine is added (injected) to form a chlorine solution.

IN-LINE FILTRATION

The addition of chemical coagulants directly to the filter inlet pipe. The chemicals are mixed by the flowing water. Flocculation and sedimentation facilities are eliminated. This pretreatment method is commonly used in pressure filter installations. Also see CONVENTIONAL FILTRATION and DIRECT FILTRATION.

INORGANIC

Used to describe material such as sand, salt, iron, calcium salts, and other mineral materials. Inorganic materials are chemical substances of mineral origin, whereas organic substances are usually of animal or plant origin. Also see ORGANIC.

INORGANIC WASTE

Waste material such as sand, salt, iron, calcium, and other mineral materials that are only slightly affected by the action of organisms. Inorganic wastes are chemical substances of mineral origin; whereas organic wastes are chemical substances usually of animal or plant origin. Also see NONVOLATILE MATTER, ORGANIC WASTE, and VOLATILE SOLIDS.

INPUT HORSEPOWER

The total power used in operating a pump and motor.

$$\text{Input Horsepower, HP} = \frac{(\text{Brake Horsepower, HP})(100\%)}{\text{Motor Efficiency, \%}}$$

INSECTICIDE

Any substance or chemical formulated to kill or control insects.

INSOLUBLE (in-SAWL-yoo-bull)

Something that cannot be dissolved.

INTEGRATOR

A device or meter that continuously measures and sums a process rate variable in cumulative fashion over a given time period. For example, total flows displayed in gallons per minute, million gallons per day, cubic feet per second, or some other unit of volume per time period. Also called a TOTALIZER.

INTERCEPTOR (INTERCEPTING) SEWER

A large sewer that receives flow from a number of sewers and conducts the wastewater to a treatment plant. Often called an interceptor. The term interceptor is sometimes used in small communities to describe a septic tank or other holding tank that serves as a temporary wastewater storage reservoir for a Septic Tank Effluent Pump (STEP) system.

INTERCONNECTOR

A sewer installed to connect two separate sewers. If one sewer becomes blocked, wastewater can back up and flow through the interconnector to the other sewer.

INTERFACE

The common boundary layer between two substances, such as water and a solid (metal); or between two fluids, such as water and a gas (air); or between a liquid (water) and another liquid (oil).

INTERLOCK

A physical device, equipment, or software routine that prevents an operation from beginning or changing function until some condition or set of conditions is fulfilled. An example would be a switch that prevents a piece of equipment from operating when a hazard exists.

INTERNAL FRICTION

Friction within a fluid (water) due to cohesive forces.

INTERSTICE (in-TUR-stuhz)

A very small open space in a rock or granular material. Also called a PORE, VOID, or void space. Also see VOID.

INVERT (IN-vert)

The lowest point of the channel inside a pipe, conduit, canal, or manhole. See FLOW LINE.

ION

An electrically charged atom, radical (such as SO_4^{2-}), or molecule formed by the loss or gain of one or more electrons.

ION EXCHANGE

A water or wastewater treatment process involving the reversible interchange (switching) of ions between the water being treated and the solid resin contained within an ion exchange unit. Undesirable ions are exchanged with acceptable ions on the resin or recoverable ions in the water being treated are exchanged with other acceptable ions on the resin.

ION EXCHANGE RESINS

Insoluble polymers, used in water or wastewater treatment, that are capable of exchanging (switching or giving) acceptable cations or anions to the water being treated for less desirable ions or for ions to be recovered.

IONIC CONCENTRATION

The concentration of any ion in solution, usually expressed in moles per liter.

IONIZATION (EYE-on-uh-ZAY-shun)

(1) The splitting or dissociation (separation) of molecules into negatively and positively charged ions.

(2) The process of adding electrons to, or removing electrons from, atoms or molecules, thereby creating ions. High temperatures, electrical discharges, and nuclear radiation can cause ionization.

J

JAR TEST

A laboratory procedure that simulates coagulation/flocculation with differing chemical doses. The purpose of the procedure is to estimate the minimum coagulant dose required to achieve certain water quality goals. Samples of water to be treated are placed in six jars. Various amounts of chemicals are added to each jar, stirred, and the settling of solids is observed. The lowest dose of chemicals that provides satisfactory settling is the dose used to treat the water.

JOGGING

The frequent starting and stopping of an electric motor.

JOULE (JOOL)

A measure of energy, work, or quantity of heat. One joule is the work done when the point of application of a force of one newton is displaced a distance of one meter in the direction of the force. Approximately equal to 0.7375 ft-lbs (0.1022 m-kg).

K

KELLY

The square section of a rod that causes the rotation of the drill bit. Torque from a drive table is applied to the square rod to cause the rotary motion. The drive table is chain- or gear-driven by an engine.

KILO

(1) Kilogram.

(2) Kilometer.

(3) A prefix meaning "thousand" used in the metric system and other scientific systems of measurement.

KINETIC ENERGY

Energy possessed by a moving body of matter, such as water, as a result of its motion.

KJELDAHL (KELL-doll) NITROGEN

Nitrogen in the form of organic proteins or their decomposition product ammonia, as measured by the Kjeldahl Method.

L

LAG TIME

The time period between the moment a process change is made and the moment such a change is finally sensed by the associated measuring instrument.

LANGELIER INDEX (LI)

An index reflecting the equilibrium pH of a water with respect to calcium and alkalinity. This index is used in stabilizing water to control both corrosion and the deposition of scale.

$$\text{Langelier Index} = pH - pH_s$$
where pH = actual pH of the water
pH_s = pH at which water having the same alkalinity and calcium content is just saturated with calcium carbonate

LAUNDERING WEIR

Sedimentation basin overflow weir. A plate with V-notches along the top to ensure a uniform flow rate and avoid short-circuiting.

LAUNDERS

Sedimentation basin and filter discharge channels consisting of overflow weir plates (in sedimentation basins) and conveying troughs.

LEAD (LEED)

A wire or conductor that can carry electric current.

LEATHERS

O-rings or gaskets used with piston pumps to provide a seal between the piston and the side wall.

LEVEL CONTROL

A float device (or pressure switch) that senses changes in a measured variable and opens or closes a switch in response to that change. In its simplest form, this control might be a floating ball connected mechanically to a switch or valve, such as is used to stop water flow into a toilet when the tank is full.

LINDANE (LYNN-dane)

A pesticide that causes adverse health effects in domestic water supplies and also is toxic to freshwater and marine aquatic life.

LINEARITY (lin-ee-AIR-it-ee)

How closely an instrument measures actual values of a variable through its effective range.

LITTORAL (LIT-or-ul) ZONE

(1) That portion of a body of fresh water extending from the shoreline lakeward to the limit of occupancy of rooted plants.

(2) The strip of land along the shoreline between the high and low water levels.

LOGARITHM (LOG-uh-rith-um) (LOG)

The exponent that indicates the power to which a number must be raised to produce a given number. For example: if $B^2 = N$, the 2 is the logarithm of N (to the base B), or $10^2 = 100$ and $\log_{10} 100 = 2$.

LOWER EXPLOSIVE LIMIT (LEL)

The lowest concentration of a gas or vapor (percent by volume in air) that explodes if an ignition source is present at ambient temperature. At temperatures above 250°F (121°C) the LEL decreases because explosibility increases with higher temperature.

M

M or MOLAR

A molar solution consists of one gram molecular weight of a compound dissolved in enough water to make one liter of solution. A gram molecular weight is the molecular weight of a compound in grams. For example, the molecular weight of sulfuric acid (H_2SO_4) is 98. A one *M* solution of sulfuric acid would consist of 98 grams of H_2SO_4 dissolved in enough distilled water to make one liter of solution.

MBAS

Methylene Blue Active Substance. Another name for surfactants or surface active agents. The determination of surfactants is accomplished by measuring the color change in a standard solution of methylene blue dye.

MCL

Maximum Contaminant Level. The largest allowable amount. MCLs for various water quality indicators are specified in the National Primary Drinking Water Regulations (NPDWR).

MCLG

Maximum Contaminant Level Goal. MCLGs are health goals based entirely on health effects. They are a preliminary standard set but not enforced by EPA. MCLs consider health effects, but also take into consideration the feasibility and cost of analysis and treatment of the regulated MCL. Although often less stringent than the corresponding MCLG, the MCL is set to protect health.

MF

See MICROFILTRATION.

MRDL

Maximum Residual Disinfectant Level. The highest level of a disinfectant allowed in drinking water without causing an unacceptable possibility of adverse health effects.

mg/*L*

See MILLIGRAMS PER LITER, mg/*L*.

MPN

MPN is the Most Probable Number of coliform-group organisms per unit volume of sample water. Expressed as a density or population of organisms per 100 m*L* of sample water.

MSDS

See MATERIAL SAFETY DATA SHEET.

MACROSCOPIC (MACK-row-SKAWP-ick) ORGANISMS

Organisms big enough to be seen by the eye without the aid of a microscope.

MAN MACHINE INTERFACE (MMI)

The device at which the operator interacts with the control system. This may be an individual instrumentation and control device or the graphic screen of a computer control system. Also called HUMAN MACHINE INTERFACE (HMI) and OPERATOR INTERFACE.

MANDREL (MAN-drill)

(1) A special tool used to push bearings in or to pull sleeves out.

(2) A testing device used to measure for excessive deflection in a flexible conduit.

MANIFOLD

A large pipe to which the ends of a series of smaller pipes are connected. Also called a HEADER.

MANOMETER (man-NAH-mut-ter)

An instrument for measuring pressure. Usually, a manometer is a glass tube filled with a liquid that is used to measure the difference in pressure across a flow measuring device, such as an orifice or a Venturi meter. The instrument used to measure blood pressure is a type of manometer.

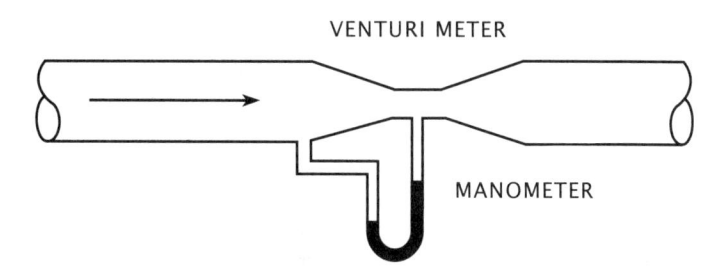

MATERIAL SAFETY DATA SHEET (MSDS)

A document that provides pertinent information and a profile of a particular hazardous substance or mixture. An MSDS is normally developed by the manufacturer or formulator of the hazardous substance or mixture. The MSDS is required to be made available to employees and operators or inspectors whenever there is the likelihood of the hazardous substance or mixture being introduced into the workplace. Some manufacturers are preparing MSDSs for products that are not considered to be hazardous to show that the product or substance is not hazardous.

MAXIMUM CONTAMINANT LEVEL (MCL)

The largest allowable amount. MCLs for various water quality indicators are specified in the National Primary Drinking Water Regulations (NPDWR).

MAXIMUM RESIDUAL DISINFECTANT LEVEL (MRDL)

The highest level of a disinfectant allowed in drinking water without causing an unacceptable possibility of adverse health effects.

MEASURED VARIABLE

A factor (flow, temperature) that is sensed and quantified (reduced to a reading of some kind) by a primary element or sensor.

MECHANICAL JOINT

A flexible device that joins pipes or fittings together by the use of lugs and bolts.

MEG

(1) Abbreviation of MEGOHM.

(2) A procedure used for checking the insulation resistance on motors, feeders, bus bar systems, grounds, and branch circuit wiring. Also see MEGGER.

MEGGER (from megohm)

An instrument used for checking the insulation resistance on motors, feeders, bus bar systems, grounds, and branch circuit wiring. A megger reads in millions of ohms. Also see MEG.

MEGOHM (MEG-ome)

Millions of ohms. Mega- is a prefix meaning one million, so 5 megohms means 5 million ohms.

MEMBRANE FOULING

The cause of a loss of flow through the membrane as a result of material being retained on the surface of the membrane or within the membrane pores. Membrane fouling may be reversible or irreversible. Loss of flow through the membrane by reversible fouling can be recovered by regular backwashing of the membrane surface. Flow loss by irreversible fouling cannot be recovered.

MENISCUS (meh-NIS-cuss)

The curved surface of a column of liquid (water, oil, mercury) in a small tube. When the liquid wets the sides of the container (as with water), the curve forms a valley. When the confining sides are not wetted (as with mercury), the curve forms a hill or upward bulge.

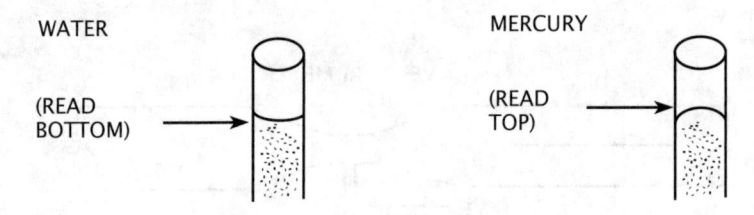

MESH

One of the openings or spaces in a screen or woven fabric. The value of the mesh is usually given as the number of openings per inch. This value does not consider the diameter of the wire or fabric; therefore, the mesh number does not always have a definite relationship to the size of the hole.

MESOTROPHIC (MESS-o-TRO-fick)

Reservoirs and lakes that contain moderate quantities of nutrients and are moderately productive in terms of aquatic animal and plant life.

METABOLISM

All of the processes or chemical changes in an organism or a single cell by which food is built up (anabolism) into living protoplasm and by which protoplasm is broken down (catabolism) into simpler compounds with the exchange of energy.

METALIMNION (met-uh-LIM-nee-on)

The middle layer in a thermally stratified lake or reservoir. In this layer there is a rapid decrease in temperature with depth. Also called THERMOCLINE.

METHOXYCHLOR (meth-OX-e-klor)

A pesticide that causes adverse health effects in domestic water supplies and is also toxic to freshwater and marine aquatic life. The chemical name for methoxychlor is 2,2-bis(p-methoxyphenol)-1,1,1-trichloroethane.

METHYL ORANGE ALKALINITY

A measure of the total alkalinity in a water sample. The alkalinity is measured by the amount of standard sulfuric acid required to lower the pH of the water to a pH level of 4.5, as indicated by the change in color of methyl orange from orange to pink. Methyl orange alkalinity is expressed as milligrams per liter equivalent calcium carbonate.

MICROBIAL (my-KRO-bee-ul) GROWTH

The activity and growth of microorganisms, such as bacteria, algae, diatoms, plankton, and fungi.

MICROFILTRATION (MF)

A pressure-driven membrane filtration process that separates particles down to approximately 0.1 μm diameter from influent water using a sieving process.

MICRON (MY-kron)

μm, Micrometer or Micron. A unit of length. One millionth of a meter or one thousandth of a millimeter. One micron equals 0.00004 of an inch.

MICROORGANISMS (MY-crow-OR-gan-is-ums)

Living organisms that can be seen individually only with the aid of a microscope.

MIL

A unit of length equal to 0.001 of an inch. The diameter of wires and tubing is measured in mils, as is the thickness of plastic sheeting.

MILLIGRAMS PER LITER, mg/L

A measure of the concentration by weight of a substance per unit volume in water or wastewater. In reporting the results of water and wastewater analysis, mg/L is preferred to the unit parts per million (ppm), to which it is approximately equivalent.

MILLIMICRON (MILL-uh-MY-kron)

A unit of length equal to 10^{-3}μ (one thousandth of a micron), 10^{-6} millimeters, or 10^{-9} meters; correctly called a nanometer, nm.

MOLAR

See M or MOLAR.

MOLARITY

A measure of concentration defined as the number of moles of solute per liter of solution. Also see M or MOLAR.

MOLE

The name for a number (6.02×10^{23}) of atoms or molecules. See MOLECULAR WEIGHT.

MOLECULAR WEIGHT

The molecular weight of a compound in grams per mole is the sum of the atomic weights of the elements in the compound. The molecular weight of sulfuric acid (H_2SO_4) in grams is 98.

Element	Atomic Weight	Number of Atoms	Molecular Weight
H	1	2	2
S	32	1	32
O	16	4	64
			98

MOLECULE

The smallest division of a compound that still retains or exhibits all the properties of the substance.

MONOMER (MON-o-mer)

A molecule of low molecular weight capable of reacting with identical or different monomers to form polymers.

MONOMICTIC (mah-no-MICK-tick)

Lakes and reservoirs that are relatively deep, do not freeze over during the winter months, and undergo a single stratification and mixing cycle during the year. These lakes and reservoirs usually become destratified during the mixing cycle, usually in the fall of the year.

MONOVALENT

Having a valence of one, such as the cuprous (copper) ion, Cu^+.

MOST PROBABLE NUMBER (MPN)

See MPN.

MOTILE (MO-till)

Capable of self-propelled movement. A term that is sometimes used to distinguish between certain types of organisms found in water.

MOTOR EFFICIENCY

The ratio of energy delivered by a motor to the energy supplied to it during a fixed period or cycle. Motor efficiency ratings will vary depending on motor manufacturer and usually will be near 90.0 percent.

MUDBALLS

Material, approximately round in shape, that forms in filters and gradually increases in size when not removed by the backwashing process. Mudballs vary from pea-sized up to golf-ball-sized or larger.

MULTISTAGE PUMP

A pump that has more than one impeller. A single-stage pump has one impeller.

N

N or NORMAL

A normal solution contains one gram equivalent weight of reactant (compound) per liter of solution. The equivalent weight of an acid is that weight that contains one gram atom of ionizable hydrogen or its chemical equivalent. For example, the equivalent weight of sulfuric acid (H_2SO_4) is 49 (98 divided by 2 because there are two replaceable hydrogen ions). A one *N* solution of sulfuric acid would consist of 49 grams of H_2SO_4 dissolved in enough water to make one liter of solution.

NESHTA (formerly NETA)

See NATIONAL ENVIRONMENTAL, SAFETY & HEALTH TRAINING ASSOCIATION.

NETA

See NATIONAL ENVIRONMENTAL, SAFETY & HEALTH TRAINING ASSOCIATION.

NF

See NANOFILTRATION.

NIOSH (NYE-osh)

The National Institute of Occupational Safety and Health is an organization that tests and approves safety equipment for particular applications. NIOSH is the primary federal agency engaged in research in the national effort to eliminate on-the-job hazards to the health and safety of working people. The NIOSH Publications Catalog, Seventh Edition, NIOSH Pub. No. 87-115, lists the NIOSH publications concerning industrial hygiene and occupational health. To obtain a copy of the catalog, write to National Technical Information Service (NTIS), 5285 Port Royal Road, Springfield, VA 22161. NTIS Stock No. PB88-175013.

NOM

See NATURAL ORGANIC MATTER.

NPDES PERMIT

National Pollutant Discharge Elimination System permit is the regulatory agency document issued by either a federal or state agency that is designed to control all discharges of potential pollutants from point sources and stormwater runoff into US waterways. NPDES permits regulate discharges into US waterways from all point sources of pollution, including industries, municipal wastewater treatment plants, sanitary landfills, large animal feedlots, and return irrigation flows.

NPDWR

National Primary Drinking Water Regulations.

NSDWR

National Secondary Drinking Water Regulations.

NSF

NSF International is a noncommercial, not-for-profit organization concerned with public health safety and environmental protection. NSF Standard 60 lists certified drinking water chemicals and NSF Standard 61 lists certified drinking water system components.

NTU

Nephelometric Turbidity Units. See TURBIDITY UNITS.

NAMEPLATE

A durable, metal plate found on equipment that lists critical operating conditions for the equipment.

NANOFILTRATION (NF)

A pressure-driven membrane filtration process that separates particles down to approximately 0.002 to 0.005 μm diameter from influent water using a sieving process.

NATIONAL ENVIRONMENTAL, SAFETY & HEALTH TRAINING ASSOCIATION (NESHTA) (formerly NATIONAL ENVIRONMENTAL TRAINING ASSOCIATION (NETA))

A professional organization devoted to serving the environmental trainer and promoting better operation of waterworks and pollution control facilities. For information on NESHTA membership and publications, contact NESHTA, PO Box 10321, Phoenix, AZ 85064-0321. Phone (602) 956-6099.

NATIONAL ENVIRONMENTAL TRAINING ASSOCIATION (NETA)

See NATIONAL ENVIRONMENTAL, SAFETY & HEALTH TRAINING ASSOCIATION.

NATIONAL INSTITUTE OF OCCUPATIONAL SAFETY AND HEALTH (NIOSH)

See NIOSH.

NATURAL ORGANIC MATTER (NOM)

Humic substances composed of humic and fulvic acids that come from decayed vegetation.

NEPHELOMETRIC (neff-el-o-MET-rick)

A means of measuring turbidity in a sample by using an instrument called a nephelometer. A nephelometer passes light through a sample and the amount of light deflected (usually at a 90-degree angle) is then measured.

NEWTON

A force that, when applied to a body having a mass of one kilogram, gives it an acceleration of one meter per second per second.

NITRIFICATION (NYE-truh-fuh-KAY-shun)

An aerobic process in which bacteria change the ammonia and organic nitrogen in water or wastewater into oxidized nitrogen (usually nitrate).

NITROGENOUS (nye-TRAH-jen-us)

A term used to describe chemical compounds (usually organic) containing nitrogen in combined forms. Proteins and nitrate are nitrogenous compounds.

NOBLE METAL

A chemically inactive metal (such as gold). A metal that does not corrode easily and is much scarcer (and more valuable) than the so-called useful or base metals. Also see BASE METAL.

NOMINAL DIAMETER

An approximate measurement of the diameter of a pipe. Although the nominal diameter is used to describe the size or diameter of a pipe, it is usually not the exact inside diameter of the pipe.

NONIONIC (NON-eye-ON-ick) POLYMER

A polymer that has no net electrical charge.

NON-PERMIT CONFINED SPACE

See CONFINED SPACE, NON-PERMIT.

NONPOINT SOURCE

A runoff or discharge from a field or similar source, in contrast to a point source, which refers to a discharge that comes out the end of a pipe or other clearly identifiable conveyance. Also see POINT SOURCE.

NONPOTABLE (non-POE-tuh-bull)

Water that may contain objectionable pollution, contamination, minerals, or infective agents and is considered unsafe or unpalatable for drinking.

NONVOLATILE MATTER

Material such as sand, salt, iron, calcium, and other mineral materials that are only slightly affected by the actions of organisms and are not lost on ignition of the dry solids at 550°C (1,022°F). Volatile materials are chemical substances usually of animal or plant origin. Also see INORGANIC WASTE and VOLATILE SOLIDS.

NORMAL

See *N* or NORMAL.

NORMALITY

The number of gram-equivalent weights of solute in one liter of solution. The equivalent weight of any material is the weight that would react with or be produced by the reaction of 8.0 grams of oxygen or 1.0 gram of hydrogen. Normality is used for certain calculations of quantitative analysis. Also see *N* or NORMAL.

NOTICE

This word calls attention to information that is especially significant in understanding and operating equipment or processes safely. Also see CAUTION, DANGER, and WARNING.

NUTRIENT

Any substance that is assimilated (taken in) by organisms and promotes growth. Nitrogen and phosphorus are nutrients that promote the growth of algae. There are other essential and trace elements that are also considered nutrients. Also see NUTRIENT CYCLE.

NUTRIENT CYCLE

The transformation or change of a nutrient from one form to another until the nutrient has returned to the original form, thus completing the cycle. The cycle may take place under either aerobic or anaerobic conditions.

O

ORP (pronounce as separate letters)

Oxidation-Reduction Potential. The electrical potential required to transfer electrons from one compound or element (the oxidant) to another compound or element (the reductant); used as a qualitative measure of the state of oxidation in water and wastewater treatment systems. ORP is measured in millivolts, with negative values indicating a tendency to reduce compounds or elements and positive values indicating a tendency to oxidize compounds or elements.

OSHA (O-shuh)

The Williams-Steiger Occupational Safety and Health Act of 1970 (OSHA) is a federal law designed to protect the health and safety of workers, including the operators of water supply and treatment systems and wastewater collection and treatment systems. The Act regulates the design, construction, operation, and maintenance of water and wastewater systems. OSHA regulations require employers to obtain and make available to workers the Material Safety Data Sheets (MSDSs) for chemicals used at industrial facilities and treatment plants. OSHA also refers to the federal and state agencies that administer the OSHA regulations.

OCCUPATIONAL SAFETY AND HEALTH ACT OF 1970 (OSHA)

See OSHA.

ODOR THRESHOLD

The minimum odor of a gas or water sample that can just be detected after successive dilutions with odorless gas or water. Also called THRESHOLD ODOR.

OFFSET

(1) The difference between the actual value and the desired value (or set point); characteristic of proportional controllers that do not incorporate reset action. Also called DRIFT.

(2) A pipe fitting in the approximate form of a reverse curve or other combination of elbows or bends that brings one section of a line of pipe out of line with, but into a line parallel with, another section.

(3) A pipe joint that has lost its bedding support, causing one of the pipe sections to drop or slip, thus creating a condition where the pipes no longer line up properly.

OHM

The unit of electrical resistance. The resistance of a conductor in which one volt produces a current of one ampere.

OLFACTORY (all-FAK-tore-ee) FATIGUE

A condition in which a person's nose, after exposure to certain odors, is no longer able to detect the odor.

OLIGOTROPHIC (ah-lig-o-TRO-fick)

Reservoirs and lakes that are nutrient poor and contain little aquatic plant or animal life.

OPERATING PRESSURE DIFFERENTIAL

The operating pressure range for a hydropneumatic system. For example, when the pressure drops below 40 psi in a system designed to operate between 40 psi and 60 psi, the pump will come on and stay on until the pressure builds up to 60 psi. When the pressure reaches 60 psi the pump will shut off. The operating pressure differential in this example is 20 psi.

OPERATING RATIO

The operating ratio is a measure of the total revenues divided by the total operating expenses.

OPERATOR INTERFACE

The device at which the operator interacts with the control system. This may be an individual instrumentation and control device or the graphic screen of a computer control system. Also called HUMAN MACHINE INTERFACE (HMI) and MAN MACHINE INTERFACE (MMI).

ORGANIC

Used to describe chemical substances that come from animal or plant sources. Organic substances always contain carbon. (Inorganic materials are chemical substances of mineral origin.) Also see INORGANIC.

ORGANICS

(1) A term used to refer to chemical compounds made from carbon molecules. These compounds may be natural materials (such as animal or plant sources) or manmade materials (such as synthetic organics). Also see ORGANIC.

(2) Any form of animal or plant life. Also see BACTERIA.

ORGANIC WASTE

Waste material that may come from animal or plant sources. Natural organic wastes generally can be consumed by bacteria and other small organisms. Manufactured or synthetic organic wastes from metal finishing, chemical manufacturing, and petroleum industries may not normally be consumed by bacteria and other organisms. Also see INORGANIC WASTE and VOLATILE SOLIDS.

ORGANISM

Any form of animal or plant life. Also see BACTERIA.

ORGANIZING

Deciding who does what work and delegating authority to the appropriate persons.

ORIFICE (OR-uh-fiss)

An opening (hole) in a plate, wall, or partition. An orifice flange or plate placed in a pipe consists of a slot or a calibrated circular hole smaller than the pipe diameter. The difference in pressure in the pipe above and at the orifice may be used to determine the flow in the pipe. In a trickling filter distributor, the wastewater passes through an orifice to the surface of the filter media.

ORTHOTOLIDINE (or-tho-TOL-uh-dine)

Orthotolidine is a colorimetric indicator of chlorine residual. If chlorine is present, a yellow-colored compound is produced. This reagent is no longer approved for chemical analysis to determine chlorine residual.

OSMOSIS (oz-MOE-sis)

The passage of a liquid from a weak solution to a more concentrated solution across a semipermeable membrane. The membrane allows the passage of the water (solvent) but not the dissolved solids (solutes). This process tends to equalize the conditions on either side of the membrane.

OUCH PRINCIPLE

This principle says that as a manager when you delegate job tasks you must be **O**bjective, **U**niform in your treatment of employees, and the tasks must be **C**onsistent with utility policies, and **H**ave job relatedness.

OVERALL EFFICIENCY, PUMP

The combined efficiency of a pump and motor together. Also called the WIRE-TO-WATER EFFICIENCY.

OVERDRAFT

The pumping of water from a groundwater basin or aquifer in excess of the supply flowing into the basin. This pumping results in a depletion or "mining" of the groundwater in the basin.

OVERFLOW RATE

One factor of the design flow of settling tanks and clarifiers in treatment plants used by operators to determine if tanks and clarifiers are hydraulically (flow) over- or underloaded. Also called SURFACE LOADING.

$$\text{Overflow Rate, GPD/sq ft} = \frac{\text{Flow, gallons/day}}{\text{Surface Area, sq ft}}$$

or

$$\text{Overflow Rate, } \frac{\text{cu m/day}}{\text{sq m}} = \frac{\text{Flow, cu m/day}}{\text{Surface Area, sq m}}$$

OVERHEAD

Indirect costs necessary for a utility to function properly. These costs are not related to the actual treatment or delivery processes but include the costs of rent, lights, office supplies, management, and administration.

OVERTURN

The almost spontaneous mixing of all layers of water in a reservoir or lake when the water temperature becomes similar from top to bottom. This may occur in the fall/winter when the surface waters cool to the same temperature as the bottom waters and also in the spring when the surface waters warm after the ice melts. This is also called turnover.

OXIDATION

Oxidation is the addition of oxygen, removal of hydrogen, or the removal of electrons from an element or compound; in the environment and in wastewater treatment processes, organic matter is oxidized to more stable substances. The opposite of REDUCTION.

OXIDATION-REDUCTION POTENTIAL (ORP)

The electrical potential required to transfer electrons from one compound or element (the oxidant) to another compound or element (the reductant); used as a qualitative measure of the state of oxidation in water and wastewater treatment systems. ORP is measured in millivolts, with negative values indicating a tendency to reduce compounds or elements and positive values indicating a tendency to oxidize compounds or elements.

OXIDIZING AGENT

Any substance, such as oxygen (O_2) or chlorine (Cl_2), that will readily add (take on) electrons. When oxygen or chlorine is added to water or wastewater, organic substances are oxidized. These oxidized organic substances are more stable and less likely to give off odors or to contain disease-causing bacteria. The opposite is a REDUCING AGENT.

OXYGEN DEFICIENCY

An atmosphere containing oxygen at a concentration of less than 19.5 percent by volume.

OXYGEN ENRICHMENT

An atmosphere containing oxygen at a concentration of more than 23.5 percent by volume.

OZONATION (O-zoe-NAY-shun)

The application of ozone to water, wastewater, or air, generally for the purposes of disinfection or odor control.

P

PCBs

Polychlorinated biphenyls. A class of organic compounds that cause adverse health effects in domestic water supplies.

pCi/*L*

picoCurie per liter. A picoCurie is a measure of radioactivity. One picoCurie of radioactivity is equivalent to 0.037 nuclear disintegrations per second.

pcu

Platinum cobalt units. A measure of color using platinum cobalt standards by visual comparison.

PMCL

Primary Maximum Contaminant Level. Primary MCLs for various water quality indicators are established to protect public health.

POU/POE

See POINT-OF-USE/POINT-OF-ENTRY.

PPM

See PARTS PER MILLION.

PSIG

Pounds per Square Inch Gauge pressure. The pressure within a closed container or pipe measured with a gauge in pounds per square inch. Also see GAUGE PRESSURE.

PACKER ASSEMBLY

An inflatable device used to seal the tremie pipe inside the well casing to prevent the grout from entering the inside of the conductor casing.

PALATABLE (PAL-uh-tuh-bull)

Water at a desirable temperature that is free from objectionable tastes, odors, colors, and turbidity. Pleasing to the senses.

PARSHALL FLUME (PAR-shul FLOOM)

A device used to measure the flow in an open channel. The flume narrows to a throat of fixed dimensions and then expands again. The rate of flow can be calculated by measuring the difference in head (pressure) before and at the throat of the flume.

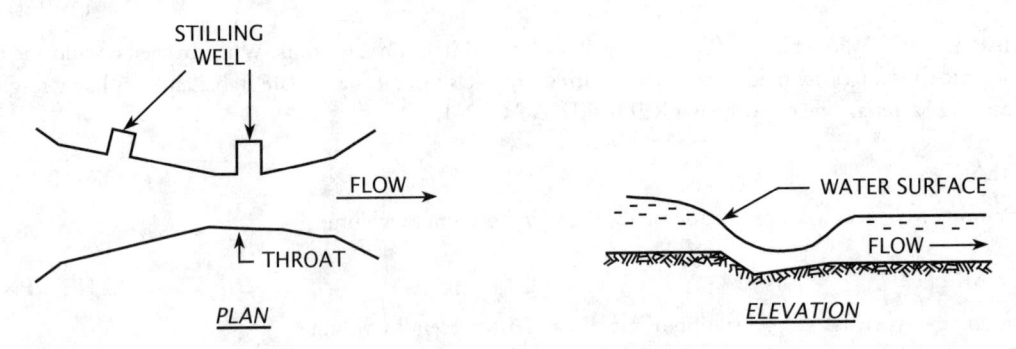

PARTICLE COUNT

The results of a microscopic examination of treated water with a special particle counter, which classifies suspended particles by number and size.

PARTICLE COUNTER

A device that counts and measures the size of individual particles in water.

PARTICLE COUNTING

A procedure for counting and measuring the size of individual particles in water. Particles are divided into size ranges and the number of particles is counted in each of these ranges. The results are reported in terms of the number of particles in different particle diameter size ranges per milliliter of water sampled.

PARTICULATE (par-TICK-yoo-let)

A very small solid suspended in water that can vary widely in size, shape, density, and electrical charge. Colloidal and dispersed particulates are artificially gathered together by the processes of coagulation and flocculation.

PARTS PER MILLION (PPM)

Parts per million parts, a measurement of concentration on a weight or volume basis. This term is equivalent to milligrams per liter (mg/L), which is the preferred term.

PASCAL (Pa)

The pressure or stress of one newton per square meter.

1 psi = 6,895 Pa = 6.895 kN/sq m = 0.0703 kg/sq cm

PATHOGENIC (path-o-JEN-ick) ORGANISMS

Organisms, including bacteria, viruses, or cysts, capable of causing diseases (such as giardiasis, cryptosporidiosis, typhoid, cholera, dysentery) in a host (such as a person). Also called PATHOGENS.

PATHOGENS (PATH-o-jens)

See PATHOGENIC ORGANISMS.

PEAK DEMAND

The maximum momentary load placed on a water treatment plant, pumping station, or distribution system. This demand is usually the maximum average load in one hour or less, but may be specified as the instantaneous load or the load during some other short time period.

PERCENT SATURATION

The amount of a substance that is dissolved in a solution compared with the amount dissolved in the solution at saturation, expressed as a percent.

$$\text{Percent Saturation, \%} = \frac{\text{Amount of Substance That Is Dissolved} \times 100\%}{\text{Amount Dissolved in Solution at Saturation}}$$

PERCOLATING (PURR-ko-lay-ting) WATER

Water that passes through soil or rocks under the force of gravity.

PERCOLATION (purr-ko-LAY-shun)

The slow passage of water through a filter medium; or, the gradual penetration of soil and rocks by water.

PERIPHYTON (pair-e-FI-tawn)

Microscopic plants and animals that are firmly attached to solid surfaces under water, such as rocks, logs, pilings, and other structures.

PERMEABILITY (PURR-me-uh-BILL-uh-tee)

The property of a material or soil that permits considerable movement of water through it when it is saturated.

PERMEATE (PURR-me-ate)

(1) To penetrate and pass through, as water penetrates and passes through soil and other porous materials.

(2) The liquid (demineralized water) produced from the reverse osmosis process that contains a low concentration of dissolved solids.

PERMIT-REQUIRED CONFINED SPACE (PERMIT SPACE)

See CONFINED SPACE, PERMIT-REQUIRED (PERMIT SPACE).

PESTICIDE

Any substance or chemical designed or formulated to kill or control animal pests. Also see INSECTICIDE and RODENTICIDE.

PET COCK

A small valve or faucet used to drain a cylinder or fitting.

pH (pronounce as separate letters)

pH is an expression of the intensity of the basic or acidic condition of a liquid. Mathematically, pH is the logarithm (base 10) of the reciprocal of the hydrogen ion activity.

$$pH = \text{Log} \frac{1}{\{H^+\}}$$

If $\{H^+\} = 10^{-6.5}$, then pH = 6.5. The pH may range from 0 to 14, where 0 is most acidic, 14 most basic, and 7 neutral.

PHENOLIC (fee-NO-lick) COMPOUNDS

Organic compounds that are derivatives of benzene. Also called phenols (FEE-nolls).

PHENOLPHTHALEIN (FEE-nol-THAY-leen) ALKALINITY

The alkalinity in a water sample measured by the amount of standard acid required to lower the pH to a level of 8.3, as indicated by the change in color of phenolphthalein from pink to clear. Phenolphthalein alkalinity is expressed as milligrams per liter of equivalent calcium carbonate.

PHOTOSYNTHESIS (foe-toe-SIN-thuh-sis)

A process in which organisms, with the aid of chlorophyll (green plant enzyme), convert carbon dioxide and inorganic substances into oxygen and additional plant material, using sunlight for energy. All green plants grow by this process.

PHYTOPLANKTON (FIE-tow-plank-ton)

Small, usually microscopic plants (such as algae), found in lakes, reservoirs, and other bodies of water.

PICO- (PEE-ko)

A prefix used in the metric system and other scientific systems of measurement which means 10^{-12} or 0.000 000 000 001.

PICOCURIE (PEE-ko-KYOOR-ee)

A measure of radioactivity. One picoCurie (pCi) of radioactivity is equivalent to 0.037 nuclear disintegrations per second.

PILOT-SCALE STUDY

A method of studying different ways of treating water or wastewater and solids or to obtain design criteria on a small scale in the field.

PIPE GAUGE

A number that defines the thickness of the sheet used to make steel pipe. The larger the number, the thinner the pipe wall.

PIPE SCHEDULE

A sizing system of numbers that specifies the ID (inside diameter) and OD (outside diameter) for each diameter pipe. The schedule number is the ratio of internal pressure in psi divided by the allowable fiber stress multiplied by 1,000. Typical schedules of iron and steel pipe are schedules 40, 80, and 160. Other forms of piping are divided into various classes with their own schedule schemes.

PITLESS ADAPTER

A fitting that allows the well casing to be extended above ground while having a discharge connection located below the frost line. Advantages of using a pitless adapter include the elimination of the need for a pit or pump house and it is a watertight design, which helps maintain a sanitary water supply.

PLAN or PLAN VIEW

A drawing or photo showing the top view of sewers, manholes, streets, or structures.

PLANKTON

(1) Small, usually microscopic, plants (phytoplankton) and animals (zooplankton) in aquatic systems.

(2) All of the smaller floating, suspended, or self-propelled organisms in a body of water.

PLANNING

Management of utilities to build the resources and financial capability to provide for future needs.

PLUG FLOW

A type of flow that occurs in tanks, basins, or reactors when a slug of water or wastewater moves through a tank without ever dispersing or mixing with the rest of the water or wastewater flowing through the tank.

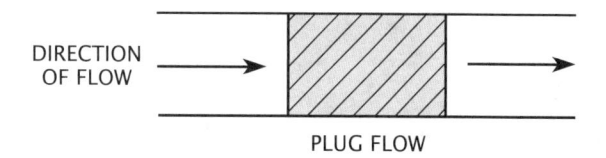

POINT SOURCE

A discharge that comes out the end of a pipe or other clearly identifiable conveyance. Examples of point source conveyances from which pollutants may be discharged include: ditches, channels, tunnels, conduits, wells, containers, rolling stock, concentrated animal feeding operations, landfill leachate collection systems, vessels, or other floating craft. A NONPOINT SOURCE refers to runoff or a discharge from a field or similar source.

POINT-OF-USE/POINT-OF-ENTRY (POU/POE)

Point-of-use applications refer to a water treatment device that treats water at the location of the end user. Point-of-entry application refers to a water treatment device that is located at the inlet to an entire building or facility.

POLARIZATION

The concentration of ions in the thin, boundary layer adjacent to a membrane or pipe wall.

POLE SHADER

A copper bar circling the laminated iron core inside the coil of a magnetic starter.

POLLUTION

The impairment (reduction) of water quality by agricultural, domestic, or industrial wastes (including thermal and radioactive wastes) to a degree that the natural water quality is changed to hinder any beneficial use of the water or render it offensive to the senses of sight, taste, or smell or when sufficient amounts of wastes create or pose a potential threat to human health or the environment.

POLYANIONIC (poly-AN-eye-ON-ick)

Characterized by many active negative charges especially active on the surface of particles.

POLYCHLORINATED BIPHENYLS (PCBs)
(POLY-KLOR-uh-nate-ed BI-FEEN-alls)

A class of organic compounds that cause adverse health effects in domestic water supplies.

POLYELECTROLYTE (POLY-ee-LECK-tro-lite)

A high-molecular-weight (relatively heavy) substance, having points of positive or negative electrical charges, that is formed by either natural or synthetic (manmade) processes. Natural polyelectrolytes may be of biological origin or obtained from starch products or cellulose derivatives. Synthetic polyelectrolytes consist of simple substances that have been made into complex, high-molecular-weight substances. Used with other chemical coagulants to aid in binding small suspended particles to larger chemical flocs for their removal from water. Often called a POLYMER.

POLYMER (POLY-mer)

A long-chain molecule formed by the union of many monomers (molecules of lower molecular weight). Polymers are used with other chemical coagulants to aid in binding small suspended particles to larger chemical flocs for their removal from water. Also see POLYELECTROLYTE.

PORE

A very small open space in a rock or granular material. Also called an INTERSTICE, VOID, or void space. Also see VOID.

POROSITY

(1) A measure of the spaces or voids in a material or aquifer.

(2) The ratio of the volume of spaces in a rock or soil to the total volume. This ratio is usually expressed as a percentage.

$$\text{Porosity, \%} = \frac{(\text{Volume of Spaces})(100\%)}{\text{Total Volume}}$$

POSITIVE BACTERIOLOGICAL SAMPLE

A water sample in which gas is produced by coliform organisms during incubation in the multiple tube fermentation test.

POSITIVE DISPLACEMENT PUMP

A type of piston, diaphragm, gear, or screw pump that delivers a constant volume with each stroke. Positive displacement pumps are used as chemical solution feeders.

POSTCHLORINATION

The addition of chlorine to the plant discharge or effluent, following plant treatment, for disinfection purposes.

POTABLE (POE-tuh-bull) WATER

Water that does not contain objectionable pollution, contamination, minerals, or infective agents and is considered satisfactory for drinking.

POWER FACTOR

The ratio of the true power passing through an electric circuit to the product of the voltage and amperage in the circuit. This is a measure of the lag or lead of the current with respect to the voltage. In alternating current, the voltage and amperes are not always in phase; therefore, the true power may be slightly less than that determined by the direct product.

PRECHLORINATION

The addition of chlorine in the collection system serving the plant or at the headworks of the plant prior to other treatment processes.

(1) For drinking water, used mainly for disinfection, control of tastes, odors, and aquatic growths, and to aid in coagulation and settling.

(2) For wastewater, used mainly for control of odors, corrosion, and foaming, and for BOD reduction and oil removal.

PRECIPITATE (pre-SIP-uh-TATE)

(1) An insoluble, finely divided substance that is a product of a chemical reaction within a liquid.

(2) The separation from solution of an insoluble substance.

PRECIPITATION

(1) The total measurable supply of water received directly from clouds as rain, snow, hail, or sleet; usually expressed as depth in a day, month, or year, and designated as daily, monthly, or annual precipitation.

(2) The process by which atmospheric moisture is discharged onto a land or water surfaces.

(3) The chemical transformation of a substance in solution into an insoluble form (precipitate).

PRECISION

The ability of an instrument to measure a process variable and repeatedly obtain the same result. The ability of an instrument to reproduce the same results.

PRECURSOR, THM (PRE-curse-or)

Natural, organic compounds found in all surface and groundwaters, which may react with halogens (such as chlorine) to form trihalomethanes (THMs); they must be present in order for THMs to form.

PRESCRIPTIVE (pre-SKRIP-tive) RIGHTS

Water rights that are acquired by diverting water and putting it to use in accordance with specified procedures. These procedures include filing a request (with a state agency) to use unused water in a stream, river, or lake.

PRESENT WORTH

The value of a long-term project expressed in today's dollars. Present worth is calculated by converting (discounting) all future benefits and costs over the life of the project to a single economic value at the start of the project. Calculating the present worth of alternative projects makes it possible to compare them and select the one with the largest positive (beneficial) present worth or minimum present cost.

PRESSURE CONTROL

A switch that operates on changes in pressure. Usually this is a diaphragm pressing against a spring. When the force on the diaphragm overcomes the spring pressure, the switch is activated.

PRESSURE HEAD

The vertical distance (in feet or meters) equal to the pressure (in psi or kPa) at a specific point. The pressure head is equal to the pressure in psi (or kPa) times 2.31 ft/psi (or 1.0 m/9.81 kPa).

PRESTRESSED

A prestressed pipe has been reinforced with wire strands (which are under tension) to give the pipe an active resistance to loads or pressures on it.

PREVENTIVE MAINTENANCE

Regularly scheduled servicing of machinery or other equipment using appropriate tools, tests, and lubricants. This type of maintenance can prolong the useful life of equipment and machinery and increase its efficiency by detecting and correcting problems before they cause a breakdown of the equipment.

PREVENTIVE MAINTENANCE UNITS

Crews assigned the task of cleaning sewers (for example, balling or high-velocity cleaning crews) to prevent stoppages and odor complaints. Also see PREVENTIVE MAINTENANCE.

PRIMARY ELEMENT

(1) A device that measures (senses) a physical condition or variable of interest. Floats and thermocouples are examples of primary elements. Also called a SENSOR.

(2) The hydraulic structure used to measure flows. In open channels, weirs and flumes are primary elements or devices. Venturi meters and orifice plates are the primary elements in pipes or pressure conduits.

PRIME

The action of filling a pump casing with water to remove the air. Most pumps must be primed before start-up or they will not pump any water.

PROCESS VARIABLE

A physical or chemical quantity that is usually measured and controlled in the operation of a water, wastewater, or industrial treatment plant. Common process variables are flow, level, pressure, temperature, turbidity, chlorine, and oxygen levels.

PRODUCT WATER

Water that has passed through a water treatment plant. All the treatment processes are completed or finished. This water is the product from the water treatment plant and is ready to be delivered to consumers. Also called FINISHED WATER.

PROFILE

A drawing showing elevation plotted against distance, such as the vertical section or side view of sewers, manholes, or a pipeline.

PROGRAMMABLE LOGIC CONTROLLER (PLC)

A microcomputer-based control device containing programmable software; used to control process variables.

PROPORTIONAL WEIR (WEER)

A specially shaped weir in which the flow through the weir is directly proportional to the head.

PRUSSIAN BLUE

A blue paste or liquid (often on a paper like carbon paper) used to show a contact area. Used to determine if gate valve seats fit properly.

PUMP BOWL

The submerged pumping unit in a well, including the shaft, impellers, and housing.

PUMPING WATER LEVEL

The vertical distance from the centerline of the pump discharge to the level of the free pool while water is being drawn from the pool.

PURVEYOR (purr-VAY-or), WATER

An agency or person that supplies water (usually potable water).

PUTREFACTION (PYOO-truh-FACK-shun)

Biological decomposition of organic matter, with the production of foul-smelling and -tasting products, associated with anaerobic (no oxygen present) conditions.

Q

QUICKLIME

A material that is mostly calcium oxide (CaO) or calcium oxide in natural association with a lesser amount of magnesium oxide. Quicklime is capable of combining with water, that is, becoming slaked. Also see HYDRATED LIME.

R

RO

See REVERSE OSMOSIS.

RADIAL TO IMPELLER

Perpendicular to the impeller shaft. Material being pumped flows at a right angle to the impeller.

RADICAL

A group of atoms that is capable of remaining unchanged during a series of chemical reactions. Such combinations (radicals) exist in the molecules of many organic compounds; sulfate (SO_4^{2-}) is an inorganic radical.

RANGE

The spread from minimum to maximum values that an instrument is designed to measure. Also see EFFECTIVE RANGE and SPAN.

RANNEY COLLECTOR

This water collector is constructed as a dug well from 12 to 16 feet (3.5 to 5 m) in diameter that has been sunk as a caisson near the bank of a river or lake. Screens are driven radially and approximately horizontally from this well into the sand and gravel deposits underlying the river.

[SEE DRAWING ON PAGE 828]

RATE OF RETURN

A value that indicates the return of funds received on the basis of the total equity capital used to finance physical facilities. Similar to the interest rate on savings accounts or loans.

RAW WATER

(1) Water in its natural state, prior to any treatment.

(2) Water entering the first treatment process of a water treatment plant.

READOUT

The reading of the value of a process variable from an indicator or recorder or on a computer screen.

REAERATION (RE-air-A-shun)

The introduction of air through forced air diffusers into the lower layers of the reservoir. As the air bubbles form and rise through the water, oxygen from the air dissolves into the water and replenishes the dissolved oxygen. The rising bubbles also cause the lower waters to rise to the surface where oxygen from the atmosphere is transferred to the water. This is sometimes called surface reaeration.

REAGENT (re-A-gent)

A pure, chemical substance that is used to make new products or is used in chemical tests to measure, detect, or examine other substances.

RECARBONATION (re-kar-bun-NAY-shun)

A process in which carbon dioxide is bubbled into the water being treated to lower the pH.

RECEIVER

A device that indicates the result of a measurement, usually using either a fixed scale and movable indicator (pointer), such as a pressure gauge, or a moving chart with a movable pen like those used on a circular flow-recording chart. Also called an INDICATOR.

RECORDER

A device that creates a permanent record, on a paper chart, magnetic tape, or in a computer, of the changes in a measured variable.

REDUCING AGENT

Any substance, such as base metal (iron) or the sulfide ion (S^{2-}), that will readily donate (give up) electrons. The opposite is an OXIDIZING AGENT.

REDUCTION (re-DUCK-shun)

Reduction is the addition of hydrogen, removal of oxygen, or the addition of electrons to an element or compound. Under anaerobic conditions (no dissolved oxygen present), sulfur compounds are reduced to odor-producing hydrogen sulfide (H_2S) and other compounds. In the treatment of metal finishing wastewaters, hexavalent chromium (Cr^{6+}) is reduced to the trivalent form (Cr^{3+}). The opposite of OXIDATION.

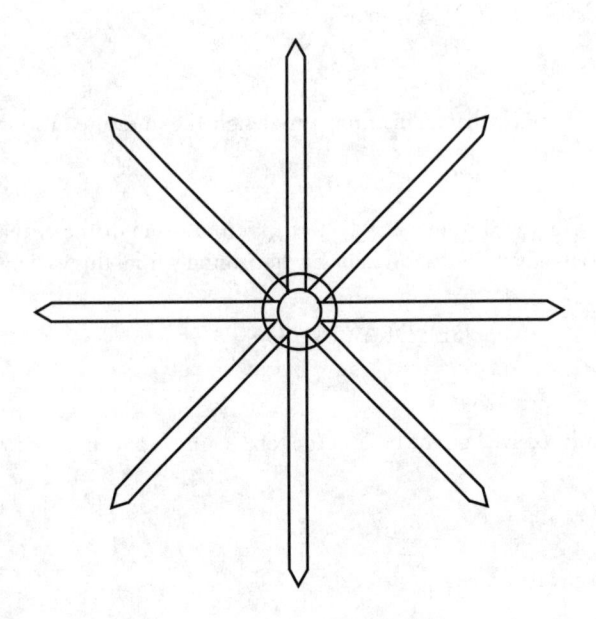

PLAN VIEW OF COLLECTOR PIPES

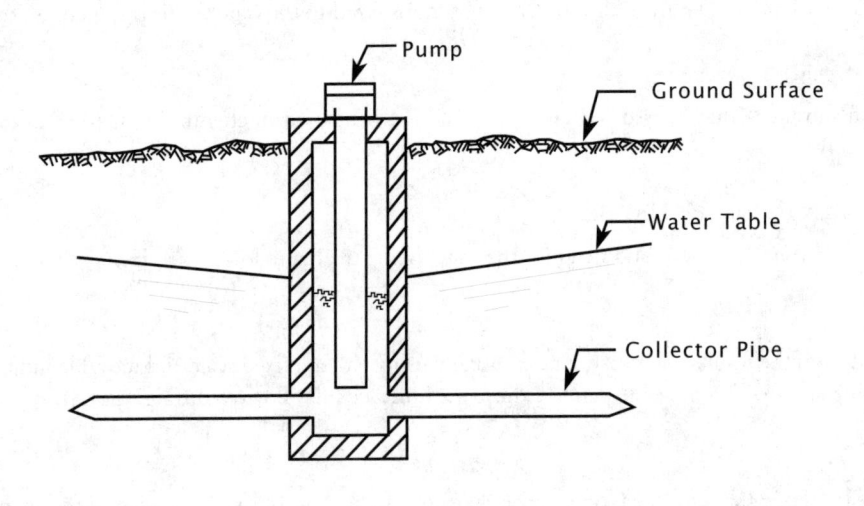

ELEVATION VIEW

Ranney collector

REFERENCE
A physical or chemical quantity whose value is known exactly, and thus is used to calibrate instruments or standardize measurements. Also called a STANDARD.

REGULATORY NEGOTIATION
A process whereby the US Environmental Protection Agency acts on an equal basis with outside parties to reach consensus on the content of a proposed rule. If the group reaches consensus, the US EPA commits to propose the rule with the agreed upon content.

RELIQUEFACTION (re-lick-we-FACK-shun)
The return of a gas to the liquid state; for example, a condensation of chlorine gas to return it to its liquid form by cooling.

REPRESENTATIVE SAMPLE
A sample portion of material, water, or wastestream that is as nearly identical in content and consistency as possible to that in the larger body being sampled.

RESIDUAL CHLORINE
The concentration of chlorine present in water after the chlorine demand has been satisfied. The concentration is expressed in terms of the total chlorine residual, which includes both the free and combined or chemically bound chlorine residuals. Also called CHLORINE RESIDUAL.

RESIDUE
The dry solids remaining after the evaporation of a sample of water or sludge. Also see TOTAL DISSOLVED SOLIDS.

RESINS
See ION EXCHANGE RESINS.

RESISTANCE
That property of a conductor or wire that opposes the passage of a current, thus causing electric energy to be transformed into heat.

RESPIRATION
The process in which an organism takes in oxygen for its life processes and gives off carbon dioxide.

RESPONSIBILITY
Answering to those above in the chain of command to explain how and why you have used your authority.

REVERSE OSMOSIS (oz-MOE-sis) (RO)
The application of pressure to a concentrated solution, which causes the passage of a liquid from the concentrated solution to a weaker solution across a semipermeable membrane. The membrane allows the passage of the water (solvent) but not the dissolved solids (solutes). In the reverse osmosis process, two liquids are produced: (1) the reject (containing high concentrations of dissolved solids), and (2) the permeate (containing low concentrations). The clean water (permeate) is not always considered to be demineralized. Also see OSMOSIS.

RIPARIAN (ri-PAIR-ee-an) RIGHTS
Water rights that are acquired together with title to the land bordering a source of surface water. The right to put to beneficial use surface water adjacent to your land.

RODENTICIDE (row-DENT-uh-side)
Any substance or chemical formulated to kill or control rodents.

ROOF LEADER
In plumbing, a pipe installed to drain water from the roof gutters or roof catchment to the storm drain or other means of disposal. Also called a CONDUCTOR, DOWNSPOUT, or roof drain.

ROTAMETER (ROTE-uh-ME-ter)

A device used to measure the flow rate of gases and liquids. The gas or liquid being measured flows vertically up a tapered, calibrated tube. Inside the tube is a small ball or bullet-shaped float (it may rotate) that rises or falls depending on the flow rate. The flow rate may be read on a scale behind or on the tube by looking at the middle of the ball or at the widest part or top of the float.

ROTOR

The rotating part of a machine. The rotor is surrounded by the stationary (nonmoving) parts (stator) of the machine.

ROUTINE SAMPLING

Sampling repeated on a regular basis.

S

SCADA (SKAY-dah) SYSTEM

Supervisory Control And Data Acquisition system. A computer-monitored alarm, response, control, and data acquisition system used to monitor and adjust treatment processes and facilities.

SCFM

Standard Cubic Feet per Minute. Cubic feet of air per minute at standard conditions of temperature, pressure, and humidity (0°C, 14.7 psia, and 50 percent relative humidity).

SDWA

See SAFE DRINKING WATER ACT.

SMCL

Secondary Maximum Contaminant Level. Secondary MCLs for various water quality indicators are established to protect public welfare.

SNARL

Suggested No Adverse Response Level. The concentration of a chemical in water that is expected not to cause an adverse health effect.

SACRIFICIAL ANODE

An easily corroded material deliberately installed in a pipe or tank. The intent of such an installation is to give up (sacrifice) this anode to corrosion while the water supply facilities remain relatively corrosion free.

SAFE DRINKING WATER ACT (SDWA)

An act passed by the US Congress in 1974. The act establishes a cooperative program among local, state, and federal agencies to ensure safe drinking water for consumers. The act has been amended several times, including the 1980, 1986, and 1996 amendments.

SAFE WATER

Water that does not contain harmful bacteria, or toxic materials or chemicals. Water may have taste and odor problems, color, and certain mineral problems and still be considered safe for drinking.

SAFE YIELD

The annual quantity of water that can be taken from a source of supply over a period of years without depleting the source permanently (beyond its ability to be replenished naturally in "wet years").

SALINITY

(1) The relative concentration of dissolved salts, usually sodium chloride, in a given water.

(2) A measure of the concentration of dissolved mineral substances in water.

SANITARY SURVEY

A detailed evaluation or inspection of a source of water supply and all conveyances, storage, treatment, and distribution facilities to ensure protection of the water supply from all pollution sources.

SAPROPHYTES (SAP-row-fights)

Organisms living on dead or decaying organic matter. They help natural decomposition of organic matter in water or wastewater.

SATURATION

The condition of a liquid (water) when it has taken into solution the maximum possible quantity of a given substance at a given temperature and pressure.

SATURATOR (SAT-yoo-ray-tor)

A device that produces a fluoride solution for the fluoridation process. The device is usually a cylindrical container with granular sodium fluoride on the bottom. Water flows either upward or downward through the sodium fluoride to produce the fluoride solution.

SCALE

(1) A combination of mineral salts and bacterial accumulation that sticks to the inside of a collection pipe under certain conditions. Scale, in extreme growth circumstances, creates additional friction loss to the flow of water. Scale may also accumulate on surfaces other than pipes.

(2) The marked plate against which an indicator or recorder reads, usually the same as the range of the measuring system. See RANGE.

SCHEDULE, PIPE

See PIPE SCHEDULE.

SCHMUTZDECKE (shmoots-DECK-ee)

A layer of trapped matter at the surface of a slow sand filter in which a dense population of microorganisms develops. These microorganisms within the film or mat feed on and break down incoming organic material trapped in the mat. In doing so, the microorganisms both remove organic matter and add mass to the mat, further developing the mat and increasing the physical straining action of the mat.

SECCHI (SECK-key) DISK

A flat, white disk lowered into the water by a rope until it is just barely visible. At this point, the depth of the disk from the water surface is the recorded Secchi disk transparency.

SEDIMENTATION (SED-uh-men-TAY-shun)

The process of settling and depositing of suspended matter carried by water or wastewater. Sedimentation usually occurs by gravity when the velocity of the liquid is reduced below the point at which it can transport the suspended material.

SEDIMENTATION (SED-uh-men-TAY-shun) BASIN

A tank or basin in which water or wastewater is held for a period of time during which the heavier solids settle to the bottom and the lighter materials float to the surface. Also called settling tank or CLARIFIER.

SEIZING or SEIZE UP

Seizing occurs when an engine overheats and a part expands to the point where the engine will not run. Also called freezing.

SENSITIVITY

The smallest change in a process variable that an instrument can sense.

SENSITIVITY (PARTICLE COUNTERS)

The smallest particle a particle counter will measure and count.

SENSOR

A device that measures (senses) a physical condition or variable of interest. Floats and thermocouples are examples of sensors. Also called a PRIMARY ELEMENT.

SEPTIC (SEP-tick) or SEPTICITY

A condition produced by bacteria when all oxygen supplies are depleted. If severe, the bottom deposits produce hydrogen sulfide, the deposits and water turn black, give off foul odors, and the water has a greatly increased oxygen and chlorine demand.

SEQUESTRATION (SEE-kwes-TRAY-shun)

A chemical complexing (forming or joining together) of metallic cations (such as iron) with certain inorganic compounds, such as phosphate. Sequestration prevents the precipitation of the metals (iron). Also see CHELATION.

SERVICE PIPE

The pipeline extending from the water main to the building served or to the consumer's system.

SET POINT

The position at which the control or controller is set. This is the same as the desired value of the process variable. For example, a thermostat is set to maintain a desired temperature.

SEWAGE

The used household water and water-carried solids that flow in sewers to a wastewater treatment plant. The preferred term is WASTEWATER.

SHEAVE

V-belt drive pulley, which is commonly made of cast iron or steel.

SHIM

Thin metal sheets that are inserted between two surfaces to align or space the surfaces correctly. Shims can be used anywhere a spacer is needed. Usually shims are 0.001 to 0.020 inch (0.025 to 0.50 mm) thick.

SHOCK LOAD

The arrival at a treatment process of water or wastewater containing unusually high concentrations of contaminants in sufficient quantity or strength to cause operating problems. Organic or hydraulic overloads also can cause a shock load.

(1) For activated sludge, possible problems include odors and bulking sludge, which will result in a high loss of solids from the secondary clarifiers into the plant effluent and a biological process upset that may require several days to a week to recover.

(2) For trickling filters, possible problems include odors and sloughing off of the growth or slime on the trickling filter media.

(3) For drinking water treatment, possible problems include filter blinding and product water with taste and odor, color, or turbidity problems.

SHORT-CIRCUITING

A condition that occurs in tanks or basins when some of the flowing water entering a tank or basin flows along a nearly direct pathway from the inlet to the outlet. This is usually undesirable since it may result in shorter contact, reaction, or settling times in comparison with the theoretical (calculated) or presumed detention times.

SIMULATE

To reproduce the action of some process, usually on a smaller scale.

SINGLE-STAGE PUMP

A pump that has only one impeller. A multistage pump has more than one impeller.

SLAKE

To mix with water so that a true chemical combination (hydration) takes place, such as in the slaking of lime.

SLAKED LIME

See HYDRATED LIME.

SLOPE

The slope or inclination of a trench bottom or a trench side wall is the ratio of the vertical distance to the horizontal distance or rise over run. Also see GRADE (2).

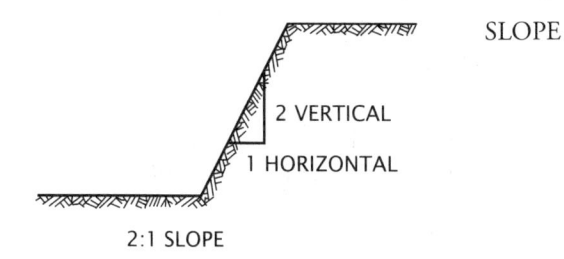

SLUDGE (SLUJ)

(1) The settleable solids separated from liquids during processing.

(2) The deposits of foreign materials on the bottoms of streams or other bodies of water or on the bottoms and edges of wastewater collection lines and appurtenances.

SLURRY

A watery mixture or suspension of insoluble (not dissolved) matter; a thin, watery mud or any substance resembling it (such as a grit slurry or a lime slurry).

SOFT WATER

Water having a low concentration of calcium and magnesium ions. According to US Geological Survey guidelines, soft water is water having a hardness of 60 milligrams per liter or less.

SOFTWARE PROGRAM

Computer program; the list of instructions that tell a computer how to perform a given task or tasks. Some software programs are designed and written to monitor and control treatment processes.

SOLENOID (SO-luh-noid)

A magnetically operated mechanical device (electric coil). Solenoids can operate small valves or electric switches.

SOLUTION

A liquid mixture of dissolved substances. In a solution it is impossible to see all the separate parts.

SOUNDING TUBE

A pipe or tube used for measuring the depths of water.

SPAN

The scale or range of values an instrument is designed to measure. Also see RANGE.

SPECIFIC CAPACITY

A measurement of well yield per unit depth of drawdown after a specific time has passed, usually 24 hours. Typically expressed as gallons per minute per foot or cubic meters per day per meter (GPM/ft or cu m/day/m).

SPECIFIC CAPACITY TEST

A testing method used to determine the adequacy of an aquifer or well by measuring the specific capacity.

SPECIFIC CONDUCTANCE

A rapid method of estimating the total dissolved solids content of a water supply. The measurement indicates the capacity of a sample of water to carry an electric current, which is related to the concentration of ionized substances in the water. Also called CONDUCTANCE.

SPECIFIC GRAVITY

(1) Weight of a particle, substance, or chemical solution in relation to the weight of an equal volume of water. Water has a specific gravity of 1.000 at 4°C (39°F). Particulates with specific gravity less than 1.0 float to the surface and particulates with specific gravity greater than 1.0 sink.

(2) Weight of a particular gas in relation to the weight of an equal volume of air at the same temperature and pressure (air has a specific gravity of 1.0). Chlorine gas has a specific gravity of 2.5.

SPECIFIC YIELD

The quantity of water that a unit volume of saturated permeable rock or soil will yield when drained by gravity. Specific yield may be expressed as a ratio or as a percentage by volume.

SPLIT SAMPLE

A single grab sample that is separated into at least two parts such that each part is representative of the original sample. Often used to compare test results between field kits and laboratories or between two laboratories.

SPOIL

Excavated material, such as soil, from the trench of a water main or sewer.

SPORE

The reproductive body of certain organisms, which is capable of giving rise to a new organism either directly or indirectly. A viable (able to live and grow) body regarded as the resting stage of an organism. A spore is usually more resistant to disinfectants and heat than most organisms. Gangrene and tetanus bacteria are common spore-forming organisms.

SPRING LINE

Theoretical center of a pipeline. Also, the guideline for laying a course of bricks.

STALE WATER

Water that has not flowed recently and may have picked up tastes and odors from distribution lines or storage facilities.

STANDARD

A physical or chemical quantity whose value is known exactly, and thus is used to calibrate instruments or standardize measurements. Also called a REFERENCE.

STANDARD DEVIATION

A measure of the spread or dispersion of data.

STANDARD METHODS

STANDARD METHODS FOR THE EXAMINATION OF WATER AND WASTEWATER, 21st Edition. A joint publication of the American Public Health Association (APHA), American Water Works Association (AWWA), and the Water Environment Federation (WEF) that outlines the accepted laboratory procedures used to analyze the impurities in water and wastewater. Available from: American Water Works Association, Bookstore, 6666 West Quincy Avenue, Denver, CO 80235. Order No. 10079. Price to members, $168.00; nonmembers, $213.00; price includes cost of shipping and handling. Also available from Water Environment Federation, Publications Order Department, 601 Wythe Street, Alexandria, VA 22314-1994. Order No. S82010. Price to members, $164.75; nonmembers, $209.75; price includes cost of shipping and handling.

STANDARD SOLUTION

A solution in which the exact concentration of a chemical or compound is known.

STANDARDIZE

To compare with a standard.

(1) In wet chemistry, to find out the exact strength of a solution by comparing it with a standard of known strength. This information is used to adjust the strength by adding more water or more of the substance dissolved.

(2) To set up an instrument or device to read a standard. This allows you to adjust the instrument so that it reads accurately, or enables you to apply a correction factor to the readings.

STARTERS (MOTOR)

Devices used to start up large motors gradually to avoid severe mechanical shock to a driven machine and to prevent disturbance to the electrical lines (causing dimming and flickering of lights).

STATIC HEAD

When water is not moving, the vertical distance (in feet or meters) from a reference point to the water surface is the static head. Also see DYNAMIC HEAD, DYNAMIC PRESSURE, and STATIC PRESSURE.

STATIC PRESSURE

When water is not moving, the vertical distance (in feet or meters) from a specific point to the water surface is the static head. The static pressure in psi (or kPa) is the static head in feet times 0.433 psi/ft (or meters × 9.81 kPa/m). Also see DYNAMIC HEAD, DYNAMIC PRESSURE, and STATIC HEAD.

STATIC WATER DEPTH

The vertical distance in feet (or meters) from the centerline of the pump discharge down to the surface level of the free pool while no water is being drawn from the pool or water table.

STATIC WATER LEVEL

(1) The elevation or level of the water table in a well when the pump is not operating.

(2) The level or elevation to which water would rise in a tube connected to an artesian aquifer, basin, or conduit under pressure.

STATOR

That portion of a machine that contains the stationary (nonmoving) parts that surround the moving parts (rotor).

STERILIZATION (STAIR-uh-luh-ZAY-shun)

The removal or destruction of all microorganisms, including pathogens and other bacteria, vegetative forms, and spores. Compare with DISINFECTION.

STETHOSCOPE

An instrument used to magnify sounds and carry them to the ear.

STORATIVITY (S)

The volume of groundwater an aquifer releases from or takes into storage per unit surface area of the aquifer per unit change in head. Also called the storage coefficient.

STORM RUNOFF

Water that flows over the ground surface directly into streams, rivers, or lakes. Also called DIRECT RUNOFF.

STORMWATER

The excess water running off from the surface of a drainage area during and immediately after a period of rain. Also see STORM RUNOFF.

STRATIFICATION (STRAT-uh-fuh-KAY-shun)

The formation of separate layers (of temperature, plant life, or animal life) in a lake or reservoir. Characteristics within each layer are similar; for instance, all water in the same layer has the same temperature. Also see THERMAL STRATIFICATION.

STRAY CURRENT CORROSION

A corrosion activity resulting from stray electric current originating from some source outside the plumbing system such as DC grounding on phone systems.

SUBMERGENCE

The distance between the water surface and the media surface in a filter.

SUBSIDENCE (sub-SIDE-ence)

The dropping or lowering of the ground surface as a result of removing excess water (overdraft or overpumping) from an aquifer. After excess water has been removed, the soil will settle, become compacted, and the ground surface will drop, which can cause the settling of underground utilities.

SUCTION HEAD

The positive pressure [in feet (meters) of water or pounds per square inch (kilograms per square centimeter) of mercury vacuum] on the suction side of a pump. The pressure can be measured from the centerline of the pump up to the elevation of the hydraulic grade line on the suction side of the pump.

SUCTION LIFT

The negative pressure [in feet (meters) of water or inches (centimeters) of mercury vacuum] on the suction side of a pump. The pressure can be measured from the centerline of the pump down to (lift) the elevation of the hydraulic grade line on the suction side of the pump.

SUPERCHLORINATION (SOO-per-KLOR-uh-NAY-shun)

Chlorination with doses that are deliberately selected to produce free or combined residuals so large as to require dechlorination.

SUPERNATANT (soo-per-NAY-tent)

Liquid removed from settled sludge. Supernatant commonly refers to the liquid between the sludge on the bottom and the scum on the surface.

SUPERSATURATED

An unstable condition of a solution (water) in which the solution contains a substance at a concentration greater than the saturation concentration for the substance.

SURFACE LOADING

One factor of the design flow of settling tanks and clarifiers in treatment plants used by operators to determine if tanks and clarifiers are hydraulically (flow) over- or underloaded. Also called OVERFLOW RATE.

$$\text{Surface Loading, GPD/sq ft} = \frac{\text{Flow, gallons/day}}{\text{Surface Area, sq ft}}$$

or

$$\text{Surface Loading, } \frac{\text{cu m/day}}{\text{sq m}} = \frac{\text{Flow, cu m/day}}{\text{Surface Area, sq m}}$$

SURFACTANT (sir-FAC-tent)

Abbreviation for surface-active agent. The active agent in detergents that possesses a high cleaning ability.

SURGE CHAMBER

A chamber or tank connected to a pipe and located at or near a valve that may quickly open or close or a pump that may suddenly start or stop. When the flow of water in a pipe starts or stops quickly, the surge chamber allows water to flow into or out of the pipe and minimize any sudden positive or negative pressure waves or surges in the pipe.

[SEE DRAWING ON PAGE 837]

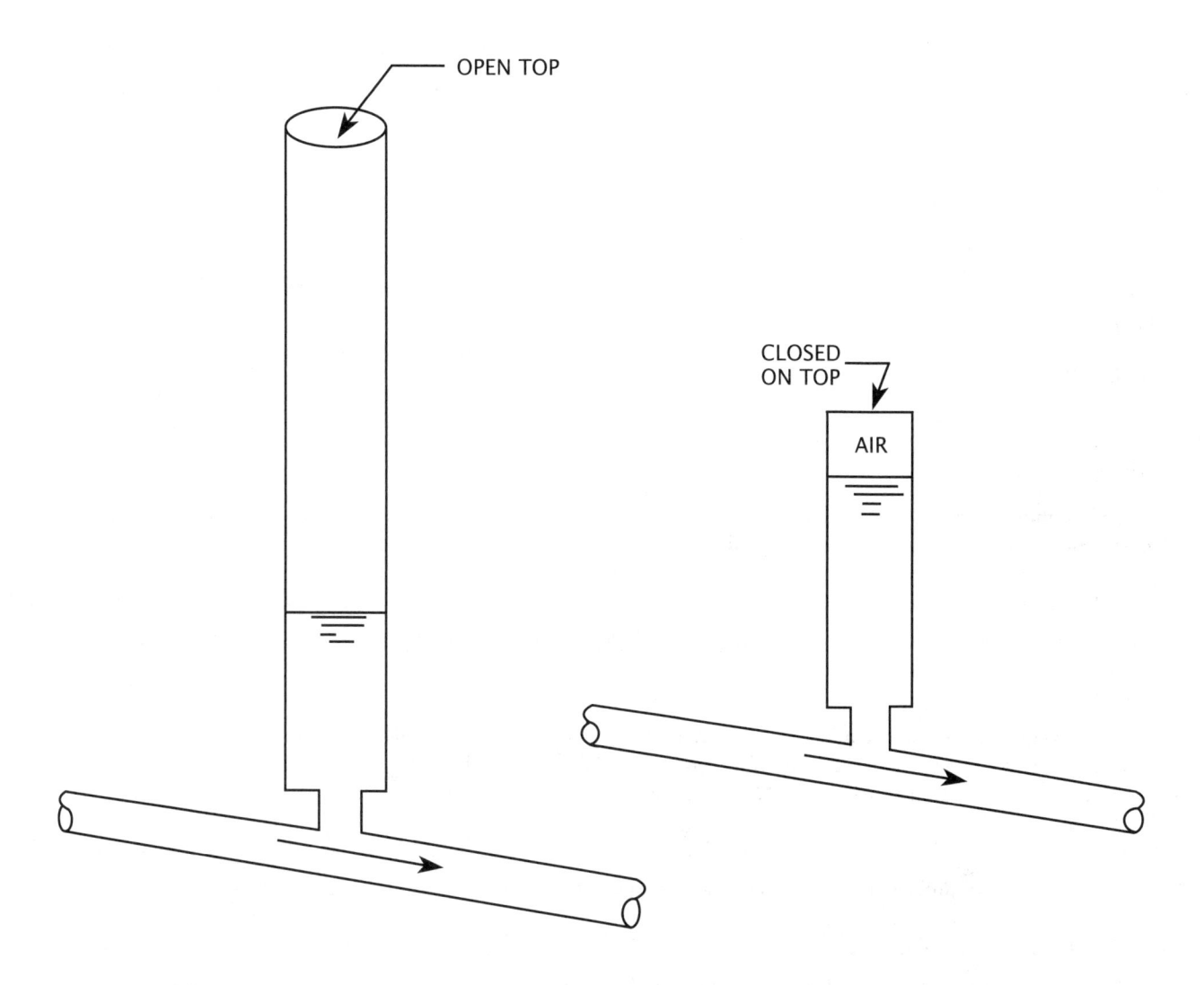

Types of surge chambers

SUSPENDED SOLIDS

(1) Solids that either float on the surface or are suspended in water, wastewater, or other liquids, and that are largely removable by laboratory filtering.

(2) The quantity of material removed from water or wastewater in a laboratory test, as prescribed in *STANDARD METHODS FOR THE EXAMINATION OF WATER AND WASTEWATER*, and referred to as Total Suspended Solids Dried at 103–105°C.

T

TCE
See TRICHLOROETHANE.

TDS
See TOTAL DISSOLVED SOLIDS.

THM
See TRIHALOMETHANES.

THM PRECURSOR
See PRECURSOR, THM.

TAILGATE SAFETY MEETING
Brief (10 to 20 minutes) safety meetings held every 7 to 10 working days. The term comes from the safety meetings regularly held by the construction industry around the tailgate of a truck.

TELEMETRY (tel-LEM-uh-tree)
The electrical link between a field transmitter and the receiver. Telephone lines are commonly used to serve as the electrical line.

TEMPERATURE SENSOR
A device that opens and closes a switch in response to changes in the temperature. This device might be a metal contact, or a thermocouple that generates a minute electric current proportional to the difference in heat, or a variable resistor whose value changes in response to changes in temperature. Also called a HEAT SENSOR.

THERMAL STRATIFICATION (strat-uh-fuh-KAY-shun)
The formation of layers of different temperatures in a lake or reservoir. Also see STRATIFICATION.

THERMOCLINE (THUR-moe-kline)
The middle layer in a thermally stratified lake or reservoir. In this layer there is a rapid decrease in temperature with depth. Also called the METALIMNION.

THERMOCOUPLE
A heat-sensing device made of two conductors of different metals joined together. An electric current is produced when there is a difference in temperature between the ends.

THICKENING
Treatment to remove water from the sludge mass to reduce the volume that must be handled.

THRESHOLD ODOR
The minimum odor of a gas or water sample that can just be detected after successive dilutions with odorless gas or water. Also called ODOR THRESHOLD.

THRESHOLD ODOR NUMBER (TON)

The greatest dilution of a sample with odor-free water that still yields a just-detectable odor.

THRUST BLOCK

A mass of concrete or similar material appropriately placed around a pipe to prevent movement when the pipe is carrying water. Usually placed at bends and valve structures.

TIDE GATE

A gate installed at the end of a drain or outlet pipe to prevent the backward flow of water or wastewater. Generally used on storm sewer outlets into streams to prevent backward flow during times of flood or high tide. Also called a BACKWATER GATE. Also see CHECK VALVE and FLAP GATE.

TIME LAG

The time required for processes and control systems to respond to a signal or to reach a desired level.

TIMER

A device for automatically starting or stopping a machine or other device at a given time.

TITRATE (TIE-trate)

To titrate a sample, a chemical solution of known strength is added drop by drop until a certain color change, precipitate, or pH change in the sample is observed (end point). Titration is the process of adding the chemical reagent in small increments (0.1–1.0 milliliter) until completion of the reaction, as signaled by the end point.

TOPOGRAPHY (toe-PAH-gruh-fee)

The arrangement of hills and valleys in a geographic area.

TOTAL CHLORINE

The total concentration of chlorine in water, including the combined chlorine (such as inorganic and organic chloramines) and the free available chlorine.

TOTAL CHLORINE RESIDUAL

The total amount of chlorine residual (including both free chlorine and chemically bound chlorine) present in a water sample after a given contact time.

TOTAL DISSOLVED SOLIDS (TDS)

All of the dissolved solids in a water. TDS is measured on a sample of water that has passed through a very fine mesh filter to remove suspended solids. The water passing through the filter is evaporated and the residue represents the total dissolved solids. Also see SPECIFIC CONDUCTANCE.

TOTAL DYNAMIC HEAD (TDH)

When a pump is lifting or pumping water, the vertical distance (in feet or meters) from the elevation of the energy grade line on the suction side of the pump to the elevation of the energy grade line on the discharge side of the pump. The total dynamic head is the static head plus pipe friction losses.

TOTAL ORGANIC CARBON (TOC)

TOC is a measure of the amount of organic carbon in water.

TOTALIZER

A device or meter that continuously measures and sums a process rate variable in cumulative fashion over a given time period. For example, total flows displayed in gallons per minute, million gallons per day, cubic feet per second, or some other unit of volume per time period. Also called an INTEGRATOR.

TOXAPHENE (TOX-uh-feen)

A chemical that causes adverse health effects in domestic water supplies and also is toxic to freshwater and marine aquatic life.

TOXIC

A substance that is poisonous to a living organism. Toxic substances may be classified in terms of their physiological action, such as irritants, asphyxiants, systemic poisons, and anesthetics and narcotics. Irritants are corrosive substances that attack the mucous membrane surfaces of the body. Asphyxiants interfere with breathing. Systemic poisons are hazardous substances that injure or destroy internal organs of the body. Anesthetics and narcotics are hazardous substances that depress the central nervous system and lead to unconsciousness.

TOXIC SUBSTANCE

See HARMFUL PHYSICAL AGENT and TOXIC.

TRANSDUCER (trans-DUE-sir)

A device that senses some varying condition measured by a primary sensor and converts it to an electrical or other signal for transmission to some other device (a receiver) for processing or decision making.

TRANSMISSION LINES

Pipelines that transport raw water from its source to a water treatment plant. After treatment, water is usually pumped into pipelines (transmission lines) that are connected to a distribution grid system.

TRANSMISSIVITY (TRANS-miss-SIV-it-tee)

A measure of the ability to transmit (as in the ability of an aquifer to transmit water).

TRANSPIRATION (TRAN-spur-RAY-shun)

The process by which water vapor is released to the atmosphere by living plants. This process is similar to people sweating. Also see EVAPOTRANSPIRATION.

TREMIE (TREH-me)

A device used to place concrete or grout under water.

TRICHLOROETHANE (TCE) (try-KLOR-o-ETH-hane)

An organic chemical used as a cleaning solvent that causes adverse health effects in domestic water supplies.

TRIHALOMETHANES (THMs) (tri-HAL-o-METH-hanes)

Derivatives of methane, CH_4, in which three halogen atoms (chlorine or bromine) are substituted for three of the hydrogen atoms. Often formed during chlorination by reactions with natural organic materials in the water. The resulting compounds (THMs) are suspected of causing cancer.

TRUE COLOR

Color of the water from which turbidity has been removed. The turbidity may be removed by double filtering the sample through a Whatman No. 40 filter when using the visual comparison method.

TUBE SETTLER

A device that uses bundles of small-bore (2 to 3 inches or 50 to 75 mm) tubes installed on an incline as an aid to sedimentation. The tubes may come in a variety of shapes including circular and rectangular. As water rises within the tubes, settling solids fall to the tube surface, and as the resulting sludge gains weight, it moves down the tubes and settles to the bottom of the basin for removal by conventional sludge collection means. Also called high-rate settlers.

TUBERCLE (TOO-burr-kull)

A crust of corrosion products (rust) that builds up over a pit caused by the loss of metal due to corrosion.

TUBERCULATION (too-BURR-kyoo-LAY-shun)

The development or formation of small mounds of corrosion products (rust) on the inside of iron pipe. These mounds (tubercles) increase the roughness of the inside of the pipe thus increasing resistance to water flow (decreases the C Factor).

TURBID

Having a cloudy or muddy appearance.

TURBIDIMETER

See TURBIDITY METER.

TURBIDITY (ter-BID-it-tee)

The cloudy appearance of water caused by the presence of suspended and colloidal matter. In the waterworks field, a turbidity measurement is used to indicate the clarity of water. Technically, turbidity is an optical property of the water based on the amount of light reflected by suspended particles. Turbidity cannot be directly equated to suspended solids because white particles reflect more light than dark-colored particles and many small particles will reflect more light than an equivalent large particle.

TURBIDITY (ter-BID-it-tee) METER

An instrument for measuring and comparing the turbidity of liquids by passing light through them and determining how much light is reflected by the particles in the liquid. The normal measuring range is 0 to 100 and is expressed as nephelometric turbidity units (NTUs).

TURBIDITY (ter-BID-it-tee) UNITS (TU)

Turbidity units are a measure of the cloudiness of water. If measured by a nephelometric (deflected light) instrumental procedure, turbidity units are expressed in nephelometric turbidity units (NTU) or simply TU. Those turbidity units obtained by visual methods are expressed in Jackson turbidity units (JTU), which are a measure of the cloudiness of water; they are used to indicate the clarity of water. There is no real connection between NTUs and JTUs. The Jackson turbidimeter is a visual method and the nephelometer is an instrumental method based on deflected light.

TURN-DOWN RATIO

The ratio of the design range to the range of acceptable accuracy or precision of an instrument. Also see EFFECTIVE RANGE.

U

UF

See ULTRAFILTRATION.

US EPA

United States Environmental Protection Agency. A regulatory agency established by the US Congress to administer the nation's environmental laws.

ULTRAFILTRATION (UF)

A pressure-driven membrane filtration process that separates particles down to approximately 0.01 μm diameter from influent water using a sieving process.

UNCONSOLIDATED FORMATION

A sediment that is loosely arranged or unstratified (not in layers) or whose particles are not cemented together (soft rock); occurring either at the ground surface or at a depth below the surface. Also see CONSOLIDATED FORMATION.

UNIFORMITY COEFFICIENT (UC)

The ratio of (1) the diameter of a grain (particle) of a size that is barely too large to pass through a sieve that allows 60 percent of the material (by weight) to pass through, to (2) the diameter of a grain (particle) of a size that is barely too large to pass through a sieve that allows 10 percent of the material (by weight) to pass through. The resulting ratio is a measure of the degree of uniformity in a granular material, such as filter media.

$$\text{Uniformity Coefficient} = \frac{\text{Particle Diameter}_{60\%}}{\text{Particle Diameter}_{10\%}}$$

UPPER EXPLOSIVE LIMIT (UEL)

The point at which, due to insufficient oxygen present, the concentration of a gas in air becomes too great to allow an explosion upon ignition.

V

VARIABLE COSTS

Costs that a utility must cover or pay that are directly associated with the actual production and delivery of service. These costs vary or fluctuate. Also see FIXED COSTS.

VARIABLE FREQUENCY DRIVE

A control system that allows the frequency of the current applied to a motor to be varied. The motor is connected to a low-frequency source while standing still; the frequency is then increased gradually until the motor and pump (or other driven machine) are operating at the desired speed.

VARIABLE, MEASURED

A factor (flow, temperature) that is sensed and quantified (reduced to a reading of some kind) by a primary element or sensor.

VARIABLE, PROCESS

A physical or chemical quantity that is usually measured and controlled in the operation of a water, wastewater, or industrial treatment plant.

VELOCITY HEAD

The energy in flowing water as determined by a vertical height (in feet or meters) equal to the square of the velocity of flowing water divided by twice the acceleration due to gravity ($V^2/2g$).

VENTURI (ven-TOOR-ee) METER

A flow measuring device placed in a pipe. The device consists of a tube whose diameter gradually decreases to a throat and then gradually expands to the diameter of the pipe. The flow is determined on the basis of the difference in pressure (caused by different velocity heads) between the entrance and throat of the Venturi meter.

NOTE: Most Venturi meters have pressure sensing taps rather than a manometer to measure the pressure difference. The upstream tap is the high pressure tap or side of the manometer.

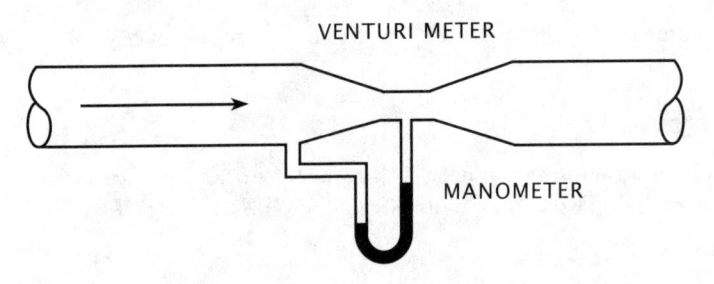

VISCOSITY (vis-KOSS-uh-tee)

A property of water, or any other fluid, that resists efforts to change its shape or flow. Syrup is more viscous (has a higher viscosity) than water. The viscosity of water increases significantly as temperatures decrease. Motor oil is rated by how thick (viscous) it is; 20 weight oil is considered relatively thin while 50 weight oil is relatively thick or viscous.

VOID

A pore or open space in rock, soil, or other granular material, not occupied by solid matter. The pore or open space may be occupied by air, water, or other gaseous or liquid material. Also called an INTERSTICE, PORE, or void space.

VOLATILE (VOL-uh-tull)

(1) A volatile substance is one that is capable of being evaporated or changed to a vapor at relatively low temperatures. Volatile substances can be partially removed from water or wastewater by the air stripping process.

(2) In terms of solids analysis, volatile refers to materials lost (including most organic matter) upon ignition in a muffle furnace for 60 minutes at 550°C (1,022°F). Natural volatile materials are chemical substances usually of animal or plant origin. Manufactured or synthetic volatile materials, such as plastics, ether, acetone, and carbon tetrachloride, are highly volatile and not of plant or animal origin. Also see NONVOLATILE MATTER.

VOLATILE ACIDS

Fatty acids produced during digestion that are soluble in water and can be steam-distilled at atmospheric pressure. Also called organic acids. Volatile acids are commonly reported as equivalent to acetic acid.

VOLATILE LIQUIDS

Liquids that easily vaporize or evaporate at room temperature.

VOLATILE SOLIDS

Those solids in water, wastewater, or other liquids that are lost on ignition of the dry solids at 550°C (1,022°F). Also called organic solids and volatile matter.

VOLTAGE

The electrical pressure available to cause a flow of current (amperage) when an electric circuit is closed. Also called ELECTROMOTIVE FORCE.

VOLUMETRIC

A measurement based on the volume of some factor. Volumetric titration is a means of measuring unknown concentrations of water quality indicators in a sample by determining the volume of titrant or liquid reagent needed to complete particular reactions.

VOLUMETRIC FEEDER

A dry chemical feeder that delivers a measured volume of chemical during a specific time period.

VORTEX

A revolving mass of water that forms a whirlpool. This whirlpool is caused by water flowing out of a small opening in the bottom of a basin or reservoir. A funnel-shaped opening is created downward from the water surface.

W

WARNING

The word *WARNING* is used to indicate a hazard level between *CAUTION* and *DANGER*. Also see CAUTION, DANGER, and NOTICE.

WASTEWATER

A community's used water and water-carried solids (including used water from industrial processes) that flow to a treatment plant. Stormwater, surface water, and groundwater infiltration also may be included in the wastewater that enters a wastewater treatment plant. The term sewage usually refers to household wastes, but this word is being replaced by the term wastewater.

WATER AUDIT

A thorough examination of the accuracy of water agency records or accounts (volumes of water) and system control equipment. Water managers can use audits to determine their water distribution system efficiency. The overall goal is to identify and verify water and revenue losses in a water system.

WATER CYCLE

The process of evaporation of water into the air and its return to earth by precipitation (rain or snow). This process also includes transpiration from plants, groundwater movement, and runoff into rivers, streams, and the ocean. Also called the HYDROLOGIC CYCLE.

WATER HAMMER

The sound like someone hammering on a pipe that occurs when a valve is opened or closed very rapidly. When a valve position is changed quickly, the water pressure in a pipe will increase and decrease back and forth very quickly. This rise and fall in pressures can cause serious damage to the system.

WATER PURVEYOR (purr-VAY-or)

An agency or person that supplies water (usually potable water).

WATER TABLE

The upper surface of the zone of saturation of groundwater in an unconfined aquifer.

WATER TABLE HEAD

See PRESSURE HEAD.

WATERSHED

The region or land area that contributes to the drainage or catchment area above a specific point on a stream or river.

WATT

A unit of power equal to one joule per second. The power of a current of one ampere flowing across a potential difference of one volt.

WEIR (WEER)

(1) A wall or plate placed in an open channel and used to measure the flow of water. The depth of the flow over the weir can be used to calculate the flow rate, or a chart or conversion table may be used to convert depth to flow. Also see PROPORTIONAL WEIR.

(2) A wall or obstruction used to control flow (from settling tanks and clarifiers) to ensure a uniform flow rate and avoid short-circuiting.

WEIR (WEER) DIAMETER

Many circular clarifiers have a circular weir within the outside edge of the clarifier. All the water leaving the clarifier flows over this weir. The diameter of the weir is the length of a line from one edge of a weir to the opposite edge and passing through the center of the circle formed by the weir.

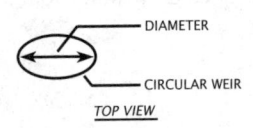

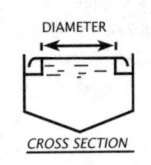

WEIR (WEER) LOADING

A guideline used to determine the length of weir needed on settling tanks and clarifiers in treatment plants. Used by operators to determine if weirs are hydraulically (flow) overloaded.

$$\text{Weir Loading, GPM/ft} = \frac{\text{Flow, GPM}}{\text{Length of Weir, ft}}$$

or

$$\text{Weir Loading, cu m/day/meter} = \frac{\text{Flow, cu m/day}}{\text{Length of Weir, m}}$$

WELL ISOLATION ZONE

A surface or zone with restricted land uses surrounding a public water system water well or well field. The zone is established to prevent contaminants from a nonpermitted land use to move toward and reach the water well or well field. Also see WELLHEAD PROTECTION AREA.

WELL LOG

A record of the thickness and characteristics of the soil, rock, and water-bearing formations encountered during the drilling (sinking) of a well.

WELLHEAD PROTECTION AREA (WHPA)

The surface and subsurface area surrounding a public water system water well or well field, through which contaminants are reasonably likely to move toward and reach such water well or well field. Also see WELL ISOLATION ZONE.

WET CHEMISTRY

Laboratory procedures used to analyze a sample of water using liquid chemical solutions (wet) instead of, or in addition to, laboratory instruments.

WHOLESOME WATER

Water that is safe and palatable for human consumption.

WIRE-TO-WATER EFFICIENCY

The combined efficiency of a pump and motor together. Also called the OVERALL EFFICIENCY.

WYE STRAINER

A screen shaped like the letter Y. Water flows through the upper parts of the Y and the debris is trapped by the screen at the fork.

X

(NO LISTINGS)

Y

YIELD

The quantity of water (expressed as a rate of flow—GPM, GPH, GPD, cu m/day, ML/day, or total quantity per year) that can be collected for a given use from surface or groundwater sources. The yield may vary with the use proposed, with the plan of development, and also with economic considerations. Also see SAFE YIELD.

Z

ZEOLITE

A type of ion exchange material used to soften water. Natural zeolites are siliceous compounds (made of silica) that remove calcium and magnesium from hard water and replace them with sodium. Synthetic or organic zeolites are ion exchange materials that remove calcium or magnesium and replace them with either sodium or hydrogen. Manganese zeolites are used to remove iron and manganese from water.

ZETA POTENTIAL

In coagulation and flocculation procedures, the difference in the electrical charge between the dense layer of ions surrounding the particle and the charge of the bulk of the suspended fluid surrounding this particle. The zeta potential is usually measured in millivolts.

ZONE OF AERATION

The comparatively dry soil or rock located between the ground surface and the top of the water table.

ZONE OF SATURATION

(1) The soil or rock located below the top of the groundwater table. By definition, the zone of saturation is saturated with water. Also see WATER TABLE.

(2) Where raw wastewater is exfiltrating from a sewer pipe, the area of soil that is moistened around the leak point is often called the zone of saturation.

ZOOPLANKTON (ZOE-uh-PLANK-ton)

Small, usually microscopic animals (such as protozoans), found in lakes and reservoirs.

SUBJECT INDEX

A

ABC (Association of Boards of Certification), 630
ADA (Americans with Disabilities Act of 1990), 638
Abbreviations, instrumentation, 395
Accident
  causes, 459
  prevention, 460, 485
  reports, 497, 498, 678, 679, 680
Accountable, 621
Accuracy, instrumentation, 400
Acetic acid (glacial), 460
Acid feed systems, 42, 207
Acids
  chemical handling, 460
  safety, 460
Action levels, lead and copper, 569
Activated alumina, 143, 146, 147
Activated carbon, 136, 138, 139, 474
Acute health effects, 480, 571, 576
Ad valorem taxes, 649
Additional reading, 23, 65, 114, 131, 140, 223, 299, 378, 499, 582, 693
Administration
  advertising positions, 626
  applications, 626
  backflow prevention, 659
  budgeting, 647
  capital improvements, 649
  certification, 630
  communication, 641
  compensation, 626, 629
  conducting meetings, 643
  confined space entry procedures, 674
  consumer complaints, 646
  consumer inquiries, 645
  contaminated water supply, 664
  contingency planning, 661
  continuing education, 630
  controlling, 621
  cross connection control program, 655–661
  directing, 619
  discipline problems, 631
  disposition of plant records, 687
  emergency response, 661–670
  employee complaints, 631
  equipment repair/replacement fund, 648
  evaluation, performance, 631
  financial assistance, 650
  financial management, 646–650
  financial stability, 646
  first aid, 673
  functions of a manager, 619
  harassment, 635
  hazard communication, 673
  interviewing job applicants, 626, 627, 628
  interviews, media, 644
  meetings, conducting, 643
  number of employees needed, 625
  operations and maintenance (O & M), 652–661
  operator certification, 630
  oral communication, 641
  organizing, 621
  orientation, new employee, 629
  paper screening, 626
  performance evaluation, 631
  personnel records, 639
  planning, 621
  planning for emergencies, 661
  plant tours, 646
  policies and procedures, 629
  probationary period, 629
  procurement of materials, 685
  public relations, 644
  public speaking, 645
  qualifications profile, 625
  rates, water, 648
  recordkeeping, 684–687
  records, personnel, 625, 639
  repair/replacement fund, equipment, 648
  SCADA systems, 654–655
  safety program, 670–683
  selection process, 628
  sexual harassment, 635
  staffing, 624–639
  supervisor, training, 630
  telephone contacts, 645
  tours, plant, 646
  training, 630
  unions, 639
  water rates, 648
  written communication, 641
Adsorption, 136, 138, 143
Adverse effects
  hardness, 77
  iron and manganese, 7

Advertising, job openings, 626
Aeration
    iron and manganese, 12
    trihalomethanes, 135–136, 140
Affirmative action, 626
Air compressor failure, membranes, 180
Air cushion chamber, 414, 417
Air gap, 655, 658, 661
Air release assembly, 109
Air supply systems, instrument, 434, 436
Air temperatures, fluoridation, 33
Air-cooled engines, 368
Alarms
    annunciator panel, 433
    fluoridation, 49
    instrumentation, 433
    reverse osmosis, 207
Algae counts, 510
Alignment, pumps, 305, 324, 331–335
Alkalinity, softening, 77, 80, 90
Allocations, water shortage, 691
Alternating current (AC), 272
Alum, 472
Alum sludge, 249
Aluminum, regulations, 587
Aluminum sulfate, 472
Amendments to the Safe Drinking Water Act (SDWA), 559
Americans with Disabilities Act of 1990 (ADA), 638
Ammeter, 277
Ammonia, 464
Amplitude, 272
Amps, 273
Analog
    chlorine residual indicator, 429
    input (AI) control, 414
    instrumentation, 401, 402
    output (AO) control, 414
Analysis
    Also see Laboratory test procedures
    iron and manganese, 8, 9
Annual report, 643, 648
Annunciator panel, alarms, 433
Antimony, 566
Applications, job, 626
Aquifer, 8
Arch, chemicals, 372, 460
Arc-mark trademark, 404
Arithmetic assignment, 23, 65, 114, 223, 378, 497
Arsenate, 142
Arsenic
    activated alumina, 143, 146, 147
    Best Available Technology (BAT), 143
    blending, 143
    chemistry, 142
    compliance, 150
    endocrine effects, 142
    engineered blending, 143
    field test kits, 150
    granular ferric hydroxide (GFH), 146
    health effects, 142
    ion exchange, 144, 145
    monitoring, 150, 151
    new source alternative, 142
    oxidation-filtration, 146
    POU/POE (point-of-use/point-of-entry) devices, 143, 146
    physical separation, 143
    problem, 142
    process control, 151
    proprietary media, 146
    recordkeeping, 151
    regulations, 566
    reporting, 151
    residuals, 149
    sampling, 150
    sources, 142
    treatment, 142
    typical plant, 148
    wastewater, 149
Arsenic Rule
    compliance, 562
    MCL, 562
    monitoring requirements, 562
    public health benefits, 562
Arsenic treatment plant
    backwash, 148
    brine, 144, 148
    chromatographic peaking, 144
    competing ions, 144
    compliance, 150
    configuration, 144, 146
    empty bed contact time (EBCT), 144, 146
    hazardous waste, 149, 151
    ion exchange run, 144
    maintenance, 146, 149
    monitoring, 150, 151
    operation, 148
    pH, 144, 146
    plans, 149
    process control, 151
    Programmable Logic Controller (PLC), 148
    records, 151
    regenerant, salt, 144
    reporting, 151
    resin, 144
    review of plans and specifications, 149
    rinse, 148
    safety, 149
    security, 149
    service, 148
    shutdown, 148
    specifications, 149
    spent media, activated alumina, 146
    start-up, 148
    troubleshooting, 149
    valves, 148, 149
Arsenite, 142

Asbestos, 566
Assessment bonds, 650
Assessment of system vulnerability, 661
Association of Boards of Certification (ABC), 630
Atmospheres, explosive, 494
Atomic absorption, 509
Atomic adsorption (AA), 150
Audience, communications, 641, 642, 644
Audiovisuals, 642
Audits, water, conservation, 689
Authority, 621
Autoclaves, 493
Automatic
   control system, 403
   controllers, 433
   valves, 360
Automation systems, computer, 399
Auxiliary electrical power, 293

B

BBP (bloodborne pathogens) program, 412
Back pressure, 655–661
Backflow, 14, 655–661
Backsiphonage, 22, 655–661
Backup controls, 399
Backwash
   filters, 15
   ion exchange, 104, 107, 148
   membranes, 167, 175
   recovery ponds, 237, 239
   wastewater, 249
Bacteria
   iron and manganese, 7, 22, 23
   regulations, 568, 573–582
Barium, 566
Bases
   chemical handling, 464
   safety, 464
Basket centrifuge, 245, 246
Batch systems, 35, 42
Batteries, 272, 297
Battery-backed power supplies, 430
Bearings, pump, 305, 311, 324, 326
Beer's Law, 509
Bellows, pressure measurement, 414, 415
Belt drives, pumps, 330
Belt filter presses, 234, 241, 244
Belts, compressors, 343
Bench-scale tests
   iron and manganese, 10, 14
   membranes, 174
   trihalomethanes, 135, 139, 140
Benefits
   employee, 626, 629
   softening, 78, 82
Benzene, 573
Bernoulli Effect, 421

Beryllium, 566
Best Available Technology (BAT), arsenic, 143
Best management practice (BMP), conservation, 688, 693
Billing inserts, 690
Black water, 16
Blending
   arsenic, 143
   engineered, 143
   ion exchange, 112
Block grants, 650
Block rate structure, 692
Bloodborne pathogens (BBP) program, 412
Blown fuse, 276
Blue baby (methemoglobinemia), 530, 567, 571
Bonds, 649, 650
Booster shots, immunization, 493
Bourdon tube, pressure measurement, 414, 415
Brackish water, 183
Breakdown maintenance, membranes, 182
Breakers, circuit, 273, 278, 281
Breakpoint chlorination, 16
Brine
   disposal, demineralization, 184
   disposal, process wastes, 234, 235, 248
   electrodialysis, 214
   ion exchange, 104, 105, 108
   recycle, arsenic treatment, 144, 148
   reverse osmosis, 194, 207
Bromide, 129, 134
Bubbler tubes, level measurement, 415, 418, 419
Budgeting, 647
Buildings, maintenance, 376, 377
Butterfly valves, 347, 350
Bypass, ion exchange, 112, 113
By-products, disinfection, 165, 568, 583

C

CAD (computer-aided design) system, 661
CCR (Consumer Confidence Report), 602
CMMS (Computer-based Maintenance Management System), 661
CPR (cardiopulmonary resuscitation), 412, 460, 469, 471, 483, 673
CPU (central processing unit), 439, 440
CT
   calculations, 760–769
   values, 580, 583, 760–769
Cadmium, 566
Cake formation fouling, 167
Calcium carbonate equivalent, 78, 79
Calcium carbonate saturation test, 528
Calcium hydroxide, 465
Calcium test procedures, 511, 512
Calculations
   chemical feeders, 374
   fluoridation, 60–65
   ion exchange softening, 110–113

Calculations *(continued)*
  lime–soda ash softening, 82, 87, 93–98
  reverse osmosis, 188–198
  trihalomethanes, 132
Calibration
  chemical feeders, 372, 373
  curve, spectrophotometer, 510
  equipment, 437
  measurement process, 401
California Urban Water Conservation Council (CUWCC), 688, 693
Call date, 650
Capital
  improvements, 649
  planning, 649
Carbon dioxide, 83, 90, 469
Carbon tetrachloride, 573
Carbonate hardness, 77, 83
Cardiopulmonary resuscitation (CPR), 412, 460, 469, 471, 483, 673
Cartridge filters, 207, 215, 218
Casing, pumps, 306, 311
Categories, instrumentation, 429–443
Cathodic protection, 377
Caustic soda
  chemical handling, 465
  softening, 84, 87
Cavitation, pumps, 308, 309
Cellulose acetate (CA) membranes, 167, 173, 186
Central processing unit (CPU), 439, 440
Centrifugal pumps, 301, 310, 338
Centrifuges, 244, 245, 246
Certification, operator, 630
Chain drives, pumps, 331
Chart recorders, 429, 431
Charts, fluoridation, 49–51
Check sampling, 591
Check valves, 324, 326, 347–360
Chemical analyzer, 427
Chemical feed rate, instrumentation, 426
Chemical feeders
  acid feed systems, 42
  batch systems, 35, 42
  calculating doses, 60–65, 85, 93–98
  calculations, 374
  calibration, 372, 373
  chemical storage, 371
  chlorinators, 375
  day tank, 35, 44
  diaphragm pumps, 35, 36
  dose, 97, 374
  drainage, 372
  dry chemical, 372
  dry feeders, 35, 39–41, 59, 372
  electronic pumps, 35, 38, 52
  feed rate, 93–97, 374
  gas, 372
  gravimetric feeders, 35, 41
  instrumentation, 426
  liquid, 372
  maintenance, 20, 58, 372
  metering, 372
  operation, 49
  peristaltic pumps, 35, 37, 52
  positive displacement pumps, 35, 36
  saturators, 35, 42, 43, 58
  shutdown, 58
  solid, 372
  solution feeders, 35, 42
  solution preparation, 49
  start-up, 49
  storage, chemical, 371, 475
  volumetric feeders, 35, 39, 40, 42
Chemical flush system, 220
Chemical handling
  acetic acid (glacial), 460
  acids, 460
  activated carbon, 474
  alum, 472
  aluminum sulfate, 472
  ammonia, 464
  bases, 464
  calcium hydroxide, 465
  carbon dioxide, 469
  caustic soda, 465
  chlorine, 467
  CHLORINE MANUAL, 468
  drains, 475
  ferric chloride, 472
  ferric sulfate, 472
  ferrous sulfate, 472
  fluoride compounds, 473
  gas mask, 469–470
  gas-detection equipment, 481
  gases, 467
  hydrochloric acid, 463
  hydrofluoric acid, 463
  hydrofluosilicic acid, 461
  hypochlorite, 466
  muriatic acid, 463
  nitric acid, 463
  potassium permanganate, 473
  powdered activated carbon, 474
  powders, 473, 474
  safety, 460
  safety shower, 461, 462, 463
  salts, 471
  self-contained breathing apparatus, 468–470, 494
  sodium aluminate, 473
  sodium carbonate, 466
  sodium hydroxide, 465
  sodium silicate, 466
  storage, 371, 475
  sulfur dioxide, 469
  sulfuric acid, 463
Chemical metering pumps, 312, 372

Chemical reactions
　lime–soda ash softening, 83–86
　THM, 129, 133
Chemical storage, 149, 371, 475
Chemicals
　labeling, 475
　laboratory, 492
Chemistry
　arsenic, 142
　softening, 79
　trihalomethanes, 129, 133
CHEMTREC [(800) 424-9300], 269, 497, 662
Chloramination, 581
Chloramines, 138, 139
Chloride
　regulations, 587
　test procedures, 512, 513
Chlorination
　iron and manganese, 10–12, 14, 18, 21
　reverse osmosis, 207
Chlorinators, 10–12, 375
Chlorine, 129, 133, 134, 136, 139, 467
Chlorine dioxide, 135, 138, 139
CHLORINE MANUAL, 468
Chlorine residual substitution, 579
Chloroform, 129, 135
"Christmas Tree" arrangement, reverse osmosis, 194, 197, 207
Chromatographic peaking, 144
Chromium, 569
Chronic health effects, 571
Circuit breakers, 273, 278, 281
Circuits, 272, 274
Circular-chart recorders, 430, 432
Civil rights, 626
Classification, fire protection, 476
Cleaning
　membranes, 167, 211
　safety, 479
　stacks, 220
　tanks, 235, 238, 239
Clean-in-place (CIP), membranes, 177
Closed-loop control system, 403, 404
Coagulants, 90
Coagulation/sedimentation/filtration, 135, 136
Code requirements, fuel storage, 370
Coliform
　Colilert™ system (ONPG-MUG), 574
　Colisure test, 574
　community water system, 565, 575
　compliance, 576
　laboratory procedures, 574
　MCL, 568, 573
　membrane filter technique (MF), 574
　monitoring frequency, 575, 576
　multiple-tube fermentation (MTF), 574
　noncommunity water system, 575
　notification requirements, 576
　presence-absence (P-A), 573, 574
　reporting requirements, 576
　routine and repeat sampling, 575
　rule, 573–577
　sampling plan, 574
　sanitary survey, 573
Colilert™ system (ONPG-MUG), 574
Colisure test, 574
Collecting samples, 8, 570, 591–592
Collection of sludges, 234
Collection systems, wastewater, 248, 249
Colloidal suspensions, 7
Colloids, 206
Color
　iron and manganese, 7
　regulations, 587
　removal, 90
　test procedures, 514
Colorimeter, 8, 9
Commercial, industrial, and institutional (CII), conservation, 691
Commodity rates, 689
Communication, 641–643
Community water system, 565
Comparator, pocket, 509
Compensation, 629
Competing ions, arsenic treatment, 144
Compilation of Quick Reference Guides, EPA, 561
Complaints
　customer, 646
　employee, 631
　procedure, 631
Compliance
　arsenic, 150
　Arsenic Rule, 562
Compounds, fluoride, 34, 59
Compressors
　belts, 343
　controls, 343
　drain, 343
　filter, 342
　fins, 342
　lubrication, 342
　maintenance, 342
　types, 342
　unloader, 342
　use, 342
　valves, 342, 344
Computer automation systems, 399
Computer control system, 430, 439, 440, 441
Computer lockup, membranes, 180
Computer recordkeeping systems, 687
Computer-aided design (CAD) system, 661
Computer-based Maintenance Management System (CMMS), 661
Concentration polarization, 194
Concrete tanks, 377
Conditioning of sludges, 235
Conducting meetings, 643

852 Water Treatment

Conductivity, 535
Conductors, 274
Configuration, arsenic treatment plants, 144, 146
Confined space
   entry permit, 411, 482, 677
   procedures, 480–483, 674
   safety hazards, instrumentation, 409–412
Conservation
   audits, water, 689
   best management practice (BMP), 688, 693
   commercial, industrial, and institutional (CII), 691
   commodity rates, 689
   coordinator, 692
   elements of programs, 688–693
   high-efficiency clothes washers, 690
   large landscape, 690
   leak detection, 689
   metering, 689
   need, 688
   pricing, 692
   public information programs, 690
   public relations, 644
   residential plumbing retrofits, 689
   residential water surveys, 688
   school education programs, 690
   system water audits, 689
   water audits, 689
   water waste prohibition, 692
   wholesale agency assistance programs, 691
Consumer
   complaints, 646
   inquiries, 645
Consumer Confidence Report (CCR), 602
Consumer Confidence Report (CCR) Rule
   information included, 564
   public health benefits, 564
   purpose, 564
Contaminants, unregulated, 561
Contaminated water supply
   contamination, 664–670
   countermeasures, 668
   Cryptosporidium, 669
   emergency contaminant limits, 664
   emergency treatment, 668, 669
   lethal dose 50 (LD 50), 668
   maximum allowable concentration (MAC), 668
   protective measures, 668
   response, 668
   toxicity, 664
   treatment, emergency, 668, 669
Contingency planning, 661
Continuing education, 630
Contract negotiations, 639
Control
   electrical, 149, 336
   iron and manganese, 7, 8
   loop, 399, 401, 404
   panels, 149, 405, 490
   pumps, 327
   room, 441, 442
   systems, 399–413, 439, 440, 441
      Also see Instrumentation
   trihalomethanes, 131, 134, 139
Controllers
   automatic, 433
   modulating, 434
   pump, instrumentation, 434, 435
Controlling, administration, 621
Controls, backup, 399
Conventional filtration, 563
Cooling systems, engines, 367–369
Coordinator, conservation, 692
Copper, 569, 587
Coprecipitation, 143
Corrective maintenance
   administration, 652
   membranes, 182
Corrosion control, lead and copper, 569
Corrosivity
   lead and copper, 569
   Lead and Copper Rule, 562
   secondary contaminants, 587
Couplings, 305, 332–335
Coverage ratio, 646, 649
Cranes, 480
Crenothrix, 588
Criteria for avoiding filtration, 578
Cross connection, 318, 655–661
Cross-flow filtration, 166, 171
Cryptosporidium, 165, 562, 563, 573, 581, 669
CT calculations, 760–769
CT values, 580, 583, 760–769
Current, electrical, 272, 273, 274, 277, 281, 489, 490
Current imbalance, 277–279
Customer
   complaints, 646
   inquiries, 645
Cuvette, 509
Cyanide, 569
Cycle, electrical, 272

D

DCS (Distributed Control System), 439
D/DBP (Disinfectants and Disinfection By-Products) Rule, 562, 563, 583
DO (dissolved oxygen)
   samples, 516, 518
   saturation table, 518
   test procedures, 515
DWCCL (Drinking Water Contaminant Candidate List), 561
Dangerous air contamination, 409
Dateometer, 330
Day tank, 35, 44

Dead-end filtration, 167, 170
Debt service, 646
Decant, 237
Dechlorination, 14
Decibel, 484
Deficiency, oxygen, 412
Deionization, 165
Delegation, 621
Demand-initiated regenerating (DIR) models, 692
Demineralization
   Also see Electrodialysis and Reverse osmosis (RO)
   brackish water, 183
   brine disposal, 184
   distillation, 184, 185
   electrodialysis, 184, 185, 213–223
   feedwater, 184
   freezing, 184, 185
   ion exchange, 184, 185
   membranes, 165
   mineralized waters, 183
   processes, 184, 185
   reverse osmosis, 184, 185, 186
   salinity, 183
   seawater, 183
   total dissolved solids, 183
Derivative, controller adjustment, 404
Desalination, 165
Desiccant, 444
Design review
   electrodialysis, 220
   fluoridation, 47
Dewatering of sludges, 234, 235, 240
Dial indicators, 333, 335
Diaphragm, pressure measurement, 414, 415, 416
Diaphragm pumps, 35, 36
Diaphragm-operated valves, 360, 361
Diarrhea, "traveler's," 573
1,2-Dichloroethane, 573
1,1-Dichloroethylene, 572
Diesel
   engines, 364–368
   fuel storage, 370
Differential pressure, flow sensing, 420, 421, 424, 425, 426
Digital, instrumentation, 401, 402
Digital readouts, 429
Direct current (DC), 272
Direct current, electrodialysis, 215, 216, 220
Direct filtration, 563
Direct hydrostatic pressure, level measurement, 415
Directing, administration, 619
Dirty water
   black water, 16
   iron and manganese, 7
   red water, 16, 23
Disciplinary problems, 631
Discrete control, 414
Discrete I/O (input/output), 414

Discriminatory hiring, 625, 626
Disinfectants and Disinfection By-Products (D/DBP) Rule, 562, 563, 583
Disinfection alternatives, 138, 581
Disinfection by-products, 165, 568, 583
Disinfection, CT values, 580, 583, 760–769
Displacers, level measurement, 415
Disposal of
   records, 687
   sludges, 234, 235, 249
   spent brine, 106, 234, 249
Disposition of plant records, 687
Dissolved organic carbon (DOC), 173, 174
Dissolved organic matter (DOM), 165, 167
Dissolved oxygen (DO)
   samples, 516, 518
   saturation table, 518
   test procedures, 515
Distillation, 184, 185
Distributed Control System (DCS), 439
Divalent, 7, 77
Documentation, administration, 625
Dose, chemical feeders, 97, 374
Downflow saturators, 45
Drainage waters, plant, 251
Draining tanks, 235, 238, 239
Drains
   chemical feeders, 372
   chemical handling, 475
   compressors, 343
Drinking Water Contaminant Candidate List (DWCCL), 561
Drinking water regulations
   See Regulations, drinking water
Drinking Water Standards, 561, 565–585, 586–590
Driving equipment, pumps, 336
Drowning, 496
Dry chemical feeders, 35, 39–41, 59, 372
Drying beds, 240, 242, 243, 249
Duplex motor control panel, 405
Dust, fluoridation, 59
Dynamic types of pumps, 302

E

EPA Compilation of Quick Reference Guides, 561
EPA Safe Drinking Water Hotline [(800) 426-4791], 561
EPA's WaterSense, 693
ESWTR (Enhanced Surface Water Treatment Rule), 562
E. coli, 573, 575, 576
Eccentric valves, 347–349
Effective range, instrumentation, 401
Effects of
   fluoride, 33
   iron and manganese, 7
   trihalomethanes, 129
Efficiency program, water, 690
Electric motors, 284–293, 327–329

Electrical
- control panels, 149
- energy consumption, 441
- hazards, 271, 404, 489
- probes, level measurement, 415, 417
- symbols, 395–398

Electrical conductivity (EC), 151

Electrical equipment
- additional reading, 299
- alternating current (AC), 272
- ammeter, 277
- amplitude, 272
- amps, 273
- auxiliary electrical power, 293
- batteries, 272, 297
- blown fuse, 276
- breakers, 273, 278, 281
- circuit breakers, 273, 278, 281
- circuits, 272, 274
- conductors, 274
- control panels, 490
- controls, 336
- current, 272, 277, 281, 489, 490
- cycle, 272
- direct current (DC), 272
- electric motors, 284–293, 327–329
- electromotive force (EMF), 272
- emergency lighting, 296, 297
- fundamentals, 272
- fuses, 272, 273, 275, 276, 281
- ground, 282
- ground-fault circuit interrupter (GFCI), 490
- hazards, 271
- hertz, 272
- instrumentation, 490
- insulation, motors, 286
- insulation resistance, 280
- insulators, 274
- kirk-key, 296
- lighting, emergency, 296, 297
- limitations, 271
- magnetic starters, 282–284
- maintenance, 267, 271, 327
- megger, 280
- megohm, 280
- meters, electrical, 274
- motor insulation, 286
- motor starters, 282, 489
- motors, 272, 273, 284–293, 327–329
- nameplate, 286
- ohm, 273
- ohmmeters, 281
- Ohm's Law, 273
- overload relays, 282
- panels, control, 490
- phase, 272
- power requirements, 274, 337
- protection devices, 281
- recordkeeping, 267, 268, 287, 293
- safety, 212, 269, 271, 299, 404, 489
- standby power generation, 293
- starters, 282, 489
- switch gear, 281, 298
- switches, 272, 273, 275
- tag, warning, 270, 319
- testers, 274
- thermal overloads, 282
- transformers, 272, 278, 282, 298, 489
- transmission, 298
- voltage testing, 274, 276, 489
- volts, 272, 273, 274, 298
- warning tag, 270, 319
- watts, 274

Electrode tab, 221

Electrodialysis
Also see Demineralization and Reverse osmosis (RO)
- additional reading, 223
- advantages, 213
- arithmetic assignment, 223
- brine, 213
- cartridge filters, 215, 218
- chemical flush system, 220
- cleaning stacks, 220
- description, 213, 214
- design, 220
- direct current, 215, 216, 220
- electrode tab, 221
- electrodialysis polarity reversal (EDR), 214, 220
- energy requirements, 213
- feedwater quality, 215
- flow diagram, 215, 218
- fouling, 213
- Langelier Index, 213
- log sheet, 221, 222
- membranes, 213–223
- multicompartment units, 215, 217
- operation, 213, 220, 221
- pH, 220
- piping, 220
- power supply, 220
- pressures, 220
- pretreatment, 215
- principles, 214
- pumping equipment, 220
- recordkeeping, 221, 222
- safety, 221
- scaling, 213, 220
- specifications, 220
- stack, 213, 215, 220
- stages, 213

Electrodialysis *(continued)*
    temperature, 220
    three-cell unit, 215, 216
Electrodialysis polarity reversal (EDR), 214, 220
Electrolyte, 296
Electromedia process, 15
Electromotive force (EMF), 272
Electronic chemical pumps, 35, 38, 52
Elements of conservation programs
    audits, water, 689
    best management practice (BMP), 688, 693
    commercial, industrial, and institutional (CII), 691
    commodity rates, 689
    coordinator, 692
    high-efficiency clothes washers, 690
    large landscape, 690
    leak detection, 689
    metering, 689
    pricing, 692
    public information programs, 690
    residential plumbing retrofits, 689
    residential water surveys, 688
    school education programs, 690
    system water audits, 689
    water audits, 689
    water waste prohibition, 692
    wholesale agency assistance programs, 691
Elements, primary, instrumentation, 429
Emergencies
    administration, 661–670
    contaminated water supply, 664–670
    phone numbers, 269
    preparation for, 497, 621, 661–670
    procedures, 269, 497, 668
    response plan, 621, 661–670
    safety, 497
    team, 269
Emergency
    arsenic reporting, 151
    lighting, 296, 297
    maintenance, 652
    response plan, 621, 661–670
    response procedures, 441
    treatment, 668, 669
Employee
    benefits, 626, 629
    evaluation, 631–633
    orientation, 629
    pride, 630
    problems, 631, 639
Employee-management relationship, 639
Employees, number, 625
Employer policies, 629, 636
Empty bed contact time (EBCT), 144, 146
End bells, 284

Endemic, 33
Endocrine effects, arsenic, 142
Energy
    conservation
        See Conservation
    consumption, electrical, 441
    requirements, electrodialysis, 213
Enforcement, regulations, 559
Engineered blending, 143
Engines
    air cooled, 368
    cooling systems, 367–369
    diesel, 364–368
    fuel storage, 370
    fuel system, 364–367
    gasoline, 362
    governor, 367
    lubrication, 362
    maintenance, 362, 368
    operation, diesel, 364
    problems, 362
    running, 362
    standby, 371
    starting, 362, 363, 368
    troubleshooting, 362, 368
    water cooled, 367
Enhanced Surface Water Treatment Rule (ESWTR), 562
Enrichment, oxygen, 412
Entry permit, confined space, 411, 482, 677
Equal Employment Opportunity Commission, 638
Equipment
    calibration, 437
    failures, membranes, 177, 180
    gas detection, 481
    laboratory safety, 492–494
    records, 687
    repair/replacement fund, 648
    safety, operator, 459, 493, 494–497, 673
    service card, 267, 268
    test, 437
Equivalent weight, 79
Establishment, regulations, 560–566
Evaluation, performance, 631–633
Evapotranspiration, 690
Explosive
    atmospheres, 494
    gas mixtures, 412
Extinguishers, fire, 476–478
Eye protection, 495

F

FMLA (Family and Medical Leave Act of 1993), 638
Fail-safe design, 404
Failure, motors, 287–293

Falls, hazards, 413
Family and Medical Leave Act of 1993 (FMLA), 638
Feasibility analysis process, 131
FEDERAL REGISTER, 129, 131, 139, 583
Feed rate, chemical, 93–97, 374
Feedback, controller, 401, 404
Feeders, chemical
   See Chemical feeders
Feedwater
   demineralization, 184
   electrodialysis, 215, 220
   reverse osmosis, 207
Fees, user, 647
Ferric chloride, 472
Ferric sulfate, 472
Ferrous sulfate, 472
Field test kits, arsenic, 150
Filter
   backwash wastewater, 249
   compressor, 342
   maintenance and inspection, membranes, 181
   membranes, uses, 175
   presses, 245, 247
Filter Backwash Recycling Rule (FBRR)
   application, 563
   public health benefits, 563
   purpose, 562
   recycle flows, 563
   thickener supernatant, 563
Filtration, iron and manganese, 14, 15, 18–21
Financial
   assistance, 650
   management, 646–650
   stability, 646
Fins, compressors, 342
Fire protection
   classification, 476
   exits, 479
   extinguishers, 476–478
   facilities, 413
   flammables, storage, 479
   hoses, 478
   prevention, 476
   storage, flammables, 479
First aid
   acetic acid, 461, 463
   alum, 472
   aluminum sulfate, 472
   ammonia, 465
   calcium hydroxide, 465
   carbon dioxide, 469
   chlorine, 467–468
   equipment, 462
   ferric chloride, 472
   ferric sulfate, 472
   ferrous sulfate, 472, 473
   fluoride, 60, 473
   hydrochloric acid, 463
   hydrofluosilicic acid, 461, 463
   hydrogen fluoride, 463
   hypochlorite, 466
   lime, 465
   nitric acid, 463
   potassium permanganate, 473
   powdered activated carbon, 474
   safety program, 673
   sodium aluminate, 473
   sodium carbonate, 466
   sodium silicate, 466
   sulfur dioxide, 471
   sulfuric acid, 463, 464
Flammables, storage, 479
Flanges, pumps, 305
Float, level measurement, 415
Flow measurements, 418–426
Fluoridation
   additional reading, 65
   air temperatures, 33
   alarms, 49
   arithmetic assignment, 65
   batch mix, 35, 42
   calculating doses, 60–65
   charts, 49–51
   chemical feeders, 35–48
      Also see Chemical feeders
   compounds, 34, 59
   day tank, 35, 44
   downflow saturators, 45
   dust, 59
   effects on teeth, 33, 569, 588
   first aid, fluoride, 60
   fluoride ion, 34
   formulas, 60
   hydrofluosilicic acid, 34, 47, 48, 50, 51, 52
   importance, 33
   log sheets, 49, 52, 53, 55, 56
   maintenance, 58
   operation, 49
   overfeeding, 47, 54
   poisoning, fluoride, 59
   programs, 33
   public notification, 58
   records, 49, 53, 55, 56
   safety, 59
   safety equipment, 49, 59
   sanitary defects, 59
   saturator, 35, 42, 43, 46, 59
   shutdown, 58
   sodium fluoride, 34, 35
   sodium fluorosilicate, 34
   sodium silicofluoride, 54
   solution preparation, 49
   specification review, 47
   start-up, 49
   systems, 35, 42
   training, 60

Fluoridation *(continued)*
  treatment charts, 49–51
  underfeeding, 58
  upflow saturators, 43, 45, 46
Fluoride
  compounds, 34, 59, 473
  fluorosis, 569, 588
  ion, 34
  levels, 569, 588
  regulations, 569, 588
  test procedures, 519
Flux
  decline, 188
  reverse osmosis, 186, 188
FmHA loans, 650
Foaming agents, 588
Foot protection, 495
Foot valves, 324, 351
Forklifts, 487, 488
Formation of THMs, 129, 133, 134, 136
Fouling
  cake formation, 167
  electrodialysis, 213
  irreversible, 167
  membrane, 167
  pore blockage, 167
  reversible, 167
Freezing, demineralization, 184, 185
Frequency of sampling, 570, 575–576, 579, 580, 584, 587, 591
Fuel storage
  code requirements, 370
  diesel, 370
  gasoline, 370
  liquid petroleum gas (LPG), 370
  natural gas, 370, 371
Fuel systems, engines, 364–367
Fueling vehicles, 485
Functions of a manager, 619
Fund, repair/replacement, 648
Fuse
  blown, 276
  puller, 276
Fuses, 273, 275, 276, 281

G

GFCI (ground-fault circuit interrupter), 483, 490
GIS (Geographic Information System), 661
GWR (Ground Water Rule), 563, 573
Gas
  chemical feeders, 372
  detection equipment, 481
  explosive mixtures, 412
  masks, 468–470
Gases
  chemical handling, 467
  safety, 467

Gasoline
  engines, 362
  fuel storage, 370
Gate valves, 344, 345
Gear train guard, 408
General obligation bonds, 649
Geobacter, 142
Geographic Information System (GIS), 661
Giardia, 165, 573, 578–582
Glassware, laboratory safety, 492
Globe valves, 346, 358, 360, 361
Gloves, 495
Goals, utility, 621
Governor, engines, 366, 367
Grant programs, 650
Granular activated carbon, 138, 140
Granular ferric hydroxide (GFH), 146
Gravimetric chemical feeders, 35, 41
Graywater use, 693
Greensand filters, 146
Greensand, iron and manganese, 14, 18–21
Grievances, 631, 639
Ground, electrical, 282
Ground Water Rule (GWR), 563, 573
Ground-fault circuit interrupter (GFCI), 483, 490
Group 1 and Group 2 THM treatment techniques, 139, 140
Groups, protected, 628

H

HMI (human machine interface), 433, 439, 440
Hand protection, 495
Handling of chemicals
  See Chemical handling
Handling process wastes, 233, 234, 235
Harassment, 635, 636
Hard hat, 496
Hardness leakage, 105
Hardness, water, 77–79, 82, 83, 529
  Also see Softening
Hazard communication program, 673
Hazardous gases, 467, 480
Hazardous waste, arsenic treatment, 149, 151
Hazards
  drowning, 496
  electrical, 271, 404, 489
  falls, 413
  gases, 467, 480
  instrumentation
    confined spaces, 409–412
    electrical, 404
    explosive gas mixtures, 412
    falls, 413
    mechanical, 408
    oxygen deficiency or enrichment, 412
    pneumatic, 408
  laboratory, 492–494
  maintenance, 479

Head protection, 496
Health effects
    arsenic, 142
    bacteria, 573
    fluoride, 33, 569, 588
    iron and manganese, 7
    nitrate, 571
    primary contaminants, 567
    secondary contaminants, 586
    trihalomethanes, 129
Health regulations, 559, 565, 567, 586
Hepatitis B vaccination, 412
Hertz, 272
Heterotrophic bacteria, 568, 574
High-efficiency clothes washers, 690
Hiring procedures, 624–629
History of drinking water laws, 559
Hollow fine fiber, reverse osmosis, 202, 204, 206
Hollow-fiber membranes, 166, 167
Homeland defense, 663–667
Horizontal centrifugal pumps, 301, 302, 310, 338
Horn, alarm, 433
Hoses, fire, 478
Hot plates, 493
Human factors, safety, 683
Human machine interface (HMI), 433, 439, 440
Hydrated lime, 83, 465
Hydrochloric acid, 463
Hydrofluosilicic acid, 34, 47, 48, 50, 52, 53, 461
Hydrogen fluoride, 463
Hydrogen sulfide, 589
Hydrolysis, 194, 196, 206
Hydrostatic pressure, level measurement, 415
Hygroscopic, 371
Hypochlorite, 466

I

ICR (Information Collection Rule), 583
IESWTR (Interim Enhanced Surface Water Treatment Rule), 581
Immediate threats to health, 571, 573
Immunization, booster shots, 493
Impeller
    flow measurement, 420, 421, 422
    pumps, 301, 303, 311
Improvements, capital, 649
Improving system management, 620
Indicators, instrumentation, 429
Inferential flow-sensing techniques, 420–426
Information Collection Rule (ICR), 583
Initial sampling, 591
Injury
    See Accident
Inorganic chemicals, regulations, 566–572
Input/output (I/O) control, 414
Inquiries, pre-employment, 627, 628

Insoluble compounds, 7
Inspection
    pumps, 340
    tanks, 376
    walk-through, membranes, 175, 180
Instrument air system, 434, 436
Instrumentation
    abbreviations, 395
    accuracy, 400
    alarms, 433
    analog, 401, 402
    annunciator panel, alarms, 433
    automatic control system, 403
    backup controls, 399
    battery-backed power supplies, 430
    Bourdon tube, 414, 415
    bubbler tubes, 415, 418, 419
    calibration
        equipment, 437
        measurement process, 401
    categories, 429–443
    chemical feed rate, 426
    closed-loop modulating control, 403
    confined spaces, 409–412
    control loop, 399, 401, 404
    control systems, 399–413
    digital, 401, 402
    discrete control, 414
    effective range, 401
    electrical
        equipment, 490
        hazards, 404
        symbols, 395–398
    fail-safe design, 404
    feedback, controller, 401, 404
    flow measurements, 418–426
    importance, 399
    indicators, 429
    inferential flow-sensing techniques, 420–426
    input/output (I/O) control, 414
    laboratory, 437
    level measurements, 414, 415
    linearity, 401
    loop, control, 399
    maintenance, preventive, 443–445
    measured variables, 414–428
    measurement, 399–413
    measurement process, 400
    mechanical hazards, 408
    motor control stations, 401, 404
    National Fire Protection Association Standard 820, 413
    ON/OFF control, 414
    operation and maintenance, 443–445
    orifice plate, 424, 425, 426
    panel instruments, 429
    power, 428, 430
    precision, 400

Index 859

Instrumentation *(continued)*
  pressure measurements, 414
  preventive maintenance, 443–445
  primary elements, 429
  process instrumentation, 426
  recorders, 430
  rotameter, 419, 420, 421
  safety, 409–412, 490
  safety hazards, 404–413
  sensitivity, 401
  sensors, types, 414–428
  signal transmitters/transducers, 428
  snubbers, 414, 417
  span, 401
  standardization, 401
  standby power generators, 430
  symbols, electrical, 395–398
  totalizers, 430
  types of sensors, 414–428
  ultrasonic devices, 415, 418, 420
  V-O-M (Volt-Ohm-Milliammeter), 437, 438
  variables, measured, 414–428
  Venturi tube, 424, 425, 426
Insulation
  motors, 286
  resistance, 280
Insulators, 274
Interface, operator, 401
Interim Enhanced Surface Water Treatment Rule (IESWTR), 581
Interim regulations, 565
Interlock, instrumentation, 408
Interview
  job, 626, 627, 628
  media, 644
Inventory, 685
Ion exchange
  arsenic, 144, 145
  demineralization, 184, 185
  iron and manganese, 12
  resin, 12, 99, 102, 104, 135, 138
  run, 144
  softeners, 99–102
  trihalomethanes, 135, 138
  wastes, 249
Ion exchange softening
  Also see Lime–soda ash softening and Softening
  air release assembly, 109
  backwash, 104, 107
  blending, 112
  brine, 104, 106, 108
  bypass, 112, 113
  calculations, 110–113
  description of process, 99–102
  disposal of spent brine, 106
  formulas, 110
  hardness leakage, 105
  iron and manganese problems, 105, 107
  limitations, 105
  maintenance, 106
  monitoring, 105
  operation, 102–105
  recordkeeping, 114
  resin, 99, 102, 104
  rinse, 103, 105, 108
  sanitary defects, brine storage tanks, 107
  service, 102, 103, 107
  shutdown, 108
  split treatment, 112
  start-up, 108
  testing, 105
  troubleshooting, 107
  wastes, 249
  zeolite, 99
Iron and manganese
  additional reading, 23
  adverse effects, 7
  aeration, 12
  analysis, 8, 9
  arithmetic assignment, 23
  bacteria, 7, 22, 23
  bench-scale tests, 10, 14
  breakpoint chlorination, 16
  chlorination, 10–12, 14, 18, 21
  collecting samples, 8
  color, 7
  colorimetric testing, 8
  control, 7, 10
  dechlorination, 14
  dirty water, 7
  divalent, 7
  effects, 7
  electromedia process, 15
  ferrous ion, 7
  filtration, 14, 15, 18–21
  formulas, 17–19, 737
  greensand, 14, 18–21
  health effects, 7
  ion exchange, 12
  ion exchange problems, 105
  jar tests, 19
  limits, 7
  maintenance, 16, 20
  measurement, 7, 8, 17
  monitoring, 16, 21
  need to control, 7
  objections, 7
  occurrence, 7
  operation, 18–21
  oxidation, 12, 14
  permanganate, 14, 18–20
  pH, 13, 18, 19, 20
  phosphate treatment, 10–12
  polyphosphate treatment, 10–12

Iron and manganese *(continued)*
  problems, ion exchange, 105
  proprietary processes, 15
  red water problems, 16, 23
  regulations, 588
  reservoirs, 7
  samples, 8
  sludge handling and disposal, 249
  stains, 7, 8
  tastes and odors, 7
  troubleshooting, 20, 21, 22
  zeolite, 14
Iron, regulations, 588
Iron (total), test procedures, 523
Iron-based adsorption, 146
Irreversible fouling, 167

J

Jar tests, softening, 93–97
Job
  duties, 619
  interview, 626, 627, 628
Jogging, 290

K

Kemmerer-type sampler, 518, 519
Kirk-key, 296

L

LAS (Linear Alkylbenzene Sulfonate), 588
LEL (lower explosive limit), 412, 413
LPG (liquid petroleum gas), 370
LSLR (lead service line replacement), 562
Labeling chemicals, 475
Laboratory instrumentation, 437
Laboratory safety
  autoclaves, 493
  biological considerations, 493
  booster shots, immunization, 493
  chemicals, 492
  equipment, 492–494
  glassware, 492
  hazards, 492–494
  hot plates, 493
  immunization, booster shots, 493
  pipet washers, 493
  radioactivity, 493
  sterilizers, 493
  water stills, 493
Laboratory test procedures
  algae counts, 510
  calcium, 511, 512
  calcium carbonate saturation test, 528
  chloride, 512, 513
  coliform, 574
  color, 514
  dissolved oxygen, 515
  fluoride, 519
  iron (total), 523
  manganese, 526
  marble test, 528
  metals, 530
  nitrate, 530
  odor, 538
  pH, 534
  specific conductance, 535
  sulfate, 535
  taste and odor, 538
  total dissolved solids, 543
  trihalomethanes, 542
Lag time, 404
Lagoons, process wastes, 237
Landfills, sanitary, 249, 250
Langelier Index, 80, 134, 213, 530
Laptop computer, membranes, 174
Large landscape conservation, 690
Lead, 569
Lead and Copper Rule
  action level (AL), 562
  corrosion control treatment (CCT), 562
  corrosivity, 562
  lead service line replacement (LSLR), 562
  public health benefits, 562
  purpose, 562
  regulations, 569–571
  water quality parameter (WQP), 562
Lead service line replacement (LSLR), 562
Leak detection, 689
Leakage, hardness, 105
Legionella, 568, 578
Lethal dose 50 (LD 50), 668
Let's Build a Pump!, 301–310
Level controls, 336
Level measurements, 414, 415
Library, maintenance, 269
Lifeline rates, 692
Lighting, emergency, 296, 297
Lime, 83, 86, 87, 88, 89
Lime sludge, 249
Lime softening, 86, 87
Lime–soda ash softening
  Also see Ion exchange softening and Softening
  alkalinity, 90
  application of lime, 89
  benefits, 82
  calculation of dosages, 85, 93–98
  carbon dioxide, 83, 90
  caustic soda softening, 84, 87
  chemical reactions, 83–86
  coagulants, 90
  color removal, 90

Lime–soda ash softening *(continued)*
   formulas, 85
   handling lime, 89
   hardness removal, 83
   hydrated lime, 83
   jar tests, 93–97
   lime, 83, 89
   lime softening, 86, 87
   lime–soda ash softening, 82, 87
   limitations, 78, 83
   marble test, 91
   National Lime Association, 90
   partial lime softening, 86
   permanent hardness, 83
   polyphosphate, 86, 91
   quicklime, 83, 91
   recarbonation, 84, 85, 86, 90
   recordkeeping, 93
   safety, 89, 91
   slake, 83, 91
   sludge, 92, 249
   split treatment, 86–89
   stability, 80, 84, 90
   storage of lime, 89
   supersaturated, 83
   temporary hardness, 83
Limitations
   ion exchange softening, 78
   lime–soda ash softening, 83
   softening, 78
Limits, iron and manganese, 7
Linear Alkylbenzene Sulfonate (LAS), 588
Linearity, instrumentation, 401
Liquid chemical feeders, 372
Liquid petroleum gas (LPG), 370
List, priority, 561
Listening skills, 641
Loan funding programs, 649, 650
Location of sampling points, 591
Lockout devices, 409
Lockout/tagout procedure, safety, 269, 490, 491
Log removals, 763
Log sheets
   electrodialysis, 221, 222
   fluoride, 49, 52, 53, 55, 56
   reverse osmosis, 209, 210
Long-term capital improvements, 649
Loop, control, 399
Lower explosive limit (LEL), 412, 413
Lubrication
   compressors, 342
   engines, 362
   maintenance, 312–317
   mechanical equipment, 312–317
   pumps, 305, 312–317
   valves, 346

M

MBAS (methylene-blue-active substances), 588
MCL (Maximum Contaminant Level)
   See Maximum Contaminant Level (MCL) (primary standards)
MCLG (Maximum Contaminant Level Goal), 564
MSDS (Material Safety Data Sheet), 180, 674–676
Magnetic
   flow sensing, 420, 423
   starters, 282–284
Maintenance
   administration, 652–661
   arsenic plant, 149
   buildings, 376, 377
   cathodic protection, 377
   chemical feeders, 20, 58, 372
   chlorinators, 375
   cleaning, 479
   compressors, 342
   concrete tanks, 377
   cooling systems, 368–370
   corrective, 652
   cranes, 479, 480
   diesel engines, 364–368
   electric motors, 284–293
   electrical equipment, 271, 327
   emergency, 652
   engines, 362, 364
   equipment service card, 267, 268
   fluoridation, 58
   gasoline engines, 362
   hazards, 479
   inspection, tanks, 376, 377
   ion exchange softeners, 106
   iron and manganese, 16, 20
   library, 269
   lubrication, 312–317
   manholes, 483
   manuals, 269
   manufacturers, 267, 268, 269
   mechanical equipment, 301
   motors, 284–293
   POU/POE (point-of-use/point-of-entry), 146
   painting, 479
   power tools, 483
   predictive, 652
   preventive, 267, 318, 652
   preventive, instrumentation, 443–445
   program
      membranes, 181, 182
      treatment plants, 267
   pumps, 301, 318–335
   recordkeeping, 267, 685
   reservoirs, 376
   safety, 479–484, 489
   service record card, 267, 268

Maintenance *(continued)*
   steel tanks, 376
   tanks, 376
   tools, power, 483
   types, 652
   valves, 344, 346, 347, 351, 484
   vehicles, 485, 486
   welding, 484
Management
   See Administration
Manager's responsibilities, 619, 624, 652
   Also see Administration
Manganese
   Also see Iron and manganese
   aeration, 12
   greensand systems, 14, 18–21
   limits, 7
   manganous ion, 8
   oxidation, 14
   regulations, 589
   test procedures, 526
Manholes, 483
Manmade radioactivity, 584
Manuals, maintenance, 269
Manufacturers, 269
Marble test, 22, 91, 528
Mass media, 644
Mass spectrometry (MS), 150
Material Safety Data Sheet (MSDS), 180, 674–676
Maximum allowable concentration (MAC), 668
Maximum Contaminant Level Goal (MCLG), 564
Maximum Contaminant Level (MCL) (primary standards)
   chlorine residual substitution, 579
   development process, 560
   disinfection by-products, 583
   fluoride, 569, 588
   inorganic chemicals, 560, 566–572
   manmade radioactivity, 584
   microbial contaminants, 560, 573–582
   natural radioactivity, 584
   organic chemicals, 559, 560, 572
   radiological contaminants, 560, 561, 584
   trihalomethanes, 129, 568, 583
   turbidity, 560, 568, 579
   types, 560
Measure, standard, 400
Measured variables, instrumentation, 414–428
Measurement
   chemical feed rate, 426
   flow, 418
   instrumentation, 399–413
   iron and manganese, 7, 8, 17
   level, 414
   pressure, 414
   process, 400
   safety, 682

Mechanical equipment
   centrifugal pumps, 301
   electric motors, 284–293
   hazards, 408
   Let's Build a Pump!, 301–310
   lubrication, 312–317
   maintenance, electric motors, 284–293
   maintenance, pumps, 301
   repair shop, 301
   seals, 321
Media, news, 644
Meetings, conducting, 643
Megger, 280
Megohm, 280
Membranes
   air compressor failure, 180
   backwashing, 167, 175
   bench-scale studies, 174
   breakdown maintenance, 182
   cellulose acetate (CA), 167, 173
   cleaning, 167
   clean-in-place (CIP), 177
   computer lockup, 180
   corrective maintenance, 182
   cross-flow filtration, 166, 171
   dead-end filtration, 167, 170
   electrodialysis, 213–223
   equipment failures, 177, 180
   filter
      maintenance and inspection, 181
      technique, 574
      uses, 175
   filtration, 166, 169, 170, 171, 172, 173
   fouling, 167
   hollow-fiber, 166, 167
   laptop computer, 174
   maintenance program, 181, 182
   membrane filter technique, 574
   monitoring, 177, 180
   operation and maintenance, 174–180
   operational procedures, 177
   operator logbook, 180
   performance, 188
   performance monitoring, 177
   pilot-scale studies, 174
   polypropylene (PP), 167
   pressure vessels, 166, 168, 174
   pretreatment, 167, 174, 175
   preventive maintenance program (PMP), 181
   process failures, 177, 180
   properties, 188
   pump station failure, 180
   recordkeeping, 180, 181
   regeneration, 177
   reverse osmosis (RO), 165, 186, 188, 202, 211
   routine maintenance, 181

Membranes *(continued)*
   SCADA system, 174, 175, 176, 180, 181
   standby engine generator, 177
   submerged flow, 166, 172, 173, 174
   threads, 166
   treatment processes, 165–182
   troubleshooting, 177, 180, 181
   walk-through inspection, 175, 180
   water quality monitoring, 180
Mercury, 571
Metals, test procedures, 530
Metering
   chemical feeders, 372
   conservation, 689
Meters, electrical, 274
Methemoglobinemia (blue baby), 530, 567, 571
Microbial
   contaminants, 560, 565, 568, 573–582
   organism fouling, 206
Microfiltration (MF), 165, 167
Microwave, level sensors, 418
Mineral rejection, 188, 189, 191
Mineralized waters, 183
Modulating
   control systems, 401
   controller, 434
Monitoring
   arsenic, 150, 151, 562
   coliform frequency, 575–576
   ion exchange softening, 105
   iron and manganese, 16, 21
   membrane performance, 177
   process wastes, 233, 252
   regulations, 570, 575, 580, 584, 587
   requirements, Arsenic Rule, 562
   SWTR, 580
   secondary contaminants, 587
   trihalomethanes, 129, 133, 139, 583
   wastes, 233, 252
   water quality, membranes, 180
Motor
   control circuit operation, 407
   control stations, 401, 404
   electric motors, 284–293
   electrical, 271, 273, 284, 490
   failure, 287–293
   induction, 292
   insulation, 286
   pump, 284, 311
   safety, 490
   starters, 282, 489
   troubleshooting, 287–293
Mottled teeth, 33
Multicompartment units, electrodialysis, 215, 217
Multimeter, 274, 437, 438
Multiple-tube fermentation, 574
Muriatic acid, 463

N

NP4, 142
NPDES (National Pollutant Discharge Elimination System) Permit, 106, 233
NTU (nephelometric turbidity unit), 579
Nameplate, 286
Nanofiltration (NF)
   membranes, 165
   reverse osmosis, 190
National Electrical Safety Code, 212
National Fire Protection Association Standard 820, 413
National Labor Relations Act, 640
National Lime Association, 90
National Safety Council, 681
National Secondary Drinking Water Standards *See* Secondary Drinking Water Standards
Natural
   gas, 371
   radioactivity, 584
Natural organic matter (NOM), 165, 167
Negotiations, contract, 639
Nephelometric turbidity unit (NTU), 579
Nessler tubes, 8, 509, 514
New employee orientation, 629
New source alternative, arsenic, 142
News media, 644
Nitrate
   health effects, 567, 571
   regulations, 567, 571
   test procedures, 530
Nitric acid, 463
Nitrite, 571
Noise protection, 484
Noisy pump, 337
Noncarbonate hardness, 83, 84
Noncommunity water system, 565
Non-permit confined space, 410
Nontransient noncommunity water system, 565
North American Industry Classification System (NAICS) code, 691
Notification, 58, 576, 592
Number of employees needed, 625

O

ORP (oxidation-reduction potential), instrumentation, 426
OSHA regulations, 410
Occupational Safety and Health Act (OSHA), 673
Odor
   regulations, 589
   test procedures, 538
Offset, instrumentation, 404
Ohm, 273
Ohmmeter, 281
Ohm's Law, 273
Olfactory fatigue, 461

ON/OFF control, 414
ONPG-MUG test (coliforms), 574
Open-loop control system, 404
Operating ratio, 646, 649
Operation
    arsenic plant, 148
    chemical feeders, 49
    diesel engines, 364
    electrodialysis, 213, 221
    fluoridation, 49
    greensand, 18–21
    instrumentation, 443–445
    ion exchange softeners, 102–105
    iron and manganese, 18–21
    motor control circuit, 407
    reverse osmosis, 205–212
    safety
        See Safety
    valves, 344, 346
Operation and maintenance, membranes, 174–180
Operational checks, instrumentation, 445
Operational procedures, membranes, 177
Operations and maintenance (O & M), 652–661
Operator certification, 630
Operator interface, 401
Operator logbook, membranes, 180
Operator protection
    Also see Safety
    atmospheres, explosive, 494
    drowning, 496
    equipment, safety, 462, 494
    explosive atmospheres, 494
    eye protection, 495
    first-aid equipment, 673
    foot protection, 495
    gas-detection equipment, 481
    gloves, 495
    hand protection, 495
    hard hat, 496
    hazardous gases, 467, 480
    head protection, 496
    noise, 484
    personal protective equipment (PPE), 412
    respiratory protection, 468–470, 494
    safety equipment, 462, 494
    safety program, 670–683
    self-contained breathing apparatus, 468–470, 494
    traffic, 485
    water safety, 496
Optimum level, fluoride, 35
Oral communication, 641
Organic chemicals, regulations, 560, 567, 572
Organizational plan, 621, 622
Organizing, administration, 619, 621
Orientation, new employee, 629
Orifice plate, instrumentation, 424, 425, 426
Osmosis, 186, 187
OUCH principle, administration, 629

Overfeeding fluoride, 47, 54
Overload relays, 282
Overrange protection, pressure, 414
Oxidation
    arsenic, 142
    iron and manganese, 12, 14
    trihalomethanes, 135, 136
Oxidation-filtration, arsenic, 146
Oxidation-reduction potential (ORP), instrumentation, 426
Oxygen
    deficiency, 412, 480, 494
    enrichment, 412
    saturation table, 518
Ozone, 135, 138, 140

P

POU/POE (point-of-use/point-of-entry)
    arsenic, 143, 146
    membranes, 165
PPE (personal protective equipment), 412
Packing
    pumps, 307, 318, 320, 321, 322
    valves, 344
Painting, 479
Panel instruments, 429
Panels, control, 490
Paper screening, 626
Para-Dichlorobenzene, 573
Partial lime softening, 86
Parts, valves, 344
Pathogenic organisms, 565
Payback time, 648, 649
Percentile, 90th, lead and copper, 571
Performance
    evaluation, 631–633
    monitoring, membranes, 177
Peristaltic pumps, 35, 37, 52
Permanent hardness, 83
Permanganate, potassium
    See Potassium permanganate
Permeability constants, 188
Permeate, 188, 202
Permeate projections, computerized, 192, 193
Permit
    confined space entry, 677
    NPDES, 106, 233
Permit-required confined space, 410
Personal protective equipment (PPE), 412
Personnel records, 625, 639
Pesticides, 567
pH
    arsenic treatment, 144, 146
    effects on
        electrodialysis, 220
        reverse osmosis, 194, 196, 206, 207
        softening, 80, 81
        trihalomethanes, 134

pH *(continued)*
   meter, 427
   regulations, 589
   test procedures, 534
Phase, electrical, 272
Phosphate treatment, 10–12
Physical separation, arsenic treatment, 143
Pilot-scale studies, membranes, 174
Pink water, 16, 20
Pipet washers, 493
Piston pumps, 310
Pitot tube, 421
Plan, organizational, 621, 622
Planning
   administration, 619, 621
   emergencies, 621, 661
Plans, arsenic plant, 149
Plant
   drainage waters, 251
   maintenance, 479
   operations data, 684
   records, 684–687
   tours, 646
Pneumatic
   bubbler tubes, level measurement, 415
   hazards, 408
   instrumentation, 418, 434
Pocket comparators, 509
Poisoning, fluoride, 59
Pole shader, 290
Policies and procedures, administration, 629
Policy statement, safety, 491, 671, 672
Polymers, 149
Polyphosphate treatment
   iron and manganese, 10–12
   reverse osmosis, 206, 207
   softening, 86, 91
Polypropylene (PP) membranes, 167
Ponds, process wastes, 237
Population decline, 621
Pore blockage fouling, 167
Positive displacement pumps, 35, 336
Potassium permanganate
   handling, 473
   iron and manganese, 14, 18–20
   safety, 473
   trihalomethanes, 135, 136, 139
Powdered activated carbon, 138, 139, 474
Powders, 473, 474
Power
   failures, 430
   instrumentation, 428, 430
   requirements, 274, 337
   supply, electrodialysis, 220
   tools, 483
Precision, instrumentation, 400
Precursors, THM, 129, 133, 134, 136
Predictive maintenance, 652

Pre-employment inquiries, 627, 628
Pre-entry checklist, confined space, 411
Preparation for emergencies, 497, 661
Presence-absence (P-A), 573, 574
Present worth, 649
Pressure measurements, 414
Pressure vessel, membranes, 166, 168, 174
Pressures, electrodialysis, 220
Pretreatment
   electrodialysis, 215
   membranes, 167, 174, 175
   reverse osmosis, 205
Prevention of fires, 476
Preventive maintenance
   administration, 652
   instrumentation, 443–445
Preventive maintenance program (PMP), membranes, 181
Pricing, conservation, 692
Pride, employee, 630
Primary Drinking Water Standards
   bacteria, 573
   disinfectants and disinfection by-products, 583
   establishment, 559–566
   inorganic chemicals, 565–572
   MCLs, 567–568
   microbial standards, 573–581, 582
   organic chemicals, 572
   radioactivity, 584
   regulations, 559
   treatment techniques, 565
Primary elements, instrumentation, 429
Primary standards, measurements, 400
Prime, pumps, 336, 338
Priority alarms, 442
Priority list, 561
Probationary period, 629
Probes, level measurement, 415, 417
Problem, arsenic, 142
Problem employees, 631
Procedures, operational, membranes, 177
Process computer control system, 439
Process control, arsenic, 151
Process failures, membranes, 177, 180
Process instrumentation, 426
Process variables, 400, 414, 426
Process wastes
   alum sludge, 249
   backwash recovery ponds, 237, 239
   backwash wastewater, 249
   basket centrifuge, 245, 246
   belt filter presses, 241, 244
   brine, 234, 235, 248
   centrifuges, 244, 245, 246
   cleaning tanks, 235, 238, 239
   collection of sludges, 234
   collection systems, 248, 249
   conditioning, 235
   decant, 237

866 Water Treatment

Process wastes *(continued)*
  dewatering of sludges, 234, 235, 240
  disposal of sludges and brines, 234, 235, 249
  draining tanks, 235, 238, 239
  drying beds, 240, 242, 243, 249
  filter backwash wastewater, 249
  filter presses, 245, 247
  handling, 233, 234, 235
  ion exchange wastes, 249
  iron sludge, 249
  lagoons, 237
  landfills, 249, 250
  lime sludge, 249
  monitoring, 233, 252
  NPDES Permit, 106, 233
  need for handling and disposal, 233
  ponds, 237
  processing alternatives, 235, 236
  Public Law 92-500, 233
  reporting, 252
  sand drying beds, 240, 242, 243, 249
  sand filter, 249
  sanitary landfills, 249, 250
  scroll centrifuge, 244, 245
  septic wastes, 252
  sewers, 248, 249
  sludge pumps, 252
  sludge volumes, 234
  solar lagoons, 237, 240
  sources, 233, 234
  supernatant, 240
  tanks, draining and cleaning, 235, 238, 239
  temperature effects, 234
  thickening, 235
  vacuum filters, 245, 248
  vacuum tank truck, 239, 240, 249
  volumes of sludges, 234
  wastewater collection system, 248, 249
  Water Pollution Control Act, 233
  Winklepress, 244
Procurement of materials, 685, 686
Procurement records, 685
Productive meetings, 643
Profile, qualifications, 625
Program, maintenance, 267
Programmable Logic Controller (PLC), 148
Progressive cavity pumps, 312, 313, 326
Propeller meters, 420, 421, 422
Propeller pumps, 326
Property taxes, 649
Proportional-band, controller adjustment, 404
Proprietary media, arsenic, 146
Proprietary processes, iron and manganese, 15
Protected groups, 628
Protection devices, electrical equipment, 281
Protective measures, water supply, 668

Prussian blue, 346
Public
  notification, fluoride, 58
  relations, 644–646
  speaking, 645
  water systems, 565
Public health benefits
  Arsenic Rule, 562
  Consumer Confidence Report (CCR) Rule, 564
  Filter Backwash Recycling Rule (FBRR), 563
  Lead and Copper Rule, 562
  Radionuclides Rule, 564
  Stage 1 Disinfectants and Disinfection By-Products Rule, 563
  Stage 2 Disinfectants and Disinfection By-Products Rule, 563
  Total Coliform Rule (TCR), 562
Public information programs, conservation, 690
Public Law 92-500, 233
Pump controllers, instrumentation, 434, 435
Pump maintenance
  alignment, 305, 324, 331–335
  bearings, 324, 326
  belt drives, 330
  chain drives, 331
  check valves, 324, 326, 347–360
  controls, 327
  couplings, 332–335
  dial indicators, 333, 335
  electric motors, 284–293, 327–329
  foot valves, 324, 351
  mechanical seals, 321
  packing, 307, 318, 321, 322
  preventive maintenance, 318
  progressive cavity pumps, 326
  propeller pumps, 326
  reciprocating pumps, 325
  shear pin, 325, 334
  shutdown, 324, 336
  variable speed belt drives, 332
  wearing rings, 324
Pump operation
  centrifugal pumps, 338–340
  discharge, 337
  driving equipment, 334, 336
  electrical controls, 334, 336
  inspection, 340
  level controls, 334, 336
  noisy pump, 337
  positive displacement pumps, 35, 340
  power requirements, 337
  prime, 334, 336, 338
  rotation, 337
  shutdown, 334, 336, 338
  starting, 334–340
  troubleshooting, 336
Pump station failure, membranes, 180
Pumping equipment, electrodialysis, 220

Pumps
    alignment, 305, 324, 331–335
    bearings, 305, 311, 324, 326
    casing, 306, 311
    cavitation, 308, 309
    centrifugal pumps, 301, 310, 338
    chemical metering, 312, 372
    couplings, 305, 332–335
    displacement, 302
    dynamic types, 302
    electric motors, 284–293
    flanges, 305
    horizontal centrifugal pumps, 301, 302, 310, 338
    impeller, 301, 303, 311
    Let's Build a Pump!, 301–310
    lubrication, 305, 312–317
    maintenance
        See Pump maintenance
    motors, 271–296
    packing, 307, 318, 321, 322
    piston type, 310
    progressive cavity, 312, 313, 326
    reciprocating, 310, 325
    rings, wearing, 307
    screw flow, 312, 313
    seal, 307
    shaft, 303, 305, 311
    sleeves, 305, 311
    stuffing boxes, 307
    suction, 305, 306, 308
    vertical centrifugal pumps, 310
    wearing rings, 307
Purchase order, 647, 685, 686

Q

Qualifications profile, 625
Questions
    management, 620
    pre-employment, 627, 628
Quick Reference Guides, Compilation of, EPA, 561
Quicklime, 83, 91, 465

R

Radio, 644
Radioactivity, 584
Radiological contaminants, 560, 561, 584
Radionuclides Rule
    public health benefits, 564
    purpose, 564
Radon, 585
Range, instrument, 401
Rate of flow, 418
Rates, water, 648
Recarbonation, 84, 85, 86, 90

Reciprocating pumps, 310, 325
Recognition, employee, 629, 630
Recorders, 430
Recordkeeping
    accident reports, 678
    arsenic, 151
    bacteriological, 602
    chemical, 602
    computer, 687
    disposition of records, 687
    electrical equipment, 267, 268, 287, 293
    electrodialysis, 221, 222
    equipment, 687
    fluoridation, 49, 53, 55, 56
    inventory, 685
    ion exchange softening, 114
    lime–soda ash softening, 93
    maintenance, 685
    membranes, 180, 181
    operations data, 684
    personnel, 625, 639
    plant, 684–687
    process wastes, 252
    procurement, 685
    purchase order, 647, 685, 686
    purpose, 684
    requisition, 647, 685, 686
    retention of records, 602, 687
    reverse osmosis, 207, 209, 210
    safety, 678
    sanitary survey, 602
    softening, 93
    types of records, 684
    variance and exemptions, 602
Recovery, reverse osmosis, 194, 207
Recycle flows, 563
Red water problems, 16, 22
Regenerant, salt, arsenic treatment, 144
Regeneration, membranes, 177
Regulations, drinking water
    arsenic, 566
    bacteria, 568, 573–582
    barium, 566
    cadmium, 566
    check sampling, 591
    chloride, 587
    chlorine residual substitution, 579
    chromium, 569
    coliforms, 568, 573–577
    color, 587
    community water system, 565
    copper, 569, 587
    Drinking Water Standards, 561, 565–585, 586–590
    establishment, 560–566
    filtration, 578–581
    fluoride, 588

Regulations, drinking water *(continued)*
  foaming agents, 588
  Giardia lamblia, 573, 578–581
  health, 559, 565, 567–568, 586
  immediate threats to health, 571, 573
  initial sampling, 591
  inorganic chemicals, 566–572
  iron, 588
  lead, 569
  maximum contaminant levels (MCLs), 560, 565–585
  membrane filter, 574
  mercury, 571
  microbiological contaminants, 560, 568, 573–581
  monitoring, 570, 575, 580, 584, 587
  multiple-tube fermentation, 574
  natural radioactivity, 584
  nitrate, 567, 571
  nontransient noncommunity water system, 565
  notification, 576, 592
  odor, 589
  organic chemicals, 560, 567, 572
  pH, 589
  Primary Drinking Water Standards, 565–585
  radioactivity, 584
  radiological contaminants, 560, 561, 584
  reporting procedures, 576, 592
  required sampling, 570, 574, 575, 576, 580, 584, 587, 591–592
  routine sampling, 591
  Safe Drinking Water Act, 591
  sampling points, 591
  sampling procedures, 591–592
  Secondary Drinking Water Standards, 586–590
  selenium, 572
  setting standards, 559
  silver, 590
  solvents, 572
  sulfate, 590
  total dissolved solids (TDS), 590
  treatment techniques, 565
  trihalomethanes, 129, 139, 583
  turbidity, 560, 565, 579
  zinc, 590
Regulations, OSHA, 410
Regulatory agencies, safety, 671
Rejection, mineral, 188, 189, 191
Repair shop, 301
Repair/replacement fund, 648
Replacement cost, 648
Report writing, 641–643
Reporting
  arsenic
    emergency, 151
    routine, 151
    special, 151
    violations, 151
  procedures, 576, 592
  safety, 497, 498, 678
  waste disposal, 252
Representative sample, 131
Required sampling, 570, 574, 576, 580, 584, 587, 591–592
Requisition system, 647, 685, 686
Reservoirs, 7, 376
Reset, controller adjustment, 404
Residential plumbing retrofits, 689
Residential water surveys, 688
Residuals, arsenic, 149
Resin
  arsenic ion exchange, 144
  ion exchange, 12, 99, 102, 104, 135, 138
Respiratory protection, 468–470, 494
Responsibilities
  manager, 619, 624
  safety, 459, 670
Responsibility, 621
Retaliation, 638
Retrofits, residential plumbing, 689
Revenue bonds, 649
Reverse osmosis (RO)
  Also see Demineralization and Electrodialysis
  acid feed system, 207
  additional reading, 223
  alarms, 207
  arithmetic assignment, 223
  brine, 194, 207
  calculations, 188–198
  cartridge filters, 207
  chlorination, 207
  "Christmas Tree" arrangement, 194, 197, 207
  cleaning membrane, 211
  colloids, 206
  components, 199
  concentration polarization, 194
  definition, 186
  feed, 207
  flow diagram, 187, 208
  flux, 186, 189
  flux decline, 188
  hollow fine fiber, 202, 204, 206
  hydrolysis, 194, 196, 206
  layout, 201
  log sheet, 209, 210
  membranes, 165, 186, 188, 202, 211
  microbial organisms, 206
  mineral rejection, 188, 189, 191
  operation, 205–212
  osmosis, 186, 187
  permeate, 188, 202
  pH effects, 194, 196, 206, 207
  polyphosphate treatment, 206, 207
  pretreatment, 205
  pumps, 199
  recordkeeping, 207, 209, 210
  recovery, 194, 207

Reverse osmosis (RO) *(continued)*
  rejection, mineral, 188, 189, 191
  safety, 211
  scalants, 206
  spiral wound, 202, 203
  suspended solids, 205
  temperature effects, 194, 195, 196, 206
  threshold treatment, 206
  troubleshooting, 211
  tubular, 202
  turbidity, 205
  types of plants, 202
Reversible fouling, 167
Review of plans and specifications, arsenic plant, 149
Right-To-Know (RTK) Laws, 475, 673
Rings, wearing, 307
Rinse, ion exchange softening, 103, 105, 108, 148
Rotameter, 419, 420, 421
Rotation of pump operation, 337
Rotor, 284
Route, sampling, 592
Routine arsenic reporting, 151
Routine maintenance, membranes, 181
Routine sampling, 570, 575, 591
Rules
  coliform, 573–577
  copper, 569
  lead, 569
  surface water treatment, 578–582

S

SCADA (Supervisory Control And Data Acquisition) systems
  computer control systems, 439
  distribution systems, 654–655
  membranes, 174, 175, 176, 180, 181
  treatment plants, 654–655
SCBA (self-contained breathing apparatus), 410, 468–470, 494
SDWA (Safe Drinking Water Act), 559, 591, 592
SRF (State Revolving Fund), 650
SURF (Small Utility Rates and Finances) software, 687
Safe Drinking Water Act (SDWA), 559, 591, 592
Safe Drinking Water Hotline [(800) 426-4791], 561
Safety
  accident
    causes, 459, 670
    prevention, 459, 485
    reports, 497, 498, 678, 679, 680
  acetic acid (glacial), 460
  acids, 460
  activated carbon, 474
  additional reading, 499
  alum, 472
  aluminum sulfate, 472
  ammonia, 464
  arsenic plant, 149
  atmospheres, explosive, 494
  autoclaves, 493
  bases, 464
  biological considerations, 493
  booster shots, immunization, 493
  calcium hydroxide, 465
  carbon dioxide, 469
  caustic soda, 465
  check, vehicles, 485, 486
  chemical handling, 211, 460
  chemical storage drains, 475
  chemicals, laboratory, 492
  CHEMTREC [(800) 424-9300], 497
  chlorine, 467
  CHLORINE MANUAL, 468
  classification, fires, 476
  cleaning, 479
  confined spaces, 409–412, 480–483, 674, 677
  control panels, 490
  costs, 682
  cranes, 480
  current, 489, 490
  dangerous air, 480
  definition, 459
  drains, 475
  drowning, 496
  education, 678
  electrical equipment, 212, 269, 271, 299, 404, 489
  electrodialysis, 221
  emergencies, 497, 673
  equipment
    fluoridation, 49, 59
    laboratory, 492–494
    operator protection, 459, 493, 494–497, 673
    personal protective equipment (PPE), 412
    shower, 461, 462, 463
  exits, 479
  explosive atmospheres, 494
  extinguishers, fire, 476–478
  eye protection, 495
  ferric chloride, 472
  ferric sulfate, 472
  ferrous sulfate, 472, 473
  fire protection, 476
  first aid, 60, 673
  flammables, storage, 479
  fluoridation, 59
  fluoride compounds, 473
  foot protection, 495
  forklifts, 488
  fueling vehicles, 485
  gas masks, 468–470
  gas-detection equipment, 481
  gases, 467
  glassware, 492
  gloves, 495
  ground-fault circuit interrupter, 483, 490
  hand protection, 495

Safety (continued)
- handling chemicals, 460
- hard hat, 496
- hazard communication program, 673
- hazardous gases, 467, 480
- hazards
  - empty digesters, 376
    - instrumentation, 404–413
    - laboratory, 492–494
    - maintenance, 479
- head protection, 496
- homeland defense, 663–667
- hot plates, 493
- human factors, 683
- hydrated lime, 465
- hydraulic equipment, 212
- hydrochloric acid, 463
- hydrofluosilicic acid, 60, 461, 473
- hydrogen fluoride, 463
- hypochlorite, 466
- immunization, booster shots, 493
- injury frequency rate, 682
- injury severity rate, 683
- instrumentation, 409–412, 490
- laboratory, 492–494
- lime–soda ash softening, 89, 91
- lockout/tagout procedure, 269, 490, 491
- maintenance, 479–484, 489
- manholes, 483
- Material Safety Data Sheet (MSDS), 674–676
- measuring, 682
- membrane, 221
- motors, 490
- muriatic acid, 463
- National Safety Council, 681
- nitric acid, 463
- noise, 484
- OSHA, 673, 674
- olfactory fatigue, 461
- operator, 459, 493, 494–497, 673
- oxygen deficiency, 480
- oxygen enrichment, 412
- painting, 479
- panels, control, 490
- permit, confined space, 482, 674, 677
- permit-required confined space, 481
- personal protective equipment (PPE), 412
- pipet washers, 493
- plant maintenance, 479
- policy statement, 491, 671, 672
- potassium permanganate, 473
- powdered activated carbon, 474
- powders, 473, 474
- power tools, 483
- program, 670–683
- quicklime, 83, 91, 465
- radioactivity, 493
- regulatory agencies, 671
- reporting, 497, 498, 678, 679, 680
- respiratory protection, 468–470, 494
- responsibilities, 459, 670
- reverse osmosis, 211
- safety check, vehicles, 485, 486
- safety shower, 461, 462, 463
- salts, 471, 472
- seat belts, 485
- self-contained breathing apparatus, 468–470, 494
- shots, booster, immunization, 493
- shower, safety, 461, 462, 463
- sodium aluminate, 473
- sodium carbonate, 466
- sodium fluoride, 60, 473
- sodium hydroxide, 465
- sodium silicate, 466
- softening, 89, 91
- standard operating procedures (SOPs), 481, 490, 491, 673
- starters, 489
- statistics, 682
- sterilizers, 493
- stills, water, 493
- storage, chemicals, 475
- storage, flammables, 479
- sulfur dioxide, 469
- sulfuric acid, 463
- supervisors, 671, 672, 679, 681
- tailgate training, 460, 681
- threat levels, homeland defense, 664–667
- tools, power, 483
- traffic, 485
- training, 629, 678
- transformers, 489
- underwater inspection, 496
- unsafe acts, 672
- valves, 484
- vehicles, 485
- voltage, 489
- water, 496
- water stills, 493
- welding, 484
- Worker Right-To-Know (RTK) Laws, 673

Salinity, 183
Salt solution characteristics, 12
Salts
- handling, 471, 472
- safety, 471, 472

Sampling
- arsenic, 150
- coliform, 574
- iron and manganese, 8
- lead and copper, 570
- points, 591
- procedures, 591–592
- schedule, 591
- trihalomethanes, 131

Sampling procedures
    check sampling, 591
    collection, 570, 591–592
    frequency, 570, 574–576, 579, 580, 584, 587, 591–592
    how often, 570, 574–576, 579, 580, 584, 587, 591–592
    initial sampling, 591
    lead and copper, 570
    location, 591
    number of samples, coliform, 574
    plan, 574
    required sampling, 570, 574, 591
    route, 592
    routine sampling, 591
    Safe Drinking Water Act, 591
Sand drying beds, 240, 242, 243, 249
Sand filter, 249
Sanitary defects
    brine storage tanks, 107
    fluoridation, 59
Sanitary landfills, 249, 250
Sanitary survey, 573, 602
Saturators, fluoridation, 35, 42, 43, 46, 59
Scaling
    electrodialysis, 213, 220
    reverse osmosis, 206
Schedule, sampling, 591
School education programs, conservation, 690
Screw flow pumps, 312, 313
Scroll centrifuge, 244, 245
Seal, pump, 307, 308
Seat belts, 485
Seats, valves, 346
Seawater, 183
Secondary Drinking Water Standards
    aluminum, 587
    chloride, 587
    color, 587
    copper, 587
    enforcement, 586
    fluoride, 588
    foaming agents, 588
    hydrogen sulfide, 589
    iron, 588
    iron and manganese, 588
    manganese, 588
    maximum contaminant levels (MCLs), 586
    monitoring, 587
    odor, 589
    pH, 589
    silver, 590
    sulfate, 590
    total dissolved solids (TDS), 590
    zinc, 590
Secondary standards, measurements, 400
Security, arsenic plant, 149
Selection, employees, 628
Selenium, 572

Self-contained breathing apparatus (SCBA), 410, 468–470, 494
Sensitivity, instrumentation, 401
Sensors, types, 414–428
Septic wastes, 251
Service
    cost recovery, 692
    ion exchange softeners, 102, 103, 107, 148
    line replacement, 571
    record card, 267, 268
Set point, 404
Sewers, 248, 249
Sexual harassment, 635, 636
Shaft, pump, 303, 305, 311
Shear pin, 325, 334
Shop steward, 639
Shots, booster, immunization, 493
Shower, safety, 461, 462, 463
Shutdown
    arsenic plant, 148
    chemical feeders, 58
    fluoridation, 58
    instrumentation systems, 444
    ion exchange softeners, 108
    pumps, 324, 334, 338
Signal
    transducers, 428
    transmitters, 428, 653
Silver, 590
Slake, 83, 91
Sleeves, pump, 305, 311
Sludge
    pumps, 251
    softening, 92
    volumes, 234
Snubbers, 414, 417
Sodium
    aluminate, 473
    carbonate, 466
    fluoride, 34, 35
    fluorosilicate, 34
    hydroxide, 465
    silicate, 466
    silicofluoride, 54
Softening
    Also see Ion exchange softening and Lime–soda ash softening
    additional reading, 114
    alkalinity, 77, 80, 90
    arithmetic assignment, 114
    basic methods, 82
    benefits, 78, 82
    calcium carbonate equivalent, 78, 79
    carbonate hardness, 77, 83
    chemical reactions, 83–86
    chemistry, 79
    hardness, water, 77, 82, 84
    importance, 78
    ion exchange softening, 99

Softening *(continued)*
   jar tests, 93–97
   Langelier Index, 80
   lime–soda ash softening, 82, 87
   limitations, 78, 83
   need, 78
   noncarbonate hardness, 83, 84
   permanent hardness, 83
   pH, 80, 81
   recordkeeping, 93
   safety, 89, 91
   sludge, 92
   stability, 80, 84, 90
   temporary hardness, 83
   total hardness, 77
   zeolite, 99
Softening membrane, RO, 190
Solar lagoons, 237, 240
Solid chemical feeders, 372
Solution feeders, 35, 42
Solution preparation, fluoridation, 49
Solvents, 572
Sonar devices, 415
Sources, arsenic, 142
Span, instrumentation, 401
Special arsenic reporting, 151
Special assessment bonds, 650
Specific conductance test procedures, 535
Specification review
   electrodialysis, 220
   fluoridation, 47
Specifications, arsenic plant, 149
Spectrophotometer
   absorbance, 509
   calibration, 509
   description, 509
   iron and manganese analysis, 8
   percent transmittance, 509
   standards, 510
   transmittance, 509
   units, 509
Spent media, activated alumina, 146
Spiral wound membrane, 202, 203
Split samples, 150, 151
Split treatment
   ion exchange softening, 112
   lime–soda ash softening, 86–89
Square root extraction, 395
Stability, water, 80, 84, 90
Stack, electrodialysis, 213, 215, 220
Staffing
   advertising positions, 626
   applications, job, 626
   certification, 630
   compensation, 626, 629
   disciplinary problems, 631
   employee orientation, 629
   employee problems, 631, 639
   employee-management relationship, 639
   employees, 624
   employer policies, 629, 636
   evaluation, performance, 631–633
   grievances, 631, 639
   groups, protected, 628
   harassment, 635, 636
   hiring procedures, 624–629
   interview questions, 627, 628
   interviewing, 626
   manager's responsibilities, 619, 624
   National Labor Relations Act, 640
   new employee orientation, 629
   number of employees, 625
   orientation, new employees, 629
   paper screening, 626
   performance evaluation, 631–633
   personnel records, 639
   policies, employer, 629, 636
   probationary period, 629
   problem employees, 631
   profile, qualifications, 625
   protected groups, 628
   qualifications profile, 625
   retaliation, 638
   selection process, 628
   sexual harassment, 635, 636
   training, 630
   unions, 639
Stage 1 Disinfectants and Disinfection By-Products Rule
   public health benefits, 563
   purpose, 563
Stage 2 Disinfectants and Disinfection By-Products Rule
   public health benefits, 563
   purpose, 563
Stages, electrodialysis, 213
Standard deviation, 541
Standard Industrial Classifications (SIC) code, 691
Standard operating procedures (SOPs), 480, 481, 490, 491, 673
Standard unit of measure, 400
Standardization, instrumentation, 401
Standardized Monitoring Framework (SMF)
   benefits, 564
   contaminants regulated, 564
   monitoring requirements, 564
   purpose, 564
Standards, drinking water, 559, 586
Standby engine generator, membranes, 177
Standby engines, 371
Standby power generation, 293
Standby power generators, 430
Starters, electrical, 282, 489
Start-up
   arsenic plant, 148
   chemical feeders, 49
   engines, 362, 363, 368
   fluoridation, 49
   instrumentation systems, 444

Start-up (continued)
  ion exchange softeners, 108
  pumps, 334–340
State Revolving Fund (SRF), 650
Stator, 284
Steel tanks, 376
Sterilizers, 493
Stethoscope, 327
Steward, shop, 639
Stills, water, 493
Storage of
  chemicals, 371, 475
  flammables, 479
  fuel, 370
  lime, 89
Strip-chart recorders, 430, 432
Stuffing boxes, pumps, 307
Submerged flow, membranes, 166, 172, 173, 174
Suction, pumps, 306, 308
Sulfate
  regulations, 590
  test procedures, 535
Sulfur dioxide, 469
Sulfuric acid, 463
Sulfurospirillum, 142
Supernatant, 240
Supersaturated, 83
Supervising inquiries, pre-employment, 627
Supervision, 630
Supervisors, safety, 671, 672, 679, 681
Supervisory Control And Data Acquisition (SCADA), 439
Surface Water Treatment Rules, 562, 578–582
Surfactant, 588
Surge protection, pressure, 414, 417
Switch gear, electrical, 281, 298
Switches, electrical, 272, 273, 275
Symbols, electrical, 395–398
Synthetic organic chemicals, 572
Synthetic resins, 135, 136
System water audits, 689

T

TCE (trichloroethylene), 572
TDR (time domain reflectometry) level sensors, 418
TON (Threshold Odor Number), 538–542, 589
Tag, warning, 270, 319
Tagged off, 409
Tagout/lockout procedure, 269, 490, 491
Tailgate training, 460, 681
Tanks
  draining and cleaning, 235, 238, 239
  steel, maintenance, 376
Target audience, 641, 642, 644
Taste rating scale, 540
Tastes and odors
  iron and manganese, 7
  test procedures, 538

Teeth, mottling and cavities, 33, 569, 588
Telemetering, instrumentation, 653
Telephone contacts, 645
Television, 644
Temperature effects
  electrodialysis, 220
  process wastes, 234
  reverse osmosis, 194, 195, 196, 206
  THM formation, 134
Temporary hardness, 83
Test and calibration equipment, 437
Test procedures
  See Laboratory test procedures
Testers, electrical, 274
Testing ion exchange softeners, 105
Thallium, 572
Thermal overloads, 282
Thickener supernatant, 563
Thickening wastes, 235
Thin film composite membrane, 188
Thread membranes, 166
Threat level categories, 665
Threshold Odor Number (TON), 538–542, 589
Threshold treatment, 206
Tier 1 and 2 violations, 576, 592
Time domain reflectometry (TDR) level sensors, 418
Titrate, 79
Tools, power, 483
Total Coliform Rule (TCR)
  MCL, 562
  public health benefits, 562
  purpose, 562
  standards, 573–577
Total dissolved solids (TDS), 183, 185, 543, 590
Total flow, 418
Total hardness, 77
Total trihalomethanes, 129, 131, 134, 139, 583
Totalizers, instrumentation, 430
Tours, plant, 646
Toxicity, 664
Traffic, 485
Training
  administration, 630
  fluoridation, 60
  safety, 630, 678
Transducers (signal), 428
Transformers, 272, 278, 282, 298, 489
Transient noncommunity water system, 565
Transmission, electrical, 298
Transmitters (signal), 428
"Traveler's" diarrhea, 573
Treatment
  arsenic, 142
  emergency, 668, 669
  processes, membranes, 165–182
Treatment charts, fluoridation, 49–51
Treatment techniques, 565
1,1,1-Trichloroethane, 572

Trichloroethylene (TCE), 572
Trihalomethanes
    activated carbon, 135, 138, 139
    additional reading, 131, 140
    adsorption, 135, 138
    aeration, 135–136, 140
    arithmetic assignment, 140
    bench-scale studies, 135, 139, 140
    bromide, 129, 134
    calculations, 132
    chemical reactions, 129, 133
    chloramines, 138, 139
    chlorine, 129, 134, 138, 139
    coagulation/sedimentation/filtration, 135, 136
    control strategies, 131, 134, 139
    disinfection alternatives, 138
    existing treatment processes, 135
    feasibility analysis process, 131
    FEDERAL REGISTER, 129, 131, 139
    formation, 129, 133, 134, 136
    formulas, 132, 743, 755
    Group 1 and Group 2 treatment techniques, 140
    health effects, 129
    ion exchange resins, 135, 138
    maximum contaminant level (MCL), 129, 140, 583
    monitoring, 129, 133, 139
    options for control, 134
    oxidation, 135, 136
    ozone, 135, 136, 138, 140
    pH, THM formation, 134
    potassium permanganate, 135, 136, 139
    precursors, THM, 129, 133, 134, 136
    problem, 129, 131
    regulations, 129, 139, 583
    resins, synthetic, 135, 136
    sampling, 131
    sources of water, 134
    synthetic resins, 135, 136
    temperature, THM formation, 134
    test procedures, 542
    total trihalomethanes, TTHMs, 129, 131, 134, 139
    ultraviolet light, 135, 136
    variance, 139
Troubleshooting
    arsenic plant, 149
    engines, 362, 368
    ion exchange softeners, 107
    iron and manganese, 20, 21, 22
    membranes, 177, 180, 181
    motors, 287–293
    pumps, 336
    reverse osmosis, 211
Tubular membranes, 202
Turbidity regulations, 560, 565, 568, 579
Types of sensors, 414–428
Typical plant, arsenic, 148

U

UEL (upper explosive limit), 412, 413
US Terrorism Alert System, 664
Ultrafiltration (UF), 165, 167
Ultra-low-flush [flow] toilet (ULFT), 689, 691, 692
Ultrasonic devices, level measurement, 415, 418, 420
Ultraviolet light, 135, 136
Unaccounted for water, 620
Unbalanced current, 277–279
Underfeeding, fluoridation, 58
Underwater inspection, 496
Unions, 639
Unity of command, 621
Unloader, compressor, 342
Unregulated contaminants, 561
Unsafe acts, 672
Upflow saturators, 43, 45, 46
Upper explosive limit (UEL), 412, 413
Uranium, 584
User fees, 647
Utility management, 619

V

VOCs (volatile organic chemicals), 567, 572
V-O-M (Volt-Ohm-Milliammeter), 437, 438
Vaccination, hepatitis B, 412
Vacuum breakers, 659, 660
Vacuum filters, 245, 248
Vacuum tank truck, 239, 240, 251
Valves
    arsenic treatment plant, 148, 149
    automatic, 360
    butterfly, 347, 350
    check, 324, 326, 347–360
    compressor, 343, 344
    diaphragm-operated, 360, 361
    eccentric, 347–349
    foot, 324, 351
    gate, 344, 345
    globe, 346, 358, 360, 361
    lubrication, 346
    maintenance, 344, 346, 347, 351
    operation, 344, 346
    packing, 344
    parts, 344
    safety, 484
    seats, 346
    types, 352
    use, 344, 347
    wafer check valve, 347, 350, 352, 357
Variable speed belt drives, 332
Variables
    measured, instrumentation, 414–428
    process, 400, 414, 426

Variance and exemptions, 139, 602
Variance, THM, 139
Vehicles
   accident prevention, 485
   forklifts, 488
   fueling, 485
   maintenance, 485
   operation, 485–487
   safety check, 485, 486
   seat belts, 485
   types, 485
Velocity flow sensing, 420, 422
Venturi tube, 424, 425, 426
Vertical centrifugal pumps, 310
Vinyl chloride, 572
Violation arsenic reporting, 151
Violations, tier 1 and 2, 592
Violence, 634
Viscosity, 149, 316
Volatile, 135
Volatile organic chemicals (VOCs), 567, 572
Voltage testing, 274, 276, 489
Volt-Ohm-Milliammeter (V-O-M), 437, 438
Volts, 272, 273, 274, 298
Volume-related charges, 692
Volumes of sludges, 234
Volumetric feeders, 35, 39, 40, 42
Vulnerability assessment, 661

W

Wafer check valve, 350, 352, 357
Walk-through inspection, membranes, 175, 180
Warning tag, 270, 319
Wastewater, arsenic, 149
Wastewater collection system, 248, 249
Water
   audits, 689
   conservation
      See Conservation
   cooled engines, 367
   efficiency program, 690
   hammer, 338
   Pollution Control Act, 233
   rates, 648
   safety, 496
   shortage allocations, 691
   stills, 493
   waste prohibition, 692
Water quality monitoring, membranes, 180
WaterSense, 693
Watts, 274
Wearing rings, pumps, 307
Welding, 484
Wholesale agency assistance programs, 691
Wilson's disease, 569
Winklepress, 244
Wood preservative, arsenic, 142
Worker Right-To-Know (RTK) Laws, 475, 673
Working conditions, 626
Written communication, 641–643

X

(NO LISTINGS)

Y

(NO LISTINGS)

Z

Zeolite
   iron and manganese, 14
   softening, 99
Zinc, 590

NOTES